D0944588

OVERDUE
25¢

Modern Genetics

The Benjamin/Cummings Series in the Life Sciences

F. J. Ayala and J. A. Kiger, Jr.
Modern Genetics (1980)

F. J. Ayala and J. W. Valentine
Evolving: The Theory and Processes of Organic Evolution (1979)

M. G. Barbour, J. H. Burk, and W. D. Pitts
Terrestrial Plant Ecology (1980)

L. L. Cavalli-Sforza
Elements of Human Genetics, second edition (1977)

R. E. Dickerson and I. Geis
The Structure and Action of Proteins, second edition (1981)

L. E. Hood, I. L. Weissman, and W. B. Wood
Immunology (1978)

L. E. Hood, J. H. Wilson, and W. B. Wood
Molecular Biology of Eucaryotic Cells (1975)

A. L. Lehninger
Bioenergetics: The Molecular Basis of Biological Energy Transformations, second edition (1971)

S. E. Luria, S. J. Gould, and S. Singer
Life: Molecules to Organisms (1981)

A. P. Spence and E. B. Mason
Human Anatomy and Physiology (1979)

J. D. Watson
Molecular Biology of the Gene, third edition (1976)

I. L. Weissman, L. E. Hood, and W. B. Wood
Essential Concepts in Immunology (1978)

N. K. Wessells
Tissue Interactions and Development (1977)

W. B. Wood, J. H. Wilson, R. M. Benbow, and L. E. Hood
Biochemistry: A Problems Approach, second edition (1981)

Modern Genetics

Francisco J. Ayala
John A. Kiger, Jr.
University of California, Davis

The Benjamin/Cummings Publishing Company, Inc.
Menlo Park, California • Reading, Massachusetts
London • Amsterdam • Don Mills, Ontario • Sydney

Sponsoring editor James W. Behnke
Manuscript editor Fred Raab
Book and cover designer Marjorie Spiegelman
Chief artist Georg Klatt; additional art by Fran Milner
Layout artist John F. Kelly, Jr.
Production coordination Madeleine Dreyfack, Fred Raab
Typesetting coordination Robin Taylor-Radford, Nancy Fontaine

Photograph on page 27 from G. Lefevre, in *The Genetics and Biology of Drosophila,* Vol. 1a, ed. by M. Ashburner and E. Novitski, Academic Press, New York, 1976. See also Figure 5.23 on page 159.

Photograph on page 337 courtesy of Prof. Barbara Hamkalo, University of California, Irvine, and Prof. Oscar L. Miller, Jr., University of Virginia. See also Figure 11.13 on page 379.

The photograph on page 533 is from F. J. Ayala. See also Figure 18.10 on page 615.

Library of Congress Cataloging in Publication Data

Ayala, Francisco José, 1934–
Modern Genetics

Bibliography: p.
Includes index.
1. Genetics. I. Kiger, John A., Jr., joint author.
II. Title.
QH430.A92 575.1 80-10192

ISBN 0-8053-0314-6
BCDEFGHIJK-HA-89876543210

The Benjamin/Cummings Publishing Company, Inc.
2727 Sand Hill Road
Menlo Park, California 94025

To Our Teachers

Table of Contents

II Expression of the Genetic Materials 336

Preface

Genetics is a rapidly advancing science. *Modern Genetics* introduces the basic concepts of this science in a way that we believe is both up-to-date and rigorous. We have maintained a historical perspective and have described classical experiments, but the emphasis is on current knowledge.

Modern Genetics is a basic genetics text intended for a one-semester course. Prior college courses in general biology and chemistry are assumed, although some essential concepts (for example, mitosis and meiosis) that the student should have learned in such courses are reviewed here. We have intentionally included more material than can be covered in most one-semester courses. However, the book is written so as to enable instructors to choose topics according to their preferences.

The molecular basis of heredity is a dominant theme of modern genetics, and of this text. A distinctive feature of the text, however, is that population genetics and evolutionary genetics are treated in greater depth than in most introductory texts. This balanced approach in *Modern Genetics* culminates in a discussion of molecular evolution (Chapter 22), which is probably unique in genetics texts.

Wherever appropriate, we have used examples from human genetics, which students find particularly interesting. We have integrated discussions of recombinant DNA technology into the text so that the student can gain an appreciation of how this technology has developed and is being applied to make—almost daily—new discoveries in genetics.

Organization

We believe that genetics is best presented in terms of the three fundamental features of genes: transmission, expression, and change. Thus, *Modern Genetics* is composed of three parts: Part I, Organization and Replication of the Genetic Materials; Part II, Expression of the Genetic Materials; and Part III, Evolution of the Genetic Materials. This format is not only logical but also, we think, of practical value in the classroom. The student should comprehend more easily why the various topics are considered and how they contribute to a full understanding of the process of heredity.

Because of this organization, certain topics are presented in a different sequence from the usual. A detailed discussion of protein structure is deferred until Part II, and mutation is not discussed as a separate topic until Part III. However, the book is designed to give the instructor maximum flexibility in devising a syllabus. The order in which the chapters are taught can be substantially altered without creating serious gaps in knowledge. We believe, however, that most instructors will agree that the subject matter within the chapters need not be rearranged in any way; each chapter has been developed directly from our lectures.

Special Features

- The book is generously illustrated to enhance the clarity of the text.
- About two dozen special topics of various kinds are set apart as "boxes," most of which can be skipped without loss of comprehension of the text. These boxes provide another element of choice in selecting the material for study.
- The problems at the end of each chapter are an integral part of the book; some contain new information that we consider subsidiary or too detailed to be included in the text proper. A solutions manual is available to the instructor.
- Students who have not taken a course in statistics will find in the appendix the concepts and methods needed to understand the text and solve the problems.
- A glossary is provided as an aid in reviewing both new and familiar concepts.
- The bibliography lists many basic references that document the material covered in the text; the credit lines for figures and tables cite additional sources of information.

We have exploited our complementary areas of expertise in writing *Modern Genetics.* F. J. A. wrote the first draft of most or all of Chapters 1–3, 10, 15, and 17–22 and the appendix; J. A. K. wrote the draft of Chapters

4–9, 11–14, and 16. We have worked together, however, on every part of the book since its conception four years ago and, prior to that, on its antecedent—a draft that we have both used in teaching genetics at our university. The flip of a coin decided the order of authorship.

Acknowledgments

We are indebted in a most fundamental way to all the scientists whose brilliant efforts have made genetics such an exciting and mature science. To many of them we are particularly thankful for their generosity in providing photographs, permission to reproduce illustrations, and other needed items. The manuscript was extensively reviewed by both specialists and generalists (they are listed below) to ensure a sound and balanced coverage of the subject. Although we alone bear the responsibility for any shortcomings, they are all the fewer because of these helpful reviews.

We thank Ms. Candy Miller and Ms. Janette Bannan, who typed the manuscript, Ms. Elizabeth Toftner, who helped with proofreading and indexing, and Ms. Lorraine Barr, also for proofreading. During a period in which the book was in production and we were overseas and hard to reach, Professor John Stubbs of San Francisco State University provided valuable advice as a consultant to the publisher. Fred Raab edited the manuscript with skill and wisdom, again and again finding ways to clarify the text. We trust that readers will agree that the book has benefited greatly from the work of Georg Klatt, who did most of the drawings.

Davis, California
January, 1980

Francisco J. Ayala
John A. Kiger, Jr.

List of Reviewers

Joan W. Bennett, Tulane University
Rowland H. Davis, University of California, Irvine
Irving Finger, Haverford College
Barbara A. Hamkalo, University of California, Irvine
Yun-Tzu Kiang, University of New Hampshire
George Lefevre, California State University, Northridge
Joyce B. Maxwell, California State University, Northridge
John R. Merriam, University of California, Los Angeles
Virginia Merriam, Loyola Marymount University
Roger Milkman, University of Iowa
Jeffry B. Mitton, University of Colorado
William H. Petri, Boston College
Henry E. Schaffer, North Carolina State University
John Stubbs, San Francisco State University

Modern Genetics

1

Introduction

The eminent geneticist Theodosius Dobzhansky stated that "Nothing in biology makes sense except in the light of evolution." It is even more certain that nothing in biology is understandable except in the light of genetics. Genetics is the core biological science; it provides the framework within which the diversity of life and its processes can be comprehended as an intellectual whole.

The foundations of genetics were discovered by Gregor Mendel in 1866, but remained generally unknown until 1900. During the first half of the twentieth century it was gradually established that genes play major roles in the function and evolution of higher organisms. The fundamental significance of these roles, however, became apparent only with the recognition that nucleic acids are the hereditary materials of all organisms. The discovery of the chemical nature of DNA laid open the principles of heredity and led to an understanding of how the genes—in the form of DNA molecules—are transmitted from generation to generation and expressed within each generation. The hereditary information is contained within the nucleotide sequence of the DNA; it is expressed through that sequence as it specifies the amino acid sequences of proteins. The unity of all living things is beautifully demonstrated by the fact that the code relating nucleotide sequences to amino acid sequences is the same in all organisms: in bacteria, in plants, in animals, in human beings. The code is universal.

Within the last ten years geneticists have discovered tools that allow them to recreate, in the laboratory, steps in the evolution of organisms.

Indeed, these tools provide the means to perform experiments that nature alone is incapable of performing. With the techniques of recombinant DNA research, geneticists have learned how to transplant genes from one organism to another, thus reshuffling the genetic materials in ways never experienced in the evolution of life on earth. Such knowledge and our ability to apply it to new purposes have profound implications for all of biology. To "life as we know it" can now be added, to a small but significant degree, "life as we make it."

The purpose of this book is to present the science of genetics in such a way that the student can appreciate its place in biology as well as the means by which we have arrived at our current state of knowledge. The genetic material, DNA, has three major features: organization, expression, and evolution. This book is organized in three parts that correspond to these three features. Part I presents the nature and organization of the hereditary materials, as well as the laws by which the information contained in these materials is transmitted from generation to generation. Part II explains how the genetic information inherited by an organism directs the organism's development and activities. Part III discusses the origin of genetic variation and the genetic basis of biological evolution.

This introductory chapter reviews some knowledge that students are assumed to have acquired in introductory biology courses. First, the various kinds of organisms are briefly considered. Then mitosis and meiosis—the two processes by which eukaryotic cells divide—are reviewed.

Viruses

The smallest things that can be considered to be living are the viruses, which are familiar as the agents responsible for diseases such as the common cold, poliomyelitis, and meningitis. Viruses were discovered in the late nineteenth century, when it was shown that some diseases (such as the mosaic disease of tobacco plants) could be transmitted by self-reproducing bodies so small that they passed through the pores of filters that retain bacteria. Viruses are obligatory parasites (i.e., they cannot exist alone) of animals, plants, and microorganisms, in which they subvert the machinery of the host cell into synthesizing new virus materials. In spite of this and other peculiarities (e.g., they can be crystallized and are unable to carry out their own metabolism), viruses are generally considered to be organisms because they are capable of self-reproduction.

Viruses vary in composition, shape, and size (see Figure 1.1). In 1935 Wendell M. Stanley (1904–1971) discovered that viruses are made up of nucleic acid and protein, the principal constituents of the chromosomes of higher organisms. Some (mainly plant viruses) contain ribonucleic acid (RNA); others (including many bacterial and animal viruses) contain deoxyribonucleic acid (DNA). They may be spherical, rod-shaped, or consist of a "head" and a "tail." The virus of the hoof-and-mouth disease

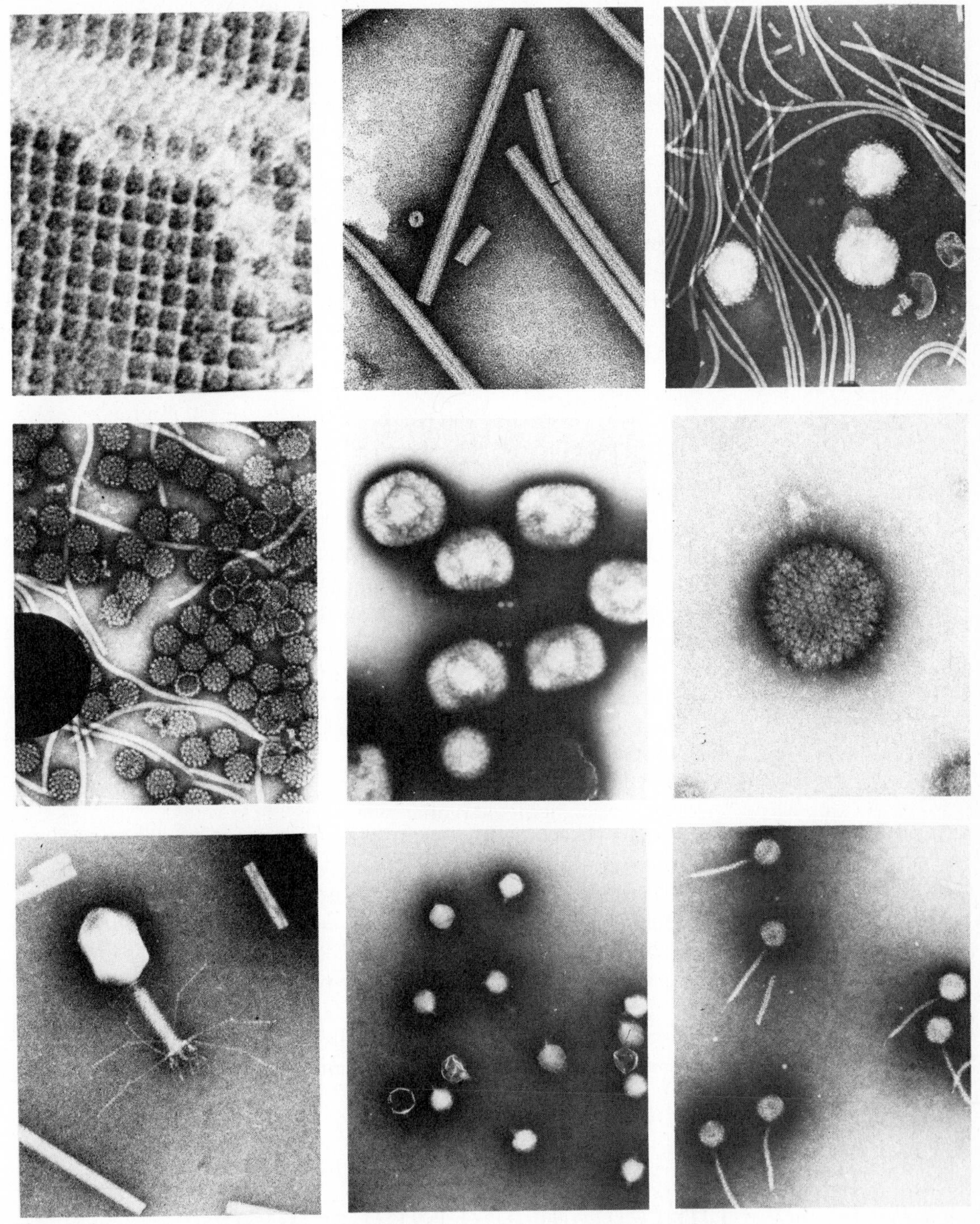

Figure 1.1
Electron micrographs of various viruses. Top row, left to right: RNA viruses of polio (×115,000), tobacco mosaic (×145,000), and Rous sarcoma (×55,000). Middle row: DNA viruses of rabbit papilloma (×65,000), vaccinia (×40,000), and herpes simplex (×140,000). Bottom row: DNA bacteriophages T4 (×110,000), T7 (×65,000), and lambda (×65,000). (Courtesy of Prof. Robley C. Williams, University of California, Berkeley, and Prof. Harold W. Fisher, University of Rhode Island.)

is a sphere with a diameter of about 10 nm (1 nm = 10^{-6} mm); the parrot fever virus is also spherical, with a diameter of 450 nm, or about 90,000 times larger in volume than the hoof-and-mouth virus. The tobacco mosaic virus is a rod some 15 nm in diameter and 300 nm long.

The viruses most widely used in genetic studies are bacterial viruses, also known as *bacteriophages* ("bacteria eaters"—a misnomer) or simply *phages* (Figure 1.2). Phages attach themselves to the surface of bacteria and inject their DNA into them, which directs the bacteria to synthesize bacteriophage components. These components are assembled into numerous new bacteriophages, which are released as each bacterium bursts open (Figure 1.3).

Prokaryotes: Bacteria and Blue-Green Algae

The simplest cellular organisms are the *prokaryotes* ("pre-nuclear"), which include the bacteria and the blue-green algae. The smallest bacteria are about 0.1 μm (100 nm) in diameter, smaller than the largest viruses, but some rod-shaped bacteria are about 6 μm wide and 60 μm long. Bacteria may be spherical, rod-shaped, or spiral (Figure 1.4). They have a rigid cell wall outside the cell membrane. Their hereditary materials are contained in a single chromosome, but bacteria lack a nuclear membrane separating the chromosome from the rest of the cell (hence the name prokaryote). Bacteria also lack mitochondria and some other organelles found in the cytoplasm of higher (eukaryotic) cells.

Some bacteria, such as *Escherichia coli,* abundant in the intestines of humans and other mammals, and *Pneumococcus pneumoniae,* responsible for bacterial pneumonia, are used extensively in genetic studies.

Blue-green algae, like bacteria, have a cell wall but lack a nuclear membrane as well as some cytoplasmic organelles. In contrast to bacteria, blue-green algae usually exist in clumps or filaments consisting of many cells.

Bacteria and blue-green algae usually reproduce by simple division, or *fission,* after duplication of their chromosome.

Unicellular and Multicellular Eukaryotes

The *eukaryotes* ("having a true nucleus") include all cellular organisms except the bacteria and blue-green algae. Eukaryotic cells have a nuclear

Figure 1.2
Morphology of several bacteriophages, showing various degrees of complexity. Two phages much used in genetic studies are λ and T4.

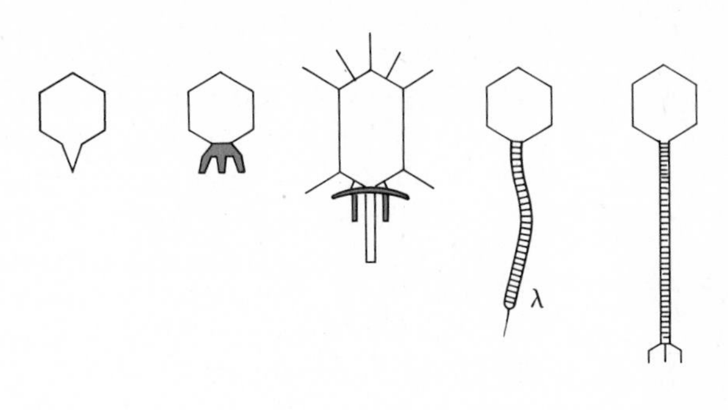

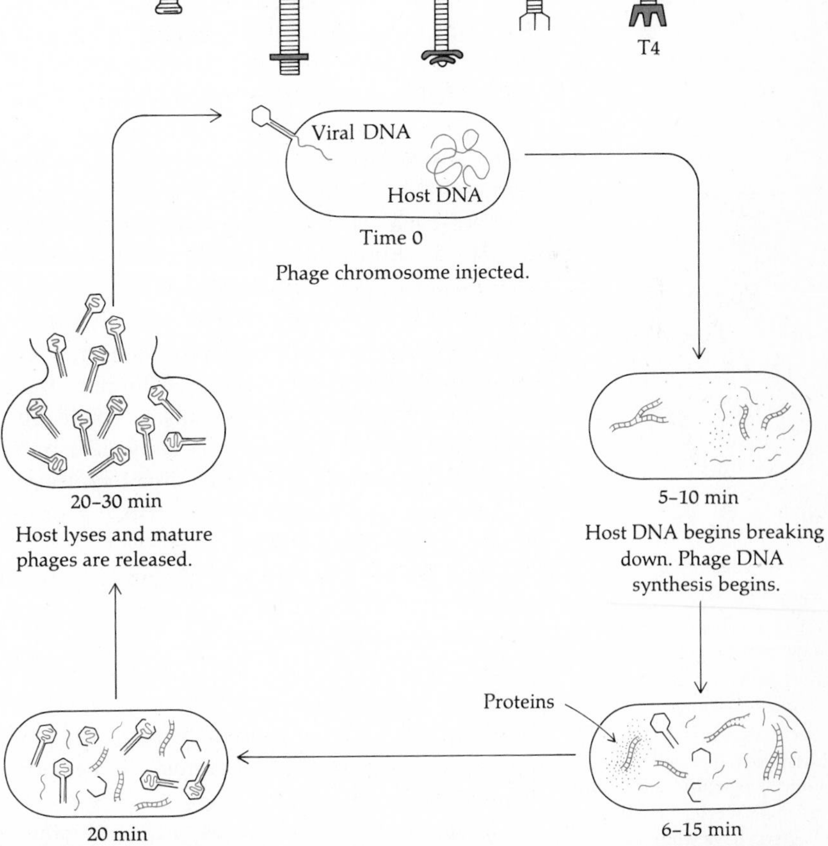

Figure 1.3
The life cycle of a bacteriophage. The virus attaches itself to the cell wall of a bacterium and injects its DNA, which then takes over the metabolic machinery of the bacterium and directs the synthesis of the viral DNA and proteins. New phages are reassembled and leave the bacterium as the cell lyses (bursts open).

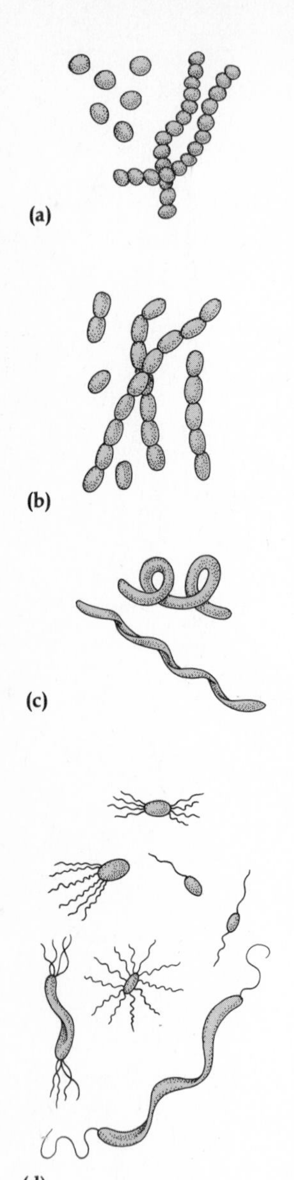

Figure 1.4
Various types of bacteria. (**a**) Cocci (singular: coccus) are spherical cells that grow singly or as long chains. (**b**) Bacilli (singular: bacillus) are rod-shaped cells. (**c**) Spirilla (singular: spirillum) are helical cells. (**d**) Many bacteria have whiplike appendages called flagella (singular: flagellum).

membrane delimiting a *nucleus,* which contains two or more chromosomes. The DNA of eukaryotic chromosomes is complexed with certain kinds of proteins called *histones* (see Chapter 4). In addition, eukaryotic cells contain certain organelles and structures that are lacking in prokaryotes, such as mitochondria, chloroplasts (in plant cells only), Golgi apparatus, endoplasmic reticulum, and vacuoles (Figure 1.5).

Eukaryotes may be either *unicellular* or *multicellular.* Unicellular eukaryotes commonly used in genetic experiments include the alga *Chlamydomonas reinhardi,* the ciliate *Paramecium aurelia,* and the yeast *Saccharomyces cerevisiae.* Some multicellular eukaryotes used in genetic research are the fungi *Neurospora crassa* and *Aspergillus nidulans,* corn (*Zea mays*), the fruit fly (*Drosophila melanogaster*), the house mouse (*Mus musculus*), and humans (*Homo sapiens*).

Reproduction in eukaryotes may be *asexual* (also called *vegetative* reproduction) or *sexual.* In asexual reproduction a single parent divides into two or more parts, each of which grows into a new individual. Asexual reproduction is common in plants—a small portion of the plant, when isolated under favorable conditions, may establish itself as a new individual. Potatoes, for example, are more easily cultivated from pieces of tubers than from seeds, and most fruit trees are propagated by cuttings. Asexual reproduction also occurs in fungi and in some simple animals, such as flatworms.

Sexual reproduction entails the union of two sex cells, or *gametes;* these form a single cell, or *zygote,* from which the new individual develops. Generally the two gametes come from two different parents, the exception being self-fertilization, in which a single parent provides both gametes.

The cycle of growth and sexual reproduction in multicellular eukaryotes is shown in Figure 1.6. The number of chromosomes characteristic of an organism is maintained constant from generation to generation because there are two kinds of cell division, one for the formation of *somatic* ("body") cells, the other for the formation of gametes. Somatic cells divide by a process called *mitosis* (plural: mitoses). The chromosomes are exactly duplicated before the onset of cell division. During mitosis the duplicated chromosomes are equally distributed into the two daughter cells. All the somatic cells of an organism therefore have the same number of chromosomes. Mitosis is also the process by which unicellular eukaryotes divide.

Gametes are formed by the process of *meiosis* (plural: meioses), during which each cell divides twice while the chromosomes are duplicated only once. The resulting gametes therefore have only half as many chromosomes as the somatic cells. Two gametes (one male sex cell and one female sex cell) join in the process called *fertilization.* The resulting zygote thus has the number of chromosomes characteristic of the somatic cells of the organism. The processes of mitosis and meiosis are described in more detail in the following sections.

If the number of chromosomes in a gamete is represented as N, a zygote will have $2N$ chromosomes, half obtained from each of the two gametes. The zygote divides mitotically, producing two cells, each with $2N$

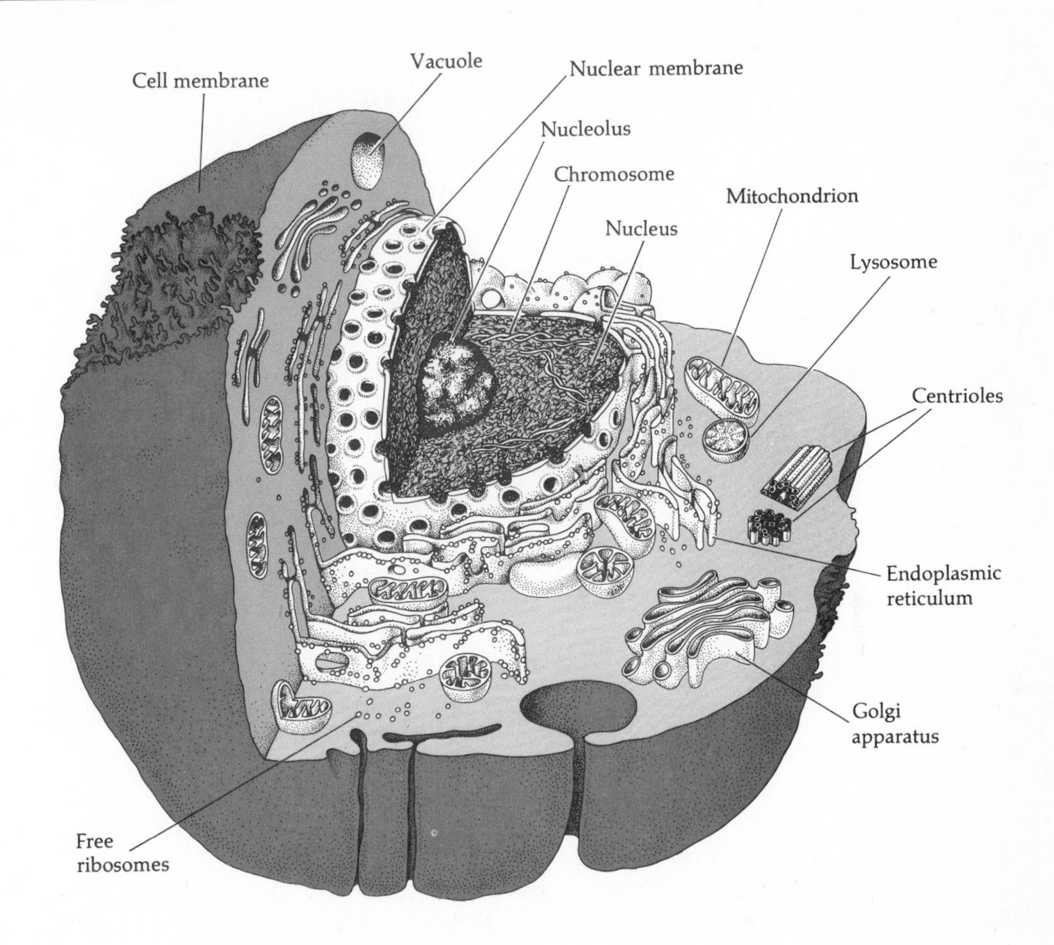

Figure 1.5
Diagram of a eukaryotic (here, mammalian) cell. A conspicuous nuclear organelle is the nucleolus.

chromosomes. These cells divide again and again during development, so that a multicellular organism consists of many cells, each with $2N$ chromosomes. The organism will also produce gametes, but these will come about by meiosis and will thus carry only N chromosomes. When two gametes unite at fertilization, the number $2N$ of chromosomes characteristic of the organism is restored and is thus maintained generation after generation. The number of chromosomes varies considerably among eukaryotes. Some species have only two; others have several hundred (see Table 1.1). Cells with two sets of chromosomes, such as somatic cells, are said to be *diploid*; cells with only one set of chromosomes, such as gametes, are called *haploid*.

In diploid organisms the two members of any pair of chromosomes are called *homologous* chromosomes; chromosomes that are not members of the same pair are called *nonhomologous*. In organisms with separate sexes, such as most animals, there is usually one pair of chromosomes involved in sex determination; these are called *sex chromosomes*. The other chromo-

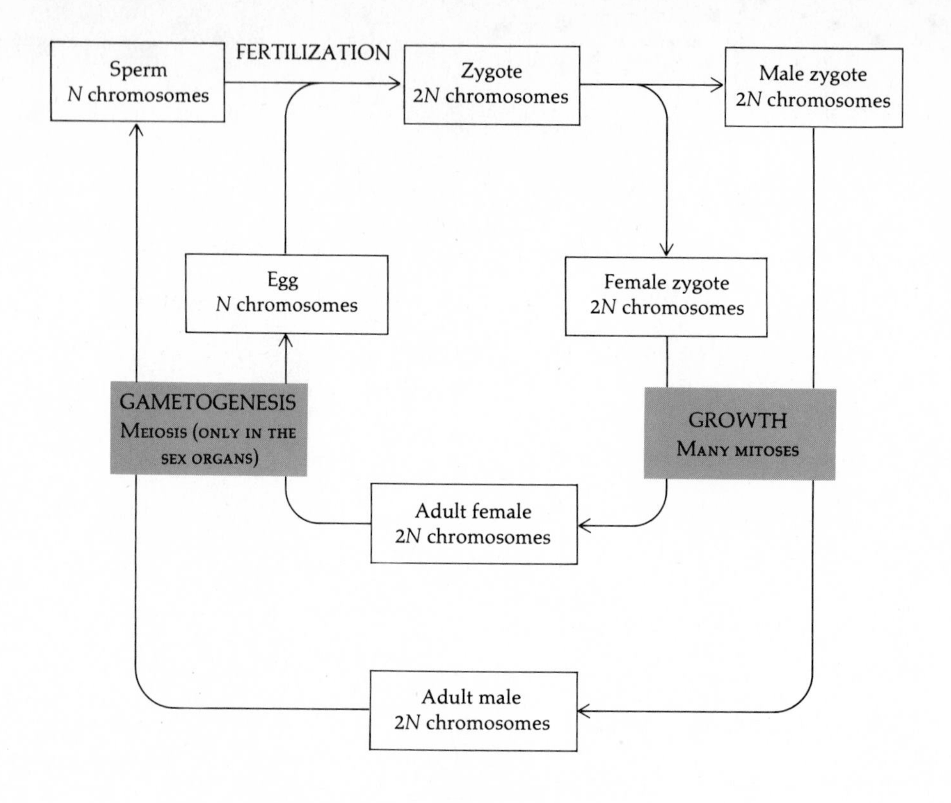

Figure 1.6
The cycle of growth and reproduction in sexual organisms. The zygote is formed by the fertilization of a female gamete by a male gamete. The zygote divides mitotically, giving rise to the multiple cells that make up the organism. The cells from which the gametes are formed are called *germ-line cells*. These multiply by mitosis but eventually undergo meiosis, during which the number of chromosomes is halved. Sexual reproduction involves the alternation of diploid and haploid (gametic) phases.

somes are called *autosomes*. The two sex chromosomes, unlike all other chromosome pairs, need not exactly match in size and shape. One of the sexes (the males in mammals and many insects, but the females in butterflies and birds) is called *heterogametic* because it has two sex chromosomes (usually designated X and Y) that are quite different from each other, while the other sex, which is *homogametic*, has two similar chromosomes (both called X). Thus in humans, mice, and *Drosophila* flies, males are XY with respect to the sex chromosomes, while females are XX (Figure 1.7). In some species the Y chromosome is altogether lacking, so the heterogametic sex is X0, while the homogametic sex is XX.

Mitosis

Mitosis is the process of nuclear division by which one cell results in two daughter cells, each with a set of chromosomes identical to that of the parental cell. The chromosomes are duplicated during a particular period prior to mitosis (although the two duplicates do not separate until later). This period is called S, for synthesis, because the DNA in the chromo-

Table 1.1
Number of chromosomes (2*N*) in various animals and plants.

Organism	Chromosomes	Organism	Chromosomes
Man, *Homo sapiens*	46	White oak, *Quercus alba*	24
Chimpanzee, *Pan troglodytes*	48	Yellow pine, *Pinus ponderosa*	24
Rhesus monkey, *Macaca mulatta*	48	Cherry, *Prunus cerasus*	32
Horse, *Equus caballus*	64	Cabbage, *Brassica oleracea*	18
Donkey, *Equus asinus*	62	Radish, *Raphanus sativus*	18
Dog, *Canis familiaris*	78	Garden pea, *Pisum sativum*	14
Cat, *Felis domesticus*	38	Sweet pea, *Lathyrus odoratus*	14
House mouse, *Mus musculus*	40	Bean, *Phaseolus vulgaris*	22
Rat, *Rattus norvegicus*	42	Cucumber, *Cucumis sativus*	14
Opossum, *Didelphys virginiana*	22	Upland cotton, *Gossypium hirsutum*	52
Chicken, *Gallus domesticus*	78	Potato, *Solanum tuberosum*	48
Turkey, *Meleagris gallopavo*	82	Tomato, *Solanum lycopersicum*	24
Frog, *Rana pipiens*	26	Tobacco, *Nicotiana tabacum*	48
Platyfish, *Platypoecilus maculatus*	48	Bread wheat, *Triticum aestivum*	42
Starfish, *Asterias forbesi*	36	Emmer wheat, *Triticum turgidum*	28
Silkworm, *Bombyx mori*	56	Barley, *Hordeum vulgare*	14
Housefly, *Musca domestica*	12	Rye, *Secale cereale*	14
Fruit fly, *Drosophila melanogaster*	8	Rice, *Oryza sativa*	24
Mosquito, *Culex pipiens*	6	Snapdragon, *Antirrhinum majus*	16
Cockroach, *Blatta germanica*	23 ♂, 24 ♀	Yeast, *Saccharomyces cerevisiae*	ca. 18
Hermit crab, *Eupagurus ochotensis*	ca. 254	Green alga, *Acetabularia mediterranea*	ca. 20

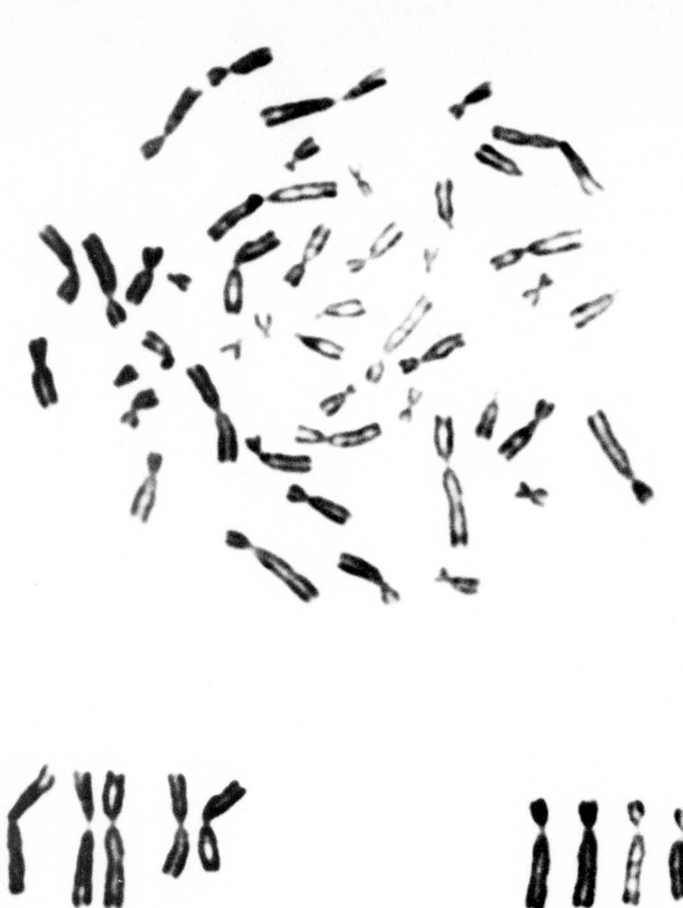

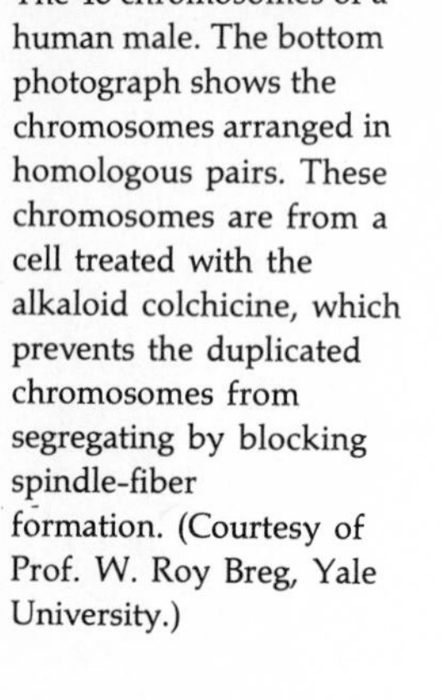

Figure 1.7
The 46 chromosomes of a human male. The bottom photograph shows the chromosomes arranged in homologous pairs. These chromosomes are from a cell treated with the alkaloid colchicine, which prevents the duplicated chromosomes from segregating by blocking spindle-fiber formation. (Courtesy of Prof. W. Roy Breg, Yale University.)

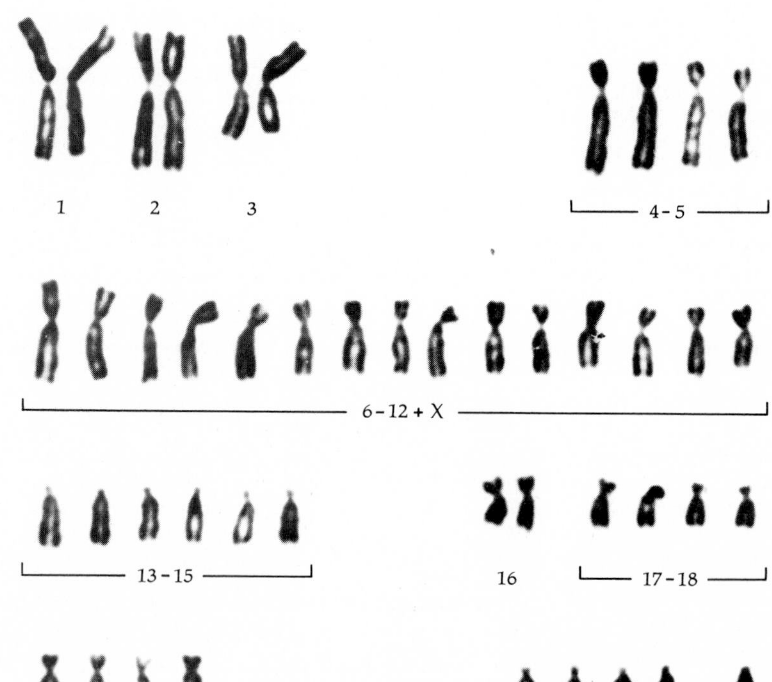

Box 1.1 Chromosomes

Chromosomes are long, threadlike bodies that, during cell division, contract into shorter, thicker bodies in which one can distinguish a *centromere* and one or two *chromosome arms.* Depending on the position of the centromere, chromosomes are classified as follows (see Figure 1.8):

1. *Metacentric,* when the two arms are nearly identical in length (i.e., the centromere is in the middle of the chromosome).
2. *Acrocentric,* when the two arms are unequal in length (i.e., the centromere is closer to one end of the chromosome than to the other).
3. *Telocentric,* when there is only one clearly distinguishable arm (i.e., the centromere is at, or very near, one end of the chromosome).

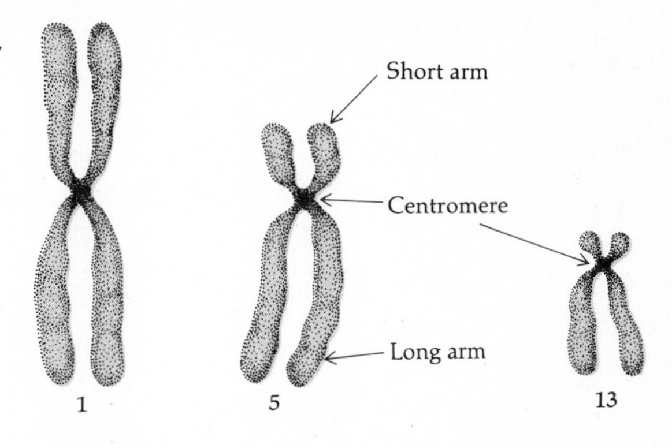

Figure 1.8
Human chromosomes 1, 5, and 13, which exemplify metacentric, acrocentric, and telocentric chromosomes, respectively.

Nonhomologous chromosomes can often be distinguished by their size and the position of the centromere.

Some chromosomal segments or even whole chromosomes are called *heterochromatic* ("differently stained") because they possess a dense, compact structure during the interphase and early prophase stages of cell division. The other chromosomal segments or whole chromosomes are called *euchromatic* ("normally stained"). Heterochromatic regions facilitate the identification of chromosomes.

somes is synthesized during this period. The S period is preceded by a G_1 (for gap) period and followed by a G_2 period. During the G_1 and G_2 periods, cells are engaged in metabolism and growth but not in chromosome replication. If we designate mitosis as M, the *cell cycle* is a sequence that can be represented as $G_1 \rightarrow S \rightarrow G_2 \rightarrow M$ (Figure 1.9). This cycle repeats itself again and again when cells are proliferating.

Although mitosis is a smooth process with no clear-cut discontinuities, certain landmark events serve to identify four stages: *prophase, metaphase, anaphase,* and *telophase* (Figures 1.10 and 1.11). A cell not undergoing mitosis is said to be in *interphase;* successive mitoses are always separated by periods of interphase during which DNA synthesis takes place.

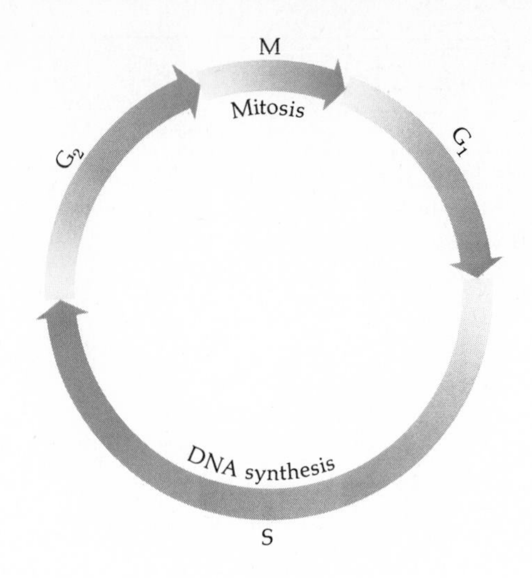

Figure 1.9
The cell cycle. DNA synthesis (S) is separated from the preceding and ensuing mitoses (M) by two gap periods, G_1 and G_2, respectively.

PROPHASE This is characterized by the gradual condensation and coiling of the chromosomes, which become visible under the microscope as threadlike structures. Each chromosome appears to consist of two duplicates lying next to each other and joined at the centromere. The two duplicates are called sister *chromatids* as long as they remain connected. Other events characteristic of prophase are the gradual disappearance of the nucleolus, whose components disperse throughout the nucleus. In most organisms the nuclear membrane starts to break down during prophase.

METAPHASE The nuclear membrane disappears, leaving the chromosomes free in the cytoplasm. The chromosomes appear to be attached by their centromeres to the *spindle* fibers. The chromosomes move about, eventually becoming arranged on a plane midway between the two poles of the spindle. This *metaphase plate* is the most distinctive characteristic of metaphase.

ANAPHASE This is usually the shortest stage of mitosis. Each centromere separates into two—by which the two chromatids become chromosomes—and the two centromeres start moving toward opposite poles of the spindle, each carrying along one of the two chromosomes.

TELOPHASE The two sets of chromosomes gather at opposite poles of the spindle, where they start to uncoil and revert to the extended form of interphase chromosomes. A nuclear membrane forms around each set of chromosomes, and the nucleoli are reconstituted. Cell division (*cytokinesis*) is also completed during this stage.

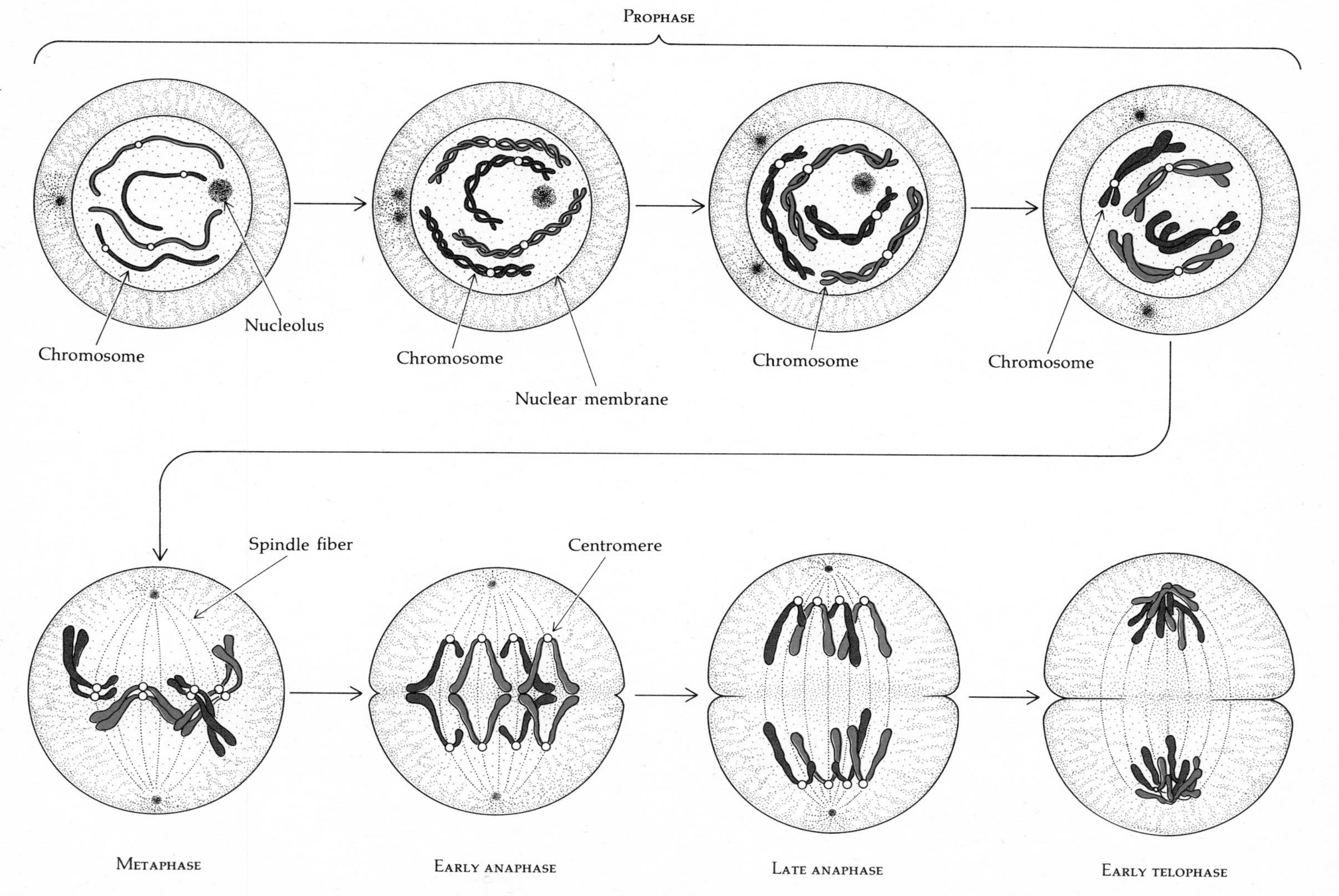

Figure 1.10
The four stages of mitosis.

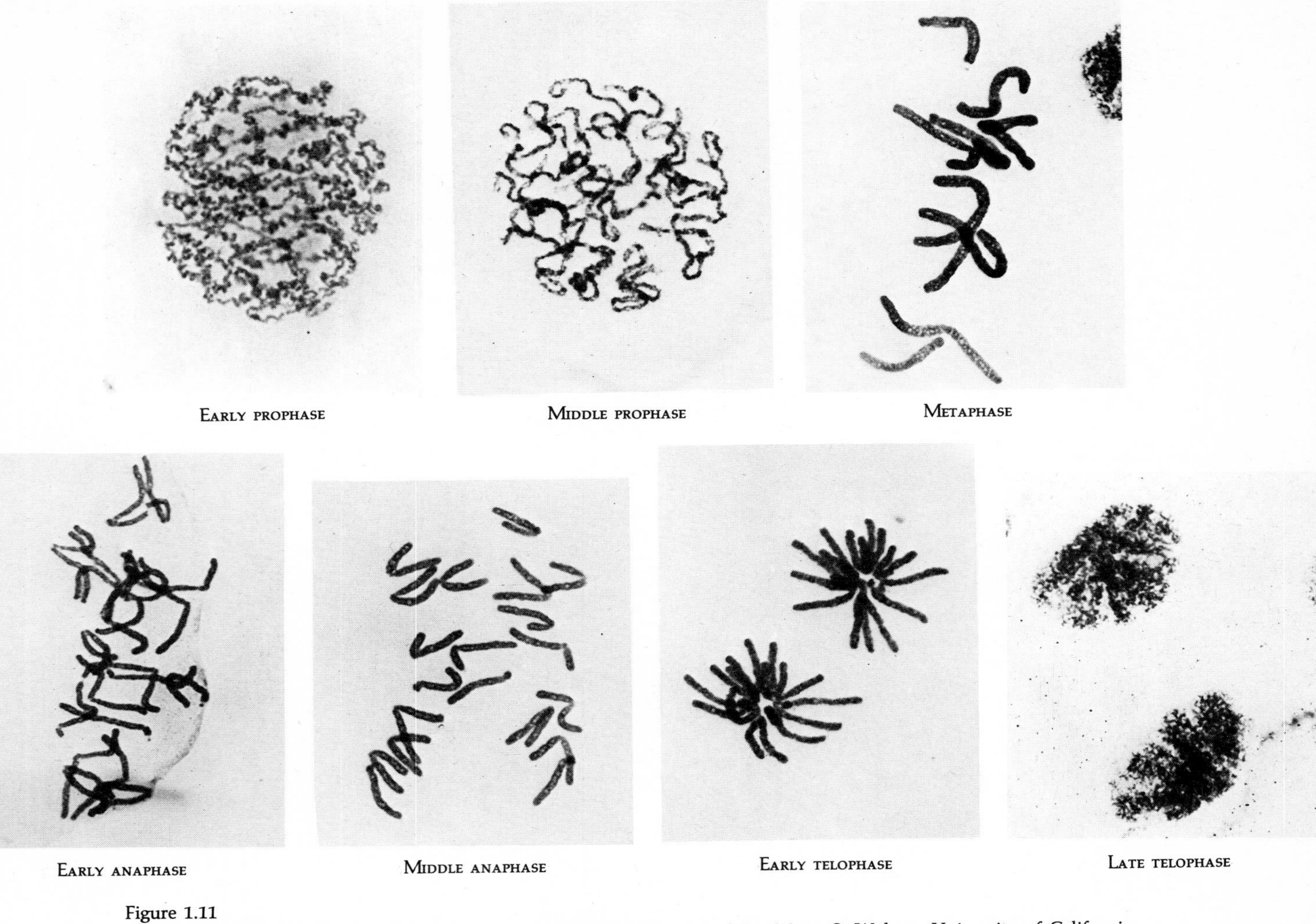

Figure 1.11
Mitosis in the peony, *Paeonia californica*, $2N = 10$ ($\times 850$). (Courtesy of Dr. Marta S. Walters, University of California, Santa Barbara, and the late Prof. Spencer W. Brown, University of California, Berkeley.)

Meiosis

Meiosis is the sequence of two nuclear divisions that leads to the formation of gametes. During meiosis each cell divides twice while the chromosomes are duplicated only once, so the resulting gametes have only half as many chromosomes as the original cell. The two cell divisions are called *meiosis I* and *meiosis II.* The four stages—prophase, metaphase, anaphase, and telophase—can usually be distinguished within each of these two meiotic divisions (Figures 1.12 and 1.13). The interphase prior to meiosis is quite similar to mitotic interphase; chromosome duplication occurs during the S period.

Meiosis I

PROPHASE I This is a particularly complex stage, which is usually divided into five substages: leptotene, zygotene, pachytene, diplotene, and diakinesis.

The *leptotene* stage is characterized by the initial condensation and coiling of the chromosomes, which appear as threadlike structures similar to those in early mitotic prophase.

During the *zygotene* stage, homologous chromosomes pair side by side in a zipperlike fashion. The lateral association of the homologous chromosomes, called *synapsis* (plural: synapses), is an important genetic event because it makes possible the exchange of homologous chromosome parts known as *crossing over.* The two synapsed chromosomes (consisting of four chromatids in total) are known as a *bivalent.*

The *pachytene* stage is characterized by a shortening and thickening of the bivalents.

During the *diplotene* stage, the two homologous chromosomes separate from each other at most places, particularly near the centromere. The sister chromatids, however, remain held together by their common centromere. Moreover, the homologous chromosomes maintain one or more zones of contact known as *chiasmata* (singular: chiasma). Each chromatid may form chiasmata with one or the other chromatid of the homologous chromosome, so that two, three, or all four chromatids in a bivalent may be involved in chiasmata, although only two chromatids are involved in each chiasma (Figure 1.14). The number of chiasmata per bivalent is variable, but two or three are typical. For example, in human (female) meiosis about two to three chiasmata per bivalent are observed, on the average, although larger chromosomes usually exhibit more chiasmata than smaller ones. Chiasmata reflect the occurrence of crossing over (exchange of parts) between chromatids.

Diakinesis is characterized by maximum condensation and coiling of the chromosomes, which appear as thick bodies. In most organisms

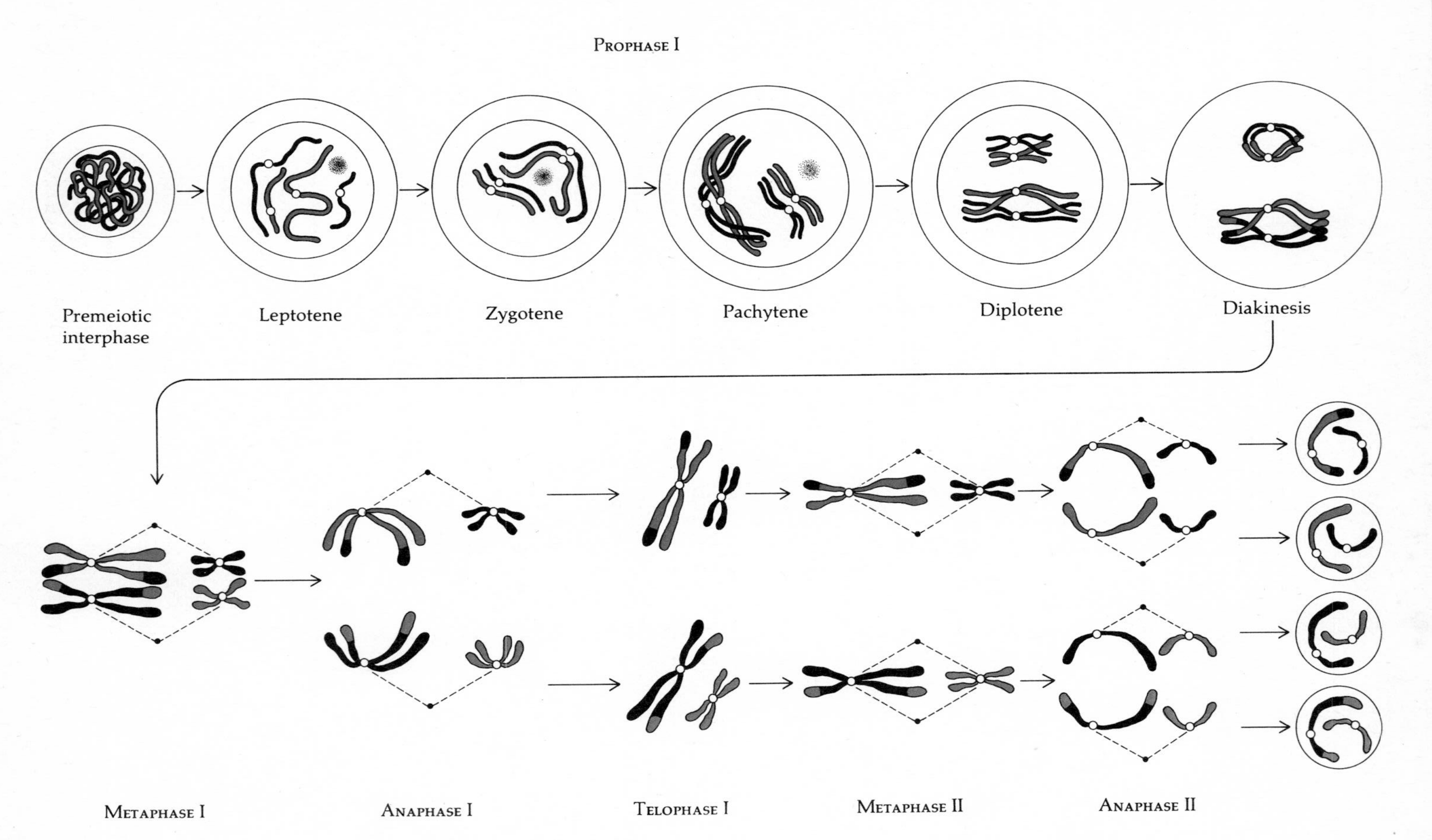

Figure 1.12
The stages of meiosis.

chiasmata appear to move away from the centromere toward the ends of the bivalents. By the end of diakinesis, the bivalents are therefore characterized by the presence of one or two terminal points of contact (Figure 1.15). As diakinesis is completed, the nuclear membrane and the nucleolus dissolve.

METAPHASE I The bivalents become attached to spindle fibers by their centromeres and move toward the metaphase plate, with the two centromeres of each pair of homologues on opposite sides of the plate. *The pairing of homologous chromosomes makes metaphase I of meiosis distinct from mitotic metaphase, where no such pairing exists.*

ANAPHASE I The centromeres of each pair of homologous chromosomes move toward opposite poles of the spindle, carrying along both chromatids of each chromosome. The chiasmata slip off the ends of the homologous chromosomes as these are pulled farther and farther apart. *An important distinction with mitotic anaphase is that in meiotic anaphase I the centromeres do not divide.*

TELOPHASE I As the anaphase migration of the chromosomes toward the spindle poles is completed, a nuclear membrane forms, in some organisms, around each set of homologous chromosomes, and the cell divides into two daughter cells.

The interphase between meiosis I and meiosis II is usually either brief or lacking altogether. It is importantly different from the interphase preceding meiosis I or mitosis in that no synthesis of new DNA takes place between meiosis I and meiosis II.

Meiosis II

The chromosomes are already duplicated when meiosis II begins, the two sister chromatids being attached at their common centromere. However, each cell contains only one set of chromosomes (N), rather than two ($2N$), as in mitosis or meiosis I. *Prophase II* is often very brief. In *metaphase II* the chromosomes are attached to spindle fibers by their centromeres and align on a metaphase plate. At the beginning of *anaphase II*, each centromere divides (for the first and only time during meiosis), and the sister chromatids thereby become chromosomes, which now move to opposite poles. *Telophase II* is completed when the nuclear membrane forms around each of the haploid nuclei.

Meiosis I starts with $2N$ *duplicated* chromosomes per cell and ends with two cells (or chromosome complexes, since cell division is not always complete), each having N *duplicated* chromosomes. Meiosis II ends with four cells, each having N *single* chromosomes. In the male reproductive organs of animals, these four cells are *spermatozoa* (singular: spermato-

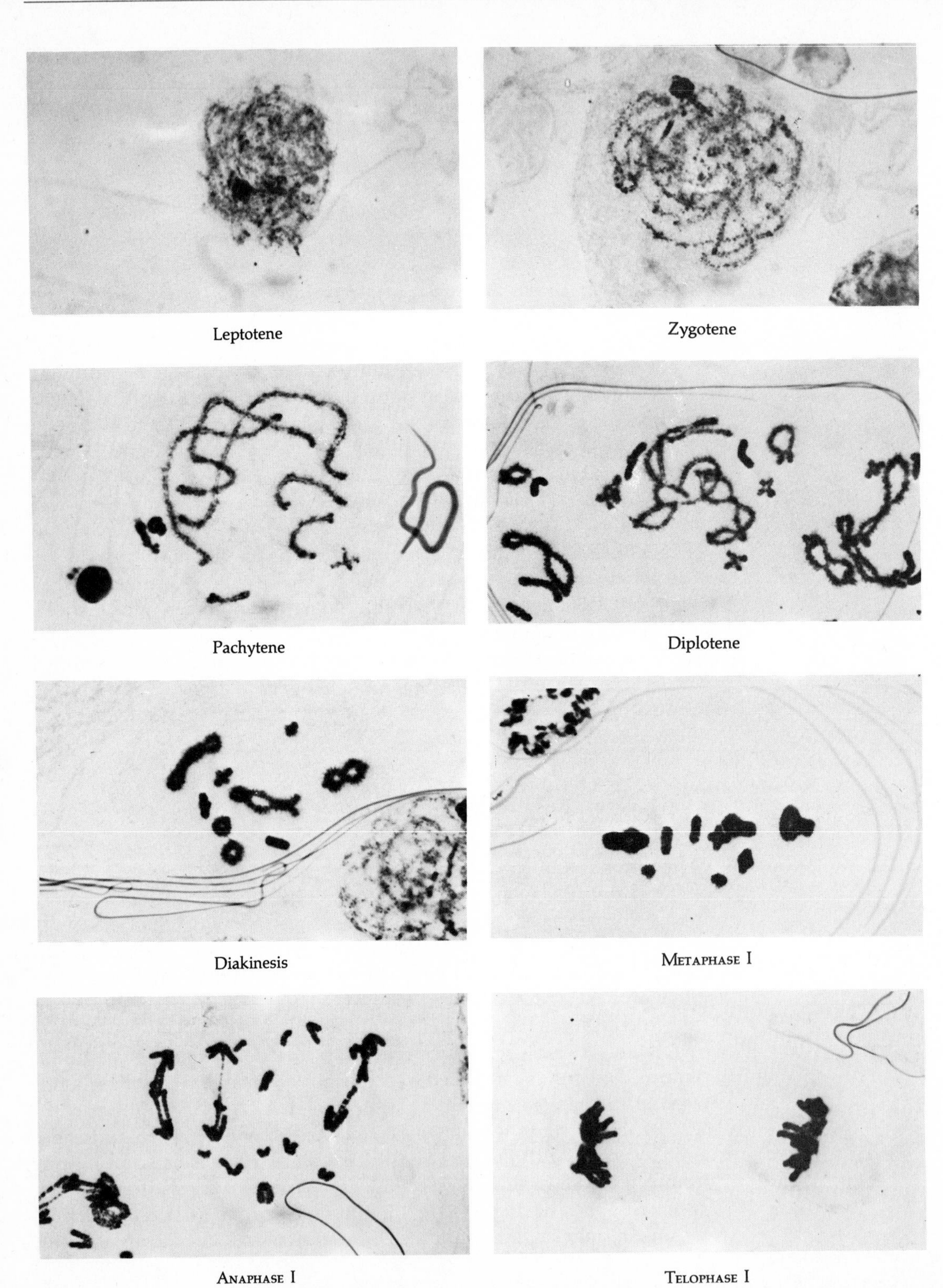

Leptotene

Zygotene

Pachytene

Diplotene

Diakinesis

Metaphase I

Anaphase I

Telophase I

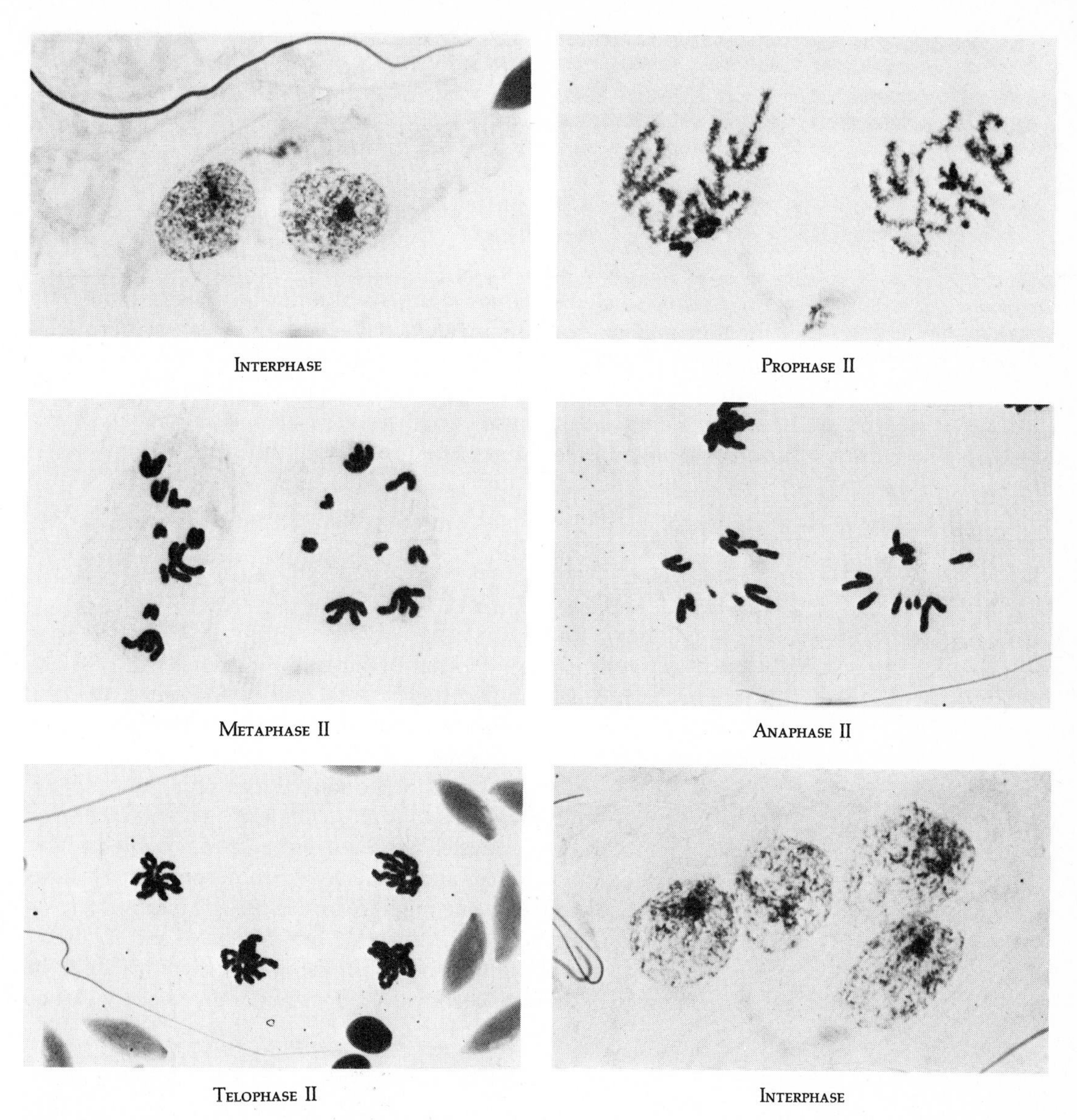

Figure 1.13
Meiosis in a male grasshopper, *Chorthippus parallelus*, $2N = 17$ (18 in the female) ($\times 1500$). Of the four resulting nuclei (see last photo in the series), two have nine chromosomes; the other two have eight because they lack the X chromosome. (Courtesy of Prof. James L. Walters, University of California, Santa Barbara.)

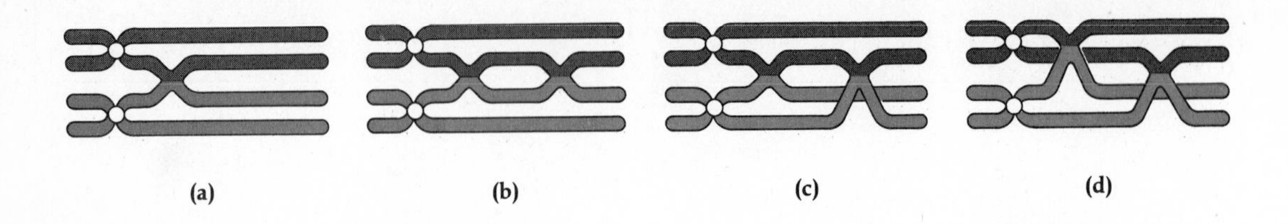

Figure 1.14
Four types of chiasmata. (**a**) A single chiasma. (**b**) Two chiasmata involving two chromatids. (**c**) Two chiasmata involving three chromatids. (**d**) Two chiasmata involving all four chromatids.

zoon). In the female reproductive organs of animals, however, only one of the four meiotic products is functional as an egg cell, or *ovum* (plural: ova), which contains most of the cytoplasmic material; the other three cells are small *polar bodies,* which do not function as gametes (Figure 1.16). In higher plants the products of meiosis are called *microspores* and *megaspores* for male and female sex cells, respectively (Figure 1.17).

Significance of Meiosis

Mitosis is an *equational* cell division because the resulting two cells are identical to each other (and to the parental cell) with respect to their chromosome content. Meiosis is different: the first meiotic division is *reductional,* the second is equational.

Meiosis I is a reductional division because the resulting nuclei have only half as many centromeres and chromosomes as the parental cells. For each pair of chromosomes, one cell has the paternal chromosome and the other, the maternal. Paternal and maternal chromosomes may have different genetic content; e.g., one might carry the genetic information for "brown eye" and "B blood group," the other for "blue eye" and "O blood group." The products of the first meiotic division may therefore differ in genetic content. This difference, however, does not apply to every part of

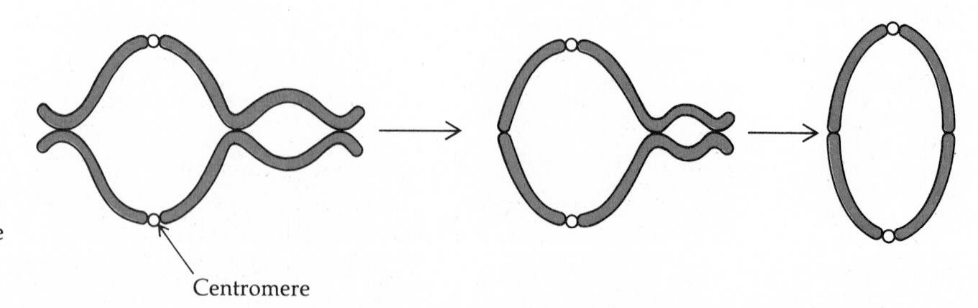

Figure 1.15
During diakinesis the chiasmata appear to move toward the ends of the bivalents.

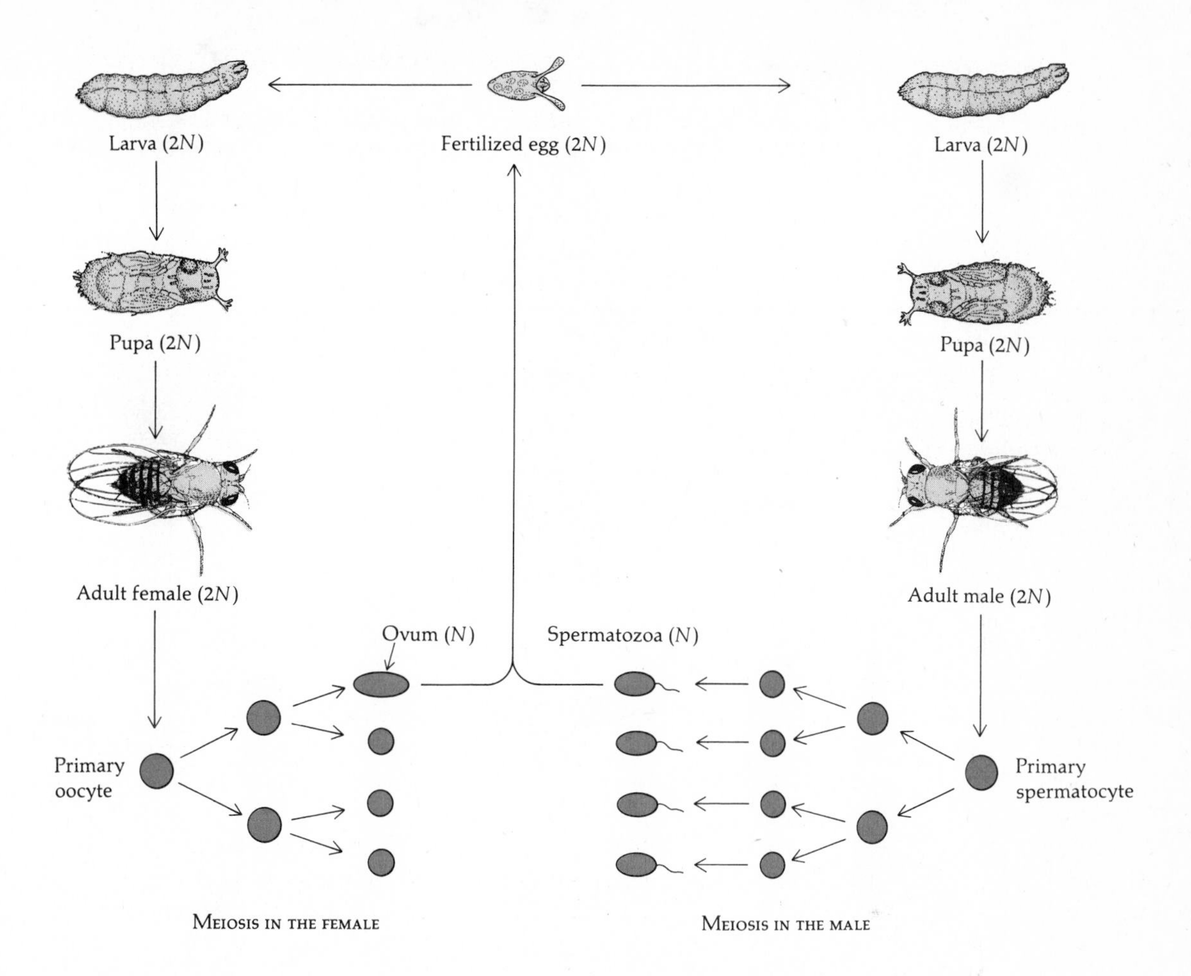

Figure 1.16
Gamete formation in *Drosophila*. Germ-line cells that multiply by mitosis eventually undergo meiosis to become oocytes and spermatocytes. Only one of the four products of female meiosis forms the nucleus of the ovum. The four products of male meiosis are called *spermatids*, which eventually differentiate into the functional spermatozoa.

the chromosomes; whenever nonsister chromatids have exchanged parts as a result of the chiasmata formed during prophase I, the two chromatids within a chromosome will be different (see Figure 1.14).

Meiosis II, on the other hand, is an equational division, entailing division of the centromeres. Sister chromatids—and therefore the nuclei resulting from meiosis II—are identical to each other (with the exception of recombined parts, as noted in the preceding paragraph). Other differences between mitosis and meiosis can be seen in Figure 1.18.

The genetic significance of meiosis can be summarized as follows:

1. Meiosis makes possible the *conservation of the number of chromosomes* from generation to generation in sexually reproducing organisms. Sexual reproduction involves fertilization—the fusion of two sex cells, or gametes. If the gametes of an organism had as many

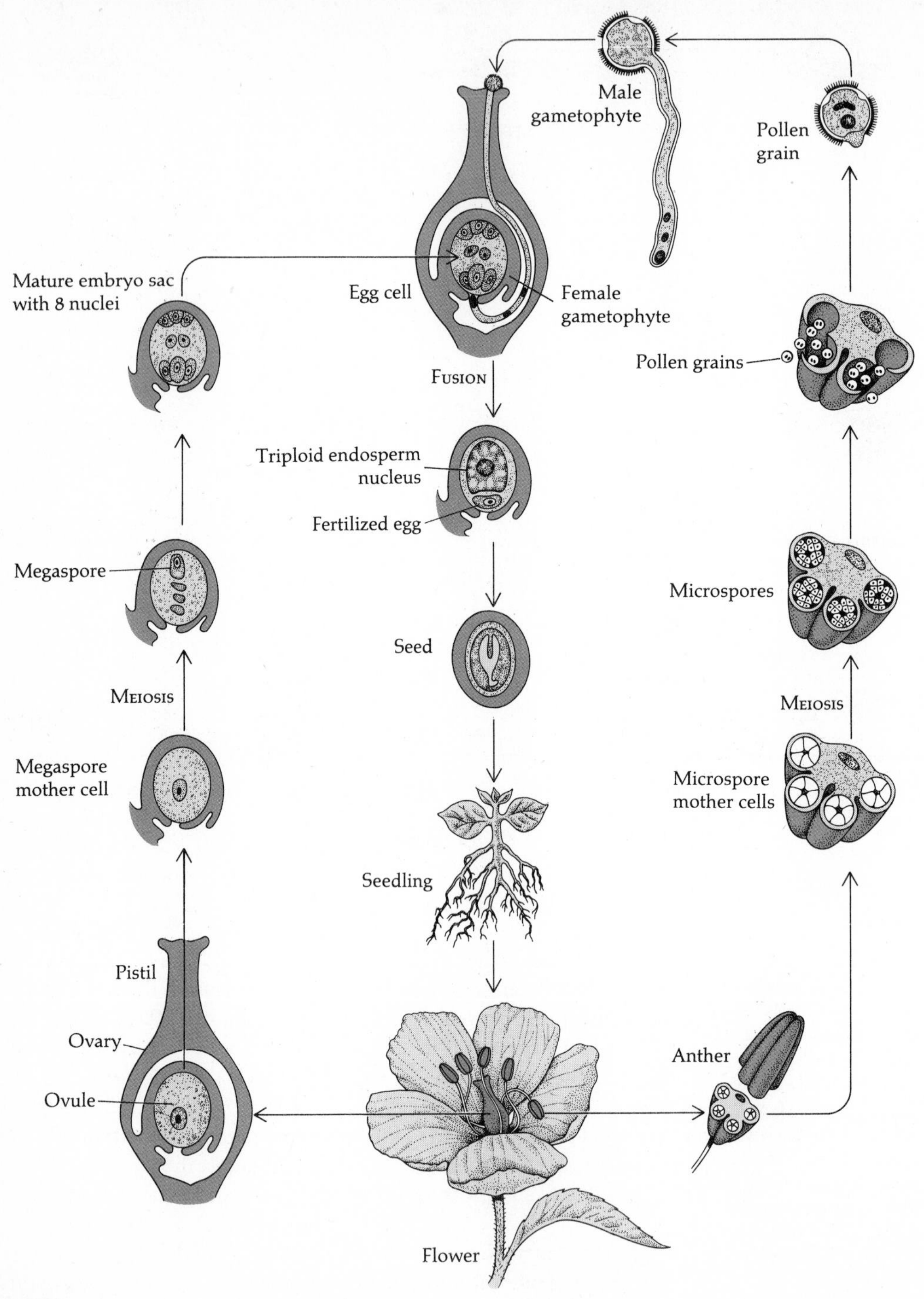

Figure 1.17
Life cycle and gamete formation in a plant. In this cycle the diploid phase, which produces the spores by meiosis, is called the *sporophyte*; the haploid phase, which involves the maturation of the gametes, is called the *gametophyte*. The haploid phase may exist as a separate plant, independently of the diploid phase. In some primitive plants, such as mosses, the gametophyte is in fact the more conspicuous component of the life cycle (what we recognize as the moss), while the sporophyte consists of a stalk living "parasitically" on the gametophyte.

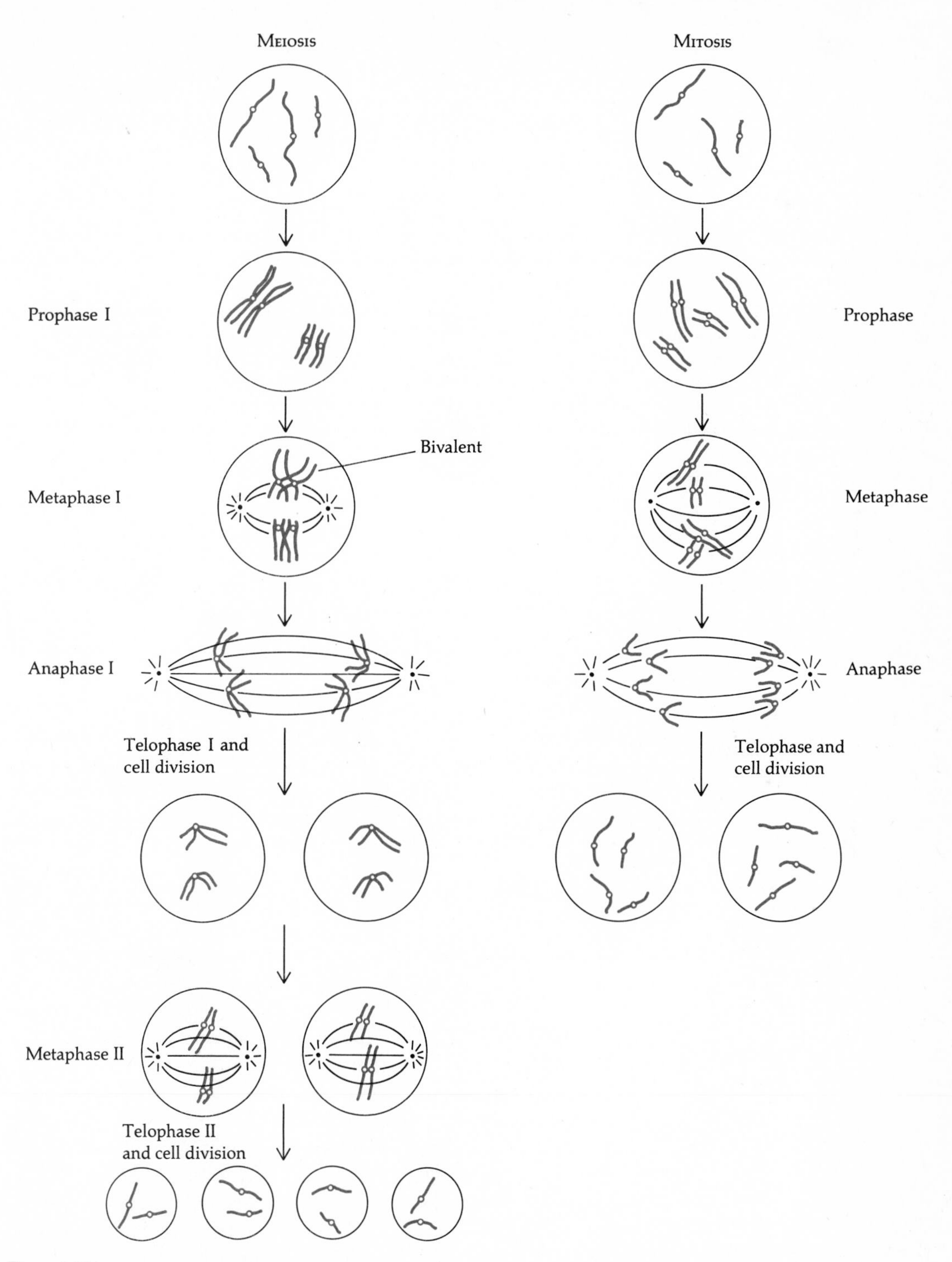

Figure 1.18
Comparison of meiosis and mitosis. The chromosomes duplicate once in each process, but meiosis entails two cell divisions, resulting in half the number of chromosomes per nucleus. Another important difference is that homologous chromosomes synapse in meiosis, but not in mitosis.

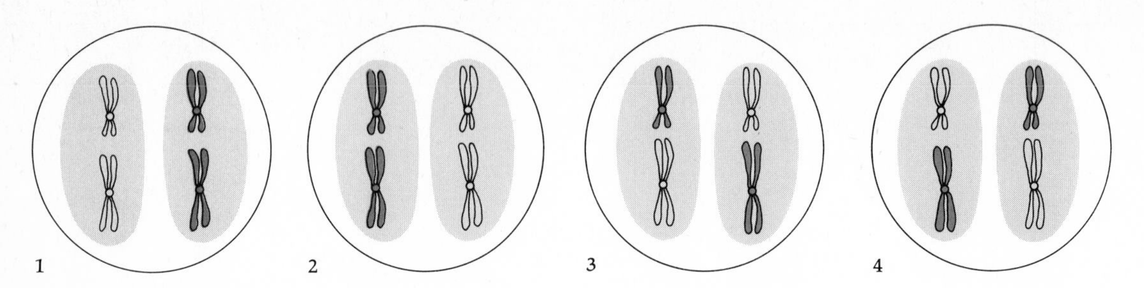

Figure 1.19
The four possible arrangements of two pairs of chromosomes on the metaphase plate. The two chromosomes going together to the same pole are shown in the same color: paternal chromosomes are shown in light color, maternal chromosomes in dark color. With one pair of chromosomes, the number of different metaphase arrangements is 2; with two pairs of chromosomes, it is $2^2 = 4$; and with n pairs of chromosomes, it is 2^n. Note that half the arrangements give the same meiotic products as the other half: arrangements 1 and 2 give the same meiotic products, and arrangements 3 and 4 give the same meiotic products. The probability that all paternal (and therefore all maternal) chromosomes will go to the same pole is $(1/2)^{n-1}$. This probability is ½ for two pairs of chromosomes ($n = 2$), but becomes smaller as the number of pairs increases.

chromosomes as the somatic cells, the number of chromosomes would double every generation.

2. In metaphase I each paternal and maternal chromosome has equal probability of falling on one or the other side of the metaphase plate. Consequently, *paternal and maternal chromosomes may be combined in each gamete.* The number of possible chromosome combinations is very large when the number of chromosomes is large, and the probability of one gamete's having chromosomes from only one parent is very small. Consider humans, for example. A normal person has 23 pairs of chromosomes. Assume that the paternal chromosome I is on one side of the metaphase plate. The probability that the paternal chromosome II is on the same side is $\frac{1}{2}$, and the same is true for the paternal chromosomes III, IV, etc. (Figure 1.19). The probability that all 22 other paternal chromosomes are on the same side as the paternal chromosome I is $(\frac{1}{2})^{22} = \frac{1}{4{,}194{,}304}$, or less than one in 4 million.
3. Crossing over between nonsister chromatids further contributes to the *recombination of paternal and maternal hereditary traits in the gametes.* Owing to nonsister chromatid exchanges, the number of different kinds of gametes is virtually infinite. Recall that in humans there are between two and three chiasmata (and therefore exchanges) per chromosome, on the average. Moreover, the exchange points vary from one meiosis to another, so that exactly the same set of exchanges is unlikely ever to occur.

Problems

1. Assume that there was no meiosis and that sexual organisms reproduced by the fertilization of two somatic cells with an unreduced number of chromosomes. If a certain organism had 8 chromosomes, how many chromosomes would its descendants have 5, 10, and 100 generations later?

2. Enumerate the similarities and differences between mitosis and meiosis.

3. In some organisms, such as honeybees, males develop from unfertilized eggs while fertilized eggs develop into females. Would you expect meiosis to occur in the formation of spermatozoa in such organisms?

4. The number of chromosomes in ordinary human cells is 46. How many chromosomes are present in human: (a) spermatids; (b) spermatozoa; (c) ova; (d) polar bodies?

5. It is not rare to find somatic cells that have a different number of chromosomes from most other somatic cells. In humans, for example, some liver cells have 92 chromosomes. How can such cells arise?

6. Where would you expect greater genetic variation, in the progeny of an asexually reproducing organism or a sexually reproducing organism? Why?

7. Assume that an organism has three pairs of chromosomes, and that each chromosome differs from its homologue by one morphological characteristic (e.g., by having or lacking a constriction near one end of the chromosome). How many different kinds of gametes with respect to such morphological differences can the organism produce?

I

Organization and Replication of the Genetic Materials

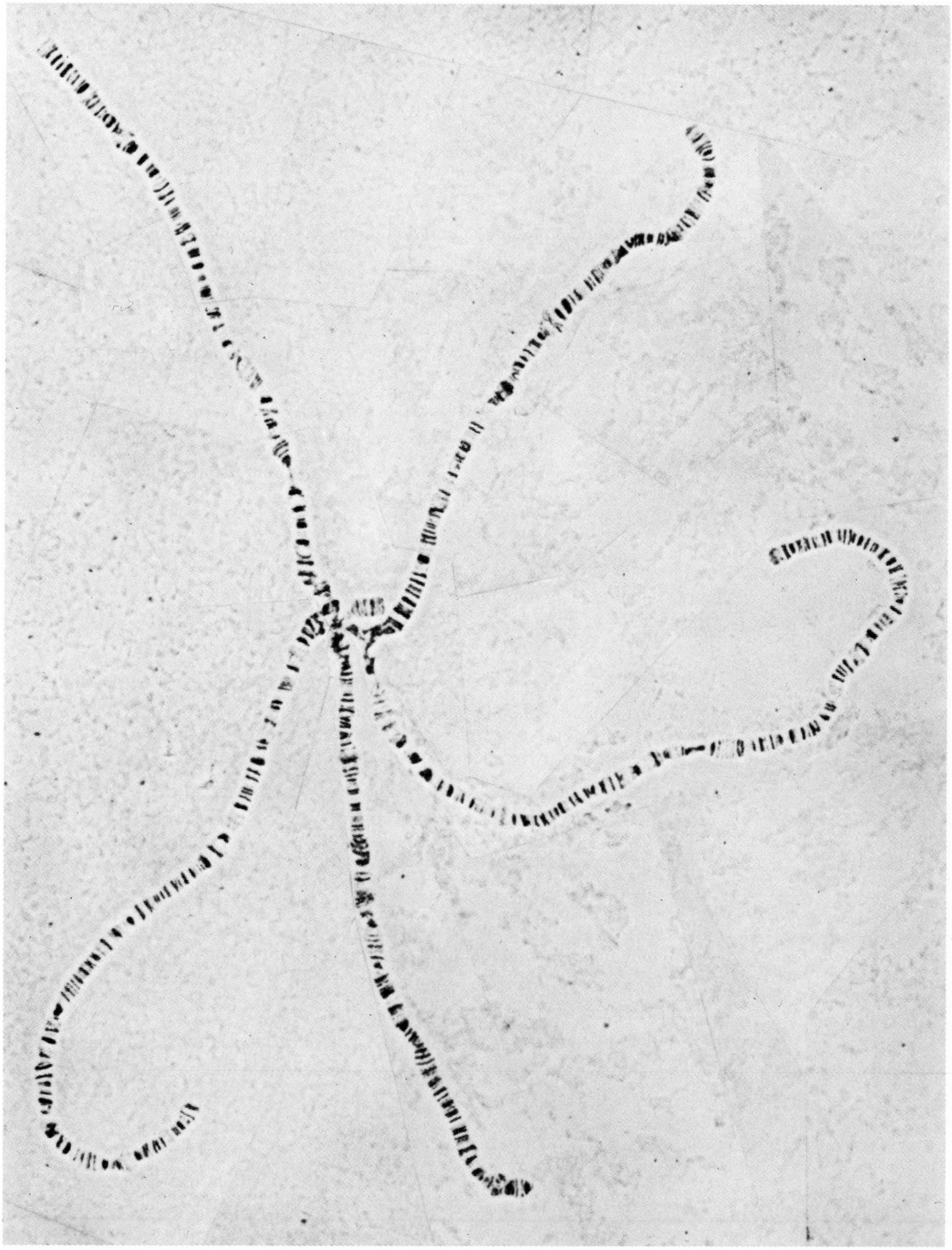

2

Mendelian Genetics

Early Ideas About Heredity

The existence of biological heredity is obvious in the resemblance of children to their parents, although the resemblance is far from exact. People have long known that in humans and animals the sexual act is involved in procreation. Hence it was natural to assume that semen was the carrier of heredity, but how this was accomplished proved difficult to establish. One explanation that prevailed for many centuries was the theory of *pangenesis,* according to which the semen was formed everywhere in the body and traveled through the blood vessels and by way of the testicles into the penis. The similarity between parents and offspring was accounted for by postulating that the semen formed in each part of the body reflected the characteristics of that part.

The theory of pangenesis was proposed by Aristotle (384–322 B.C.) and other ancient Greeks, and prevailed into the nineteenth century. Jean Baptiste de Lamarck (1744–1829) considered it the fundamental mechanism of evolutionary change. For Lamarck, evolution was the result of acquired characteristics accumulated over many generations: body modifications acquired by use and disuse, such as muscle development in an athlete, could be transmitted to the offspring if the semen formed throughout the body would reflect such modifications. The theory of pangenesis was accepted by other great biologists of the nineteenth century, including Charles Darwin (1809–1882).

The first serious challenge to the theory of pangenesis was made by August Weismann (1834–1914), who proposed instead the *germ-plasm*

Figure 2.1
Gregor Mendel, who discovered the fundamental principles of heredity. (The Bettmann Archive.)

theory. He distinguished between the germ plasm—i.e., the sex cells and the cells from which they arise—and the *somatoplasm,* which comprises all other parts of the body. According to Weismann, the germ plasm perpetuates itself in reproduction for generation after generation, while the somatoplasm is produced almost incidentally by the germ plasm only as a means to protect and help reproduce itself. This view contrasts with the theory of pangenesis, according to which the semen is made up of representative particles secreted by the somatoplasm. Weismann supported his theory with experiments that seem crude today but had considerable influence on subsequent ideas about heredity. He cut off the tails of mice for many generations and observed that the progeny continued having tails of normal length. He concluded that inheritance of the tail did not depend on particles formed in the tail itself; rather, it was determined by the germ-plasm cells, which remained unaffected when the tails were cut.

Discovery of the Laws of Heredity

The fundamental principles of heredity were discovered by Gregor Mendel (1822–1884), who was a monk in the Augustinian monastery in Brünn, Austria (now Brno, Czechoslovakia). Starting around 1856, he used garden peas (*Pisum sativum*) to investigate how individual traits are inherited. His experiments and publications are masterly examples of scientific research, even by today's standards. He published his results in 1866 in the

Proceedings of the Natural History Society of Brünn, but his publication received little attention from scientists anywhere.

Mendel's achievements were rediscovered in 1900 by three scientists who had obtained similar results and recognized his precedence. These three, who had been working independently, were Hugo de Vries in Holland, Carl Correns in Germany, and Erich von Tschermak in Austria. The significance of Mendel's principles now became apparent: the riddle of heredity was open to solution. Many biologists became interested in genetics. The first step was to show that Mendel's principles also apply to animals. This was done in the early years of the twentieth century, principally by Lucien Cuénot in France, William Bateson in England, and William E. Castle in the United States. Other important discoveries would soon follow.

Mendel's Methods

Many scientists before Mendel had tried to elucidate how biological characteristics are inherited. They had crossed plants or animals and had looked at the overall similarities between the offspring and their parents. The results were confusing: the offspring resembled one parent in some traits, the other parent in other traits, and apparently neither one in still other traits. No precise regularities could be discovered.

Mendel succeeded where earlier investigators had failed, owing to his brilliant insight and methodology. He saw the need to pay attention to *a single trait* at a time—the shape of the seeds, for example—rather than to the whole plant. For this purpose he selected characters for which he had plants that differed in clear-cut ways (Figure 2.2). Before starting crosses between plants, he also made sure they were *true-breeding*: he obtained many pea varieties from seedsmen and bred them for two years in order to select for his experiments only those strains in which the offspring always resembled their parents in a given trait. Another important feature of Mendel's work was his *quantitative* approach: he counted the number of progeny of each kind in order to ascertain whether carriers of alternate traits always appeared in the same proportions.

Mendel's method of genetic analysis—counting the number of individuals of each class in the progeny of appropriate crosses—is still in use. In fact, this was the *only* method of genetic analysis until the discoveries of molecular genetics in the 1950s. Besides this successful methodology, what made Mendel a scientific genius was his ingenuity in formulating a theory that accounted for his experimental results and in devising appropriate experimental tests to confirm it. Although properly presented as a hypothesis, Mendel's theory was formulated with an air of completeness. Time has shown that it is fundamentally complete and correct.

Let us now review Mendel's experiments, the basic laws of heredity derived from the experiments, and the theory that explains these laws and accounts for the experimental results.

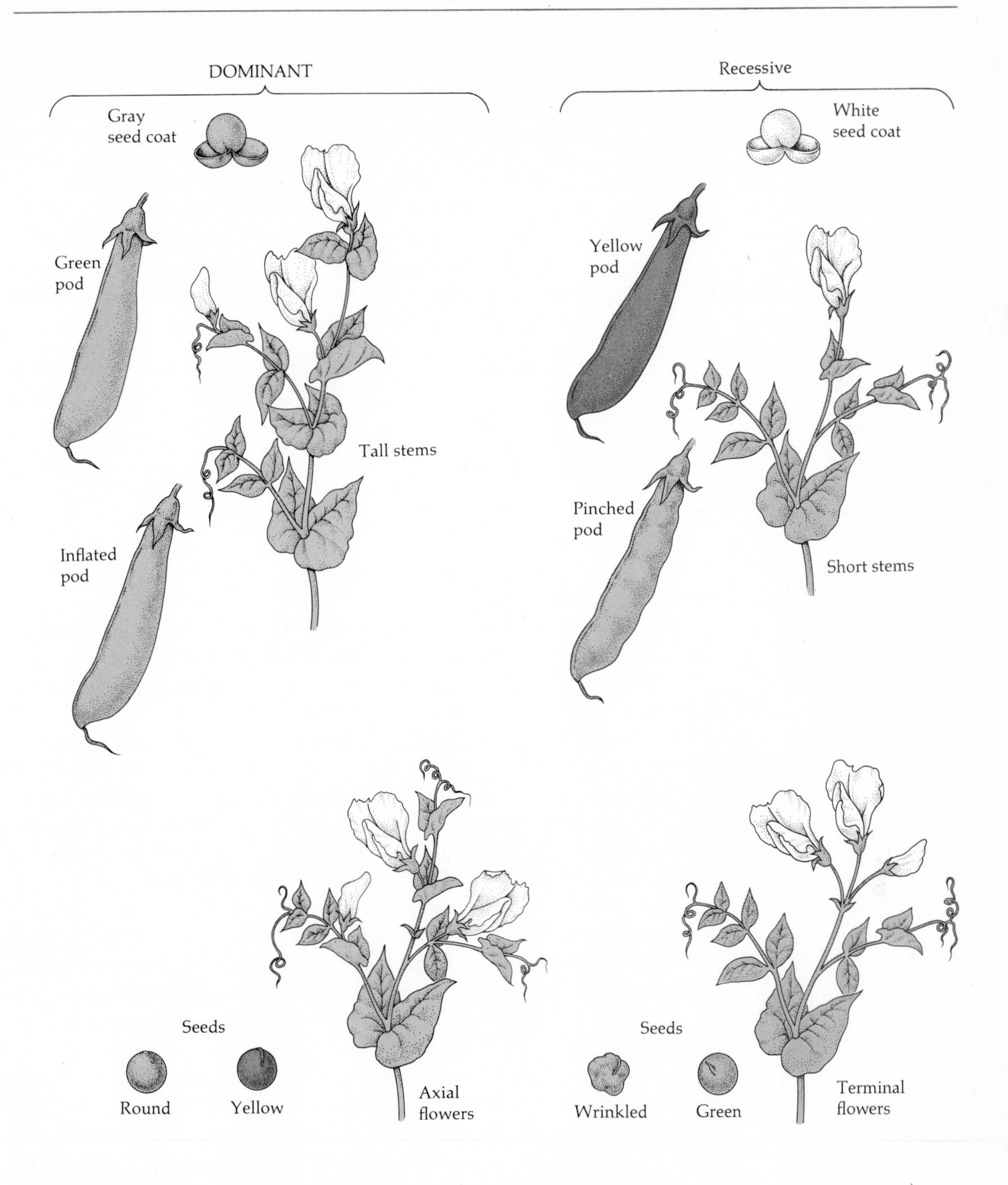

Figure 2.2
The seven characters studied by Mendel in the garden pea, *Pisum sativum.* Mendel obtained true-breeding plants that differed in clear-cut ways with respect to any one character.

Dominance and Recessivity

The garden pea reproduces by self-fertilization: the plants are so constructed that pollen from a flower normally falls on the stigma of the same flower and fertilizes it (Figure 2.3). However, it is relatively simple to obtain cross-fertilization. Mendel would open a flower bud and remove

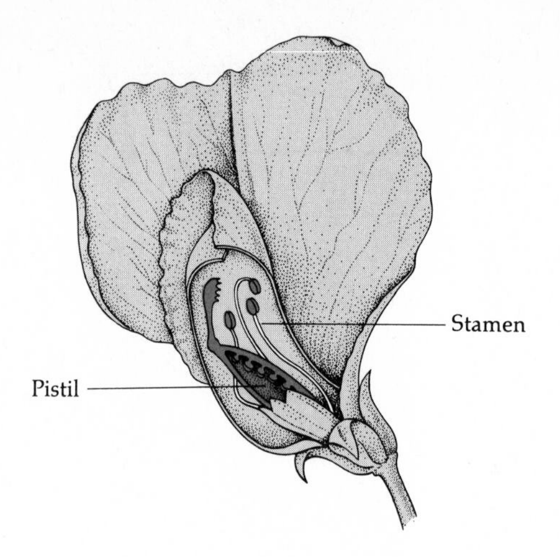

Figure 2.3
Cross-section of a *Pisum sativum* flower, showing the female (pistil) and male (stamen) reproductive organs.

the stamens before any pollen had been shed, thus preventing self-fertilization; then he used pollen from another flower to fertilize the first one.

In one experiment Mendel studied the inheritance of seed shape by crossing plants yielding round seeds with plants yielding wrinkled seeds. The results were clear-cut: all hybrid plants of the first filial generation (F_1) produced round seeds, regardless of whether the round-seed plant had been the female parent or the male parent. Wrinkling seemed to be suppressed by the dominance of roundness (Figure 2.4). Mendel found that all seven characters he had selected for study behaved in this way; in each case only one of the two contrasting traits appeared in the F_1 hybrids. Mendel called such traits (round seeds, yellow peas, axial flowers, etc.) *dominant,* and their alternatives (wrinkled seeds, green peas, terminal flowers, etc.) he called *recessive* (Figure 2.2).

Scientists later found that dominance of one trait over another is a common but not universal phenomenon. In some cases there is *incomplete dominance*: the F_1 hybrids are intermediate between the two parents. In the snapdragon, for example, a plant with crimson flowers crossed to a plant with white flowers produces F_1 hybrids all with pink flowers. This is simply because pink flowers possess less red pigment than crimson flowers, and white flowers have none. There are also cases in which the traits of the parents are both exhibited in their F_1 progeny; this is called *codominance.* In humans, for example, the characteristics of blood group A and blood group B are expressed equally in individuals who have inherited the two, one from each parent. The substances (*antigens*) characteristic of blood group A and blood group B are both present in the blood of such individuals and can be identified by an appropriate (antigenic) reaction.

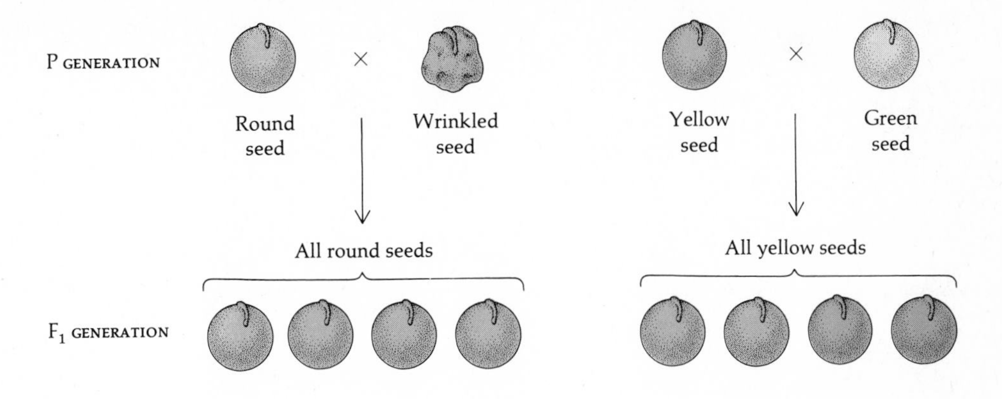

Figure 2.4
The F_1 generation in two of Mendel's crosses. The first-generation hybrids exhibited the trait of one of the parents (dominant), while the alternative trait (recessive) seemed to be suppressed. The results were the same whether the plant having the dominant trait was the male parent or the female parent.

Segregation

Mendel planted seeds from the F_1 hybrid progeny and allowed the plants that grew from these seeds to fertilize themselves. Both round and wrinkled seeds appeared side-by-side in the same pods in the second generation (F_2) of the cross between round-seed and wrinkled-seed plants. He counted the seeds: 5474 were round and 1850 were wrinkled (Figure 2.5). The ratio was very close to 3:1 (in fact, 2.96:1). A similar ratio appeared in all the other crosses; in every case the dominant trait was about three times as common in the F_2 as the recessive trait (Table 2.1).

Mendel was now ready to investigate whether the round seeds and the wrinkled seeds of the F_2 generation bred true. He planted the F_2 seeds and allowed the plants that grew from them to fertilize themselves. The wrinkled seeds all developed into plants that produced only wrinkled peas. But the round seeds behaved quite differently. Although indistinguishable in appearance, these round seeds were of two kinds: about one-third of them developed into plants with only round seeds; the other two-thirds developed into plants with round and wrinkled seeds in the ratio 3:1. That is, one-third of the F_2 round seeds (or one-quarter of all F_2 seeds) bred true, while the other two-thirds (or one-half of all F_2 seeds) were like the F_1 hybrids: they developed into plants with round and wrinkled seeds in the ratio 3:1.

These results were the same for other characters. In every case the F_2 plants showing the recessive trait bred true: their seeds produced F_3 plants identical to their parents. The F_2 plants showing the dominant trait were of two kinds, however, one-third bred true, while the other two-thirds produced F_3 progenies in which the dominant and recessive traits appeared in the ratio 3:1.

Mendel's results have been proved to be generally valid for plants, as well as for animals, including humans. In cases of codominance or in-

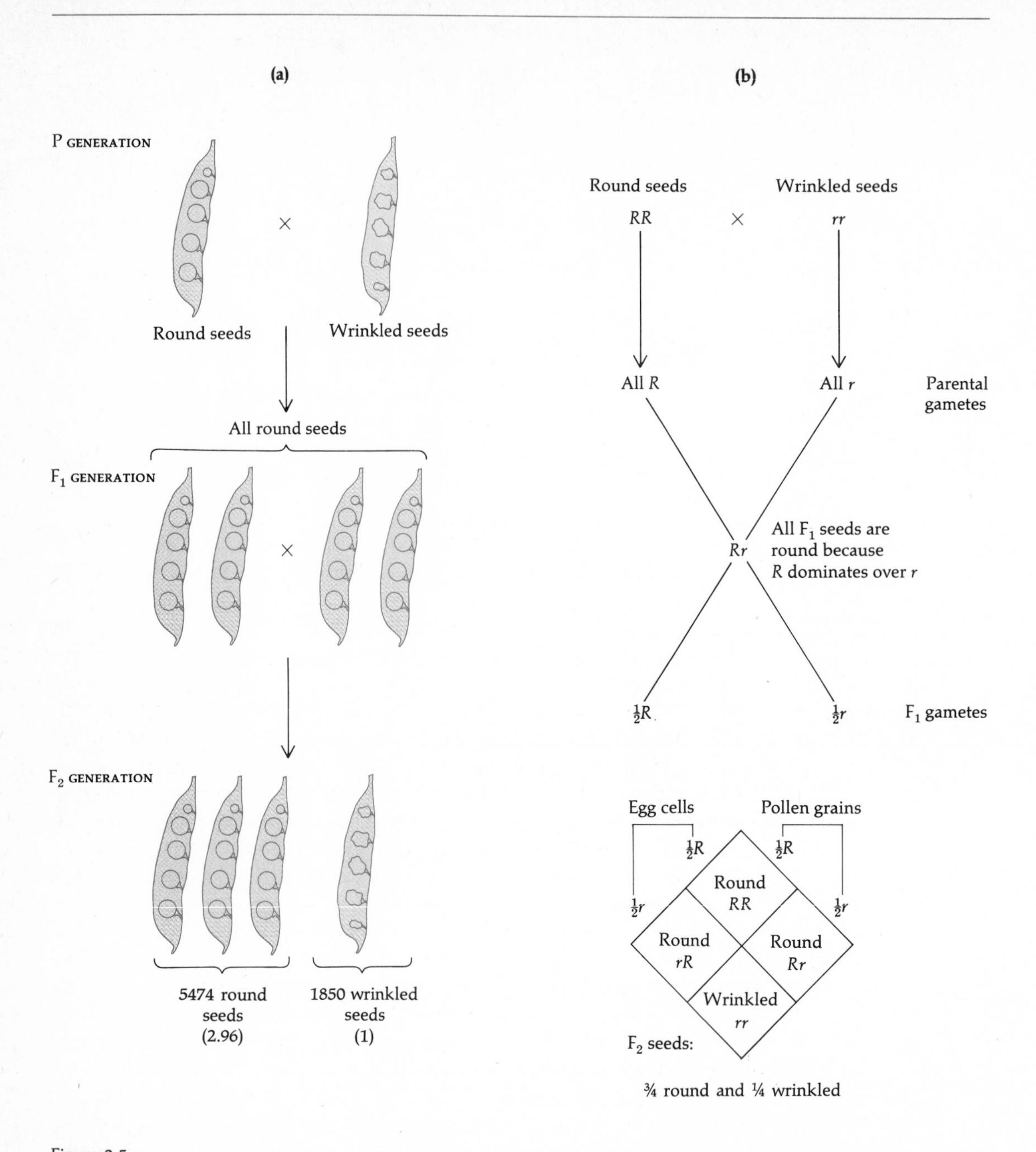

Figure 2.5
The F_2 progeny of a cross between a round-seed and a wrinkled-seed pea plant. **(a)** When the round-seed plants of the F_1 generation are self-fertilized or crossed to each other, the seeds of the F_2 generation are, approximately, three-quarters round and one-quarter wrinkled. **(b)** The explanation proposed by Mendel. *R* and *r* represent the alternative factors (alleles) determining roundness and wrinkling, respectively. The probability with which each kind of plant is expected is obtained by multiplying the probabilities of the types of gametes that must unite in order to form that kind of plant. Thus, one-quarter of the F_2 plants are *RR* because the probability of *R* is ½ among F_1 male gametes and also ½ among F_1 female gametes: (½)(½) = ¼.

Table 2.1
Mendel's results in crosses between plants differing in one of seven characters.

Character*	F_1	F_2 (Number) Dominant	F_2 (Number) Recessive	F_2 (Number) Total	F_2 (Percent) Dominant	F_2 (Percent) Recessive
Seeds: round vs. wrinkled	All round	5,474	1,850	7,324	74.7	25.3
Seeds: yellow vs. green	All yellow	6,022	2,001	8,023	75.1	24.9
Flowers: purple vs. white	All purple	705	224	929	75.9	24.1
Flowers: axial vs. terminal	All axial	651	207	858	75.9	24.1
Pods: inflated vs. pinched	All inflated	882	299	1,181	74.7	25.3
Pods: green vs. yellow	All green	428	152	580	73.8	26.2
Stem: tall vs. short	All tall	787	277	1,064	74.0	26.0
Total or average		14,949	5,010	19,959	74.9	25.1

*The dominant trait is always written first.

complete dominance, the F_2 generation consists of three classes: one-quarter of the organisms exhibit the trait of one parent, one-quarter exhibit the trait of the other parent, and one-half are like the F_1 hybrids. The individuals that are like the parents breed true, while the hybrid ones produce progenies in the F_3 that are again one-half hybrid and one-quarter like each of the parents (Figures 2.6 and 2.7).

Genes, the Carriers of Heredity

Mendel advanced the following hypothesis to explain the results of his experiments with peas. Contrasting traits, such as the roundness or wrinkling of peas, are determined by "factors" (now called *genes*) that are transmitted from parents to offspring through the gametes; each factor may exist in alternative forms (now called *alleles*) responsible for the alternative forms that the character may take. For each character, every pea plant has two genes, one inherited from the male parent and the other inherited from the female parent. Thus, each pea plant has two genes for seed shape; these genes may exist in the form that determines round seeds (allele for roundness) or in the form that determines wrinkled seeds (allele for wrinkling).

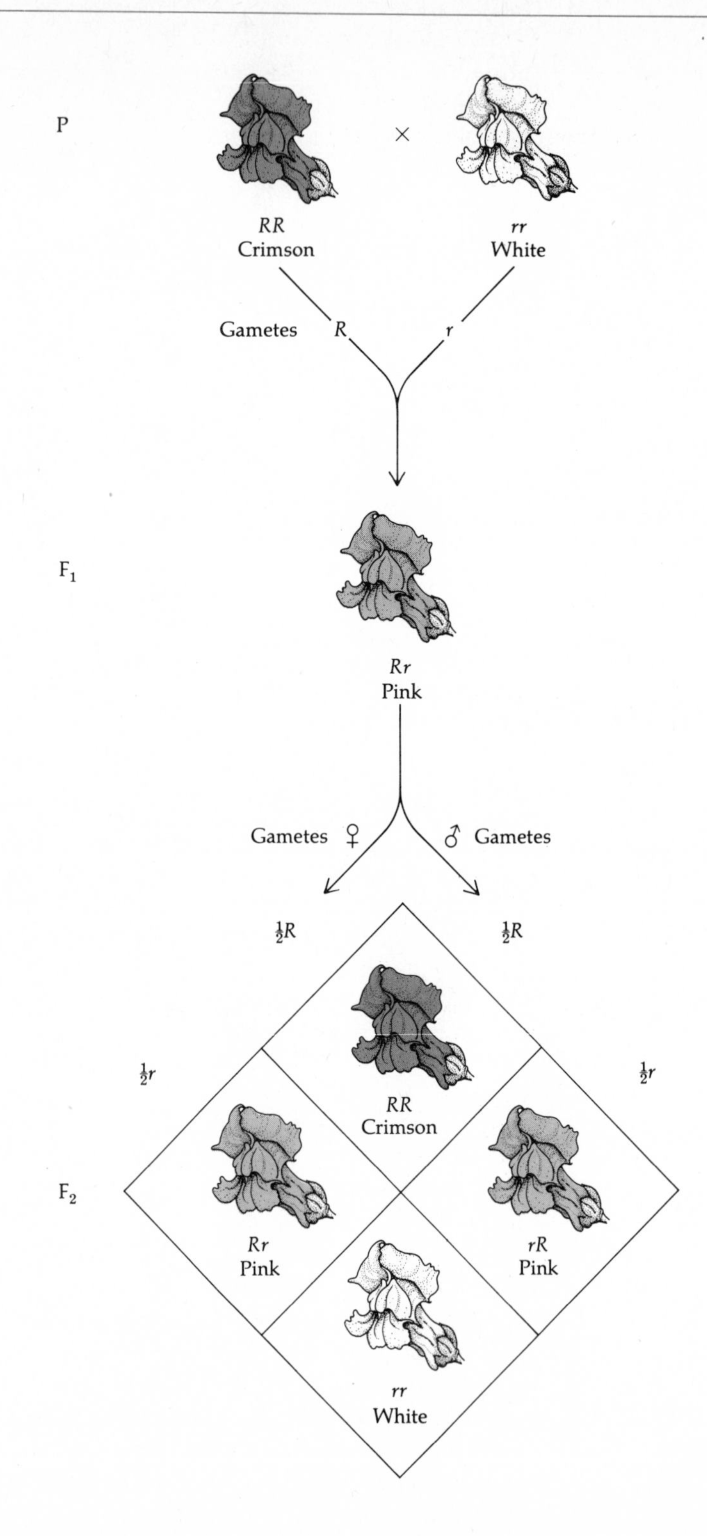

Figure 2.6
Cross between a crimson-flowered and a white-flowered snapdragon. The F_1 flowers are pink, showing incomplete dominance of crimson over white. The F_2 flowers are crimson, pink, and white in the proportions 1:2:1.

Figure 2.7
Inheritance of plumage color in Andalusian fowls. The F_1 progeny are blue-gray, intermediate between the black and white parents. In the F_2 generation, black, blue-gray, and white appear in the proportions 1:2:1. The black and white F_2 fowls are true-breeding, producing only black and only white F_3 progenies, respectively. The blue-gray F_2 fowls crossed with each other produce F_3 progenies consisting of black, blue-gray, and white progenies, again in the proportions 1:2:1.

It is appropriate at this point to introduce two other terms of the genetics vocabulary. A *homozygote* (adjective: homozygous) is an individual in which the two genes for a given character are identical, i.e., an individual with two identical alleles. A *heterozygote* (adjective: heterozygous) is an individual in which the two genes for a given character are different, i.e., an individual with two different alleles. Thus, the true-breeding round-seed plants are homozygous for roundness, the true-breeding wrinkled-seed plants are homozygous for wrinkling, and the F_1 hybrids from the cross round × wrinkled are heterozygous for roundness and wrinkling (Figure 2.8).

From the phenomenon of dominance, Mendel inferred that in heterozygous individuals one allele is dominant and the other is recessive. From the reappearance of the two parental traits in the progeny of hybrids (heterozygotes), he concluded that *the two factors (genes) for each trait do not fuse or blend in any way, but remain distinct throughout the life of the individual and segregate in the formation of gametes,* so that half the gametes carry one gene and the other half carry the other gene. This conclusion is known as Mendel's *law of segregation.*

Pairs of genes are often symbolized by letters, the dominant allele by a capital italic letter, the recessive allele by the same letter in lowercase italic. For example, the allele for roundness is generally represented as *R* and the allele for wrinkling as *r*. The genetic makeup of Mendel's plants is then as follows (Figure 2.5). The true-breeding round plants are *RR*, and the true-breeding wrinkled plants are *rr*. The F_1 hybrids are *Rr* and produce two kinds of gametes, *R* and *r*, in equal amounts. When *Rr* plants fertilize themselves (or are crossed with other *Rr* plants), they produce three kinds of progeny: (1) one-quarter are *RR*, namely, the true-breeding round-seed plants; (2) one-half are *Rr*, namely, the round-seed plants that, when self-fertilized, produce both round and wrinkled seeds in the F_3 generation; (3) one-quarter are *rr*, namely, the wrinkled-seed plants, which are true-breeding because they are homozygous and thus produce only one kind of gamete.

Mendel tested his hypothesis in various ways. One way, much employed by later geneticists, is called a *testcross* (Figure 2.9). It consists of crossing F_1 hybrid individuals to the recessive parent. If Mendel's hypothesis were correct, the offspring of such a cross should consist of recessive and dominant individuals in approximately equal proportions. This is precisely what he observed.

Independent Assortment

Mendel's experiments described so far concern the inheritance of alternative expressions of a single character. What happens when two characters are considered simultaneously? Mendel formulated the *law of independent*

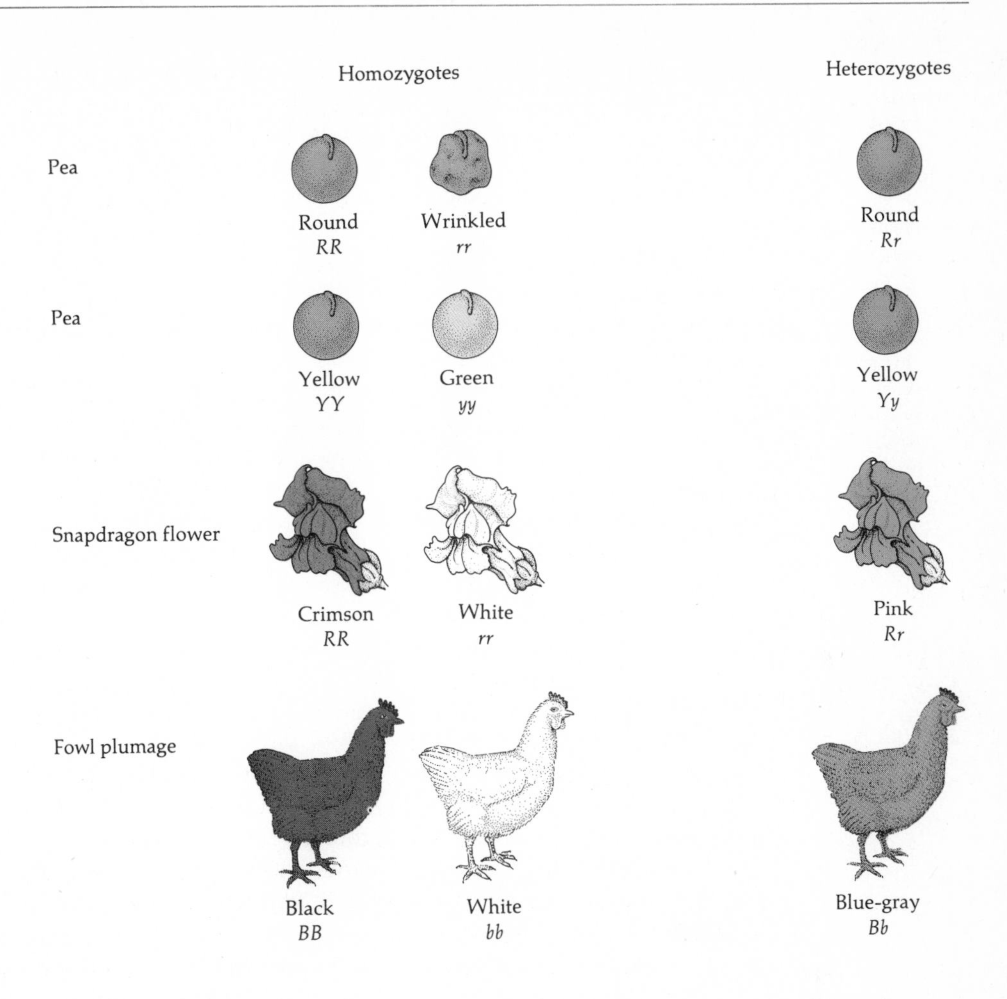

Figure 2.8
Homozygous and heterozygous individuals in various kinds of organisms. Homozygotes have two identical alleles of the gene determining the character under consideration; heterozygotes have two different alleles.

assortment, which says that *genes for different characters are inherited independently of one another.* (This law, however, would later be shown to apply only to genes on different chromosomes.)

Mendel derived this law from the results of crosses between plants that were different with respect to two separate characters (a *dihybrid cross*). In one experiment, plants having round and yellow seeds were crossed with plants having wrinkled and green seeds. As expected, all peas in the F_1 generation were round and yellow, but the interesting results came in the F_2 generation. Mendel had already considered two possibilities: (1) that traits derived from one parent are transmitted together, and (2) that they are transmitted independently of each other. With characteristic lucidity, he formulated the expectations from these alternatives. If (1) were true, there would be only two kinds of seeds in the F_2 generation: round-

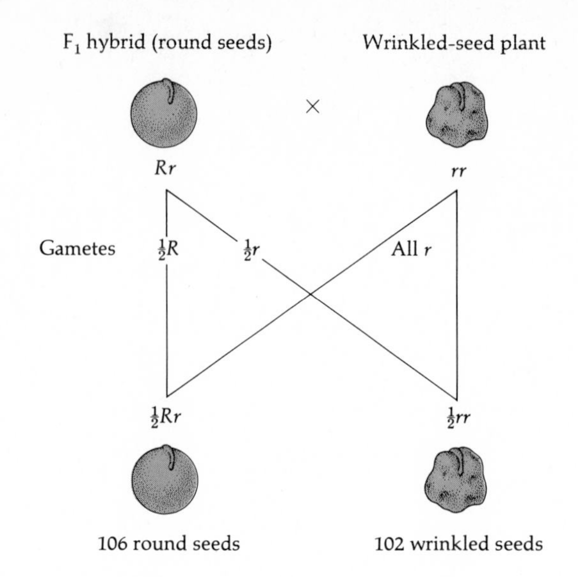

Figure 2.9
A testcross made by crossing a hybrid F_1 individual with the recessive parent. In the testcross shown, an F_1 hybrid between round and wrinkled (*Rr*) is crossed to a wrinkled individual (*rr*). Mendel made this cross and observed that—as predicted by his hypothesis—approximately half the progeny were round, like the hybrid parent, and half were wrinkled, like the recessive parent.

yellow and wrinkled-green, and they should appear in the ratio 3:1 according to the law of segregation. If (2) were correct, however, there should be four kinds of seeds: round-yellow (two dominant traits), round-green (dominant-recessive), wrinkled-yellow (recessive-dominant), and wrinkled-green (two recessive traits), which should appear in the proportions 9:3:3:1 (Figure 2.10).

Mendel found that the peas in the F_2 generation were of four kinds: 315 round-yellow, 108 round-green, 101 wrinkled-yellow, and 32 wrinkled-green. This was reasonably close to the proportions 9:3:3:1 expected from the second alternative, and Mendel concluded that the genes determining different characters are transmitted independently. (Note that the results of this experiment also confirm the law of segregation, since the expected 3:1 ratio was obtained for each character separately. In the F_2 generation there are 423 round seeds versus 133 wrinkled seeds, and 416 yellow seeds versus 140 green seeds.)

Trihybrid Crosses

Mendel confirmed the law of independent assortment for various combinations of two characters. He also confirmed it with an experiment in which the parents differed simultaneously with respect to three characters, which is called a *trihybrid cross*.

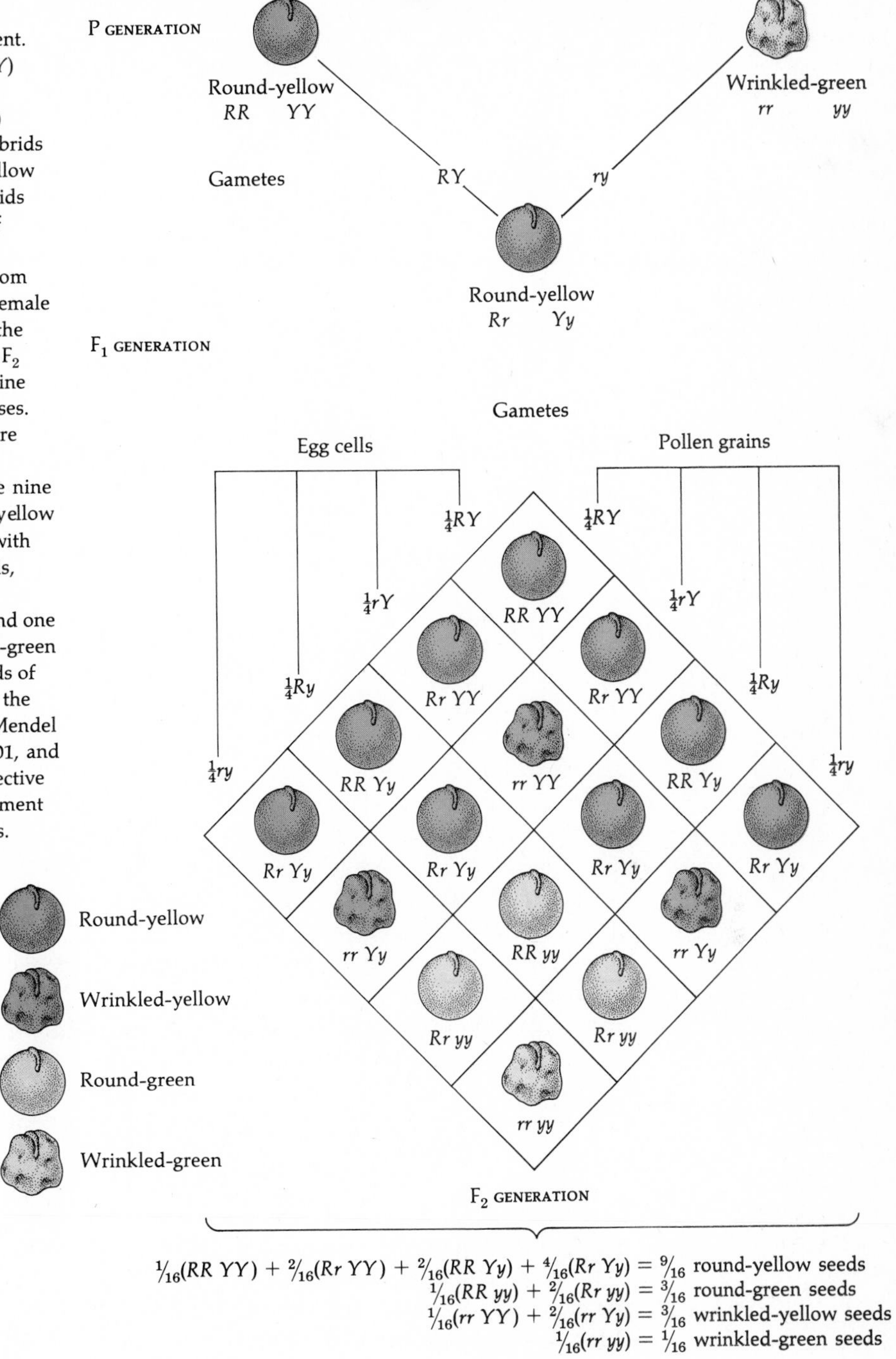

Figure 2.10
Independent assortment. Round-yellow (*RR YY*) plants crossed with wrinkled-green (*rr yy*) plants produce F_1 hybrids that are all round-yellow (*Rr Yy*). The F_1 hybrids produce four kinds of gametes, each with a frequency of ¼. Random association between female and male gametes of the four kinds produces F_2 seeds belonging to nine different genetic classes. When the F_2 seeds are grouped by their appearance, there are nine squares with round-yellow seeds, three squares with wrinkled-yellow seeds, three squares with round-green seeds, and one square with wrinkled-green seeds; these four kinds of seeds are expected in the proportions 9:3:3:1. Mendel obtained 315, 108, 101, and 32 seeds of the respective kinds, in good agreement with the expectations.

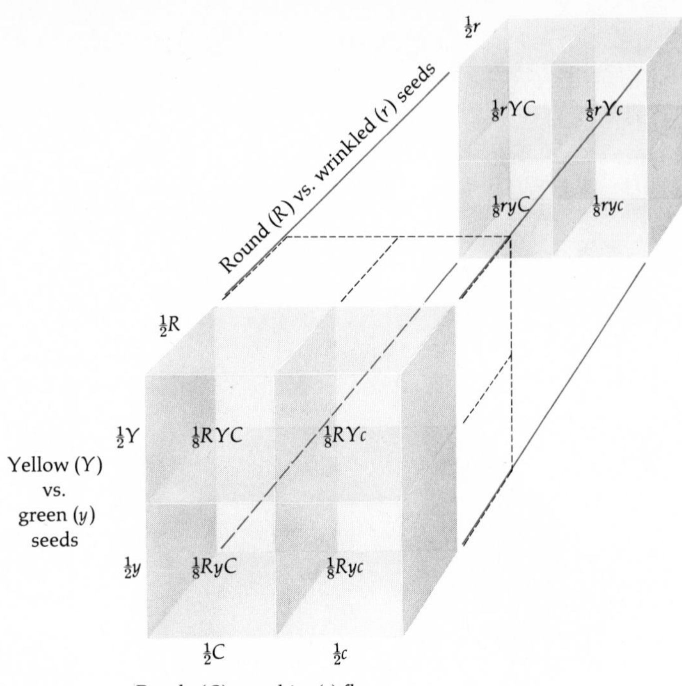

Figure 2.11
Gametes produced by a trihybrid individual. For each gene, the probability of each of the two kinds of gametes is ½. When all three genes are taken into account, eight kinds of gametes are possible. If the three genes assort independently, the probability of each kind of gamete is (½)(½)(½) = ⅛.

Consider, for example, a cross between the following two pea plants:

Female parent	Male parent
Round seeds (*RR*)	Wrinkled seeds (*rr*)
Yellow seeds (*YY*)	Green seeds (*yy*)
Purple flowers (*CC*)	White flowers (*cc*)

The gametes of the female parent will all be *RYC*; the gametes of the male parent will all be *ryc*; therefore the F_1 hybrid will be a triple heterozygote, or trihybrid, of the genetic class *Rr Yy Cc*. Because of dominance, it will have round-yellow seeds and purple flowers. If the three gene pairs assort independently, the trihybrid plant will produce eight kinds of gametes, all with equal probability (Figure 2.11).

The random union among the 8 different gametes from the 2 parents results in 27 different genetic classes (Figure 2.12). Because of dominance, these 27 genetic classes reduce to 8 different kinds of plants, which are

F$_1$ TRIHYBRID

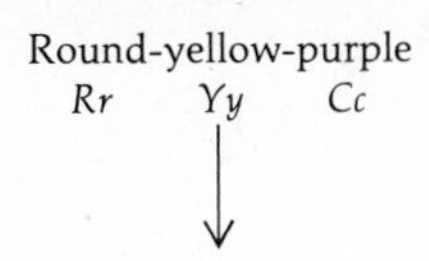

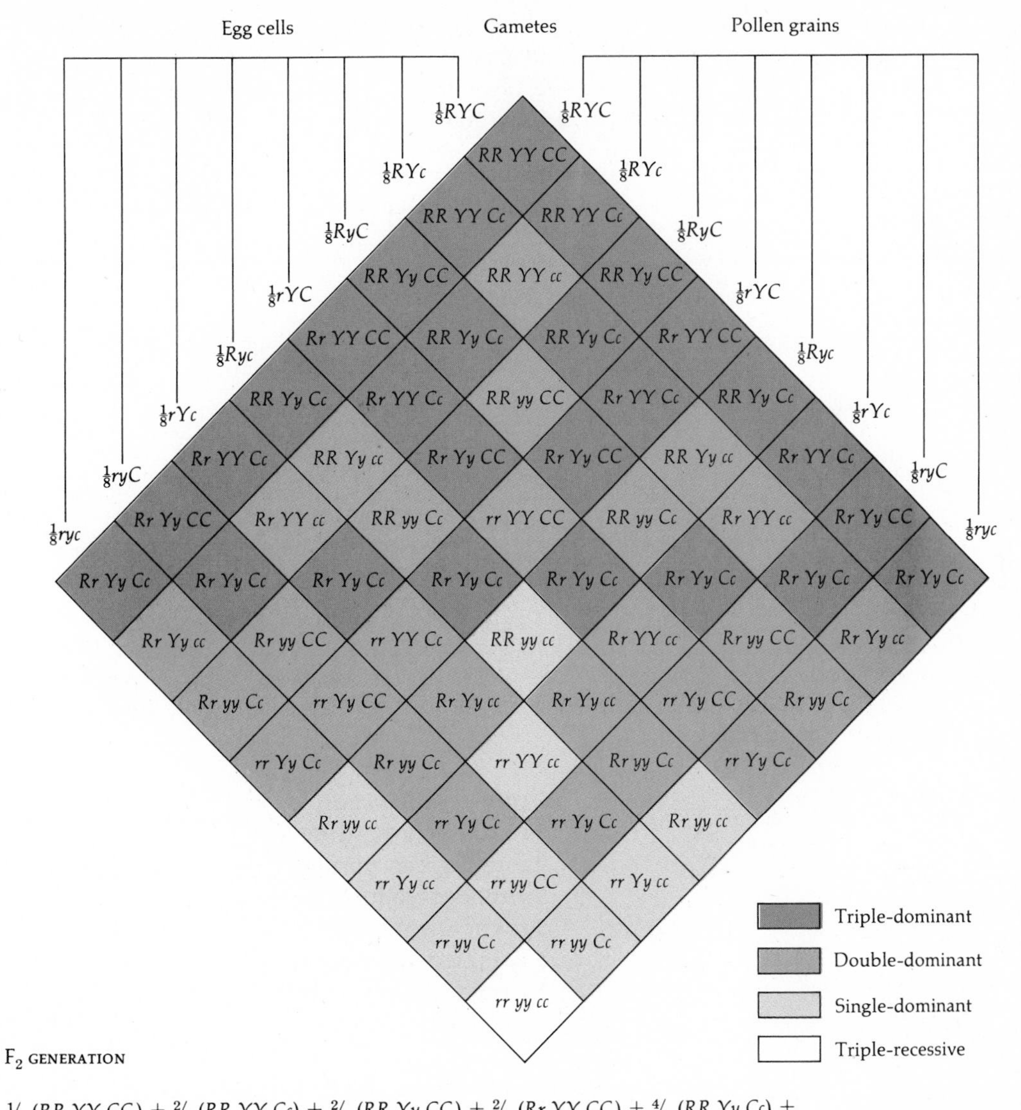

$\frac{1}{64}(RR\ YY\ CC) + \frac{2}{64}(RR\ YY\ Cc) + \frac{2}{64}(RR\ Yy\ CC) + \frac{2}{64}(Rr\ YY\ CC) + \frac{4}{64}(RR\ Yy\ Cc) +$
$\frac{4}{64}(Rr\ YY\ Cc) + \frac{4}{64}(Rr\ Yy\ CC) + \frac{8}{64}(Rr\ Yy\ Cc) = \frac{27}{64}$ round-yellow-purple
$\frac{1}{64}(RR\ YY\ cc) + \frac{2}{64}(RR\ Yy\ cc) + \frac{2}{64}(Rr\ YY\ cc) + \frac{4}{64}(Rr\ Yy\ cc) = \frac{9}{64}$ round-yellow-white
$\frac{1}{64}(RR\ yy\ CC) + \frac{2}{64}(RR\ yy\ Cc) + \frac{2}{64}(Rr\ yy\ CC) + \frac{4}{64}(Rr\ yy\ Cc) = \frac{9}{64}$ round-green-purple
$\frac{1}{64}(rr\ YY\ CC) + \frac{2}{64}(rr\ Yy\ CC) + \frac{2}{64}(rr\ YY\ Cc) + \frac{4}{64}(rr\ Yy\ Cc) = \frac{9}{64}$ wrinkled-yellow-purple
$\frac{1}{64}(RR\ yy\ cc) + \frac{2}{64}(Rr\ yy\ cc) = \frac{3}{64}$ round-green-white
$\frac{1}{64}(rr\ YY\ cc) + \frac{2}{64}(rr\ Yy\ cc) = \frac{3}{64}$ wrinkled-yellow-white
$\frac{1}{64}(rr\ yy\ CC) + \frac{2}{64}(rr\ yy\ Cc) = \frac{3}{64}$ wrinkled-green-purple
$\frac{1}{64}(rr\ yy\ cc) = \frac{1}{64}$ wrinkled-green-white

Figure 2.12
Genotypes produced in the progeny of trihybrid individuals self-fertilized or crossed to each other. There are 64 combinations of 8 paternal and 8 maternal gametes, but these correspond to only 27 different genotypes. If there is dominance, as in the example shown, the 27 genotypes reduce to 8 different phenotypes. The example shown corresponds to Mendel's peas trihybrid for round and wrinkled, yellow and green, and purple and white.

Table 2.2
The number of different gametes produced by the F_1, and of different genotypes and phenotypes in the F_2, of a cross between parents differing in a given number of genes.

Gene Pairs	Kinds of Gametes	Kinds of Genotypes	Kinds of Phenotypes*
1	2	3	2
2	4	9	4
3	8	27	8
4	16	81	16
n	2^n	3^n	2^n

*For dominance; if there is no dominance, there are as many kinds of phenotypes as of genotypes.

expected in the following proportions (the dominant traits are printed in bold type):

27 **round-yellow-purple**
9 **round-yellow-**white
9 **round-**green-**purple**
9 wrinkled-**yellow-purple**
3 **round-**green-white
3 wrinkled-**yellow-**white
3 wrinkled-green-**purple**
1 wrinkled-green-white

We can now introduce some general rules that apply to the progenies of hybrids between parents that differ with respect to a given number of genes (Table 2.2). In general, every added gene multiplies the number of different gametes by 2 and the number of genetic classes (*genotypes*) by 3. An individual heterozygous for n gene pairs may therefore produce 2^n kinds of gametes and 3^n kinds of genotypes. The number of kinds of organisms with respect to their appearance (*phenotype*) is the same as the number of different gametes if there is dominance, but the same as the number of different genotypes if there is no dominance.

A simple procedure can also be applied to calculate the frequency of a given genotype in the progeny of parents that differ with respect to a number of independently assorting gene pairs. The procedure consists of

Table 2.3
Genetic determination of coat color in rabbits.

Allele	Genotype	Phenotype
c^+	c^+c^+, c^+c^{ch}, c^+c^h, c^+c^a	Wild type
c^{ch}	$c^{ch}c^{ch}$	Chinchilla
	$c^{ch}c^h$, $c^{ch}c^a$	Light gray
c^h	c^hc^h, c^hc^a	Himalayan
c^a	c^ac^a	Albino

calculating the probability of the genotype separately for each gene pair and then multiplying these probabilities. Assume that we want to calculate the expected frequency of the genotype *Rr yy Cc* in the progeny of the cross *Rr Yy cc* × *Rr Yy Cc*. The probability of an *Rr* individual in the progeny of *Rr* × *Rr* is ½; the probability of *yy* in the progeny of *Yy* × *Yy* is ¼; and the probability of *Cc* in the progeny of *cc* × *Cc* is ½. Therefore the probability of the *Rr yy Cc* genotype is (½)(¼)(½) = 1/16.

Multiple Alleles

The examples of Mendelian segregation discussed so far in this chapter, including Mendel's own experiments, involve only two alleles of each gene. However, many, and possibly all, genes have *multiple alleles,* i.e., they exist in more than two allelic forms, although any one diploid organism can carry no more than two alleles.

Many examples of multiple alleles are known, some of which will be mentioned throughout this book. One series of multiple alleles is known in rabbits for a gene determining coat color; four of these alleles are listed in Table 2.3. The wild-type allele, c^+, is dominant over the other three alleles: individuals homozygous for the c^+allele, or heterozygous for c^+ and some other allele, exhibit the normal wild-type gray (agouti) coat color. Individuals homozygous for the c^{ch} allele have the chinchilla phenotype, which is less intense than the wild-type gray. Heterozygotes for the c^{ch} allele and either c^h or c^a are light gray (intermediate between chinchilla and albino); the c^{ch} allele, therefore, exhibits incomplete dominance with respect to either c^h or c^a. Homozygotes c^hc^h and heterozygotes c^hc^a have the Himalayan phenotype, which is similar to the albino phenotype

Box 2.1 A Word on Genetic Notation

Genetic notation has developed without firm rules, and inconsistencies have often arisen as knowledge of genetics has expanded. Moreover, little effort has been made by geneticists working with one organism to develop a notation consistent with that used by geneticists working with other organisms. The principles followed in this text with regard to allele designations and genotype designations are described here.

If only two alleles of a gene are known, it is convenient to designate the dominant allele with a capital italic letter and the recessive allele with a lowercase italic letter; e.g., the three diploid genotypes possible for the gene pair *A*, *a* are *AA*, *Aa*, and *aa*.

However, a different notation is generally used when multiple alleles are to be considered or when independent mutations of a gene to the same mutant phenotype are known. A lowercase italic letter (or group of letters) designates the gene, or locus, and superscripts are used to designate different alleles. For example, *c* may be used to indicate the coat-color gene of rabbits. The normal, or *wild-type*, allele (which is often the most dominant allele in a series of multiple alleles) is designated c^+, and the other alleles are designated c^{ch}, c^h, c^a, etc. Frequently c^+ is abbreviated to +, and the locus in question is identified by the recessive allele present.

In writing diploid genotypes, a slash is generally used to indicate that the two alleles of a gene are present on each of the two homologous chromosomes, e.g., c^{ch}/c^+ or $c^{ch}/+$. When genes located at separate loci on the same chromosome are considered (see Chapter 5), the utility of the slash becomes more obvious. For example, consider the recessive *scarlet* (*st*) mutation of *Drosophila melanogaster*, which has a bright red eye color when homozygous (*st*/*st*), while the dominant wild-type allele (st^+) confers a dark red eye color. The recessive *ebony* (*e*) mutation has a black body color when homozygous (*e*/*e*), while the dominant wild-type allele (e^+) confers a brownish yellow body color. The *scarlet* and *ebony* loci are on the same chromosome, and two kinds of doubly heterozygous genotypes are possible: $st^+e^+/st\ e$ and $st^+e/st\ e^+$. In the former arrangement st^+ and e^+ are said to be *coupled* (they reside on the same homologue), and in the latter arrangement st^+ and e^+ are said to be in *repulsion* (they reside on different homologues).

Infrequently, mutation of a gene to a dominant mutant phenotype may occur. Such mutations are often designated with a capital italic letter. For example, the *Bar* (*B*) eye-shape mutation in *Drosophila* is dominant over the normal wild-type allele, which is designated B^+.

The ABO human blood groups represent a hybrid type of notation that is idiosyncratic: I^A and I^B designate the codominant alleles, and *i* designates the recessive allele for the O phenotype.

The notation described above is generally used when dealing with either diploid or haploid eukaryotic organisms. However, the notation developed for bacteria follows different rules and will be presented in Chapter 7.

Figure 2.13
Four phenotypes due to allelic variation at a gene for coat color in rabbits.

except for the feet, tail, ears, and the tip of the nose, which are pigmented. Homozygotes for the c^a allele exhibit the typical albino phenotype of white fur and pink eyes (Figure 2.13).

Another series of multiple alleles underlies the ABO blood groups, discovered by Karl Landsteiner (1868–1943) in 1900. Blood groups must be considered when matching donors with recipients for blood transfusions, to prevent the donor's red blood cells from becoming agglutinated in the recipient's blood stream (Figure 2.14).

There are four common blood groups within the ABO system: O, A, B, and AB. These are determined by three alleles: I^A, I^B, and i. Alleles I^A and I^B are dominant over allele i, but codominant with each other. With three alleles, six genotypes are possible, but these reduce to four blood groups because of the recessivity of the i allele (Table 2.4).

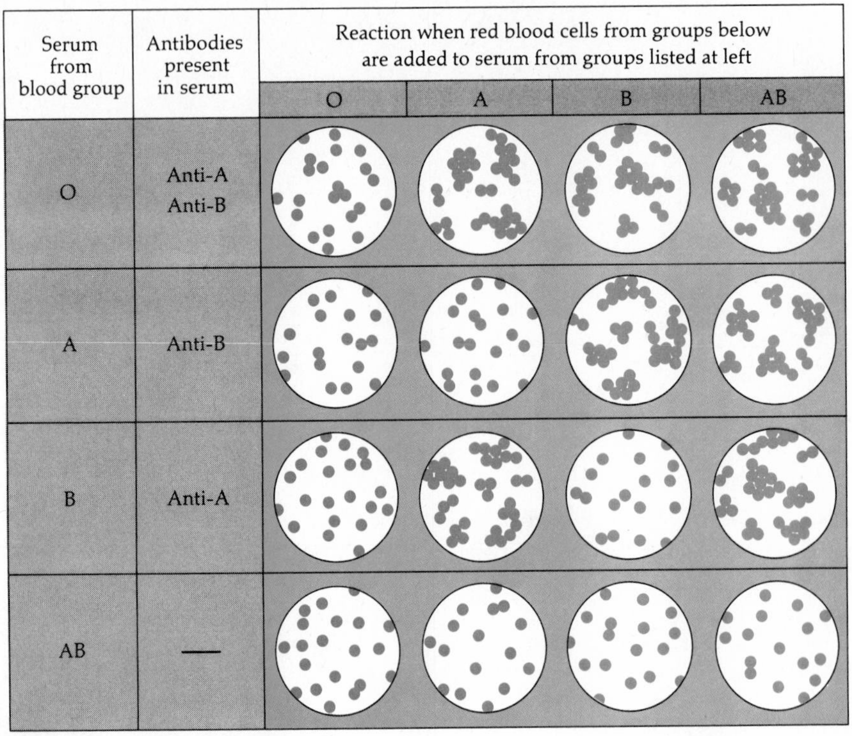

Serum from blood group	Antibodies present in serum	Reaction when red blood cells from groups below are added to serum from groups listed at left			
		O	A	B	AB
O	Anti-A Anti-B				
A	Anti-B				
B	Anti-A				
AB	—				

Figure 2.14
Antigenic reactions used to determine the ABO blood groups. Sera from each of the four blood groups are used as tester solutions. When a drop of blood is mixed with the tester, the expected reactions are as shown. For example, blood from an O individual is not agglutinated by any of the four kinds of sera, whereas blood from an A individual is agglutinated by O and B sera. Cell agglutination is detected when the tester solution becomes turbid.

The number of possible genotypes in a multiple-allele series depends on the number of alleles. With only one allele, A, only one genotype is possible, AA. With two alleles, A_1 and A_2, three genotypes are possible: the two homozygotes A_1A_1 and A_2A_2 and one heterozygote, A_1A_2. With three alleles, A_1, A_2, and A_3, six genotypes are possible: the three homozygotes A_1A_1, A_2A_2, and A_3A_3 and three heterozygotes, A_1A_2, A_1A_3, and A_2A_3. In general, given n alleles, there are $n(n + 1)/2$ genotypes, of which n are homozygotes and $n(n - 1)/2$ are heterozygotes (Table 2.5).

Genotype and Phenotype

In 1909 Wilhelm Johannsen (1857–1927) introduced the important distinction between phenotype and genotype. The phenotype of an organism is its appearance—what we can observe: its morphology, physiology, and behavior. The genotype is the genetic constitution it has inherited. During the lifetime of an individual, the phenotype may change; the genotype, however, remains constant.

The distinction between phenotype and genotype must be kept in mind because the relation between the two is not fixed. This is because the

Table 2.4
The ABO blood groups.

Allele	Genotype	Phenotype (Blood Group)
I^A	I^AI^A, I^Ai	A
I^B	I^BI^B, I^Bi	B
	I^AI^B	AB
i	ii	O

Table 2.5
Number of different genotypes, given a certain number of alleles.

Alleles	Kinds of Genotypes	Kinds of Homozygotes	Kinds of Heterozygotes
1	1	1	0
2	3	2	1
3	6	3	3
4	10	4	6
5	15	5	10
n	$\frac{n(n+1)}{2}$	n	$\frac{n(n-1)}{2}$

phenotype results from complex networks of interactions between different genes and between the genes and the environment.

In general, individuals do not have identical phenotypes, although the phenotypes may be similar when only one or a few traits are considered. Moreover, organisms having similar phenotypes with respect to a given trait do not necessarily have identical genotypes. For example, yellow peas can be either homozygous for the yellow allele or heterozygous for the yellow and green alleles.

Genetic diversity is widespread in nature. Sexually reproducing organisms do not have identical *total* genotypes except in the case of identical (monozygous) twins, which develop from a single fertilized egg (see Chapter 18), but they may have identical genotypes with respect to a given

Figure 2.15
Three sets of monozygous twins, showing some differences in appearance. (Photo credits, clockwise from top left: Ken Heyman; Joanne Leonard; Charles Harbutt/Magnum Photos.)

gene. On the other hand, products of the asexual reproduction of a given individual are genetically identical to each other and to their parent. Even individuals with identical genotypes may have different phenotypes, however, owing to different interactions with the environment. Monozygous twins, for example, may differ in height, weight, and longevity because of different life experiences (Figure 2.15 and Table 2.6).

A good illustration of environmental effects on the phenotype is shown in Figure 2.16. Three plants of the cinquefoil, *Potentilla glandulosa,*

Table 2.6
Percent concordance and discordance with respect to certain traits in identical twins. In concordance, both twins exhibit the same trait in all cases in which at least one twin is affected.

Trait	Number of Twin Pairs Studied	Concordance (percent)	Discordance (percent)
Measles	189	95	5
Scarlet fever	31	64	36
Tuberculosis	190	74	26
Specific kinds of tumors	62	58	42
Diabetes mellitus	63	84	16
Feeble-mindedness	126	91	9

From various sources.

were collected in California—one on the coast at about 100 feet above sea level, the second at about 4600 feet, and the third in the Alpine zone of the Sierra Nevada at about 10,000 feet. Each plant was cut into three parts and the parts were planted in three experimental gardens at different altitudes, the same gardens for all three plants. The division of one plant insured that all three parts planted at different altitudes had the same genotype.

Comparison of the plants in any row shows how a given genotype gives rise to different phenotypes in different environments. Besides the obvious divergence in appearance, there are differences in fertility, growth rates, etc. Comparison of the plants in any column shows that in a given environment different genotypes result in different phenotypes. An important observation derived from this experiment is that there is no single genotype that is "best" in all environments. For example, the plant from near sea level that is so prosperous there fails to develop at 10,000 feet, and the plant collected at 10,000 feet prospers at that altitude, but withers at sea level.

The interaction between the genotype and the environment is further illustrated in Figure 2.17. Two strains of rats were selected, one for brightness at finding their way through a maze and the other for dullness. Selection was done in the bright strain by using the brightest rats of each generation to breed the following generation, and in the dull strain by breeding the dullest rats every generation. After many generations of selection, bright rats made only about 120 errors running through the maze, whereas dull rats averaged 165 errors. However, the differences

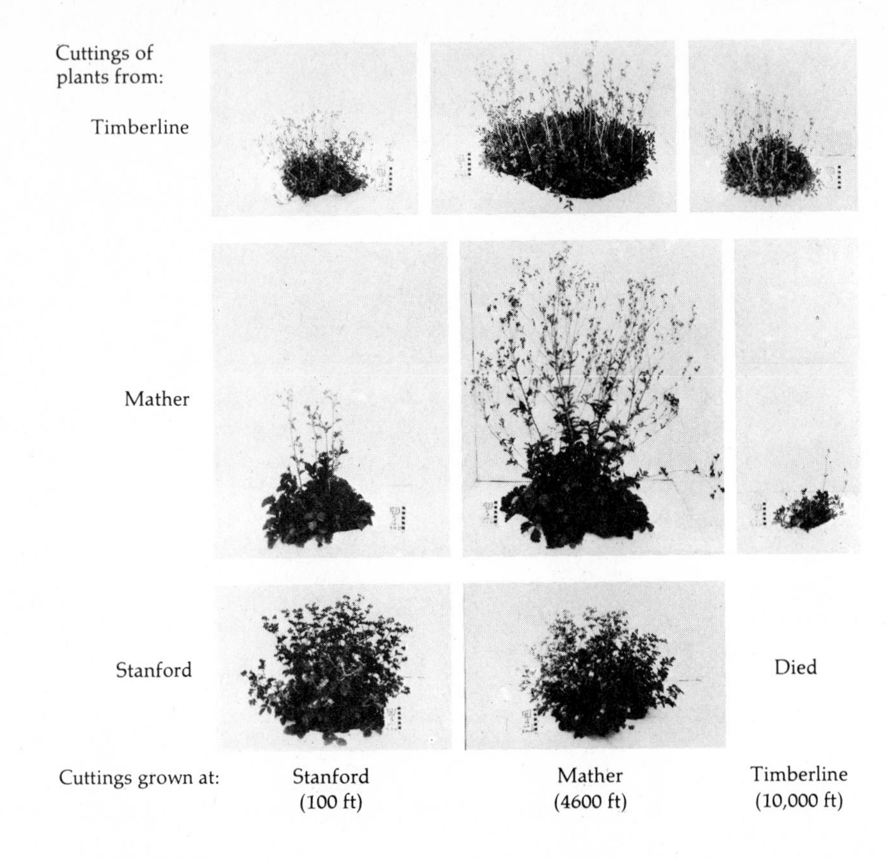

Figure 2.16
Effects of the genotype and the environment on the phenotype. Cuttings from *Potentilla glandulosa* plants collected at different altitudes were planted together in three different experimental gardens. Plants in the same row are genetically identical because they have been grown from cuttings of a single plant; plants in the same column are genetically different but have been grown in the same environment. Genetically identical plants (for example, those in the bottom row) may prosper or die, depending on the environmental conditions. Genetically different plants (for example, those in the first column) may have quite different phenotypes, even when grown in the same environment. (Photographs courtesy of Dr. William M. Hiesey, Carnegie Institution of Washington, Palo Alto, Cal.)

between the strains disappeared when rats of both strains were raised in an unfavorable environment of severe deprivation, and they nearly disappeared when the rats were raised with abundant food and other unusually favorable conditions. As with the cinquefoil plants, we see (1) that a given genotype gives rise to different phenotypes in different environments and (2) that the *differences* in phenotype between two genotypes change from one environment to another—the genotype that is best in one environment may not be best in another.

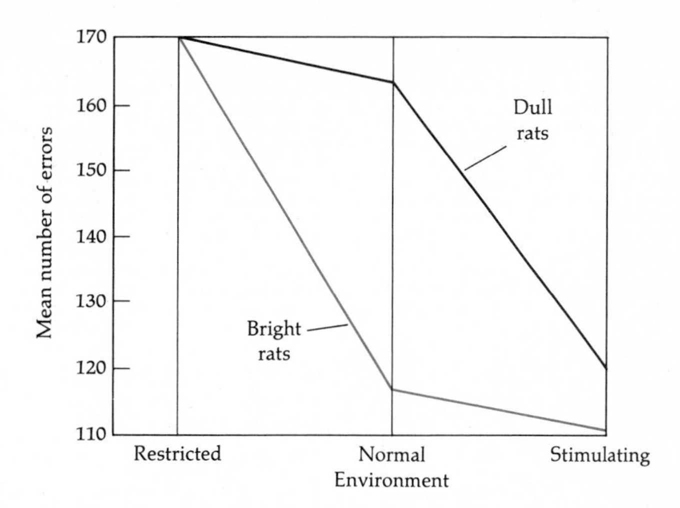

Figure 2.17
Results of an experiment with two strains of rats, one selected for brightness, the other for dullness. When raised in the same environment as that in which the selection was practiced ("normal"), bright rats made about 45 fewer errors than dull rats in the maze used for the tests. However, when they were raised in an impoverished ("restricted") environment, bright and dull rats made the same number of errors. When raised in an abundant ("stimulating") environment, the two strains performed nearly equally well. [After R. M. Cooper and J. P. Zubek, *Can. J. Psychol.* *12*:159 (1958).]

Because of the variable interactions it may have with different environments, the genotype of an organism does not unambiguously specify its phenotype. Rather, the genotype determines the range of phenotypes that *may* develop; this is called the *range of reaction,* or norm of reaction, of the genotype. Which phenotype will be realized depends on the environment in which development takes place. For this reason, the entire range of reaction of a genotype is never known, because this would require that individuals with that genotype be exposed to all possible kinds of environments, which are virtually infinite.

Problems

Note: The chi-square method of testing a hypothesis is described in the Statistical Appendix, page 781.

1. Mendel discovered that in peas axial position of the flowers is dominant over terminal position. Let the allele for axial be represented by A and the

allele for terminal by *a*. Determine the kinds and proportions of gametes and of progeny produced by each of the following crosses: $AA \times aa$, $AA \times Aa$, $Aa \times aa$, $Aa \times Aa$.

2. A geneticist working with guinea pigs made the cross black × albino twice, using different animals. In the first cross he obtained 12 black progeny, but in the second he obtained 6 black and 5 albino progeny. What are the likely genotypes of the parents in each cross?

3. The ovaries of an albino guinea pig were surgically replaced with those from a homozygous black female. When mated with an albino male, the albino female produced two black offspring. Is this result consistent with the theory of pangenesis? With Mendelian inheritance?

4. Two black female rats were crossed with a brown male and several litters were obtained from each female. The first female produced 36 black rats, the second produced 14 blacks and 10 browns. What is the mode of inheritance of black and brown coat color in rats? What are the genotypes of the parents? Use the chi-square method to test your hypothesis.

5. A geneticist inbred six green corn plants of a special stock and raised several seeds from each plant. The seeds produced green and albino (lacking chlorophyll) plants in the following ratios:

Parent	Green offspring	White offspring
1	38	11
2	26	11
3	42	12
4	30	9
5	36	14
6	48	12

What is the likely mode of inheritance of albinism in corn? What was the genotype of the parent plants? Use the chi-square method to test your explanation for each parent.

6. A corn plant from the same stock as in the previous problem was fertilized with pollen from a plant from a different stock. Of the seeds raised, 20 produced green plants and 10 produced albino plants. Is this result com-

patible with a 3:1 ratio or with a 1:1 ratio? What was the likely genotype of the male parent?

7. In poultry, rose comb is dominant over single comb. A farmer believes that some of his rose-combed Wyandotte fowls may carry an allele for single comb. How could he find out which fowls are heterozygous?

8. A form of peroneal muscular atrophy consists of a gradual wasting of the distal muscles of the limbs, starting between the ages of 10 and 20. Pedigrees show that a person never has this kind of peroneal muscular atrophy unless one of the parents also had it. What is the likely mode of inheritance of this condition?

9. A second kind of peroneal muscular atrophy is known that is more severe than the one referred to in the previous problem. This second kind is found almost exclusively among children whose parents are first cousins and do not suffer from the condition. How is this kind of peroneal atrophy inherited?

10. A man of blood group A marries a woman of blood group B and they have a child of blood group O. What are the genotypes of these three people? What other genotypes, and in which frequencies, would you expect among the children of such marriages?

11. Table 2.1 gives the F_2 numbers for each of the seven crosses made by Mendel. Use the chi-square method to test whether or not the numbers obtained in each case are consistent with the 3:1 ratio predicted by his hypothesis.

12. In guinea pigs, black coat *B* is dominant over albino *b*, and rough coat *R* is dominant over smooth coat *r*. *R* and *B* are independent genes. When a cross is made between a homozygous black rough animal and an albino smooth one, what will be the appearance of the F_1? Of the F_2? Of the offspring of a cross between the F_1 and an albino smooth parent?

13. A black rough guinea pig bred with an albino rough one (see previous problem) produces the following progeny: 13 black rough, 16 albino rough, 6 black smooth, and 5 albino smooth. Identify the genotypes of the parents, using the chi-square method to test your explanation.

14. In *Drosophila melanogaster* a recessive allele causes short vestigial wings (vg), whereas the dominant allele (vg^+) results in normal long wings. An allele from a different gene pair is recessive and produces scarlet eyes (st), whereas the dominant allele (st^+) produces normal red eyes. The progenies obtained in three different experiments are listed below. Determine

the genotypes of the parents, using the chi-square method to test your hypothesis.

Parents	Progeny			
	Long wings, red eye	Long wings, scarlet eye	Vestigial wings, red eye	Vestigial wings, scarlet eye
1. Long wings, red eye × vestigial wings, scarlet eye	178	164	142	140
2. Long wings, red eye × long wings, red eye	364	0	107	0
3. Long wings, red eye × long wings, red eye	309	107	95	29

15. In sesame, the one-pod condition is dominant over three-pod, and normal leaf is dominant over wrinkled leaf. The two characters are inherited independently. What are the genotypes of the two parents in each of the five experiments listed below?

Parents	Progeny			
	One-pod, normal	One-pod, wrinkled	Three-pod, normal	Three-pod, wrinkled
1. One-pod, normal × three-pod, normal	362	118	0	0
2. One-pod, normal × three-pod, wrinkled	211	0	205	0
3. One-pod, wrinkled × three-pod, normal	78	90	84	88
4. One-pod, normal × three-pod, normal	318	98	323	104
5. One-pod, normal × one-pod, wrinkled	110	113	33	38

16. How many different kinds of gametes, of genotypes, and of phenotypes (assuming dominance) are expected in the progeny of a self-fertilizing plant heterozygous for three, five, or seven different genes?

17. A plant heterozygous for four independently assorting pairs of genes (*Aa Bb Cc Dd*) is self-fertilized. Determine the expected frequency of the following genotypes in the progeny of such a plant: (1) *aa bb cc dd;* (2) *aa bb Cc Dd;* (3) *Aa Bb Cc Dd.*

18. In a species of *Drosophila,* seven different alleles are known of the gene coding for the enzyme acid phosphatase. How many different genotypes are possible for this gene?

19. What proportion of all possible genotypes are homozygotes when the number of different alleles for a certain gene is three? Five? Six?

20. In rats a cross of wild color (agouti, solid, dark-eyes, undilute color) × black, spotted, ruby-eyed, dilute produced F_1 individuals all having wild color. Crosses of F_1 wild individuals × black, spotted, ruby-eyed, dilute produced the following:

Wild color	41	Spotted, dilute agouti	33
Black	50	Spotted, dilute black	34
Spotted agouti	29	Ruby-eyed, dilute agouti	39
Spotted black	30	Ruby-eyed, dilute black	34
Ruby-eyed, agouti	41	Spotted, ruby-eyed, agouti	35
Ruby-eyed, black	38	Spotted, ruby-eyed, black	32
Dilute agouti	36	Spotted, ruby-eyed, dilute agouti	25
Dilute black	29	Spotted, ruby-eyed, dilute black	30

Explain the results. In what proportions will these phenotypes appear in the progeny of the cross $F_1 \times F_1$?

3

The Chromosomal Basis of Heredity

Genes and Chromosomes

The discovery that two seemingly separate branches of science are in fact causally interrelated contributes to the advancement of science, because the knowledge gained in each field provides explanations for the other field. Such an advance took place when it was shown that Mendelian genetics and the processes of meiosis and mitosis are related. In 1902 two investigators—Walter S. Sutton in the United States and Theodor Boveri in Germany—independently suggested that genes are contained in chromosomes, an idea that came to be known as the *chromosome theory of heredity.* Their argument was based on the parallel behavior, at meiosis and at fertilization, between chromosomes on the one hand and genes on the other hand. The existence of two alleles for a given character, one inherited from each parent, parallels the existence of two chromosomes, also derived one from each parent. The two alleles for a character segregate in the formation of the gametes because the two chromosomes of each pair pass into different gametes during meiosis (Figure 3.1). Some genes for different characters assort independently because they are in nonhomologous chromosomes, and these chromosomes assort themselves in the gametes independently of the parent from which they came (Figure 3.2).

The parallel behavior of the chromosomes and the genes in the formation of gametes and in fertilization strongly suggested that genes were located in chromosomes. More compelling evidence supporting the

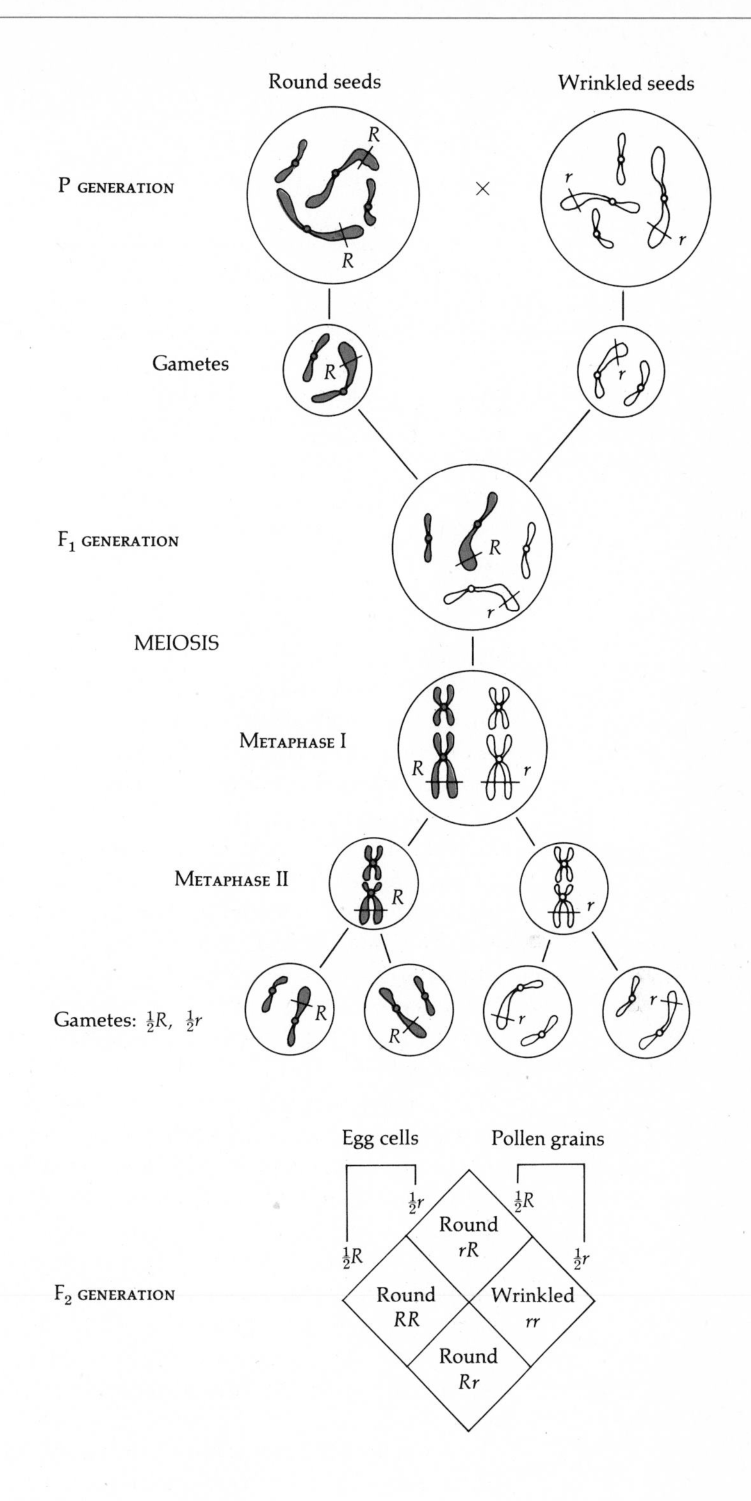

Figure 3.1
Chromosomal basis of Mendel's law of segregation. The example is a cross between a round-seed plant and a wrinkled-seed plant; only two pairs of chromosomes are shown, one of them carrying the allele for either roundness or wrinkling. Meiosis results in gametes all carrying the allele for roundness (*R*) in round-seed plants, but for wrinkling (*r*) in wrinkled-seed plants. The F_1 plants have one chromosome carrying *R* and the other carrying *r*; meiosis in F_1 plants produces gametes with *R* and gametes with *r* in equal proportions. Random union of these gametes at fertilization produces the observed 3:1 ratio of round versus wrinkled seeds.

chromosome theory of heredity came from demonstrations of the association between specific genes and specific chromosomes. An early demonstration was provided in two steps through experiments performed by Nobel laureate Thomas Hunt Morgan in 1910 and by his student and colleague Calvin B. Bridges in 1916. These experiments were done with *Drosophila melanogaster,* the small, yellowish brown fruit fly that hovers around fallen fruit in summer and fall (see Figures 3.3 and 3.4).

Sex-Linked Heredity

Like so many other great scientific achievements, the discoveries of Morgan and Bridges came from the study of exceptions to the expected results. Morgan had a strain of *D. melanogaster* flies that had white, rather than the normal red, eyes. The strain was true-breeding: the offspring of white-eyed flies also had white eyes. However, when white-eyed flies were crossed to red-eyed flies, the progenies were not as expected according to Mendelian heredity.

When a red-eyed female and a white-eyed male are crossed (Figure 3.5), the F_1 flies are all red-eyed, exactly as expected if red-eye is dominant over white-eye. When the F_1 red-eyed flies are bred together, they produce F_2 flies that are three-quarters red-eyed and one-quarter white-eyed, also as expected if red-eye is dominant and white-eye recessive. However, *all the F_2 females are red-eyed, whereas half the males are red-eyed and half are white-eyed.* This is *not* expected according to the principles of Mendelian heredity. Other unexpected results appear when the F_2 flies are bred. The males are all true-breeding: the red-eyed males carry only red-eye genes, and the white-eyed males carry only white-eye genes. The F_2 females are of two kinds: half produce only red-eyed offspring, while the other half produce male offspring half of which are red-eyed and half white-eyed.

The results are different when white-eyed females are crossed with red-eyed males (Figure 3.6). The F_1 offspring are not all red-eyed, as would be expected according to Mendelian principles if red-eye is dominant over white-eye. Rather, half the offspring are red-eyed and half are white-eyed; moreover, all the red-eyed flies are females and all the white-eyed flies are males. When these are bred together, the F_2 offspring consist of half (rather than one-quarter) white-eyed flies and half red-eyed flies, this time in equal numbers in both sexes.

Morgan saw that the results could be explained if (1) the eye-color gene was in the sex (X) chromosome and (2) the male sex chromosome (Y) contained *no* gene for eye color. As noted in Chapter 1, chromosomes occur in pairs. However, males and females differ with respect to the one pair of chromosomes that are associated with sex determination. *Drosophila* females carry two identical X chromosomes, but males carry two different chromosomes, one X and one Y (Figure 3.4). Females inherit one

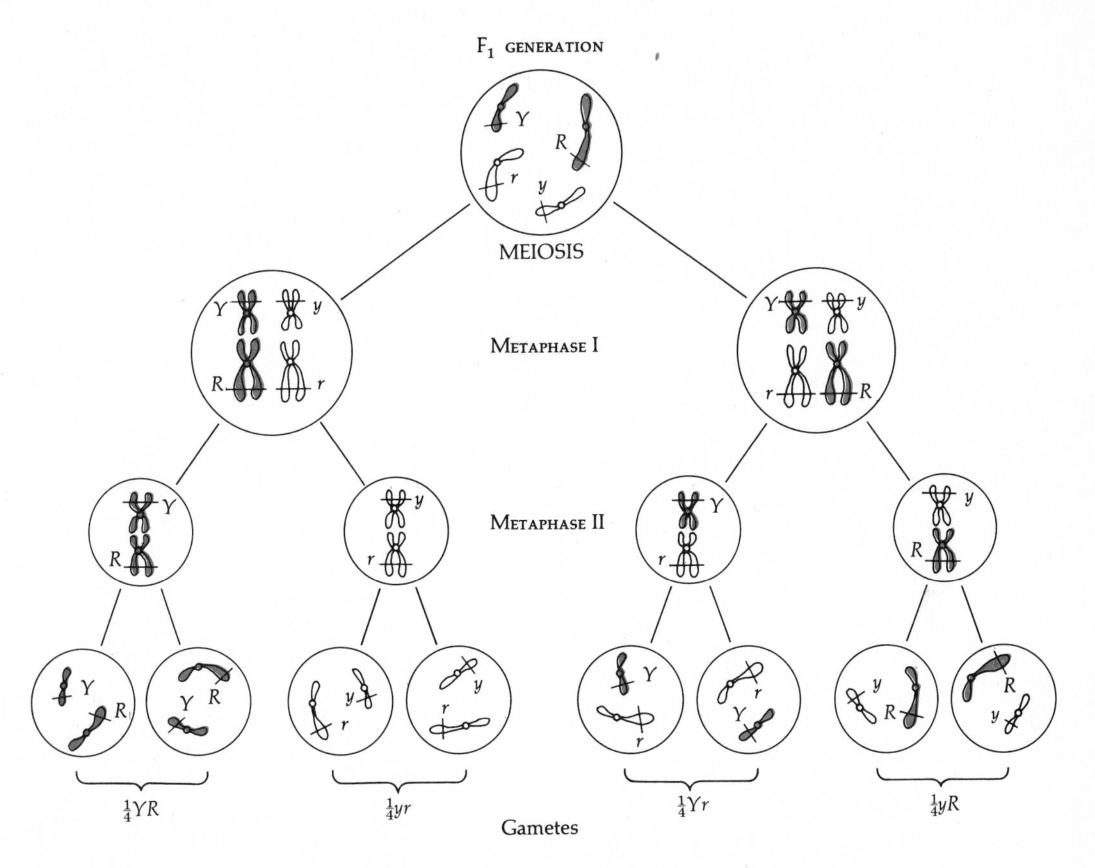

Figure 3.2
The law of independent assortment as a consequence of independent association of nonhomologous chromosomes at meiosis. The cross round-yellow × wrinkled-green produces hybrids (*Rr Yy*) carrying either *R* or *r* in two homologous chromosomes and *Y* or *y* in the other two homologous chromosomes. In metaphase I of meiosis, the chromosomes coming from the same parent may align, with equal probability, either on the same side (as on the left of the diagram) or on different sides (as on the right). In the first case, the resulting gametes have the same gene combinations (*YR* and *yr*) as were present in the parental generation; in the second case, the alternative combinations appear (*Yr* and *yR*). The final results are four kinds of gametes, each with a frequency of ¼; random combination among these four kinds of gametes produces the proportions 9:3:3:1, as observed by Mendel.

X chromosome from their father and one from their mother, and transmit their X chromosomes to their sons as well as to their daughters. Males, on the other hand, receive their only X chromosome from their mother and transmit it only to their daughters (Figures 3.5 and 3.6). Consequently, a trait determined by a gene located in the X chromosome will follow a crisscross inheritance: the male transmits such traits to his grandsons through his daughters, never to or through his sons.

Morgan concluded that the eye-color gene is *sex-linked,* i.e., it is carried in the X chromosome. His demonstration that the behavior of one specific gene corresponded to the behavior of one specific chromosome visible in

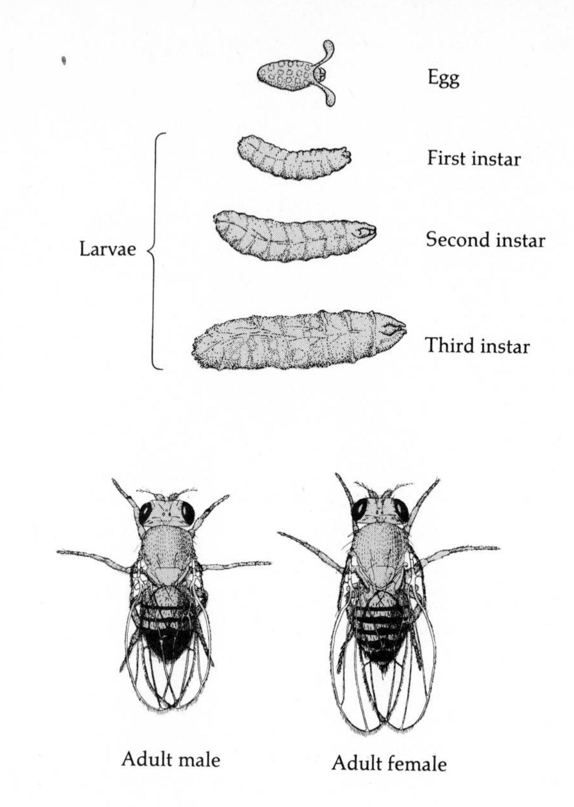

Figure 3.3
Drosophila melanogaster. Usually called fruit fly, *D. melanogaster* breeds in decaying fruit and fermented matter. It is a small animal, about 2 mm in length and 1 mg in weight, with a yellowish brown body and red eyes. Its extensive use in genetic studies is facilitated by a short generation time (two weeks), large number of progeny (several hundred per mating pair), and ease and low cost of rearing in the laboratory.

microscopic preparations was strong evidence in favor of the chromosome theory of heredity. Definite proof of the theory was provided six years later by Bridges.

Nondisjunction of X Chromosomes

As shown by Morgan, white-eyed *Drosophila* females crossed with red-eyed males produce red-eyed daughters and white-eyed sons (Figure 3.6). There are, however, occasional exceptions to this rule as well. About 1 in 2000 flies in the F_1 offspring have an unexpected eye color, either white in females or red in males. Bridges proposed that these exceptional flies are produced because of X-chromosome *nondisjunction,* i.e., the two X chromosomes fail to separate at meiosis, so both go to the same pole, producing eggs with either two X chromosomes or no X chromosome (Figure 3.7).

When a white-eyed female produces an egg with two X chromosomes and this is fertilized by a sperm carrying a Y chromosome, the resulting zygote has two X and one Y chromosomes, but both X chromosomes carry

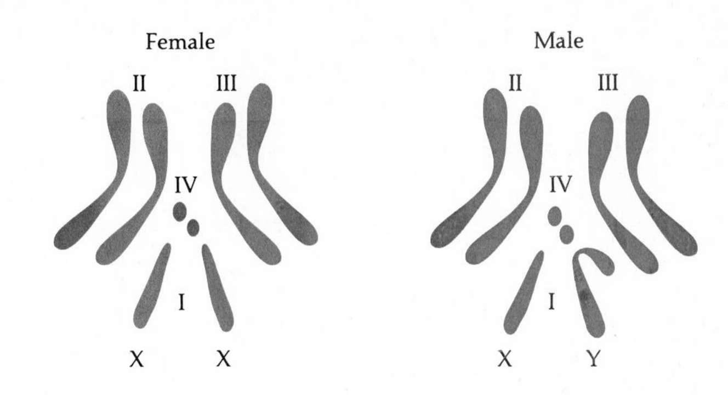

Figure 3.4
The chromosomes of *Drosophila melanogaster.* Chromosome I is telocentric in the female (X) and acrocentric in the male (Y); chromosomes II and III are metacentric; chromosome IV is small and dotlike.

the white-eye gene. Bridges hypothesized that the exceptional white-eyed females develop from such zygotes. When an egg with no X chromosome is fertilized by an X-carrying sperm from a red-eyed male, the zygote has one X but no Y chromosome, and the X chromosome carries the red-eye gene. Bridges proposed that the exceptional red-eyed males develop from such zygotes. In other words, Bridges' hypothesis was that the exceptional females inherit both X chromosomes from their mother (and a Y chromosome from their father) and that the exceptional males inherit their only X chromosome from their father (they therefore lack the Y chromosome).

Bridges' hypothesis was imaginative and lent itself to experimental test by microscopic examination of the chromosomes of the exceptional flies. Bridges indeed confirmed that the exceptional white-eyed females had two X and one Y chromosomes. Moreover, the exceptional red-eyed males were shown to have one X but no Y chromosome. This proof of the chromosome theory of heredity was considered definitive. A specific gene had been shown beyond reasonable doubt to be associated with a specific chromosome.

Secondary Nondisjunction

Drosophila males without a Y chromosome are normal in appearance, but they are sterile. Females with two X and one Y chromosomes are normal and fertile. Bridges crossed the exceptional white-eyed females (XXY) to normal red-eyed males (XY). He found that about 4% of the female progeny had white eyes and about 4% of the male progeny had red eyes; the other 96% of each sex consisted of red-eyed females and white-eyed males. Bridges hypothesized that the exceptional females and males again resulted from nondisjunction of the X chromosomes in female meiosis. He called this *secondary nondisjunction* because it happened in the progenies of females that were themselves produced by *primary nondisjunction* (and thus had two X and one Y chromosomes) (Figure 3.8). Secondary nondisjunc-

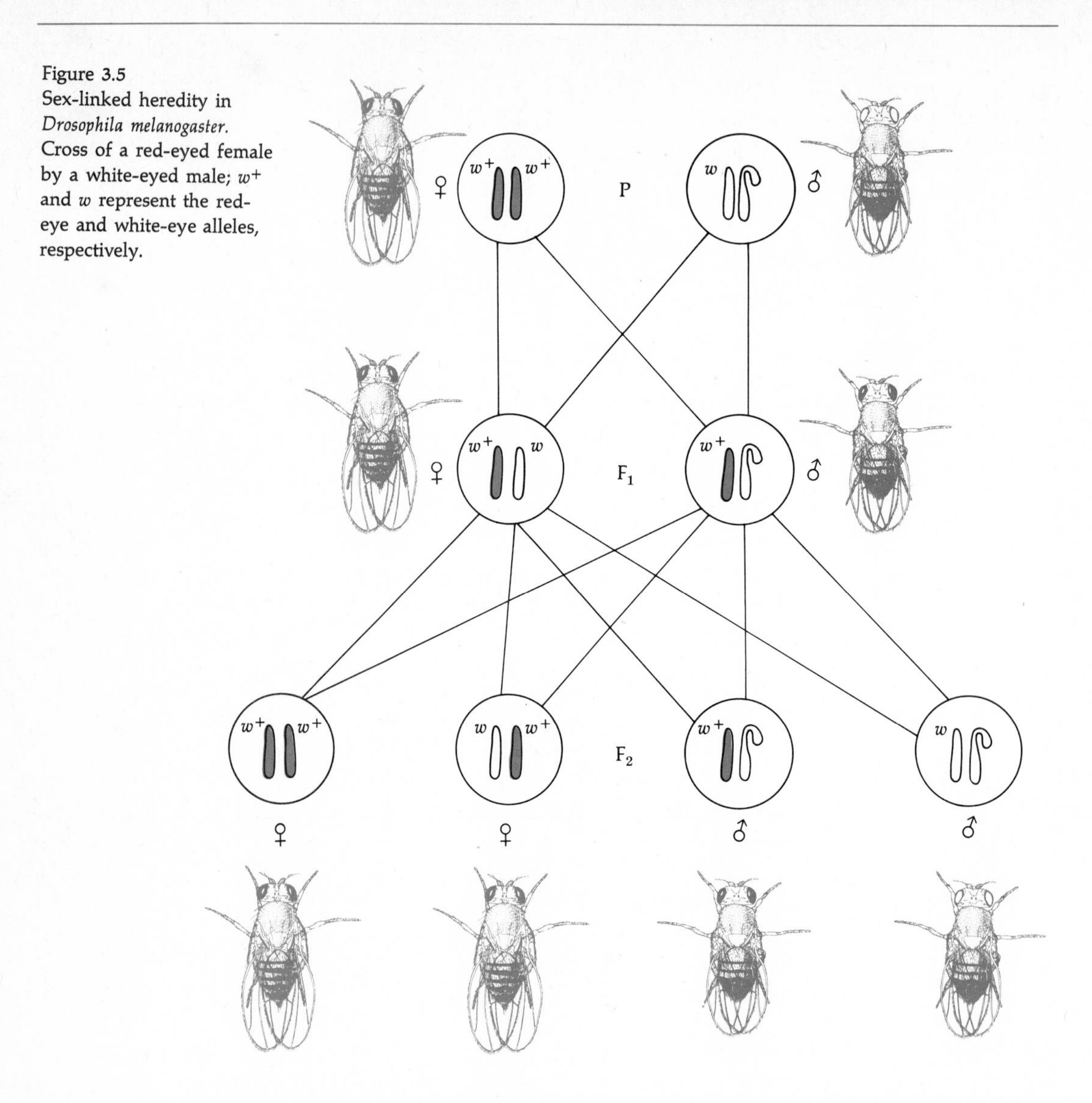

Figure 3.5
Sex-linked heredity in *Drosophila melanogaster.* Cross of a red-eyed female by a white-eyed male; w^+ and w represent the red-eye and white-eye alleles, respectively.

tion occurred with a frequency (about 1 in 25 offspring) about 100 times greater than that of primary nondisjunction (about 1 in 2000 offspring).

Nondisjunction may result from the physical attachment of two X chromosomes, a phenomenon that leads to 100% nondisjunction; this was discovered in 1922 by Lilian V. Morgan (T. H. Morgan's wife) (Figure 3.9). Yellow body color in *Drosophila melanogaster* is due to a sex-linked allele (y) that is recessive to the wild-type allele (y^+). Yellow-bodied *Drosophila* have pale yellow bodies rather than the yellowish brown characteristic of normal flies. L. V. Morgan found an exceptional yellow female that, when crossed to a normal male, produced only yellow daughters and normal sons, rather than normal daughters and yellow sons. The yellow daughters repeated the exceptional performance of their mother: they produced only yellow female and normal male offspring. Cytological examination proved

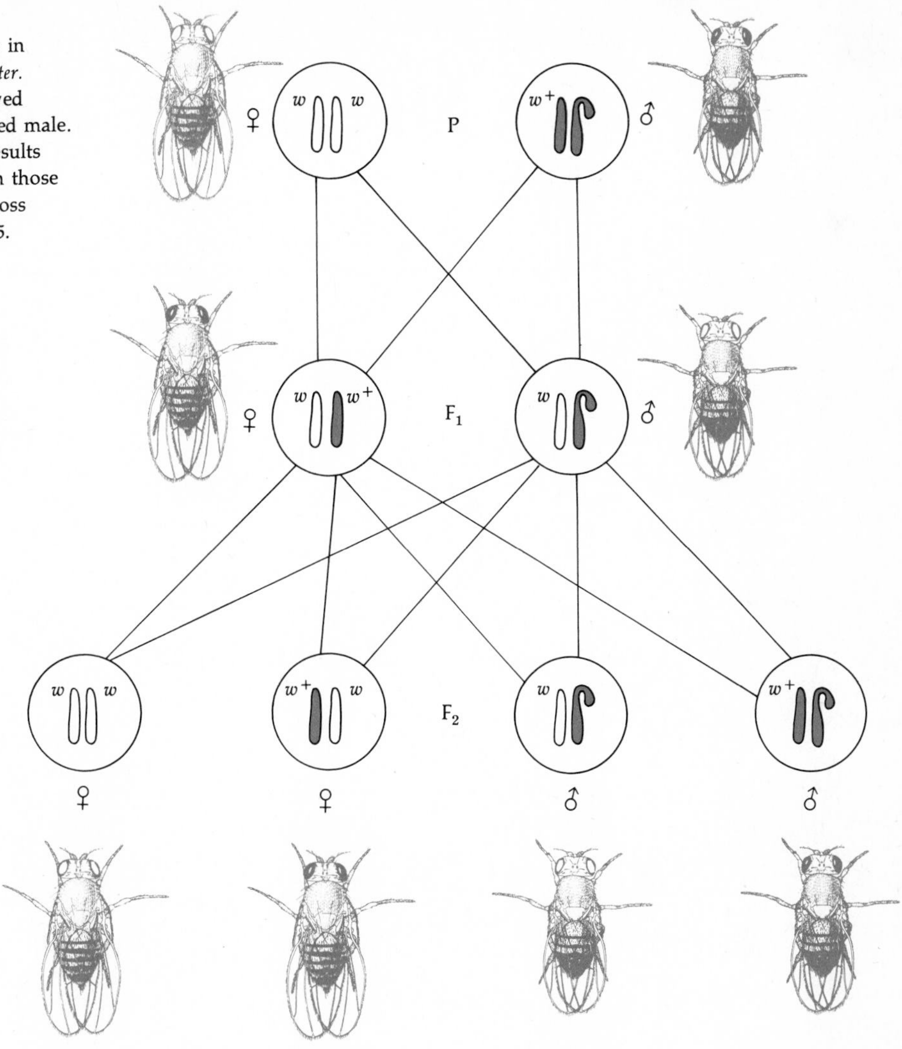

Figure 3.6
Sex-linked heredity in *Drosophila melanogaster.* Cross of a white-eyed female by a red-eyed male. This cross yields results quite different from those of the reciprocal cross shown in Figure 3.5.

that all the exceptional yellow females had their two X chromosomes attached to a single centromere and carried a Y chromosome as well.

Sex-Linked Heredity in Humans and Other Organisms

The pattern of sex-linked inheritance described for *Drosophila* is also observed in all animals and plants in which the males are the heterogametic sex. The males are said to be *hemizygous* with respect to the genes in the

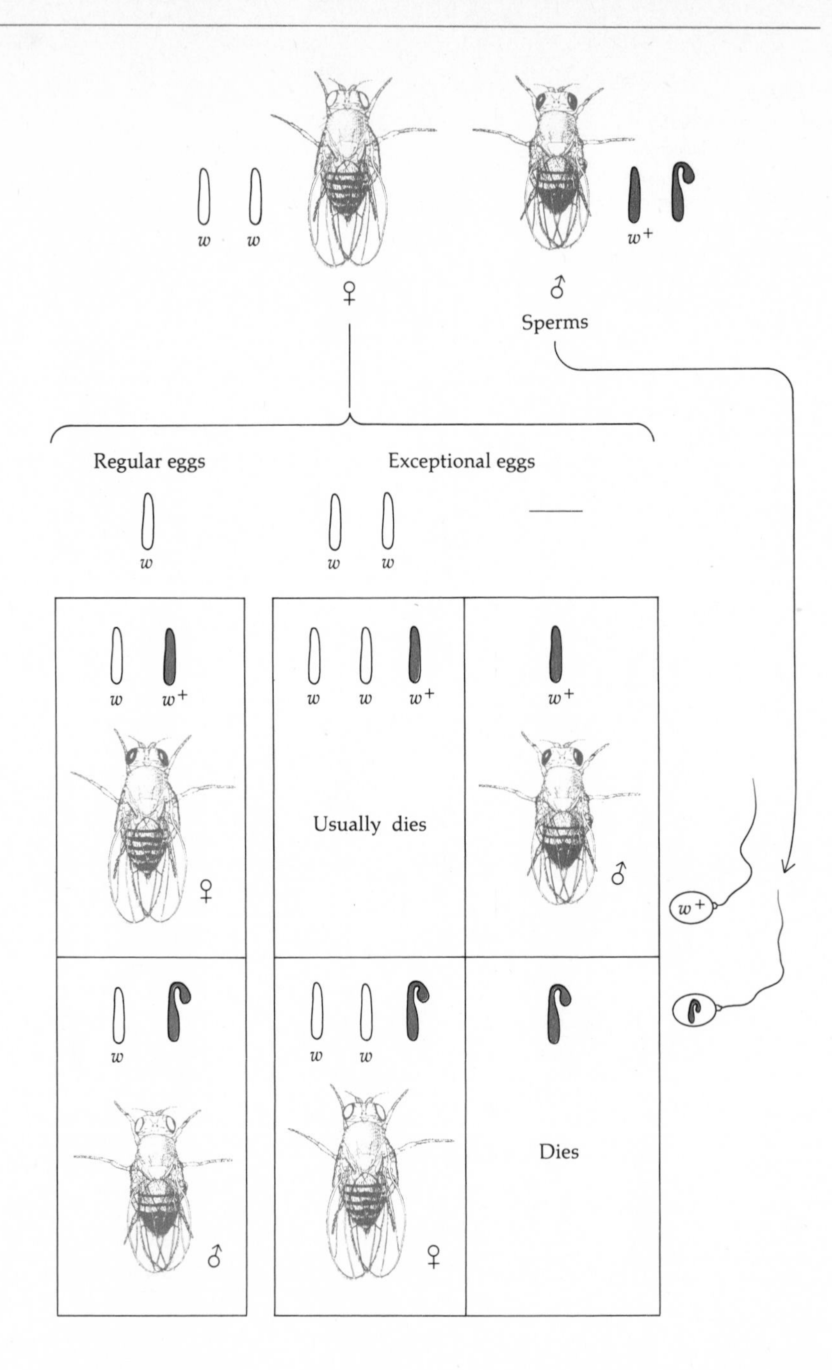

Figure 3.7
Primary nondisjunction in *Drosophila melanogaster.* Normal results are shown on the left for comparison.

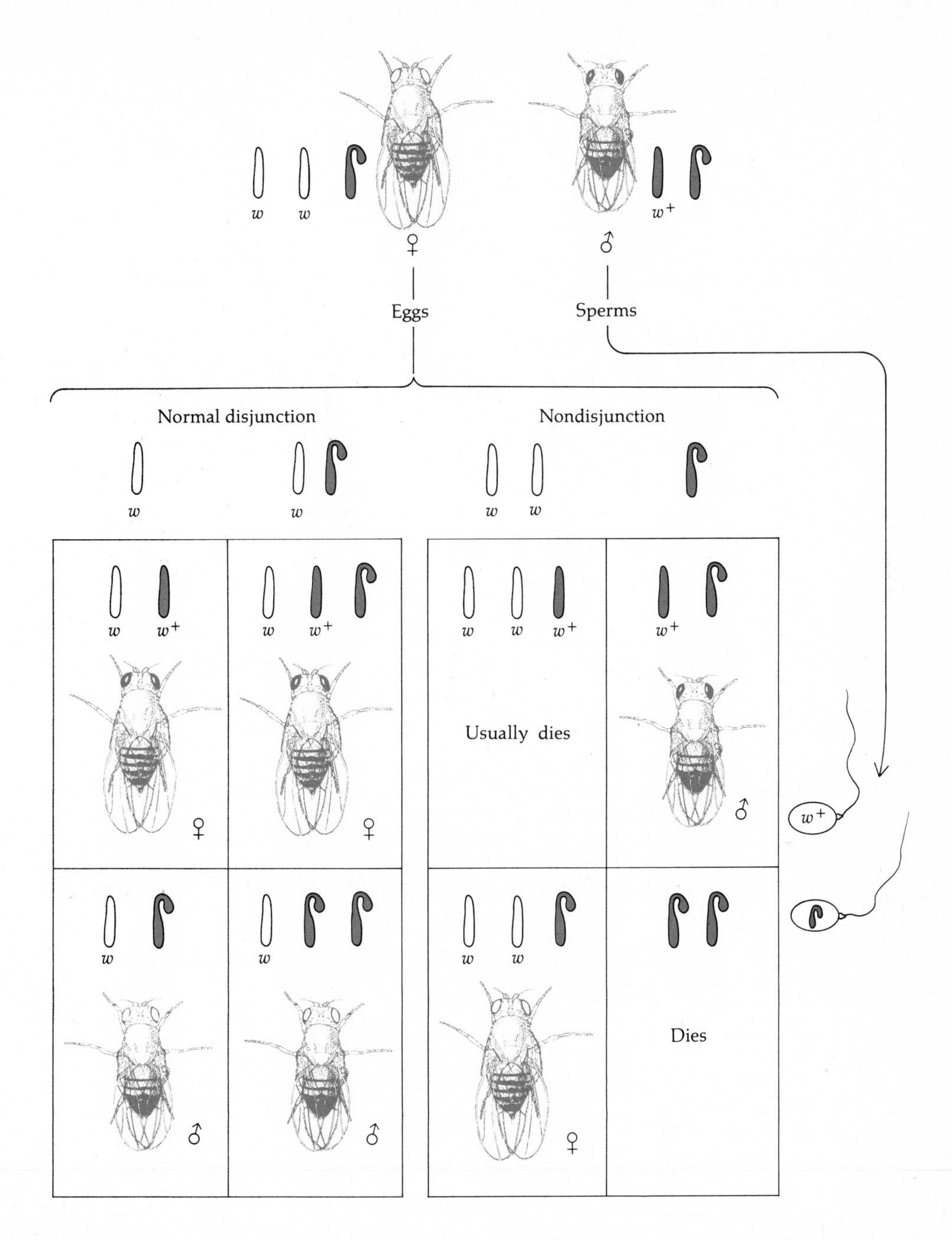

Figure 3.8
Secondary nondisjunction in *Drosophila melanogaster.* Bridges hypothesized that segregation in the meiosis of females with two X and one Y chromosomes can occur in two ways: (1) The two X chromosomes each go to a different pole, one of them accompanied by a Y chromosome—this happens about 96% of the time. (2) The two X chromosomes go to the same pole and the Y chromosome to the other—this happens about 4% of the time. Bridges verified his hypothesis by microscopic examination of the chromosomes of the various kinds of flies.

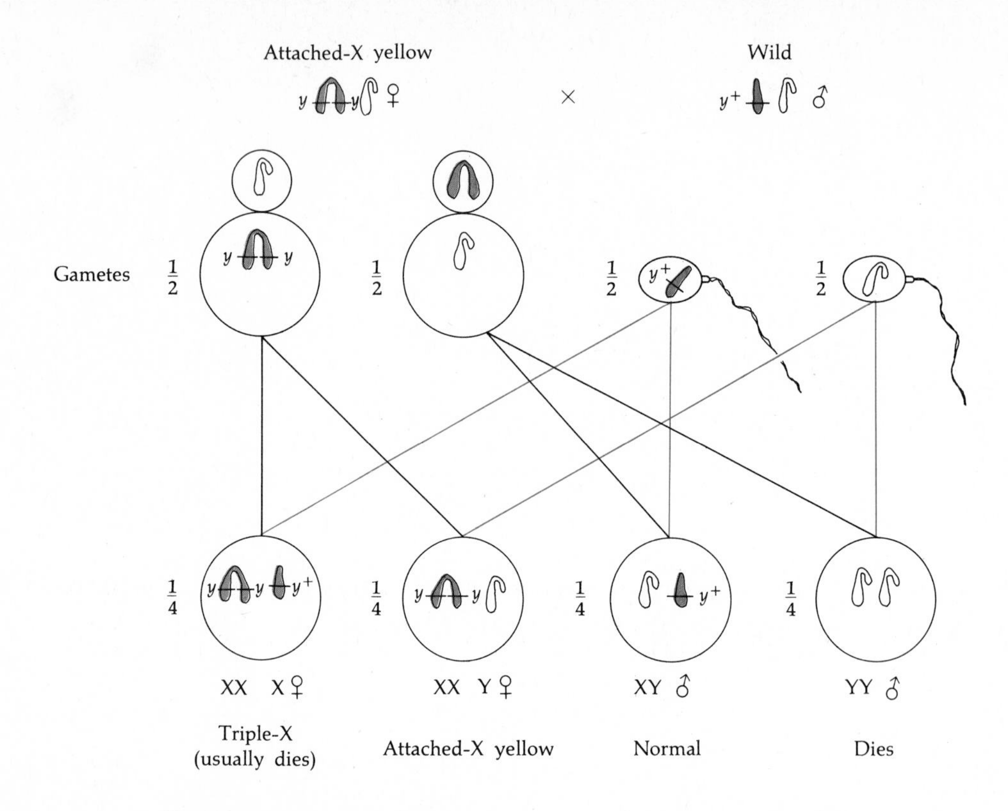

Figure 3.9
Nondisjunction as a consequence of attached-X chromosomes. X chromosomes are shown in color, Y chromosomes in white. An attached-X yellow female produces only two kinds of eggs, because the two attached-X chromosomes remain attached during meiosis and must, therefore, go together to the same pole. When fertilized by an X-carrying sperm, these eggs produce zygotes either with three X chromosomes (which usually die) or with one X and one Y chromosome, like normal males. When the two kinds of eggs are fertilized by a Y-carrying sperm, they produce zygotes either with two attached-X chromosomes and one Y chromosome or with two Y but no X chromosomes (which always die).

X chromosome; indeed, they are neither homozygous nor heterozygous with respect to such genes.

About 150 sex-linked hereditary traits are known in humans. The mode of transmission of red-green color blindness—from the mother's family to her sons—has been known for more than one hundred years. In 1911 Edmund B. Wilson pointed out that all the facts about the heredity of red-green color blindness in humans could be explained if the allele responsible for this condition were a sex-linked recessive and males were the heterogametic sex (Figure 3.10).

Another example of human sex-linked inheritance is hemophilia, a serious disease characterized by the inability of blood to clot. In normal persons, bleeding after a moderate injury is limited by the clotting of

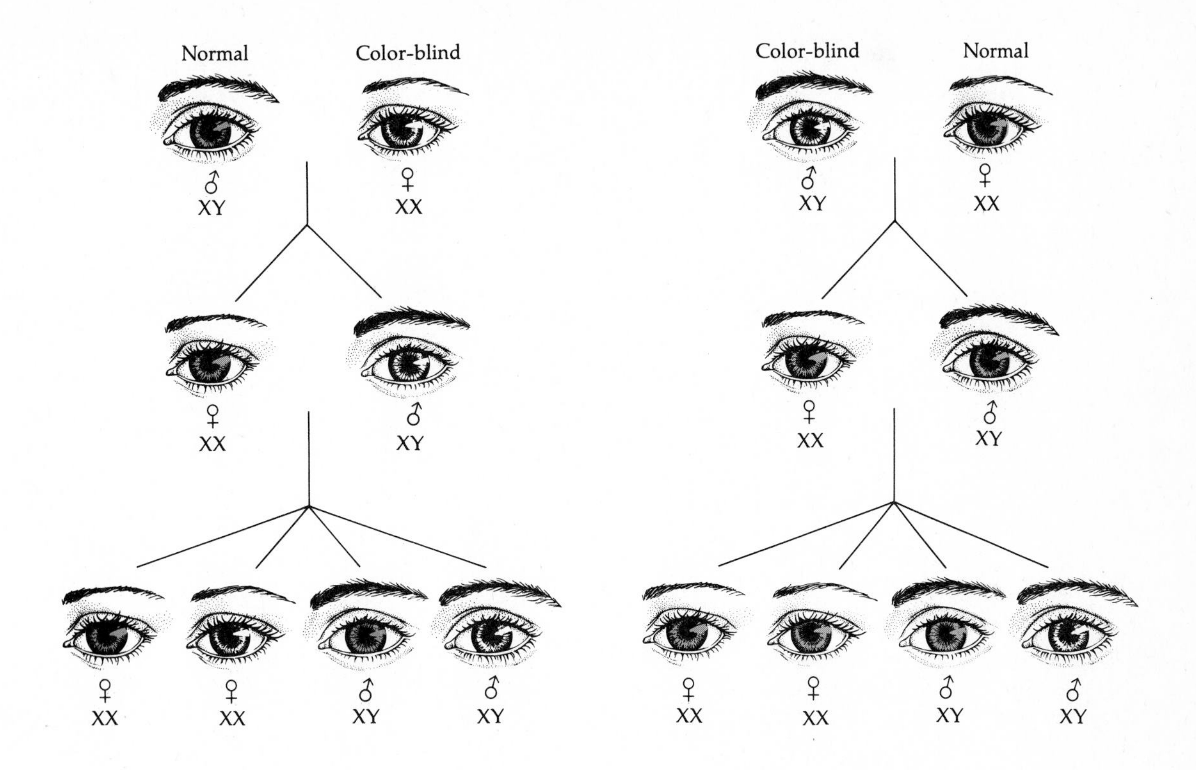

Figure 3.10
Inheritance of red-green color blindness in humans. The sons of a color-blind woman are all color-blind; her daughters are all carriers. A color-blind man transmits the gene for color blindness to all his daughters and, through them, to half the daughters' sons.

blood. In hemophiliacs, even a small injury can lead to death from bleeding. There are at least three kinds of hemophilia, two of them caused by recessive sex-linked genes and a very rare kind caused by an autosomal recessive gene. Each affects a different factor normally required for blood to clot. A famous case of sex-linked hemophilia that has affected some royal houses of Europe can be traced to Queen Victoria (Figure 3.11). Since no hemophilia is known in her ancestors, it is likely that the hemophilia allele appeared by mutation in one of the gametes from which she was conceived.

The pattern of inheritance of sex-linked traits described for *Drosophila* and humans is reversed when females are the heterogametic sex, as is true in birds, moths, butterflies, and some fish. In these organisms the females are hemizygous with respect to sex-linked traits and transmit such traits only to their sons, whereas the males transmit sex-linked traits to both their sons and their daughters (Figures 3.12 and 3.13).

The Y Chromosome

The sex-linked genes discussed in the previous sections of this chapter are contained in the X chromosome but have no corresponding allele in the Y

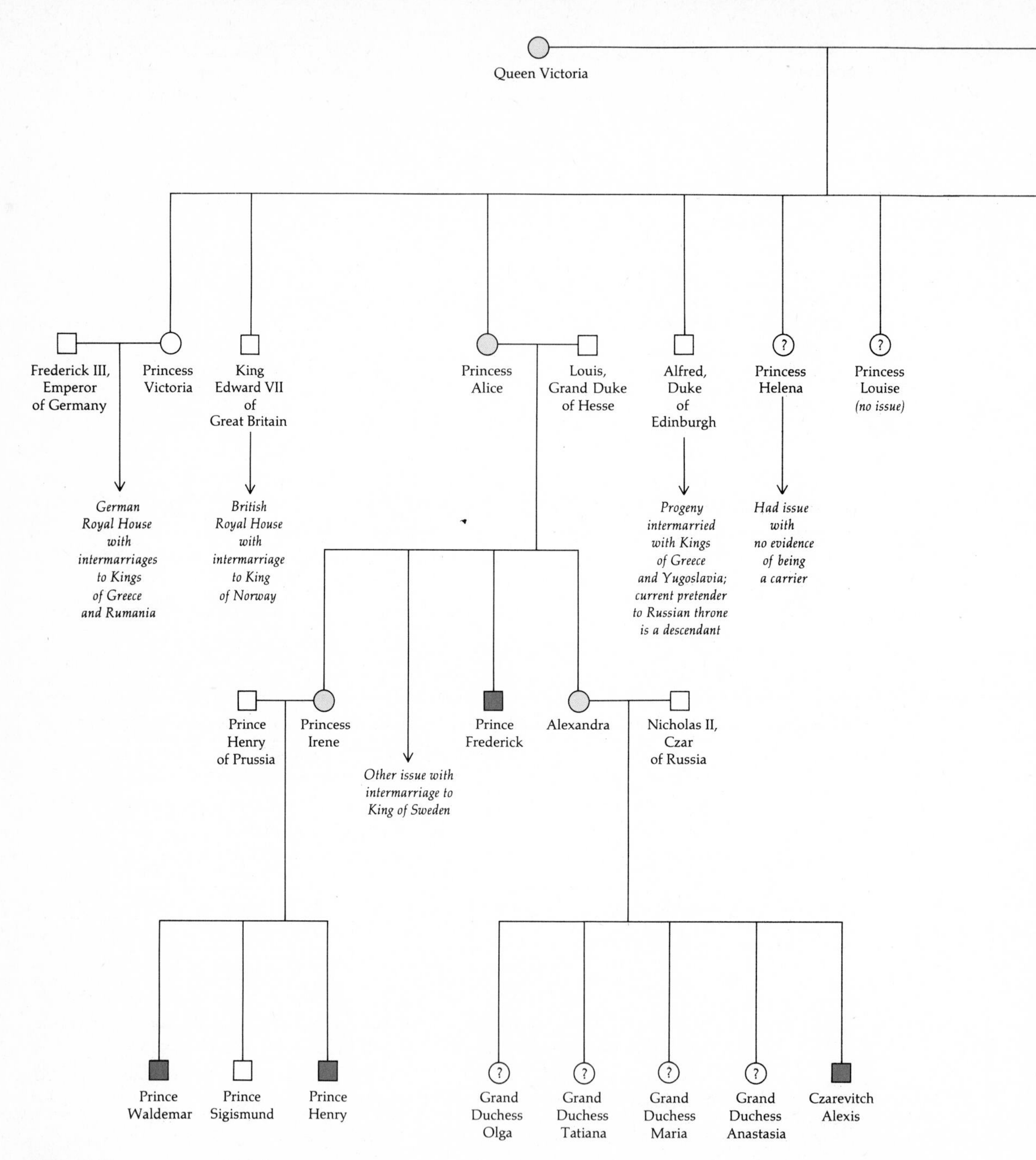

Figure 3.11
Queen Victoria's descendants, showing the transmission of hemophilia. As is customary in human pedigrees, females are represented with circles, males with squares. A horizontal line connecting two individuals indicates marriage; vertical lines from a marriage line lead to the progeny of that marriage.

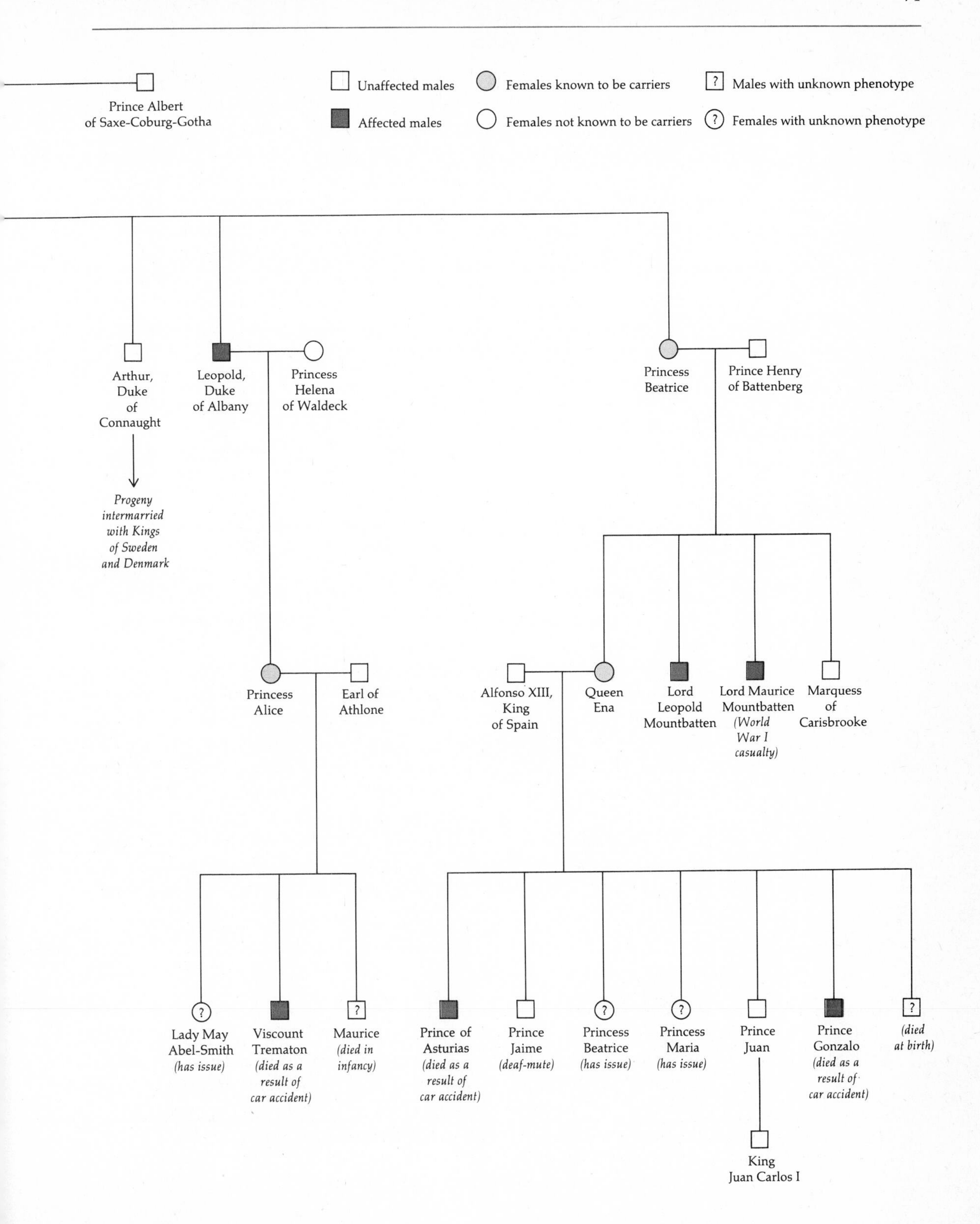

Prince Albert
of Saxe-Coburg-Gotha
Unaffected males
Affected males
Females known to be carriers
Females not known to be carriers
Males with unknown phenotype
Females with unknown phenotype
Arthur, Duke of Connaught
Progeny intermarried with Kings of Sweden and Denmark
Leopold, Duke of Albany
Princess Helena of Waldeck
Princess Beatrice
Prince Henry of Battenberg
Princess Alice
Earl of Athlone
Alfonso XIII, King of Spain
Queen Ena
Lord Leopold Mountbatten
Lord Maurice Mountbatten (World War I casualty)
Marquess of Carisbrooke
Lady May Abel-Smith (has issue)
Viscount Trematon (died as a result of car accident)
Maurice (died in infancy)
Prince of Asturias (died as a result of car accident)
Prince Jaime (deaf-mute)
Princess Beatrice (has issue)
Princess Maria (has issue)
Prince Juan
Prince Gonzalo (died as a result of car accident)
(died at birth)
King Juan Carlos I

chromosome. This is why the males (or the heterogametic sex) are hemizygous with respect to sex-linked genes and always exhibit the phenotype of all their sex-linked alleles, even when these are recessive.

Does the Y chromosome contain any genes at all? The Y chromosome lacks counterparts of most genes present in the X chromosome, yet it is not completely empty of genetic information. We can distinguish two kinds of genetic information: (1) genes that are present only in the Y chromosome and (2) genes that are present in the Y as well as the X chromosome.

In organisms in which males are the heterogametic sex, the Y chromosome is transmitted from the father to all his sons, and only to them. Hence genes present only in the Y chromosome will exhibit *holandric inheritance*—they will be transferred only from father to sons—and will be manifested only in males.

In humans the Y chromosome is necessary for maleness: individuals with one X but no Y chromosome are phenotypically female, although most are sterile (their characteristics are known as Turner's syndrome; see Chapter 17, page 588). In general, the Y chromosome appears to be necessary in animals either for maleness (or femaleness, when the females are the heterogametic sex), or at least for fertility. In *D. melanogaster*, individuals with one X and no Y chromosome are phenotypically male, although sterile; the Y chromosome contains genes necessary for spermatogenesis.

An instance of a gene's being carried in the Y as well as the X chromosome is known in *D. melanogaster*. The mutant bobbed phenotype is characterized by shorter, more slender bristles than those of normal flies. The *bobbed* allele (*bb*) is recessive to its wild-type allele (bb^+), which is associated with the nucleolus organizer (the chromosomal region that forms the nucleolus in the interphase cell). Atypical results are expected in the cross between a female homozygous for the recessive allele ($X^{bb}X^{bb}$) and a heterozygous male. If the male carries the dominant allele in the X chromosome ($X^{bb^+}Y^{bb}$), all the females in the F_1 offspring will be normal ($X^{bb}X^{bb^+}$) and all the males will be bobbed ($X^{bb}Y^{bb}$). If the heterozygous father carries the dominant allele in the Y chromosome ($X^{bb}Y^{bb^+}$), all the F_1 females will be bobbed ($X^{bb}X^{bb}$) and all the F_1 males will be normal ($X^{bb}Y^{bb^+}$).

Sex Determination

Sex is an important aspect of an individual's phenotype. Throughout this chapter it has become apparent that males and females usually have different chromosome constitutions. We now consider more explicitly the role that chromosomes play in sex determination.

In humans as well as in *Drosophila*, females have two X chromosomes and males have one X and one Y chromosome. This type of difference

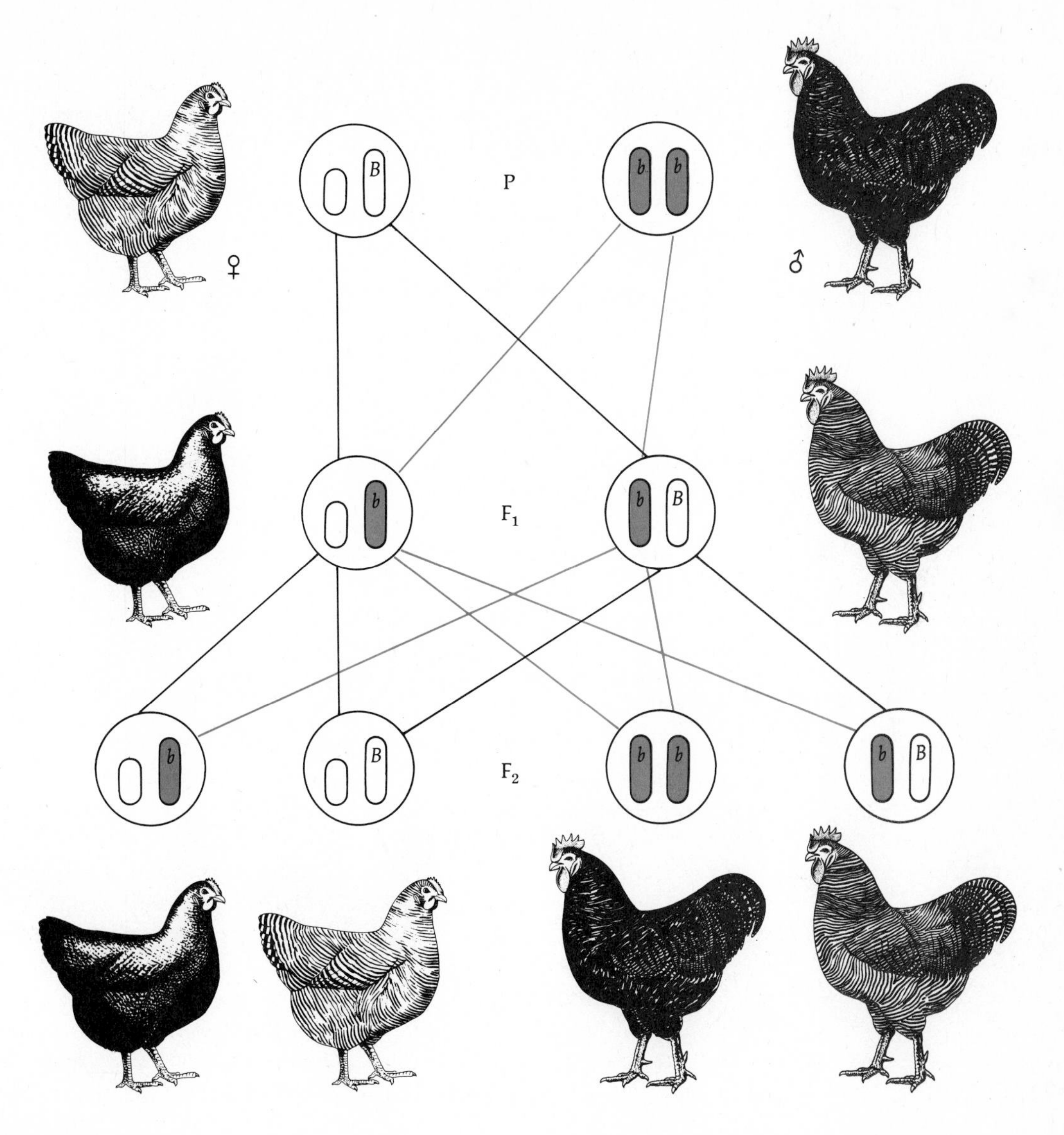

Figure 3.12
Sex-linked inheritance in poultry: cross of a barred female with a nonbarred male. Barred (B) plumage is dominant over nonbarred (b) plumage. In poultry, females are the heterogametic sex and therefore transmit the X chromosome only to their sons, whereas daughters receive their only X chromosome from their father. Hence, daughters exhibit the recessive phenotype present in the father, whereas sons have the dominant phenotype of the mother. This pattern contrasts with the one shown in Figure 3.5, where the dominant phenotype of the mother reappears in both the sons and the daughters.

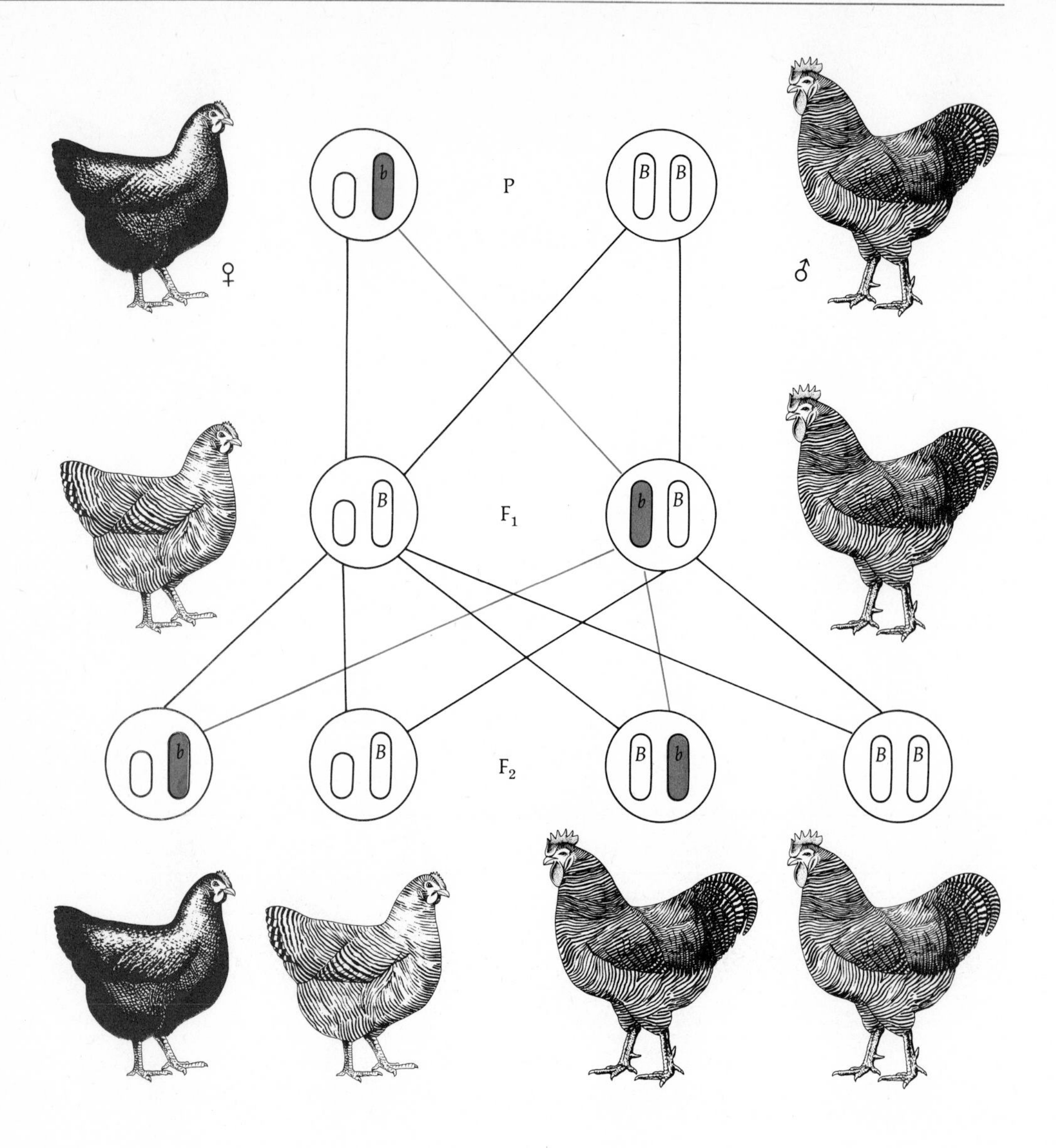

Figure 3.13
Sex-linked inheritance in poultry: cross of a non-barred female with a barred male. Sons and daughters all exhibit the dominant phenotype of the father. This contrasts with Figure 3.6, where only the daughters manifest the dominant phenotype of the father.

between the sexes is characteristic of most vertebrates, many insects and other invertebrates, and many *dioecious* plants (plants in which each individual is either male or female, but not both). But the genetic basis of sex determination is not the same in all these organisms.

In *D. melanogaster* an individual with one X and no Y chromosome has a generally normal male phenotype (although it is sterile). The sex phenotype of *D. melanogaster* flies is determined by the balance between the number of X chromosomes and the number of sets of autosomes (A). When the ratio X/A = 1, the individual is a female; when the ratio is 0.5,

Table 3.1
Sex determination in *Drosophila melanogaster.*

Number of X Chromosomes	Sets of Autosomes (A)	Ratio X/A	Sex Phenotype
3	2	1.5	Metafemale*
2	2	1	Normal female
2	3	0.67	Intersex
1	2	0.5	Normal male
1	3	0.33	Metamale*

*These are sometimes called superfemale and supermale, but there is nothing "super" about them. Flies with these genotypes either die before maturing or are very weak.

the individual is a male; when the ratio is between 0.5 and 1, the individual has an "intersex" phenotype, intermediate between male and female (Table 3.1). The mechanism whereby the ratio of X chromosomes to sets of autosomes is translated into the development of one or the other sex phenotype is unknown. However, specific autosomal genes affecting the sex phenotypes are known in *D. melanogaster.* The mutant gene *tra* (for transformer) in the homozygous condition causes XX individuals that would otherwise be females to be transformed into phenotypic males (although they are sterile because they lack the fertility genes present in the Y chromosome).

With respect to sex determination, humans differ from *Drosophila* in that the Y chromosome is necessary for the development of a normal male phenotype. Sex determination in mammals is discussed in Chapter 14.

In birds, butterflies, and moths, the males are the homogametic sex (XX) and the females are the heterogametic sex (XY or X0). The sex chromosomes of these organisms are sometimes represented as Z and W to distinguish this mode of sex determination from the one with heterogametic males. The males are then represented as ZZ and the females as ZW or Z0 (see Figure 3.14).

A completely different mode of sex determination, called *haplo-diploidy,* is common in bees and ants. There are no sex chromosomes in these organisms: diploid individuals are female and haploid individuals are male (drones). Females develop from fertilized eggs, while unfertilized eggs develop into drones. The drones, therefore, have no fathers, although they have grandfathers on the maternal side, and there is no meiotic

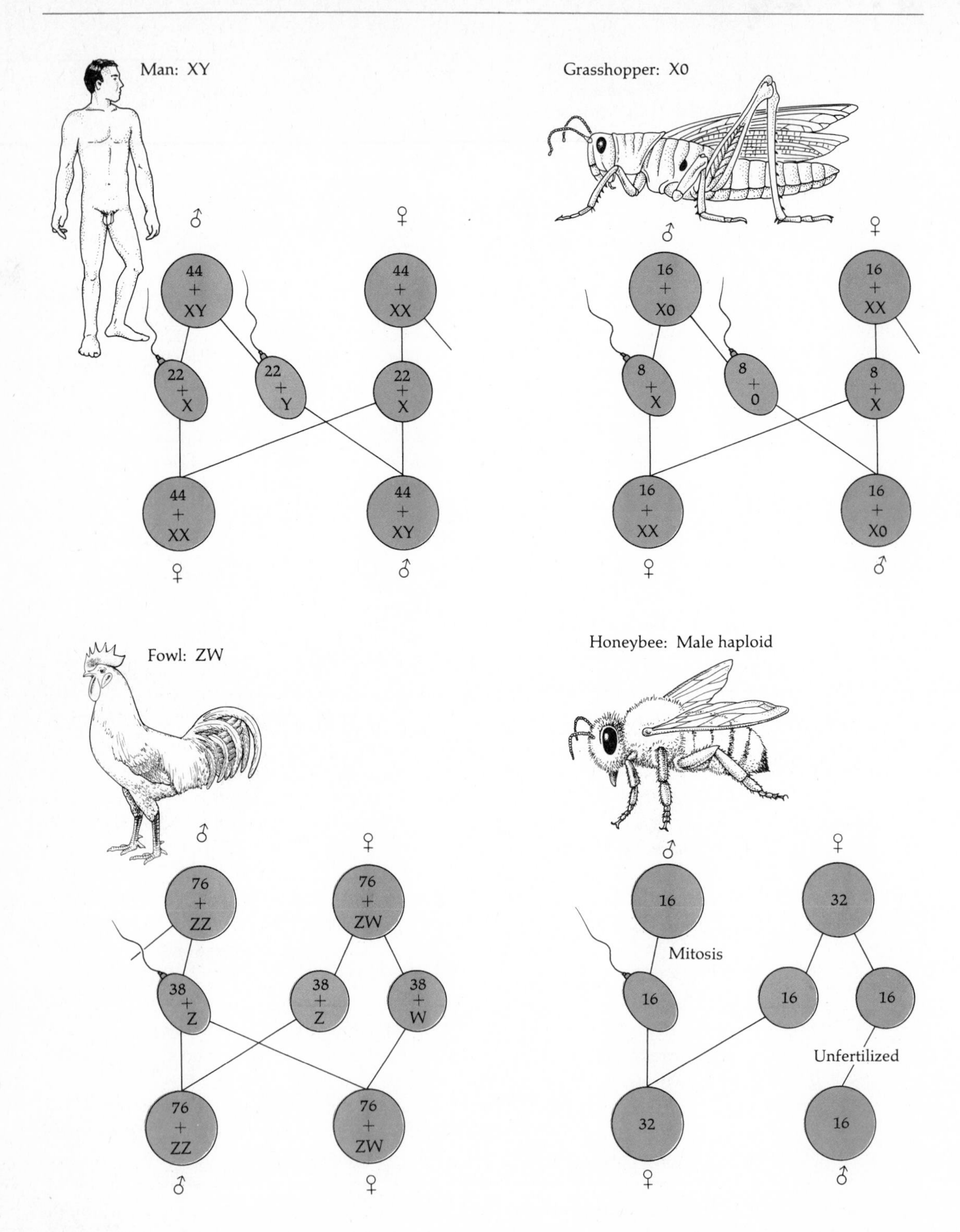

Figure 3.14
Four types of sex determination.

reduction of the chromosome number during spermatogenesis in the drones. The females may develop either into queens, which are large, fertile females, or into sterile female workers, depending on the nutrition supplied to the larvae by the workers.

Most plants and a few animals are *hermaphroditic,* with both sexual capabilities present in the same individual. Most hermaphrodites reproduce by self-fertilization, although in some animals and fewer plants the construction of the sexual organs facilitates cross-fertilization.

Some plants may be either *monoecious* (hermaphroditic) or dioecious. Fernando Galán has shown in *Ecballium elaterium,* a plant of the gourd family, that sex determination is due to a series of three alleles, a^D, a^+, and a^d, where a^D is dominant over the other two alleles, and a^+ is dominant over a^d; a^D is the allele for maleness, a^+ for hermaphroditism, and a^d for femaleness. The sexual phenotypes determined by the five possible genotypes are shown in Figure 3.15. (The homozygotes a^Da^D do not exist because they could only arise from a cross between two males.)

Finally, in some animals sex determination depends on environmental circumstances. In the marine worm *Boniella,* individuals that remain free-swimming throughout their entire larval stage become females, while larvae that attach themselves to the bodies of mature females are turned into males by a masculinizing hormone secreted by the females. In the coral-reef fish *Labroides dimidiatus,* individuals live in groups consisting of one male and several females. When the male dies, the most dominant female repels males approaching her group and, if successful, starts courting the females and otherwise behaving like a male; in about two weeks she (now he) becomes able to produce fertile sperm. The sex-determination mechanisms of *Boniella* and *Labroides* are economic in that there are no surplus males in search of mates.

The Sex Ratio

The sex ratio, i.e., the number of males divided by the number of females, is approximately 1 in most organisms with separate sexes. This results from segregation of the sex chromosomes in the heterogametic sex (Figure 3.16). Females (or whichever is the homogametic sex) produce only one kind of gametes, X; males produce two kinds of gametes in equal proportions, X and Y. The progeny, therefore, normally consist of XX and XY individuals in the ratio 1:1.

In bees and related insects, the ratio of males to females depends on the proportion of eggs that are fertilized; not surprisingly, the sex ratio is not 1, but is usually much smaller. In organisms with environmental sex determination, the sex ratio is usually different from 1, and often considerably smaller, as in *Labroides dimidiatus.*

Figure 3.15
Sex determination in a plant of the gourd family, *Ecballium elaterium,* is due to a series of three alleles, a^D for maleness, a^+ for hermaphroditism, and a^d for femaleness. Allele a^D is dominant to a^+ and a^d, while allele a^+ is dominant to a^d. (Drawings by Carmen Simón, provided by Prof. Fernando Galán, University of Salamanca.)

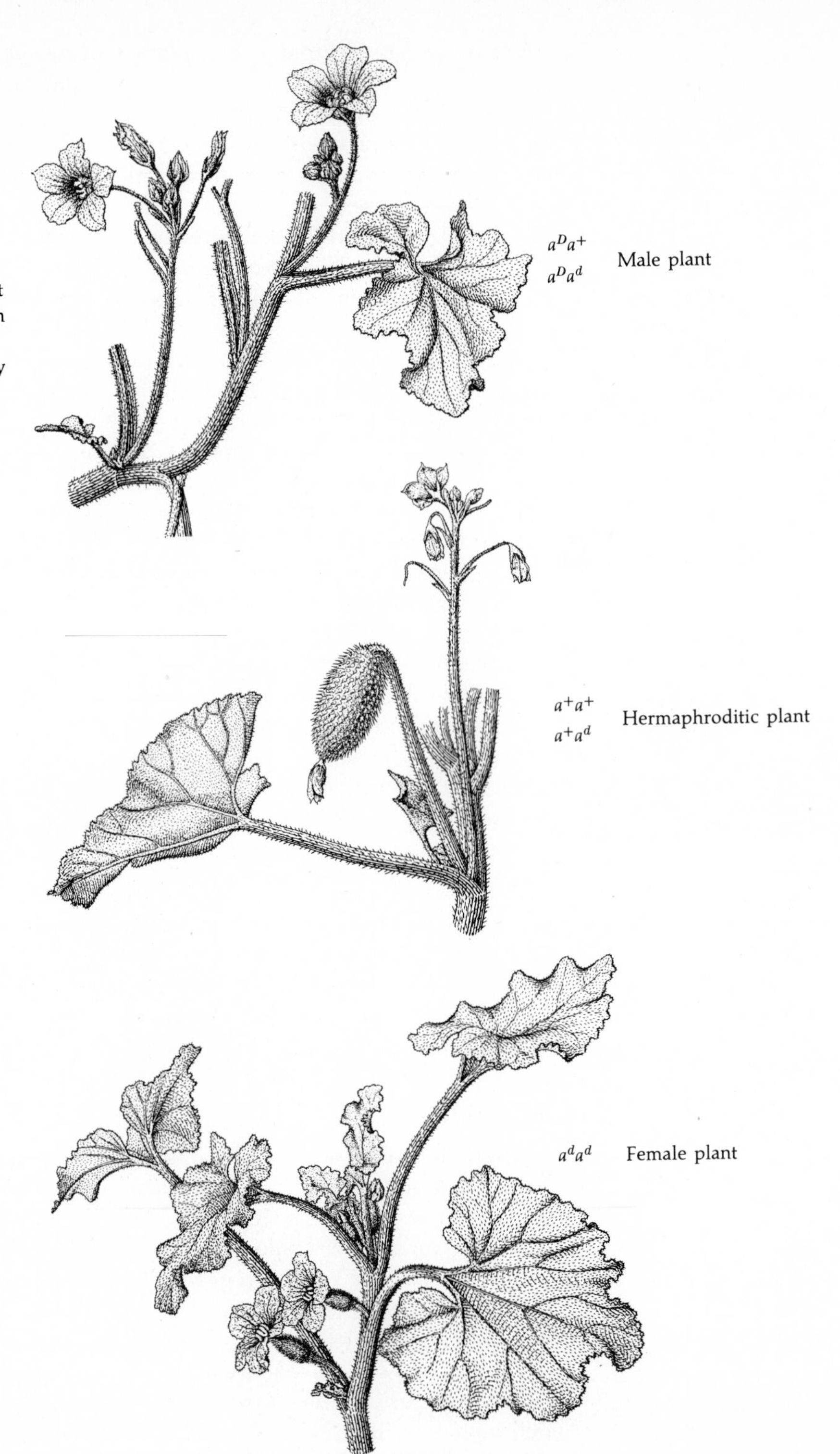

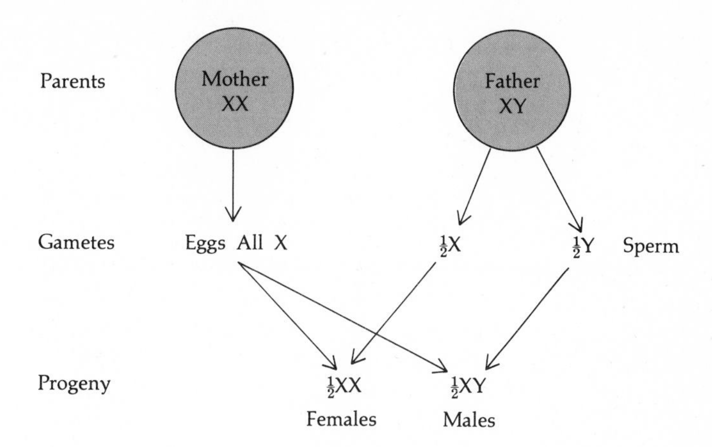

Figure 3.16
Determination of the sex ratio. The numbers of males and of females are approximately identical in organisms in which one sex is homogametic and the other heterogametic. With respect to sex, homogametic individuals produce only one kind of gamete, whereas heterogametic individuals produce two kinds of gametes in equal numbers.

Certain genes modify the sex ratio. Recessive *lethal* genes (i.e., genes causing death before sexual maturity) in the X chromosome of *Drosophila,* for example, kill all the males but not the heterozygous females. In *Drosophila pseudoobscura* there are stocks, known as sex-ratio strains, that produce almost exclusively female progeny, presumably because most Y-bearing gametes of sex-ratio males fail to develop or to function.

Problems

1. Two people with normal vision have four children: the first is a daughter of normal vision who has three sons, two of them color-blind; the second is a daughter of normal vision who has five sons, all with normal vision; the third is a color-blind son who has two daughters and two sons, all with normal vision; the fourth is a son with normal vision who has four sons, all with normal vision. What are the probable genotypes of the two original parents, their children and the children's spouses, and their grandchildren?

2. If a woman whose father suffered from hemophilia marries a normal man, what is the chance of hemophilia in the children of this marriage? Assume now that the husband's father also had hemophilia; what will then be the chance of hemophilia in the children?

3. In *Drosophila, vestigial* wings (vg) is recessive to normal wings (vg^+), and the gene for this character is not in the X chromosome; *yellow* body (y) is recessive to normal-color body (y^+), and this gene is sex-linked. If a homozygous yellow vestigial female is crossed to a normal male, what will the appearance of the F_1 and the F_2 offspring be?

4. Assume that the yellow vestigial female in the previous problem is $\widehat{XX}$ (attached-X). What offspring will you expect when it is crossed to a normal male?

5. A rooster is heterozygous for a sex-linked recessive lethal. What ratio of males to females will be produce when crossed to normal hens?

6. Occasionally the ovary of a hen fails to develop or loses its function, and testes develop instead. Some sex-reversed "hens" are known to have fathered chicks. What types of progeny are expected when a sex-reversed hen is crossed to a normal hen? If WW eggs fail to develop, what will be the sex ratio?

7. A barred hen is crossed to a nonbarred rooster. The F_2 progeny are allowed to interbreed freely. What will be the appearance of the F_3 with respect to barring?

8. What kinds of offspring, and in what proportions, will be produced when a nonbarred hen is crossed to a heterozygous barred rooster?

9. In *Ecballium elaterium,* monoecious (hermaphroditic) plants are classified in the variety *elaterium,* and dioecious (male or female) plants in the variety *dioicum.* The usual genetic constitutions are a^+a^+ for *E. elaterium* var. *elaterium,* a^Da^d for *E. e.* var. *dioicum* males, and a^da^d for *E. e.* var. *dioicum* females. Assuming that wild plants have the usual genetic constitutions, what kinds of plants, and in what proportions, will the following crosses produce: (a) var. *dioicum* ♀ × var. *dioicum* ♂; (b) var. *elaterium* × var. *elaterium;* (c) var. *elaterium* × var. *dioicum* ♂; (d) var. *dioicum* ♀ × var. *elaterium?*

10. Assume that the progeny of cross (c) in the previous problem are allowed to interbreed freely. What kinds of plants, and in what proportions, will the various types of matings produce in the following generation? What kinds of plants will be obtained if we allow the progeny of cross (d) to interbreed?

11. Two *Ecballium elaterium* plants are crossed and produce only hermaphroditic plants. When these are intercrossed, one-fourth of the offspring are female dioecious and three-fourths are monoecious. What are the genotypes of the various kinds of plants?

12. Female honeybees, *Apis mellifera,* are diploid, whereas drones are haploid. There is, however, a sex locus with multiple alleles that determines the sex of diploid individuals. All diploid individuals heterozygous at the sex locus are female, but homozygotes for any one allele are male. Diploid males are not encountered as adults because homozygotes are usually

eaten by workers in the larval stage within 72 hours after the male homozygotes hatch from the eggs. Although homozygotes are therefore "behavioral" lethals, they can develop into viable, fertile males if removed from the hive so that they are not eaten. About 20 different alleles exist at the sex locus, which can be represented as $A_1, A_2, A_3, \ldots, A_{20}$. A queen fertilized by a haploid drone produces 50% viable brood (this is noticed because half the larval cells of the comb are empty, and the empty cells are randomly distributed). What can you infer about the genetic constitution of the queen and the drone?

13. Assume that a diploid male fertilizes a queen that carries one allele identical to that of the male. What proportion of the offspring will you expect to be viable if they are left in the hive?

14. Assuming that in a population there are exactly 20 different alleles at the sex locus, how many genetically different kinds are possible of haploid males? Diploid males? Females?

15. The first demonstration of a gene located in the Y chromosome was made by Antonio de Zulueta in 1925. The beetle *Phytodecta variabilis* has four phenotypes: lined, yellow, red, and black, distinguished by the color and pattern of the elytra (wing covers). The four phenotypes are determined by four alleles of a single gene, which can be represented as e^l for *lined* elytra, e^y for *yellow*, e^r for *red*, and e^b for *black*. The dominance relationships can be represented as $e^b > e^r > e^y > e^l$, where $>$ means "dominant to." The gene is in the sex chromosomes, but all four alleles exist in the X as well as in the Y chromosome. Zulueta observed that the lined phenotype was very rare among males (0.5% of all males) but common among females (59% of all females). How can you explain this disparity?

16. A cross of a *Phytodecta variabilis* lined female to a yellow male produced in the F_1 13 lined females and 11 yellow males. In the F_2 all 31 females were also lined, and all 29 males were yellow. What are the genotypes of the parents, the F_1, and the F_2 individuals? Use a chi-square to test your hypothesis.

17. Assume that you did not know the sex of the F_1 and F_2 individuals in the previous problem. Could the results given there have been obtained if the gene for elytra color were not sex-linked? Now, taking sex into account, use a chi-square test to ascertain whether the hypothesis that the gene is not sex-linked can be accepted.

18. A cross of a *Phytodecta variabilis* red female with a red male produced 15 yellow females, 15 red females, and 34 red males. Crosses of individual F_1 yellow females to individual F_1 red males produced diverse results: about half the crosses produced males and females all red, whereas the

other half produced only red males but the females were lined and yellow in about equal numbers. What are the probable genotypes of the parental cross?

19. In some tropical fish, such as the platyfish and the guppies, the heterogametic sex is the male in some strains but the female in others. Wild strains often have XX females and XY males; some domestic strains have ZZ males and ZW females. When the two kinds are intercrossed, males can be obtained with chromosome complements ZZ, XZ, XY, ZY, or YY, and females with complements XX, XW, ZW, or YW. What is the expected sex ratio in the following crosses: (a) XX females with ZZ males; (b) ZW females with XZ males; (c) XW females with XZ males?

4

The Nature of the Genetic Materials

The association of Mendelian genes with the chromosomes of cells was firmly established by Bridges' work in 1916 (Chapter 3). Several general properties of genes themselves had been established. First, their ability to generate copies of themselves (self-replication) during chromosome duplication prior to meiosis was evident. Second, genes were known to mutate to different allelic forms that also possess the property of self-replication. The infrequent occurrence of mutation, however, indicated that genes are basically very stable entities capable of exact duplication. Third, genes affect the phenotypes of organisms in specific ways. The expression of alternative traits, as first noted by Mendel (e.g., tall versus short plants or round versus wrinkled seeds), is essential in identifying genes by observation of the segregation of alleles in matings. The stable transmission of traits from generation to generation, altered only by mutation, raised the questions: How are such traits determined? What is the physical substance composing the gene that is capable of self-replication, mutation, and phenotypic expression?

Answers to basic questions such as these appear to be quite complex when they are considered in terms of the growth, development, and reproduction of higher organisms, such as humans, or even less complex organisms, such as pea plants and flies. Simpler organisms have provided less complex systems for the study of such questions. Our initial knowledge of the physical and chemical basis of heredity was deduced from the study of microorganisms.

Bacteria, initially studied as agents of disease in humans and domestic animals, have proved to be useful organisms for studying heredity and the fundamental nature of the gene. Viruses, which are less complex entities than bacteria, have proved even more useful. Viruses are capable of reproduction only through infection of cells. Bacteriophages infect bacterial cells; other types of viruses infect plant or animal cells, and many are agents of disease.

As early as 1922, the geneticist H. J. Muller pointed out two essential similarities between bacteriophages and genes: Both are capable of producing exact replicas of themselves, and both are capable of mutating to new forms. Muller wrote:*

> *On the other hand, if these d'Hérelle bodies [bacteriophages] were really genes, fundamentally like our chromosome genes, they would give us an utterly new angle from which to attack the gene problem. . . . It would be very rash to call these bodies genes, and yet at present we must confess that there is no distinction known between the genes and them. Hence we cannot categorically deny that perhaps we may be able to grind genes in a mortar and cook them in a beaker after all. Must we geneticists become bacteriologists, physiological chemists and physicists, simultaneously with being zoologists and botanists? Let us hope so.*

The Mendelian genes studied by geneticists in 1922 were known to exist only in the nuclei of cells. Bacteriophages were obviously much smaller than bacterial cells. Could it be that they were "liberated" genes, free and capable of replicating when they happened to infect a bacterial cell? This speculation led to many studies of bacteriophages and other viruses that partly confirmed Muller's idea.

In this chapter we will see how the study of bacteria and viruses led to the chemical identification of the genetic material and how its physical structure and properties were deduced. In later chapters we will see how microorganisms have contributed to our knowledge of the organization and expression of the genetic material.

Bacteria as Experimental Organisms

The most intensively studied species of bacterium is *Escherichia coli,* an inhabitant of the human intestine. This bacterium can be easily grown in a liquid medium containing a few salts (NaCl, NH_4Cl, KH_2PO_4, $CaSO_4$) and a simple carbon source, such as the sugar glucose. From these compounds, *E. coli* is able to synthesize all of the complex organic molecules constituting the cell (such bacteria are called *prototrophs*). In such a medium, *E. coli* populations as high as 2–3 $\times$ 10^9 cells/ml can be produced from only a few thousand cells in a few hours. Many other bacteria, particularly pathogenic species, require a more complex medium containing organic molecules that they are unable to synthesize themselves (such bacteria are called *auxotrophs*).

*H. J. Muller, *American Naturalist* 56:32 (1922).

The number of living cells in a bacterial culture can be determined by appropriately diluting a sample of the culture and spreading a measured volume on a nutrient medium solidified with agar in a Petri dish (Figure 4.1). If the culture was diluted sufficiently that individual cells are separate from one another when spread on the solid medium (plate), incubation of the plate will allow each bacterium to grow and produce a visible colony of descendants. The number of colonies on the plate is then counted, and the number of cells in the original culture is determined by multiplying that number by the dilution factor employed, as shown in Figure 4.1. Moreover, the descendants of a single cell in the original culture can be obtained for study from any one of the individual colonies on the plate. A pure culture of any mutant bacterium that may have been present in the original culture when it was diluted for plating on the agar can be obtained in this way.

The contributions that bacteria have made to our understanding of hereditary mechanisms have come relatively late to the field of genetics. Early studies of heredity concentrated on higher organisms whose sexual life cycles are shared by humans to some degree. Bacterial reproduction is mainly asexual. It was not until 1946 that bacteria were found to have any form of sexuality. Only gradually, therefore, did scientists come to recognize that man and bacteria might share fundamentally similar genetic or hereditary processes.

Experimental Study of Bacteriophages

Phages have made indispensable contributions to our understanding of heredity. Enormous populations of a phage can be easily obtained—up to 10^{11} phages/ml or more in infected bacterial cultures. They can be detected by their lethal effects on the cells they infect.

The number of phages in a culture is determined by appropriately diluting the suspension, mixing a measured volume with a culture of living bacterial cells in melted agar, and pouring this mixture into a Petri dish containing nutrient agar, where the melted agar solidifies. Upon incubation, the living bacteria multiply to produce a dense and visible *lawn* of bacteria. Wherever a virus particle has initiated an infection, however, a hole in the lawn is visible in which the bacterial cells have been killed by the infection. Such a hole is called a *plaque* (Figure 4.2). If the number of phage particles originally used to infect the plated bacterial culture was relatively small, each plaque observed in a lawn of bacteria will contain the descendants of a single ancestral phage particle. The concentration of phage particles in the original suspension can be determined by counting the number of plaques on the lawn and multiplying that number by the dilution factor employed, as shown in Figure 4.2. Pure strains of a phage can be obtained by using the contents of a single plaque to infect a fresh culture of bacteria.

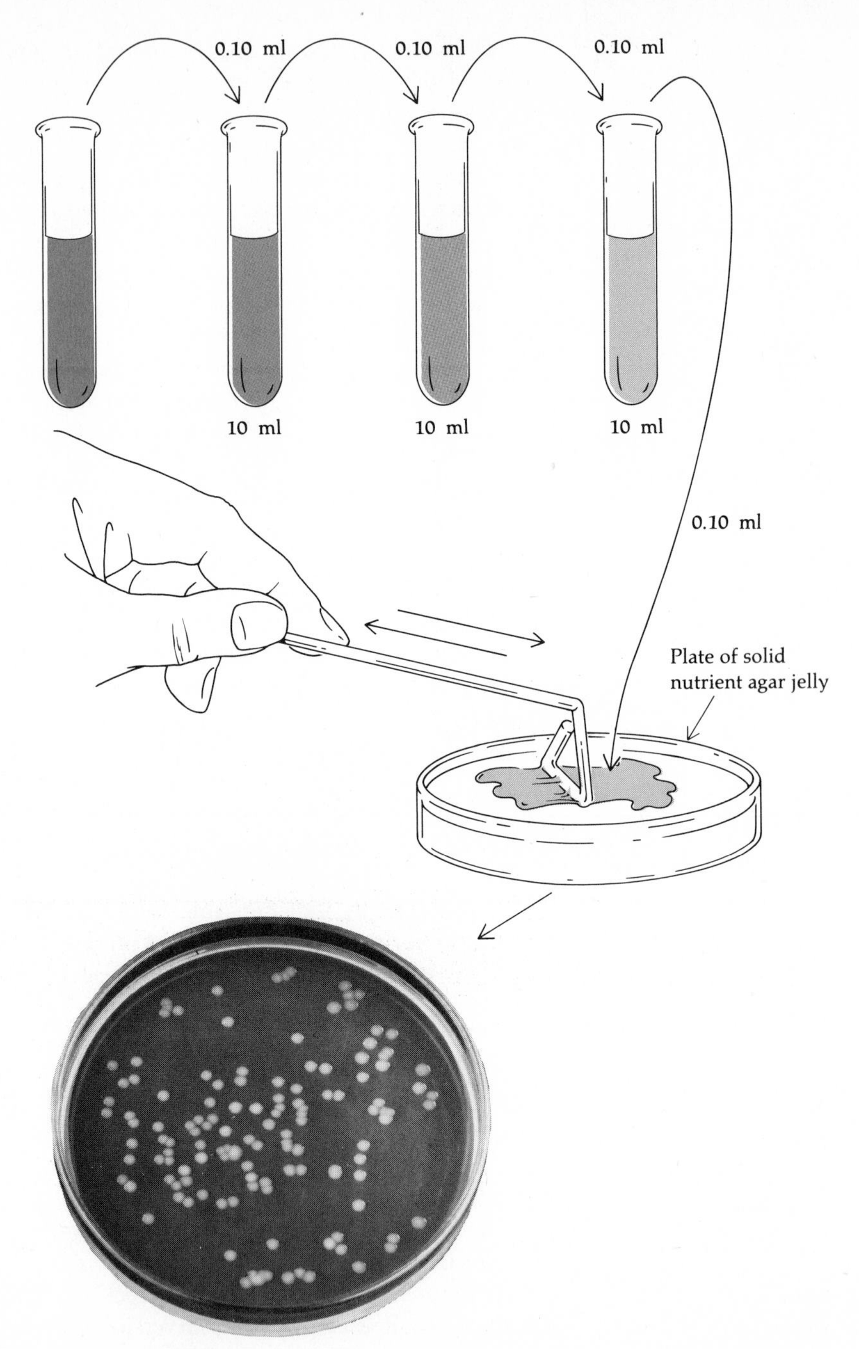

Figure 4.1
The number of living bacterial cells in a culture is determined by serially diluting a measured volume of the culture and spreading a measured volume of the diluted culture on a nutrient agar medium. Upon incubation, each bacterial cell grows to produce a visible colony of descendants. The number of colonies counted on the medium reflects the number of cells in the diluted culture. For example, if 100 colonies are counted, there were 100 cells in the 0.10 ml of diluted culture. The number of cells in the original culture was then (100 cells/0.10 ml) $\times\ 10^2 \times 10^2 \times 10^2 = 10^9$ cells/ml. (Photograph from G. S. Stent and R. Calendar, *Molecular Genetics,* 2nd ed., W. H. Freeman, San Francisco, 1978.)

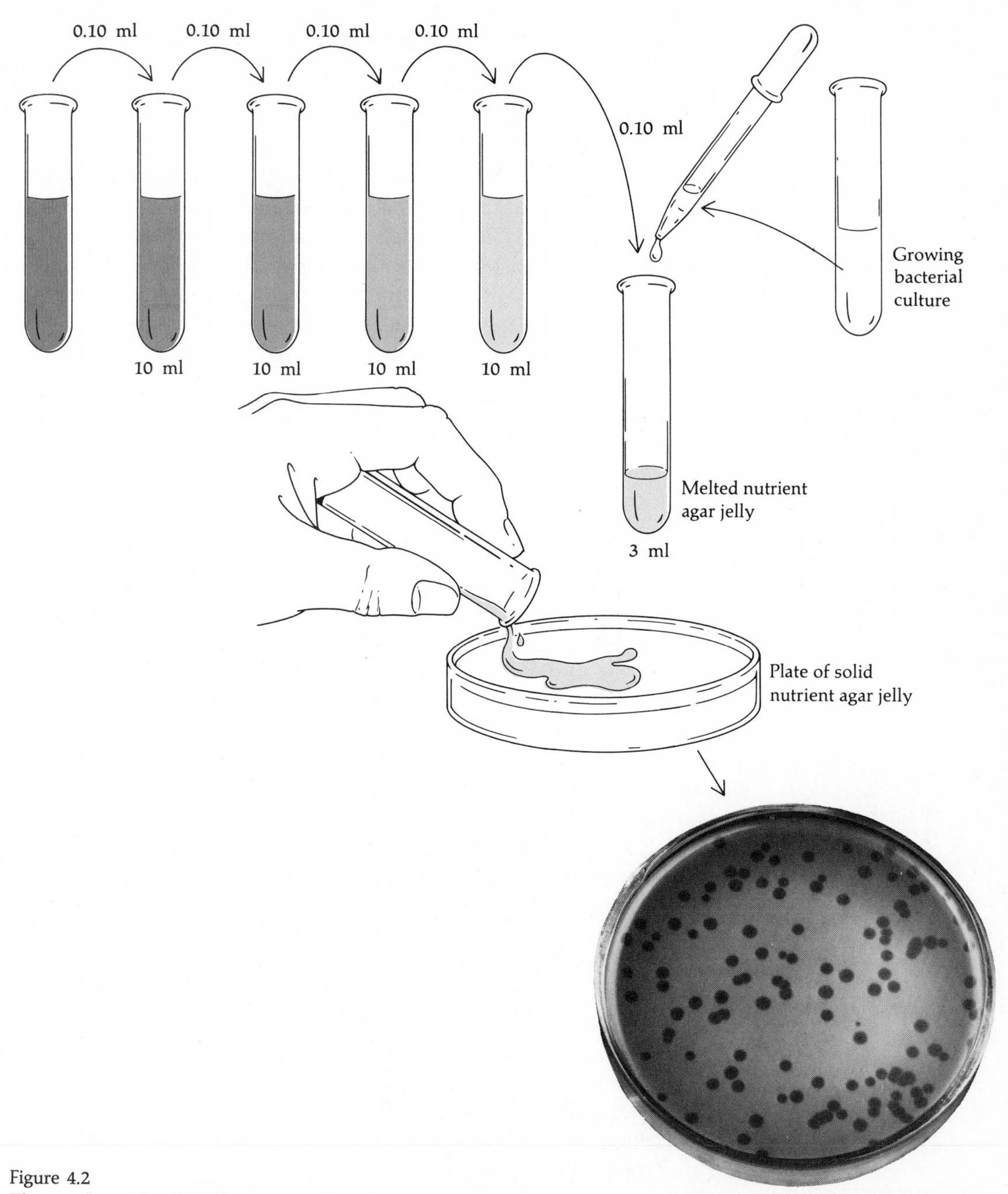

Figure 4.2
The number of bacteriophages in a culture is determined by serially diluting a measured volume of the culture. A measured volume of the diluted culture is added to a few milliliters of melted nutrient agar, along with a few drops of growing bacterial culture ($\approx 10^6$ cells). The melted agar is then poured onto solidified nutrient medium in a Petri dish, where it hardens. Upon incubation, the bacteria grow to form a dense lawn of cells except where phages have initiated infections that kill the cells and create holes in the lawn. The number of holes, or plaques, reflects the number of phages added to the soft agar. If 100 plaques are counted, there were 100 phages in the 0.10 ml of diluted culture. The number of phages in the original culture was then (100 phages/0.10 ml) $\times 10^2 \times 10^2 \times 10^2 \times 10^2 = 10^{11}$ phages/ml. (Photograph from G. S. Stent and R. Calendar, *Molecular Genetics,* 2nd ed., W. H. Freeman, San Francisco, 1978.)

DNA—the *Pneumococcus* Transforming Factor

Deoxyribonucleic acid (DNA) is now known to be the hereditary material of all prokaryotes and eukaryotes. This was first demonstrated with bacteria by studies of strains of the bacterium *Pneumococcus,* some of which cause a fatal pneumonia in mammals. The pathogenic strains synthesize a capsular polysaccharide that forms a mucous coat, protecting the bacterial cells from the phagocytes of the immune system of the infected animal. Different strains of *Pneumococcus* can be distinguished from one another by the types of antibodies induced in an infected animal by the different polysaccharides and proteins of each bacterial strain. Antibodies synthesized in response to the capsular polysaccharides of different strains serve to distinguish one pathogenic strain from another; these are designated type I, type II, type III, etc.

Pathogenic strains, grown on nutrient agar in the laboratory, produce shiny, smooth colonies, owing to the mucous coats synthesized by the bacteria. Occasionally, mutant cells arise that have lost an enzyme activity required to synthesize a mucous coat. These mutant cells produce colonies that have a rough surface; they are designated R to distinguish them from the parent S (smooth) strains (Figure 4.3). R strains propagate themselves as faithfully as do S strains, but some R strains occasionally give rise to cells, by reverse mutation, that have regained the ability to synthesize a mucous coat. The type of capsular polysaccharide synthesized by these revertant strains is always that produced by the original parent S strain; that is, IIS $\rightleftarrows$ IIR, IIIS $\rightleftarrows$ IIIR, etc. Therefore R strains are not identical: each is specific to the S strain from which it arose.

The R strains are nonpathogenic. When they are injected into a test animal such as a mouse, the mouse generally survives the infection by producing antibodies that lead to the phagocytosis and death of the bacteria. However, when bacteria of an S strain are injected into a mouse, the mouse invariably succumbs to the pneumonial infection, owing to the protective mucous coats synthesized by the bacterial cells. In 1928 Frederick Griffith showed that, if cells of a IIR strain of *Pneumococcus* are injected into a mouse along with heat-killed cells of a IIIS strain, the mouse will succumb to an infection shown by autopsy to be caused by cells of a IIIS type. Control experiments showed that injection of mice with only live IIR cells or with only heat-killed IIIS cells never led to a fatal infection. The fact that the cells causing the infection synthesized a mucous coat characteristic of a type III strain rather than of a type II strain was evidence that the pathogenic cells had not arisen by reversion of live IIR cells (IIR $\rightarrow$ IIS). Griffith concluded that nonpathogenic IIR cells are capable of being *transformed* to pathogenic cells by heat-killed IIIS cells. The process of heat killing had to be carefully controlled: too high a temperature destroyed the transforming factor and too low a temperature failed to destroy enzyme activities that would also destroy the transforming factor. A tem-

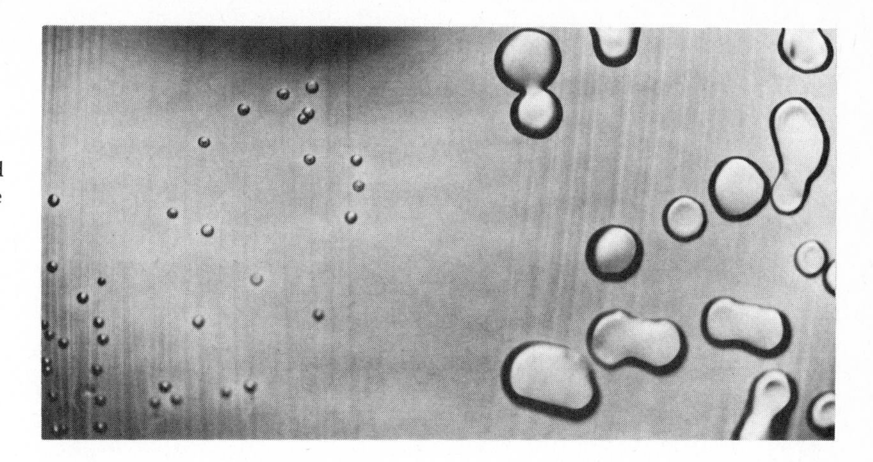

Figure 4.3
Colonies of *Pneumococcus* growing on nutrient medium. The small colonies are type IIR, and the large, glistening mucoid colonies are ones that have been transformed to type IIIS. (Courtesy of Prof. Maclyn McCarty, Rockefeller University.)

perature of 65°C was generally found to destroy enzyme activity and maintain transforming activity.

Later, other investigators found that transformation of nonpathogenic strains of *Pneumococcus* to pathogenic strains can be carried out in test-tube cultures of *Pneumococcus.* The rare transformed cells can be easily selected from nontransformed cells by their resistance to agglutination by serum containing antibodies directed against IIR cells. Agglutinated IIR cells clump at the bottom of a culture tube, while transformed cells are protected from clumping and grow to produce a cloudy suspension of IIIS cells (Figure 4.4). The development of this *in vitro* ("in glass") assay for the detection of transformed cells was an important advance, for it provided the means by which the nature of the transforming factor in heat-killed IIIS cells could be directly investigated without recourse to injecting mice and waiting to see whether they died.

Oswald Avery, Colin MacLeod, and Maclyn McCarty used this assay to identify the substance responsible for the transforming activity in heat-killed IIIS cells. The results of their studies, published in 1944, demonstrated that the transforming factor was deoxyribonucleic acid (DNA). When added to a growing culture of IIR cells agglutinated by antiserum, purified DNA from IIIS *Pneumococcus* was shown to be sufficient to cause some IIR cells to acquire the ability to synthesize the capsular polysaccharide characteristic of IIIS cells. Avery and his co-workers then demonstrated that the transforming factor could be destroyed by enzymes that degrade DNA (deoxyribonucleases). These experiments showed that a cell of strain IIR transformed by DNA from strain IIIS acquires the ability to transmit this newly acquired biosynthetic capacity to its progeny.

This new function, acquired from the DNA of another strain, thus becomes part of the heredity of the transformed cell, just as that function was a part of the heredity of the IIIS strain from which the DNA was obtained. It is noteworthy that DNA is chemically quite different from the

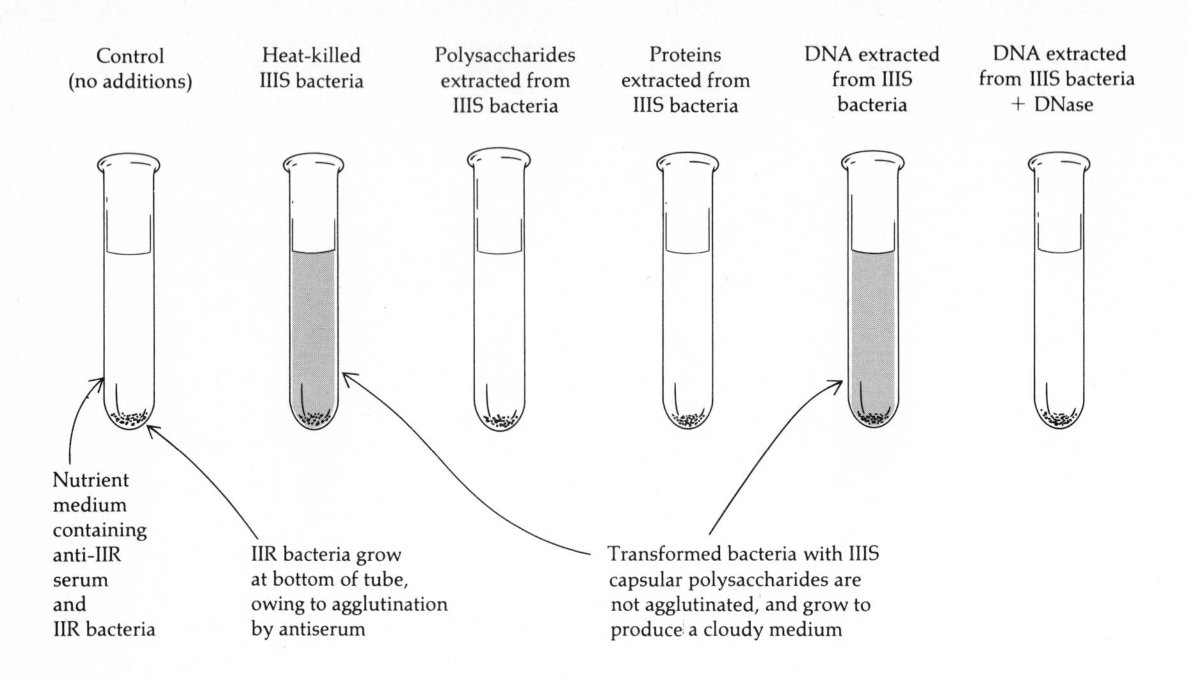

Figure 4.4
In vitro assay for transforming activity. The active factor present in heat-killed IIIS cells can be purified by fractionation of the heat-killed cells and assay of each fraction for transforming activity, using this assay. The purified factor turned out to be a highly viscous preparation of native *Pneumococcus* DNA.

polysaccharide composing the mucous coat synthesized by the IIIS strain; i.e., the transforming factor is not the polysaccharide that confers the phenotypic character of a mucous coat. The startling discovery made by Avery, MacLeod, and McCarty was that the heritable ability of cells to perform a biosynthetic function can be transferred from one cell to another by purified DNA. Subsequent studies of the mechanism of transformation showed that DNA fragments (some carrying the gene required for polysaccharide synthesis) are released from the heat-killed cells and taken into the R cells, where, by recombination, they may be incorporated into the R cell's DNA (Figure 4.5).

The general importance of this discovery was not immediately comprehended, however, for two reasons. The prevailing knowledge of the chemical structure of DNA was very incomplete and incorrectly interpreted; DNA was believed to be a chemically simple substance not complex enough to contain the amount of information needed to direct the development of plants or animals. On the other hand, proteins, which were known to be complex substances, were suspected by many scientists to be the chemical substance of the genes. Moreover, the principles of heredity in bacteria were just beginning to be studied in 1944, and it had not yet been clearly established that bacteria had genes that were in any way similar to those being studied in higher organisms. The appearance of R cells and their reversion to S cells were not firmly established as being due to mutation. DNA might be the hereditary substance in bacteria, but what about other, more complex organisms?

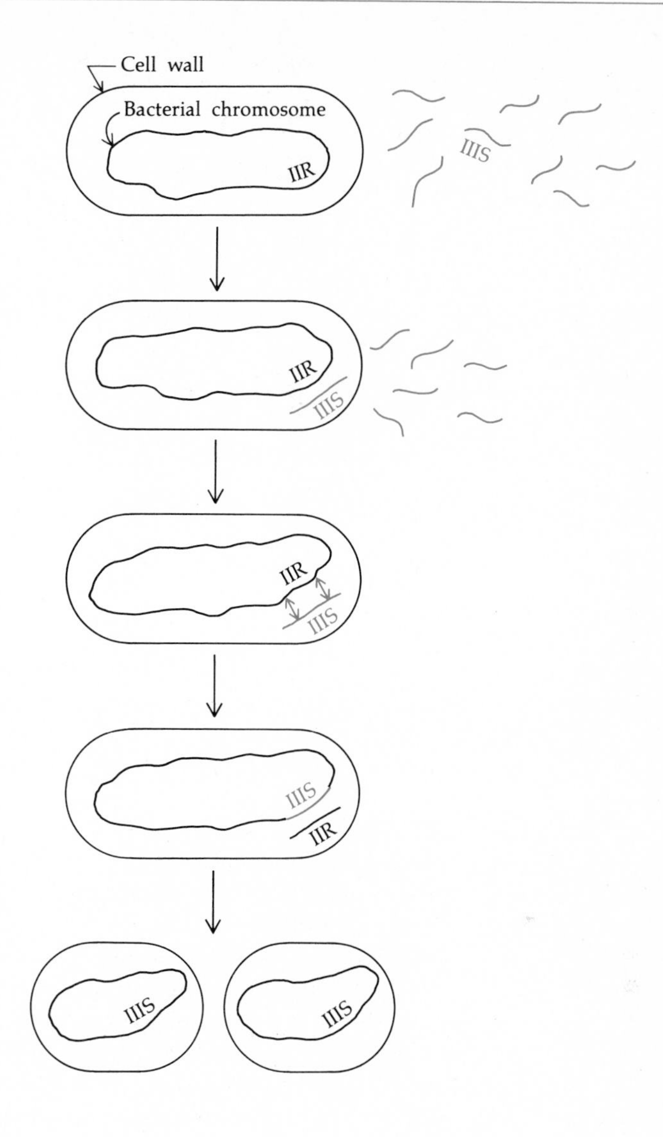

Figure 4.5
Transformation of a IIR cell requires the entrance of IIIS DNA into the cell and incorporation of the DNA into the chromosome of the IIR cell. Expression of the gene causing the synthesis of the polysaccharide coat characteristic of IIIS cells then occurs. This gene is transmitted to progeny as part of the heredity of the transformed IIR cell.

Nucleic Acids—the Genetic Materials of Viruses

Because bacteriophages became objects of genetic study much earlier than did bacteria (owing in part to Muller's insight), the demonstration in 1952 that DNA is the genetic material of the phage T2 was greeted with great excitement by geneticists, and served to recall to general attention the similar demonstration made with *Pneumococcus* several years earlier. The bacteriophage T2 (Figure 4.6) is one of the most thoroughly studied phages that infect the bacterium *E. coli.* This virus contains DNA within a

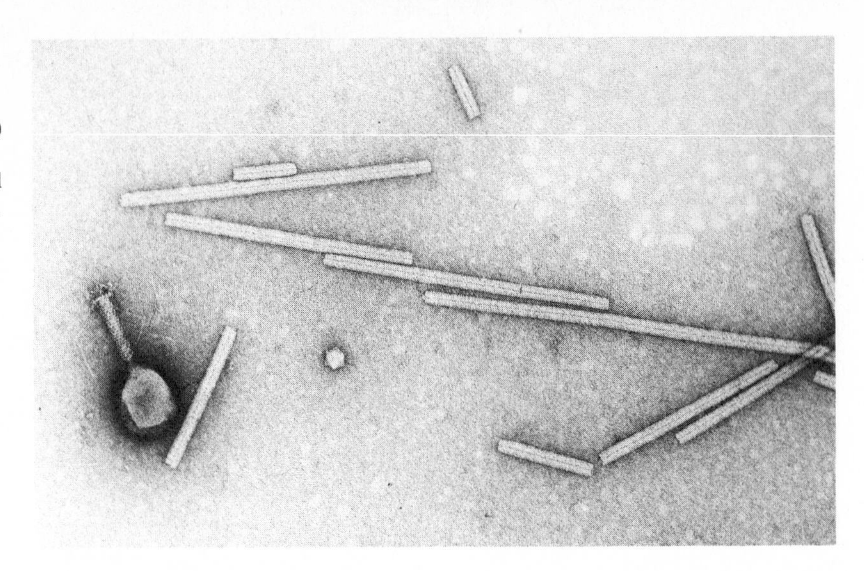

Figure 4.6
An electron micrograph showing long, rodlike tobacco mosaic virus (TMV) and some broken TMV particles, the tadpole-shaped bacteriophage T4 (identical in appearance to T2), and the small, spherical bacteriophage ϕX174. (Courtesy of Prof. Fred Eiserling, University of California, Los Angeles.)

coat of protein. In 1952 Alfred D. Hershey and Martha Chase determined the roles of each of these two components in the production of progeny phage.

Only the protein component of T2 contains sulfur (in the amino acids methionine and cysteine); it can be specifically labeled by growing T2 on host bacteria growing in medium containing the radioactive isotope ^{35}S. The DNA component contains at least 99% of all the phosphorus in T2. This component can be specifically labeled by growth in a medium containing the radioactive isotope ^{32}P. These radioactive elements are sensitive tracers for the T2 protein and DNA during the course of infection.

The initial step in the process of infection is the adsorption of the phage to the bacterial cells (Figure 4.7). This step can be observed in the electron microscope and is confirmed by the observation that both ^{35}S- and ^{32}P-labeled phages sediment with the bacterial cells when the infected cells are centrifuged. Hershey and Chase found that, once the infection had been initiated, most of the ^{35}S-labeled protein could be sheared from the bacterial cells by agitation in a kitchen blender, but that most of the ^{32}P-labeled DNA remained with (presumably inside) the cells. Removing the empty protein coats, called *ghosts*, did not affect the course of the infection; the bacteria lysed and released progeny phage just as if the ghosts had remained associated with the cells (Figure 4.7). Hershey and Chase concluded that only the DNA of the parental phage was essential for the production, in the infected bacteria, of a progeny phage, which consists of both DNA and protein components. The protein component of the phage was therefore assumed only to provide protection to the DNA from degradative enzymes and to provide efficient entry of the DNA into the bacterium, whereas the DNA was the hereditary material itself.

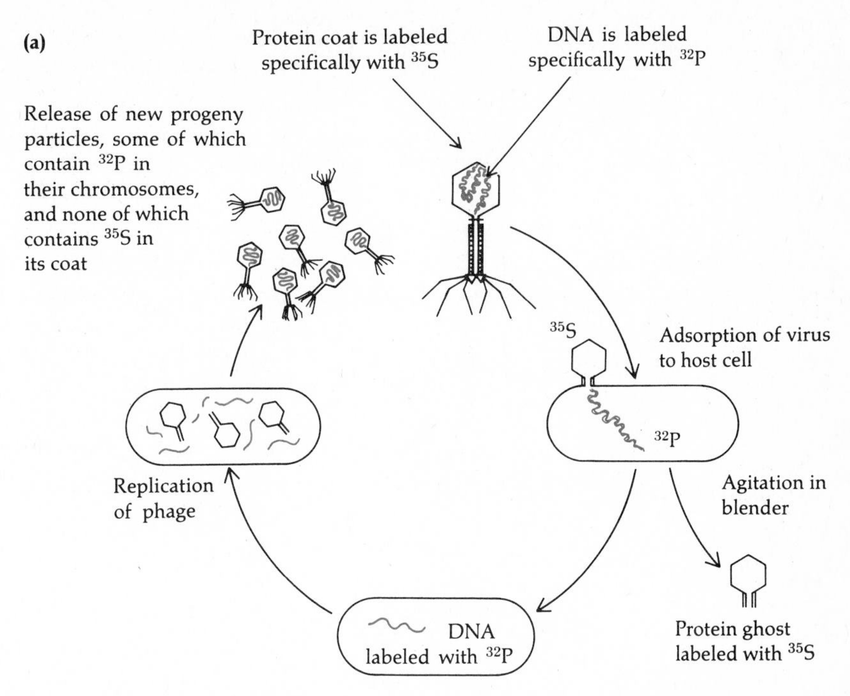

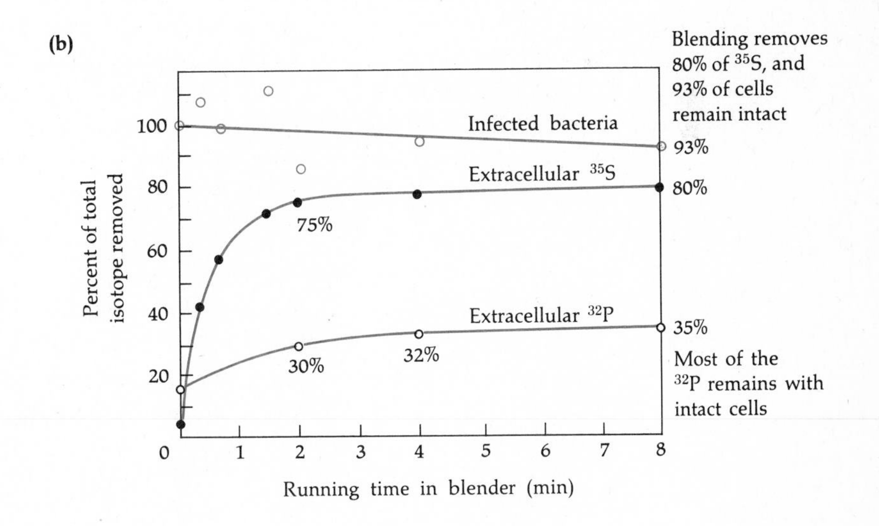

Figure 4.7
(a) Diagram of the Hershey-Chase experiment, which demonstrated that T2 DNA is the phage component essential to the production of progeny T2 phage in an infected bacterial cell. **(b)** The fate of parental T2 DNA is observed by measuring the radioactivity of ^{32}P, which remains associated with the infectecd cell, while the parental T2 protein, labeled with radioactive ^{35}S, can be sheared from the infected cell without interrupting the infection.

The Hershey-Chase experiment marked the general acceptance of an important genetic function for DNA. There are two reasons why this experiment was immediately accepted as a demonstration of the genetic role of DNA, whereas the transformation experiment of Avery, MacLeod, and McCarty had not been. First, it was carried out with a bacteriophage that (in contrast to *Pneumococcus*) was recognized as clearly possessing properties of heredity that are similar to those identified in higher organisms; spontaneous mutation had been demonstrated in T2, and mutant genes of T2 had been observed to recombine with one another, just as Mendelian genes in higher organisms do. Second, chemical studies of the composition of DNA from many different organisms, carried out between 1944 and 1952, had ruled out the widely held belief that DNA consisted of a simple polymer of one tetranucleotide repeated many times in all DNA molecules. These studies, by Erwin Chargaff (see next section), showed that nucleic acids do possess the chemical complexity necessary for genetic material.

Confirming evidence that it is the nucleic acid component of viruses, and not the protein component, that is the carrier of the hereditary information came from the direct demonstration that viral proteins of tobacco mosaic virus (TMV) do not play a genetic role in infection. Like most plant viruses, TMV consists of a protein component and a ribonucleic acid (RNA) component (Figure 4.6). RNA is similar to DNA in structure, as will be discussed below. Each virus particle consists of a single RNA molecule of about 6400 nucleotides wrapped in a protein coat. The protein coat is composed of about 2130 identical subunits, each a polypeptide chain of 158 amino acids ordered in a certain sequence.

Chemical techniques permit the RNA and protein components of the intact TMV virus to be separated from each other (Figure 4.8). Usually a purified preparation of TMV RNA retains no more than 0.1% of the infectivity of the intact virus preparation. Under the proper conditions, however, intact virus can be reconstituted in the test tube from a mixture of purified RNA and purified protein. The protein subunits aggregate with each other and with the RNA to form an intact virus with normal infectivity.

A number of varieties of TMV are known that are distinguishable by the host plants they infect and by the virulence of the infection produced in a particular host. Detectable differences in the amino acid compositions of the protein exist between some of these varieties. For example, the standard TMV strain lacks the amino acids histidine and methionine in its coat, while a strain known as HR contains these amino acids. Experiments have been carried out in which, for example, purified HR protein and purified standard RNA are used to reconstitute infective hybrid viruses. When these viruses are used to infect plants, the amino acid composition of the progeny viruses always corresponds to the strain whose RNA was employed in forming the parental hybrid. The nature of the protein coat of the parental hybrid virus is not transmitted to the progeny viruses; only

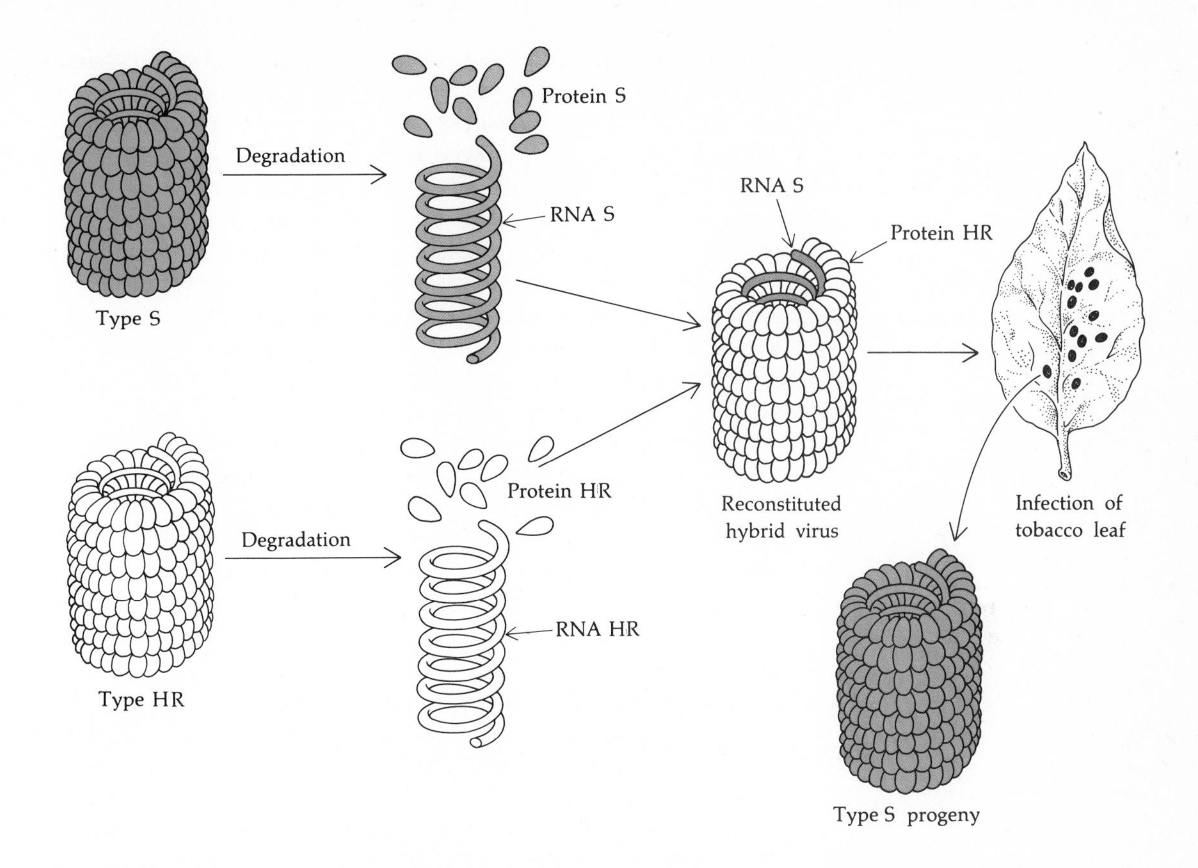

Figure 4.8
Degradation of TMV particles, yielding RNA and protein subunits. Hybrid TMV particles can be reconstituted from the RNA of one strain and the protein subunits of another strain. Infection by the hybrid particles yields progeny consisting of both RNA and protein of the type specified by the RNA of the hybrid TMV. The character of the protein of the hybrid TMV is not transmitted to the progeny.

the RNA component is observed to possess the hereditary function necessary to specify the character of the protein component as well as the RNA of the progeny virus (Figure 4.8).

The hereditary function of viral nucleic acids is conclusively demonstrated by the infective properties of the purified nucleic acids themselves. As pointed out above, purified TMV RNA preparations retain a residual infectivity. This infectivity was at first believed to be due to intact TMV viruses that might be contaminating the RNA preparations. Subsequent studies, however, showed that the infectivity of TMV RNA preparations is destroyed by a purified enzyme from mammalian pancrease, called ribonuclease, which hydrolyzes unprotected RNA but does not affect the infectivity of the TMV virus itself. The reduced infectivity of TMV RNA preparations, in comparison to TMV virus preparations, is due to the absence of the protection to hydrolysis conferred on the RNA by the protein coat. Plant ribonucleases destroy most free RNA molecules before they can initiate infection. However, careful studies have shown that a single, intact TMV RNA molecule is capable of initiating an infection leading to the production of complete TMV viruses.

Subsequently, the purified DNAs of some bacteriophages, ϕX174 and λ, were shown to be infective in the absence of their protein coats. Free DNA molecules do not easily pass through the bacterial cell wall. However, by treating *E. coli* with an enzyme, egg-white lysozyme, holes in the cell wall can be produced. Bacterial cells whose cell walls have been damaged in this manner are called *spheroplasts* (owing to the spherical shape assumed by the normally rod-shaped bacteria) and are incapable of subsequent normal growth. However, spheroplasts are capable of being infected by DNA molecules from ϕX174 and λ, yielding normal bacteriophages as progeny of the infecting DNA molecules. Such experiments confirm that DNA is the hereditary component of bacteriophages, while the protein component of the phage plays no genetic role.

Since the early studies just described, it has become clear that nucleic acids are the hereditary material in *all* organisms. Two kinds of nucleic acids, DNA and RNA, play genetic roles in all prokaryotic and eukaryotic cells. Viruses, however, contain only one or the other of these nucleic acids.

The Chemical Composition and Structure of Nucleic Acids

The basic unit of nucleic acid structure is the *nucleotide.* The nucleotide is composed of three distinct chemical parts that are joined by covalent bonds (Figure 4.9). One part is a pentose sugar: deoxyribose in DNA and ribose in RNA. The second part is a nitrogenous base of either purine or pyrimidine derivation, covalently bonded to the 1′ carbon of the pentose sugar to form a *nucleoside.* DNA contains the purine bases adenine (A) and guanine (G) and the pyrimidine bases cytosine (C) and thymine (T); the corresponding nucleosides are deoxyadenosine, deoxyguanosine, deoxycytidine, and deoxythymidine. RNA contains the same purine bases as DNA, as well as the pyrimidine cytosine, but contains uracil (U) rather than thymine; the corresponding nucleosides are adenosine, guanosine, cytidine, and uridine. The third part of the nucleotide is a phosphate group; in a nucleic acid polymer, this group joins two nucleosides to each other by forming a phosphodiester bridge between the 5′ carbon of one sugar moiety and the 3′ carbon of another. Nucleotides are nucleosides with one or more phosphate groups esterified to either the 5′ or 3′ carbon of the sugar. Nucleotides are the precursors in the synthesis of nucleic acids and are also the products of chemical or enzymatic hydrolysis of nucleic acids.

Nucleic acids are very large polymers of mononucleotides joined to one another by 5′–3′ phosphodiester bonds. Intact natural RNA molecules range in size from about 100 to 100,000 or more nucleotides. (Further

Figure 4.9
The nucleotide building blocks of nucleic acids. Note both the similarities and differences between DNA and RNA polymers. Note the numbering system for the carbon and nitrogen atoms in the purine and pyrimidine rings. The carbon atoms of the sugar moiety are numbered separately, using primed numbers.

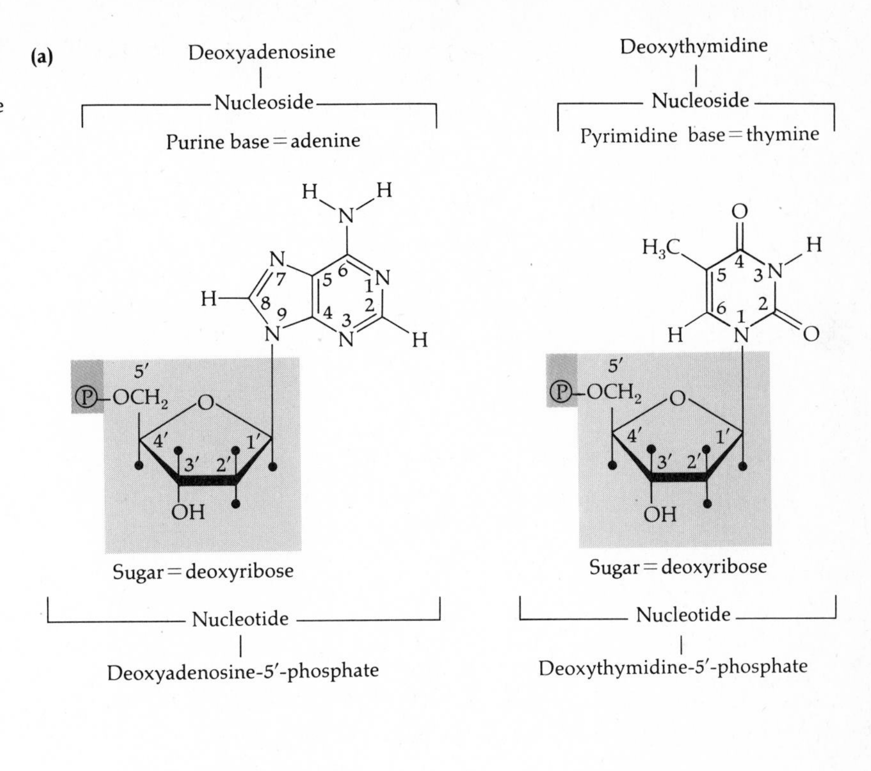

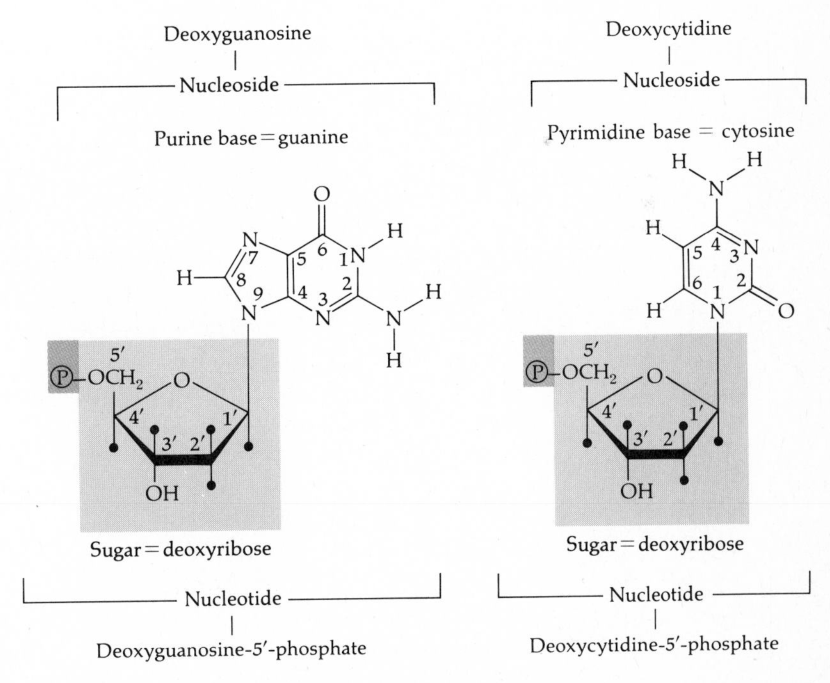

(continued)

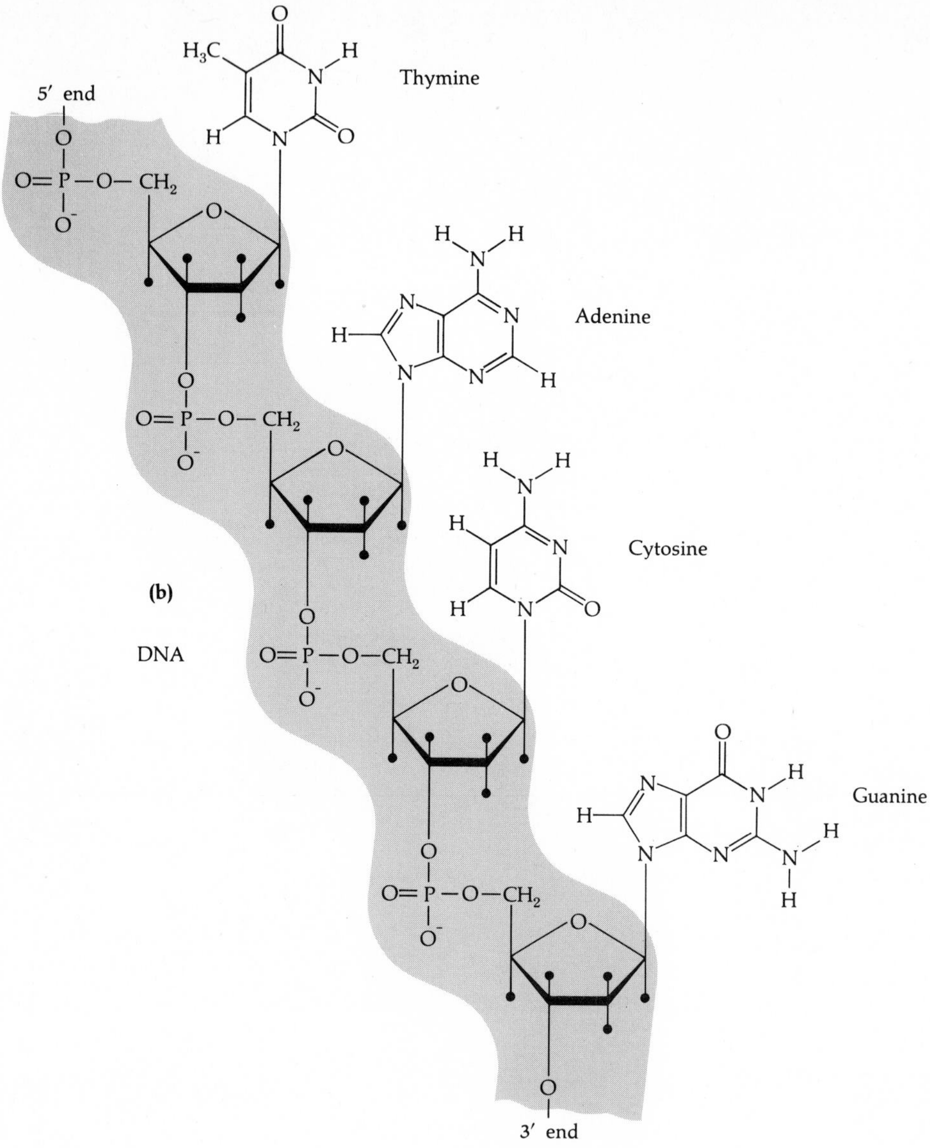

5′ end

Thymine

Adenine

Cytosine

Guanine

(b)

DNA

3′ end

Figure 4.9 (*continued*)

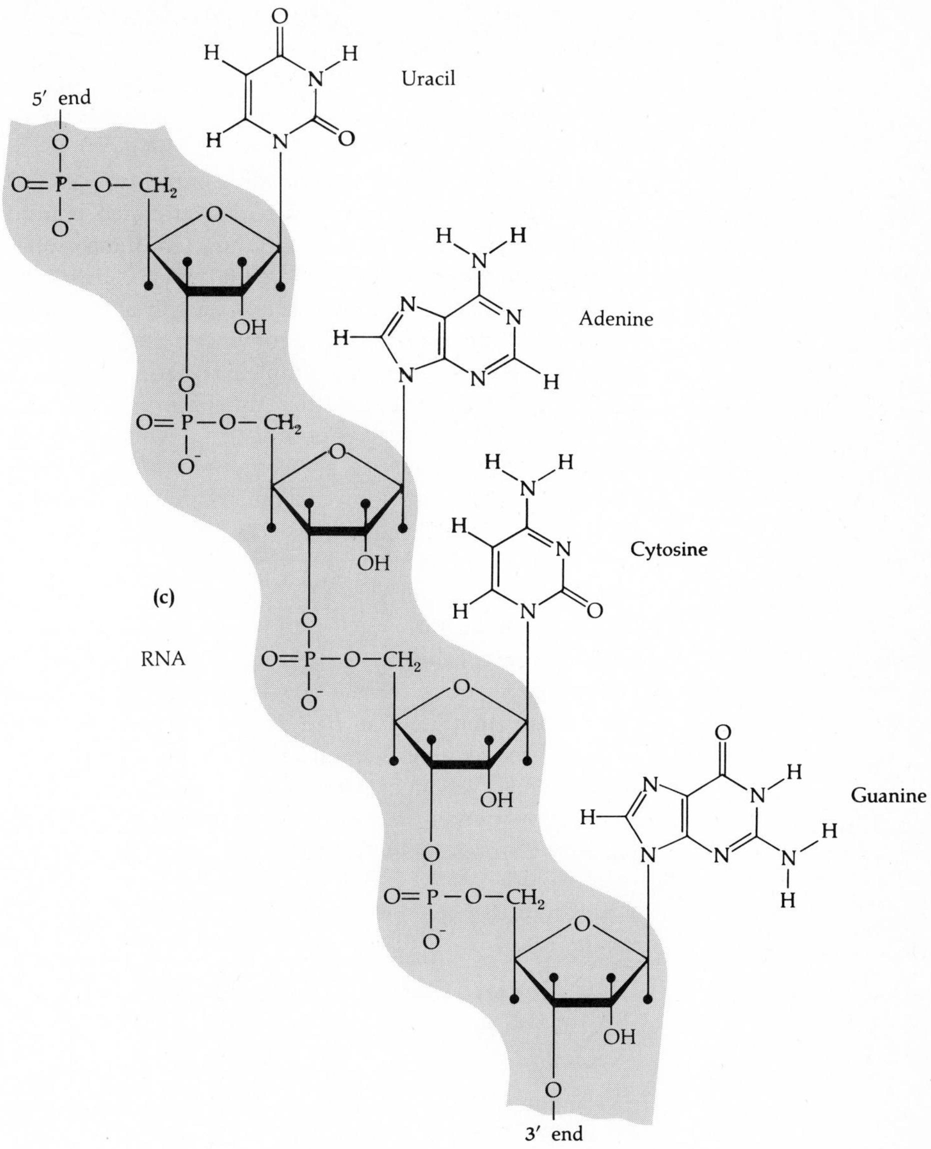

5′ end

Uracil

Adenine

Cytosine

(c)

RNA

Guanine

3′ end

discussion of the structure of RNA molecules will be deferred to Chapter 11.) Intact natural DNA molecules range in size from a few thousand to many millions of nucleotides, depending on the species of organism. At the time that Avery, MacLeod, and McCarty carried out their experiments with *Pneumococcus* DNA, it was generally believed that the structure of nucleic acid molecules was relatively simple and consisted of a tetranucleotide sequence, e.g., $pApCpGpT_{OH}$, reiterated many times to constitute the polymer $(pApCpGpT)_n$. Thus DNA did not seem to possess sufficient complexity to be the hereditary material.

Subsequent chemical studies by Chargaff of the composition of DNA from many different organisms convinced the scientific world that DNA does possess the chemical complexity necessary for genetic material. Chargaff's studies showed that the base compositions of DNA are different in different species of organisms. This observation ruled out the possibility that all DNAs are composed of identical tetranucleotides. Chargaff's studies also showed a remarkably constant feature of all DNA: the molar quantity of adenine is equal to that of thymine, and the molar quantity of guanine is equal to that of cytosine. These equivalences are called *Chargaff's rules:* [A] = [T]; [G] = [C]; [purines] = [pyrimidines]. It is only the molar ratio ([A] + [T])/([G] + [C]) that is observed to vary among species (Table 4.1).

Chargaff's observations suggested that DNA molecules might be much more complex than had previously been thought, because many different base sequences could be present in DNA and still accord with his data. Indeed, Watson and Crick soon demonstrated that Chargaff's rules place no restriction on the possible number of different base sequences that may be present in DNA.

The Watson-Crick Model for DNA Structure

In 1953 James Watson and Francis Crick proposed a model for the structure of DNA that has withstood many tests and is now known to be correct. The model proposed was based on four pieces of evidence (Figure 4.10). (1) A DNA polymer consists of nucleotides linked by 3′–5′ phosphodiester bonds. (2) DNA base compositions follow Chargaff's rules. (3) Heating gently purified "native" DNA can produce a marked change in its physical properties, such as viscosity, to produce "denatured" DNA, without the cleavage of covalent bonds. (4) X-ray diffraction patterns of DNA fibers indicate that the molecules have a helical structure. To account for these observations, Watson and Crick proposed that the native DNA molecule consists of two polymeric chains of nucleotides paired in a double helix. The pairing between the two strands is effected by specific

Figure 4.10
Observations used by Watson and Crick in constructing their model for the structure of DNA. **(a)** Nucleotides are linked by 3′-5′ phosphodiester bonds. **(b)** Chargaff's rules. **(c)** Heating causes a change in the physical properties of native DNA without breaking covalent bonds. **(d)** X-ray diffraction pattern produced by a DNA fiber. (Photograph courtesy of Prof. Maurice H. F. Wilkins, Kings College, London.)

(a)

5′ end

O=P—O—CH_2 (5′) — Base

O^-

O (3′)

O=P—O—CH_2 (5′) — Base

O^-

OH (3′)

3′ end

(b) Moles A = moles T
Moles G = moles C

(c) Viscous DNA solution $\xrightarrow{\text{Heat}}$ nonviscous DNA solution

(d)

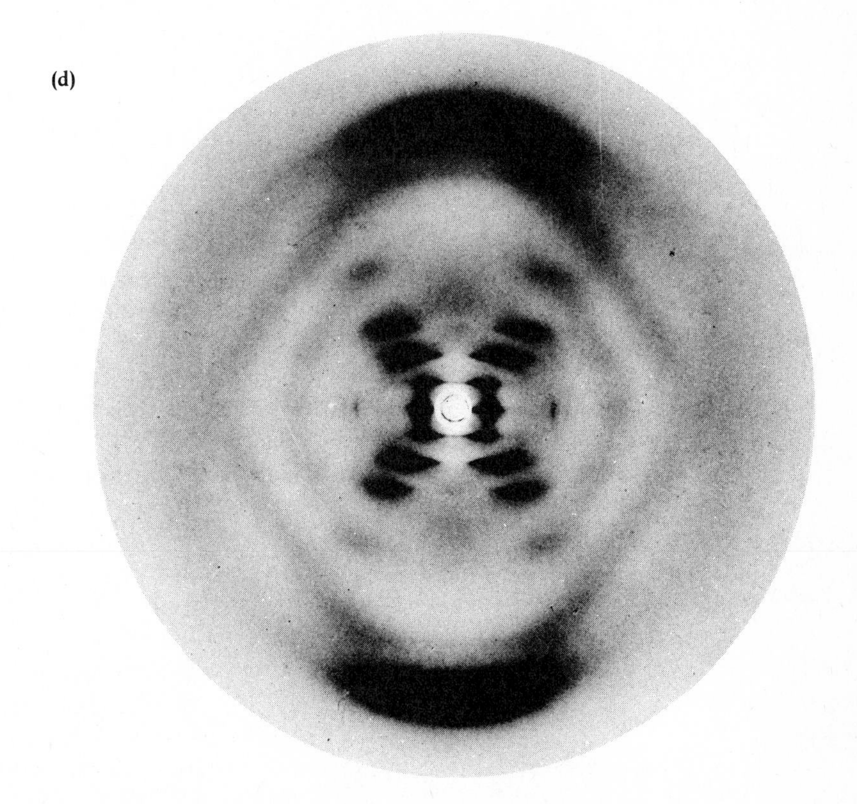

Table 4.1
Base compositions of DNA from different organisms.

Organism	Base Composition (mole percent)				Base Ratios			Asymmetry
	A	G	C	T	A/T	G/C	Pu/Py	$\frac{A+T}{G+C}$
Animals								
Man	30.9	19.9	19.8	29.4	1.05	1.00	1.04	1.52
Sheep	29.3	21.4	21.0	28.3	1.03	1.02	1.03	1.36
Hen	28.8	20.5	21.5	29.2	1.02	0.95	0.97	1.38
Turtle	29.7	22.0	21.3	27.9	1.05	1.03	1.00	1.31
Salmon	29.7	20.8	20.4	29.1	1.02	1.02	1.02	1.43
Marine crab	47.3	2.7	2.7	47.3	1.00	1.00	1.00	17.5
Sea urchin	32.8	17.7	17.3	32.1	1.02	1.02	1.02	1.58
Locust	29.3	20.5	20.7	29.3	1.00	1.00	1.00	1.41
Plants								
Wheat germ	27.3	22.7	22.8	27.1	1.01	1.00	1.00	1.19
Yeast	31.3	18.7	17.1	32.9	0.95	1.09	1.00	1.79
Aspergillus niger (mold)	25.0	25.1	25.0	24.9	1.00	1.00	1.00	1.00

Bacteria								
Escherichia coli	24.7	26.0	25.7	23.6	1.04	1.01	1.03	0.93
Staphylococcus aureus	30.8	21.0	19.0	29.2	1.05	1.11	1.07	1.50
Clostridium perfringens	36.9	14.0	12.8	36.3	1.01	1.09	1.04	2.70
Brucella abortus	21.0	29.0	28.9	21.1	1.00	1.00	1.00	0.72
Sarcina lutea	13.4	37.1	37.1	12.4	1.08	1.00	1.04	0.35
Bacteriophages								
T7	26.0	24.0	24.0	26.0	1.00	1.00	1.00	1.08
λ	21.3	28.6	27.2	22.9	0.92	1.05	1.00	0.79
ϕX174, viral	24.6	24.1	18.5	32.7	0.75	1.30	0.95	1.34
ϕX174, replicative	26.3	22.3	22.3	26.4	1.00	1.00	1.00	1.18

After A. L. Lehninger, *Biochemistry*, 2nd ed., Worth Publishers, New York, 1975.

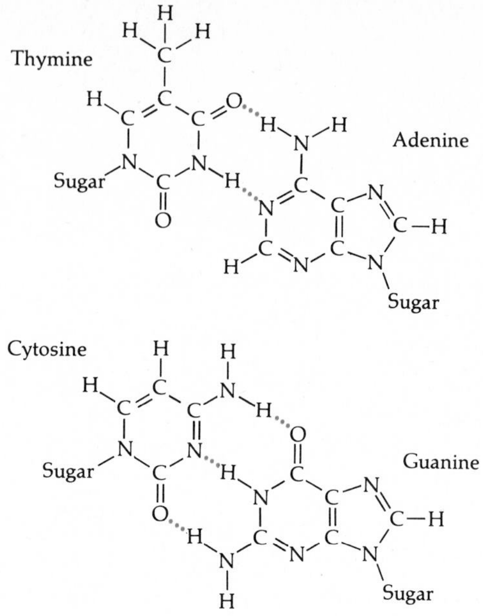

Figure 4.11
The hydrogen-bonded base pairs in DNA. Adenine pairs with thymine through two hydrogen bonds, and guanine pairs with cytosine through three hydrogen bonds. Note that adenine and guanine, which are purines, pair with the pyrimidines thymine and cytosine. The hydrogen bonds are much weaker than the covalent bonds joining the atoms of each nucleotide, but are strong enough to insure the specificity of the pairings A–T and G–C.

hydrogen bonds between adenine in one strand and thymine in the other strand, and between guanine in one strand and cytosine in the other strand (Figure 4.11). These pairing properties, which render A the complement of T and G the complement of C, were deduced by constructing molecular models showing the precise spatial arrangements of the atoms of the bases. Construction of a molecular model of the proposed double-stranded helical structure further requires that the orientation of the two strands be antiparallel, as shown in Figure 4.12.

The proposed specificity of hydrogen bonding between adenine and thymine and between guanine and cytosine accounts for the constancies in base composition observed by Chargaff, and renders the two strands of the double helix complementary. Further, any sequence of base pairs along the length of the double helix is possible. The precise order of A-T and G-C base pairs may be a species-dependent character that does not affect the double helical nature of the DNA molecule. The possible number of different sequences of base pairs in a DNA molecule is virtually infinite and is capable of encoding an enormous amount of information. This fact made the hypothesis that DNA might be the genetic material of all organisms very appealing. It is also evident from the model that the physical structure of the native molecule might be greatly altered, without breaking covalent bonds, if heating caused the hydrogen bonds to break, allowing the two strands to separate. A space-filling molecular model of

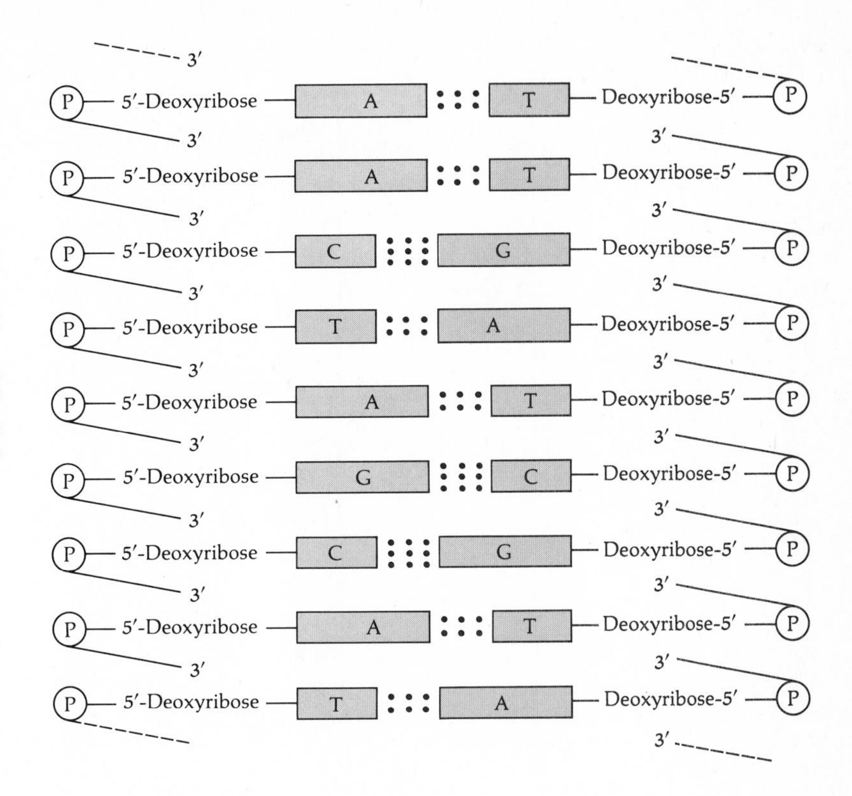

Figure 4.12
The two strands of the DNA double helix pair with the 5′→3′ polarity of the strands in opposite directions. Note that A-T and G-C base pairs occupy equal spaces between the phosphate-sugar backbones of the two strands. Any sequence of base pairs along the length of the molecule is possible without changing the overall conformation of the DNA double helix. The helical twists of the strands are not shown.

the double helix is shown in Figure 4.13; note that separation of the two strands requires that they be unwound from each other rather than pulled directly apart.

The Watson-Crick model suggests how the native DNA molecule might replicate to produce two identical daughter molecules. Because of their complementarity, if the two strands were separated, each could serve as a template for the synthesis of a new complementary strand. The base sequence of a newly synthesized strand would be dictated by specific hydrogen bond formation between bases in the template strand and bases in the newly forming strand (Figure 4.14). Thus the genetic information that might be encoded in the sequence of base pairs in a parental molecule could be faithfully transmitted to two daughter molecules. Moreover, were an error in base pairing to occur during DNA replication, the insertion of an incorrect nucleotide into the newly forming strand might change the information content of the molecule and could be expected to be transmitted to progeny DNA molecules. Such an alteration in a nucleotide pair could possess the properties of genetic mutations. Thus the Watson-Crick model of DNA structure, if proved to be true, could account for both the replicational and informational properties of genes.

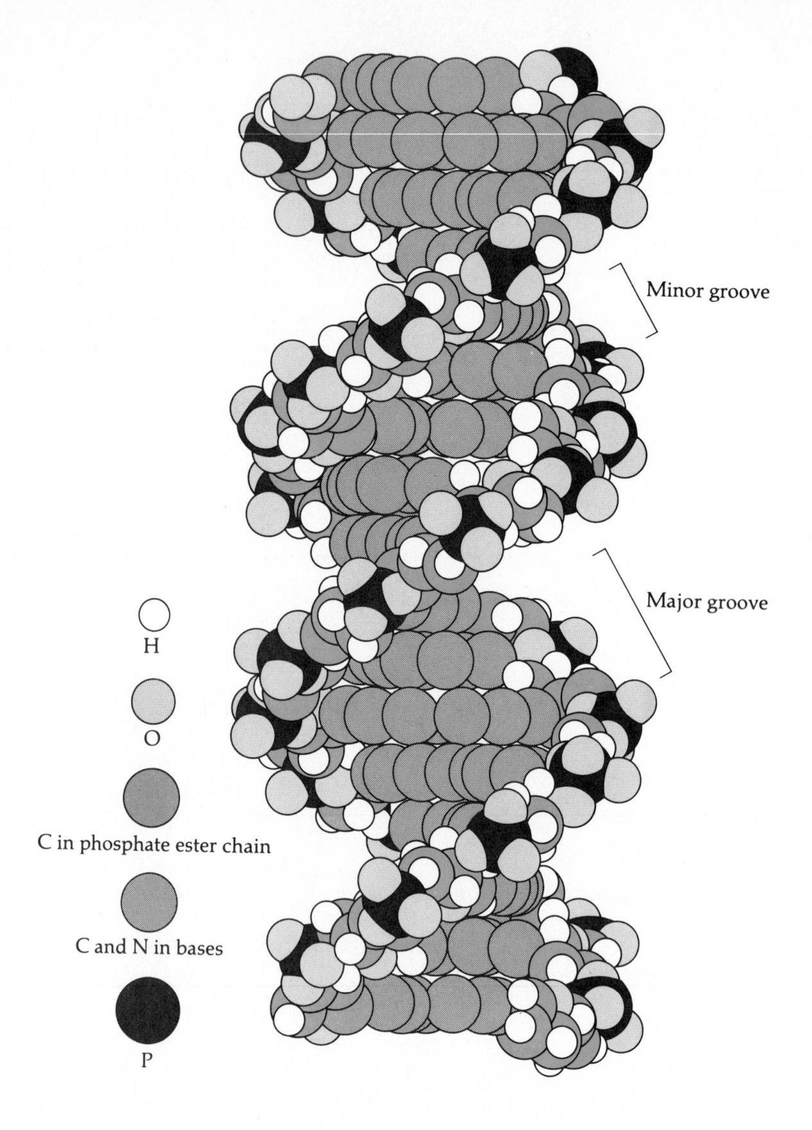

Figure 4.13
A space-filling model of the DNA double helix, showing the intertwining of the two strands. Note the close stacking of the flat base pairs within the double helix.

Tests of the Watson-Crick Model

The double helix model for DNA was proposed in 1953 and marks the beginning of the science of molecular biology. It was not until five years later, however, that the first convincing experimental confirmation of the Watson-Crick model was presented, by Matthew Meselson and Franklin Stahl. An explicit prediction of the double helix model, confirmed by their experiments, is that DNA replication is *semiconservative*: each daughter

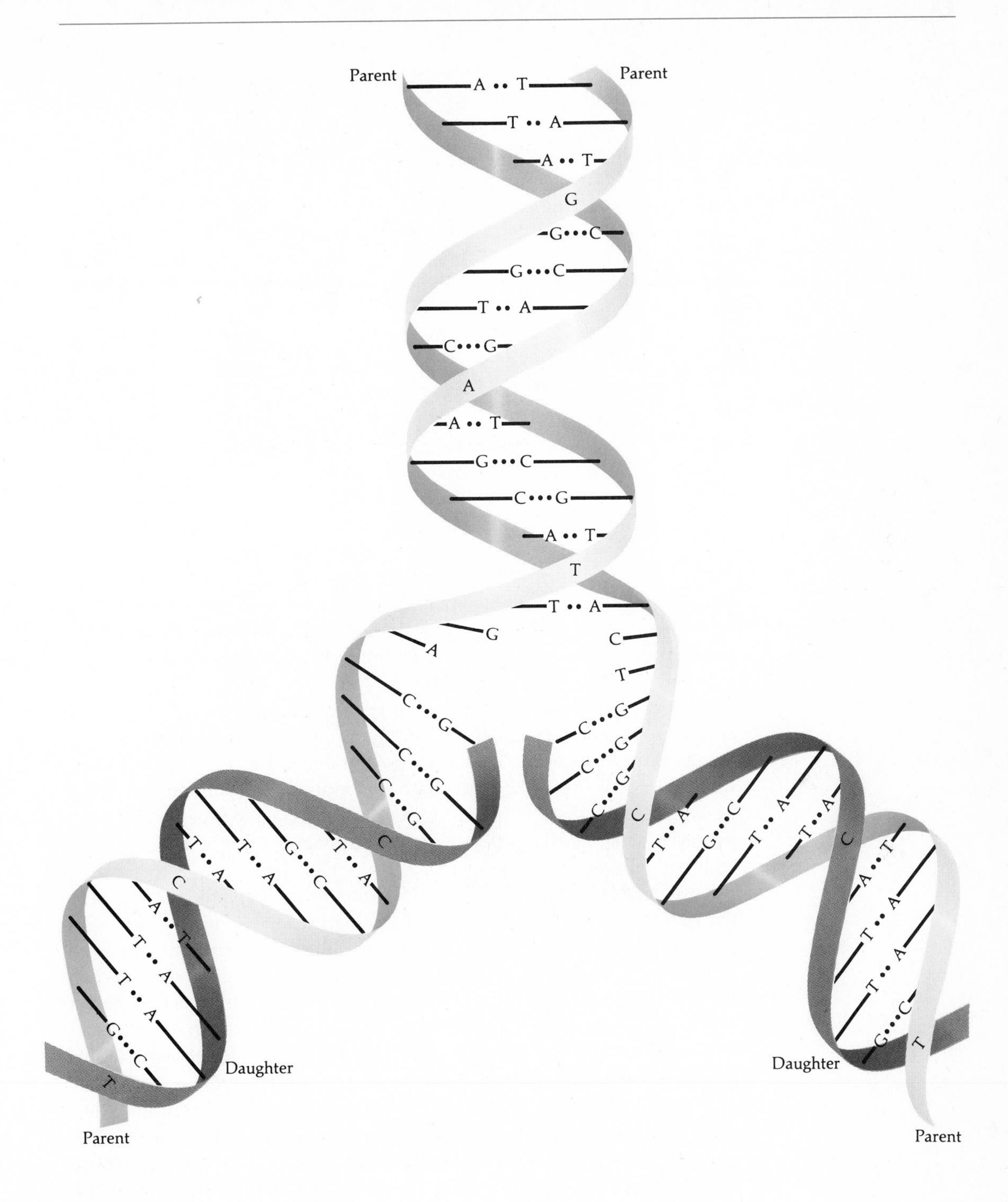

Figure 4.14
The model proposed by Watson and Crick for the replication of the DNA double helix.

DNA molecule will consist of one intact (conserved) strand from the parental double-stranded helix and one newly synthesized complementary strand.

Other possible modes of DNA replication that are *not* predicted by the double helix model are: (1) a conserved mode, in which parental DNA is fully conserved and daughter DNA molecules consist wholly of newly synthesized DNA, and (2) a dispersive mode, in which both daughter DNA molecules consist of newly synthesized DNA and the parental molecule is degraded to nucleotides, which may or may not form part of the daughter molecules.

Demonstration of the mode of DNA replication depends on the ability to distinguish parental from daughter DNA molecules. Meselson and Stahl used a heavy isotope of nitrogen, ^{15}N, to label the DNA of *E. coli* by growing cells in a medium in which the only nitrogen source was $^{15}NH_4Cl$. A dozen generations of growth in such a medium is sufficient to label uniformly all of the bacterial DNA with ^{15}N. DNA molecules containing ^{15}N can be distinguished from DNA molecules containing the lighter, common isotope, ^{14}N, on the basis of their densities, because DNA containing ^{15}N has a higher mass per nucleotide than DNA containing ^{14}N. DNA molecules of different densities can be separated from one another by centrifugation in a density gradient of a cesium chloride solution (Figure 4.15). If the CsCl solution is sufficiently concentrated, the gradient of density created by centrifugation of this heavy salt solution will possess a value at which DNA molecules are buoyant. During centrifugation, DNA molecules will come to rest at a level where their density is equal to that of the surrounding solution. The DNA of *E. coli* cells grown in a medium containing ^{15}N has a buoyant density of 1.724 g/cm^3, while that from cells grown in a medium containing the common ^{14}N has a buoyant density of 1.710 g/cm^3. A mixture of these two types of DNA can be easily separated into two density components by this centrifugation technique (Figure 4.15).

An analogy that illustrates the principle of density-gradient centrifugation is that of a submerged submarine. By admitting the right amount of water to its ballast tanks, the submarine can float at a fixed depth where its density is equal to that of the surrounding water. When more water is admitted, the density of the submarine increases, and it sinks until it encounters colder water with a greater density than that of the warmer water above. When it reaches water of a density equal to its own, the submarine will again float at the new depth.

Meselson and Stahl performed a density-shift experiment, in which cells growing for many generations in a medium containing ^{15}N are abruptly shifted to a ^{14}N medium. Portions of this culture are removed at progressive times following the shift, and the buoyant density of the DNA in each portion is determined. The results are shown in Figure 4.16. After one generation of growth in the ^{14}N medium, all of the DNA in the culture is observed to have a buoyant density exactly intermediate between those

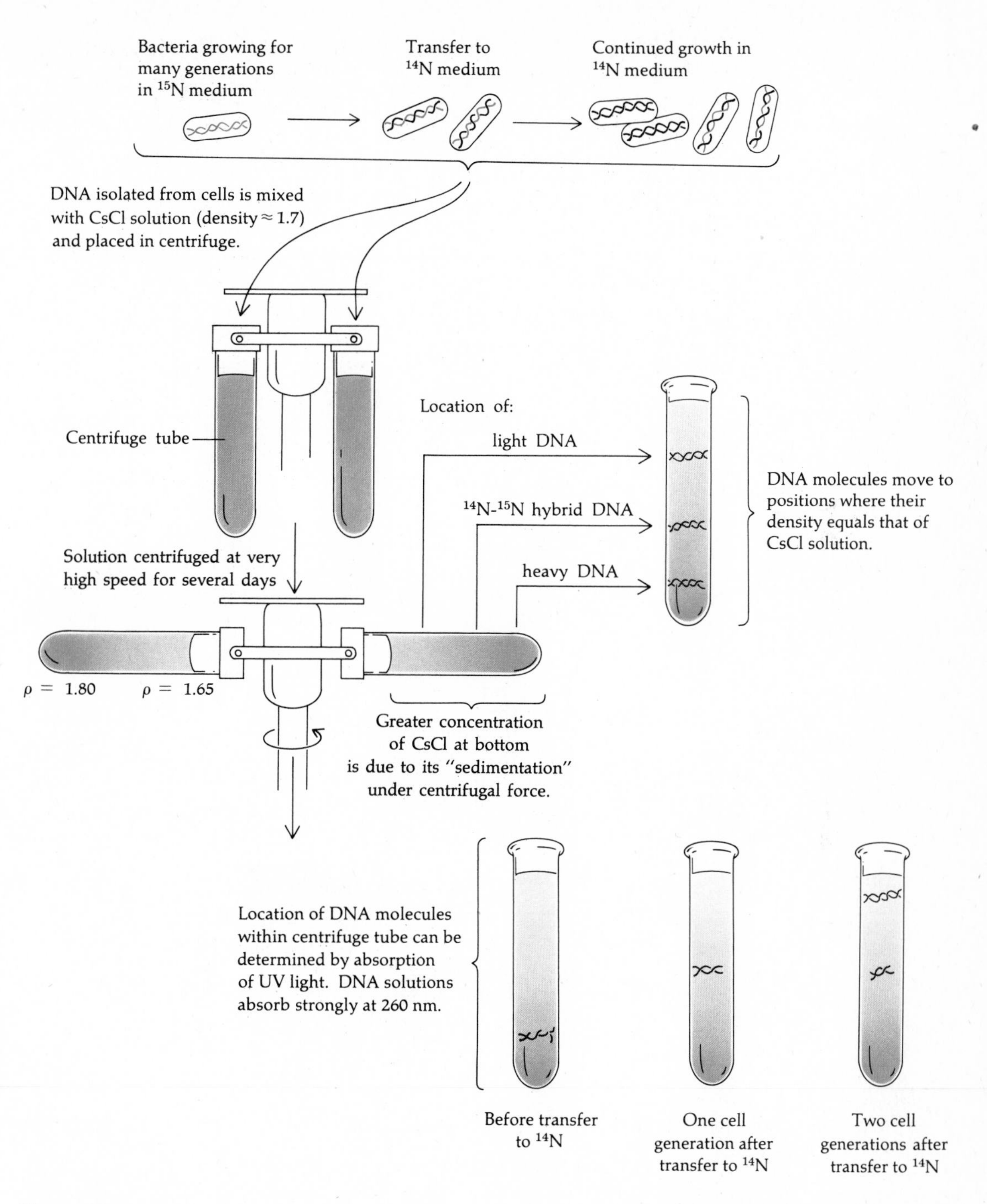

Figure 4.15
Separation of the DNA molecules of different densities by centrifugation in a cesium chloride density gradient. The steps followed by Meselson and Stahl in demonstrating the semiconservative replication of *E. coli* DNA are diagrammed. (After J. D. Watson, *Molecular Biology of the Gene,* 3rd ed., W. A. Benjamin, Menlo Park, Cal., 1976.)

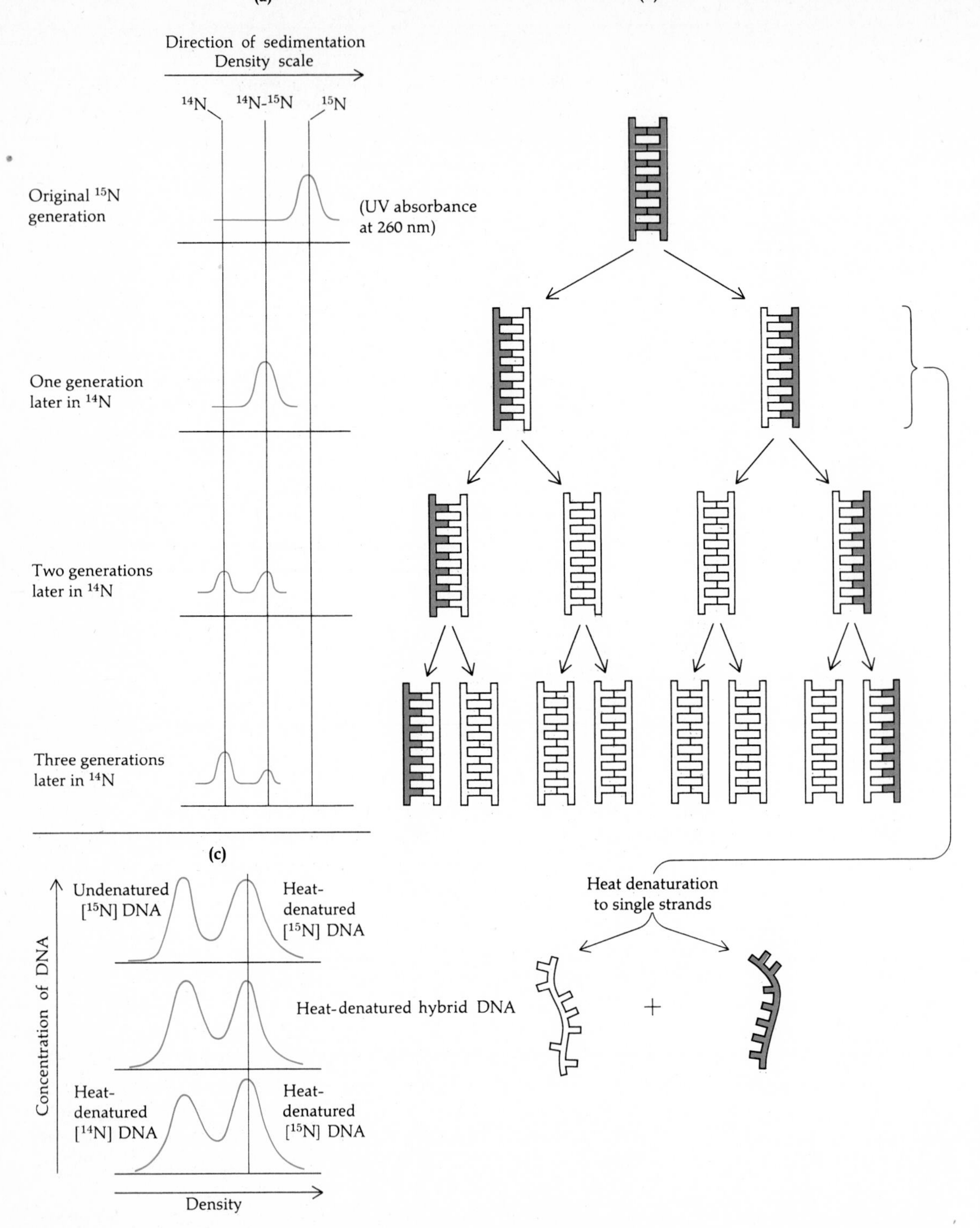

Figure 4.16
(a) The density distribution of DNA molecules observed by Meselson and Stahl at different times following the shift of *E. coli* cells growing in heavy medium to light medium. **(b)** Schematic interpretation of the observations in part a. The ^{15}N-labeled DNA and the newly synthesized ^{14}N-labeled DNA are shown in different colors. **(c)** Heating the DNA of intermediate density converts it to single-stranded DNA containing two density components, one with the density of heated ^{15}N-labeled DNA and the other with the density of heated ^{14}N-labeled DNA.

of [^{15}N]DNA and [^{14}N]DNA. After a second generation of growth in ^{14}N medium, one-half of the DNA is observed to have a density equal to that of [^{14}N]DNA, and one-half retains the intermediate density observed after one generation of growth in ^{14}N medium. After a third generation of growth in ^{14}N medium, three-quarters of the DNA has a density equal to that of [^{14}N]DNA, and one-quarter retains the intermediate density. This relationship between the number of generations of growth and the distribution of densities of the DNA is exactly that expected for a semiconservative mode of replication, as predicted by the Watson-Crick model (Figure 4.16).

The model further predicts that the DNA of intermediate density should consist of one fully heavy (^{15}N) single strand and one fully light (^{14}N) single strand. Meselson and Stahl heated the DNA of intermediate density at 100°C for 30 minutes (a procedure known to alter the physical properties of DNA without breaking covalent bonds) and found that it was converted to a DNA with two density components in equal amounts. One of these components had a density equal to that of heated fully light DNA, and the other had a density equal to that of heated fully heavy DNA (Figure 4.16). They concluded that the DNA of intermediate density, formed after one generation of growth in light medium, is a hybrid molecule consisting of one fully heavy, conserved parental strand and one fully light, newly synthesized daughter strand, exactly as predicted by the Watson-Crick model.

Similar experiments have been performed with replicating DNA from a number of prokaryotic and eukaryotic organisms, and in each case the mode of replication has been found to be semiconservative. The Meselson-Stahl experiments were the first, and are still the strongest, proof of the Watson-Crick model. Many other types of experiments have been carried out since, all of which support the double helix model in detail. The model is now so fully accepted as being correct that it is the foundation of modern genetics.

General Features of DNA Replication

Native DNA molecules are very large and subject to enzymatic or physical breakage when extracted from the organism for study. The replicating *E. coli* DNA studied by Meselson and Stahl was extensively fragmented, and their experiment provides information only concerning the state of the DNA before and after replication. The entire replicating *E. coli* chromosome was first observed by John Cairns, who developed a technique for breaking *E. coli* cells gently to avoid shearing the intact chromosome by fluid motion in the solution. Cairns was thus able to liberate intact *E. coli* chromosomes, which he had labeled with radioactive [^{3}H]thymidine. The chromosomes were allowed to settle gently from solution onto a solid surface, and the dried surface was then covered in the dark with a photographic emulsion and stored for several weeks. During this time, the emission of electrons from the radioactive DNA induced the formation of silver grains in the photographic emulsion adjacent to the DNA molecules. Subsequent development of the emulsion provided an *autoradiograph* of the radioactive chromosome, in which a track of silver grains traced the conformation of the DNA molecule. These autoradiographs led to the surprising finding that the *E. coli* chromosome is circular (Figure 4.17). Circularity has subsequently been found to be a prominent feature of the DNA molecules of prokaryotic organisms, of viruses, and of organelles in eukaryotic organisms. Not all DNA molecules are circular, however; the chromosomes of eukaryotic organisms and of many viruses contain linear DNA molecules.

Three general conformations and modes of DNA replications are now known for DNA molecules. Circular DNA molecules, such as the replicating form of phage λ DNA, can replicate in the manner shown by autoradiographs of the *E. coli* chromosome (Figure 4.17). The replication of a circular DNA molecule is initiated at a specific point on the circle and leads to the formation of a replication "bubble," which grows in size as replication proceeds in two directions around the chromosome (Figure 4.17). This mode of DNA replication produces intermediates with a structure similar to that of the Greek letter theta, Θ. The *theta* mode of replication converts a circular parental chromosome to two circular daughter chromosomes, in each of which one strand of the parental DNA molecule is conserved and a complementary strand is newly synthesized.

As we shall see later, the life cycles of certain organisms require the conversion of a circular chromosome to a linear chromosome. This conversion is effected by a different mode of DNA replication, known as the *sigma* (Greek letter σ) or "rolling circle" mode. Initiation of sigma DNA replication begins with the cleavage of a phosphodiester bond in one strand of the parental circular molecule to produce a nick with 3′-OH and 5′-PO_4 ends on that strand. The complementary circular strand then serves as a template for the synthesis of a new strand, which is covalently attached to the 3′-OH end of the nicked parental strand. As this strand grows at the 3′-OH end, the 5′-PO_4 end of the same strand is displaced to form a "tail" on the circle. Synthesis of a complement to the strand forming the tail then occurs (Figure 4.18). Intermediates in this mode of replication have a sigma (σ) conformation. As replication proceeds, a circular parental molecule is converted to two daughter molecules, one circular and the other linear. The sigma mode of replication is necessary to the life cycle of some bacteriophages, such as λ and ϕX174. It is also involved in sexual conjugation in bacteria, and occurs during oogenesis in certain eukaryotic organisms.

The chromosomes of some viruses and of all eukaryotic organisms are linear DNA molecules. The replication of linear molecules is initiated at specific points by the formation of replication bubbles. Small viral molecules may have only one point of initiation per molecule. Large DNA molecules, such as those found in eukaryotic chromosomes, may have hundreds of initiation points per molecule (Figure 4.19). Once a replication bubble forms, it grows in size as DNA replication proceeds in both

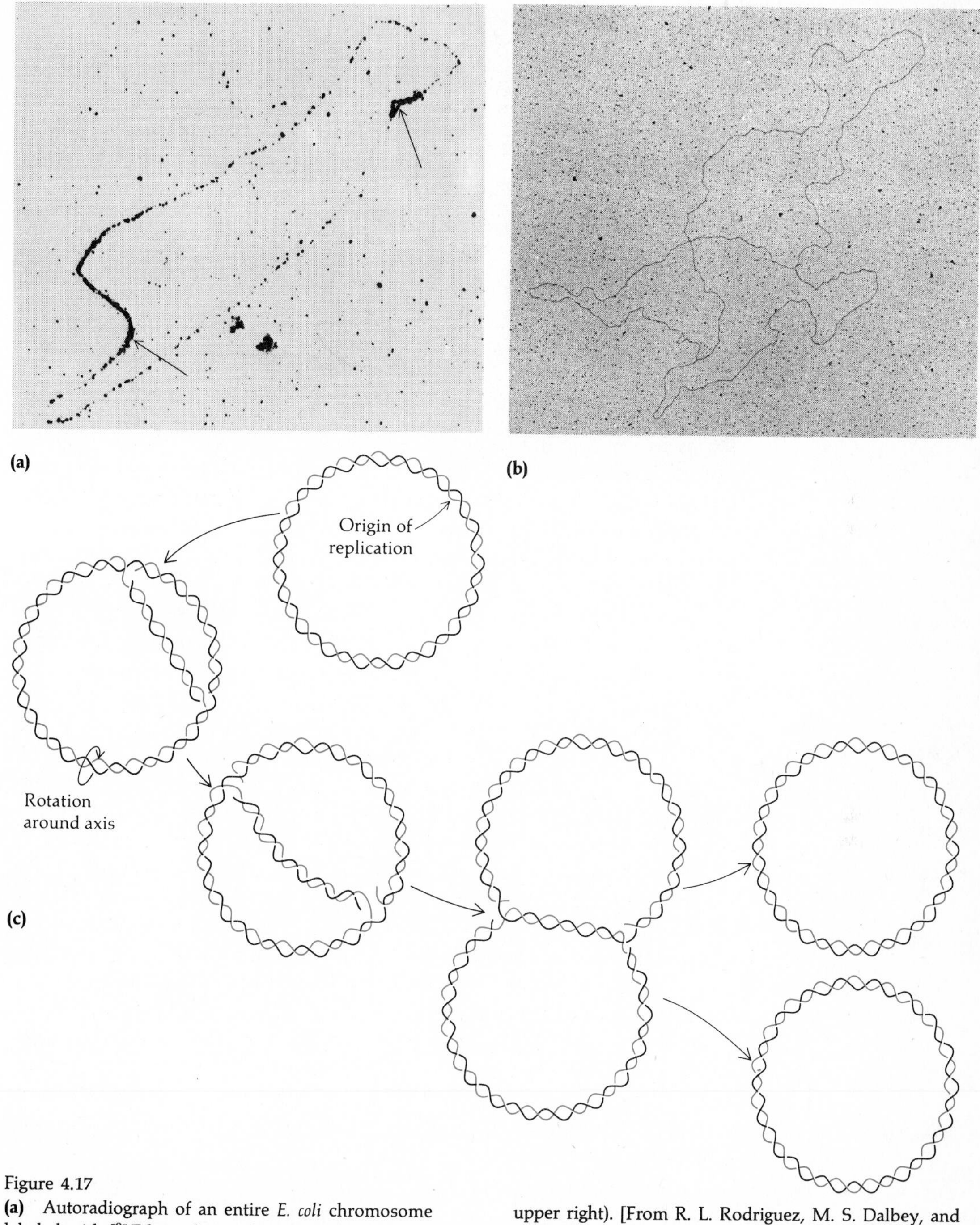

Figure 4.17
(a) Autoradiograph of an entire *E. coli* chromosome labeled with [^{3}H]thymidine at low specific activity for most of a round of DNA replication, concluded by a pulse of high-specific-activity [^{3}H]thymidine. The high-specific-activity pulse (dense grains) labels the terminus of replication of this molecule (lower left) and the initiation of a new round of replication (bubble at upper right). [From R. L. Rodriguez, M. S. Dalbey, and C. I. Davern, *J. Mol. Biol.* 74:599 (1973).] **(b)** θ-shaped replicating bacteriophage λ DNA during early parental DNA synthesis. (Courtesy of Dr. David Dressler and Dr. John Wolfson, Harvard University.) **(c)** Diagram of bidirectional replication of circular DNA molecules.

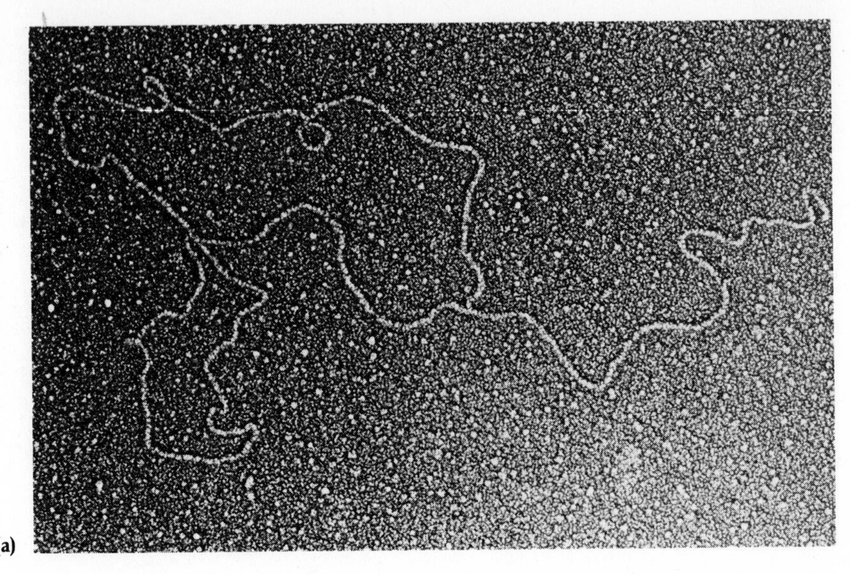

Figure 4.18
(a) Electron micrograph of bacteriophage λ DNA replicating by the sigma (σ) mode during synthesis of linear progeny viral DNA. [From J. A. Kiger, Jr., and R. L. Sinsheimer, *Proc. Natl. Acad. Sci. USA* 68:112 (1971).] **(b)** Diagram of the sigma mode of DNA replication.

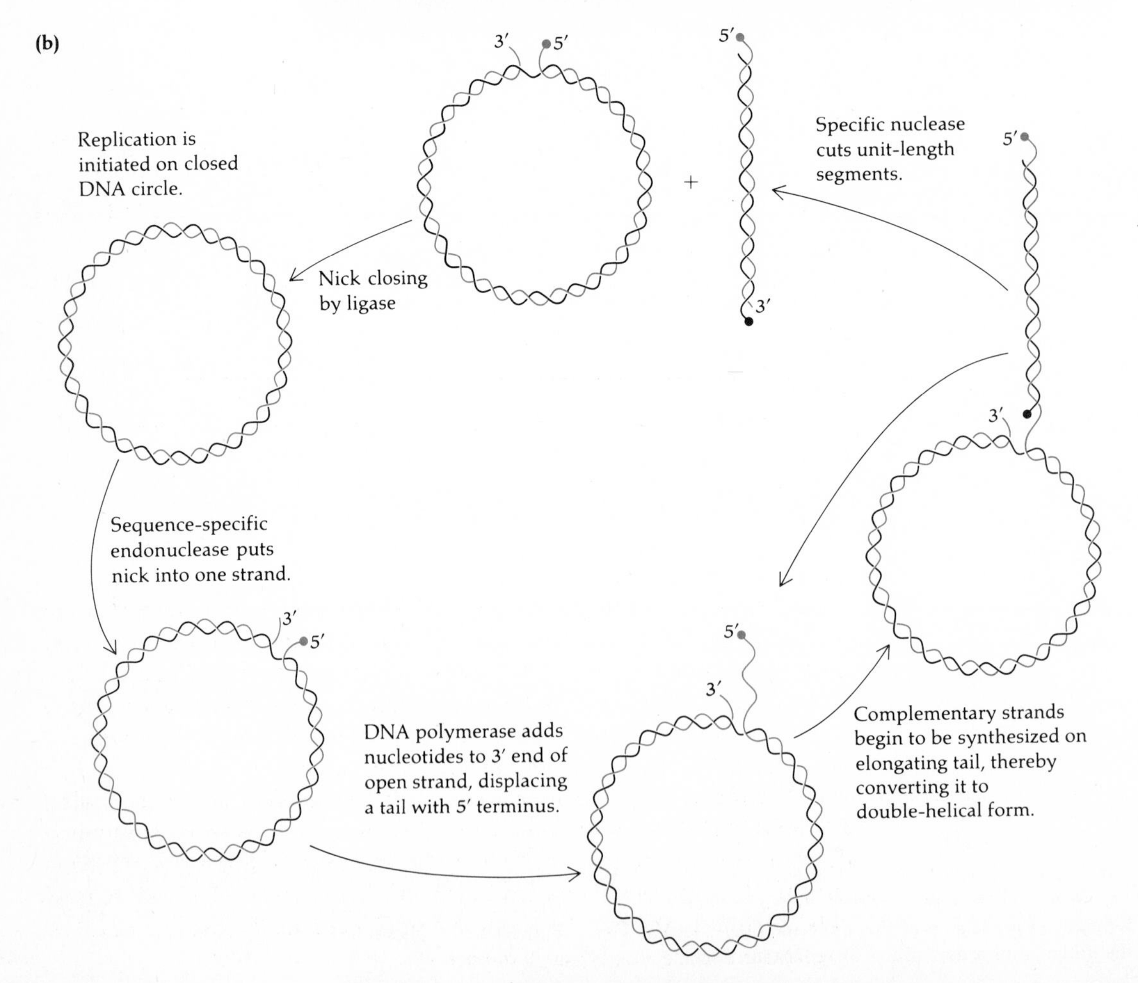

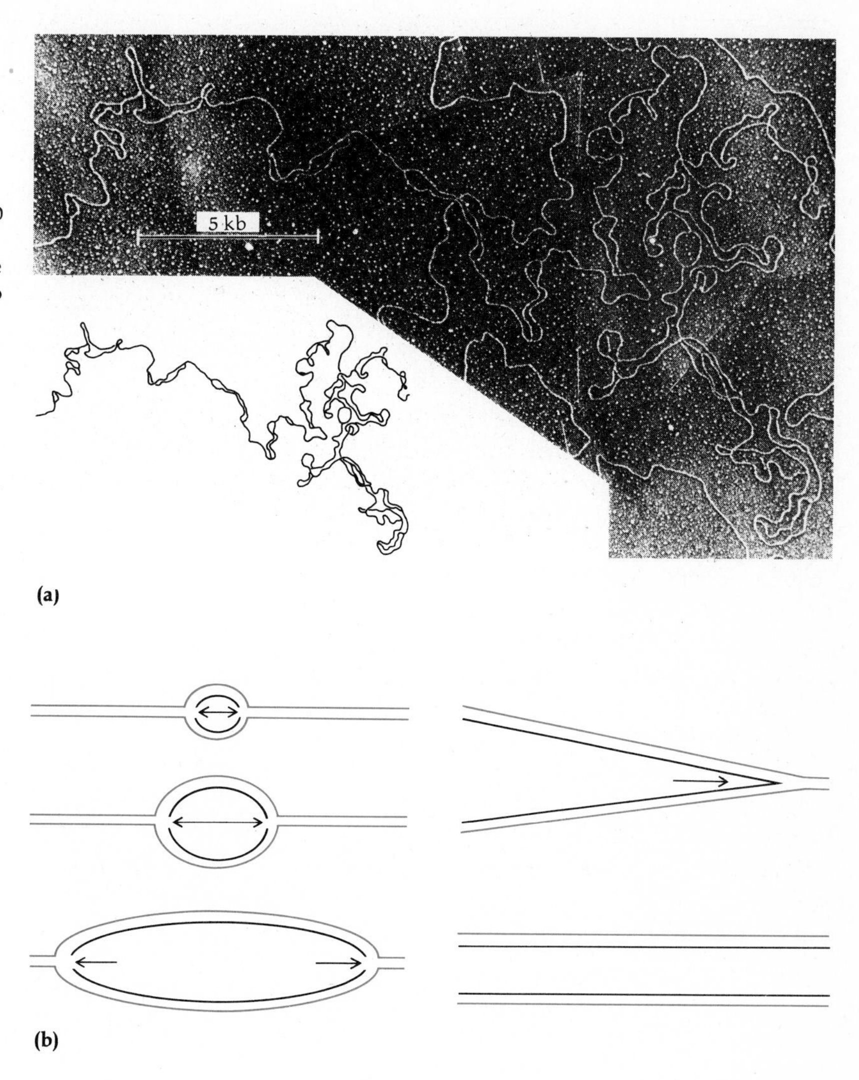

Figure 4.19
(a) Replicating *Drosophila melanogaster* embryo DNA. Note the many replication bubbles along the DNA strand. The inset is a tracing of a fragment of chromosomal DNA 119,000 nucleotide pairs long and containing 23 bubbles. The bar is a length equivalent to 5000 nucleotide pairs (5 kb = 5 kilobase pairs). [From H. J. Kriegstein and D. S. Hogness, *Proc. Natl. Acad. Sci. USA 71*:135 (1974).] **(b)** Diagram of bidirectional replication of a linear DNA molecule.

directions from the point of initiation. As replication continues, adjacent bubbles fuse to form larger bubbles, and Y-shaped intermediates may form as a bubble reaches the end of a molecule. When replication is complete, two linear double helical daughter molecules, each containing a conserved strand from the parental molecule and a newly synthesized strand, have descended from a single linear parental molecule.

The process of DNA replication plays a vital role in the transmission of the hereditary information, contained in the sequence of base pairs, from parent to progeny DNA molecule, from parent to progeny somatic cell, and from parent to progeny organism. This chapter serves as a general introduction to this vital process. We shall examine the process of DNA replication more closely in Chapter 9.

Box 4.1 Renaturation Kinetics of DNA

The steps involved in the denaturation and renaturation of DNA fragments are diagrammed in Figure 4.20. Renaturation of complementary single strands to produce fully double-stranded molecules is a two-step process: (1) nucleation and (2) "zippering." Because it involves two single-stranded DNA molecules, nucleation occurs at a rate proportional to the square of the DNA concentration (measured in molarity of nucleotides) and is therefore a second-order reaction. In contrast, "zippering" to produce a fully double-stranded molecule is a unimolecular reaction and is expected to be first-order with respect to DNA concentration. Empirically, renaturation of DNA is observed to be a second-order reaction; therefore, nucleation rather than zippering must be the rate-limiting step.

If we let c = concentration of single-stranded DNA at time t, and c_0 = concentration of single-stranded DNA at $t = 0$, then the second-order rate equation describing the loss of single-stranded DNA is

$$-\frac{dc}{dt} = k_2c^2 \qquad (1)$$

where k_2 is the second-order rate constant. Rearranging the terms gives us

$$-\frac{dc}{c^2} = k_2dt$$

and integrating this equation gives

$$\frac{1}{c} = k_2t + \text{Constant}$$

At $t = 0$, $c = c_0$, and therefore the Constant = $1/c_0$. Substituting, we get

$$\frac{1}{c} = k_2t + \frac{1}{c_0}$$

Rearranging, we get the *cot equation* (the quantity c_0t is called "cot"—hence the name):

$$\frac{c}{c_0} = \frac{1}{1 + k_2c_0t} \qquad (2)$$

At half-renaturation, $c/c_0 = 0.5$, and $t = t_{1/2}$. Substituting in the cot equation, we get

$$0.5 = \frac{1}{1 + k_2c_0t_{1/2}}$$

which rearranges to give

$$c_0t_{1/2} = \frac{1}{k_2} \qquad (3)$$

Figure 4.21 shows a plot of c/c_0 as a function of c_0t; this is called a *cot plot.*

James Wetmur and Norman Davidson have shown that for DNA renaturation the second-order rate constant, k_2, is inversely proportional to N, the number of units in a nonrepeating sequence of nucleotide pairs in a haploid nucleus or prokaryotic genome. Their relationship is given by the equation

$$k_2 = \frac{\alpha\beta^3L^{0.5}}{N} \qquad (4)$$

where L = the average number of nucleotides in a single-strand fragment, β = the average density of nucleation sites in a fragment, and α = a proportionality constant.

Combining equations 3 and 4, we see that $c_0t_{1/2}$ is directly proportional to N (a is the proportionality constant):

$$c_0t_{1/2} = aN \qquad (5)$$

The validity of equation 5 is demonstrated in Figure 4.22 by the observed renaturation curves of DNA fragments from DNAs of known complexity.

For DNA of known L and N values, the proportionality constant a can be evaluated from the observed $c_0t_{1/2}$ value under a given set of experimental conditions (temperature and ionic strength). One of the most important uses of cot analysis then emerges: namely, under the same set of experimental conditions, a can be used to calculate N for an unknown DNA from the observed $c_0t_{1/2}$ value of that DNA.

A cot plot of the renaturation of a bacterial DNA is compared with that for a eukaryotic DNA in Figure 4.23. It is apparent from the shape of the curve that the bacterial DNA renatures at a rate consistent with a single value for k_2 as given by the cot equation. In contrast, the cot plot for the renaturation of the eukaryotic DNA shows that the value of k_2 must vary. Cot plots for eukaryotic DNAs such as that in Figure 4.23 are interpreted to mean that a number of different classes of nucleotide sequences of different complexity (N) and frequency of repetition are present in the DNA of the eukaryotic cell. That fraction of the DNA that renatures most slowly (has the highest $c_0t_{1/2}$ value) represents unique sequences that are present only once per genome. Those fractions that renature more rapidly represent families of identical or very similar sequences, each characterized by a particular rate constant reflecting the complexity (N) of that family. The fraction of the genome occupied by each type of sequence can be estimated from the ordinate in Figure 4.23.

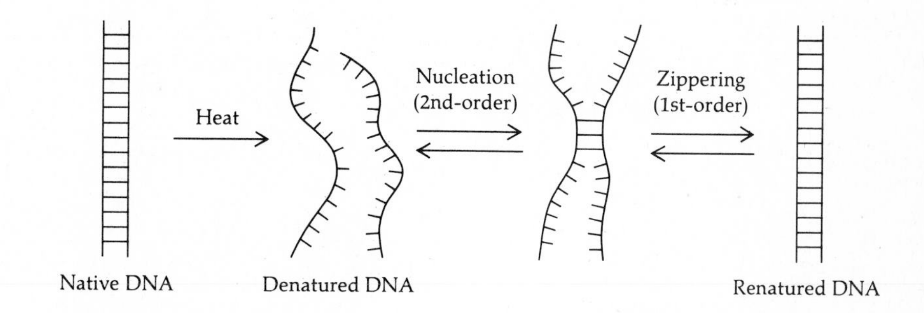

Figure 4.20
Steps in the denaturation and renaturation of DNA fragments. Renaturation requires two separate reactions: (1) In the nucleation reaction, hydrogen bonds form between two complementary single strands; this is a bimolecular, second-order reaction. (2) In the zippering reaction, hydrogen bonds form between all the bases in the complementary strands; this is a unimolecular, first-order reaction.

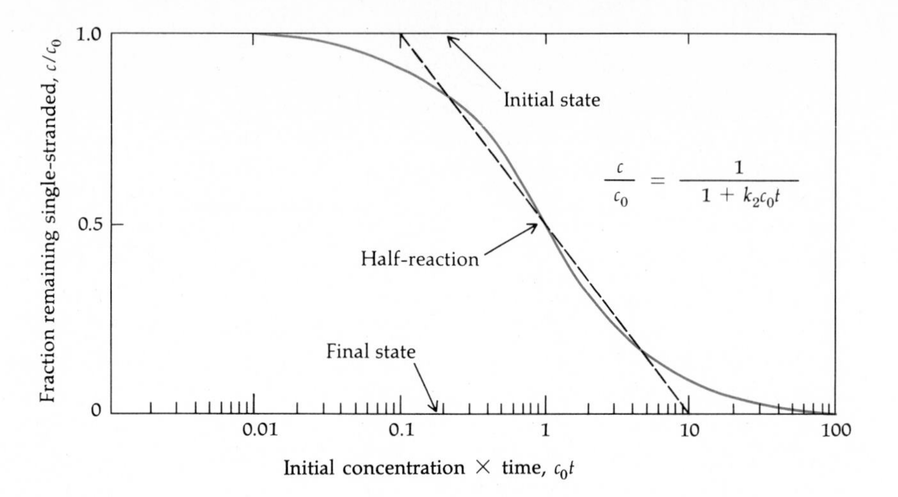

Figure 4.21
Time course of an ideal second-order reaction as given by the cot equation when $k_2 = 1$. The ordinate represents the fraction of single strands remaining after t seconds of reaction. The abscissa represents c_0t, where c_0 is the initial concentration of single strands of DNA, measured in moles of nucleotides per liter.

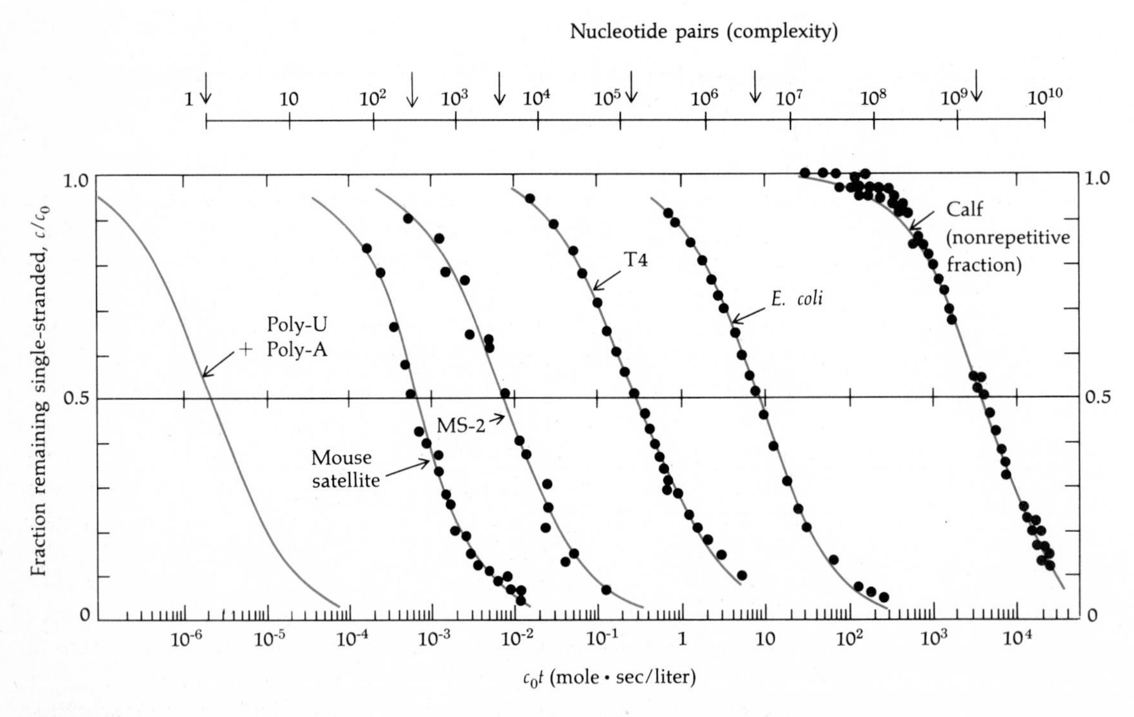

Figure 4.22
The direct relation between the $c_0t_{1/2}$ value, determined experimentally, and the complexity (N) of the DNA involved is indicated here. For a fixed value of c_0, the time t required to reach half-renaturation is proportional to N. [After R. J. Britten and D. E. Kohne, *Science 161*:529 (1968).]

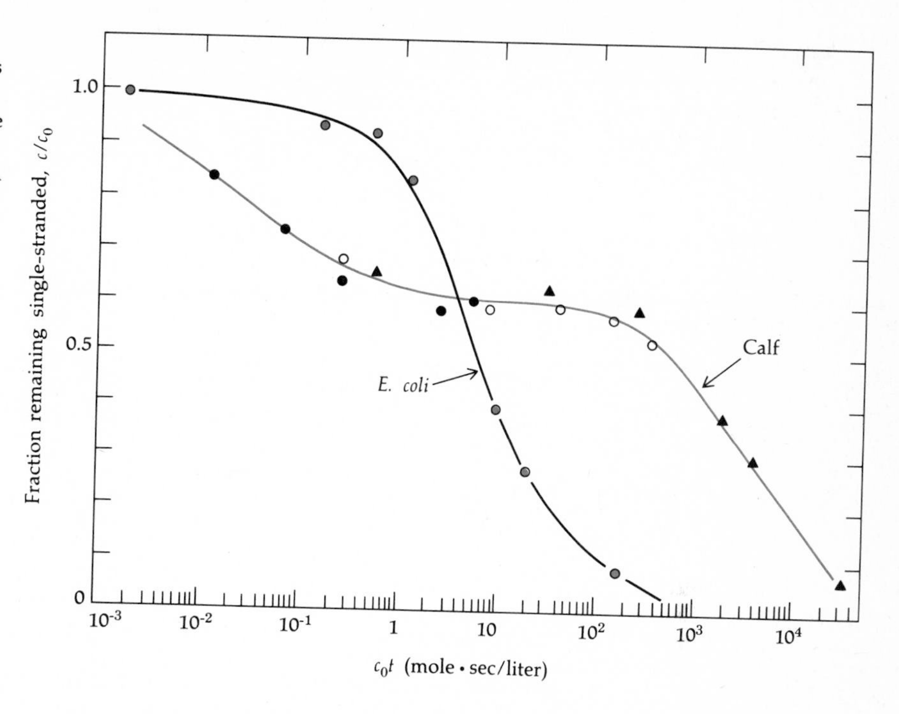

Figure 4.23
Comparison of the kinetics of reassociation of *E. coli* DNA fragments and bovine (calf) DNA fragments. The cot plot for the *E. coli* DNA has a shape very similar to that of the theoretical plot in Figure 4.21, indicating that this DNA renatures at a rate consistent with a single, unique value for k_2. In contrast, the cot plot for bovine DNA has a shape very different than that in Figure 4.21. The bovine DNA contains some DNA sequences that renature much more rapidly than *E. coli* DNA sequences and some that renature much more slowly. [After R. J. Britten and D. E. Kohne, *Science 161*:529 (1968).]

Organization of Nucleotide Sequences in DNA

The distinguishing features of prokaryotes and eukaryotes, the two major groups of living organisms, were described in Chapter 1 in terms of their cellular organization. A further distinction between prokaryotic and eukaryotic organisms is also quite evident in the nature of the DNA sequences themselves. The nucleotide sequences of eukaryotic DNAs are internally more complex and more complicated in organization than those of prokaryotic DNAs. Both eukaryotic and prokaryotic DNAs contain many nucleotide sequences that are present only once per genome. The characteristic feature of eukaryotic DNAs, however, is that they also contain groups of very similar sequences, called *families,* that are present more than once in each haploid genome. The evidence for this comes primarily from kinetic studies of the renaturation of single-stranded DNA molecules obtained by fragmentation of the chromosomal DNA (see Box 4.1).

The results of DNA renaturation studies can be seen in a qualitative way by understanding that the more abundant a particular sequence is in the genome, the greater is the concentration of that single-strand sequence and its complementary sequence in a solution of single-strand fragments. Therefore the probability that an abundant sequence will find its complement and anneal to form double-stranded DNA is greater for that sequence than for a unique sequence in the genome.

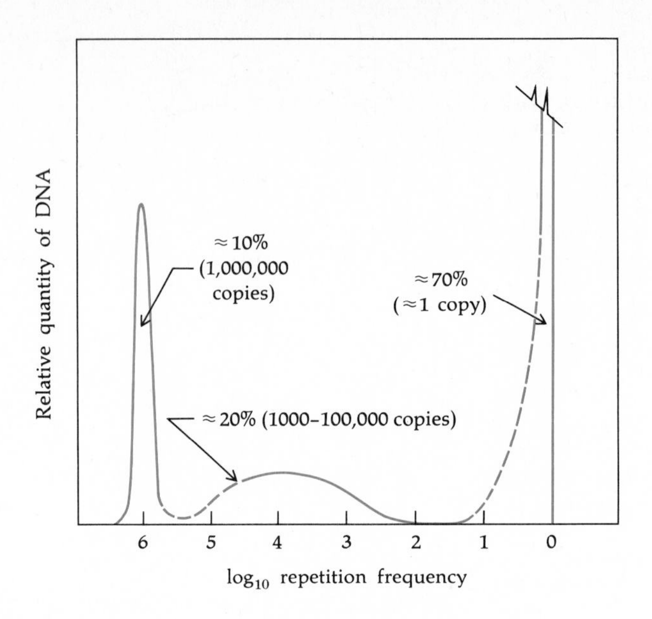

Figure 4.24
The frequency of repetition of nucleotide sequences in the DNA of the mouse is represented here as a spectrogram. The number of copies of each type of sequence is estimated from the cot plot for mouse DNA. [After R. J. Britten and D. E. Kohne, *Science* *161*:529 (1968).]

The cot equation (described in Box 4.1) provides a quantitative means for evaluating the complexity of different abundance classes of sequences. Figure 4.24 shows the relative abundance of different sequences in the mouse genome, as determined from its cot plot. About 70% of the DNA of the mouse represents sequences present about once per genome. At the other end of the spectrum, about 10% of the DNA represents highly repetitive sequences occurring about 10^6 times per genome. The remaining 20% of mouse DNA represents families of repetitive sequences occurring about 10^3–10^5 times per genome. Data on the genome composition of some representative eukaryotes are given in Table 4.2.

There is some biochemical and cytological evidence for the organization and possible function of these different classes of nucleotide sequences. The highly repetitive DNAs of mouse and *Drosophila* consist of a tandem repetition of a small number (about 10) of nucleotides. These DNAs are localized preferentially in the centromeric heterochromatin of the chromosomes. The great majority of these sequences appear to have no coding function, since they are not transcribed into RNA; their role may be a structural one. Most of the moderately repetitive sequences are distributed throughout the genome and are interspersed with unique sequences. At least a portion of both the moderately repetitive and unique sequences are transcribed into large, heterogeneous size classes of RNA, called hnRNA, in nuclei (see Chapter 11). However, most moderately repetitive sequences are not found as RNA transcripts in the cytoplasm. Most messenger RNA (mRNA) molecules are transcribed from unique sequences, with the known exception of those coding for histone proteins. The histone genes and genes for ribosomal RNAs (see Chapter 11) form classes of clustered, rather than interspersed, repetitive sequences.

Table 4.2
Sequence composition of some eukaryotic genomes.

Species	Fraction in Each Frequency Class			
	Nonrepetitive	Moderately repetitive*	Repetitive†	Highly repetitive‡
Nassaria obsoleta (snail)	0.38	>0.12	>0.15	0.18
Calf	0.55	–	0.38	0.05
Xenopus laevis (toad)	0.54	0.06	0.31	0.09
Strongylocentrotus purpuratus (sea urchin)	0.38	0.25	0.27	0.10
Drosophila	0.75	–	0.15	0.10

*20–50 copies.
†250–60,000 copies.
‡Up to 10^6 copies, including satellite DNAs.

From P. Grant, *Biology of Developing Systems*, Holt, Rinehart & Winston, New York, 1978.

Organization of DNA in Chromosomes

The circular *E. coli* DNA molecule is 20 Å in diameter and about 10^7 Å in length. This enormous length of DNA (about 1 mm) is contained in a cell only 2×10^4 Å in length and about 8×10^3 Å in diameter. The DNA of *E. coli* can be isolated from intact cells, employing very gentle procedures, as a compact *nucleoid* in which the DNA strand is highly folded and coiled in an ordered way. A model of the condensed *E. coli* chromosome is detailed in Figure 4.25. The chromosome is folded into a number of loops held in place by RNA molecules, and within each loop the DNA is further condensed by twisting or supercoiling. These two types of compacting processes have been revealed by treating intact nucleoids with either ribonuclease or deoxyribonuclease, as shown in Figure 4.25., and observing the structural changes induced by each enzyme.

The DNA molecules in eukaryotic chromosomes are much larger than those in bacterial chromosomes. Autoradiography of ^{3}H-labeled *Drosophila* DNA has revealed intact DNA molecules up to 1.2 cm in length. It is generally believed that eukaryotic chromosomes each contain a single continuous DNA molecule. Condensation of such enormously long

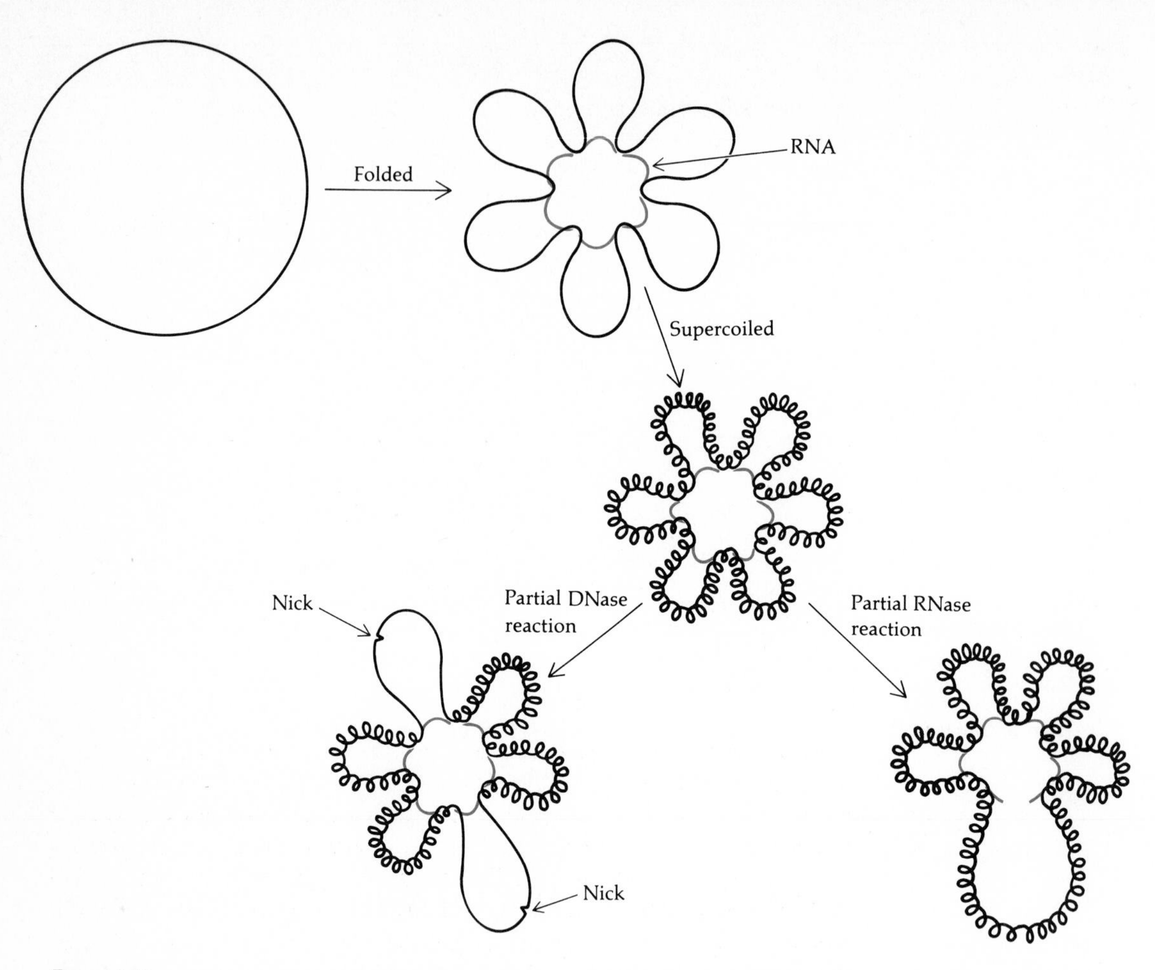

Figure 4.25
A model of the condensed *E. coli* chromosome, or nucleoid. RNA molecules bind the DNA molecule into folds. Within each fold, supercoiling further condenses the DNA. Partial digestion by ribonuclease (RNase) releases some of the folds but does not relax the supercoiling. Partial digestion by deoxyribonuclease (DNase) nicks one or another strand of the double helix, permitting rotation and relaxation of the supercoiled twists, but does not release the folds. [After D. E. Pettijohn and R. Hecht, *Cold Spring Harbor Symp. Quant. Biol. 38*:39 (1974).]

molecules within the nuclei of cells is the major function of the histone proteins, which are one of the distinctive features of eukaryotic cells.

The basic structural unit of the eukaryotic chromosome is the *nucleosome.* The nucleosome is composed of two each of four histone proteins—H2A, H2B, H3, and H4—which are combined to form an octamer. Each octamer is associated with about 200 nucleotide pairs of DNA, 700 Å in length. The precise arrangement of histone and DNA in the nucleosome is not known, but the DNA is believed to be wrapped around

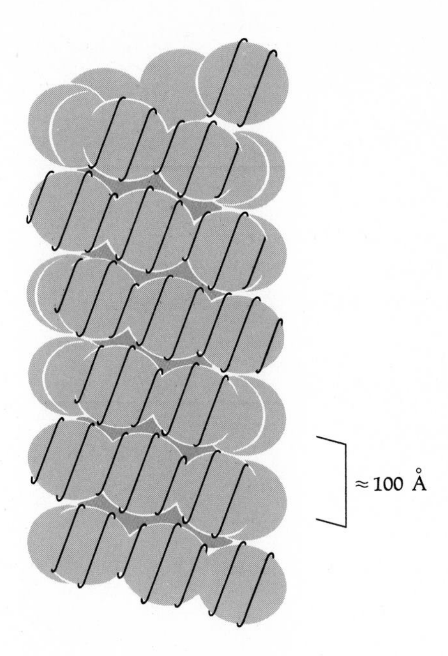

Figure 4.26
Diagram of a portion of a solenoid, showing the helically coiled nucleosomes (spheres), around which is wound the DNA (lines).

the protein in some way. The nucleosome has a diameter of about 100 Å, and thus the coiling of the DNA in the nucleosome reduces its length by a factor of 7. Another histone, H1, stabilizes the association of nucleosomes to form a helical coil (Figure 4.26). The diameter of this coil (called a *solenoid*) is measured by some to be 300 Å and by others to be 500 Å. The difference is probably due to the methods of preparing the material for electron microscopy. The condensation of DNA length created by solenoid formation is a factor of 6 over that of the condensation present in the nucleosome, if 300 Å is taken as the diameter of the solenoid. The basic structure of the solenoid is revealed by partial digestion of chromosomes with micrococcal nuclease; this enzyme cuts the DNA strand between nucleosomes, producing monomers, dimers, trimers, and so on, of nucleosomes (Figure 4.27). In interphase chromosomes, the solenoid is coiled once more, forming a hollow tube about 2000 Å in diameter, and condensing the length of the DNA by another factor of about 18 (Figure 4.28).

The transition from interphase chromosome to metaphase chromatid is probably accomplished by one more coiling of the 2000-Å tubes to form a larger coil of diameter about 6000 Å (Figure 4.28). This general scheme for the organization of DNA in the nucleus ignores the differential degrees of coiling that almost certainly exist between those regions of the chromosomes that are involved in RNA synthesis and DNA replication and those that are not. Moreover, heterochromatic regions of chromosomes are more compact than euchromatic regions. In any event, the DNA in the nuclei of eukaryotic cells is almost certainly organized in a hierarchy of coils in which the nucleosome is the basic structural unit.

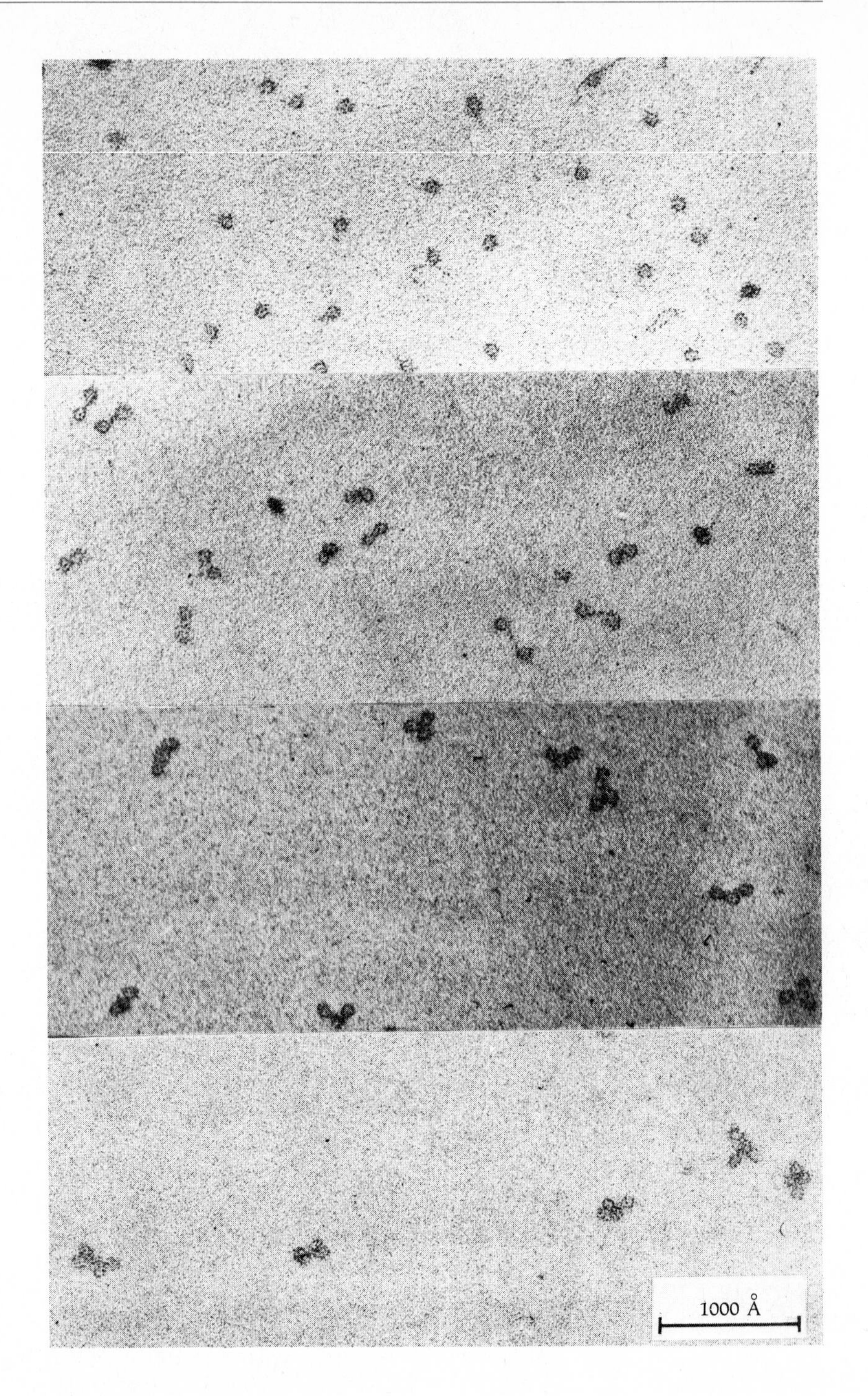

Figure 4.27
Electron micrographs of (from top to bottom) nucleosome monomers, dimers, trimers, and tetramers obtained by partial digestion of rat-liver nuclei by micrococcal nuclease. [From J. T. Finch, M. Noll, and R. D. Kornberg, *Proc. Natl. Acad. Sci. USA* 72:3321 (1975).]

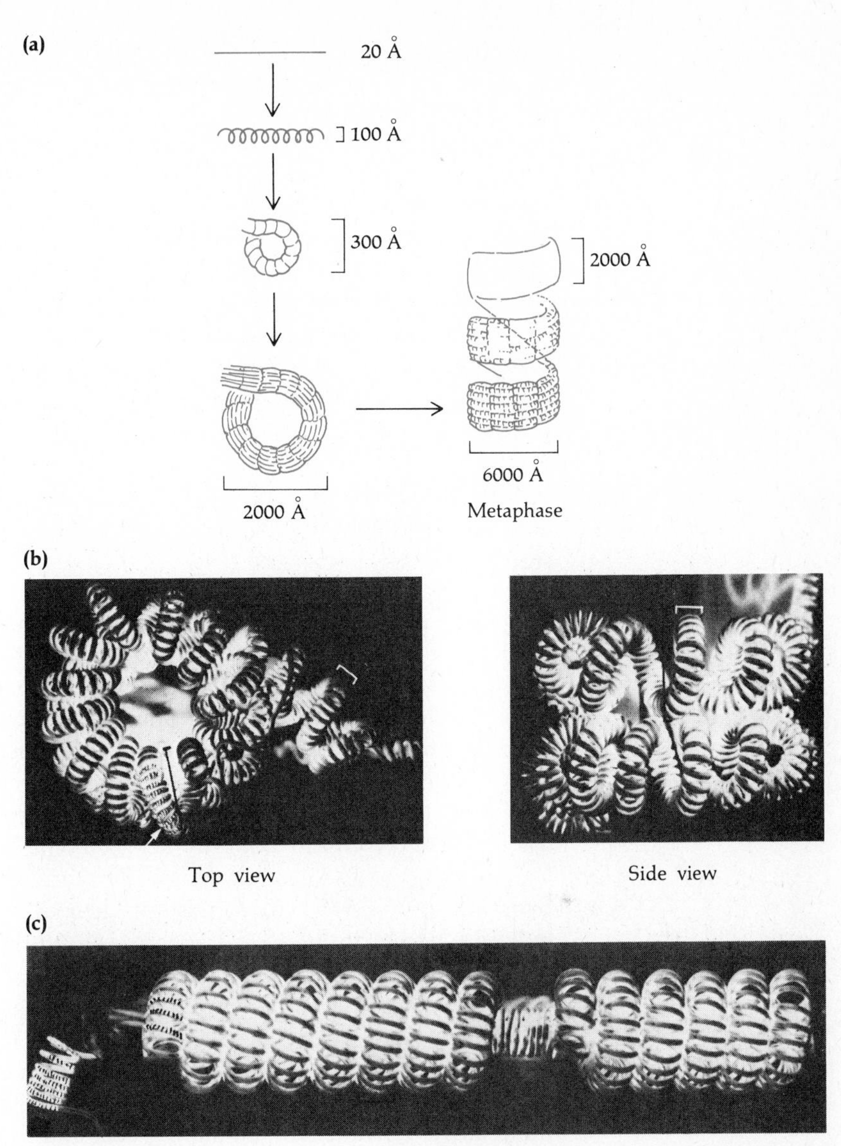

Figure 4.28
Scale models of interphase and metaphase eukaryotic chromosomes. **(a)** Diagram of the hierarchy of coiling, going from the 20-Å-diameter Watson-Crick helix through the 100-Å-diameter nucleosome, the 300-Å-diameter solenoid, and the 2000-Å-diameter tube to the 6000-Å-diameter metaphase chromatid. **(b)** Scale model of the metaphase chromatid, showing the final two coilings with wire. The finest white lines drawn on the wire (white arrow) represent the 20-Å Watson-Crick helix, the white bar shows the 300-Å coil, and the black bar shows the 2000-Å tube. **(c)** A metaphase chromosome model on a smaller scale than in part b. The finest lines drawn on the wire represent the 100-Å nucleosomes coiled into 300-Å solenoids (wire), which in turn coil into 2000-Å tubes and finally coil to form the length of the metaphase chromatid. The centromere is shown as just a 2000-Å tube connecting the two arms of this metacentric chromosome. [Photographs from J. Sedat and L. Manuelidis, *Cold Spring Harbor Symp. Quant. Biol.* 42:346 (1978).]

Problems

1. A culture of growing bacteria is diluted serially by 1/100, 1/100, and 1/50, and 0.2-ml portions of the final dilution are spread on three nutrient plates. After incubation the three plates contain 105, 84, and 98 bacterial colonies. What was the approximate concentration of bacteria in the culture at the time of sampling?

   At the same time, the same culture is diluted serially by 1/100 and 1/10, and 0.2-ml portions of the final dilution are spread on three nutrient plates seeded heavily with T2 phage. After incubation the three plates contain 40, 25, and 30 bacterial colonies. What was the frequency of T2-resistant cells in the original culture?

2. What would be the pattern of DNA densities observed by Meselson and Stahl if DNA replication were conservative? What would it be if it were dispersive?

3. Some phages with DNA genomes, such as ϕX174, contain a single strand of DNA in a ring form. How do you suppose these genomes replicate? The DNA isolated from ϕX174 phage is infective to spheroplasts of *E. coli;* however, cleavage of only one phosphodiester bond in the DNA molecule by deoxyribonuclease (which cleaves single phosphodiester bonds) destroys the infective property of the molecule. In contrast, the infective replicating ϕX174 DNA molecules isolated from infected *E. coli* cells are much more resistant to inactivation by deoxyribonuclease. Why?

4. Synthetic poly-dA/dT (d stands for deoxyribose), a double-stranded molecule in which one strand is poly-dA and the other is poly-dT, undergoes strand separation at a characteristic temperature called the *melting temperature,* T_m. This is the temperature at which one-half of the nucleotides are present in double-stranded form and the other half in single-stranded form. Poly-dA/dT has a lower T_m than does poly-dG/dC. Why might you expect this to be so? From the data in Table 4.1, what would be the relative T_m values for human, *E. coli,* and *Sarcina lutea* DNAs?

5. The mitotic chromosomes of the root-tip cells of the bean *Vicia faba* can be labeled by incubation in a medium containing [^{3}H]thymidine, and the radioactive chromosomes can be detected by autoradiography. Figure 4.29 shows the labeled mitotic chromosomes fixed for examination after incubation in [^{3}H]thymidine for less than one cell generation time. If the same cells are transferred to nonradioactive medium for an additional cell generation before fixation, the labeled mitotic chromosomes shown in Figure 4.29 are observed. Interpret these observations.

Figure 4.29
Mitotic chromosomes of *Vicia faba.* **(a)** Autoradiography after radioactive labeling for less than one cell generation. **(b)** Autoradiography after an additional generation of growth in nonradioactive medium. (From J. H. Taylor, in *Molecular Genetics,* Part I, ed. by J. H. Taylor, Academic Press, New York, 1963.)

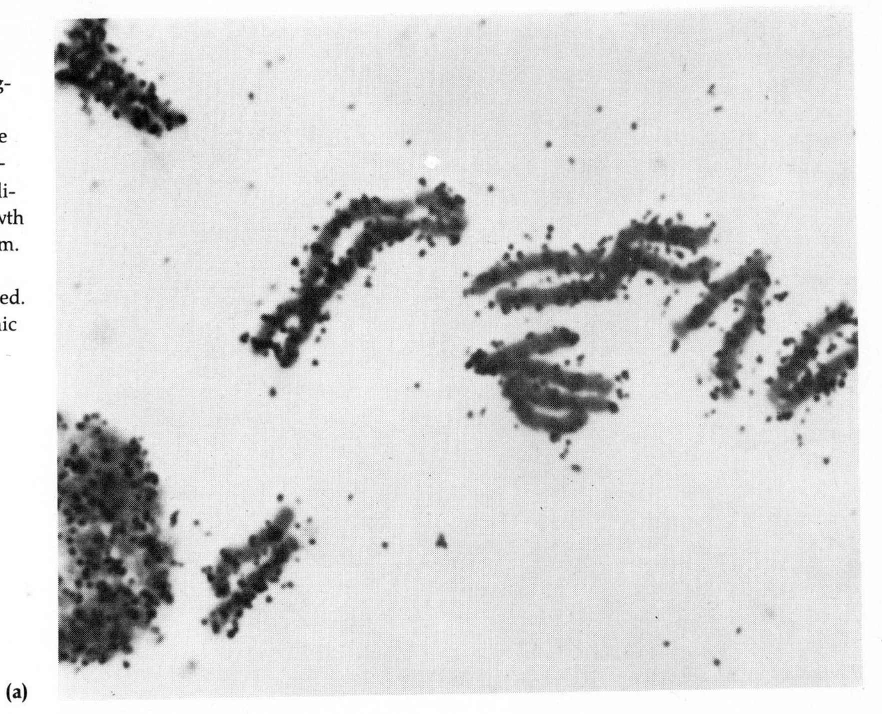

(a)

(b)

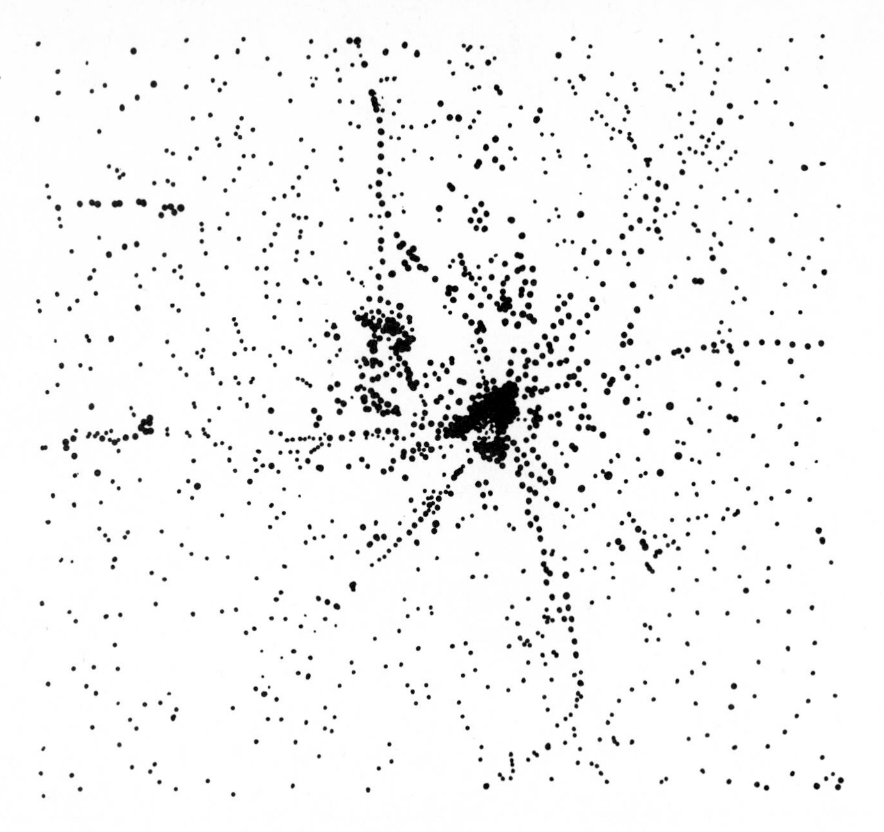

Figure 4.30
A "star" created by radioactive decay of the ^{32}P atoms in a phage particle embedded in a photographic emulsion.

6. Large DNA molecules are very fragile: shear forces, such as those created by pipetting, break them into smaller DNA molecules. The T2 phage head contains a quantity of DNA equal to 120×10^6 daltons, yet the DNA molecules purified from T2 phage by early investigators had an average molecular weight of only 12×10^6 daltons. Such observations suggested that T2 might contain as many as 10 distinct DNA molecules or chromosomes. The first evidence suggesting that T2 phage heads contain a single DNA molecule came from quantitative autoradiographic studies of T2 phage labeled with ^{32}P. When ^{32}P-labeled phages are embedded in a photographic emulsion and allowed to stand for 1–2 weeks (the half-life of ^{32}P is 14 days), development of the emulsion reveals "stars" with rays created by beta particles emitted by each decayed ^{32}P atom (one ray per decay), as shown in Figure 4.30. The center of each star identifies the position of a single phage particle, and the number of rays reflects the amount of ^{32}P present in each phage. How would you use this autoradiographic technique to prove that each phage contains a single DNA molecule? *Hint:* Assume that the mean number of rays per star is 13.3 for the intact phage.

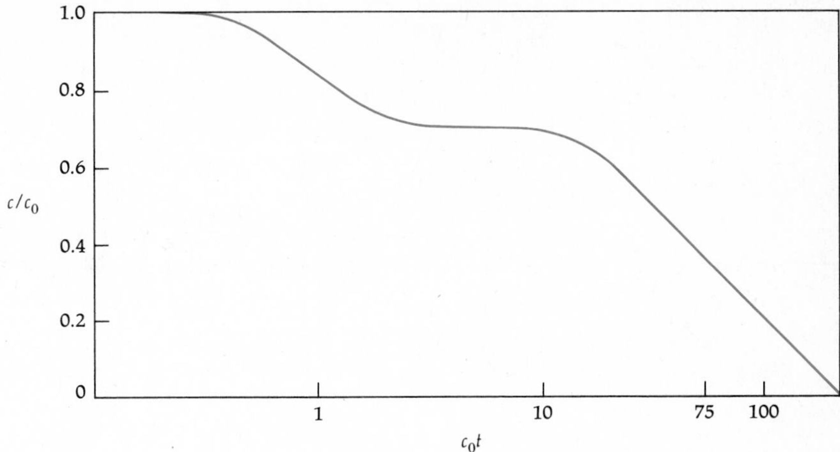

Figure 4.31
A cot plot detailing the renaturation of fragmented Troglodyte DNA.

7. The cot plot shown in Figure 4.31 was obtained from fragmented DNA of a simple Troglodyte. Tell as much as you can about the genome of this creature, given the following data: The *E. coli* genome consists of 3.2×10^6 nucleotide pairs and has a molecular weight of 2.5×10^9 daltons. Under the same conditions used to determine the cot plot for the Troglodyte DNA, *E. coli* DNA has the value $c_0t_{1/2} = 5$. The DNA content of a sperm from a male Troglodyte is 37.5×10^9 daltons.

8. Considerable controversy has surrounded the question of whether eukaryotic chromosomes are *unineme* or *polyneme,* i.e., whether they contain one or more than one identical complements of DNA. How could a study of renaturation kinetics help to resolve this question?

5

The Eukaryotic Genome

The gene is a functional entity composed of a sequence of nucleotides in a DNA molecule. A complete description of the structure and organization of an organism's genes requires a description of the sequence of nucleotides in that organism's DNA. However, the determination of the complete nucleotide sequence of even the smallest viral DNA molecule is a formidable task (see Chapter 6), and it is virtually impossible to accomplish for the entire DNA of a higher organism. Indeed, the genetic variability present in all species of organisms means that there is no single nucleotide sequence for the entire genome that is unique and invariant in all individuals of the species. The *E. coli* genome is composed of approximately 3.2×10^6 nucleotide pairs (np). It is obvious that an enormous number of different nucleotide sequences are possible for even a small genome such as that of *E. coli.* For each nucleotide pair there are four possibilities (AT, TA, GC, CG), and the number of different possible sequences is therefore $4^{3.2 \times 10^6} = 10^{1.93 \times 10^6}$. The number of possible sequences in the DNA of the human genome is clearly much larger than even this enormous number. The haploid DNA content of various eukaryotic organisms is shown in Figure 5.1 in terms of multiples of the amount of DNA in the *E. coli* genome.

It is obvious, therefore, why most of our current understanding of the genetic organization of DNA is based upon *genetic analysis* and not upon chemical sequencing of nucleotides in DNA. Genetic analysis has permitted the construction of detailed models (maps) showing the genetic organization of chromosomes. For a number of organisms it has been

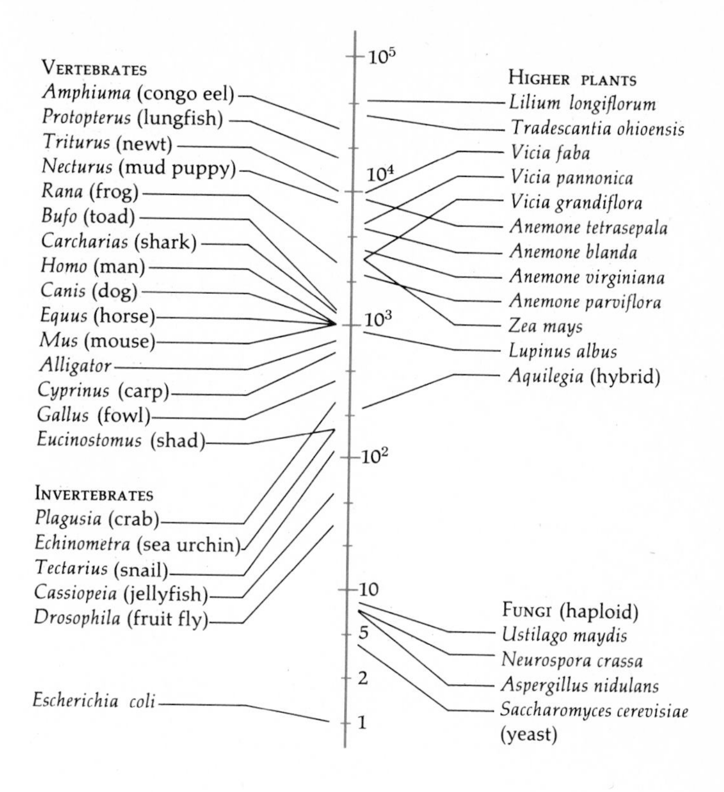

Figure 5.1
Haploid amounts of DNA in cells of various organisms, expressed in multiples of the amount of DNA in the *E. coli* genome (3.2×10^6 np). (From J. D. Watson, *Molecular Biology of the Gene*, 3rd ed., W. A. Benjamin, Menlo Park, Cal., 1976.)

possible to make very precise correlations between such genetic maps and the physical organization of the DNA itself. Such genetic analysis began with Mendel (Chapter 2). The independent assortment of segregating alleles observed by Mendel suggests that particular genes may reside on different chromosomes and, therefore, in different DNA molecules. In this chapter genetic analysis is extended to the study of genetic organization within the DNA of single chromosomes and organelles.

The methods of genetic analysis were developed to explore the genetics of diploid eukaryotic organisms. Because these methods were first developed for meiotic organisms, they are introduced in that context here. In succeeding chapters we shall see how the methodology of genetic analysis has been adapted to study genetic organization in viruses and bacteria, which do not exhibit meiosis.

In order to acquire a thorough understanding of genetic organization, geneticists have focused on a relatively few organisms that offer unique advantages for genetic analysis. The most intensively studied eukaryotic organism is the fruit fly, *Drosophila melanogaster*. *E. coli* has been the bacterial organism of choice, and the bacteriophages T2, T4, λ, and ϕX174 have been the most intensively studied viruses. The studies of these ge-

Figure 5.2
Progeny genotypes from a testcross involving alleles at two genetic loci. The observed numbers of each progeny genotype are those expected among 100 progeny if (1) loci a and b assort independently or (2) they are completely linked.

Parental cross: $\frac{a^+b^+}{a^+b^+} \times \frac{a\ b}{a\ b}$

Testcross: $\frac{a^+b^+}{a\ b} \times \frac{a\ b}{a\ b}$

		Genotype	Class	Expected number
(1) Independent assortment	Progeny types:	$\frac{a^+b^+}{a\ b}$	Parental types	25
		$\frac{a\ b}{a\ b}$		25
		$\frac{a^+b}{a\ b}$	Recombinant types	25
		$\frac{a\ b^+}{a\ b}$		25
(2) Complete linkage	Progeny types:	$\frac{a^+b^+}{a\ b}$	Parental types	50
		$\frac{a\ b}{a\ b}$		50
		$\frac{a^+b}{a\ b}$	Recombinant types	0
		$\frac{a\ b^+}{a\ b}$		0

nomes serve as paradigms for study of genetic organization as it exists in other organisms.

Recombination of Linked Genes

In Chapter 2 we saw how Mendel postulated two possible outcomes for a dihybrid cross, involving alleles of two different genes: (1) independent assortment of alleles of the two genes or (2) complete association (linkage) of alleles of the two genes. In the first case he expected, and indeed observed, among the progeny of a testcross, four genotypes in equal numbers (Figure 5.2). Two classes of progeny exhibit the same association of alleles as observed in the parents in the parental cross (a^+b^+ and $a\ b$), and the other two classes of progeny exhibit new (recombinant) association of alleles (a^+b and $a\ b^+$). If parental-type and recombinant-type progeny occur in equal numbers, the genes a and b assort independently at meiosis in the heterozygous parent of the testcross and are said to be *unlinked.* The recombination frequency between the two genes is defined as the summed frequency of recombinant types among the total progeny (50/100

			Observed number
Parental cross:	$\frac{c^+d^+}{c^+d^+} \times \frac{c\ d}{c\ d}$		
Testcross:	$\frac{c^+d^+}{c\ d} \times \frac{c\ d}{c\ d}$		
Progeny types:	$\frac{c^+d^+}{c\ d}$	Parental types	40
	$\frac{c\ d}{c\ d}$		40
	$\frac{c^+d}{c\ d}$	Recombinant types	10
	$\frac{c\ d^+}{c\ d}$		10

Figure 5.3
Progeny genotypes from a testcross involving alleles at two linked loci. The observed numbers of each progeny genotype differ significantly from those expected for independent assortment, as determined by the χ^2 test (see Appendix).

= 0.50 in the example of Figure 5.2). Independent assortment is characterized by a recombination frequency of 50% and may indicate that two genes reside on different chromosomes. As we shall see below, however, the latter is not invariably true.

In the second case postulated by Mendel, on the other hand, complete association of alleles of two genes leads to the expectation of only two genotypes among the progeny of a testcross, these genotypes exhibiting the association of alleles observed in the parents in the parental cross (Figure 5.2). Such complete linkage is strong evidence that the two genes are located on the same chromosome. When two genes do reside on the same chromosome, alleles of these genes do not usually assort independently of each other, and yet four progeny genotypes are observed. A preponderance of parental-type progeny in a testcross is evidence of such linkage (Figure 5.3); this type of observation is intermediate between the two alternative postulates of Mendel shown in Figure 5.2. In the example shown in Figure 5.3, the testcross progeny exhibit a recombination frequency, calculated as the summed frequency of recombinant types among total progeny, of 20% (20/100 = 0.20). When a recombination frequency significantly less than 50% is observed between alleles of two genes, those genes are said to be *linked* and to reside on the same chromosome. The recombinant gametes giving rise to the recombinant-type progeny are formed by *crossing over* between homologous chromatids during meiosis in the heterozygous parent. The chiasmata observed at the first meiotic division are physical evidence that two chromatids have undergone crossing over (see Chapter 1).

Linkage is demonstrated in the testcross between tomato plants doubly heterozygous for the mutants *dwarf* (d) and *pubescent* fruit (p) (Figure 5.4), which are alleles of the normal *tall* (d^+) and *smooth* fruit (p^+). Note that the two largest classes among the progeny types can be identi-

Figure 5.4
Progeny genotypes from a testcross involving the mutant alleles *dwarf* (d) and *pubescent* (p) in the tomato.

Parental cross: $\frac{d^+p^+}{d^+p^+} \times \frac{d\ p}{d\ p}$

Testcross: $\frac{d^+p^+}{d\ p} \times \frac{d\ p}{d\ p}$

	Genotype	Phenotype	Number
Progeny types:	$\frac{d^+p^+}{d\ p}$	Tall, smooth	161
	$\frac{d\ p}{d\ p}$	Dwarf, pubescent	118
	$\frac{d^+p}{d\ p}$	Tall, pubescent	5
	$\frac{d\ p^+}{d\ p}$	Dwarf, smooth	5
		Total:	289

Box 5.1 Doubly Homozygous Recessive Strains

It is unusual for doubly homozygous recessive strains to be found; they must usually be created by the geneticist for the purpose of making a testcross. The manner in which this is done is diagrammed in Figure 5.5, where doubly homozygous individuals are found in the F_2 generation as the union of two recombinant, doubly recessive gametes.

Parental cross (P) $\frac{d\ p^+}{d\ p^+} \times \frac{d^+p}{d^+p}$

First filial generation (F_1) $\frac{d\ p^+}{d^+p} \times \frac{d\ p^+}{d^+p}$

♀ gametes	♂ gametes: $d\ p^+$	d^+p	d^+p^+	$d\ p$
$d\ p^+$	$\frac{d\ p^+}{d\ p^+}$	$\frac{d\ p^+}{d^+p}$	$\frac{d\ p^+}{d^+p^+}$	$\frac{d\ p^+}{d\ p}$
d^+p	$\frac{d^+p}{d\ p^+}$	$\frac{d^+p}{d^+p}$	$\frac{d^+p}{d^+p^+}$	$\frac{d^+p}{d\ p}$
d^+p^+	$\frac{d^+p^+}{d\ p^+}$	$\frac{d^+p^+}{d^+p}$	$\frac{d^+p^+}{d^+p^+}$	$\frac{d^+p^+}{d\ p}$
$d\ p$	$\frac{d\ p}{d\ p^+}$	$\frac{d\ p}{d^+p}$	$\frac{d\ p}{d^+p^+}$	$\frac{d\ p}{d\ p}$

Second filial generation (F_2)

Figure 5.5
F_2 progeny genotypes derived from a cross between *dwarf* and *pubescent* tomato plants. The doubly homozygous recessive strain appears at the lower right.

Figure 5.6
Progeny genotypes in the F_2 generation from a cross between *yellow*-body (y) males and *white*-eye (w) females in *Drosophila melanogaster.* The hemizygosity of the X chromosome in the male progeny permits genotypes of the male progeny to be determined from their phenotypes, just as in a testcross involving autosomal loci. The genotypes of F_2 females cannot be determined from their phenotypes (see Figure 5.8 for testcross data).

Parental cross (P) $\frac{y^+w}{y^+w}$ ♀ × $\frac{y\ w^+}{Y}$ ♂

F_1 $\frac{y^+w}{y\ w^+}$ ♀ × $\frac{y^+w}{Y}$ ♂

F_2 genotype	Class	Observed number
$\frac{y^+w}{Y}$	♂ parental types	190
$\frac{y\ w^+}{Y}$		196
$\frac{y^+w^+}{Y}$	♂ recombinant types	3
$\frac{y\ w}{Y}$		1
$\frac{y^+w}{-\ w}$, $\frac{y^+w}{-\ w^+}$	♀ parental and recombinant types (Genotypes at the *yellow* locus cannot be distinguished because all females received the dominant y^+ allele from their fathers.)	403

fied as the parental types. The data from this cross demonstrate a recombination frequency of about 3.5% (10/289 = 0.0346).

An example of linkage between two sex-linked X-chromosome genes in *Drosophila melanogaster* is presented in Figure 5.6. All male progeny receive their only X chromosome from their mother, and that X chromosome may be either a parental type or a recombinant type. In this example a testcross of a doubly heterozygous female to a male carrying the recessive alleles of the two genes is not necessary, since the recombination frequency can be determined from the phenotypes of hemizygous male progeny. The Y chromosome does not carry alleles of these X-chromosome genes. The two mutant genes in this cross are *yellow* body color (y) and *white* eye (w), and the dominant wild-type alleles are *brownish* body color (y^+) and *red* eye (w^+). The recombination frequency observed in this cross between the *yellow* and *white* genes is $4/390 \approx 0.010$.

The recombination frequency observed in crosses depends on which genes are involved in the cross. The recombination frequencies observed for some other sex-linked genes in *Drosophila melanogaster* are shown in Table 5.1. Alfred H. Sturtevant, in 1913, was the first to realize that data on the frequency of recombination between linked genes could be used to construct a linear, one-dimensional map of the chromosome. Such a

Table 5.1
Recombination frequencies for some sex-linked mutations in *Drosophila melanogaster*.

Genes			Recombination Frequency
yellow (y)	&	*white (w)*	0.010
yellow (y)	&	*vermilion (v)*	0.322
yellow (y)	&	*miniature (m)*	0.355
vermilion (v)	&	*miniature (m)*	0.030
white (w)	&	*vermilion (v)*	0.300
white (w)	&	*miniature (m)*	0.327
white (w)	&	*rudimentary (r)*	0.450
vermilion (v)	&	*rudimentary (r)*	0.269

From A. H. Sturtevant, *J. Exp. Zool.* 14:43 (1913).

map of the *Drosophila* X chromosome, based on the data in Table 5.1, is shown in Figure 5.7. Sturtevant's insight was based on the observation that, under the microscope, chromosomes have a threadlike form. He reasoned that, if genes were distributed along a chromosome, then the farther apart two genes lay, the greater the likelihood that a recombination event, or crossover, between homologous chromosomes would occur between the two genes. Thus, the frequency of recombination between widely separated genes should be greater than that between closely situated genes. This was a major discovery because it demonstrated that genetic analysis is capable of producing a coherent model of the organi-

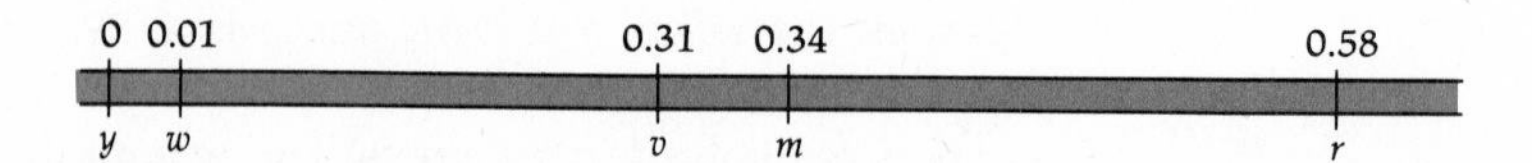

Figure 5.7
The first genetic map of the X chromosome of *D. melanogaster*, showing the relative map positions of the genes *yellow, white, vermilion, miniature,* and *rudimentary*. The *yellow* gene was arbitrarily chosen as zero on the genetic map.

Figure 5.8
Progeny genotypes from a testcross involving three loci on the X chromosome of *D. melanogaster.*

Parents in testcross: $\frac{y^+w^+m^+}{y\ w\ m}$ ♀ × $\frac{y\ w\ m}{Y}$ ♂

Progeny: Brothers and sisters are classified together, and only the allelic compositions of the maternal X chromosomes are indicated.

Genotype	Class	Number
$y\ w\ m$ $y^+w^+m^+$	Parental	6972
$y\ w\ m^+$ y^+w^+m	w—m recombinants	3454
$y\ w^+m^+$ $y^+w\ m$	y—w recombinants	60
$y^+w\ m^+$ $y\ w^+m$	Double recombinants	9

zation of the genetic material. This linear model of the chromosome provided a framework for all future work in genetics and presaged the discovery of the linear nature of the DNA molecule.

Three-Factor Crosses

Sturtevant's proof of the linearity of the genetic map was based on the analysis of data from three-factor crosses (crosses in which alleles of three genes segregate). The measurement of recombination frequencies in a cross involving three mutant genes permits the detection of chromosomes that have participated simultaneously in two recombination events (double crossovers) and consequently permits the order of the three genes on the chromosome to be established. Consider the cross shown in Figure 5.8, based on some of the data analyzed by Sturtevant from testcrosses that allowed the genotypes of *both* male and female progeny to be determined.

Genetic mapping is based on one principle: the closer together two *loci* (singular: locus—the position of a gene), the less probable it is that these loci will be separated by a recombination event. Assume, for the sake of discussion, that recombination events occur at random along the length of a chromosome. If two loci are very close, the probability that a random

Table 5.2
Three-factor cross data treated as three separate two-factor crosses.

Genes	Recombination Frequency	Map Distance
w—m	$\frac{3454 + 9}{10{,}495} = 0.330$	33.0
y—w	$\frac{60 + 9}{10{,}495} = 0.007$	0.7
y—m	$\frac{3454 + 60}{10{,}495} = 0.335$	33.5

recombination event will separate them is low. As the separation between two loci increases, the probability that a random recombination event will separate them also increases. As the separation becomes large, the probability that two or more recombination events will occur between them also increases, but is always less than that for one recombination event. For example, if the probability of a single crossover is 0.2, then the probability of two independent crossovers is $0.2 \times 0.2 = 0.04$. The fact that double recombination events are always less frequent than single recombination events makes it possible to deduce the order of three loci in a linear sequence.

First let us treat the data in Figure 5.8 as if they were derived from three two-factor crosses, as shown in Table 5.2. Clearly *y—m* and *w—m* are farther apart than *y—w*, based on the observed recombination frequencies. But is the order *y—w—m* or *w—y—m*? In considering the data from two-factor crosses, a choice between these two possible orders depends, among other things, on the statistical significance of the data and environmental influences on their reproducibility. In many cases the choice may not be as clear as the data in Table 5.2 suggest, i.e., *y—w—m*. However, consideration of these data in the context of the three-factor cross in which environmental influences (temperature, parental age, etc.) are identical indicates unequivocally that the order must be *y—w—m*, because a double recombination event between *y* and *m* must be expected to occur less frequently than a single recombination event. The presence of *w* allows this double event to be observed and proves that *w* is between *y* and *m* rather than *y* between *w* and *m*. The three possible orders for the three genes are shown in Figure 5.9. Only order I predicts that the least frequent recombinant classes will be *y*—+—*m* and +—*w*—+, in agreement with the data in Figure 5.8.

Thus three-factor crosses provide both the order of the three genes involved and the recombination frequencies between them. In construct-

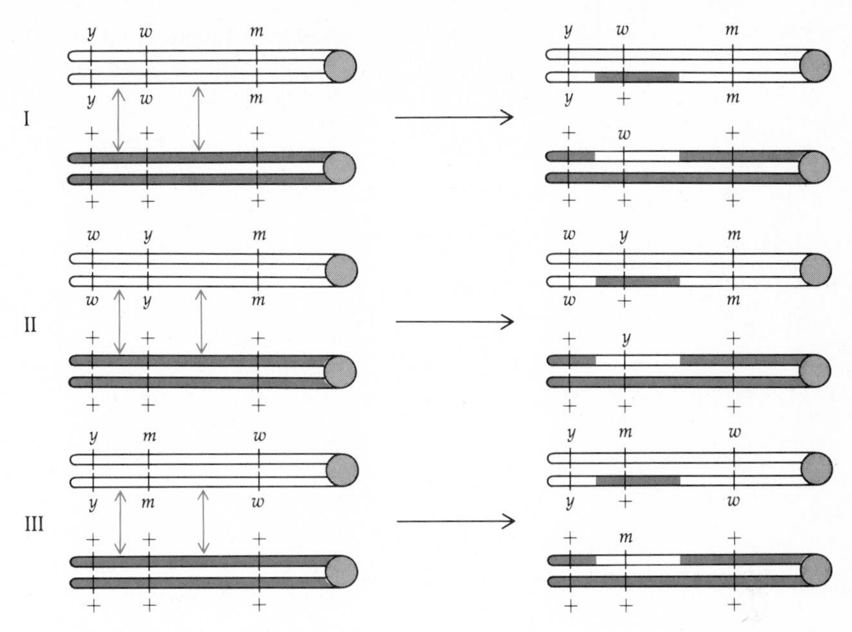

Figure 5.9
Chromatids before and after postulated double recombination events prior to the first meiotic division in triply heterozygous mothers. Three possible orders (I, II, and III) for the *yellow*, *white*, and *miniature* loci are indicated. Only order I is compatible with the data in Figure 5.8.

ing chromosome maps, such as that in Figure 5.7, the recombination frequencies between near loci are used to define the map distance between them, since these are more accurate than those between distant loci. The farther apart two genes are, in the absence of intervening markers, the greater is the underestimation of genetic map distance due to failure to observe double recombination events. *In fact,* two loci far enough apart on a chromosome can appear to be unlinked (recombination frequency = 50%) even though they reside on the same chromosome.

In constructing chromosome maps, a 1% recombination frequency is defined as 1 *map unit,* or one centimorgan. (The *morgan,* named in honor of T. H. Morgan, is 100 map units.) Thus the map distance between *y* and *m* is 33.7 centimorgans (0.7 + 33.0), while the observed recombination frequency between *y* and *m* is 0.335. In the absence of the *white* mutation, we would be forced to evaluate the map distance between *y* and *m* as 33.5 centimorgans. This lower value would be the result of failure to detect the double recombination events that occur and that *are* detected because of the presence of the *w* mutation. This point is illustrated more forcefully by comparing the map distance of 57 centimorgans between *w* and *r* (Figure 5.7) with the observed recombination frequency of 0.450 between *w* and *r* in a two-factor cross (Table 5.1). Because genetic maps are constructed by adding map units between the nearest genes, the maps of some chromosomes exceed 100 centimorgans, even though the maximum recombination frequency that can be observed is 50%. These methods, originally set forth by Sturtevant, form the basis of all genetic mapping employing recombination frequencies. The genetic maps of several eukaryotic organisms of scientific or agricultural importance are shown in Figure 5.10, 5.11, 5.12, and 5.13.

Genetic Interference

When genes are not too far apart, recombination frequency can be viewed as an estimate of the probability that a recombination event will occur between them. This estimate can be used to determine whether recombination events occurring within a chromosome are independent of one another. If two recombination events occur independently, the frequency of a double recombination event should be the product of the frequencies of the two single events. From the data in Figure 5.8, the expected frequency of double crossovers is $0.330 \times 0.007 = 0.00231$ (see Table 5.2). The observed frequency of double crossovers is, however, $9/10{,}495 = 0.00086$, which is significantly lower than the expected frequency. Thus, the occurrence of one recombination event between *y* and *m* affects the occurrence of a second event. This phenomenon is called *interference* (I). The ratio of observed to expected double recombination events is called the *coefficient of coincidence* (c). By definition, $I = 1 - c$. In the above example,

$$I = 1 - 0.00086/0.00231 = 1 - 0.374 = 0.626$$

The value observed for I depends strongly on which loci are involved in a cross. If the loci are very far apart or are separated by a centromere, I can be zero. As the two loci become closer to one another, the value of I increases. For close loci, $I = 1$, and for this reason the recombination frequency between close markers is used to construct genetic maps. Evidently, recombination events do not occur independently of one another. A crossover at one point on a chromatid can hinder the occurrence of a second crossover nearby. This fact does not affect the usefulness of genetic maps or their linear character.

The Time of Crossing Over

Crossing over occurs during the four-strand, or tetrad, stage of meiosis, at which time each chromosome consists of two sister chromatids (see Chapter 1). This is demonstrated quite clearly by the genotypes of female progeny of *Drosophila* derived from mothers heterozygous for a mutant attached-X chromosome.

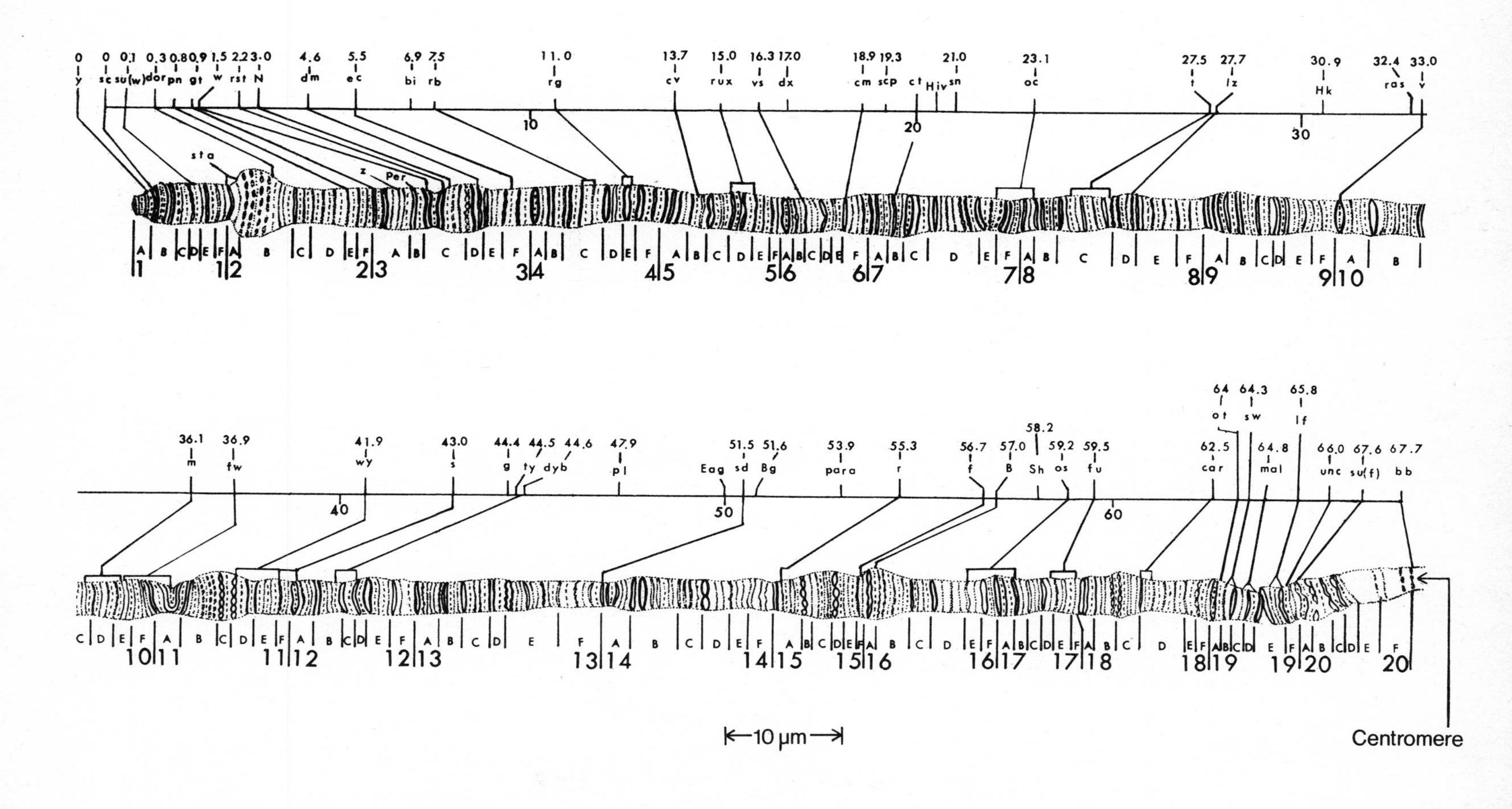

Figure 5.10
Genetic map of some of the known mutations on the X chromosome of *D. melanogaster*. Cytological correlation of the genetic map and polytene salivary-gland chromosomes are indicated where known. (From *Handbook of Genetics*, Vol. 3, ed. by R. C. King, Plenum Press, New York, 1975.)

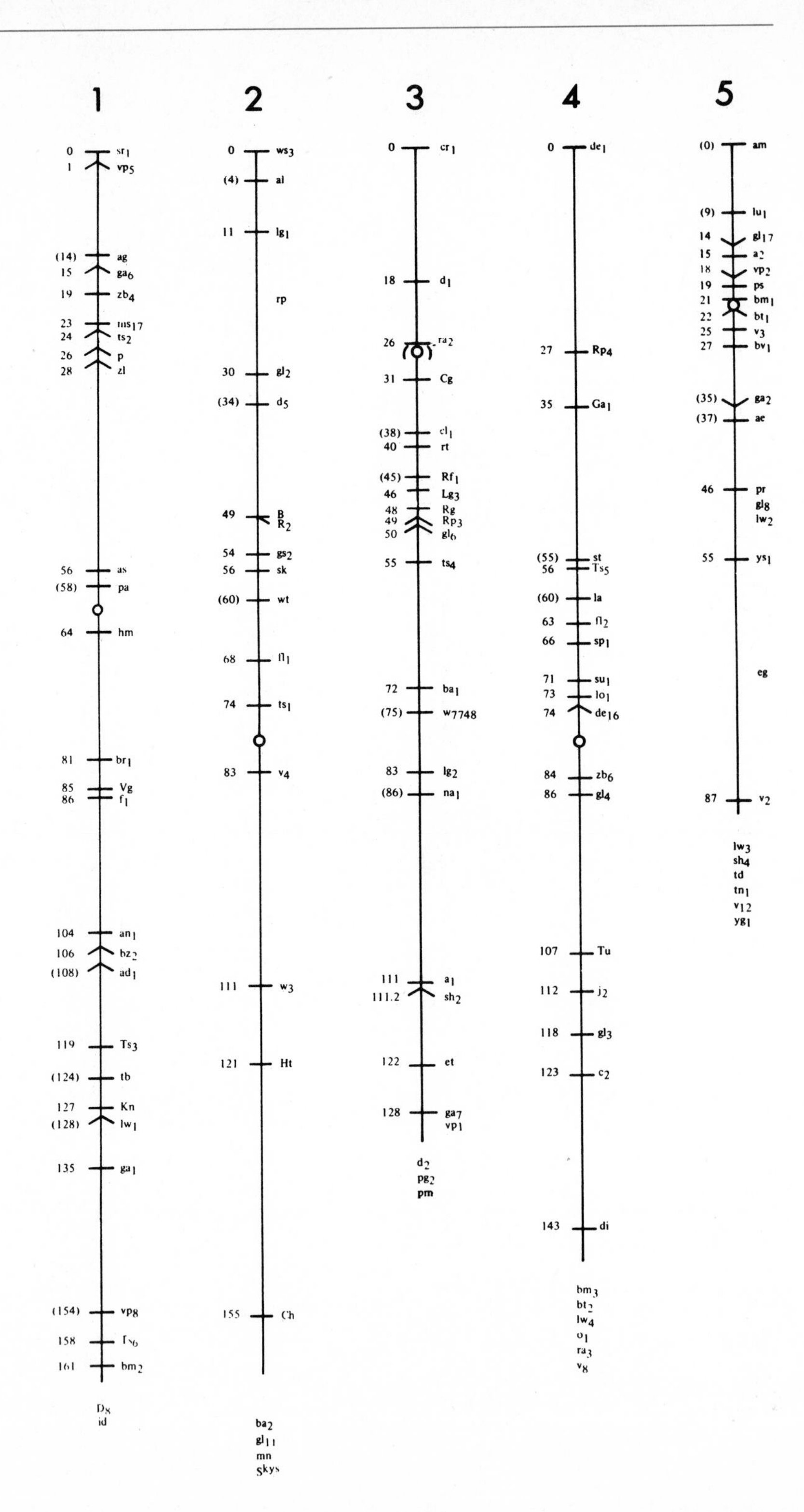

Figure 5.11
Genetic map of corn (maize), *Zea mays*. (From *Handbook of Biochemistry: Selected Data for Molecular Biology*, 2nd ed., ed. by H. A. Sober, CRC Press, Boca Raton, Fla., 1970.)

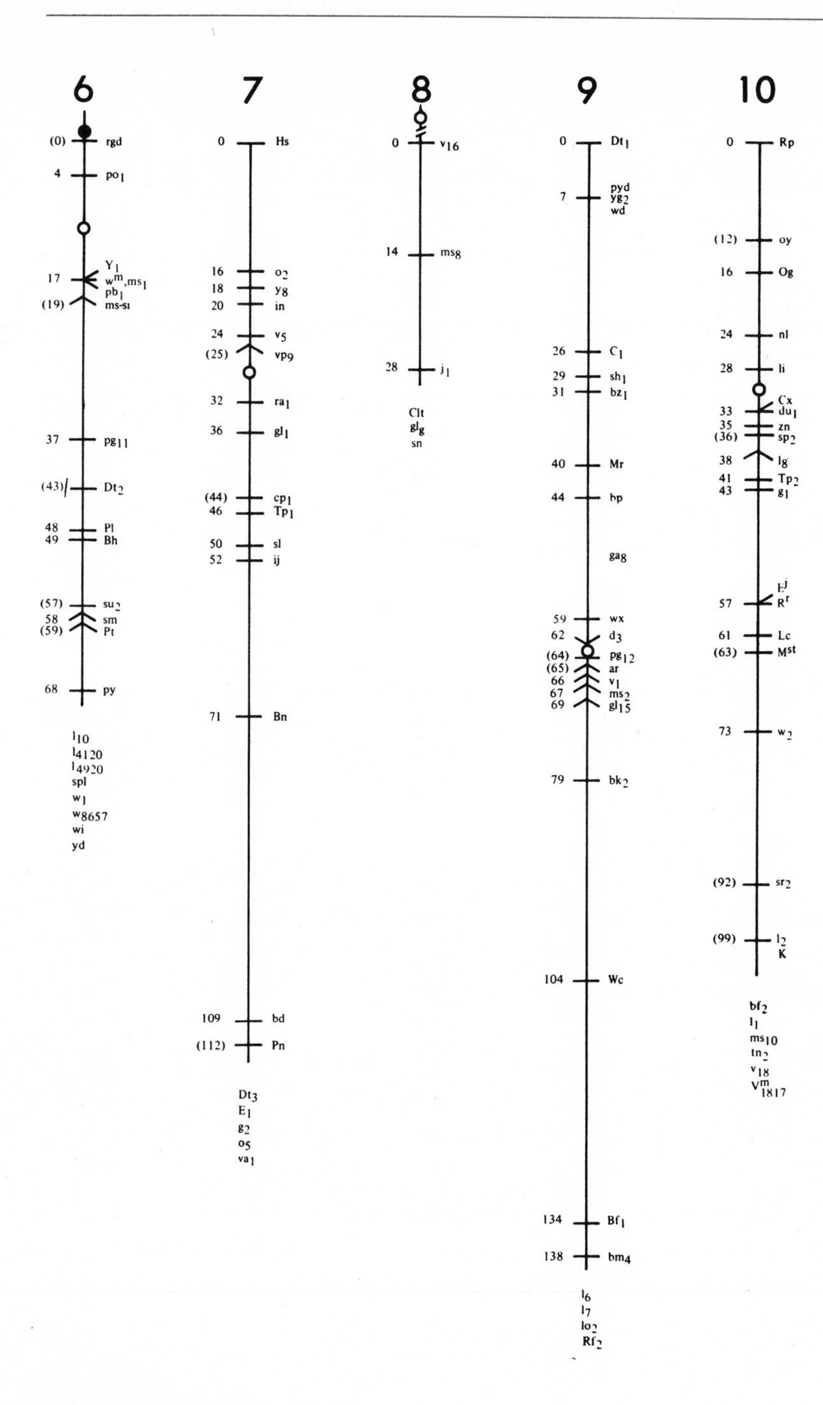

( ) Indicates probable position, based on insufficient data
o Indicates centromere position
● Indicates organizer

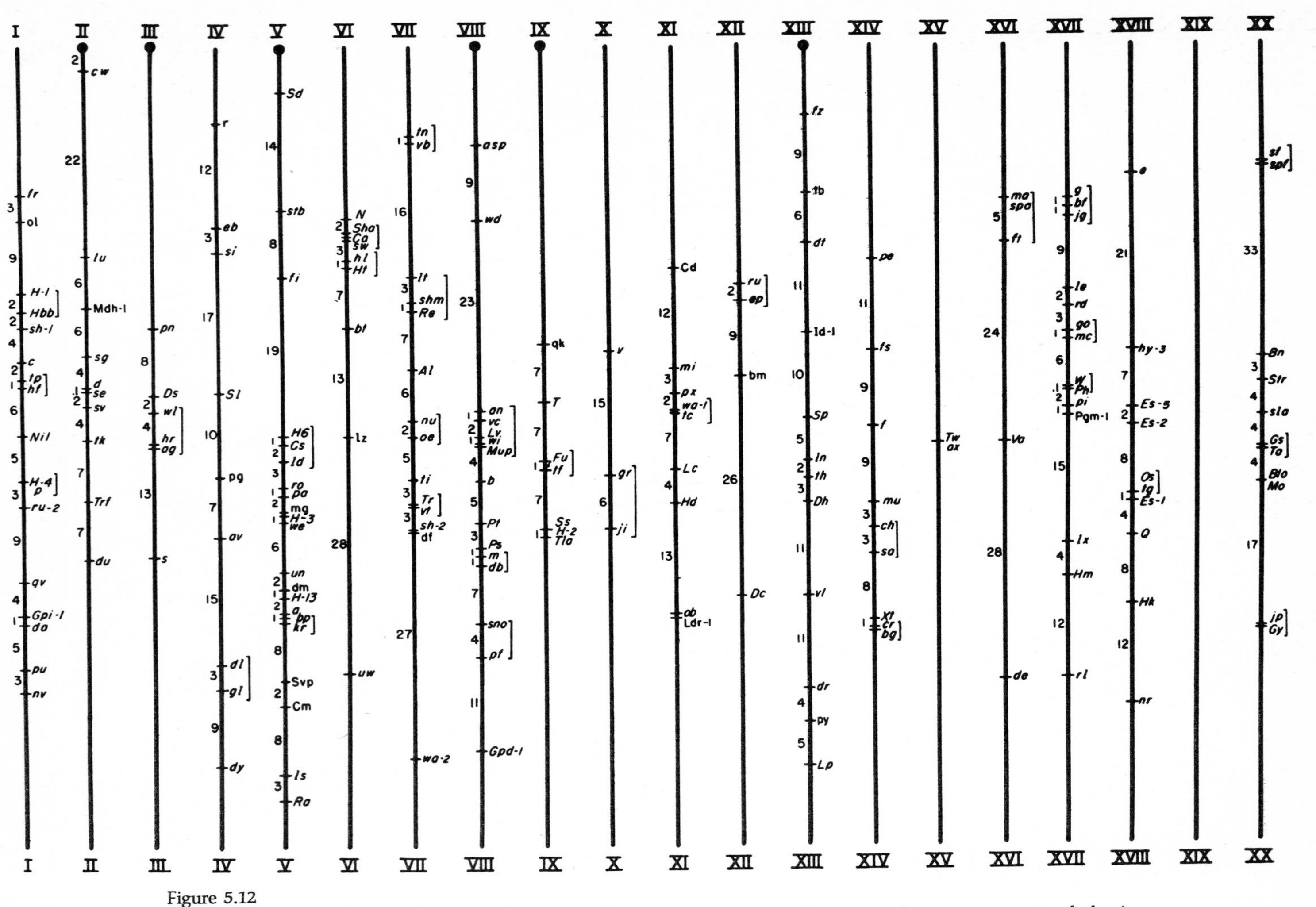

Figure 5.12
Genetic map of the laboratory mouse, *Mus musculus*. Loci whose order is uncertain are shown in roman symbols. A bracket indicates that the order within the bracketed group has not been established. Closed circles represent centromeres. (From *Handbook of Biochemistry: Selected Data for Molecular Biology*, 2nd ed., ed. by H. A. Sober, CRC Press, Boca Raton, Fla., 1970.)

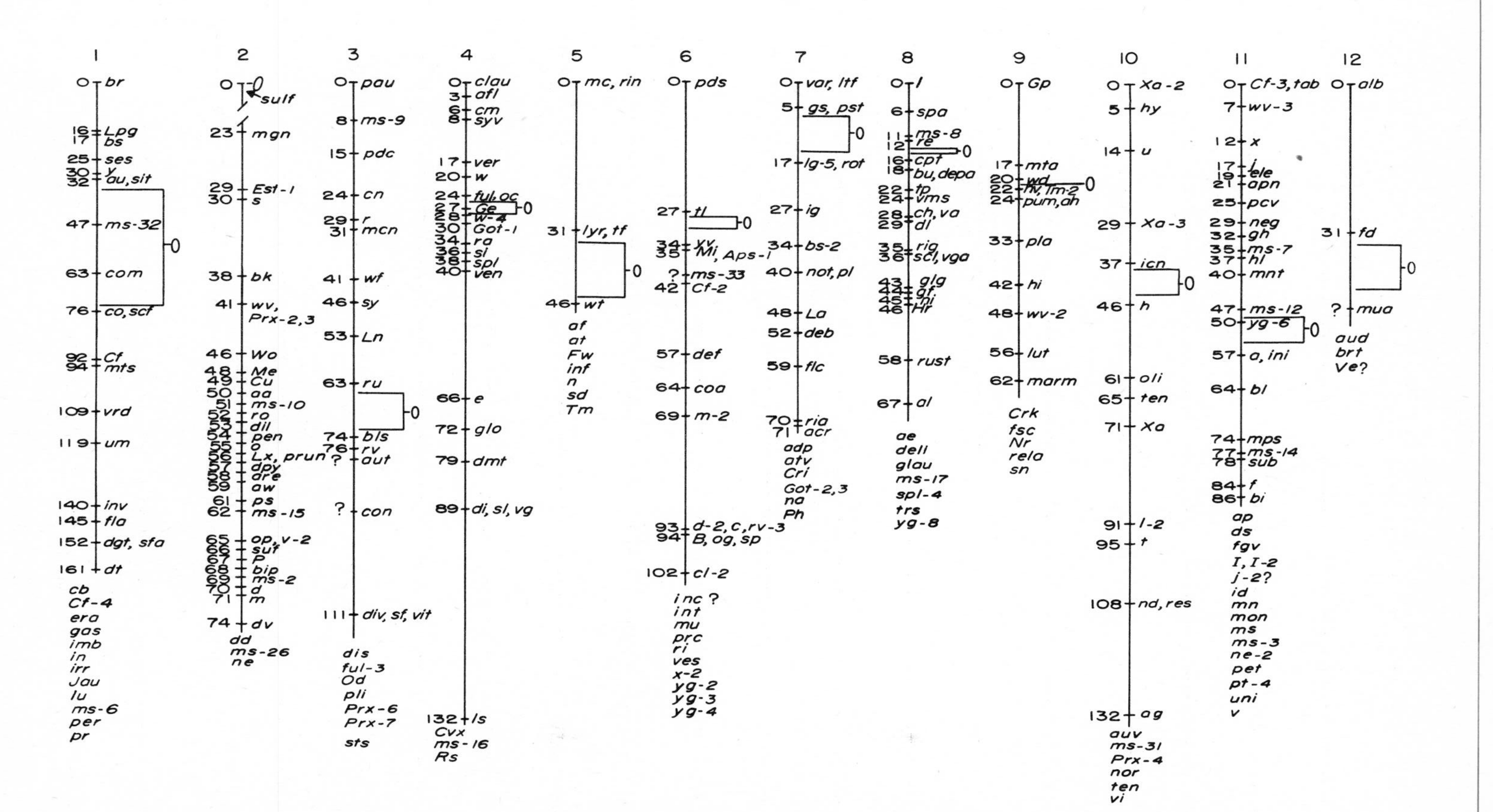

Figure 5.13
Genetic map of the tomato, *Lycopersicon esculentum*. (Courtesy of Prof. Charles W. Rick, University of California, Davis.)

The normal *Drosophila* X chromosome is a telocentric chromosome (the centromere is located at the end of the chromosome). An attached-X chromosome (X•X) is one in which two X chromosomes share a common centromere and form a metacentric chromosome in which each arm is a virtually complete X chromosome (see Chapter 3). *Drosophila* attached-X females generally carry a Y chromosome as well (X•X/Y). The presence of a Y chromosome in *Drosophila* females has no effect on their physiology, since sex is determined by the ratio of X chromosomes to autosomes. The Y chromosome is required only for male fertility. Attached-X females exhibit an abnormal segregation of sex-linked genes, as shown in Figure 3.9. Only half of the zygotes from such females develop normally to produce female progeny (carrying their mother's attached-X chromosome and their father's Y chromosome) and male progeny (carrying their father's X chromosome and their mother's Y chromosome).

Females with an attached-X chromosome, in which the two homologues are heterozygous for recessive mutations, produce some female progeny that are homozygous for one or more of the recessive mutations. The frequency with which a given mutation appears in a homozygous state in an attached-X chromosome increases with increasing map distance from the centromere. Thus crossing over is responsible for the appearance of these new attached-X genotypes, and they can only arise from crossing over at the tetrad state of meiosis, as shown in Figure 5.14. Crossing over at any stage prior to the tetrad stage, e.g., before chromosome duplication, is ruled out.

In some organisms, particularly fungi, all of the meiotic products of a tetrad may be recovered and their genotypes determined. In the haploid fungus *Neurospora crassa*, meiosis occurs immediately after fusion of nuclei of opposite mating types within a cell called an *ascus* (plural: asci). (Sex, or mating type, in this fungus is determined by the mating-type allele of a single gene.) The two meiotic divisions are followed by a mitotic division to produce eight spores within the ascus. The spindles of these nuclear divisions do not overlap one another, and so the order of spores in the ascus reflects the order of segregation of centromeres at the first and second meiotic divisions, as shown in Figure 5.15. The ordered development of spores from the tetrad makes it possible to detect crossing over between the centromere and mutant genes, as well as between mutant genes (Figure 5.16). In the absence of crossing over, alleles segregate from each other at the first meiotic division and are found in spores at opposite ends of the ascus. If a single crossover occurs between the centromere and different alleles of a gene, the different alleles do not segregate until the second meiotic division, producing an ascus in which parental genotypes are found at both ends of the ascus. A single crossover between two genes produces an ascus in which alleles proximal to the crossover segregate at the first division, and alleles distal to the crossover segregate at the second division. Such asci contain half recombinant-type spores and half parental-type spores. The frequency of asci showing second-division segregation is a measure of the genetic distance between the centromere and

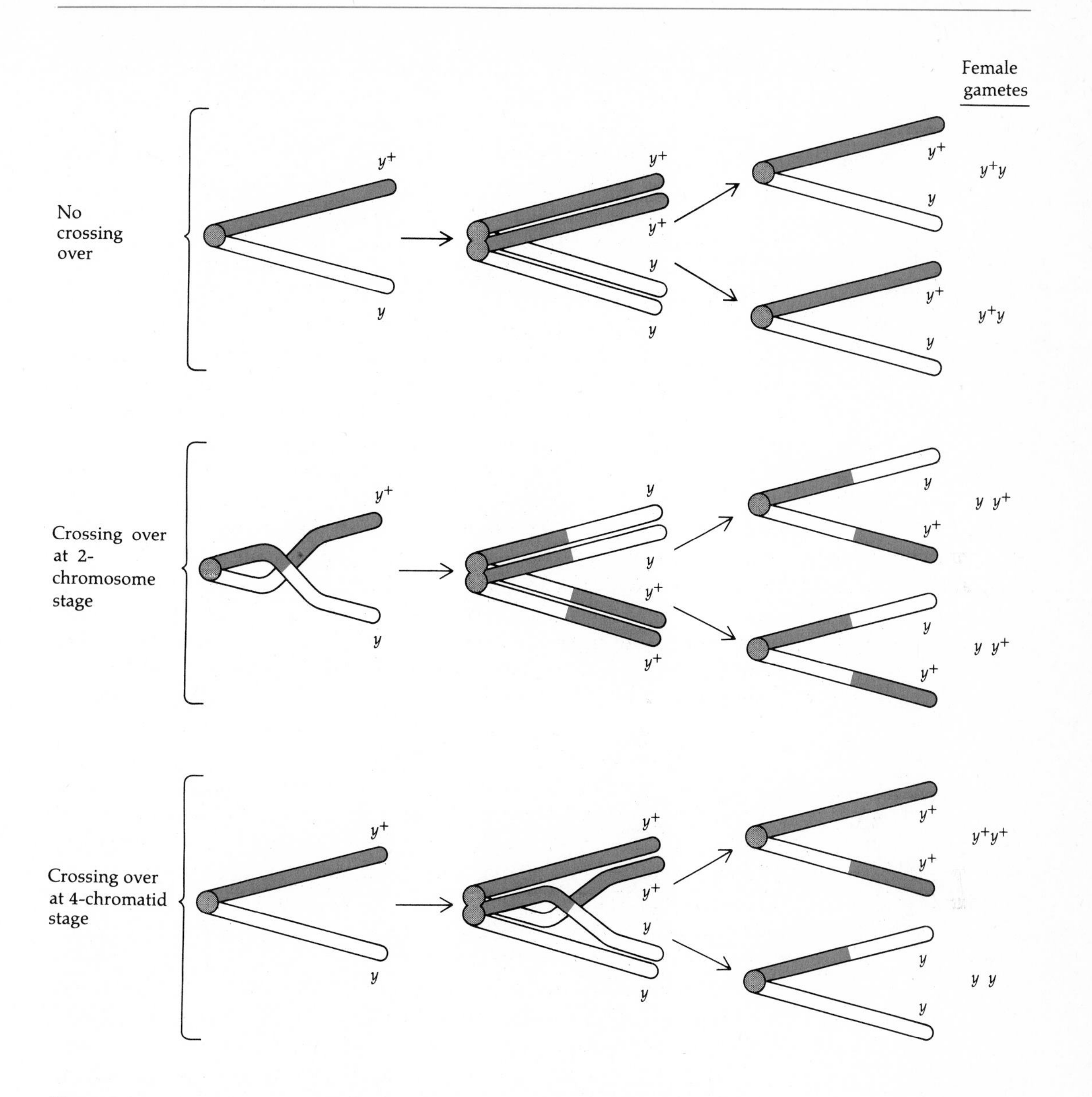

Figure 5.14
Crossing over must occur at the four-chromatid stage in meiosis to produce a homozygous attached-X chromosome from a heterozygous attached-X chromosome. The crossover must occur between the centromere and the mutant locus.

the gene locus. However, since only half of the chromatids in a tetrad are recombinant, the map distance between gene and centromere is half of the frequency of asci showing second-division segregation. The ordered ascospores of *Neurospora* thus permit the centromere to be mapped with respect to mutant alleles of genes on the same chromosome.

Four types of double crossover events can be observed in *Neurospora* tetrads, as shown in Figure 5.17. The occurrence of crossover events involving three and four chromatids, revealed by the ordered ascospores, proves that crossing over occurs at the tetrad stage, when four chromatids are present, rather than at a two-strand stage prior to replication of

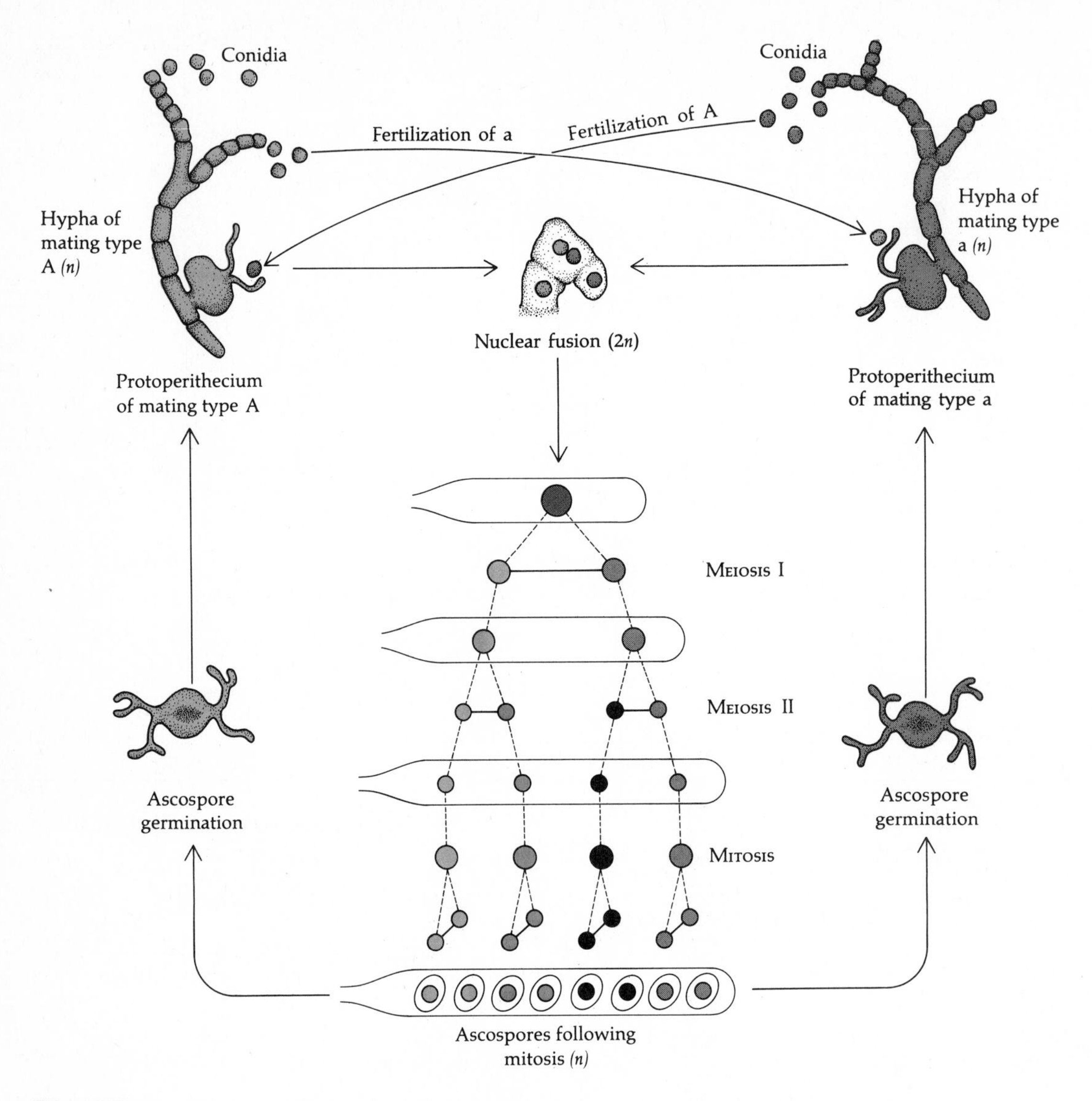

Figure 5.15
Life cycle of *Neurospora,* showing the alignment of meiotic nuclei in the ascus. The spindles of the second meiotic division do not overlap, so nuclei in each half of the ascus bear centromeres that segregated at the first meiotic division. Similarly, the spindles of the following mitotic division do not overlap, so nuclei in each quarter-ascus bear mitotic sister chromatids of centromeres that segregated at the second meiotic division. Thus ascospores in the ascus are ordered with respect to first- and second-division segregation of centromeres.

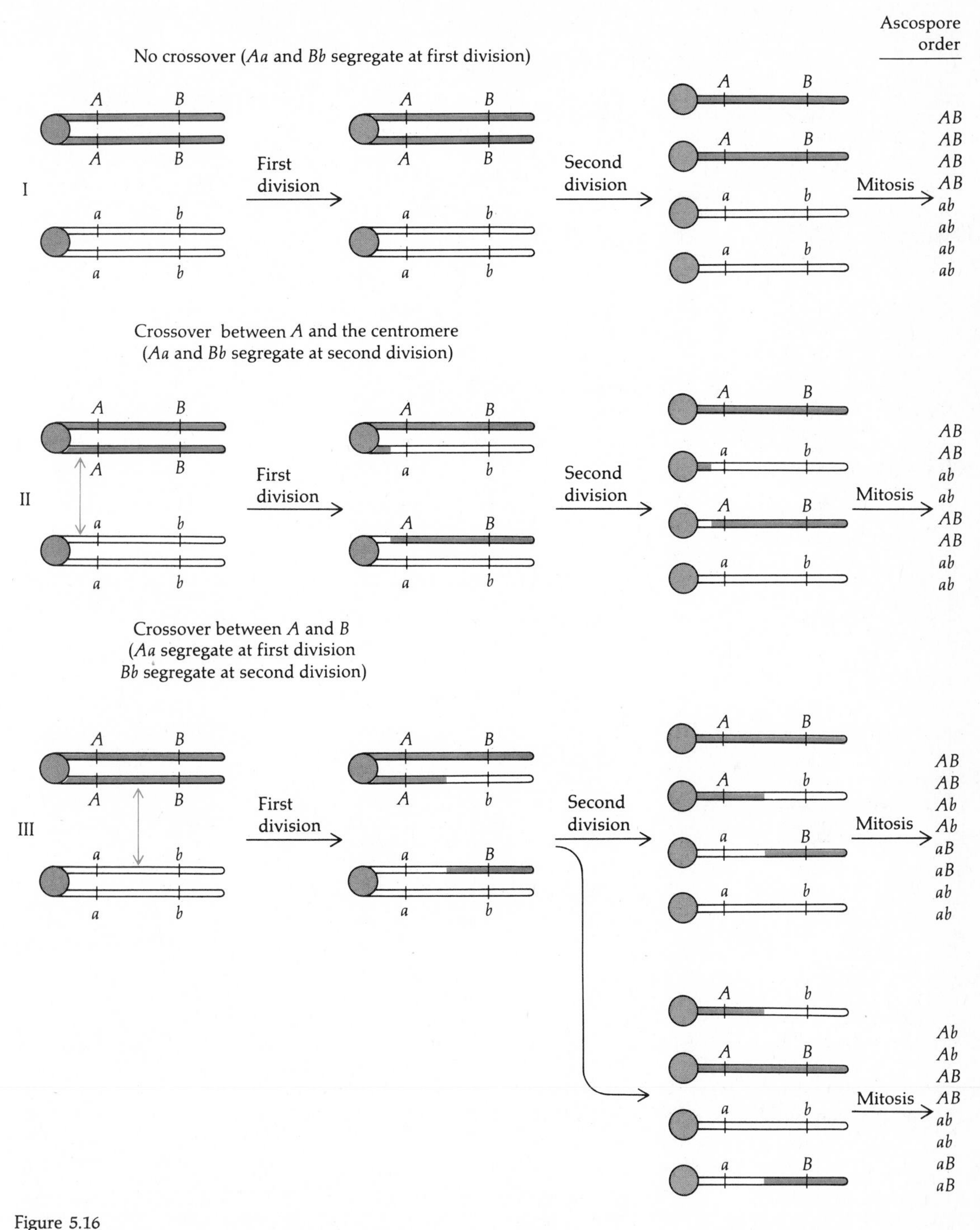

Figure 5.16
Observation of first- and second-division segregation of alleles by position in the ascus of each genotype. Ascospores are haploid, and a spore's phenotype reflects its genotype. A crossover between the centromere and *Aa* results in second-division segregation of all markers distal (*Aa*, *Bb*) to the crossover (II). A crossover between *Aa* and *Bb* results in first-division segregation of markers proximal (*Aa*) to the crossover and second-division segregation of markers distal (*Bb*) to the crossover (III).

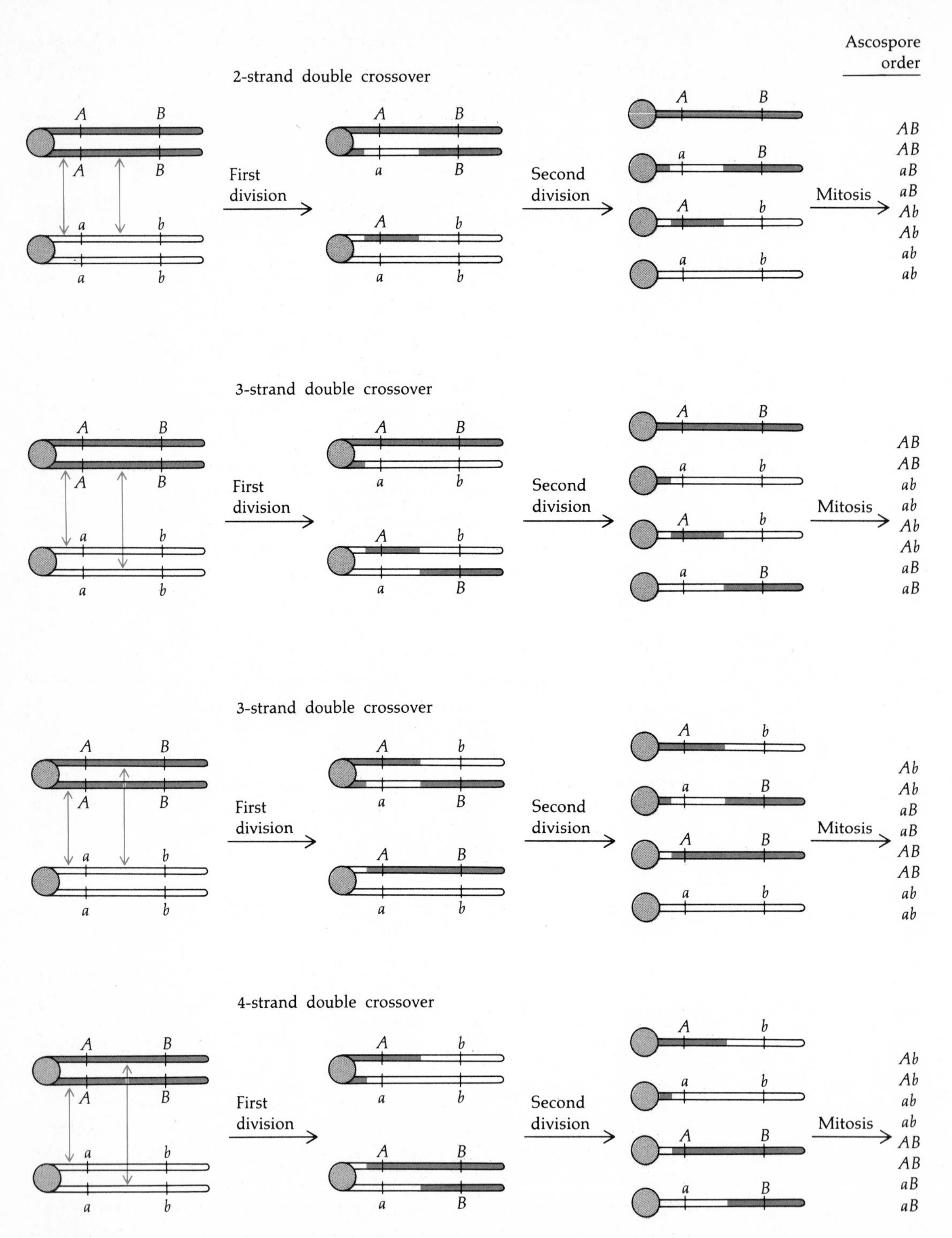

Figure 5.17
Four types of double crossover events can be detected in ordered asci.

the DNA of the chromosomes. Moreover, such data also demonstrate a very important point, namely, that recombination events occurring during a single meiosis yield *reciprocal* recombinant types.

Cytological Observation of Crossing Over

Thus far we have implicitly assumed that recombination events involve a physical exchange, or crossing over, of reciprocal pieces of chromosome from one broken chromatid to another. Evidence that this is indeed true was first presented in 1931 by Harriet B. Creighton and Barbara McClintock, studying corn, and by Curt Stern, studying *Drosophila.* The evidence for physical exchange in both organisms consists of a correlation of genetic exchange with exchange of cytologically visible chromosome segments. The evidence for physical exchange in corn is as follows. Corn plants heterozygous for alleles of two genetic markers, *colorless* (c) versus *colored* (c^+) and *waxy* (wx) versus *starchy* (wx^+), and two cytological markers, a heterochromatic knob and a translocated piece of another chromosome, were bred. Recombinant corn kernels of genotypes $c^+wx^+/c\ wx^+$ and $c\ wx/c\ wx$ were selected from the progeny of these corn plants and grown, and their chromosomes were examined cytologically. The parental and recombinant chromosomes are shown in Figure 5.18. The results, which show a correlation between genetic crossing over and exchange of the cytological markers, indicate that genetic exchange of alleles is accompanied by physical exchange of chromosome segments. Only those plants that were genetically recombinant were also cytologically recombinant.

Allelism and Complementation

Genetic analysis is the experimental study of the relationships that exist between mutant genes. Two experimental procedures or tests employed to determine the allelic nature of these relationships are the *recombination test* and the *complementation test;* each provides quite different information to the geneticist.

It is important to distinguish clearly between the recombination test and the complementation test. The recombination test for allelism is performed by noting the *genotypes of progeny* derived from a mating in which one or both of the parents are doubly heterozygous for different mutants. In contrast, the complementation test for allelism between recessive mutations is performed by observing the *phenotype of the doubly heterozygous diploid* (the parent employed in testing for recombination). If the observed

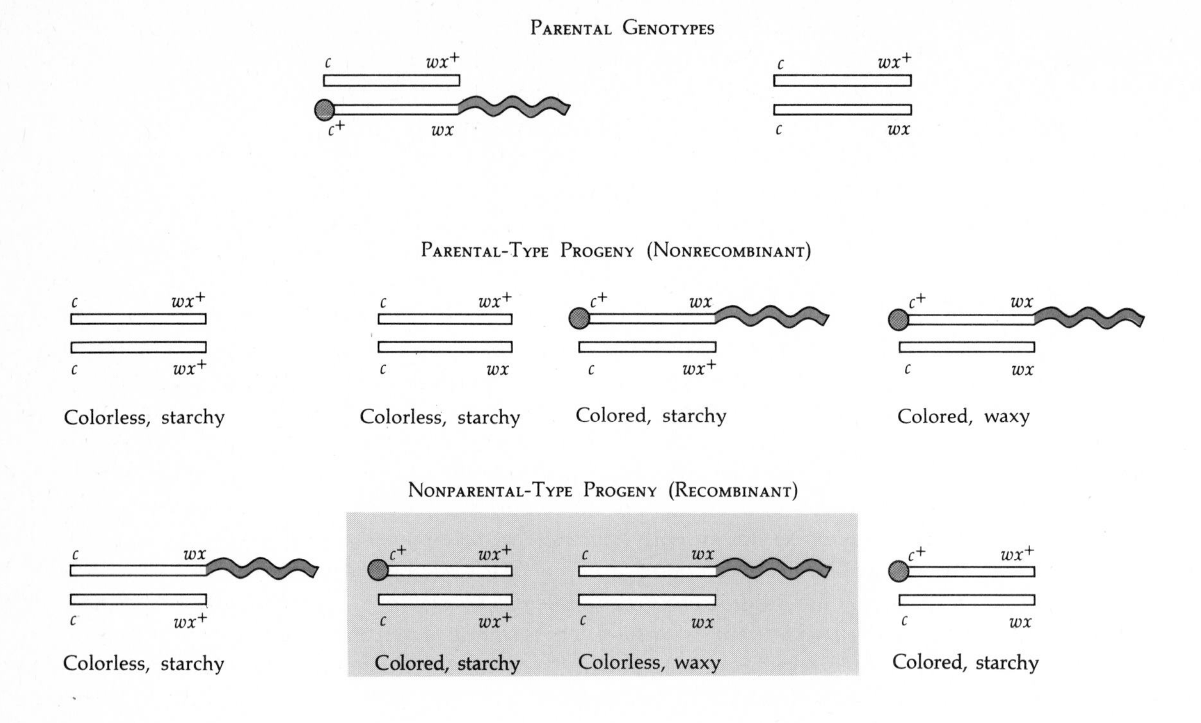

Figure 5.18
Correlation of genetic exchange with physical exchange of cytological markers; *c* and *wx* are the genetic markers, and the knob (circle) and translocated piece (wavy segment) are the cytological markers. The two shaded genotypes are distinguished as kernels of a recombinant genetic type, the kernels are grown, and the chromosomes are examined for cytological markers. wx^+/wx^+ can be distinguished from wx^+/wx by pollen phenotypes of progeny plants; all pollen grains of wx^+/wx^+ plants stain for starch, whereas only half the pollen grains of wx^+/wx stain for starch.

phenotype is mutant, then both recessive mutants must affect the same genetic function; if it is normal (wild type), then both mutants must affect different genetic functions.

The recombination test determines whether or not two alleles are segregational alleles (alleles that obey Mendel's law of segregation and always segregate from each other at meiosis). If two alleles recombine with each other, they must reside at different positions on the chromosome and can usually be considered to be alleles of different genes. The exceptions to this statement will be explored in depth in Chapter 8. Several examples of the application of the recombination test have been given in this chapter. For example, the alleles *dwarf* and *pubescent* in the tomato are alleles of different genes because they recombine with each other.

The complementation test determines whether or not two recessive alleles are functional alleles, i.e., whether or not they affect the same genetic function. *A priori* we might expect the recessive mutation *dwarf*, which causes short plant height when homozygous (d/d), to affect a different genetic and physiological function than *pubescent*, which causes hairy fruit when homozygous (p/p). When homozygous *dwarf* plants are crossed with homozygous *pubescent* plants (Figure 5.5), the double heterozygote dp^+/d^+p in the F_1 is wild-type—tall plant, smooth fruit. Each of the homologous chromosomes supplies the wild-type function that the other lacks (each *complements* the other). Therefore *d* and *p* are not functional alleles of the same gene, as we suspected.

In another example, consider two independent, recessive *Drosophila* mutations discovered by different geneticists and named independently *raspberry* (*ras*) and *prune* (*pn*). Both are sex-linked and, when homozygous or hemizygous, exhibit virtually identical mutant phenotypes: the eyes are a dark ruby color. The complementation test can be employed to determine if these two mutations are identical. When *ras*/*ras* females are mated with *pn*/Y males, the female progeny have wild-type eyes. Therefore these two mutations must affect different genetic functions. The *ras* chromosome must also carry pn^+ and the *pn* chromosome must carry ras^+. The genotype of the F_1 females is therefore *pn* +/+ *ras*, and these two mutations are said to complement each other. The recombination test also confirms that *ras* and *pn* are not identical. When *pn* +/+ *ras* females are mated, some recombinant wild-type male progeny are observed (+ +/Y), as well as doubly mutant male progeny (*pn ras*/Y). Subsequent mapping experiments showed that *pn* is located at position 0.8 and *ras* at position 32.8 on the X-chromosome map.

In contrast, the recessive, sex-linked eye-color mutations *white* (*w*), *apricot* (*a*), *coffee* (*cf*), and *buff* (*bf*) exhibit distinctly different eye colors when homozygous or hemizygous. However, females heterozygous for any two of these mutations do not exhibit wild-type eye color. Therefore each of these mutations must affect the same genetic function, but each in a different way: they fail to complement one another. The failure of these mutations to complement one another indicates that they are alleles of the same gene (called the *white* gene); this is indicated by the notation w^a, w^{cf}, and w^{bf}. These different mutations of the w^+ gene are an example of a multiple-allelic series (see Chapter 2). Recombination studies of these mutations will be discussed in Chapter 8.

The complementation test is particularly useful for determining whether recessive lethal mutations occur in identical genes. Complementation analysis has been essential in studies of mutations at the *T* locus in the mouse, because many of the mutations at this locus suppress recombination in the region of the chromosome in which they occur, making recombinational analysis impossible. The dominant mutation referred to as *Brachyury* (*T*) arose spontaneously in a laboratory strain and was observed because of a short-tail phenotype in heterozygotes (*T*/+). Homozygotes (*T*/*T*) die as embryos. Shortly after the discovery of *T*, several apparently normal wild-type stocks were observed to carry recessive mutations, designated *t*, which caused tailless progeny when crossed to *T*/+, i.e., *T*/*t* exhibits a tailless phenotype (Figure 5.19). These *t* mutations were also recessive lethals, as shown in Figure 5.20. The cross in Figure 5.20 creates a true-breeding stock of *T*/*t* mice because only progeny of this genotype survive to breed a new generation, and, since *t* mutations usually suppress recombination, this genotype is maintained from generation to generation. True-breeding stocks of this type are called *balanced*

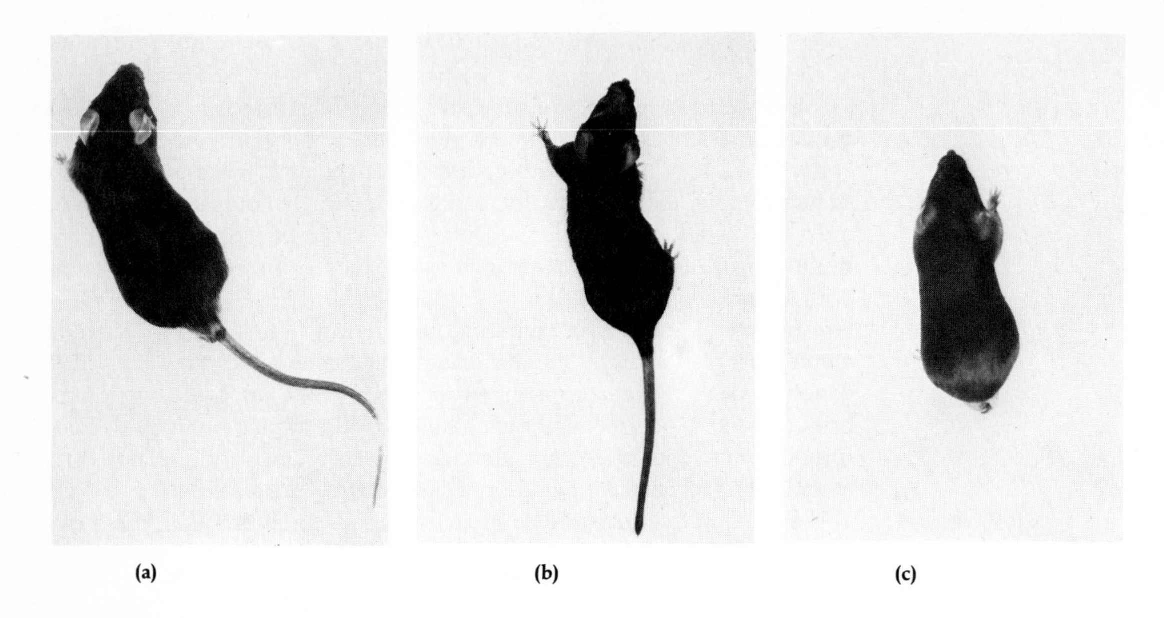

Figure 5.19
Mice of different *T*-allele genotypes. **(a)** $+/+$. **(b)** $T/+$. **(c)** T/t. (Courtesy of Dr. Keiko Yanagisawa, Mitsubishi-Kasei Institute of Life Sciences, Tokyo.)

lethal stocks and are very useful to geneticists because the stock need not be selected every generation to preserve the two mutations for future study.

Large numbers of *t* mutations have been found in natural mouse populations that are geographically widely separated. Their interesting developmental properties will be considered in Chapter 14. Complementation analysis of recessive lethal *t* mutations made by crossing different balanced stocks (Figure 5.20) reveals that some combinations are viable and have normal tails, while others die as embryos. The known lethal *t* mutations fall into six complementation groups (Table 5.3). Members of each group are lethal in heterozygous combination with one another, whereas double heterozygotes involving members of two different complementation groups are normal. As indicated in Table 5.3, some other *t* mutations are only semilethal or are viable as homozygotes.

Correlation of the Genetic and Cytological Maps of *Drosophila* Chromosomes

Most eukaryotic chromosomes, including those of *Drosophila*, are composed of genetically active regions called *euchromatin* and genetically inactive regions called *heterochromatin* (see Box 1.1). Most of the known genes in *Drosophila* are in euchromatic regions. *Drosophila* species and other dipteran species possess giant chromosomes in the somatic cells of some

(a) Parents: $\frac{T}{t} \times \frac{T}{t}$

F_1: $\frac{T}{T}$ Die as embryos; $\frac{T}{t}$ Tailless; $\frac{t}{t}$ Die as embryos

(b) Parents: $\frac{T}{t^x} \times \frac{T}{t^y}$

F_1: $\frac{T}{T}$ Die as embryos; $\frac{T}{t^x}$ $\frac{T}{t^y}$ Tailless; $\frac{t^x}{t^y}$ Morphologically normal or die as embryos

Figure 5.20
Genetic properties of the *T* locus in the mouse. **(a)** The dominant mutation *T* causes a short tail in the genotype T/t^+. Recessive *t* alleles cause taillessness in the genotypes T/t. As shown by the cross, both *T* and *t* are recessive lethal genes; progeny homozygous for these genes die as embryos. **(b)** The functional identity of the different *t* alleles *x* and *y* can be determined from the indicated cross by observing the phenotypes of t^x/t^y genotypes.

tissues, especially those of the larval salivary glands. The giant chromosomes are the product of a number of chromosome replications, without nuclear division, that are confined to the euchromatic portion of the chromosomes. Depending on the species and the stage of development, each euchromatic chromosome strand may be represented more than 1000 times in each nucleus, and all the strands of both homologues are precisely paired along their length to create a ropelike interphase polytene chromosome. A schematic model of a polytene X chromosome is shown in Figure 5.21, and electron micrographs of nuclei from *Drosophila* salivary-gland cells are shown in Figure 5.22. Differences in protein content and DNA coiling along the length of the interphase chromosomes are magnified by polytenization to produce the banded structures seen in Figures 5.22 and 5.23.

Many chromosomal aberrations are known in *Drosophila* in which pieces of chromosome have been lost (deficiencies: symbol *Df*) or are duplicated (symbol *Dp*) elsewhere in the genome (see Chapter 17). The endpoints of deficiencies and duplications can generally be located quite precisely in cytological preparations of the polytene chromosomes of individuals carrying these chromosomal aberrations. The extents of some deficiencies and duplications involving a portion of the X, or first, chromosome are shown in Figure 5.24. The genetic loss associated with deficiencies can be determined using the complementation test. For example, the heterozygotes $Df(1)N^{8}/w$, $Df(1)N^{St}/w$, and $Df(1)w^{-67k30}/w$ exhibit

Table 5.3
Properties of recessive t alleles in the mouse.

Type of Allele	Complementation Group	Member Alleles	T/t Phenotype	Recombination	Transmission Ratio (%)*
Homozygous lethal	t^{0}	t^{0}, t^{6}, t^{30}, t^{h16}	Tailless	Suppressed	$80^{\dagger}$
		t^{h7}, t^{h13}	Normal	Suppressed	Normal
		t^{h18}	Short-tailed	Enhanced	Normal
	t^{9}	t^{4}, t^{9}, t^{w18}, t^{w30}, t^{w52}	Tailless	Normal	$\text{Normal}^{\ddagger}$
	t^{12}	t^{12}, t^{w32}	Tailless	Suppressed	$75^{\dagger}$
	t^{w1}	t^{w1}, t^{w3}, t^{w12}, t^{w20}, t^{w21}, t^{w71}, t^{w72}	Tailless	Suppressed	95
	t^{w5}	t^{w5}, t^{w6}, t^{w10}, t^{w11}, t^{w13}, t^{w14}, t^{w15}, t^{w16}, t^{w17}, t^{w37}, t^{w38}, t^{w39}, t^{w41}, t^{w46}, t^{w47}, t^{w74}, t^{w75}, t^{w80}, t^{w81}	Tailless	Suppressed	90
	t^{w73}	t^{w73}	Tailless	Suppressed	95

Homozygous semilethal	t^{w2} (51% viability)	Tailless	Suppressed	95
	t^{w8} (12% viability)	Tailless	Suppressed	76
	t^{w36} (20% viability)	Tailless	Suppressed	97
	t^{w49} (2% viability)	Tailless	Suppressed	95
Homozygous viable	21 independent t^{v}'s	Tailless	Normal	10–40
	43 independent t^{v}'s	Tailless	Normal	40–60 (normal)
	1 independent t^{v}	Tailless	Normal	60–80
	t^{AE5}	Tailless	(No data)	(No data)

*See Chapter 14 for explanation of this column.

†This represents an average ratio; however, some individual males of these genotypes may have ratios as low as 30%.

‡This represents an average ratio, but individual males have very different ratios, some being as low as 25% and others as high as 90%. This indicates some abnormality in transmission of these mutants through sperm, since, for example, $T/+$ males do not show this kind of variability.

From D. Bennett, *Cell* 6:441 (1975).

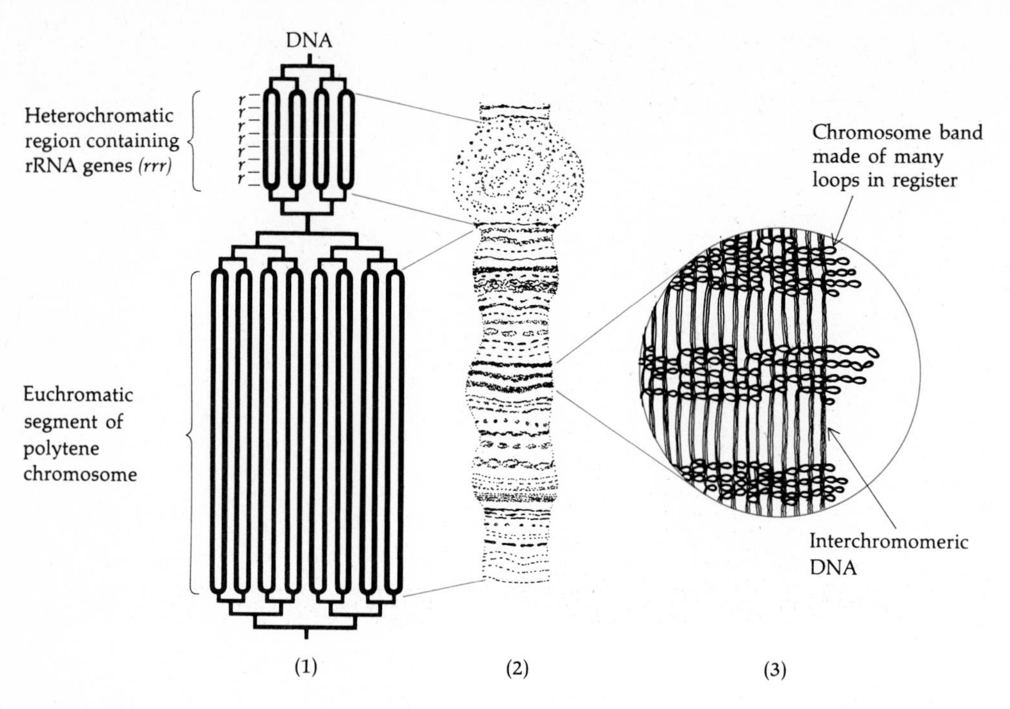

Figure 5.21
Diagram of the possible structure of a polytene X chromosome from a *Drosophila* salivary-gland cell. Replication of the euchromatic portions of the chromosome to give as many as 1000 copies is shown (1). Pairing of homologous regions of the chromosomal strands amplifies inhomogeneities along the strands (3) to give the observed banding (2). The ribosomal RNA (rRNA) genes located in the X heterochromatin are also partially amplified but do not form discrete bands, for unknown reasons.

white eyes, indicating that a w^+ gene is absent in these three deficiency chromosomes and that the physical location of the w^+ gene is within the missing portion of the X chromosome that is common to the three deficiencies—that is, within the bands designated 3C2–3C6. The heterozygote $Df(1)N^{64i16}/w$ exhibits red eyes, indicating that the w^+ gene is present in this chromosome. Thus the w^+ gene must be located between the left breakpoints of $Df(1)N^{64i16}$ and $Df(1)N^{St}$ or $Df(1)w^{-67k30}$ and must be in band 3C2. Consistent with the assignment of w^+ to 3C2 is the observation that the three duplications shown in Figure 5.24 all include band 3C2 in portions of the X chromosome that have been translocated to the second autosome, the third autosome, or the Y chromosome, respectively. For example, males of genotype w/w^+Y exhibit red eyes, indicating that the w^+ gene is contained in the portion of the X chromosome translocated to the Y chromosome. Such cytological assignments have been made for many genes located on the different chromosomes of the *Drosophila* genome, and they demonstrate conclusively that the genetic map of each chromosome derived from recombination data is colinear with the physi-

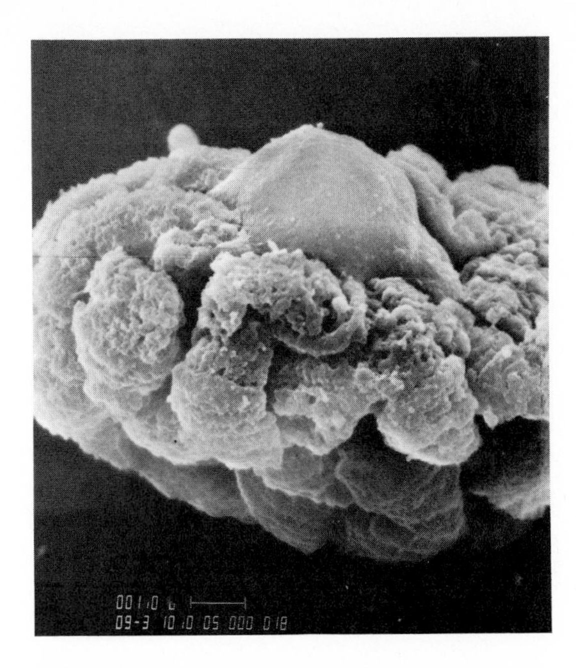

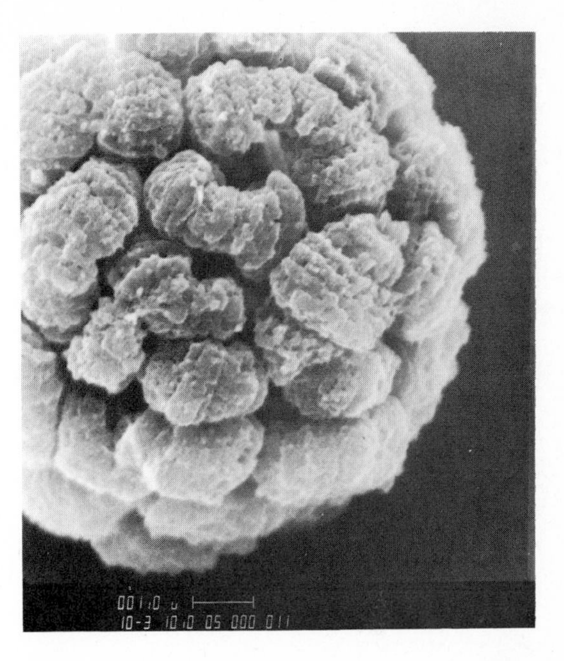

Figure 5.22
Scanning electron micrographs of intact *Drosophila* salivary-gland cell nuclei from which the nuclear membrane has been removed. The banding is apparent. The large globular structure is the nucleolus. (Courtesy of Dr. John W. Sedat, University of California, San Francisco.)

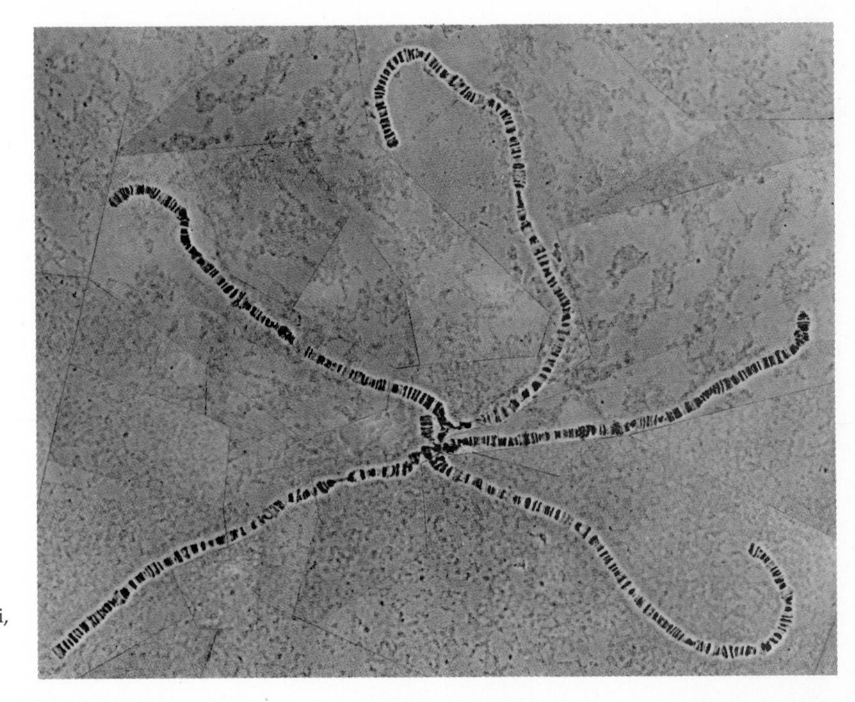

Figure 5.23
Stained light-microscope composite of *Drosophila* salivary-gland polytene chromosomes. (From G. Lefevre, in *The Genetics and Biology of Drosophila*, Vol. 1a, ed. by M. Ashburner and E. Novitski, Academic Press, New York, 1976.)

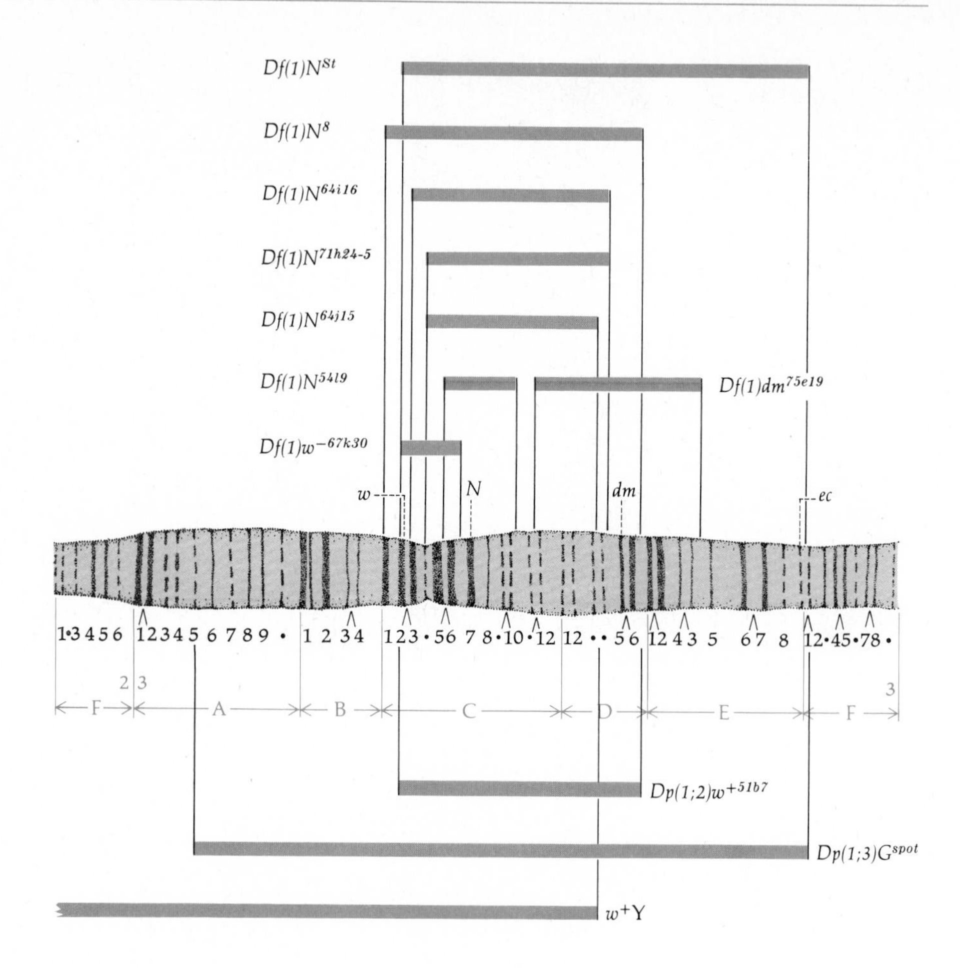

Figure 5.24
Drawing of a segment of the X polytene chromosome of *Drosophila*, showing the cytological extent of some deficiencies and duplications discussed in the text.
[From J. A. Kiger, Jr., and E. Golanty, *Genetics* 85:609 (1977).]

cal chromosome. The corresponding genetic and cytological maps are shown in Figure 5.10. The lack of a consistent quantitative correspondence between genetic map distances and physical length of the polytene chromosome may be due to differential stretching of portions of the chromosome during preparation. It is possible, however, that some portions of the X chromosome may participate in genetic recombination more frequently than others and therefore show a local expansion of the genetic map in comparison to the physical map.

Extranuclear Inheritance

The principles of genetic analysis have been applied to the study of many eukaryotic organisms: higher animals and plants, and simpler eukaryotes such as fungi, algae, and protozoa. Detailed chromosome maps have been

Figure 5.25
A common houseplant shows variegation of the leaves. (Courtesy of Constance M. Kiger.)

constructed for a number of organisms of particular interest in research or of economic importance. These maps detail the way in which genes are organized in the nucleus of the cell. Not all eukaryotic genes are located on nuclear chromosomes, however. Soon after the rediscovery of Mendel's laws of inheritance, it became apparent that some types of variations do not obey these laws. In 1909 C. Correns presented studies on the inheritance of variegation in the four-o'clock plant, *Mirabilis jalapa*, which demonstrated a non-Mendelian mode of inheritance.

Many varieties of ornamental plants, including four-o'clocks, exhibit variegation—the appearance of yellow or white patches in an otherwise green plant (Figure 5.25). The yellow patches may be small and give the leaves of the plant a mottled or striped appearance, or sometimes whole branches may be yellow and others green or variegated. Correns used pollen from flowers on yellow, variegated, or green branches to pollinate the stigmas of flowers (from which he had removed the anthers) on yellow, variegated, or green branches. The properties of the seeds produced showed that the character of the embryo within the seed was determined only by the character of the flower bearing the seeds and not by the character of the flower producing the pollen (Table 5.4). These results are an example of extranuclear, or cytoplasmic, inheritance.

We now know that the green character is due to the presence of chloroplasts, which contain the photosynthetic pigment chlorophyll. Green chloroplasts develop from self-replicating organelles called *plastids*, which are located in the cytoplasm of plant cells. The cells of yellow regions in variegated plants contain plastids that are unable to develop into normal green chloroplasts. The progeny produced by flowers on

Table 5.4
Progeny resulting from crosses between the different types of flowers on a variegated four-o'clock plant.

Pollen from Branch of Type:	Pollenated Flowers on Branch of Type:	Progeny Grown from Seed
Yellow	Yellow	Yellow
	Variegated	Yellow, green, and variegated
	Green	Green
Variegated	Yellow	Yellow
	Variegated	Yellow, green, and variegated
	Green	Green
Green	Yellow	Yellow
	Variegated	Yellow, green, and variegated
	Green	Green

After A. M. Srb, R. D. Owen, and R. S. Edgar, *General Genetics,* 2nd ed., W. H. Freeman, San Francisco, 1965.

yellow branches (Table 5.4) are unable to carry out photosynthesis, and they soon die. The variegated character in plants is due to the presence of both types of self-replicating plastids in the cytoplasm of a plant embryo. A relatively small number of plastids is contained in each cell of a plant. As the plant grows by cell division, some daughter cells, by chance, will obtain all normal plastids, others will obtain all abnormal plastids, and some will retain a mixture of the two. These cells will then give rise to branches that are all green, all yellow, or variegated. The results in Table 5.4 show that the plastids of a seed are inherited from the maternal cytoplasm and not from the pollen, a conclusion that has been confirmed by microscopic observation.

Extranuclear DNA

The self-replicating property exhibited by abnormal plastids (and, by inference, normal plastids) is attributable to their DNA, which encodes genetic functions necessary for normal plastid activity. DNA has been

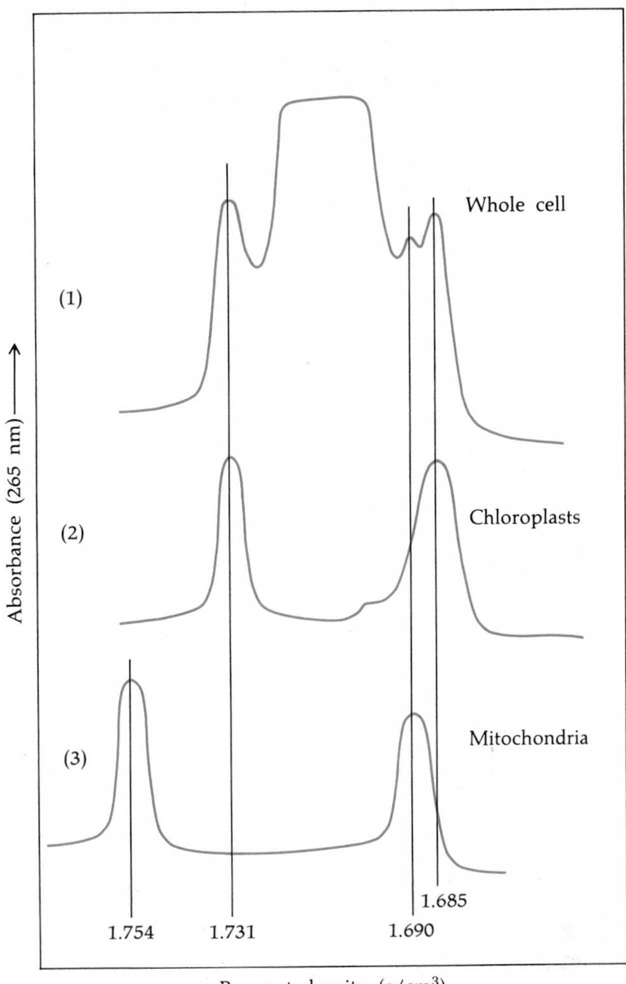

Figure 5.26
Cesium chloride density-gradient centrifugation of DNA of *Euglena gracilis* extracted from (1) whole cells, (2) purified chloroplasts, and (3) purified mitochondria. DNAs of density 1.754 and 1.731 are reference markers added prior to centrifugation. [From J. E. Manning, D. R. Wolstenholme, R. S. Ryan, J. A. Hunter, and O. C. Richards, *Proc. Natl. Acad. Sci. USA* 68:1169 (1971).]

identified in the chloroplasts of many species of plants. Its isolation has often been facilitated by the fact that many chloroplast DNAs have a different base composition than nuclear DNA and therefore exhibit a different buoyant density during cesium chloride density-gradient centrifugation. For example, as shown in Figure 5.26, the chloroplast DNA of the photosynthetic protozoan *Euglena gracilis* has a buoyant density of 1.685 g/cm^3 and can be identified in the DNA extracted from whole cells as a satellite of the nuclear DNA, which has a higher density. A strain of the same species of *Euglena* that lacks chloroplasts also lacks DNA of this buoyant density. The chloroplast DNA isolated from *Euglena gracilis* is circular and has a contour length expected for a molecule with molecular weight of 8.3×10^7 daltons.

Chloroplasts contain their own protein-synthesizing system (see Chapter 11), which is distinctly different from that in the cell's cytoplasm. They also contain the pigments, enzymes, and other proteins necessary for converting the energy of sunlight and CO_2 into carbon compounds (photosynthesis). Some of the genetic functions necessary for chloroplast protein synthesis and for photosynthesis are carried out by the chloroplast DNA, and others by the nuclear DNA. An active area of research on the mechanisms of photosynthesis and chloroplast function employs genetic analysis of mutations that affect these functions. Some of these mutations exhibit Mendelian segregation in crosses and therefore occur in genes that reside in the nucleus. Other mutations affecting these functions exhibit non-Mendelian inheritance and therefore occur in genes that reside in the chloroplast.

The DNA extracted from *Euglena* exhibits a second satellite (Figure 5.26), which is also present in cells lacking chloroplasts. This species of DNA is found in mitochondria, which are organelles responsible for the cell's respiratory activity; they produce most of the cell's ATP (adenosine triphosphate) via the citric acid cycle and oxidative phosphorylation. ATP is the cell's source of energy for all metabolic reactions. Mitochondria are also self-replicating organelles. Their DNA encodes a number of functions necessary for normal respiratory activity.

Genetic Analysis of Mitochondrial Mutations

Mutations that affect mitochondrial functions can occur in either nuclear genes or mitochondrial genes, and demonstrate either Mendelian or non-Mendelian inheritance as a result. The most intensive genetic studies of mitochondrial function have been carried out with the baker's yeast *Saccharomyces cerevisiae,* which can grow either aerobically or anaerobically. In the former case the ATP needed for growth is supplied by the mitochondria, employing a carbon source such as glycerol in the medium, whereas in the latter case ATP is supplied by the fermentation of glucose to ethyl alcohol without the need for mitochondria. Yeast cultures grown on glucose accumulate respiratory-deficient mutants called *petites,* owing to the small size of the colonies formed on agar, in comparison to normal colonies. Petite mutants are unable to grow on media containing carbon sources that require cellular respiration for growth, because they lack functional mitochondria. Such mutations would be lethal in organisms that cannot meet their energy requirements by fermentation.

A minority of independent petite mutants exhibit Mendelian inheritance and are called *segregational petites* to indicate this fact (Figure 5.27). The majority of such mutants, however, exhibit non-Mendelian inheri-

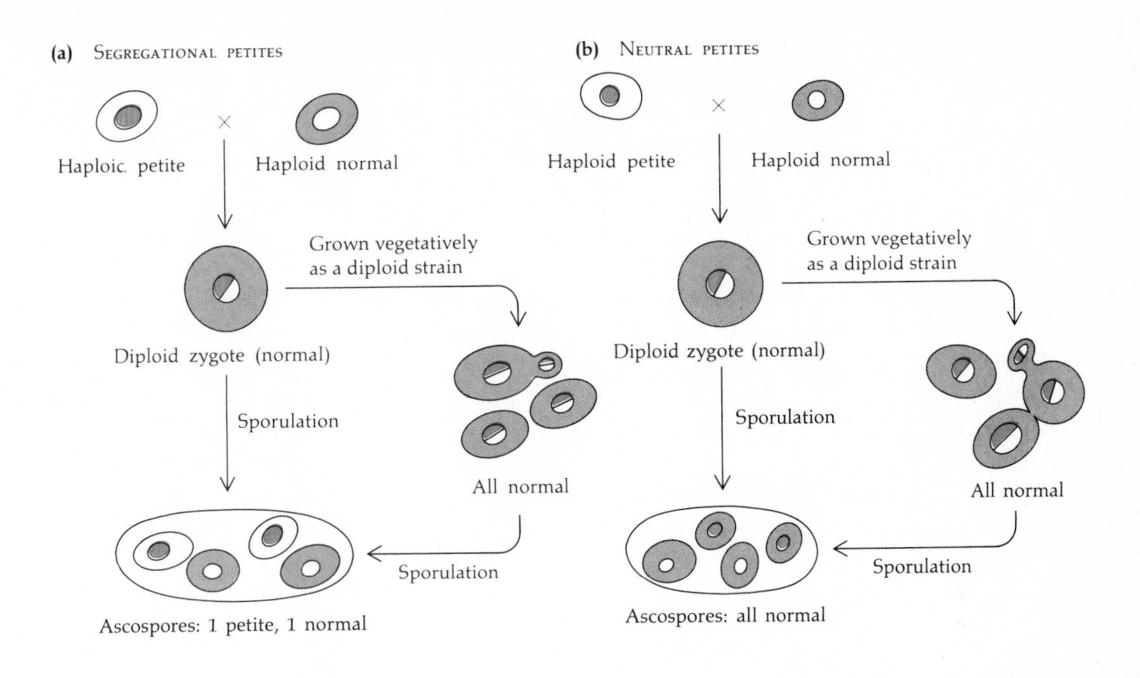

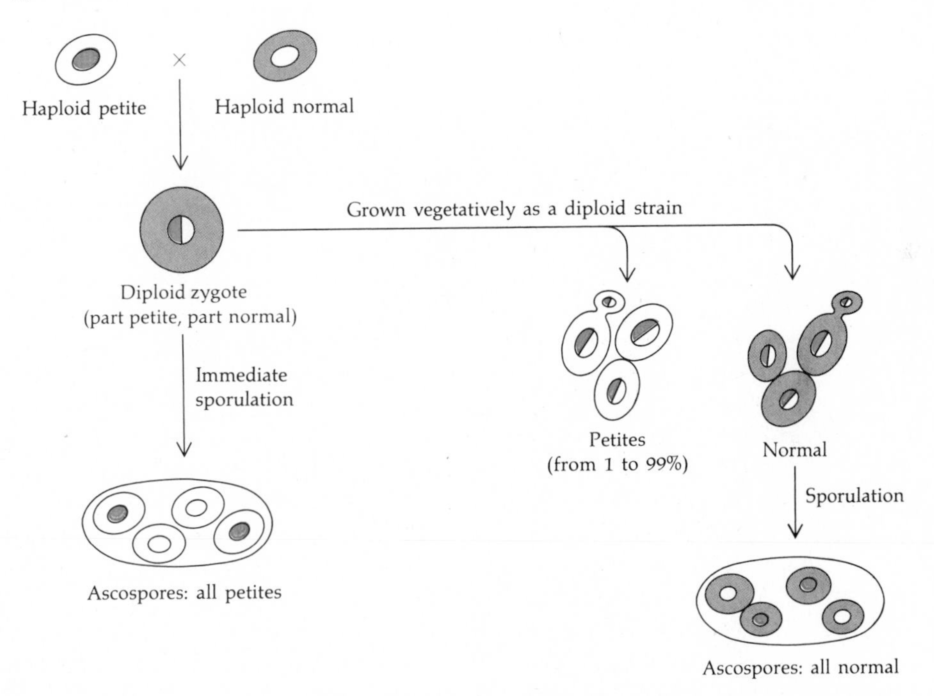

Figure 5.27
Results of crosses between different kinds of petite and normal strains of *S. cerevisiae.* (After M. W. Strickberger, *Genetics,* 2nd ed., Macmillan, New York, 1976.)

Table 5.5
Genetic characteristics of mitochondrial genomes of normal and petite strains of yeast.

Strain	Mitochondrial Genotype*					
	cap1	*ery1*	*oli1*	*ana1*	O_{II}	*par1*
Wild-type strain						
U	R	R	R	R	S	S
Y	R	R	R	R	S	R
Z	R	R	S	R	R	R
Petite strain						
U4	R	R	R	R	S	0
U5.1	R	R	R	0	0	0
Y1.2	0	0	R	R	S	0
Y1.4	0	0	0	R	S	R
Y1.5	R	R	R	0	0	0
Y2	0	0	R	R	–	R
Y7	R	R	0	0	0	R
Y8	R	R	0	0	0	R
987–19	R	R	0	0	0	R
Z3.3	0	0	–	0	R	R
Z3.31	0	0	–	0	R	R
Z3.32	0	0	–	0	R	0
Z3.33	0	0	–	0	0	R

*The symbols R, S, and 0 refer to the presence of the resistance allele, the presence of the sensitivity allele, and loss of the locus, respectively. A dash means not analyzed for the particular locus.

From K.-B. Choo et al., *Mol. Gen. Genet.* *153*:279 (1977).

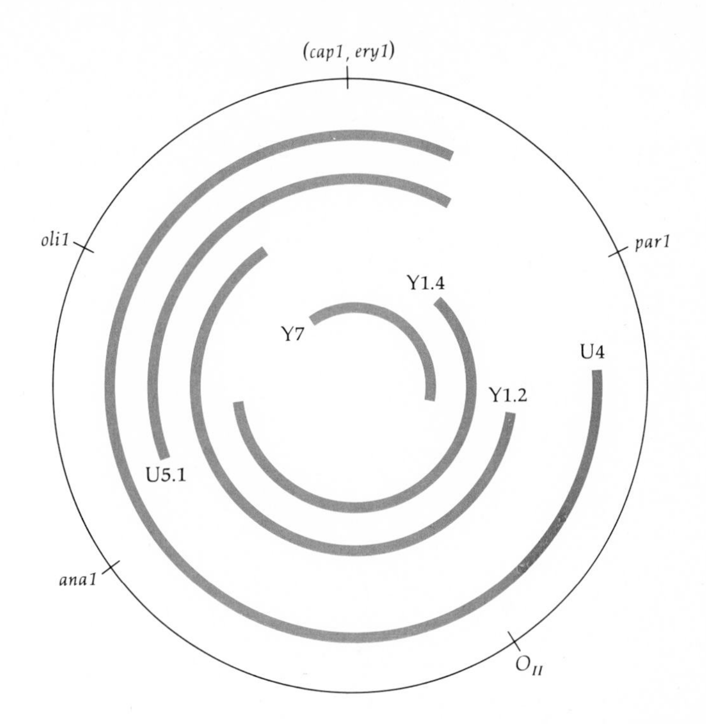

Figure 5.28
Deletion map of the mitochondrial genome of *S. cerevisiae* derived from the data in Table 5.5.

tance when crossed to the wild type and are called *vegetative petites.* Two types of vegetative petites can be distinguished by their behavior in crosses to the wild type: (1) *Neutral petites* are not transmitted through meiosis to progeny spores; they survive only in a vegetative state, reproducing asexually. (2) *Suppressive petites* are transmitted through meiosis to *all* of the progeny spores (non-Mendelian inheritance).

Yeast strains are normally haploid, but they can also be propagated in a diploid state. Diploid strains formed by fusion of haploid normal and haploid suppressive petite cells possess a cytoplasm formed equally by the two strains. Yet they exhibit an abnormal behavior, in that during vegetative growth of the diploids, the mitotic progeny are gradually converted to petite cells—hence the designation *suppressive.*

The cytoplasmic factor responsible for the non-Mendelian inheritance exhibited by vegetative petite mutants has been shown to be mitochondrial DNA. The evidence for this is as follows. First, the dye ethidium bromide, which is known to intercalate between the base pairs of DNA, induces petite mutations with almost 100% efficiency. Ethidium bromide also induces nuclear mutations, but not as efficiently as it does petite mutations. Prolonged exposure to ethidium bromide results in the complete loss of mitochondrial DNA; such mutations are neutral petites. Second, suppressive petites, induced by ethidium bromide or of spontaneous origin, contain abnormal mitochondrial DNAs containing only some portion of the base sequences present in normal mitochondrial DNA. Thus, mutant mitochondrial DNA is responsible for the respiratory

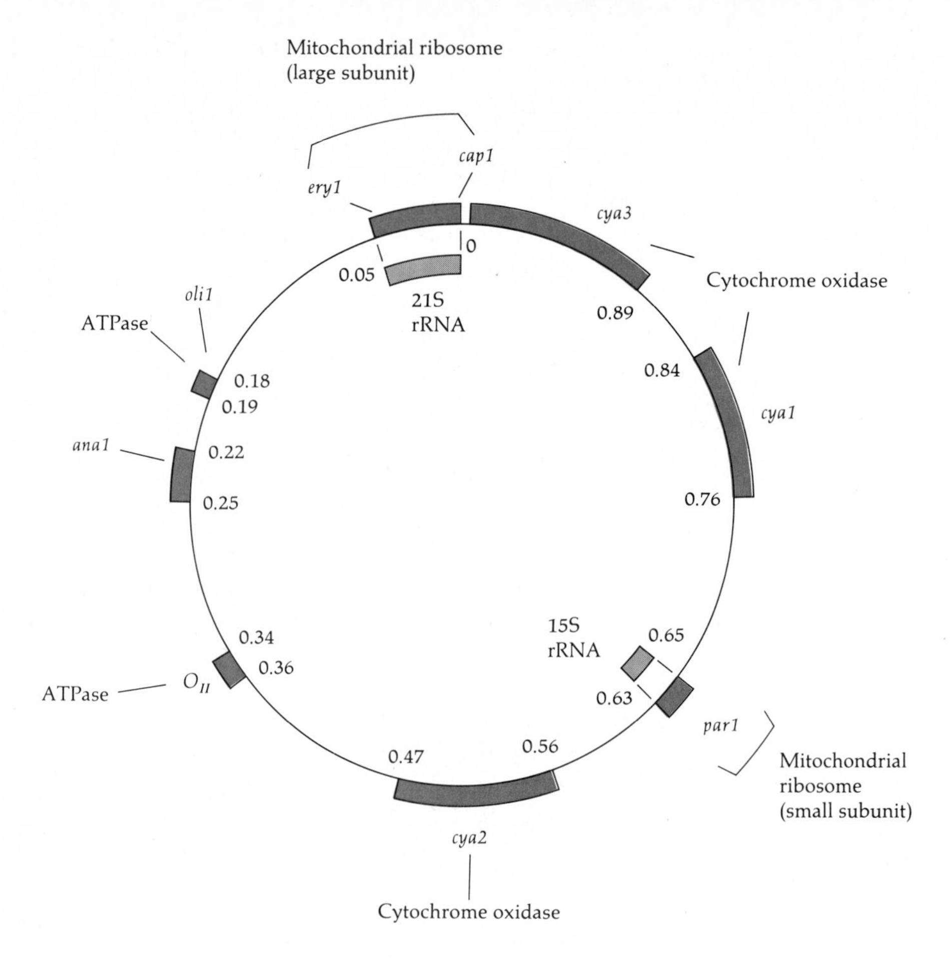

Figure 5.29
Physical map of the mitochondrial genome of *S. cerevisiae.* Inner numbers running counterclockwise indicate physical map units in percent of total DNA. Regions coding for two mitochondrial enzymes—cytochrome oxidase and ATPase—and the mitochondrial ribosomal RNAs are indicated. S, the svedberg unit, is a measure of the rate at which a particle sediments in a centrifugal field. In general, the larger the mass, the greater the observed sedimentation rate; hence 21S rRNA is larger than 15S rRNA.

deficiency exhibited by suppressive petites, and the entire loss of mitochondrial DNA is responsible for that of neutral petites.

Many of the functions necessary for normal mitochondrial activity are dependent on the mitochondrial protein-synthesizing system or on the enzymes required for oxidative phosphorylation, and some of these functions are coded by genes in the mitochondrial DNA. The mitochondrial localization of many genetic functions has been confirmed by the study of mutations that affect them. For example, mutations conferring resistance to the antibiotics chloramphenicol, erythromycin, oligomycin, antimycin A, and paromomycin, which prevent respiratory growth of yeast, all exhibit non-Mendelian inheritance. The location of these antibiotic-resistance mutations in mitochondrial DNA has been proved by the observation that some induced petite mutations causing a partial loss of mitochondrial DNA have also lost a particular antibiotic-resistance mutation.

The study of the retention and loss of particular antibiotic-resistance markers in petites of independent origin provides a unique way of mapping these mutations with respect to one another on the mitochondrial DNA. The pattern of retention of six different resistance markers in a number of different petite strains is shown in Table 5.5. These data allow the construction of a deficiency map of the mitochondrial genome if we assume that each petite strain results from the loss of a single contiguous portion of the mitochondrial DNA, an assumption that has been verified in a number of ways. Strain U4 has lost only the *par1* resistance marker, while strain U5.1 has lost *par1*, O_{II}, and *ana1*; thus these three markers must be adjacent. The pattern of loss exhibited by other petites in Table 5.5 indicates that the order of these three markers must be *par1*–O_{II}–*ana1*. Strain Y1.2 has lost *par1*, *cap1*, and *ery1*; thus these three markers must be adjacent. Strain Y1.4 has lost *cap1*, *ery1*, and *oli1*. Strains Y7 and Y8 have lost *oli1*, *ana1*, and O_{II}, indicating that these markers are adjacent and the order must be *par1*–(*cap1*, *ery1*)–*oli1*–*ana1*–O_{II}. Only a circular genetic map can accommodate the order of markers indicated by these data, as shown in Figure 5.28.

Studies of the amount of DNA retained by each petite strain and of the degree of overlap in DNA content among petite strains have allowed a physical map of the mitochondrial genome to be constructed (Figure 5.29). This map should be compared with the restriction enzyme map presented in Figure 9.26.

Problems

1. The homozygous mutant c^e, an allele of the albinism gene in the mouse, has a white coat and dark eyes. The homozygous mutant *p* has pink eyes. Mice of genotype $c^e\ +/+\ p$ are testcrossed with mice homozygous for the two mutations. The progeny obtained are shown below. Calculate the recombination frequency and map distance between the two genes.

Phenotype of progeny	Number
White coat, dark eyes	240
White coat, pink eyes	31
Dark coat, dark eyes	34
Dark coat, pink eyes	474

2. In the rabbit the dominant gene *B* specifies black hair pigment and its recessive allele *b* specifies brown hair pigment. Alleles of the albino locus, *chinchilla* (c^{ch}) and *Himalayan* (c^{h}), affect the distribution of pigment in the coat. True-breeding (homozygous) black chinchilla rabbits are crossed with true-breeding brown Himalayan rabbits. The progeny, which are black chinchilla, are crossed to brown Himalayan rabbits; the phenotypes of their progeny are given below. What is the linkage between these two genes?

Phenotype of progeny	Number
Black chinchilla	244
Brown chinchilla	134
Black Himalayan	109
Brown Himalayan	233

3. Human albinos are the result of homozygosity for a recessive mutant autosomal gene. In a case observed in England, a marriage between two albinos produced three children, all of whom exhibited normal pigmentation. This observation can be explained in at least two ways. What are they?

4. *Antennapedia* (*Antp*) is a dominant third-chromosome mutation in *Drosophila melanogaster* that transforms the antennae to legs; *Stubble* (*Sb*) is a dominant mutation causing short bristles; and *rosy* (*ry*) is a recessive eye-color mutation. *Antp*/*ry Sb* ♀♀ are crossed to *ry*/*ry* ♂♂, and the progeny classes shown below are observed. Determine the order, recombination

Phenotype of progeny	Number
Antennapedia	1140
Stubble, rosy	1078
Wild type	58
Antennapedia, stubble	70
Rosy	110
Stubble	2
Antennapedia, stubble, rosy	43

frequencies, and map distances between these mutant genes. Calculate the coefficient of coincidence and the interference value from the data.

5. *Lyra* (*Ly*) is a dominant third-chromosome mutation in *D. melanogaster* that causes incised wings (it is across the centromere from *Antp*). *Antp*/*Ly Sb* ♀♀ are crossed to wild-type ♂♂, and the progeny classes shown below are observed. Determine the order, recombination frequencies, and map distances between these mutant genes. Calculate the coefficient of coincidence and the interference value from the data. Compare the interference observed in this cross with that observed in the cross in the previous problem. How might you account for these observations? (See Chapter 8 for a possible explanation.)

Phenotype of progeny	Number
Antennapedia	851
Lyra, stubble	823
Antennapedia, stubble	73
Stubble	42
Lyra	77
Antennapedia, lyra	18
Antennapedia, lyra, stubble	6
Wild type	6

6. The *Drosophila* mutants *bithoraxoid* (*bxd*) and *ebony* (*e*) are recessive third-chromosome mutations. The dominant mutant *Contrabithoraxoid* (*Cbx*) is a pseudoallele of *bxd* and consequently almost always segregates from *bxd* at meiosis. X-irradiation of a stock of *bxd e*/*bxd e* induces a linked dominant suppressor of *bxd*, *Su*(*bxd*), which has no other phenotype. A male that is

Phenotype of Cbx^+ progeny	Number
Bithoraxoid, ebony	156
Bithoraxoid	72
Ebony	256
Wild type	9

phenotypically ebony and that is homozygous for *bxd* and *Su(bxd)* is crossed to a female homozygous for *Cbx*. Daughters from this cross are testcrossed to *bxd e*/*bxd e* males and the Cbx^+ progeny, which must be homozygous for *bxd*, are scored (see table, page 171). Determine the order, recombination frequencies, and map distances for the three mutant genes. Why are only four progeny classes scored rather than eight?

7. *Drosophila melanogaster* and *Drosophila simulans* are closely related species that can mate and produce viable but sterile progeny. In each species, mutant genes are known that have identical phenotypes and fail to complement each other in hybrid progeny produced by crossing pheno-

+ *H Dl* +/*st* + + *p* ♀♀ × *st* + + *p* ♂♂

Phenotype of progeny	Number *melanogaster*	Number *simulans*
H Dl	400	320
st p	423	330
p	18	69
st H Dl	17	75
st	0	181
H Dl p	1	181
H p	2	7
st Dl	2	11
Dl	10	3
st H p	14	0
Dl p	0	0
st H	0	1
Wild type	85	41
st H Dl p	91	29
H	0	3
st Dl p	1	1
	1070	1252

typically identical mutants of the two species. Four mutant genes found in both species are the recessives *scarlet* (*st*) and *peach* (*p*), and the dominants *Hairless* (*H*) and *Delta* (*Dl*); the latter two, however, are recessive lethals in both species. The relative map positions of these four linked genes in each species can be determined from the testcross data given on the previous page. Construct maps of the homologous chromosomes in the two species. What do your results mean?

8. Why must crossing over be suppressed if a balanced lethal stock is to be maintained for any period of time?

9. From the data in Figure 5.4, calculate the expected frequencies of the genotypes present in the F_2 generation represented in Figure 5.5. What are the expected phenotypic frequencies?

10. Crossing over occurs during meiosis only in *Drosophila* females; it does not occur in males. Why do you think this fact is of great usefulness to geneticists?

11. Balanced lethal stocks have been of great importance in the development of *Drosophila* genetics. They are made possible by the construction of *balancer chromosomes* containing multiple inversions, which suppress crossing over (see Chapter 17), and recessive lethal or female sterilizing genes, as well as a dominant mutation that allows the presence of the chromosome to be observed in heterozygotes. For example, the X chromosome designated FM7 is a balancer containing inversions, a female sterilizing mutation, and the dominant eye-shape mutation *Bar* (*B*). Show why FM7 permits recessive X-chromosome lethal mutations or other deleterious mutations to be saved indefinitely for future study.

12. *Minute* (*M*) is a class of mutations in *Drosophila* that all show retarded development, thin bristles, and recessive lethality. In a study of *Minute* mutations on the second autosome, males homozygous for the eye-color mutation *cinnabar* (*cn*), which causes bright red eyes, were X-rayed and crossed to females carrying the second-chromosome balancer SM1 [marked with the dominant mutation *Curly* wings (*Cy*) and a recessive lethal mutation] and a homologue marked with a dominant *Brown* eye mutation (*Bw*). Progeny with curly wings that also exhibit thin bristles are each mated as shown below to establish a stock of the induced putative *Minute* mutation.

$$\text{P:}\quad \frac{cn\ M}{Cy}\ ♀ \times \frac{Bw}{Cy}\ ♂ \qquad \text{or} \qquad \frac{cn\ M}{Cy}\ ♂ \times \frac{Bw}{Cy}\ ♀$$

$$\text{F}_1\text{:}\quad \frac{cn\ M}{Cy}\ ♀♀ \times \frac{cn\ M}{Cy}\ ♂♂$$

The F_1 cross establishes a stock of each putative *Minute* mutation. True *Minutes* are recessive lethals, and stocks of them are identified by the absence of homozygous *cn* progeny from the F_1 cross. In the experiment, twelve new induced *Minute* mutations were recovered, each in a balanced lethal stock. In order to determine if these new *Minute* mutations have occurred in one or more genes, complementation tests are performed by crossing the different stocks (numbered 1–12) with one another and observing whether or not homozygous *cn* progeny survive. The results of these crosses are given in the table below; a + sign indicates that homozygous *cn* progeny are produced, a − sign that they are not. Assign these 12 mutations to complementation groups.

	1	2	3	4	5	6	7	8	9	10	11	12
1	−	+	+	+	−	+	+	+	+	+	−	−
2		−	+	−	+	+	+	+	−	+	+	+
3			−	+	+	+	+	−	+	+	+	+
4				−	+	+	+	+	−	+	+	+
5					−	+	+	+	+	+	−	−
6						−	+	+	+	+	+	+
7							−	+	+	+	+	+
8								−	+	+	+	+
9									−	+	+	+
10										−	+	+
11											−	−
12												−

13. Two mating types *A* and *a* are known in the haploid fungus *Neurospora crassa*. Ascus formation occurs only between nuclei of different mating types (Figure 5.16). During meiosis, mating types behave like alleles and segregate from one another. From the following ordered tetrad data, calculate the map distance from the centromere to the mating-type locus.

Position of spore in ascus								
1	2	3	4	5	6	7	8	Number of asci
A	*A*	*A*	*A*	*a*	*a*	*a*	*a*	105
a	*a*	*a*	*a*	*A*	*A*	*A*	*A*	129
A	*A*	*a*	*a*	*A*	*A*	*a*	*a*	9
a	*a*	*A*	*A*	*a*	*a*	*A*	*A*	5
A	*A*	*a*	*a*	*a*	*a*	*A*	*A*	11
a	*a*	*A*	*A*	*A*	*A*	*a*	*a*	14

14. The haploid unicellular green alga *Chlamydomonas reinhardi* possesses mating types mt^+ and mt^-, and fusion between cells of opposite types occurs to give a diploid cell that undergoes meiosis. Four zoospores result from meiosis and form an unordered tetrad. The mutant *yellow* (y) does not form chlorophyll in the dark, producing yellow cells, whereas y^+ does, producing green cells. The mutation sr^2 is resistant to a concentration of the antibiotic streptomycin, which kills *sensitive* (ss) cells. The following classes of tetrads were observed from the cross $sr^2mt^+y^+ \times ss\ mt^-y$:

Tetrad class				Number
$sr^2mt^+y^+$;	$sr^2mt^+y^+$;	sr^2mt^-y;	sr^2mt^-y	9
sr^2mt^+y;	sr^2mt^+y;	$sr^2mt^-y^+$;	$sr^2mt^-y^+$	15
$sr^2mt^+y^+$;	sr^2mt^+y;	$sr^2mt^-y^+$;	sr^2mt^-y	25

Backcrosses of sr^2 spore progeny to ss of opposite mating type give the following types of tetrads:

	Number	
Backcross	All spores sr^2	All spores ss
$sr^2mt^+ \times ss\ mt^-$	30	0
$sr^2mt^- \times ss\ mt^+$	0	80

What can you conclude about alleles of the three genes participating in the cross?

15. How do unobserved double crossovers affect estimates of map distances?

16. The *Neurospora* mutations *arg* and *thi* prevent the biosynthesis of the amino acid arginine and the vitamin thiamine, respectively. The following asci were observed from a cross in which these mutant alleles were segregating. What is the relation between these genes, their centromeres, and each other?

Position of spore in ascus				
1 & 2	3 & 4	5 & 6	7 & 8	Number of asci
arg thi	*arg thi*	+ +	+ +	63
+ *thi*	+ *thi*	*arg* +	*arg* +	60
+ +	+ +	*arg thi*	*arg thi*	58
arg +	*arg* +	+ *thi*	+ *thi*	63

17. *Black* body (*b*) and *purple* eye (*pr*) are recessive autosomal mutations in *Drosophila*. Bridges crossed *b*/*b* females with *pr*/*pr* males and in the F_2 observed: 684 wild-type, 371 black-bodied, and 300 purple-eyed flies. Do these results indicate that the *b* and *pr* genes are closely linked? Explain. Remember that there is no crossing over in male *Drosophila*.

18. Bridges then crossed *b pr*/*b pr* females with wild-type males and in the F_2 observed: 147 wild-type, 3 black-bodied, 3 purple-eyed, and 47 black-bodied purple-eyed flies. What is the map distance between *b* and *pr*?

19. Considering that only euchromatic portions of the *Drosophila* genome are polytenized in salivary-gland cells, draw cot plots expected for DNA extracted from salivary glands and from whole embryos. Only qualitative cot plots showing the differences expected are required.

6

The Viral Genome

The study of viruses, particularly bacterial viruses, has made an enormous contribution to our knowledge of genetic phenomena. The rapid multiplication of bacteriophages makes it possible to perform genetic crosses with the progeny of two successive generations within a 24-hour period. Similar genetic crosses with *Drosophila* require 3½ weeks, and those with corn, at least a year. Moreover, the enormous populations of phage that can be produced in only a few milliliters of culture medium make it possible to observe genetic events that occur only very rarely. Furthermore, the small size of many phage genomes compared to that of a bacterium such as *E. coli* makes it possible to identify all or most of the genes and to understand the genetic organization and regulation of the entire genome in great detail. The genome of ϕX174 (see Figure 6.1) comprises only nine genes and that of phage λ less than 60, whereas the genome of *E. coli* probably comprises several thousand genes. These obvious advantages have combined to make the genomes of certain bacteriophages the best understood of all "organisms." They provide model systems for attempts to understand much more complex genomes.

Mutant Bacteriophages

Virtually all genetic experiments with phages are performed without direct observation of either parents or progeny. This means that rather indirect methods are employed to determine the phenotypes of the genetic

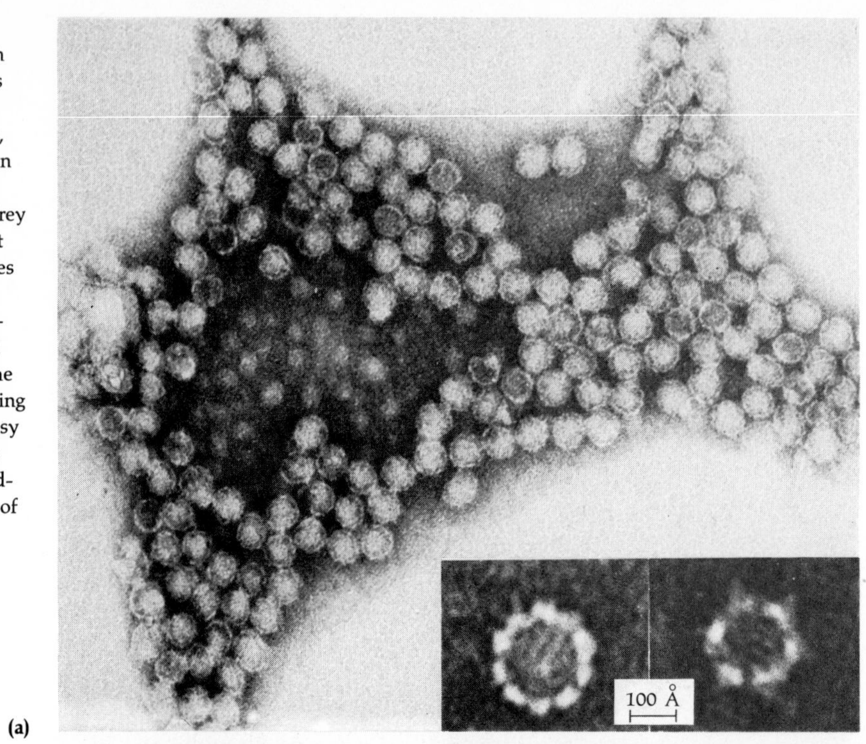

(a)

(b)

Figure 6.1
(a) Electron micrograph of ϕX174 phage particles ($\times$190,000). Inset shows highly magnified phages, revealing the organization of the capsid subunits. (Courtesy of Dr. W. Jeffrey Tromans and Dr. Robert W. Horne, the John Innes Institute, Norwich, England. **(b)** Electron micrograph of single-stranded ϕX174 viral DNA and the double-stranded replicating form ($\times$21,000). (Courtesy of Dr. Richard Junghans and Prof. Norman Davidson, California Institute of Technology.)

variants studied. As discussed in Chapter 4, phages are usually detected through their ability to kill their host bacteria in the course of an infection. The presence of a phage is observed as a plaque in the lawn of bacterial cells in a Petri dish. The plaque itself is obviously not the phage that initiated the infection, but is the result of host-cell death caused by the progeny of that phage. Before we consider the genetics of phages, therefore, we must first examine the methods used to describe the phenotypes of mutant phages.

Mutations that affect the morphology of plaques are the most directly observed. Some mutations affect the size of the plaque. Others produce plaques that are either clear or turbid (a result of the infection's not killing all the infected cells). Phages such as T2, T4, and φX174, which kill all infected cells, making clear plaques, are called *virulent* phages. Phages such as λ and P1, which do not invariably kill their host, making turbid plaques, are called *temperate* phages.

Phages infect bacterial cells by first adsorbing to specific receptors on the cell surface. The nature of these receptors is genetically controlled by the cell, and it is possible to isolate mutant bacterial strains that possess altered receptors to which the phage cannot attach. Phage mutations that extend the host range of the mutant phage to include strains to which the wild type cannot adsorb identify the genes responsible for phage adsorption. Plaque-morphology mutations and extended host-range mutations were among the first phage mutations to be studied; however, they occur in only a small number of all the genes present in the phage genome.

The majority of phage genes provide functions that are essential for replication and for the production of progeny phages. Mutations in these genes prevent the production of progeny and are lethal—no plaques are formed. Except under special circumstances, lethal mutations cannot be propagated, as recessive lethals are in many eukaryotes, because phages are haploid. *Conditional lethal mutations* are those that are lethal under one set of conditions (*restrictive* conditions) but that can be propagated under another set of conditions (*permissive* conditions). These mutations permit most phage genes to be identified and studied.

Mutations that cause the production of progeny phage to be sensitive to temperature are an example of a type of conditional lethal mutation. Most phages can infect their host and produce progeny over a wide range of temperature. *Temperature-sensitive* (*ts*) mutants of many phages that infect *E. coli* are usually propagated at 30°C (the permissive condition) and exhibit their mutant phenotype at 40–42°C (the restrictive condition) by failing to make plaques. Cold-sensitive mutations are also known. Temperature sensitivity is almost invariably the result of the mutant's coding for a protein with an amino acid substitution that renders it unstable at the restrictive temperature, thereby causing loss of activity (see Chapter 10).

Suppressor-sensitive (*sus*) mutations are another type of conditional lethal mutation. A phage carrying a *sus* mutation can produce progeny when it infects one strain of host cell [carrying a suppressor gene (Su^+), the

Table 6.1
Phenotypes of *sus* phage mutations in bacterial hosts of different genotypes.

Genotype of Phage	Genotype of Bacterial Host*			
	Su^-	Su^+_{amber}	Su^+_{ochre}	Su^+_{opal}
Wild type	+	+	+	+
sus_{amber}	−	+	+	−
sus_{ochre}	−	−	+	−
sus_{opal}	−	−	−	+

*+ = progeny produced; − = no progeny produced.

permissive condition], but fails to produce progeny when it infects another strain of host cell [lacking a suppressor gene (Su^-), the restrictive condition]. The wild-type phage can produce progeny in both strains of host cell (Table 6.1). Suppressor-sensitive phage mutants do not affect the adsorption of the phage to the host, as do the host-range mutations mentioned above. The phage adsorbs normally, injects its DNA, and may kill the host cell, but produces no progeny. There are three classes of suppressor-sensitive mutations, called *amber* (*am*), *ochre* (*och*), and *opal* (*op*). We will discuss them here simply as genetic markers. They can occur in all genes that code for a protein product. The mutation prevents the synthesis of a normal phage protein by the protein-synthesizing system of an Su^- host but not by that of an Su^+ host. The biochemical nature of *sus* mutations and their mode of suppression will be discussed in Chapter 12.

With this introduction to phage mutants, we will now turn to the genetic analysis of mutations in several different phages that have contributed to our knowledge of the organization of the viral genome.

Complementation Analysis of Conditional Lethal Mutations—ϕX174

Conditional lethal mutations in all essential genes have identical phenotypes: the mutants fail to produce progeny when the infection is carried out under restrictive conditions. The complementation test described in

Chapter 5 can be used to determine whether two conditional lethal mutations of independent origin affect different genetic functions or the same genetic function. Bacterial cells are infected simultaneously with phages of both mutant types under restrictive conditions, e.g., 42°C if the mutants are both temperature-sensitive. If the doubly infected bacterial cells yield progeny phages then each mutant phage must have supplied the function that could not be supplied by the other (Figure 6.2). The two mutations are then said to complement each other and to reside in different complementation groups, or *cistrons* (a cistron is the unit of genetic function defined by the complementation test; see Chapter 8 for the origin of this term, which has become a synonym for gene). The complementation test carried out in this way is quite analogous to that described in Chapter 5. A "diploid" infected cell is created in which each phage chromosome carries a mutation, and the phenotype of that "diploid" is noted—i.e., whether progeny phages are produced or not. Note that the genotypes of the progeny are irrelevant in determining the outcome of the complementation test.

Thirty-nine conditional lethal mutations of the small phage φX174 are listed in Table 6.2. Pairwise complementation tests of these mutants show that they reside in eight complementation groups. For example, infection of a restrictive host by both *am10* (cistron *D*) and *am9* (cistron *G*) is observed to yield progeny phage when the host lyses. In contrast, infection of a restrictive host by both *am9* and *am32* fails to yield progeny, and the two mutations are assigned to the same complementation group (cistron *G*). Assuming that all complementation groups are equally mutable, the fact that several mutations exist in most complementation groups suggests that these mutations define all the essential complementation groups in the φX174 genome. In other words, enough independent mutations have been studied to saturate the genetic map (see Chapter 16). We shall see below that this inference accurately reflects the biochemical expression of these functions, with one exception.

Complementation tests between phage mutants are easily carried out employing the *spot test*. For example, a culture of Su^- bacteria is infected with one phage *sus* mutant at an infection multiplicity of about 0.1 (1 phage/10 bacterial cells), mixed with warm agar, and spread on a nutrient plate. When the agar has solidified, a drop of medium containing another phage *sus* mutant is spotted on the agar. Within the area of the drop, some Su^- bacteria will be infected by both mutants. If complementation occurs, the infected cells will release progeny phages of both parental genotypes (and of recombinant genotypes as well). These progeny will infect all the remaining bacteria within the area of the drop and kill them. Thus complementation will be evidenced by the absence of growing bacteria over the entire area of the spot (Figure 6.3).

When the spot test is carried out with two noncomplementing mutants, the uninfected bacteria grow to cover the area of the spot and blend with the background bacterial growth. Occasionally, plaques may appear within the area of the spot, owing to recombination between two non-

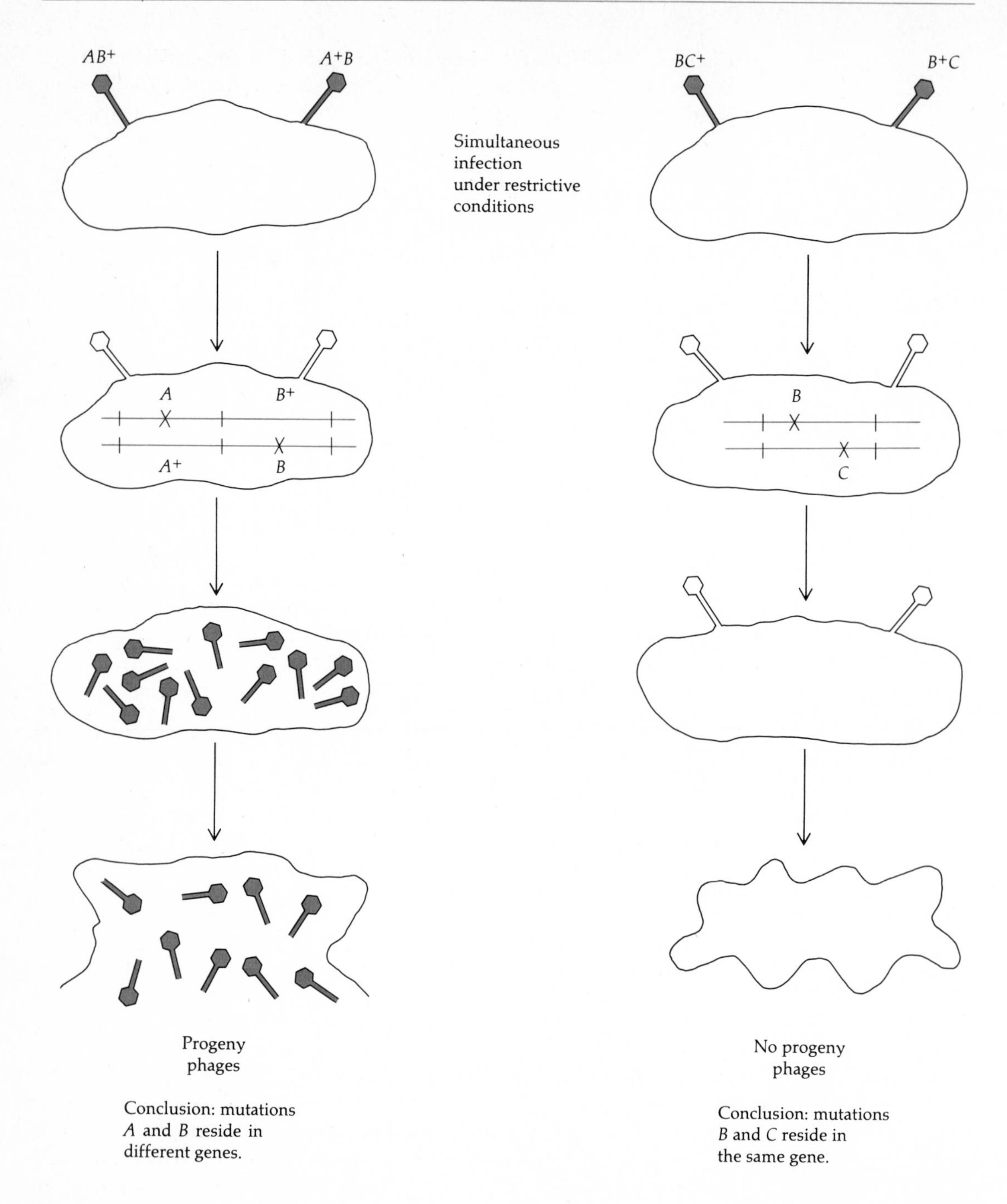

Figure 6.2
Schematic representation of positive (mutants *A* and *B*) and negative (mutants *B* and *C*) complementation tests between phage mutants.

Table 6.2
Classification of ϕX mutants.

Cistron	Mutant
A	*am8, am18, am30, am33, am35, am50, am86, ts128*
B	*am14, am16, och5, ts9, ts116, och1, och8, och11*
C	*och6*
D	*am10, amH81*
E	*am3, am6, am27*
F	*am87, am88, am89, amH57, op6, op9, tsh6, ts41D*
G	*am9, am32, tsγ, ts79*
H	*amN1, am23, am80, am90, ts4*

From R. M. Benbow et al., *J. Virol.* 7:549 (1971).

complementing mutants within a single cell; this produces a wild-type recombinant phage genome that is able to multiply in the Su^- cell. The appearance of these occasional plaques can generally be distinguished from a positive complementation response, unless the plaques are so dense that they cover the entire area of the spot. Recombination between mutants in the same cistron is infrequent enough, however, that this is not a serious problem in interpreting spot tests.

Recombination Analysis of Phage Mutants—ϕX174

When phages with different mutant genotypes infect a cell in which both can produce progeny, the progeny phages are found to contain both parental and recombinant genotypes. A cross between two different *ts* phage mutants is performed at the permissive temperature, e.g., 30°C (Figure 6.4), or, in the case of *sus* phage mutants, in an Su^+ bacterial strain. The total number of progeny (of all genotypes) is determined by plating an aliquot of the culture under permissive conditions. The number of recombinant wild-type progeny is easily determined by plating an aliquot

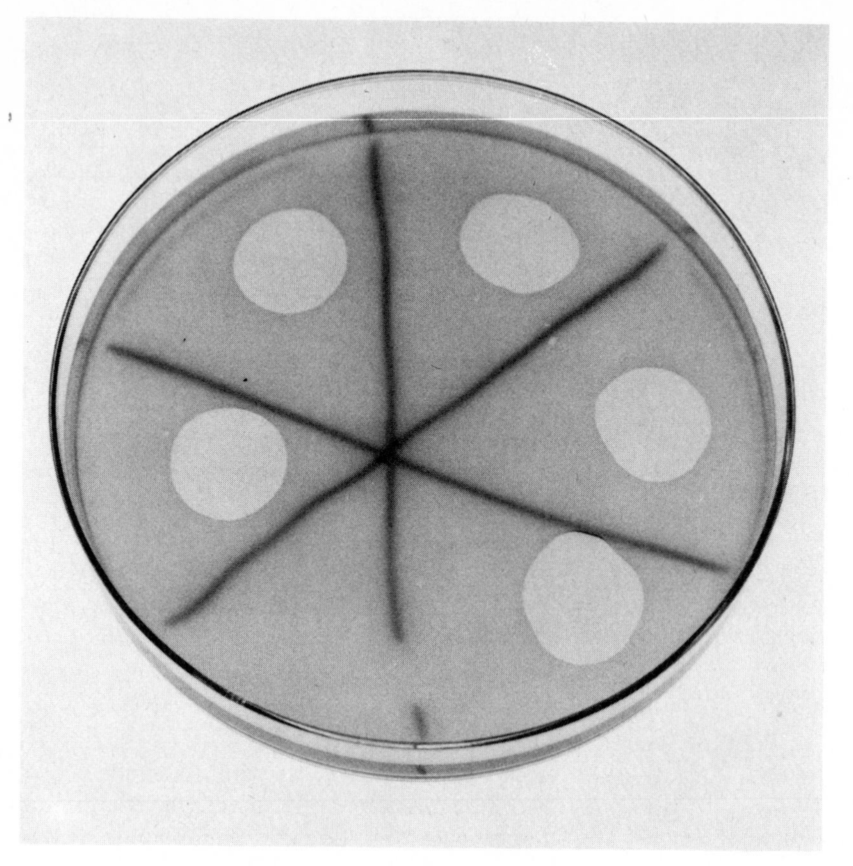

Figure 6.3
Petri dish showing spot test for complementation between pairs of T4 phage mutants. Five spots show positive complementation (clearing), and the sixth shows negative complementation.

under restrictive conditions where *only* progeny with the wild-type genotype are able to make plaques.

Phage crosses are not strictly analogous to crosses involving eukaryotic organisms. When two eukaryotic parents mate, we know that (1) the genetic contribution of each parent to the progeny is equal, owing to the meiotic process, and that (2) when genetic markers are unlinked, equal numbers of parental and recombinant genotypes appear in the progeny. In phage crosses, however, the relative genetic contribution of parents to the progeny is found to depend on the relative numbers of each parent phage that infects a given bacterial cell. For example, if the parental ratio of genotypes A and B is $A/B = 10/1$, the number of recombinant progeny produced is often found to be larger than the number of parental type-B progeny. In a phage cross, the number of parental genotypes is not limited to two. If phages with three different genotypes, A, B, and C, all infect the same cell, some recombinant progeny of genotype ABC appear. Clearly, the dynamics of a phage cross are more akin to a problem in population

Figure 6.4
Schematic representation of a cross between two mutant phages, A and B. The cross is made by infecting the host under permissive conditions that allow both mutant strains to propagate.

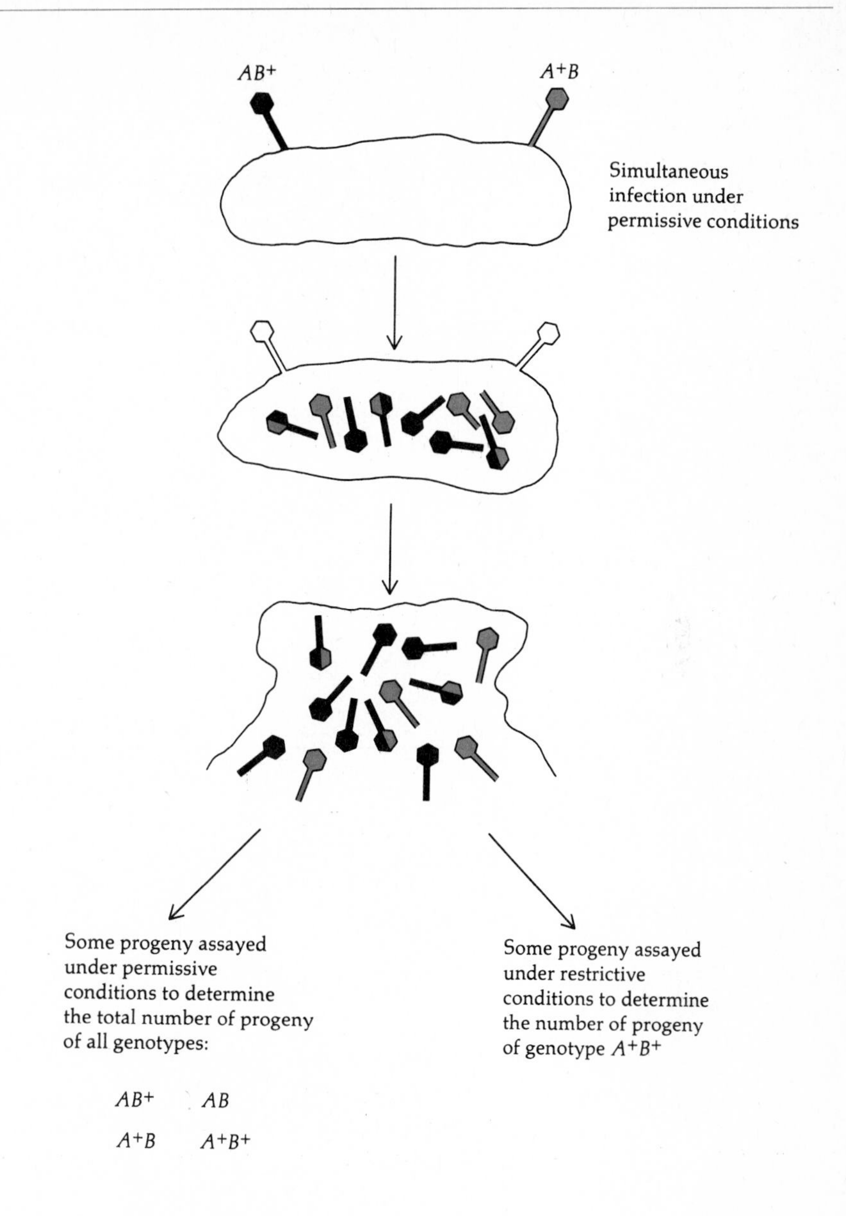

biology than to a genetic cross between meiotic organisms. The infecting parental genomes represent the founder population. These genomes replicate and recombine with other genomes of like or unlike genotype; recombinant genotypes replicate along with parental genotypes; and when the infected cell lyses, the progeny phages represent a sample of genomes from the mating pool present in the cell at the time the DNA molecules were encapsulated in the phage coats. Moreover, not all replicating genomes of one parental type may have an equal opportunity to recombine

Table 6.3
Two-factor recombination frequencies observed in crosses between ϕX174 mutants.*

	A					B		C	D	E		F		G	H	
Mutant	*am18* (*A*)	*am33* (*A*)	*am35* (*A*)	*am50* (*A*)	*am86* (*A*)	*am14* (*B*)	*am16* (*B*)	*och6* (*C*)	*am10* (*D*)	*am3* (*E*)	*am6* (*E*)	*am88* (*F*)	*op6* (*F*)	*am9* (*G*)	*am23* (*H*)	*amN1* (*H*)
A																
am18(A)																
am33(A)	21.9 ±3.2 (2)[†]															
am35(A)	0.4 ±0.1	21.2 ±2.4[†]														
am50(A)	7.9 ±0.5[†]	4.5 ±0.6[†]	13.9 ±1.0[†]													
am86(A)	11.7 ±2.4[†]	5.5 ±2.0 (2)[†]	16.8 ±1.5[†]	0.5 ±0.1 (2)												
B																
am14(B)	2.8 ±0.3	11.3 ±0.4[†]	2.8 ±0.3	2.6 ±0.5 (2)	4.0 ±0.2											
am16(B)	2.0 ±0.1	9.1 ±2.8[†]	2.7 ±0.1	6.4 ±0.3	7.3 ±1.2	1.3 ±0.3										
C																
och6(C)	1.3 ±0.2[‡]	3.9 ±0.4[‡]	0.7 ±0.1[‡]	2.2 ±1.3[‡]	1.1 ±0.1[‡]	1.0 ±0.2	1.3 ±0.3									
D																
am10(D)	1.4 ±0.5	6.2 ±1.0	2.0 ±0.5	4.0 ±0.6 (2)	4.2 ±0.2	1.8 ±0.2 (2)	2.3 ±0.4	2.0 ±0.2								

E															
am3(E)	4.1 ±0.6 (2)	10.8 ±0.8 (3)	4.2 ±1.0 (3)	10.2 ±1.6 (3)	8.3 ±0.9 (2)	3.4 ±0.5 (4)	4.6 ±0.9 (5)	1.3 ±0.1 (2)‡	1.5 ±0.2 (3)						
am6(E)	6.6 ±1.0	5.7 ±0.8	6.3 ±1.2	8.3 ±2.0	6.5 ±2.0‡ (2)	2.4 ±0.7	3.5 ±0.4	0.2 ±0.7‡	0.2 ±0.4	0.2 ±0.4 (6)					
F															
am88(F)	10.8 ±1.2	12.4 ±2.1	11.9 ±0.3	8.0 ±1.0	5.3 ±0.9	10.3 ±0.7	9.3 ±3.2	11.1 ±2.2	14.4 ±0.6	4.4 ±0.8 (2)‡	7.1 ±0.7 (2)				
op6(F)	6.5 ±0.2		6.0 ±0.2	4.2 ±1.5‡	1.3 ±0.6‡	4.8 ±0.4	5.3 ±0.7		2.2 ±0.1	1.2 ±0.1	2.5 ±0.4				
G															
am9(G)	5.8 ±1.4 (2)	11.5 ±0.8	8.0 ±1.0 (2)	8.2 ±0.8	6.8 ±0.4	2.9 ±0.9	5.4 ±1.2	1.2 ±0.1‡	5.9 ±2.1	6.8 ±0.8 (9)	4.7 ±1.0	1.3 ±0.2			
H															
am23(H)	1.7 ±0.8 (2)‡	2.0 ±0.4‡	4.7 ±0.5‡	1.2 ±0.1‡	0.4 ±0.1‡	1.8 ±0.5 (3)‡	2.1 ±0.4 (2)‡		2.6 ±0.3‡	2.2 ±0.6 (2)‡	3.4 ±0.9‡	3.4 ±0.4	9.2 ±1.4	2.1 ±0.3‡	
amN1(H)	3.0 ±0.3	7.5 ±1.2	3.1 ±0.2	2.0 ±0.3 (2)	2.1 ±0.3 (2)	3.0 ±0.5	2.8 ±0.5	1.4 ±0.2‡	4.6 ±0.3 (2)	8.1 ±1.3 (3)	6.2 ±0.9 (2)	4.1 ±0.6	8.1 ±0.5	3.1 ±0.8 (2)	0.26 ±0.3 (2)

*Values are expressed times 10^4.

†These cistron-*A* recombination frequencies were used in the construction of Figure 9.15 only. Numbers in parentheses represent the number of independent determinations used to calculate the average recombination frequency.

‡Exact correspondence between map distance in Figure 6.6 or 9.15 and recombination frequency is not obtained, for the reason noted in Chapter 9 (page 313).

From R. M. Benbow et al., *J. Virol.* 7:549 (1971).

with genomes of the other parental type—e.g., if a phage of genotype *A* adsorbs to one end of a long bacterium and a phage of genotype *B* adsorbs to the other end. The exclusion of some genomes from the mating pool is evidenced by *low negative interference* (*c* is slightly greater than 1), which is observed in most phage crosses (see Chapter 8).

Despite the dissimilarity of mechanisms between phage crosses and crosses involving eukaryotic organisms, one *can* apply the formalism of eukaryotic genetics to a phage cross. To do so, it is necessary to control the input of each of the two parental phage genotypes and to control the time permitted for replication and recombination to occur. When these two factors are controlled, by adopting certain standard conditions for performing crosses, linkage of genetic markers can be detected, and reproducible recombination frequencies can be measured and used to create genetic maps.

The φX174 mutations listed in Table 6.2 have been analyzed in both two-factor and three-factor crosses. Two-factor crosses between suppressor-sensitive mutations are performed by infecting a culture of Su^+ bacteria with each of two mutant phage strains, using five of each type of phage per bacterial cell to assure that all cells are infected with phages of both parental genotypes. Upon lysis of the infected bacteria, the total number of progeny phages is determined by plaque assay, employing a bacterial host strain (indicator) carrying the appropriate Su^+ genes. In crosses between sus_{amber} phage mutants, an Su^+_{amber} strain is used as a host; in crosses between sus_{amber} and sus_{ochre} phage mutants or between sus_{amber} and sus_{opal} phage mutants, an indicator is used that carries both Su^+ genes. The number of wild-type recombinant progeny produced in the cross is determined by plaque assay employing an Su^- indicator, on which only wild-type phage will make plaques. The recombination frequencies found in crosses between these φX174 mutants are shown in Table 6.3.

Three-factor crosses have been used to deduce and confirm the order of these mutations on a genetic map. The method used to deduce the order of three markers differs from that presented in Chapter 5 for three-factor crosses. Consider the cross *amA tsC* × *amB*, shown in Figure 6.5. If *amA* and *amB* are more closely linked to each other than either is to *tsC*, as determined by two-factor crosses, *tsC* is treated as an unselected marker by plating the progeny from the cross on Su^- indicator at a permissive temperature. Progeny that are wild-type recombinants with respect to the two *amber* mutations will produce plaques whether they also carry *tsC* or its wild-type allele. The proportion of these recombinant phages that are *tsC* or tsC^+ determines whether the order is *amA amB tsC* or *tsC amA amB*. If the progeny are predominantly wild type, the order is deduced to be *tsC amA amB*. If they are predominantly temperature-sensitive, the order is deduced to be *amA amB tsC*. The order assigned from such a cross is then confirmed by carrying out the reciprocal cross, as indicated in Figure 6.5. The results of these three-factor crosses are shown in Table 6.4.

The data presented in Tables 6.3 and 6.4 demonstrate unequivocally that the genetic map of φX174 is circular. This map, shown in Figure 6.6,

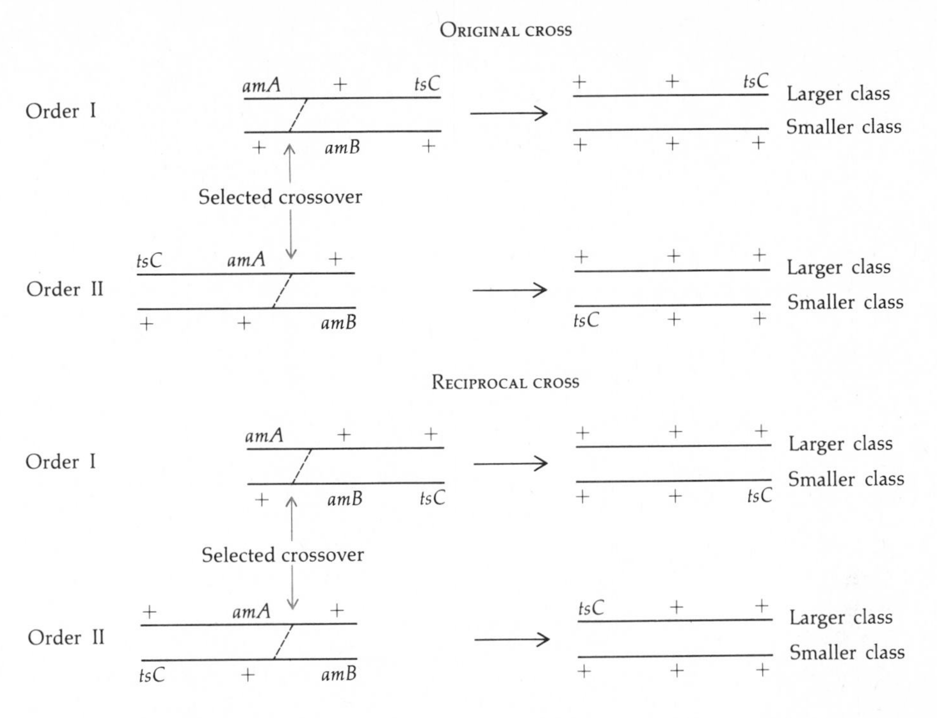

Figure 6.5
Example of the use of a distantly linked, unselected marker, *tsC*, to determine the order of markers in a three-factor phage cross. I and II represent the two possible orders of the three markers. Wild-type recombinants between *amA* and *amB* are selected, and the genotype with respect to the unselected marker is then determined to deduce the correct order. For example, if the order is I, the majority of the selected wild-type recombinants in the original cross will be + + *tsC*. Recombinants that are + + + require a second crossover and will be less frequent. If the order is II, the frequency of these two recombinant genotypes will be reversed. The reciprocal cross is made in order to confirm the order deduced in the first cross.

reflects the fact that the φX174 genome is a circular DNA molecule. The map indicates that noncomplementing mutations occupy contiguous portions of the genome, demonstrating that the functional units identified by the complementation test are also structural units of the genome. The boundaries between genes are shown in an arbitrary manner on the map, since the mutations used to construct the map do not identify these boundaries.

The Nucleotide Sequence of φX174

The genome of φX174 comprises 5375 nucleotides in a single strand of circular DNA. The feat of determining the nucleotide sequence in the φX174 DNA molecule was accomplished in 1977 by Frederick Sanger and his colleagues. The sequence is shown in Figure 6.7. The correlation

Table 6.4
Three-factor recombination frequencies observed in crosses between ϕX174 mutants.*

Cross	Recombination Frequency (Selected Markers)	Percent *wt* (Unselected Markers)	Predominant Genotype (Unselected Markers)	Order Deduced or Confirmed (*H G F E D C B A H G*)
1. *amN1 ts*γ × *ts79*	1.2 ± 0.4	33	*am*	N1—79—γ
2. *amN1 ts79* × *ts*γ	0.8 ± 0.3	90	*wt*	N1—79—γ
3. *am88 ts*γ × *ts79*	1.0 ± 0.7	97	*wt*	79—γ–88
4. *am88 ts79* × *ts4*	1.8 ± 0.4	96	*wt*	4–79—88
5. *am88 ts79* × *op6*	13.4 ± 3.7	91	*wt*	79—88—6
6. *am88 ts79* × *am9*	1.4 ± 0.3	38	*ts*	79—9–88
7. *am3 ts79* × *ts*γ	0.7 ± 0.1	25	*am*	79—γ—3
8. *am3 ts*γ × *ts79*	0.9 ± 0.2	88	*wt*	79—γ—3
9. *am88 ts79* × *am3*	3.0 ± 0.3	99	*wt*	79—88—3
10. *am3 ts*γ × *am88*	4.3 ± 0.5	46	*ts*	γ–88—3
11. *am3 ts*γ × *op6*	1.2 ± 0.1	43	*ts*	γ—6–3
12. *am3 ts*γ × *am27*	0.3 ± 0.2	95	*wt*	γ—3–27
13. *am3 ts9* × *am27*	0.3 ± 0.2	37	*ts*	3–27—9
14. *am3 ts9* × *ts116*	2.3 ± 0.4	16	*am*	3—116—9
15. *am33 ts116* × *ts9*	1.5 ± 0.3	39	*am*	116—9–33
16. *am33 ts116* × *am50*	4.5 ± 0.6	89	*wt*	116—33–50
17. *am33 ts116* × *am86*	7.1 ± 2.2	77	*wt*	116—33–86
18. *am33 ts4* × *am16*	9.7 ± 1.5	95	*wt*	16–33—4
19. *am33 ts79* × *ts*γ	1.0 ± 0.4	79	*wt*	33—79—γ
20. *am33 ts*γ × *amN1*	8.3 ± 1.1	34	*ts*	33—N1—γ
21. *am9 ts128* × *amN1*	3.0 ± 0.3	7	*ts*	128—N1—9
22. *am33 ts*γ × *am86*	6.0 ± 0.8	43	*ts*	33–86—γ
23. *am9 ts128* × *am88*	1.6 ± 0.3	99	*wt*	128—9—88

24. *am88 ts79* × *och6*	10.8 ± 2.3	93	*wt*	79——88——6
25. *am3 tsγ* × *och6*	1.4 ± 0.3	95	*wt*	γ——3——6
26. *am3 ts9* × *och6*	1.0 ± 0.4	45	*ts*	3——6——9
27. *am3 och8* × *och6*	Low burst†	4	*am*	3——6——8
28. *am3 och11* × *och6*	Low burst	6	*am*	3——6——11
29. *am3 och6* × *och1*	Low burst	95	*wt*	3——6——1
30. *am3 och6* × *och5*	Low burst	98	*wt*	3——6——5
31. *am33 ts116* × *och6*	3.9 ± 0.4	21	*ts*	6——116——33
32. *am33 ts4* × *och6*	3.0 ± 1.2	97	*wt*	6——33——4
33. *am88 ts79* × *am10*	14.0 ± 0.5	99	*wt*	79——88——10
34. *am3 ts79* × *am10*	1.4 ± 0.7	90	*wt*	79——3——10
35. *am3 tsγ* × *am10*	1.3 ± 0.2	99	*wt*	γ——3——10
36. *am3 ts9* × *am10*	1.6 ± 0.1	42	*ts*	3——10——9
37. *am33 ts116* × *am35*	18.4 ± 1.2	55	Intermediate	116——(35, 33)
38. *am33 ts116* × *am18*	21.9 ± 6.8	56	Intermediate	116——(18, 33)

*Values are expressed × 10^4.
†Low burst means that very few progeny were produced in the cross.

From R. M. Benbow et al., *J. Virol.* 7:549 (1971).

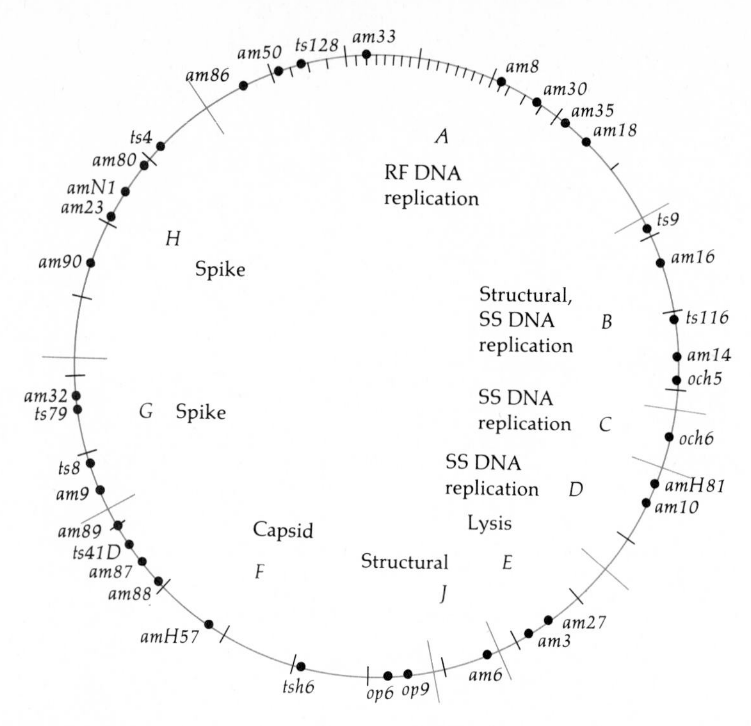

Figure 6.6
Genetic map of ϕX174. The frequency of wild-type recombinants in two-factor crosses is represented schematically. The distance between tick marks represents one map unit, defined as 10^{-4} wild-type recombinants per total progeny phage. Within cistron *A*, the tick marks on the outside of the circle represent recombination frequencies across cistron *A* measured by recombination of markers in other cistrons; tick marks on the inside of the circle represent recombination frequencies between cistron-*A* markers (see text for explanation). The boundaries between cistrons (colored lines) are arbitrary. The function of each cistron is indicated. [From R. M. Benbow et al., *J. Virol. 13*:898 (1974).]

Figure 6.7 (*opposite*)
The nucleotide sequence of the DNA of ϕX174 *cs70* and (above it) the amino acid sequences of the proteins for which it codes. The sequence of the circular DNA molecule is written starting after the termination codon of the H protein, but the numbering starts at the unique *Pst* I cleavage site. The letters shown in the left margin indicate the protein whose amino acid sequence is shown in the corresponding line. Restriction-enzyme recognition sites are indicated by underlining; they are defined as follows:

A	*Alu* I	H	*Hha* I	P	*Pst* I	R	*Hin* dII	Y	*Hap* II
F	*Hin* fI	M	*Mbo* II	Q	*Hph* I	T	*Taq* I	Z	*Hae* III

The *cs70* mutation is cold-sensitive and is believed to be a point mutation in the F protein. The K protein nucleotide sequence, shown nested within the A and C protein sequences, is indicated here through homology with the closely related phage G4. The K protein has been identified in cells infected by G4, but its presence in cells infected by ϕX174 has not been confirmed. G4 also produces proteins that are homologous with all of those identified in cells infected with ϕX174. [From F. Sanger et al., *J. Mol. Biol. 125*:225 (1978).]

```
CCGTCAGGATTGACACCCTCCCAATTGTATGTTTTCATGCCTCCAAATCTTGGAGGCTTT
     3927      3937      3947      3957      3967      3977
                                         ↑ mRNA start

A   MET VAL ARG SER TYR TYR PRO SER GLU CYS HIS ALA ASP TYR PHE ASP PHE GLU ARG
TTTATGGTTCGTTCTTATTACCCTTCTGAATGTCACGCTGATTATTTTGACTTTGAGCGT
     3987      3997      4007      4017      4027      4037
   ↑ mRNA end

A ILE GLU ALA LEU LYS PRO ALA ILE GLU ALA CYS GLY ILE SER THR LEU SER GLN SER PRO
ATCGAGGCTCTTAAACCTGCTATTGAGGCTTGTGGCATTTCTACTCTTTCTCAATCCCCA
     4047      4057      4067      4077      4087      4097
 T1/6 (TCGA)

A MET LEU GLY PHE HIS LYS GLN MET ASP ASN ARG ILE LYS LEU LEU GLU GLU ILE LEU SER
ATGCTTGGCTTCCATAAGCAGATGGATAACCGCATCAAGCTCTTGGAAGAGATTCTGTCT
     4107      4117      4127      4137      4147      4157
 A7b/7a (AGCT)   M5/8 (GAAGA)   F5c/3 (GATTC)

A PHE ARG MET GLN GLY VAL GLU PHE ASP ASN GLY ASP MET TYR VAL ASP GLY HIS LYS ALA
TTTCGTATGCAGGGCGTTGAGTTCGATAATGGTGATATGTATGTTGACGGCCATAAGGCT
     4167      4177      4187      4197      4207      4217
 T6/2 (TCGA)   Q2/3a (GGTGA)   R4/3 (GTTGAC)   Z2/6b (GGCC)

A ALA SER ASP VAL ARG ASP GLU PHE VAL SER VAL THR GLU LYS LEU MET ASP GLU LEU ALA
GCTTCTGACGTTCGTGATGAGTTTGTATCTGTTACTGAGAAGTTAATGGATGAATTGGCA
     4227      4237      4247      4257      4267      4277

A GLN CYS TYR ASN VAL LEU PRO GLN LEU ASP ILE ASN ASN THR ILE ASP HIS ARG PRO GLU
CAATGCTACAATGTGCTCCCCCAACTTGATATTAATAACACTATAGACCACCGCCCCGAA
     4287      4297      4307      4317      4327      4337
            Origin of viral strand replication ↑

A GLY ASP GLU LYS TRP PHE LEU GLU ASN GLU LYS THR VAL THR GLN PHE CYS ARG LYS LEU
GGGGACGAAAAATGGTTTTTAGAGAACGAGAAGACGGTTACGCAGTTTTGCCGCAAGCTG
     4347      4357      4367      4377      4387      4397
 M8/6 (GAAGA)   A7a/4 (AGCT)

A ALA ALA GLU ARG PRO LEU LYS ASP ILE ARG ASP GLU TYR ASN TYR PRO LYS LYS LYS GLY
GCTGCTGAACGCCCTCTTAAGGATATTCGCGATGAGTATAATTACCCCAAAAAGAAAGGT
     4407      4417      4427      4437      4447      4457

A ILE LYS ASP GLU CYS SER ARG LEU LEU GLU ALA SER THR MET LYS SER ARG ARG GLY PHE
ATTAAGGATGAGTGTTCAAGATTGCTGGAGGCCTCCACTATGAAATCGCGTAGAGGCTTT
     4467      4477      4487      4497      4507      4517
 Z6b/6a (GGCC)

A ALA ILE GLN ARG LEU MET ASN ALA MET ARG GLN ALA HIS ALA ASP GLY TRP PHE ILE VAL
GCTATTCAGCGTTTGATGAATGCAATGCGACAGGCTCATGCTGATGGTTGGTTTATCGTT
     4527      4537      4547      4557      4567      4577
```

(continued)

Figure 6.7 (*continued*)

```
A    PHE ASP THR LEU THR LEU ALA ASP ASP ARG LEU GLU ALA PHE TYR ASP ASN PRO ASN ALA
     TTTGACACTCTCACGTTGGCTGACGACCGATTAGAGGCGTTTTATGATAATCCCAATGCT
     4587  4597  4607  4617  4627  4637

A    LEU ARG ASP TYR PHE ARG ASP ILE GLY ARG MET VAL LEU ALA ALA GLU GLY ARG LYS ALA
     TTGCGTGACTATTTTCGTGATATTGGTCGTATGGTTCTTGCTGCCGAGGGTCGCAAGGCT
     4647  4657  4667  4677  4687  4697

A    ASN ASP SER HIS ALA ASP CYS TYR GLN TYR PHE CYS VAL PRO GLU TYR GLY THR ALA ASN
     AATGATTCACACGCCGACTGCTATCAGTATTTTTGTGTGCCTGAGTATGGTACAGCTAAT
     F3/5a  4707  4717  4727  4737  4747  A4/11  4757

A    GLY ARG LEU HIS PHE HIS ALA VAL HIS PHE MET ARG THR LEU PRO THR GLY SER VAL ASP
     GGCCGTCTTCATTTCCATGCGGTGCACTTTATGCGGACACTTCCTACAGGTAGCGTTGAC
     Z6a/9  M6/1 4767  4777  4787  4797  4807  R3/8 4817

A    PRO ASN PHE GLY ARG ARG VAL ARG ASN ARG ARG GLN LEU ASN SER LEU GLN ASN THR TRP
     CCTAATTTTGGTCGTCGGGTACGCAATCGCCGCCAGTTAAATAGCTTGCAAAATACGTGG
     4827  4837  4847  4857  A11/10  4867  4877  Z9/10

A    PRO TYR GLY TYR SER MET PRO ILE ALA VAL ARG TYR THR GLN ASP ALA PHE SER ARG SER
     CCTTATGGTTACAGTATGCCC↑ATCGCAGTTCGCTACACGCAGGACGCTTTTTCACGTTCT
     4887  4897  4907  4917  4927  4937
     mRNA start

A    GLY TRP LEU TRP PRO VAL ASP ALA LYS GLY GLU PRO LEU LYS ALA THR SER TYR MET ALA
     GGTTGGTTGTGGCCTGTTGATGCTAAAGGTGAGCCGCTTAAAGCTACCAGTTATATGGCT
     4947  Z10/3  4957  4967  Q3a/1  4977  A10/12b  4987  4997

A    VAL GLY PHE TYR VAL ALA LYS TYR VAL ASN LYS LYS SER ASP MET ASP LEU ALA ALA LYS
     GTTGGTTTCTATGTGGCTAAATACGTTAACAAAAAGTCAGATATGGACCTTGCTGCTAAA
     5007  5017  5027  R8/5  5037  5047  5057

A    GLY LEU GLY ALA LYS GLU TRP ASN ASN SER LEU LYS THR LYS LEU SER LEU LEU PRO LYS
B                            MET GLU GLN LEU THR LYS ASN GLN ALA VAL ALA THR SER GLN GLU
     GGTCTAGGAGCTAAAGAATGGAACAACTCACTAAAAACCAAGCTGTCGCTACTTCCCAAG
     5067  A12b/15b  5077  5087  5097  A15b/17  5107  5117

A    LYS LEU PHE ARG ILE ARG MET SER ARG ASN PHE GLY MET LYS MET LEU THR MET THR ASN
B      ALA VAL GLN ASN GLN ASN GLU PRO GLN LEU ARG ASP GLU ASN ALA HIS ASN ASP LYS SER
     AAGCTGTTCAGAATCAGAATGAGCCGCAACTTCGGGATGAAAATGCTCACAATGACAAAT
     A17/12a  5127  F5a/6  5137  5147  5157  5167  5177
```

```
A     LEU   SER   THR   GLU   CYS   LEU   ILE   GLN   LEU   THR   LYS   LEU   GLY   TYR   ASP   ALA   THR   PRO   PHE   ASN
B        VAL   HIS   GLY   VAL   LEU   ASN   PRO   THR   TYR   GLN   ALA   GLY   LEU   ARG   ARG   ASP   ALA   VAL   GLN   PRO
      C T G T C C A C G G A G T G C T T A A T C C A A C T T A C C A A G C T G G G T T A C G A C G C G A C G C C G T T C A A C
                5187                5197                5207      A12a/5    5217                5227                5237

A     GLN   ILE   LEU   LYS   GLN   ASN   ALA   LYS   ARG   GLU   MET   ARG   LEU   ARG   LEU   GLY   LYS   VAL   THR   VAL
B        ASP   ILE   GLU   ALA   GLU   ARG   LYS   LYS   ARG   ASP   GLU   ILE   GLU   ALA   GLY   LYS   SER   TYR   CYS   SER
      C A G A T A T T G A A G C A G A A C G C A A A A A G A G A G A T G A G A T T G A G G C T G G G A A A A G T T A C T G T A
                5247                5257                5267                5277                5287                5297

A     ALA   ASP   VAL   LEU   ALA   ALA   GLN   PRO   VAL   THR   THR   ASN   LEU   LEU   LYS   PHE   MET   ARG   ALA   SER
B        ARG   ARG   PHE   GLY   GLY   ALA   THR   CYS   ASP   ASP   LYS   SER   ALA   GLN   ILE   TYR   ALA   ARG   PHE   ASP
      G C C G A C G T T T G G C G G C G C A A C C T G T G A C G A C A A A T C T G C T C A A A T T T A T G C G C G C T T C G
                5307          H1/15 5317                5327                5337                5347    H15/8b     5357 T2/7

A     ILE   LYS   MET   ILE   GLY   VAL   SER   ASN   LEU   GLN   SER   PHE   ILE   ALA   SER   MET   THR   GLN   LYS   LEU
B        LYS   ASN   ASP   TRP   ARG   ILE   GLN   PRO   ALA   GLU   PHE   TYR   ARG   PHE   HIS   ASP   ALA   GLU   VAL   ASN
      A T A A A A A T G A T T G G C G T A T C C A A C C T G C A G A G T T T T A T C G C T T C C A T G A C G C A G A A G T T A
                5367                5377          P1/1    1                  11                  21          R5/7b    31

A     THR   LEU   SER   ASP   ILE   SER   ASP   GLU   SER   LYS   ASN   TYR   LEU   ASP   LYS   ALA   GLY   ILE   THR   THR
K                                        MET   SER   ARG   LYS   ILE   ILE   LEU   ILE   LYS   GLN   GLU   LEU   LEU   LEU
B        THR   PHE   GLY   TYR   PHE   ***
      A C A C T T T C G G A T A T T T C T G A T G A G T C G A A A A A T T A T C T T G A T A A A G C A G G A A T T A C T A C T
                 41                  51   F6/9  T7/8   61                  71                  81                  91

A     ALA   CYS   LEU   ARG   ILE   LYS   SER   LYS   TRP   THR   ALA   GLY   GLY   LYS   ***
K        LEU   VAL   TYR   GLU   LEU   ASN   ARG   SER   GLY   LEU   LEU   ALA   GLU   ASN   GLU   LYS   ILE   ARG   PRO   ILE
C                                                                                  MET   ARG   LYS   PHE   ASP   LEU   SER
      G C T T G T T T A C G A A T T A A A T C G A A G T G G A C T G C T G G C G G A A A A T G A G A A A A T T C G A C C T A T
                 101                 111 T8/9            121                 131                 141   T9/10       151

K        LEU   ALA   GLN   LEU   GLU   LYS   LEU   LEU   LEU   CYS   ASP   LEU   SER   PRO   SER   THR   ASN   ASP   SER   VAL
C        LEU   ARG   SER   SER   ARG   SER   SER   TYR   PHE   ALA   THR   PHE   ARG   HIS   GLN   LEU   THR   ILE   LEU   SER
      C C T T G C G C A G C T C G A G A A G C T C T T A C T T T G C G A C C T T T C G C C A T C A A C T A A C G A T T C T G T
          H8b/4  161 A5/18  T10/4  171 A18/6            181                 191                 201    F9/13      211

K        LYS   ASN   ***
C        LYS   THR   ASP   ALA   LEU   ASP   GLU   GLU   LYS   TRP   LEU   ASN   MET   LEU   GLY   THR   PHE   VAL   LYS   ASP
      C A A A A A C T G A C G C G T T G G A T G A G G A G A A G T G G C T T A A T A T G C T T G G C A C G T T C G T C A A G G
                 221                 231                 241                 251                 261                 271

C        TRP   PHE   ARG   TYR   GLU   SER   HIS   PHE   VAL   HIS   GLY   ARG   ASP   SER   LEU   VAL   ASP   ILE   LEU   LYS
      A C T G G T T T A G A T A T G A G T C A C A T T T T G T T C A T G G T A G A G A T T C T C T T G T T G A C A T T T T A A
                 281     F13/17  291                 301                 311 F17/16a         321 R7b/6c          331
```

(continued)

Figure 6.7 (*continued*)

```
D                                                                                                  MET
C      GLU  ARG  GLY  LEU  LEU  SER  GLU  SER  ASP  ALA  VAL  GLN  PRO  LEU  ILE  GLY  LYS  LYS  SER  ***
    A A G A G C G T G G A T T A C T A T C T G A G T C C G A T G C T G T T C A A C C A C T A A T A G G T A A G A A A T C A T
               341                 351  F16a/16b ↑    361                 371                 381                 391
                                        mRNA start

D     SER  GLN  VAL  THR  GLU  GLN  SER  VAL  ARG  PHE  GLN  THR  ALA  LEU  ALA  SER  ILE  LYS  LEU  ILE
    G A G T C A A G T T A C T G A A C A A T C C G T A C G T T T C C A G A C C G C T T T G G C C T C T A T T A A G C T C A T
    F16b/1     401                 411                 421                 431   Z3/7      441        A6/1     451

D     GLN  ALA  SER  ALA  VAL  LEU  ASP  LEU  THR  GLU  ASP  ASP  PHE  ASP  PHE  LEU  THR  SER  ASN  LYS
    T C A G G C T T C T G C C G T T T T G G A T T T A A C C G A A G A T G A T T T C G A T T T T C T G A C G A G T A A C A A
               461                 471                 481 M1/7            491                 501                 511
                                                                           T4/5

D     VAL  TRP  ILE  ALA  THR  ASP  ARG  SER  ARG  ALA  ARG  ARG  CYS  VAL  GLU  ALA  CYS  VAL  TYR  GLY
E                                                                                                MET  VAL
    A G T T T G G A T T G C T A C T G A C C G C T C T C G T G C T C G T C G C T G C G T T G A G G C T T G C G T T T A T G G
               521                 531                 541                 551                 561                 571

D     THR  LEU  ASP  PHE  VAL  GLY  TYR  PRO  ARG  PHE  PRO  ALA  PRO  VAL  GLU  PHE  ILE  ALA  ALA  VAL
E       ARG  TRP  THR  LEU  TRP  ASP  THR  LEU  ALA  PHE  LEU  LEU  LEU  LEU  SER  LEU  LEU  LEU  PRO  SER
    T A C G C T G G A C T T T G T G G G A T A C C C T C G C T T T C C T G C T C C T G T T G A G T T T A T T G C T G C C G T
               581                 591                 601                 611                 621                 631

D     ILE  ALA  TYR  TYR  VAL  HIS  PRO  VAL  ASN  ILE  GLN  THR  ALA  CYS  LEU  ILE  MET  GLU  GLY  ALA
E       LEU  LEU  ILE  MET  PHE  ILE  PRO  SER  THR  PHE  LYS  ARG  PRO  VAL  SER  SER  TRP  LYS  ALA  LEU
    C A T T G C T T A T T A T G T T C A T C C C G T C A A C A T T C A A A C G G C C T G T C T C A T C A T G G A A G G C G C
               641                 651     R6c/7a      661                671                 681                 691
                                                                       Z7/5                                    H4/13

D     GLU  PHE  THR  GLU  ASN  ILE  ILE  ASN  GLY  VAL  GLU  ARG  PRO  VAL  LYS  ALA  ALA  GLU  LEU  PHE
E       ASN  LEU  ARG  LYS  THR  LEU  LEU  MET  ALA  SER  SER  VAL  ARG  LEU  LYS  PRO  LEU  ASN  CYS  SER
    T G A A T T T A C G G A A A A C A T T A T T A A T G G C G T C G A G C G T C C G G T T A A A G C C G C T G A A T T G T T
               701                 711                 721                 731                 741                 751
                                                         T5/3                Y1/3

D     ALA  PHE  THR  LEU  ARG  VAL  ARG  ALA  GLY  ASN  THR  ASP  VAL  LEU  THR  ASP  ALA  GLU  GLU  ASN
E       ARG  LEU  PRO  CYS  VAL  TYR  ALA  GLN  GLU  THR  LEU  THR  PHE  LEU  LEU  THR  GLN  LYS  LYS  THR
    C G C G T T T A C C T T G C G T G T A C G C G C A G G A A A C A C T G A C G T T C T T A C T G A C G C A G A A G A A A A
               761                 771 H13/11          781                 791                 801        M7/3     811

J                                                                      MET  .SER  LYS  GLY  LYS  LYS  ARG  SER
D     VAL  ARG  GLN  LYS  LEU  ARG  ALA  GLU  GLY  VAL  MET  ***
E       CYS  VAL  LYS  ASN  TYR  VAL  ARG  LYS  GLU  ***
    C G T G C G T C A A A A A T T A C G T G C G G A A G G A G T G A T G T A A T G T C T A A A G G T A A A A A A C G T T C T
               821                 831                 841                 851                 861                 871
```

```
J   GLY   ALA   ARG   PRO   GLY   ARG   PRO   GLN   PRO   LEU   ARG   GLY   THR   LYS   GLY   LYS   ARG   LYS   GLY   ALA
    G G C G C T C G C C C T G G T C G T C C G C A G C C G T T G C G A G G T A C T A A A G G C A A G C G T A A A G G C G C T
     H11/14     881                 891                 901                 911                 921          H14/12  931

J   ARG   LEU   TRP   TYR   VAL   GLY   GLY   GLN   GLN   PHE   ***
    C G T C T T T G G T A T G T A G G T G G T C A A C A A T T T T A A T T G C A G G G G C T T C G G C C C C T T A C T T G A
                941                 951 R7a/6b          961                 971                 981                 991
                                                                                                   Z5/8     Minor mRNA end

F                     MET   SER   ASN   ILE   GLN   THR   GLY   ALA   GLU   ARG   MET   PRO   HIS   ASP   LEU   SER   HIS
    G G A T A A A T T A T G T C T A A T A T T C A A A C T G G C G C C G A G C G T A T G C C G C A T G A C C T T T C C C A T
                1001                1011                1021                1031                1041                1051
                                                         H12/10

F   LEU   GLY   PHE   LEU   ALA   GLY   GLN   ILE   GLY   ARG   LEU   ILE   THR   ILE   SER   THR   THR   PRO   VAL   ILE
    C T T G G C T T C C T T G C T G G T C A G A T T G G T C G T C T T A T T A C C A T T T C A A C T A C T C C G G T T A T C
                1061                1071                1081                1091                1101     Y3/2       1111

F   ALA   GLY   ASP   SER   PHE   GLU   MET   ASP   ALA   VAL   GLY   ALA   LEU   ARG   LEU   SER   PRO   LEU   ARG   ARG
    G C T G G C G A C T C C T T C G A G A T G G A C G C C G T T G G C G C T C T C C G T C T T T C T C C A T T G C G T C G T
                1121                1131                1141                1151                1161                1171
                F1/14b        T3/1                            H10/7

F   GLY   LEU   ALA   ILE   ASP   SER   THR   VAL   ASP   ILE   PHE   THR   PHE   TYR   VAL   PRO   HIS   ARG   HIS   VAL
    G G C C T T G C T A T T G A C T C T A C T G T A G A C A T T T T T A C T T T T T A T G T C C C T C A T C G T C A C G T T
     Z8/4       1181          F14b/2  1191                1201                1211                1221                1231

F   TYR   GLY   GLU   GLN   TRP   ILE   LYS   PHE   MET   LYS   ASP   GLY   VAL   ASN   ALA   THR   PRO   LEU   PRO   THR
    T A T G G T G A A C A G T G G A T T A A G T T C A T G A A G G A T G G T G T T A A T G C C A C T C C T C T C C C G A C T
             Q1/3c      1241                1251                1261                1271                1281                1291

F   VAL   ASN   THR   THR   GLY   TYR   ILE   ASP   HIS   ALA   ALA   PHE   LEU   GLY   THR   ILE   ASN   PRO   ASP   THR
    G T T A A C A C T A C T G G T T A T A T T G A C C A T G C C G C T T T T C T T G G C A C G A T T A A C C C T G A T A C C
     R6b/1      1301                1311                1321                1331                1341                1351

F   ASN   LYS   ILE   PRO   LYS   HIS   LEU   PHE   GLN   GLY   TYR   LEU   ASN   ILE   TYR   ASN   ASN   TYR   PHE   LYS
    A A T A A A A T C C C T A A G C A T T T G T T T C A G G G T T A T T T G A A T A T C T A T A A C A A C T A T T T T A A A
                1361                1371                1381                1391                1401                1411

F   ALA   PRO   TRP   MET   PRO   ASP   ARG   THR   GLU   ALA   ASN   PRO   ASN   GLU   LEU   ASN   GLN   ASP   ASP   ALA
    G C G C C G T G G A T G C C T G A C C G T A C C G A G G C T A A C C C T A A T G A G C T T A A T C A A G A T G A T G C T
     H7/5       1421                1431                1441                1451   A1/12c       1461                1471

F   ARG   TYR   GLY   PHE   ARG   CYS   CYS   HIS   LEU   LYS   ASN   ILE   TRP   THR   ALA   PRO   LEU   PRO   PRO   GLU
    C G T T A T G G T T T C C G T T G C T G C C A T C T C A A A A A C A T T T G G A C T G C T C C G C T T C C T C C T G A G
                1481                1491                1501                1511                1521                1531
```

(continued)

Figure 6.7 (*continued*)

```
F   THR   GLU   LEU   SER   ARG   GLN   MET   THR   THR   SER   THR   THR   SER   ILE   ASP   ILE   MET   GLY   LEU   GLN
    A C T G A G C T T T C T C G C C A A A T G A C G A C T T C T A C C A C A T C T A T T G A C A T T A T G G G T C T G C A A
       A12c/13 1541                 1551                1561                1571                1581                1591

F   ALA   ALA   TYR   ALA   ASN   LEU   HIS   THR   ASP   GLN   GLU   ARG   ASP   TYR   PHE   MET   GLN   ARG   TYR   HIS
    G C T G C T T A T G C T A A T T T G C A T A C T G A C C A A G A A C G T G A T T A C T T C A T G C A G C G T T A C C A T
    A13/2         1601                 1611                1621                1631                1641                1651

F   ASP   VAL   ILE   SER   SER   PHE   GLY   GLY   LYS   THR   SER   TYR   ASP   ALA   ASP   ASN   ARG   PRO   LEU   LEU
    G A T G T T A T T T C T T C A T T T G G A G G T A A A A C C T C T T A T G A C G C T G A C A A C C G T C C T T T A C T T
                1661 M3/4           1671                1681                1691                1701                1711

F   VAL   MET   ARG   SER   ASN   LEU   TRP   ALA   SER   GLY   TYR   ASP   VAL   ASP   GLY   THR   ASP   GLN   THR   SER
    G T C A T G C G C T C T A A T C T C T G G G C A T C T G G C T A T G A T G T T G A T G G A A C T G A C C A A A C G T C G
          H5/9a 1721                1731                1741                1751                1761                1771

F   LEU   GLY   GLN   PHE   SER   GLY   ARG   VAL   GLN   GLN   THR   TYR   LYS   HIS   SER   VAL   PRO   ARG   PHE   PHE
    T T A G G C C A G T T T T C T G G T C G T G T T C A A C A G A C C T A T A A A C A T T C T G T G C C G C G T T T C T T T
          Z4/1    1781                1791                1801                1811                1821                1831

F   VAL   PRO   GLU   HIS   GLY   THR   MET   PHE   THR   LEU   ALA   LEU   VAL   ARG   PHE   PRO   PRO   THR   ALA   THR
    G T T C C T G A G C A T G G C A C T A T G T T T A C T C T T G C G C T T G T T C G T T T T C C G C C T A C T G C G A C T
                  1841                1851                1861 H9a/8a         1871                1881                1891

F   LYS   GLU   ILE   GLN   TYR   LEU   ASN   ALA   LYS   GLY   ALA   LEU   THR   TYR   THR   ASP   ILE   ALA   GLY   ASP
    A A A G A G A T T C A G T A C C T T A A C G C T A A A G G T G C T T T G A C T T A T A C C G A T A T T G C T G G C G A C
                  1901                1911                1921                1931                1941                1951
                F2/11

F   PRO   VAL   LEU   TYR   GLY   ASN   LEU   PRO   PRO   ARG   GLU   ILE   SER   MET   LYS   ASP   VAL   PHE   ARG   SER
    C C T G T T T T G T A T G G C A A C T T G C C G C C G C G T G A A A T T T C T A T G A A G G A T G T T T T C C G T T C T
                  1961                1971                1981                1991                2001                2011

F   GLY   ASP   SER   SER   LYS   LYS   PHE   LYS   ILE   ALA   GLU   GLY   GLN   TRP   TYR   ARG   TYR   ALA   PRO   SER
    G G T G A T T C G T C T A A G A A G T T T A A G A T T G C T G A G G G T C A G T G G T A T C G T T A T G C G C C T T C G
     Q3c/6  F11/7   2021                2031                2041                2051                2061  H8a/6         2071

F   TYR   VAL   SER   PRO   ALA   TYR   HIS   LEU   LEU   GLU   GLY   PHE   PRO   PHE   ILE   GLN   GLU   PRO   PRO   SER
    T A T G T T T C T C C T G C T T A T C A C C T T C T T G A A G G C T T C C C A T T C A T T C A G G A A C C G C C T T C T
                  2081                2091                2101                2111                2121                2131
                                      Q6/5

F   GLY   ASP   LEU   GLN   GLU   ARG   VAL   LEU   ILE   ARG   HIS   HIS   ASP   TYR   ASP   GLN   CYS   PHE   GLN   SER
    G G T G A T T T G C A A G A A C G C G T A C T T A T T C G C C A C C A T G A T T A T G A C C A G T G T T T C C A G T C C
     Q5/3b        2141                2151                2161                2171                2181                2191
```

```
F    VAL   GLN   LEU   LEU   GLN   TRP   ASN   SER   GLN   VAL   LYS   PHE   ASN   VAL   THR   VAL   TYR   ARG   ASN   LEU
     G T T C A G T T G T T G C A G T G G A A T A G T C A G G T T A A A T T T A A T G T G A C C G T T T A T C G C A A T C T G
           2201                2211                2221                2231                2241                2251

F    PRO   THR   THR   ARG   ASP   SER   ILE   MET   THR   SER   ***
     C C G A C C A C T C G C G A T T C A A T C A T G A C T T C G T G A T A A A A G A T T G A G T G T G A G G T T A T A A C G
           2261      F7/5b     2271                2281                2291                2301                2311

     C C G A A G C G G T A A A A A T T T T A A T T T T T G C C G C T G A G G G G T T G A C C A A G C G A A G C G C G G T A G
           2321                2331                2341                2351 R1/9           2361   H6/3         2371

G                                                  MET   PHE   GLN   THR   PHE   ILE   SER   ARG   HIS   ASN   SER   ASN   PHE
     G T T T T C T G C T T A G G A G T T T A A T C A T G T T T C A G A C T T T T A T T T C T C G C C A T A A T T C A A A C T
           2381                2391                2401                2411                2421                2431

G       PHE   SER   ASP   LYS   LEU   VAL   LEU   THR   SER   VAL   THR   PRO   ALA   SER   SER   ALA   PRO   VAL   LEU   GLN
     T T T T T T C T G A T A A G C T G G T T C T C A C T T C T G T T A C T C C A G C T T C T T C G G C A C C T G T T T T A C
           2441      A2/16     2451                2461                2471                2481                2491
                                                                       A16/15a   M4/10

G       THR   PRO   LYS   ALA   THR   SER   SER   THR   LEU   TYR   PHE   ASP   SER   LEU   THR   VAL   ASN   ALA   GLY   ASN
     A G A C A C C T A A A G C T A C A T C G T C A A C G T T A T A T T T T G A T A G T T T G A C G G T T A A T G C T G G T A
           2501   A15a/3       2511   R9/10        2521                2531                2541                2551

G       GLY   GLY   PHE   LEU   HIS   CYS   ILE   GLN   MET   ASP   THR   SER   VAL   ASN   ALA   ALA   ASN   GLN   VAL   VAL
     A T G G T G G T T T T C T T C A T T G C A T T C A G A T G G A T A C A T C T G T C A A C G C C G C T A A T C A G G T T G
           2561      M10/9     2571                2581                2591 R10/2          2601                2611

G       SER   VAL   GLY   ALA   ASP   ILE   ALA   PHE   ASP   ALA   ASP   PRO   LYS   PHE   PHE   ALA   CYS   LEU   VAL   ARG
     T T T C T G T T G G T G C T G A T A T T G C T T T T G A T G C C G A C C C T A A A T T T T T T G C C T G T T T G G T T C
           2621                2631                2641                2651                2661                2671

G       PHE   GLU   SER   SER   SER   VAL   PRO   THR   THR   LEU   PRO   THR   ALA   TYR   ASP   VAL   TYR   PRO   LEU   ASN
     G C T T T G A G T C T T C T T C G G T T C C G A C T A C C C T C C C G A C T G C C T A T G A T G T T T A T C C T T T G A
           2681      M9/2      2691                2701                2711                2721                2731
         F5b/8

G       GLY   ARG   HIS   ASP   GLY   GLY   TYR   TYR   THR   VAL   LYS   ASP   CYS   VAL   THR   ILE   ASP   VAL   LEU   PRO
     A T G G T C G C C A T G A T G G T G G T T A T T A T A C C G T C A A G G A C T G T G T G A C T A T T G A C G T C C T T C
           2741                2751                2761                2771                2781                2791

G       ARG   THR   PRO   GLY   ASN   ASN   VAL   TYR   VAL   GLY   PHE   MET   VAL   TRP   SER   ASN   PHE   THR   ALA   THR
     C C C G T A C G C C G G G C A A T A A C G T T T A T G T T G G T T T C A T G G T T T G G T C T A A C T T T A C C G C T A
           2801                2811                2821                2831                2841                2851
             Y2/5
```

(continued)

Figure 6.7 (*continued*)

```
G      LYS  CYS  ARG  GLY  LEU  VAL  SER  LEU  ASN  GLN  VAL  ILE  LYS  GLU  ILE  ILE  CYS  LEU  GLN  PRO
     C T A A A T G C C G C G G A T T G G T T T C G C T G A A T C A G G T T A T T A A A G A G A T T A T T T G T C T C C A G C
                   2861                2871                2881                2891                2901                2911
                                                         F8/4

H                                                  MET  PHE  GLY  ALA  ILE  ALA  GLY  GLY  ILE  ALA  SER  ALA  LEU  ALA
G      LEU  LYS  ***
     C A C T T A A G T G A G G T G A T T T A T G T T T G G T G C T A T T G C T G G C G G T A T T G C T T C T G C T C T T G C
                   2921                2931                2941                2951                2961                2971
                         Q3b/4

H      GLY  GLY  ALA  MET  SER  LYS  LEU  PHE  GLY  GLY  GLY  GLN  LYS  ALA  ALA  SER  GLY  GLY  ILE  GLN
     T G G T G G C G C C A T G T C T A A A T T G T T T G G A G G C G G T C A A A A A G C C G C C T C C G G T G G C A T T C A
                   2981                2991                3001                3011                3021                3031
                 H3/2                                                                          Y5/4

H      GLY  ASP  VAL  LEU  ALA  THR  ASP  ASN  ASN  THR  VAL  GLY  MET  GLY  ASP  ALA  GLY  ILE  LYS  SER
     A G G T G A T G T G C T T G C T A C C G A T A A C A A T A C T G T A G G C A T G G G T G A T G C T G G T A T T A A A T C
                   3041                3051                3061                3071                3081                3091
        Q6/7                                                                  Q7/2

H      ALA  ILE  GLN  GLY  SER  ASN  VAL  PRO  ASN  PRO  ASP  GLU  ALA  ALA  PRO  SER  PHE  VAL  SER  GLY
     T G C C A T T C A A G G C T C T A A T G T T C C T A A C C C T G A T G A G G C C G C C C C T A G T T T T G T T T C T G G
                   3101                3111                3121                3131                3141                3151
                                                                            Z1/2

H      ALA  MET  ALA  LYS  ALA  GLY  LYS  GLY  LEU  LEU  GLU  GLY  THR  LEU  GLN  ALA  GLY  THR  SER  ALA
     T G C T A T G G C T A A A G C T G G T A A A G G A C T T C T T G A A G G T A C G T T G C A G G C T G G C A C T T C T G C
                   3161                3171                3181                3191                3201                3211
                         A3/9

H      VAL  SER  ASP  LYS  LEU  LEU  ASP  LEU  VAL  GLY  LEU  GLY  GLY  LYS  SER  ALA  ALA  ASP  LYS  GLY
     C G T T T C T G A T A A G T T G C T T G A T T T G G T T G G A C T T G G T G G C A A G T C T G C C G C T G A T A A A G G
                   3221                3231                3241                3251                3261                3271

H      LYS  ASP  THR  ARG  ASP  TYR  LEU  ALA  ALA  ALA  PHE  PRO  GLU  LEU  ASN  ALA  TRP  GLU  ARG  ALA
     A A A G G A T A C T C G T G A T T A T C T T G C T G C T G C A T T T C C T G A G C T T A A T G C T T G G G A G C G T G C
                   3281                3291                3301                3311                3321                3331
                                                                               A9/12d

H      GLY  ALA  ASP  ALA  SER  SER  ALA  GLY  MET  VAL  ASP  ALA  GLY  PHE  GLU  ASN  GLN  LYS  GLU  LEU
     T G G T G C T G A T G C T T C C T C T G C T G G T A T G G T T G A C G C C G G A T T T G A G A A T C A A A A A G A G C T
                   3341                3351                3361                3371                3381                3391
                                                            R2/6a        Y4/1             F4/14a              A12d/7c
```

```
H    THR LYS MET GLN LEU ASP ASN GLN LYS GLU ILE ALA GLU MET GLN ASN GLU THR GLN LYS
     T A C T A A A A T G C A A C T G G A C A A T C A G A A A G A G A T T G C C G A G A T G C A A A A T G A G A C T C A A A A
                   3401          3411          3421          3431          3441   F14a/12  3451

H    GLU ILE ALA GLY ILE GLN SER ALA THR SER ARG GLN ASN THR LYS ASP GLN VAL TYR ALA
     A G A G A T T G C T G G C A T T C A G T C G G C G A C T T C A C G C C A G A A T A C G A A A G A C C A G G T A T A T G C
                   3461          3471          3481          3491          3501          3511

H    GLN ASN GLU MET LEU ALA TYR GLN GLN LYS GLU SER THR ALA ARG VAL ALA SER ILE MET
     A C A A A A T G A G A T G C T T G C T T A T C A A C A G A A G G A G T C T A C T G C T C G C G T T G C G T C T A T T A T
                   3521          3531          3541   F12/1o  3551          3561          3571

H    GLU ASN THR ASN LEU SER LYS GLN GLN GLN VAL SER GLU ILE MET ARG GLN MET LEU THR
     G G A A A A C A C C A A T C T T T C C A A G C A A C A G C A G G T T T C C G A G A T T A T G C G C C A A A T G C T T A C
                   3581          3591          3601          3611     3621          3631
                                                                    H2/9b

H    GLN ALA GLN THR ALA GLY GLN TYR PHE THR ASN ASP GLN ILE LYS GLU MET THR ARG LYS
     T C A A G C T C A A A C G G C T G G T C A G T A T T T T A C C A A T G A C C A A A T C A A A G A A A T G A C T C G C A A
           A7c/8   3641          3651          3661          3671          3681   F10/15  3691

H    VAL SER ALA GLU VAL ASP LEU VAL HIS GLN GLN THR GLN ASN GLN ARG TYR GLY SER SER
     G G T T A G T G C T G A G G T T G A C T T A G T T C A T C A G C A A A C G C A G A A T C A G C G G T A T G G C T C T T C
                   3701   R6a/4  3711          3721          3731  F15/5c       3741        M2/5  3751

H    HIS ILE GLY ALA THR ALA LYS ASP ILE SER ASN VAL VAL THR ASP ALA ALA SER GLY VAL
     T C A T A T T G G C G C T A C T G C A A A G G A T A T T T C T A A T G T C G T C A C T G A T G C T G C T T C T G G T G T
                   3761          3771          3781          3791          3801          3811
                   H9b/1

H    VAL ASP ILE PHE HIS GLY ILE ASP LYS ALA VAL ALA ASP THR TRP ASN ASN PHE TRP LYS
     G G T T G A T A T T T T T C A T G G T A T T G A T A A A G C T G T T G C C G A T A C T T G G A A C A A T T T C T G G A A
                   3821          3831          3841          3851          3861          3871
                                               A8/14

H    ASP GLY LYS ALA ASP GLY ILE GLY SER ASN LEU SER ARG LYS ***
     A G A C G G T A A A G C T G A T G G T A T T G G C T C T A A T T T G T C T A G G A A A T A A
                   3881  A14/7b  3891          3901          3911
```

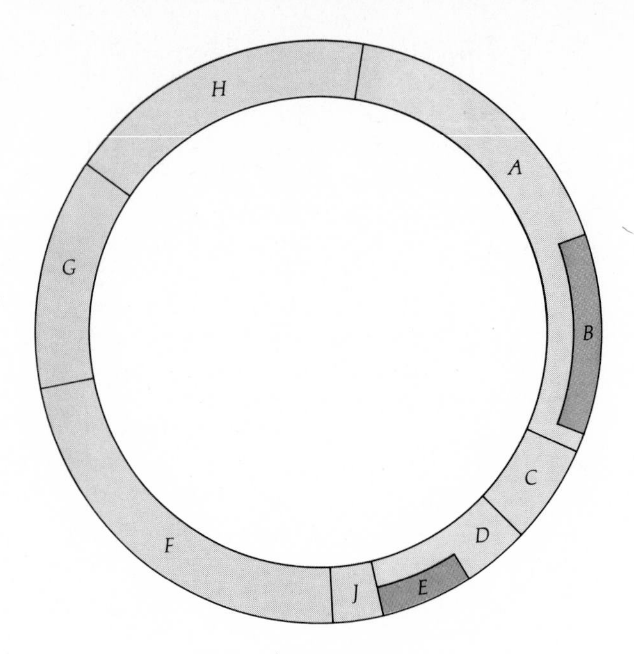

Figure 6.8
The physical genetic map of ϕX174. Note the nesting of cistron *B* in cistron *A*, and of *E* in *D*.

between the genetic map in Figure 6.6 and the chemical map in Figure 6.7 is good, with two exceptions: first, recombination frequencies do not correlate directly with physical distance, especially within the *A* gene; second, the boundaries between genes, in two cases, are not those that might be expected from the genetic map. The mutations used to construct the genetic map are located on the physical map in positions that fail to reveal that gene *B* is physically nested within the nucleotide sequence of gene *A*, and that gene *E* is likewise nested in gene *D* (Figure 6.8). The functional aspects of this remarkable finding will be discussed in Chapter 12. The answer to the question of whether "overlapping" genes is a general phenomenon or one restricted to genomes of very small size, such as that of ϕX174, awaits the sequencing of other DNA molecules. The lack of correlation between genetic map distances and physical distances will be discussed in Chapter 9. Finally, these and other studies have shown that the mutation *am6*, shown to be in gene *E* in Tables 6.2 and 6.3, is actually in a different, additional gene, *J*.

The Temperate Bacteriophage λ

Studies of the temperate bacteriophage λ have made major contributions to genetics. The λ genome consists of approximately 49,000 np; it is thus almost ten times the size of the ϕX174 genome. Phage λ is of great interest because of the complexity of its genetic regulatory mechanisms. When a temperate bacteriophage, such as λ, infects a sensitive bacterial cell, it can either follow a lytic course in which it replicates, produces many progeny, and destroys the host cell, or it can establish a lysogenic state in which the viral genome is integrated into the bacterial chromosome [*E. coli* (λ)] and becomes a largely passive portion of the bacterial genome. The viral genome resides there as a *prophage,* or *provirus,* and is replicated as a part of the host genome and passed on to daughter cells. The many phage genes, which are potentially lethal to the host, are inactive, or *repressed.* Thus, λ serves as a model genetic system for understanding virus-host interactions, and it has served as a paradigm for those mammalian tumor viruses capable of integration into the mammalian genome, such as polyoma and SV40. In this chapter the types of mutations found in λ will be discussed, along with the construction of genetic and physical maps of the genome. The expression and regulation of the λ genome will be deferred to Chapter 13.

λ Genes

As with other bacteriophages, plaque assay is the chief tool for identifying mutants of λ. The λ genes can be classified in two groups: essential and nonessential. The essential genes are identified by mutations that cause conditional lethality—either *ts* mutations or *sus* mutations, as discussed earlier. Complementation analysis of these mutations shows that they reside in 25 different complementation groups. Seven of these cistrons are required for normal phage heads to be formed, and eleven are required for the formation of normal phage tails. These genes either code for the structural proteins of the phage particle or are necessary for the proper assembly of the phage particle. Two genes are required for cell lysis and the release of progeny phage; two genes are necessary for λ DNA replication; the remaining three genes have essential regulatory roles.

Nonessential genes are identified by plaque morphology or by deletion of regions of the genome in which they reside. Normal λ phages make turbid plaques when plated with a sensitive indicator because some of the infected cells become lysogenic. Cells carrying the λ prophage are immune to further infection by λ phage and therefore grow within the plaque, giving it a turbid appearance. Some mutations (*clear, c*) cause the affected phages to produce clear plaques. These mutants are unable to lysogenize and must follow the lytic course of infection. *Clear* mutations occur in four different genes; *cI, cII, cIII,* and *cy.* Other nonessential genes are identified in phage mutants that carry large deletions and yet are capable of making plaques. Deletion mutations are easily detected in λ because phage particles carrying deletions have less than a normal amount of DNA in their heads and therefore have a different buoyant density in cesium chloride

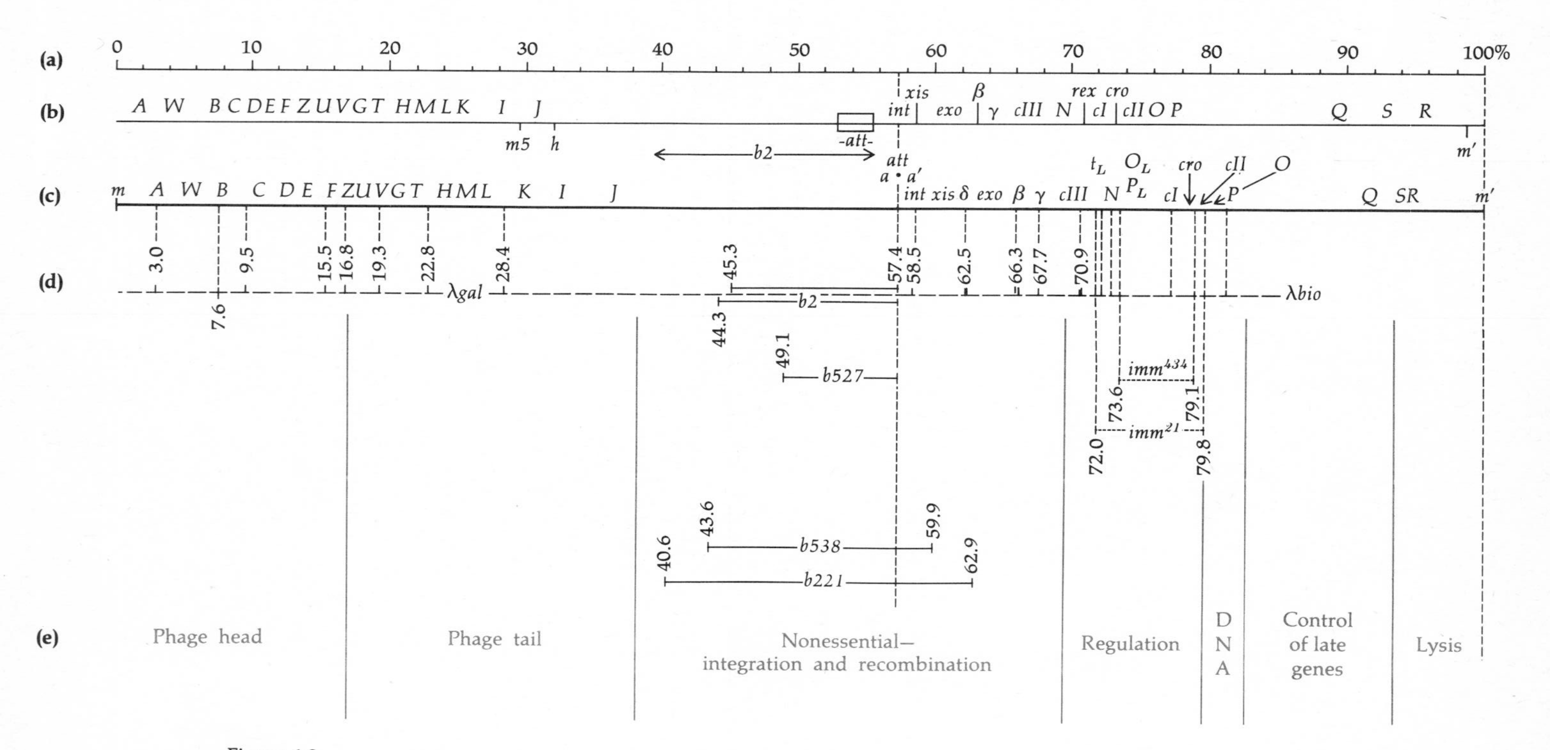

Figure 6.9
The λ genome. **(a)** Percentage scale of molecular distance. **(b)** The genetic map based on recombination frequencies. **(c)** The physical map based on heteroduplex formation. **(d)** Some of the *gal* and *bio* substitutions, *b* deletions, and lambdoid immunity substitutions used in heteroduplex mapping. **(e)** Distribution of genetic functions in the genome.

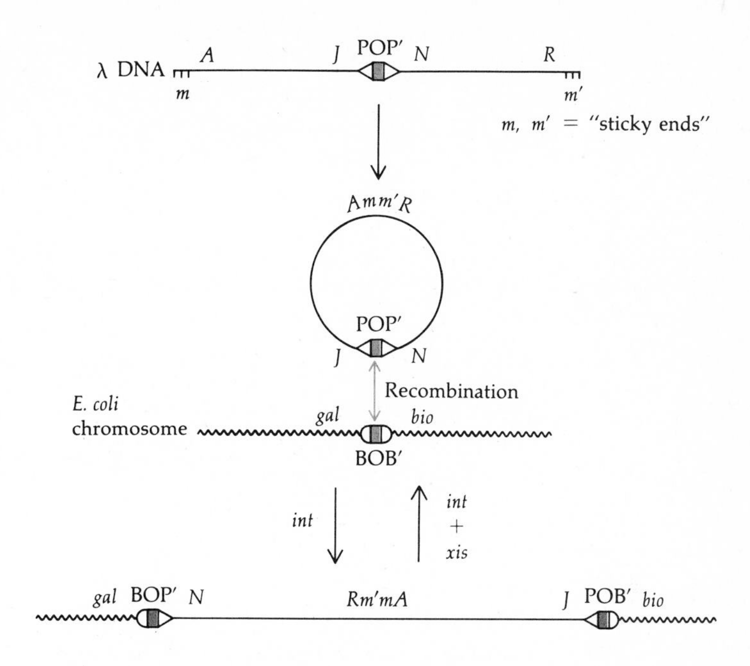

Figure 6.10
The lytic and prophage maps of λ, showing the mode of integraton of the circular λ genome into the host chromosome. The *attP* site is represented as POP′ and the *attB* site as BOB′.

density gradients. Such buoyant mutations are designated *b*. As much as 22% of the center of the λ DNA molecule may be lost without affecting the ability of the phage to make plaques. However, these deletion mutations are unable to lysogenize or to carry out genetic recombination, owing to the loss of a number of functions that are nonessential for plaque formation.

The genetic map constructed from recombination frequencies between mutants undergoing lytic phage reproduction is shown in Figure 6.9. The most striking feature of this map is that genes responsible for related physiological functions are grouped together, e.g., the genes for head formation, for tail formation, and for lysis. Another important feature of the λ genetic map is its linearity: the DNA isolated from λ phage particles is a linear double-stranded DNA molecule.

The λ Prophage

Mutant genes carried on a λ prophage integrated into the bacterial genome can be mapped, employing the techniques of bacterial genetics presented in the next chapter. These studies have shown that the λ prophage is integrated between the *E. coli* genes *gal* and *bio*, which control utilization of the sugar galactose and synthesis of the vitamin biotin, respectively. The order of the genes on the prophage genetic map is compared with that on the phage genetic map in Figure 6.10. Surprisingly, they are not identical:

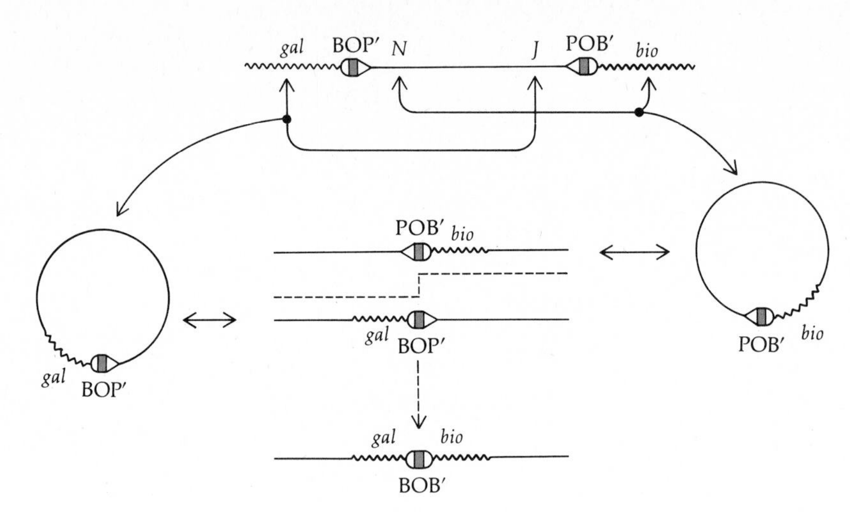

Figure 6.11
Formation of λ*gal* and λ*bio* transducing phages by aberrant excision. The *int*-mediated recombination between λ*gal* and λ*bio* transducing phages produces a hybrid λ*gal bio* phage carrying BOB′. POP′, POB′, BOP′, and BOB′ have been sequenced (see Figure 9.21).

one is a circular permutation of the other. This difference in gene order arises from the manner in which the λ DNA becomes inserted into the bacterial chromosome.

The linear DNA molecules present in phage particles possess complementary single-stranded ends consisting of twelve nucleotides. Upon entering the bacterial host cell, these "sticky ends" anneal, and the single-strand nicks present in the circularized molecule are closed covalently by the bacterial enzyme *DNA ligase* (also called *polynucleotide ligase*). The circular λ genome can then be inserted into the bacterial chromosome through the action of the phage *int* gene, which codes for a protein catalyzing a site-specific recombination event between the phage attachment site (*attP*) and the bacterial attachment site (*attB* or *att*λ). This event breaks the λ chromosome at the *attP* site (POP′), which is located in the middle of the phage genetic map, and thereby changes the order of the genes in the prophage, as indicated in Figure 6.10.

Once integrated, all but two λ genes remain inactive and under the control of a *repressor* protein coded by the *cI* gene. The λ prophage is replicated as part of the host chromosome and segregated to daugher bacterial cells. In a population of lysogenic bacterial cells, perhaps 1 in 10^6 cells per generation will enter a lytic cycle (owing to the spontaneous *induction* of the prophage), produce progeny λ phage, and lyse. This ability of a bacterial culture carrying a prophage to produce phage spontaneously is the origin of the term *lysogenic.* The cause of spontaneous induction is not clear. However, it is known that injury to the host's ability to replicate its own DNA will cause induction. For example, ultraviolet radiation will cause most of the cells in a culture to experience prophage induction. Temperature-sensitive mutants of the *cI* gene are known that cause induction at high temperature. These mutations provide a useful experimental means for the study of events accompanying induction.

Table 6.5
Recombination between some λ*dgal* strains and representative *sus* mutations in cistrons of the left arm of the λ map.*

	λ*dgal1*	λ*dgal2*	λ*dgal3*	λ*dgal4*	λ*dgal5*
λ*susA*	−	+	+	+	+
λ*susB*	−	−	+	+	+
λ*susE*	−	−	−	+	+
λ*susG*	−	−	−	−	+
λ*susH*	−	−	−	−	+
λ*susM*	−	−	−	−	−

*+ = wild-type recombinants observed; − = none observed.

Data modified from A. Campbell, *Virology* 9:293 (1959).

An early event associated with induction is the excision of the prophage from the host chromosome by a mechanism similar to its insertion. The excision process is controlled by two λ genes, *int* and *xis*, and results in the formation of a circular λ genome that first replicates through a θ mode to produce a number of progeny genomes. Later the mode of DNA replication changes to a σ mode to produce linear progeny-phage genomes that are encapsulated in head proteins (see Chapter 4).

The λ prophage is almost always excised in a precise manner, creating an intact λ genome in its circular form. Occasionally, however, an improper excision event occurs that leads to the creation of a circular DNA molecule in which a portion of the λ genome at one end of the prophage map is lost and is replaced by a portion of the *E. coli* chromosome contiguous to the other end of the prophage map (Figure 6.11). Depending upon whether essential genes have been lost from the λ genome, these hybrid molecules may or may not produce viable phage particles. Those molecules that possess the *E. coli bio* gene (λ*bio*) are generally viable since, usually, only nonessential genes between *attP* and gene *N* are lost. Those molecules that possess the *E. coli gal* gene (λ*gal*), on the other hand, are defective, owing to the loss of essential genes between *attP* and the left end of the chromosome. The defective λ*dgal* phage or λ*dbio* phage (if the essential gene *N* has been lost) can be propagated in the presence of normal λ phage, which complements the lost functions of these phages. λ*dgal* and λ*bio* phages are easily detected by their ability to transduce the gal^{+} or bio^{+} gene into gal^{-} or bio^{-} bacteria. These transducing phages can also form lysogens, e.g., *E. coli* (λ*dgal*).

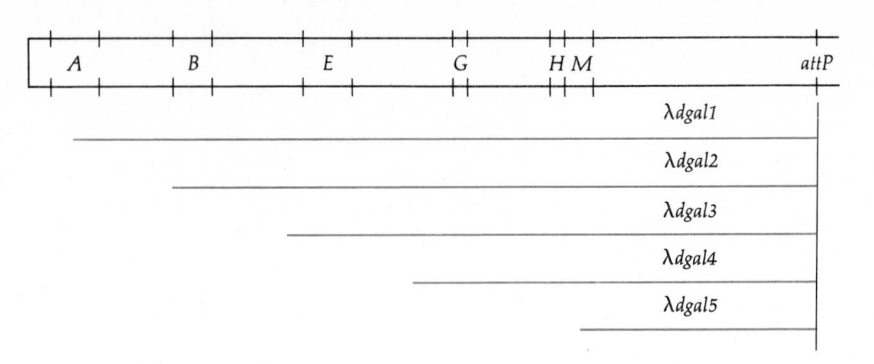

Figure 6.12
Placing the endpoints of *gal* substitutions on the λ genetic map is accomplished by seeking wild-type recombinants between λ*dgal* strains and known *sus* mutations. This map was deduced from the data in Table 6.5.

The extent of the substituted portion of the λ genome in individual λ*dgal* or λ*bio* strains can be ascertained by either recombination or complementation tests in which, for example, a λ*dgal* phage is tested for its ability to form wild-type recombinants when crossed to a number of different mutant phage strains, or to complement with known phage mutants in a mixed infection. The data in Table 6.5 score the presence or absence of wild-type phage recombinants formed between certain *sus* mutations in a number of cistrons and several λ*dgal* phages. If the λ*dgal* strain is deficient for the wild-type gene in question, no wild-type recombinants can be produced. These data allow the left endpoints of the *gal* substitutions to be placed with respect to the map positions of the *sus* mutations on the genetic map, as shown in Figure 6.12. The right endpoint of each substitution is always at the integration site, *attP.*

Conversely, when a set of mapped λ*dgal* phage strains is available, point mutations of unknown location can be rapidly mapped with respect to the left endpoints of the *gal* substitutions, simply by determining whether any wild-type recombinants can form between the unknown mutant and the λ*dgal* strains. (Such a mapping technique is analogous to the deletion mapping of *rII* mutations; see Chapter 8.)

Correlation of the Genetic and Physical Maps of λ

Electron microscopy has permitted precise correlations to be made between the genetic map of λ, constructed from recombination data, and the DNA molecule constituting the λ chromosome. This involves the use of the deletion (λ*b*) and substitution (λ*dgal*, λ*bio*) mutations described above. Closely related phages in the lambdoid family (e.g., phages 434, 82, 21), which share some genes with λ but which also contain regions

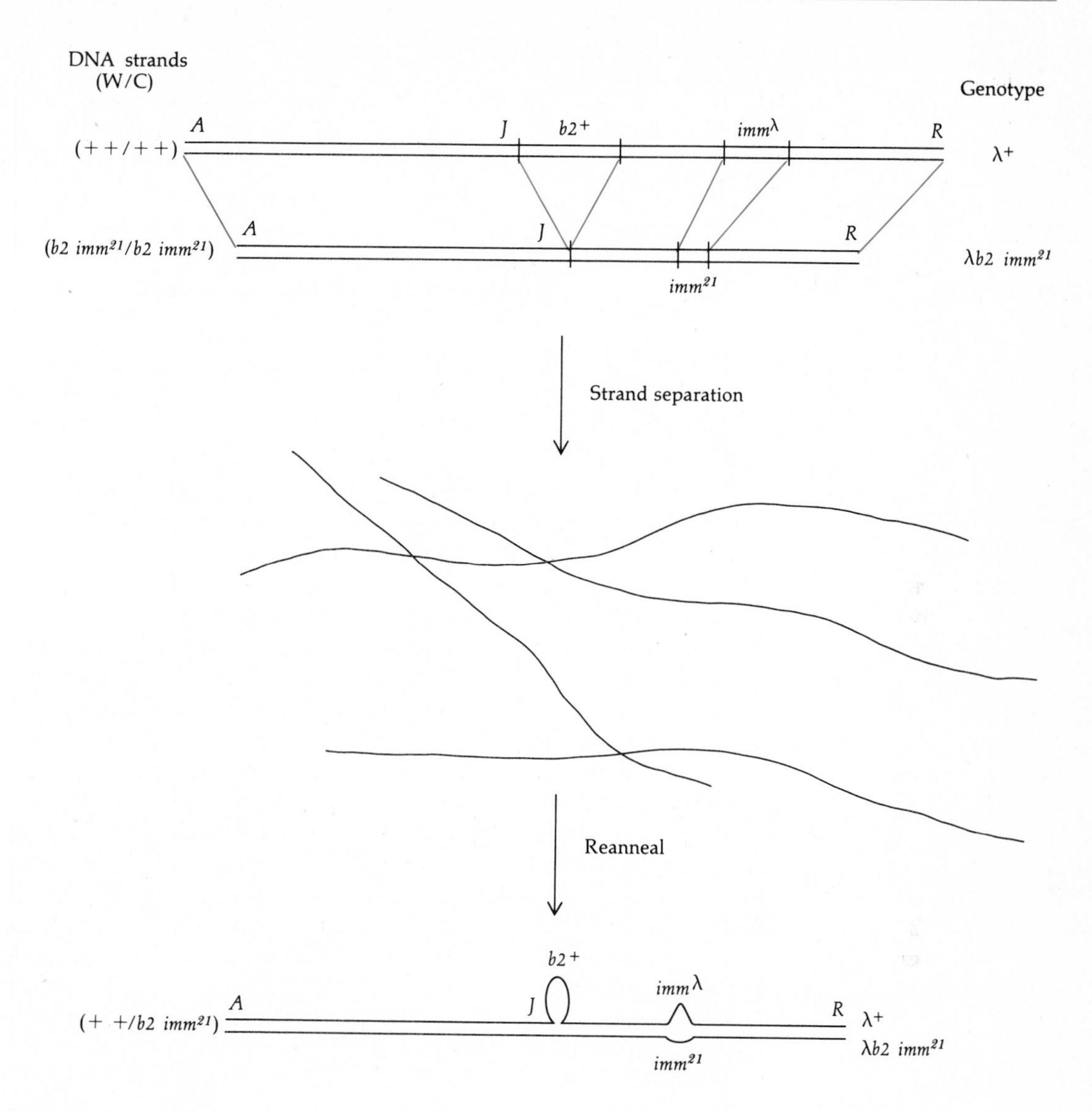

Figure 6.13
Schematic representation of heteroduplex formation between DNA single strands of two different λ DNA molecules differing in deletions and substitutions. The genome carrying the imm^{21} region substituted for the imm^{λ} region is shorter than the λ genome, since the imm^{21} region is shorter than the imm^{λ} region it replaces; this is called the *b5* deletion.

of genetic nonhomology, have also been useful. The formation of heteroduplex DNA molecules, in which each of the complementary strands has a different genetic origin, is diagrammed in Figure 6.13. Examination of heteroduplex molecules with the electron microscope permits the identification of complementary and noncomplementary regions of the molecules (Figure 6.14). Accurate, reproducible measurements of the lengths of the double-stranded portions of these heteroduplex DNA molecules allow the positions of substitution and deletion endpoints to be determined on the molecule. The correspondence of the ends of the DNA molecule with the right and left ends of the genetic map can be determined by heteroduplex formation between DNA single strands each carrying a

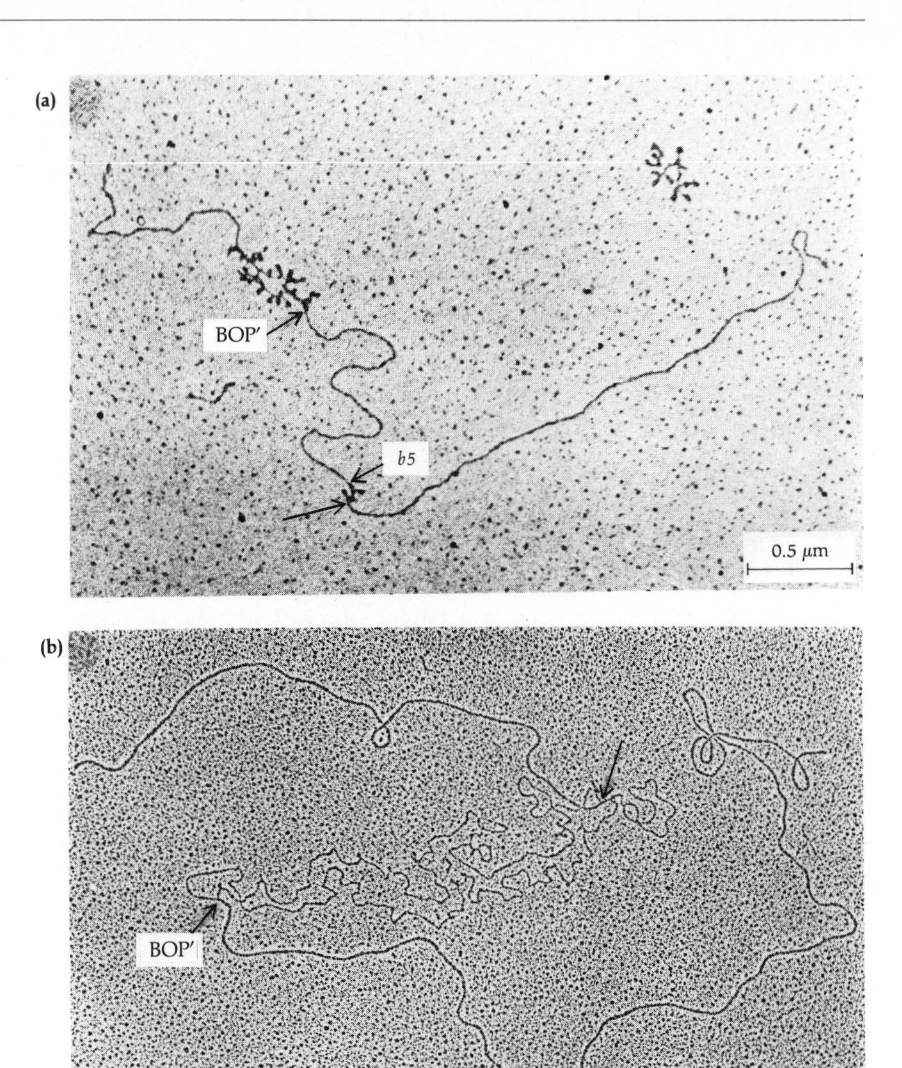

Figure 6.14
Electron micrographs of λ*gal*/λ heteroduplexes. **(a)** Heteroduplex of λ*dgalP73*/λ*b5* prepared in a way that shows single-stranded regions as blobs (see Figure 6.13 caption for a description of *b5*). **(b)** Heteroduplex of λ*dgalP74*/λ^+ prepared in a way that permits single-stranded regions to be observed clearly. Measurement of double-stranded regions provides physical distances of deletion endpoints from the ends of the molecule. The presence of the *b5* substitution in part a indicates that the right end of the molecule shown there is the right end of the genetic map, because *b5* maps to the right of the *gal* substitution. [From R. W. Davis and J. S. Parkinson, *J. Mol. Biol.* *56*:403 (1971).]

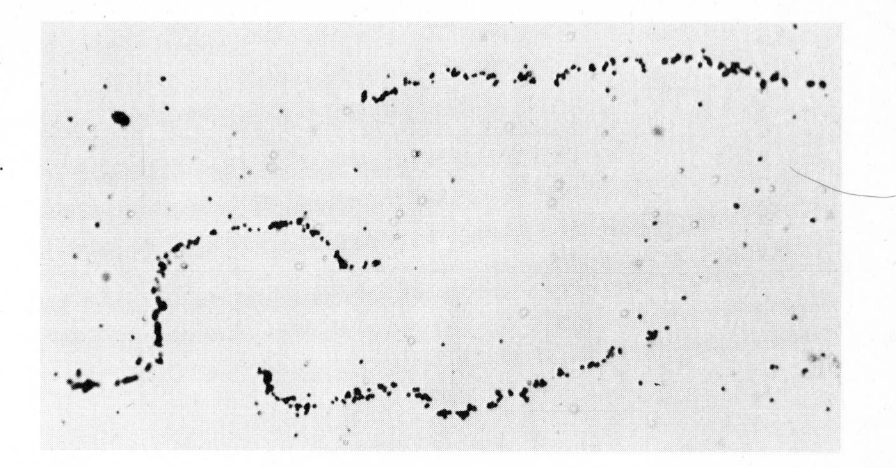

Figure 6.15
Autoradiograph of several T2 chromosomes. These molecules measure 52 μm and represent whole genomes. (Courtesy of Dr. John Cairns, Imperial Cancer Research Fund, London.)

genetically mapped deletion or substitution (Figure 6.13). Heteroduplex mapping has resulted in the physical map shown (along with the genetic map) in Figure 6.9. Since the λ DNA molecule contains approximately 49,000 np, the approximate size of each gene can be determined from the map.

The use of restriction enzymes (see Chapter 9) to cleave λ DNA at discrete sites, followed by purification of the discrete fragments thus produced, provides DNA molecules of a size that can be sequenced by present methods.

The Genomic Organization of Phages T2 and T4

The large phages T2 and T4 are very closely related, having most genes in common and having identical genetic organization. The autoradiograph in Figure 6.15 demonstrates that the T2 chromosome is a linear DNA molecule. It has a molecular weight of 120×10^6 daltons and consists of 182,000 np.

Phage T4 has been the subject of intensive genetic analysis, and mutations in many complementation groups are known. Recombination analysis employing three- and four-factor crosses has shown that the genetic map of T4 is circular (Figure 6.16). The paradox of a linear DNA molecule's possessing a circular genetic map has been resolved by genetic and physical experiments that show that the linear DNA molecules isolated from T2 phages are terminally redundant and possess circularly permuted gene orders.

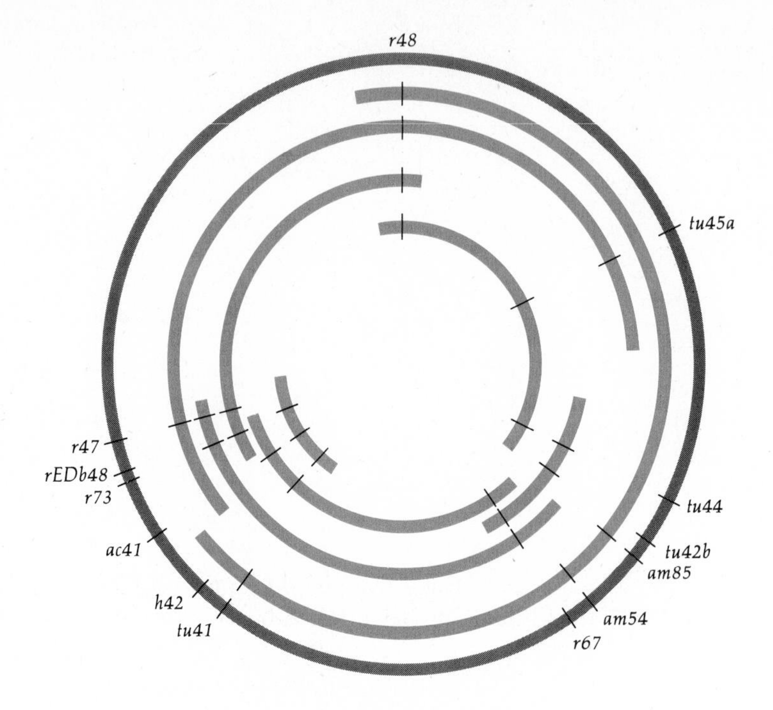

Figure 6.16
The circular genetic map of T4. The markers used in three- or four-factor crosses are connected by arcs.

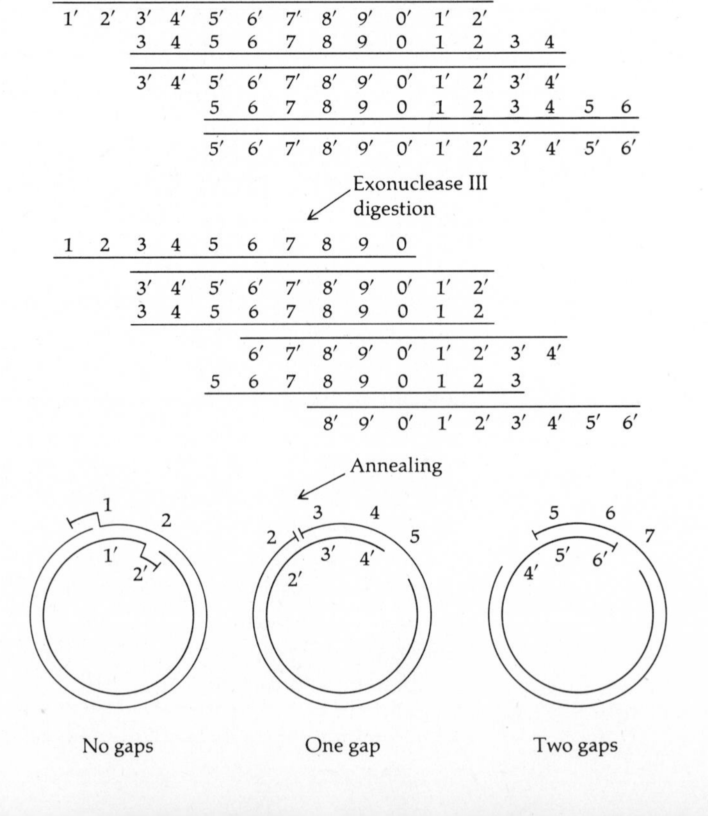

Figure 6.17
Schematic diagram of the demonstration of terminal redundancy in the T2 genome by exonuclease III digestion and circle formation. Each molecule is shown having a different circular permutation of a common sequence and a terminal repetition of its first sequence. Exonuclease III exposes the complementary 5′-ended chains at the ends, and circle formation takes place upon annealing. If the degradation proceeds beyond the limits of the terminal repetition, then two single-chain "gaps" will bracket a duplex segment within a circle, the length of which is the length of the terminal repetition (see Figure 6.18).

Figure 6.18
Electron micrographs of circularized T2 DNA molecules prepared as diagrammed in Figure 6.17. **(a)** A circular molecule measuring 54.9 μm. **(b)** A circular molecule showing two closely spaced single-chain regions separated by a duplex segment 1.7 μm long (see arrows). This duplex segment is the terminally redundant portion of this molecule. [From L. A. MacHattie et al., *J. Mol. Biol.* 23:355 (1967).]

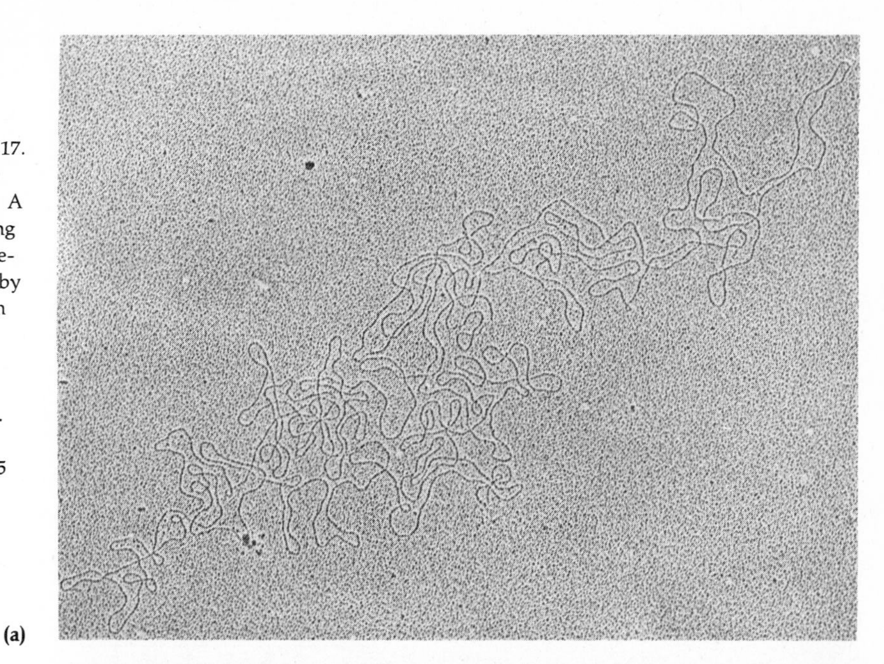

(a)

(b)

Physical evidence for terminally redundant (duplicated) base sequences comes from experiments in which T2 DNA is partially digested by the enzyme exonuclease III, which sequentially cleaves nucleotides from the 3′-OH ends of DNA strands, producing single-stranded 5′-PO_4 ends on double-stranded molecules (Figure 6.17). Incubation of these partially digested molecules under conditions that permit complementary

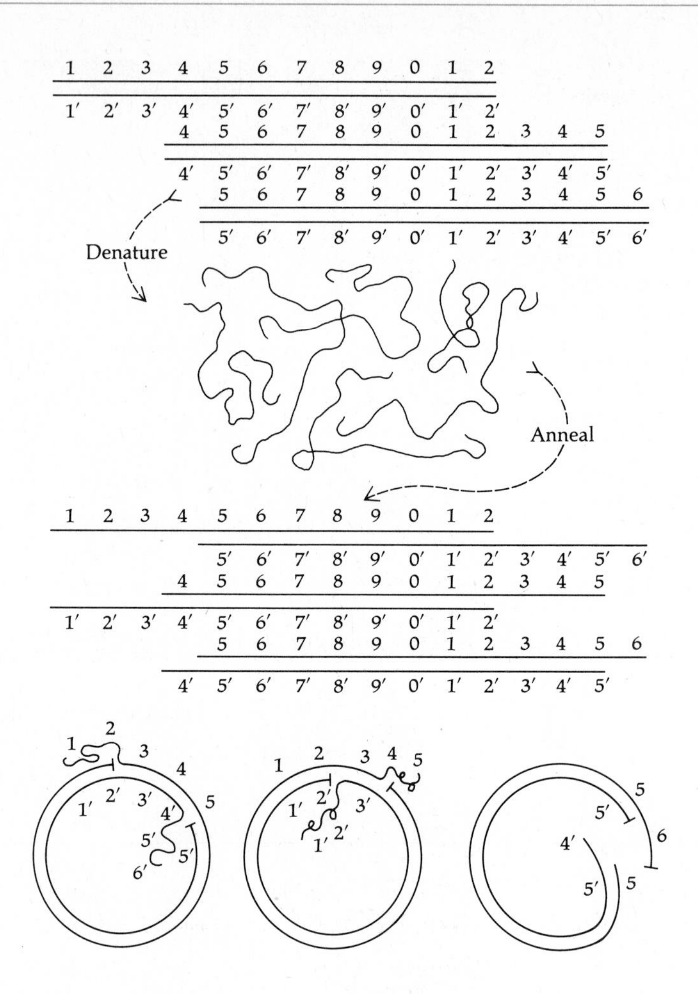

Figure 6.19
Schematic diagram of the demonstration of circular permutation of sequences in a population of T2 genomes. Each permutation is also terminally repetitious. The repetitious ends cannot find complementary partners and are left out of the circular duplexes formed by annealing. The separation of the repetitious single-stranded ends depends on the relative permutations of the complementary chains.

single-strand sequences to form hydrogen bonds results in the formation of circular molecules (Figure 6.18). This would be possible only if the original double-stranded T2 DNA molecules possessed identical base sequences at each end, as diagrammed in Figure 6.17. Before it is possible to form circles, it is necessary to digest about 2% of the T2 DNA which indicates that about 1% of the genome is present in the terminal redundancy.

Genetic evidence for terminal redundancy in the genomes of T4 and T2 is provided by the existence of heterozygous phage particles. These phage heterozygotes are products of recombination between phages of different genotypes. They are identified as single-phage particles that produce plaques containing progeny phages of both genotypes. A single-phage genome is capable of being heterozygous for only a few closely linked genes. In a population of phage particles, however, heterozygotes for all genes can be detected regardless of the map position of the gene in question. The size of the region that is terminally redundant and heterozygous is increased in deletion mutations as a consequence of the "headful" packaging mechanism discussed below. (Deletion mutations of

Figure 6.20
Electron micrographs of annealed circular T2 duplexes formed by the procedure in Figure 6.19. The circular molecules bear two single-stranded "bushes" formed by the terminal repetitions (arrows). [From L. A. MacHattie et al., *J. Mol. Biol.* *23*:355 (1967).]

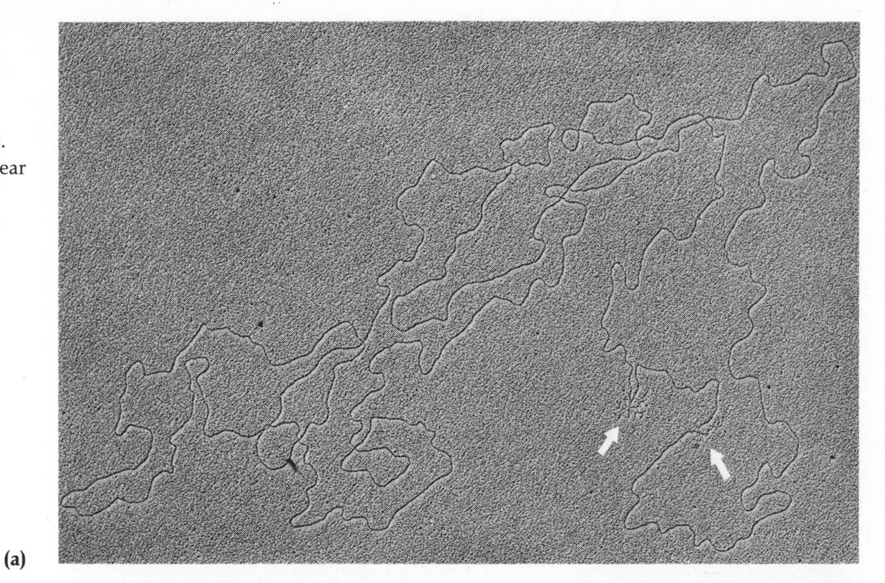

(a)

(b)

T2 and T4 do not reduce the amount of DNA contained in a phage head, as occurs with λ deletions.) This observation permits terminal redundancy heterozygotes to be distinguished from heteroduplex heterozygotes, which are discussed in Chapter 9. The logical conclusion to be drawn from these observations is that there is *not* a unique order to the sequence of genes along the DNA molecule, i.e., it is possible for any gene to be terminally redundant in some phage particle in a population.

Evidence for this bizarre conclusion is provided by the following experiment. A preparation of T2 DNA is denatured by heating to separate the complementary strands of each double-helical molecule. The

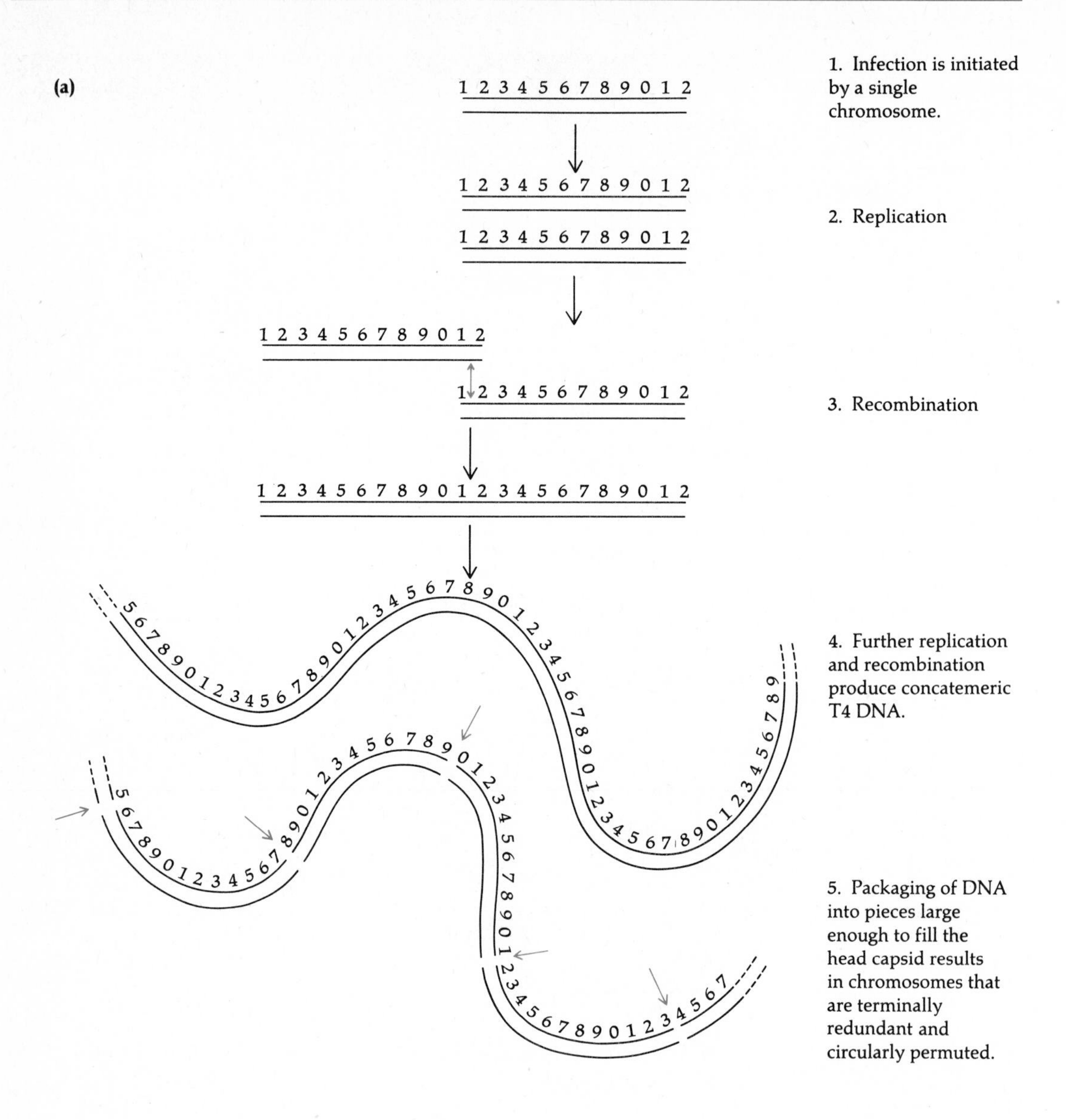

Figure 6.21
(a) Concatemeric intermediates in T4 DNA synthesis are matured to create terminally repetitious and circularly permuted progeny DNA molecules.

mixture of single strands is then incubated under conditions that permit the reformation of hydrogen-bonded double-helical molecules between complementary base sequences. In the mixture of single strands from a large population of molecules, most single strands will reform double-stranded molecules with a single strand originating from a different native DNA molecule (Figure 6.19). When viewed in the electron microscope, many circular double-stranded molecules are observed (Figure 6.20). This would be possible only if the original DNA preparation contained a population of molecules in which the gene order of any one molecule was

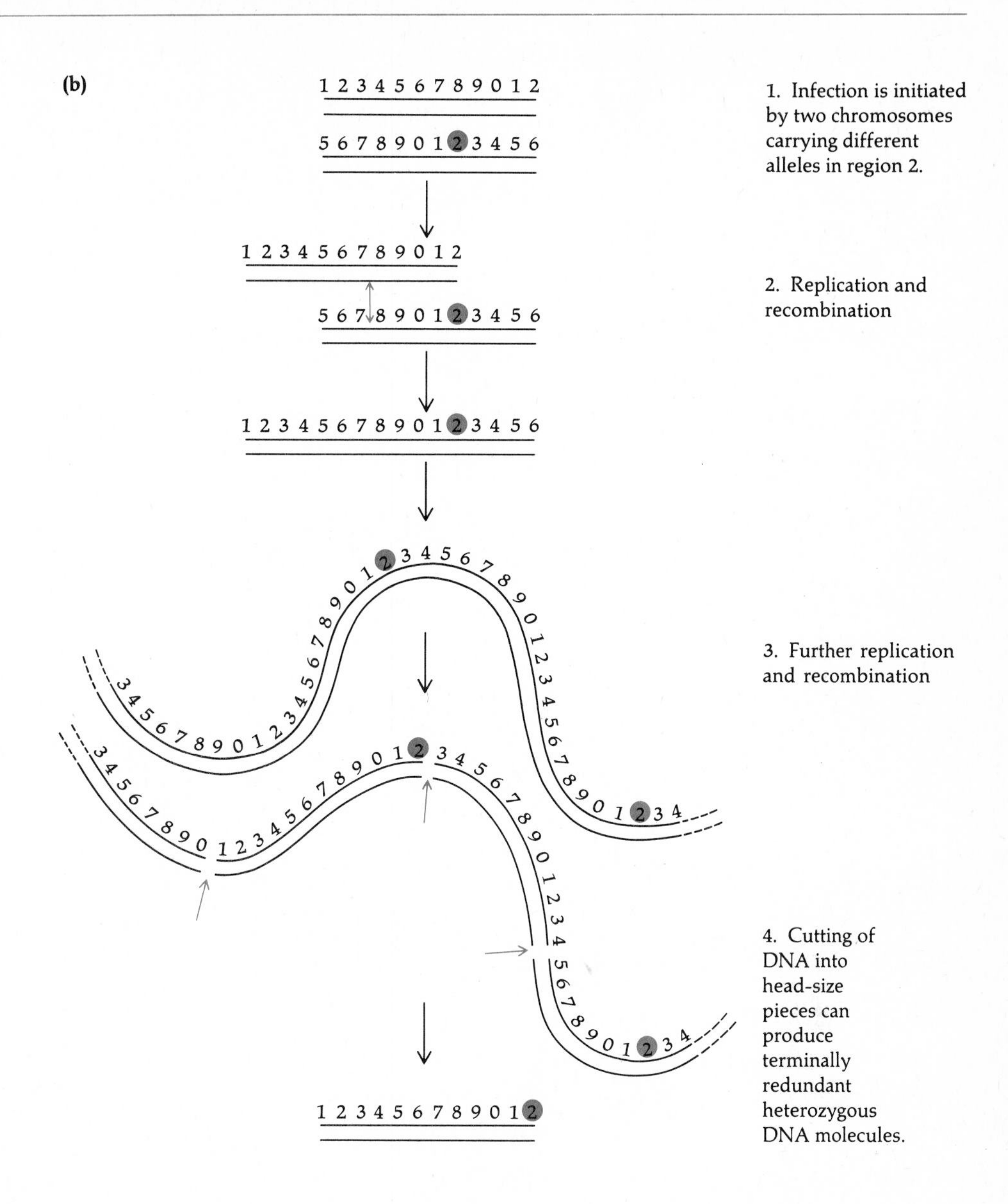

Figure 6.21 (*continued*)
(b) Recombination between genomes with different alleles in, for example, region 2 can lead to terminally redundant heterozygous progeny genomes.

a circular permutation of other gene orders in the population. It is this circular permutation and terminal redundancy of individual T2 and T4 DNA molecules that produces a circular genetic map, which is a representation of linkage relationships between genes in a population of individual DNA molecules.

The production of progeny phage, with circularly permuted and terminally redundant genomes, from a single parental phage is a consequence of the manner of DNA replication and packaging of progeny DNA molecules in the phage heads of T2 and T4. During the early stages

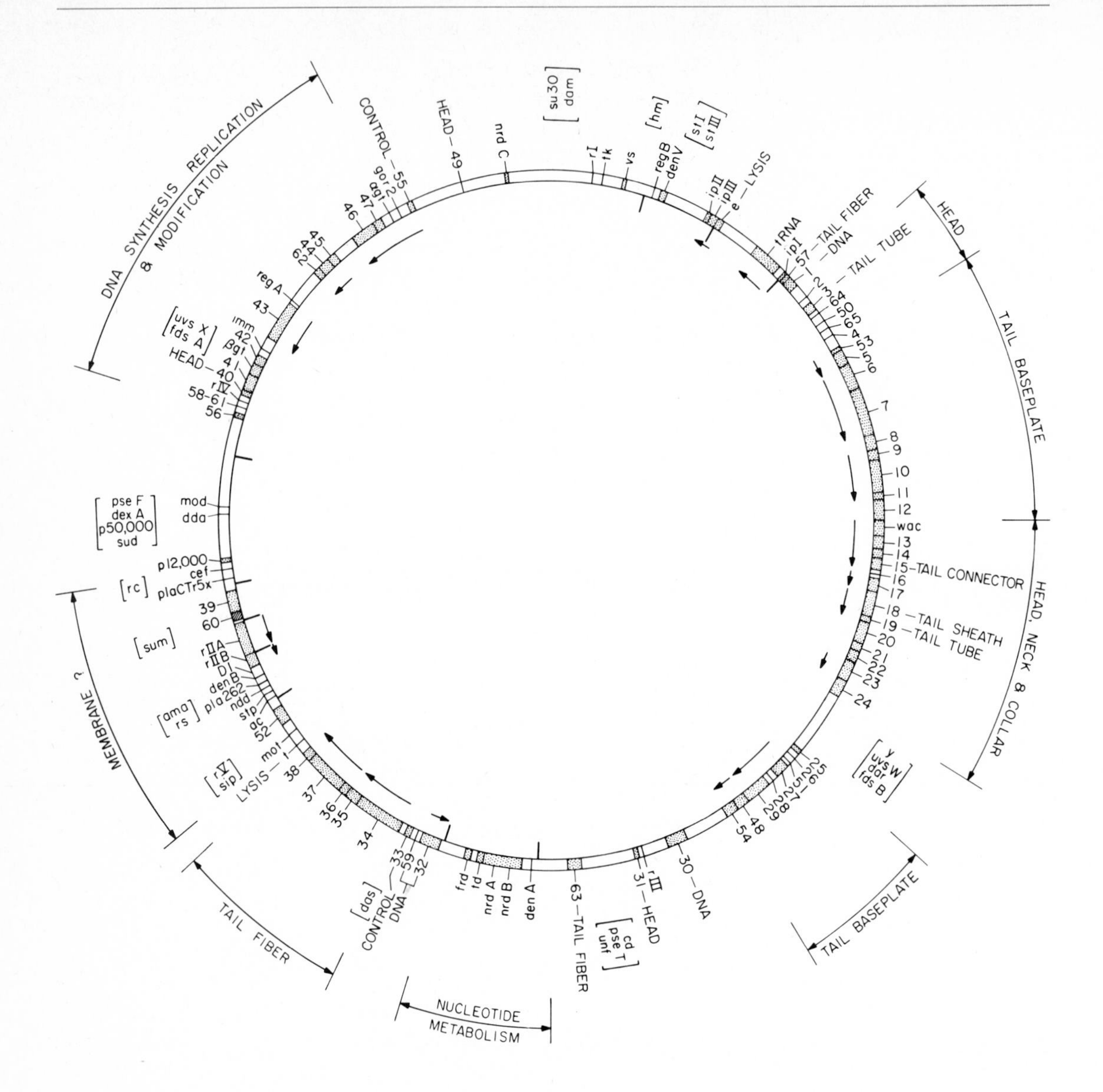

Figure 6.22
Genetic map of T4, showing the clustering of genes essential for head formation, tail formation, and DNA synthesis. The arrows inside the circle indicate transcriptional units. [From W. B. Wood and H. R. Revel, *Bact. Rev.* *40*:847 (1976).]

of infection, the linear parental DNA molecule undergoes several rounds of DNA replication to produce daughter molecules of unit length plus the terminal redundancy. Then recombination between the redundant ends of the daughter molecules produces long *concatemers* (tandemly repeated sequences) of the T4 genome (Figure 6.21), which themselves replicate and recombine to form longer concatemers. During the latter part of the infection, progeny DNA molecules begin to be packaged into phage heads. The length of DNA incorporated into each phage head is determined by the head itself. The phage head holds slightly more DNA than is present in the unit genome of T4—just enough more to provide the terminally redundant portion of the genome. Sequential packaging of "headfuls" of DNA cut from a long concatemer results in the circular permutation of gene order in different progeny particles, as diagrammed in Figure 6.21. The creation of terminal redundancy heterozygotes in doubly infected cells is also a consequence of this mode of encapsulating T4 genomes (Figure 6.21).

The genetic map of the T4 genome (Figure 6.22) demonstrates the clustering of genes responsible for related physiological functions. This is similar to the organization of genes with related functions seen in phage λ. This type of organization of genetic functions is important for their regulation, as will be discussed in Chapter 13.

Problems

1. In a phage cross involving the following combinations of markers, list the permissive and restrictive conditions for each marker and the markers that you would choose as the selective markers in three-factor crosses.
   (a) *amber, amber,* temperature-sensitive
   (b) *opal,* temperature-sensitive, temperature-sensitive
   (c) *ochre, ochre, amber*
   (d) plaque morphology, *amber,* temperature-sensitive
   (e) *ochre, opal,* plaque morphology
   (f) plaque size, plaque morphology, *amber*
   (g) host range, host range, plaque size

2. *E. coli* C is the normal host of wild-type ϕX174. The mutant strain *E. coli* C_1 has an altered cell surface, to which ϕX cannot adsorb; therefore, ϕX cannot grow and produce progeny on *E. coli* C_1. A double-mutant strain of ϕX designated H_aH_b has an altered phage protein that allows ϕXH_aH_b to adsorb to *E. coli* C_1 (as well as *E. coli* C), infect it, and produce progeny phage. When wild-type and H_aH_b phages simultaneously infect *E. coli* C cells, some of the progeny phages have the property of infecting *E. coli* C_1, but the progeny produced in *E. coli* C_1 by these phages cannot reinfect *E. coli* C_1. Explain.

3. ϕ105 is a temperate phage whose host is *Bacillus subtilis.* The wild-type phage makes turbid plaques at both 30°C and 40°C. A number of temperature-sensitive mutations have been mapped in two-factor crosses, and the percent of wild-type recombinants assayed at 40°C is recorded for the crosses in the following table. Construct a map from these data.

	N10	N15	N22	N29	N30	N31	N34
N9	3.9	13.4	5.2	10.4	—	13.2	—
N10		16.9	6.6	12.5	3.9	5.3	1.7
N15			5.1	2.2	12.3	16.1	15.8
N22				2.6	11.2	13.7	10.4
N29					—	13.8	—
N30						1.5	4.0
N31							4.1

The mutant *cc1* makes clear plaques at both temperatures, and *c4* makes turbid plaques at 30°C and clear plaques at 40°C. From the data given below, construct a map of the ϕ105 genome incorporating the map constructed above.

Cross	Clear plaques / Turbid plaques (40°C)
+ + × N15, *cc1*	149/1304
+ + × N22, *cc1*	213/1441
+ + × N10, *cc1*	251/1459
+ + × N31, *cc1*	104/1548
cc1 × *c4*	2339/8
c4 × N15	980/123
c4 × N22	846/159
c4 × N9	1010/151
c4 × N31	1248/68

Data from L. Rutberg, *J. Virol.* 3:38 (1969).

4. The following three-factor crosses with mutants of a small phage were made, and the predominant genotype for the unselected marker in each cross was recorded. Selected markers employed always mapped closer to each other than either did to the unselected marker, as determined

in two-factor crosses. Construct a map showing the order of the mutant genes.

Cross	Predominant genotype
amA ts1 × *ts5*	*am*
amA ts5 × *ts1*	*wt*
amC ts3 × *ts9*	*am*
amC ts9 × *ts3*	*wt*
amN ts9 × *ts3*	*am*
amN ts3 × *ts9*	*wt*
amN ts1 × *ts5*	*wt*
amN ts5 × *ts1*	*am*
amC ts1 × *amA*	*ts*
amC × *amA ts1*	*wt*

5. *E. coli* cells that are lysogenic for λ are immune to further infection by λ and cannot be lysogenized by a second λ phage. However, simultaneous infection of a sensitive cell by λ phages of two different genotypes frequently produces lysogens that carry both prophages. Double infection by λ and λ*cI* gives rise to lysogens carrying both prophages, whereas λ*cI* alone never forms a single lysogen, even in the presence of a λcI^+ helper during the infection. Diagram a possible sequence of events leading to such a double lysogen. What can you conclude about the nature of the cI^+ gene product?

6. Early genetic studies of λ were confined to mutations that altered the wild-type plaque morphology or size. Some of these mutations are: *s* (*small* plaques); *c* (*clear* rather than turbid plaques, later designated *cI*); *co* (*cocarde*, clear except for a ring of lysogenic colonies in the center; the ring is more dense for co_2 than for co_1, later designated *cIII* and *cII*, respectively); and *mi* (*minute* plaques smaller than *s*). The progeny genotypes from three-factor crosses involving these mutations are given overleaf. Construct a map from the data, showing map distances between adjacent genes.

Cross	Progeny								
	+ + +	*s co mi*	*s* + +	+ co_1 *mi*	*s* co_1 +	+ + *mi*	*s* + *mi*	+ *co* +	Total
s co_1 *mi* × + + +	975	924	30	32	61	51	5	13	2,091
s + *mi* × + co_1 +	38	23	273	318	112	121	6389	5050	12,324
	+ + +	+ + *mi*							
c + + × + co_1 *mi*	8	1							6,600
co_2 + + × + *c mi*	28	13							5,800

Data from A. Kaiser, *Virology* 3:42 (1957).

7. λ*cII* and λ*cIII* are capable of forming stable single lysogens if each infects a sensitive *E. coli* in the presence of a helper λ^+ phage or a helper λ*cI* phage. Recalling your answers to problems 5 and 6, what can you conclude about *cI*, *cII*, and *cIII*?

8. The lambdoid family of temperate phages all possess the same cohesive single-strand ends on their chromosomes. They integrate at different sites on the host chromosome and have the immunity properties shown in the table below, where + indicates a fruitful infection and − indicates immunity to infection.

Phage	K12	K12(λ)	K12(434)	K12(82)	K12(21)
λ	+	−	+	+	+
434	+	+	−	+	+
82	+	+	+	−	+
21	+	+	+	+	−

Different lambdoid phages are capable of exchanging some of their genes with one another during combined infections. The following table records the occurrence of wild-type recombinants with λ immunity obtained in crosses of λ mutants with wild-type 434, 21, and 82 phages. Which phages are more closely related to λ? What can you conclude about the genetic control of immunity and of lysogenization from these observations and those of problems 5, 6, and 7?

Cross	λ mutant				
	m_5	co_2	c	co_1	mi
434 × λ	+	+	−	+	+
21 × λ	+	+	−	−	+
82 × λ	+	−	−	(+)	+

(+) = more turbid than co_1 but not wild type.

9. Recombination between λ and phage 434 can produce a hybrid phage in which the λ chromosome has exchanged its immunity with that of phage 434. Such a recombinant is called λimm^{434}; it will grow on hosts carrying a λ prophage but not on hosts carrying a 434 prophage. The *cy* gene lies between the immunity region and the *O* gene. The following two crosses were made to order the mutations *2001* and *42* in the *cy* gene. Progeny were plated on *E. coli* Su⁻ (λimm^{434} *Oam 29*). What is the order of the mutations?

Cross	Recombinants
λimm^{434} *cy2001* × λ*cy42 Oam29*	cy^+O^+/O^+ = 40/1728 = 2.3%
λimm^{434} *cy42* × λ*cy2001 Oam29*	cy^+O^+/O^+ = 10/5300 = 0.19%

Data from M. O. Jones and I. Herskowitz, *Virology 88*:199 (1978).

10. The λ mutation *can1* permits lysogenization by λ*cIII* mutants, i.e., λ*cIII can1* make turbid plaques. λcan^+ can be distinguished from λ*can1* if they are plated on *E. coli* strain WA8067, where λcan^+ makes slightly clearer plaques than λ*can1*. From the results of the following crosses, locate *can1* with respect to the *cy* mutations mapped in problem 9. Progeny are plated on *E. coli* Su⁻ (λimm^{434} *Oam29*).

Cross	Recombinants
λimm^{434} *can1* × λ*cy2001 Oam29*	$can^+cy^+O^+/O^+$ = 1/1856 = 0.05%
λimm^{434} *cy2001* × λ*can1 Oam29*	$can^+cy^+O^+/O^+$ = 18/2236 = 0.81%
λimm^{434} *can1* × λ*cy42 Oam29*	$can^+cy^+O^+/O^+$ = 12/3933 = 0.31%
λimm^{434} *cy42* × λ*can1 Oam29*	$can^+cy^+O^+/O^+$ = 0/1221 $<$ 0.08%

Data from M. O. Jones and I. Herskowitz, *Virology 88*:199 (1978).

11. Bacteriophage T7 is a relatively small phage possessing a linear double-stranded DNA molecule of approximately 38,000 np. Nineteen essential genes have been identified and mapped among more than 800 conditional lethal *amber* mutations. One nonessential gene specifies a T7 polynucleotide ligase. Mutant HA13 inactivates T7 ligase, but this mutant grows normally in *E. coli*, which synthesizes its own ligase; the *E. coli* ligase meets the requirements of the replicating T7 DNA for ligase function (see Chapter 9). However, HA13 does not grow in the ligase-deficient *E. coli* strain N1252, which does support growth of wild-type T7. Thus N1252, which is also Su^-, acts as a restrictive host for HA13 as well as for T7 *amber* mutations in essential genes, permitting HA13 to be mapped with respect to essential genes. The table below gives the percent of wild-type progeny observed in two-factor crosses between HA13 and *amber* mutations in genes 1 and 2. Construct the most reasonable map possible from these data.

	1–27	1–193	2–64
HA13	10.0	12.9	8.7
1–27		9.8	15.1
1–193			20.2

Three-factor crosses were performed to confirm the gene order suggested by the data in the table. The percent of wild-type recombinant progeny observed in the crosses below is recorded. From the data in the table, calculate the expected frequencies of wild-type recombinants in these crosses and compare them with the observed frequencies. What do you conclude?

Three-factor cross	Recombinants observed (%)
1–193 × 1–27, HA13	8.3
1–27 × 1–193, HA13	1.8
HA13 × 1–193, 2–64	1.5
HA13 × 1–27, 2–64	1.4
2–64 × 1–193, HA13	6.0
2–64 × 1–27, HA13	11.8

Data from F. W. Studier, *J. Mol. Biol.* 79:227 (1973).

12. An important factor in demonstrating that the T4 genetic map is circular was the early interruption of mating by artificial lysis of the infected cells, when perhaps 10 progeny phages are present per cell. If lysis is allowed to occur normally, when perhaps 200 progeny phages are present per cell, linkage of more distant markers, required for the proof of circularity, cannot be observed. Based on the text discussion of the dynamics of phage crosses, form a hypothesis to account for these observations. On the basis of your hypothesis, what effect would the prevention of normal cell lysis have on observed recombination frequencies? Cell lysis can be prevented if both parental phages carry mutations in the lysozyme gene (*e*). Progeny phages can be recovered if the cells are lysed artificially later.

7

The Bacterial Genome

Genetic studies of the organization of prokaryotic bacterial genomes began soon after the demonstration that DNA is the genetic material of *Pneumococcus*. Like viruses, bacteria enable the geneticist to work with enormous populations over a short time period. Selective techniques, which will be described in this chapter, make it possible to detect and study very rare genetic events. The most intensively studied bacterial species is *Escherichia coli*, which will be the focus of this chapter. The genetic properties of *E. coli* are not unique to this species; the genetic methodology applied to *E. coli* has formed the basis for investigation of other species as well.

Genetic studies of bacteria have made at least two major contributions to human knowledge. The first is a growing appreciation of the *variety* of genetic processes in which a single species of organism can engage in nature. This appreciation is enlightening studies of how the human genome may interact with viral genomes, and it is causing a reevaluation of many observed, but unexplained, genetic phenomena in other eukaryotic organisms. The second contribution, which will be examined in Part II, is to a basic understanding of the regulation and expression of gene activity in prokaryotic organisms. The insights gained into these processes in the less complex prokaryotes have provided models for understanding these processes in the more complex eukaryotes.

As discussed in Chapter 6, the genetic analysis of viral genomes progressed along lines that were formally analogous to those applied to the study of meiotic organisms. When this approach was applied to the analysis of mutations in *E. coli*, however, it led to much confusion until geneticists realized that any analogy between sexuality in meiotic organisms and in bacteria was impossible. Our present understanding of genetic organization in bacteria rests on the discovery that bacteria contain a number of genetic elements, more or less independent of one another, that interact in ways that bear no formal analogy to the meiotic process. The discovery of episomes (a class of genetic elements), particularly the F episome, and of transducing phages has made it possible to apply the principles of genetic analysis successfully to bacteria and to describe the organization of the bacterial genome in great detail. Although genetics has played an essential role in understanding the biology of bacteria, our understanding of genetic mechanisms in bacteria is still incomplete.

Mutants of *E. coli*

Before the genetics of bacteria can be discussed, we must introduce the types of mutations that are studied and the notation used to describe them. The wild-type *E. coli* grows on a very simple laboratory medium in which the only organic constituent is a carbon source, often the sugar glucose. Wild-type strains are prototrophs ("basic nutrition"; see Chapter 4): they possess the genetic functions necessary to synthesize all the complex organic molecules required for their metabolism and growth. These biosynthetic capabilities (*anabolic* functions) require the expression of many essential genes. Many mutations that prevent the expression of essential biosynthetic functions are conditional lethal mutations (see Chapter 6), since bacteria possessing them can be rescued by providing the necessary organic molecule in the laboratory medium. Such biosynthetic mutants are auxotrophs ("supplementary nutrition"). In studying the organization of bacterial genes, we will consider auxotrophs simply as genetic variants; they will be discussed in more detail in Chapter 10. The phenotypes of auxotrophic bacteria will be designated by roman symbols that indicate their growth requirements. For example, Met^-, Thi^-, and Pur^- designate mutant strains that require methionine, thiamine, and purine, respectively; the corresponding prototrophic (wild-type) phenotypes are designated Met^+, Thi^+, and Pur^+.

Mutations in a number of different genes may exhibit identical auxotrophic phenotypes, and the genotypes of bacteria possessing them will be designated by italic symbols. For example, the mutations *metA* and *metB* are mutant alleles of the wild-type genes $metA^+$ and $metB^+$, and each mutant exhibits a Met^- phenotype. As we shall see later, each wild-type gene provides an essential function in the biosynthesis of methionine.

E. coli is able to utilize many different carbon sources more complex than glucose, because it possesses functions that convert the more complex sugar molecules to glucose or to other simple sugars, or that degrade other types of complex molecules, such as amino acids or fatty acids, to acetate or to intermediates of the tricarboxylic acid cycle. These degradative functions are called *catabolic* functions. Mutations that abolish such functions restrict the types of carbon sources that can be utilized by the mutant bacterium. For example, a mutant with a Lac^- phenotype is unable to grow when lactose is the only available carbon source, whereas the wild-type Lac^+ bacterium can utilize lactose. The Lac^- phenotype may be due to mutation in the $lacZ^+$ or $lacY^+$ gene, producing strains of genotype *lacZ* or *lacY*, respectively. Note that, in order to understand whether a given mutant strain will grow on a given medium, it is necessary to know whether the mutant affects an anabolic or catabolic function. For example, a Met^- mutant requires methionine-supplemented medium in order to grow, whereas a Lac^- mutant cannot grow with lactose as the sole carbon source and must be provided with another carbon source. Both types of mutants are useful conditional lethal genetic markers.

Temperature-sensitive lethal mutations are also important in bacterial genetics. Genes performing essential functions that cannot be identified by auxotrophic mutation can generally be identified by mutation to temperature sensitivity. Examples of such essential functions are those associated with the synthesis of proteins or nucleic acids from their precursor molecules—amino acids or nucleotides (see Chapter 9 for an extensive discussion of mutations affecting DNA synthesis).

Bacteria are also capable of mutation to resistance to particular bacteriophages or to antibiotics. The former mutations usually affect the ability of the particular phage to adsorb to the mutant bacterium by changing a bacterial membrane protein in some way. Resistance mutants are easily selected by spreading mutagenized cells directly onto a selective medium containing a given phage or antibiotic: colony formation indicates a resistant mutant. Mutations that confer resistance to particular antibiotics are well known because they represent serious public health problems. Resistance phenotypes are designated $T1^S$ or $T1^R$ (sensitive or resistant to phage T1), or Str^S or Str^R (sensitive or resistant to the antibiotic streptomycin), whereas the genes conferring these phenotypes are designated *ton* and *str*, ton^+ being $T1^S$ and str^+ being Str^S.

Mutant strains of bacteria are easily produced by treating a wild-type strain with a mutagenic agent, such as X rays, ultraviolet light, or a chemical mutagen (see Chapter 16), and spreading a diluted sample from the culture on a permissive medium. For example, if amino acid auxotrophs are to be selected, the mutagenized bacteria are spread on permissive medium containing all twenty amino acids. When colonies have appeared, they are *replica-plated* onto a minimal medium (Figure 7.1) to identify those colonies that are unable to grow without any amino acids in the medium. Each mutant bacterial colony (selected from the original permissive plate) is then tested on medium supplemented separately with one of the twenty amino acids in order to determine precisely which one is required. Temperature-sensitive mutants can be selected by spreading mutagenized cells on a plate that is incubated at 30°C and then replica-plating the colonies onto a plate that is incubated at 42°C. Failure of a

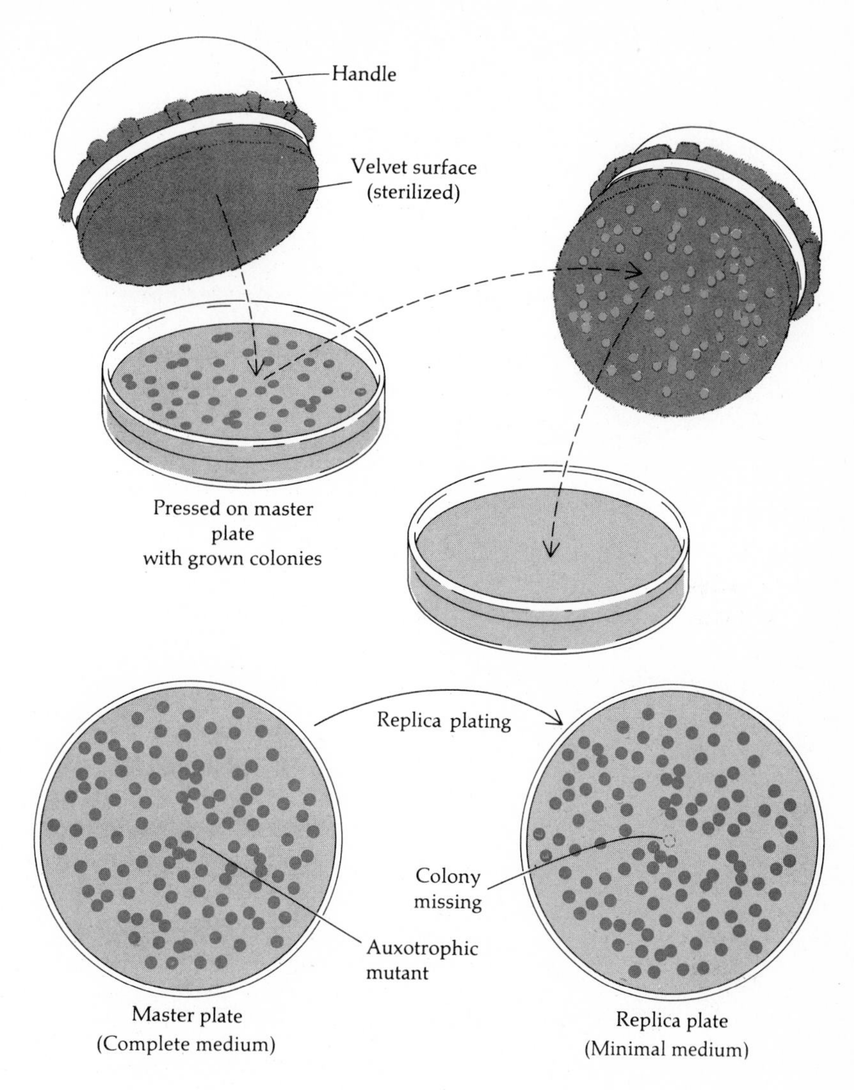

Figure 7.1
Replica plating is a technique for transferring bacteria quickly and easily from colonies on a master plate (Petri dish) to a different medium in another plate. The procedure diagrammed here permits the identification of auxotrophic mutants, which form colonies on the supplemented medium of the master plate but are unable to grow on the minimal medium of the replica plate. (After G. S. Stent and R. Calendar, *Molecular Genetics*, 2nd ed., W. H. Freeman, San Francisco, 1978.)

colony to grow at 42°C identifies the colony grown at the permissive temperature as a *ts* mutant. Bacterial strains carrying multiple mutations can be created by successive mutagenesis and selection for each of the desired phenotypes.

The Genetic Elements of *E. coli*

E. coli cells may contain a number of distinct genetic entities, each capable of self-replication. The bacterial chromosome itself is a large, circular DNA molecule of molecular weight approximately 2.5×10^9 daltons,

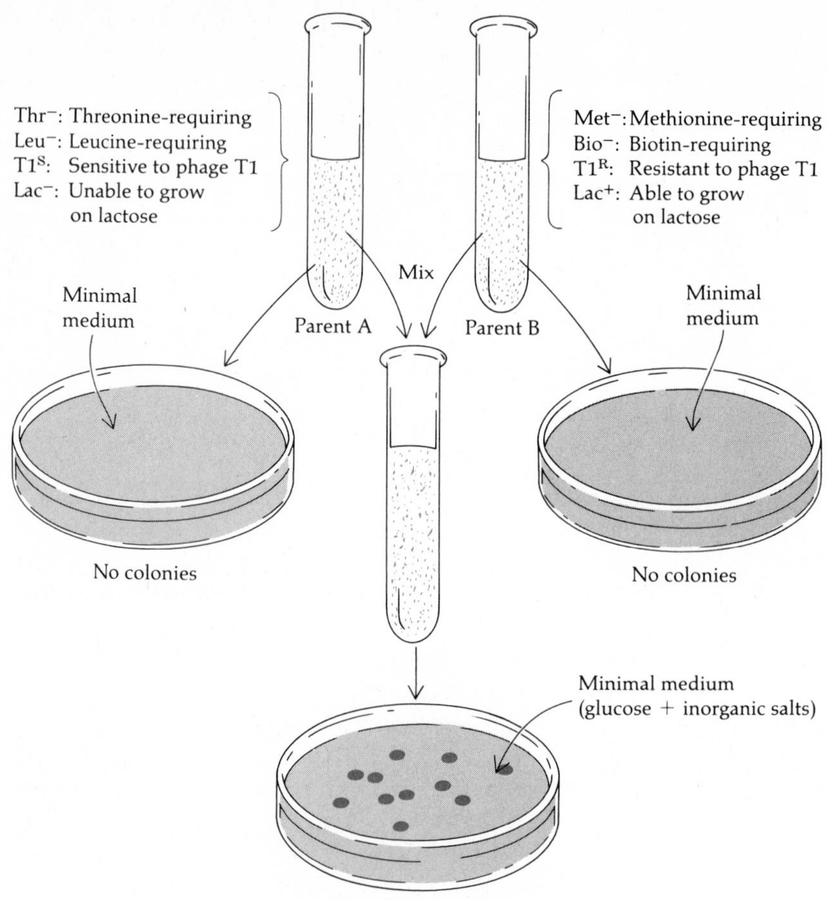

Figure 7.2
The use of multiply auxotrophic strains to demonstrate mating by *E. coli* cells. In this cross, met^+, bio^+, thr^+, and leu^+ are the selected markers (see page 236), and *ton* and *lac* are unselected markers. (After J. D. Watson, *Molecular Biology of the Gene*, 3rd ed., W. A. Benjamin, Menlo Park, Cal., 1976.)

containing about 3.2×10^6 np. As discussed in Chapter 4, the bacterial chromosome possesses a site at which DNA replication is initiated and from which it then proceeds in both directions in a theta replication mode. The genetic functions for controlling the replication of the bacterial chromosome reside on the chromosome; together with the initiation site, they form what has been called by François Jacob, Sydney Brenner, and François Cuzin a *replicon*—a self-replicating genetic molecule.

E. coli cells can harbor other replicons as well, which may exist separately from the bacterial chromosome. These are called *episomes* and *plasmids* and are circular DNA molecules of quite diverse sizes, ranging from 1.5 × 10^6 daltons up to almost one-third the size of the bacterial chromosome itself. One of the most interesting and thoroughly studied episomes is the F (fertility) factor, which confers sexuality on *E. coli.* Episomes are genetic elements that may exist either as a replicon separate from the bacterial chromosome or integrated into the bacterial chromosome, where they are a part of the bacterial replicon. The bacteriophage λ is an episome—one that, as a phage, has an extracellular form and can exist in bacteria as either its own replicon (during lytic replication) or as a prophage and part of the bacterial replicon. In contrast to episomes, plasmids do not integrate into other replicons but always remain as free (autonomous) replicons. The temperate phage P1 in its prophage state usually exists as a plasmid separate from the bacterial replicon. Most plasmids, however, do not have an extracellular form.

Some episomes are infective because they possess the ability to transfer a copy of themselves from the bacterial cell in which they reside to a cell that does not contain a resident episome of the same type. The genetic functions necessary for replication, for infectivity, and for exclusion of other episomes are encoded in the episomal DNA. Many episomes and plasmids also carry genes specifying functions that are not essential to their existence. For example, some infective episomes carry genes causing resistance to certain antibiotics. Bacteria harboring such *resistance transfer factors* become resistant and are serious menaces to public health because these factors can quickly spread antibiotic resistance to other strains and species of bacteria, some of which may be pathogenic. A particularly dangerous property of resistance transfer factors (one that obviously has a high selective advantage in antibiotic-rich societies) is their ability to accumulate resistance genes for several different antibiotics and to transfer these cumulative resistances simultaneously to other bacteria that are not resistant.

The F Factor—a Dispensable Sex Element

The F factor possesses the fascinating property of conferring sex on indifferent bacteria. Its existence was first detected when geneticists sought to determine if mating occurs between different strains of *E. coli.* The production of genetic recombinants between mutant strains was sought as possible evidence for the existence of a mating process. Multiply auxotrophic strains, produced by successive steps of mutagenesis and selection, were mixed and spread on minimal medium to detect wild-type recombinants, as diagrammed in Figure 7.2.

The use of multiply auxotrophic strains assures that those colonies that appear on minimal medium are due to the formation of wild-type

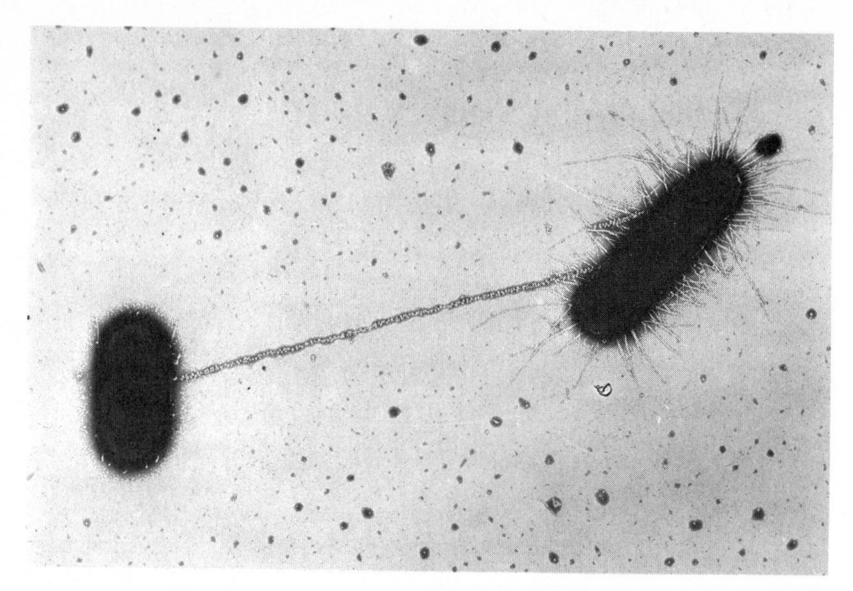

Figure 7.3
Electron micrograph of F^+ (right) and F^- (left) *E. coli* cells. The F pili can be distinguished from other pili, which are not F-specific, by their covering of male-specific bacteriophages. A number of male-specific bacteriophages are known, all containing RNA as their genetic material, that infect the cells by attachment to the F pili. The phage RNA enters the cell through the pili. (Courtesy of Prof. Charles C. Brinton, Jr., and Judith Carnahan, University of Pittsburgh.)

recombinants and are not the result of reverse mutation to wild type. If a single auxotrophic mutant reverts to wild type at a frequency of 1 cell in 10^6 cells, the revertant cells might obscure the detection of any recombinant cells formed in a mating. However, when two auxotrophic mutants are present in a strain, the frequency of simultaneous reversion of each is expected to be $10^{-6} \times 10^{-6} = 10^{-12}$. Some pairwise combinations of the mutant strains employed in these experiments yielded wild-type colonies on minimal medium at frequencies as high as 1 cell in 10^5 cells, well above that expected for reversion. These experiments led to the identification of two mating types of *E. coli*, designated $F^+$and F^-. The F^+ mating type was subsequently shown to harbor the F factor, or episome, and the F^- mating type was shown to lack it.

Further study of these two mating types revealed a fascinating (and totally unexpected) behavior of the F factor: it was found to be infective. When streptomycin-sensitive cells that carry an F factor (Str^S, F^+) are mixed with streptomycin-resistant cells (Str^R, F^-) and spread on medium containing streptomycin, only cells of the Str^R strain survive and produce colonies (the *str* alleles are carried on the bacterial chromosome). Most of these Str^R colonies are no longer F^-, but F^+. The F factor carries a number of genes that make it infective. Several of these genes code for proteins that form F pili, hairlike structures that cover the surface of F^+ cells (Figure 7.3). The F pili bind cell-surface receptors on F^- cells, leading to the formation of a cytoplasmic bridge between the two cells. During the growth of F^+ cells, the F factor replicates by the θ mode of DNA replica-

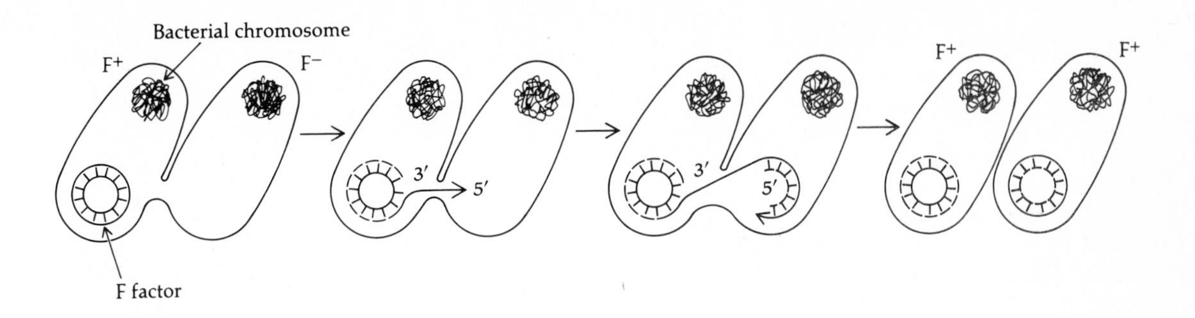

Figure 7.4
Diagram showing the manner in which a copy of the F-factor DNA is transferred from an F+ cell to an F− cell via rolling circle replication.

tion, as does the bacterial chromosome. However, when the cytoplasmic bridge is established between F+ and F− cells, the F factor begins replicating by the rolling circle, or σ mode (see Chapter 4). The 5′-PO_4 single-stranded tail of the replicating F factor passes into the F− cell, where a complementary strand is synthesized. A round of DNA replication by the F factor effects the transfer of a copy of the F factor to the F− cell, converting it to F+ (Figure 7.4).

As mentioned above, the F factor is an episome, which can exist either independently of, or integrated into, the bacterial replicon. When integrated, it can effect the transfer of the bacterial chromosome to an F− cell. The frequency with which wild-type recombinants for genes on the bacterial chromosome arise in crosses between F+ and F− strains is very low (approximately 10^{-5} per parental cell) because very few of the cells in an F+ culture engage in recombinant formation, even though there is a high frequency of infection by the F factor. However, from an F+ culture it is possible to isolate strains that give a much higher frequency of recombinant formation when mixed with F− strains. These strains are designated Hfr (high-frequency recombination) and no longer possess autonomous F factors. In an Hfr strain the F factor is integrated into the bacterial chromosome. When an Hfr cell establishes contact with an F− cell via the F pilus, a bridge called a *conjugation tube* forms, and the integrated F factor initiates rolling circle replication at the site in the bacterial chromosome where it resides. This mobilizes the transfer of the bacterial chromosome into the F− cell (Figure 7.5).

As a copy of the bacterial chromosome of the Hfr cell enters the F− cell, those genetic markers that have entered become available for recombination with the chromosome of the F− cell (Figure 7.5). During mating, disruption of the conjugation tube between the Hfr and F− cells often occurs spontaneously, with concomitant breakage of the entering Hfr chromosome. Consequently, the entire Hfr chromosome is rarely transferred to the F− cell.

Figure 7.5
Mobilization of the bacterial chromosome by an integrated F factor in an Hfr *E. coli* cell. Rolling circle replication begins within the sequence of the integrated F factor when conjugation commences. A portion of the F-factor DNA at the 5′ tail leads the bacterial DNA into the F^- cell. If the conjugal attachment is broken before the complete Hfr chromosome is transferred, the exconjugant cell will remain F^-.

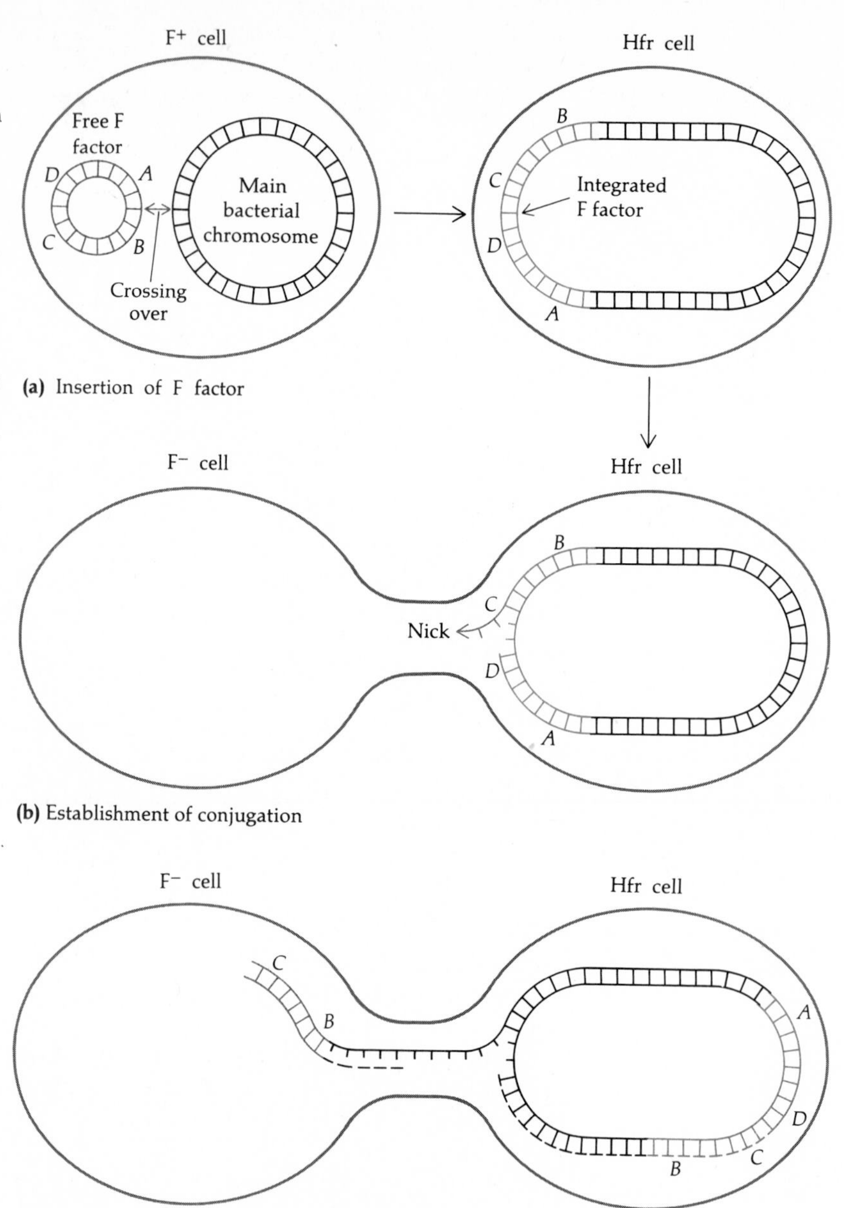

(a) Insertion of F factor

(b) Establishment of conjugation

(c) Single-strand transfer via rolling circle mechanism

Once mating is initiated, the production of wild-type recombinants depends only on the survival of the F^- parent, as is shown by the following reciprocal crosses plated on minimal medium containing streptomycin:

Cross 1: Hfr, Thr^+ Leu^+ $Str^S \times F^-$, Thr^- Leu^- Str^R
Result: Thr^+ Leu^+ Str^R recombinants form the only colonies.

Cross 2: Hfr, Thr^- Leu^- $Str^R \times F^-$, Thr^+ Leu^+ Str^S
Result: No colonies of any type are recovered.

Cross 1 demonstrates the use of streptomycin to eliminate colonies that would otherwise be formed by the Hfr parent. Only wild-type recombinants for the selected genes are able to form colonies, because the F^- parent does not grow in the absence of threonine and leucine in the selective medium. Cross 2 demonstrates that formation of recombinants requires the survival of the F^- parent but not the Hfr parent. Str^R recombinant colonies are not formed, because the str^R gene is located quite far from the site at which chromosome transfer is initiated and almost never enters the F^- cell.

The oriented transfer of bacterial genes, beginning at the site of F-factor integration into the bacterial chromosome, provides a means for physically mapping the organization of bacterial genes in the bacterial chromosome based on the order of their transfer to the F^- cell, as will be discussed in detail below. As shown in Figure 7.5, the origin of chromosome transfer is within the nucleotide sequence of the integrated F factor. Only a portion of the F factor is initially transferred to the F^- cell. Transfer of the remainder of the F factor requires transfer of the entire Hfr chromosome; since this rarely occurs, owing to spontaneous breakage of the conjugation tube, most recombinants that are selected remain F^-.

Physical Mapping of Bacterial Genes by Interrupted Mating

The ordered transfer of genes from an Hfr cell into an F^-cell provides a means for mapping genes with respect to their order and time of entry into the F^- cell. For example, consider cross 3:

Cross 3: Hfr, Thr^+ Leu^+ Az^S $T1^S$ Lac^+ Gal^+, Str^S
$\times$
F^-, Thr^- Leu^- Az^R $T1^R$ Lac^- Gal^-, Str^R

Mating between the indicated Hfr and F^- strains is initiated by mixing the two cultures at time $t = 0$. At successive time intervals, a portion of the mixed culture is agitated in a kitchen blender to break the conjugation

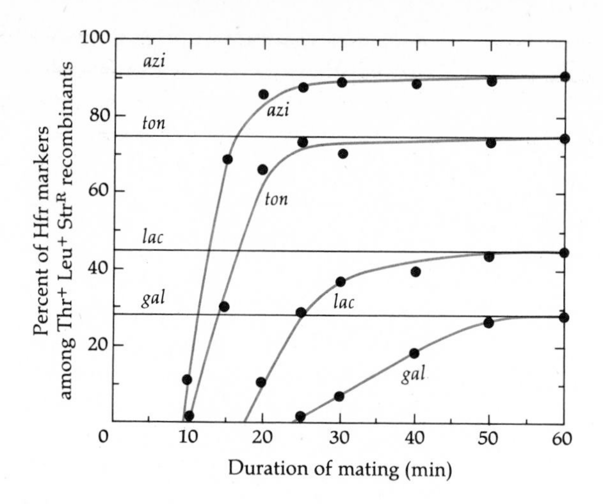

Figure 7.6
The frequencies of unselected markers among Thr$^+$ Leu$^+$ Str^R selected recombinants of cross 3 are plotted as a function of duration of mating. Mating is interrupted prior to plating of parents on the selective medium. Extrapolation of the frequency of each unselected marker to zero indicates the earliest time at which markers become available for recombination with the chromosome of the F$^-$ cell.

tubes joining the mating cells and is then plated on medium containing streptomycin with glucose as a carbon source. On this medium, Thr$^+$ Leu$^+$ Str^R recombinants are selected. The corresponding three markers, *thr*$^+$, *leu*$^+$, and *str*^{R}, are referred to as *selected* markers. Azide sensitivity (*azi*), phage T1 sensitivity (*ton*), lactose utilization (*lac*), and galactose utilization (*gal*) are referred to as *unselected* markers because the medium on which the cells are plated does not discriminate between the different alleles of these loci that are present in the cross. The Thr$^+$ Leu$^+$ Str^R recombinants that form colonies are then scored for the alleles of the unselected markers that are present in the selected recombinants by replica plating to selective media. The frequency of each unselected marker in the selected recombinants, plotted as a function of the duration of mating, is shown in Figure 7.6. The data show that the different unselected Hfr markers become available for recombinant formation with the chromosome of the F$^-$ cell at different times following the initiation of mating (Figure 7.7). Extrapolation of the frequency of each marker to zero indicates a time that characterizes the entry of each marker into the F$^-$ cell (Figure 7.6). These data permit the ordering of genes along the Hfr chromosome with respect to the point of F-factor integration and the assignment of physical distances between them, in units of minutes elapsed from the initiation of mating at $t = 0$.

Circularity of the *E. coli* Genome

A single F$^+$ strain gives rise to many different Hfr strains, each characterized by a unique point of insertion and orientation of the F factor in the bacterial chromosome (Figure 7.8). Thus each Hfr strain demonstrates a different origin and orientation of chromosome transfer in an interrupted mating experiment, as just described. The order of marker transfer by a

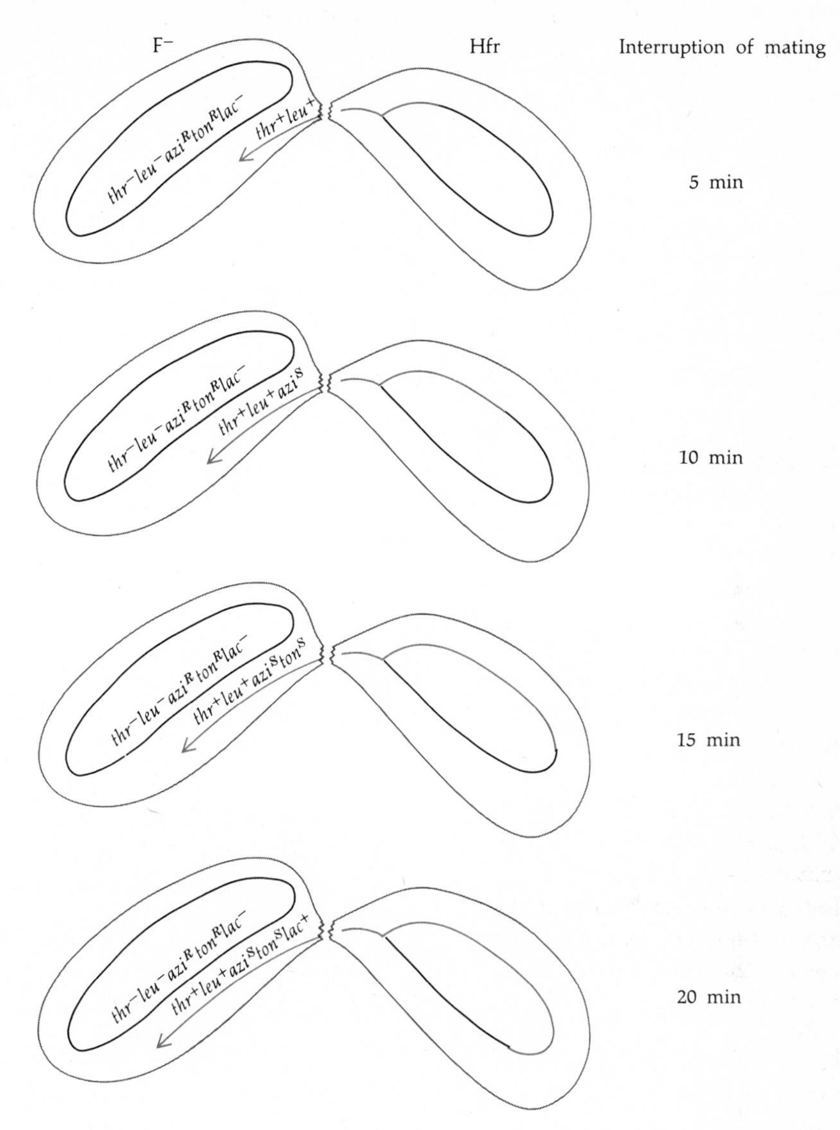

Figure 7.7
Diagram showing polarized transfer of the Hfr chromosome in an interrupted mating experiment. Only the Hfr genes present in the F⁻ recipient at the time of interruption of mating are available for recombination with the chromosome of the F⁻ cell.

number of different Hfr strains permits the construction of a physical map of the entire *E. coli* chromosome. As shown in Figure 7.9, this map is circular, in concordance with the circularity of the DNA itself.

F′ Strains and Partial Diploids

The integrated F factor in an Hfr strain often undergoes spontaneous excision to produce an F⁺ cell. Occasionally an aberrant excision event can

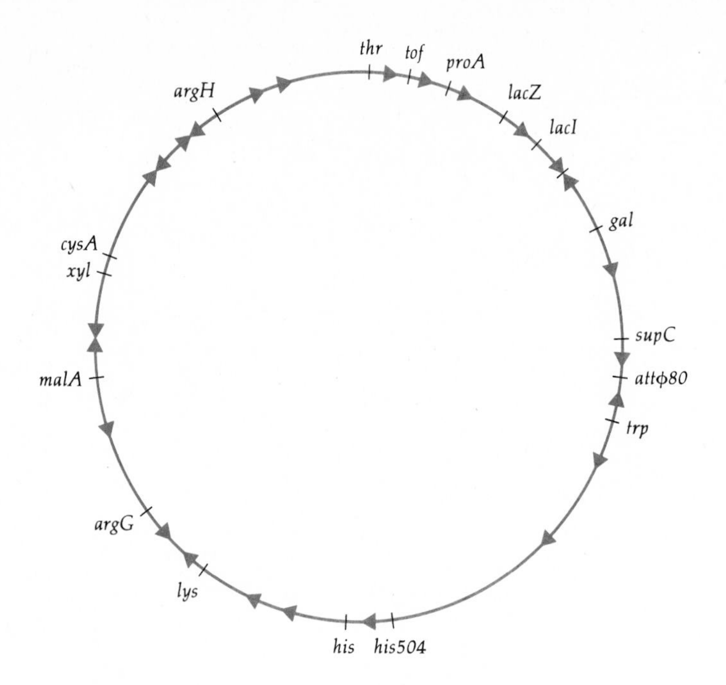

Figure 7.8
Sites of F-factor integration in the chromosome of several Hfr *E. coli* strains. Each point of insertion is marked by an arrowhead pointing in the direction of chromosome transfer during conjugation. The gene behind an arrowhead is the first gene to be transferred by that Hfr strain. Each Hfr strain has only one F factor present. Note that some Hfr strains transfer genes from the same initiation point but in opposite directions.

result in an F factor that carries an adjacent block of bacterial chromosome and that may have lost a block of F-specific DNA (Figure 7.10). The circular F factor with inserted bacterial genes is a replicon called an F′ element. The F′ element is usually infective, as is the F factor itself, and readily transfers a copy of itself to an F^- cell. F′ strains can be distinguished from Hfr strains by their behavior in crosses to F^- strains. Compare the results of crosses 4 and 5, in which the mixed cultures are spread on minimal medium containing streptomycin:

Cross 4: Hfr, Thr^+ Leu^+ Str^S × F^-, Thr^- Leu^- Str^R
Result: F^-, Thr^+ Leu^+ Str^R recombinants are selected.

Cross 5: F′, Thr^+ Leu^+ Str^S × F^-, Thr^- Leu^- Str^R
Result: F′, Thr^+ Leu^+ Str^R cells are selected.

The selected colonies in cross 4 will all be F^- and unable to transfer Thr^+ and Leu^+ to an F^- strain. In contrast, the selected colonies in cross 5 will carry an active F factor, enabling them to mate with other F^- strains and to transfer Thr^+ and Leu^+ to them. The selected progeny of these matings will also be phenotypically F^+, owing to the F′ element.

The existence of F′ elements provides a means of creating *partial diploids* by transfer of the F′ element to an F^- cell. Such partial diploids, in

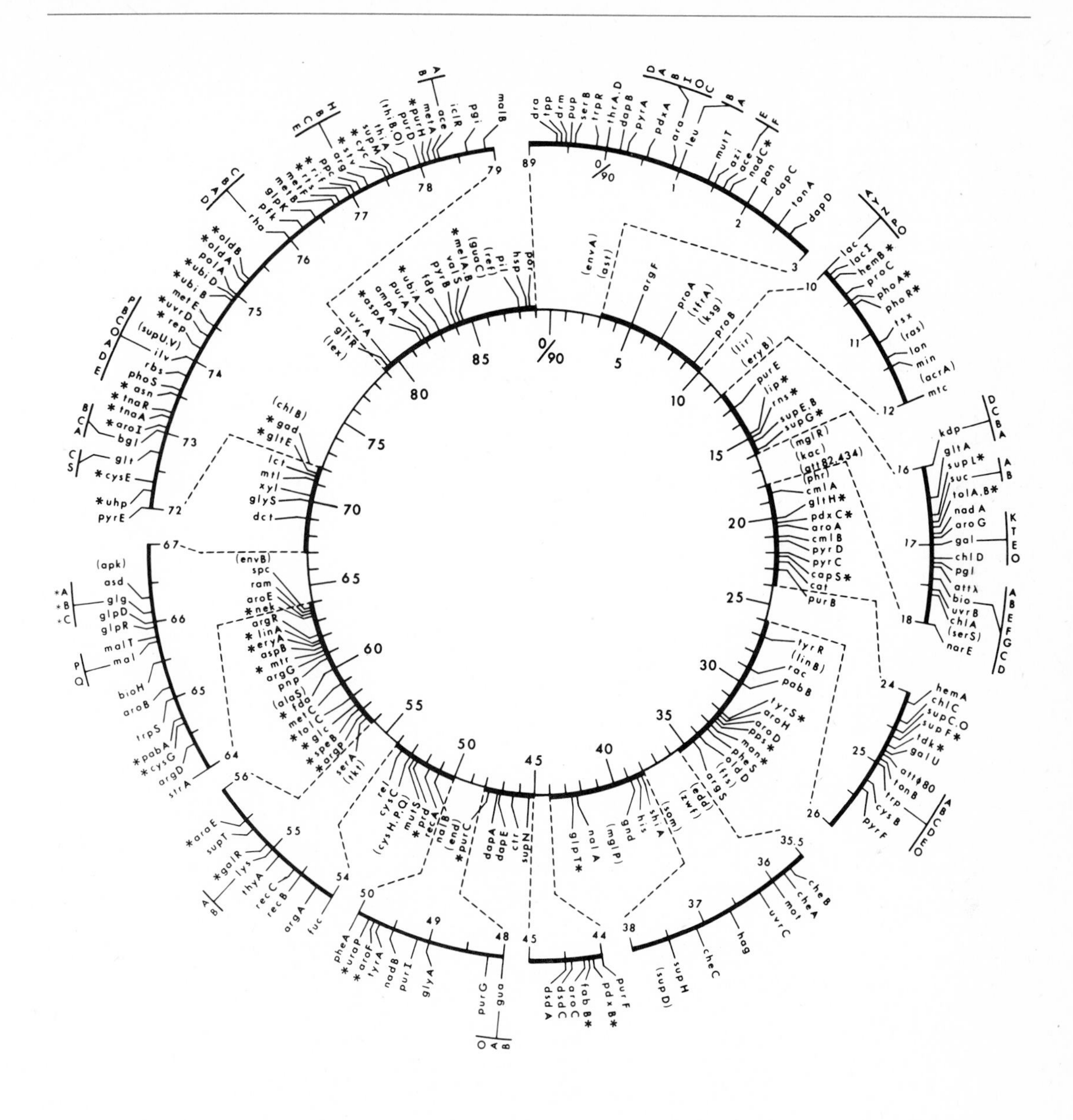

Figure 7.9
The genetic map of *E. coli* constructed from interrupted mating experiments and recombinational mapping experiments. [From A. L. Taylor, *Bact. Rev.* 34:155 (1970).]

turn, provide a means of carrying out complementation analysis of different mutants and of assessing the dominance or recessiveness of different alleles of particular genes. (Examples of these applications will be presented in Chapter 13.)

Recombination may occur in a partial diploid between bacterial genes carried on an F′ element and the homologous region of the bacterial chromosome. Depending on whether a single or double crossover occurs, recombination may lead to an Hfr strain carrying a duplication for the

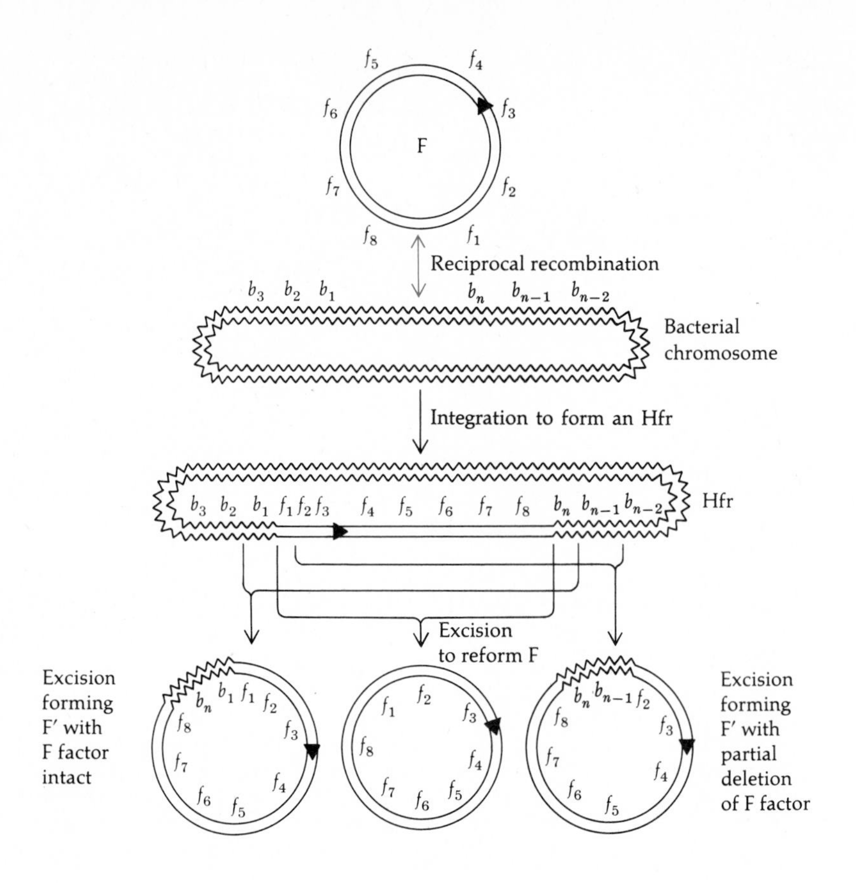

Figure 7.10
Formation of an Hfr strain by integration of an F factor and subsequent excision to form either F+ or F′ strains. The *f*'s and *b*'s represent arbitrary sites on the F factor and the bacterial chromosome, respectively. Two types of aberrant excisions may occur, as indicated, to form an F′ element with an intact F factor or an F′ element with a partially deleted F factor.

genes present on the F′ element or to an F′ strain in which markers have been exchanged between the bacterial chromosome and the F′ element.

Transposable Genetic Elements

Ideas concerning the stability of the organization of DNA molecules over time were profoundly altered during the 1970s by the study of insertional genetic elements in bacteria. The first examples of transposable elements to be recognized in bacteria are now called *insertion sequences* (IS). They were detected as the cause of a class of spontaneous mutations of *E. coli* occurring in genes in operons (see Chapter 13). These mutations abolish the expression of the gene in which they occur and show strong polar effects on the expression of genes distal from the operon promoter.

Heteroduplex molecules formed between mutant and wild-type DNAs reveal that these mutants have an inserted DNA sequence within

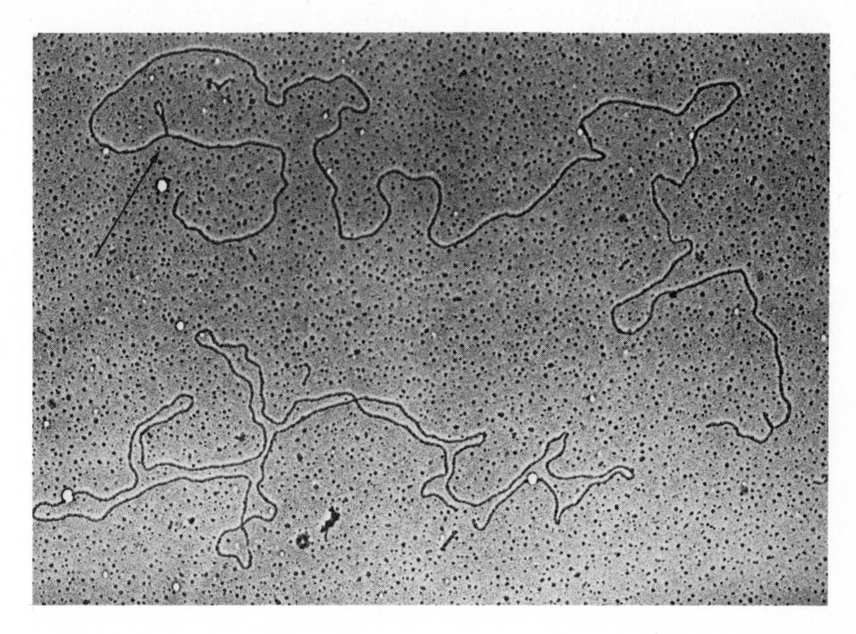

Figure 7.11
Electron micrograph of $\lambda gal^{+}/\lambda gal3$ heteroduplex DNA molecules. The single-stranded loop (arrow) is the IS2 insertion in the gal^{+} gene. [From A. Ahmed and D. Scraba, *Mol. Gen. Genet.* *136*:233 (1975).]

the wild-type DNA sequence (Figure 7.11). Four such sequences, IS1, IS2, IS3, and IS4, have been found repeatedly as the cause of mutation in many genes. These IS elements contain about 800, 1300, 1200, and 1400 np, respectively. Mutations caused by IS elements spontaneously revert to wild type; reversion is accompanied by precise removal of the IS element from the gene in which it was present. The wild-type *E. coli* chromosome is known to contain eight copies of IS1 and five copies of IS2, but the number of other insertion sequences present has not been determined. The positions of these sequences appear not to be fixed, however. They move from place to place at a frequency of 10^{-6} to 10^{-8} per cell. Usually when an IS lands within the nucleotide sequence of a gene, the gene is inactivated. Reversion of such a mutation may occur when the IS element moves again.

A physical map of the *E. coli* F factor is shown in Figure 7.12a. Within its 94,500 np it carries a number of genes essential for conjugative transfer of the chromosome (*tra* genes) and other genes essential for its own replication. In addition, four insertion sequences are present—two IS3's, one IS2, and another designated $\gamma\delta$. These sequences are the sites at which the F factor recombines with homologous sequences within the bacterial chromosome to form an Hfr strain, just as phage λ integrates by recombination between *attP* and *attB* (see Chapter 6). Evidence for this is provided by studies of the structure of F′ elements using heteroduplex mapping. In the F′ DNA, the bacterial DNA integrated into the F factor is separated from the F-factor DNA at each end by identical insertion sequences. The ability of the F factor to integrate at many places around

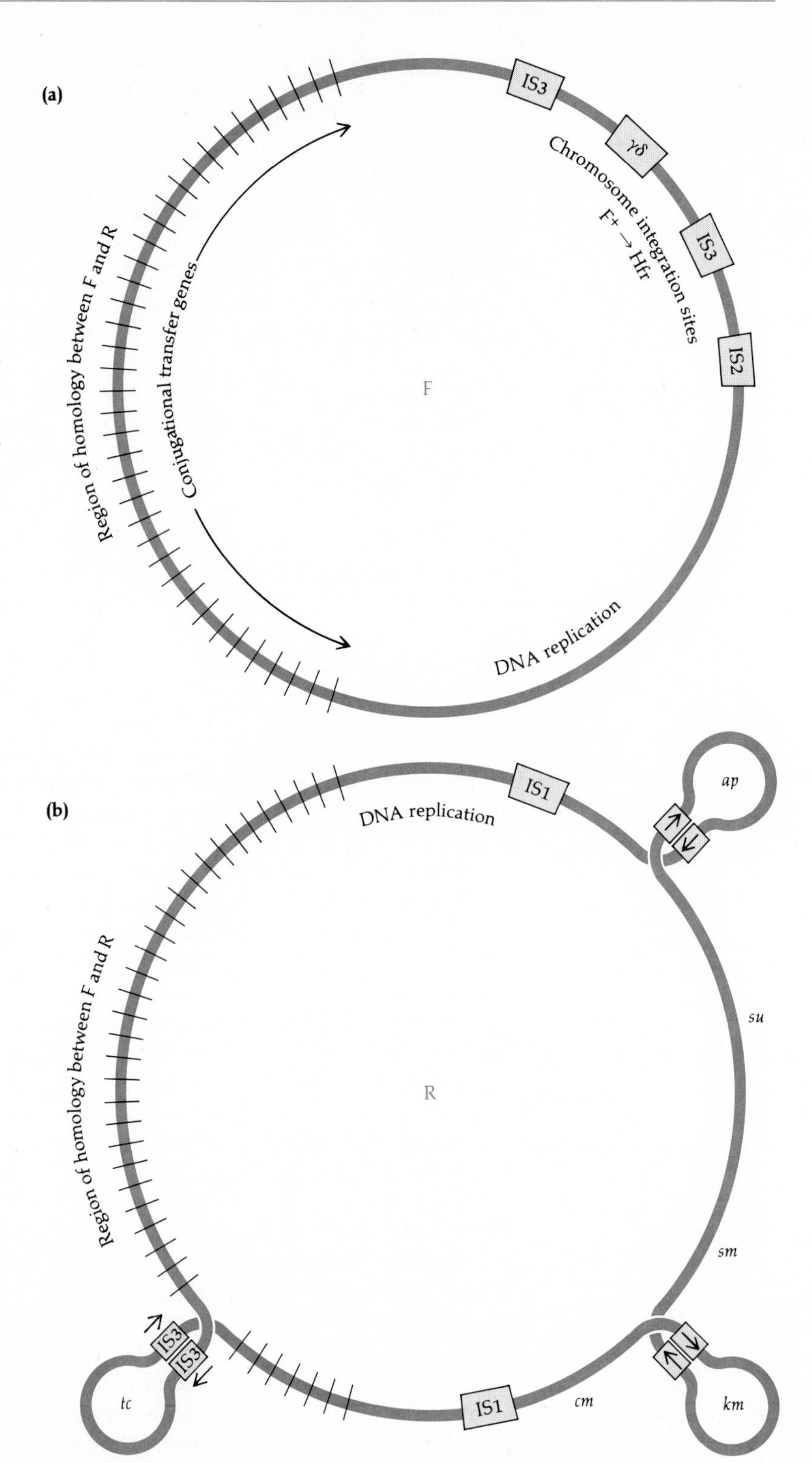

Figure 7.12
(a) Physical map of the *E. coli* F factor, showing the locations of the genes required for transfer and of the insertion elements responsible for integration into the bacterial chromosome. **(b)** Physical map of a generalized resistance transfer factor, showing regions of homology with the F factor and sites of some antibiotic-resistance genes, some bounded by insertion sequences, forming transposons.

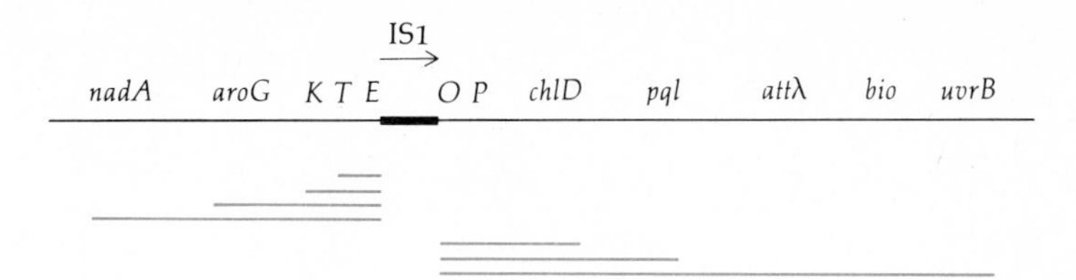

Figure 7.13
Extents of deletions induced by the IS1 insertion in the *gal* operon of *E. coli* are indicated by solid lines below the genetic map of the region. [After P. Nevers and H. Saedler, *Nature 268*:109 (1977).]

the *E. coli* chromosome, in either orientation, to form Hfr strains is due to the presence of identical insertion sequences shared by F-factor DNA and bacterial chromosomes.

Antibiotic resistance transfer factors have the general structure shown in Figure 7.12b. An evolutionary relationship to the F factor is indicated by the presence of DNA homology (determined by heteroduplex DNA formation) throughout the region of the *tra* genes. Inserted into the region of F-factor homology, bounded by IS3 elements, is a DNA sequence coding for tetracycline resistance, *tc*. Within the region of nonhomology, genes coding for ampicillin resistance (*ap*), sulfonamide resistance (*su*), streptomycin resistance (*sm* here, rather than *str*), chloramphenicol resistance (*cm*), and kanamycin resistance (*km*) can be found. These resistance genes are bounded individually or in groups by designated IS elements or by other inverted repeat sequences (indicated by arrows). The DNA sequences carrying individual resistance markers such as *tc* or *ap* are observed to jump occasionally to other episomes or plasmids, to phage chromosomes, or to bacterial chromosomes. The ability to jump is conferred by the presence of IS, or other special sequences, bounding the antibiotic-resistance genes. A transposable DNA sequence bounded by insertion sequences is called a *transposon.*

The mechanism of transposition (jumping) exhibited by both insertion sequences and transposons is poorly understood. Transposition occurs in *recA* mutants of *E. coli*, which are recombination-deficient, indicating that a homologous or general recombination mechanism is not involved (see Chapter 9). At least one transposon is believed to code for an enzyme required for its own transposition.

Insertion sequences are also known to be sites at which spontaneous deletions of bacterial genes arise. Deletions induced by the presence of IS1 in the *gal* operon have either their left or right endpoints at the site of the IS1 element (Figure 7.13).

The temperate phage mu is a transposon that also has an extracellular form as a phage. Mu can integrate anywhere in the *E. coli* genome; it must be integrated into a host chromosome in order to replicate. Even mu DNA isolated from phage has short random sequences of host DNA at each end of the linear phage DNA molecule. Mu can also cause transposition of

Figure 7.14
A nightmare induced by recent discoveries in bacterial genetics. (From J. A. Shapiro, *Trends in Biochemical Sciences*, August, 1977, p. 178.)

pieces of the bacterial chromosome. When this occurs, a mu prophage resides at each end of the transposed piece of DNA, just as IS sequences do in a transposon (see Figure 7.14).

Genetic Mapping in *E. coli*

Interrupted mating is useful for physically mapping genes separated by large distances but is not at all useful for mapping markers separated by only one or two minutes on the chromosome. However, loci separated by no more than two or three minutes on the physical map can be ordered and mapped effectively by recombination analysis, based on the principles employed in three-factor crosses, as discussed in Chapters 5 and 6.

Recombinational mapping requires a recipient cell, with its circular chromosomal DNA, and DNA from a donor cell. The DNA from the donor cell can be introduced by transfer of a portion of an Hfr chromosome by conjugation, by a phage vector that had incorporated host DNA

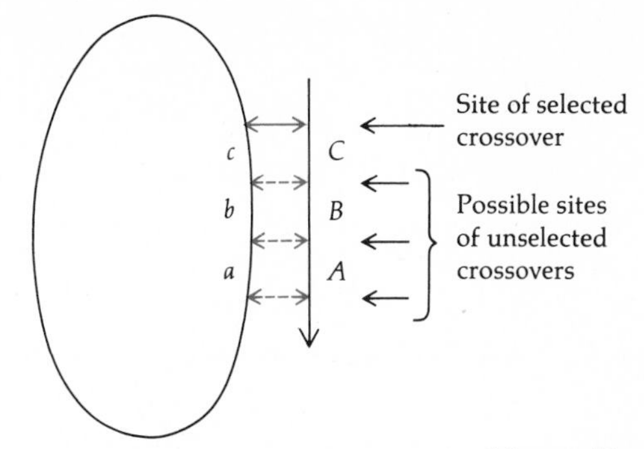

$$\text{Recombination frequency between } B \text{ and } C = \frac{bC \text{ recombinants}}{C \text{ recombinants}}$$

Figure 7.15
Genetic mapping by recombinational analysis in a merozygote created by conjugation. Selection is for the origin-distal marker *C*. The map-unit distance between *B* and *C* is the number of *bC* recombinants divided by the number of selected *C* recombinants, times 100. Because a reciprocal recombinant class is not produced in the mating, as it is in a cross between meiotic organisms, map units obtained in this way do not correspond to map units determined for meiotic organisms.

rather than phage DNA into its coat (*transduction*), or by direct assimilation of DNA (*transformation*), as discussed in Chapter 4 for *Pneumococcus*. The partially diploid cell thus created is called a *merozygote*. Merozygotes are unstable because the donor DNA is only a fragment of a complete replicon. In order for genetic markers carried by the donor DNA to replicate and be retained by descendants, they must become incorporated into the replicon of the recipient cell by recombination. As diagrammed in Figure 7.15, two (or an even number) separate crossovers are required both to insert donor DNA into the recipient chromosome and to maintain the circular integrity of the recipient chromosome.

Conjugational Mapping

When mapping is done with F^- merozygotes created by conjugation with an Hfr strain, knowledge of the order of transfer of flanking markers is helpful. Mapping is accomplished by selecting for the origin-distal marker *C* (Figure 7.15). This assures that all selected cells will actually have received the markers of interest during conjugation. The frequency with which unselected markers appear in the selected recombinants can be used to obtain the map distance of the unselected markers from *C*, as indicated in Figure 7.15. For example, closely linked mutations that affect the ability to utilize lactose (*lacY*, *lacZ*, *lacI*) as a carbon source are trans-

Table 7.1
Frequencies of unselected markers among ade^+ str^R recombinants produced in merozygotes.

Cross 6

Hfr	Y_R^-	Z_4^+	I_3^+	ade^+	str^S
F⁻	Y_R^+	Z_4^-	I_3^-	ade^-	str^R
Region:	1	2	3	4	

Among recombinants ade^+ str^R (crossover in 4):

22% are $Y_R^+ Z_4^-$ (crossover in 2 or 3)

2.3 are $Y_R^+ Z_4^+ I_3^+$ (crossover in 1)

0.20% are $Y_R^- Z_4^+ I_3^-$ (crossover in 2 and 3)

0.048% are $Y_R^+ Z_4^+ I_3^-$ (crossover in 1, 2, and 3)

Cross 7

Hfr	Y_R^+	Z_4^-	I_3^-	ade^+	str^S
F⁻	Y_R^-	Z_4^+	I_3^+	ade^-	str^R
Region:	1	2	3	4	

Among recombinants ade^+ str^R (crossover in 4):

26% are $Y_R^- Z_4^+$ (crossover in 2 or 3)

1.9% are $Y_R^+ Z_4^+$ (crossover in 2 or 3 and 1)

1.4% are $Y_R^- Z_4^+ I_3^-$ (crossover in 2)

0.22% are $Y_R^+ Z_4^+ I_3^-$ (crossover in 1 and 2)

After F. Jacob and E. L. Wollman, *Sexuality and the Genetics of Bacteria,* Academic Press, New York, 1961, p. 229.

ferred by a particular Hfr strain after pro^+ and before ade^+ in the two reciprocal crosses 6 and 7:

$$\text{Cross 6: Hfr, } Y_R^- Z_4^+ I_3^+ \, ade^+ \, str^S \times \text{F}^-, \; Y_R^+ Z_4^- I_3^- \, ade^- \, str^R$$

$$\text{Cross 7: Hfr, } Y_R^+ Z_4^- I_3^- \, ade^+ \, str^S \times \text{F}^-, \; Y_R^- Z_4^+ I_3^+ \, ade^- \, str^R$$

The ade^+ str^R recombinants are selected by plating on glucose medium containing streptomycin. The data in Table 7.1 show the genetic constitution of more than 10,000 such recombinant colonies analyzed by replica plating onto media that allow scoring of the Lac phenotypes. The largest classes of recombinants in the two crosses establish the map distance between *ade* and *lacZ*. In cross 6 this distance is 22 units. The reciprocal cross 7 exhibits a distance of 26 units, in reasonable agreement with cross 6. Figure 7.16 diagrams the crossovers required to produce these and the other classes of recombinants recorded in Table 7.1. Cross 6 recombinant classes B and C (Figure 7.16) establish that *lacI* lies between *lacZ* and *ade* because a quadruple crossover (class C) must be less frequent than a double crossover (class B). Comparison of classes C and D with class B indicates that *lacZ* must lie between *lacI* and *lacY* by the same reasoning. The reciprocal cross 7 confirms the order of these genes deduced in cross 6. The observation of quadruple crossovers at the frequencies seen in these data is due to high negative interference (see Chapter 8). Figure 7.17 shows the recombinational map obtained from the data in Table 7.1 and other data involving alleles at the *pro* locus.

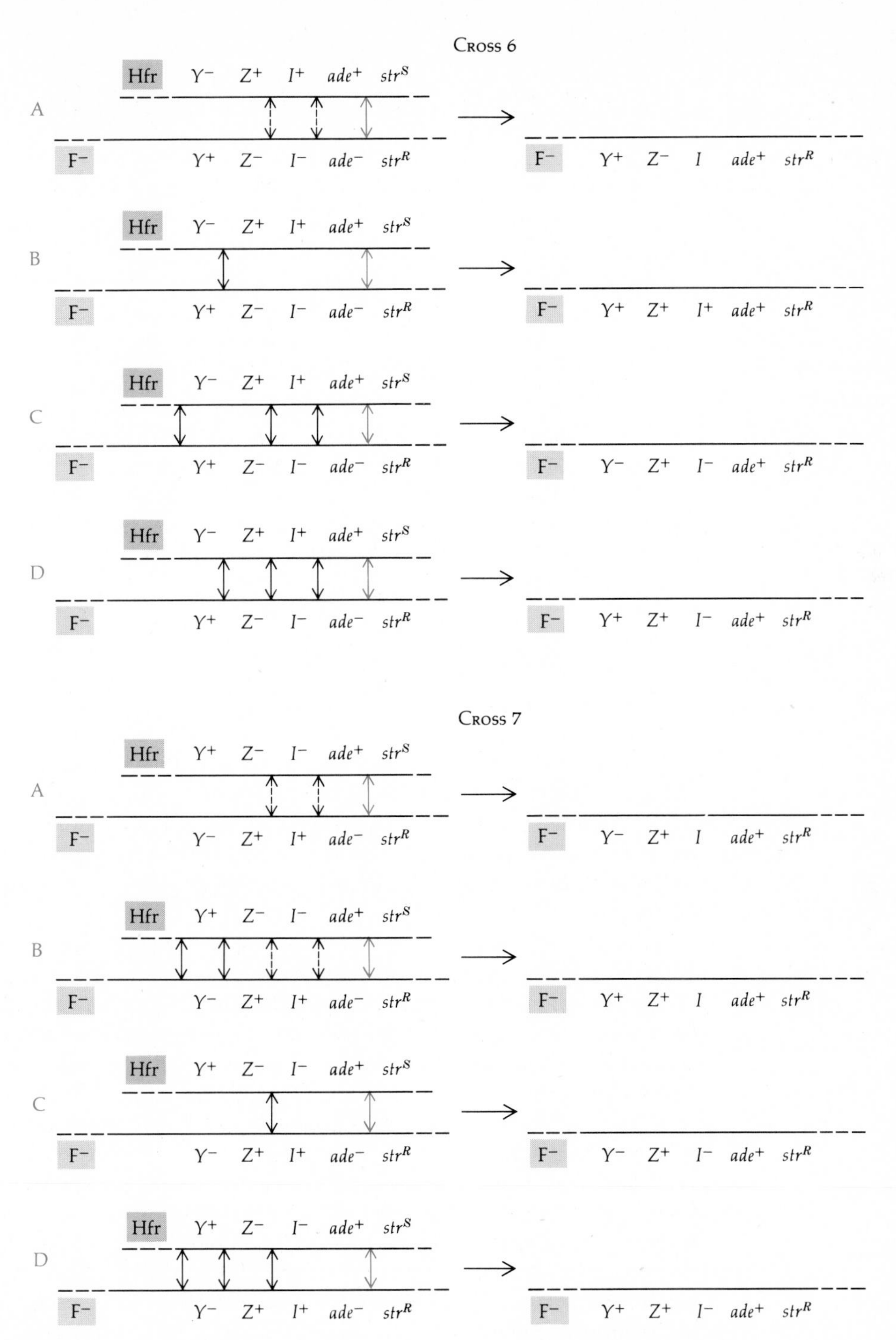

Figure 7.16
Origin of the selected and unselected recombinant genotypes recorded in Table 7.1 for crosses 6 and 7. All recombinants are selected for the presence of *ade*$^+$, the origin-distal marker, dictating the crossover shown in color; unselected crossovers are shown in black. Dotted lines indicate crossovers whose positions with respect to the *lacI* gene are not determined.

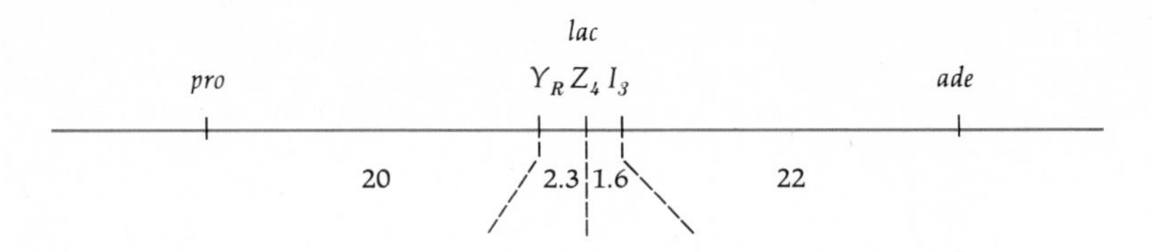

Figure 7.17
Genetic map of the *lac* region of *E. coli*.

Correlation of the physical map obtained by interrupted mating with the genetic map constructed from recombination data indicates that 1 minute = 20 map units. The entire *E. coli* chromosome has a length of 90 minutes, or 1800 map units. Since the *E. coli* chromosome contains about 3.2×10^6 np, 1 map unit corresponds to approximately 1750 np. Recombinational mapping is clearly useful only over short distances, because markers separated by no more than 3 minutes will assort independently (appear unlinked).

Transductional Mapping

It is desirable to confirm recombination mapping in merozygotes by performing reciprocal crosses, such as crosses 6 and 7. Constructing the Hfr and F^- strains employed in reciprocal crosses is often not a simple matter, and most recombination mapping in *E. coli* is therefore done by another method, using merozygotes created by transduction with the temperate bacteriophage P1.

When P1 infects a sensitive *E. coli* cell, it may follow a lytic course, resulting in the production of progeny phage, or it may lysogenize the cell. During the production of progeny phage, the circularly permuted P1 DNA (approximately 10^5 np) is packaged into the phage head by a headful mechanism, as discussed in Chapter 6 for T4. Occasionally, however, a fragment of the host-cell chromosome, which is broken down during lytic infection, is packaged into a head instead of the P1 DNA. The frequency of such defective phage particles is about 10^{-5} to 10^{-7} of the progeny phage produced. Defective P1 phages bearing *E. coli* DNA can be detected by the genetic markers present in that DNA. For example, if a thr^- cell is "infected" by a phage carrying a piece of *E. coli* DNA with a thr^+ gene, the thr^+ gene may be recombined into the bacterial chromosome to yield a prototrophic recombinant detectable by growth in the absence of threonine (Figure 7.18).

When P1 is grown on a thr^+ leu^+ azi^R host and then used to infect a thr^- leu^- azi^S recipient, only 3% of the selected Thr^+ recombinants are also Leu^+, and none are Az^R. However, if Leu^+ recombinants are selected, about 50% of these are also Az^R. Thus leu^+ is more closely linked to azi^R

Figure 7.18
Transfer of DNA from a donor bacterial cell to a recipient cell by encapsulation in the head of the transducing phage P1. The entire genome of the donor bacterial strain may be found, albeit in small pieces, in a large population of P1 phage. Only selection procedures permit the phage carrying the desired piece of host DNA to be identified through the bacterial recombinants produced as a result of its infection of a recipient cell.

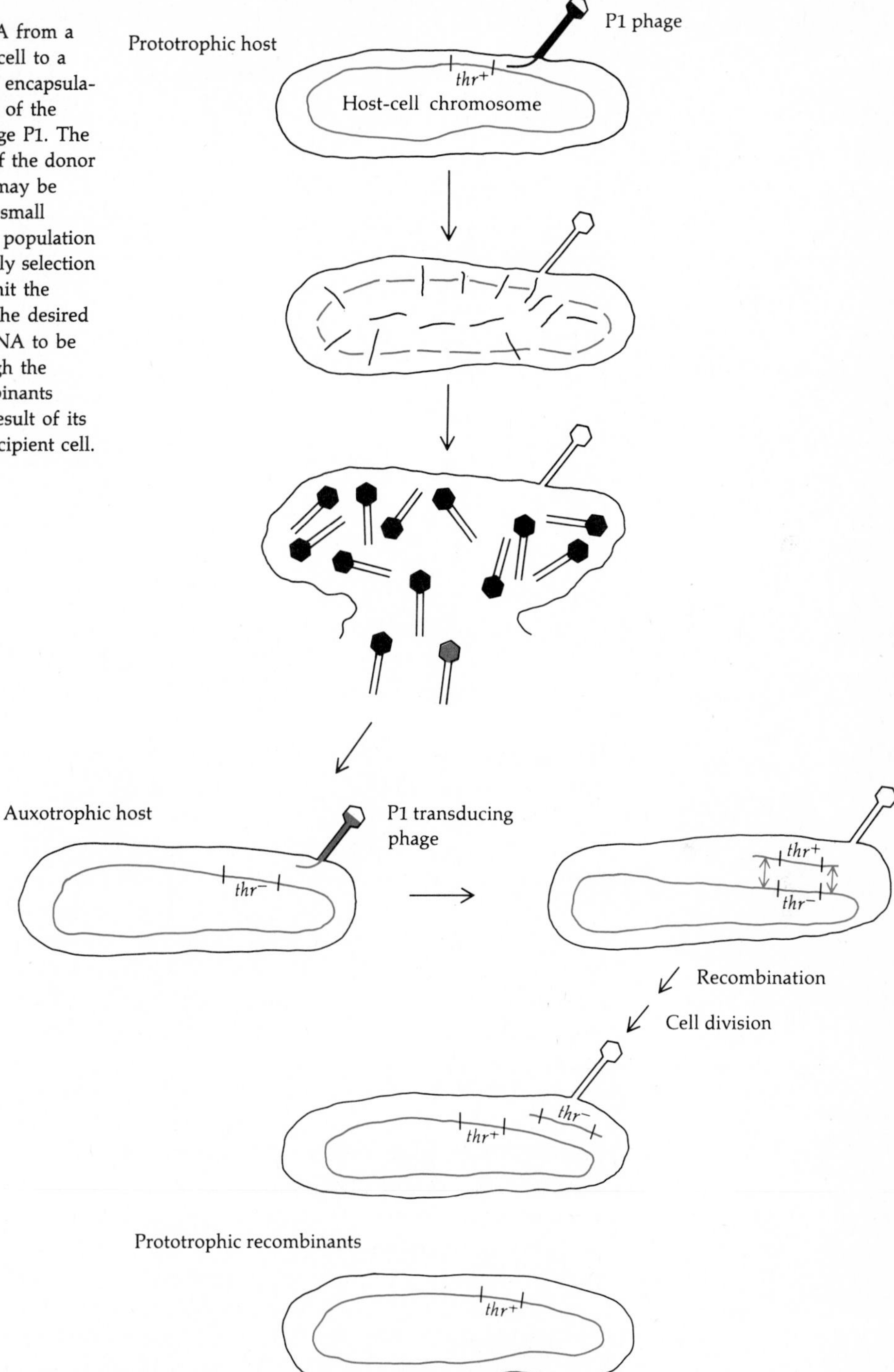

Table 7.2
Cotransduction of tryptophan auxotroph mutations with *cys*$^+$.

Donor Genotype	Recipient Genotype	Selected Marker	Unselected Marker	% Cotransduction of Unselected Marker with *cys*$^+$
cys$^+$ *trpE*	*cys*$^-$ *trpE*$^+$	*cys*$^+$	*trpE*	63
cys$^+$ *trpC*	*cys*$^-$ *trpC*$^+$	*cys*$^+$	*trpC*	53
cys$^+$ *trpA*	*cys*$^-$ *trpA*$^+$	*cys*$^+$	*trpA*	46
cys$^+$ *trpB*	*cys*$^-$ *trpB*$^+$	*cys*$^+$	*trpB*	47

Data modified from C. Yanofsky and E. S. Lennox, *Virology* 8:425 (1959).

than it is to *thr*$^+$ and the suggested order is *thr*$^+$ *leu*$^+$ *azi*^{R}. The actual frequencies of cotransduction of markers can be used to measure their degree of linkage. The observation that only 3% of *thr*$^+$ transducing phages also carry *leu*$^+$ indicates that these two genes are so far apart that they are rarely both included in a DNA fragment that fills the P1 head. The physical map produced by interrupted mating experiments shows these markers to be separated by a distance that is approximately 1/50 the length of the whole bacterial chromosome, or 6.4×10^4 np. This is in good agreement with the observation that the P1 phage head carries a DNA molecule of slightly less than 10^5 np.

Reciprocal three-factor crosses can be carried out easily by growing P1 on each bacterial strain and using the progeny phage to transduce markers into the other strain. Reciprocal three-factor transduction experiments allow the order of mutant genes on the chromosome to be established when cotransduction experiments fail to do so, as in the following example, which establishes the order of mutant genes causing tryptophan auxotrophy in *E. coli*. The data in Table 7.2 show the cotransduction frequencies of *cys*$^+$ and four closely linked Trp$^-$ auxotrophs. The frequency of cotransduction is the frequency with which recipient cells that acquire *cys*$^+$ (prototrophic for cysteine) also acquire the Trp$^-$ phenotype (auxotrophic for tryptophan). Thus, merozygotes are plated on minimal medium containing tryptophan. Colonies are then replica-plated to minimal medium to determine which ones are not prototrophs, i.e., which are Trp$^-$. The frequency of cotransduction suggests the order *cys–trpE–trpC–(trpA, trpB)*, but does not permit the relative order of *trpA* and *trpB* to be determined with any assurance, since the difference between 46% and 47% is, in this case, statistically insignificant.

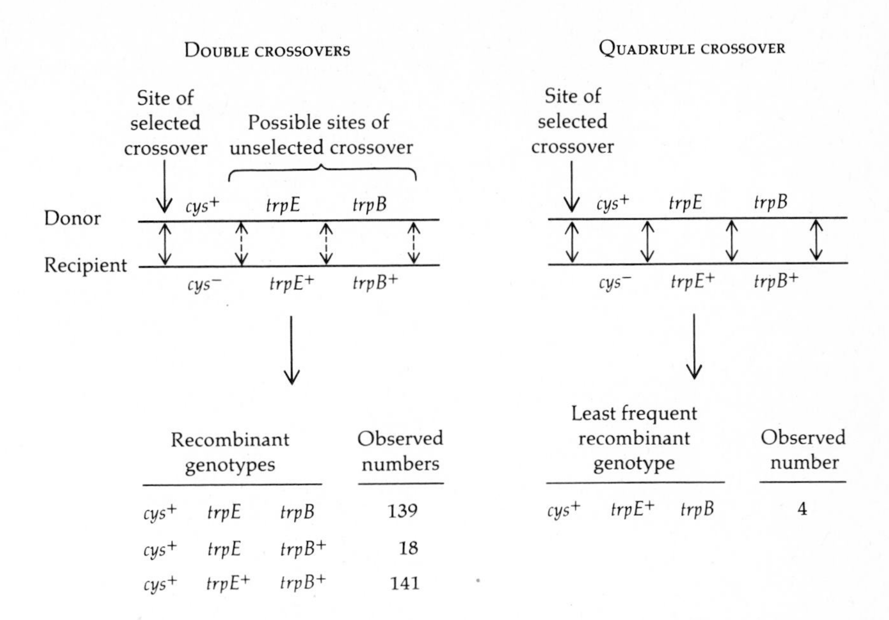

Figure 7.19
Genetic mapping by recombinational analysis in a merozygote created by transduction. Selection of the marker *cys*$^+$, which is known to be closely linked to *trp*, identifies a class of merozygotes that may also have received *trp* markers from the donor cell on the same piece of transferred DNA. The order of the markers can then be deduced from the least frequent recombinant class, which must result from a quadruple crossover.

A three-factor transduction cross involving *cys, trpE,* and *trpB,* diagrammed in Figure 7.19, confirms the order *cys–trpE–trpB.* This order is deduced from the recombinant genotype that is found in lowest frequency and must be the result of a quadruple crossover event. (Compare the frequency of genotype *cys*$^+$ *trpE trpB* with that of *cys*$^+$ *trpE*$^+$ *trpB*.) The scoring of the recombinant classes *trpE trpB*$^+$, *trpE*$^+$ *trpB*, and *trpE trpB,* all of which are Trp$^-$ in phenotype, is possible with selective media containing one or another intermediate in the tryptophan biosynthetic pathway. Biosynthetic pathways and auxotrophic mutations that interrupt them are discussed in detail in Chapter 10; here we view these mutations simply as genetic markers that can be scored.

Three-factor transduction experiments permit all four tryptophan auxotrophs, *E, C, B,* and *A,* to be ordered on a genetic map. These reciprocal three-factor crosses are recorded in Table 7.3. In these crosses the choice of selected markers dictates the selection of one crossover event. The position of the second of the required two crossover events is revealed by the assortment of the unselected markers in the cross. Some prior knowledge (from cotransduction experiments) of marker order is necessary in order that selected markers be adjacent, rather than flanking the

Table 7.3
Ordering of tryptophan auxotrophic mutations by three-factor transduction crosses.

Experiment	Donor Genotype	Recipient Genotype	Selected Markers	Recombinant Classes of Unselected Markers		Possible Order	Order Dictated by Relative Unselected Marker Frequencies
1	$E\ A\ C^+$	E^+A^+C	A^+, C^+	E	427	E–C–A	E–C–A
				E^+	120	E–A–C	
2	E^+A^+C	$E\ A\ C^+$	A^+, C^+	E	241	E–C–A	E–C–A
				E^+	65	E–A–C	
3	$E\ B\ C^+$	E^+B^+C	B^+, C^+	E	398	E–C–B	E–C–B
				E^+	83	E–B–C	
4	E^+B^+C	$E\ B\ C^+$	B^+	E^+C^+	5	E–C–B	E–C–B
				$E\ C^+$	87		
				E^+C, $E\ C$	1561	E–B–C	
5	E^+B^+A	$E\ B\ A^+$	B^+	E^+A^+	27	E–A–B	
				$E\ A^+$	6		
				E^+A, $E\ A$	1913	E–B–A	E–B–A
6	$E\ B\ A^+$	E^+B^+A	A^+, B^+	E	15	E–B–A	E–B–A
				E^+	107	E–A–B	

Data modified from C. Yanofsky and E. S. Lennox, *Virology* 8:425 (1959).

unselected marker on both sides. The nature of the crossover events leading to the recombinants formed in experiments 1 and 2 of Table 7.3 are diagrammed in Figure 7.20. The choice of one of the two possible gene orders in a cross is determined by the following principles: (1) the closer the unselected marker lies to the selected markers, the less frequently will the unselected crossover fall between them; (2) quadruple crossover events must occur less frequently than double crossover events. The choice between the two possible gene orders is dictated by the predominant genotype for the unselected marker in a cross. Thus, in experiment 1 of Figure 7.20, the genotype EC^+A^+ is found more frequently than

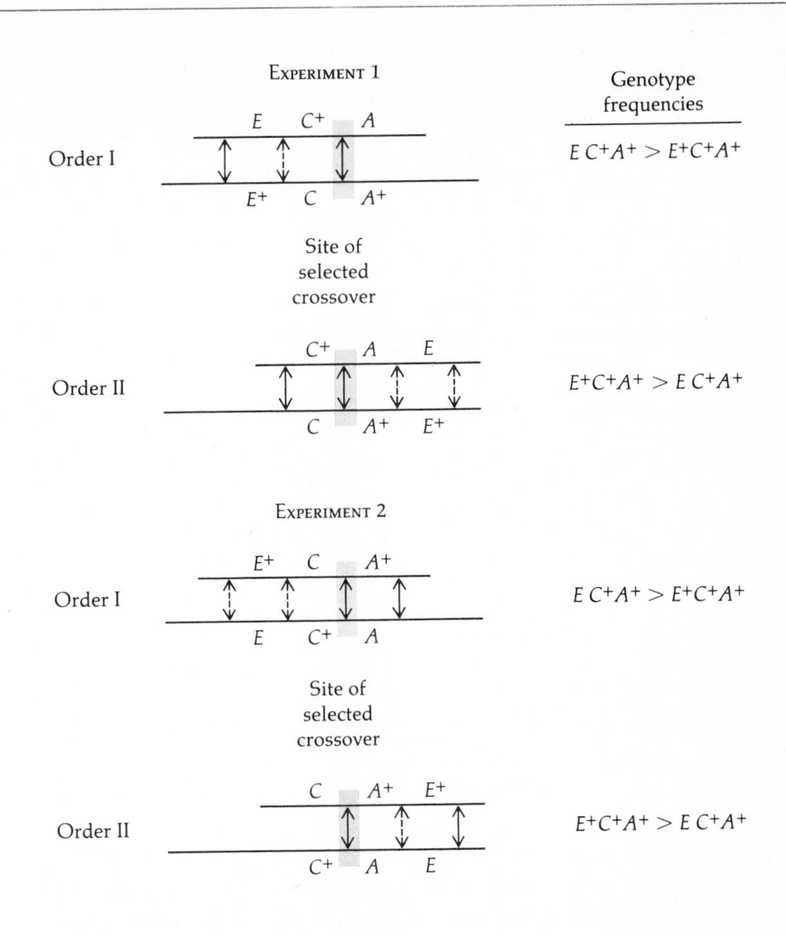

Figure 7.20
Example of the use of reciprocal crosses to deduce and confirm the order of markers present in three-factor transduction crosses. The relative frequencies of the two recombinant genotypes indicated are used to deduce the gene order.

$E^+C^+A^+$, suggesting the gene order E–C–A. This order is confirmed by experiment 2, the reciprocal cross, in which the most frequent genotype is again $E\ C^+A^+$.

The tryptophan auxotrophs mapped here constitute an operon in which the order of these genes happens to reflect the order of the biochemical steps leading to the synthesis of tryptophan. We have already seen that mutations that affect the utilization of lactose are located very close to one another on the chromosome (Figure 7.17). Such clustering of related genetic functions is one of the most important facts revealed by the study of genetic organization in bacteria. Recall also that clustering of related functions is observed in bacteriophages λ and T4 (Chapter 6). This genetic organization is not fortuitous but reflects the underlying mechanisms by which genetic functions are regulated in prokaryotic organisms. The methodology of genetic analysis exhibited in Chapters 5, 6, and 7 provides evidence of the organization of functional units in the genomes of both prokaryotic and eukaryotic organisms. The functional significance of this organization will be explored in Part II.

Problems

1. In *E. coli* the mutations *thr* and *leu* are auxotrophic mutations requiring threonine and leucine for growth. The mutant *ara3* renders the cell unable to grow with arabinose as the sole carbon source. The table below gives the frequency of cotransduction of these genes by phage P1. What selective medium is used in each case, and what is the order of the genes?

Recipient genotype	Selected marker	% Selected colonies containing the unselected marker: leu^+	thr^+	ara^+
ara3 leu thr	thr^+	4.1	—	6.7
	leu^+	—	1.9	55.4
	thr^+ leu^+	—	—	80.0
	ara^+	72.6	4.3	—

2. The mutations *ara1* and *ara2* are very closely linked to *ara3* and also render *E. coli* cells unable to use arabinose as the sole carbon source. The reciprocal transduction crosses carried out to order these mutations give the results in the table below. In all cases Ara$^+$ recombinants are selected and the fraction of these carrying the unselected markers leu^+ or thr^+ is recorded. What is the order of these markers?

Recipient	Donor	% Selected colonies containing the unselected marker: leu^+	thr^+
thr ara1 leu	*ara2*	64.4	1.2
thr ara2 leu	*ara1*	17.4	7.4
thr ara1 leu	*ara3*	26.1	6.4
thr ara3 leu	*ara1*	52.4	2.4
thr ara2 leu	*ara3*	14.3	9.5
thr ara3 leu	*ara2*	65.8	2.8

After J. Gross and E. Englesberg, *Virology* 9:314 (1959).

3. When *ara3* cells are used as a recipient and P1 is grown on wild type, *ara1*, or *ara2* as donors, and selection is carried out for Ara^+ transductants, numerous minute colonies, invisible to the naked eye, appear on the selection plates in addition to the normal-sized Ara^+ colonies. Form a hypothesis to account for these minute colonies.

4. Some strains of *Salmonella paratyphi* are motile because they possess flagella; others lack flagella and are nonmotile. The *Salmonella* general transducing phage P22 is grown on a motile strain and used to infect a nonmotile strain. When the infected nonmotile cells are deposited on the surface of a column of soft nutrient agar in a test tube and incubated, trails of small bacterial colonies are found to pass throughout the soft agar (Figure 7.21). Subculture of these colonies reveals that all the cells in the colonies are nonmotile. Explain.

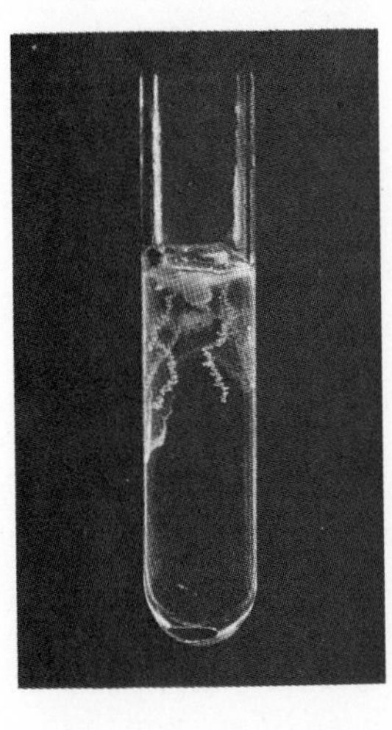

Figure 7.21
Trails of bacterial colonies (*Salmonella paratyphi*) produced in a tube of soft nutrient agar, as described in problem 4. [From J. Lederberg, *Genetics* 41:845 (1956).]

5. Different Hfr strains exhibit different origins and directions of transfer of the bacterial chromosome in mating to F^- strains. The order of transfer of genes close to the origin (*O*) of several different Hfr strains is given below. Use the data to construct a physical map of the bacterial chromosome.

Hfr strain	Order of transfer of genes
H	*O—thr—leu—azi—ton—pro—lac—ade*
4	*O—thi—met—ile—mtl—xyl—mal—str*
6	*O—ile—met—thi—thr—leu—azi—ton*
AB311	*O—his—trp—gal—ade—lac—pro—ton*
AB313	*O—mtl—xyl—mal—str—his*

6. An interrupted mating experiment is performed to locate the *arg7*$^+$ gene on the *E. coli* map, employing a prototrophic Str^S Hfr 4 strain (see problem 5) mated to F$^-$, Met$^-$ Mtl$^-$ Str^R, which is *arg7*$^-$. Following interruption of mating, portions of the culture are spread on plates containing glucose as a carbon source, streptomycin, and either arginine or methionine. From the table below determine the location of the *arg7*$^+$ gene.

Time of interruption of mating (min)	Number of recombinants	
	Arg$^+$ Str^R	Met$^+$ Str^R
0	0	0
6	0	0
8	3	0
10	19	4
11.5	56	28
13	126	62
14.5	217	183

7. A genetics student wishes to locate the map position of an auxotrophic mutant that he believes (from cotransduction experiments) is closely linked to *his*, by interrupted mating employing an Hfr AB313 strain that is Str^S, Ade$^-$ T1^S (see problem 5). He mates it to the F$^-$ auxotroph that is Str^R, Ade$^+$ T1^R and plates portions of the culture, following interruption of mating, on medium containing glucose and streptomycin. Very few recombinants are recovered, and those few that *are* recovered do not provide enough data to establish close linkage. He explains his difficulties to his genetics professor. She tells him to omit streptomycin from his selective medium and redo the experiment. He does, and obtains good data proving that the gene he wished to locate enters 2 minutes before *his* enters. Explain what was wrong with the first experiment, and why it worked the second time. Propose another selection scheme that would also work.

8. In some crosses involving lysogenic strains, prophages such as λ can be treated just as if they were bacterial genetic markers. The table below shows the results obtained when an Hfr H, Str^S nonlysogenic [(ly)$^-$] strain is crossed to F$^-$, Thr$^-$ Leu$^-$ Gal$^-$ Str^R strains lysogenic for one of a number of different prophages, e.g., F$^-$, Thr$^-$ Leu$^-$ Gal$^-$ Str^R (λ). Either Thr$^+$ Leu$^+$ Str^R recombinants or Gal$^+$ Str^R recombinants are selected, and the fre-

quency of the (ly)$^-$ Hfr marker is determined by sensitivity to infection by the phage in question, i.e., loss of the prophage from the recombinant is treated as an unselected marker. From the data, order the prophages on the physical map of *E. coli* with respect to each other and Thr$^+$ Leu$^+$ and Gal$^+$.

Prophage	Fraction of recombinants that are nonlysogenic: Thr$^+$ Leu$^+$ Str^R	Gal$^+$ Str^R
λ	0.16	0.82
21	0.01	0.10
82	0.21	0.89
434	0.11	0.68
424	<0.01	0.03
381	0.025	0.15

9. When an Hfr strain that is lysogenic for λ is crossed to an F$^-$ strain that is nonlysogenic, a phenomenon called *zygotic induction* can occur. The effects of zygotic induction can be seen by comparing the recombinants obtained in reciprocal crosses involving the prophage between Hfr H, Str^S and F$^-$, Thr$^-$ Leu$^-$ Az^R T1^R Lac$^-$ Gal$^-$ Str^R strains in the table below. Propose an explanation for zygotic induction that will account for the observed effect on recombinant formation, and propose an experiment that will support your explanation.

		Selected and unselected markers in recombinants:									
		thr$^+$ *leu*$^+$ *str*^{R} (selected)					*gal*$^+$ *str*^{R} (selected)				
	Ratio	Unselected marker frequency					Unselected marker frequency				
Cross	Thr$^+$ Leu$^+$ Str^R / Gal$^+$ Str^R	*azi*^{S}	*ton*^{S}	*lac*$^+$	*gal*$^+$	(λ)	*thr*$^+$ *leu*$^+$	*azi*^{S}	*ton*^{S}	*lac*$^+$	(λ)
Hfr H (λ)$^-$ × F$^-$ (λ)$^+$	3.7	0.92	0.73	0.49	0.31	0.15	0.75	0.75	0.74	0.74	0.84
Hfr H (λ)$^+$ × F$^-$ (λ)$^-$	54	0.86	0.60	0.21	0.025	0.001	0.82	0.79	0.78	0.74	0.01

10. Diagram possible recombinants that might be segregated from a newly formed F′ strain in which the F′ element carries the bacterial genes thr^+ leu^- and the bacterial chromosome is thr^- leu^+. Designate the location of the sex factor in each recombinant.

11. Propose two ways to determine whether a recombinant colony selected from a cross between Hfr and F^- strains is F′ or F^-.

8

The Fine Structure of the Gene

With the acceptance of the chromosome theory of inheritance, genes came to be thought of as beads on a chromosomal string. Mutant alleles of a single gene were considered as beads of different colors, with only one bead of a particular color on each string. Recombination was believed to involve breaking and rejoining of the string at positions between the beads, while recombination within a gene was thought not to occur. The gene was regarded as an indivisible entity, and it was even defined as the basic unit of recombination as well as the unit of function and mutation. The bead theory of the gene held sway until about 1940, owing to the lack of resolving power of the genetic systems being studied.

Then in two favorable cases in *Drosophila*, wild-type recombinants were obtained between mutations that were believed to be allelic on the basis of phenotypic criteria. The term *pseudoallele* was adopted to define such mutations, which phenotypically appear to be alleles but which undergo recombination with one another. Not until the introduction of microbial genetic systems, with their great resolving power, were questions concerning the possible substructure of genes fully settled. These latter studies sharply defined the basic units of mutation, recombination, and genetic function. The genetic investigations that split the gene into its subunits began at about the same time that the Watson-Crick model for the structure of DNA was proposed. The elucidation of the fine structure of the gene bridged the gap between the genetic map and the physical structure of the genetic material itself. These studies destroyed the indivisible bead theory and led to the concept of the gene as a sequence of nucleotide pairs in a DNA molecule. They revolutionized genetics just as

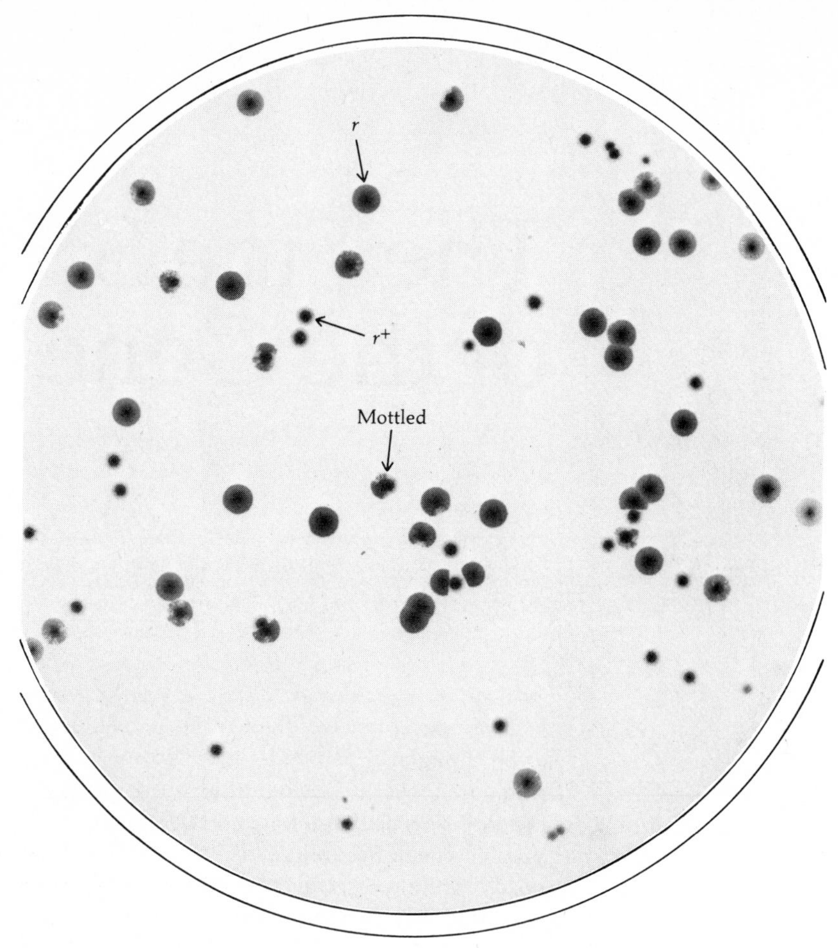

Figure 8.1
Bacterial lawn of *E. coli* B containing plaques made by T4 r^+ and T4 *rII* phages. Several mottled plaques formed by the presence of both genotypes at a focus are also present. (From G. S. Stent and R. Calendar, *Molecular Genetics,* 2nd ed., W. H. Freeman, San Francisco, 1978.)

the discovery of the divisibility of the atom had revolutionized physics at the turn of the century.

The *rII* System of Bacteriophage T4

The classical fine-structure analysis of the gene was carried out by Seymour Benzer with *rII* mutants of the phage T4 during the 1950s. Benzer exploited two selective advantages that these mutants provide. First, *rII* mutants can easily be selected in large numbers by their distinctive plaque morphology when plated on *E. coli* strain B. The *r* stands for *rapid lysis* of the infected host, which results in the formation of a much larger plaque than is made by a wild-type phage (Figure 8.1). One *rII* plaque among 2000

Table 8.1
Phenotypes exhibited by T4 wild type and *rII* mutants.

Phage	Phenotype when Plated on:	
	E. coli B	*E. coli* K(λ)
T4 rII^+	Small plaques	Small plaques
T4 *rII*	Large plaques	No plaques (lethal)

rII^+ plaques can be easily detected on a single Petri plate. The use of chemical mutagens greatly increases the mutation rate to *rII* above the spontaneous rate; consequently, hundreds of independent *rII* mutants can be obtained. Second, Benzer discovered that *rII* mutants are unable to produce progeny when infecting strain *E. coli* K(λ), which is lysogenic for bacteriophage λ. The *rII* mutants adsorb to *E. coli* K(λ), and inject their DNA, but the infected cells die without yielding progeny phage. The rII^+ wild-type T4, however, reproduces normally on *E. coli* K(λ). The physiological cause of this *rII* lethality in *E. coli* K(λ) is not well understood; it is known, however, that the product of the λrex^+ gene is responsible. This gene and the *cI* gene coding for the λ repressor are the only two λ genes expressed by the prophage. This system was the first conditional lethal system to be exploited in phage genetics (Table 8.1). The inability of *rII* mutants to grow on *E. coli* K(λ) provides a selective screen whereby a single rII^+ phage among as many as 10^6 *rII* phages can be detected on a single Petri plate. Thus, rare wild-type recombinants between two different *rII* mutants can be detected easily.

The Nature of Mutations in the *rII* Region

The Watson-Crick model for DNA suggests that alteration of a single nucleotide pair in the normal nucleotide sequence of a gene might result in a mutant phenotype. A change of a single nucleotide pair would be expected to exhibit the following properties: (1) the mutation should revert to a normal phenotype at some frequency approximating that at which the original mutation arose; (2) it should occupy a single point on the genetic map; (3) it should recombine with all other point mutations except those that are due to an independent alteration of the same nucleotide pair.

Table 8.2
Initial characterizations of eight T4 *rII* mutants.

Mutant	Map Position	Relative Reversion Rate
r47	0	$<0.01 \times 10^{-6}$
r104	1.3	$<1 \times 10^{-6}$
r101	2.3	4.5×10^{-6}
r103	2.9	$<0.2 \times 10^{-6}$
r105	3.4	1.8×10^{-6}
r106	4.9	$<1 \times 10^{-6}$
r51	6.7	170×10^{-6}
r102	8.3	$<0.01 \times 10^{-6}$

After S. Benzer, *Proc. Natl. Acad. Sci. USA* 41:344 (1955).

Some of the *rII* mutants studied by Benzer exhibit these properties, but others do not. The data in Table 8.2 show that a wide range of reversion rates is exhibited by *rII* mutations that do recombine with one another. Some of these mutations are quite stable and do not revert [make plaques on *E. coli* K(λ)] at an observable rate; others do revert to wild type at measurable and characteristic rates. A genetic map of the *rII* mutants in Table 8.2 is shown in Figure 8.2. The map distances are calculated from the frequency of recombinant rII^+ progeny, using the relation

$$\text{Map distance} = \frac{2 \times \text{number of } rII^+ \text{ progeny}}{\text{total number of progeny}} \times 100 \qquad (8.1)$$

As discussed in Chapter 6, recombination experiments are carried out by infecting permissive cells (strain B) with both mutant T4 strains. The total number of progeny phages is assayed by plating progeny on strain B, and the number of rII^+ recombinants is determined by plating on the restrictive host, strain K(λ). Since the reciprocal recombinants, which are double *rII* mutants, are not detected, they are assumed to be as frequent as wild-type recombinants. Thus the number of wild-type recombinants is multiplied by two in calculating map distances.

Recombination analysis of a larger set of 60 independent *rII* mutants produced the map in Figure 8.3. Careful examination of this map will show

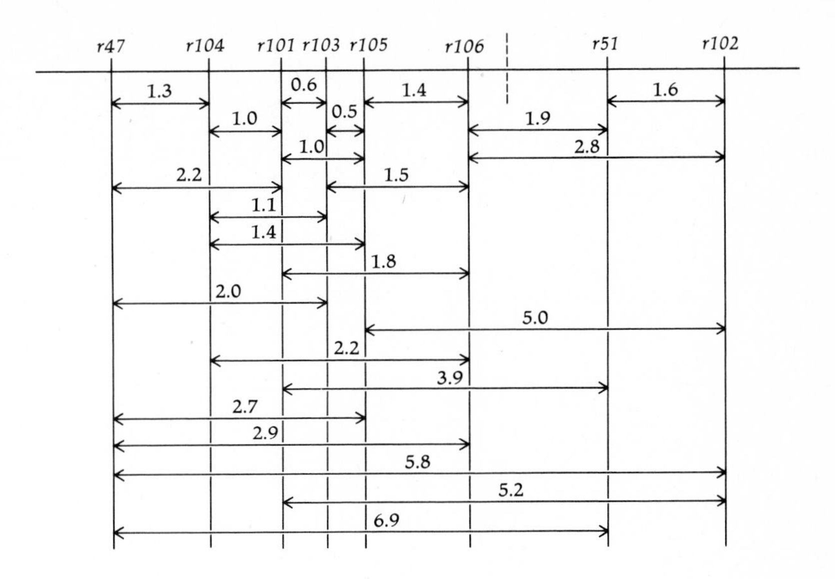

Figure 8.2
Genetic map of the T4 *rII* mutations listed in Table 8.2. The figures between arrowheads are map distances. [From S. Benzer, *Proc. Natl. Acad. Sci. USA* *41*:344 (1955).]

that some *rII* mutants (those that do not revert at a measurable frequency) do not recombine to yield rII^+ progeny with all other *rII* mutants, but only with some *rII* mutants. Consequently these mutations occupy a portion of the genetic map larger than that of a point mutation (the size depending on the particular mutation) and are represented as a bar overlapping the mutations with which they do not recombine. For example, consider the mutations mapping within region *a* of Figure 8.3. Mutant *168* recombines with *295* and *312*, but none of these three mutants yields wild-type recombinants with *47*. All four of these mutants yield recombinants with *145*, *282*, and *228*. If mutants *168* and *295* are single-point mutations, then mutant *47* must be due to alteration of more than one point on the genetic map (more than a single nucleotide pair).

Mutants such as *47*, which lack all of the properties expected of single-nucleotide-pair alterations, have been shown to be deletions of a number of contiguous nucleotide pairs. Recombination studies with double *rII* mutations demonstrate that such mutations contract the genetic map between flanking genetic markers, as would be expected if a portion of the DNA between the flanking markers were deleted (Figure 8.4).

An important conclusion of these studies is that *rII* mutations, which are phenotypically identical, may result from either the alteration of a single nucleotide pair or the deletion of some number of nucleotide pairs. The properties exhibited by deletion mutations are not unexpected. The loss of a number of nucleotide pairs from a gene would not be expected to be a reversible process, since both the number and proper sequence of the nucleotide pairs would have to be restored. Likewise, wild-type recombinants should not be produced in a cross between a deletion mutation and a point mutation located within the region deleted in the other ge-

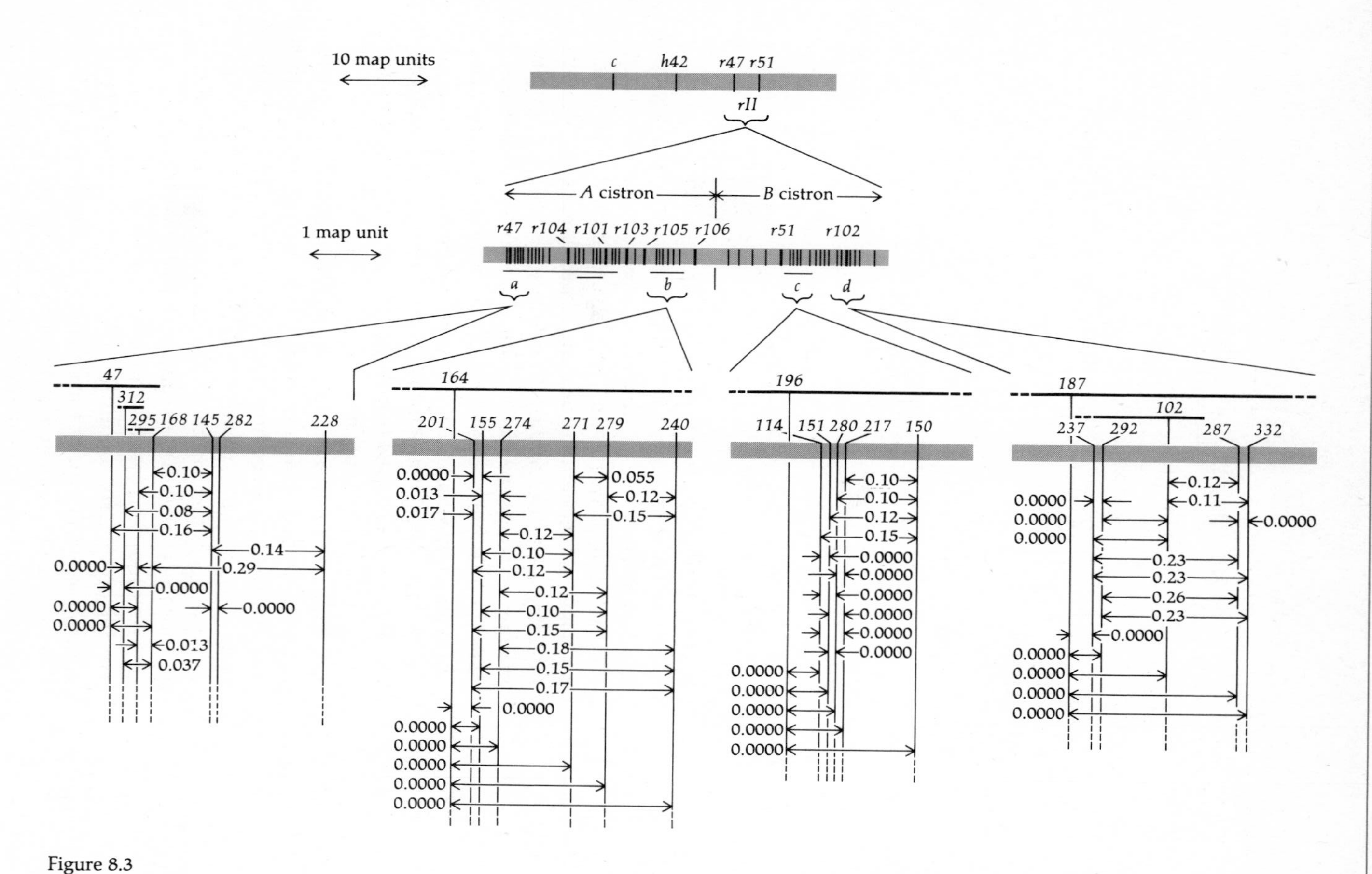

Figure 8.3
A fine-structure map of portions of the T4 *rII* region. Figures between arrowheads in the expanded maps of microclusters *a*, *b*, *c*, and *d* are the percent of wild-type recombinants observed between pairs of mutants in two-factor crosses. [From S. Benzer, *Proc. Natl. Acad. Sci. USA* 41:344 (1955).]

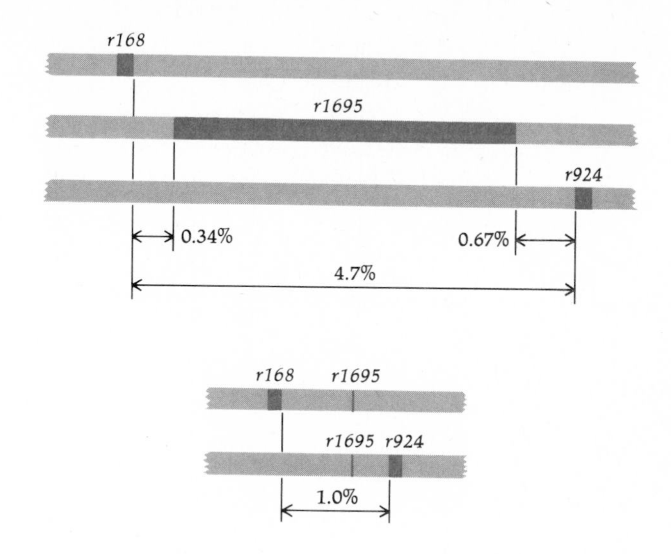

Figure 8.4
Schematic illustration of how an *rII* mutation, which occupies a segment of the genetic map, contracts the map distance between flanking *rII* mutations when it is present in both parents in the cross. Contraction of the genetic map is evidence that such a mutation is a deletion. [After M. Nomura and S. Benzer, *J. Mol. Biol.* 3:684 (1961).]

nome. Neither genome involved in such a cross contains the correct nucleotide pair at the site of the point mutation; therefore, restoration of a wild-type nucleotide sequence by recombination is impossible.

The functional properties of phenotypically identical *rII* mutations have been examined by complementation analysis as well as recombination analysis. Different *rII* mutants are used to infect the restrictive host *E. coli* K(λ) in pairwise combinations, as described in Chapter 6. If the doubly infected cells yield progeny, the two mutations must affect different genetic functions. The results of complementation analysis demonstrate that *rII* mutations, with the exception of some deletion mutations, fall into one of two complementation groups, designated *A* and *B* (Figure 8.3). Thus, the *rII* phenotype can be caused by the loss of either of two genetic functions, *A* or *B*. For example, mutation *104* in complementation group *A* does not yield progeny in double infections with mutations *47, 101,* or *106* (Figure 8.3). Mutation *104* does yield progeny in double infections with mutations in complementation group *B,* e.g., *51* and *102.* The exceptional deletions are ones that overlap the boundary between the two complementation groups and have lost both functions.

The complementation test (discussed more fully in Chapters 5 and 6) employed by Benzer examines the functional relationship between mutations in a *trans* configuration (see Figure 8.5). If the two mutations being tested are in the same complementation group and in the *trans* configuration, they will fail to complement and progeny phage will not be produced, whereas if the two mutations are in the *cis* configuration, progeny phage will be produced. For example, if a double *rII* mutant and a wild-type T4 both infect a cell of strain K(λ), progeny phages of *both* genotypes

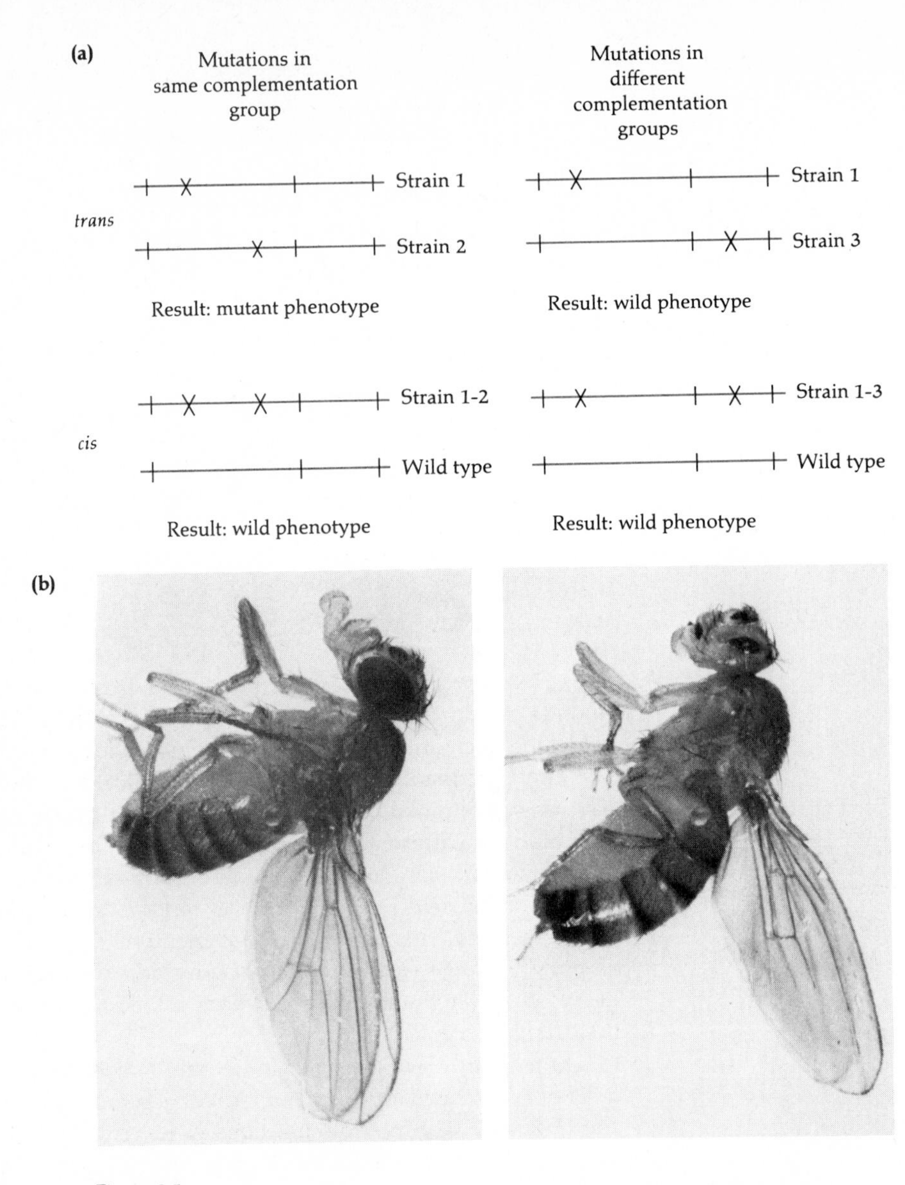

Figure 8.5
(a) The *cis-trans* comparison test, an example of the complementation test, applied to *rII* mutations. If two mutations under study show different phenotypes in *cis* and *trans*, they affect the same genetic function. If they show identical phenotypes, they affect different genetic functions.
(b) The *cis-trans* test applied to two *Drosophila* mutations affecting the eye: *Star* (*S*) and *asteroid* (*ast*). The mutant *S*/+ exhibits a dominant phenotype owing to a loss of function in the diploid cell; the single wild-type gene does not supply sufficient function to provide a wild phenotype. The fly on the left is *S ast*/+ +, and that on the right is *S* +/+ *ast*. Because the phenotypes of these two genotypes are different, the two mutations are believed to affect the same function in different ways. [Photographs from E. B. Lewis, *Cold Spring Harbor Symp. Quant. Biol.* *16*:159 (1951).]

are produced. The wild-type genome provides functions required for the successful replication of both mutant and wild-type genomes. In contrast, if the two mutations being tested are in different complementation groups, both the *trans* and *cis* configurations will exhibit the same phenotype, i.e., both will yield progeny phage.

These two alternative phenotypes of a doubly infected complex, possible when two mutations are in the *cis* or *trans* configuration, provide a genetic test for function that is analogous to the *cis-trans* test, first employed by Edward B. Lewis, to study the functional interaction of pseudoalleles at a number of loci in *Drosophila*. For example, the mutations *Star* (*S*) and *asteroid* (*ast*) affect the shape and structure of the *Drosophila* eye. These mutations were once thought to be alleles because they were observed to segregate from each other during meiosis. The *ast* mutation was first designated *star*-recessive (*s*) to indicate its allelism to *Star*-dominant (*S*). However, extensive observation of the progeny of females of genotype *al S ho*/ + *s* + backcrossed to *al s ho*/*al s ho* males showed that occasionally s^+ (wild-type) progeny could be found. Such wild-type progeny were found at a frequency of about 1 in 10,000, and all were observed to be recombinant for the flanking markers *al* (*aristaless*) and *ho* (*h*eld-*o*ut wings). Their genotype was + + *ho*/*al s ho*. The existence of such recombinant progeny suggested that *S* and *s* must be mutant at different sites in the chromosome, *S* to the left of *s*. The *star*-recessive mutation was then renamed *asteroid* to reflect this fact. However, since both *S* and *ast* affect the phenotype of the eye in similar fashion, the question of their functional relationship was still unanswered. Genotypes with these mutations in the *cis* (*S ast*/+ +) and *trans* (*S* +/+ *ast*) configurations were then constructed to investigate their functional (phenotypic) interaction. The *cis* and *trans* configurations of these mutations exhibit very different phenotypes (Figure 8.5), leading to the conclusion that they both affect the same genetic function. The term pseudoallele was then introduced to indicate this functional relationship between mutations exhibiting very close linkage.

In recognition of the utility of the *cis-trans* test for defining functional genetic units, Benzer coined the term *cistron* for the complementation groups in which the *rII* mutations are found. The use of the term cistron as a synonym for complementation group or gene has since become common.

Deletion Mapping of *rII* Mutations

The fact that a mutation will not yield wild-type recombinants when crossed with a deletion that overlaps that mutation on the genetic map provides a powerful analytical tool, which Benzer exploited to easily map thousands of independent *rII* mutations to subregions of the *rII* map. The judiciously chosen set of deletion mutations shown in Figure 8.6 divides the *rII* region into 47 segments defined by the adjacent endpoints of pairs of deletions. In as few as a dozen crosses, in which only the presence or absence of wild-type recombinants is scored by plating on *E. coli* K(λ), a new *rII* mutation can be mapped to one of the 47 segments of the region. The procedure employed is diagrammed in Figure 8.7. Once a number of independent *rII* mutations have been mapped to a specific segment, these mutations need only be mapped with respect to one another by further crosses, without the need also to map them with respect to other independent mutations in other segments.

The fine-structure map of spontaneous *rII* mutations produced by this method (Figure 8.8) shows that these mutations are scattered more or less randomly over the extent of the genetic map, with the exception of two "hot spots" that appear to be highly mutable regions of the DNA (see Chapter 16 for further discussion of hot spots). This map demonstrates the important fact that many sites within the *rII* cistrons are capable of mutation to a loss of function, conferring identical phenotypes. The number of possible *rII* alleles is clearly very large. Further differentiation of the genetic fine structure is indicated by the fact that spontaneous point mutations mapping at different positions revert to wild type at characteristically different rates. Therefore the nature of these different spontaneous mutations of identical phenotype cannot be identical. A comparison of the spontaneous mutation map with similar maps constructed with mutations induced by specific chemical mutagens shows quite different and characteristic patterns of mutagenic activity (Figure 8.8), further revealing the fine structure of the genetic material. Mapped mutable sites, identified by mutations from all sources, total 200 for the *rIIA* cistron and 108 for the *rIIB* cistron. Heteroduplex mapping of *rII* deletions by electron microscopy has subsequently shown that *rIIA* comprises 1800 np and *rIIB*, 850 np. The paucity of mutants at many sites and the failure to identify as many sites as there are nucleotide pairs suggest that the *rII* region has not been saturated. Further, many potentially mutable sites may remain undetected because their alteration does not produce a detectable phenotype. Owing to the degeneracy of the genetic code (see Chapter 12), a nucleotide change does not invariably cause an amino acid change in the protein coded by a cistron. Clearly, however, the number of mutable sites within the *rII* region capable of causing a mutant phenotype is a significant portion of the nucleotide pairs present.

The Ultimate Resolution of Recombinational Analysis

The *rII* mutations provide a means for testing the ultimate resolution of recombinational analysis. As mentioned earlier, even one rII^+ recombi-

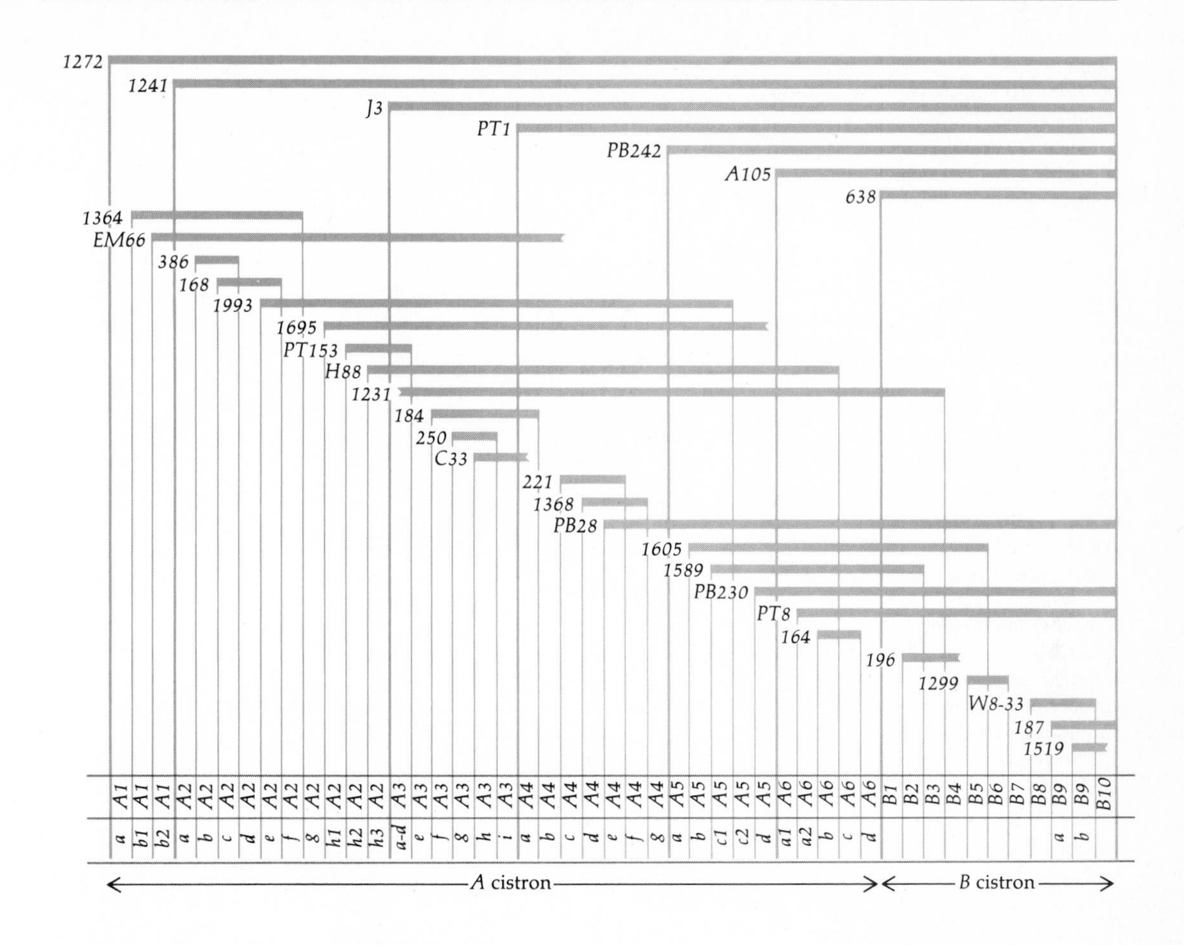

Figure 8.6
Deletions whose adjacent endpoints divide the *rII* region into 47 smaller segments. The seven large deletions at the top divide the region into seven smaller segments. The remaining deletions divide each of these segments into still smaller segments. Those ends shown fluted are not used to define a region. The extent of the *A* and *B* cistrons is indicated. [From S. Benzer, *Proc. Natl. Acad. Sci. USA* 47:403 (1961).]

nant phage in 10^6 *rII* phages can be detected by plating the phage on strain K(λ). Thus, by Equation 8.1, a resolution of 0.0002 map unit between two *rII* mutations is theoretically possible ($2 \times 10^2/10^6 = 2 \times 10^{-4}$). However, the smallest observed map distances between various *rII* mutations are only about 0.02 map unit (see Figure 8.3). Smaller map distances are not found, despite the great sensitivity of the system. This estimate of the minimum possible observable map distance is close to that which would be expected, from a rough calculation, if adjacent nucleotide pairs were able to recombine. The T4 chromosome contains 1.8×10^5 np and is 1500 map units in length. Thus 0.02 map unit corresponds to approximately 2 np [$(0.02/1500) \times 1.8 \times 10^5$ np $= 2.4$ np]. As this is only a rather crude estimate that ignores such factors as negative interference (see below), it is reasonable to conclude that mutations at adjacent nucleotide positions within a gene are capable of recombining. This has been shown conclusively in the fine-structure map of the *trpA* gene (see Figure 11.20). Thus, those point mutations (Figure 8.3) that yield no rII^+ recombinants when crossed are very likely due to independent alterations of the same nucleotide pair.

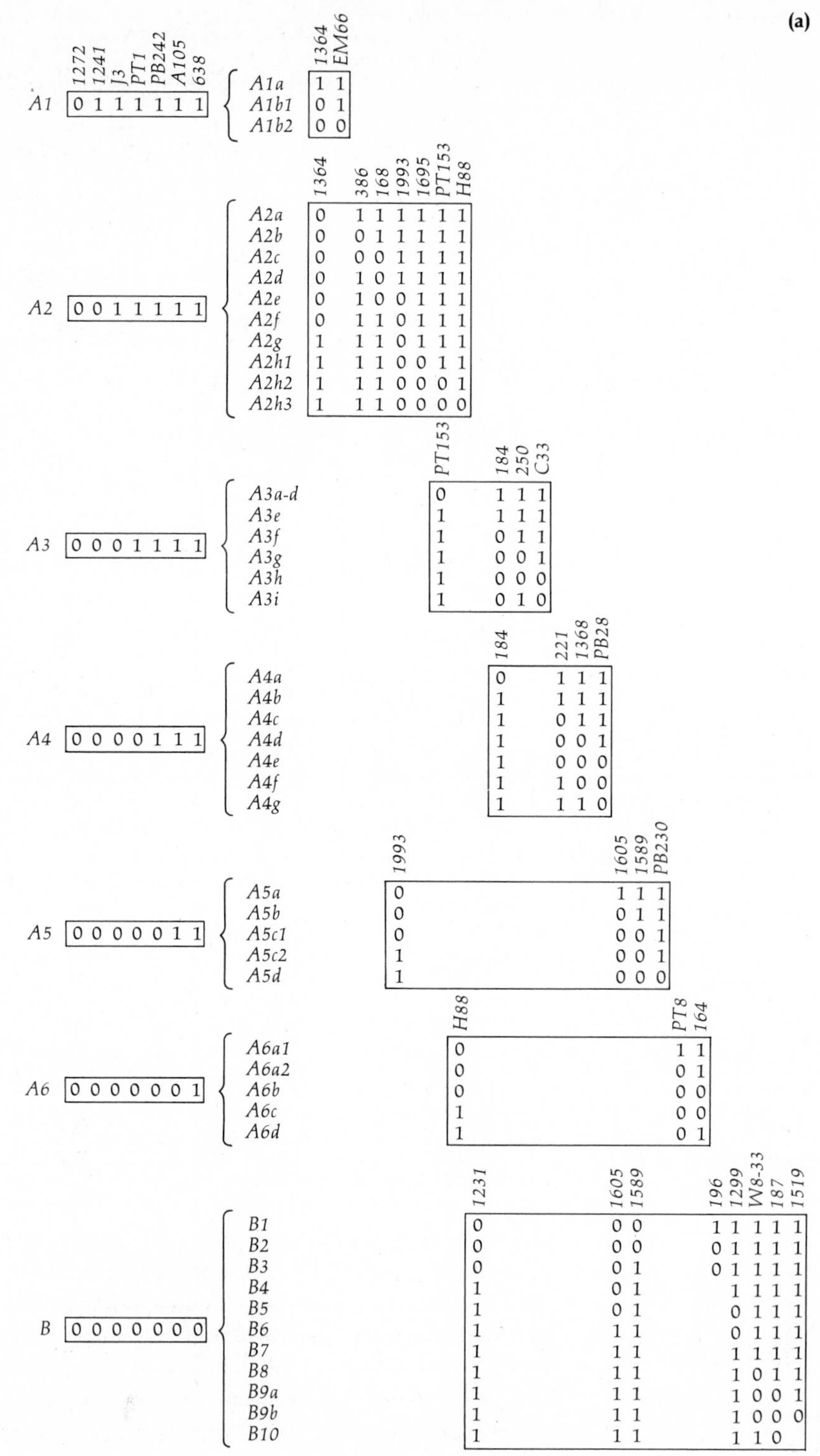

Figure 8.7
(a) Two-step localization of an unmapped *rII* mutation to one of 47 small regions in the *rII* map. The mutant is first crossed with each of the "big seven" deletions in Figure 8.6. The number 0 in the left set of boxes signifies no detectable recombination, and 1 signifies some recombination. The number of zeros defines the major segment in which the mutant resides. Next the mutant is crossed with each of the deletions that define smaller segments of the major segment in which the mutant is located. The results obtained with a number of different *rII* mutations are recorded in the right set of boxes. The number of zeros again defines the subsegment in which the mutation resides.

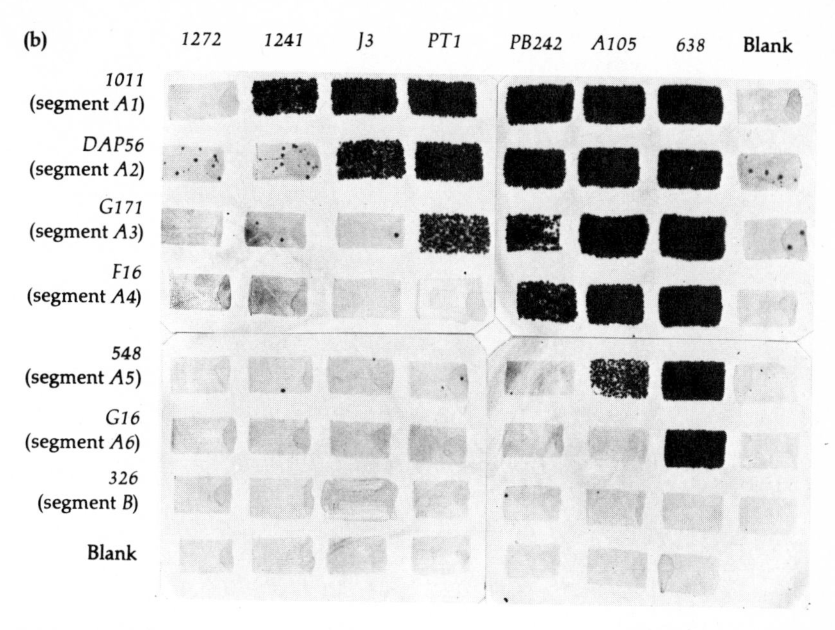

Figure 8.7 (*continued*)
(b) A sample of some plates showing the results of an abbreviated test for production of wild-type recombinants devised by Benzer. 0.5-ml aliquots of a culture of *E. coli* B (permissive host) are inoculated with both a standard deletion mutation and the *rII* mutation being tested, at a multiplicity of about five of each phage per cell. After time is allowed for adsorption of the phage, a drop of the culture is picked up on a strip of sterile paper and laid briefly on a plate inoculated with *E. coli* K(λ) (restrictive host). If wild-type recombinants are produced in the infected cells, they will infect the lawn of *E. coli* K(λ), producing clearing of the area covered by the paper strip. A negative result signifies that the proportion of recombinants is less than about 10^{-3}% of the progeny. The appearance of only a few plaques in the blank is due to reversion of the point mutation. [From S. Benzer, *Proc. Natl. Acad. Sci. USA* 47:403 (1961).]

A Redefinition of Genetic Terminology

The fine-structure analysis of the *rII* region provides precise definitions for genetic entities all of which had previously been referred to under the broad term *gene*. As discussed above, the unit of genetic *function* is designated by Benzer as the *cistron* and is defined by the *cis-trans* test; *cistron* is synonymous with *complementation group* and *gene*. The unit of genetic *mutation*, which Benzer designates a *muton*, is the smallest unit of the cistron capable of mutation; a muton is certainly equivalent to a nucleotide pair in

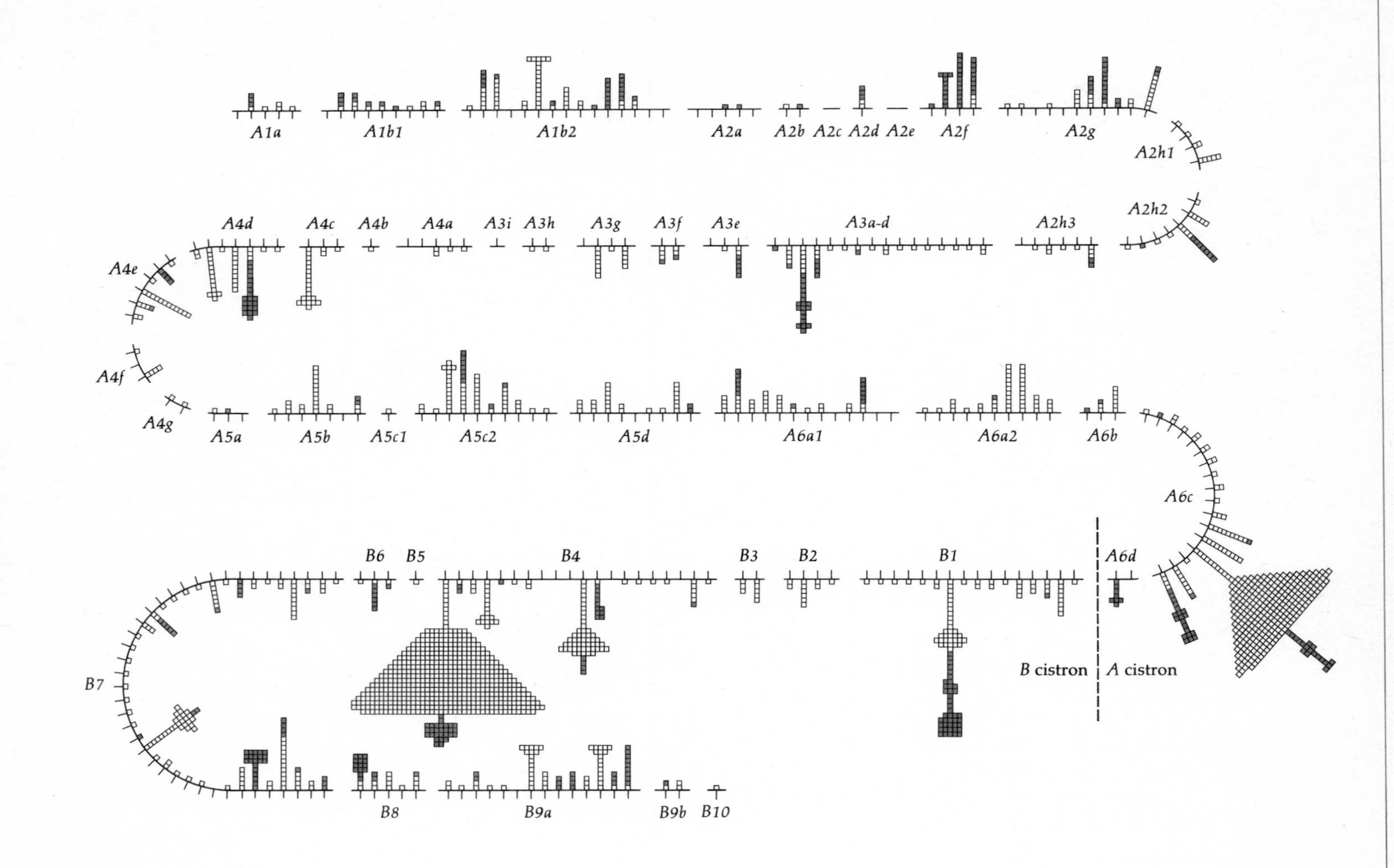
A1a
A1b1
A1b2
A2a
A2b
A2c
A2d
A2e
A2f
A2g
A2h1
A2h2
A2h3
A3a-d
A3e
A3f
A3g
A3h
A3i
A4a
A4b
A4c
A4d
A4e
A4f
A4g
A5a
A5b
A5c1
A5c2
A5d
A6a1
A6a2
A6b
A6c
A6d
B cistron
A cistron
B1
B2
B3
B4
B5
B6
B7
B8
B9a
B9b
B10

Figure 8.8 (*opposite*)
The fine-structure map of spontaneous *rII* mutations produced by the method described in Figure 8.7. Each white square represents an independent spontaneous mutation. The separation of mutations within each of the 47 small segments is based on further crosses between these mutations. Two major "hot spots" for spontaneous mutations are evident. Each gray square represents a mutation induced by the action of nitrous acid, and each colored square, by 5-bromouracil (5-BU). [After S. Benzer, *Proc. Natl. Acad. Sci. USA* 47:403 (1961).]

the DNA. The unit of genetic *recombination*, which Benzer designates a *recon*, is also an individual nucleotide pair in the DNA.

The term *allele* can now also be defined more precisely. Thus at each site in the DNA of a gene, four true alleles of the gene are possible: AT, TA, GC, and CG. These alleles can never recombine with one another (one of them may be defined as the wild type). If different nucleotide pairs within a gene are changed by mutation, the two forms of the gene are termed *heteroalleles* (or pseudoalleles). Heteroalleles can recombine to produce new recombinant alleles (one of which may be defined as the wild type) and can be distinguished from one another by recombination analysis. Complementation analysis, however, which resolves only functional units, will not usually distinguish between heteroalleles and true alleles (see Chapter 10 for a discussion of exceptions to this generalization). The total number of true alleles and heteroalleles possible for any gene is a function of the size of that gene; it is 4^n, where n is the number of nucleotide pairs constituting the gene.

Fine-Structure Analysis in a Higher Eukaryote—*Drosophila*

Fine-structure mapping of heteroalleles in higher eukaryotic organisms has been confined almost exclusively to *D. melanogaster* because of its short generation time and ease of culture. It is generally a long, tedious enterprise to rear and determine the genotypes of the thousands of progeny that are required to detect crossing over of heteroalleles. In a few cases, chemical selection techniques for the detection of rare recombinants have been developed that permit a resolution comparable to that obtained with conditional lethal systems in microorganisms.

The most extensively analyzed cistron in *Drosophila* is the *rosy* (*ry*) locus, which is the structural gene for the enzyme xanthine dehydrogenase (XDH). Flies that lack XDH activity (null-activity mutants) are easily identified by their rosy eye color, caused by the absence of the pigment isoxanthopterin. Larvae that lack XDH are also very sensitive to the toxic effect of purine (Figure 8.9) in their food; purine-supplemented medium thus provides a means for screening for rare wild-type recombinants between different XDH null heteroalleles. Females heterozygous for two *rosy* heteroalleles (ry^x/ry^y) that are to be tested for recombination are crossed in large numbers to males heterozygous for two other *rosy* heteroalleles (ry^A/ry^B). Since recombination does not occur in male *D. melanogaster*, the only heterozygous wild-type progeny produced will be due to recombination between ry^x and ry^y and will have one of two possible genotypes: ry^+/ry^A or ry^+/ry^B. Larvae heterozygous for ry^+ have enough XDH activity to survive purine in their food. All other progeny will be

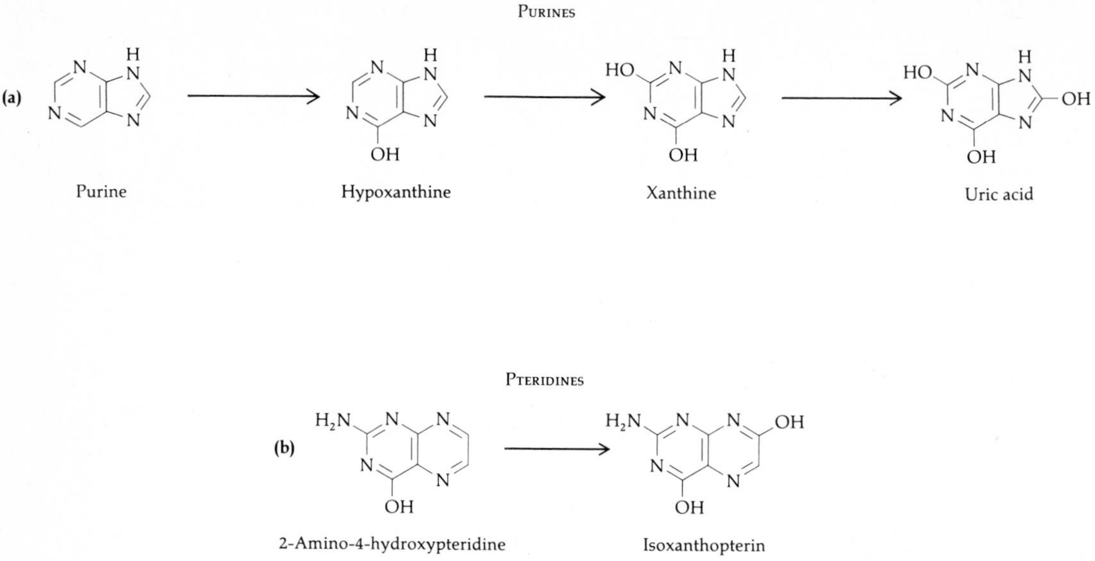

PURINES

(a) Purine → Hypoxanthine → Xanthine → Uric acid

PTERIDINES

(b) 2-Amino-4-hydroxypteridine → Isoxanthopterin

Figure 8.9
Chemical conversions catalyzed by xanthine dehydrogenase (XDH). **(a)** The elimination of toxic purine. **(b)** The production of isoxanthopterin, an eye pigment in *Drosophila*.

heterozygous for *rosy* heteroalleles, lack XDH activity, and die from the purine, which is added to the culture bottles after the parents have been removed. A typical recombination experiment may involve treating several hundred identical culture bottles with purine to select for wild-type recombinants. The total number of progeny screened can be estimated by omitting purine from one or two of the culture bottles and counting all the progeny that emerge. The fine-structure map obtained for XDH null mutants is shown in Figure 8.10. In addition, the map positions of several purine-sensitive mutants with low XDH activity but wild-type eye color are shown. The map positions of several electrophoretic variants of the XDH enzyme that have an altered amino acid but that retain enzyme activity are also shown (see Chapters 10 and 18 for discussions of electrophoretic variants of enzymes).

The map positions of the sites causing the electrophoretic variants cannot be directly mapped employing the purine-selection system, because they do have XDH activity. XDH null mutations are first induced in these variant genes and selected on the basis of rosy eye color; the sites at which these null mutations reside are different from the electrophoretic-variant sites. The electrophoretic sites are then treated as unselected markers in crosses between two null mutations in which ry^+ recombinants are selected using purine. This procedure is identical to that discussed for three-factor phage crosses in Chapter 6.

The XDH enzyme consists of two identical polypeptide subunits of 160,000 daltons each. A length of DNA comprising approximately 4500 np is required to code for a polypeptide of this size (see Chapter 12). If we as-

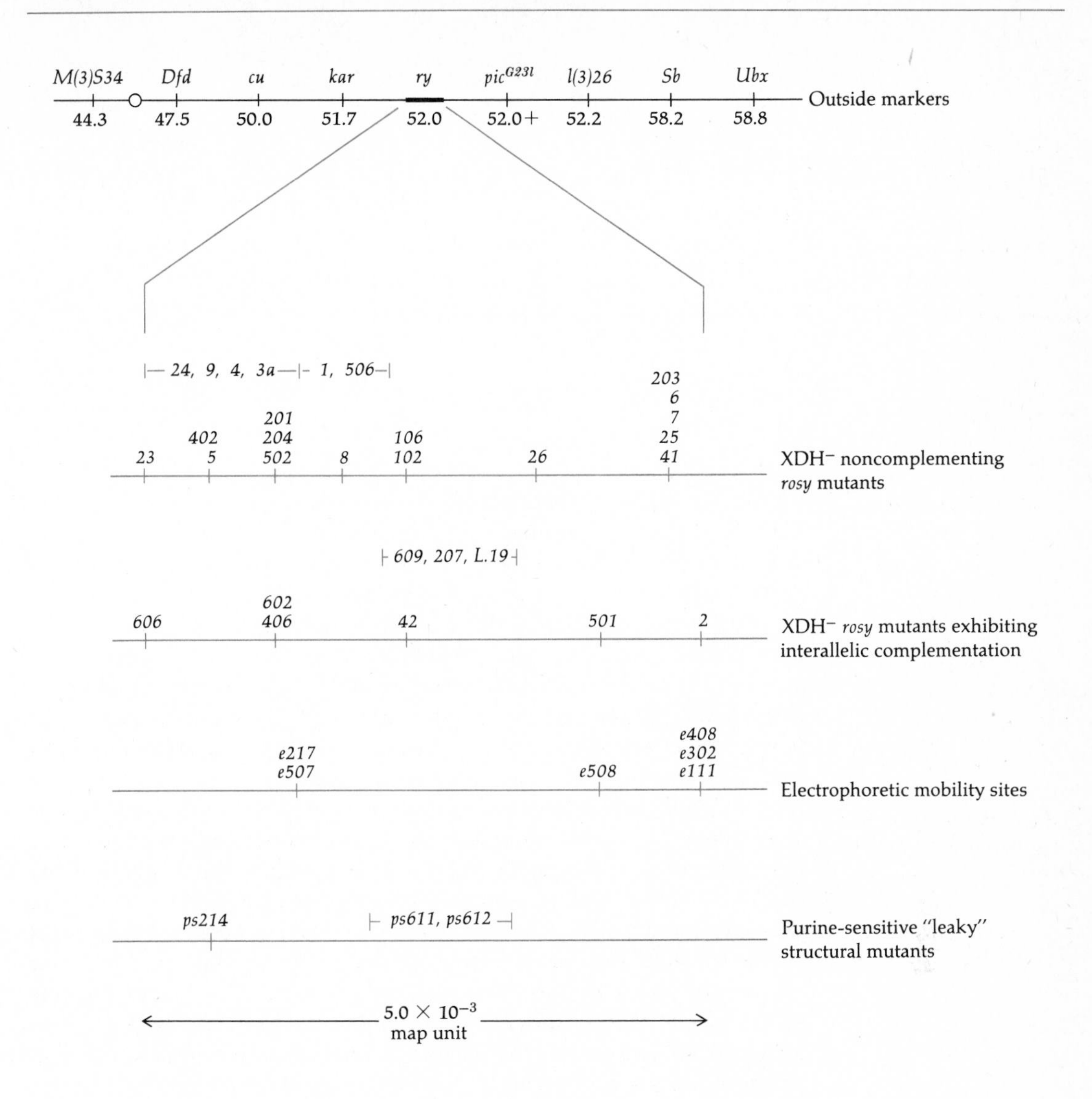

Figure 8.10
Fine-structure map of the *rosy* gene of *Drosophila melanogaster.* [After W. Gelbart et al., *Genetics* 84:211 (1976).]

sume that the XDH mutants define the entire length of the XDH structural gene (0.005 centimorgan), then *0.01 centimorgan = 9000 np* of DNA. The euchromatic portion of the haploid genome, within which most *Drosophila* genes map, contains approximately 1.5×10^8 np in 275 centimorgans; thus *0.01 centimorgan = 5400 np.* The closeness of these two independent estimates of the relation between map units and DNA structure suggests that fine-structure genetic analysis in *Drosophila* has a resolution comparable to that which is possible in microorganisms.

Fine-structure analysis has been carried out for a number of other *Drosophila* genes for which special selection systems are not available, but

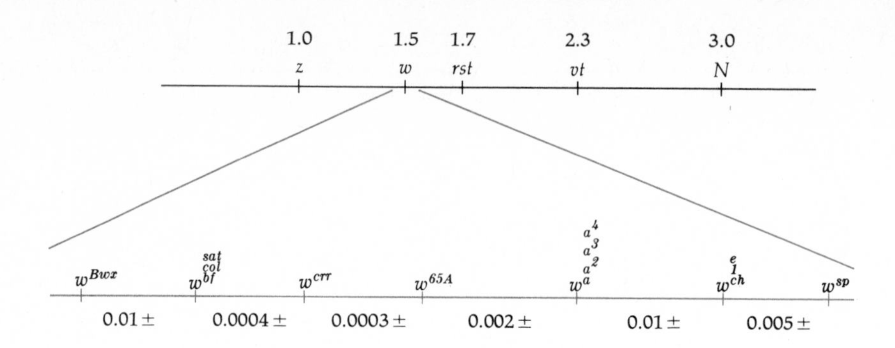

Figure 8.11
Fine-structure map of the *white* gene of *Drosophila melanogaster*. [After H. M. LeFever, *Dros. Inform. Serv.* 50:109 (1973).]

for which persistent geneticists are. The *white* eye-color gene has long been famous for the large number of its mutant alleles, each of which produces a different (non-wild-type) eye color when homozygous or when heterozygous with another mutant allele. The allele w^1, originally described by Morgan, is exceptional in that it causes a completely white eye; most other mutant alleles cause eye colors such as carrot, coffee, blood, or apricot. All of these alleles are functional alleles as determined by the *cis-trans* test. Recombinational analysis has produced the map shown in Figure 8.11, in which seven sites are found, with many alleles clustering at three of these sites. Constructing this map required the visual examination of millions of progeny flies. Although complementation tests indicate that mutation at all of these sites affects a function needed for the appearance of a normal eye color, some functional diversity also exists within the *white* gene. Mutations at the two right-most sites act as dominant suppressors of the eye-color mutant *zeste*, whereas mutations at the remaining five sites do not. The mechanism of this suppression is unknown.

Another gene that has been extensively studied is the *lozenge* (*lz*) gene. Mutant alleles of this gene can exhibit *pleiotropic effects* (independent effects on apparently unrelated functions) on a number of *Drosophila* organs: the shape, structure, and pigment distribution of the eye are affected; the tarsal claws are affected; and the fertility of the female and the development of her reproductive structures are affected. The functional alleles that have been analyzed by recombination are found to cluster at four sites, as shown in Figure 8.12.

The fine-structure maps of the *white* and *lozenge* genes differ from the map of the *rosy* gene (Figure 8.10) in two respects. First, the size of the *white* gene is 0.025 centimorgan and that of the *lozenge* gene is 0.140 centimorgan, as compared to 0.005 centimorgan for the *rosy* gene. Second, the mutant sites in the *rosy* gene are more or less equally distributed throughout the map, while this is clearly not true of *white* or *lozenge* mutants. The latter two maps contain gaps larger than the whole *rosy* map. The natures of the w^+ and lz^+ functions are not known, but Melvin M. Green has suggested that

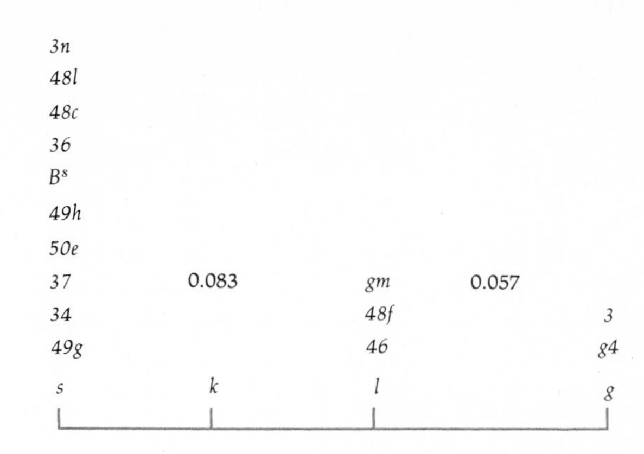

Figure 8.12
Fine-structure map of the *lozenge* gene of *Drosophila melanogaster*. [After D. Lindsley and E. Grell, *Genetic Variations of Drosophila melanogaster*, Carnegie Institution of Washington Publ. No. 627 (1968).]

the organization of the *white* and *lozenge* genes must be fundamentally different from that of the *rosy* gene, which appears to correspond closely to the cistronic organization present, for example, in T4.

If we assume that genetic map distances more or less reflect the amount of DNA present, then two quite different models can be proposed for the organization of the *white* and *lozenge* genes. The first model proposes that these genes code for enormously large polypeptides and that certain sites in these genes (hot spots) are more highly mutable than others, as reflected by the clustering of mutant sites on the maps. The second model proposes that these genes contain long stretches of nonfunctional DNA separating functional stretches that code for a polypeptide. Recombination frequencies would then reflect physical distances, and the clustering of mutant sites would be only *apparent*—it would reflect a lack of sufficient resolving power, owing to the absence of a selective screen such as is used to resolve *rosy* mutations. The recent discoveries of eukaryotic structural genes containing large intervening noncoding DNA sequences provide some support for this second model (see Chapter 11).

High Negative Interference

The concept of interference (I) was introduced in Chapter 5 and defined as $I = 1 - c$, where c, the coefficient of coincidence, is the ratio of observed to expected double crossovers in a three-factor cross. For most crosses involving markers in three different linked genes (x–y–z), a crossover in the interval between x and y is observed to reduce the probability of a second crossover in the interval between y and z. Thus c is less than 1 in such crosses, and I is a positive number that denotes the magnitude of the interference observed (see Chapter 5).

Recombinational analysis of very closely linked markers (usually heteroalleles) in both prokaryotic and eukaryotic organisms frequently discloses values of c much larger than 1, which give large negative values to I; this is called *high negative interference.* In other words, in fine-structure studies double and even multiple crossing over occurs more frequently than expected. This fact has been evident in some of the examples of mapping data presented in Chapters 6 and 7 for phages and bacteria. High negative interference (sometimes called *localized* negative interference) is exhibited, for example, in a number of three-factor crosses between mutants of phage λ. The value of c for each cross is graphed in Figure 8.13 as a function of the sum of the map distances in the two intervals between each set of three mutations. For outside markers separated by less than one map unit, the value of c is observed to exceed 70.

Four- or five-factor crosses, which permit triple and quadruple crossovers to be observed, reveal that these events are also observed much more frequently than would be expected if they were independent events. A naive interpretation of a high value of c is that double crossovers occur much more frequently within a small region of DNA than would be expected on the basis of the frequency of crossing over between each pair of markers. However, as will be discussed in the next section, high negative interference does not arise from multiple crossing over within small regions of the chromosome, as the definition of interference would imply. Instead, it is a result of events that occur within the region of a single crossover.

It should also be noted in Figure 8.13 that, even for very distantly separated outside markers, the value of c is never less than 1, as it is in the crosses discussed in Chapter 5. This *low negative interference* is usually observed in all phage crosses. It is a result of some parental phages, being excluded from the mating pool of phage genomes, as mentioned in Chapter 6.

Gene Conversion

The cause of the high negative interference observed in phage crosses remained a mystery until high negative interference came to be studied in fungi and was associated with a phenomenon called gene conversion.

In fungi, all the products of meiosis are recovered as a group of spores contained within an ascus, as discussed in Chapter 5. When two fungal parent strains carrying different alleles of a particular gene mate, each contributes equally to the diploid cell that is formed and that then undergoes meiosis. The diploid cell is heterozygous: it carries the two alleles of the gene in question. Following meiosis, the spores within an ascus are expected to contain equal numbers of each of the two alleles involved in the cross. Alleles of a gene responsible for spore pigmentation in the fungus *Sordaria fimicola* permit this expectation to be observed quite easily.

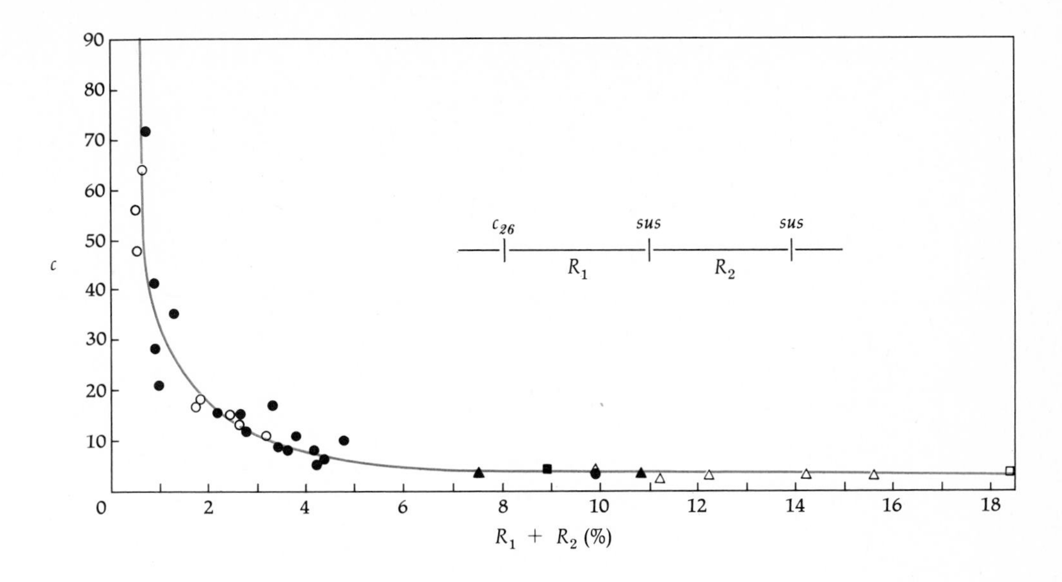

Figure 8.13
Coefficient of coincidence, c, is a function of the sum of the map distances R_1 and R_2 between λ markers in a number of different three-factor crosses, as indicated in the inset. [From P. Amati and M. Meselson, *Genetics* 51:369 (1965).]

The segregation patterns of a gray-spored mutant allele and the black-spored wild-type allele of the gene are drawn in Figure 8.14. Depending on whether the alleles segregate at the first or second meiotic division, one of the two patterns on the left is usually observed: each ascus contains four wild-type black spores and four mutant gray spores (4 + :4m). However, when many asci are scored, a few aberrant segregation patterns can be found. Five types of aberrant asci are shown in Figure 8.14 along with the actual numbers of each type observed.

A priori we might expect that these aberrant asci result from the mutation of wild type to gray or of gray to wild type during meiosis. Extensive studies have shown that mutation is not the cause of such aberrant asci, however. They result instead from a process associated with recombination that is called *gene conversion.* Gene conversion has been observed in many organisms, including *Drosophila,* from which all, or half, of the products of meiosis can be recovered. It is observable in *Drosophila* because the use, for example, of attached-X chromosomes permits half of the tetrad to be recovered.

Gene conversion has been extensively studied in the yeast *Saccharomyces cerevisiae,* whose asci contain four spores that are not ordered as are those of *Sordaria* or *Neurospora.* Thus normal asci produced by a heterozygous yeast diploid contain alleles in 2:2 ratios, and aberrant asci exhibit 3:1 or 1:3 ratios. Seymour Fogel, Robert Mortimer, and their colleagues have made intensive studies of the conversion of heteroalleles of the *arg*4 cistron of *S. cerevisiae,* which codes for the enzyme argininosuccinase (Figure 8.15).

Three important observations concerning gene conversion have come from the studies of Fogel and Mortimer. First, half of all conversion events

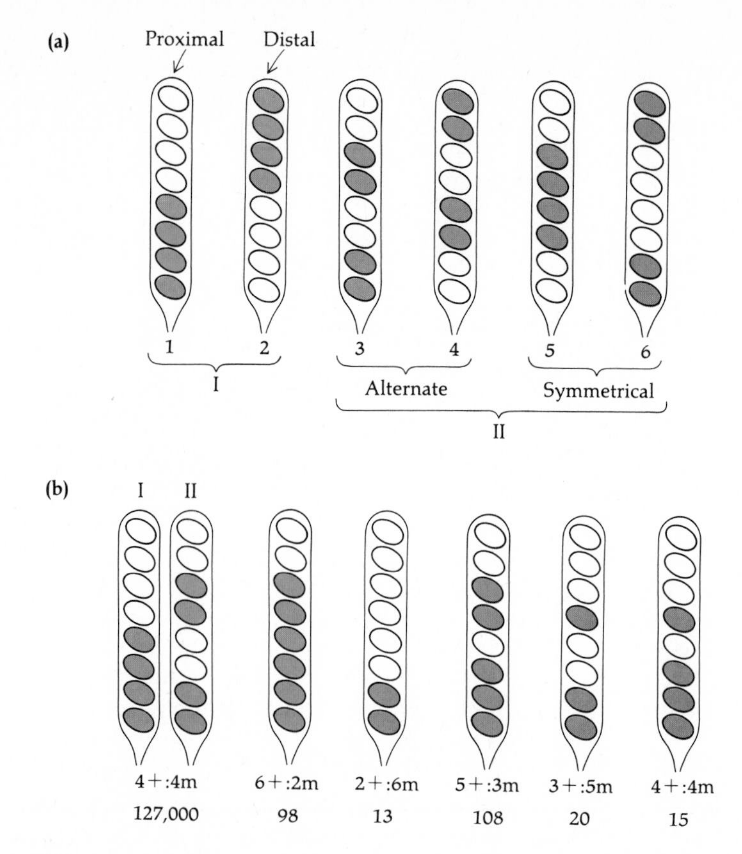

Figure 8.14
Normal and aberrant asci observed in a cross between a gray-spored mutant and its wild-type black-spored allele in the fungus *Sordaria fimicola.* **(a)** Normal segregation patterns are drawn and grouped: I. Proximal and distal patterns from first-division segregation. II. Alternate and symmetrical patterns from second-division segregation. **(b)** Normal and aberrant asci are grouped without regard to segregation pattern. The observed number of each type of ascus is given in the bottom line. [After Y. Kitani et al., *Am. J. Bot. 49*:697 (1962).]

are associated with recombination of flanking markers. This suggests that conversion is not a random mutational event, since there is no a priori reason to believe that recombination and mutation should be associated. Second, conversion is specific. The wild-type heteroallele of *arg4-4* is converted to *arg4-4*, and the wild-type heteroallele of *arg4-17* is converted to *arg4-17*. Random mutational events do not exhibit such specificity (see Chapter 16). Third, *coconversion* (simultaneous conversion of two heteroalleles) occurs much more frequently than would be expected for independent events. This suggests that conversion involves a *region* of the chromosome, rather than single sites. Moreover, the frequency of coconversion for closely linked heteroalleles (*2* and *17*) is greater than for more distantly linked ones (*4* and *17*). This also suggests that conversion affects

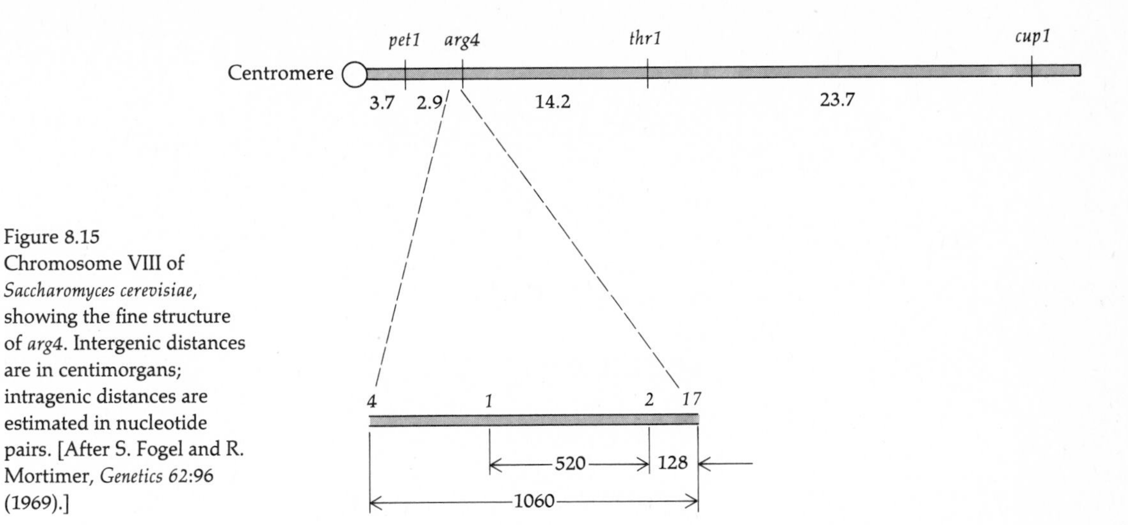

Figure 8.15
Chromosome VIII of *Saccharomyces cerevisiae*, showing the fine structure of *arg4*. Intergenic distances are in centimorgans; intragenic distances are estimated in nucleotide pairs. [After S. Fogel and R. Mortimer, *Genetics 62*:96 (1969).]

a region of the chromosome rather than single sites, because an event occurring in a region of the chromosome is more likely to affect two sites if they are close to each other rather than distantly separated. A sample of some of the data leading to these conclusions is presented in Box 8.1.

Gene conversion is a consequence of two factors that are discussed in detail in Chapter 9. First, the mechanism of recombination creates partially heteroduplex DNA molecules. Heteroduplex DNA may be formed with or without the recombination of outside flanking markers (Figure 8.16). If a region of heteroduplex DNA overlaps a region containing heterozygosity, mispaired nucleotides will be present, as diagrammed in Figure 8.16. Second, cells possess mechanisms for detecting and correcting mispaired nucleotides by removing a mispaired nucleotide from one of the two strands of the helix and employing the other strand as a template to effect repair synthesis. As far as is known, the repair of a mispaired nucleotide lacks any strand specificity; consequently a given mismatch may be repaired in one of two ways, or it may not be repaired at all if it fails to be detected prior to the next replication of the DNA molecule. Some possible patterns of repair of mismatched nucleotides and their genetic consequences are diagrammed in Figure 8.16. Depending upon which mismatched nucleotides are repaired, single-site conversion, co-conversion, or no conversion may occur. Some patterns of repair mimic the occurrence of double or triple crossing over, as indicated in Figure 8.16. The creation of *apparent* double- or triple-crossover chromatids leads to a high coefficient of coincidence and the observation of high negative interference when these chromatids, or DNA molecules, are part of a randomly sampled population of progeny molecules. That is, if random spores, rather than whole asci, were used to study the frequency of recombination between the markers diagrammed in Figure 8.16, the occurrence of gene conversion would be missed, and only the multiple-

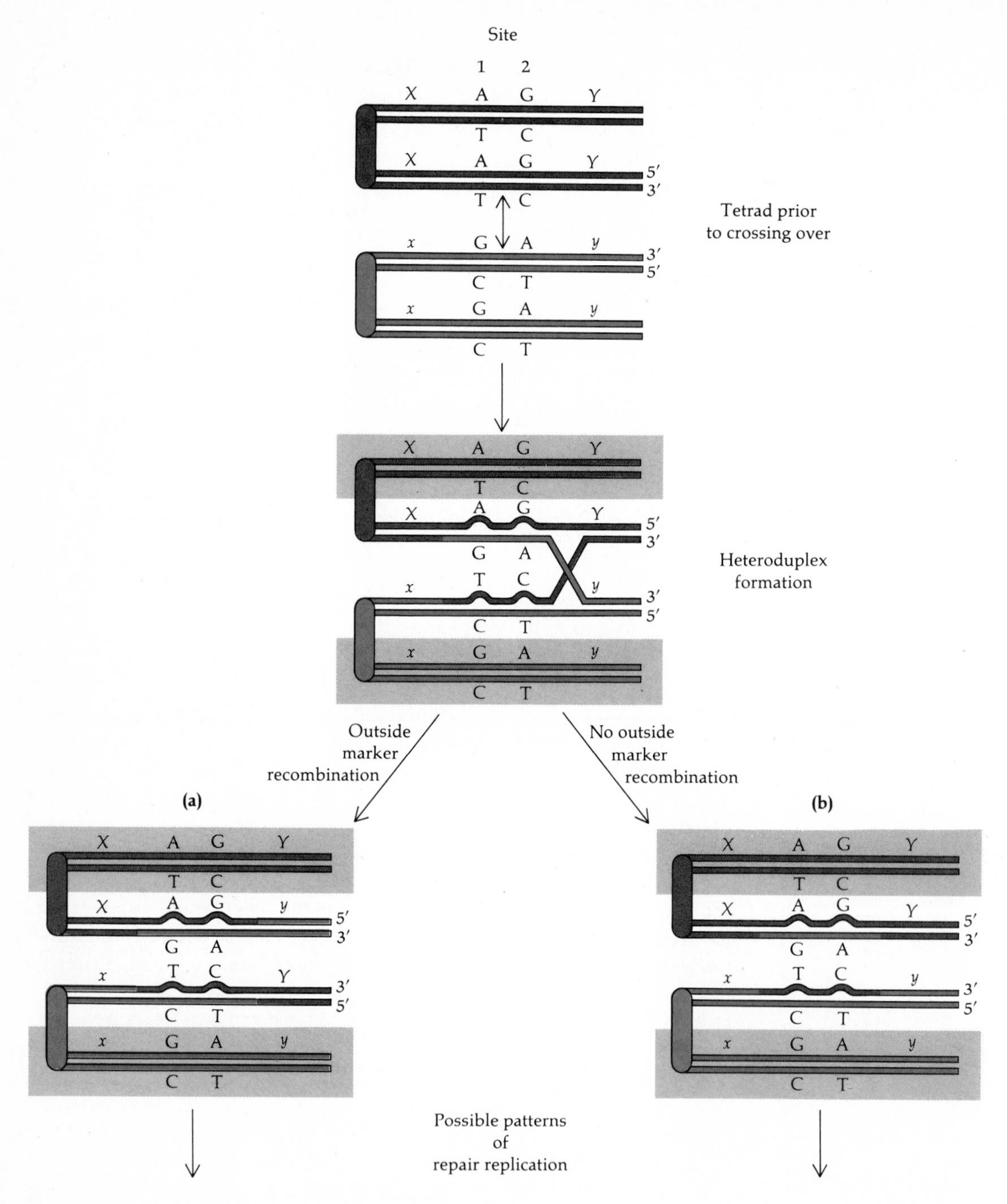

Figure 8.16
Heteroduplex DNA overlapping a region of heterozygosity of genetic markers is a prerequisite for gene conversion. Heteroduplex formation may occur **(a)** with or **(b)** without recombination of outside flanking markers. Repair of base mismatches can produce the various types of tetrads diagrammed.

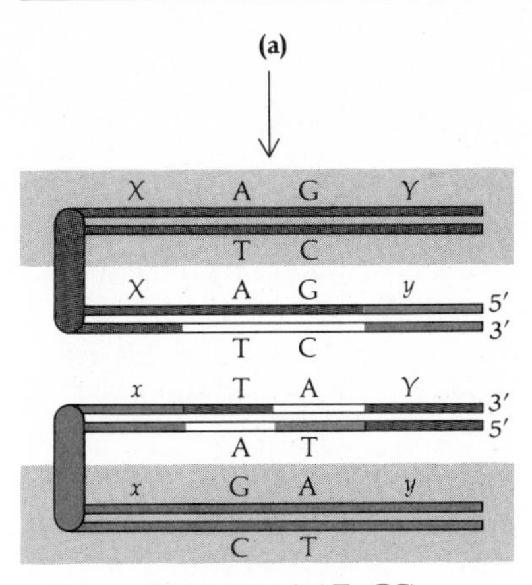

Conversion at site 1 (3AT:1GC) with generation of an apparent triple-crossover chromatid and retention of one single-crossover chromatid

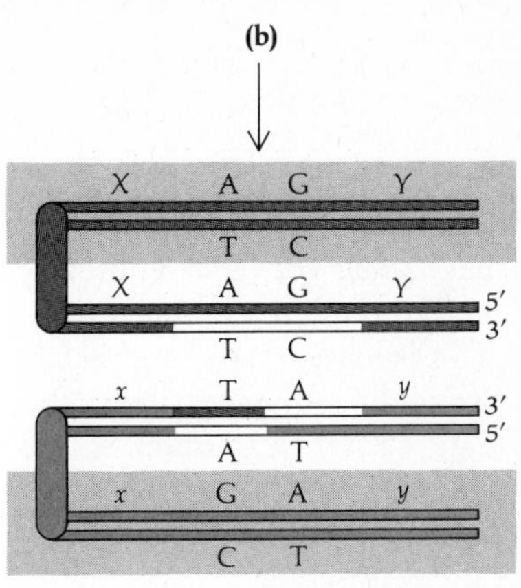

Conversion at site 1 (3AT:1GC) with generation of an apparent double-crossover chromatid

Coconversion at sites 1 (3AT:1GC) and 2 (3GC:1AT) and retention of two single-crossover chromatids

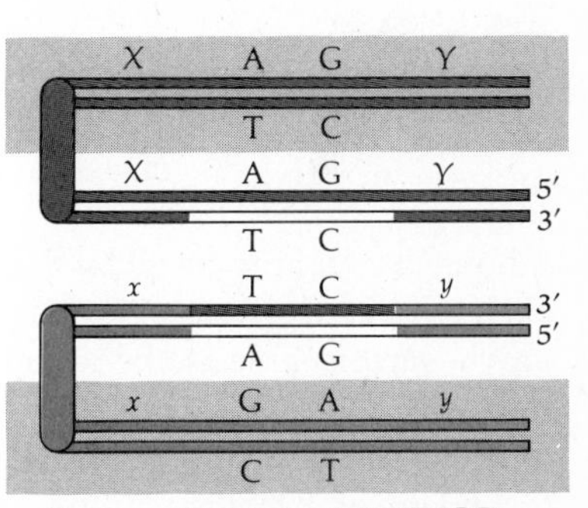

Coconversion at sites 1 (3AT:1GC) and 2 (3GC:1AT) with generation of an apparent double-crossover chromatid

No conversion at either site (2AT:2GC) and retention of two single-crossover chromatids

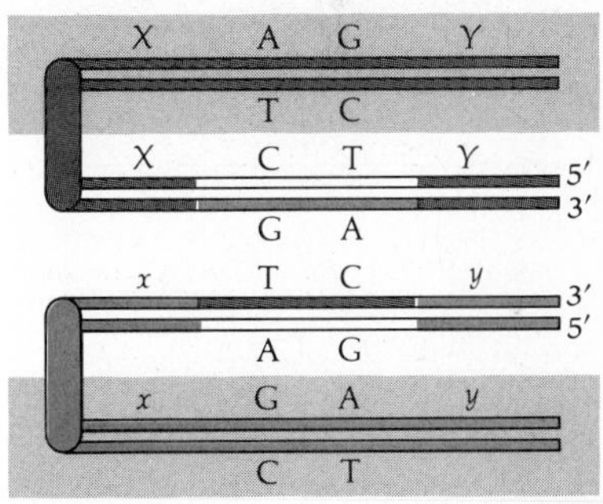

No conversion at either site (2AT:2GC) with generation of two apparent double-crossover chromatids

Box 8.1 Heteroallele Conversions at the *arg4* Locus of Yeast

The data in Table 8.3 show the patterns of heteroallele conversion observed in asci produced by four different diploid genotypes of *Saccharomyces cerevisiae*, each heterozygous for two of the four heteroalleles shown in the map (Figure 8.15). In addition, each diploid is heterozygous for flanking markers, e.g., *pet1* and *thr1*, whose recombination during meiosis can be observed.

For example, the diploid BZ34 carried *arg4-4* and *thr1* on one homologue and *pet1* and *arg4-17* on the other:

$$\frac{+ \quad arg4\text{-}4 \quad + \quad thr1}{pet1 \quad + \quad arg4\text{-}17 \quad +}$$

Among 690 asci produced by this diploid, there were eight conversions of *arg4-4*, seven of which were observed to be recombined for *pet1* and *arg4-17*, which flank *arg4-4*. The data recorded do not distinguish between conversion of *arg4-4* to wild type and wild type to *arg4-4*, but both are observed as either 3:1 or 1:3 asci. In addition, there were 42 conversions of *arg4-17*, of which 19 were recombined for *arg4-4* and *thr1*, which flank *arg4-17*. Finally, there were five double-site conversions in which both *arg4-4* and *arg4-17* were converted; of these, three were observed to be recombined for the flanking markers *pet1* and *thr1*. The results obtained with three other diploids are presented in Table 8.3 in similar fashion.

Three important observations come from the data in Table 8.3. First, half (73/140) of all conversion events are associated with recombination of the flanking markers. Second, conversion is the substitution of one true allele in a cross for another. Thus, the wild-type allele of *arg4-4* is converted to *arg4-4*, and *arg4-4* is converted to its wild-type allele. Conversion is not a random event in the sense that one heteroallele can convert to any other. Third, coconversion of two heteroalleles in a cross is more frequent than would be expected if conversions were independent events. For example, the frequency of conversion of *arg4-4* is 0.012 (8/690), and that of *arg4-17* is 0.061 (42/690). If conversions were independent events, we would expect coconversion to exhibit a frequency of 0.00073 (0.012×0.061), whereas the observed frequency is 0.0072 (5/690). This and similar calculations from the data in Table 8.3 for the other pairs of heteroalleles of *arg4* are given in Table 8.4.

"crossover" chromatids would be detected. These byproducts of gene conversion would be interpreted as evidence of multiple crossing over—their true origin would go undetected.

Overview of the Results of Genetic Analysis

We have seen in Chapters 5, 6, and 7 how genetic analysis can be employed to deduce the overall organization of the genetic material in euka-

Table 8.3
The association of gene conversion with recombination in the *arg4* gene of yeast.

Diploid	*arg4* Genotype	Number of Asci	Allele(s) Converted	Number of Conversions 3:1 plus 1:3	Interval	Number of Conversions with Recombination in Interval
BZ34	*4 +* / *+ 17*	690	4	8	*pet1–17*	7
			17	42	*4–thr1*	19
			4-17	5	*pet1–thr1*	3
BZ140	*2 +* / *+ 17*	544	2	1	*pet1–17*	1
			17	7	*2–thr1*	3
			2-17	28	*pet1–thr1*	14
X841	*1 +* / *+ 2*	367	1	3	*pet1–2*	1
			2	14	*1–thr1*	8
			1-2	19	*pet1–thr1*	8
X901	*1 +* / *+ 2*	116	1	2	*c–2*	2
			2	6	*1–thr1*	3
			1-2	5	*c–thr1*	4
Total:		1717		140		73

After D. Hurst, S. Fogel, and R. Mortimer, *Proc. Natl. Acad. Sci. USA 69*:101 (1972).

Table 8.4
Frequency of expected and observed coconversion events in the *arg4* gene of yeast.

Heteroallele Pairs	Frequency of Single-Site Conversion	Frequency of Coconversion	
		Expected	Observed
4 & 17	(*4*) 0.012 (*17*) 0.061	0.00073	0.0072
1 & 2	(*1*) 0.010 (*2*) 0.041	0.00041	0.050
2 & 17	(*2*) 0.002 (*17*) 0.013	0.00003	0.051

ryotic and prokaryotic organisms and their viruses. The complementation test permits mutations to be assigned to functional units. Recombination analysis permits these functional units to be mapped with respect to one another and leads to the creation of genetic maps that are models for the chromosomal organization of functional genetic units. We have seen that these genetic models of the hereditary material correspond quite closely to the actual physical organization of the DNA itself. In this chapter we have seen how genetic analysis has been extended to examine alterations in the fine structure of the gene—alterations as small as the nucleotide pairs themselves.

In the final chapter of Part I, the details of our knowledge of replication and recombination of DNA molecules will be explored. The mechanisms employed by cells to maintain the integrity of their genetic information and to transmit it unaltered to the next generation will be introduced. Then, having established an understanding of the physical and genetic organization of genomes, we will turn in Part II to an examination of how genes are expressed and how they determine the phenotypes of organisms.

Problems

1. A new T4 *rII* mutation yields wild-type recombinants when crossed with *J3* (see Figure 8.6) but not when crossed with *1241*. In other crosses it yields wild-type recombinants when crossed to *1993, 1695, PT153,* or *H88,* but not when crossed to *386* or *168*. In what section of the *rII* map does it reside?

2. An *rII* mutation selected after prolonged growth on a plate containing 5-bromouracil (5-BU) in the medium is found not to yield wild-type recombinants when crossed to *638* (see Figure 8.6). In other crosses it does not yield wild-type recombinants when crossed to *196, W8-33, 187,* or *1519,* but does when crossed to *1589* or *1299*. Explain.

3. Fine-structure mapping employing deletions provides a means for ordering mutations unequivocally on a linear map. Problems of interpretation, which may occur when high negative interference is observed in crosses between point mutations, do not arise. Only the presence (or absence) of wild-type recombinants need be noted. In the table below, the results of pairwise crosses of six deletion mutations are noted; 1 indicates that wild-type recombinants are observed in a cross, and 0 indicates that none are observed. Draw a genetic map indicating the relative extent of each deletion. Which of the six mutations could be point mutations and still give the same results as recorded in the table?

	1	2	3	4	5	6
1	0	0	0	1	0	0
2		0	0	0	1	0
3			0	0	1	1
4				0	1	1
5					0	1
6						0

4. The T4 *rII* deletions chosen by Benzer to delineate the *rII* region into 47 subregions (Figure 8.6) were chosen from a much larger set of *rII* mutations. The deletions can be ordered with respect to one another by pairwise crosses in which only the presence or absence of wild-type recombinants is scored, as in problem 3. The table below records the results of such crosses, employing 15 of Benzer's nonreverting *rII* mutations. Order

	184	215	221	250	347	455	506	749	782	882	A103	C4	C33	C51	H23
184	0	0	1	0	1	0	1	1	0	1	1	1	0	0	0
215		0	1	1	1	1	1	1	1	1	1	1	1	1	0
221			0	1	0	1	0	0	0	0	0	0	0	1	0
250				0	1	1	1	1	1	1	1	1	0	0	0
347					0	1	1	1	0	1	1	1	1	1	0
455						0	1	1	1	1	1	1	1	1	0
506							0	1	0	1	1	1	1	1	0
749								0	0	1	1	0	1	1	0
782									0	0	0	0	0	1	0
882										0	1	0	1	1	0
A103											0	1	1	1	0
C4												0	1	1	0
C33													0	1	0
C51														0	0
H23															0

these mutations on a genetic map. Assume that each mutation deletes only one contiguous region, i.e., there are no double mutations in the group. *Hint:* Order the larger deletions first.

5. The mutations am^2 and am^3 in *Neurospora crassa* are amination-deficient. Spores carrying these mutations will not grow when germinated on a medium containing 0.02 *M* glycine; this provides a selective screen for wild-type recombinants formed in matings of the two mutant strains. In such matings, the results of which are given in the table below for random spores, *inos* (inositol-requiring) and *sp* (a morphological mutation called *spray*) are employed as flanking markers. What can you deduce from the results of these crosses?

Cross	Observed number of following genotypes of flanking markers in am^+ ascospores:				Number of am^+ ascospores	Number of live ascospores scored (thousands)	Frequency of am^+ ascospores per 10^6 live ascospores
	sp^+ $inos^+$	sp $inos$	sp^+ $inos$	sp $inos^+$			
$am^2 \times am^2$	—	—	—	—	0	465	—
$am^3 \times am^3$	—	—	—	—	0	585	—
$am^2 \times am^3$	—	—	—	—	14	651	21.54
sp am^2 $inos^+$ × sp^+ am^2 $inos$	0	0	0	0	0	209	—
sp am^2 $inos^+$ × sp^+ am^3 $inos$	0	6	0	17	23	1549	14.85
sp^+ am^2 $inos$ × sp am^3 $inos^+$	0	8	7	9	24	1214	19.77
sp am^2 $inos$ × sp^+ am^3 $inos^+$	6	9	6	5	26	881	29.45
sp^+ am^2 $inos^+$ × sp am^3 $inos$	7	4	2	4	17	1148	14.80
Total:					104	5443	19.11

After J. A. Pateman, *Genetics* 45:839 (1960).

6. Mutations that affect the ability of *E. coli* to utilize lactose as a carbon source occur in the cistrons $lacY^+$, $lacZ^+$, and $lacI^+$, which map in that order (Figure 7.17). The $lacZ^+$ cistron specifies the enzyme β-galactosidase. Two types of regulatory mutations affect the expression of the $lacZ^+$ gene.

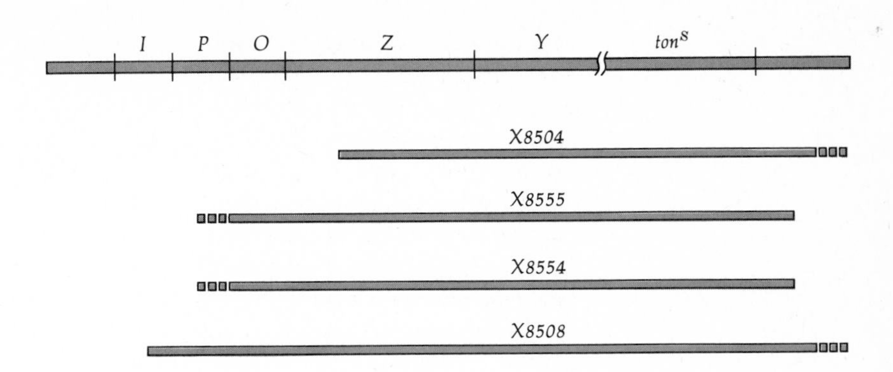

Figure 8.17
The extent of some deletions of the ton^S gene extending into the *lac* operon of *E. coli*.

Normal *E. coli* cells synthesize β-galactosidase only when lactose is present in the medium; O^c (*operator-constitutive*) mutations permit the synthesis of the enzyme in the absence of lactose. On the other hand, P^- (*promoter-deficient*) mutations drastically reduce the amount of enzyme synthesized when lactose is present. Employing a strain of *E. coli* in which the *lac* region is transposed to a position adjacent to the ton^S gene, deletions of all or most of the *lac* region can be selected by screening for $T1^R$ cells (deletion of the ton^S gene causes resistance to phage T1), which simultaneously lose the ability to utilize lactose and fail to synthesize β-galactosidase. Some of these deletions are shown in Figure 8.17.

The P^- or O^c mutations are present on F′ episomes carrying the *lac* region and are introduced into F^- cells carrying one or another of the deletions shown in Figure 8.17. The frequencies of wild-type recombinants (synthesizing normal amounts of β-galactosidase only in the presence of lactose) are recorded in the table below (+ indicates that wild-type recombinants are observed). What are the map positions and order of the P^- and O^c mutations?

	Deletions			
Point mutants	X8508	X8504	X8555	X8554
P^- *L8*	0	+	0.065	0.087
P^- *L29*	0	+	+	+
P^- *L37*	0	0.79	0.086	0.086
O^c *307*	<0.00008	0.52	0.0022	<0.00008
O^c *15*	<0.00007	0.23	0.0015	0.00017

After J. H. Miller et al., *J. Mol. Biol.* 38:413 (1968).

7. *Notch* (*N*) is a dominant X-chromosome mutation in *Drosophila melanogaster* that causes a number of effects: notched wings, thickened wing veins, and minor bristle abnormalities. *N* mutations are also recessive lethal mutations. The recessive mutations *facet* (*fa*) and *split* (*spl*) cause roughened eye surfaces but are distinguishable from each other. The genotypes *fa*/*N* and *spl*/*N* exhibit the phenotype of the respective recessive mutation as well as that of *N*/+, i.e., *N*-mutant chromosomes do not carry fa^+ or spl^+. The recessive mutations *facet-notchoid* (fa^{no}) and *notchoid* (*nd*) affect the wings in a manner similar to that of *N*. The genotype *nd*/*N* has severely notched wings and fa^{no}/N is almost completely lethal; *N*-mutant chromosomes do not carry nd^+ or fa^{no+}. From the data given below, construct a map of this pseudoallelic locus. Data for only two of the many known *N* alleles are included.

Cross	Male recombinants	Total males
$\frac{y\ w\ spl\ sn}{+\ +\ fa^{no}\ +}$ ♀ × fa^{no} ♂	4 *y w* 2 fa^{no} *spl sn*	9,400
$\frac{y\ w\ fa^{no}\ sn}{+\ +\ fa\ +}$ ♀ × *fa* ♂	2 *y w* 1 *fa* fa^{no} *sn*	8,100
$\frac{y\ w\ spl\ sn}{+\ +\ fa\ +}$ ♀ × *y w spl sn* ♂	4 *y w* *fa spl* (cannot be distinguished from parental types)	6,100
$\frac{+\ nd\ rb}{w^a\ spl\ +}$	6 w^a *spl nd rb* 5 + + + +	38,900
$\frac{w^a\ N^{N1c}\ +}{+\ spl\ rb}$	3 + + + +	42,482
$\frac{y\ N^{g11}\ +}{+\ spl\ rb}$	12 *y* + + *rb*	41,041

Data from W. Welshons, *Proc. Natl. Acad. Sci. USA* 44:254 (1958); and from W. Welshons and E. Van Halle, *Genetics* 47:743 (1962).

8. Mutations at the *pan* locus of *Neurospora crassa* confer dependence on pantothenic acid, and those at the *trp* locus confer dependence on tryptophan. The *ylo* mutation causes spores to be yellow rather than black. From the data in the table below, construct a map showing the relation between the *ylo* and *trp* genes and the different mutations at the *pan* locus. Ascospores from the four crosses were plated on minimal medium supplemented only with tryptophan. The genotypes of the unselected markers among the pan^+ recombinants are recorded for each cross.

Cross	Frequency of unselected markers			
	ylo trp$^+$	*ylo*$^+$ *trp*	*ylo trp*	*ylo*$^+$ *trp*$^+$
ylo$^+$ *pan18 trp* × *ylo pan20 trp*$^+$	0.084	0.053	0.067	0.796
ylo$^+$ *pan20 trp* × *ylo pan18 trp*$^+$	0.033	0.052	0.710	0.105
ylo$^+$ *pan20 trp* × *ylo pan25 trp*$^+$	0.160	0.200	0.100	0.54
ylo$^+$ *pan25 trp* × *ylo pan20 trp*$^+$	0.080	0.270	0.520	0.130

9. How would you measure the recombination frequency between *r168* and *r924* in the cross *r168 r1695* × *r1695 r924*, diagrammed in Figure 8.4? *Hint*: Employ the easy and rapid method of doing crosses outlined in Figure 8.7 at some point in your answer. You have available strains of each of the single *rII* mutants as well as the double mutants.

9

DNA Replication, Repair, and Recombination

The beautiful simplicity with which the double helical structure and complementarity of nucleotides in the two strands of the DNA molecule lend themselves to a semiconservative mode of replication (see Chapter 4) conceals an extremely complex biochemical mechanism for effecting that replication. The biochemical complexity of the replication apparatus has almost certainly evolved to assure the extreme fidelity of information transfer that is observed between parental and daughter DNA molecules. It is estimated that errors of replication leading to the insertion of an incorrect nucleotide into a DNA molecule of *E. coli* occur at a frequency of only one in 10^9–10^{10} nucleotides. Yet prokaryotic DNA strands are synthesized at the rate of about 1000 nucleotides per second at a replication fork. Eukaryotic DNA strands are synthesized more slowly—only about 100 nucleotides per second. The slower rate of eukaryotic DNA synthesis at a replication fork is probably due to the close association of eukaryotic DNA with histone proteins in the nucleosomes, which must be dissociated as the replication fork progresses along the DNA. Yet errors are as infrequent in eukaryotes as in prokaryotes.

The enzymes and other proteins involved in semiconservative replication represent only a portion of all the proteins involved in the metabolism of DNA molecules. Other enzymatic systems exist to repair incorrect or damaged nucleotides located away from replication forks. Some of these enzymes also play roles in recombination, in association with other enzymes whose roles are confined to recombination. Many of the enzymatic functions associated with DNA metabolism are shared by prokaryotic and eukaryotic organisms. The topics of DNA replication, repair, and recombination will be discussed in terms of enzymatic activities in order to emphasize the biochemical functions employed by both prokaryotes and eukaryotes. These functions can be assigned to specific enzymes in some instances. However, in some organisms a particular function may be performed by more than one enzyme, and sometimes more than one function may be performed by a single enzyme. Evolution has apparently created a number of mechanisms for carrying out the functions necessary for DNA metabolism and for maintaining the hereditary information encoded in the DNA.

A list of the major enzymatic activities involved in DNA synthesis is found in Table 9.1. In this chapter these activities will be discussed first from the point of view of semiconservative DNA replication and then from the point of view of their involvement in repair processes and in mechanisms of recombination.

Genetic analysis plays a central role in studying the biochemical complexities of DNA metabolism. Mutants that affect a function associated with DNA metabolism by altering or eliminating a particular enzyme can demonstrate the *in vivo* role played by that enzyme. If such a mutant is lethal or conditionally lethal, then the role played by that enzyme can be assigned an essential function in the process, rather than a peripheral function.

DNA Polymerization at a Replication Fork

The parental strands of the Watson-Crick double helix, as they unwind and separate, provide templates on which growing complementary daughter strands are synthesized by DNA polymerase, employing deoxyribonucleoside triphosphates (dATP, dGTP, dTTP, and dCTP) as substrates. In order to carry out the template-directed polymerization of these substrates, all DNA polymerases require a 3′-OH end as a primer for the addition of the next nucleotide to the growing chain of nucleotides. The reaction catalyzed by DNA polymerase is

$$(\mathrm{dNp})_n\, \mathrm{dN_{OH}} + \mathrm{dNTP} \rightarrow (\mathrm{dNp})_{n+1}\, \mathrm{dN_{OH}} + \mathrm{P\text{-}P}$$

where $(\mathrm{dNp})_n\mathrm{dN_{OH}}$ is a deoxyribonucleoside phosphate polymer with a terminal 3′-OH group, dNTP is a single deoxyribonucleoside triphosphate molecule, and P-P is a pyrophosphate molecule. The new nucleotide is added to the 3′ end of the growing strand, with the elimination of a pyrophosphate molecule. The 3′-OH group of the newly added nucleotide then provides a primer for the addition of the next nucleotide, etc. The

Table 9.1
Enzymatic activities involved in DNA synthesis.

Enzymatic activity
Helix-unwinding protein (ATP-dependent)
Helix-destabilizing protein (HDP)
Helix-relaxing protein (DNA gyrase)
DNA polymerase (deoxyribonucleoside triphosphate polymerization)
RNA polymerase (RNA primer synthesis)
5′→3′ exonuclease
3′→5′ exonuclease
DNA ligase
Endonuclease

strands of the Watson-Crick helix are antiparallel, and therefore both 3′ and 5′ ends of the newly synthesized strands are present at a replication fork, as shown in Figure 9.1a.

Because DNA polymerases require a 3′-OH primer, only one of the two growing strands (the *leading* strand) may be synthesized in a continuous manner. The other strand (the *lagging* strand) is synthesized in a discontinuous manner. Evidence for discontinuous synthesis of at least one of the two growing strands at a replication fork comes from *pulse-labeling* experiments, in which cells are permitted to incorporate [^{3}H]thymidine into newly synthesized DNA. If the radioactive substrate is present for only a few seconds before DNA is extracted from the cells, much of the newly synthesized DNA (identified by its radioactivity) is found to be in small, single-stranded pieces only 1000–2000 nucleotides in length. These fragments are called *Okazaki fragments* after their discoverer, Reiji Okazaki. If the pulse of [^{3}H]thymidine is terminated by transferring the growing cells to a nonradioactive medium and growth is allowed to continue for a few minutes (a *pulse-chase* experiment), the radioactivity is then found in very large DNA strands (Figure 9.2).

The observation that synthesis of the lagging strand is discontinuous was initially puzzling because the origin of new primers was not apparent; all DNA polymerases require a 3′-OH primer and are inactive in its absence. For example, purified DNA polymerase, when provided with a single-stranded circular template of ϕX174 phage DNA, is completely inactive. RNA polymerases, however, do possess the ability to initiate RNA synthesis on a single-stranded template in the absence of a primer.

(a)

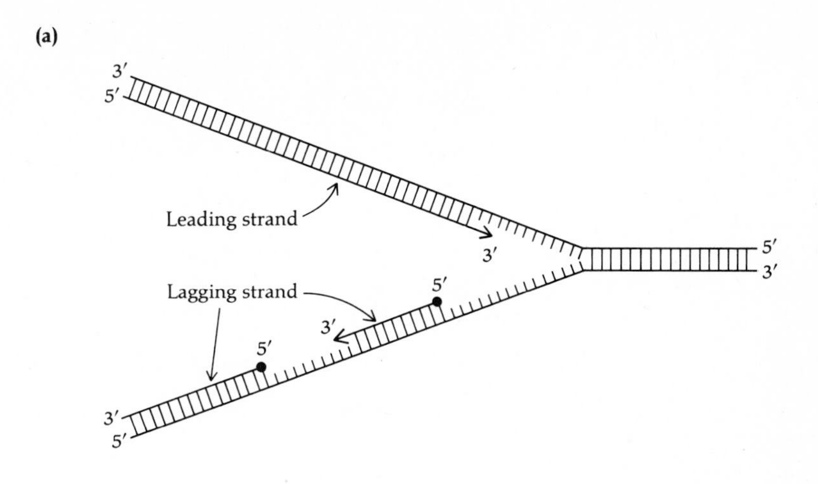

(b)

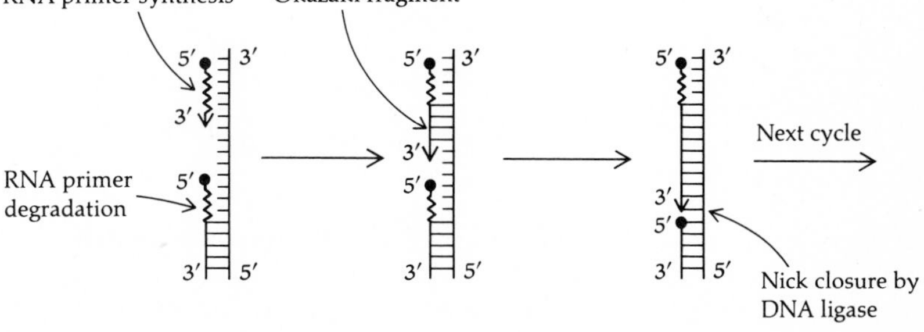

(c)

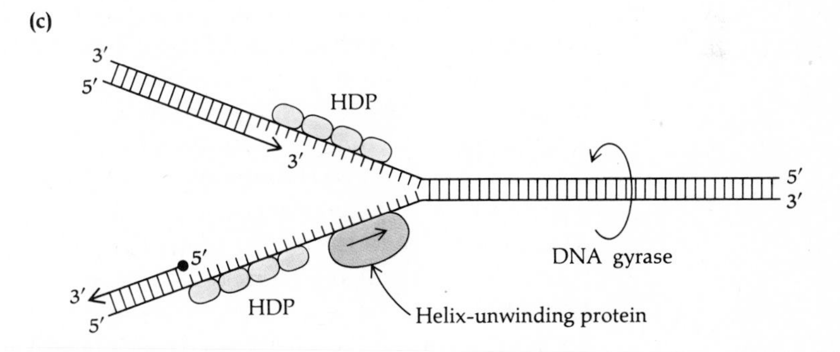

Figure 9.1
(a) A DNA replication fork, at which the polarities of the two newly synthesized strands differ. The strand possessing a 3′-OH primer is the leading strand. The strand whose synthesis must be primed *de novo* is the lagging strand. **(b)** Lagging-strand synthesis proceeds through synthesis of a short RNA strand that primes the synthesis of a DNA fragment (Okazaki fragment). The RNA primer of one DNA fragment is removed as the synthesis of the next fragment is concluded. The nick remaining between the two completed DNA fragments is closed by DNA ligase. **(c)** Participation of a helix-unwinding protein, a helix-destabilizing protein (HDP), and DNA gyrase in opening the parental DNA helix ahead of the replication fork. [After B. Alberts and R. Sternglanz, *Nature 269*:655 (1977).]

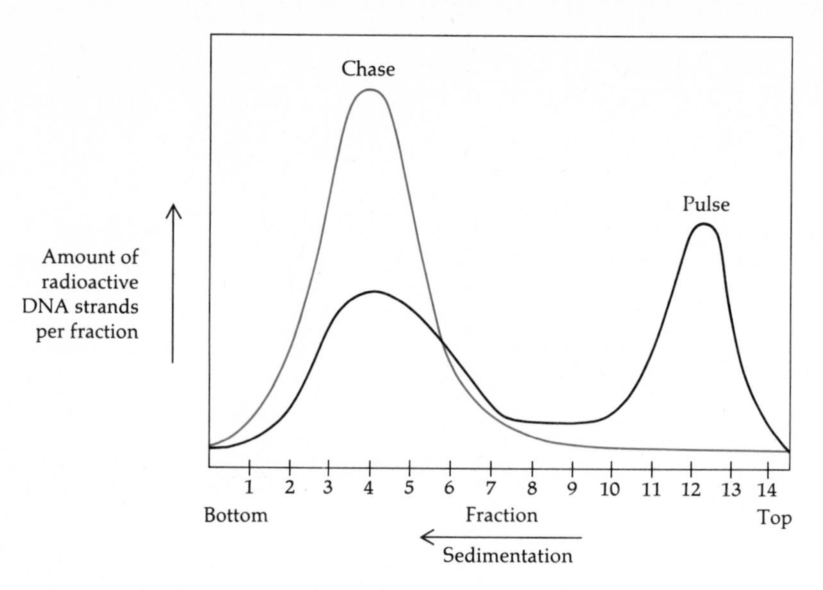

Figure 9.2
Demonstration that a portion of newly synthesized *E. coli* DNA is in short single strands that are later incorporated into long single strands. The black curve shows the size distribution of pulse-labeled single strands of DNA, sized by sedimentation through a sucrose gradient at pH 12 to prevent hydrogen-bond formation between complementary strands. The small fragments near the top of the centrifuge tube become incorporated into larger single strands during a "chase" period of growth in nonradioactive medium (colored curve). The larger single strands sediment farther through the sucrose gradient during centrifugation than the smaller single strands.

This is evidenced by observations that newly synthesized RNA molecules possess a triphosphate moiety at their 5′ ends. The subsequent finding that Okazaki fragments possess RNA at *their* 5′ ends provides firm evidence for primer synthesis by an RNA polymerase.

RNA primer strands do not remain within completed DNA strands. Once they have served to prime the synthesis of a DNA strand, they are removed by a 5′→3′ exonuclease activity, as diagrammed in Figures 9.1b and 9.3. When the RNA primer has been removed and replaced by a DNA strand that was initiated on an RNA primer located closer to the replication fork, a *nick* (absence of a phosphodiester bond) remains between the two discontinuously synthesized DNA strands. The nick is the absence of a covalent bond between a 3′-OH end and a 5′-PO_4 end. It is closed by the enzyme DNA ligase, which creates a covalent phosphodiester bond, as diagrammed in Figure 9.4.

Leading-strand and lagging-strand syntheses occur as the replication fork proceeds along the parental DNA helix (see Figure 9.1). The unwinding of the parental double helix to create two single-strand templates entails the participation of three types of proteins (see Table 9.1) and the

Figure 9.3
Lagging-strand synthesis is initiated by an RNA polymerase that creates a short RNA molecule complementary to the template strand of DNA. The RNA molecule's 3′-OH end serves as a primer for synthesis of a DNA strand by DNA polymerase III. DNA polymerase I removes the RNA primer, beginning at the 5′-triphosphate end of the RNA, employing the 3′-OH end of the next DNA fragment as a primer. As it removes the RNA, it extends the DNA molecule that served as its primer. When the entire RNA is removed, a single-strand nick remains between the two DNA fragments and is closed by DNA ligase.

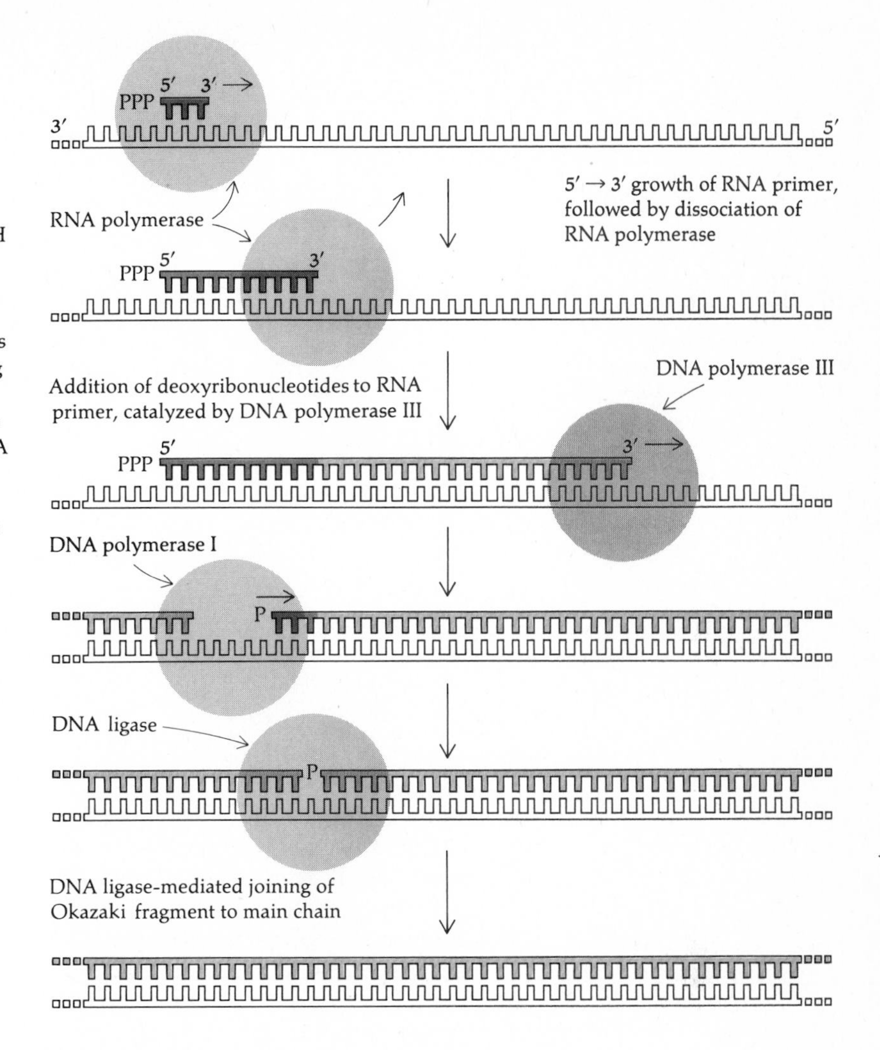

expenditure of considerable chemical energy derived from the hydrolysis of ATP. The first of these proteins is one that unwinds the parental strands to create template strands. This unwinding reaction requires energy and is accompanied by ATP hydrolysis. The second protein is a helix-destabilizing protein (HDP), which binds to single-stranded DNA at the fork and prevents the separated parental strands from reannealing. Unwinding of the parental double helix without rotation creates supercoiled twists in it, just as the abrupt separation of the twisted threads of a string creates a knot in the string at the point of separation. The third protein, DNA gyrase, relaxes the tension of these supercoiled twists by breaking a phosphodiester bond in one of the parental strands ahead of the replication fork; this creates a swivel point on the opposite strand, around which

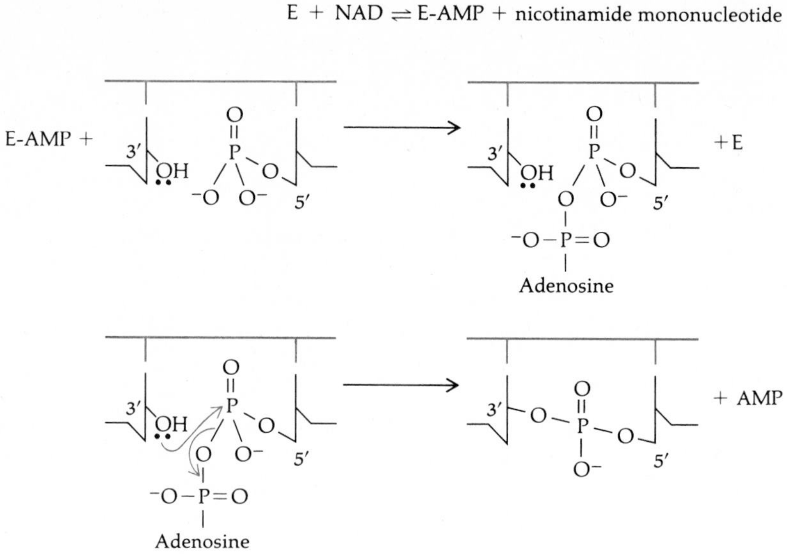

Figure 9.4
Nick closure by *E. coli* DNA ligase (E) employs the energy created by the hydrolysis of NAD (nicotinamide adenine dinucleotide). The reaction proceeds through a ligase-AMP intermediate. The AMP is transferred to the 5′-PO_4 of the nick. The diphosphate thus created is hydrolyzed by attack of the 3′-OH of the nick, creating a covalent phosphodiester bond. The T4 phage-induced DNA ligase employs ATP rather than NAD to form a ligase-AMP intermediate.

$$\text{E} + \text{NAD} \rightleftharpoons \text{E-AMP} + \text{nicotinamide mononucleotide}$$

rotation occurs to remove the supercoiling created by strand separation. When the tension has been removed from the parental helix, the DNA gyrase restores the broken phosphodiester bond. The participation of these three types of proteins is diagrammed in Figure 9.1c.

Combined Biochemical and Genetic Analyses of DNA Replication

The details of DNA synthesis at the replication fork have been elucidated by combined biochemical and genetic analyses. Biochemical analysis proceeds by the purification of enzymes and other proteins that carry out one or more steps in the replication process *in vitro.* The development of *in vitro* replication systems has been greatly facilitated by the use of small, well-characterized viral genomes, such as ϕX174, as templates because the products of *in vitro* replication can be equally well characterized.

Genetic analysis proceeds by the isolation of mutants that lack the ability to carry out particular functions or that have impaired functions. Some of these mutations may be lethal, or conditionally lethal, if they affect a vital step in DNA replication. The first evidence that DNA replication is complicated and involves the interaction of a number of different essential functions was provided by the selection of conditional lethal (temperature-sensitive) mutations that could be assigned to a number of different complementation groups and map positions on the *E. coli* chromosome. At the restrictive temperature, these mutations exhibit a phenotype of either the immediate cessation of DNA synthesis or the inability to initiate a new round of genome replication. The map positions of a number of these mutations and other mutations affecting DNA metabolism in *E. coli* are shown in Figure 9.5. The biochemical determination of which replication function (which protein) is affected by each of these mutations allows the *in vivo* roles of proteins studied *in vitro* to be determined.

E. coli possesses three distinct DNA polymerase enzymes. DNA polymerase III (Pol III) is coded by the *dnaE* gene, identified by conditional lethal mutations in that gene. It functions at the replication fork to extend DNA synthesis on both the leading and lagging strands. On the lagging strand it utilizes a primer synthesized by an RNA polymerase that is coded by another vital gene, *dnaG*.

DNA polymerase I (Pol I) is the most abundant DNA polymerase in *E. coli* and the most thoroughly studied of these enzymes. The functional aspects of this enzyme are shown schematically in Figure 9.6. In addition to its ability to polymerize in the 5′→3′ direction when provided with a template strand and a primer, it possesses a 5′→3′ exonucleolytic activity at a different site on its surface. The combination of these two activities in the same enzyme molecule enables Pol I to carry out *nick translation*. When purified Pol I is provided *in vitro* with a circular double-stranded ϕX174 DNA molecule containing a 5′-PO_4/3′-OH nick in one strand and the four deoxyribonucleoside triphosphates, the combined action of polymerization on the 3′-OH primer and exonucleolytic removal of the adjacent 5′-nucleotide moves the nick around the circular genome (Figure 9.7). This nick translation will continue indefinitely until either the pool of deoxyribonucleoside triphosphates is exhausted or DNA ligase and NAD are added to close the nick and eliminate the 3′-OH primer. (Nick translation provides an extremely useful means for labeling DNA radioactively *in vitro*; see Chapter 11.)

Pol I was believed to be the only DNA polymerase enzyme in *E. coli* until a mutant (*polIA*) was discovered that abolishes this polymerase activity (by at least 98%) but does not noticeably impair the ability of the mutant strain to grow. This discovery at first suggested that Pol I was nonessential. The *polIA* mutant provided biochemists with a strain in which the other DNA polymerase enzymes, Pol III and Pol II (Pol II has not yet been identified with any *in vivo* function and will not be discussed further), could be identified and purified as minor contributors to the total DNA polymerase activity of normal *E. coli*. Further study, however, proved that the 5′→3′ exonuclease activity of Pol I was not impaired in the *polIA* mutant. Subsequently, temperature-sensitive mutations in the *polI* gene

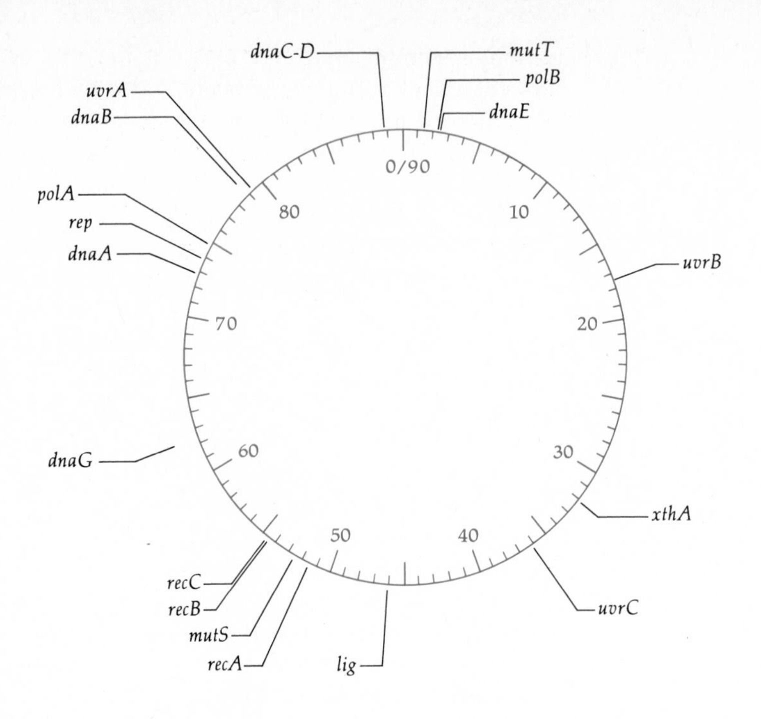

Locus	Minutes	Comments
mutT	1.5	Mutator, AT → CG transversions
polB	2.5	Pol II
dnaE	2.5	Pol III
uvrB	17.5	Thymine-dimer excision
xthA	32	Exonuclease III, endonuclease II
uvrC	36	Thymine-dimer excision
lig	46	DNA ligase
recA	51.5	Recombination
mutS	52.5	Mutator
recB	54.5	*recB,C* nuclease
recC	54.5	*recB,C* nuclease
dnaG	60–65	DNA replication—primer synthesis
dnaA	73	DNA replication
rep	74	DNA unwinding protein
polA	75	Pol I
dnaB	≈79	DNA replication
uvrA	79.5	Thymine-dimer excision
dnaC-D	89	DNA replication

Figure 9.5
Map positions of genes involved in DNA metabolism in *E. coli.* (After A. Kornberg, *DNA Synthesis,* W. H. Freeman, San Francisco, 1974.)

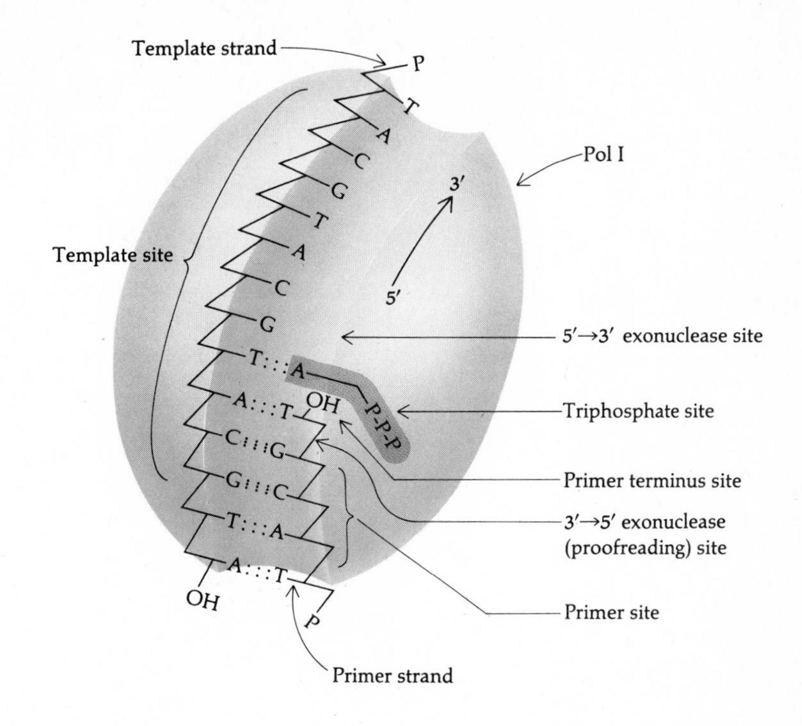

Figure 9.6
Schematic diagram of DNA polymerase I, showing its functional sites in relation to template and primer DNA strands.

were isolated that abolished the 5′→3′ exonuclease activity, and these mutations were found to be lethal at the restrictive temperature. These studies led to the conclusion that the 5′→3′ exonuclease activity of Pol I is essential and is used to remove the RNA primers that are necessary for discontinuous DNA synthesis on the lagging strand (Figure 9.3). The fact that the mutant *polIA* grows normally but that Okazaki fragments are retarded in joining to larger molecules suggests that Pol III may be able to fill the gaps created by the removal of RNA primers from the lagging strand by the 5′→3′ exonuclease.

Initiation of DNA Synthesis at the Origin of Replication

DNA replication is initiated at specific nucleotide sequences. Most prokaryotic genomes have only one origin per genome, but eukaryotic chromosomes certainly have a number of such sequences (see Figure 4.19). Conditional lethal mutations have identified a number of complementation groups, whose functions are necessary for the initiation of DNA synthesis, that are distinct from those necessary for synthesis at a

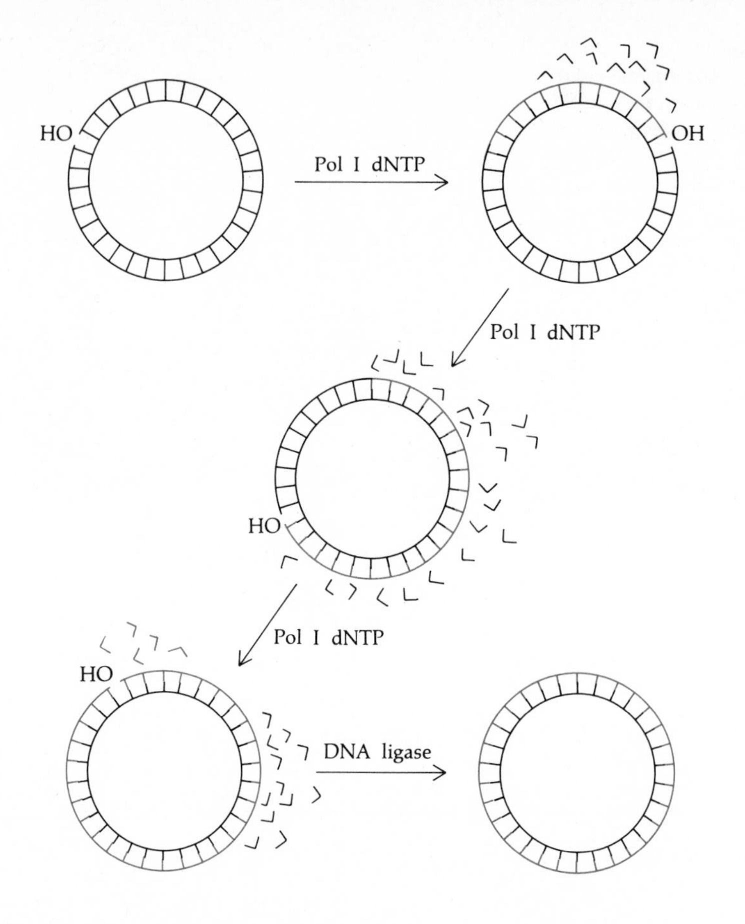

Figure 9.7
Nick translation on ϕX174 replicative form (RF) DNA by DNA polymerase I; dNTP represents any deoxyribonucleoside triphosphate.

replication fork. The biochemical mechanism involved has not yet been elucidated in *E. coli*, however.

The most detailed study of initiation has been carried out with ϕX174. The *A*-cistron protein (cis A) of ϕX174 is essential for its DNA replication. The origin of ϕX174 replication is within the *A* cistron. The cis A protein is an endonuclease that nicks a specific site in the nucleotide sequence of one of the strands of the covalently closed double-stranded genome. The 3′-OH group at this nick serves as a primer for the initiation of leading-strand synthesis by Pol III while the cis A protein remains covalently attached to the 5′ end of the nicked strand. A helix-unwinding protein and a helix-destabilizing protein are necessary for polymerization to proceed, as detailed in Figure 9.8. This modified rolling circle mode of replication gives rise to a double-stranded circular daughter molecule with the cis A protein still attached and a single-stranded viral ring complexed with helix-destabilizing protein. The cis A protein also exhibits a special ligase function that covalently joins the ends of the single-stranded daughter molecule to form a ring.

The bacteriophage λ *O* gene codes for a protein required for the initiation of λ DNA replication. The *O* gene also contains the site of

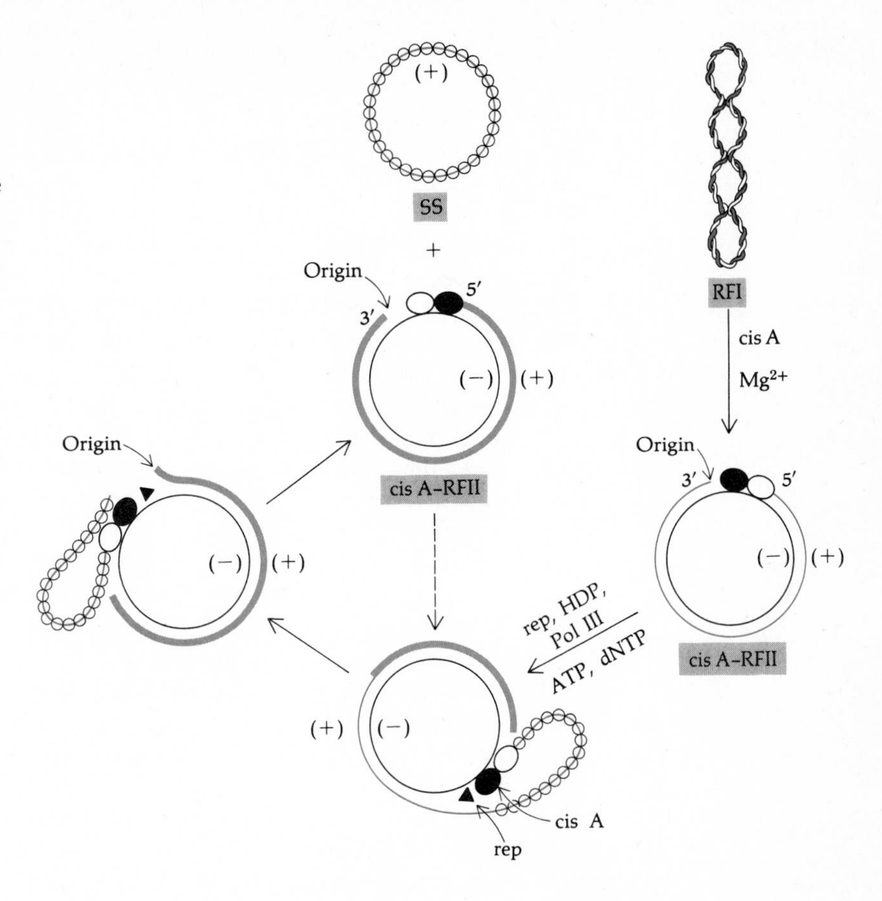

Figure 9.8
The role of cis A protein in the initiation and termination of ϕX174 viral DNA synthesis and RF replication *in vitro*. Note the "dimeric" nature of the cis A protein (one "monomer" open, the other closed).The substance rep is a helix-unwinding protein requiring ATP for its action; HDP is a helix-destabilizing protein. SS is a single-stranded circular daughter molecule. [After S. Eisenberg et al., *Proc. Natl. Acad. Sci. USA* 74:3198 (1977).]

initiation of λ replication. This association of an essential replication protein that recognizes an initiation sequence within its own structural gene may be a general phenomenon in prokaryotes.

Error Correction and Repair Mechanisms

As discussed above, the rate at which incorrect nucleotides are incorporated into newly replicated strands of DNA is extremely low (10^{-9}–10^{-10}). For some time it was believed that the fidelity of DNA replication was due primarily to the specificity of hydrogen-bond formation between a template base and the base of the incoming deoxyribonucleoside triphosphate. However, physicochemical estimates of the specificity of base pairing suggest that the error rate ought to be no less than 10^{-5} errors per nucleotide incorporated. Two properties that all three DNA polymerases of *E. coli* possess make the high fidelity of DNA replication understandable. First, each enzyme possesses a 3′→5′ exonucleolytic activity. Second, in order for the 3′-OH group of the terminal nucleotide of a growing strand to serve as a primer, the correct hydrogen bonding must occur between this nucleotide and its mate in the template strand. If an incorrect nucleotide is inserted by the polymerase, further polymerization is stopped because of the absence of correct base pairing. The polymerase backs up, excises the incorrect nucleotide by employing the 3′→5′ exonucleolytic activity, and then proceeds with further polymerization. Thus, each nucleotide inserted into a growing strand is tested twice for its correctness. This *proofreading* function of the enzyme is evidently essential for maintaining the low error rate of one mistake in 10^9–10^{10} nucleotide pairs. If the thermodynamic error rate is as great as 10^{-5}, then testing each nucleotide twice ($10^{-5} \times 10^{-5} = 10^{-10}$) will permit the observed error rate of 10^{-9}–10^{-10} to be achieved.

It has been suggested that the evolution of this proofreading mechanism, which requires a correctly base-paired primer, is responsible for the role of an RNA polymerase in priming discontinuous DNA synthesis on the lagging strand. The ability exhibited by RNA polymerase to create a primer *de novo* cannot be as error-free as is the proofreading mechanism employed by DNA polymerase, owing to the thermodynamic estimates discussed above. The removal of the error-prone RNA primer by DNA polymerase I and its replacement by DNA subject to proofreading eliminates the "bad copy" that may be present in primers initiated *de novo*.

Other mechanisms exist to maintain the integrity of the hereditary information encoded in the nucleotide sequence of DNA. These mechanisms can correct the presence of incorrectly paired bases in DNA that arise either through an error of the proofreading mechanism or through heteroduplex DNA formation during recombination in heterozygotes (see Chapter 8). Moreover, DNA is subject to damage from environmental agents, such as solar ultraviolet light and chemical agents of biological or industrial origin. Many chemical agents that induce cancers (carcinogens) act by damaging DNA. Mechanisms for repairing many types of such damage and restoring undamaged nucleotides to DNA are present in all cells.

A major type of damage induced in DNA by ultraviolet light is *thymine dimerization* between adjacent thymine bases in a DNA strand (Figure 9.9). These dimers distort the conformation of the double helix and interfere with normal DNA replication, leading either to the eventual death of the cell or to the creation of mutations. Thymine dimers are recognized by a specific repair endonuclease that creates a nick in the DNA strand near the dimer. A nick to the 5′ side of the dimer provides a substrate for a 5′→3′ exonuclease that removes a portion of the strand, including the dimer, thereby creating a gap. This gap is then filled in by a repair polymerase, using the exposed 3′-OH as a primer. The nick that remains is closed by

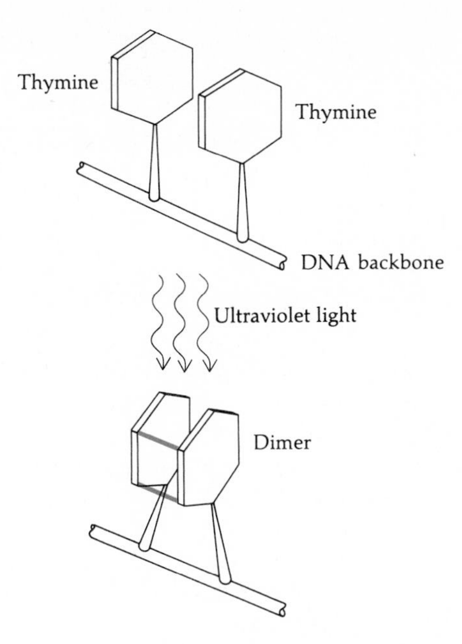

Figure 9.9
Illustration of thymine dimerization between adjacent thymine bases in a single strand of DNA. (After J. D. Watson, *Molecular Biology of the Gene,* 3rd ed., W. A. Benjamin, Menlo Park, Cal., 1976.)

DNA ligase (Figure 9.10). The finding that the *polIA* mutant is unable to close such gaps and is thus very sensitive to the lethal effects of ultraviolet light revealed that Pol I serves this function in *E. coli*. Indeed, the 5′→3′ exonuclease activity of Pol I is also capable of excising thymine dimers.

Humans who are homozygous for a mutant gene causing *xeroderma pigmentosum* are very sensitive to sunlight, freckle easily, and have a high incidence of skin cancer. Skin cells from those patients having one of several forms of this genetic disease are unable to excise thymine dimers. At least one form of the disease appears to be due to a deficiency of an endonuclease that recognizes the dimers and introduces an adjacent nick into the damaged DNA strand.

Another type of repair pathway that is common to both bacterial and mammalian cells employs a type of enzyme called *N*-glycosidase, which hydrolyzes the bond linking a damaged base to the deoxyribose in the backbone of the DNA strand. One such enzyme recognizes the improperly hydrogen-bonded base pair dG/dU, which occurs by deamination of dC at a dG/dC site. Deamination of dC could cause a dG/dC → dA/dT nucleotide-pair mutation upon replication, owing to the hydrogen-bonding properties of uracil, which are similar to those of thymine. For example, a T4 *rII* mutation that is the result of a dA/dT → dG/dC substitution reverts spontaneously at a frequency of 2×10^{-7} by

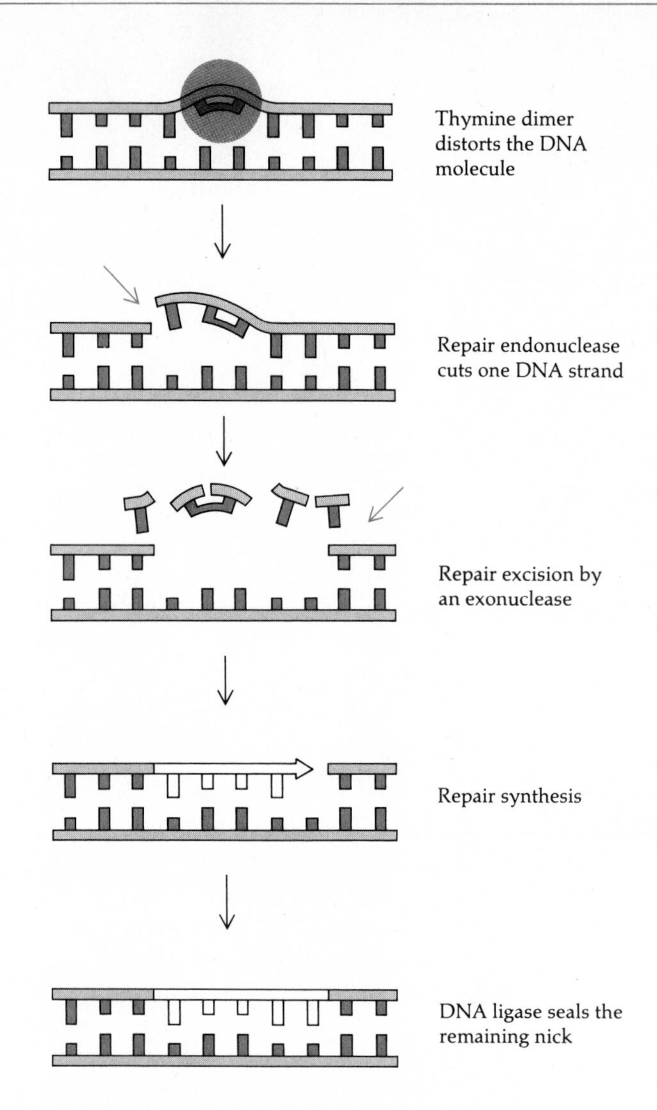

Figure 9.10
Steps in the excision and repair of a damaged region of a DNA molecule.

this mechanism. Conditions that increase the frequency of dC deamination (37°C, pH 5.0) inactivate 50% of the phages and increase the reversion rate of this mutation 40-fold among the survivors. Another enzyme of this type recognizes chemically modified (alkylated) bases produced by the actions of some mutagenic or carcinogenic compounds called *alkylating agents.* The action of an *N*-glycosidase produces an *AP site* (apyrimidinic or apurinic site) in one strand; this site is recognized first by an endonuclease and then by an exonuclease, which act to introduce a gap in the DNA strand, as indicated in Figure 9.11. A repair polymerase and DNA ligase then close the gap by inserting the proper nucleotide opposite the template strand.

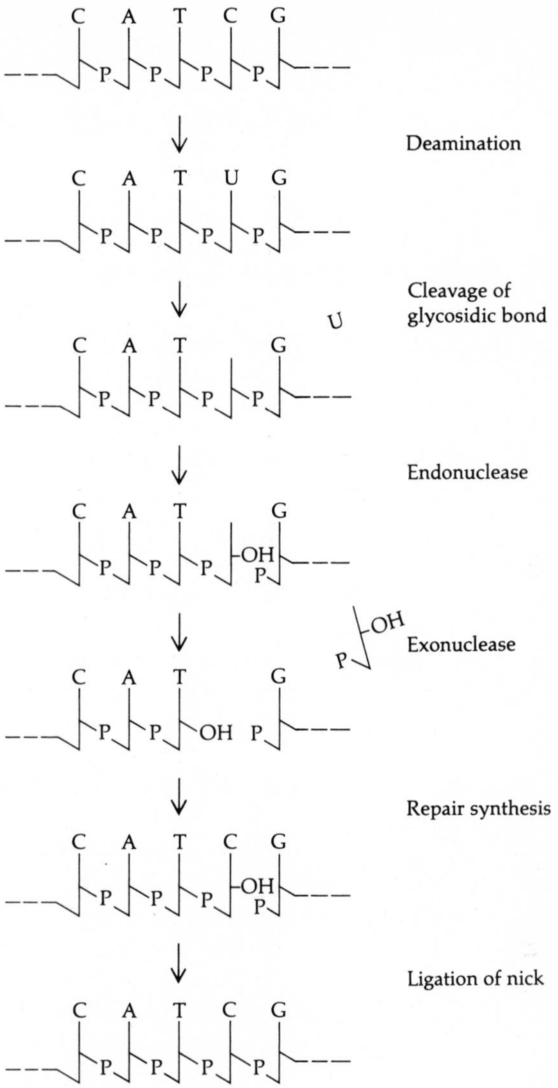

Figure 9.11
Repair of deaminated or alkylated DNA in *E. coli.* Deamination of a cytosine to uracil is followed by removal of the uracil through action of an *N*-glycosidase. Endonuclease II then cleaves the adjacent phosphodiester bond. The 3′-OH created serves as a site for the action of exonuclease III. Repair replication using the complementary strand (not shown) as template then occurs, and the remaining nick is closed by DNA ligase.

C A T C G

Deamination

C A T U G

Cleavage of glycosidic bond

U

C A T G

Endonuclease

C A T G

OH

Exonuclease

P OH

C A T G

OH P

Repair synthesis

C A T C G

OH

Ligation of nick

C A T C G

The alkylating agent dimethylnitrosamine is a potent carcinogen. When it is fed to rats it is metabolized to a form that actively methylates DNA at a number of different sites. One DNA product of the methylating activity is O^6-methylguanine formed at dG/dC sites. Dimethylnitrosamine specifically induces tumors of the kidney in rats, even though O^6-methylguanine is produced in the DNA of other organs, such as the liver, as well as in DNA of the kidney. Both kidney and liver possess enzyme systems that remove O^6-methylguanine from DNA and repair the

DNA. The susceptibility of the kidney to tumor induction by dimethylnitrosamine in comparison to the liver is correlated with a decreased efficiency of the kidney repair system for removing O^6-methylguanine from its DNA at high doses of dimethylnitrosamine. The persistence of this damage in the DNA of kidney leads to the induction of tumors, presumably by the induction of mutations when the DNA is replicated.

Evidence for a large number of different DNA repair pathways in both prokaryotic and eukaryotic cells is rapidly accumulating. It is clear that the evolution of these pathways is due to the high selective value of a metabolism that assures that the hereditary information is maintained intact and is transmitted in an unaltered form to progeny cells. No information-transferral system can be perfect, however, and mutations do arise (see Chapter 16) to provide the raw material upon which natural selection acts.

The Mechanism of General Recombination

Recombination of mutant genes to provide new associations of genes among progeny DNA molecules is an extremely important factor in evolution. Mechanisms to assure that recombination occurs have been of high selective value. These mechanisms can be divided into two categories: general, or homologous, recombination and site-specific recombination.

General recombination, which occurs between homologous DNA molecules or homologous chromatids at meiosis, has been extensively discussed in previous chapters, since it forms the basis of genetic mapping techniques. Recombination between homologous DNA molecules occurs with great precision, indicating that precise pairing must occur between homologues. Only very rarely do mutations arise that appear to be due to mispairing prior to a recombination event (see Chapter 17). The precision of the pairing between homologous parental DNA molecules is mediated by base pairing between complementary nucleotide-pair sequences in two parental molecules during the process of recombination.

The general sequence of events that leads to the formation of recombinant DNA molecules from two parental DNA molecules is diagrammed in Figure 9.12. The diagram shows that first an exchange of single strands between the parental DNA molecules takes place, resulting in a cross-bridged structure involving regions of heteroduplex DNA in each double-stranded molecule. The existence of this cross-bridged structure, or *Holliday structure*, was first predicted by Robin Holliday in 1964 on genetic grounds, from studies of gene conversion. The regions of heteroduplex DNA shown in the diagram provide for the precise alignment, just mentioned between homologous parental DNA molecules. Rotation around

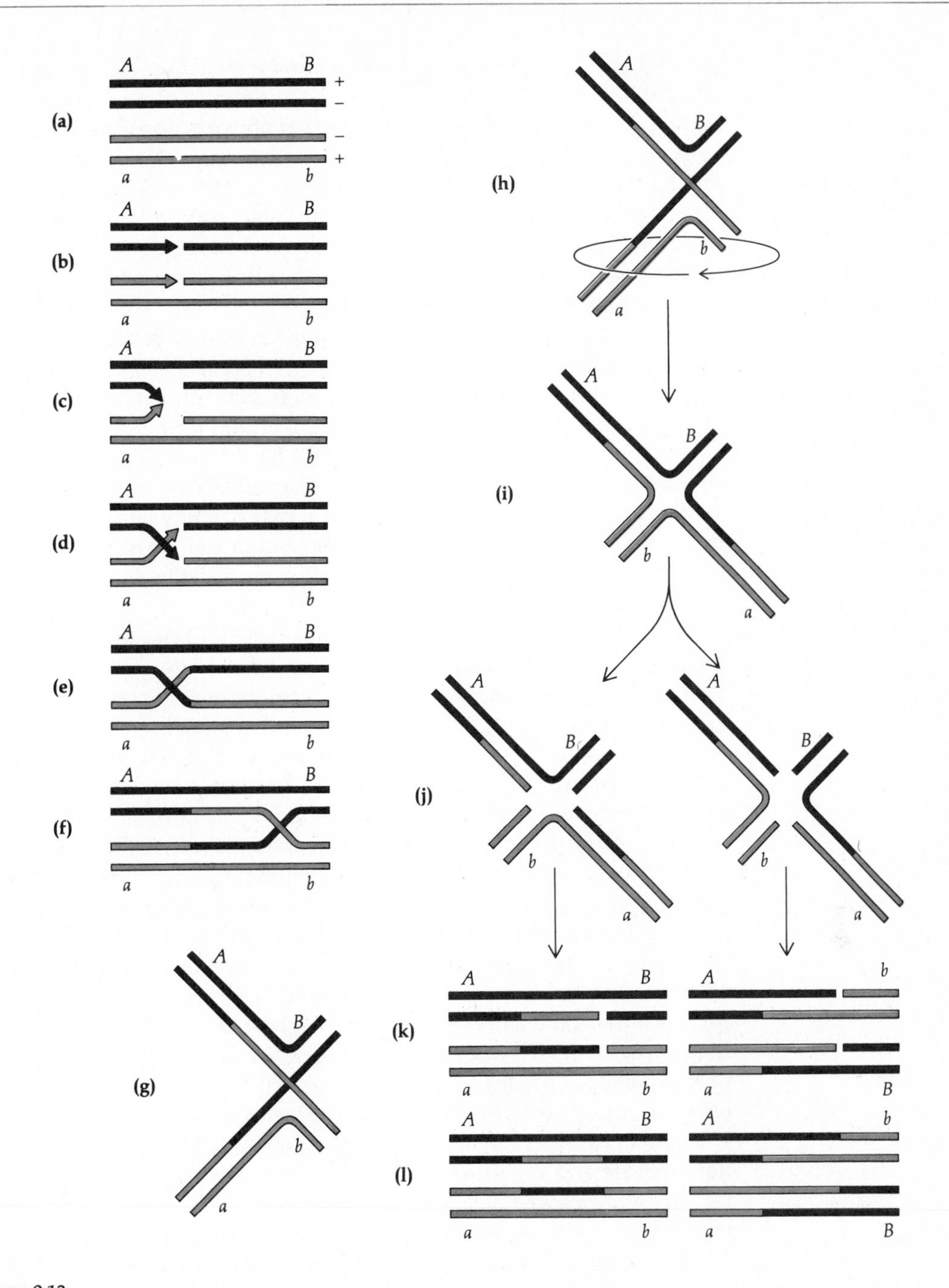

Figure 9.12
The Holliday model for general recombination. Note that migration of the cross-bridge can create large regions of heteroduplex DNA. Two types of intermediates in recombination are shown that are related by a 180° rotation. The cross-bridge may be broken in either of two ways; one results in recombination of markers flanking the region of heteroduplex DNA, while the other does not. [After H. Potter and D. Dressler, *Proc. Natl. Acad. Sci. USA* *73*:3000 (1976).]

the cross-bridge strand may occur to produce another type of Holliday structure, as shown in the diagram. Resolution of the cross-bridge by cutting of single strands can occur by either of two types of cuts, as shown, to restore two linear DNA molecules. One type of cut yields linear molecules that are recombinant for parental genetic markers on either side of a region of heteroduplex DNA. The other type of cut yields linear molecules that are not recombinant for the parental genetic markers flanking the position of the cross-bridge but that do each contain a region of heteroduplex DNA.

Our current understanding of the mechanism of genetic recombination as diagrammed in Figure 9.12 is the product of many years of genetic and biochemical research carried out with both prokaryotic and eukaryotic organisms. Evidence for the steps featured in this mechanism will now be presented. Much of this evidence has been obtained by physical and genetic studies of bacteriophage or plasmid DNA molecules, since the small size of these molecules makes them easy to study intact. The genetic predictions of this mechanism are also borne out by studies on gene conversion in fungi and observations of high negative interference in a number of organisms.

A central feature of the model in Figure 9.12 is that recombinant DNA molecules are formed by the breaking and rejoining of intact DNA molecules of parental genotype in a process independent of semiconservative DNA replication. This was first shown in studies with phage λ.

Phages grown on an *E. coli* host in a medium containing the heavy isotopes ^{15}N and ^{13}C of nitrogen and carbon are easily separated from phages grown on a host in a medium containing the normal, light isotopes, ^{14}N and ^{12}C (Figure 9.13a). Figure 9.13b shows the density distribution of the progeny phages produced by heavy parental phages when the heavy phages ($\approx$1 phage/host cell) infect a host growing in light medium. Most of the progeny contain fully light DNA in which both strands of the double helix are newly synthesized. A few of the progeny, however, contain DNA that is semiconserved parental DNA, consisting of a heavy parental strand and a newly synthesized light strand. These phages have a higher density than the rest. Figure 9.13c shows the density distribution of progeny phages produced by heavy parental phages when many heavy phages ($\approx$20 phages/host cell) infect a host growing in light medium. When many phages infect a single cell, some of the injected genomes do not replicate and are simply repackaged in new phage heads during the maturation of progeny DNA molecules. These unreplicated genomes can be detected as a third density component among the progeny phages.

When heavy phages of two genotypes, *a* + and + *b*, simultaneously infect a host cell in large numbers growing in light medium, examination of the genotypes of those phage particles that have not replicated reveals that some are of + + genotype, as shown in Figure 9.13d. This observation shows that recombination of parental genotypes can occur independently of DNA replication.

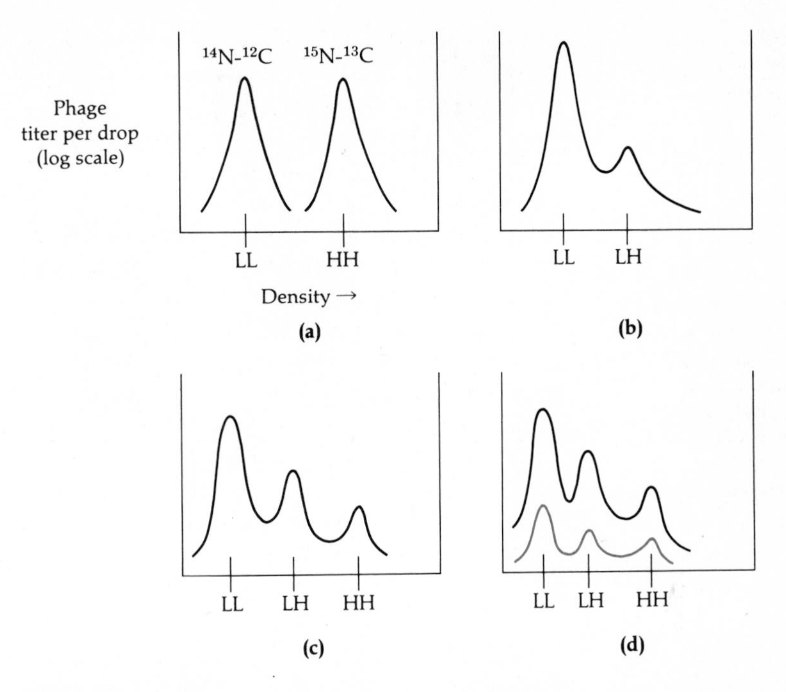

Figure 9.13
(a) Separation of fully light (LL) and fully heavy (HH) λ phages in a CsCl density gradient. Fractions are collected from the centrifuge tube, and each fraction is assayed for the number of phages by plaque assay. **(b)** Density distribution of progeny phages observed when fully heavy parental phages infect host cells in light medium at a low multiplicity. **(c)** Density distribution of progeny phages observed when fully heavy parental phages infect host cells in light medium at a high multiplicity. **(d)** Density distribution of progeny phages observed when fully heavy parental phages of two different genotypes infect host cells in light medium at a high multiplicity of each genotype. Parental genotypes are indicated in black, and wild-type recombinants in color.

The formation of recombinant DNA molecules by the breaking and rejoining of DNA molecules is shown in Figure 9.12 to be accompanied by the formation of a heteroduplex region of DNA. Indeed, a heteroduplex region of DNA is produced whether or not the cutting of the Holliday structure leads to recombination of flanking genetic markers. Genetic evidence for the existence of heteroduplex DNA molecules is provided by the existence of heterozygous λ phage particles that, upon single infection (1 phage/host cell), yield progeny in which two alleles segregate. In an experiment similar to that described for Figure 9.13d, in which one parent is heavy λc^+ and the other is heavy λc, single genomes are formed that carry both the c^+ and c alleles. These phage particles are distinguished by the creation of mottled plaques, which contain clear and turbid areas, from the clear plaques produced by λc and the turbid plaques produced

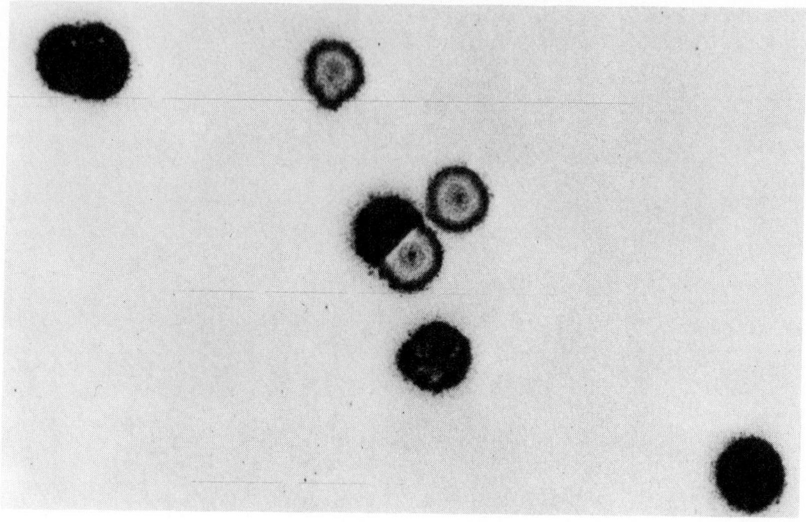

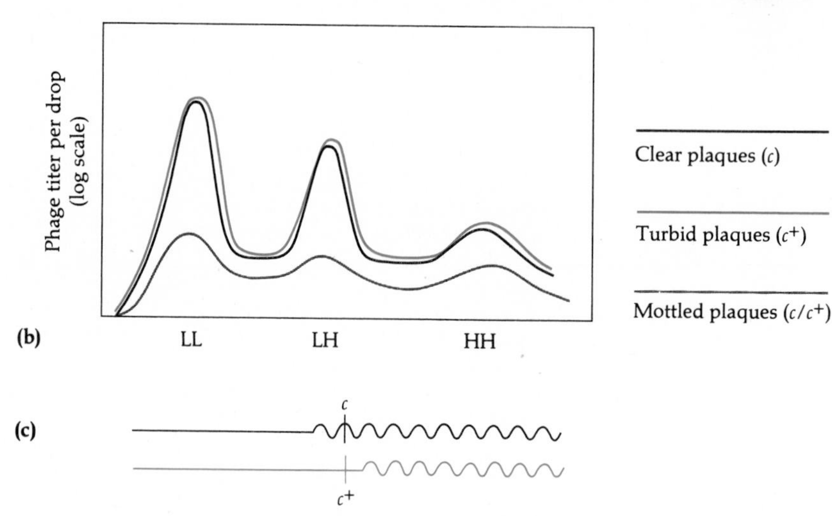

Figure 9.14
(a) Photograph of a Petri dish showing plaques made by λ*cI* (clear), λcI^+ (turbid), and a λc^+/c heterozygote (mottled). (Courtesy of Dr. Matthew Meselson, Harvard University.) **(b)** Density distribution of progeny phages observed when fully heavy parental phages of genotypes *cI* and cI^+ infect host cells in light medium at a high multiplicity of each genotype. **(c)** DNA structure of a λc/c^+ heterozygote.

by λc^+ (Figure 9.14a). The density distribution of mottled, clear, and turbid plaque-forming particles from such an experiment is shown in Figure 9.14b. These c/c^+ heterozygous phage genomes must have the structure diagrammed in Figure 9.14c. The base-pair mismatch that is present in such DNA molecules must have escaped detection by repair enzymes prior to maturation of the DNA and formation of the phage particle.

Some of the initial steps in the formation of the Holliday structure shown in Figure 9.12 have been identified. The first step is the introduction of single-strand nicks into one or both of the double-stranded molecules participating in the exchange. Two mutant genes of *E. coli* have been

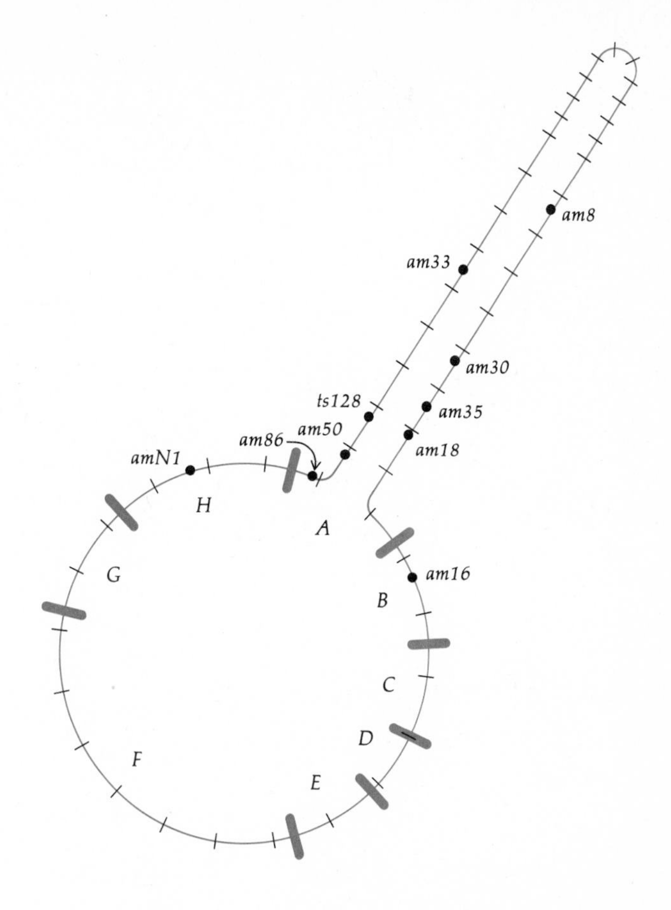

Figure 9.15
Genetic map of ϕX174 based upon recombination frequencies only. Note the map expansion present in cistron *A*. Compare with Figure 6.6 and with the nucleotide map of Figure 6.7. [From R. M. Benbow et al., *J. Virol.* 7:549 (1971).]

identified, *recB* and *recC*, that greatly reduce the frequency of genetic recombination in *E. coli.* These genes code for two polypeptides that constitute a powerful nuclease.

In ϕX174, recombination occurs much more frequently within cistron *A* than it does within other cistrons; this leads to the recombination map shown in Figure 9.15, which is distorted in comparison with the actual physical map of the genome. As discussed above, the *A* cistron is the site where a nick is always introduced to prime a new round of DNA replication. Thus, nicks occur more frequently in cistron *A* than in other cistrons. Apparently any nick can serve to initiate a recombination event, whether it is introduced by a special recombination nuclease or by a nuclease associated directly with DNA replication. The formation of a Holliday structure (which appears to be a *figure-eight* structure because of the circularity of the ϕX174 genome) between two genomes is diagrammed in Figure 9.16. The cross-bridge is created by the unwinding of a single strand of DNA from a nicked molecule, followed by its insertion between

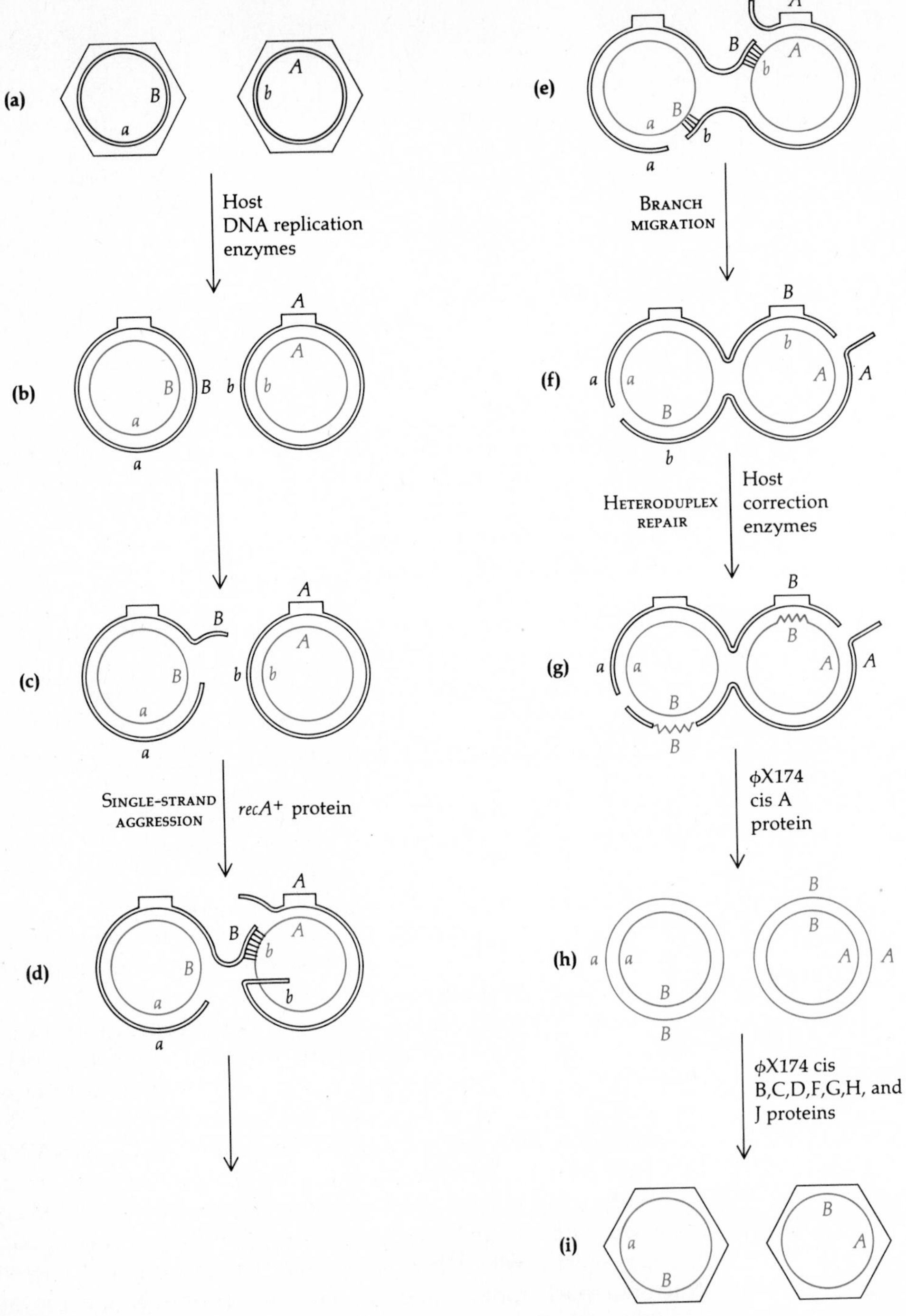

Figure 9.16
A model for ϕX174 recombinant formation. The parental DNA strands and genetic markers are shown in black. All subsequently synthesized DNA strands and genetic markers are shown in color. Heteroduplex mismatch repair synthesis is indicated by a jagged line. Enzymes or proteins known to be involved in specific steps are indicated. The hexagons represent the phage capsids. (After R. M. Benbow et al., in *Mechanisms in Recombination*, ed. by R. F. Grell, Plenum Press, New York, 1974.)

the strands of the second DNA molecule. This process has been termed *single-strand aggression.* The creation of ϕX174 figure-eight molecules requires the participation of a host function, *recA*$^+$, which is also required for recombination by *E. coli.* The *recA*$^+$ protein is believed to mediate single-strand aggression, as indicated in the diagram. The helix-unwinding protein and helix-destabilizing proteins that participate in DNA replication probably also play roles in the process of single-strand aggression.

During meiosis in both rat (or mouse) and lily, the stage-specific appearance of both helix-destabilizing proteins and DNA nicking enzymes is observed to occur at those meiotic stages at which crossing over occurs (Figure 9.17). Similar enzymatic mechanisms for general recombination in both prokaryotes and eukaryotes probably exist. The nucleosome structure of chromatin in eukaryotes (see Chapter 4) and its absence in prokaryotes, however, mean that the mechanisms cannot be completely identical. Pairing between homologous chromatids during meiosis is mediated by an organized protein structure called the *synaptinemal complex,* which draws homologous chromosomal regions together prior to the initiation of crossing over and the formation of Holliday structures at the sites of crossing over (Figure 9.18). The requirement for prepairing of homologous chromosomal regions in the synaptinemal complex is probably dictated by the nucleosomes, which must impede single-strand aggression and annealing of complementary regions of DNA, as well as by the great length of the DNA molecules present in most eukaryotic chromosomes.

Once a cross-bridge between two parental DNA molecules has formed, its position can "diffuse" along the DNA molecules by a zipper-like action in which the pairing of individual nucleotides changes from one strand to another in forming complementary hydrogen bonds. This process creates long regions of heteroduplex DNA in both molecules participating in the Holliday structure (Figure 9.12f). The "arms" of the Holliday structure (Figure 9.12g) may isomerize by rotation to give an equivalent structure (Figure 9.12i). An electron micrograph of such forms created by recombination between two plasmid DNA molecules is seen in Figure 9.19, where the single strands of the cross-bridge are revealed.

The final steps in the recombination process are the cutting of the cross-bridge strands and the ligation of the resulting nicks to yield recombinant molecules.

Two types of cuts can remove the cross-bridge, as indicated in Figure 9.12j. An "east-west" cut yields products that are nonrecombinant for flanking genetic markers and contain regions of heteroduplex DNA, whereas a "north-south" cut yields recombinants for flanking markers that also contain regions of heteroduplex DNA. If east-west and north-south cuts are equally probable ways of resolving a Holliday structure, then one-half of all genetic exchanges should result in recombination of flanking markers. Recall from Chapter 8 that one-half of all gene-conversion events show recombination of the flanking markers. The fact that one-half of such exchanges result in recombination of flanking markers provides genetic evidence that isomerization of a Holliday structure oc-

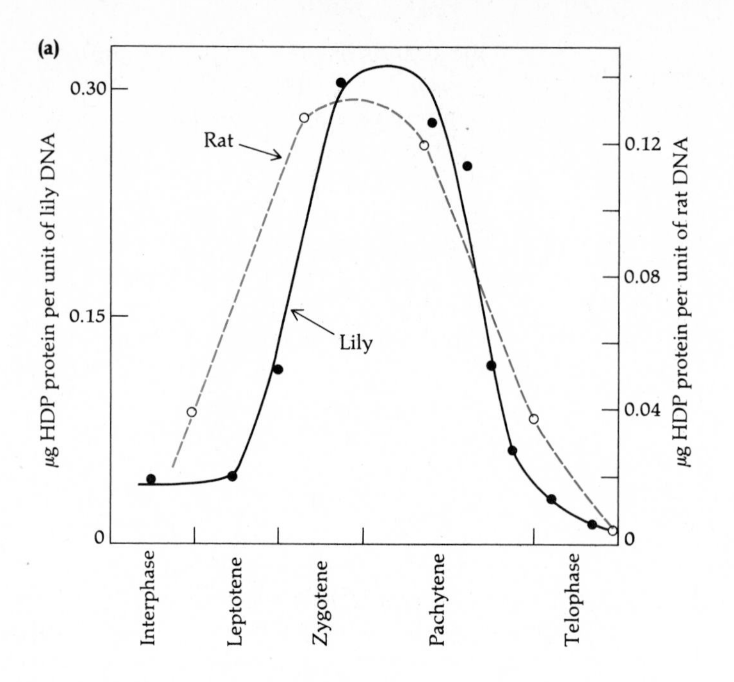

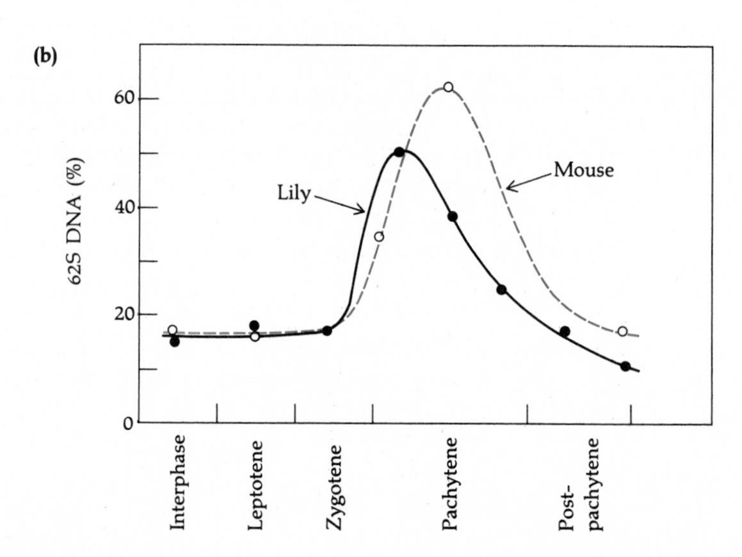

Figure 9.17
(a) Changes in activity of a helix-destabilizing protein during meiosis in lily microsporocytes and rat spermatocytes. **(b)** Changes in sedimentation behavior of single-stranded DNA prepared from lily microsporocytes and mouse spermatocytes at different stages during meiosis. The increase in the amount of smaller (62S) DNA single strands reflects an increase in nicking activity (most of the DNA single strands sediment faster than 100S; see Figure 5.29 caption). [After Y. Hotta, A. C. Chandley, and H. Stern, *Nature 269*:240 (1977).]

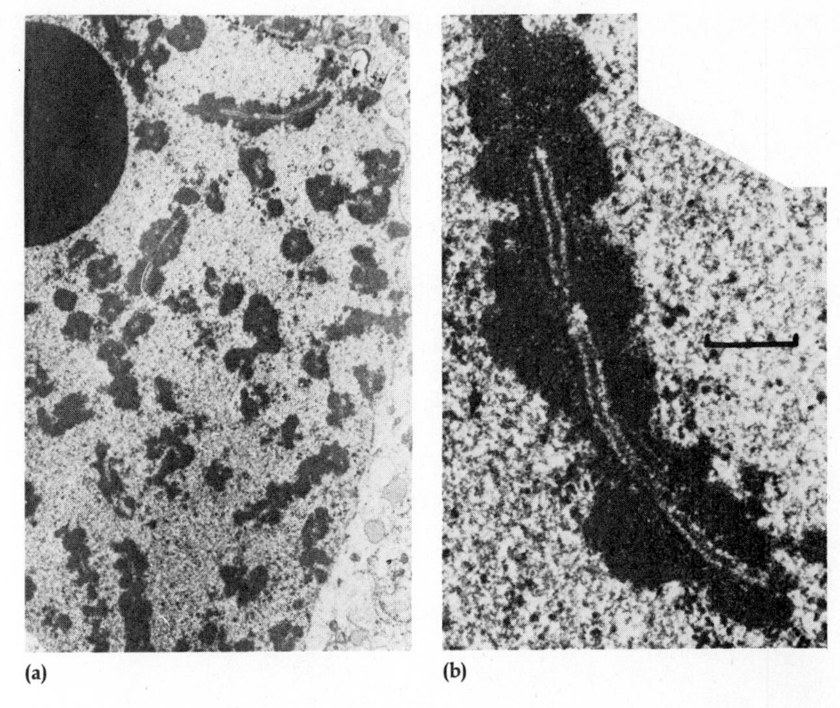

Figure 9.18
Synaptinemal complexes in lily occupy the spaces between the paired homologues in these electron micrographs of a sectioned nucleus.
(a) ×2900. **(b)** ×15,500.
[From P. B. Moens, *Chromosoma 23*:418 (1968).]

curs, and it substantiates the existence of Holliday structures (such as those shown in Figure 9.19) as intermediates in genetic recombination in eukaryotes as well as in prokaryotes. Gene conversion results from the formation of heteroduplex DNA in which allelic differences (base-pair mismatches) reside, as discussed in Chapter 8. The frequency with which asci show conversion of a particular marker is then a measure of genetic exchange in that region.

Another way of representing the events of Figure 9.12 is diagrammed in Figure 9.20. This figure shows isomerization of a cross-bridge by a double rotation of the participating double-helical chromatids, which is a more likely event for the constrained chromatids within a synaptinemal complex. If the two isomeric forms are equally probable, then cutting the cross-bridge in either form will, on the average, result in exchange events one-half of which show recombination of flanking markers.

Site-Specific Recombination

As the name implies, the second category of recombination events comprises those whose occurrence is restricted to specific sites on a DNA molecule. This is in contrast to the general recombination events discussed

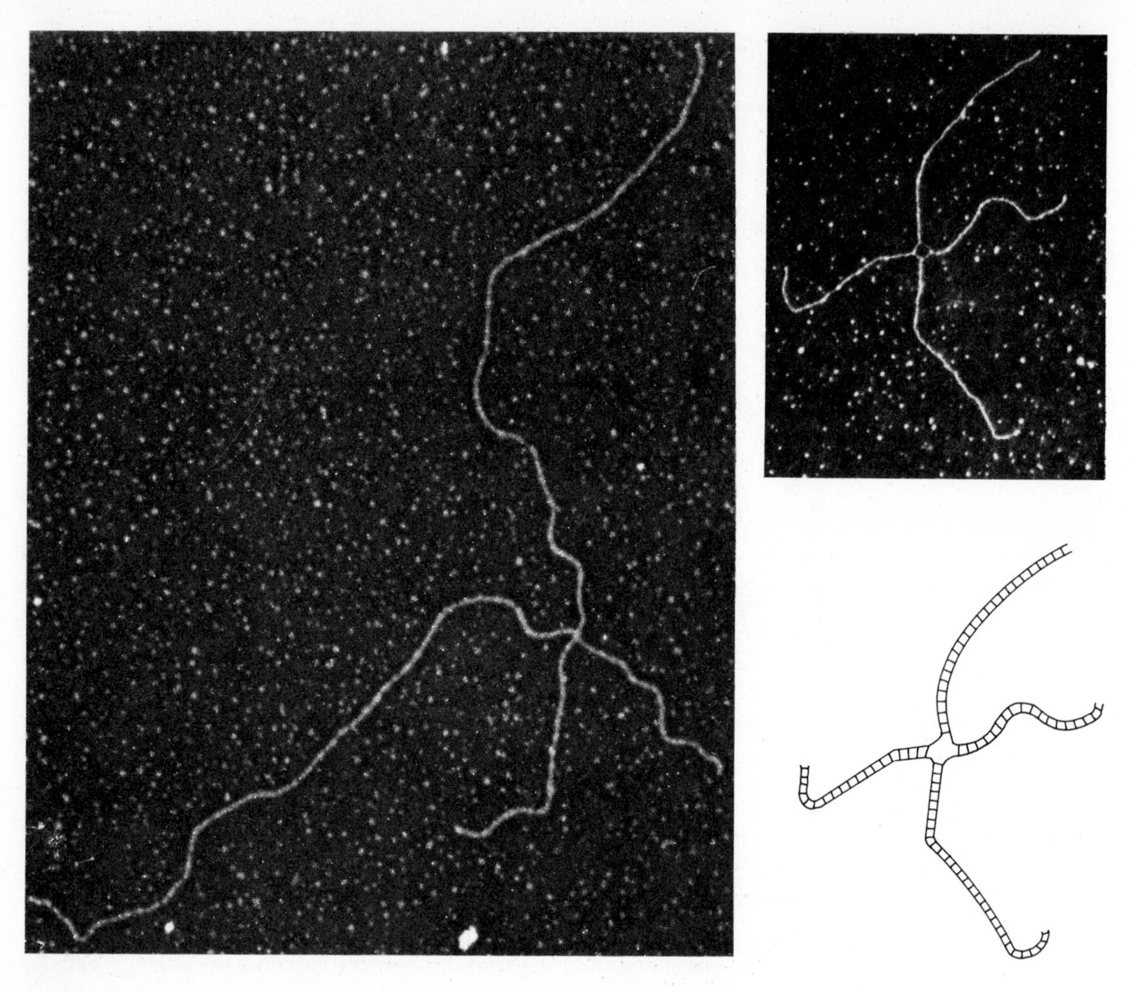

Figure 9.19
Electron micrographs of Holliday structures formed by recombining circular plasmid DNA molecules in *E. coli*. The circular molecules have been opened by cleavage with restriction enzyme *Eco* RI (this plasmid contains only one *Eco* RI site). The two molecules shown are the two isomeric forms of the Holliday structure. Separation of single strands at the cross-bridge can be seen in the form on the right. [From H. Potter and D. Dressler, *Proc. Natl. Acad. Sci. USA* 73:3000 (1976); 74:4168 (1977).]

above, in which heteroduplex DNA is formed and the site of crossing over may occur at the junction between any two adjacent nucleotide pairs. In *recA*, *recB*, and *recC* mutants of *E. coli* that are defective in general recombination, site-specific recombination is not affected, thus demonstrating that the two categories of recombination events are functionally distinct.

The integration of phage lambda into the *E. coli* chromosome is the most intensively studied case of site-specific recombination. Lambda is inserted into the bacterial chromosome by recombination between a site, *attB*, on the host chromosome and a site, *attP*, on the phage chromosome. Each of the two chromosomes contains a single *att* site. Lambda is a member of a family of temperate phages called lambdoid phages. Each species of lambdoid phage (λ, 434, 21, ϕ80) possesses a unique insertion site that is different from all the others. Recombination between the λ *attP* and *attB* sites requires a protein encoded by the λ*int* (integrase) gene. The specificity of site recognition resides in the ability of the *int* protein to recognize its own sites of action on both the phage and bacterial chromosomes. Each species of lambdoid phage possesses its own distinct integration enzyme. The integration and excision of the λ prophage are diagrammed in Figure 6.10.

The nucleotide sequences of *attP* (POP′) and *attB* (BOB′) and of the junctions between prophage and host chromosome, POB′ and BOP′, are shown in Figure 9.21. The POP′ and BOB′ sites share a common core of 15 nucleotide pairs, which is also present in POB′ and BOP′. The site-specific recombination event takes place within this common core, but precisely where each of the two strands of the core region is cut and rejoined is unknown. As far as is known, cutting and rejoining occur as a single event—the host ligase is not required to close any nicks that may exist transiently. Two models for site-specific recombination have been proposed. In model I (Figure 9.22a), staggered cuts in the complementary strands create "sticky ends" that cross-anneal to insert the phage DNA into the host chromosome. In model II (Figure 9.22b), cuts made at a single nucleotide pair in each DNA molecule create blunt ends that are then covalently rejoined to effect the recombination event.

Many other examples of site-specific recombination are known in prokaryotes, and a few have recently been found in eukaryotes. The integration of the *E. coli* F factor, which occurs at a number of different sites in the *E. coli* chromosome, is another example of site-specific recombination. The mechanisms involved in these other cases may be similar to that employed by lambda.

DNA Restriction and Modification Enzymes

Virtually all bacterial species synthesize one or more types of site-specific endonucleases called *restriction enzymes*, which make double-stranded cuts in DNA. They are called restriction enzymes because they appear to function primarily to restrict the entrance of foreign DNA into bacterial cells. The DNA of a cell that synthesizes a restriction enzyme is protected from the action of that enzyme because such cells also synthesize a *modification enzyme*, which alters the structure of the DNA sites that are recognized by the restriction enzyme. If a cell with an active restriction-

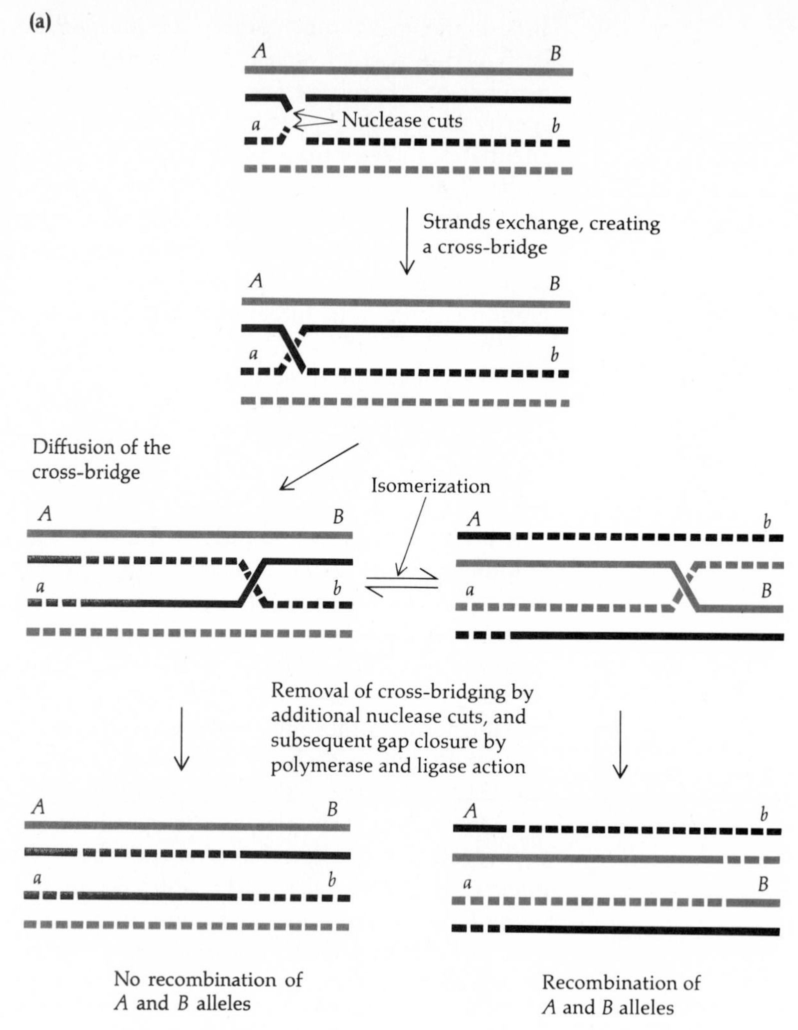

Figure 9.20
(a) Isomerization of a cross-bridge can occur without physical rotation of an entire DNA strand (as is illustrated in Figure 9.12). Isomerization must occur if recombination of outside markers is to result. (After J. D. Watson, *Molecular Biology of the Gene*, 3rd ed., W. A. Benjamin, Menlo Park, Cal., 1976.)

modification system is infected by a bacteriophage whose DNA has not previously been modified, the probability of the phage DNA's initiating an infection is several orders of magnitude less than if the phage DNA were modified. The unmodified DNA is broken into fragments, the number of fragments being a function of the number of sensitive sites it possesses, and the fragments are then degraded by exonucleases. Rarely, the infecting phage DNA is modified by the host modification enzymes *before* it is attacked by the restriction enzymes, and it then initiates a lytic infection. All of the progeny phages produced will then possess modified DNA and will be able to infect other bacterial cells (with the same restriction-modification system) with high efficiency.

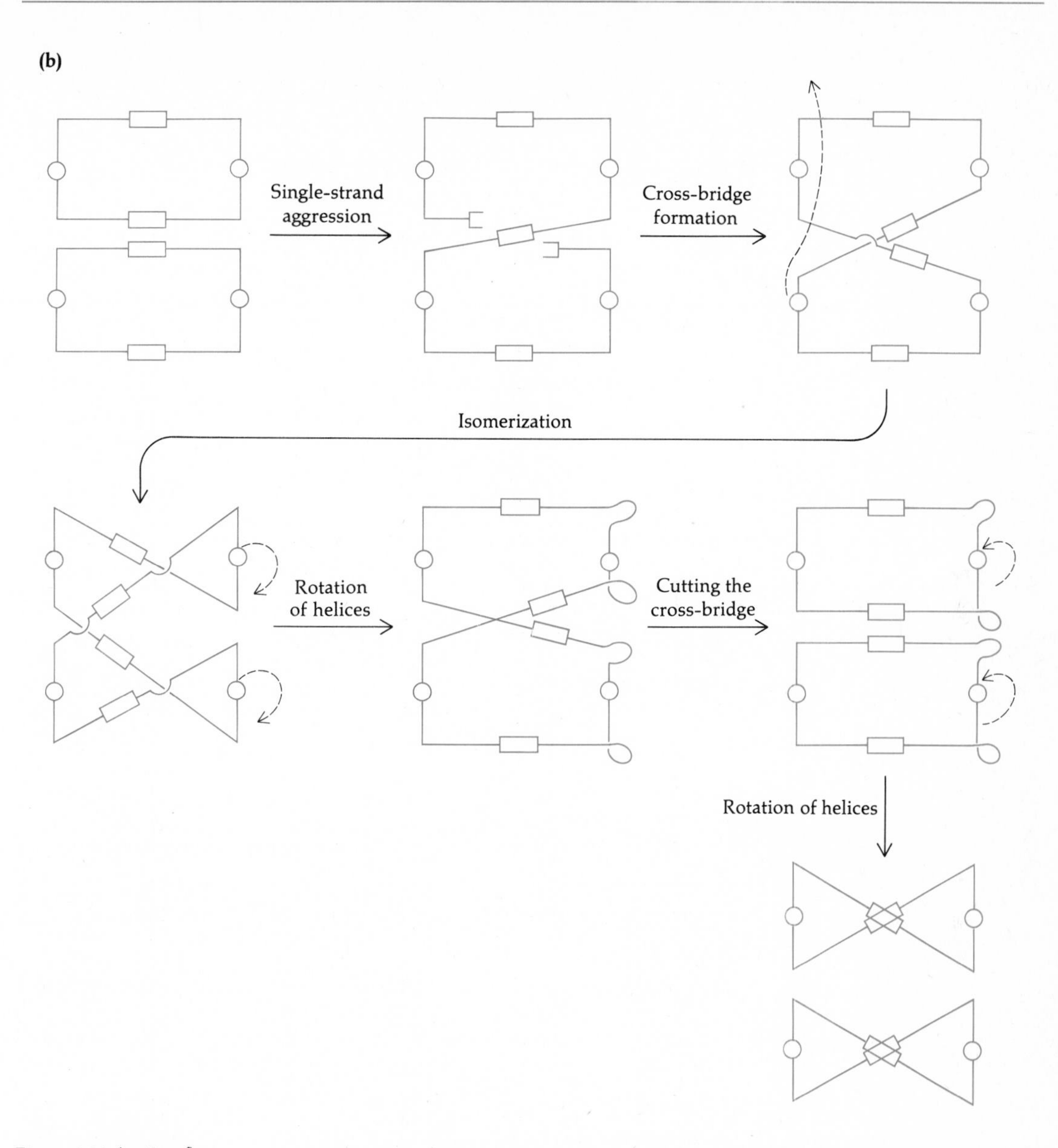

Figure 9.20 (*continued*)
(b) The isomerization illustrated in part (a) must be seen to be appreciated. The steps of the "Recombination Shuffle" illustrated here should be performed by students in the class. Any number of pairs can participate in forming a "mating pool."

Two major types of restriction enzymes are known. Type I restriction endonucleases recognize a specific site and then make a double-strand cut somewhere nearby, but without any specificity as to the nucleotide sequence that is cut. Type II restriction endonucleases recognize a site and cleave specifically at the recognition site. Our discussion of restriction enzymes will be confined to type II because the special properties of these enzymes have spawned a new field of scientific endeavor during the 1970s—that of genetic engineering by *in vitro* recombinant DNA technology. Equally important is that type II restriction enzymes provide powerful tools for generating unique fragments from very large DNA

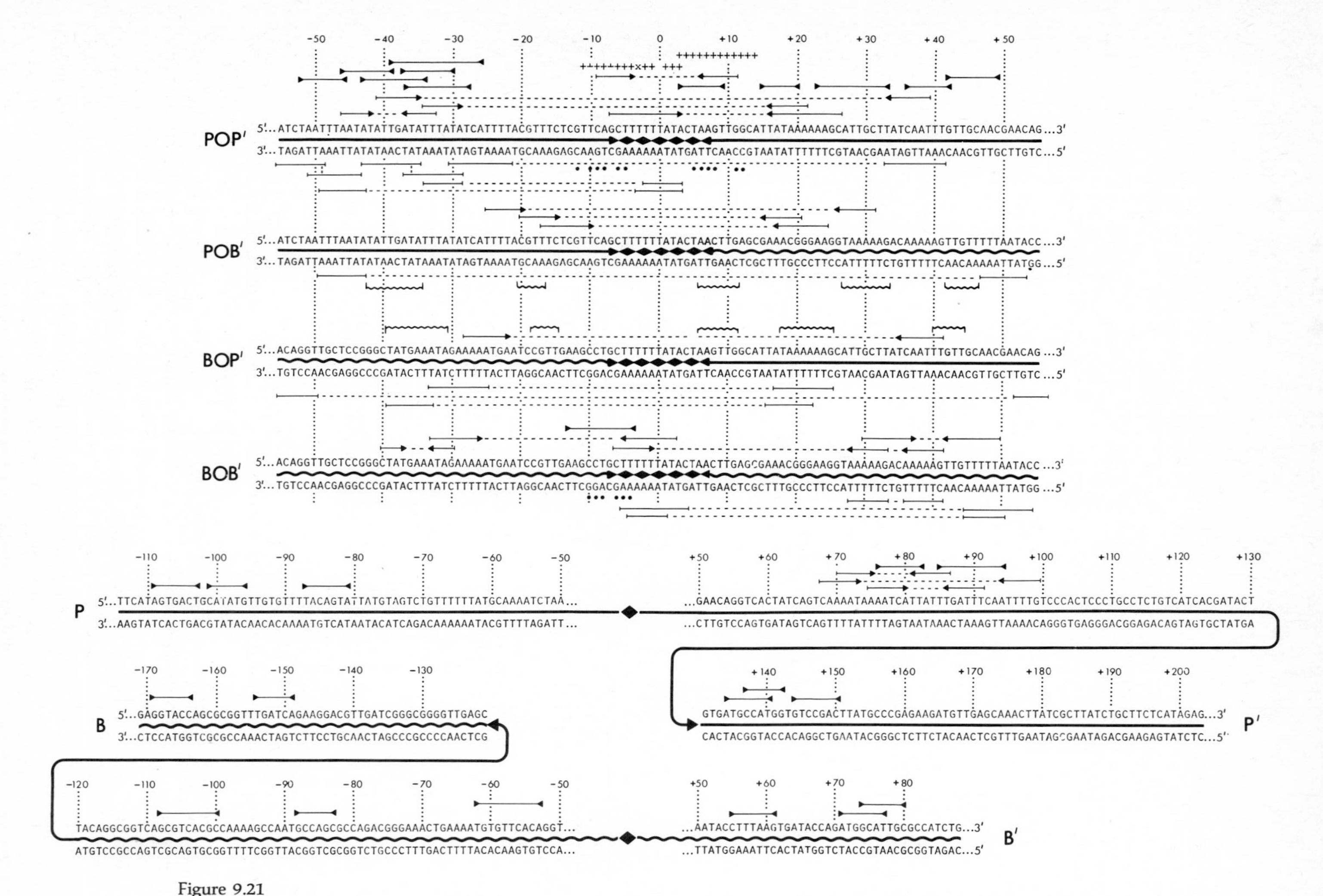

Figure 9.21
Nucleotide sequences of the attachment sites of λ phage DNA (POP′) and the *E. coli* DNA (BOB′), as well as of the junction between phage and bacterial DNAs (POB′ and BOP′). The common core region is indicated. [From A. Landy and W. Ross, *Science 197*:1147 (1977). Copyright © 1977 by the American Association for the Advancement of Science.]

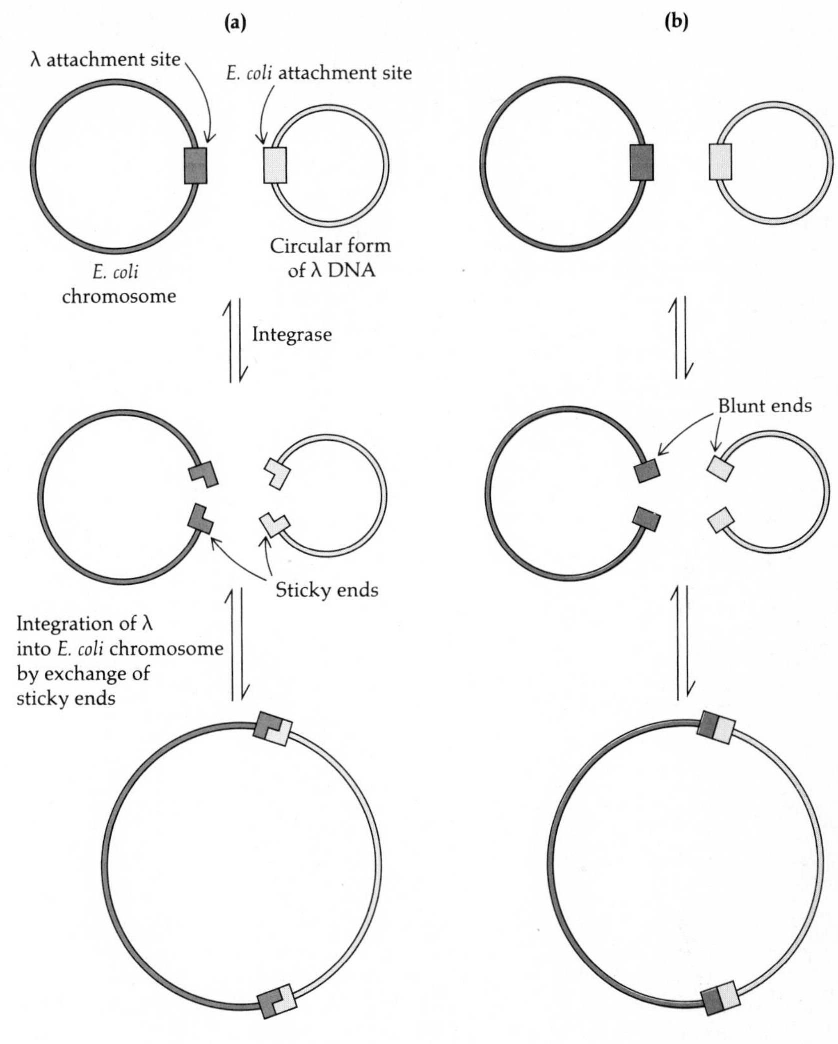

Figure 9.22
Two possible models for the *int*-mediated integration of λ DNA into the host chromosome, differing only in the nature of the cuts made and the configurations of the resulting ends of the strands.

molecules from both prokaryotes and eukaryotes, making it possible to physically map these large molecules and to sequence unique fragments of special interest; for an example, see Figure 9.21.

The recognition sites of restriction enzymes are palindromic sequences—sequences that have a central axis of symmetry and read identically in both directions from that axis. The recognition sequences of several restriction enzymes are shown in Table 9.2. As indicated by the arrows, the double-strand cuts may be either staggered or aligned along the axis of symmetry. In the former case, the staggered cuts create sticky ends, which may reanneal with each other and be covalently closed by DNA ligase. Protection from cleavage is conferred by modification en-

Table 9.2
Recognition sequences of several restriction endonucleases.

Microbial Origin	Enzyme	Recognition Site*	Number of Recognition Sites per Viral Genome†		
			SV40	λ	Adenovirus 2
Escherichia coli KY13	*Eco* RI	5′ G↓A A* T T C 3′ C T T A* A↑G	1	5	5
Hemophilus influenzae R_d	*Hin* dII	5′ G T Py↘ Pu A* C 3′ C A* Pu ↖Py T G	7	34	>20
	Hin dIII	5′ A*↓A G C T T 3′ T T C G A↑A*	6	6	11
Hemophilus parainfluenzae	*Hpa* I	5′ G T T↘ A A C 3′ C A A ↖T T G	5	11	6
	Hpa II	5′ C↓C* G G 3′ G G C*↑C	1	>50	>50
Hemophilus aegyptius	*Hae* III	5′ G G↘ C C 3′ C C ↖G G	18	>50	>50

*The asterisks mark bases that can be methylated by modification enzymes. The small arrows mark the cuts made by the type II restriction endonuclease. The long vertical arrow is the axis of symmetry. Pu = purine; Py = pyrimidine.

†The molecular weights of these genomes are as follows: SV40 = 3×10^6 daltons; $\lambda = 32 \times 10^6$ daltons; adenovirus 2 = 25×10^6 daltons.

zymes that methylate certain bases within the recognition sites; these bases are indicated with an asterisk. Methylation by the modification enzymes occurs after a particular nucleotide has been incorporated into a DNA strand. These enzymes act either on completely unmethylated sites or on semimethylated sites created by semiconservative replication of a fully methylated site.

More than 30 different restriction enzymes from different species and strains of bacteria have been purified and their cleavage sites determined. Some of these enzymes almost certainly do serve the function of restrict-

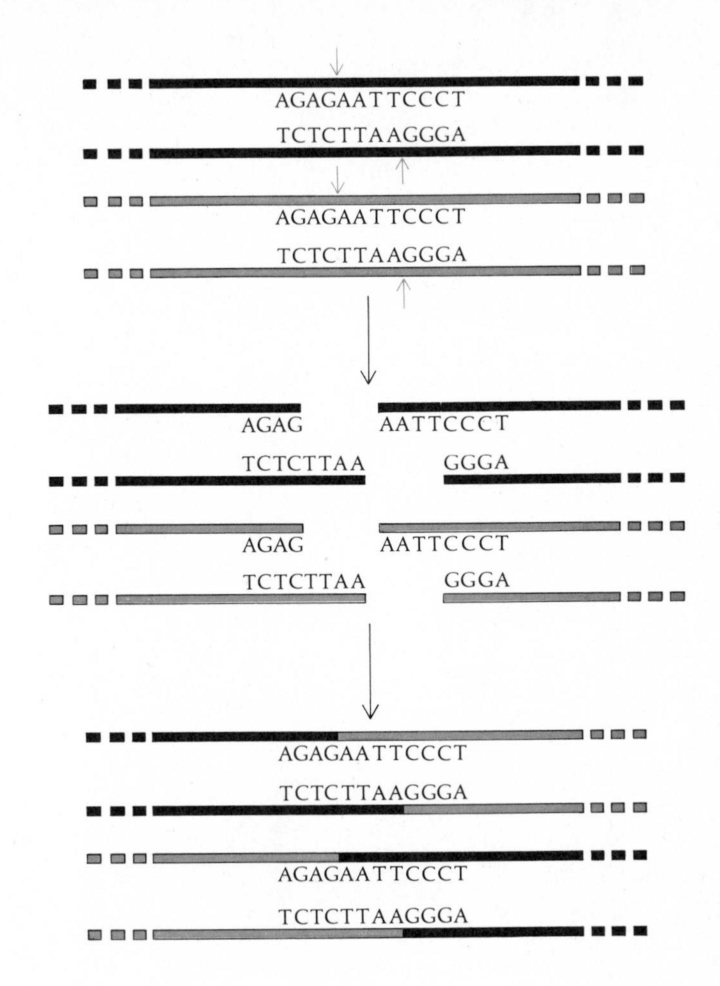

Figure 9.23
Recombination of different DNA molecules can occur by site-specific cleavage of the molecules, creating complementary single-stranded ends that dissociate and reanneal. DNA ligase bonds the recombinant molecules covalently.

ing the entrance of unmodified DNAs into the cell. Some species of bacteria, however, contain "restriction" endonucleases but exhibit no detectable restriction of foreign DNA *in vivo.* Such observations suggest that these enzymes may serve other functions, such as carrying out site-specific recombination. An endonuclease that creates sticky ends might cleave at two or more sites that have escaped modification and provide an opportunity for the sticky ends to undergo rearrangement or recombination (Figure 9.23). The recombined molecules would subsequently be ligated and modified by the ligase and modification enzymes.

Shing Chang and Stanley N. Cohen have shown in a number of experiments that the restriction enzyme *Eco* RI can promote site-specific recombination of plasmid DNAs in *E. coli*. A simple demonstration of this action by *Eco* RI, which also serves to outline the principle of *in vitro* recombinant DNA formation, is detailed in Figure 9.24. The resistance

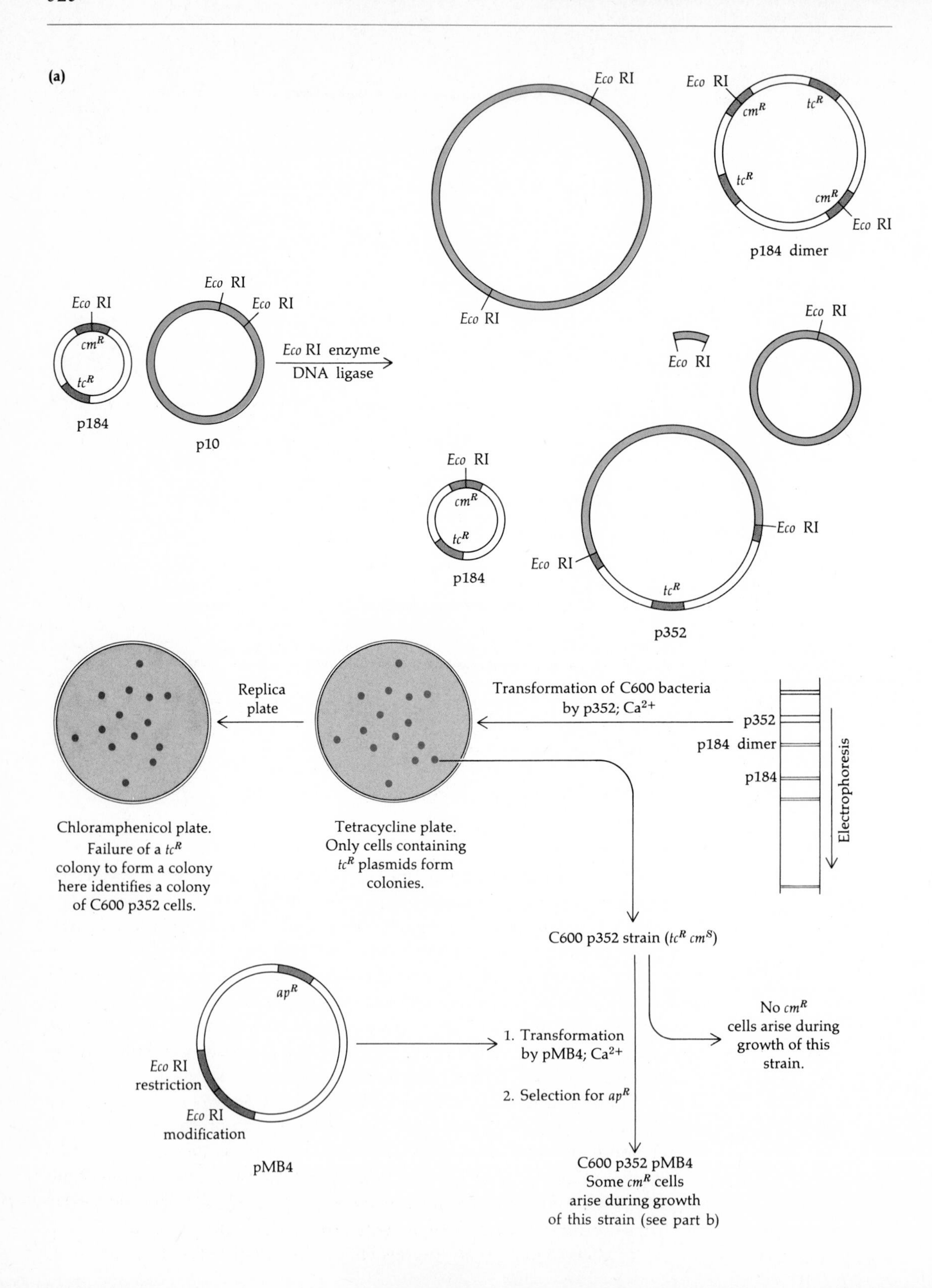

(a)
Eco RI
cm^R
tc^R
p184
Eco RI
Eco RI
p10
Eco RI enzyme
DNA ligase
Eco RI
Eco RI
Eco RI
cm^R
tc^R
tc^R
cm^R
Eco RI
p184 dimer
Eco RI
Eco RI
Eco RI
cm^R
tc^R
p184
Eco RI
Eco RI
tc^R
p352
Replica plate
Transformation of C600 bacteria by p352; Ca2+
p352
p184 dimer
p184
Electrophoresis
Chloramphenicol plate. Failure of a tc^R colony to form a colony here identifies a colony of C600 p352 cells.
Tetracycline plate. Only cells containing tc^R plasmids form colonies.
C600 p352 strain (tc^R cm^S)
ap^R
Eco RI restriction
Eco RI modification
pMB4
1. Transformation by pMB4; Ca2+
2. Selection for ap^R
No cm^R cells arise during growth of this strain.
C600 p352 pMB4
Some cm^R cells arise during growth of this strain (see part b)

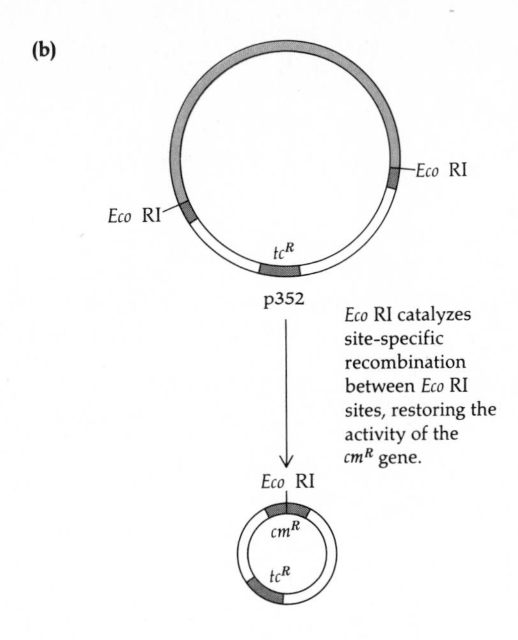

Figure 9.24 (*begins opposite*)
In vitro formation of a recombinant plasmid, p352, from p184 by insertion of an *Eco* RI fragment of p10 into the cm^R gene. The inserted fragment inactivates the cm^R gene. The p352 plasmid DNA is purified from other recombinant and nonrecombinant DNA molecules by electrophoresis, which separates DNA molecules of different sizes. This plasmid is cloned by transformation of antibiotic-sensitive C600 cells. *E. coli* cells carrying p352 are resistant to tetracycline. The tc^R transformants are tested for sensitivity to chloramphenicol to screen out transformants that may have received a contaminating p184 plasmid instead of p352. Demonstration of *in vivo* recombination mediated by *Eco* RI is made by transforming C600 p352 cells with pMB4. See text for details.

transfer plasmid p184 carries genes conferring resistance to the antibiotics tetracycline (tc^R) and chloramphenicol (cm^R); it has one recognition site for *Eco* RI, located within the cm^R gene. The purified p184 DNA is mixed with purified DNA from another plasmid (which carries no genes of interest for this experiment) also possessing *Eco* RI sites. The mixture is cleaved with pure *Eco* RI enzyme at a temperature that permits the sticky ends to come apart and reanneal with each other, either to reconstitute the original plasmids or to form larger hybrid plasmids. The addition of DNA ligase covalently closes the nicks present and produces a mixture of the original plasmids and a hybrid plasmid, p352, in which the cm^R gene of p184 is split and inactivated by insertion of another DNA. This hybrid plasmid p352 is separated from the smaller parental plasmids by electrophoresis and is used to transform *E. coli* cells of strain C600, which is sensitive to all antibiotics. The transformation of the *E. coli* cells can be promoted very efficiently by "shocking" them with a transient high concentration of Ca^{2+}, which renders them permeable to DNA molecules. The cells that

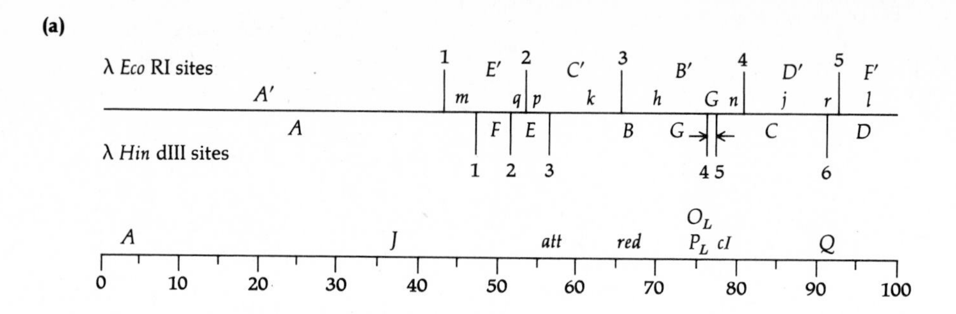

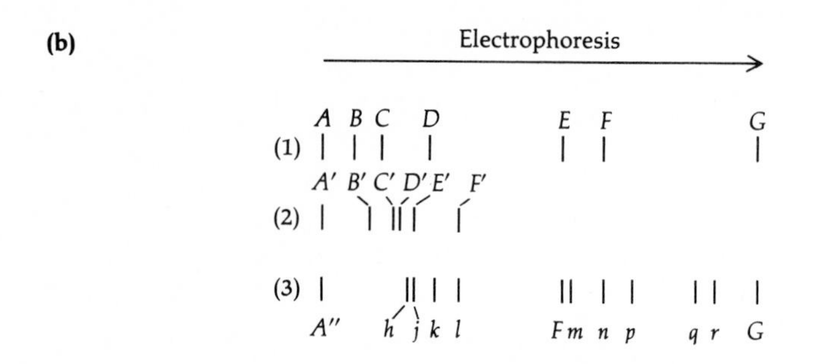

Figure 9.25
(a) A physical map of λ, showing the positions of *Eco* RI and *Hin* dIII restriction sites. **(b)** The electrophoretic size distributions of fragments: (1) *Hin* dIII cleavage; (2) *Eco* RI cleavage; (3) *Hin* dIII and *Eco* RI cleavage. [After K. Murray and N. Murray, *J. Mol. Biol.* 98:551 (1975).]

take up p352 DNA are easily detected by plating on a medium containing tetracycline, because of the tc^R gene carried by the plasmid. The strains of tetracycline-resistant cells derived by this procedure are all sensitive to chloramphenicol because the extraneous DNA insert in the cm^R gene destroys the function of that gene. The C600 p352 strains can grow for many generations and remain tetracycline-resistant and chloramphenicol-sensitive. Chang and Cohen tested 10^{11} cells that were the progeny of single transformed cells without detecting any chloramphenicol-resistant cells.

Another plasmid, called pMB4, carries genes coding for both the *Eco* RI restriction enzyme and the modification enzyme, as well as a gene for ampicillin resistance (ap^R). The pMB4 plasmid is introduced into C600 p352 cells using the Ca^{2+} shock technique, and C600 p352 pMB4 cells are selected by their newly acquired resistance to ampicillin. When strains of C600 p352 pMB4 are grown to produce large numbers of progeny, some of the progeny cells (≈ 20 of 10^{10} cells) are found to be chloramphenicol-resistant, as well as tetracycline- and ampicillin-resistant. These chloramphenicol-resistant cells have been shown to contain a new plasmid, which appears to be identical to the original parent p184 plasmid, except that the *Eco* RI site in the restored cm^R gene is modified by methylation, rendering it resistant to *Eco* RI restriction. This result provides sound evidence that the *Eco* RI enzyme is able to promote a site-specific recombination between the two *Eco* RI sites in p352, causing the deletion of the extraneous DNA inserted into the cm^R gene by *in vitro* recombination.

Regardless of the roles played by restriction enzymes *in vivo*, they provide powerful tools for the biochemical analysis of genomes. The DNA sequencing techniques that have provided the structure of the ϕX174 genome (Figure 6.7), the SV40 genome (a mammalian tumor virus), and those portions of the λ genome that have been sequenced (e.g., Figure 9.21) require purified DNA fragments of unique nucleotide sequence no more than a few hundred nucleotides in length. The discovery and purification of a battery of restriction enzymes have provided the means to obtain such unique DNA fragments for sequencing purposes.

Restriction enzymes can also be used to construct physical maps of genomes based on the order of restriction sites and the molecular distances between them. The locations of *Eco* RI and *Hin* dIII restriction sites on wild-type λ DNA shown in Figure 9.25 provide, in conjunction with the physical maps in Figure 6.9, a means of assigning particular genes to specific DNA fragments produced by cleavage with *Eco* RI and *Hin* dIII, either singly or together. A restriction map of the yeast mitochondrial genome with the positions of the genes discussed in Chapter 5 is shown in Figure 9.26 (compare with Figure 5.29).

Recombinant DNA Technology

Restriction enzymes offer tools with which any two DNAs can be cleaved and recombined *in vitro*, regardless of the origins of the DNAs. The employment of plasmids as cloning vehicles makes it possible to insert any foreign DNA fragment into a plasmid and to amplify that foreign DNA in cloned bacterial strains carrying the recombinant plasmid. The implications of this fact for science and technology are only now becoming clear. *In vitro* recombination and molecular cloning of eukaryotic genes, and even of synthetic genes, offer heretofore unknown possibilities, and also some uncertainties.

The techniques are simple, and follow the events diagrammed in Figure 9.23. An *Eco* RI fragment derived from DNA (human, animal, or plant) is inserted into an antibiotic-resistance gene of a plasmid that also carries a second antibiotic-resistance gene lacking an included *Eco* RI site. The plasmids are then introduced into an antibiotic-sensitive strain of *E. coli*, and strains of *E. coli* resistant to the second antibiotic are selected (in Figure 9.24, tetracycline-resistant strains are selected). Replica plating to medium containing the first antibiotic allows identification of clones that are sensitive to that antibiotic and that therefore carry an insertion of foreign DNA in that resistance gene. Techniques have also been developed for inserting shear-fragmented DNAs into plasmids, making entire eukaryotic genomes subject to fractionation and cloning in plasmids without regard to particular restriction sites.

If sheared fragments of the *Drosophila melanogaster* genome (comprising a total of about 1.5×10^8 np) of average size $15\text{–}20 \times 10^3$ np

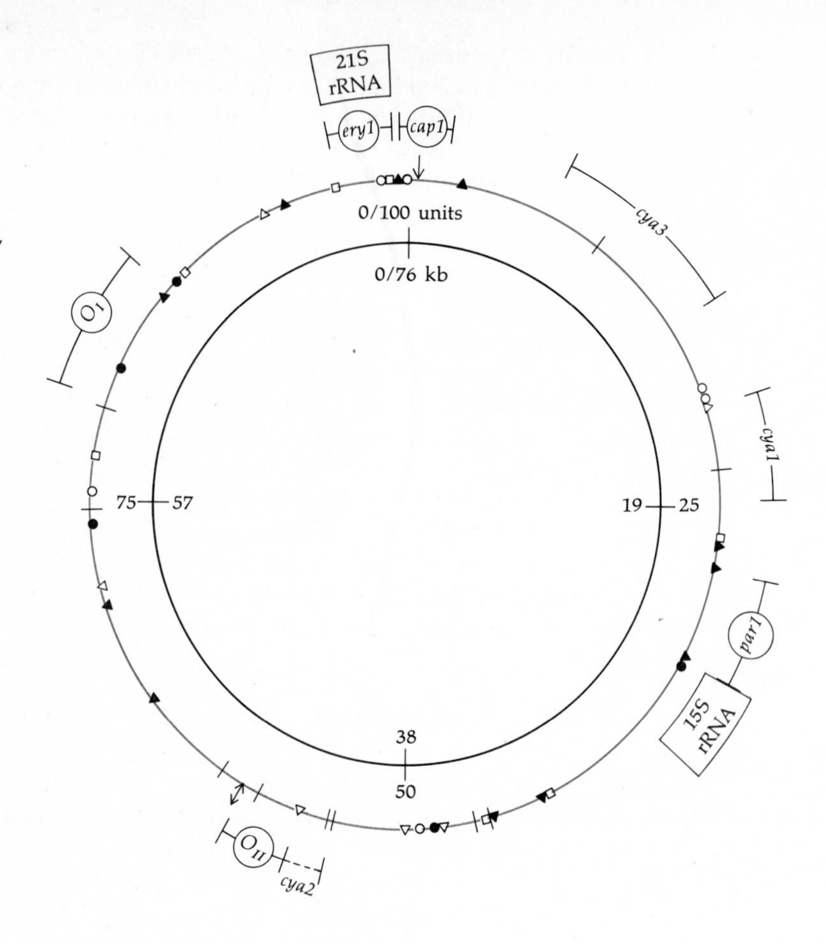

Figure 9.26
A physical map of yeast mitochondrial DNA based on restriction-enzyme cleavage. 1 kb = 1 kilobase pair = 1000 np. See also Figure 5.29. (Courtesy of Prof. Murray Rabinowitz, University of Chicago.)

are cloned in a plasmid vector, about 40,000 independent selected clones (called a *library*) are sufficient to ensure that the entire *D. melanogaster* genome is represented at least once in the collection; about 800,000 clones are required for a library of the human genome. Such libraries are available for a number of eukaryotic genomes (yeast, *Neurospora, Drosophila,* and human) and for some bacterial genomes.

These libraries are of potential value for the detailed study of the structure and organization of any eukaryotic gene that can be selected from the library. Their existence poses the question whether some eukaryotic gene might represent a hazard for other forms of life if it is selected and amplified during growth of the bacterial strain containing it and if that strain should inadvertently escape to the environment or infect other living organisms.

In order to avoid such hypothetical hazards, public health authorities in most countries where such research is conducted have established rules to restrict the types of DNAs that may be cloned in *E. coli* and to specify the containment conditions that must be used for different types of experiments. The general rule is that the more closely a species of organism is evolutionarily related to humans, the more stringent the containment conditions must be. The cloning of genes coding for obviously hazardous products such as toxins or potential human tumor viruses is forbidden at present.

Recombinant DNA technology has an industrial potential as well, especially for the production of human proteins or hormones of medicinal value. A gene coding for the mammalian hormone somatostatin, a peptide of 14 amino acid residues, has been chemically synthesized and cloned in a plasmid. The *E. coli* cells carrying this plasmid synthesize this peptide hormone in a form that can be recovered and chemically converted *in vitro* to the active hormone. The rat insulin gene has also been cloned. It is not unlikely that within a few years bacteria capable of synthesizing a number of important, currently expensive, rare human proteins will be available. It is not impossible to foresee the entire chemical engineering industry revolutionized by human-designed enzymes produced in bacterial cells by synthetic genes cloned in plasmids.

Problems

1. The plasmid pBR322 is a useful vehicle for the cloning of any DNA sequence in *E. coli*. The physical restriction map of pBR322 is shown in Figure 9.27. The plasmid carries genes conferring ampicillin and tetracycline resistance, as shown. Which restriction enzyme would you choose for inserting a foreign DNA fragment into this plasmid for cloning? How would you select an *E. coli* strain that carries the fragment that you wish?

2. In the experiment diagrammed in Figure 9.24, why must the plasmid pMB4 carry genes for both *Eco* RI restriction and modification?

3. DNA polymerase I will synthesize radioactively labeled DNA *in vitro* when provided with nicked double-stranded DNA from any source. If [α-^{32}P]dYTP (5′-P-P-^{32}P-Y_{OH}-3′) is the only radioactive substrate provided, subsequent degradation of the radioactive DNA by micrococcal DNase and spleen diesterase will yield 3′-deoxyribonucleotides, as shown

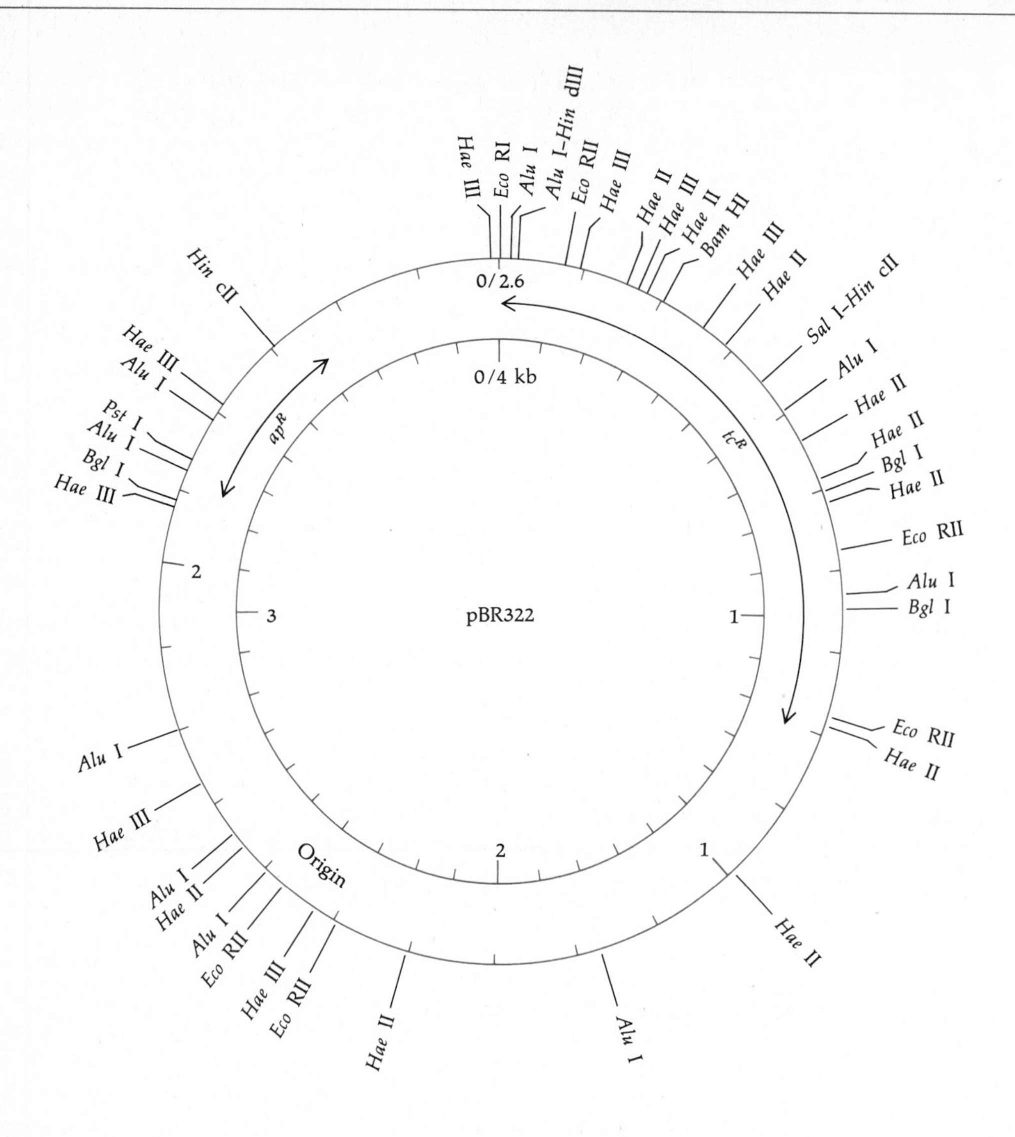

Figure 9.27
Physical map of the pBR322 cloning vehicle. (From F. Bolivar et al., in *DNA Insertion Elements, Plasmids and Episomes,* ed. by A. I. Bukhari, J. A. Shapiro, and S. L. Adhya, Cold Spring Harbor Laboratory, Cold Spring Harbor, N.Y., 1977.)

in Figure 9.28. Assume that you have available all four [α-^{32}P]dNTPs and all four nonradioactive dNTPs. Also assume that you can separate and identify all four 3′-deoxyribonucleotides and determine the amount of ^{32}P associated with each one. This will give the frequency of the nearest neighboring nucleotides to nucleotide Y. How could you prove that the strands of the Watson-Crick double helix are antiparallel?

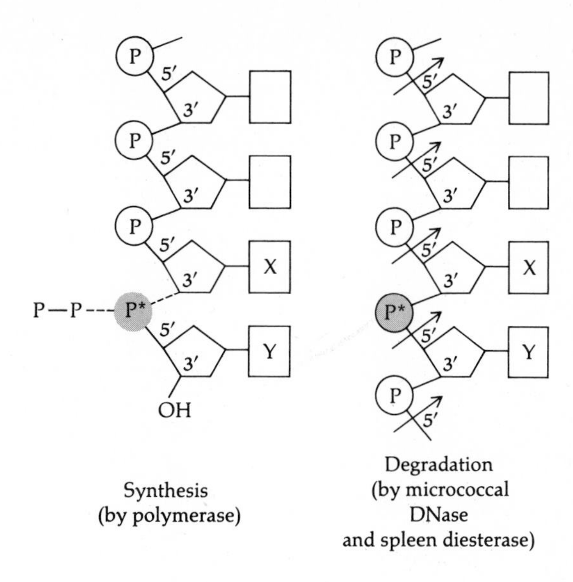

Figure 9.28
Synthesis of a ^{32}P-labeled DNA strand by Pol I and its subsequent degradation by micrococcal DNase and spleen diesterase. Note that the radioactive ^{32}P, originally linked to nucleotide Y, is transferred to nucleotide X.

4. Phage λ is grown on *E. coli* strain K12, and the concentration of progeny phage in the lysed culture is determined by plaque assay on *E. coli* K12 to be 6.0×10^{10} λ•K/ml. (λ•K identifies phage produced in K12.) When this same culture is assayed by plating on *E. coli* K12(P1), the titer is observed to be only 1×10^6/ml, i.e., the efficiency of plaque formation by λ•K on strain K12(P1) is only 0.0017% of that on strain K12. When phages from a plaque formed on K12(P1) [designated λ•K(P1)] are assayed on either of the two *E. coli* strains, they form plaques with 100% efficiency on each strain. When λ•K(P1) infect strain K12, however, the progeny, which are λ•K, again exhibit a very low efficiency of plaque formation on *E. coli* K12(P1). Explain.

5. The cross-bridged plasmid DNA molecules shown in Figure 9.19 were purified from *E. coli* as figure-eight molecules consisting of two unit-circumference circles and were cleaved with *Eco* RI just prior to fixation for electron microscopy. These cleaved molecules are unstable at room temperature and quickly convert to linear double-strand molecules of unit

length. On the other hand, the figure-eight molecules are quite stable. Explain.

6. Mutants of T4 with temperature-sensitive ligase exhibit an increased frequency of recombination in crosses between various markers performed at a permissive temperature. Ultraviolet irradiation of parental T4 phage also increases the recombination frequency observed between various markers. Explain.

7. A strain of *E. coli* that is *lac*Z$^-$ and carries an F′ Lac$^+$ element is mutagenized and plated on lactose EMB nutrient agar at 30°C. (On this agar, Lac$^+$ colonies are red and Lac$^-$colonies are white.) Colonies are replica-plated to lactose EMB nutrient agar and incubated at 42°C. Some white colonies are found on the 42°C plate that formed red colonies at 30°C on the original plate. Some of the white colonies are the result of temperature-sensitive *lac* mutants induced in the F′ Lac$^+$ element; others are temperature-sensitive mutants in which the replication of the F factor is specifically blocked. How would you distinguish these two classes of mutants from each other? Among those mutants that lose the F′ Lac$^+$ element during growth at 42°C, some carry the temperature-sensitive mutation blocking F-factor replication on the bacterial chromosome, and others carry it on the F′ element itself. How would you distinguish these mutations from one another? What types of genes might these mutations affect? (*Note*: Replication of F factors requires attachment of the F factor to certain sites on the cell membrane.)

8. Three temperature-sensitive *dna*$^-$ mutations, *508*, *46*, and *177*, isolated in *E. coli* are cotransducible with *ilv* (see table below). Calculate the linkage of these mutations to the *ilv* locus. What is your guess as to the relationship of these three mutations to one another?

Donor	Recipient	Selected marker (number tested)	Unselected marker (number)
ilv$^-$ *dna*$^+$	*ilv*$^+$ *dna*$^-$ *508*	*dna*$^+$ (100)	*ilv*$^-$ (4)
ilv$^+$ *dna*$^-$ *508*	*ilv*$^-$ *dna*$^+$	*ilv*$^+$ (100)	*dna*$^-$ (11)
ilv$^-$ *dna*$^+$	*ilv*$^+$ *dna*$^-$ *177*	*dna*$^+$ (100)	*ilv*$^-$ (13)
ilv$^+$ *dna*$^-$ *177*	*ilv*$^-$ *dna*$^+$	*ilv*$^+$ (100)	*dna*$^-$ (17)
ilv$^+$ *dna*$^+$	*ilv*$^-$ *dna*$^-$ *46*	*ilv*$^+$ (100)	*dna*$^+$ (7)
ilv$^+$ *dna*$^+$	*ilv*$^-$ *dna*$^-$ *46*	*dna*$^+$ (100)	*ilv*$^+$ (8)

After J. A. Wechsler and J. D. Gross, *Mol. Gen. Genet.* *113*:273 (1971).

9. Diagram all the possible steps leading to recombination of genetic markers between two circular parental λ genomes. Recall that λ progeny-phage DNA molecules are synthesized by a rolling circle mode of replication.

10. Phage T7 contains a linear, completely double-stranded DNA. Partial exonuclease I digestion and annealing show that the T7 DNA contains a terminal redundancy, but the gene order is not circularly permuted (see Chapter 6). Upon infection, T7 DNA replication begins but the DNA does not form circles. Late in the infection, very long concatemers of T7 DNA are observed. Propose a model for T7 DNA replication, showing why concatemers are a necessary step in the complete replication of the linear T7 chromosome.

II

Expression of the Genetic Materials

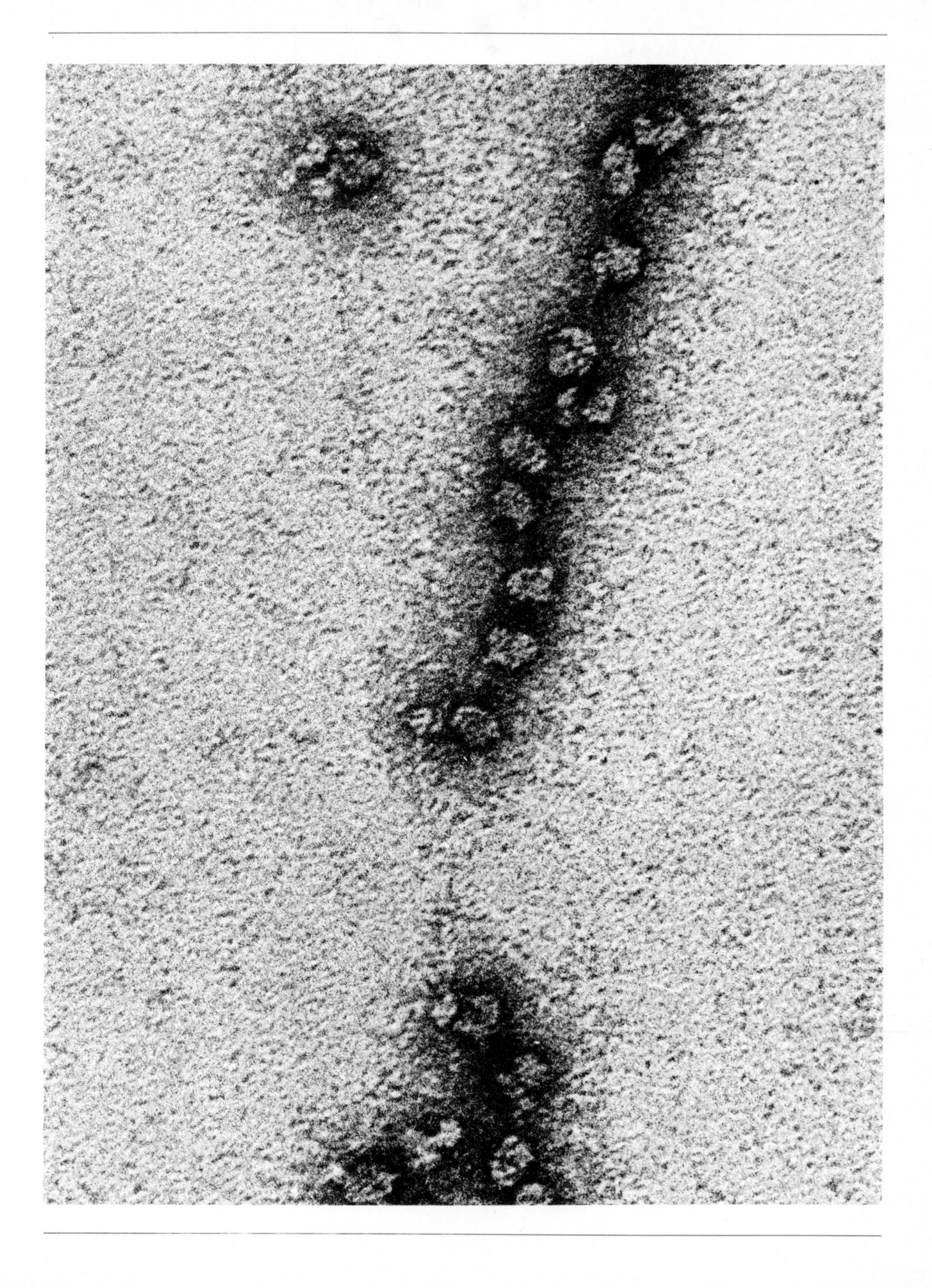

10

The Genetic Function

Preformation and Epigenesis

As we have seen in previous chapters, the genetic material, DNA, is capable of self-replication. Moreover, the DNA of organisms contains the information that directs their development and functioning. The second part of this book investigates the problems of gene expression. How, for example, does the tiny amount of material contained in a human zygote direct the transformation of that single initial cell into a human being? The changes involved are staggering. First, there is *cell proliferation,* or *growth*—the initial cell multiplies to about 1 trillion (10^{12}) cells by the time of birth, and to 20 trillion cells in the mature adult. Second, there is *differentiation*—a human being has arms, legs, eyes, liver, and kidneys made of cells with diverse shapes and functions.

The problem of cell proliferation seems simple, a matter of multiplication by cell division. But how does a single cell give rise to progeny cells of different types? How is it that cell division leads from the zygote to a nerve cell in one case, to a muscle cell in another, and to a bone cell in yet another?

In the eighteenth century, the *preformation* theory provided a radical solution to the problem of differentiation by simply denying it. In the late seventeenth century, an observer using a bad microscope and a lively imagination had claimed to see a *homunculus,* a miniature figure of a man, inside the human spermatozoon. The Dutch naturalist Jan Swammerdam

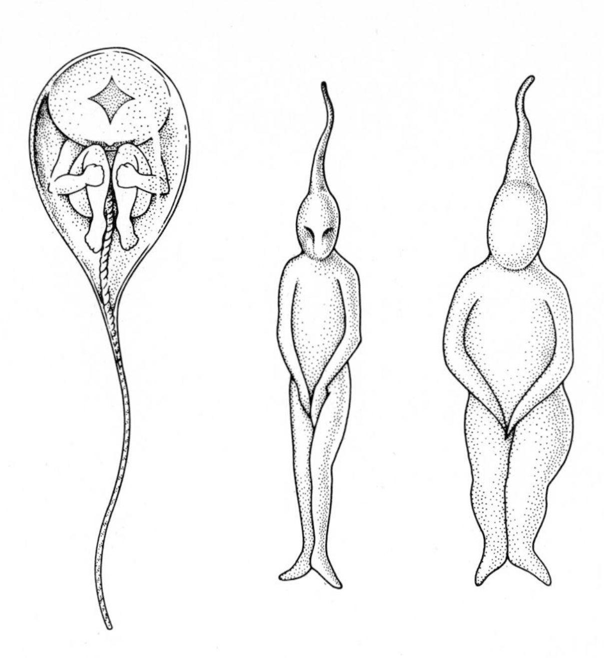

Figure 10.1
Homunculi claimed to have been seen in human sperm.

(the discoverer of the eggs of insects and of red blood corpuscles) and the Swiss naturalist Charles Bonnet developed this spurious discovery, which had been "confirmed" by other wishful thinkers, into a theory of preformation. The human body, according to some preformationists, was already *preformed* in the spermatozoon; development was simply a matter of growth of the tiny homunculus contained in the sperm into a full-sized human being (Figure 10.1). This was the theory of "spermatists," but then there were "ovists," who claimed that the egg, rather than the spermatozoon, contained the homunculus. Bonnet asserted that a woman contains all the "germs" of her descendants, immediate as well as remote. The implications are amusing. Eve (or Adam, depending on one's preferences) had in her sexual organs the germs of all humans to come, stored like Chinese boxes within boxes. And which one of your two grandmothers had your body preformed in her ovaries?

Kaspar Friedrich Wolff in the eighteenth century, and Karl Ernst von Baer in the nineteenth, proposed instead the theory of *epigenesis.* According to this theory, the sex cells are largely homogeneous bits of organic matter containing nothing whatsoever resembling the body that will develop from them. Development is differentiation as well as growth. The various tissues and organs gradually form from the zygote through a series of radical transformations. Von Baer provided a fairly accurate description of

Table 10.1
Possible combinations of four different bases.

Number of Bases Combined	Combinations	Total
1	A C G T	4
2	AA AC AG AT CA CC CG CT GA GC GG GT TA TC TG TT	16
3	AAA ACA AGA ATA CAA CCA CGA CTA GAA GCA GGA GTA TAA TCA TGA TTA AAC ACC AGC ATC CAC CCC CGC CTC GAC GCC GGC GTC TAC TCC TGC TTC AAG ACG AGG ATG CAG CCG CGG CTG GAG GCG GGG GTG TAG TCG TGG TTG AAT ACT AGT ATT CAT CCT CGT CTT GAT GCT GGT GTT TAT TCT TGT TTT	64
n	$4 \times 4 \times \cdots \times 4$ (n times)	4^n

the embryonic development of a chick. An orderly process of gradual change leads from the egg to the fetus and finally to the adult body.

We now know that the theory of epigenesis was more nearly correct than that of preformation. The organism is not preformed in the zygote. What is transmitted from parents to offspring is a set of "instructions"—the genetic information contained in the DNA—which, interacting with the environment, direct the development of the organism. Development is, in a way, similar to the process of building a skyscraper, following the instructions written by an architect. It is not simply a matter of starting with a tiny skyscraper and having its parts grow to their full size.

Genetic Information

The genetic information is encoded in the sequence of bases in nucleic acids in a manner analogous to that in which the sequence of letters in a book contains information. Genes can be thought of as molecular "sentences" in which the "words" consist of certain sequences of nucleotide "letters." The genetic endowment contained in the zygote of an organism can then be considered as a "book" made up of molecular sentences.

Since there are four kinds of bases in DNA, the number of different possible sequences in a gene with n nucleotides is 4^n (see Table 10.1). This is a staggeringly large number when n is in the hundreds, as is generally true for individual genes. However, as we shall see in the following chapters, the units of information for specifying the amino acid sequences of proteins are not the individual bases, but discrete groups of three consecutive bases, called *triplets*. Because there are four different bases, there are $4^3 = 64$ different triplets, but some of these are synonymous in the sense that they code for the same information, so there are only 21 different units of information among the 64 triplets. A polynucleotide chain with $n = 600$ nucleotides has $n/3 = 200$ nonoverlapping groups of three bases. The number of different messages that could be contained in chains of that length is thus $21^{n/3} = 21^{200} \approx 10^{264}$, a number incomparably greater than the number of protons and neutrons in the known universe (estimated to be about 10^{76}). Thus there is virtually no limit to the number of different messages that can be encoded in long nucleic acid chains.

Garrod's "Inborn Errors of Metabolism"

The question of how genes direct the development and functioning of organisms has now been answered. Genes control cell metabolism by specifying the structure of enzymes and other proteins, and enzymes are the catalysts that modulate all chemical activity in living organisms. The English physician Archibald Garrod was the first to suggest a specific connection between genes and enzymes. In 1902 he described the first clearcut case of a disease, *alkaptonuria*, that is inherited according to Mendel's laws.

Alkaptonurics suffer from arthritis and produce urine that turns black on exposure to air. Garrod suggested that alkaptonuria is due to a biochemical block in a metabolic process. He managed to identify the location of the block by observing that alkaptonurics excrete in the urine all the homogentisic acid fed to them. Normal individuals can metabolize homogentisic acid to its breakdown products, but alkaptonurics cannot. Therefore, Garrod suggested, alkaptonurics must lack the enzyme that metabolizes homogentisic acid. Thus he was the first to recognize the relation between a gene and an enzyme (Figure 10.2). Garrod proposed a similar explanation for three other human conditions, including albinism, which he suggested must be due to a metabolic block in the pathway leading from tyrosine to melanin.

Garrod called these biochemical hereditary diseases *inborn errors of metabolism*. His discoveries were made possible because he realized that the position of a metabolic block can be identified by the accumulation in the body of the substance immediately preceding the blocked step. This simple principle would become extremely important in the investigation of metabolic pathways.

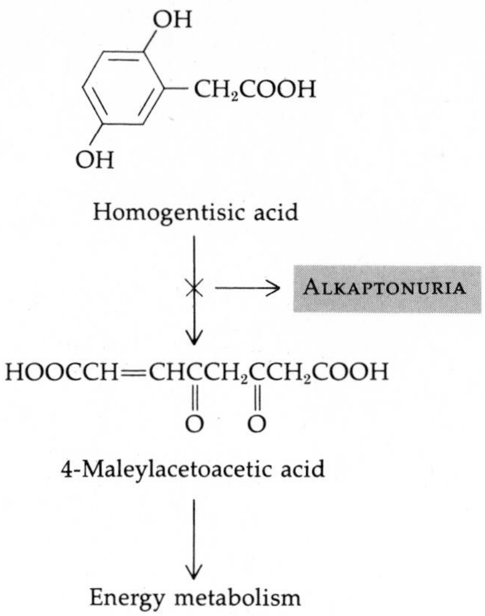

OH
CH_2COOH
OH

Homogentisic acid

→ ALKAPTONURIA

$HOOCCH{=}CHCCH_2CCH_2COOH$ (with C=O at the two C positions)

4-Maleylacetoacetic acid

Energy metabolism

Figure 10.2
The metabolic step that is blocked in alkaptonuria. The lack of the enzyme (homogentisic acid oxidase) that catalyzes the reaction of homogentisic acid to 4-maleylacetoacetic acid results in the accumulation of homogentisic acid, which is then excreted in the urine.

The One Gene–One Enzyme Hypothesis

The precise relationship between genes and enzymes was formulated by George W. Beadle and Edward L. Tatum in 1941 as the *one gene–one enzyme hypothesis,* a discovery for which they received the Nobel Prize in 1958. In order to analyze the biochemical effects of genes, Beadle and Tatum worked with the common bread mold, *Neurospora crassa* (see Figure 5.15). Normally this fungus can be grown on a defined minimal medium consisting of sugar, some inorganic acids and salts, a nitrogen source (ammonium compounds), and the vitamin biotin.

Beadle and Tatum induced mutations by means of X irradiation and isolated mutants that required specific supplementary compounds in order to grow (Figure 10.3). Such mutants are auxotrophs; stocks able to grow on minimal medium are prototrophs. (Nutritional mutations in bacteria are discussed in Chapter 7.) The method used by Beadle and Tatum to confirm that auxotrophs are truly hereditary mutations is shown in Figure 10.4.

The first three mutant strains isolated by Beadle and Tatum were labeled *pab, pdx,* and *thi*; in order to grow, they needed minimal medium supplemented, respectively, with *p*-aminobenzoic acid, pyridoxine, and

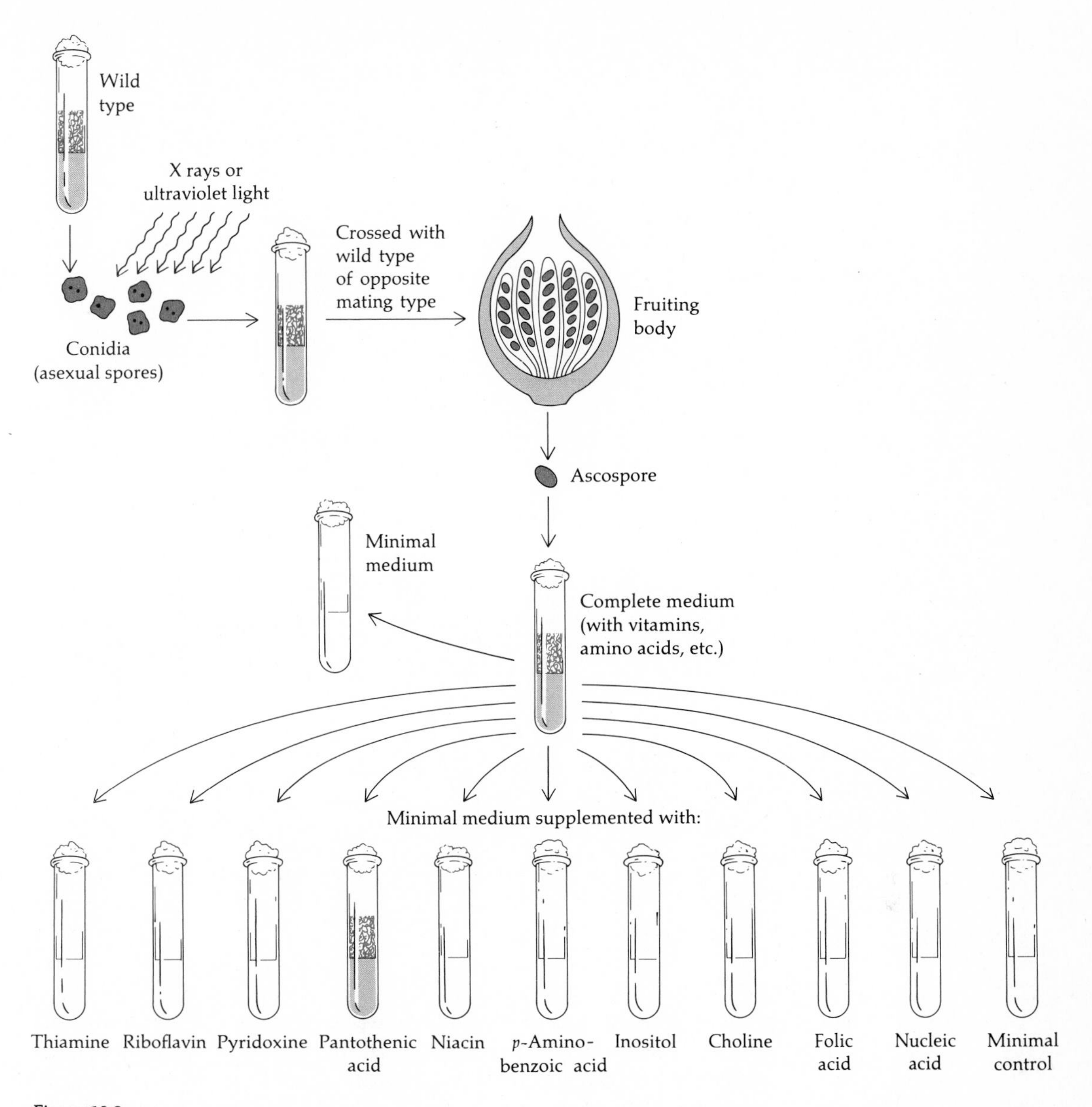

Figure 10.3
Method for detecting nutritional mutations in *Neurospora*. Conidia are exposed to mutagens (such as X rays or ultraviolet light) and crossed to wild type. Haploid spores are then grown on complete medium, and cultures from this growth are grown on minimal medium. Failure to grow on minimal medium indicates a growth defect. The nature of the defect is investigated by growing the deficient strain on minimal medium supplemented with various nutrients, such as amino acids and vitamins. In the example shown, the strain is able to grow on minimal medium supplemented with pantothenic acid.

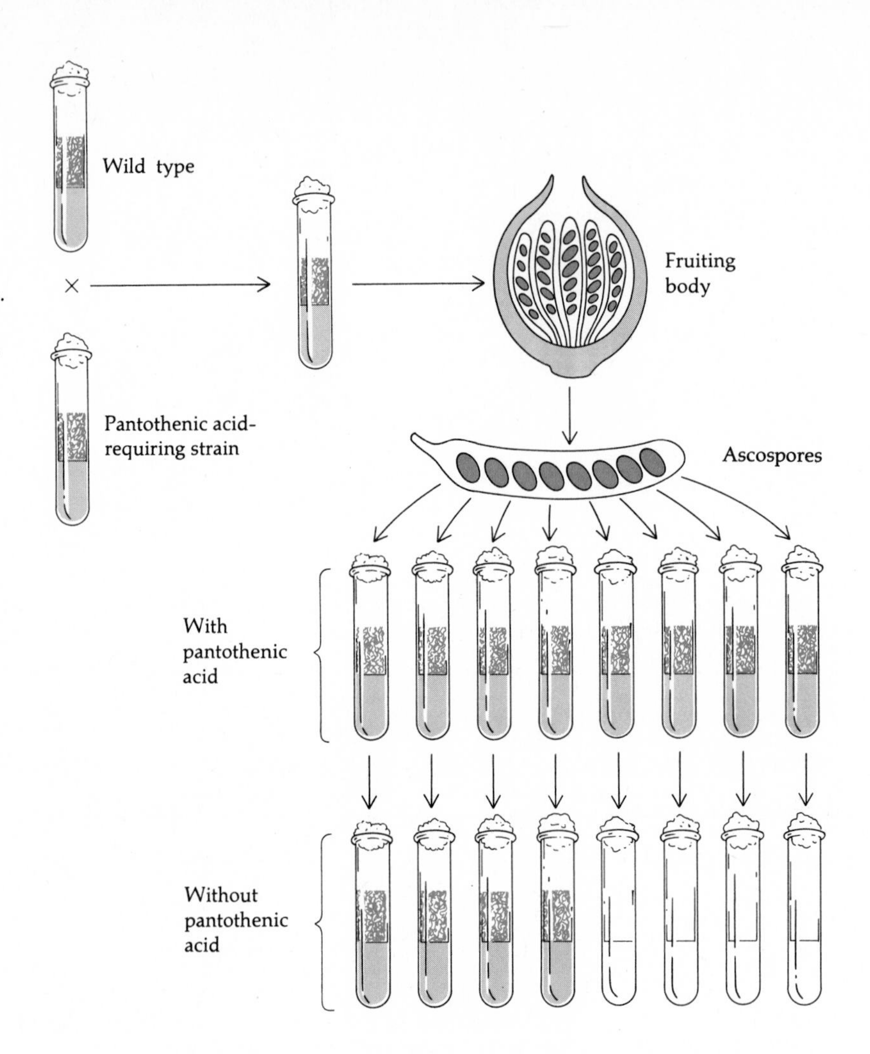

Figure 10.4
Method for confirming the genetic character of a growth defect. A growth-deficient strain isolated by the method shown in Figure 10.3 is crossed to wild type. The haploid spores are then placed on minimal medium. Observation that four spores are able to grow and four are unable to grow confirms that the defect is due to a genetic mutation.

thiamine. In each case a metabolic step leading to the synthesis of a specific compound had been blocked. There was a one-to-one correspondence between a genetic mutation and the lack of a specific enzyme required in a biochemical pathway. Beadle and Tatum then proposed the one gene–one enzyme hypothesis: Each gene specifies the synthesis of one enzyme (Figure 10.5). This hypothesis, in its modified form of "one gene–one polypeptide," was thoroughly confirmed by subsequent studies.

Biochemical Pathways

Different auxotrophic strains able to grow when the same compound is added to minimal medium are not necessarily mutants of the same gene. The three *Neurospora* mutants shown in Table 10.2 can all grow when

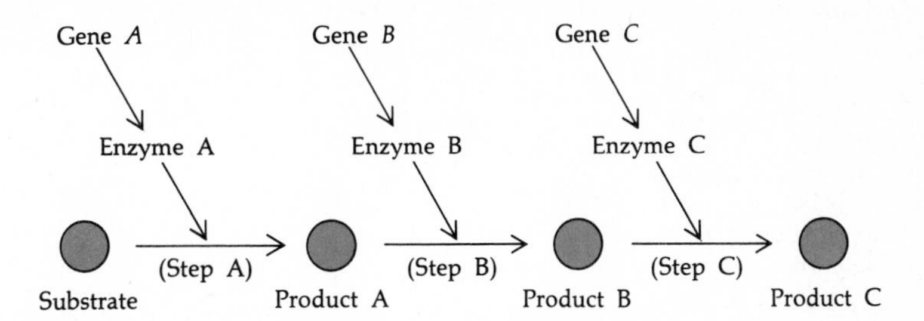

Figure 10.5
The one gene–one enzyme hypothesis. Each gene specifies the synthesis of one enzyme. A mutant gene produces a nonfunctional enzyme, blocking the metabolic pathway. A mutation in gene *B* would block the step from product A to product B. Product A would accumulate, while products B, C, etc., would not be formed. The accumulation of product A serves to identify the location of the blocked step in the pathway.

minimal medium is supplemented with arginine. However, the second will also grow when supplied with citrulline, and the first can grow with arginine, citrulline, or ornithine. The second one can, therefore, produce arginine by itself if supplied with citrulline, while the first can also produce arginine from ornithine. The production of the three compounds is part of the same biochemical pathway, having the sequence shown in Figure 10.6. (We will ignore the argininosuccinate intermediate in the present discussion.)

Mutation of gene *argE* blocks the pathway prior to the synthesis of ornithine; the effect of the mutation can be bypassed by supplying ornithine (from which the organism can produce citrulline and then arginine), citrulline, or arginine. A mutant of gene *argF* blocks the step from ornithine to citrulline and cannot be bypassed by providing ornithine, but the mutant will grow with either citrulline or arginine. A mutant of gene *argH* will grow when supplied with arginine but not with either of the other two compounds. The important inference is that different mutants

Table 10.2
Three arginine-requiring mutant strains of *Neurospora crassa*.

Mutant	Compound Required
1	Arginine, citrulline, or ornithine
2	Arginine or citrulline
3	Arginine

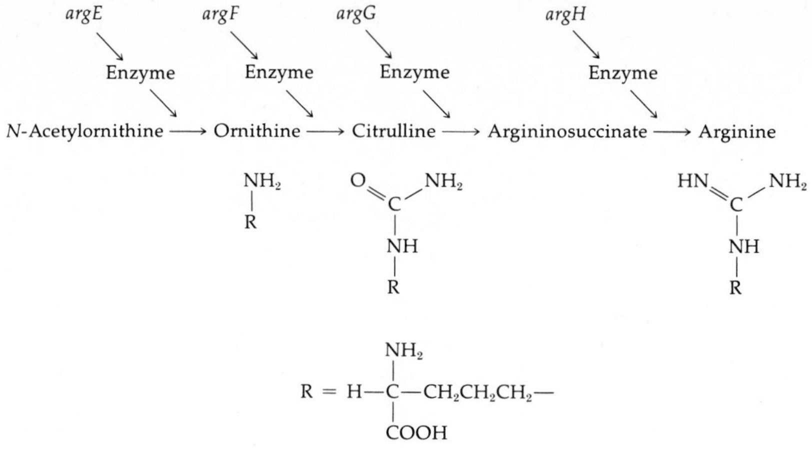

Figure 10.6
Biochemical pathway leading to the synthesis of arginine. A mutation in gene *argE* may block the synthesis of ornithine, but this block can be bypassed by supplying ornithine or any of the ensuing products (citrulline, argininosuccinate, or arginine) in the pathway. A mutation in *argF* can be bypassed by supplying citrulline or any of the ensuing products in the pathway, but not by supplying ornithine. A mutation in *argH* can be bypassed only by supplying arginine.

argE → Enzyme; *argF* → Enzyme; *argG* → Enzyme; *argH* → Enzyme

N-Acetylornithine ⟶ Ornithine ⟶ Citrulline ⟶ Argininosuccinate ⟶ Arginine

Ornithine: NH_2—R

Citrulline: $O=C(NH_2)$—NH—R

Arginine: $HN=C(NH_2)$—NH—R

$R = H—C(NH_2)(COOH)—CH_2CH_2CH_2—$

able to grow when supplied with the same substance may have blocks at different steps in the biochemical pathway leading to the synthesis of that substance.

The involvement of different genes in a given biochemical sequence can be further established by the complementation test discussed in previous chapters. A modified version of the complementation test is used with *Neurospora.* When hyphae from different strains grow in mutual contact, cell fusion can occur, forming a *heterokaryon*—a hybrid cell containing nuclei from two strains in a common cytoplasm (Figure 10.7). A heterokaryon will grow on minimal medium when the two nuclei contain mutants affecting different genes. When the mutants are allelic, however, the complementation test yields no "wild-type" heterokaryons able to grow on minimal medium. Complementation tests of many arginine-requiring *Neurospora* mutants have shown that seven different genes are involved in the biochemical pathway leading to the synthesis of arginine.

The growth requirements of several tryptophan mutants of the bacterium *Salmonella typhimurium* are shown in Table 10.3. None of the mutants can grow on minimal medium, and all can grow when supplied with tryptophan. Mutant *trp8*, however, can also grow when supplied with anthranilic acid (Ant), indole glycerol phosphate (IGP), or indole (I). Mutants *trp2* and *trp4* cannot grow when supplied with Ant, but they will grow when supplied with either IGP or I; moreover, these mutants accumulate Ant. Mutant *trp3* grows only when supplied with I (or tryptophan), and it accumulates IGP. The other mutants can grow only when supplemented with tryptophan, and they accumulate IGP and I. The sequence of biochemical synthesis is obviously Ant→IGP→I→tryptophan (Figure 10.8).

An early study of a biochemical pathway involved the synthesis of eye pigment in *Drosophila.* It was carried out by G. W. Beadle and Boris Ephrussi, using an experimental approach that was later extended to *Neurospora.* In *Drosophila* and other Diptera, many of the adult structures,

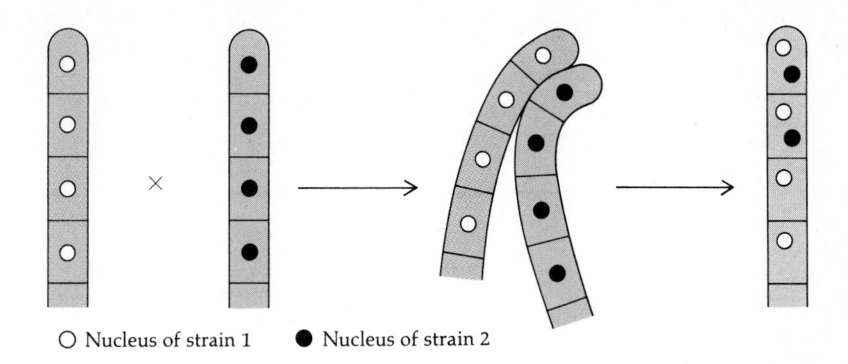

Figure 10.7
Heterokaryon formation in *Neurospora*. Cell fusion without nuclear fusion takes place when hyphae from two different strains are grown in contact with each other.

such as wings, legs, and eyes, develop from separate groups of cells (an *anlage*, or *imaginal disk*) that remain in an undifferentiated state during the larval stages (see Chapter 14). The imaginal disk of the compound eye can be transplanted from one larva into the abdomen of another, where it will complete its development. After metamorphosis of the host, the characteristics of the implanted eyes can be examined (Figure 10.9). The mutants *vermilion* (*v*) and *cinnabar* (*cn*) are recessive alleles of different genes that result in bright red eyes. When eye disks from vermilion or cinnabar larvae are transplanted into the abdomens of wild-type larvae, the implants develop into eyes with the normal wild-type eye color. Obviously some diffusible substance from the host enters the eye-tissue transplants, permitting normal pigment development. Vermilion disks transplanted into cinnabar larvae also develop into wild-type eyes. However, cinnabar disks transplanted into vermilion larvae develop into cinnabar eyes. From

Table 10.3
Growth requirements of tryptophan mutants in *Salmonella typhimurium.*

		Supplemented Substance*				
Tryptophan Mutant	Minimal Medium	Anthranilic acid	Indole glycerol phosphate	Indole	Tryptophan	Accumulated Substance
8	−	+	+	+	+	—
2, 4	−	−	+	+	+	Anthranilic acid
3	−	−	−	+	+	Indole glycerol phosphate
1, 6, 7, 9, 10, 11	−	−	−	−	+	Indole

*+ = growth; − = no growth.

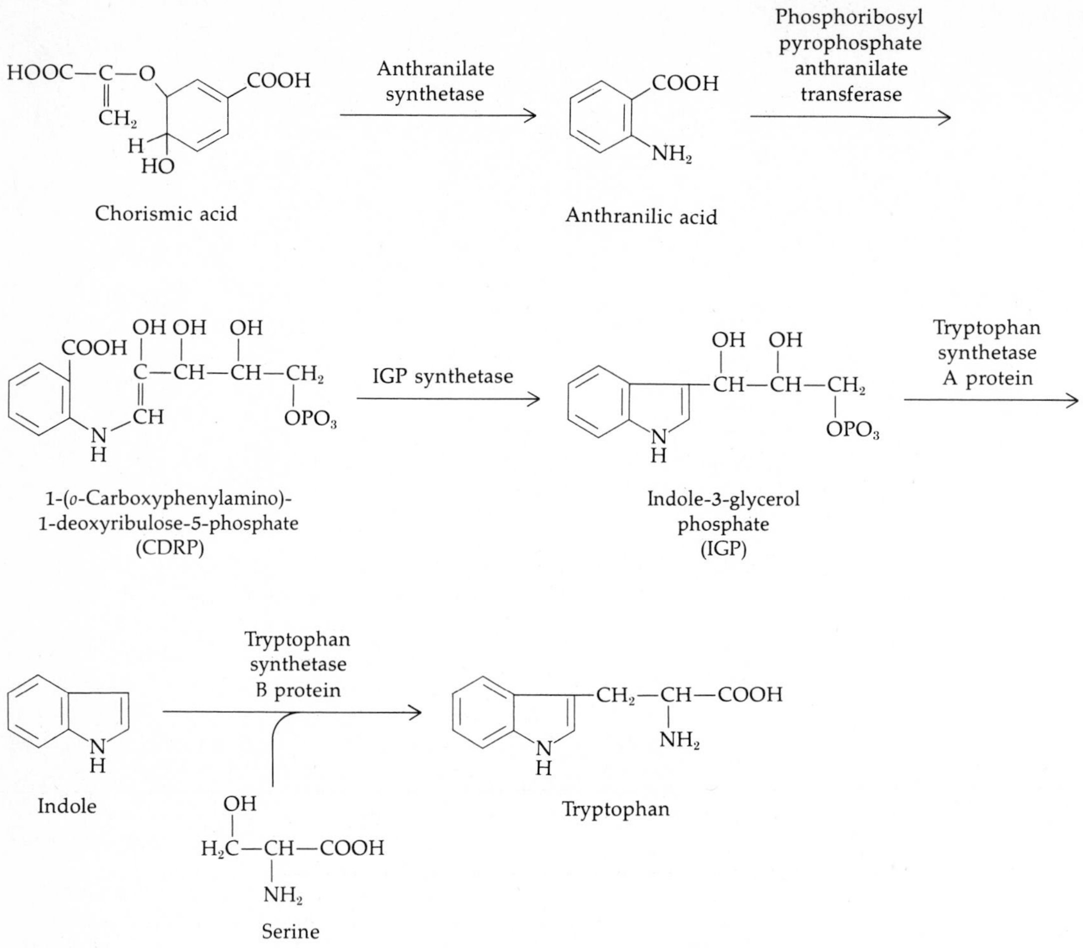

Chorismic acid → Anthranilate synthetase → Anthranilic acid → Phosphoribosyl pyrophosphate anthranilate transferase →

1-(*o*-Carboxyphenylamino)-1-deoxyribulose-5-phosphate (CDRP) → IGP synthetase → Indole-3-glycerol phosphate (IGP) → Tryptophan synthetase A protein →

Indole + Serine → Tryptophan synthetase B protein → Tryptophan

Figure 10.8
Part of the pathway of tryptophan biosynthesis in bacteria. The tryptophan synthetase enzyme consists of A and B subunits each of which is an enzyme in its own right.

these results it can be inferred that the *v* mutant causes an earlier block than the *cn* mutant in the biochemical pathway leading to the synthesis of the eye pigment, because cinnabar larvae can provide vermilion larvae with the substance missing in them, but not vice versa (Figure 10.10).

In summary, the studies of biochemical mutants in a variety of organisms confirm Garrod's hypothesis that genes control enzyme activities and that these enzymes are, in turn, associated with metabolism.

Genes and Proteins

The work of Beadle and Tatum suggested that each gene determines the presence of one enzyme. The nature of the relationship between genes and proteins was first suggested by the study of *sickle-cell anemia*, a severe human disease. Most people suffering from sickle-cell anemia die before

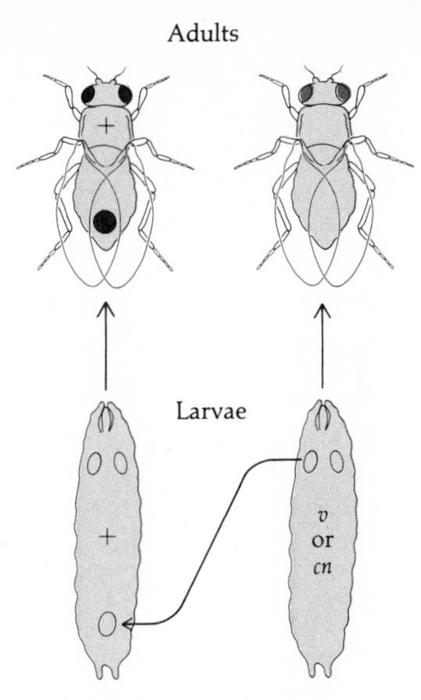

Nonautonomous development

(a)

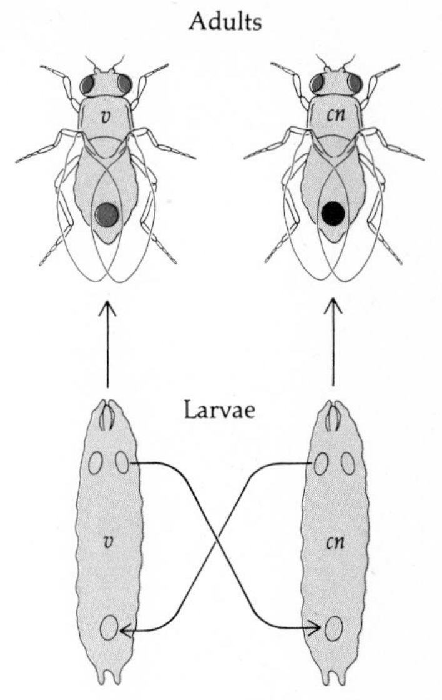

Nonautonomous development for *v*, autonomous development for *cn*

(b)

Figure 10.9
Transplantation experiments in *Drosophila melanogaster*. Imaginal disks of the eye are transplanted from one larva into the abdomen of another. (**a**) Implants from vermilion (*v*) or cinnabar (*cn*) larvae into wild-type larvae develop into eyes with wild-type color. In this case, the development of eye color is *nonautonomous:* the implant does not develop the characteristics of the donor, but is influenced by the host. (**b**) Implants from *v* larvae into *cn* larvae develop into wild-type eyes, again a case of nonautonomous development. However, implants from *cn* larvae into *v* larvae develop into *cn* eyes, i.e., *cn* eyes develop *autonomously* in *v* larvae. Thus, *cn* larvae provide *v* imaginal disks with some substance needed to produce wild-type pigment, but *v* larvae do not provide *cn* imaginal disks with this substance. Hence the step blocked by *v* occurs prior to the step blocked by *cn* in the synthesis of wild-type eye pigment.

adulthood. Under low oxygen tensions, the red blood cells of afflicted individuals exhibit a characteristic sickle shape that inspired the name of the disease. Parents of patients exhibit the sickle-cell trait—a moderate sickling of their red blood cells—although they do not suffer from severe anemia.

In 1949 James V. Neel and E. A. Beet independently suggested that sickling was caused by a mutant gene that was homozygous in individuals with sickle-cell anemia but heterozygous in people with the sickle-cell trait. In the same year, the Nobel laureate Linus Pauling and three co-workers observed that the hemoglobins of normal individuals and of sickle-cell anemics could be clearly differentiated by their different behavior in an electric field (Figure 10.11). The hemoglobin of people carrying the sickle-cell trait consisted of a mixture of normal and sickle-

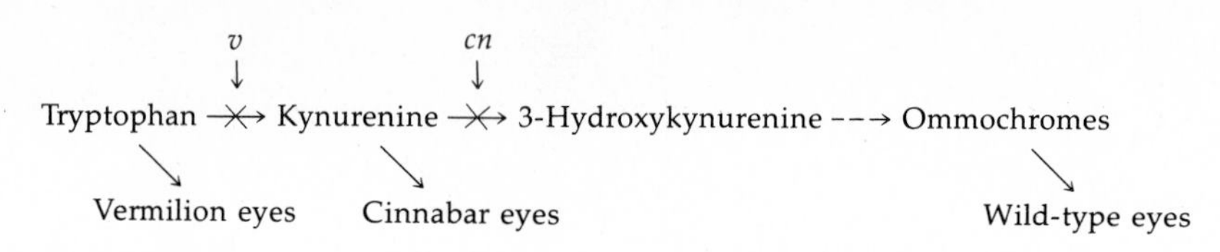

Figure 10.10
Locations of the *v* and *cn* blocks in the synthesis of eye pigments (ommochromes) in *Drosophila*.

cell hemoglobin in approximately equal amounts. The inference was that the sickle-cell mutant alters the chemical structure of the hemoglobin molecules. Homozygotes for the mutant have only the altered form of the molecule; heterozygotes have the normal as well as the mutant molecule.

Hemoglobin A, the most common form of hemoglobin in human adults, consists of four polypeptide chains, two identical alpha chains and two identical beta chains ($\alpha_2\beta_2$). Vernon M. Ingram showed in 1957 that normal and sickle-cell hemoglobins have identical alpha chains but different beta chains. At the sixth amino acid site of the beta chain, normal hemoglobin has glutamic acid, whereas sickle-cell hemoglobin has valine (Figure 10.12). Thus the difference between the normal and sickle-cell alleles was reflected in a single amino acid difference in a protein. It became apparent that genes must somehow specify the amino acid sequence of proteins.

The alpha and beta polypeptide chains are specified by separate genes. Many other proteins and enzymes (though not all) consist of two or more polypeptide chains encoded by different genes. Ingram therefore

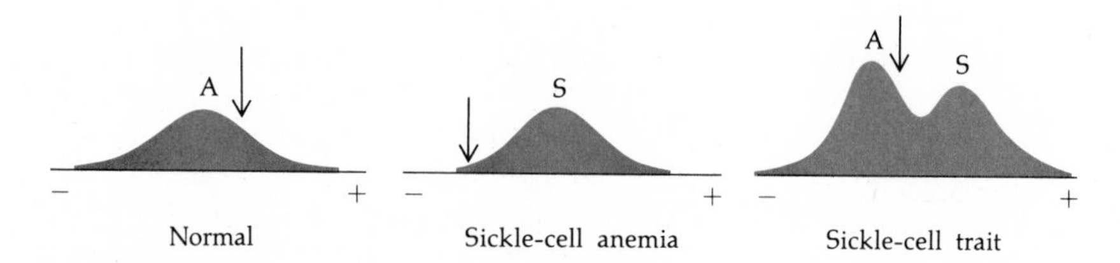

Figure 10.11
Electrophoretic migration of hemoglobin from normal individuals (left), sickle-cell anemia patients (center), and sickle-cell trait carriers (right). Hemoglobin is placed in an electric field at the position indicated by the arrows. Normal hemoglobin (A) migrates toward the cathode, while sickle-cell anemia hemoglobin (S) migrates toward the anode. Individuals with the sickle-cell trait show both kinds of hemoglobin. [After L. Pauling, H. A. Itano, S. J. Singer, and I. C. Wells, *Science* *110*:543 (1949).]

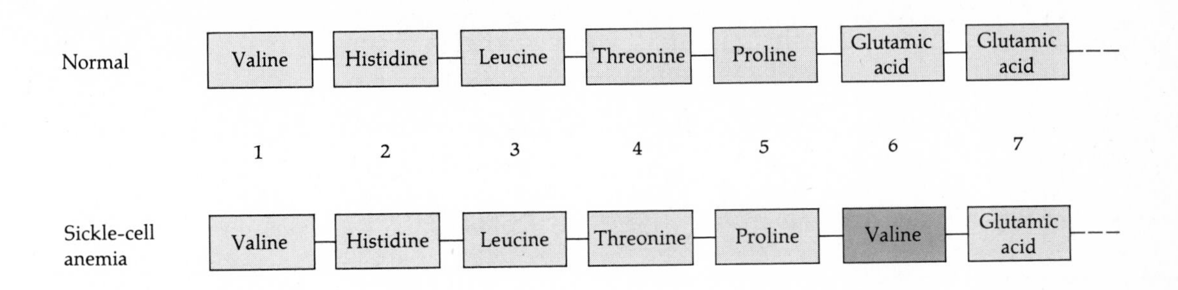

Figure 10.12
The first seven amino acids of the β chain of human hemoglobin; the β chain consists of 146 amino acids. A substitution of valine for glutamic acid at the sixth position is responsible for the severe disease known as sickle-cell anemia.

proposed that the one gene–one enzyme hypothesis be more appropriately renamed the one gene–one polypeptide hypothesis.

The Structure of Proteins

Proteins play a vital role in organisms. The diversity and complexity of life are largely a reflection of the diversity and complexity of proteins. Each protein has a unique function that depends on its particular structure and chemical properties. Some proteins are *enzymes,* which catalyze biochemical reactions in organisms. Every biochemical reaction in a living being is catalyzed by a specific enzyme. Without enzymes these reactions would not occur, or they would occur only at rates far too slow to permit life to exist. Other proteins function as *structural* building blocks of the organism, either by themselves (e.g., collagen) or in association with nucleic acids (*nucleoproteins*), polysaccharides (*glycoproteins*), or lipids (*lipoproteins*). Some proteins, such as the myoglobins and hemoglobins involved in oxygen storage and transport, are associated with metal-containing organic molecules (iron-containing *heme* groups in myoglobin and hemoglobin).

Proteins are large polymer molecules made up of amino acid monomers. Twenty different kinds of amino acids are found in proteins. All but one of them share the same basic structure:

$$\begin{array}{c} NH_2 \\ | \\ H{-}C{-}COOH \\ | \\ R \end{array} \quad \leftarrow \alpha\text{-carbon}$$

These amino acids are characterized by an *amino group* ($-NH_2$) and a *carboxyl group* ($-COOH$), both linked to the same carbon atom (the α-carbon). The remaining portion of the molecule consists of an H atom and the *R group* (R stands for radical). See Box 10.1.

Box 10.1 Chemical Structure of the Amino Acids

The 20 amino acids commonly found in proteins can be classified in a variety of ways. The classification described below and shown in Figure 10.13 highlights some biologically important properties of amino acids in proteins. *These properties are determined exclusively by the R groups.*

Eight amino acids are *nonpolar*, or *hydrophobic* (water-repelling). Note that one of these, proline, has a structure substantially different from those of the other 19 amino acids. Of the twelve *polar*, or *hydrophilic* (water-attracting), amino acids, seven are *neutral* under physiological conditions, three are *basic* (and hence tend to become positively charged), and the other two are *acidic* (and hence tend to become negatively charged). The two acidic amino acids (aspartic acid and glutamic acid) have a carboxyl group (—COOH) in their R groups; two of the basic amino acids (lysine and arginine) have an amino group ($-NH_2$) in their R groups. These carboxyl and amino groups are largely ionized at physiological pH: a —COOH group tends to lose its proton, becoming $-COO^-$, while an $-NH_2$ group tends to gain a proton, becoming $-NH_3^+$.

Two amino acids, methionine (nonpolar) and cysteine (polar), contain a sulfur atom; *disulfide bridges* (—S—S—) can form between two cysteines because the H atom of the —SH group is easily removed.

Three amino acids are *aromatic*, having unsaturated carbon rings in their R groups: phenylalanine is nonpolar, and tyrosine and tryptophan are polar. (Tryptophan is so weakly polar, however, that it is sometimes classified as a nonpolar amino acid.)

In proteins, amino acids are linked to each other by a *peptide bond*, a strong covalent bond resulting from the interaction of the amino group of one amino acid with the carboxyl group of the next amino acid (Figure 10.14). When two or a few amino acids are thus linked, the resulting polymer is called a *peptide*; a dipeptide consists of two amino acids; a tripeptide consists of three amino acids, etc. (In any peptide, the amino acids are often referred to as amino acid *residues*.) All peptides have a zigzag backbone of nitrogen and carbon atoms (Figure 10.15). The R groups project outward from the backbone in an alternating fashion. Peptides have a free amino group at one end (the *amino terminal*) and a carboxyl group at the other end (the *carboxyl terminal*). The amino acid at the amino terminal is called the *N-terminal amino acid*, and the one at the carboxyl terminal is called the *C-terminal amino acid*.

A *polypeptide* is a peptide consisting of many (up to a thousand or more) amino acids. The *primary structure* of a polypeptide is simply the sequence of its amino acids. Polypeptide chains spontaneously adopt a *secondary structure* dictated by the nature of the R groups along the backbone. A common secondary structure found in at least portions of many

NEUTRAL—NONPOLAR

Glycine (Gly)

Alanine (Ala)

Valine (Val)

Leucine (Leu)

Isoleucine (Ile)

Phenylalanine (Phe)

Proline (Pro)

Methionine (Met)

NEUTRAL—POLAR

Serine (Ser)

Threonine (Thr)

Tyrosine (Tyr)

Tryptophan (Trp)

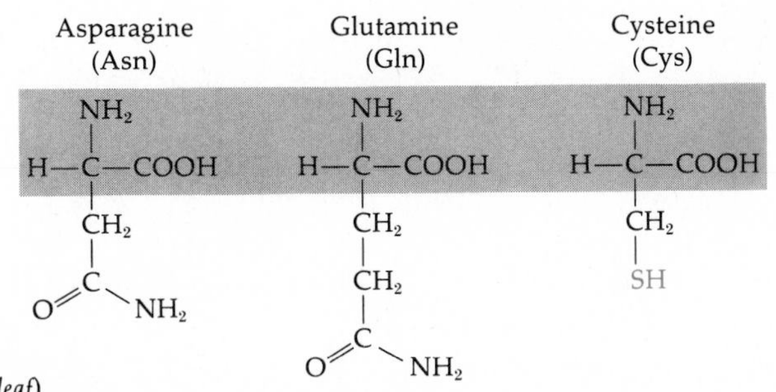

Figure 10.13 (*continues overleaf*)
The 20 amino acids commonly found in proteins. The structures below the gray bands are the R groups. The three color-shaded R groups are aromatic.

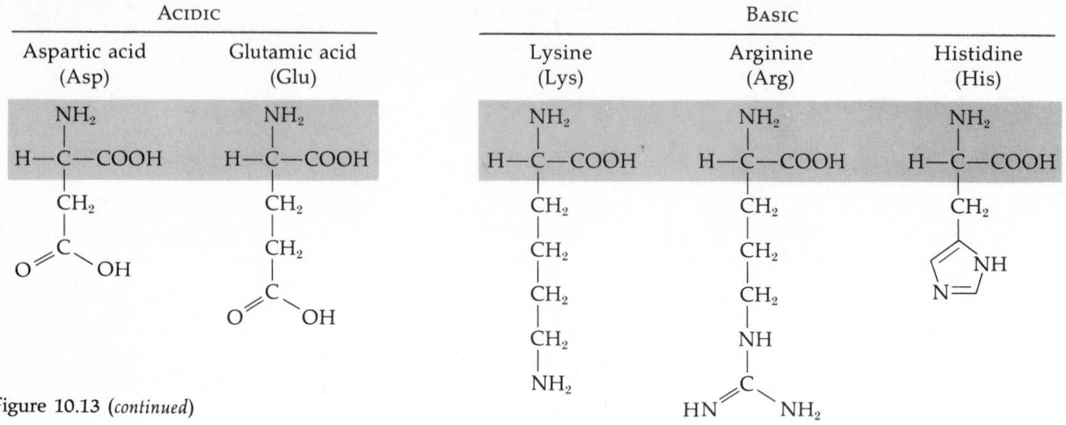

ACIDIC		BASIC		
Aspartic acid (Asp)	Glutamic acid (Glu)	Lysine (Lys)	Arginine (Arg)	Histidine (His)

Figure 10.13 (*continued*)

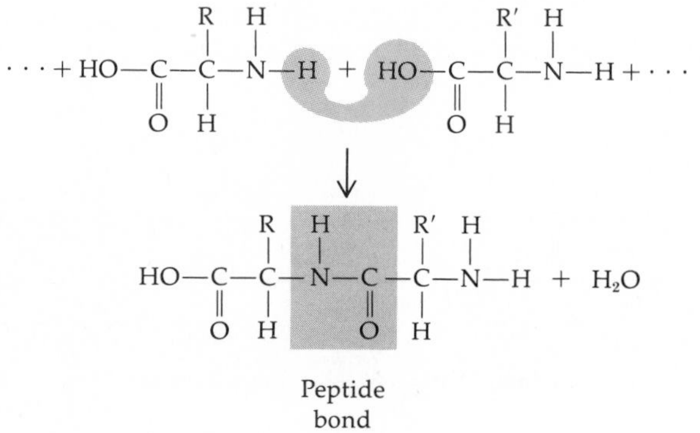

Peptide bond

Figure 10.14
The formation of a peptide bond between two amino acids.

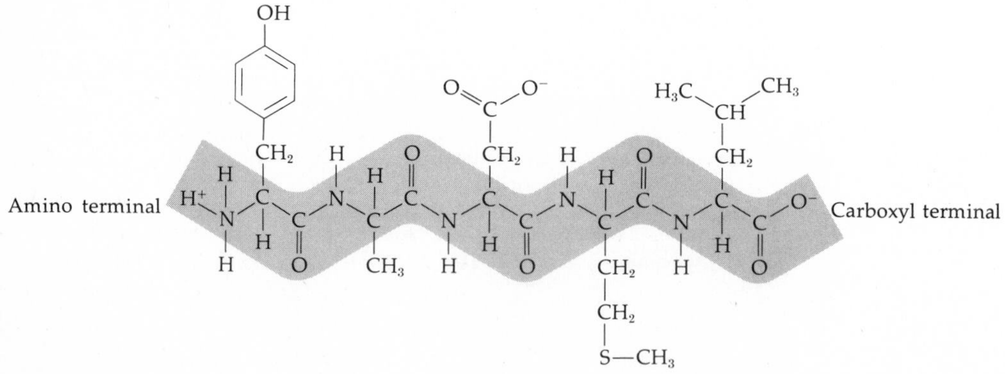

Amino terminal

Carboxyl terminal

Figure 10.15
A pentapeptide. The five amino acids are, from left to right, tyrosine, alanine, aspartic acid, methionine, and leucine. The backbone, which contains the peptide bonds, is shaded. The acidic and basic groups are shown in their ionized forms, as they exist at physiological pH.

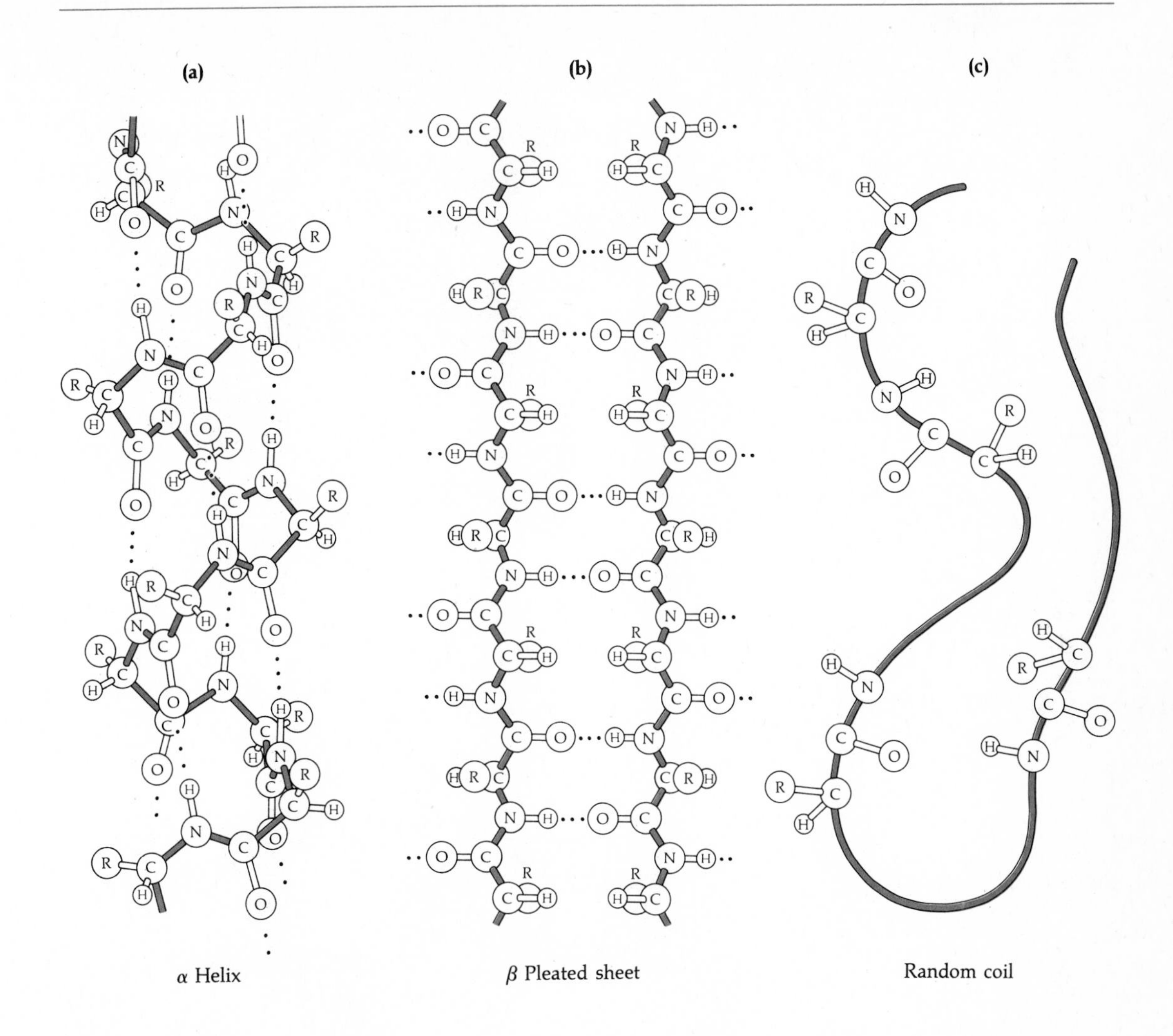

Figure 10.16
(**a**) The α helix configuration of a polypeptide. Note the hydrogen bonds (dotted lines), which stabilize this configuration. (**b**) The β pleated sheet configuration. Note the hydrogen bonding between adjacent polypeptide chains. (**c**) A random coil; the backbone of the polypeptide is drawn, but only a few amino acid residues are shown.

polypeptides is known as the *alpha helix.* The backbone forms a coil containing 3.6 amino acid residues per turn, in a right-handed helix with the R groups projecting outward; the structure is stabilized by intramolecular hydrogen bonds (Figure 10.16a). Another type of secondary structure assumed by some polypeptides is the *beta pleated sheet* configuration, which consists of rows of polypeptide chains that are hydrogen bonded to each other; the R groups project above and below the plane of the resulting pleated sheet structure (Figure 10.16b). Both of these structures were discovered by Pauling. When a polypeptide contains adjacent bulky residues such as isoleucine, or charged residues such as glutamic acid and aspartic acid, repulsion between these groups causes the polypeptide to assume a *random coil* configuration (Figure 10.16c).

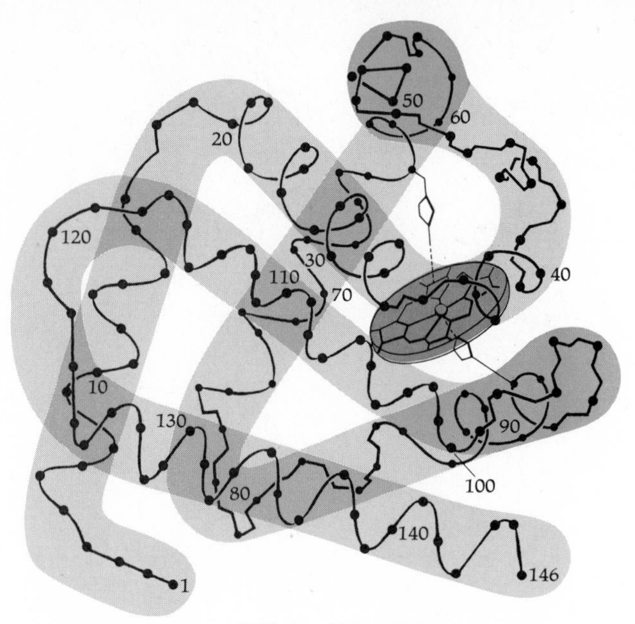

β Chain of hemoglobin

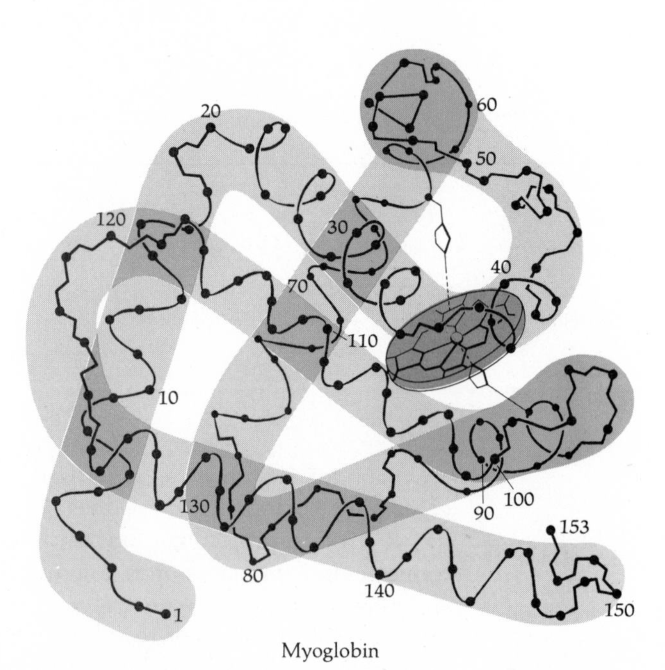

Myoglobin

Figure 10.17
Tertiary structures of the β polypeptide chain of human hemoglobin and of the distantly related oxygen-storage protein myoglobin. Note the similarities in their conformations. Alpha-helical portions of the molecules are shaded. The iron-containing heme groups are shown as disks. (After M. F. Perutz, *Scientific American*, November, 1964, p. 64.)

Figure 10.18
A model of human hemoglobin, showing how two α chains (color) and two β chains (gray) are associated to form the quaternary structure of the protein. (After M. F. Perutz, *Scientific American*, November, 1964, p. 64.)

Thus, the R groups distributed along the polypeptide backbone determine the secondary structure adopted by different portions of the polypeptide (α helix or β pleated sheet), or the lack of a well-defined structure (random coil). Proline, because of its unique structure, causes a kink in the polypeptide backbone. All of these structural factors lead to an overall three-dimensional configuration known as the *tertiary structure* of the polypeptide, which is caused by the folding of the chain (Figure 10.17).

A *protein* may consist of one or more polypeptides. Proteins consisting of two or more polypeptides are called *oligomeric* and have a *quaternary structure*, which is the overall configuration of the protein resulting from the association of the component polypeptides (Figure 10.18).

The ultimate configuration (secondary, tertiary, and quaternary structures) of a protein is determined by the primary structure of the polypeptide chain(s) and depends on the R groups of the amino acids. Nonpolar (hydrophobic) R groups, for example, tend to orient toward the interior of a protein as the polypeptide chain folds, whereas polar (hydrophilic) R groups tend to orient toward the outside surface of the protein. The three-dimensional configuration of a protein is stabilized by chemical interactions, including hydrophobic interactions, hydrogen

(a)

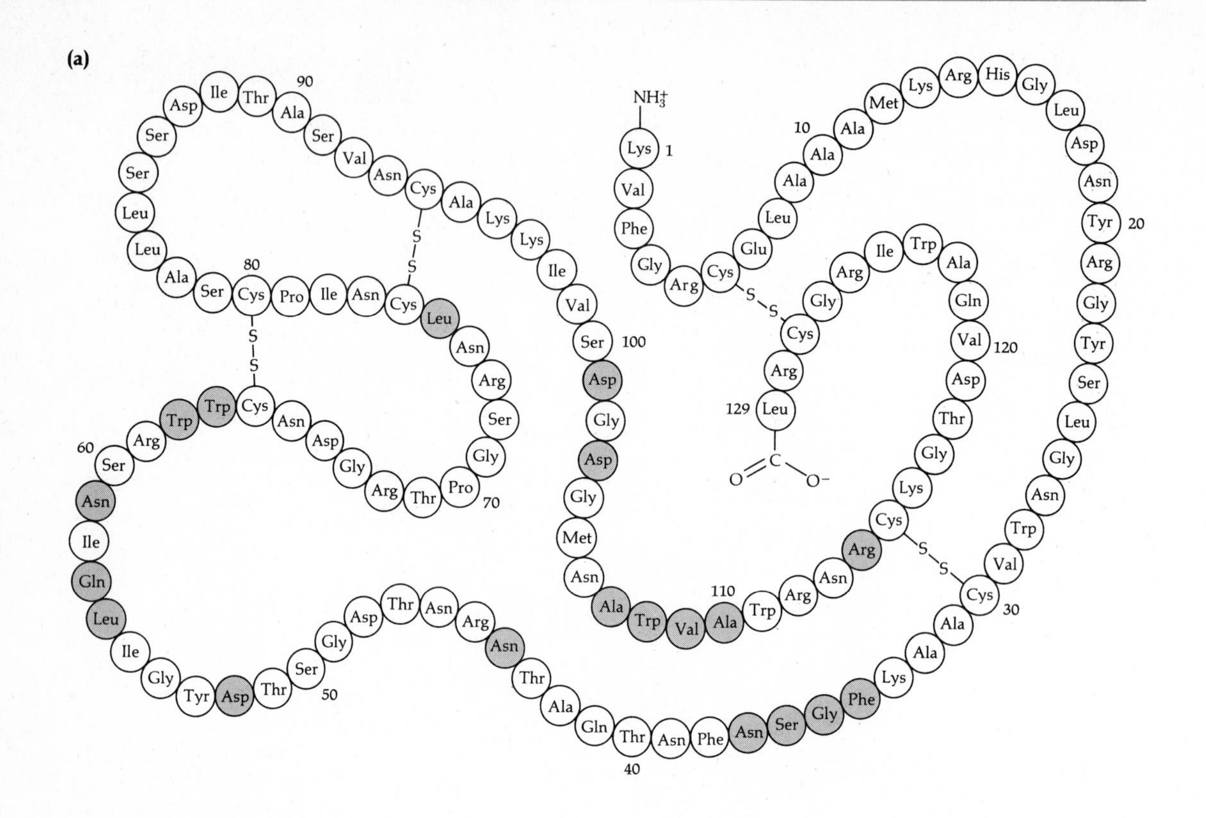

(b)

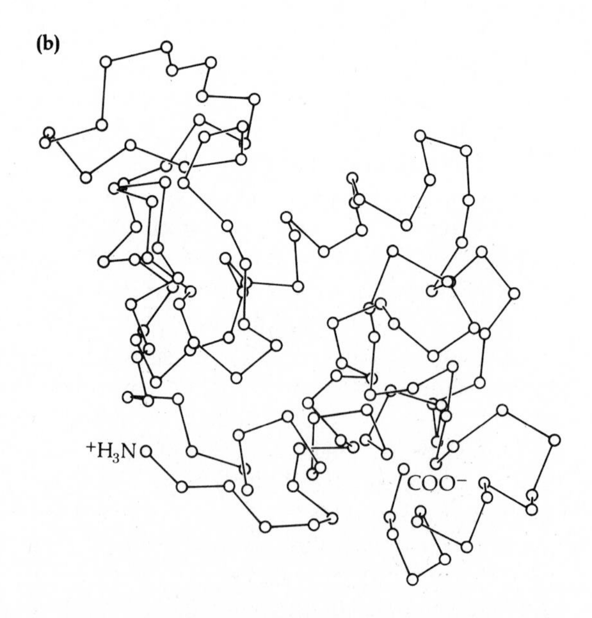

Figure 10.19
(**a**) Amino acid sequence of an enzyme, egg-white lysozyme from chicken. Residues that are part of the active site are colored. Note the disulfide bonds. (**b**) Diagram of the three-dimensional structure of lysozyme. Only the α-carbon atoms of the backbone are shown. (After L. Stryer, *Biochemistry*, W. H. Freeman, San Francisco, 1975.)

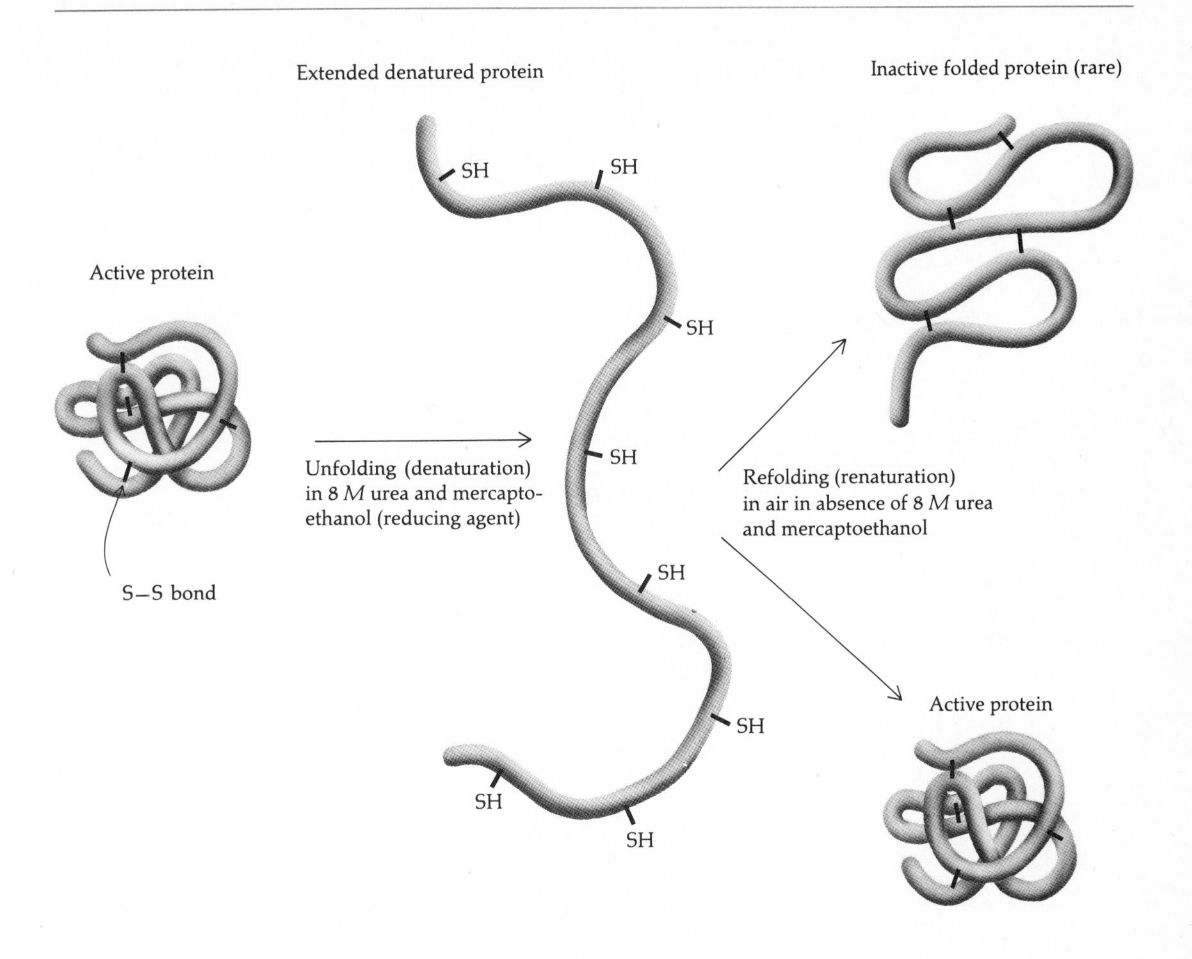

Figure 10.20
Proteins that are denatured usually reform their original structure when the proper chemical environment is restored. (After J. D. Watson, *Molecular Biology of the Gene*, 3rd ed., W. A. Benjamin, Menlo Park, Cal., 1976.)

bonds, and disulfide bridges (—S—S—); the latter often form when cysteines come into contact in the folding process (Figure 10.19).

A polypeptide adopts its characteristic three-dimensional configuration spontaneously. When the stabilizing hydrogen bonds and disulfide bridges of a polypeptide are ruptured, the polypeptide is *denatured*, but it may spontaneously reacquire its three-dimensional configuration if appropriate chemical conditions are restored (Figure 10.20).

The functional properties of a protein—for example, the active site of an enzyme—are determined by the three-dimensional configuration of the molecule, which depends in turn on the primary structure of the polypeptide(s). Some genetic mutations prevent the synthesis of all or part of a polypeptide. Others affect organisms by causing amino acid substitutions in the polypeptides coded by the mutant gene (Figure 10.12). If the substitution occurs in a functionally important part of the polypeptide, the three-dimensional configuration is altered. Depending upon the extent of the alteration, the enzyme activity or other functions may be either reduced or abolished entirely. Temperature-sensitive mutations may be caused by an amino acid substitution at a site that is important in

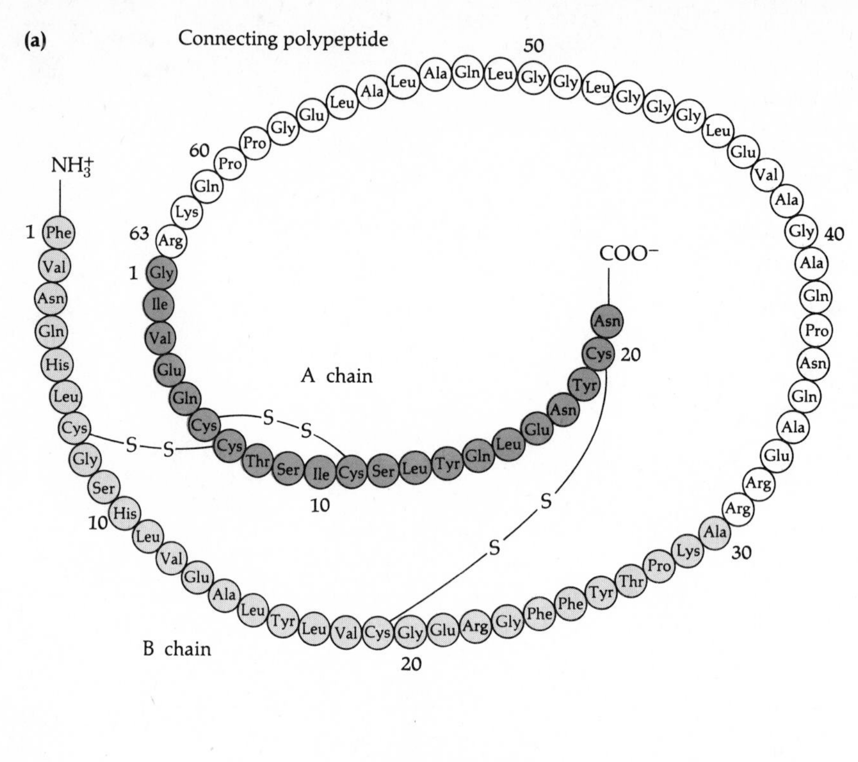

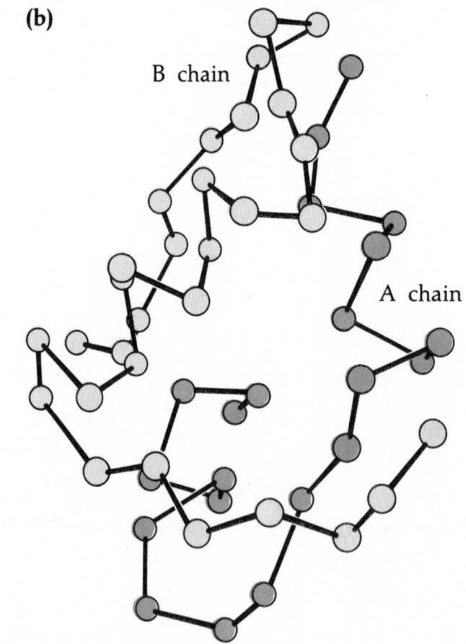

Figure 10.21

(**a**) Amino acid sequence of pig proinsulin. Insulin is composed of two polypeptides joined by disulfide bonds. The two polypeptides are found in proinsulin joined by the connecting polypeptide, which is removed by proteases. [After B. Chance et al., *Science 161*:165 (1968).]

(**b**) Three-dimensional structure of insulin. Only the α-carbon atoms of the backbones are shown: the A chain is in dark color, the B chain in light color. (After L. Stryer, *Biochemistry*, W. H. Freeman, San Francisco, 1975.)

Figure 10.22
Schematic representation of how two different mutant polypeptides may interact in the cytoplasm of a doubly heterozygous cell to exhibit intragenic complementation. (After U. Goodenough, *Genetics*, 2nd ed., Holt, Rinehart and Winston, New York, 1978.)

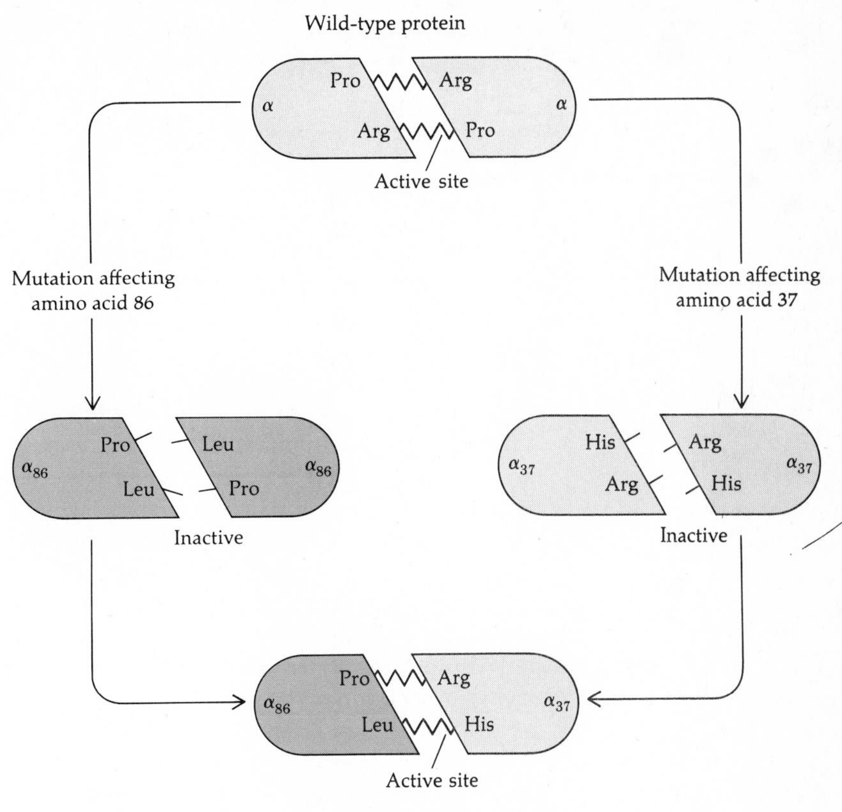

stabilizing the secondary or tertiary structure of a polypeptide. The normal conformation of the polypeptide is lost at the restrictive temperature.

Functional proteins are not always synthesized in their active form. Many important digestive enzymes are synthesized in an inactive form called a *zymogen;* the zymogens are activated at certain sites by *proteolysis,* i.e., by cutting out part of their amino acid sequence. The hormone insulin is another example of a protein synthesized in an inactive form (proinsulin), that is activated by proteolysis (Figure 10.21).

Intragenic Complementation

It is obvious from the discussions of the complementation test in Part I of this book (see page 180) that organisms homozygous for either one of two different mutant genes coding for proteins will lack one or the other normal protein. On the other hand, organisms heterozygous for two different mutant genes will possess both functional proteins, each coded by the wild-type allele of each gene, although only half the normal amount of activity of each may be present.

In some instances, however, complementation between mutant alleles of the same gene coding for a single polypeptide may be observed; this is *intragenic complementation* (see Chapter 8 for a discussion of heteroalleles). Intragenic complementation is exhibited only by mutants of genes coding for polypeptides that form dimeric or multimeric proteins. Such proteins contain two or more identical polypeptides associated through hydrogen bonds, disulfide bridges, or hydrophobic interactions. The function of the protein is dependent on the three-dimensional structure of the dimer or multimer. Amino acid substitutions in mutant (heteroallelic) polypeptides that disrupt the proper subunit polypeptide interactions cause loss of activity. Some such mutants, when present in double heterozygotes ($m_1 +/+ m_2$), exhibit restored enzyme activity, although the activity may be only partially restored. Restoration of activity results from a more nearly normal protein structure formed by interaction of subunits coded by the different heteroalleles (Figure 10.22).

Some heteroalleles of a gene may show intragenic complementation, and others not. This is probably because some mutants affect portions of polypeptides that interact with one another as subunits. Other mutants may alter a portion of the polypeptide essential for activity but may not be involved in subunit interaction.

Intragenic complementation can generally be distinguished from intergenic complementation by careful study. Where intragenic complementation is observed, often only a very low level of enzyme activity is present, compared to the normal level. The proteins formed as a result of intragenic complementation are often not normal in some other ways, such as temperature stability or pH dependence.

Problems

1. Calculate the numbers of different proteins (different in amino acid sequence) that can be encoded by three genes consisting of 30, 300, and 3000 nucleotide pairs coding for amino acids.

2. A peptide is known to consist of the six amino acids alanine, glycine, histidine, lysine, methionine, and tryptophan, but their order is unknown. Upon chemical degradation of the peptide, a biochemist is able to identify three kinds of tripeptides: Met-His-Trp, Lys-Ala-Gly, and Gly-Met-His. What is the amino acid sequence of the peptide?

3. Transplants made by Beadle and Ephrussi between *Drosophila melanogaster* mutant (and wild-type) strains gave the following results with respect to autonomous (aut) and nonautonomous (non) development:

Implant	Host +	bw	ca	cn	v
+	aut	aut	non	aut	aut
bw	aut	aut			
ca	aut		aut		
cn	non	non	aut	aut	aut
v	non	non	aut	non	aut

Interpret these results.

4. The mutant strain *susL* of bacteriophage λ fails to produce infective progeny phages when grown in an Su^- host (see Chapter 6). The infected host cells lyse, however, and examination of the lysate with the electron microscope reveals the presence of normal phage heads lacking tails. The λ mutant strain *susA* also fails to produce infective progeny when grown in an Su^- host, and electron microscopy shows the presence of phage tails, but no heads, in the lysate. If cell-free lysates produced by λ*susL* and λ*susA* infections are mixed and incubated for several hours at 37°C, infective phages are formed by the union of heads from the λ*susL* lysate and tails from the λ*susA* lysate. These phages can be assayed by their plaque-forming ability on an Su^+ host. This is an *in vitro* example of __________ __________. What will be the result if cells of an Su^- host are simultaneously infected with both λ*susA* and λ*susL*?

5. The table below shows the number of infective phages produced when lysates of several different λ*sus* mutants (*W, C, Z, U, F, V*), produced by infection of Su^- host cells, are mixed with lysates of λ*susA* or λ*susL*, as described in problem 4. Determine the nature of the defect present in each *sus* mutant.

Mutant lysate	Infective phages produced ($\times 10^{-7}$) *susL* lysate	*susA* lysate
susW	250	<0.001
susC	410	<0.001
susZ	<0.01	350
susU	<0.001	560
susF	200	<0.01
susV	<0.001	310

6. Lysates of $\lambda susB$ and $\lambda susE$ mutants produced by infection of Su^- host cells fail to form infective phages when mixed with a lysate of $\lambda susA$ (see problem 4). How could you determine if these three mutant strains are defective for the same genetic function?

7. When blood from patients suffering from X-linked hemophilia is drawn into a glass tube, it is observed to clot much more slowly than blood drawn from normal individuals. A boy named Stephen Christmas, who suffered from an X-linked hemophilia and whose blood exhibited a prolonged clotting time in a glass tube, was believed to be afflicted with classical hemophilia until his blood was mixed with blood from another hemophiliac and the mixture was observed to clot rapidly, with a clotting time indistinguishable from that of normal blood. Explain these observations.

8. The nucleotide cyclic AMP is converted to 5′-AMP by the enzyme cyclic AMP phosphodiesterase (PDE). The PDE activities present in *Drosophila* flies heterozygous for some of the deficiency chromosomes drawn in Figure 5.24 have been measured and compared with those present in nondeficiency flies. The ratio of activities in deficiency flies to activities in nondeficiency flies is recorded below.

Deficiency	Ratio of enzyme activities
$Df(1)N^{St}$	0.685 ± 0.066
$Df(1)N^{71h24\text{-}5}$	0.638 ± 0.077
$Df(1)N^{64j15}$	0.815 ± 0.063
$Df(1)N^{5419}$	0.996 ± 0.089
$Df(1)w^{-67k30}$	1.12 ± 0.054
$Df(1)dm^{75e19}$	0.662 ± 0.028

Similarly, PDE activities have been measured in males and females carrying the duplication chromosomes drawn in Figure 5.24 and compared to the PDE activities in males and females without duplications. The ratio of activities in duplication flies to activities in nonduplication flies is given below.

	Ratio of enzyme activities	
Duplication	Females	Males
$Dp(1;3)G^{spot}$	1.44 ± 0.040	1.77 ± 0.15
$Dp(1;2)w^{+51b7}$	1.58 ± 0.26	
w^+Y		0.89 ± 0.20

What is the simplest hypothesis that can account for these results?

9. Three independent methionine auxotrophs of *Neurospora* have been studied to determine which related compounds might substitute for their methionine requirement. In the table below, + indicates growth on minimal medium supplemented with the indicated compound, and − indicates no growth.

Mutant	Methionine	Homoserine	Homocysteine	Cystathionine
1	+	−	+	−
2	+	−	−	−
3	+	−	+	+

Give the order of these three compounds in the biosynthetic pathway of methionine.

10. Another *Neurospora* mutation that segregates as a single Mendelian gene confers auxotrophy for *both* methionine and threonine. Both of these growth requirements are satisfied by adding homoserine to the medium. Explain.

11

Information Transfer in Cells

The discovery that every protein in an organism has its primary structure specified by a particular gene has profound implications. Proteins contain 20 different amino acids, whereas only four different nucleotides are contained in DNA. Thus, a one-to-one correspondence between nucleotides and amino acids cannot exist, for then only four amino acids would be found in proteins. Instead, there must be a code in which some number of nucleotides stands for a single amino acid. If each code word, called a *codon*, comprised two nucleotides, then $(4^2) = 16$ codons would be possible—still not enough to specify 20 different amino acids, however. The minimum codon size is therefore three nucleotides, which would provide $(4^3) = 64$ codons—more than are necessary.

The concept of a genetic code is an important one because it suggests an information transfer system. The hereditary information for the specification of a cell's proteins is encoded in the nucleotide sequence of the cell's DNA in a 4-letter alphabet (an appropriate term because an alphabet is a set of symbols used to convey information). That information is also contained in the amino acid sequence of the proteins in a 20-letter alphabet. The genetic code is the key that relates, in Crick's words, ". . . the two great polymer languages, the nucleic acid language and the protein language."

The concept of hereditary information is the basis of what Crick has called the *Central Dogma* of molecular biology, which details the flow of information in biological systems (Figure 11.1). The transfer of information falls into three categories: (1) general transfers (those that can occur in

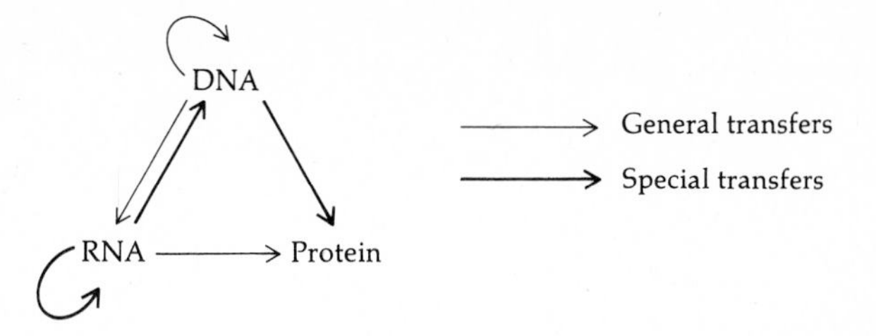

Figure 11.1
Transfers of genetic information in biological systems (the Central Dogma of molecular biology). Forbidden transfers are not diagrammed.

all cells), (2) special transfers (those that occur in cells only under special circumstances), and (3) forbidden transfers (those that have never been detected experimentally or predicted theoretically).

General Information Transfers

DNA synthesis

Three general transfers occurring in all cells are shown in Figure 11.1. The first, transfer of information from DNA to DNA, occurs during semiconservative DNA replication and is mediated by the complementarity between bases in the two strands of the DNA helix. Complex enzyme systems effect the transfer of genetic information from parental to progeny DNA molecules and assure that base-pair complementarity is maintained, as was discussed in detail in Chapter 9.

RNA synthesis

The second general transfer of genetic information is from DNA to RNA. This transfer is also mediated by a complementarity between bases, identical to that for DNA, with the exception that deoxyriboadenosine in the DNA template strand is the complement of ribouridine in an RNA strand. Information transfer from a double-stranded DNA molecule to a single-stranded RNA molecule is called *transcription* and is carried out by enzymes called RNA transcriptases or RNA polymerases. (*RNA polymerase* is the common term for these enzymes. However, *RNA transcriptase* is used

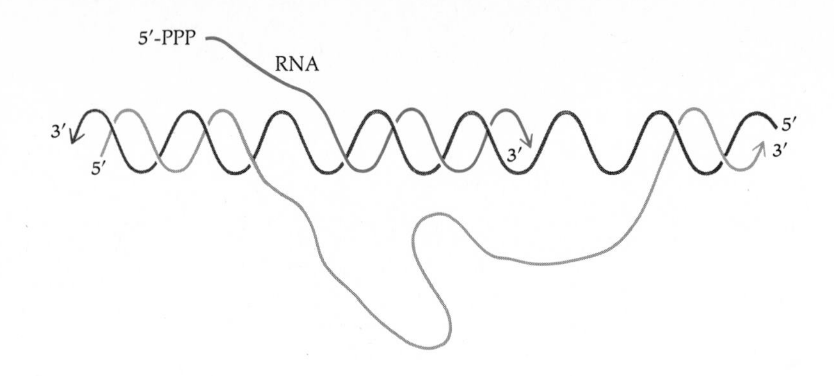

Figure 11.2
Schematic diagram of RNA transcription from a template DNA strand. RNA transcription requires a localized separation of the DNA strands to expose the template strand. Initiation of RNA synthesis occurs without the need for a primer; the nascent RNA possesses a 5′-triphosphate terminus.

here to emphasize the informational role of these enzymes and to distinguish them from the RNA polymerase involved in DNA replication.) The RNA molecules synthesized are single-stranded and complementary to only one of the two strands of the double helix. RNA transcriptases employ the ribonucleotide triphosphates as substrates. RNA synthesis proceeds from the 5′ end to a growing 3′ end of a nascent RNA molecule. The 5′ end of the nascent RNA carries a 5′-triphosphate group, evidence that RNA transcriptases and polymerases initiate a new strand without the requirement for a primer, which is needed by DNA polymerases (Figure 11.2). The DNA helix is unwound only near the point of RNA synthesis to provide a template strand, which is read in the direction 3′→5′. The selection of transcription initiation sites, called *promoters,* and of template DNA strands is determined by the RNA transcriptase and the nucleotide-pair sequence of the promoter site. The nucleotide sequences of several template strands, showing the RNA transcriptase binding sites and initial base transcribed from each, are shown in Figure 11.3. The variability in nucleotide sequences among these sites reflects, in part, the efficiency with which each is recognized by the transcriptases. The efficiency of promoter recognition by RNA transcriptase is an important control on RNA synthesis in cells.

The *E. coli* RNA transcriptase has been studied in great detail. It consists of a core enzyme containing four polypeptide subunits, two of one type (α) and one each of two other types (β and β'). Another polypeptide, called sigma factor, associates with this core enzyme, forming the complete enzyme, or *holoenzyme* (Figure 11.4). The presence of sigma factor permits the holoenzyme to recognize the correct template strand at a promoter site. Once RNA synthesis has begun, sigma factor dissociates from the core enzyme, which continues transcription until it reaches and transcribes a transcription termination site on the DNA. In *E. coli* a protein termination factor, called rho (ρ), then acts to prevent further transcription (Figure 11.5).

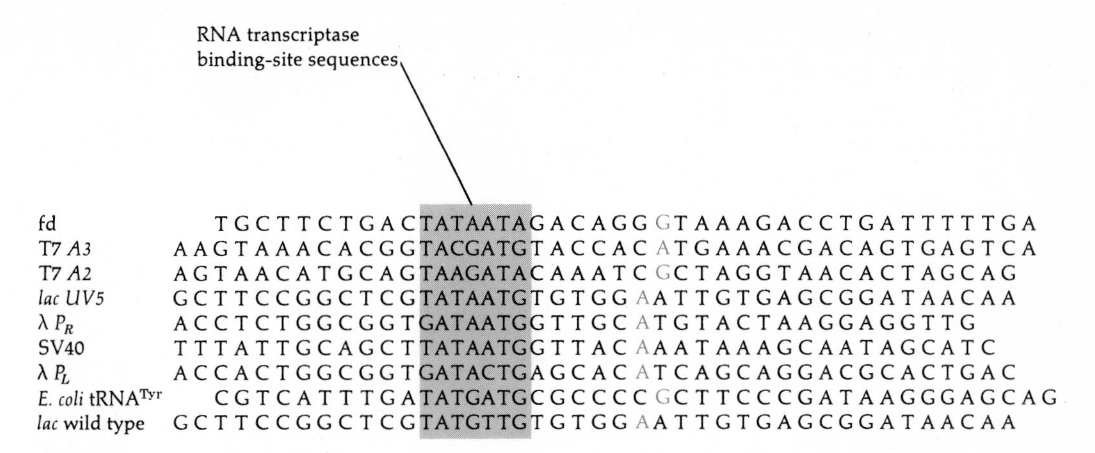

Figure 11.3
Nucleotide sequences of several DNA template strands, showing the initial positions of binding by RNA transcriptase (promoter) and the first base to be transcribed (color). (From J. D. Watson, *Molecular Biology of the Gene*, 3rd ed., W. A. Benjamin, Menlo Park, Cal., 1976.)

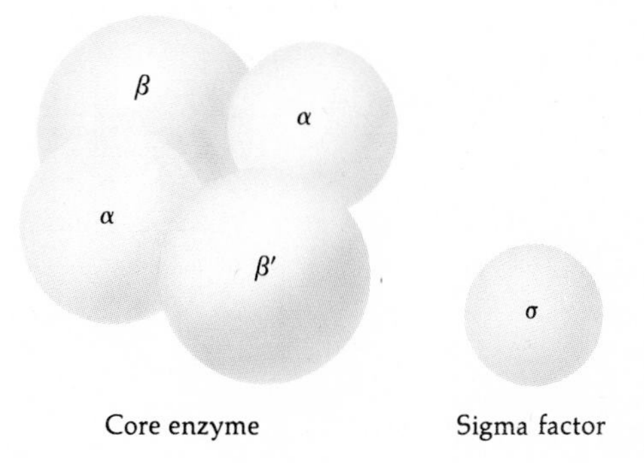

Figure 11.4
Schematic drawing of *E. coli* RNA transcriptase, showing its subunit structure. The core enzyme is $\alpha_2\beta\beta'$, and the holoenzyme is $\alpha_2\beta\beta'\sigma$.

Transcription of DNA leads to the production of three classes of RNA that are found in all cells: ribosomal RNA (rRNA), transfer RNA (tRNA), and messenger RNA (mRNA). These classes of RNA participate in the third general transfer of genetic information, that from RNA to protein.

Eukaryotic cells generally possess three distinct transcriptases that recognize different classes of promoters. Eukaryotic RNA polymerase I is found in nucleoli and is responsible for the transcription of ribosomal DNA (rDNA), which contains the sequences for ribosomal RNA. RNA polymerase III transcribes transfer RNA and 5S RNA (a component of ribosomes). RNA polymerase II transcribes all other RNAs, including messenger RNA. Enzymes II and III reside in the nucleoplasm.

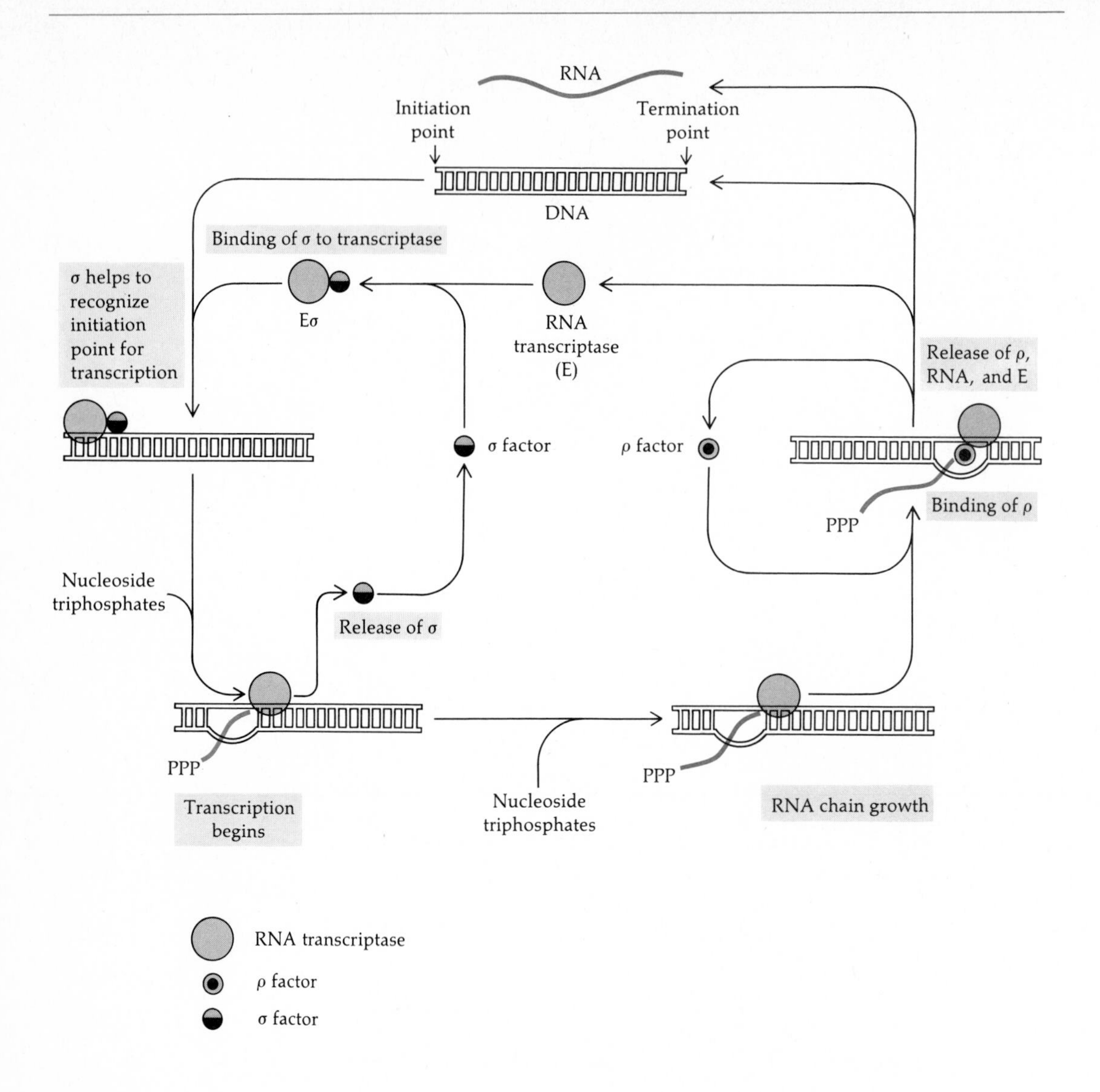

Figure 11.5
Overall outline of RNA synthesis in *E. coli*, showing the roles of σ factor, core transcriptase, and ρ factor. (After J. D. Watson, *Molecular Biology of the Gene*, 3rd ed., W. A. Benjamin, Menlo Park, Cal., 1976.)

Protein synthesis

The transfer of information encoded in the nucleotide sequence of a messenger RNA molecule to a specific amino acid sequence in a protein is complex, because it requires *translation* from a 4-letter alphabet to a 20-letter alphabet and the correct ordering of amino acids in a sequence. Translation involves the participation of both tRNA and rRNA, as well as mRNA. The information for the amino acid sequence of a protein is contained in a sequence of codons in the mRNA. Each codon consists of

Figure 11.6
Structures of some of the rare bases found in tRNA molecules as a result of post-transcriptional modifications carried out by special enzymes. These bases give tRNAs some of their special recognition functions.

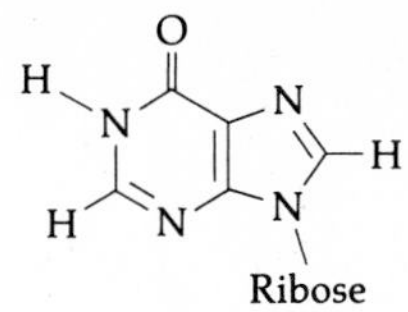

Inosine (I)

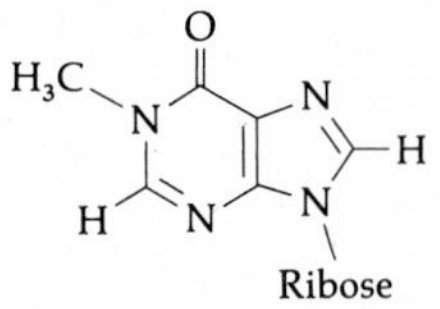

1-Methylinosine (I^m)

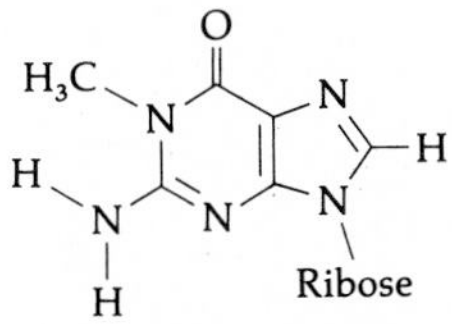

1-Methylguanosine (G^m)

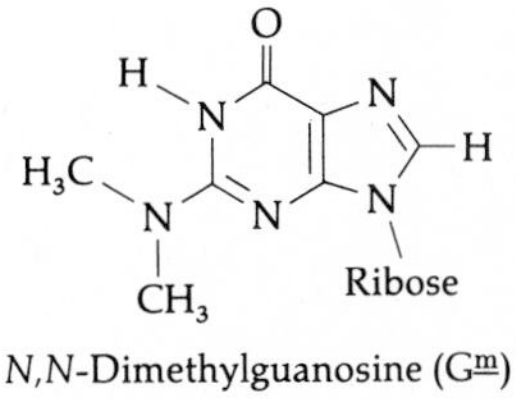

N,N-Dimethylguanosine ($G^{\underline{m}}$)

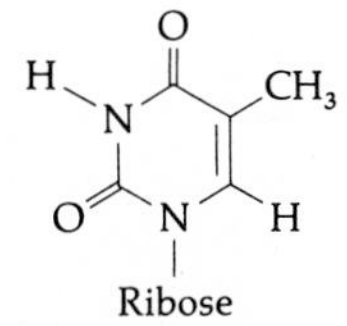

Ribothymidine (T)

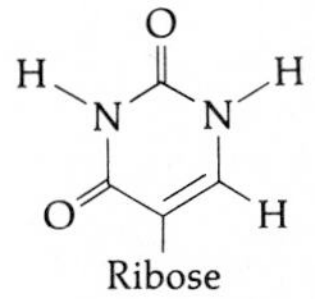

Pseudouridine (Ψ)

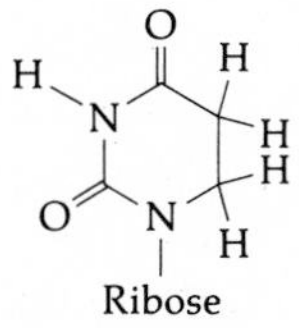

Dihydrouridine ($U^{\underline{h}}$)

three nucleotides read in a nonoverlapping manner, as will be discussed in detail in Chapter 12.

TRANSFER RNA. There is no direct interaction between a codon and the amino acid it represents. Instead, their association is mediated by transfer RNA molecules. There is at least one unique species of tRNA for each of the 20 amino acids. They are designated $tRNA^{Phe}$, $tRNA^{Ser}$, etc. The various transfer RNAs differ from one another in nucleotide sequence and in their content of certain rare nucleotides (Figure 11.6). These rare nucleotides are found only in tRNA and are formed by post-transcrip-

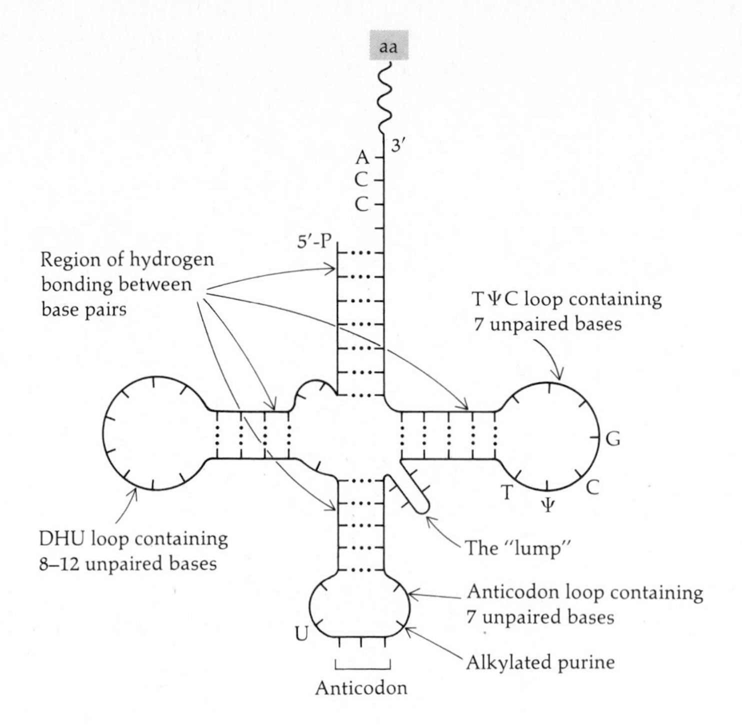

Figure 11.7
The general structure of tRNA molecules as represented in the cloverleaf configuration. The DHU loop contains several dihydrouridine bases. The point of attachment of the amino acid is indicated by the symbol aa.

tional enzyme modifications of the standard bases. All prokaryotic and eukaryotic tRNA molecules are about 80 nucleotides in length and have very similar secondary and tertiary structures, which can be drawn in the cloverleaf configuration shown in Figure 11.7. The actual 3-dimensional configuration, however, is probably that shown in Figure 11.8.

All tRNA molecules share two important properties: (1) at the 3′ end of the molecule, they possess the terminal sequence $-pCpCpA_{OH}$, which serves as a covalent attachment site for an amino acid, and (2) internally, each molecule bears a loop containing three nucleotides complementary to a codon; these nucleotides are the *anticodon.*

The function of a tRNA molecule is to provide a recognition signal between a codon and the amino acid for which the codon is specific (its *cognate*). The purpose of this recognition signal is to assure the accuracy of translation, as will be pointed out below. There are two important steps in the transfer of an amino acid from its free state to its position in a nascent polypeptide at which this recognition signal functions. Both of these steps are important in the accurate translation of mRNA into a product whose amino acid sequence reflects the sequence of the nucleotides in the DNA.

AMINOACYL-tRNA. The covalent attachment of an amino acid to its cognate tRNA is catalyzed by a specific *aminoacyl-tRNA ligase* or *synthetase,*

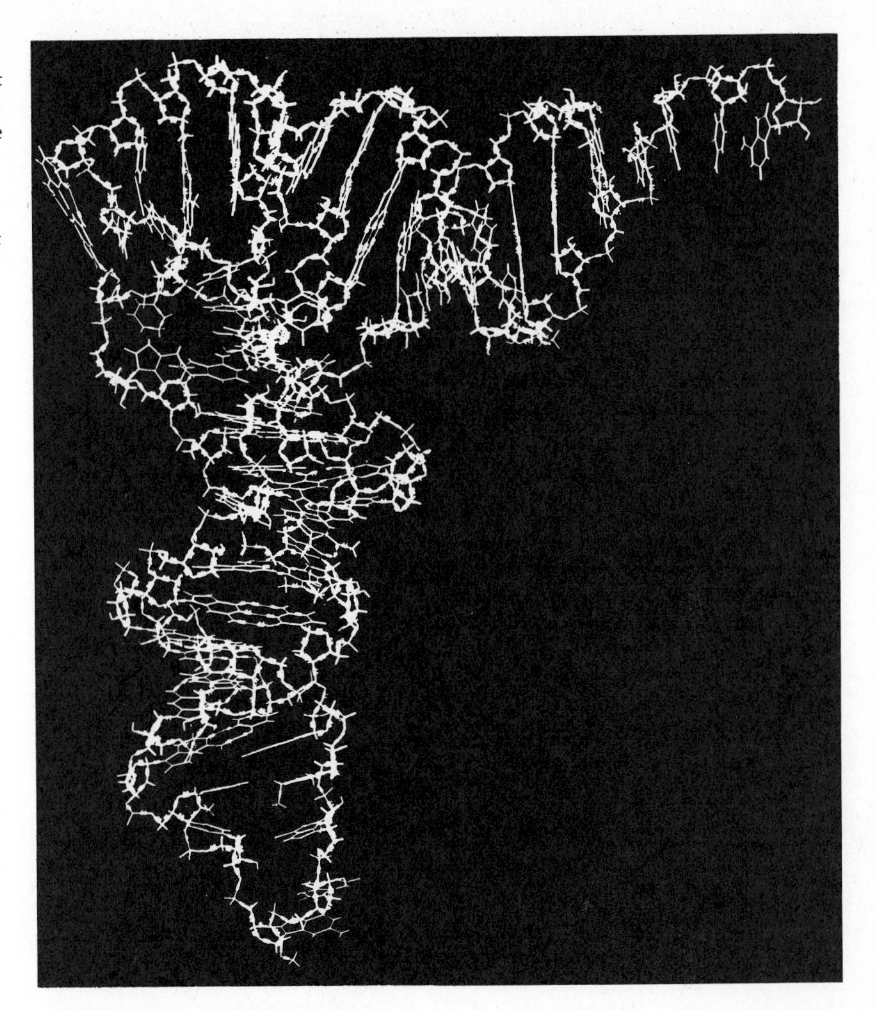

Figure 11.8
A molecular model of yeast $tRNA^{Phe}$ deduced by X-ray diffraction of crystals of the molecule. The 3′-CCA end is at the upper right, and the anticodon loop is at the bottom. [From S. H. Kim et al., *Science 185*:435 (1974). Copyright © 1974 by the American Association for the Advancement of Science.]

of which there are 20 types (one for each of the 20 amino acids). These are complex enzymes. They recognize both their cognate amino acid and cognate tRNA in order to form the covalent linkage of the carboxyl end of the amino acid to the 3′-OH of adenosine at the $-CpCpA_{OH}$ end of the tRNA. This process occurs in two steps, as can be shown using the example of serine.

The first step is amino acid activation (Figure 11.9):

Seryl-tRNA ligase + serine + ATP $\rightleftharpoons$
[seryladenylate]**seryl-tRNA ligase** + P-P

(recall that P-P represents a pyrophosphate molecule). The second step is the transfer of the activated amino acid to the 3′-OH of its cognate tRNA:

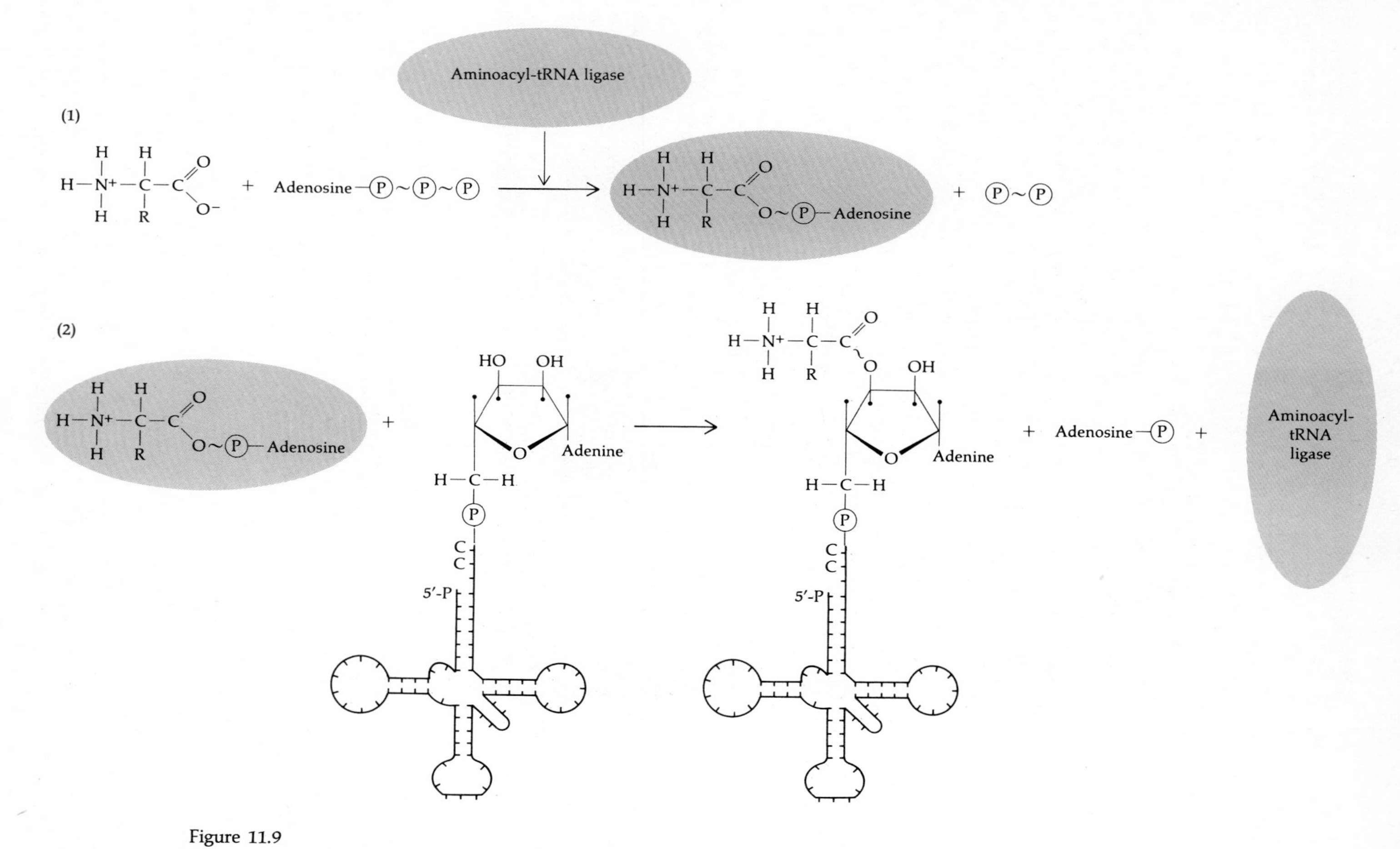

Figure 11.9
The activation of an amino acid by ATP and its transfer to the 3′-CCA end of its cognate tRNA are mediated by aminoacyl-tRNA ligase in two steps. The wavy lines denote high-energy bonds.

[Seryladenylate]**seryl-tRNA ligase** + $tRNA^{Ser} \rightleftharpoons$
seryl-$tRNA^{Ser}$ + **seryl-tRNA ligase** + AMP

Note in Figure 11.9 that the attachment is at the carboxyl end of the amino acid; the amino end is free for participation in peptide bond formation, which will be described below.

The aminoacyl-tRNA ligases play essential roles in assuring that the genetic information, which is so carefully transmitted intact from generation to generation (see Chapter 9) is also accurately expressed during translation. The accuracy of translation is obviously dependent on the accuracy with which an aminoacyl-tRNA ligase distinguishes between amino acids and attaches the correct one to its cognate tRNA. Accuracy is greatly increased by a *verification* reaction also performed by aminoacyl-tRNA ligases. For example, isoleucyl-tRNA ligase occasionally recognizes the wrong tRNA and incorrectly forms isoleucyl-$tRNA^{Phe}$ rather than isoleucyl-$tRNA^{Ile}$. This could result in the substitution of an isoleucine residue for phenylalanine during protein synthesis. However, phenylalanyl-tRNA ligase recognizes the incorrect association of isoleucine with its own cognate, $tRNA^{Phe}$, and hydrolyzes the isoleucyl-$tRNA^{Phe}$ to isoleucine and $tRNA^{Phe}$.

A *proofreading* function of aminoacyl-tRNA ligase, distinct from the verification function, is also known. Isoleucyl-tRNA ligase occasionally mistakes valine for isoleucine in the first step of amino acid activation:

Isoleucyl-tRNA ligase + valine + ATP $\rightarrow$
[valyladenylate]**isoleucyl-tRNA ligase** + P-P

The association of $tRNA^{Ile}$ with the [valyladenylate]isoleucyl-tRNA ligase complex in step 2 causes predominantly (180:1) hydrolysis of valyladenylate to valine and AMP, rather than formation of valyl-$tRNA^{Ile}$:

[Valyladenylate]**isoleucyl-tRNA ligase** + $tRNA^{Ile} \rightarrow$
valine + $tRNA^{Ile}$ + **isoleucyl-tRNA ligase** + AMP

The aminoacyl-tRNA ligases play the essential role in the translation of the genetic code: they associate specific amino acids with cognate anticodons. They also possess the functions of verification and proofreading, which assure accurate translation by subjecting each association of amino acid and anticodon to at least one check. Using the example given above, if the error rate in the first association made by isoleucyl-tRNA ligase is 1 valine per 100 isoleucines and the proofreading has an error rate of 1 in 180, the overall error rate is $1/100 \times 1/180 = 1/18{,}000$.

PEPTIDE BOND FORMATION. Following the formation of aminoacyl-tRNAs, the next step is the correct ordering of amino acids to form a polypeptide. This is accomplished by ribosomes, using the sequence of

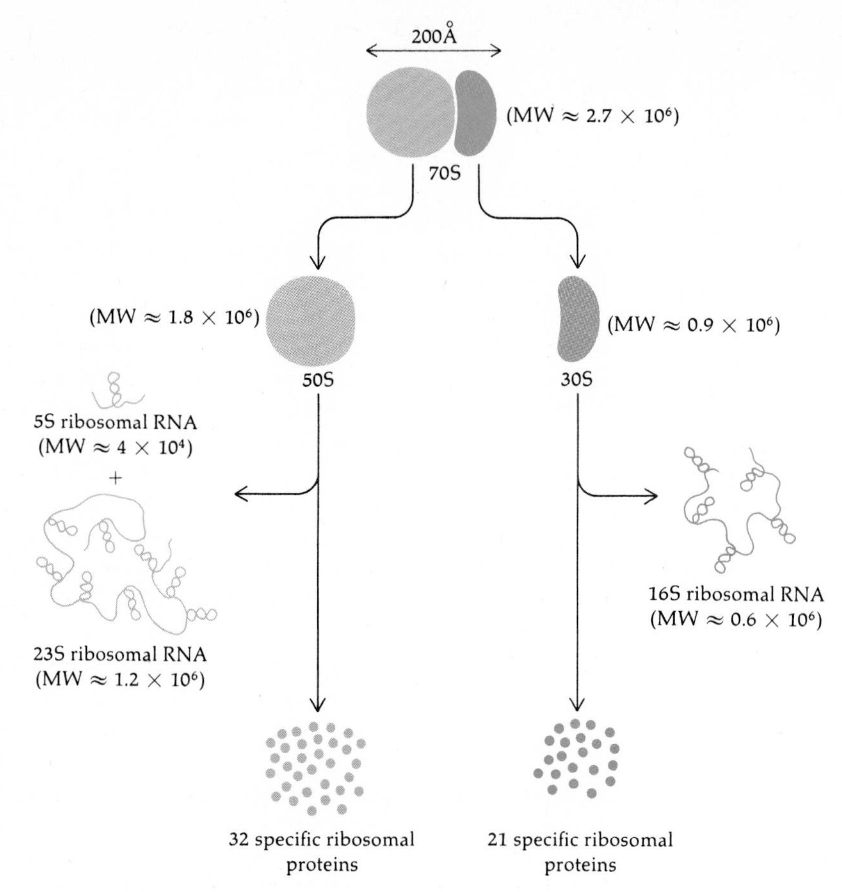

Figure 11.10
The structure of the *E. coli* ribosome. (After J. D. Watson, *Molecular Biology of the Gene*, 3rd ed., W. A. Benjamin, Menlo Park, Cal., 1976.)

codons present in mRNA. Ribosomes are complex structures consisting of ribosomal RNA complexed with a number of specific ribosomal proteins (Figure 11.10). In bacteria there are two ribosomal subunits, designated 30S and 50S (S, the svedberg unit, is a measure of size determined by the rate at which a particle in solution sediments in a centrifugal field). The analogous ribosomal subunits in eukaryotes are generally larger, 40S and 60S. The association of a large and a small subunit forms a ribosome of size 70S (bacteria) and 80S (eukaryotes). (Note that sedimentation rates are not strictly additive.) Chloroplasts and mitochondria in eukaryotic cells possess ribosomes that are more similar in size to those of prokaryotic ribosomes than to the eukaryotic ribosomes found in the cytoplasm. Much of the DNA found in chloroplasts and mitochondria (see Chapter 5) codes for their distinct protein-synthesizing apparatus.

All polypeptides are synthesized beginning at the NH_2 terminal and progressing to the COOH terminal, and all are initiated with methionine, though the N-terminal methionine residue is cleaved from certain polypeptides prior to the completion of synthesis. A specific type of $tRNA^{Met}$, designated tRNA$_i^{Met}$, always serves to initiate polypeptide synthesis. In bacteria the initiating methionine always has the NH_2 terminal blocked by a formyl residue (Figure 11.11) to form *N*-formylmethionyl-tRNA$_i^{Met}$. The formyl residue is cleaved from the N-terminal methionine residue once polypeptide synthesis has begun. In eukaryotes, however, the initiating methionyl-tRNA$_i^{Met}$ is not formylated. Methionine residues situated within a polypeptide are inserted by a second type of tRNA^{Met} in both prokaryotes and eukaryotes. Both tRNA^{Met} and tRNA$_i^{Met}$ have the same anticodon; there is only one codon, AUG, for methionine (see Chapter 12).

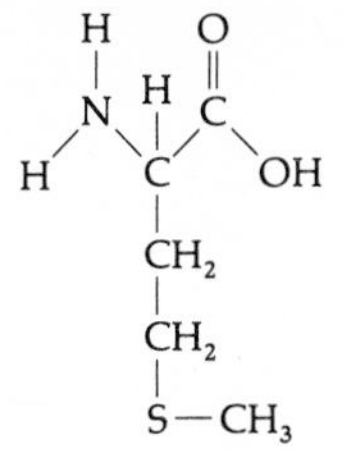

Methionine (Met)

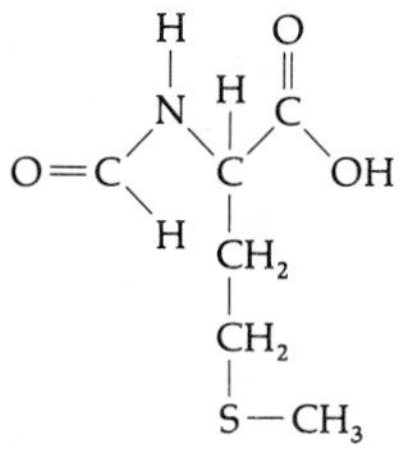

N-Formylmethionine (fMet)

Figure 11.11
The structures of methionine and *N*-formylmethionine.

The initiation of polypeptide synthesis begins with the attachment of the small ribosomal subunit to a binding site on mRNA that includes the initiating AUG codon. In prokaryotes the initiating *N*-formylmethionyl-tRNA$_i^{Met}$ then pairs with the AUG codon through its anticodon (Figure 11.12), followed by attachment of the large ribosomal subunit to form the complete ribosome. Only tRNA$_i^{Met}$, not tRNA^{Met}, can participate in the formation of this initiation complex.

The large subunit possesses two tRNA binding sites: a peptidyl-tRNA site (P site) and an aminoacyl-tRNA site (A site). The initiating *N*-formylmethionyl-tRNA$_i^{Met}$ is found in the P site, and the mRNA codon following the initial AUG codon is found opposite the A site, where it is available for pairing with an aminoacyl-tRNA having the appropriate anticodon. When the A site is occupied, the first peptide bond is formed by *peptidyl transferase*, an enzyme that is part of the larger ribosomal subunit; this enzyme shifts the *N*-formylmethionyl residue from its attachment to tRNA$_i^{Met}$ to the NH_2 group of the amino acid residue attached to its tRNA in the A site, forming a peptidyl-tRNA molecule (Figure 11.12). The ribosome then moves along the mRNA by one codon, placing the peptidyl-tRNA in the P site and freeing the A site. The A site is then occupied by the next aminoacyl-tRNA molecule, which has an anticodon appropriate for the mRNA codon now opposite that site. The deacylated tRNA$_i^{Met}$ is dissociated from the ribosome by the movement just described. The cyclic action of peptidyl transferase followed by movement of the ribosome along the mRNA, one codon at a time, leads to progressive polypeptide synthesis, the amino acid sequence being specified by the order of codons along the mRNA.

As the first ribosome moves down the mRNA, the ribosomal binding site is freed, and a second ribosome forms from subunits and begins polypeptide synthesis. The association of a number of active ribosomes with a single mRNA molecule forms a *polyribosome*, or *polysome*, as seen in Figure 11.13. Polypeptide synthesis continues until a termination codon arrives opposite the A site and is recognized by a termination protein, which enters the A site. The presence of the termination protein in the A site causes the next movement of the ribosome to induce hydrolysis of the polypeptidyl-tRNA, creating a free polypeptide, and dissociation of the ribosomal subunits to free subunits. The free subunits can then engage in another round of polypeptide synthesis (Figure 11.14).

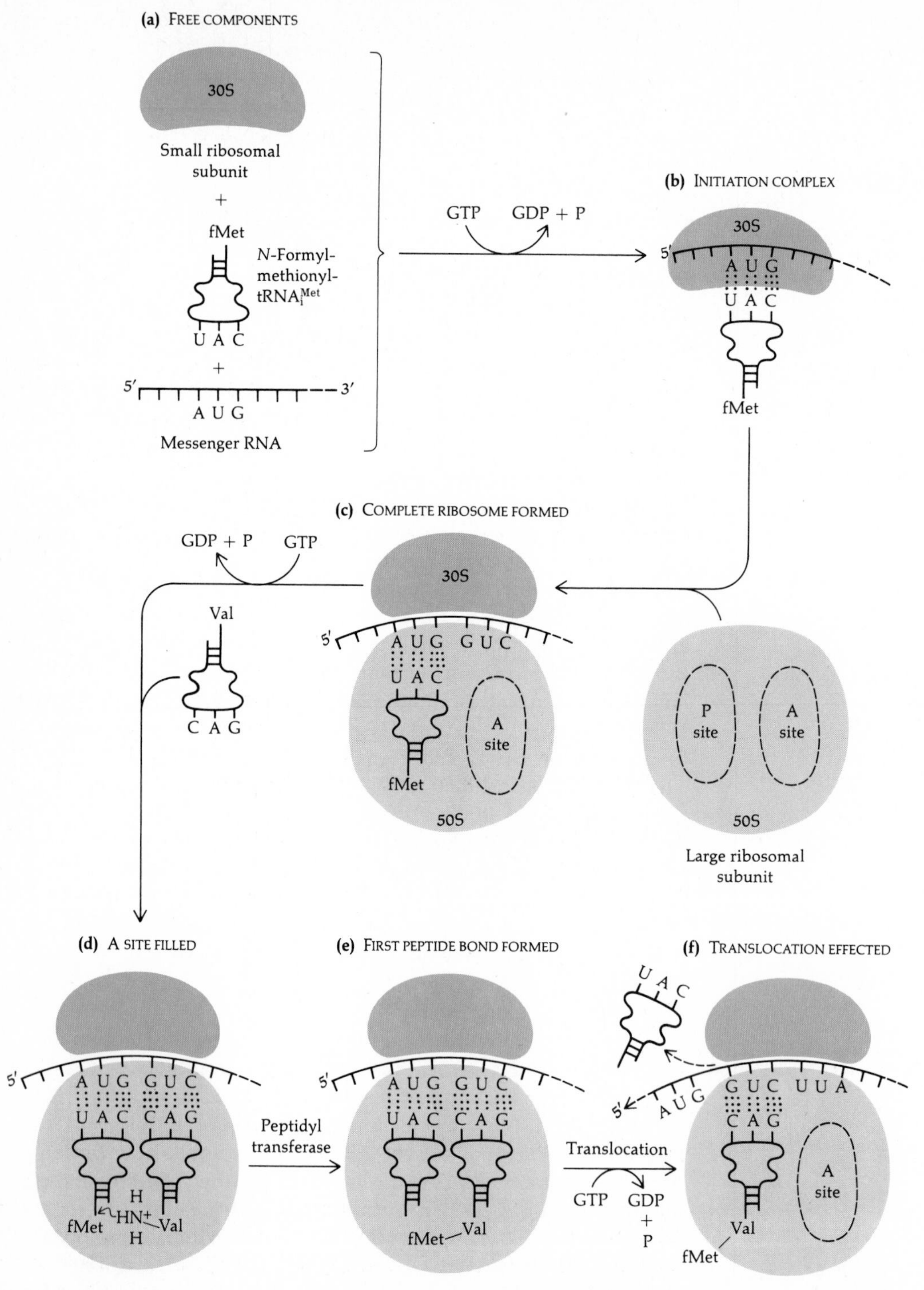

Figure 11.12
Formation of the initiation complex and of the first peptide bond during polypeptide synthesis. The cleavage of GTP provides energy for some of the steps.

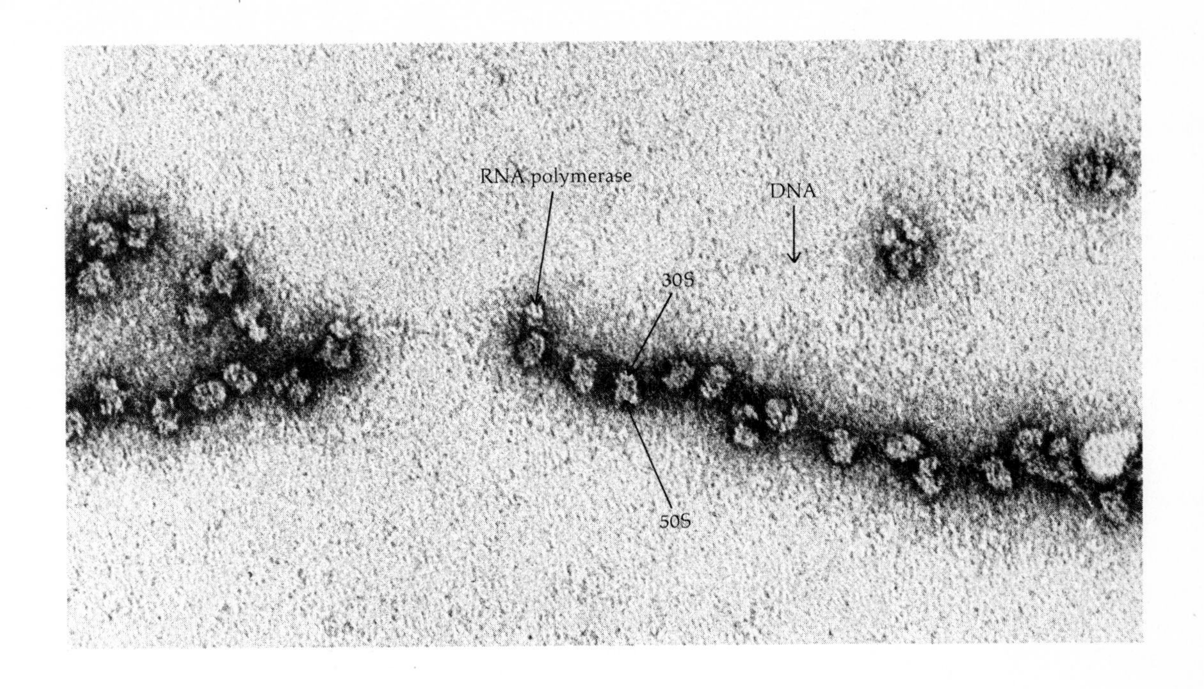

Figure 11.13
A group of *E. coli* 70S ribosomes moving along a nascent mRNA molecule (as the mRNA is being synthesized) forms a polysome. (Courtesy of Prof. Barbara Hamkalo, University of California, Irvine, and Prof. Oscar L. Miller, Jr., University of Virginia.)

The accuracy of protein synthesis depends in part on the precision with which the anticodon and the codon form hydrogen bonds. The pairing of anticodon and codon is monitored at least once by the ribosomes to assure that the correct aminoacyl-tRNA is present before the peptide bond is formed. Direct evidence for the active role of ribosomes in monitoring anticodon-codon pairing is provided by mutations that alter ribosomal proteins and affect the accuracy of protein synthesis (see Chapter 12), but our knowledge of the detailed mechanism is very incomplete.

Special Information Transfers

RNA replication

Three special transfers of genetic information are known (Figure 11.1). The first of these, from RNA to RNA, is known to occur only in cells infected by viruses whose genetic material is RNA, e.g., TMV and many other plant viruses, RNA bacteriophages, and animal RNA viruses such as polio virus. These viral RNAs, which may be either single-stranded or (in reovirus) double-stranded, carry genes encoding specific *RNA replicase* enzymes, which employ the viral RNA as a template to synthesize complementary RNA molecules. These, in turn, serve as templates for the

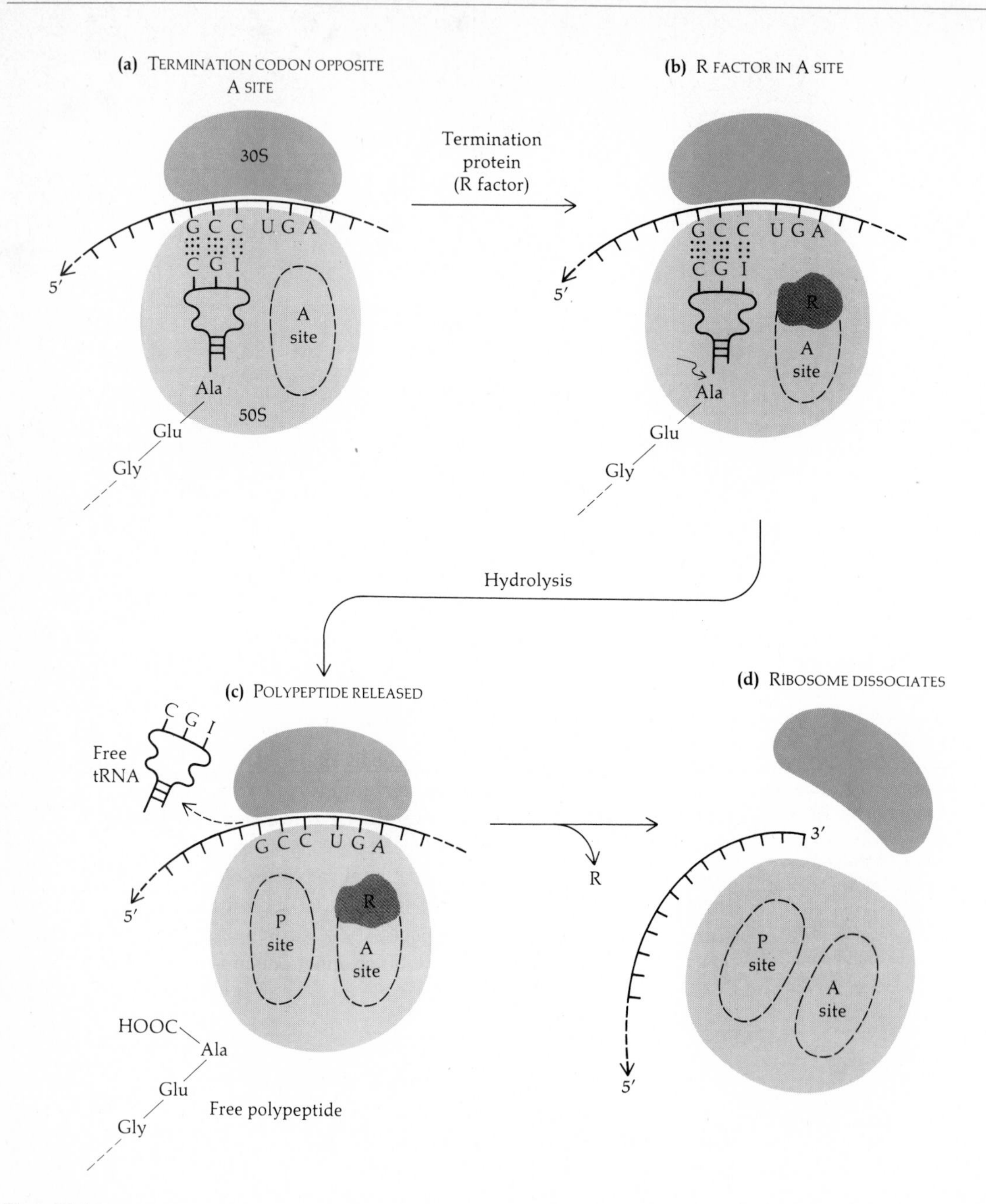

Figure 11.14
Steps in the termination of polypeptide synthesis when a termination codon (here UGA) enters the aminoacyl site on the ribosome.

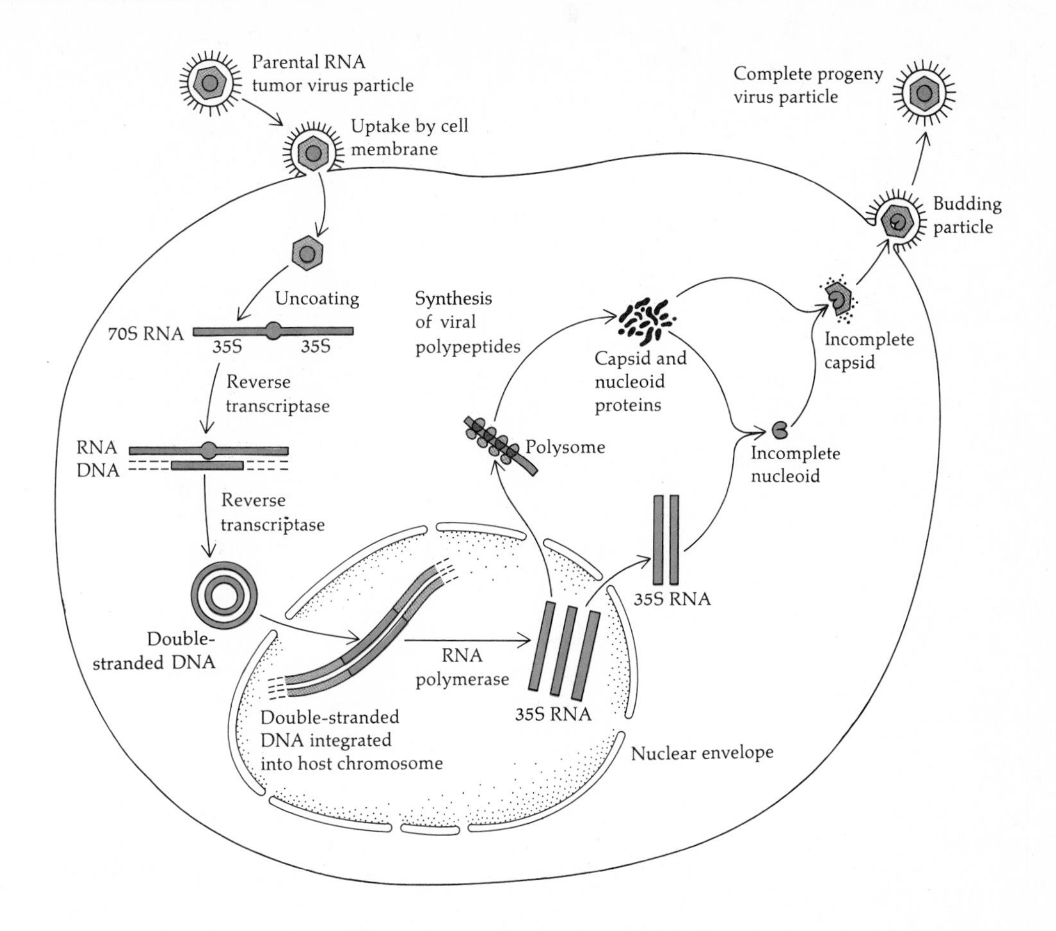

Figure 11.15
The role of reverse transcriptase in the infection of cells by an RNA tumor virus. (After J. D. Watson, *Molecular Biology of the Gene*, 3rd ed., W. A. Benjamin, Menlo Park, Cal., 1976.)

synthesis of progeny RNA molecules. The transfer of genetic information from RNA to RNA is mediated by the complementarity between bases in template and nascent RNA strands.

Reverse transcription

The second special information transfer, from RNA to DNA, occurs only in animal cells that are infected by some single-stranded RNA viruses. All viruses of this type can cause tumors in the host organism by transforming normal infected cells to cancerous cells (Figure 11.15). RNA tumor viruses code for a *reverse transcriptase*, which uses the single-stranded viral RNA as a template for the synthesis of a complementary DNA molecule. The DNA strand then serves as a template for the synthesis of a complementary DNA strand, also by reverse transcriptase, to produce a double-stranded DNA molecule containing the information originally encoded in the viral RNA. Reverse transcriptase is carried inside the RNA tumor virus particle as a normal component of this type of virus. The double-stranded DNA molecule, synthesized as the first step in infection (Figure 11.15), becomes integrated into the DNA of the host-cell chromosome, forming a provirus, perhaps by a mechanism similar to that employed by bacteriophage λ to integrate into the bacterial chromosome (see Chapter 9).

Once integrated, the provirus is transmitted to daughter cells by replication along with the host DNA, so that the progeny are also transformed to a cancerous state. These cancerous cells have lost the growth control typical of normal cells, and so they proliferate to form a tumor. Transcription of the proviral DNA produces both mRNA, which is translated to give virus-specific proteins, including reverse transcriptase, and progeny RNAs, which are encapsulated to form new tumor viruses. These bud from the cell surface without killing the transformed cells.

Aside from the deleterious effects of RNA tumor viruses, these viruses have been enormously useful in contemporary genetic engineering studies. Purified RNA tumor viruses are an excellent source of relatively pure reverse transcriptase, which is an important tool in many studies involving the cloning of recombinant DNA molecules (see Chapter 9). For example, purified mRNA coding for proteins of interest can often be obtained more easily than can the DNA of the gene coding for the protein. Reverse transcriptase is then used to synthesize a DNA copy of the mRNA, which is inserted into a DNA plasmid vehicle for cloning and production of quantities of the pure DNA. This subject will be discussed further below.

DNA translation

The third special information transfer, DNA to protein, has only been observed *in vitro* in the laboratory. Certain antibiotics, such as streptomycin and neomycin, which interact with ribosomes, can distort their messenger selectivity to enable them to accept single-stranded DNA molecules in place of mRNA. The single-stranded DNA then directs the translation of its nucleotide sequence to an amino acid sequence of a polypeptide. This special transfer probably never occurs as a normal process in cells.

Forbidden (Unknown) Information Transfers

The transfer of information from protein to DNA or from protein to RNA would involve the specification of an RNA or DNA nucleotide sequence by the amino acid sequence of a protein. This type of transfer would require translation from a 20-letter alphabet to a 4-letter alphabet, and would certainly require a mechanism as complicated as that which translates the information encoded in mRNA into protein. There is no evidence to suggest the existence of such a mechanism, and therefore these two transfers are unknown and presumably never occur.

The third forbidden information transfer, protein to protein (amino acid sequence to amino acid sequence), is also unknown. There is no evidence to suggest that proteins can replicate. Self-replication seems to be confined exclusively to nucleic acids.

Colinearity of Gene and Polypeptide—Prokaryotes

The colinearity of gene and polypeptide is a logical inference from the hypothesis that hereditary information is encoded in a linear sequence of base pairs in DNA and in a linear sequence of amino acids in polypeptides. This relation is implicit because DNA and polypeptides are both linear polymers. However, the demonstration in 1964 that gene and polypeptide *are* colinear was a confirmation of more than a decade of reasoning based on the sequence hypothesis.

Tryptophan synthetase, an enzyme of *E. coli*, consists of two polypeptides, A and B, encoded by the $trpA^+$ and $trpB^+$ genes (Figure 11.16; see also Chapter 7). Many mutants lacking tryptophan synthetase activity have been mapped within the *trpA* gene; they can be ordered on a linear genetic map, using the techniques discussed in Chapter 7.

The A polypeptide of the normal enzyme has been sequenced (Figure 11.17), as have the inactive polypeptides synthesized by a number of *trpA* mutants. Each mutation produces an alteration of one or another of the amino acids in the wild-type A polypeptide. The relative position of each altered amino acid and the genetic map position of each mutation causing that alteration are shown in Figure 11.18. The concordance of the genetic map and the positions of the altered amino acids in the protein sequence is evident. For example, *trpA3*, which alters the amino acid at position 49 from the N-terminal end (Glu→Val), maps to the left of *trpA446*, which alters the amino acid at position 175 (Tyr→Cys); *trpA446*, in turn, maps to the left of *trpA58*, which alters the amino acid at position 234 (Gly→Asp), and so on.

Note that at position 234, two different mutations cause different amino acid substitutions: *trpA58* (Gly→Asp) and *trpA78* (Gly→Cys). These two mutations recombine with each other and yet cause different amino acids to replace the normal amino acid. Therefore, they must reside within the same codon but affect different sites. As can be seen from the genetic code table (Table 12.1, page 407; this is discussed further in Chapter 16), they affect adjacent nucleotides in the same codon, providing conclusive evidence that recombination can occur between adjacent nucleotide pairs in the DNA. This is further substantiated by mutations affecting position 211: *trpA23* (Gly→Arg) and *trpA46* (Gly→Glu). These two mutations also recombine with each other to restore the wild-type enzyme activity and, presumably, the wild-type nucleotide sequence.

Colinearity of the genetic map and the coded polypeptide has also been demonstrated for the capsid-protein gene of phage T4 (discussed in

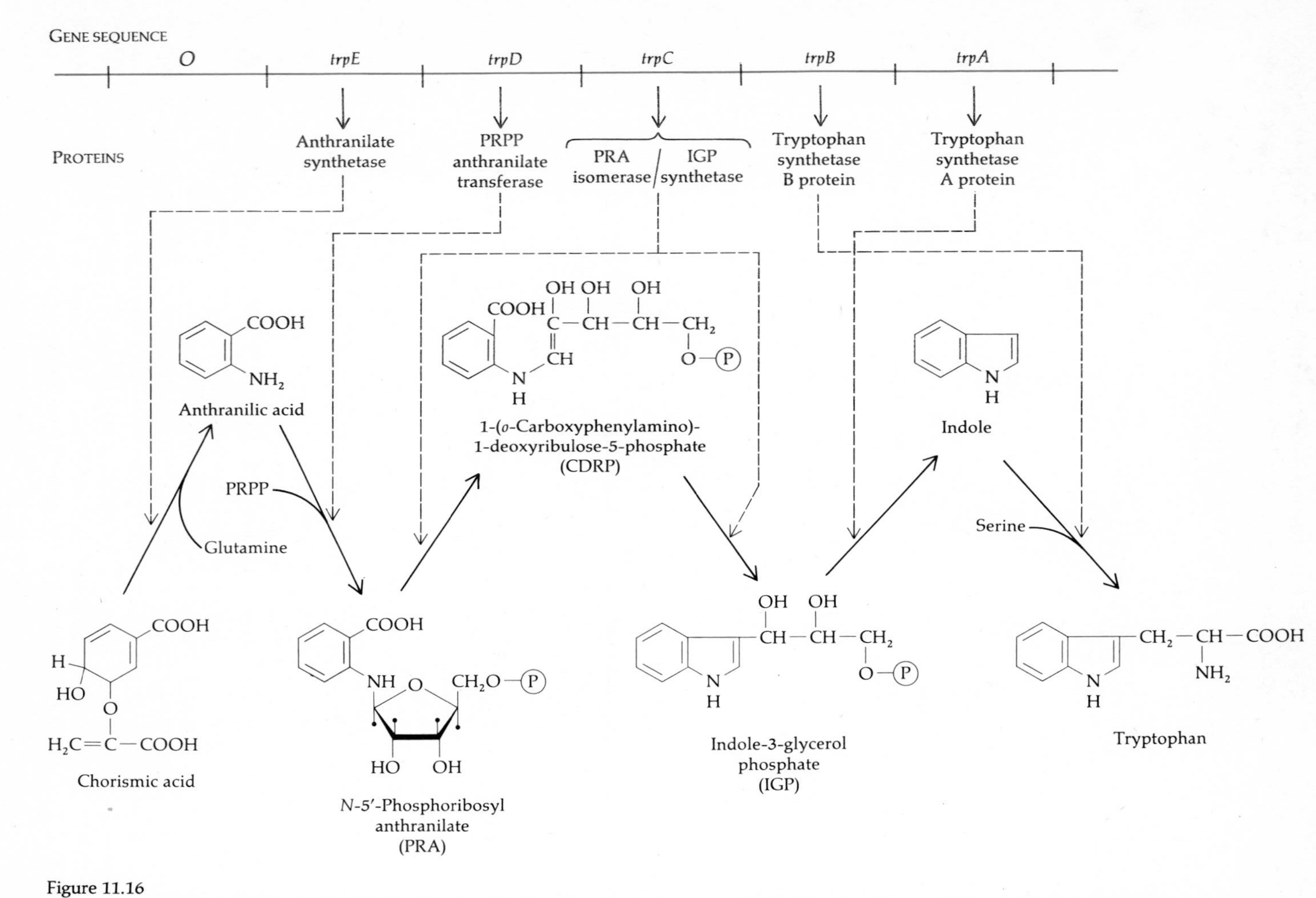

Figure 11.16
The genes coding for the enzymes of the tryptophan biosynthetic pathway are closely linked. PRPP = phosphoribosyl pyrophosphate; PRA = *N*-5′-phosphoribosyl anthranilate; IGP = indole-3-glycerol phosphate. The *trpC* gene codes for an unusual type of enzyme that catalyzes two different reactions; this bifunctional polypeptide acts as both PRA isomerase and IGP synthetase. (See also Figure 10.8, which shows the same reaction sequence except that the step involving PRA and PRA isomerase is omitted there.)

1 10 20
Met–Gln–Arg–Tyr–Glu–Ser–Leu–Phe–Ala–Gln–Leu–Lys–Glu–Arg–Lys–Glu–Gly–Ala–Phe–Val–

21 30 40
Pro–Phe–Val–Thr–Leu–Gly–Asp–Pro–Gly–Ile–Glu–Gln–Ser–Leu–Lys–Ile–Ile–Asp–Thr–Leu–

41 50 60
Ile–Glu–Ala–Gly–Ala–Asp–Ala–Leu–Glu–Leu–Gly–Ile–Pro–Phe–Ser–Asp–Pro–Leu–Ala–Asp–

61 70 80
Gly–Pro–Thr–Ile–Gln–Asn–Ala–Thr–Leu–Arg–Ala–Phe–Ala–Ala–Gly–Val–Thr–Pro–Ala–Gln–

81 90 100
Cys–Phe–Glu–Met–Leu–Ala–Leu–Ile–Arg–Gln–Lys–His–Pro–Thr–Ile–Pro–Ile–Gly–Leu–Leu–

101 110 120
Met–Tyr–Ala–Asn–Leu–Val–Phe–Asn–Lys–Gly–Ile–Asp–Glu–Phe–Tyr–Ala–Gln–Cys–Glu–Lys–

121 130 140
Val–Gly–Val–Asp–Ser–Val–Leu–Val–Ala–Asp–Val–Pro–Val–Gln–Glu–Ser–Ala–Pro–Phe–Arg–

141 150 160
Gln–Ala–Ala–Leu–Arg–His–Asn–Val–Ala–Pro–Ile–Phe–Ile–Cys–Pro–Pro–Asn–Ala–Asp–Asp–

161 170 180
Asp–Leu–Leu–Arg–Gln–Ile–Ala–Ser–Tyr–Gly–Arg–Gly–Tyr–Thr–Tyr–Leu–Leu–Ser–Arg–Ala–

181 190 200
Gly–Val–Thr–Gly–Ala–Glu–Asn–Arg–Ala–Ala–Leu–Pro–Leu–Asn–His–Leu–Val–Ala–Lys–Leu–

201 210 220
Lys–Glu–Tyr–Asn–Ala–Ala–Pro–Pro–Leu–Gln–Gly–Phe–Gly–Ile–Ser–Ala–Pro–Asp–Gln–Val–

221 230 240
Lys–Ala–Ala–Ile–Asp–Ala–Gly–Ala–Ala–Gly–Ala–Ile–Ser–Gly–Ser–Ala–Ile–Val–Lys–Ile–

241 250 260
Ile–Glu–Gln–His–Asn–Ile–Glu–Pro–Glu–Lys–Met–Leu–Ala–Ala–Leu–Lys–Val–Phe–Val–Gln–

261 268
Pro–Met–Lys–Ala–Ala–Thr–Arg–Ser

Figure 11.17
The amino acid sequence of the A polypeptide of *E. coli* tryptophan synthetase.

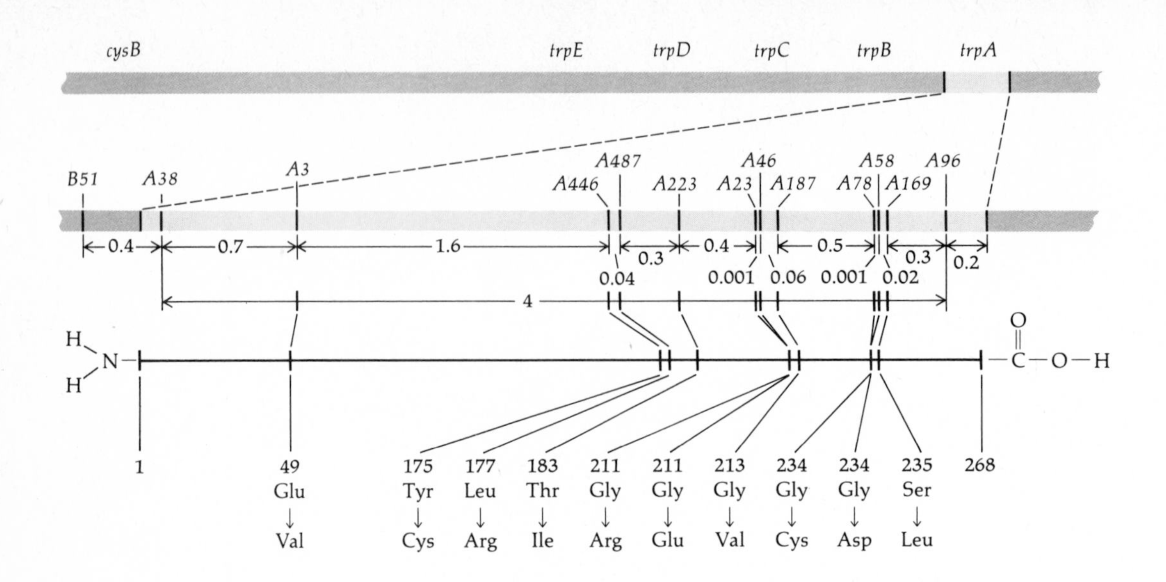

Figure 11.18
Correlation of the genetic map of the *trpA* gene and amino acid substitutions in the A polypeptide of *E. coli* tryptophan synthetase. (After C. Yanofsky, *Scientific American*, May, 1967, p. 80.)

Chapter 12) and the *E. coli* β-galactosidase gene, *lacZ*$^+$. Colinearity of nucleotide sequence and amino acid sequence is also apparent from comparison of the nucleotide sequence of ϕX174 genes (Chapter 6) with those ϕX174 proteins whose amino acid sequences are known. The genetic code table (Table 12.1) permits the amino acid sequence of the proteins to be read directly from the nucleotide sequence of the DNA.

Electron micrographs of broken *E. coli* cells provide visual evidence of how closely translation follows transcription of mRNA by RNA polymerase (Figure 11.13). Ribosomes synthesizing protein on nascent mRNA are seen to follow immediately upon the heels of the RNA polymerase transcribing the DNA. Because RNA is synthesized in a $5' \rightarrow 3'$ direction, the growing point being at the 3′-OH end of the strand, ribosomes can initiate translation at the 5′ end of the nascent strand. The ribosome closest to the RNA polymerase must be the first to have initiated translation on the messenger shown in Figure 11.13. Such micrographs substantiate the conclusions, drawn from more rigorous experiments, that gene and polypeptide are colinear in the order of their subunits.

Colinearity of Gene and Polypeptide—Eukaryotes

Recently developed recombinant DNA technology has made it possible to compare specific mRNA sequences for known proteins of eukaryotic organisms with the corresponding sequences in the chromosomal DNA.

Table 11.1
Known restriction recognition sites in the plasmid pOV230 carrying chicken ovalbumin cDNA (residues 13–1859 of Figure 11.20).

Enzyme*	Residue Numbers†	Recognition Sequence
Pst I	477	CTGCA/G
Hae III	818	GG/CC
Hin fI	258, 566, 1458, 1732	G/ANTC
Hga I	1125	GCGTC
Hph I	60, 336, 1605	TCACC
Alu I	15, 38, 117, 175, 277, 475, 495, 557, 759, 828, 986, 1042, 1235, 1332, 1338, 1554	AG/CT
Mbo II	147, 308, 389, 603, 638, 748, 887, 895, 922, 989, 1030, 1133, 1161, 1197, 1229, 1669	GAAGA or TCTTC
Eco RII	193, 213, 254, 507, 754	/CC$\binom{A}{T}$GG
Mbo I	482, 751, 898	/GATC
Taq I	41	T/CGA
Xba I	1345	T/CTAGA
Sst I	116, 494	GAGCT/C
Pvu II	474	CAG/CTG
Asu I	867	G/GACC
Mnl I	325, 444, 447, 892, 914, 929, 943, 1014, 1085, 1103, 1159	CCTC or GAGG

*No sites for *Hae* II, *Hin* dII, *Hin* dIII, *Hha* I, *Hpa* I, *Hpa* II, *Eco* RI, *Ava* I, *Bam* I, *Bal* I, *Bgl* II, *Sma* I, *Xho* I, and *Sal* I.

†The number given is the first (5′) nucleotide of the recognition sequence, not the point of cutting. Where the latter is within the recognition sequence, it is indicated by a slash.

From L. McReynolds et al., *Nature 273:*723 (1978).

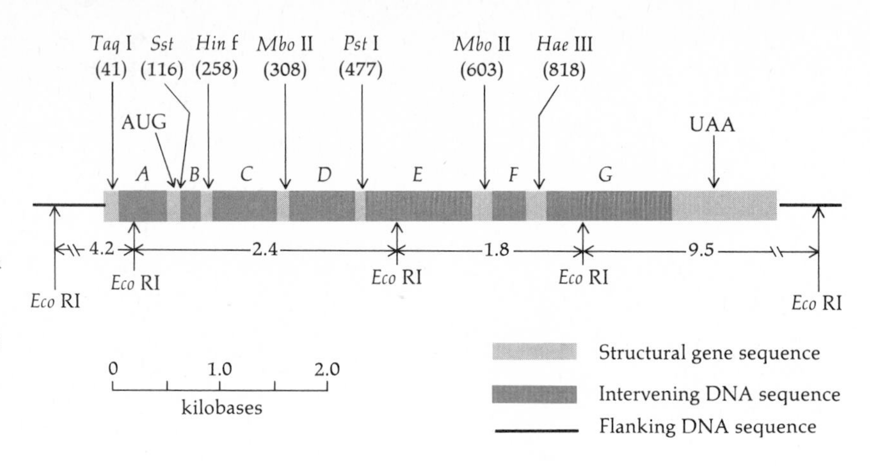

Figure 11.19
A map of the organization of structural and intervening DNA sequences constituting the chicken ovalbumin gene. Each of the noncontiguous structural segments is identified by one restriction endonuclease site. The positions of the messenger initiation codon (AUG) and termination codon (UAA) are indicated. [After A. Dugaiczyk et al., *Nature* 274:328 (1978).]

In 1977 these techniques led to the astonishing discovery that some genes contain nontranslated, intervening nucleotide sequences within the coding region of the gene. Noncoding, intervening nucleotide sequences have been found, for example, in the structural genes for both rabbit and mouse β globin chains of hemoglobin, for mouse immunoglobin light chains, and for chicken ovalbumin (egg white). The map of the organization of mRNA coding sequences and intervening sequences in the ovalbumin gene shows that seven large intervening sequences separate the mRNA sequences into eight noncontiguous sequences (Figure 11.19).

The methods employed to deduce the gene organization in Figure 11.19 will be briefly outlined. First, mRNA molecules are isolated from polysomes obtained from chicken oviduct, in which ovalbumin is the major protein being synthesized and ovalbumin mRNA is very abundant. This mRNA of 1859 nucleotides is significantly larger than is required to code for ovalbumin; it possesses a translated portion of 1158 nucleotides between untranslated regions at both 5′ and 3′ ends, as well as a tail of poly-A at the 3′ end. (A poly-A tail at the 3′ end is added to most eukaryotic mRNA molecules after transcription.) A double-stranded DNA copy (cDNA) of the mRNA is synthesized *in vitro* by reverse transcriptase and cloned in *E. coli* after insertion into a plasmid cloning vehicle (see Chapter 9). A cloned plasmid called pOV230 isolated by this procedure contained cDNA specifying mRNA residues 13–1859. The chicken cDNA portion of this hybrid plasmid has been sequenced and its restriction enzyme sites mapped. The corresponding mRNA sequence deduced from the cDNA sequence is shown in Figure 11.20, and the sites of restriction enzyme cleavage of the cDNA are given in Table 11.1. It is important to note that no site recognized by *Eco* RI or *Hin* dIII is present in the cDNA and that one *Hae* III site is present at residue 818.

When purified plasmid pOV230 is digested with *Hin* dIII and *Hae* III, the chicken cDNA can be purified from the plasmid DNA in two unique

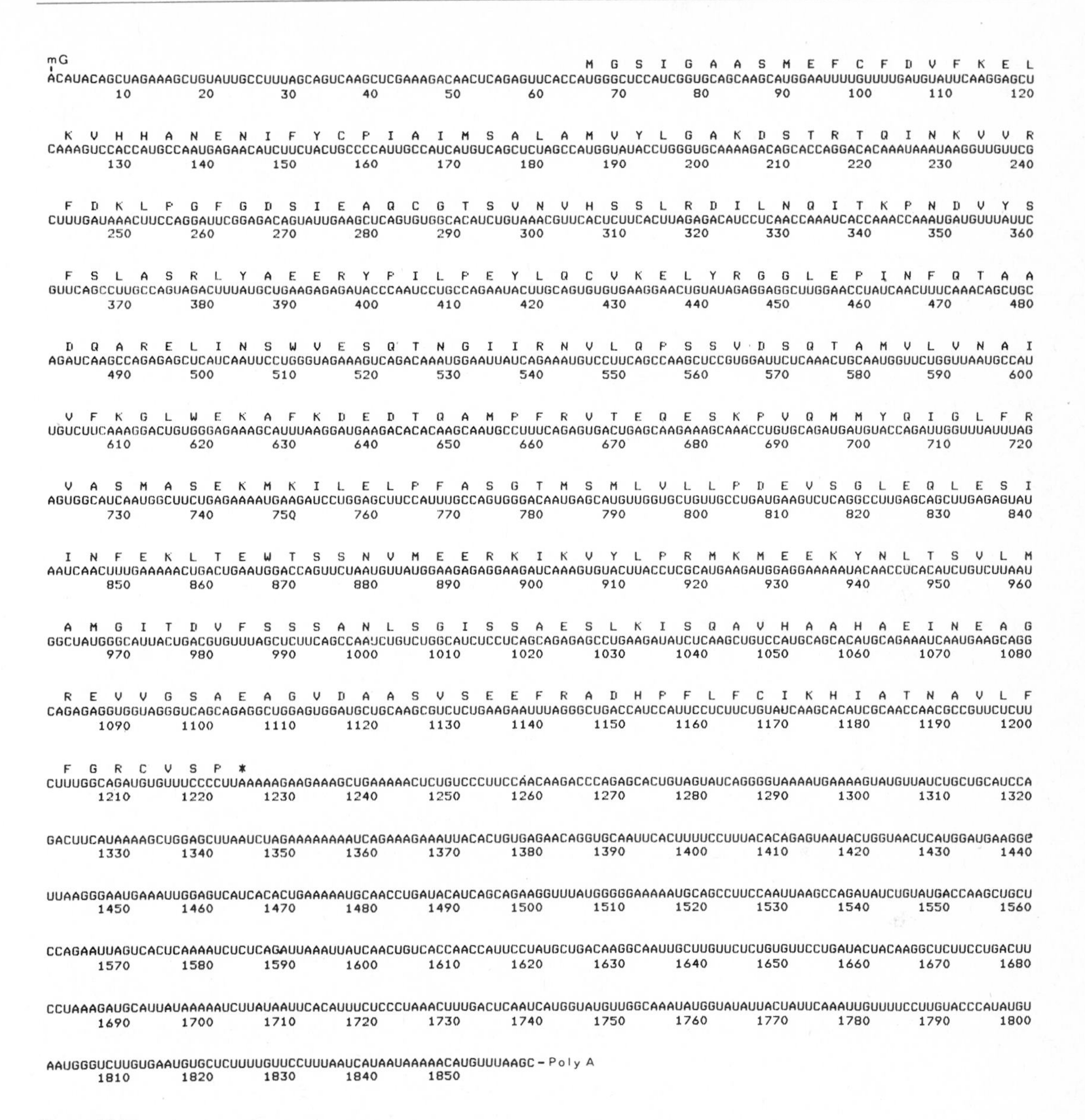

Figure 11.20
Sequence of chicken ovalbumin mRNA. Residues 13–1859 were present in the plasmid pOV230; residues 1–12 and 1731–1859 were sequenced directly from the mRNA. The ovalbumin termination codon is marked (*). Mature ovalbumin lacks the N-terminal methionine. The amino acid composition, excluding the initiating methionine, is: Ala (A), 35; Arg (R), 15; Asn (N), 17; Asp (D), 14; Cys (C), 6; Gln (Q), 15; Glu (E), 33; Gly (G), 19; His (H), 7; Ile (I), 25; Leu (L), 32; Lys (K), 20; Met (M), 16; Phe (F), 20; Pro (P), 14; Ser (S), 38; Thr (T), 15; Trp (W), 3; Tyr (Y), 10; and Val (V), 31. Total: 385 residues. [From L. McReynolds et al., *Nature* *273*:723 (1978).]

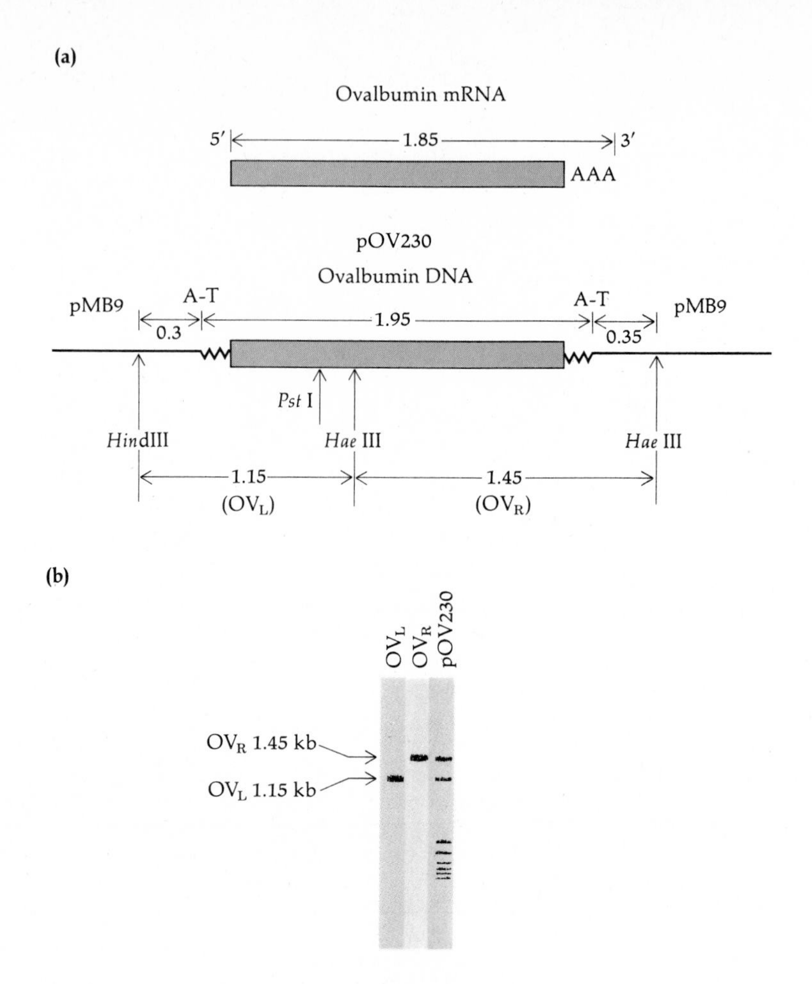

Figure 11.21
(**a**) Restriction-site map of the ovalbumin cDNA portion of pOV230 and its alignment with the ovalbumin mRNA. (**b**) Following cleavage with *Hin* dIII and *Hae* III restriction enzymes, the OV_L and OV_R fragments can be purified by electrophoresis through an agarose gel. The plasmid DNA sequence (pMB9) of pOV230 is broken into many small fragments that move faster than the OV_L and OV_R fragments. The OV_L and OV_R fragments are obtained in pure form by cutting them from the gel. [After E. Lai et al., *Proc. Natl. Acad. Sci. USA* 75:2205 (1978).]

fragments of 1150 and 1450 nucleotides each (Figure 11.21). The smaller fragment includes the cDNA containing the 5′ end of the mRNA sequence, and the larger fragment includes that for the 3′ end. Each fragment is then radioactively labeled *in vitro* to a very high specific activity by nick translation, employing *E. coli* DNA polymerase I (see Chapter 9) and α-^{32}P-deoxyribonucleoside triphosphates. The radioactive cDNA fragments, when converted to single strands by heating, serve as probes to identify complementary sequences in denatured whole chicken DNA.

Because the cDNA, representing the mRNA sequence, contains no *Eco* RI site, it is expected that digestion of whole chicken DNA with *Eco* RI should produce only one DNA fragment containing sequences complementary to the cDNA (since there are many *Eco* RI sites in chicken DNA, there are certain to be *Eco* RI sites on each side of the ovalbumin gene). This fragment of whole chicken DNA can be identified by size

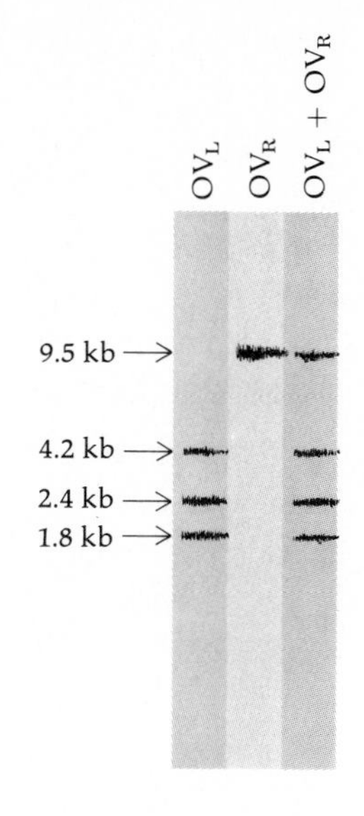

Figure 11.22
Drawing of autoradiogram of total chicken DNA after cleavage with *Eco* RI, electrophoresis through agarose gel, denaturation of the DNA fragments in the gel, and hybridization with radioactive probes of OV_L alone, OV_R alone, and OV_L plus OV_R. [After E. Lai et al., *Proc. Natl. Acad. Sci. USA* 75:2205 (1978).]

fractionation of *Eco* RI-digested DNA fragments on an electrophoretic gel, denaturation of the fragments, and hybridization to the single-strand radioactive cDNA probe (Figure 11.22). When this experiment was carried out, the unexpected observation was made that four fragments of different sizes contain sequences complementary to one or the other of the two probes (Figure 11.22). This astonishing result can only mean that the natural gene for ovalbumin contains DNA sequences possessing *Eco* RI sites that are interspersed with the sequences present in the mRNA. The four *Eco* RI fragments can be identified on the map in Figure 11.19, which is the product of a number of similar experiments carried out with different restriction enzymes, employing both whole chicken DNA and cloned fragments derived from whole chicken DNA.

The difference in structural organization of the eukaryotic gene and its mRNA transcript may be a consequence of the fact that transcription and translation in eukaryotes are not coupled as they are in prokaryotes (Figure 11.13). The nuclear membrane, a distinguishing feature of eukaryotes, physically separates DNA transcription within the nucleus from translation occurring on ribosomes in the cytoplasm. Therefore, opportunity exists prior to translation for structural alteration of the primary RNA transcript. The origin of cytoplasmic mRNA in eukaryotes has been very unclear until recently. Biochemical studies of nuclear RNA had shown that it generally consists of molecules much larger than the mRNA molecules found in the cytoplasm. This heterogeneous nuclear RNA is designated hnRNA. Most of the hnRNA synthesized in the nucleus never reaches the cytoplasm and is degraded to nucleotides. Biochemical studies had not determined whether cytoplasmic mRNA is synthesized as such in the nucleus. Its presence in the nucleus might be obscured by the large amount of hnRNA also synthesized there. On the other hand, mRNA might be synthesized as part of the hnRNA molecules and then processed by nucleases to mRNA as it is transported to the cytoplasm. The discovery of intervening sequences in structural genes strongly suggests that the latter possibility is generally true.

The highly radioactive cDNA probes for ovalbumin mRNA sequences discussed above have been used to measure the size of the ovalbumin gene transcripts present in the nuclei of oviduct cells. Nuclei are found to contain large RNA molecules possessing these sequences, which represent transcripts of the entire ovalbumin gene, including the intervening sequences. Radioactive probes for the intervening sequences show that these sequences are also transcribed into large molecules. Prior to production of mature cytoplasmic mRNA, therefore, the intervening sequences must be removed from an ovalbumin hnRNA transcript by endonuclease cuts and ligation of the coding sequences.

Studies of mouse β globin mRNA and mRNA precursor have also shown that the precursor mRNA contains the entire intervening sequence present in the β globin gene and that this sequence is absent in the mature mRNA (Figure 11.23).

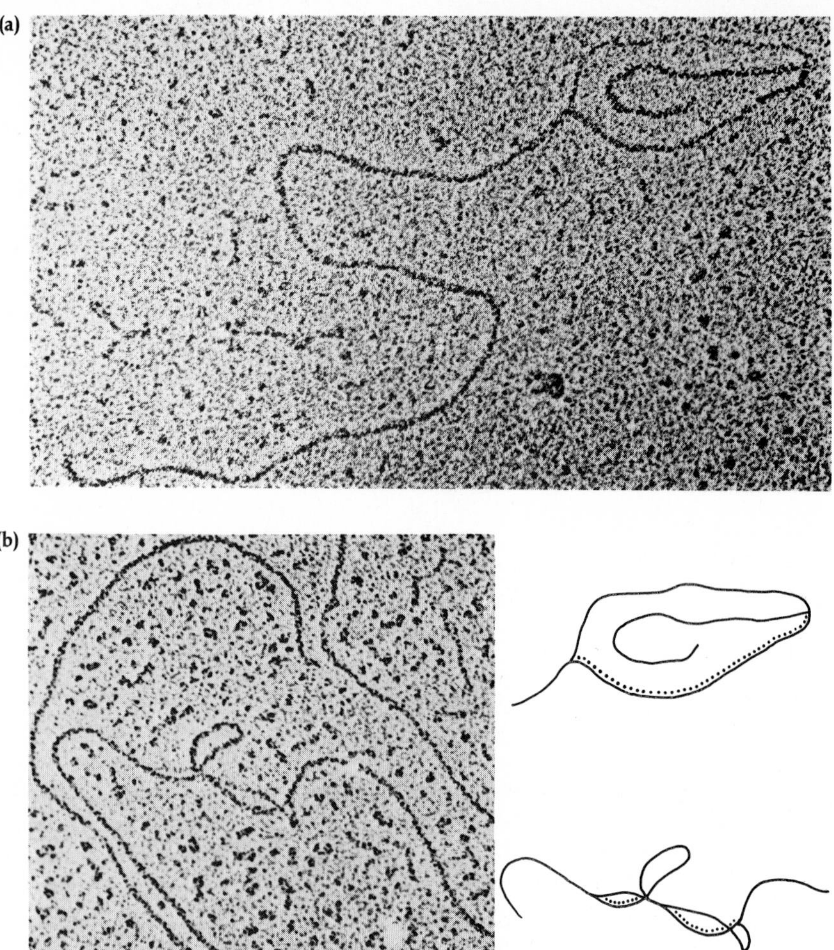

Figure 11.23
Electron micrographs of R-loop structures formed between a cloned mouse β globin gene and (**a**) a 15S globin hnRNA and (**b**) a 10S globin mRNA. The loop in part b shows that the β globin gene contains an intervening DNA sequence, about 550 np in length, that is not present in the complementary mRNA sequence. The absence of a loop in part a demonstrates that the intervening sequence is present in the hnRNA precursor to the mRNA. R-loop formation occurs when a double-stranded DNA molecule is incubated under partial denaturing conditions with a complementary RNA molecule. The RNA-DNA hybrid is more stable under such conditions than is the double-stranded DNA molecule. [From S. M. Tilghman et al., *Proc. Natl. Acad. Sci. USA* 75:1309 (1978).]

The DNA coding sequences for ovalbumin are colinear with the ovalbumin polypeptide. The ovalbumin gene map in Figure 11.19 indicates that the mRNA sequences, read from left to right, are colinear (subtracting intervening sequences) with the mRNA that contains the coded ovalbumin amino acid sequences shown in Figure 11.20. Since mRNA synthesis entails cutting and splicing of pieces of a larger hnRNA molecule, the overall colinearity of gene and polypeptide in eukaryotes remains to be proven. It is possible that some genes and polypeptides may not be colinear in their whole as well as their parts. DNA coding sequences A, B, C, D, E may be separated by intervening sequences but might be assembled into a messenger RNA in some other order, such as A, E, C, D, B. Much work remains to be done before the colinearity of gene and polypeptide is demonstrated to be invariably true in eukaryotes.

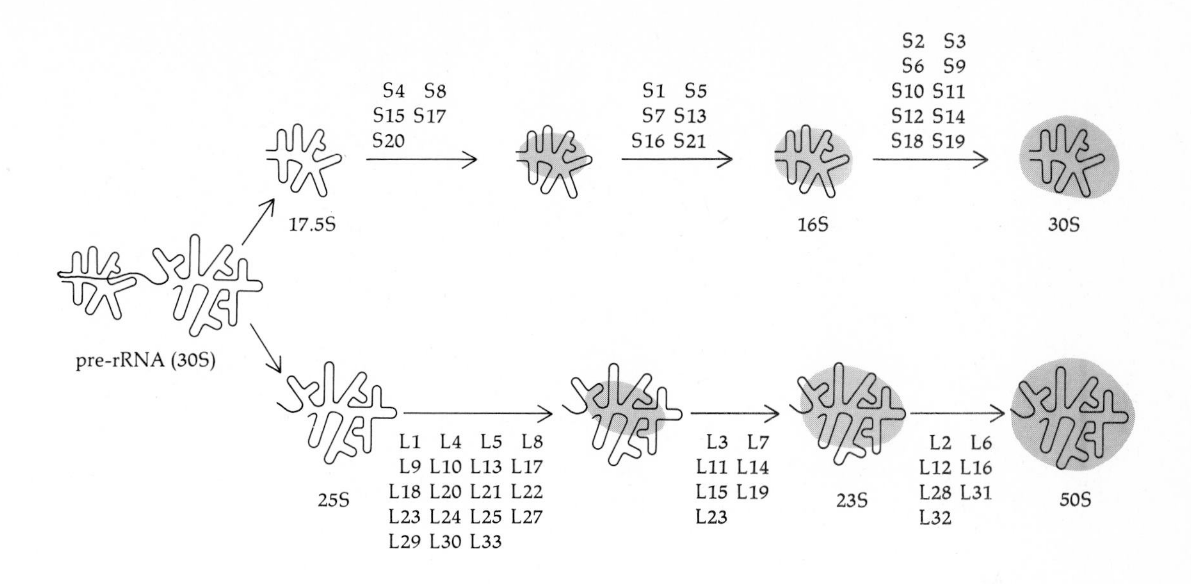

Figure 11.24
Assembly pathways for the *E. coli* 30S and 50S ribosomes. The numbers refer to different ribosomal proteins that bind to the RNAs during assembly. (After J. D. Watson, *Molecular Biology of the Gene*, 3rd ed., W. A. Benjamin, Menlo Park, Cal., 1976.)

Synthesis of Ribosomal and Transfer RNAs

In both prokaryotes and eukaryotes, post-transcriptional processing of transcripts of the ribosomal and transfer RNA genes plays a prominent role in the production of these mature RNA molecules. These RNA molecules play essential roles in the translation of mRNA and the synthesis of proteins, as discussed above.

In *E. coli* the genes for 16S, 23S, and 5S rRNA are closely linked and are transcribed from a common promoter as a single unit, or pre-rRNA. The 5′ end of the pre-rRNA contains the 16S rRNA, followed by the 23S rRNA and the 5S rRNA. As transcription proceeds, the nascent pre-rRNA is modified by sequence-specific methylation of certain bases and is subsequently cut by processing enzymes to yield the mature ribosomal RNA molecules. During the maturation process, each rRNA molecule becomes associated with specific proteins that are part of the mature ribosomes (Figure 11.24).

Transfer RNA molecules are also derived from larger transcripts. For example, the *E. coli* pre-tRNA^{Tyr} shown in Figure 11.25 is processed by removal of nucleotides from both the 3′ and 5′ ends. Some primary transcripts contain the sequences of more than one mature tRNA (Figure 11.25).

In eukaryotic organisms, the 18S and 28S rRNA molecules are also transcribed from a common promoter to form a pre-rRNA that is processed by methylation enzymes and nucleases to create the mature

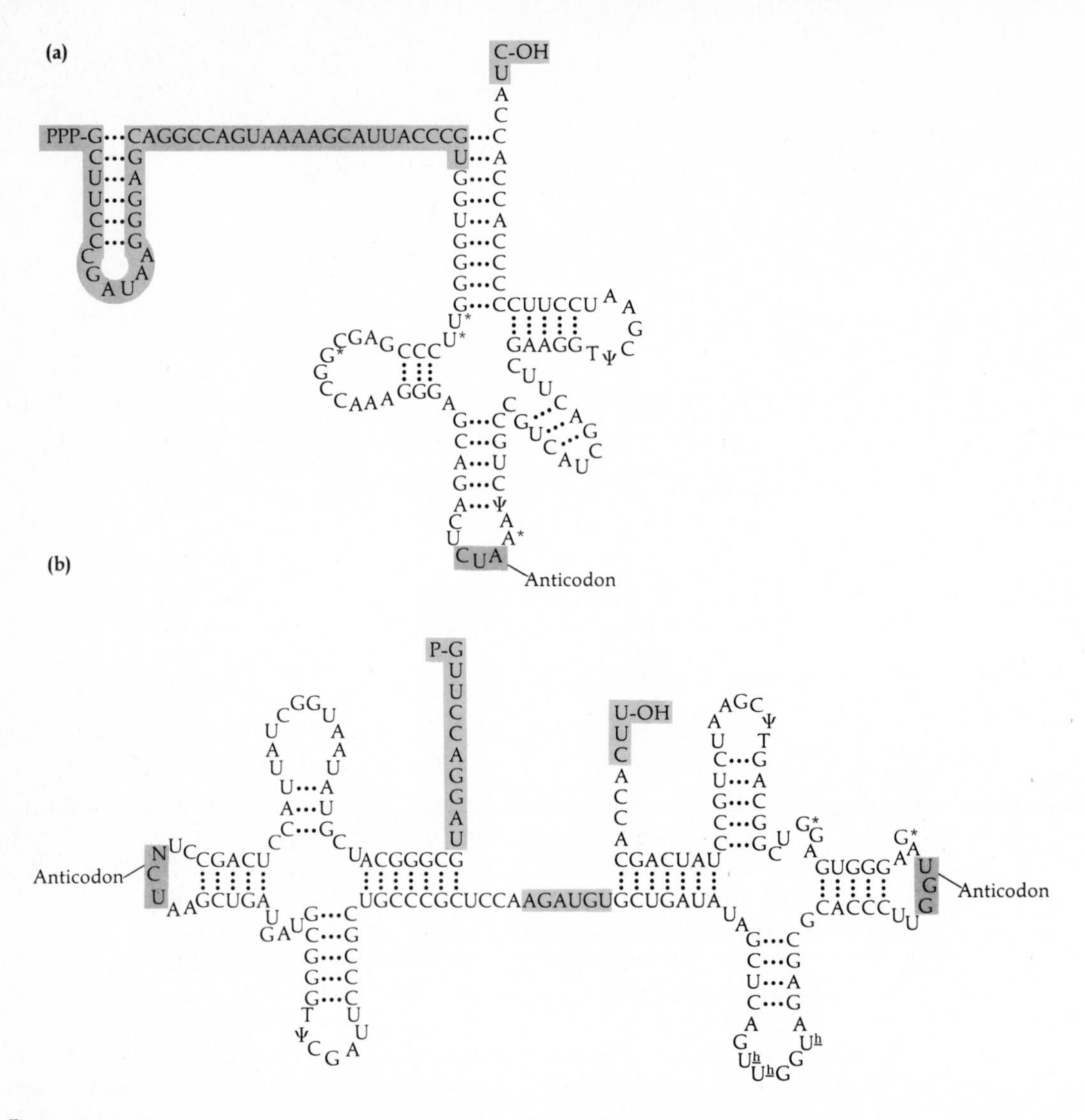

Figure 11.25
(**a**) The nucleotide sequence of *E. coli* pre-tRNA^{Tyr}, showing the residues (in color) removed during processing. (**b**) The nucleotide sequence of the *E. coli* tRNA^{Ser}-tRNA^{Thr} precursor, showing the residues (in color) removed during processing. Note the absence of a 5′-triphosphate on this precursor RNA molecule. Bases marked with asterisks carry modifications; N = nucleotide. (After J. D. Watson, *Molecular Biology of the Gene*, 3rd ed., W. A. Benjamin, Menlo Park, Cal., 1976.)

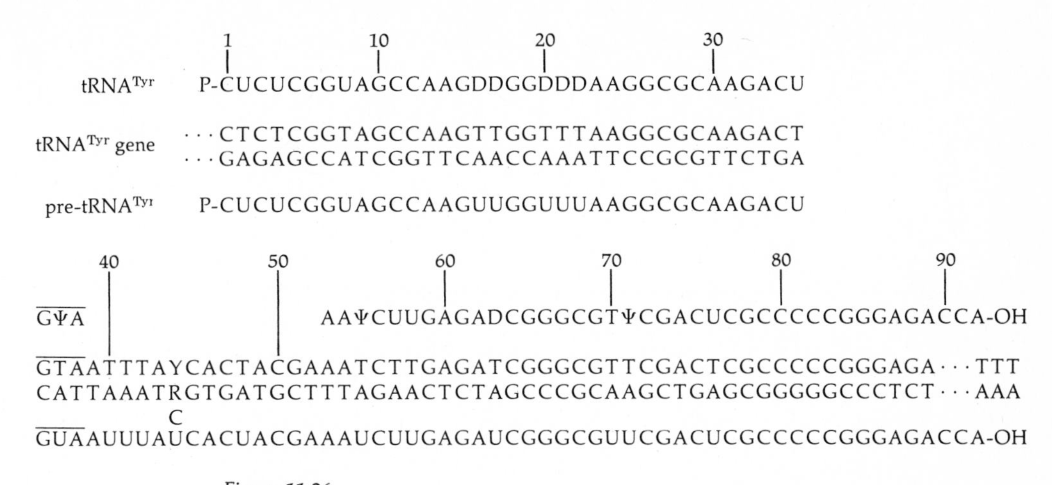

Figure 11.26
The nucleotide sequences of yeast $tRNA^{Tyr}$, the $tRNA^{Tyr}$ gene, and pre-$tRNA^{Tyr}$. Note the presence of an intervening 14-nucleotide sequence, adjacent to the anticodon (marked with an overbar), in both the gene and the pre-$tRNA^{Tyr}$, and the absence of genetic coding for the CCA-3′ end of the tRNA. D = dihydrouridine; Y = pyrimidine; R = purine. [After P. Z. O'Farrell et al., *Nature 274*:438 (1978).]

ribosomal RNAs. The 5S rRNA of eukaryotes is transcribed separately from its genes, which may not be linked to the rRNA genes. In *Drosophila* some of the genes for the 28S rRNA contain an intervening sequence that may be removed from the transcript during processing, perhaps by the same type of mechanism that removes intervening sequences from hnRNA.

Eukaryotic tRNA genes may also possess intervening sequences. The yeast $tRNA^{Tyr}$ gene has been cloned and sequenced, and the sequence of the pre-$tRNA^{Tyr}$ transcript and mature $tRNA^{Tyr}$ are shown in Figure 11.26. This eukaryotic $tRNA^{Tyr}$ gene carries an intervening sequence of 14 nucleotides adjacent to the anticodon sequence, which is also present in the pre-$tRNA^{Tyr}$ transcript. In addition, the $tRNA^{Tyr}$ gene lacks nucleotides coding for the 3′-terminal $-pCpCpA_{OH}$, which is added post-transcriptionally, as is true for most tRNAs. Maturation of the pre-$tRNA^{Tyr}$ molecule involves excision of the intervening 14 nucleotides and covalent joining of the cut ends.

The Central Dogma

In summary, the Central Dogma outlined by Crick provides a concise view of the relationships between the informational polymers found in biological systems. The hereditary information, encoded in the DNA, flows to RNA molecules and then on through a translation step to protein molecules.

This general pattern of information transfer, which occurs in all cells, is elaborated on or reversed under the special circumstances of infections by RNA viruses. During the infection, information may pass from parental RNA molecules to progeny RNA molecules or from RNA to DNA molecules.

The hereditary information flows from nucleotide sequences into amino acid sequences. This transfer, which involves a translation step, does not appear to be reversible. Protein molecules represent a "sink" in the scheme of information flow. The evolution of this system must have occurred very early in the history of life on earth. What the steps in this evolution may have been provides fascinating food for thought. The testing of any hypotheses, however, presents formidable challenges.

Problems

1. Purified *E. coli* RNA transcriptase, when incubated with pure double-stranded DNA and nucleoside triphosphates under appropriate conditions, synthesizes RNA. The amount of RNA synthesized is measured by including ^{3}H-labeled nucleoside triphosphates (the ^{3}H label is in the base) and determining the amount of ^{3}H incorporated into RNA (the RNA is precipitated with acid to remove its ^{3}H from that of unincorporated nucleoside triphosphates). The data in Table 11.2 show the relative amounts of γ-^{32}P-nucleoside triphosphates incorporated into RNA in equivalent reactions. In each reaction, only one of the four nucleoside triphosphates is labeled with ^{32}P in the γ-phosphate position. Measurement of the amount of ^{32}P and ^{3}H incorporated into RNA in each reaction can be made by counting the RNA precipitates in a scintillation spectrometer, which permits ^{32}P and ^{3}H to be counted separately, based on the difference in the energies of their radioactive emissions. What do you conclude from these data?

2. Recall that the ϕX174 DNA purified from phage is single-stranded and that, upon infection, the phage DNA is converted to a double-stranded replicating form (RF). During ϕX174 infection, RNA that is synthesized is labeled by incubating the cells with [^{3}H]uridine and is then purified from the infected cells. Incubation of the [^{3}H]RNA with ϕX single-stranded phage DNA or denatured ϕX RF DNA leads to the formation of RNA-DNA hybrid molecules between complementary RNA and DNA sequences. The amount of [^{3}H]RNA found in hybrid molecules is indicated in Table 11.3. What do you conclude from the data?

Table 11.2
Incorporation of γ-^{32}P-nucleoside triphosphates into RNA with different DNA preparations.

DNA	RNA Synthesis (pmoles)*	γ-^{32}P-Nucleotide Incorporated (pmoles)			
		ATP	GTP	UTP	CTP
T2	4800	2.4	1.2	0.12	0.10
T5	4000	1.80	1.4	0.41	0.23
SP3	5480	1.25	1.0	0.39	0.12
Clostridium perfringens	2800	1.60	2.1	0.28	0.25
Escherichia coli	2660	0.43	1.4	0.13	0.10
Micrococcus lysodeikticus	2560	0.36	2.5	0.10	0.12
Calf thymus	3560	0.77	1.3	0.33	0.18
dAT copolymer	rAU = 7200	4.40	–	0.20	–
dGdC homopolymer	rG = 1350	–	4.8	–	0.30
	rC = 120				

*1 pmole = 10^{-12} mole.
From U. Maitra and J. Hurwitz, *Proc. Natl. Acad. Sci. USA* 54:815 (1965).

Table 11.3
Hybridization of [^{3}H]RNA with RF DNA and single-stranded DNA.

RNA/DNA Pulse-Labeling Range (min)	RNA/DNA Hybrid Formed (counts/min of ^{3}H)	
	RF DNA	Single-stranded DNA*
5–6.5	253	<60
35–36.5	3261	<160
50–51.5	3473	<160

*These values represent upper limits.
From M. Hayashi et al., *Proc. Natl. Acad. Sci. USA* 50:664 (1963).

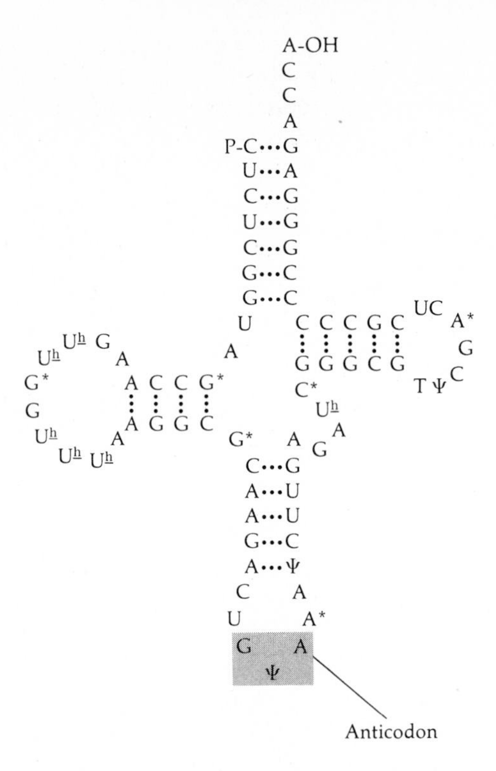

Figure 11.27
The structure of yeast tRNA^{Tyr}. See problem 6. (Bases marked with asterisks carry modifications.)

3. Mammalian cell lines grown in culture are amenable to genetic studies employing techniques used for genetic studies of microorganisms. For example, temperature-sensitive lethal mutations can be selected that will grow at 34°C but not at 40°C. Mutagenesis of a Chinese hamster ovarian cell line has produced a number of mutant strains with the temperature-sensitive phenotype. Some of these mutant cell strains do not grow at 40°C because they cannot synthesize proteins. This defect is also exhibited *in vitro* at 40°C. However, the mutant cells will grow at 40°C if supplied with a 10–100-fold *excess* concentration of a particular amino acid (normal medium always contains each of the 20 amino acids). Each particular mutant strain is rescued by an excess of only one or another of the 20 amino acids. From your knowledge of the details of protein synthesis and protein function, answer the following questions:

(a) What is the probable function of the genes in which these mutations have occurred?

(b) In one sentence, state how you would confirm your prediction.

Table 11.4
Reconstitution of RNA transcriptase *in vitro* from subunits.

Subunits				[^{3}H]UTP Incorporated into RNA at Restrictive Temperature (nmoles)*
Wild type	XH56	R120	A2R7	
$\alpha_2\beta\beta'\sigma$	–	–	–	3.5
σ	$\alpha_2\beta\beta'$	–	–	0.2
$\beta'\sigma$	$\alpha_2\beta$	–	–	4.3
$\beta\sigma$	$\alpha_2\beta'$	–	–	0.17
σ	–	$\alpha_2\beta\beta'$	–	0.09
$\beta'\sigma$	–	$\alpha_2\beta$	–	2.5
$\beta\sigma$	–	$\alpha_2\beta'$	–	0.06
σ	–	–	$\alpha_2\beta\beta'$	0.47
$\beta'\sigma$	–	–	$\alpha_2\beta$	0.52
$\beta\sigma$	–	–	$\alpha_2\beta'$	1.40

*1 nmole = 10^{-9} mole.

Data modified from J. Miller et al., in *RNA Polymerase,* ed. by R. Losick and M. Chamberlin, Cold Spring Harbor Laboratory, Cold Spring Harbor, N.Y., 1976, p. 519.

4. When intervening sequences, not present in mRNA, were discovered to be present in a number of eukaryotic genes, it was proposed that during mRNA synthesis the RNA transcriptase jumps over the intervening DNA sequences during transcription. What arguments for or against this hypothesis can you muster?

5. What explanation can you supply for the absence of a 5′-triphosphate in the *E. coli* tRNA^{Ser}-tRNA^{Thr} precursor in Figure 11.25?

6. The temperature-sensitive lethal mutation *ts136* of yeast is observed to accumulate a number of precursor tRNAs at the restrictive temperature. This mutant provided the precursor tRNA^{Tyr}, sequenced in Figure 11.26. The secondary structure of mature yeast tRNA^{Tyr} is shown in Figure 11.27.

Table 11.5
Mixed reconstitution of RNA polymerase from separated subunits of rifampicin-sensitive and rifampicin-resistant enzyme.

Original Enzymes, and Subunits Used in Reconstitution*	Relative Polymerase Activity		% Resistant Activity
	(−) Rifampicin (100 μg/ml)	(+) Rifampicin (100 μg/ml)	
Original sensitive enzyme	242	1.5	0.6
Original resistant enzyme	124	120	97
$\alpha + \beta + \beta' + \sigma$	52	1.4	2.0
$\alpha_r + \beta_r + \beta'_r + \sigma$	27	25.6	95
$\alpha_r + \beta + \beta' + \sigma$	40	0.6	1.5
$\alpha + \beta_r + \beta' + \sigma$	88	69	78
$\alpha + \beta + \beta'_r + \sigma$	17.5	1.4	8.0

*The subscript r (rifampicin) represents subunits from the resistant strain.

From A. Heil and W. Zillig, *FEBS Lett.* *11*:165 (1970).

The single-strand-specific endonuclease S_1 digests the non-hydrogen-bonded nucleotides present in this RNA, including the anticodon. In contrast, the anticodon of the pre-tRNA^{Tyr} molecule is not digested by S_1, while the majority of the intervening sequence is. The pre-tRNA^{Tyr} cannot be aminoacylated *in vitro,* in contrast to mature tRNA^{Tyr}. Draw a secondary structure for pre-tRNA^{Tyr} consistent with these observations. Consult Figures 11.26 and 11.27.

7. The genes for the β and β' subunits of *E. coli* RNA transcriptase map very close to each other and are transcribed into a common polycistronic mRNA. A number of temperature-sensitive lethal mutations that map to the $\beta\beta'$ region contain RNA transcriptase that is defective at the restrictive temperature after purification. The subunits of purified transcriptase can be separated and recombined to reconstitute the active enzyme. Reconstitution experiments were carried out to determine which subunit of the RNA transcriptase is defective in each of three mutant strains (XH56,

Table 11.6
Mixed reconstitution of RNA polymerase from separated subunits of streptolydigin-sensitive and streptolydigin-resistant enzyme.

Subunits used in Reconstitution*	Relative Polymerase Activity (−) Streptolydigin (1 mM)	Relative Polymerase Activity (+) Streptolydigin (1 mM)	% Resistant Activity
$\alpha+\beta+\beta'+\sigma$	84	1.8	2.1
$\alpha_s+\beta_s+\beta'_s+\sigma$	123	35.2	29.0
$\alpha_s+\beta+\beta'+\sigma$	36	1.1	3.0
$\alpha+\beta_s+\beta'+\sigma$	86	26	30.2
$\alpha+\beta+\beta'_s+\sigma$	113	4.2	3.6
$\alpha+\beta_s+\beta'_s+\sigma$	132	38	29.0
$\alpha_s+\beta+\beta'_s+\sigma$	75	1.2	1.6
$\alpha_s+\beta_s+\beta'+\sigma$	31	11.5	36.6

*The subscript s (streptolydigin) represents subunits from the resistant strain.

From A. Heil and W. Zillig, *FEBS Lett.* *11*:165 (1970).

R120, and A2R7). From the data in Table 11.4, state which subunit is defective in each strain.

8. The antibiotics rifampicin and streptolydigin exert their killing effect by inhibiting RNA synthesis. *In vitro* studies with purified RNA transcriptase show that rifampicin blocks initiation of RNA chains, whereas streptolydigin blocks translocation of the enzyme along the DNA, interrupting the synthesis of already initiated RNA chains. Mutant strains of *E. coli* resistant to one or the other antibiotic yield purified RNA transcriptase that is resistant to the same antibiotic *in vitro*. Reconstitution experiments, similar to those described in problem 7, yield the data in Tables 11.5 and 11.6. How do these antibiotics affect RNA synthesis?

9. The nucleolus organizers of *Drosophila melanogaster* are found in the heterochromatic portion of both the X and Y chromosomes. Chromosomal

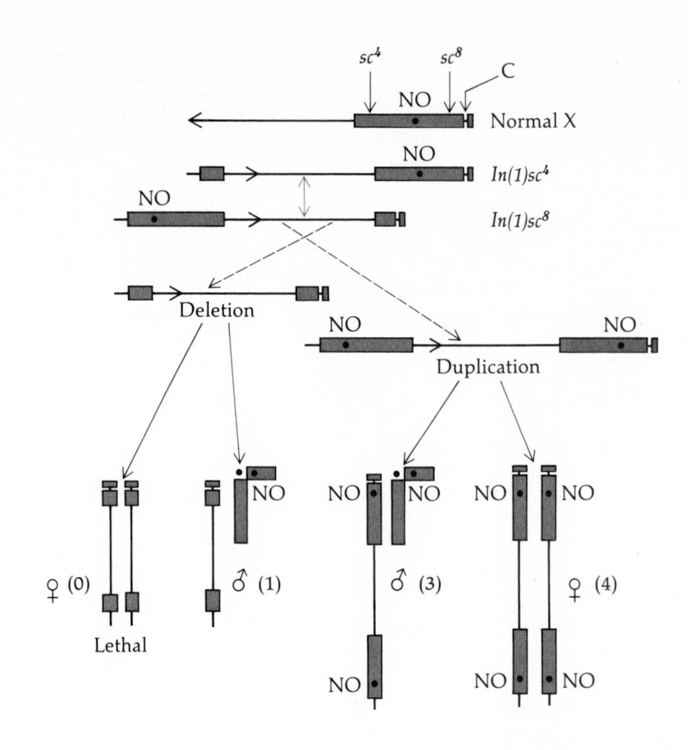

Figure 11.28
Origin and construction of X chromosomes described in problem 9. Symbols identify the nucleolus organizer (NO), the centromere (C), and the inversions *scute-4* (sc^4) and *scute-8* (sc^8). A cross between the two inversions yields the desired deletion and duplication, from which the various genotypes can be developed. [After F. Ritossa and S. Spiegelman, *Proc. Natl. Acad. Sci. USA* 53:743 (1965).]

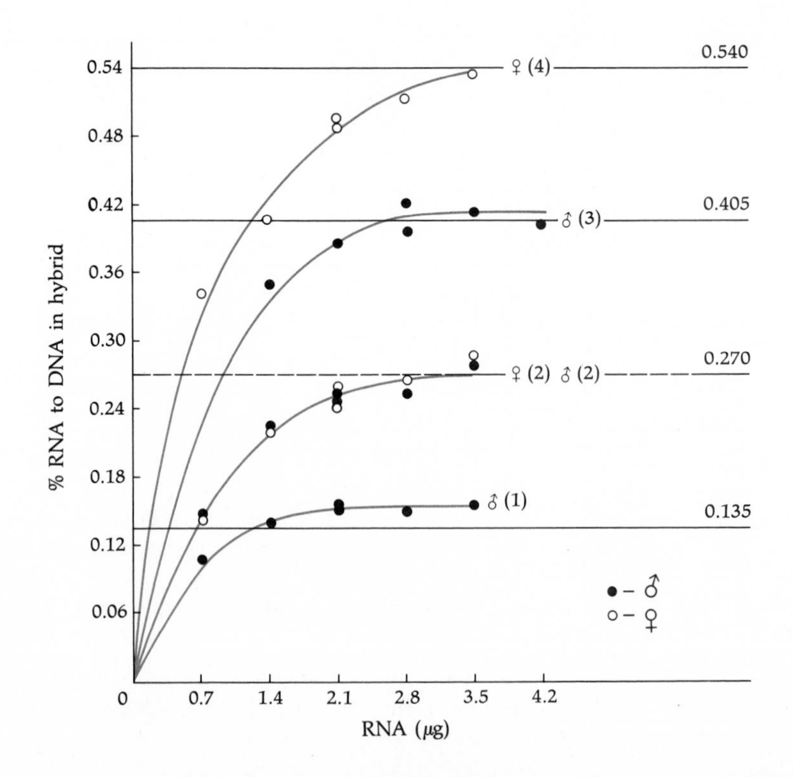

Figure 11.29
Saturation levels of DNA containing various doses of nucleolus organizer (NO) region. The doses of NO are indicated by the numbers in parentheses. The dashed horizontal line at 0.270 is assumed to be a correct estimate for a dose of 2, and the solid horizontal lines represent the predicted plateaus for doses of 1, 3, and 4, respectively. The numerical values of the plateaus are shown on the right. [After F. Ritossa and S. Spiegelman, *Proc. Natl. Acad. Sci. USA* 53:743 (1965).]

engineering, using inversions and the selection of recombinant chromosomes, as diagrammed in Figure 11.28, produces X chromosomes that are duplicated and deficient for the nucleolus organizer. These chromosomes can be used to produce flies carrying 1, 2, 3, or 4 doses of the nucleolus organizer. Quantitative hybridization experiments using purified radioactive rRNA and measured amounts of DNA from flies with 1, 2, 3, or 4 doses of the nucleolus organizer yield the data shown in Figure 11.29. How would you interpret these observations?

12

The Genetic Code

The genetic code was mentioned frequently in Chapter 11 because its nature is important to an understanding of protein synthesis. Our current knowledge of the genetic code developed hand-in-hand with detailed knowledge of the processes of transcription and mutation. These subjects, dealt with separately, are firmly entwined. We have seen that the genetic code consists of nucleotide triplets read in a nonoverlapping sequence along the mRNA molecule from the 5′ end to the 3′ end. Translation starts at a specific initiation codon, AUG, setting the frame in which succeeding codons are read, and proceeds until a termination codon is read. In this process a polypeptide of unique amino acid sequence is assembled sequentially from the NH_2 end to the COOH end. The most striking single fact about the genetic code is its universality: the codons in chicken ovalbumin mRNA, which specify the amino acid sequence of chicken ovalbumin (Figure 11.20), are identical in meaning to those that specify the amino acid sequences of proteins in *E. coli* and in ϕX174, in tobacco plants and in humans.*

Once the Watson-Crick model for DNA had been accepted, along with its corollary that the sequence of nucleotide pairs along the DNA encodes the sequence of amino acids in protein, the exact nature of the code became the focus of effort of many molecular biologists. Since 20 different amino acids must be specified, it was clear that each codon must

---

*Recent discoveries require some modification of this assertion. See page 426 for details.

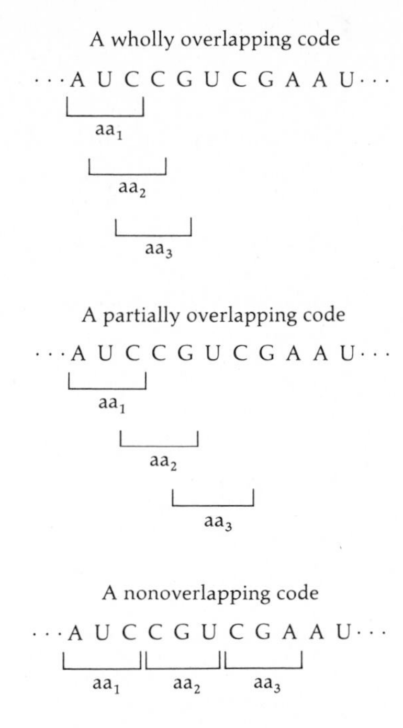

Figure 12.1
Three ways in which a nucleotide triplet code might be read to specify the sequence of three adjacent amino acids (aa) in a polypeptide.

consist of at least three nucleotides, because doublets of nucleotides can specify at most 16 (4^2) different codons. Triplets of nucleotides, however, can specify 64 (4^3) different codons.

Several types of triplet codes are possible. A code may be *overlapping* (Figure 12.1). Overlapping codes have the property that point mutations, which change single nucleotide pairs, should change two or three adjacent amino acids in the mutant protein. Sequence analysis of mutant proteins demonstrates, however, that virtually all of them have only one altered amino acid for each mutation. Overlapping codes also place restrictions upon which amino acids can be adjacent in a protein. Sequence analysis of proteins demonstrates, however, that any amino acid can be adjacent to any other amino acid. These considerations rule out overlapping codes and prove that the genetic code is *nonoverlapping*.

A nonoverlapping triplet code may be of two types. Since only 20 codons are required to specify 20 amino acids, and 64 different codons are possible in a triplet code, the remaining 44 codons may be nonsense, coding for nothing at all; such a code is called a *nondegenerate* code, i.e., each amino acid is specified by a single, unique codon. On the other hand, most or all of the 64 possible codons of the genetic code may specify amino acids; such a code is called a *degenerate* code, i.e., each amino acid may be specified by one or more codons. (The term *degenerate* in this context was borrowed from quantum mechanics by the converted physicists who moved into the new field of molecular biology.) The deduction of the actual nature of the genetic code, followed by the determination of the meaning of each codon, is one of the outstanding achievements of science. This feat was accomplished in only a few years, in a series of fascinating and quite marvelous genetic and biochemical experiments, which will be described in this chapter. Elucidation of the genetic code (Table 12.1), which was accomplished in 1966, confirmed the sequence hypothesis and provided a firm foundation for the Central Dogma, whose tenets were introduced in Chapter 11.

Genetic Evidence for the Nature of the Code

Much of the genetic evidence for the nature of the code was provided by studies of certain T4 *rII* mutations presented by Francis Crick and his colleagues in 1961. They studied *rII* mutations that had been induced by the mutagen proflavin. Proflavin-induced mutations were of interest because they were believed to be the result of an alteration to DNA that does *not* involve a nucleotide-pair substitution.

The reason for this belief was based on the properties of the mutations induced by proflavin, as well as by those induced by other types of mutagens. The base-analogue mutagens 2-aminopurine and 5-bromouracil induce mutations when they are incorporated into DNA in place of a normal nucleotide. The mutations induced by these two compounds were believed (correctly) to be the result of base substitutions. Mutations induced by these compounds were known to be reverted to wild type by the same compounds, and this was taken as evidence that the mutations were due to base substitutions (see Chapter 16). Virtually no proflavin-induced mutations, however, are reverted to wild type by base-analogue mutagens (Table 12.2). Proflavin-induced mutations do revert spontaneously to wild type and can be induced to revert at high frequency by proflavin itself. The ability to revert to wild type suggests that proflavin-induced mutations are point mutations rather than deletions, as discussed in Chapter 8. Crick guessed, correctly, as it turned out (see Chapter 16), that proflavin induces either the *insertion* or *deletion* of a single nucleotide pair in the DNA. The following experiments supported Crick's hypothesis and led to important deductions concerning the nature of the genetic code.

We shall follow the story of one proflavin-induced *rII* mutation, studied by Crick and designated *FC0,* that maps in the *B* cistron near the boundary with the *A* cistron (Figure 8.3). Growth of *FC0* in the permissive host *E. coli* B results in the spontaneous appearance, at a low frequency, of

Table 12.1
The genetic code.*

	Second Position: U		C		A		G		
First Position									**Third Position**
U	UUU	Phe	UCU	Ser	UAU	Tyr	UGU	Cys	U
	UUC		UCC		UAC		UGC		C
	UUA	Leu	UCA		UAA	*Ochre*	UGA	*Opal*	A
	UUG		UCG		UAG	*Amber*	UGG	Trp	G
C	CUU	Leu	CCU	Pro	CAU	His	CGU	Arg	U
	CUC		CCC		CAC		CGC		C
	CUA		CCA		CAA	Gln	CGA		A
	CUG		CCG		CAG		CGG		G
A	AUU	Ile	ACU	Thr	AAU	Asn	AGU	Ser	U
	AUC		ACC		AAC		AGC		C
	AUA		ACA		AAA	Lys	AGA	Arg	A
	AUG	Met	ACG		AAG		AGG		G
G	GUU	Val	GCU	Ala	GAU	Asp	GGU	Gly	U
	GUC		GCC		GAC		GGC		C
	GUA		GCA		GAA	Glu	GGA		A
	GUG		GCG		GAG		GGG		G

*The first position of a codon (5′ end) is given in the column on the left. See Figure 10.13 for amino acid abbreviations.

wild-type revertants that are detected by their ability to form plaques on the restrictive host *E. coli* K(λ) ($rIIFC0 \rightarrow rII^{"+"}$). Most of these "revertants," however, were found not to be true revertants, in the following way. Simultaneous infection of permissive cells with a "revertant" phage and a wild-type phage (cross 1, below) yields mutant *rII* progeny at a frequency significantly higher than the spontaneous mutation rate to *rII*:

Cross 1
Parents: $rII^{"+"} \times rII^{+}$
Progeny: rII^{+} and rII^{α} and rII^{β}

Table 12.2
Induced reverse mutation of a set of spontaneously reverting T4 *rII* mutants.

Mutagen Used for Inducing *rII* Mutation	Number of *rII* Mutants Tested	% of *rII* Mutants Found to Revert in Presence of 2-Aminopurine or 5-Bromouracil
2-Aminopurine	98	98
5-Bromouracil	64	95
Hydroxylamine	36	94
Nitrous acid	47	87
Ethyl ethane sulfonate	47	70
Proflavin	55	2
Spontaneous	110	14

After E. Freese, 5th International Congress of Biochemistry, Moscow, 1961.

A study of the *rII* progeny arising in cross 1 shows them to be of two types, designated here as α and β. Cross 2, below, demonstrates that α is the original *FC0* mutation, because α and *FC0* do not recombine to yield wild-type progeny. By contrast with cross 2, cross 3 shows that rII^{β} is a new *rII* mutant, which we shall now call *FC1*.

Cross 2: $rII^{\alpha} \times rIIFC0$
Result: No rII^{+} recombinants produced
Cross 3: $rII^{\beta} \times rIIFC0$
Result: rII^{+} recombinants produced

The original "revertant," $rII^{''+''}$, is shown by these crosses not to be a true revertant, but instead a double *rII* mutant, *FC0 FC1*, which exhibits a wild phenotype by making plaques on *E. coli* K(λ). *FC1* maps in the *B* cistron very close to *FC0*; these two mutations are mutual *intragenic suppressors* of each other.

The behavior of the new *rII* mutation, *FC1*, was studied in an identical manner. Growth of *FC1* in a permissive host results in the spontaneous appearance of "revertants" that are also found to be double *rII* mutants, by the procedure outlined in crosses 1–3. The "revertant" of *FC1* is the double mutant *FC1 FC2*, where *FC2* is the newly arisen *rII* mutant and is a mutual

Table 12.3
Phenotypes of T4 mutants carrying two *rII* mutations.

Double Mutant	Phenotype	Double Mutant	Phenotype
FC2 FC3	+	*FC4 FC5*	+
FC1 FC3	*rII*	*FC3 FC5*	*rII*
FC0 FC3	+	*FC2 FC5*	+
FC3 FC4	+	*FC1 FC5*	*rII*
FC2 FC4	*rII*	*FC0 FC5*	+
FC1 FC4	+		
FC0 FC4	*rII*		

suppressor with *FC1*. Proceeding in this iterative manner, Crick generated a series of *rII* mutants: *FC3, FC4, FC5,* etc. Each of these mutations, *FC(n)*, in isolation is phenotypically an *rII* mutant, and each *FC(n)* mutation is a mutual intragenic suppressor of *FC(n* − 1) and *FC(n* +1). *FC(n* − 1) is the *rII* mutant that "reverted" by the spontaneous appearance of *FC(n)*, and *FC(n* + 1) is the *rII* mutant that caused the "reversion" of the isolated *FC(n)* mutant. Recombination between different mutations in the *FC* series permits different pairs of doubly mutant chromosomes to be produced. In general, constructed double mutants of the series exhibit the phenotypes shown in Table 12.3. Note that pairwise combinations of odd- *and* even-numbered mutations are phenotypically wild type (mutual suppressors), whereas pairwise combinations of odd- *or* even-numbered mutations have a mutant phenotype (no mutual suppression).

The latter observation—that odd- or even-numbered mutations are not mutual intragenic suppressors (e.g., *FC1 FC3*) but that each is a mutual intragenic suppressor of the other type (e.g., *FC2 FC3*)—is important in view of the phenotype exhibited by constructed triple mutants. The combination of three odd-numbered *or* three even-numbered mutations exhibits a wild-type phenotype, e.g., *FC0 FC2 FC4* and *FC1 FC3 FC5*. That is, three mutations, each of which does *not* exhibit mutual suppression in pairwise combination with the other two, *do* show mutual suppression in triple combination. In contrast, triple mutants of two even-numbered and one odd-numbered types or one even-numbered and two odd-numbered types exhibit the mutant phenotype.

Crick made the following postulates to account for these observations:

1. The *FC0* mutation was either a deletion or insertion of one nucleotide pair. An intragenic suppressor of *FC0* is an insertion of one nucleotide pair (if *FC0* is a deletion) or a deletion of one nucleotide pair (if *FC0* is an insertion).
2. The reading of the genetic code begins at a fixed point in the gene and proceeds sequentially with no punctuation between codons. The insertion or deletion of a single nucleotide pair consequently shifts the frame in which succeeding nucleotides are read as codons. Thus, proflavin is often called a *frameshift* mutagen. (When this work was done, mRNA had not yet been discovered, nor its role in protein synthesis predicted.)
3. The bases are read three at a time to construct a polypeptide; codons are triplets.
4. All or most of the 64 possible triplets code for some amino acid. Thus the code is degenerate (more than one triplet can code for a given amino acid).

The following example demonstrates how these postulates account for the properties demonstrated by *FC0* and its derivatives. Suppose the following hypothetical mRNA sequence codes for the region of the *B* cistron in which *FC0* and its suppressors map:

CAU CAU CAU CAU CAU CAU CAU CAU CAU CAU
rII+

The leftmost triplet sets the reading frame for the message, which then codes His–His–His · · · . Now suppose *FC0* is a deletion of the second A:

CAU CUC AUC AUC AUC AUC AUC AUC AUC AU
rIIFC0

Then after the first triplet, all subsequent codons to be read will specify a completely different polypeptide, if the new codons created by the frameshift are not nonsense, i.e., His–Leu–Ile–Ile · · · . Were the code not degenerate, the new codons would probably be nonsense, however. Now suppose *FC1* is an insertion of U at the fourth codon, restoring the reading frame to the right:

CAU <u>CUC AUC AUU</u> CAU CAU CAU CAU CAU CAU
rIIFC0FC1

The resulting polypeptide would differ from the wild type by three consecutive amino acids, i.e., His–Leu–Ile–Ile–His · · · . If this substitution occurs in a relatively nonfunctional portion of the polypeptide, as is true with this portion of the *B* cistron, it may retain a relatively normal function. Separation of *FC1* and *FC0* by recombination yields a messenger for *FC1*,

CAU CAU CAU CAU UCA UCA UCA UCA UCA UCA U
rIIFC1

that again specifies a completely different polypeptide, compared to wild type, throughout its whole length, i.e., His–His–His–His–Ser–Ser · · · . The combination of three single nucleotide deletions or three insertions, if they occur close enough to one another, would not disturb the reading frame for the entire message. For example, deletion of the second, third, and fourth A has the following result,

CAU CUC UCU CAU CAU CAU CAU CAU CAU · · ·

which may code for a functional polypeptide, i.e., His–Leu–Ser–His–His · · · .

The attempted deduction of the nature of the code from these genetic experiments provides an internally consistent model. Crick's four postulates accounted nicely for all the genetic data, but their independent confirmation seemed to be far in the future. To everyone's surprise, it took only five years to prove biochemically that all four postulates were correct and to reveal the entire code.

Genetic Evidence for Termination Codons

Study of the *rII* system has also provided genetic evidence for the existence of signals for the termination of polypeptide synthesis. Complementation analysis shows that cistron *A* and cistron *B* code for separate genetic functions. Therefore, some genetic punctuation must exist at the boundary between the two cistrons. Deletions that cross this boundary fuse the undeleted portions of *A* and *B* into one cistron. Deletion *1589* (Figure 12.2) is an *rIIA* mutant that possesses a functional *B* cistron coding for an active B protein in spite of the partial deletion of the *B* cistron. The portion of the *B* cistron missing in *1589* contains the region in which *FC0* and its derivatives map, substantiating the conclusion reached above that this portion of the B polypeptide is not essential for its activity.

The introduction of frameshift mutations into the inactive *A* portion of the fused cistrons of *1589* destroys the active *B* function present in the chromosome. Therefore the direction of translation of the fused mRNA must be $A \rightarrow B$, and a single fused polypeptide must be synthesized. Base analogues, which do not induce frameshift mutations, can also induce mutations in the *A* portion of the fused cistrons that destroy *B* function. Some of these mutations, which must be base-substitution mutations, are of the *amber* class of suppressible conditional lethals, discussed in Chapter 6. A possible explanation for the effect on *B* function of an *amber* mutation

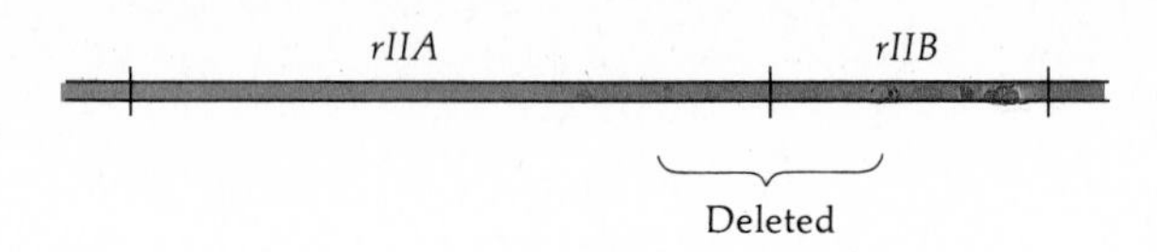

Figure 12.2
A deletion of the boundary between *rII* cistrons *A* and *B* removes sequences necessary for the independent expression of these two genetic functions, creating a single fused *A-B* cistron. The general region missing in deletion *1589* is indicated.

in *A* is that the mutation has created a termination codon that prevents the translation of the *B* portion of the fused mRNA.

Direct evidence that *amber* mutations cause premature termination of protein synthesis comes from study of the synthesis of T4 head protein. Ten *amber* mutations, mapping throughout the gene for the head protein, have been examined in infection of the restrictive host to determine how much of the head-protein mRNA is translated. The length of the head polypeptide synthesized in each abortive infection is directly correlated with the map position of each mutant in the gene (Figure 12.3). This result shows that *amber* mutations do cause termination; moreover, it demonstrates again the colinearity of gene and polypeptide.

Biochemical Elucidation of the Code

The first major biochemical contribution to breaking the genetic code came from the development of an *in vitro* cell-free protein-synthesizing system from *E. coli,* which was dependent on added mRNA for activity. Polypeptide synthesis is detected and measured by incorporation of radioactive amino acids into peptides. Usually, in a given experiment, only one amino acid is radioactive and the other 19 are not, so that only the ability of the system to utilize the radioactive amino acid is measured.

During studies of the activities of various RNA preparations as messengers in a cell-free protein-synthesizing system, Marshall W. Nirenberg and J. Heinrich Matthaei employed synthetic polyuridylic acid (poly-U) as a control, expecting it to exhibit little or no messenger activity. To their astonishment, they found that poly-U directed the synthesis of polyphenylalanine. Moreover, *only* polyphenylalanine was synthesized. This immediately suggested that the triplet UUU is a codon for phenylalanine. Poly-C was soon found to direct the synthesis of polyproline, and poly-A to direct that of polylysine, suggesting CCC = proline and AAA = lysine.

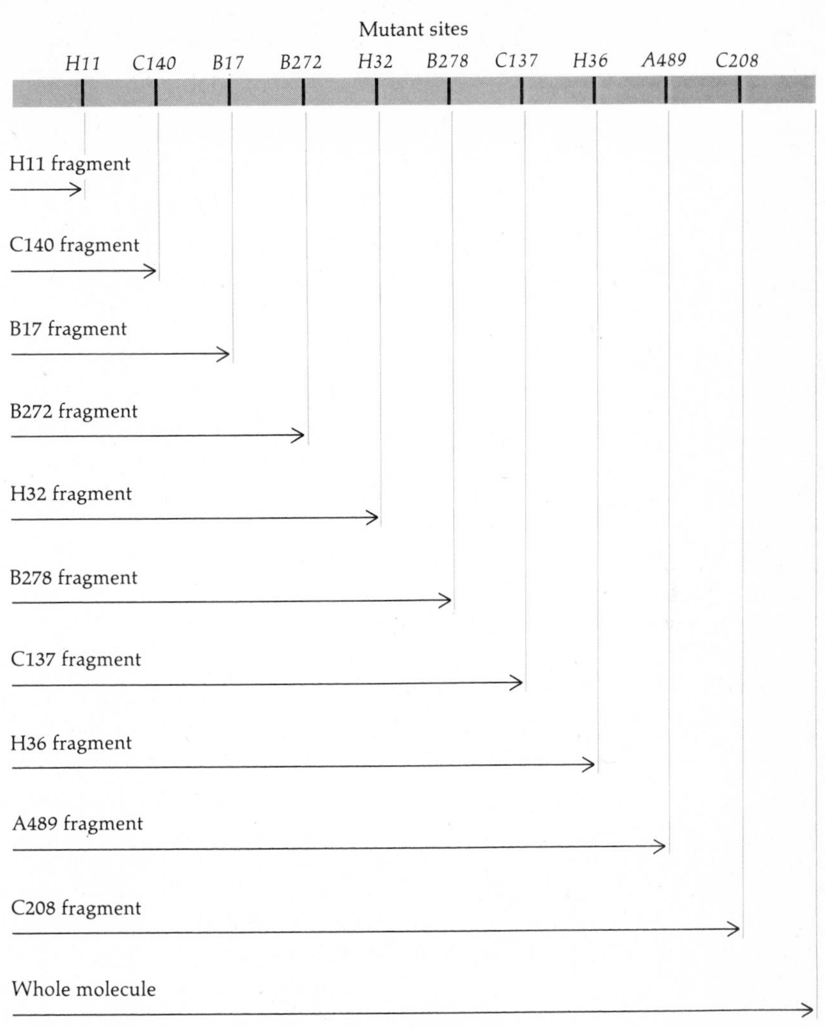

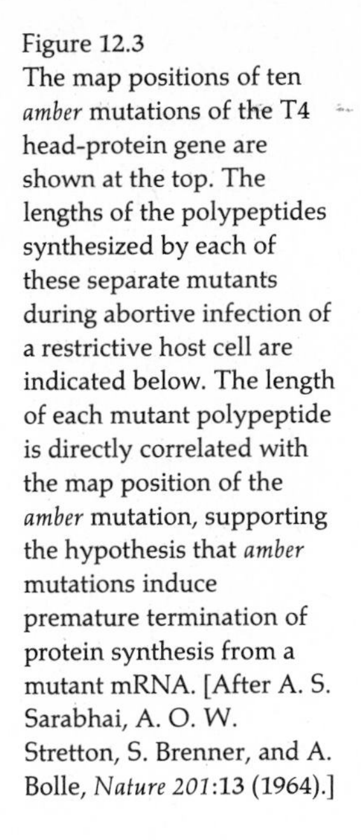
Figure 12.3
The map positions of ten *amber* mutations of the T4 head-protein gene are shown at the top. The lengths of the polypeptides synthesized by each of these separate mutants during abortive infection of a restrictive host cell are indicated below. The length of each mutant polypeptide is directly correlated with the map position of the *amber* mutation, supporting the hypothesis that *amber* mutations induce premature termination of protein synthesis from a mutant mRNA. [After A. S. Sarabhai, A. O. W. Stretton, S. Brenner, and A. Bolle, *Nature 201*:13 (1964).]

Fortunately, the cell-free protein-synthesizing systems used for these experiments employed a magnesium concentration so high that specific initiation of protein synthesis at the initiation codon AUG was not required (see Chapter 11), and abnormal initiation occurred on these synthetic messengers. This lucky accident permitted the sequence of the first codons to be determined.

Next, studies focused on the coding properties of random RNA copolymers in cell-free systems. Random copolymers are synthesized *in vitro* by the enzyme polynucleotide phosphorylase, employing 5′-ribonucleoside diphosphates as substrate. No primer or template strands are required by this polymerase, since it synthesizes polymers randomly by

Table 12.4
Calculated frequencies of codons present in a random copolymer of composition A:C = 5:1.

Codon Composition	Calculated Frequency	Normalized Relative Frequency
3A	$\left(\frac{5}{6}\right)^3 = \frac{125}{216}$	100
2A1C	$\left(\frac{5}{6}\right)^2\left(\frac{1}{6}\right) \times 3 = \frac{75}{216}$	60
1A2C	$\left(\frac{5}{6}\right)\left(\frac{1}{6}\right)^2 \times 3 = \frac{15}{216}$	12
3C	$\left(\frac{1}{6}\right)^3 = \frac{1}{216}$	0.8

addition to 3′ ends, as determined by the availability of each particular 5′-ribonucleoside diphosphate. For example, if supplied with ADP and CDP in a ratio of 5:1, the resulting RNA copolymer will contain A and C in the ratio 5:1, and the order of bases will be random and proportional to their frequency.

The frequency of codons of different composition in a random copolymer of composition A:C = 5:1 can be calculated (Table 12.4). It can be seen that, for every 100 AAA codons, the copolymer contains 60 codons with 2A1C (AAC, ACA, CAA), and so on. Owing to the 5′-3′ polarity of RNA molecules, the codons AAC and CAA are not identical. By convention, the 5′ end is always written on the left, and the 3′ end is written on the right. The codon sequences written above are abbreviations of 5′-pApApC$_{OH}$-3′, and so on.

The relative incorporations of specific amino acids into polypeptides directed by an A:C = 5:1 copolymer are given in Table 12.5A. Only six amino acids are incorporated. The most frequent, lysine, we already know to be coded by AAA, and the least frequent, proline, we know to be coded by CCC. Employing the calculations in Table 12.4 as a guide, the best assignments of amino acids to triplets of a given composition permitted by the data are shown. The reciprocal experiment, employing a copolymer of composition A:C = 1:5, is shown in Table 12.5B. Such experiments permit only the assignment of a codon composition to particular amino acids; they give no information as to codon sequence.

One approach to the determination of codon sequence is to use synthetic messengers of known sequence. For example, pancreatic ribonuclease (an endonuclease specific for pyrimidine residues) will cleave a copolymer of composition A:C = 25:1 only on the 3′ side of C to produce

Table 12.5
Assignment of codon compositions to particular amino acids, based on relative incorporation of amino acids into polypeptides under the direction of AC copolymers.

Amino Acid	Relative Amino Acid Incorporation	Best Assignment of Triplets				Sum of Triplet Frequencies for Each Amino Acid
		3A	2A1C	1A2C	3C	
A A:C = 5:1						
Asparagine	24.2		20			20
Glutamine	23.7		20			20
Histidine	6.5			4.0		4.0
Lysine	100	100				100
Proline	7.2			4.0	0.8	4.8
Threonine	26.5		20	4.0		24.0
Normalized relative frequency:		100	60	12	0.8	
B A:C = 1:5						
Asparagine	5.3		3.3			3.3
Glutamine	5.2		3.3			3.3
Histidine	23.4			16.7		16.7
Lysine	1.0	0.7				0.7
Proline	100			16.7	83.3	100
Threonine	20.8		3.3	16.7		20
Normalized relative frequency:		0.8	12	60	100	

After J. F. Speyer et al., *Cold Spring Harbor Symp. Quant. Biol.* 28:559 (1963).

RNA molecules of the approximate structure $A_{25}C$. This messenger directs the synthesis of oligolysine with a carboxyl-terminal asparagine. This proves that AAC codes for asparagine; it also demonstrates that the messenger is read in the direction $5' \rightarrow 3'$, because if it were read in the opposite direction, oligolysine with amino-terminal glutamine (CAA) would be synthesized.

A more systematic approach to the determination of codon sequences was taken by the organic chemist Har Gobind Khorana, who developed techniques for chemically synthesizing messengers of known sequence, as

Table 12.6
Specificity of aminoacyl-tRNAs in trinucleotide-stimulated binding to *E. coli* ribosomes.*

	Aminoacyl-tRNA Bound to Ribosomes (pmoles)			
Codon	(−) Codon	(+) Codon	(−) Codon	(+) Codon
	[14C]Arginyl-tRNA I (39.4 pmoles)		[3H]Arginyl-tRNA II (19.3 pmoles)	
CGU	0.19	0.54	0.16	2.61
CGC	0.19	0.29	0.16	1.29
CGA	0.19	0.21	0.16	2.21
CGG	0.19	1.24	0.16	0.14
AGA	0.19	0.22	0.16	0.14
AGG	0.19	0.19	0.16	0.30
	[14C]Glycyl-tRNA I (32.1 pmoles)		[14C]Glycyl-tRNA II (47.1 pmoles)	
GGU	0.83	1.11	1.18	4.16
GGC	0.83	1.39	1.18	4.94
GGA	0.83	4.38	1.18	1.49
GGG	0.83	4.46	1.18	2.79
	[14C]Isoleucyl-tRNA II (7.0 pmoles)		[3H]Isoleucyl-tRNA I (5.3 pmoles)	
AUU	0.18	0.96	0.05	0.32
AUC	0.18	1.20	0.05	0.43
AUA	0.18	0.16	0.05	0.05
	[14C]Phenylalanyl-tRNA I (13.4 pmoles)		[14C]Phenylalanyl-tRNA II (11.9 pmoles)	
UUU	0.26	0.50	0.33	0.75
UUC	0.26	0.58	0.33	0.92

Table 12.6—*Continued*
Specificity of aminoacyl-tRNAs in trinucleotide-stimulated binding to *E. coli* ribosomes.*

	Aminoacyl-tRNA Bound to Ribosomes (pmoles)			
Codon	(−) Codon	(+) Codon	(−) Codon	(+) Codon
	[14C]Prolyl-tRNA I (3.2 pmoles)		[14C]Prolyl-tRNA II (5.4 pmoles)	
CCU	0.39	0.48	0.07	0.09
CCC	0.39	0.42	0.07	0.09
CCA	0.39	0.71	0.07	0.15
CCG	0.39	0.81	0.07	0.30

*Compare the trinucleotide-specific binding observed here with the codon assignments of Table 12.1. Aminoacyl-tRNA I and II represent chemically distinct forms of the molecules.

From D. Söll, J. Cherayil, and R. Bock, *J. Mol. Biol. 29*:97 (1967).

well as all 64 possible RNA triplets. These synthetic molecules provided the means to complete the determination of all the code words. The attainment of this goal was speeded by the discovery that some RNA triplets will cause their cognate aminoacyl-tRNAs to bind specifically to ribosomes. The binding reaction between a synthetic codon and the anticodon of a cognate aminoacyl-tRNA, mediated by ribosomes, mimics the recognition step in protein synthesis (Chapter 11). This technique permits many, but not all, codons to be assigned to specific amino acids (Table 12.6).

The combination of all of these techniques leads to the genetic code shown in Table 12.1, in which 61 of the 64 possible triplets code for amino acids. These codon assignments have also been confirmed by observed amino acid substitutions in mutant proteins, such as TMV coat protein and human hemoglobins, employing the assumption that a single mutation involves change in only one nucleotide in a triplet. Examples for mutant human β globin chains are seen in Figure 12.4. The final result demonstrates that the code is almost totally degenerate. Only two amino acids, methionine and tryptophan, are represented by a single codon. The one for methionine, AUG, signals polypeptide initiation. As we shall see, the remaining three triplets, UAA, UAG, and UGA, act as termination signals.

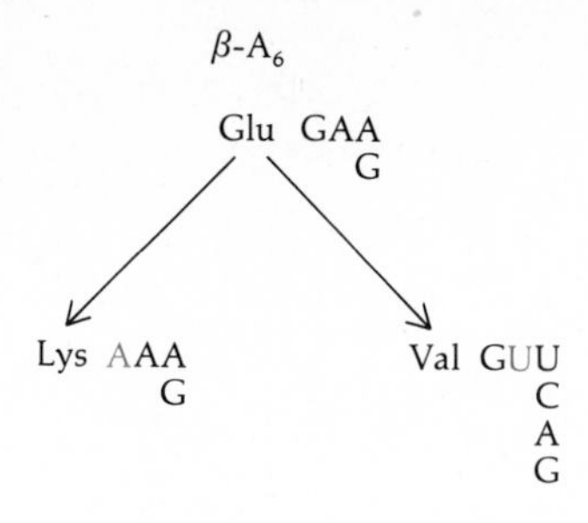

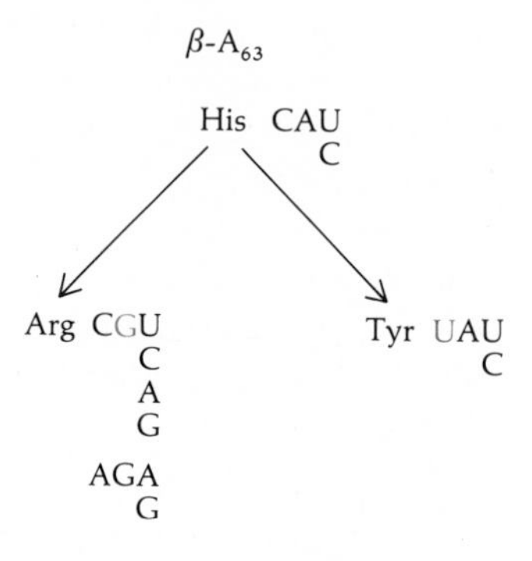

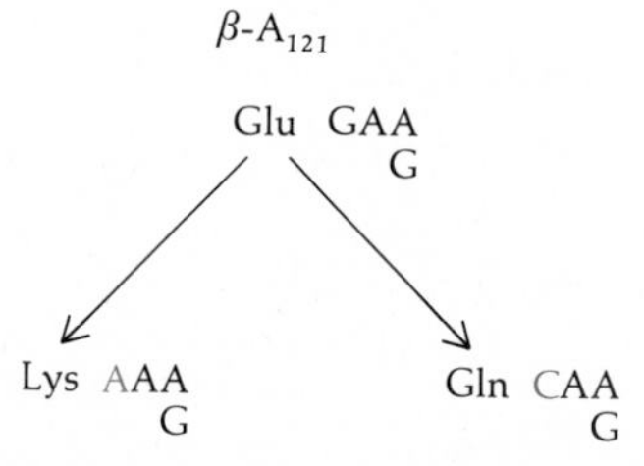

Figure 12.4
Amino acid replacements observed in six different mutant β chains of human hemoglobin A occurring at positions 6, 63, and 121. All possible codons for the wild-type and mutant amino acids are indicated. Note that only a single nucleotide substitution (color) is necessary in each case to account for the observed amino acid replacement.

Pattern in the Genetic Code

An inspection of the code words in Table 12.1 will quickly reveal a pattern to the degeneracy of the code. In many instances the first two codon positions are of more importance to the meaning than the third position. Eight amino acids possess codons in which the base in the third position is apparently irrelevant. For example, serine has six codons, four of which are UCN (N is any of the four possible nucleotides); valine has four codons, GUN; glycine has four codons, GGN. Other amino acids possess codons in which the degeneracy of the third position is restricted to purines (G and A) or pyrimidines (C and U). For example, phenylalanine has the codons UU^{U}_{C}, and tyrosine, UA^{U}_{C}.

More than 20 species of tRNA are present in all organisms that have been studied. Some amino acids have as many as six different species of specific tRNA molecules, although it is unlikely that each of these species has a unique anticodon. The data in Table 12.6 show that different species of some tRNAs do recognize different codons for the same amino acid.

Crick has proposed a set of anticodon-codon pairing rules for the third codon position, called the *wobble rules* (Table 12.7). These rules have been generally useful for relating multiple species of tRNA, each with a specific anticodon, to the reading of specific codons. For example, the wobble rules predict that the glutamine codons CA^{A}_{G} would pair with a single anticodon, UUG (codon-anticodon pairing is antiparallel; by convention, the 5′ end is written to the left, and therefore the third codon position is the first anticodon position). However, as pointed out in Chapter 11 and again below, the specificity of anticodon-codon recognition is not solely a function of their hydrogen bonding. The entire structure of the aminoacyl-tRNA may be involved in recognition, along with a number of ribosomal proteins and other proteins.

Role of the Initiation Codon

The codon AUG hydrogen bonds with formylmethionyl-$tRNA_i^{Met}$ when it occurs within a ribosomal binding site on a prokaryotic mRNA, and with methionyl-$tRNA^{Met}$ when it occurs internally within the translated portion of an mRNA. As the initiation codon, AUG performs the important role of setting the reading frame for the rest of the messenger. Binding experiments with synthetic messengers and ribosomes demonstrate that AUG stimulates reading of the next codon in the correct reading frame and suppresses reading of other codons that are out of frame. This phasing function depends on magnesium and occurs only at a low magnesium concentration (Figure 12.5). At higher Mg^{2+} concentrations, the reading frame is chosen at random. A comparision of the binding specificities of two synthetic messengers is made in Figure 12.5: AUGUUU · · · and AUGGUUU · · · . The first messenger contains a phenylalanine codon in phase with the initiating formylmethionine codon and an out-of-phase valine codon, GUU. This messenger becomes positioned on the ribosome with the AUG codon in the P site and UUU in the A site: binding of phenylalanyl-$tRNA^{Phe}$ is stimulated and binding of valyl-$tRNA^{Val}$ is suppressed. The second messenger contains a valine codon in phase with the formylmethionine codon and an out-of-phase phenylalanine codon: binding of valyl-$tRNA^{Val}$ is stimulated and binding of phenylalanyl-$tRNA^{Phe}$ is suppressed.

Table 12.7
Wobble rules for anticodon-codon pairing.

1st Position of Anticodon	3rd Position of Codon
U	A G
C	G
A	U
G	U C
I*	U C A

*Inosine—a derivative of guanosine created by post-transcription modification.

The frame-setting of the initiator codon makes it possible to understand how the nested ϕX174 genes can be expressed. Recall (Chapter 6) that the nucleotide sequence of gene *B* is nested within that of gene *A* and that the sequence of gene *E* is nested within that of gene *D*. Each of these genes codes for a unique protein. An examination of the nucleotide sequence of ϕX174 (Figure 6.7) shows that the initiating codons for the nested genes are not in the same reading frame. Therefore a common mRNA transcribed from the nested genes is read in two different frames and translated into two different proteins, each frame set by its own initiator codon.

Termination Codons

Amber mutations cause premature termination of protein synthesis (Figure 12.3). The nature of the termination signal created by an *amber* mutation has been deduced by amino acid sequence studies of the *E. coli* enzyme alkaline phosphatase and of the T4 head protein. An *E. coli* strain carrying an *amber* mutation that inactivates alkaline phosphatase is mutagenized, and strains with restored or partially restored enzyme activity are isolated. The amino acid sequences of the alkaline phosphatase enzymes in these

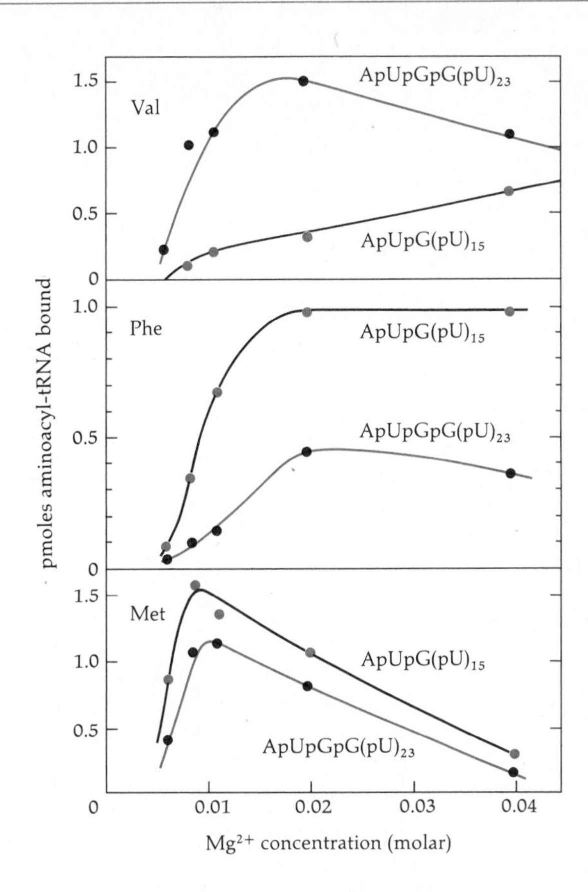

Figure 12.5
Ribosomal binding of indicated aminoacyl-tRNAs stimulated by the synthetic messengers ApUpG$(pU)_{15}$ and ApUpGpG$(pU)_{23}$. At low magnesium concentrations, the phasing role of the AUG codon is evidenced by the enhanced binding of methionyl-$tRNA^{Met}$ to both messengers and by the messenger-dependent binding of valyl-$tRNA^{Val}$ and phenylalanyl-$tRNA^{Phe}$ (see text for discussion). (1 pmole $= 10^{-12}$ mole.) [After T. A. Sundararajan and R. E. Thach, *J. Mol. Biol. 19*:74 (1966).]

reverted strains are compared with each other and with the wild-type sequence. The amino acid replacements observed at one site in the enzyme are shown in Figure 12.6. The original *amber* mutation occurred in a tryptophan codon. At this site, different revertant strains possess the amino acids tryptophan (true revertants), lysine, glutamine, glutamic acid, serine, tyrosine, and leucine (pseudorevertants). Assuming that all mutational events involve only a single nucleotide-pair substitution for each event, examination of all the possible codons for these amino acids shows that only the triplet UAG differs from each by a single nucleotide (Figure 12.6). UAG is one of the three codons not assigned to an amino acid. The relationships shown in Figure 12.6 have been found at five different sites

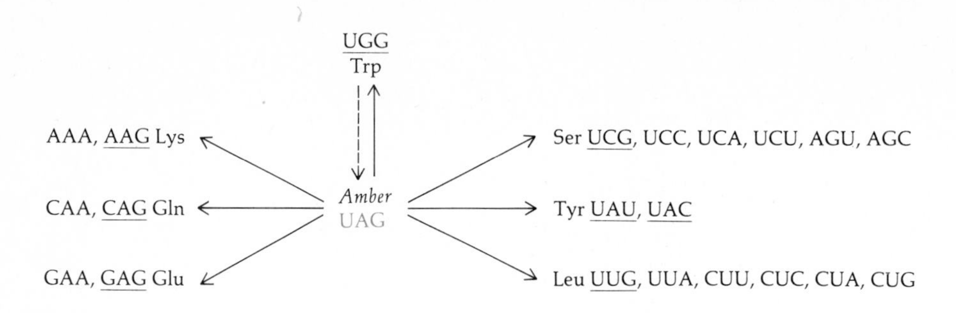

Figure 12.6
Amino acid replacements observed in revertants of *amber* mutations in *E. coli* alkaline phosphatase. The *amber* mutation occurred in a wild-type tryptophan codon (broken arrow). All possible codons for the observed amino acid substitutions are indicated (solid arrows). The *amber* codon UAG is the only codon to differ from the underlined codons by only a single nucleotide substitution. [After A. Garen, *Science 160*:149 (1968).]

in alkaline phosphatase and at eight different sites in the T4 head protein. Therefore the only termination signal created by an *amber* mutation is UAG.

Amber mutations are a type of conditional lethal mutation originally characterized in bacteriophages. They are lethal in some host strains but viable in other host strains, which carry a suppressor gene, Su^{+}_{amber} (Chapter 6, Table 6.1). Two other types of suppressible conditional lethal mutations were identified, *ochre* and *opal*. Experiments similar to those outlined for *amber* show that *ochre* mutations create a UAA termination signal and *opal* mutations create a UGA termination signal. These studies give meaning to all of the 64 possible codons: 61 code for amino acids, and 3 are termination codons. Sequence analyses of natural mRNA molecules and of DNA molecules demonstrate that these three codons are natural termination signals for protein synthesis. Because termination codons and mutations that create termination codons (e.g., *amber*) do not code for an amino acid, they are frequently called *nonsense codons* or *nonsense mutations*. Because they do have meaning, however, we have refrained from using this common terminology until now.

Mutations that Alter Translation of the Code

A complex biochemical machinery exists to translate the nucleotide sequence of DNA into the amino acid sequence of protein. This machinery itself is encoded in the DNA: genes for tRNAs, ribosomal RNA, ribosomal proteins, etc. These genes are subject to mutation just as any other genes are, and mutations that affect translation are therefore to be expected. Mutations that affect translation in a drastic way are, of course, lethal, but mutations that exert only a subtle effect on translation also occur and are frequently observed to be suppressors of other mutations.

Table 12.8
Chemical and genetic properties of five different *amber* suppressors.

amber Suppressor	Amino Acid Inserted	% Protein Synthesized to Wild Type	*ochre* Suppressor
$su1^+$	Ser	28	–
$su2^+$	Gln	14	–
$su3^+$	Tyr	55	–
$su4^+$	Tyr	16	+
$su5^+$	Lys	5	+

After A. Garen, *Science 160*:149 (1968).

We have already encountered *intragenic* suppressors of frameshift mutations, which are themselves frameshift mutations. Another example of intragenic suppressor mutations is the pseudorevertants discussed above (Figure 12.6), which occur within the same codon as the original mutation and cause a different but functional amino acid to replace the original amino acid in a protein.

Mutations that affect the functioning of the translation apparatus itself occur in different genes than do the mutations they suppress, and are called *intergenic* suppressors. In general, intergenic suppressors are allele-specific but not gene-specific in their action on other mutations. *Amber, ochre,* and *opal* mutations, which create termination codons, are conditional lethal mutations; they are distinguished by their nonlethality in hosts that carry specific suppressor genes (Table 6.1). *Amber* suppressors were the first to be understood. They can be produced in Su^- strains by mutagenesis with agents that cause nucleotide-pair substitutions. Their effect is to prevent premature termination at the site of a suppressed *amber* mutation by insertion of a characteristic amino acid. A number of different *amber* suppressors are known, and each inserts a particular amino acid to prevent premature termination (Table 12.8).

Nonsense suppressor mutations occur in tRNA genes. For example, the *amber* suppressor $su3^+$, which inserts tyrosine at UAG codons, is due to a mutation in the anticodon of the gene for $tRNA^{Tyr}$. The normal anticodon of $tRNA^{Tyr}$ is GUA, which, by the wobble rules, reads the tyrosine codons UA^U_C. The $tRNA^{Tyr}$ from the $su3^+$ bacteria has the anticodon CUA, which reads the *amber* codon UAG and inserts tyrosine to prevent termination (Figure 12.7). Not all nonsense suppressors are due to mutation

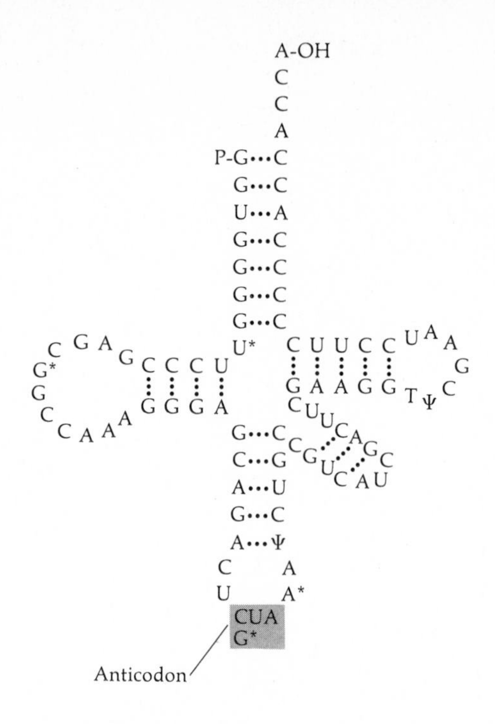

Figure 12.7
The structure of *E. coli* tRNA^{Tyr}. The sequence of the *amber* suppressor tRNA^{Tyr} is drawn and has the anticodon CUA. The normal tyrosine anticodon is $\overset{*}{\text{G}}$UA (the * means that a post-transcriptional modification is present). [After H. Goodman et al., *Nature 217*:1019 (1968).]

in the anticodon of a tRNA gene, however. The *opal* suppressor, which reads UGA codons and inserts tryptophan, is due to a mutation in the dihydrouracil loop of tRNA^{Trp} and possesses the normal anticodon CCA, which is the complement of the tryptophan codon UGG (Figure 12.8). Thus, other regions of the tRNA molecule besides the anticodon are important to the correct translation of the code.

One may wonder how cells carrying a nonsense suppressor are able to survive if the normal function of a necessary tRNA molecule is altered in a way that prevents the normal termination of protein synthesis as well as the ability to recognize the cognate codon. The answer seems to be two-fold. First, nonsense suppressor strains do not grow at a normal rate. They appear to survive because the mutant tRNAs are not efficient suppressors of termination (Table 12.8); they suppress termination infrequently and yet provide sufficient enzyme activity (perhaps only as little as 5% of normal) for growth. Second, in the case of *su3*$^+$, the strain does not lack normal tRNA^{Tyr}. The *su3*$^+$ mutation occurred in a cell carrying a duplication for the tRNA^{Tyr} gene, and thus both normal and mutant tRNA^{Tyr} species are present.

Intergenic suppressors of *missense* mutations are also known, and these too result from mutations that affect the ability of a tRNA to translate codons properly. For example, the *trpA36* mutation of *E. coli* tryptophan synthetase is due to the substitution of arginine for the normal glycine

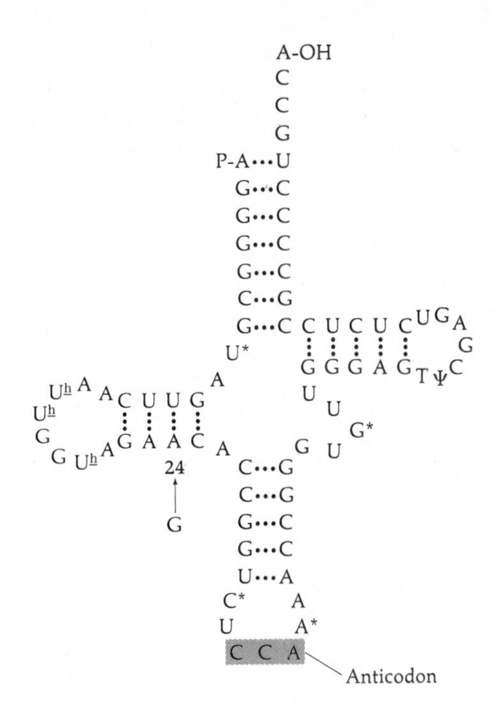

Figure 12.8
The structure of the $tRNA^{Trp}$ from the *opal* suppressor strain of *E. coli.* It differs from that of the nonsuppressor strain at position 24, where A replaces the normal G. (Bases marked with asterisks carry modifications.) [After D. Hirsh, *J. Mol. Biol. 58*:439 (1971).]

residue at position 211 (Figure 11.18); the normal glycine codon GGA is changed to AGA (arginine). The suppressor mutation *suA36* resides in the gene coding for a $tRNA^{Gly}$ with the normal anticodon UCC; *suA36* $tRNA^{Gly}$ has the anticodon UCU, which translates the AGA codon of *trpA36* as glycine rather than arginine.

Mutations that affect the structure of the ribosome also have effects on the fidelity of translation. Nonsense mutations are "leaky" to varying degrees—that is, premature termination does not occur with 100% efficiency. Low levels of streptomycin in the media of Str^Scells carrying a nonsense mutation enhance the leakiness of all three nonsense codons. Streptomycin exerts its effects on bacteria by binding to a ribosomal protein; at sufficiently high concentrations, it blocks protein synthesis altogether. Str^R mutants have an altered ribosomal protein. Nonsense mutations are observed to be much less leaky in Str^R cells; thus, this altered protein affects the frequency of misreading of termination codons. Other mutations called *ram* (ribosomal ambiguity mutation) also increase the leakiness and enhance the efficiency of both nonsense and missense suppressors. These mutations affect a 30S ribosomal subunit protein.

Nonsense suppressors that affect tRNA genes are also known in yeast, and a number of allele-specific intergenic suppressor mutations have been studied in *Drosophila,* although the molecular nature of these *Drosophila* suppressors has not yet been determined.

Despite the fact that mutations that alter the meaning of codons have been documented in both prokaryotes and eukaryotes, the code remains universal, as far as can be determined. Codons have not evolved new meanings during the evolution of all the organisms now on Earth. The universality of the genetic code is certainly a striking fact, and one of the strongest pieces of evidence for the evolution of all organisms from a common ancestor.

Exceptions to the Universality of the Genetic Code

The oft-repeated assertion regarding the universality of the genetic code, which is based upon many documented examples in prokaryotic genes and eukaryotic nuclear genes, must be modified. As *Modern Genetics* is about to go to press, recombinant DNA technology has led to the discovery of a *different* genetic code employed by the mitochondrial protein-synthesizing system. The nucleotide sequences of several cloned restriction fragments of human and yeast mitochondrial DNAs have recently been determined.

The sequences of the human mitochondrial gene for cytochrome oxidase subunit II (COII) and of the homologous protein from beef-heart mitochondria have been compared. Much of the amino acid sequence of the beef protein can be read from the human nucleotide sequence, indicating that the human protein has a very similar amino acid sequence to the beef protein. However, within the regions of homology shared by the beef and human genes are three UGA codons within the proper reading frame of the protein. Of the five tryptophan residues in the protein, three occur opposite these UGA codons, which are *opal* terminators in the genetic code of Table 12.1. The UGG codon for tryptophan in the code of Table 12.1 is also used for tryptophan in the COII gene.

Sequencing of other restriction fragments of human mitochondrial DNA has revealed a mitochondrial $tRNA^{Trp}$ gene with the anticodon UCA. By the wobble pairing rules of Table 12.7, this anticodon would be expected to read both UGA and UGG codons.

Similar comparisons of the sequences of the human mitochondrial COII gene and of the beef protein suggest that AUA is used as a methionine codon, rather than an isoleucine codon, as in the code of Table 12.1. By the wobble rules, a $tRNA^{Met}$ with a UAU anticodon might therefore be expected to read both AUA and AUG as methionine.

Similar studies have also shown that UGA codes for tryptophan in yeast mitochondria. Thus a different genetic code is probably a general feature of mitochondria.

As discussed in Chapter 5, mitochondria have a genetic and protein-synthesizing system distinct from that of the nucleus and its cytoplasmic protein-synthesizing system. A theory (the *endosymbiotic theory*) to account for two distinct genetic systems in eukaryotic cells is that eukaryotes have evolved from ancestors in which a symbiotic relationship existed between a nucleated cell and a prokaryotic cell dwelling in its cytoplasm. The different mitochondrial code may therefore represent an ancient form of the code, preserved by its cytoplasmic form of heredity from the evolutionary changes that have given rise to the code of Table 12.1. Conversely, the mitochondrial code may have evolved following the symbiotic union of the two genetic systems. In the latter case, mitochondria from different organisms might possess different codes, which have evolved independently of one another. It is now important to learn whether mitochondria in all eukaryotes use an identical code.

Problems

1. Why are mutual intragenic frameshift suppressor mutations not expected to occur at all positions in a gene?

2. An RNA copolymer is synthesized from UDP and CDP in a molar ratio of 1:4. What are the relative frequencies of the different codons present in this copolymer? What amino acids would be incorporated in a cell-free protein-synthesizing system using this messenger?

3. The base compositions of DNAs from many organisms, especially microorganisms, vary widely. G + C constitutes 72% of the DNA bases of *Micrococcus lysodeikticus* and 35% of those of *Bacillus cereus*. Only 25% of the DNA bases of the protozoan *Tetrahymena pyriformis* are G + C. Yet the amino acid compositions of the proteins from organisms having very different DNA base compositions are very similar. What explanations can you suggest for this observation? How would you test your hypotheses?

4. An intergenic suppressor of certain frameshift mutations maps to the structural gene for a $tRNA^{Gly}$. Suggest a possible type of alteration that might be present in this mutant $tRNA^{Gly}$gene. How could you test your hypothesis without sequencing the mutant and normal tRNAs?

5. A new mutation, *NG813*, was found in the *Z* gene coding for β-galactosidase in *E. coli*. *NG813* causes premature termination of polypeptide synthesis and is not suppressed by F′ elements carrying *amber* or *ochre* suppressor genes. It maps extremely close to an *amber* mutation that causes premature termination at the same amino acid as *NG813*. It mutates to form an *ochre*-suppressible UAA codon by a single base change. The wild-type gene mutates by a single base change either to *amber* or to mutations identical to *NG813*. What is the nature of the *NG813* mutation?

6. All *ochre* suppressor genes are also *amber* suppressors, but not all *amber* suppressor genes are *ochre* suppressors. Why?

7. How would you distinguish intragenic suppression from intragenic complementation (see Chapter 10)?

8. How could you determine if the single-stranded RNA purified from an RNA virus is a natural mRNA or the complement of the natural mRNA, which is synthesized following infection of the host?

9. Temperature-sensitive mutations of the *E. coli* $su3^+$ gene are known. Design a procedure for screening for such mutations. By analogy with temperature-sensitive proteins, what is the probable nature of such temperature-sensitive $tRNA^{Tyr}$ mutants?

13

Regulation of Gene Expression in Prokaryotes

The environments in which organisms live are not constant. Natural selection has favored those organisms that have been able to regulate their genetic activity to meet changing environmental situations. Regulation of gene expression permits flexibility in the way in which resources can be utilized to maintain the maximum rate of reproduction and to survive environmental stress. For example, bacteria growing in a rich medium containing a simple carbon source, such as glucose, and all 20 amino acids can divide faster if they do not synthesize the enzymes required to utilize more complex carbon sources or synthesize the amino acids themselves. These metabolic functions are needed only when the bacteria do not find the required substances in their environment. The regulation of genetic activity is most economically controlled at the level of gene transcription, and many examples of such regulation have been uncovered in *E. coli* and other bacteria.

Bacteriophages also provide interesting examples of genetic regulation. In phages, the primary level of gene regulation is also at transcription. The small size of phage genomes compared to bacterial genomes has been an important factor in advancing our knowledge of the genetics of regulatory mechanisms, because all or most of the genes can be identified. It is now clear that the same types of regulatory elements are employed to control the expression of phage and bacterial genes. These regulatory elements are certainly shared by all prokaryotes.

An example of the evidence for transcriptional control of gene activity is provided by studies of the gene coding for the enzyme β-galactosidase,

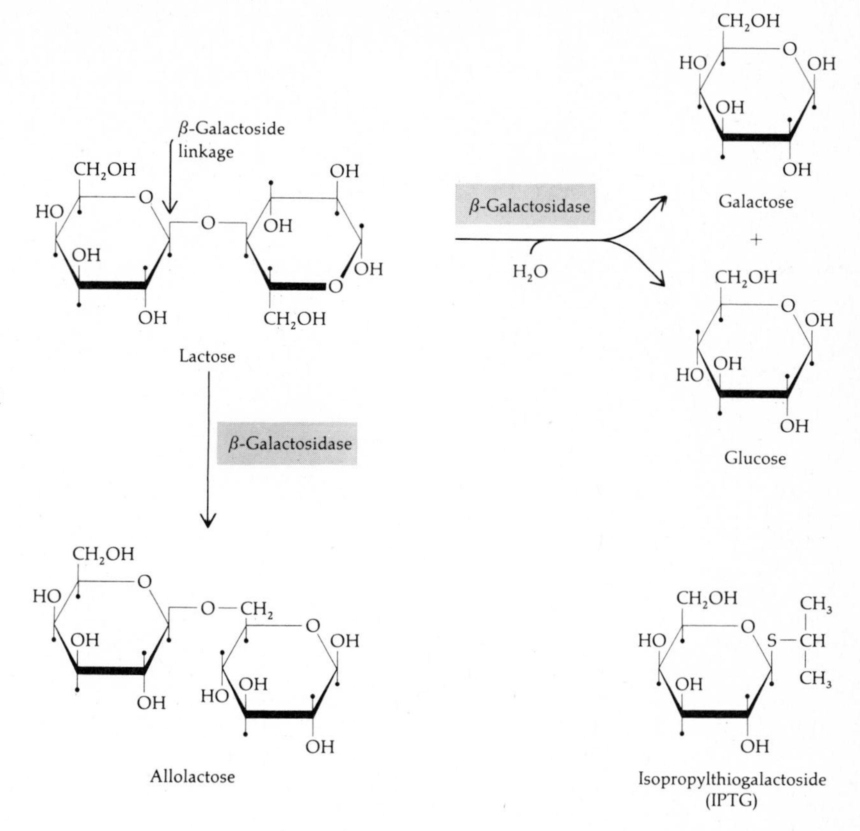

Figure 13.1
Chemical reactions catalyzed by β-galactosidase are the hydrolysis of lactose and the isomerization of lactose to allolactose. Allolactose is the natural inducer of β-galactosidase synthesis. The nonmetabolized, artificial inducer IPTG is also shown; it is often used experimentally to induce β-galactosidase synthesis because it is not hydrolyzed by β-galactosidase.

which degrades the disaccharide lactose to provide galactose and glucose (Figure 13.1). The synthesis of this enzyme is *induced* when lactose is present in the medium and glucose, the preferred carbon source, is absent. When the β-galactosidase gene is induced, mRNA transcribed from the gene *lacZ*$^+$ appears in the cells prior to the enzyme activity itself (Figure 13.2). Following induction, the rate of β-galactosidase synthesis increases to 1000 times its value before induction and is maintained as long as the inducing substance is present.

Lactose and a number of structurally similar compounds act as *inducers* of β-galactosidase synthesis. Allolactose, a metabolite of lactose shown in Figure 13.1, is actually the active inducer. The low basal level of β-galactosidase present in noninduced cells converts lactose to allolactose.

When the inducing substance is removed from the culture, the amount of mRNA rapidly decreases, with a half-life of only a few minutes, as it is degraded to nucleotides by ribonucleases. Maintenance of induction requires the continued synthesis of mRNA to balance its degradation. Virtually all prokaryotic mRNA is metabolically unstable. This mRNA

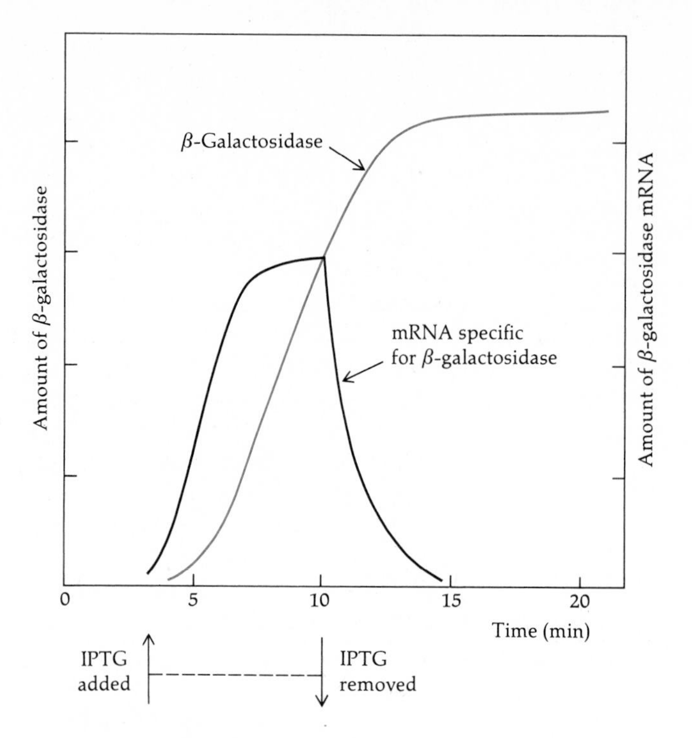

Figure 13.2
The increased synthesis of mRNA coding for β-galactosidase precedes the increased synthesis of β-galactosidase itself. The presence of an inducer is necessary to maintain the presence of the mRNA, which is degraded rapidly, in contrast to the relative stability of the enzyme. The basal level of enzyme activity present in the absence of inducer (necessary for the synthesis of allolactose, the natural inducer) is due to the synthesis of a single mRNA molecule, initiated when the DNA replication fork passes through the gene.

instability, coupled with transcriptional regulation, assures that only required proteins will be synthesized by the cell.

Regulatory Elements Controlling Transcription

RNA is synthesized by RNA transcriptase, which binds to a DNA promoter sequence, initiates RNA synthesis, and transcribes the template strand of DNA until a termination sequence is reached (see Chapter 11). Cessation of transcription and release of the RNA and the polymerase at the termination site require the action of a protein factor called ρ. Thus the control over the presence of mRNA in the cell must reside at the site of initiation—the promoter—but it may also reside at the site of termination—the terminator—if a terminator is found between the promoter and the gene itself. Examples of both types of control of mRNA synthesis will be discussed in the following sections.

A primary factor in regulating gene expression is obviously the rate at which RNA transcriptase initiates RNA synthesis at a promoter. The more efficiently this takes place, the higher the rate of RNA synthesis from a promoter. Some aspects of genetic regulation are the direct result of alteration in the recognition of different types of promoters by RNA transcriptase. Sets of genes whose functions are required simultaneously by the cell have similar sets of promoters. For example, when *E. coli* cells grow in a rich medium containing glucose and all 20 amino acids, about half of all RNA synthesis is for ribosomal RNA. Maintenance of a large pool of ribosomes permits a high rate of growth and cell division. Shifting rapidly growing cells to a medium that is deficient in amino acids quickly causes a shift in the pattern of RNA synthesis. Ribosomal RNA genes cease to be transcribed, and many genes for catabolic functions are induced. This change in gene expression is called the *stringent response*, i.e., a response to stringent or poor growth conditions. The stringent response is mediated by the increased synthesis of the nucleotide ppGpp (guanosine-5′-diphosphate-3′-diphosphate) by a ribosomal protein that is activated on ribosomes starved of aminoacyl-tRNAs. Mutations that abolish the stringent response (str^-) are said to have a *relaxed* phenotype, and fail to synthesize ppGpp.

The nucleotide ppGpp is an example of an *effector* molecule, a regulatory element reflecting the state of the cell's environment—in this case, amino acid starvation. In the presence of ppGpp, RNA transcriptase experiences a change in promoter preference, causing the shift in the pattern of genes that are transcribed. The σ factor plays some role, not completely understood, in the change in promoter preference caused by ppGpp. Some temperature-sensitive mutations that mimic the stringent response at high temperature without amino acid starvation map to the structural gene for the σ polypeptide of the RNA transcriptase holoenzyme (see Chapter 11).

Another example of the direct participation of RNA transcriptase in regulation of gene expression is provided by the phage T7. Some T7 genes are expressed early in infection (early genes), and others, late (late genes). The early gene promoters are recognized by the host-cell RNA transcriptase, but not the late gene promoters. The switch from early to late gene transcription is due to the synthesis of a new T7 RNA transcriptase (the product of an early gene) that recognizes only late gene promoters. Following the switch from early to late genes, host-gene transcription, as well as early T7 gene transcription, is turned off by inactivation of the host transcriptase by another T7 gene product.

Examples of more specific mechanisms of gene regulation have been provided by combined genetic and biochemical analyses of particular genetic functions, such as the ability to utilize specific carbon sources or to synthesize particular amino acids or other metabolites. Recall from Chapters 6 and 7 that genetic mapping of mutant genes that affect particular functions shows that such genes are often clustered in the genome. Clustering of functionally related genes, whose transcription is initiated at a single promoter, permits coordinate control of the expression of the genes. Transcription of a *polycistronic* mRNA from the cluster of genes assures that all of the proteins required for a specific function appear simultaneously in the cell upon induction.

In the examples of gene regulation to be discussed below, genetic analysis reveals the existence of three distinct regulatory elements. They are:

1. *Regulatory proteins*—proteins that affect RNA transcriptase initiation or termination. These proteins may exert either a positive or negative effect on initiation or termination. Their activities are often controlled by specific binding of effector molecules.
2. *Effector molecules*—small nonprotein molecules whose concentrations reflect the environment of the cell (ppGpp, allolactose, etc).
3. *Regulatory sites*—nucleotide sequences that are acted upon by regulatory proteins to affect RNA synthesis (promoters, terminators, etc.).

The *lac* Operon

Genetic analyses of lactose utilization by *E. coli*, carried out by François Jacob and Jacques Monod, provided the first major insights into bacterial gene regulation and led to the formulation of the *operon* model. The products of two closely linked cistrons are necessary for lactose utilization. The $lacZ^+$ gene codes for β-galactosidase (Figure 13.1), and the $lacY^+$ gene codes for a permease that actively transports lactose into the cell. A third gene, $lacA^+$, codes for the enzyme thiogalactoside transacetylase, which is coordinately controlled with the first two enzymes but plays no role in lactose utilization. Mutants of these genes map in the order Z–Y–A.

All three proteins are induced coordinately, being synthesized from a single polycistronic mRNA. Nonsense (*amber*) mutations in the Z gene abolish β-galactosidase activity, of course, but they often abolish permease and transacetylase activity in Y^+A^+ genotypes as well. Nonsense mutations in the Y gene lack permease activity but also often abolish transacetylase activity in A^+ genotypes. However, these Y-gene mutations never affect β-galactosidase activity. This *polar effect* of nonsense mutations was correctly interpreted by Jacob and Monod to indicate transcription

Table 13.1
Phenotypes of heterozygous constitutive mutations in partial diploids that are also heterozygous for β-galactosidase-negative alleles.

Genotype	Phenotype
$\frac{I^-O^+Z^-}{I^+O^+Z^+}$	Inducible control of β-galactosidase synthesis
$\frac{I^-O^+Z^+}{I^+O^+Z^-}$	Inducible control of β-galactosidase synthesis
$\frac{I^+O^cZ^-}{I^+O^+Z^+}$	Inducible control of β-galactosidase synthesis
$\frac{I^+O^cZ^+}{I^+O^+Z^-}$	Constitutive synthesis of β-galactosidase

and translation of a polycistronic mRNA in the direction $Z \rightarrow A$, the single promoter being to the left of the Z gene. Polar effects resulting from nonsense mutations are frequently observed in systems with polycistronic messengers. The mechanism by which nonsense mutations in one gene exert a polar effect on the expression of an adjacent wild-type gene will be discussed later.

Normally, the *lac* genes are transcribed only at a very low rate in the absence of an inducer, and cells contain only very low levels of β-galactosidase and permease. However, mutants that synthesize the normal induced levels of these proteins (constitutive mutants) are easily selected. These constitutive mutants map very near the Z gene and distant from Y and A. They are of two types, as distinguished by complementation analysis: I^- (inducible-negative) and O^c (operator-constitutive). Partial diploids involving these mutations can be constructed using F′ *lac* elements or conjugational merozygotes (see Chapter 7). The phenotypes of these partial diploids show that I^- and O^c mutations affect the synthesis of β-galactosidase in fundamentally different ways. The first two partially diploid genotypes listed in Table 13.1 exhibit normal, inducible phenotypes, indicating that the I^- mutation is recessive. The I^+ gene exerts a normal control over the Z^+ gene in both partially diploid genotypes, regardless of whether the Z^+ gene is on the same chromosome as the I^+ gene. The last two partially diploid genotypes in Table 13.1, on the other hand, show that the O^+ gene and the mutant O^c gene each exerts control only over the Z^+ gene on the same chromosome (these genes are said to be

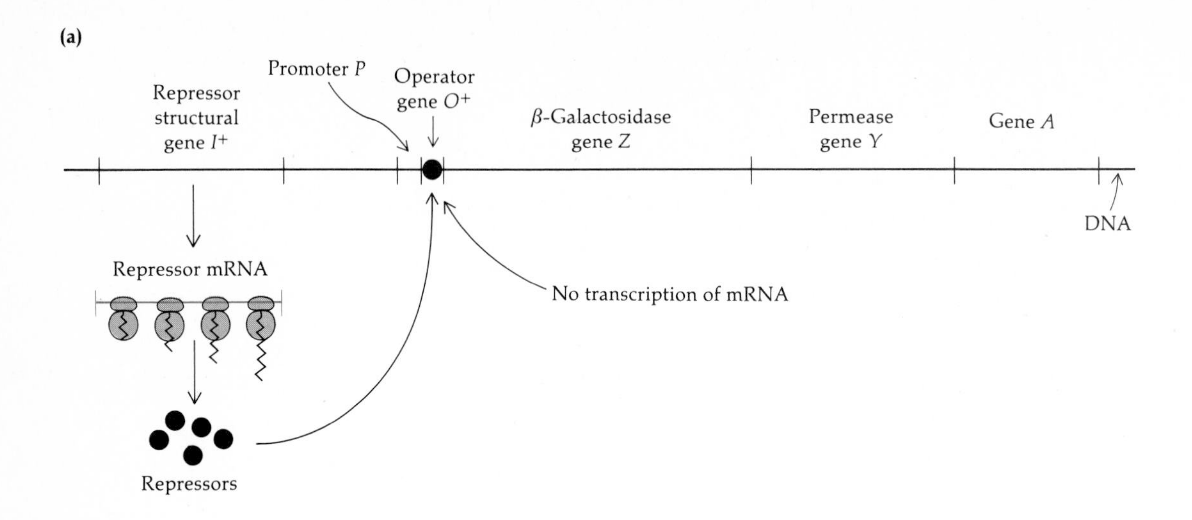

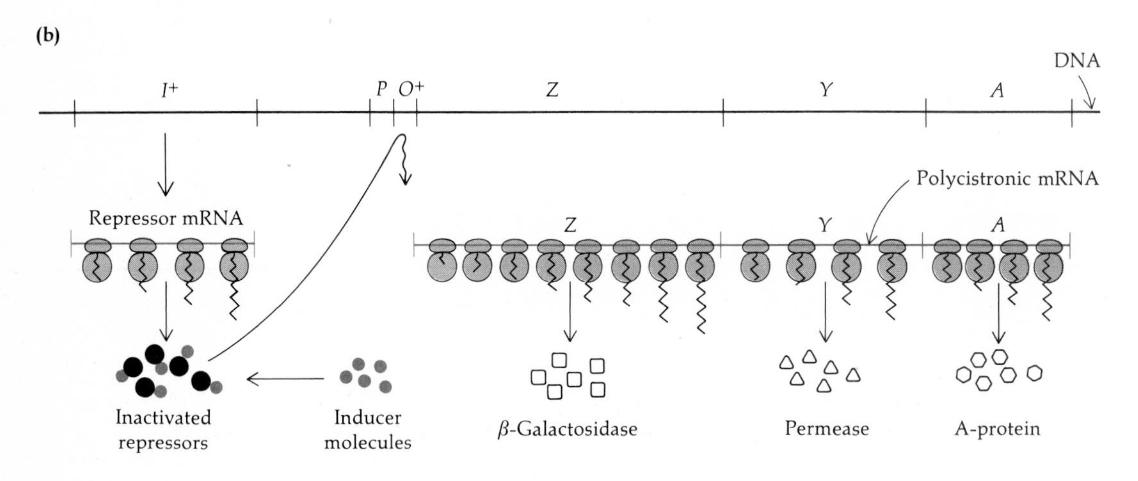

Figure 13.3
The interactions of the repressor, inducer, and operator in *E. coli* control transcription of the genes of the lactose operon. (**a**) In the absence of inducer, the repressor binds to the operator, excluding RNA transcriptase from the promoter adjacent to the operator. (**b**) The inducer binds to the repressor, inactivating it and permitting RNA transcriptase to initiate transcription at the promoter.

cis-dominant). If the O^+ gene is on the chromosome carrying the Z^+ gene, inducible enzyme synthesis is observed; if the O^c gene is on the chromosome carrying the Z^+ gene, however, constitutive synthesis is observed.

To account for these observations, Jacob and Monod proposed in 1961 that transcription of the *Z*, *Y*, and *A* genes is under the control of a regulatory site called the *operator*, or O^+ gene. They further proposed that regulation of transcription at the operator site is controlled by a *repressor*, which is the product of the I^+ gene. The repressor was postulated to have two functions. One function was to bind to the operator and thus prevent transcription. The other function was to bind to the inducer. Binding of an inducer molecule by the repressor was postulated to cause the repressor to dissociate from the operator, thereby permitting transcription (Figure 13.3). Thus the *lac*, *Z*, *Y*, and *A* genes, along with their transcriptional control sites, promoter and operator, were postulated to form an *operon*,

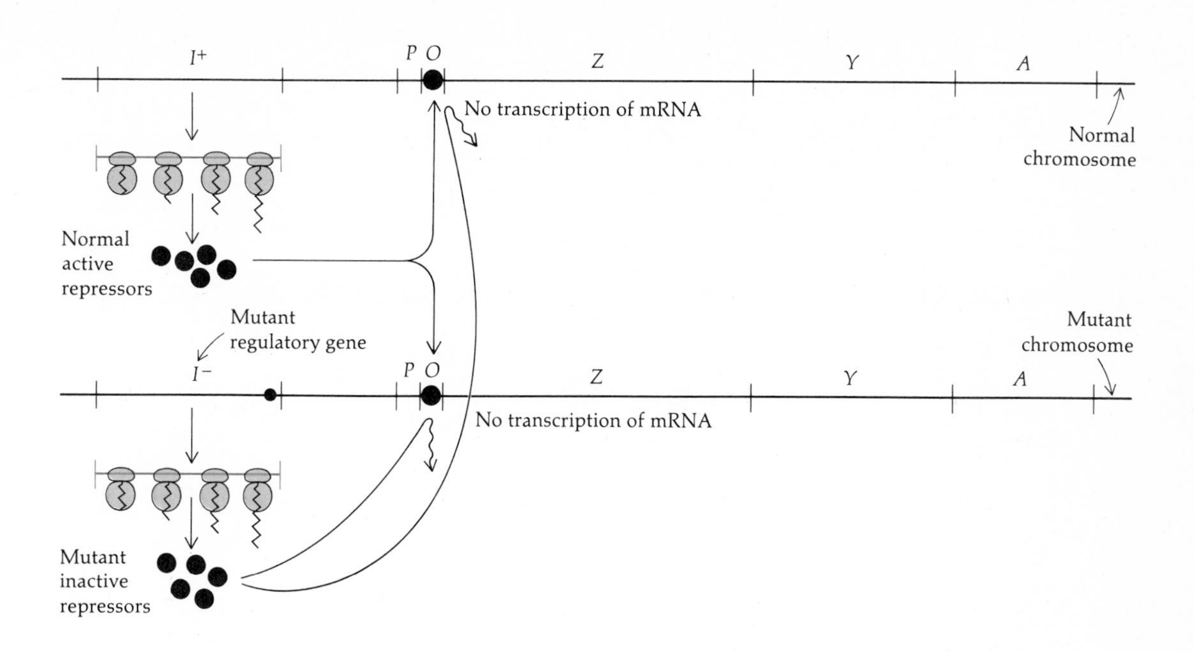

Figure 13.4
The dominance of the I^+ gene. In the partially diploid I^-/I^+ cell, the I^+ gene provides enough active repressor to bind to the operators on both chromosomes, preventing transcription of both of the lactose operons. Hence, β-galactosidase synthesis is inducible rather than constitutive in I^-/I^+ partial diploids.

which was under the control of a repressor, whose function was subject to regulation by an inducer, or effector molecule. This hypothesis explained the recessive nature of I^- mutations and the cis-dominance of the *O* gene, as outlined in Figures 13.4 and 13.5.

The hypothesis has since been verified in great detail. The repressor has been purified and shown to be a protein. Its amino acid sequence has been determined, and this, in turn, has been confirmed by determination of the nucleotide sequence of the *I* gene itself. The nucleotide sequence of the *lac* control region containing the promoter and operator has been determined (Figure 13.6), and the relative positions of the promoter and operator have been mapped genetically (see problem 6 in Chapter 8). The purified repressor has been shown to bind tightly to the purified operator DNA in the absence of inducer. It also binds tightly to the inducer and, when bound to inducer, releases its association with the operator DNA.

The *lac* Repressor

The *lac* repressor has provided a model system for the study of a protein-DNA interaction. The interaction of regulatory proteins with specific DNA sequences is the most fundamental aspect of the control of gene expression in all organisms. The *lac* repressor recognizes and binds tightly to one operator sequence of only 24 np (Figure 13.7) among all the sequences present in the *E. coli* genome, which consists of 3.2×10^6 np!

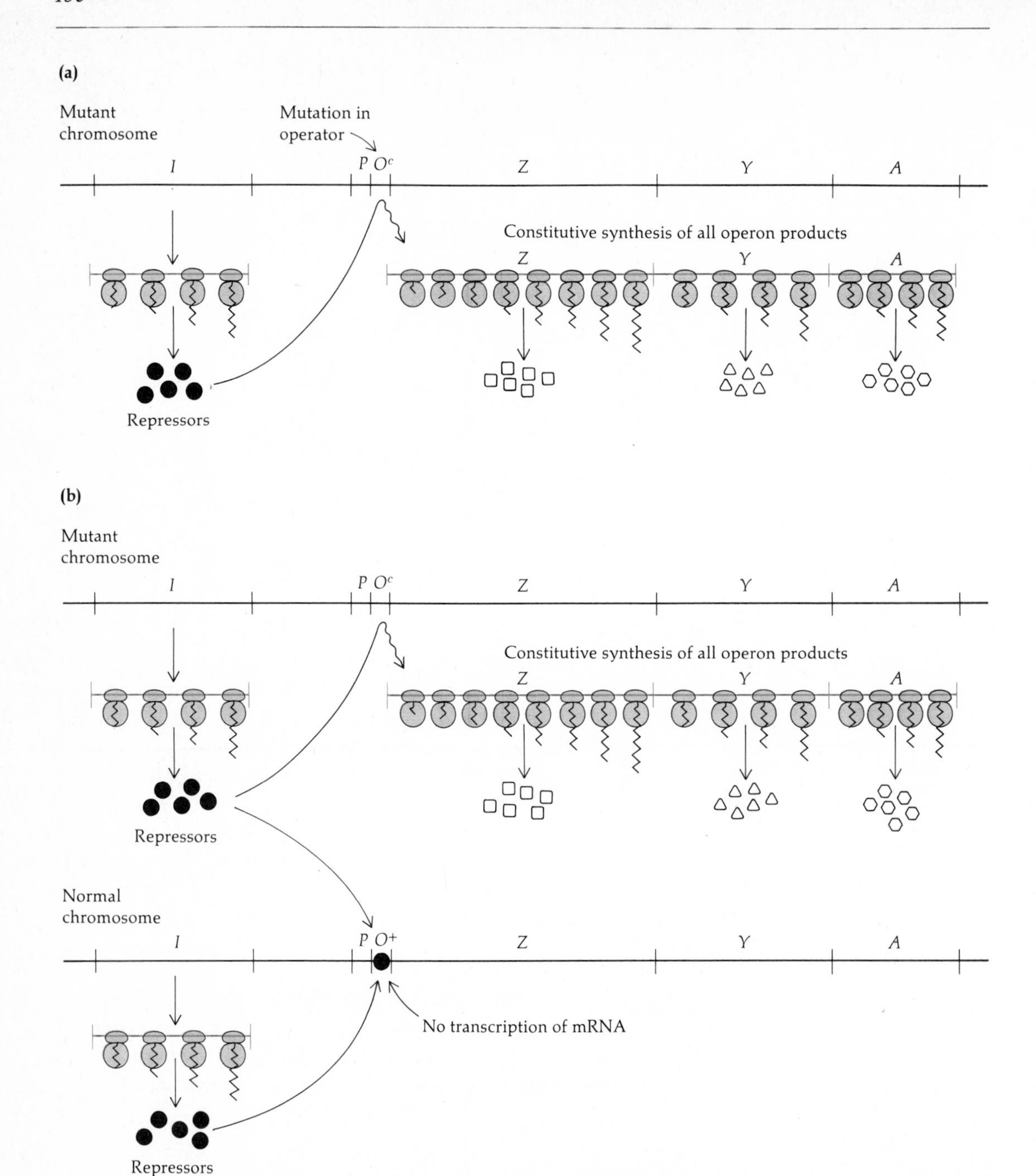

Figure 13.5
Cis-dominance of *O* alleles is a consequence of their effect on only their own chromosome's transcription. The *O* gene does not code for a product that can diffuse throughout the cell and interact with other chromosomes, as does the *I* gene. **(a)** In a haploid cell containing the mutant operator O^c, the O^c mutation causes the operator *not* to bind the repressor, permitting free transcription from the promoter adjacent to the O^c operator. **(b)** In a heterozygous partial diploid (O^+/O^c), the O^c is dominant over O^+; transcription occurs from the O^c chromosome, but not from the O^+ chromosome, in the absence of inducer. Hence, the phenotype of haploid cells and of partially diploid cells carrying an O^c mutation is constitutive.

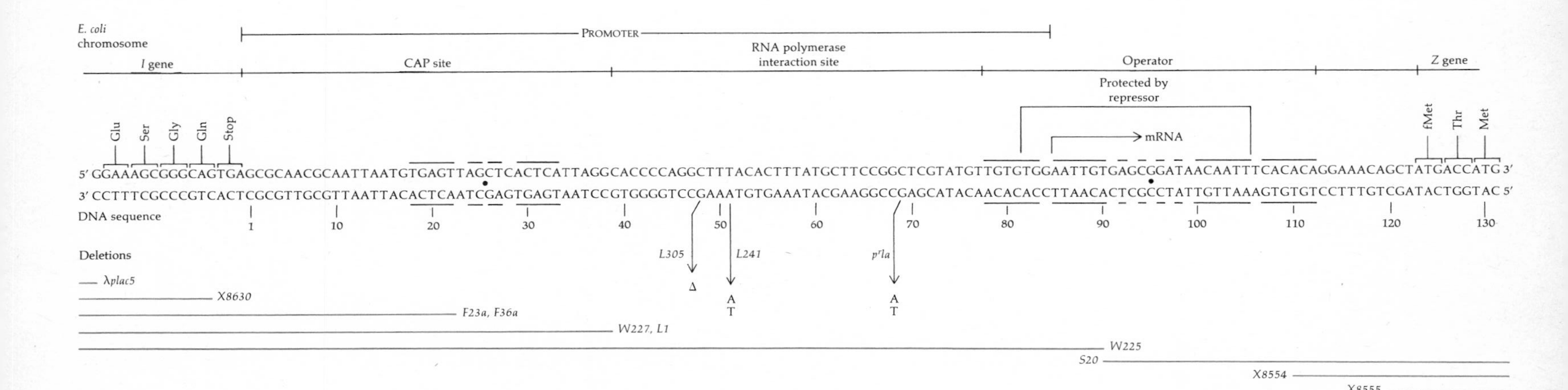

Figure 13.6
Nucleotide sequence of the *lac* control region in *E. coli*. The known amino acid sequences of the repressor and of β-galactosidase permit the coding regions of the *I* gene and the *Z* gene to be identified. The operator and promoter regions are identified by determining which sequence is protected from enzymatic degradation by bound repressor and which sequence is protected by bound RNA polymerase. Deletions or point mutations that affect one or the other function alter only certain sequences, permitting further assignment of function to region. [After R. C. Dickson et al., *Science* *187*:27 (1975).]

Sixteen of these nucleotide pairs lie in a 2-fold symmetrical pattern, or *palindrome,* which reads identically (observing polarity) from either the left or right end of the sequence.

The *lac* repressor is a tetramer of four identical polypeptides coded by the *lacI* gene. Each polypeptide consists of 360 amino acids. The N-terminal sequence of about 50 amino acids of each polypeptide plays an essential role in the binding of the tetrameric repressor to DNA. I^{-D} constitutive mutations abolish the ability of the repressor to bind to the operator, and all of these mutations map within the region of the I gene coding for the N-terminal end of the polypeptide. These mutations are dominant in partial diploids; I^{-D}/I^{+} genotypes have a constitutive phenotype. This dominance must be due to the formation of mixed tetramers. One mutant polypeptide in the tetramer somehow prevents the proper functioning of the three normal polypeptides. On the other hand, the repressor present in I^{-D} cells binds inducer normally. *Superrepressed* mutations, I^{S}, of the *lac* repressor are known in which the mutant repressor binds to the operator and is insensitive to the presence of an inducer. These mutations map to the region of the I gene that does not code for the N-terminal 50 amino acids.

Thus the two functional aspects of the I-gene polypeptide, binding of DNA and binding of inducer, reside in different portions of the polypeptide. This has been confirmed biochemically. When purified repressor is treated with the proteolytic enzyme trypsin, the N-terminal ends of the polypeptides are cleaved from the tetrameric core of the repressor. The core can still bind inducer, however, but can no longer bind to DNA. The N-terminal 50 or so amino acids probably protrude from the core in a way that permits their insertion into the major or minor grooves of the Watson-Crick double helix, where they can recognize the operator sequence.

The details of how the tetrameric repressor interacts with the symmetrical nucleotide sequence of the operator, and of how the inducer alters the interaction, are not known precisely. The dissociation constant K is a measure of the tightness of binding and is given by the mass action law,

$$K = \frac{[\mathrm{O}][\mathrm{R}]}{[\mathrm{OR}]}$$

where [O], [R], and [OR] are the molar concentrations of operator DNA, repressor, and operator-repressor complex, respectively. The observed values of K are of the order of 10^{-13} molar, which indicates extremely tight binding of repressor and operator. Such tight binding explains why only about ten repressor molecules per cell are sufficient to keep the *lac* operon fully repressed.

The repressor also has considerable affinity for nonoperator DNA. For example, it binds with poly-d(AT) with a dissociation constant of about 10^{-8} molar. The binding to nonoperator DNA is not affected by the

5′ TGGAATTGTGAGCGGATAACAATT 3′
3′ ACCTTAACACTCGCCTATTGTTAA 5′

Figure 13.7
Nucleotide sequence of the DNA bound by the *lac* repressor. Palindromic sequences, symmetrical about the rotation axis (colored dot), are shaded.

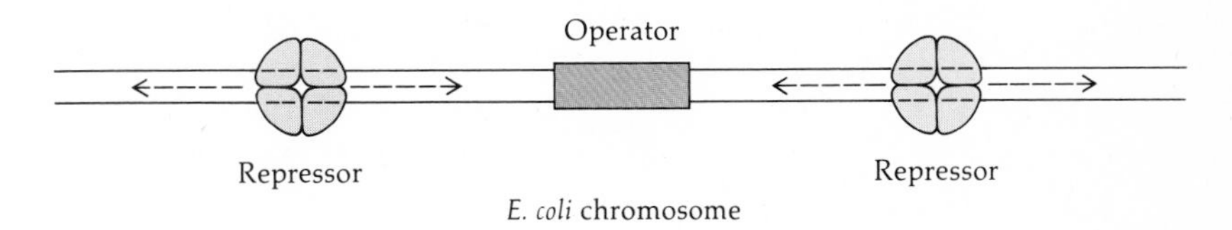

Figure 13.8
The *lac* repressor may search for the operator while bound to nonoperator DNA by diffusion along the chromosome. It may thus approach the operator from either of two possible directions.

presence of inducer. This suggests that newly synthesized repressor molecules and repressor molecules associated with inducer are probably always complexed with DNA in the cell. They probably find the operator sequence by a linear diffusion along the chromosome rather than by a three-dimensional diffusion through the cytoplasm (Figure 13.8). Diffusion of the repressor along the chromosome may explain why the operator possesses 2-fold symmetry. The appearance of the operator would then be the same, regardless of the direction with which the repressor approaches the operator as it moves along the chromosome. Palindromes are components of DNA recognition sites in general. The [CAP-cAMP] binding site in the *lac* promoter possesses one (see next section and Figure 13.6), as do sites involved in site-specific recombination and restriction (Chapter 9). The RNA transcriptase binding site is an exception to this generality, most probably because the selection of the template DNA strand requires asymmetry in the recognition site (Figure 13.6).

Catabolite Repression

The *lac* repressor is an example of a negative regulatory protein, whose action prevents expression of the genes under its control and whose function is controlled by an effector molecule—in this case, allolactose.

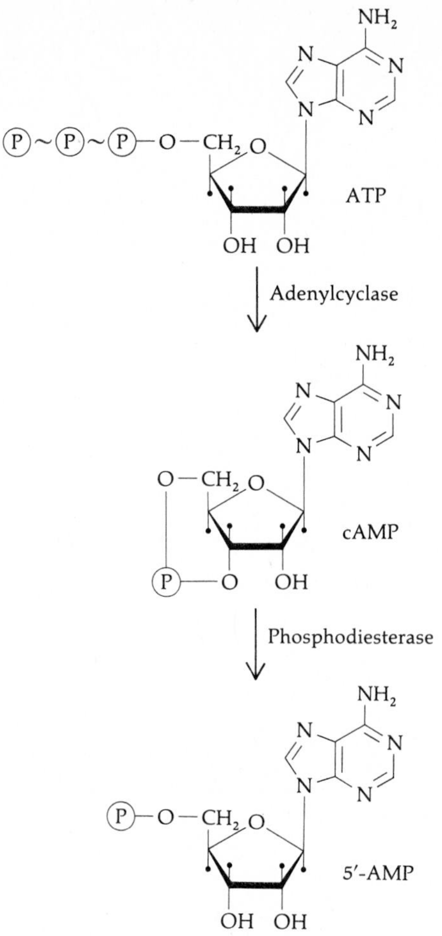

Figure 13.9
Control of the cyclic AMP (cAMP) level in cells may be either by synthesis from ATP or by degradation to 5′-AMP. Which step is affected by glucose to regulate the cAMP level in bacteria is unknown.

The *lac* operon is under the control of a positive regulatory protein that is involved in the control of a number of different catabolic systems in *E. coli*. The action of this positive regulatory protein is controlled indirectly by the preferred carbon source, glucose. Glucose prevents transcription of the *lac* operon, even in the presence of lactose; it does this in I^- and O^c mutants as well as in wild type, indicating that glucose does not function via the repressor-operator interaction. The effect of glucose is mediated by a nucleotide, *cyclic AMP* (cAMP).

The intracellular concentration of cAMP is controlled in a balanced way through its synthesis by adenylcyclase and its degradation by phosphodiesterase (Figure 13.9). The level of cAMP is high in the absence of glucose and low in its presence. The mechanism whereby the glucose concentration controls the cAMP level is not known. However, it is clear that cAMP serves as an effector molecule reflecting this aspect of the cell's environment.

Studies of mutants that are unable to activate catabolic functions have elucidated the mechanism of cAMP control on *lac* transcription. Two types of such mutants are found. The first type inactivates the adenylcyclase enzyme (Figure 13.9). The second type inactivates a protein that binds cAMP, a *catabolite activator protein* (CAP). Such mutants can be selected on the basis of their simultaneous inability to utilize two different sugars, e.g., lactose and galactose, as carbon sources. By selection for the simultaneous loss of both functions, mutations are detected in genes, other than those of either operon, that are required for the expression of both functions and for the utilization of both lactose and galactose.

In vitro studies of *lac* transcription, employing the *lac* operon carried in the DNA of a λ transducing phage, purified RNA transcriptase, and purifed CAP, demonstrate that CAP is active in enhancing *lac* transcription only when it is bound with cAMP. The site of [CAP-cAMP] action is at the *lac* promoter; this is demonstrated by the fact that a deletion mutation, *L1*,extending from the *I* gene into the promoter (Figure 13.6) abolishes the effect of cAMP on *lac* transcription without abolishing the ability of RNA transcriptase to bind to this promoter. Thus, the *lac* promoter contains both an RNA transcriptase binding site and a [CAP-cAMP] binding site. A model for the initiation of *lac* transcription is diagrammed in Figure 13.10.

In summary, regulation of expression of the *lac*-operon structural genes is under two sets of controls, both of which are governed by environmental factors. The repressor-operator interaction provides an "all or none" level of control. There are only about ten repressor molecules per cell, and these are readily inactivated by low concentrations of inducer when lactose is present. The [CAP-cAMP]-promoter interaction provides a modulatory control of the rate of initiation of mRNA molecules. This rate is low when cAMP levels are low and most CAP protein is inactive, and the rate is high when cAMP levels are high and [CAP-cAMP] is abundant.

Regulation of Transcription by Translation

Several operons coding for amino acid synthetic-pathway enzymes are controlled by the availability of that particular amino acid in the environment. The control of operon expression by the amino acid occurs at two levels, one transcriptional and the other translational.

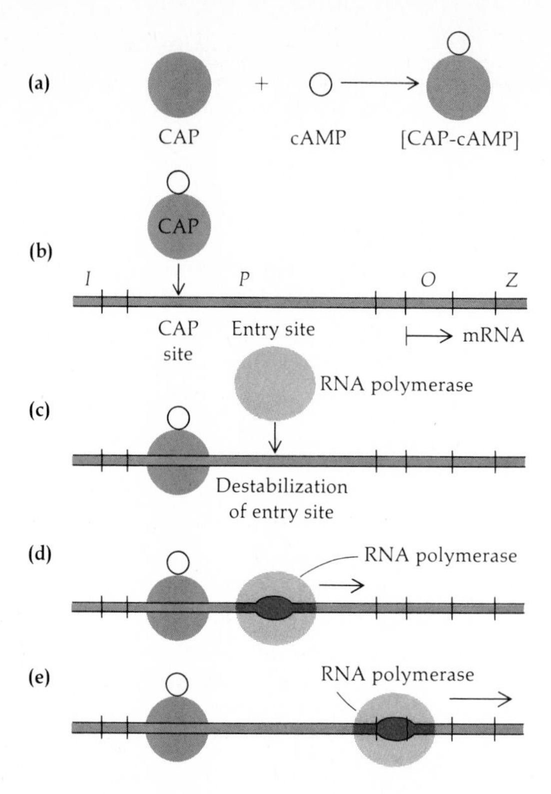

Figure 13.10
The catabolite activator protein (CAP) is inactive until bound with cAMP. The [CAP-cAMP] complex binds to a region of the promoter for the *lac* operon, stimulating the initiation of transcription by RNA polymerase.

The *E. coli tryptophan* operon is shown in Figure 13.11. Transcription is initiated at the promoter (*P*) and is under the control of a repressor protein and tryptophan. When the repressor is bound with tryptophan, it is also capable of binding the operator (*O*) and blocking RNA transcriptase from the promoter. When tryptophan is in low supply, the repressor loses its bound tryptophan molecule and releases the operator, allowing transcription to be initiated. The role of tryptophan is that of a *corepressor*, since both tryptophan and repressor are necessary for repression.

The second level of control over the amount of mRNA coding for the enzymes of the tryptophan pathway was discovered when mutations causing an overproduction of these enzymes were studied. These mutations were found to be deletions that map between the operator and the first structural gene of the operon, *trpE*. The repressor-operator interaction is normal in these mutations, suggesting that the deletions do not act on the promoter to increase, in some way, the frequency of mRNA initiation.

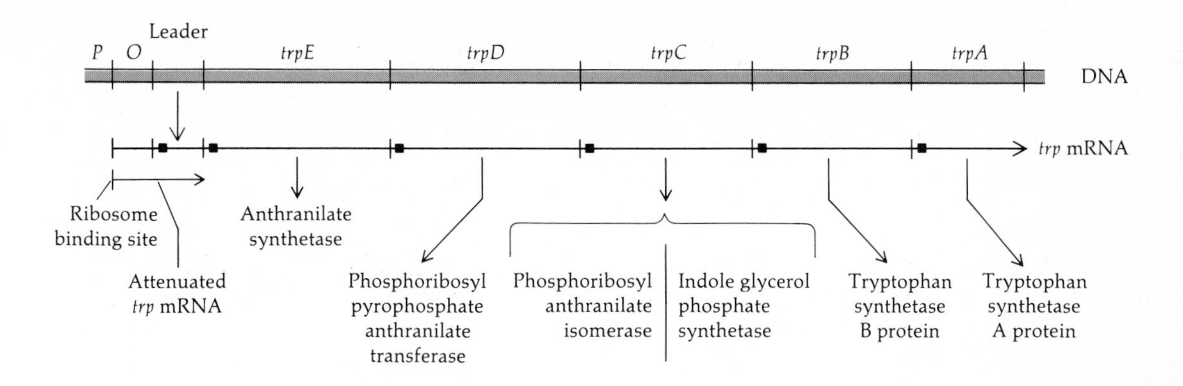

Figure 13.11
The tryptophan operon of *E. coli*, showing the relation of the regulatory region to the structural genes. Two transcription products are indicated: a polycistronic mRNA and the attenuated mRNA produced by transcription termination in the leader region. (See also Figures 10.8 and 11.16.)

Instead, these deletions remove a signal that normally causes RNA transcriptase to cease transcription before reaching the *trpE* gene, so that only the *leader* portion of the mRNA is synthesized. Normally, only about 10% of the *trp* operon mRNA molecules initiated at the promoter are completed by transcription of the structural genes of the operon. The region that, when deleted, allows larger numbers of mRNA molecules to be completed, is called an *attenuator.*

The nucleotide sequence of the leader region of the *trp* mRNA is shown in Figure 13.12. It contains two surprising features. The first is a ribosomal binding site and the codons for a 14-amino-acid peptide containing two adjacent tryptophan residues. The second is a region of nucleotides capable of forming the secondary structures indicated in the figure. The generality of these unusual features is shown by the nucleotide sequences of the leader regions of the *E. coli phenylalanine* operon and of the *E. coli* and *Salmonella histidine* operons. The *phe*-operon leader codes for a 15-amino-acid peptide containing seven phenylalanine residues, and also has a region of secondary structure similar to that in the *trp*-operon leader (Figure 13.13). The *his*-operon leader codes for a 16-amino-acid peptide containing seven adjacent histidine residues, and also has a region of secondary structure prior to the site of transcription termination.

Apparently, translation of the message for these leader peptides in some way causes transcription to cease just beyond the region of secondary structure present in the leader sequence. However, when the appropriate aminoacyl-tRNA (e.g., tryptophanyl-$tRNA^{Trp}$) is limiting, owing to starvation for that amino acid, translation of the leader peptide is slowed or prevented, thereby allowing RNA transcriptase to proceed with transcription of the entire operon. The mechanism of termination is not known, but it is possible that a ribosome that passes through the peptide region of the leader disrupts the secondary structure of the leader RNA, exposing an RNA sequence recognized by the termination factor, ρ. Certain mutations in a gene that is believed to be the structural gene for ρ decrease the efficiency of attenuation, indicating that ρ is probably involved.

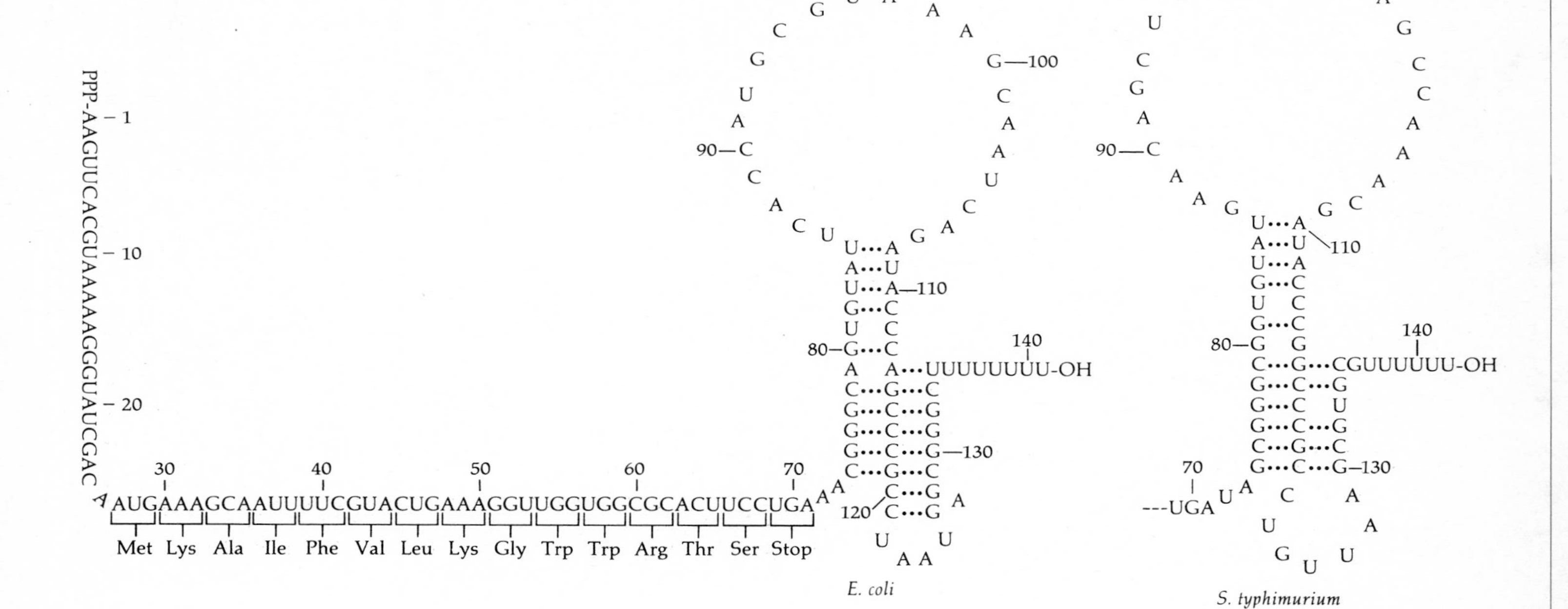

Figure 13.12
Nucleotide sequence of the attenuated *trp* mRNA of *E. coli*. The coding region for a 14-amino-acid polypeptide containing two adjacent tryptophan residues is indicated. The last half of the homologous nucleotide sequence of the attenuated *trp* mRNA of *Salmonella typhimurium* is shown for comparison. In both, termination occurs after a stretch of uridine residues, possibly preceded by a base-paired loop. A second, larger base-paired loop is also possible for both mRNAs. Complete versus incomplete translation of the leader coding region may influence which loop actually forms and determine whether or not attenuation of the mRNA occurs. [After F. Lee and C. Yanofsky, *Proc. Natl. Acad. Sci. USA* 74:4365 (1977).]

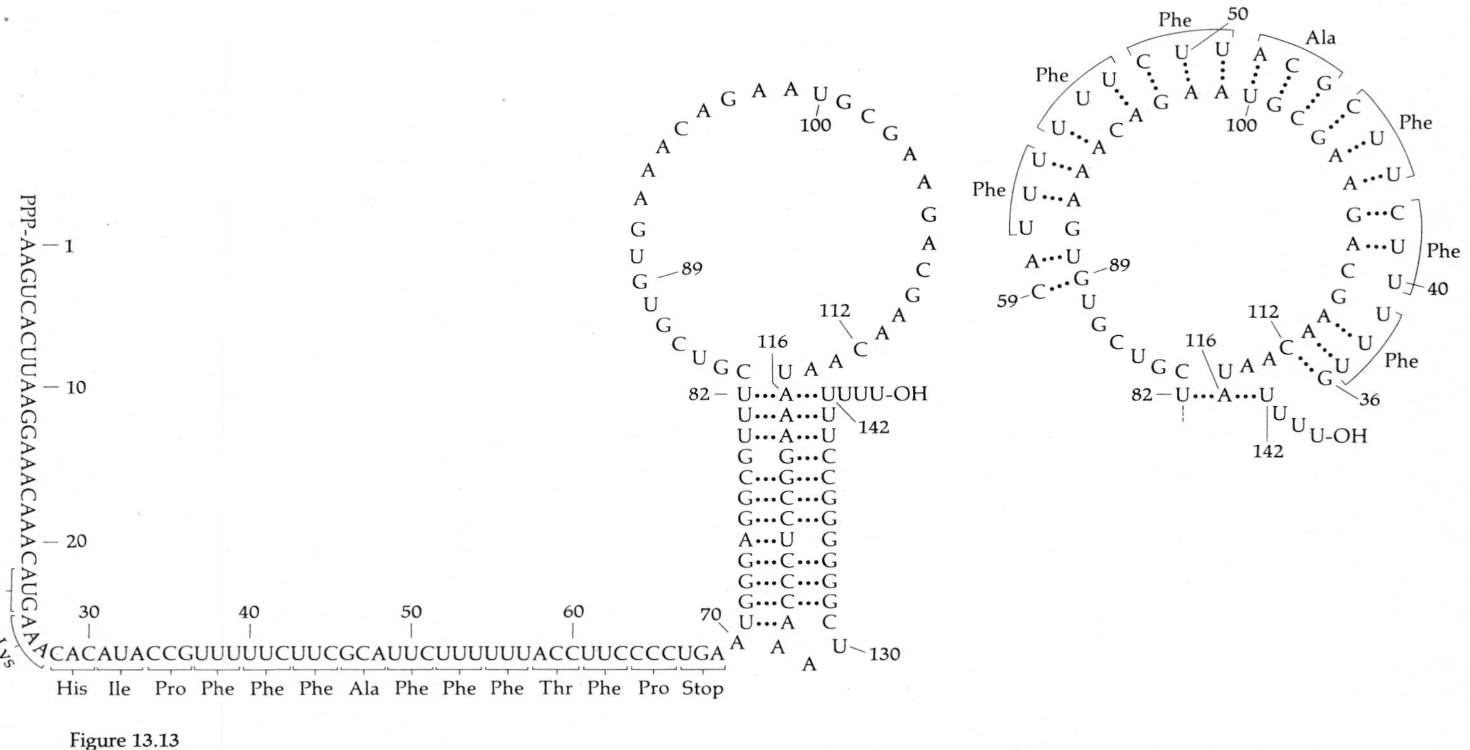

Figure 13.13
Nucleotide sequence of the attenuated *phe* mRNA of *E. coli*. The coding region for a 15-amino-acid polypeptide containing seven phenylalanine residues is indicated. Base-paired loops similar to those in the attenuated *trp* mRNA are shown. Another possible secondary structure, involving base-pairing between the leader coding region and the larger loop, is drawn on the right. It is possible that such secondary structures play regulatory roles. [After G. Zurawski et al., *Proc. Natl. Acad. Sci. USA* 75:4271 (1978).]

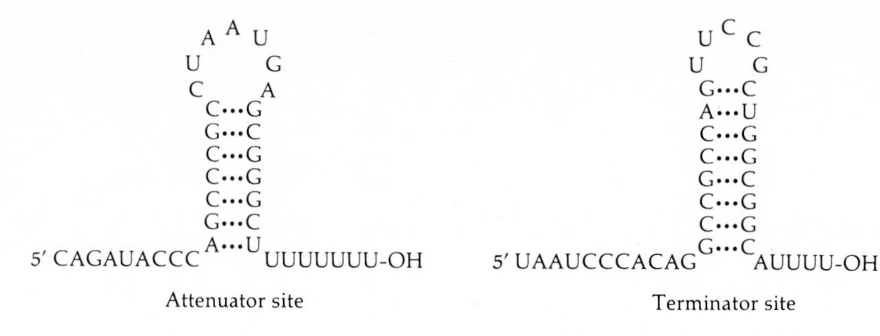

Figure 13.14
Nucleotide sequences of the *E. coli trp*-operon attenuator and terminator sites. [After A. Wu and T. Platt, *Proc. Natl. Acad. Sci. USA* 75:5442 (1978).]

The site of transcriptional termination of the complete *trp* mRNA of *E. coli* has been determined by sequencing of the 3′-OH end of the complete mRNA. This terminator is also ρ-dependent and has the structure indicated in Figure 13.14. The structural similarities between this terminator and the attenuator sites in Figures 13.12 and 13.13 are evident. All of these RNA molecules terminate after a stretch of uridine residues preceded by a base-paired loop.

Further evidence of a coupling between transcription and translation, as exhibited by attenuators, is provided by the polar effect of nonsense mutations within structural genes of an operon (see above). Polar nonsense mutations, which cause premature termination of polypeptide synthesis, also cause premature termination of mRNA synthesis. The failure of wild-type genes to be expressed distal to the gene containing a nonsense mutation is due to the absence of mRNA sequences for those genes. A suppressor mutation (*suppressor A*) that suppresses the polarity effect of nonsense mutations in many operons is known. This mutation, however, does not suppress the nonsense mutation itself, i.e., it is not a mutant tRNA gene. Instead, *supA* maps to the gene that is believed to be the structural gene for ρ, again suggesting that ρ somehow mediates the coupling of transcription and translation.

Bacteriophage λ

The genes of phage λ are regulated temporally during lytic infection to permit controlled DNA replication, recombination, synthesis, and assembly of progeny phage. A different pattern of gene expression, however, pertains in λ lysogens, in which almost all of the prophage genes are repressed. A fascinating question is: How are the λ genes involved in making the decision to either follow the lytic pathway of progeny production or follow the lysogenic pathway?

An understanding of λ gene regulation developed simultaneously with that of the *lac* operon because both were found to be controlled by repressors. In the prophage state, all of the genes required for progeny

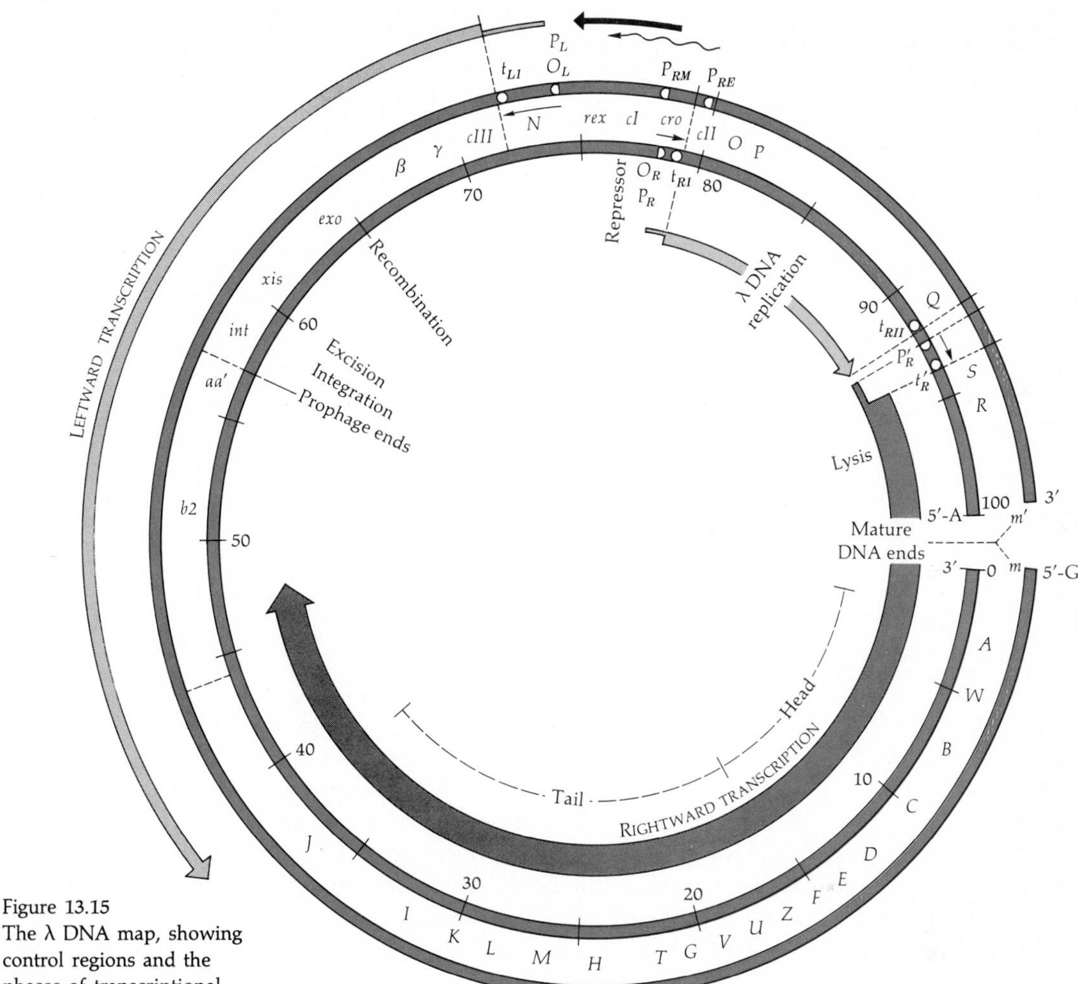

Figure 13.15
The λ DNA map, showing control regions and the phases of transcriptional activity. The three short arrows represent stage-I transcription originating at promoters P_L, P_R, and P'_R and ending at terminators t_{L1}, t_{R1}, and t'_R. The light gray arrows represent stage-II transcription in the presence of the antiterminator *N* protein, and the dark gray arrow, stage-III transcription in the presence of the antiterminator *Q* protein. The wavy arrow represents transcription of the *cI* gene from the promoter P_{RE}, and the heavy arrow, transcription of *cI* from P_{RM}.

production are repressed. The *cI* gene codes for a repressor protein that acts through binding of two operators: O_R, to the right of the *cI* gene, and O_L, to the left of the *cI* gene (Figure 13.15). These operators have been identified genetically through cis-dominant mutations that inactivate them. Phages with mutations in both O_R and O_L are *virulent* and cannot lysogenize host cells, nor are λ lysogens immune to superinfection by them, with subsequent production of virulent progeny. Some temperature-sensitive *cI* mutations have active repressor in lysogens at 30°C but cause prophage induction at 40°C; they are useful for studying the events that occur when repression of the prophage is lifted by a shift to 40°C.

In the absence of repressor, the first stage of λ mRNA transcription begins with RNA transcriptase initiation at two promoters on either side of the *cI* gene: P_R and P_L. These promoters overlap the O_R and O_L operators recognized by the *cI* repressor. Only two structural-gene transcripts result from stage-I transcription: the *N* gene and the *cro* gene. Transcription terminates at ρ-dependent terminator sequences following *N* and *cro*, as indicated in Figure 13.15.

The second stage of mRNA transcription requires the product of the *N* gene that is produced by translation of stage-I mRNA. The *N* protein prevents termination by RNA transcriptase at the ρ-dependent sites just mentioned. It does so by interacting with RNA transcriptase in some manner as stage-II transcription is initiated at P_R and P_L. The presence of *N* protein then allows transcription to proceed through genes *cIII* to *int* on the left and through genes *cII* to *Q* on the right. The proteins synthesized by translation of stage-II mRNAs provide functions necessary for the lytic pathway leading to DNA replication (genes *O* and *P*), for recombination (genes *exo* and *β*), and for the lysogenic pathway (genes *int*, *cII*, and *cIII*).

The third stage of mRNA transcription requires the product of the *Q* gene and yields mRNA for the structural proteins of the phage particle, for assembly of complete phage particles, and for host-cell lysis (Figure 13.15). Like *N* protein, *Q* protein acts as an antiterminator of mRNA transcription, which is initiated at a promoter located between genes *Q* and *S*. The RNA transcriptase initiates transcription at this promoter, but it synthesizes only a short (201-nucleotide) RNA before reaching a terminator sequence prior to the *S* gene. In the presence of *Q* protein, RNA transcriptase proceeds past the terminator to transcribe the stage-III genes. It should be noted that both the *N* and *Q* proteins are positive regulatory elements that possess antiterminator functions.

The lysogenic pathway of development requires the insertion of λ into the host genome (the *int* function), as well as the activation of repressor synthesis (*cI* function). The *cII* and *cIII* genes code for polypeptide subunits that together form a protein that activates *cI* transcription. This protein stimulates the host RNA transcriptase to initiate at a promoter located between *cro* and *cII* (Figure 13.15). This promoter is called P_{RE}, the promoter for *establishment of repression,* to distinguish it from the promoter P_{RM}, which later functions to *maintain repressor* synthesis in a lysogenic cell. In a lysogenic cell, the repressor prevents transcription rightward from P_R and leftward from P_L, and the products of the *cII* and *cIII* genes are not present. The P_{RM} promoter then serves to provide *cI* mRNA synthesis to maintain repressor concentration as the cells grow and divide.

The λ repressor exhibits the properties both of repressing transcription from P_R and P_L and of regulating its own synthesis. Nucleotide sequencing of the O_R and O_L operators shows that each contains three repressor binding sites (Figure 13.16), which overlap the promoters P_{RM}, P_R, and P_L (Figure 13.17). Binding studies with purified repressor show that the relative affinity for repressor of each binding site in O_R is $O_{R1} > O_{R2} > O_{R3}$. At low repressor concentration, O_{R1} is the first site to bind repressor, which prevents transcription of the *cro* gene from P_R but *stimulates* transcription of the *cI* gene from P_{RM}. Stimulation of *cI* transcription increases the repressor concentration, leading to binding of O_{R2} and O_{R3}.

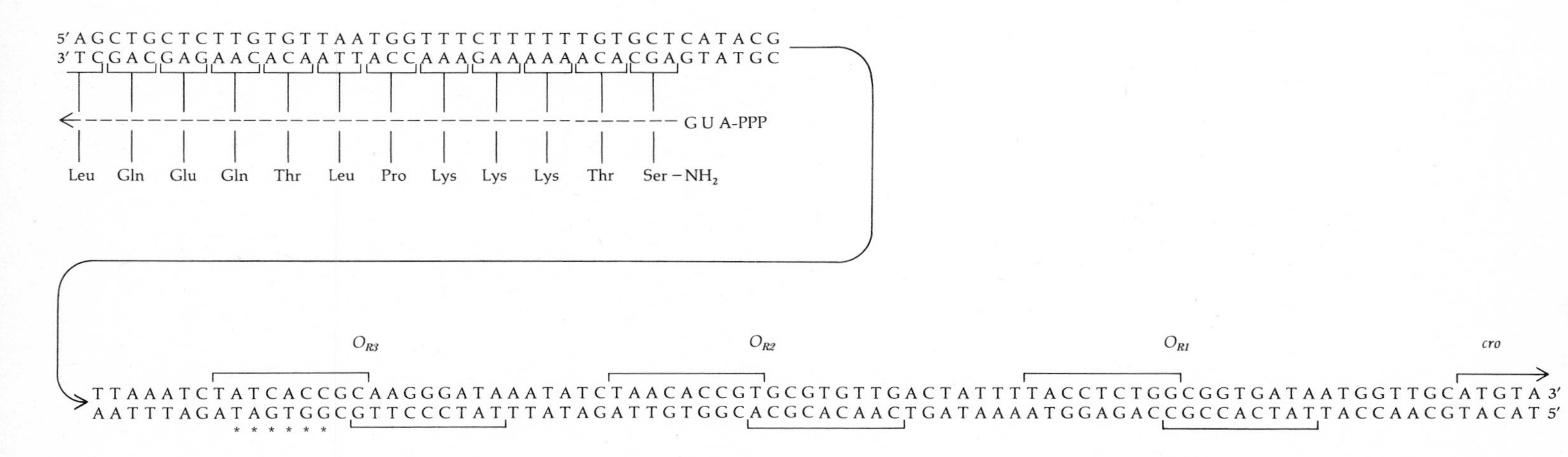

Figure 13.16
DNA, RNA, and protein sequences in and around the O_R and O_L operators in phage λ. The six repressor binding sites are indicated by brackets. The starting points of transcription of genes *N*, *cro*, and *cI* are indicated. Six bases on O_{R3} believed to code for a strong ribosomal binding site on the *cI* messenger are marked with asterisks. The O_L sequence is reversed from its orientation in Figure 13.15 to show homology with O_R. [After M. Ptashne et al., *Science* 194:156 (1976).]

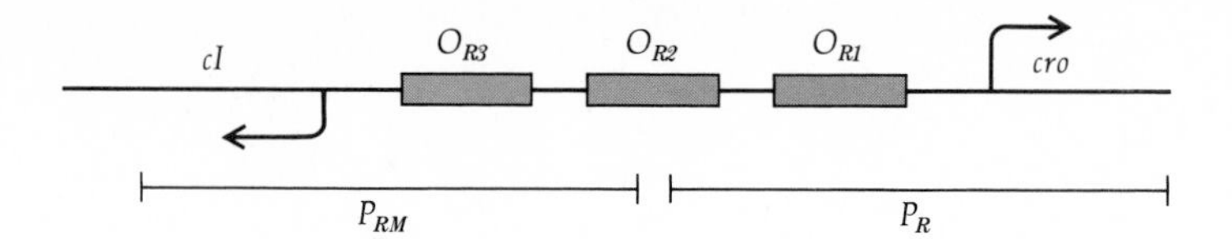

Figure 13.17
The promoters P_{RM} and P_R in phage λ overlap the repressor binding sites in O_R. Transcription starting points are indicated by arrows. At low repressor concentrations, O_{R1} is the first site to bind repressor. The binding of repressor at O_{R1} prevents transcription of the *cro* gene and stimulates transcription of the *cI* gene, perhaps in the same manner in which the [CAP-cAMP] complex stimulates transcription of the *lac* operon.

Binding of repressor to O_{R3} prevents further transcription from P_{RM}, limiting the synthesis of repressor. Thus the concentration of repressor controls the rate of synthesis of its own mRNA.

As discussed above, stage-II transcription provides functions necessary for both lytic and lysogenic pathways. The decision of which pathway is actually to be followed in an infected host cell depends, among other things, on the *cro*-gene product produced by both stage-I and stage-II transcription. The *cro* protein is a regulatory protein necessary for the adoption of the lytic pathway; *cro*$^-$ mutants always lysogenize infected cells and never produce plaques. *In vitro* studies show that the *cro* protein, like repressor, binds to O_R and O_L. However, the relative affinity of *cro* protein for the O_R binding sites is $O_{R3} > (O_{R2}, O_{R1})$. At low concentration, the *cro* protein binds preferentially to O_{R3}, which shuts off transcription of the *cI* gene from P_{RM} and *stimulates* transcription from P_R. At higher concentrations, *cro* protein binds to O_{R2} and O_{R1}, shutting off its own transcription. Thus *cro* protein has the mirror-image properties of the *cI* repressor (Figure 13.18).

Why the lytic pathway is followed in some infected cells and the lysogenic pathway in others is not known. Clearly, *cI* and *cro* gene expression provide for two mutually exclusive patterns of gene transcription, depending on whether *cro* protein or the *cI* repressor is more active, as diagrammed in Figure 13.18. The actual outcome is in some way a function of the physiological state of the infected cell. The identity and nature of any effector molecules that may activate or inhibit the activity of these two proteins are unknown.

Regulation of Transcription by Site-Specific Inversion

In *Salmonella,* the control of the synthesis of flagellin (the protein forming the flagella) involves a mechanism that alters the structure of the DNA itself. *Salmonella* bacteria possess two genes coding for different flagellins

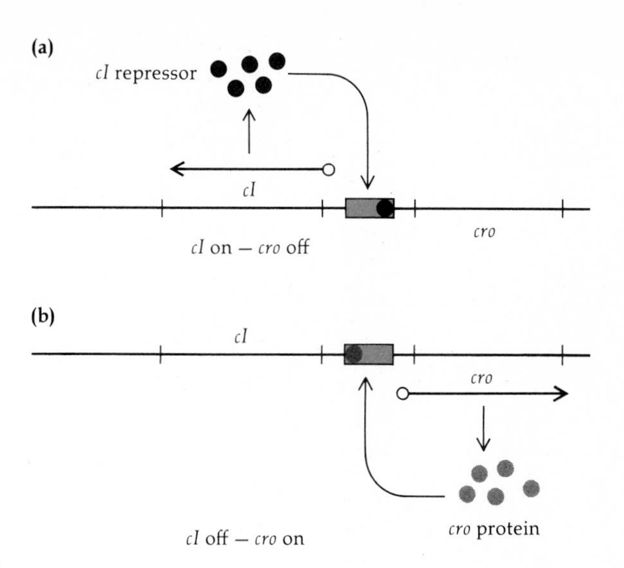

Figure 13.18
Two mutually exclusive patterns of gene transcription are possible at O_R. Depending on the relative activities (amounts) of *cI* repressor and *cro* protein, either pattern may become established and would be expected to remain stable. (**a**) This pattern would lead to lysogeny. (**b**) This pattern would lead to progeny synthesis and lysis.

(distinguishable by specific antibodies). These two proteins are designated H1 and H2; the structural genes for the two are not closely linked, and only one or the other gene is expressed in a cell. The *H2* gene is closely linked and coordinately expressed with another gene, *rH1*, which codes for a repressor of the *H1* gene. When *H2* and *rH1* are expressed, only H2 flagellin is synthesized, because the *rH1* repressor prevents expression of the *H1* gene. Alternatively, if *H2* and *rH1* are turned off, H1 flagellin is synthesized. The alternative expression of either the *H1* or *H2* gene is called *phase variation*. When a cell in either phase is grown to produce a large culture, the culture is found to contain cells in *both* phases. The rate at which cells undergo phase variation varies with different *Salmonella* strains from 1 shift per 10^5 cells per generation to 1 shift per 10^3 cells per generation. Phase variation may be one way in which the *Salmonella* bacteria avoid immunological destruction during an infection.

The control of phase variation depends on the state of the promoter for transcription of the *H2* and *rH1* genes. This has been shown by cloning of the control region in a plasmid vector, using recombinant DNA techniques described in Chapters 9 and 11. A single plasmid molecule, with the control region inserted, is grown in an *E. coli* host to produce a large population of molecules. The plasmid DNA is purified and made linear by cleavage with *Eco* RI restriction enzyme (Figure 13.19). Denaturation and reannealing of the single strands of DNA permit the formation of het-

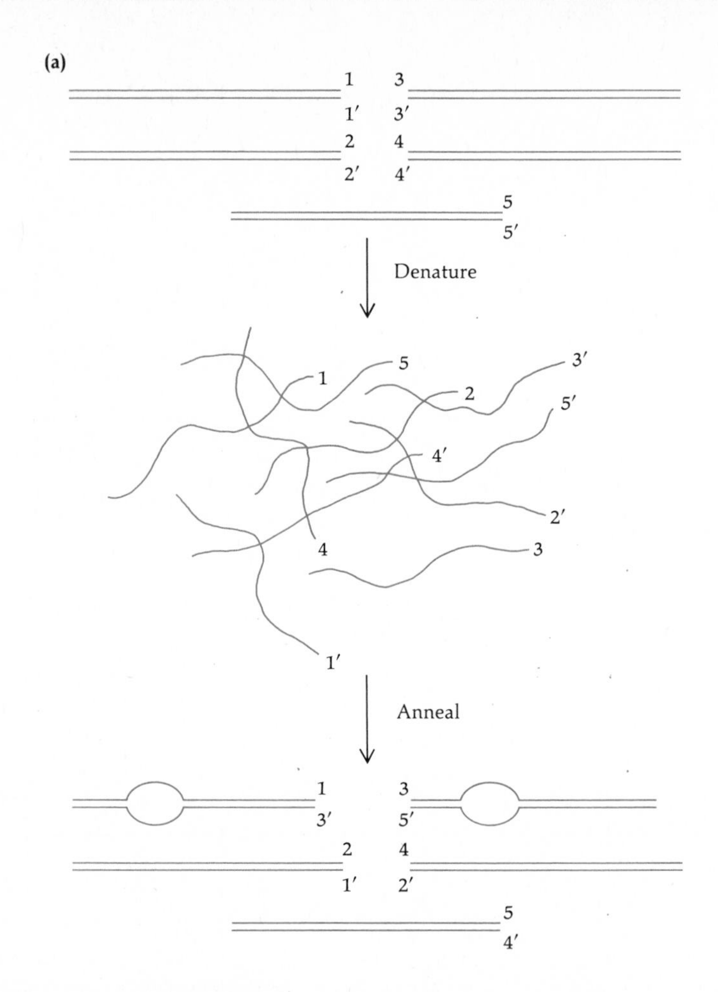

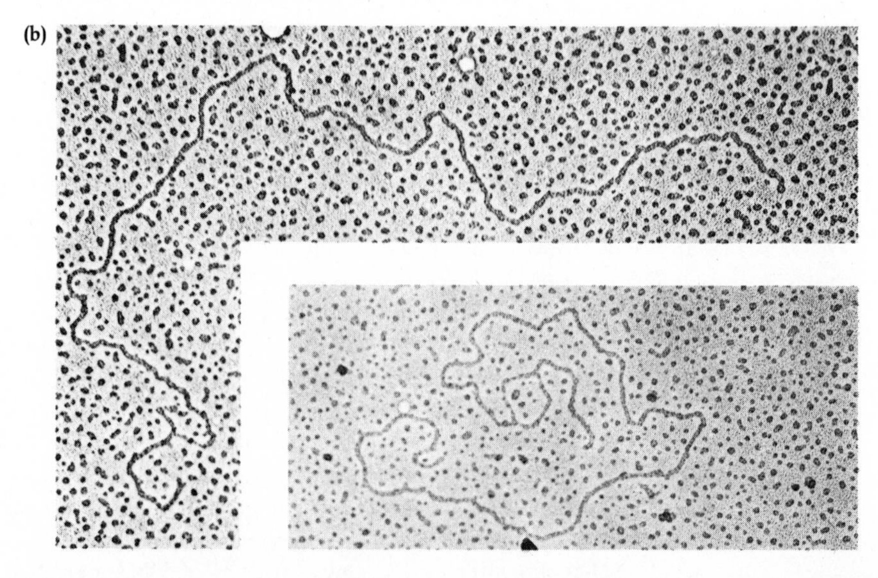

Figure 13.19
Evidence for a spontaneously invertible nucleotide sequence in the control region of the *H2-rH1* operon. (**a**) A cloned plasmid containing the control region as an insert is grown to produce a large population of plasmids whose DNA is made linear by cleavage with restriction enzyme *Eco* RI (the entire plasmid contains a single *Eco* RI site). The population of DNA molecules is then denatured and reannealed, and (**b**) the heteroduplex molecules are viewed by electron microscopy. The bubble present in some molecules is evidence for inversion occurring during growth of the plasmid population. [Electron micrographs from J. Zieg et al., *Science 196*:170 (1977). Copyright © 1977 by the American Association for the Advancement of Science.]

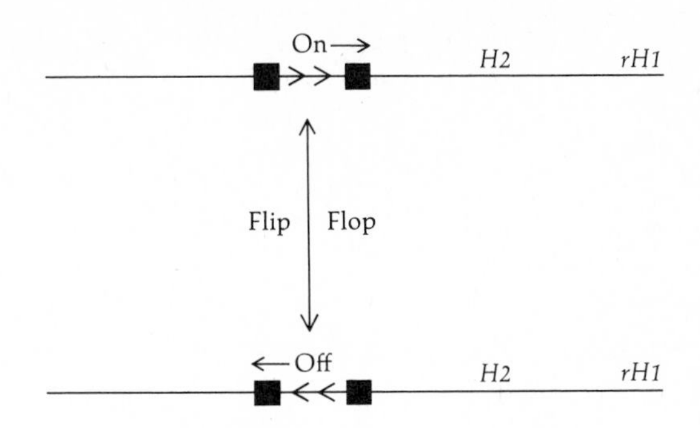

Figure 13.20
A model for phase variation by inversion of a nucleotide sequence containing the promoter for the *H2-rH1* operon. [After J. Zieg et al., *Science 196*:170 (1977).]

eroduplex DNA molecules between single strands from different plasmid molecules. Some of these heteroduplex molecules show a "bubble" 800 nucleotide pairs in size (Figure 13.19). The bubble is due to renaturation of single strands in which the 800-np region is in the opposite orientation. This observation has led to a model for phase variation in which the promoter for the *H2* and *rH1* genes is postulated to exist in one of two orientations: flip or flop. In one orientation, transcription initiated at the promoter leads to transcription of the *H2* and *rH1* genes; in the other, it does not (Figure 13.20).

Invertible nucleotide sequences about 3000 np in length have also been found in the genomes of phages P1 and Mu. In Mu the orientation of this segment during lytic growth affects the adsorption properties of the phage produced by the cell. A site-specific recombination mechanism, perhaps similar to those employed by insertion sequences or transposons, must be involved in site-specific inversion (see Chapters 7 and 9). This type of gene regulation differs from those discussed above in the role of the environment. Rather than responding to environmental change by a change in cellular gene expression, site-specific inversion prepares a population of cells for an altered environment by increasing the genetic variability of the population.

Problems

1. Prokaryotic organisms contain the following types of regulatory elements: repressors, activators, corepressors, promoters, operators, attenuators, and terminators. Which of these elements would you expect to be involved in all transcription units? All operons? All attenuated operons?

2. State the phenotypes of the following *lac* partial diploids of *E. coli* in terms of enzyme activities:

(a) $\dfrac{I^+ P^- O^+ Z^- Y^+}{I^+ P^+ O^c Z^+ Y^+}$ (d) $\dfrac{I^+ P^- O^+ Z^+ Y^+}{I^- P^+ O^+ Z^+ Y^-}$

(b) $\dfrac{I^+ P^- O^c Z^+ Y^+}{I^+ P^+ O^+ Z^- Y^+}$ (e) $\dfrac{I^+ P^+ O^c Z^+ Y^-}{I^- P^+ O^+ Z^- Y^+}$

(c) $\dfrac{I^- P^+ O^+ Z^- Y^+}{I^- P^+ O^+ Z^+ Y^-}$ (f) $\dfrac{I^+ P^+ O^+ Z^+ Y^+}{I^+ P^+ O^c Z^- Y^-}$

3. Classify the following as positive or negative regulatory elements (from the point of view of transcription):

(a) λ*N* gene product
(b) λ*cro* gene product
(c) Tryptophanyl-tRNA ligase
(d) Tryptophan
(e) λ*cII* gene product
(f) *lacI* gene product
(g) Phenylalanyl-$tRNA^{Phe}$
(h) $tRNA^{Phe}$
(i) ρ factor
(j) σ factor
(k) ppGpp
(l) β-Galactosidase
(m) cAMP
(n) λ*Q* gene product
(o) λ*cI* gene product
(p) *lacO* gene
(q) λO_R gene
(r) *trp* leader

4. *E. coli* cells carrying a lambda prophage with a temperature-sensitive *cI* mutation [*E. coli* (λcI^{ts})] will survive as lysogenic cells at 30°C, but at 42°C the cells lyse and release phage. The strain *E. coli* ($\lambda cI^{ts} N^- O^-$) grows as a normal lysogen at 30°C and is therefore immune to infection by λ phage. When this strain is grown at 42°C for several generations and is then tested at 30°C for immunity to λ infection, it is found that virtually all of the cells (99.9%) are now sensitive to λ infection and are killed, releasing progeny λ phage of the infecting type. However, cells of the strain *E. coli* ($\lambda cI^{ts} N^- O^-$) that have been grown at 42°C and then returned to growth at 30°C spontaneously reacquire immunity to λ infection during growth at 30°C, with a probability of about 1% per generation. From your knowledge of λ regulatory circuits, propose a hypothesis to account for these observations. How would you test your hypothesis?

5. Of the various mechanisms for regulating the transcription of genes discussed in this chapter, which one seems most likely to be confined exclusively to prokaryotic organisms?

6. Some *E. coli* mutations mapping in the *lacI* gene abolish the ability of the cells to utilize lactose as a carbon source. In partial diploids with *lacI*$^+$,

these mutations, designated *lacI*^{s} (for superrepressed), are dominant (the mutant phenotype is expressed rather than the normal phenotype). Propose a hypothesis accounting for the properties of *lacI*^{s} mutations. Your hypothesis should suggest testable predictions. What are your predictions and how would you test them?

7. The initial observations of Jacob and Monod that *lacI*$^-$ mutations are recessive is consistent with the hypothesis that the *lacI*$^+$ gene specifies a repressor RNA molecule that acts on the operator to prevent transcription, either by forming an R loop with the operator DNA sequence or by annealing to the transcribed operator mRNA and preventing further transcription of the operon. What *genetic* experiments (not *biochemical* experiments) can you propose that would help decide whether the *lac* repressor is an RNA or protein molecule?

8. Phage λ with R^- mutations is unable to lyse the host bacterium upon infection, even though progeny phages are produced in the infected cell. However, when λR^- phage infects cells that are lysogenic for the heteroimmune lambdoid phage 434 (see Chapter 6), the cells lyse and release λR^- progeny (no 434 phages are produced). Form a hypothesis to account for this observation, and state how you would test your hypothesis.

9. Suppose Jacob and Monod had originally worked with I^{-D} mutations instead of the I^- mutations discussed in this chapter. Would their conclusions about the nature of β-galactosidase regulation have been the same? How might they have differed?

14

Regulation of Gene Expression in Eukaryotes

Most eukaryotes are multicellular organisms in which groups of different cells are specialized to carry out different functions. Two aspects of eukaryotic gene expression must be distinguished: that related to the maintenance of cell function ("housekeeping" functions) and that related to the expression of specialized cell functions (differentiated functions). The former functions are found in all cells; the latter functions, being cell-specific, require that the cells select which functions to express, based on their location in the organism.

The number of proteins coded by an organism's DNA is a reasonable estimate of the number of genetic functions required for the development and reproduction of the organism. The sequence *complexity* (N) of an organism's DNA places an upper limit on the number of proteins that can be coded (see Chapter 4). However, probably only a fairly small fraction of eukaryotic DNA sequences actually code for proteins. Many sequences that are transcribed into hnRNA are never transported to the cytoplasm to form mRNA that is translated. End sequences and intervening sequences are removed from hnRNA during maturation to mRNA (Chapter 11). Only about 10% of the sequences transcribed actually form mRNA. If we assume that all single-copy DNA in a eukaryotic genome is transcribed in one cell or another during development, we can estimate the maximum number of proteins that constitute an organism. The human genome possesses about 2.9×10^9 np, 70% of which exist as single-copy DNA. Therefore about 2×10^8 nucleotides is the complexity of the mRNAs. An average-sized protein is coded by an mRNA of about 1800 nucleotides.

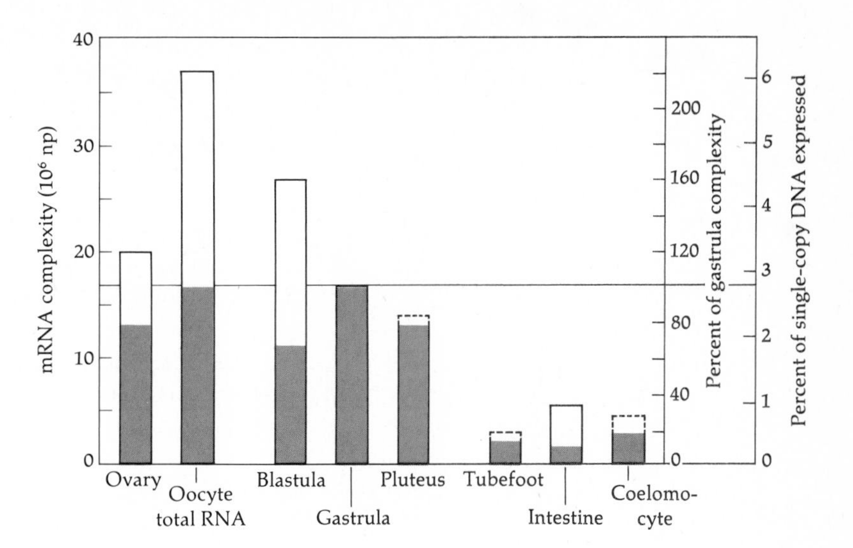

Figure 14.1
Complexity of mRNA molecules found in adult tissues, embryonic stages, and eggs of the sea urchin. The solid portion of each bar indicates the amount of single-copy nucleotide sequences shared between gastrula mRNA and the mRNA from other sources. The open portion shows the amount of single-copy nucleotide sequences present in the various mRNAs indicated but absent from gastrula mRNA. [From G. Galau et al., *Cell* 7:487 (1976).]

Therefore the maximum number of different human proteins is about 110,000. Similar calculations provide maximum estimates of 17,000 proteins for the sea urchin and 7250 proteins for *Drosophila melanogaster*. In the case of *Drosophila*, this estimate is in reasonably good agreement with the number of complementation groups estimated from mutational studies and with the number of bands in the polytene chromosomes (5000 to 6000).

A comparison of the sets of mRNA molecules present in several adult tissues, embryonic stages, and eggs of the sea urchin is shown in Figure 14.1. The gastrula stage of embryogenesis contains mRNA sequences of a complexity of 1.7×10^7 nucleotides, as determined by the kinetics of RNA-DNA hybrid formation. This quantity represents less than 3% of the unique-sequence DNA in the sea urchin genome, but is still enough to code for approximately 10,000 different average-sized proteins. The solid portions of the bars in the figure show the fraction of these sequences present in other tissues and embryonic stages. The open portions of the bars show the complexity of mRNA sequences present in the other sets of mRNA species but not present in the gastrula. There are three noteworthy features of the data in Figure 14.1. First, a small fraction of these mRNA species are shared by all the cells (about ½ of 1% of the unique sequences present in the genome). These mRNA molecules may code for "housekeeping" proteins that are responsible for the metabolic processes common to all cells. Second, all of the mRNA sequences present in the gastrula are also present in the mRNA sequences of the egg; they are put there during the process of oogenesis. Third, egg and gastrula contain many more mRNA sequences than do the differentiated adult tissues, indicating that many more genetic functions are being expressed in these cells.

Biochemical observations indicate that different cells in a eukaryotic organism share the synthesis of some sets of proteins but are distinguished from one another by the synthesis of cell-specific sets of proteins. Moreover, the amount of each type of protein that is synthesized varies from cell type to cell type and from one stage to another during development. Hemoglobin molecules of different types are synthesized by blood-forming cells at different stages during human development. Crystalline proteins forming the lens of the eye are synthesized only by those ectodermal cells of the head that come into contact with the developing retina. These examples of the expression of characteristic sets of mRNAs and proteins by different cells obviously reflect genetic regulation. Furthermore, the decisions made by mitotic sister cells to become blood-forming, retinal, or lens cells, based on their relative positions in the developing embryo, also reflect genetic regulation.

Exploration of the mechanisms of gene regulation in eukaryotes is one of the most active endeavors of modern genetics. This chapter outlines some of the fundamental observations of genetic regulation during development and differentiation.

Differential Gene Expression

Most eukaryotic organisms begin their development as a zygote with one diploid nucleus. Ensuing mitoses produce many progeny nuclei, contained within cells produced by cleavage of the zygote. The cells produced during these early cleavage divisions are destined to give rise to different and specialized types of cells upon further development of the embryo. What mechanisms cause sister nuclei, each descended from a common zygotic nucleus, to express different sets of genes and to channel their progeny nuclei into different developmental pathways? Most of the cells of an animal are somatic cells, destined to die and never contribute to future generations of organisms. Only the germ cells, which are segregated early in embryogenesis, demonstrate *totipotency* (the ability to repeat all of the steps of development and give rise to all cell types). Does cellular differentiation following the selection of one specific developmental pathway result in the selective elimination or permanent inactivation of those genes that are not to be expressed?

Several types of experiments have suggested that developmental processes do not of necessity lead to irreversible changes in the cell's nucleus. These experiments argue that *differential gene expression* occurs in somatic cells during development.

One of the most direct experiments is the demonstration of totipotency by transplantation of a nucleus from a differentiated somatic cell to an enucleated zygote. John Gurdon, using the African clawed toad, *Xenopus laevis*, has produced adult toads in this way. Unfertilized eggs are functionally enucleated with a massive dose of UV radiation, and a single nucleus from a differentiated tadpole cell is injected into each egg to initiate development. In a number of cases the eggs with the injected nuclei developed into mature toads (Figure 14.2). In order to prove that it was the transplanted nucleus that carried out development rather than the original egg nucleus, which might have survived the UV radiation, a genetic marker was employed. The eggs came from a strain exhibiting two nucleoli per nucleus, one for each nucleolus organizer on two homologous chromosomes. The somatic-cell nucleus came from an individual heterozygous for a deletion of the nucleolus organizer and exhibiting only one nucleolus per nucleus. All the nuclei in the cells of the toad that developed from the transplanted nucleus had only one nucleolus. Thus, in these cases, all of the genetic information necessary for normal development is present in the nucleus of a differentiated cell and capable of reactivation to repeat the developmental process.

Maternal Effects on Development

Regardless of the nature of the genetic control systems that operate during development, the cytoplasm of the egg plays a central role in selecting which groups of genes will be activated first during development. Many maternal-effect mutations are known that cause the mutant mother to produce eggs that fail to develop normally and cause the embryo to die, regardless of its own genotype.

Ova-deficient (*o*) is a maternal-effect mutation in the Mexican axolotl. Homozygous *o*/*o* females produce embryos that fail to gastrulate and whose embryonic nuclei never begin to synthesize RNA. Injection of only 2% of the whole cytoplasm of normal mature eggs, or nucleoplasm from the normal oocyte nucleus, into the eggs of *o*/*o* females provides enough o^+ function to permit the normal development of the eggs (Figure 14.3). This substance may be required to activate those genes responsible for gastrulation and the beginning of organogenesis.

The recessive *Drosophila* mutation *bicaudal* causes homozygous females to produce eggs that develop into double-abdomen monsters. The anterior half of these embryos fails to develop normally; instead, it forms a second abdomen in mirror-image symmetry with the normal larval abdomen in the posterior half of the embryo.

Maternal-effect mutations demonstrate that development cannot proceed normally in the absence of a properly prepared egg cytoplasm, produced during oogenesis under control of the maternal genome. It is estimated that the *Drosophila* X chromosome (20% of the entire genome) contains about 160 cistrons capable of mutation to maternal-effect lethality (see Chapter 16). This is a significant fraction ($\approx$10%) of the estimated number of cistrons on the chromosome. The nature of the oo-

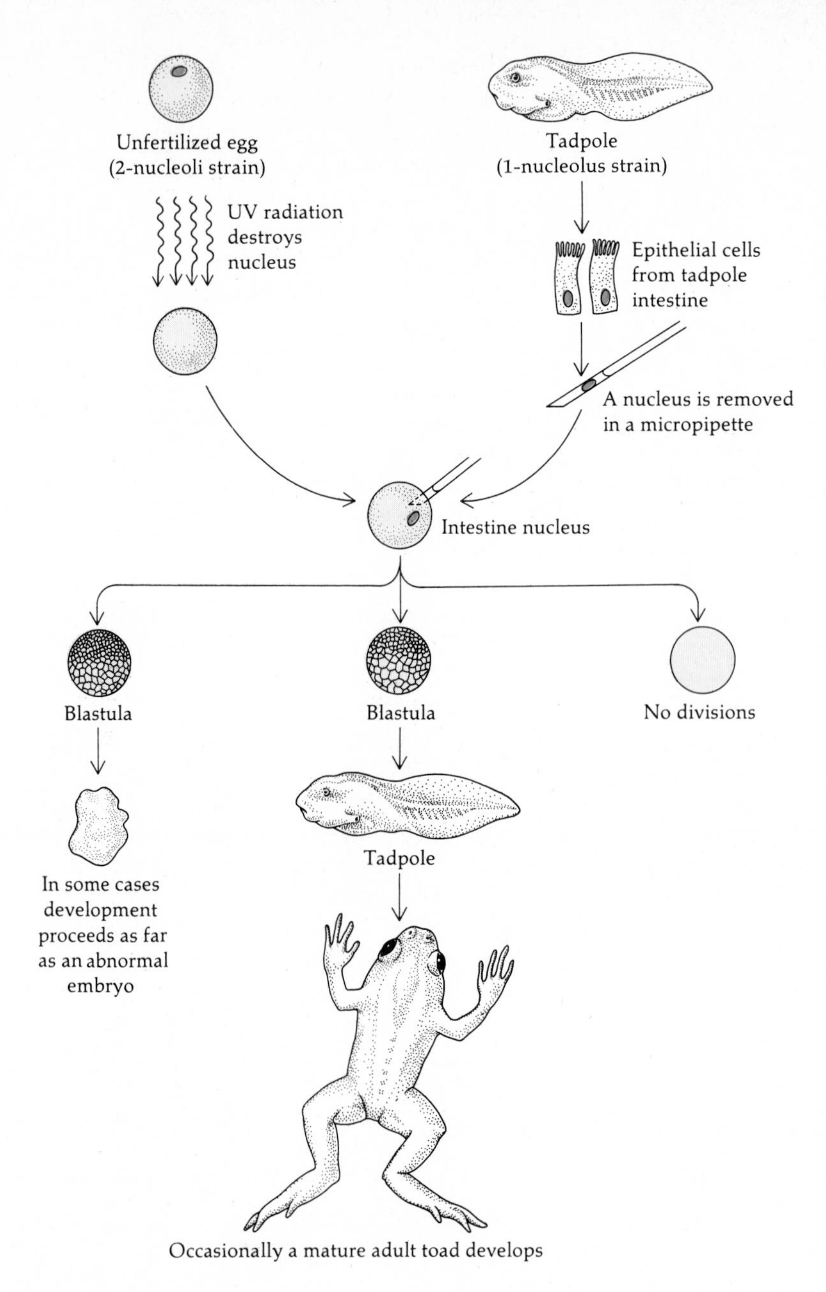

Figure 14.2
Procedure employed to demonstrate the totipotency of the nucleus from a differentiated cell. (After J. B. Gurdon, *Scientific American,* December, 1968, p. 24.)

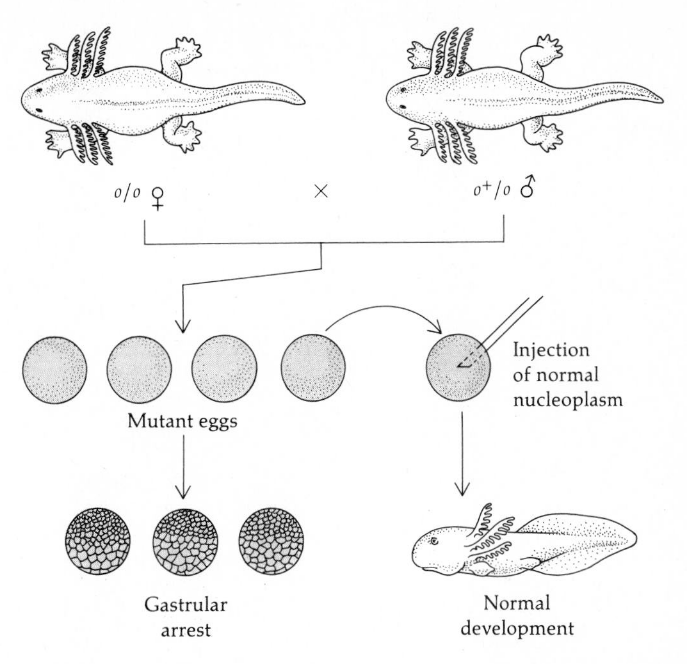

Figure 14.3
Evidence that factors essential for embryonic gene expression are controlled by maternal genes is provided in part by studies of the Mexican axolotl (*Ambystoma mexicanum*). Females homozygous for the mutant gene *o* produce eggs that cease development at gastrulation and fail to synthesize RNA, regardless of the genotype of the embryo (o/o or o^+/o). Injection of nucleoplasm from normal oocytes into fertilized eggs from the mutant females permits normal development. The active substance representing the o^+ function may be an acidic protein. [After A. Brothers, *Nature 260*:112 (1976).]

plasmic defects caused by these mutations is known only in a very few cases, and from present information it is not possible to determine whether most of these mutations cause general metabolic defects or whether they identify functions necessary for differential gene activation in specific embryonic cells.

Germ-cell determination in *Drosophila* is due to localized cytoplasmic determinants that are present in the posterior ooplasm. The germ cells are derived from the pole cells (Figure 14.4), which form prior to the cellular blastoderm (see below) when some early cleavage nuclei reach the posterior end of the egg. The presence of pole-cell determinants has been demonstrated by the injection of posterior ooplasm into the anterior end of other eggs, where pole cells then form, producing an embryo with pole cells at both anterior and posterior ends. The functionality of these anterior supernumerary pole cells is demonstrated by their transplantation to the posterior end of a third, preblastoderm embryo, where they integrate

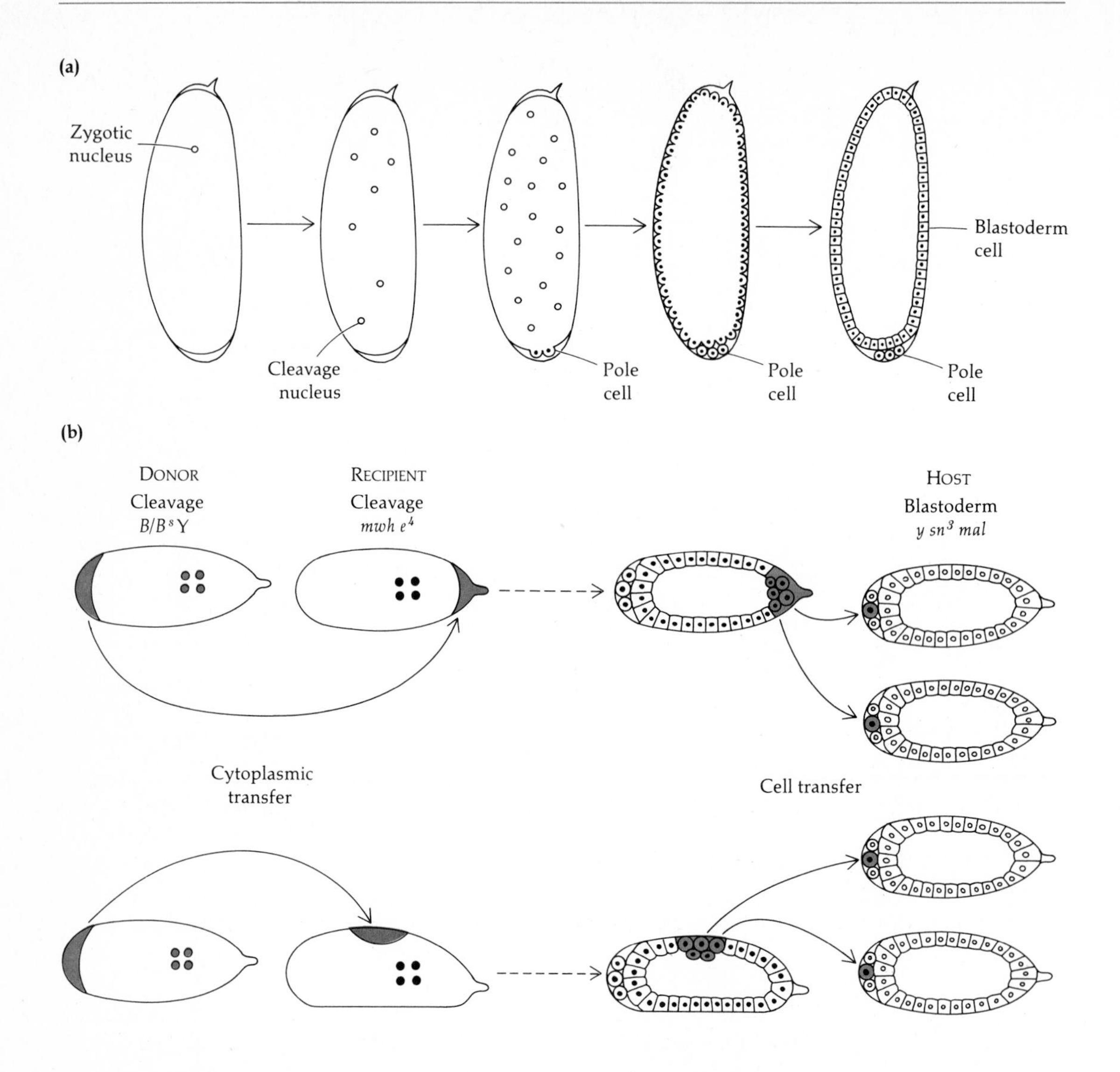

Figure 14.4

(a) Pole cells at the posterior end of the *Drosophila* embryo form prior to the cellular blastoderm. At least some of the pole cells later migrate to the somatic gonad to form the primordial germ cells. (After W. Gehring, in *Genetic Mechanisms of Development*, ed. by F. H. Ruddle, Academic Press, New York, 1973, p. 107.) **(b)** The presence of pole-cell determinants in the posterior ooplasm of *Drosophila* eggs is demonstrated by transplantation of the ooplasm to regions of other eggs where pole cells do not normally form. The cytoplasmic donor shown here is from a stock in which both X and Y chromosomes are marked by the dominant mutations *Bar* (B and B^s) in order to detect any co-transfer of donor nuclei together with the donor polar cytoplasm. The recipient eggs are from a stock homozygous for the 3rd-chromosome mutations *multiple wing hair* (*mwh*) and *ebony* (e^4) body color. The transplanted polar plasm induces the formation of pole cells in the recipient egg at the site of injection. The functional nature of the induced pole cells is demonstrated by their transplantation to the posterior end of host embryos whose X chromosome is marked with the recessive mutations *yellow* (*y*) body color, *singed* (sn^3) bristles, and *maroonlike* (*mal*) eye, where they participate in germ-cell formation. The adult flies that develop from the host embryos produce some (but not all) gametes carrying the *mwh* e^4 markers rather than the *y* sn^3*mal* markers. Thus, the induced pole cells whose genotype is that of the recipient egg are true pole cells. (After K. Illmensee, in *Insect Development*, ed. by P. Lawrence, Blackwell, London, 1976, p. 86.)

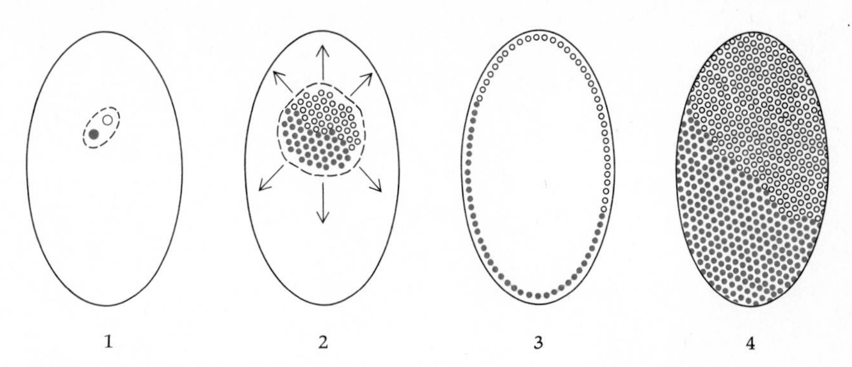

Figure 14.5
Diagram of the cleavage divisions giving rise to the syncitial blastoderm of *Drosophila*. The three sections on the left show: (1) the two nuclei present after the first cleavage division, in which one (white) has lost one X chromosome, (2) the egg following several more cleavage divisions to produce a cluster of nuclei that are half XX (colored) and half X0 (white), and (3) the egg following nuclear migration to the cortex. The section on the right (4) shows the surface of the syncitial blastoderm, consisting of half X0 and half XX nuclei, prior to cell-membrane formation. Note the sex-dividing line formed at the interface between X0 and XX nuclei. (After Y. Hotta and S. Benzer, in *Genetic Mechanisms of Development*, ed. by F. H. Ruddle, Academic Press, New York, 1973, p. 129.)

with the pole cells of that embryo and participate in gamete formation in the adult (Figure 14.4). Some maternal-effect mutations, for example, *grandchildless,* affect the migration of nuclei into the posterior polar plasm; then pole cells do not form, and the resulting adults are sterile (grandchildless with regard to the mutationally affected mother).

Fate Map of the *Drosophila* Blastoderm

Following fertilization of the *Drosophila* egg, the zygotic nucleus and ensuing daughter nuclei undergo nine synchronous divisions, forming a cluster of nuclei in a common ooplasm. Following the ninth division, the vast majority of these nuclei migrate to the egg cortex, undergo three more divisions, and form a *syncitial* (nuclei with a common cytoplasm) blastoderm (Figure 14.5). Shortly thereafter, membranes form, enclosing each nucleus within a cell and producing a cellular blastoderm. Gastrulation ensues, followed by organogenesis. From the egg hatches a larva composed of two types of cells: larval and imaginal. The larval cells make up the body of the larva. As the larva feeds, grows larger, and periodically molts, the larval cells do not divide, but increase in size. The imaginal cells, however, generally increase in number by cell division during larval growth to produce the imaginal disks (Figure 14.6). Upon pupation, the

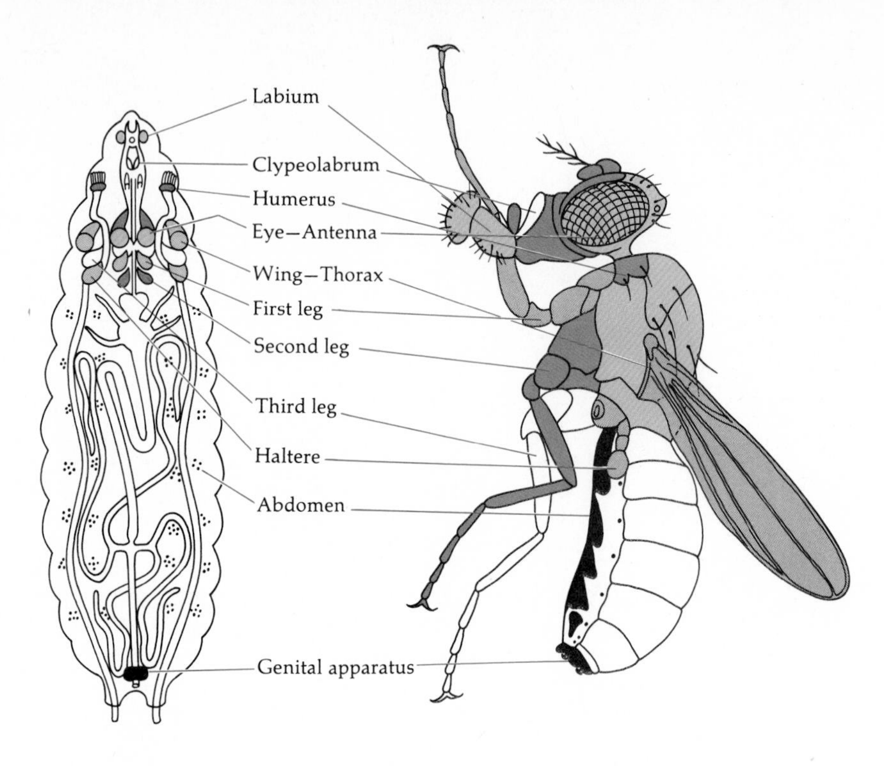

Figure 14.6
Imaginal disks in a mature larva of *Drosophila*. The abdominal tergites and sternites develop from smaller primordia called histoblasts, shown dotted.

imaginal disks differentiate, producing specific portions of the adult body (Figure 14.7).

Imaginal disks can be cultured as implants in the abdomens of adult females, where hormonal conditions permit cell proliferation but prevent differentiation (see Figure 10.9). The imaginal disks can be made to differentiate by transplantation to the body cavity of a larva just prior to that larva's pupation. The hormonal changes occurring in the pupating larva induce the transplants to differentiate, thus expressing their determined state; e.g., eye imaginal cells differentiate into eye structures, and so on.

Several lines of evidence indicate that the primordial cells for larval and imaginal organs are determined at or soon after the time of cellular blastoderm formation. First, cells isolated from the anterior portion of blastoderm embryos give rise only to anterior imaginal structures upon *in vivo* culture in adults, prior to their transfer to pupating larvae. Conversely, cells isolated from the posterior blastoderm give rise only to posterior imaginal structures.

Second, the removal of cells from specific areas of the blastoderm surface produces specific damage to the adults that eventually develop from the damaged embryo. These results suggest that the position of cells

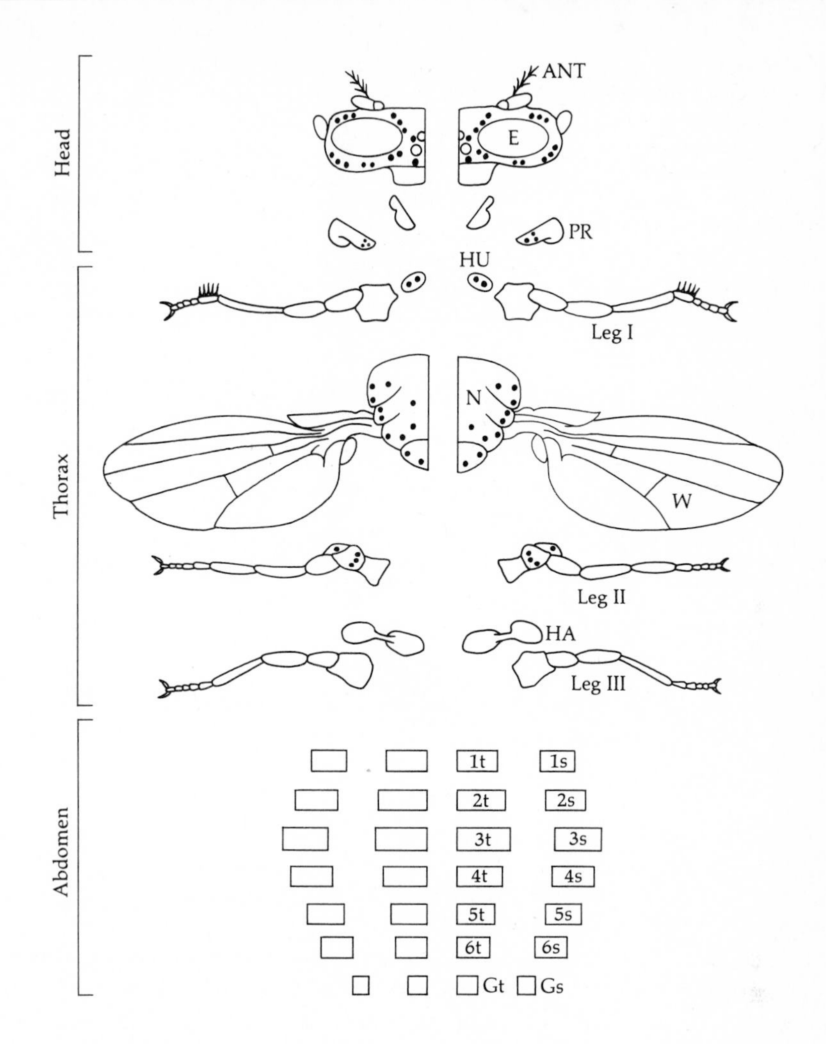

Figure 14.7
Exploded view of the external body parts of the adult *Drosophila* fly. Each part develops from a separate primordium or imaginal disk during pupation. The parts are sutured together prior to emergence of the fly from the pupal case. Abbreviations: ANT, antenna; E, eye; PR, proboscis; HU, humerus; N, notum; W, wing; HA, haltere; t, tergite; s, sternite; G, genital. (After Y. Hotta and S. Benzer, in *Genetic Mechanisms of Development*, ed. by F. H. Ruddle, Academic Press, New York, 1973, p. 129.)

in the blastoderm determines the fate of those cells throughout the course of development. Finally, studies of gynandromorphs, to be described below, suggest that the cleavage nuclei that migrate to the egg cortex are totipotent and become fated depending upon their position in the egg cortex.

Sexual differentiation in *Drosophila* is cell-autonomous and is specified directly by the chromosomal composition of each cell. As was mentioned in Chapter 3, the ratio of X chromosomes to autosomes determines sex; X0 zygotes develop as males that are, however, sterile because of the absence of Y-chromosome genes that are necessary for spermatogenesis.

Zygotes with two X chromosomes sometimes lose an X chromosome from one of the two nuclei that are formed after division of the zygotic

Figure 14.8
Typical gynandromorphs of *D. melanogaster*. Body parts derived from XX cells (color) are female; parts derived from X0 cells (white) are male and express recessive mutant genes that are uncovered by the loss of the other X chromosome, bearing the wild-type alleles. (After Y. Hotta and S. Benzer, in *Genetic Mechanisms of Development*, ed. by F. H. Ruddle, Academic Press, New York, 1973, p. 129.)

nucleus. Certain mutations, or rearrangements of the X chromosome to a ring form, greatly enhance the frequency of X elimination during the early cleavage divisions. The X0/XX mosaics that result give rise to adult flies that are approximately half male and half female in phenotype; these are called *gynandromorphs*. Gynandromorphs do not consist of a disorganized mixture of female and male cells. A clear dividing line between cells of different sexes is always present; it is accentuated if the male cells contain an X chromosome bearing recessive mutations affecting cuticular color, eye color, or bristle structure, while the female cells are heterozygous for the wild-type alleles. The sex-dividing line may divide the adult fly in any of a large number of ways, some of which are indicated in Figure 14.8.

In 1929 A. H. Sturtevant described a method for mapping the primordia of adult imaginal structures by noting the frequency, in a population of gynandromorphs, with which the sex-dividing line separates two adult structures. (Recall that Sturtevant also showed how recombination data could be used to construct a map of the chromosome—see Chapter 5.) This method was eventually (1969) used by Antonio Garcia-Bellido and John Merriam to construct a fate map of the blastoderm, showing the physical relationship between the primordia.

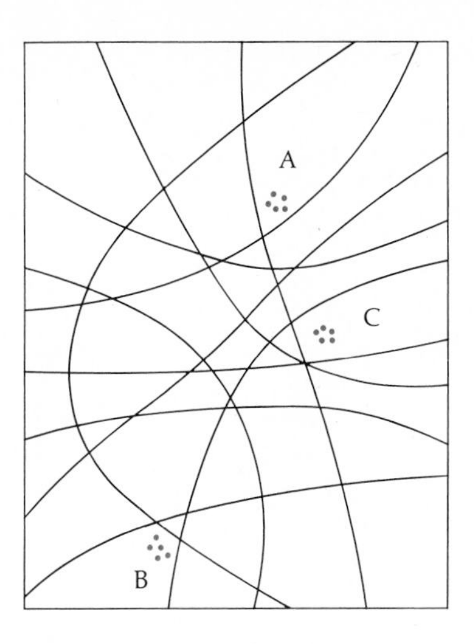

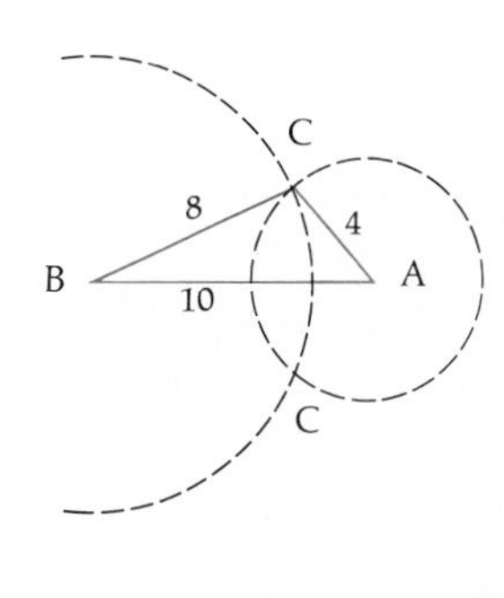

Figure 14.9
Principles of fate mapping of organ primordia. The two-dimensional closed blastoderm surface is a sheet of cells, which is shown opened into a plane on the left. Three primordia (A, B, and C) are indicated. A number of possible sex-dividing lines cutting the blastoderm surface in two (male and female) are drawn; some fall between the primordia. The greater the separation of two primordia, the greater is the chance that the sex-dividing line will fall between them in any particular gynandromorph. In that event, the corresponding adult body parts derived from them will be of opposite sex and genotype. The proportion of gynandromorphs in which two body parts differ in sex and genotype is a measure of the distance between their primordia or sites on the blastoderm surface. One *sturt unit* is defined as a distance such that the sex-dividing line passes between two sites in 1% of all gynandromorphs. In the right diagram, sites A and B are 10 sturts apart; if site C is 4 sturts from A and 8 sturts from B, then C is located at one of the two intersections shown. Choosing between these two possible sites requires mapping of the three sites with respect to additional sites. (After Y. Hotta and S. Benzer, in *Genetic Mechanisms of Development*, ed. by F. H. Ruddle, Academic Press, New York, 1973, p. 129.)

The principles of both fate mapping and gene mapping are very similar, except that in fate mapping, sites on the two-dimensional surface of the blastoderm are mapped, rather than sites on the one-dimensional length of the chromosome. In fate mapping, the farther apart two primordia are, the more frequently the sex-dividing line will fall between them. For example, if 10% of all gynandromorphs possess first and second legs of different sex, then the primordia for these two legs are said to be separated by 10 *sturt units* on the fate map (see Figure 14.9).

The fate map produced from the detailed quantitative study of gynandromorphs bears an uncanny relationship along the anterior-posterior axis to the organization of the adult fly (Figure 14.10). The egg itself possesses distinct anterior-posterior and dorsal-ventral axes, which reflect those of the female that produced the egg. The construction of a coherent fate map of the blastoderm would be possible only if the *position* to which a

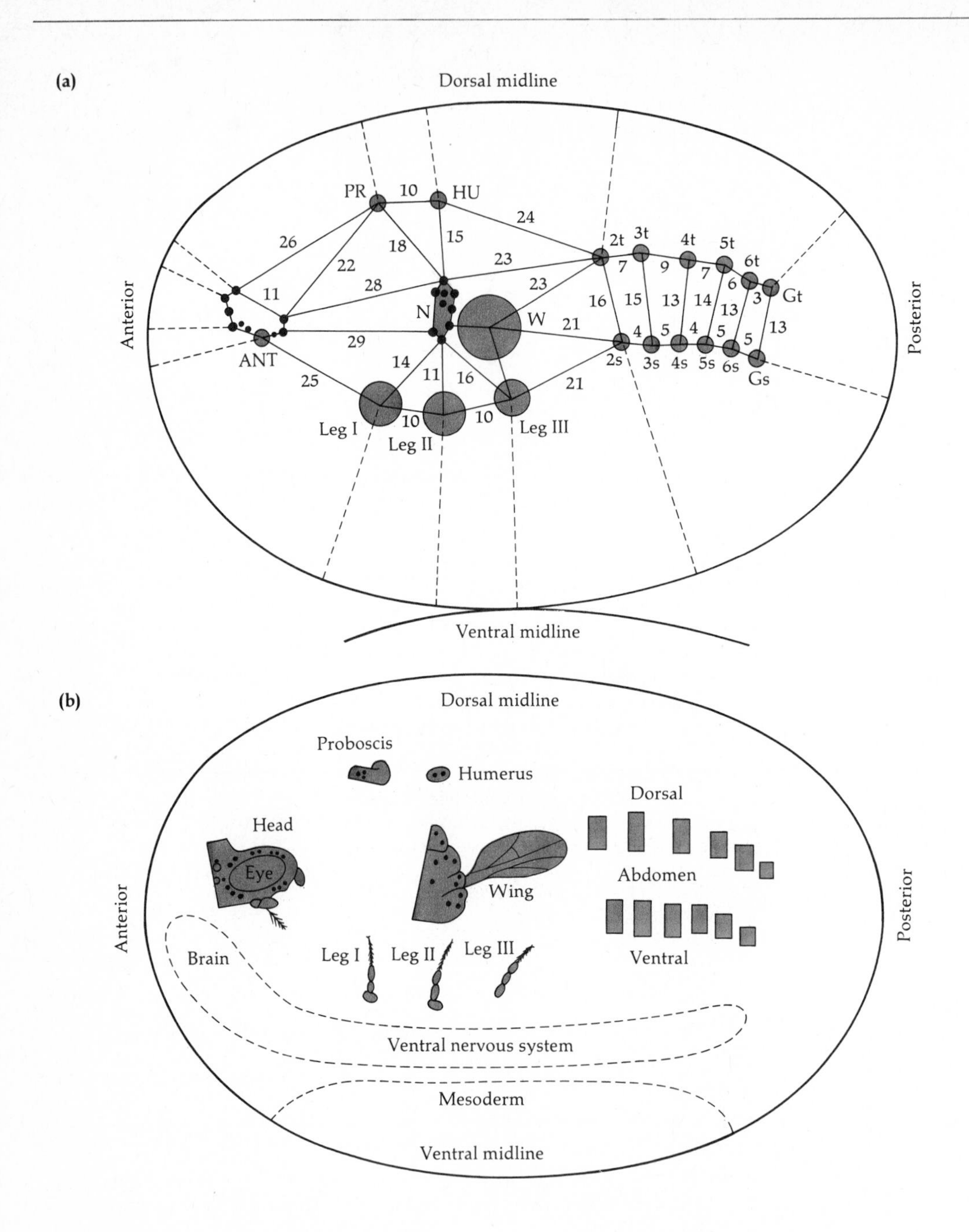

Figure 14.10

(a) Fate map of *D. melanogaster* blastoderm, constructed as described in Figure 14.9, showing sites of cells that will eventually develop into the indicated external body parts of the adult fly. Distances are given in sturts. Dashed lines indicate distances to nearest midline of the fly. The map is viewed from the inside of the hollow blastoderm, looking outward. See Figure 14.7 caption for abbreviations. **(b)** External parts of the adult fly depicted on blastoderm surface. Dashed lines indicate areas that give rise to the nervous system and the mesoderm, according to embryological studies. (After Y. Hotta and S. Benzer, in *Genetic Mechanisms of Development*, ed. by F. H. Ruddle, Academic Press, New York, 1973, p. 129.)

totipotent cleavage nucleus moved during blastoderm formation determined the fate of that nucleus. This is a consequence of the essentially random orientation of the first cleavage division (Figure 14.5), as indicated by the different orientations of the sex-dividing line in different gynandromorphs. Thus, within the egg cortex some type of *positional information* must exist that requires cells formed in different positions to follow specific developmental pathways.

Homeotic Mutations of *Drosophila*

The adoption of particular developmental pathways by blastoderm cells (cell determination) involves the specification of position in the egg cortex. This is a maternal function, as indicated by the maternal nature of the *bicaudal* mutation described above. For cell determination to be effected, however, the positional information must be correctly interpreted by the blastoderm cells in order to select the appropriate genes for activation. Thus, there should be mutations that interfere with the correct interpretation of positional information; these mutations should be expressed by the embryonic genome rather than the maternal genome. *Homeotic mutations,* which cause specific cells to substitute one developmental pathway for another, represent this type of mutation.

A number of homeotic mutations are known. The mutation *Antennapedia* causes legs to develop in place of antennae; *ophthalmoptera* causes wing tissue to replace eye tissue (Figure 14.11); *proboscipedia* causes the proboscis to develop either as a leg or as the distal part of an antenna, depending on the temperature during development; and *tumorous head* causes head tissue to be replaced by several different types of tissue, including genital structures.

Insects are segmented organisms exhibiting various degrees of specialization of each segment; they have presumably evolved from primitive, wormlike creatures in which little differentiation between segments was present (see Figures 14.7 and 14.12). Flies almost certainly evolved from insects with four wings instead of two, and insects are believed to have developed from ancestors with many legs instead of six. E. B. Lewis has postulated that during the evolution of the fly, two major groups of genes must have evolved: "leg-suppressing" genes, which removed legs from abdominal segments of millipede-like ancestors, and "halterepromoting" genes, which suppressed the posterior pair of wings of four-winged ancestors.

The most spectacular homeotic transformations are effected by mutations of the *bithorax* gene complex. From studies of a variety of mutations and deficiencies of this complex, the wild-type functions of the complex have been inferred to be the selection of segment-specific developmental pathways. The map positions and locations in the salivary-gland polytene chromosomes of these mutations are shown in Figure 14.13. The most extreme mutation studied is the homozygous deficiency for the entire

Figure 14.11
The head of the *Drosophila melanogaster* mutant *ophthalmoptera* produces wings at 29°C but is normal at 17°C. Wing tissue, identified by wing hairs and the rows of dark bristles along the anterior edges of the wings, replaces normal eye cells. The wings here appear abnormal because the upper and lower wing epithelia are ballooned by lymph fluid rather than being in contact, as in the normal flat wing on the thorax. [From J. H. Postlethwait, *Develop. Biol.* *36*:212 (1974).]

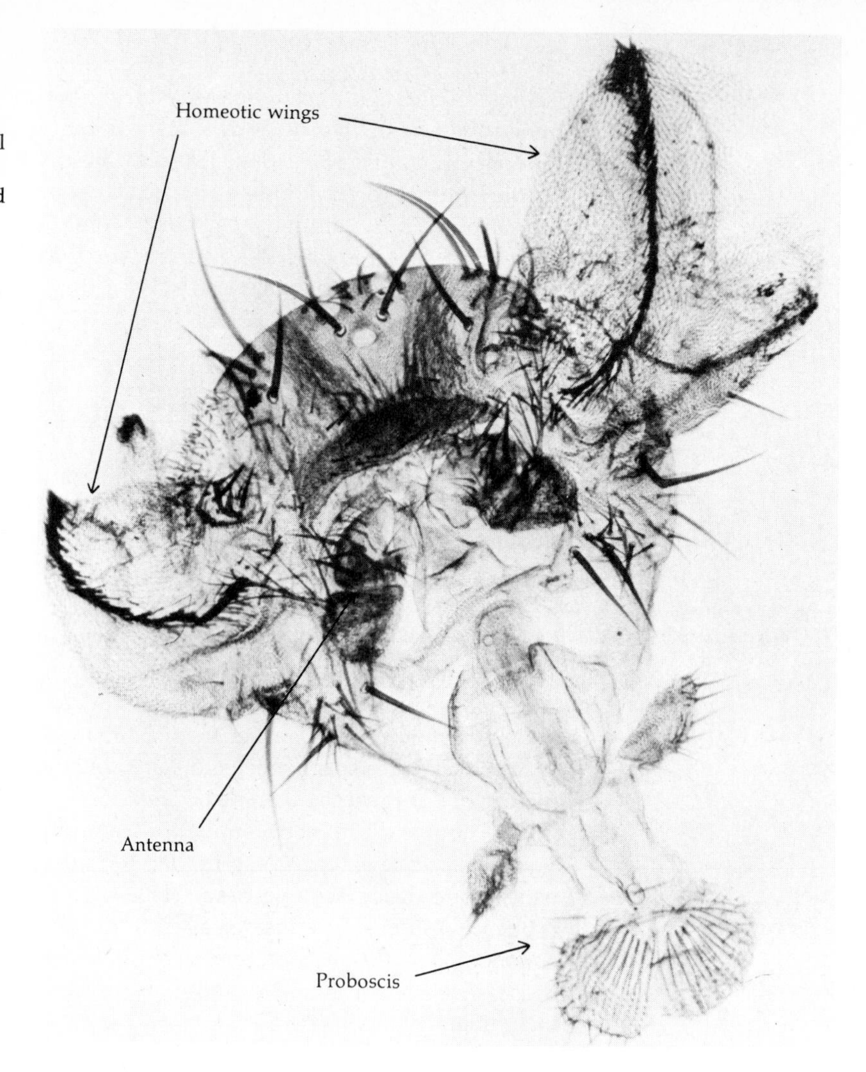

Figure 14.12 (*opposite*)
Genetic control of segmental levels of development in thoracic and abdominal segments of *Drosophila melanogaster*. **(a)** Anatomy of third-instar larva. *Head*: dorsal sense organs (DSO) and protruding portion of mandibular hooks (MH). *Thoracic segments*: each bears a pair of Keilen's organs (KO), three sensory hairs in a depression; also ventral pits (VP) and ventral setal bands (VSB) with fine, toothlike setae. *Abdominal segments*: VSB with coarse, toothlike setae; KO and VP lacking; each dorsal longitudinal trunk (DLT) terminates in an anterior spiracle (ASP) and a posterior spiracle (PSP). **(b)** Anatomy of adult female. *Thoracic segments*: coxae (C1, C2, and C3) of prothoracic legs (PRL), mesothoracic legs (MSL), and metathoracic legs (MTL); wing (W) and haltere (H), being dorsal, are shown in dotted outline. *Abdominal segments*: AB1 to AB7 bear sternital sensillae (SS); AB2 to AB7 bear bristled sternites (ST); AB2 has a raised lump, or Wheeler's organ (WO). **(c)** Levels of development (L), achieved in wild-type segments (vertical array of connected boxes) and approached in mutant segments, are depicted in panels containing genotype, phenotype, and status with respect to relative amounts of the hypothetical BX-C substances S_0, S_3, S_4, and S_5, which are products of the wild-type alleles of *Ultrabithorax* (*Ubx*), *bithoraxoid* (*bxd*), *infra-abdominal-2* (*iab-2*), and *infra-abdominal-3* (*iab-3*), respectively. Panels to the left of wild type depict hemizygous genotypes for hypomorphic or amorphic mutants, the designated chromosome being heterozygous with *Df-P9* (see Figure 14.13) in all cases. Panels to the right of wild type involve heterozygotes for dominant-constitutive mutants, *Contrabithorax* (*Cbx*), *Hyperabdominal* (*Hab*) and *Ultra-abdominal-5* (*Uab*^{5}). Panels in dotted outline represent theoretically possible L_{AB1} transformations, for which definitive mutant genotypes are not available. Panels in dashed outline are genotypes dying in the first-instar larva. All other genotypes shown survive to the adult stage except *Ubx* hemizygotes, which die in either the third-instar or early pupal stage. [From E. B. Lewis, *Nature 276*:565 (1978).]

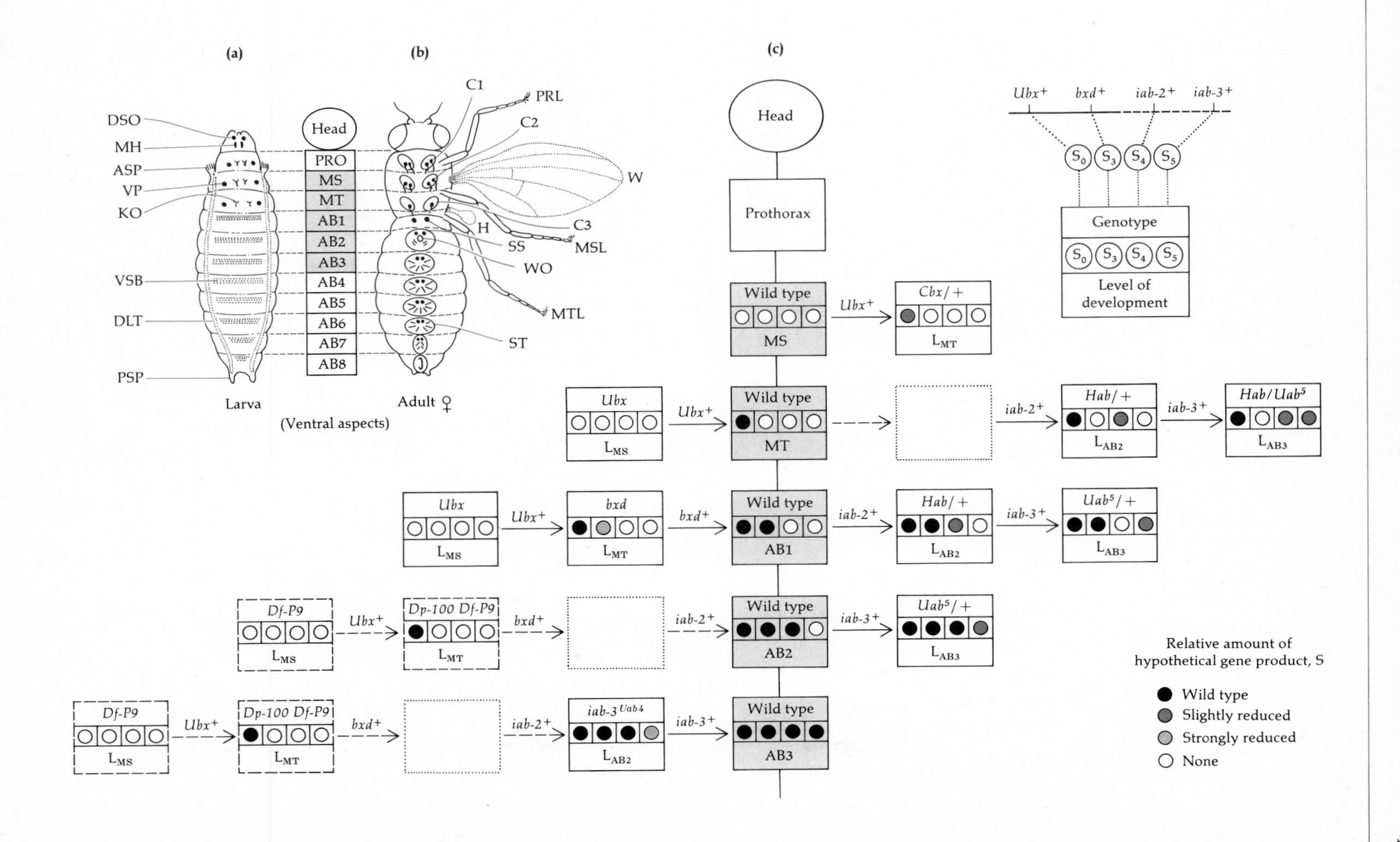

(a)
(b)
(c)
DSO
MH
ASP
VP
KO
VSB
DLT
PSP
Larva
Head
PRO
MS
MT
AB1
AB2
AB3
AB4
AB5
AB6
AB7
AB8
(Ventral aspects)
C1
PRL
C2
W
H
C3
SS
MSL
WO
MTL
ST
Adult ♀
Head
Prothorax
Ubx+
bxd+
iab-2+
iab-3+
Genotype
Level of development
Wild type
MS
Cbx/+
Ubx
MT
Hab/+
Hab/Uab5
bxd
AB1
Uab5/+
Df-P9
Dp-100 Df-P9
AB2
iab-3 Uab4
AB3
Relative amount of hypothetical gene product, S
Wild type
Slightly reduced
Strongly reduced
None

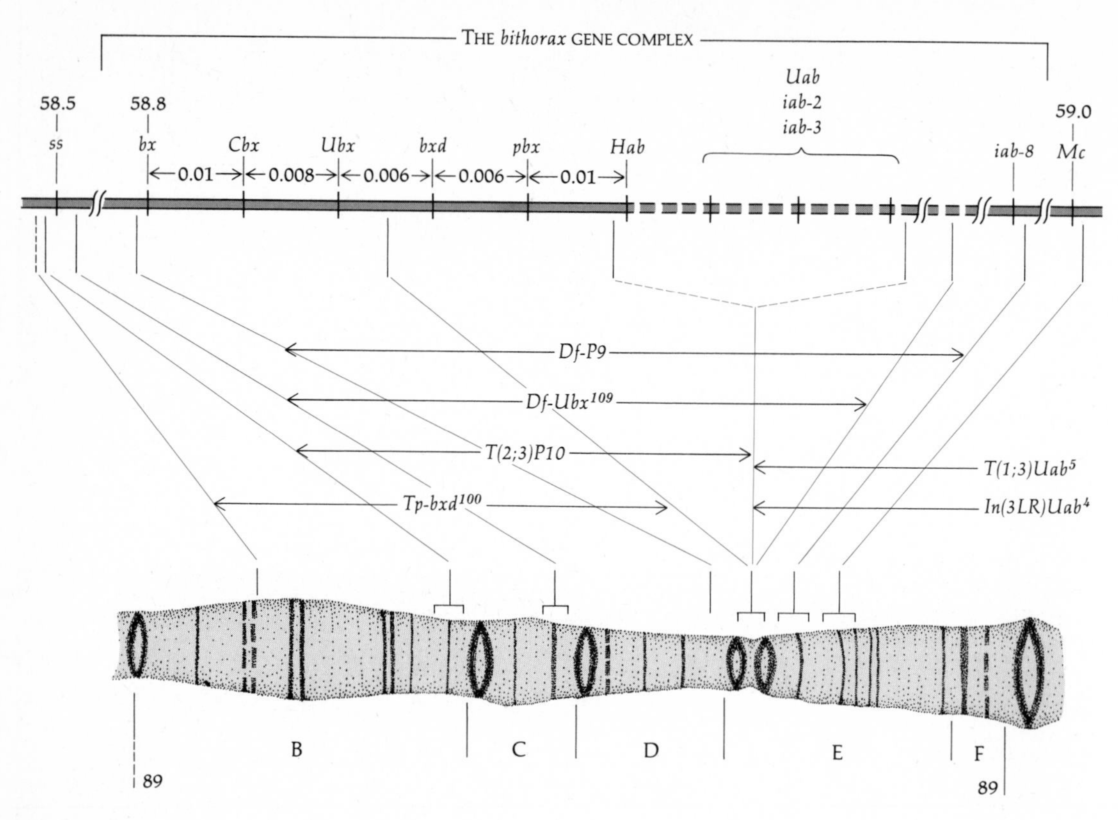

Figure 14.13
Correlation of the genetic and salivary-gland chromosome maps of the *bithorax* gene complex and the immediately surrounding regions in the right arm of the third chromosome of *D. melanogaster*. The solid portion of the linkage map is based on recombination studies. The dashed portion is based on cytogenetic studies of the designated chromosomal rearrangements; the relative order of *Hab*, *Uab*, *iab-2*, and *iab-3* with respect to each other is uncertain. The complex spans approximately 0.05–0.1 centimorgan and lies within the region of two heavy doublet structures and possibly an adjoining faint band of chromosome section 89E. *Df-P9* in its interactions with *bithorax*-complex mutants and rearrangements acts as if it were deleted for all of the genes of the complex. [From E. B. Lewis, *Nature* 276:565 (1978).]

complex. These embryos develop, but die just at the time of hatching from the egg, exhibiting transformation of all of the abdominal segments to a basic thoracic segment, the mesothorax (MS) (Figure 14.14), which is believed to be the most primitive and least specialized of the larval segments. Evidently the wild-type *bithorax* genes, missing in the homozygous deficiency, are required for the selection of developmental pathways other than that for mesothorax. The particular segmental developmental pathway selected depends on the position of the cells in the blastoderm.

Mutations that retain some of the wild-type functions of the *bithorax* complex transform only some segments to abnormal developmental pathways. Some of these homozygous mutations actually survive as adults. Figure 14.12 shows that the metathoracic (MT) and first three abdominal

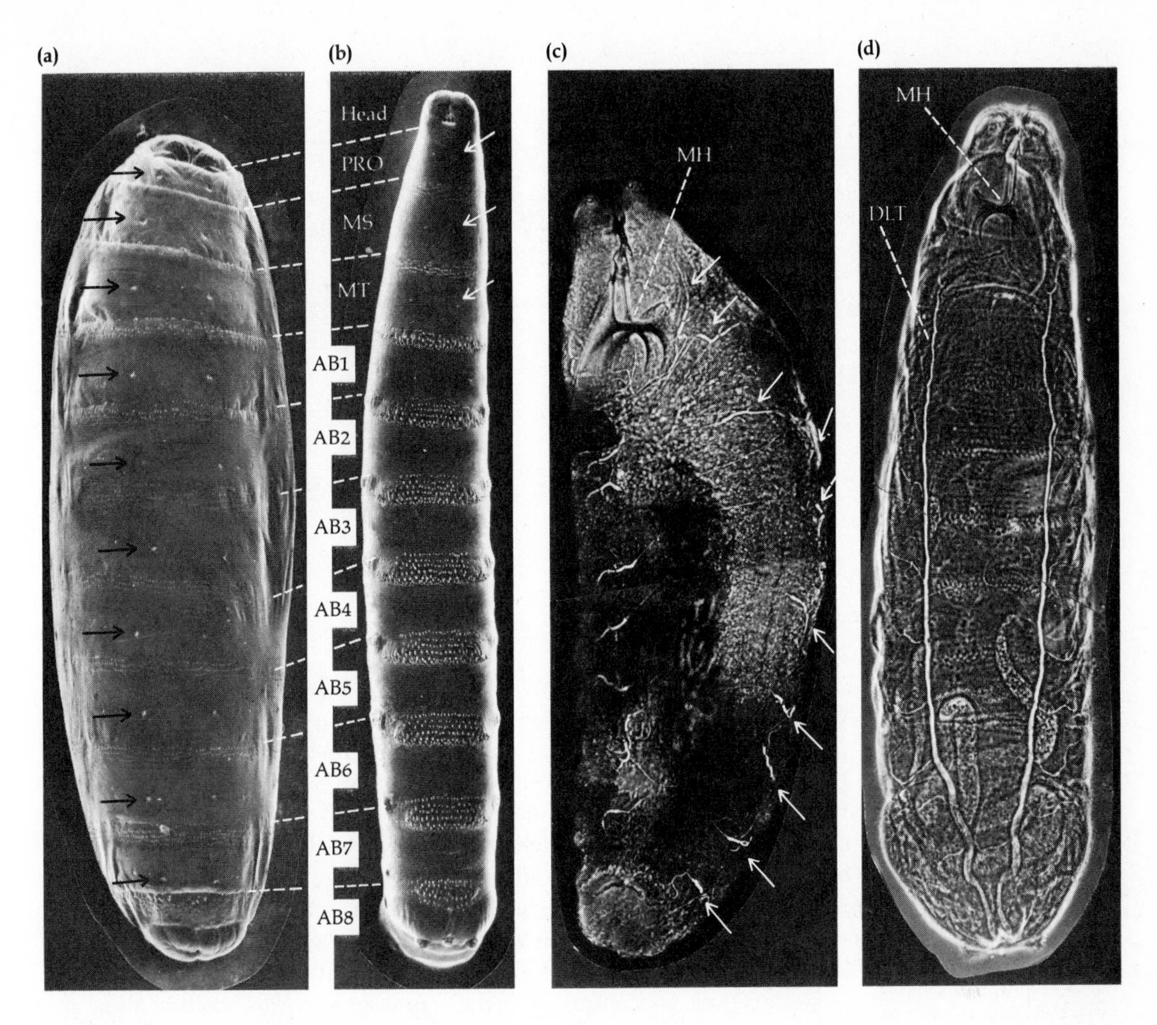

Figure 14.14
Comparison of normal, newly hatched larvae (parts b and d) with embryos that are homozygous deficient for the *bithorax*-complex genes (*Df-P9/Df-P9*) (parts a and c). Scanning electron micrographs are on the left [(a) ×110; (b) ×40], and phase-contrast light micrographs are on the right [(c and d) ×85]. In (a), note the thoracic-like VSB and KO (arrows) on segments AB1 through AB7, as well as on thoracic segments; these are also present on segment AB8 but are not visible here (see Figure 14.12 caption for abbreviations). On the normal larva (b), note the more prominent VSB on the abdominal versus the thoracic segments. KO (arrows) are restricted to thoracic segments. In (c), note that separate tracheal sections (arrows) of DLT occur in each segment from MS through AB8. Compare (c) with the normal larva in (d), which has continuous DLT. [From E. B. Lewis, *Nature 276*:565 (1978).]

segments (AB1, AB2, and AB3) can achieve almost all segmental levels of development, from the mesothorax level (L_{MS}) to the third-abdominal-segment level (L_{AB3}), depending upon which gene functions are lost. Figure 14.15 shows an adult fly of genotype bx^3pbx/bx^3pbx, which has the metathorax (MT) completely transformed to a mesothorax (MS), restoring the primitive level of development of the four-winged insect ancestor of the fly.

Figure 14.15
Adult *D. melanogaster* of genotype bx^3pbx/bx^3pbx. The metathorax is completely transformed to a mesothorax. (From E. B. Lewis, *Engineering & Science*, California Institute of Technology, Pasadena, Cal., November, 1957.)

The studies of E. B. Lewis suggest that at least eight genes are present in the *bithorax* complex. They are believed to produce the normal segmentation pattern by controlling other genes that are actually responsible for coding for segment-specific proteins and cellular organization. The functions of four of the postulated products of the genes in this complex in controlling the level of development are indicated in Figure 14.12. The state of repression of the *bithorax* complex genes is postulated to be mediated by cis-dominant regulatory genes that are acted upon by a repressor coded elsewhere in the genome. It is suggested that the repressor is present in a gradient of decreasing concentration from anterior to posterior; thus, concentration of the repressor specifies position along the anterior-posterior axis. The cis-dominant regulatory genes are postulated to have different affinities for the repressor, the affinities increasing from left to right in Figure 14.13. The Ubx^+ gene, having a cis-dominant regulatory gene with lowest affinity for repressor, is turned on at the highest repressor concentration in the gradient, raising cells to the MT state. At a lower concentration of repressor, the bxd^+ function is also turned on, raising cells to the AB1 state, and so on. Thus the *bithorax* complex genes are postulated to translate positional information into differential gene activity in different cells along the anterior-posterior axis of the blastoderm.

Early Embryogenesis of the Mouse

Following fertilization of the mouse egg, the zygote undergoes a series of complete cleavage divisions as the embryo passes through states of 2, 4, 8,

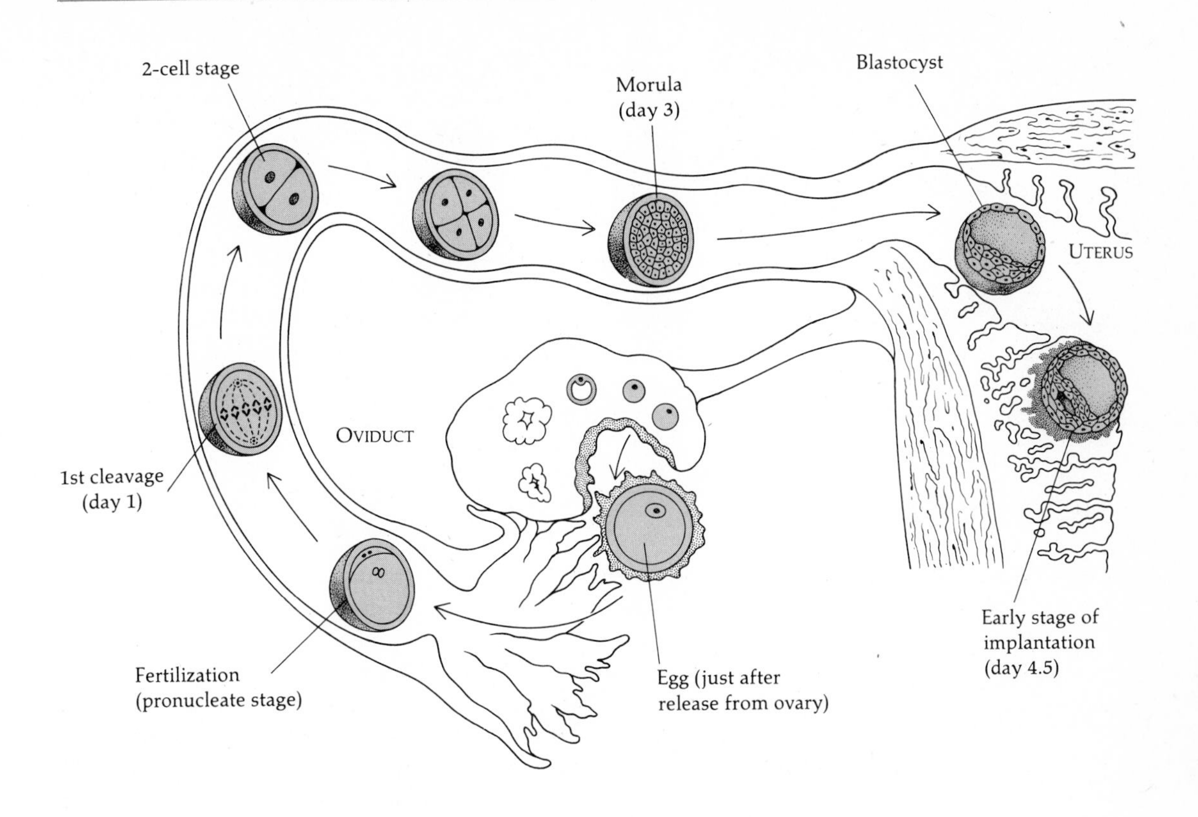

Figure 14.16
Early developmental stages in mouse embryogenesis.

16, 32, and 64 cells prior to implantation in the wall of the uterus (Figure 14.16). At the 8-cell stage, the cells are totipotent: each can give rise to a complete embryo by further cell division. By the 64-cell blastocyst stage, however, the embryo is clearly separated into two cell types: the outer cells, or *trophoblast*, and the cells of the *inner cell mass* (ICM). The cells of the blastocyst are no longer totipotent.

It is generally believed that the first signal that commits sister cells to different developmental pathways occurs at the 16-cell stage, at which, for geometric reasons, some cells find themselves surrounded on all sides by other cells. The *inside-outside theory* holds that the inner cells experience a different environment than the outer cells, which are still in contact with the oviduct, and that this positional signal induces the first step in cell determination. By the blastocyst stage, this commitment is evidenced by the division of cells into trophoblast and inner cell mass. The fetus itself develops from a few cells in the inner cell mass; the rest of the cells produce extraembryonic tissues, such as the placenta (Figure 14.17).

The genetic control of the steps in early mouse embryogenesis that are shown in Figure 14.17 has only begun to be investigated. The *T* locus of the mouse may play an important role in understanding some of these steps. The *T* locus has long fascinated geneticists for the diversity of its mutations (see Chapter 5). Many recessive *t* alleles are known that form balanced lethal systems with the dominant mutant, *T*. Complementation

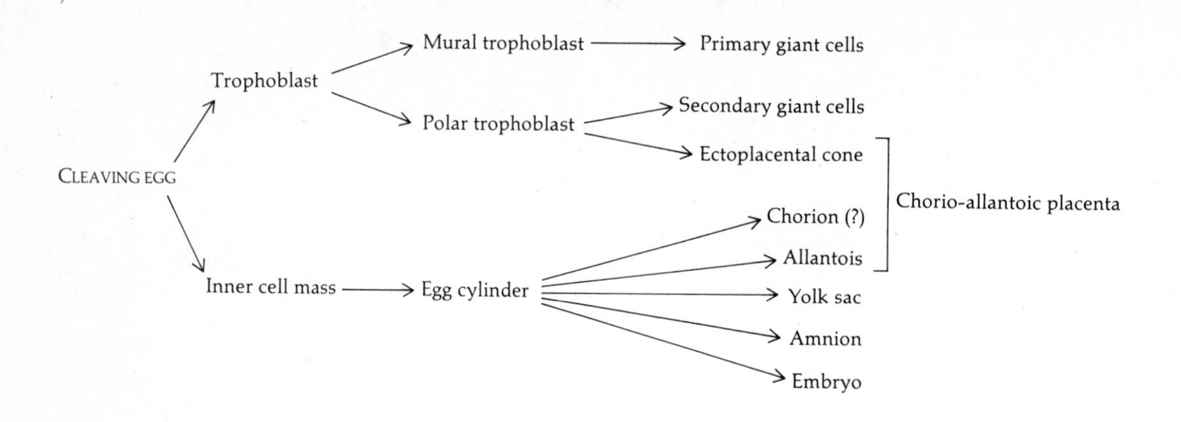

Figure 14.17
Fate map of trophoblast and inner cell mass of the 3½-day mouse blastocyst. (From R. L. Gardiner, in *The Developmental Biology of Reproduction*, ed. by C. Markert and J. Papaconstantinou, Academic Press, New York, 1975.)

analysis among the various *t* alleles is one of the primary genetic methods for distinguishing these mutations, since most of them also suppress recombination in their region of the chromosome. The recessive *t* alleles affect cell membranes on the sperm, as evidenced by the creation of new antigenic properties. Heterozygous males, $t/+$, produce two antigenically distinct populations of sperm. The *t* alleles are widely spread in mouse populations, owing to the greater efficiency of *t*-carrying sperm in fertilization (see Chapter 5, Table 5.3, Transmission Ratio). Thus, these otherwise deleterious mutations have a selective advantage over t^+ sperm and are not eliminated from the population.

Noncomplementing, lethal *t* alleles have been shown to interrupt embryonic development at characteristic stages, as diagrammed in Figure 14.18. Homozygous t^{12}embryos progress to the 16-cell stage, or *morula*, but die before developing to the blastocyst stage just prior to implantation in the uterine wall. Homozygous t^{w73} embryos begin to implant, but fail to do so properly and die. Other homozygous *t* alleles block subsequent steps in embryonic development. Since each allele seems to block development at a stage when new cell-cell interactions occur, the hypothesis that the normal *T* locus controls early development by altering cell-surface properties and thus causing new cellular associations to occur is appealing, and is supported by the effects of *t* alleles on sperm-surface antigens.

Sex Determination in Mammals

The mechanism of sex determination in mammals also involves sex-specific differences of the cell surface. Two aspects of sex determination in mammals must be distinguished: primary, or gonadal, determination and secondary, or somatic, determination. Gonadal sex determination, leading to the differentiation of testes or ovaries from an early, indifferent gonad (Figure 14.19) is effected by male-determining genes carried on the

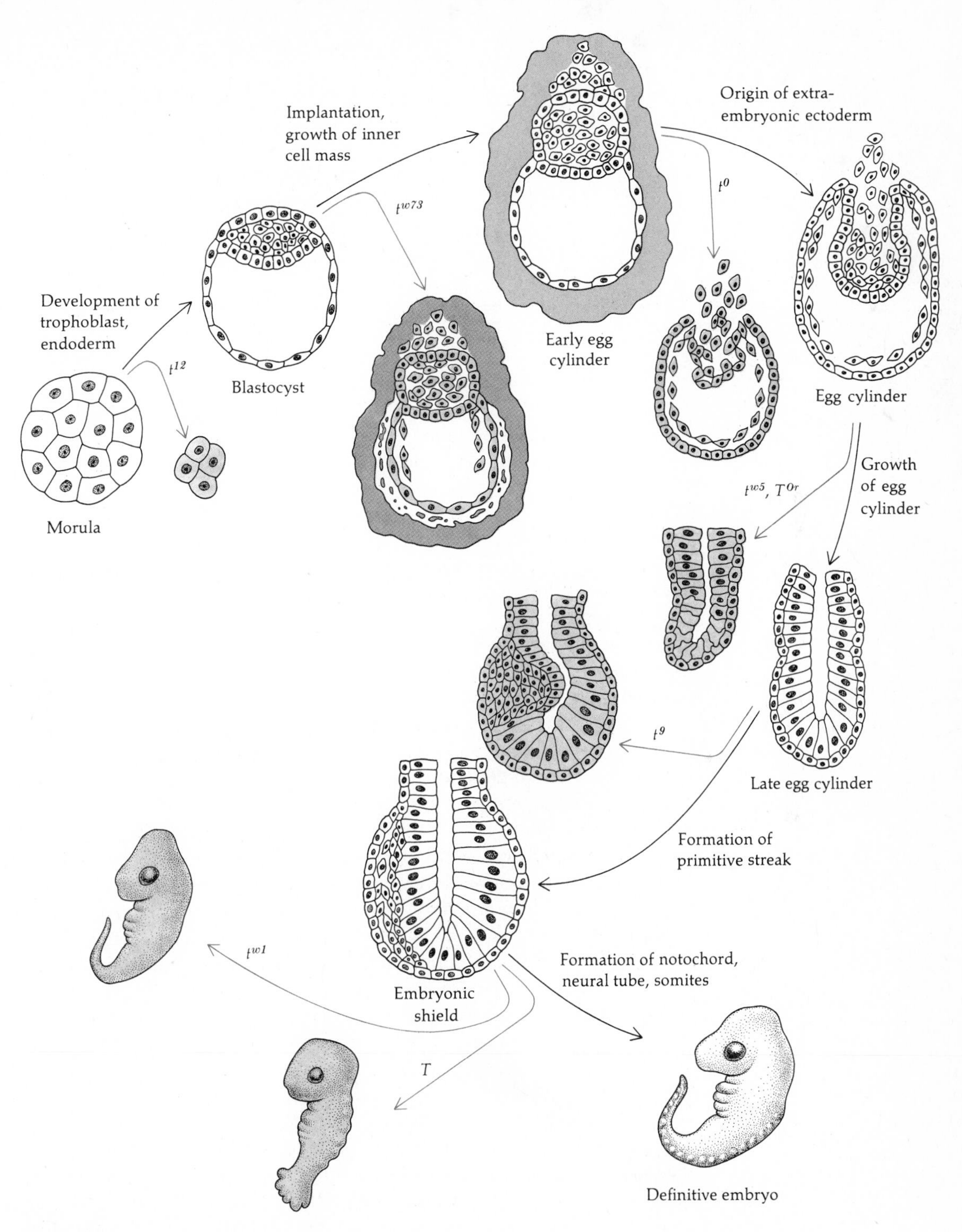

Figure 14.18
Schematic representation of early mouse embryogenesis, showing developmental arrests observed in embryos homozygous for certain mutant alleles of the *T* locus. [After D. Bennett, *Cell* 6:441 (1975).]

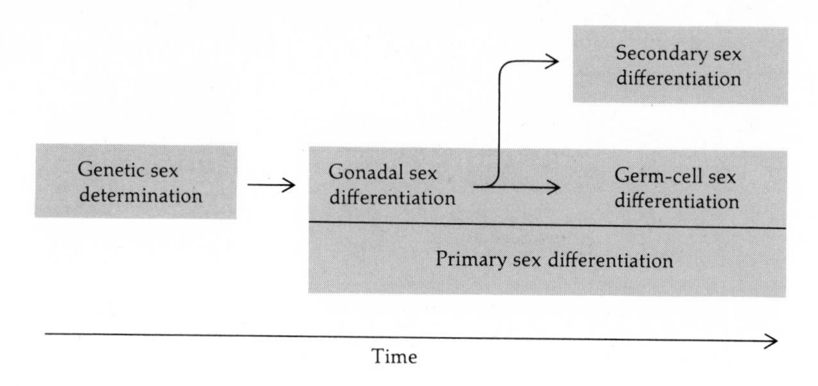

Figure 14.19
Temporal relation of events in mammalian sexual development.

Y chromosome. These genes specify a protein called the *H-Y antigen,* which is present on the surface of all Y-chromosome-bearing cells.

The existence of a male-specific cell-surface antigen was originally detected in skin-graft studies among individuals within highly inbred lines of mice. Skin grafts between female mice and between male mice within a given line were readily accepted. Female skin grafts to males were also accepted, but male skin grafts to females were rejected. Thus, males possess some antigen that is not present in otherwise isogenic females. Since the only genetic difference between these males and females is the presence of the Y chromosome in the males, it is concluded that this antigen is specified by genes on the Y chromosome. More than one gene specifying the H-Y antigen is believed to be on the Y chromosome, because of some chromosomal rearrangements involving the Y that cause intersexual phenotypes and yet express the H-Y antigen.

The H-Y antigen genes are expressed very early in development, since the antigen is present in 8-cell embryos of the mouse. Even though all cells of the male body express this antigen, its only function is believed to be testis induction in the indifferent gonad. In its absence, the indifferent gonad develops into an ovary. Antibodies prepared against mouse H-Y antigen cross-react with male cells from all mammalian species tested, including humans, suggesting that this protein's structure has been highly conserved during evolution.

Secondary sex determination is a consequence of the nature of the gonad that is induced to develop by the chromosomal constitution of the individual. The developing testes secrete the steroid hormone testosterone, which circulates to all cells of the embryo, initiating male development in the somatic cells, including the somatic cells of the gonad. In the absence of this hormonal signal, female development occurs. Male development is under the control of a single X-linked gene (Tfm^+) specifying a testosterone-binding protein that is present in the cytoplasm of all cells, male and female. This protein is a regulatory protein, activated by binding of testosterone (an effector molecule), which then enters the nucleus and activates the genes required for male differentiation. Mutations of the *Tfm* gene are known in several species, including humans, causing the syndrome called *testicular feminization*. Cells of mutant *Tfm*/Y embryos are completely insensitive to the masculinizing effect of testosterone; consequently, the fetus develops all the external secondary sexual characteristics of a female rather than a male. Internally, testes develop rather than ovaries, however; the testes suppress the development of Fallopian tubes and uterus by secreting another male hormone, known as factor chi (χ), resulting in a blind vagina (Figure 14.20).

Like XY cells, XX cells produce the testosterone receptor and respond to testosterone. Administration of testosterone to XX embryos or castrated XY embryos induces the development of all the external secondary sexual characteristics of a male. Internally, however, both male ducts and female ducts develop, owing to the absence of factor χ, producing a hermaphrodite.

Dosage Compensation

In organisms whose sex determination is associated with a heteromorphic pair of sex chromosomes, it may be asked whether the two sexes exhibit different levels of expression of sex-linked genes. In *Drosophila* and in mammals, males have one X chromosome and females have two. However, most genes located on the X chromosome are expressed equally in male and female somatic cells. These two distantly related types of organisms have developed quite different means of compensating for the sex-dependent dosage of X-linked genes.

In *Drosophila*, autoradiographic studies of RNA synthesis at specific X-chromosome loci in salivary-gland polytene nuclei show that equal amounts of RNA are synthesized in males and females. Both of the polytene X chromosomes in female cells are active, and each exhibits half the amount of RNA synthesis observed for the single polytene X chromosome in male cells. Moreover, the levels of X-linked enzyme activities are equal in male and female flies. Thus the X chromosome in males is twice as active as each of the X chromosomes in females. Superfemales (3X:2A) exhibit enzyme activities equal to those of normal females and males; thus each X chromosome in superfemales appears to be ⅔ as active as those of normal females (Figure 14.21). Triploid females (3X:3A) have 1.5 times as much enzyme activity as diploid females (2X:2A). Comparison of enzyme activities in intersexes (2X:3A) and metamales (XY:3A) with triploid females (3X:3A) shows that they are equivalent and that dosage compensation is acting to maintain a constant ratio of X and autosomal gene activities (Figure14.21). Thus, five levels of activity are

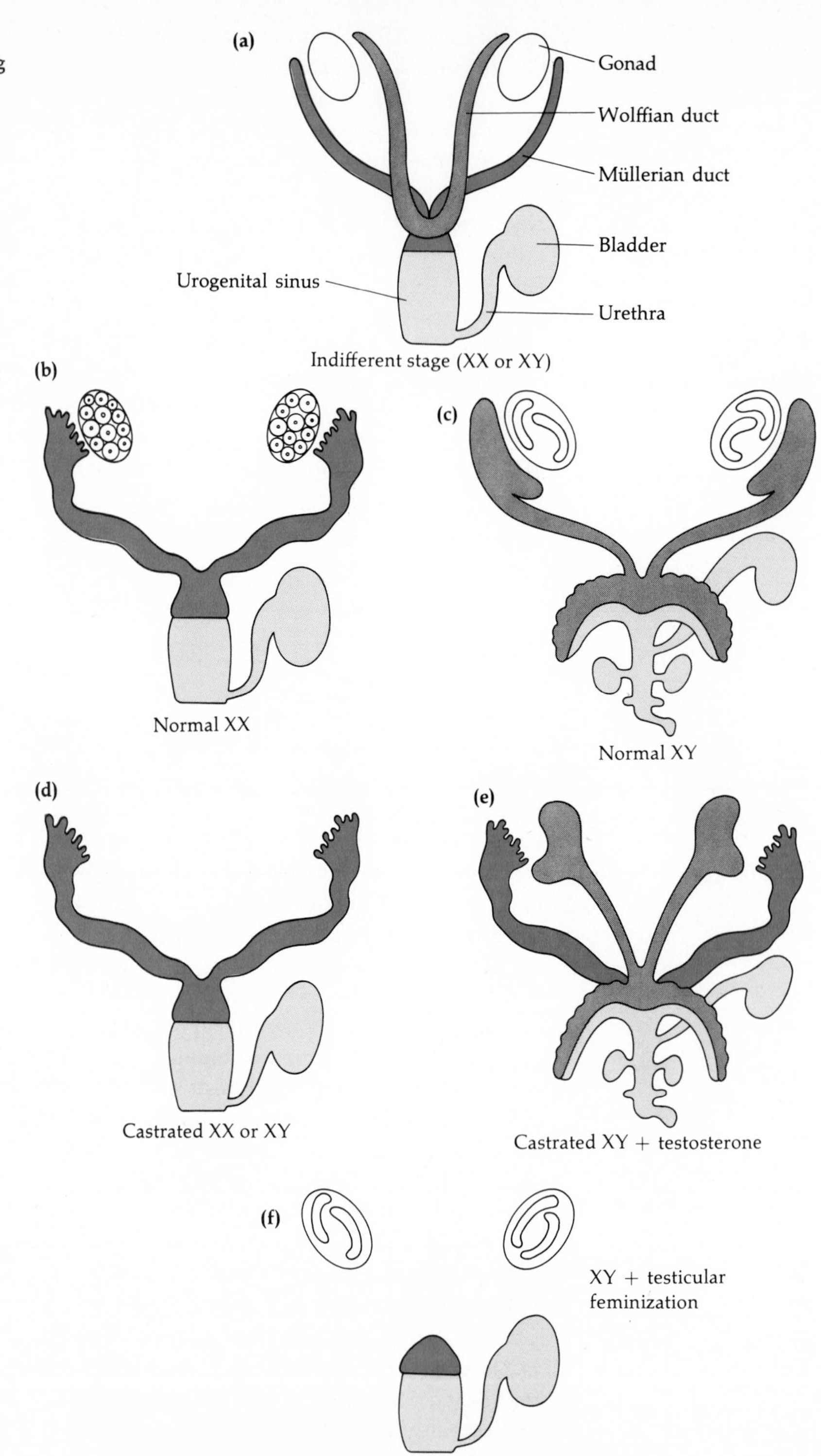

Figure 14.20
Schematic diagram showing the role of testosterone in secondary sex determination of the human reproductive organs. The Wolffian (male) duct is shown in gray, and the Müllerian (female) duct in dark color; the urogenital sinus, which includes bladder and urethra, is shown in light color. The pair of oval bodies are the gonads: indifferent gonads (empty), ovaries (filled with small circles), and testes (containing tubules). [After S. Ohno, *Nature* *234*:134 (1971).]

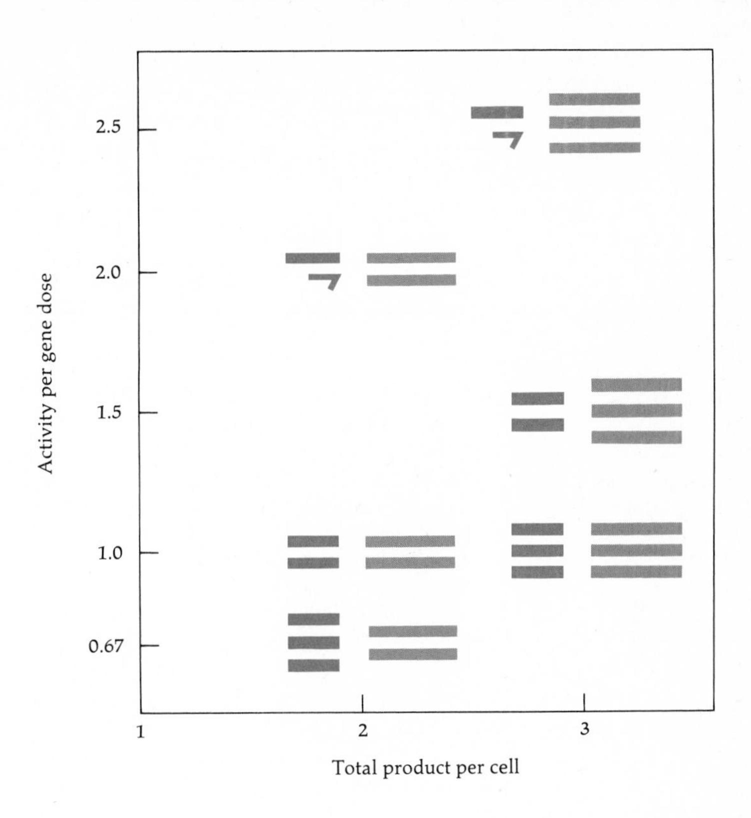

Figure 14.21
Enzyme activities of X-linked structural genes exhibited by *Drosophila melanogaster* flies with various chromosomal constitutions. The dark bands represent X chromosomes, the light bands represent autosome sets, and the hook-shaped figures represent Y chromosomes. Metamales (XY:3A) are observed to be substantially, but not fully, compensated, exhibiting only 2.5 rather than 3.0 times the level of activity of a diploid female X chromosome. [From J. C. Lucchesi, *Am. Zool.* *17*:685 (1977).]

possible for X-linked genes, depending on the number of autosomes and the ratio of X chromosomes to autosomes.

Dosage compensation does not act at the level of the entire X chromosome; if an X-linked gene is translocated to an autosome, it continues to exhibit dosage compensation, whereas an autosomal gene translocated to the X chromosome does not exhibit dosage compensation. It therefore appears that X-linked genes are regulated individually by the dosage-compensation mechanism, possibly by closely linked regulatory genes.

Two models for dosage compensation in *Drosophila* have been proposed. The first postulates an X-linked compensator gene, which itself is not compensated, synthesizing an inhibitor of X-linked gene transcription. The more X chromosomes per nucleus, the greater the concentration of the inhibitor and the greater the reduction of the X-linked gene activity per dose (Figure 14.22). The second model postulates an autosomal compensator gene synthesizing an activator of X-linked gene transcription. The amount of this activator synthesized would be proportional to the compensator gene dosage. The level of X-linked gene transcription would then be regulated by the availability of the activator substance (Figure 14.23).

Figure 14.22
A model for dosage compensation in *Drosophila* that postulates an X-linked compensator gene, which itself is not compensated. The product of this gene acts on X-chromosome regulator genes that act to reduce the level of expression (indicated by minus signs) of cis-linked structural genes (upper diagram). Zygotes carrying a duplication of the compensator gene would contain twice the amount of the postulated repressor, reducing the level of expression of all X-linked structural genes relative to those on the autosomes. This model predicts that such zygotes would fail to develop normally, owing to the imbalance in gene activities, and would die (lower diagram). [From J. C. Lucchesi, *Am. Zool.* 17:685 (1977).]

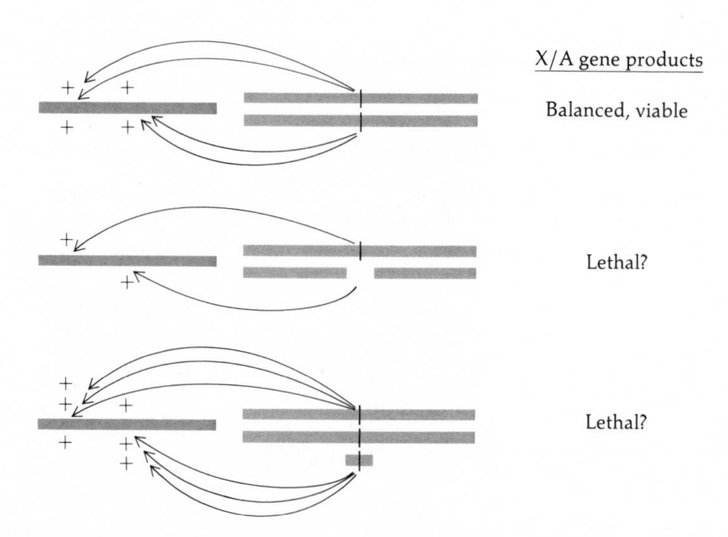

Figure 14.23
A model for dosage compensation in *Drosophila* that postulates an autosomal compensator gene synthesizing an activator of X-linked structural genes (indicated by plus signs) and acting on cis-dominant X-linked regulator genes (upper diagram). Zygotes carrying a duplication or deficiency for this compensator gene would exhibit an imbalance in X versus autosomal structural gene expression and are postulated to die as a result (lower two diagrams). [From J. C. Lucchesi, *Am. Zool.* 17:685 (1977).]

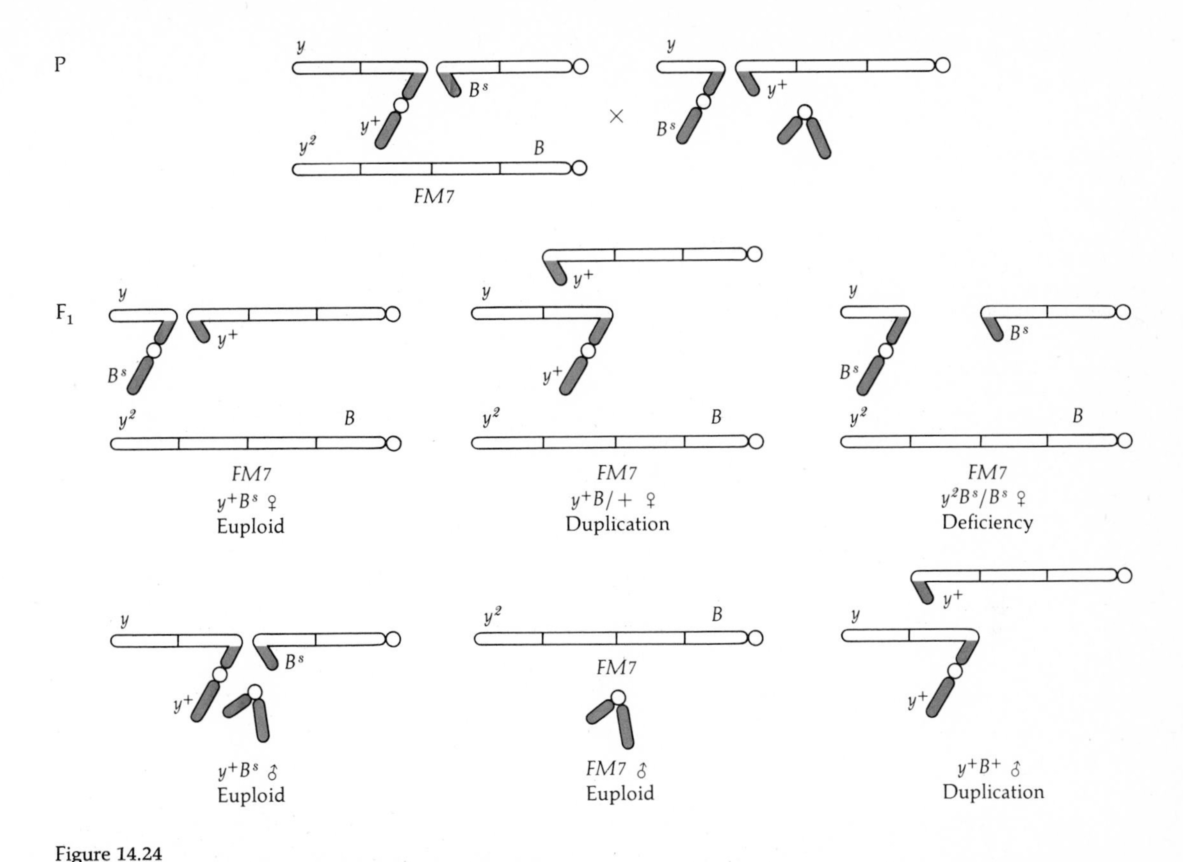

Figure 14.24
X-chromosome segmental duplications (and deficiencies) can be constructed by crossing two strains carrying X;Y translocations in which the X chromosomes are broken at different positions. Several dozen different X-chromosome breakpoints scattered throughout the length of the chromosome are available that permit sequential segmental duplications to be produced for the entire chromosome. Duplication- or deficiency-bearing progeny can be identified phenotypically by the genetic markers shown. The two arms of the Y chromosome are identified by the dominant markers y^+ and B^s previously translocated to the arms of the Y from their original locations on the X chromosome (these translocations involved different X chromosomes from those diagrammed here). The two strains employed in each cross must have Y-chromosome breakpoints in different arms of the Y chromosome, as indicated in the diagram. The *FM7* X chromosome in the mother is a balancer chromosome carrying multiple inversions to suppress recombination and is marked with y^2 (distinguishable from y) and B (distinguishable from B^s). Flies with two doses of B^s (B^s/B^s) are also distinguishable from those with only one dose of B^s. The cross diagrammed here creates progeny duplicated and deficient for the quarter of the X chromosome next to that most distal from the centromere (the small circle). Hemizygous deficient male progeny (not shown) do not survive the loss of this much genetic information, and die. All other progeny survive. [After B. Stewart and J. Merriam, *Genetics* 79:635 (1975).]

The possibility that a duplication or deficiency for the postulated compensator gene might lead to an imbalance in X versus autosomal gene expression and thus might be lethal has led to experiments to test this hypothesis. Duplications and deficiencies involving any portion of the X chromosome can be systematically produced by crossing different X;Y translocation strains, as diagrammed in Figure 14.24 (see also Chapter 17).

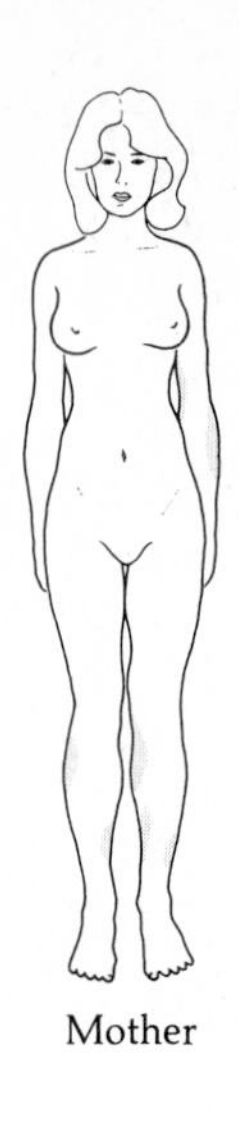

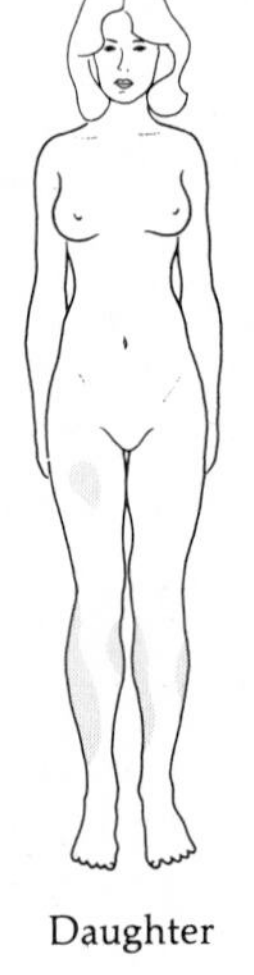

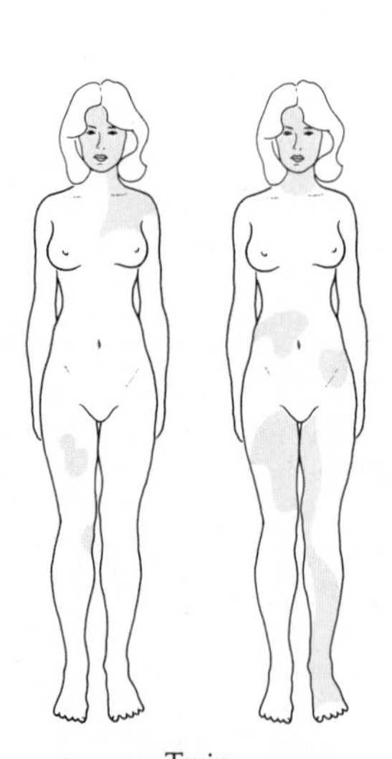

Figure 14.25
Three generations of women in a family, the last being identical twins, are heterozygous for the mutant gene causing anhidrotic ectodermal dysplasia. They exhibit patches of skin lacking sweat glands (shaded), in which the wild-type allele is inactivated, and normal skin, in which the mutant allele is inactivated. (After E. Novitski, *Human Genetics*, Macmillan, New York, 1977.)

The results of these experiments have shown that there is no single locus on the X chromosome that causes lethality when duplicated, nor one that inhibits the expression of X-linked structural genes (Figure 14.22). Y;autosome translocation strains allow segmental duplications and deficiencies of the autosomes to be produced in like manner (see Box 17.2, page 580). These experiments show that there is one region of the third chromosome, between polytene chromosome bands 83D and 83E, that causes lethality when present in either one or three doses (Figure 14.23). These results are consistent with, but do not prove, the hypothesis that there is an autosomal compensator gene.

In contrast to *Drosophila*, dosage compensation in mammals involves the entire inactivation of one of the two X chromosomes in female cells. The inactive X chromosome is observed in interphase cells as a condensed heterochromatin body, called a *Barr body* after its discoverer, Murray Barr (see Chapter 17). X inactivation occurs near the time of implantation of the early embryo in the wall of the uterus. It occurs only in somatic cells, not in germ-line cells, a fact that is believed to be the cause of sterility in XO females and XXY males (see Chapter 17), because of abnormal levels of X-chromosome activity in the germ cells. In placental mammals, such as humans and mice, maternal and paternal X chromosomes are inactivated at random in different cells. Once a particular X chromosome has been inactivated, it remains inactivated in daughter cells. Consequently, females that are heterozygous for X-chromosome genes are *mosaics*: they contain patches of cells expressing one or the other heterozygous allele. For example, cats with tortoiseshell coats (calico cats) have patches of black and yellow fur and are almost invariably females. Male kittens of tortoiseshell females are either yellow or black; *yellow coat* (C^Y) is an allele of *black coat* (C^B). The patchy tortoiseshell coat of the mother cat (genotype C^Y/C^B) is due to the random inactivation of the C^B-bearing X chromosome in some cells early in development. The clonal descendants of cells in which the C^B chromosome is inactivated produce patches of yellow fur, and those in which the C^Y chromosome is inactivated produce patches of black fur. Occasionally, tortoiseshell males are found, and these invariably have an XXY chromosomal constitution.

Human females who are heterozygous for X-linked genes are mosaics. For example, the X-linked mutation called *anhidrotic ectodermal dysplasia*, responsible for the "toothless men of Sind" (West Pakistan), shows mosaicism in heterozygous women. These women have jaw areas with and without teeth and exhibit patches of skin with and without sweat glands (Figure 14.25). Women heterozygous for an X-linked hemophilia gene show wide variability in their amount of clotting factor, which ranges from 20% to 100% of normal and perhaps depends on the random variability of X inactivation in the relatively few cells forming the blood primordia.

Genetic Regulatory Mechanisms

The molecular mechanisms underlying the examples of genetic regulation described in this chapter are largely unknown. Until recently there was a growing assumption that the molecular details of gene regulation in eukaryotes would turn out to be very similar to those in prokaryotes—only more complicated. The recent introduction of recombinant DNA technology now provides the necessary tools for studying the molecular structure and organization of eukaryotic genes, a prerequisite for understanding how they function. Already, however, it is becoming certain that

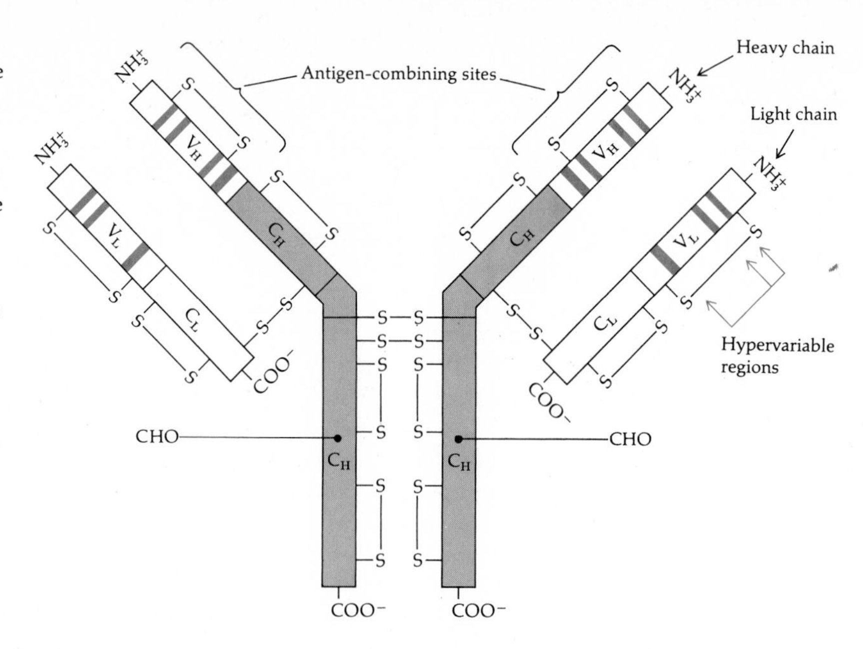

Figure 14.26
A schematic structure of the human immunoglobin (antibody) molecule. V and C indicate variable and constant regions of the amino acid sequences of the heavy (H) and light (L) chains. Each molecule is a dimer of two identical subunits joined by disulfide bonds. Each subunit consists of a light and a heavy polypeptide chain joined by disulfide bonds. Intrachain disulfide bonds contributing to the tertiary structure of the molecule are also shown. CHO represents carbohydrate groups attached to the heavy chains.

the mechanisms for the control of gene expression in eukaryotes are quite different from those in prokaryotes. The unexpected existence of intervening DNA sequences in eukaryotic structural genes is a case in point (see Chapter 11). The directed rearrangement of genes during lymphocyte differentiation—a commitment to produce a specific type of antibody—is another example, suggesting that yet other unsuspected mechanisms of eukaryotic gene regulation await our discovery.

Directed Gene Rearrangement During Lymphocyte Differentiation

The higher vertebrates possess immune systems that have evolved to combat infection and eliminate aberrant cells that might become cancerous. The immune system acts by producing antibodies, which are induced in response to foreign molecules (antigens) synthesized by the infecting or cancerous cells. Mammals are capable of producing at least 10^6 different specific antibodies. Antibodies against synthetic antigens, to which the organism has never been exposed in its evolutionary history, are elicited as easily as are antibodies against more common natural antigens. Whether all antibody diversity is encoded in the germ line or whether

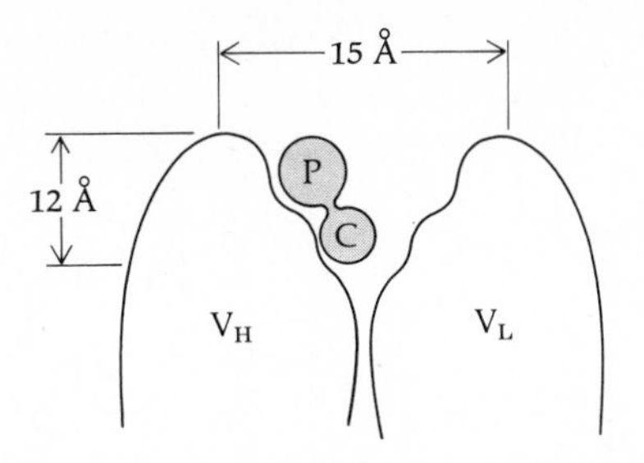

Figure 14.27
Schematic view of the binding of an antigen in the cleft between the variable regions of the heavy and light chains of an antibody subunit. The antigen is phosphorylcholine, and the antibody is a mouse myeloma protein. The third dimension, not shown, is 20 Å. (After E. Padlan et al., in *The Immune System: Genes, Receptors, Signals*, ed. by E. Sercarz, A. Williamson, and C. Fox, Academic Press, New York, 1975.)

some of it is generated during somatic development is a question attracting much current study. The vast diversity of antibody types that can be synthesized, coupled with our knowledge of the gene-protein relationship, raises questions as to how the genome can encode all of the different amino acid sequences required.

Antibody molecules are dimers, each monomer being composed of two distinct polypeptides that are designated *heavy* (H) and *light* (L); these are joined by disulfide bonds (Figure 14.26). The amino acid sequences of many different heavy and light chains have been determined; each chain was obtained from a mouse or human suffering from the cancer *multiple myeloma*, in which a particular antibody-producing plasma cell vastly overproliferates. These amino acid sequence studies have shown that each heavy and light chain contains a region of *variable* sequence (V) and another of *constant* sequence (C). Within the variable regions, V_H and V_L, of the two types of chains, are *hypervariable* regions, which confer the antigen-binding specificity that characterizes a particular heavy or light chain.

Each antibody dimer contains the same heavy and light chains. The two antigen binding sites are formed by a pocket between the V_H and V_L portions of the subunits (Figure 14.27).

The most intensively studied type of antibody, IgG, shows a great deal of internal amino acid sequence homology (Figure 14.28), suggesting that these molecules have evolved by tandem gene duplication. Much duplication and subsequent divergence of the genes coding for the variable region of light chains has also occurred, forming a family of related genes. Two major subgroups of V_L genes in humans are designated λ and κ; each subgroup contains further subgroups that show finer degrees of sequence homology (Figure 14.29).

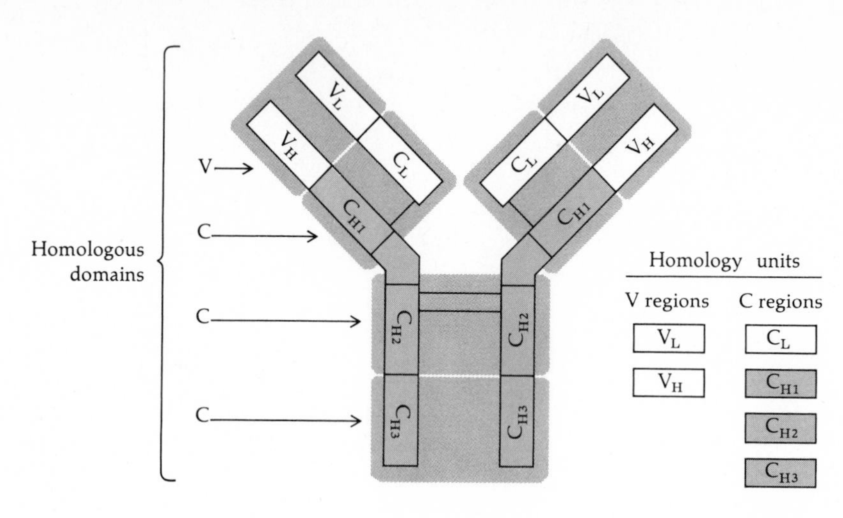

Figure 14.28
Representation of the homology units of the IgG molecule. The coding regions of the gene for each chain are separated by intervening sequences into sequences coding for homologous globular regions of the IgG protein, called *domains.* The intervening sequences are removed during maturation of the mRNA. (After L. E. Hood, I. L. Weissman, and W. B. Wood, *Immunology,* Benjamin/Cummings, Menlo Park, Cal., 1978.)

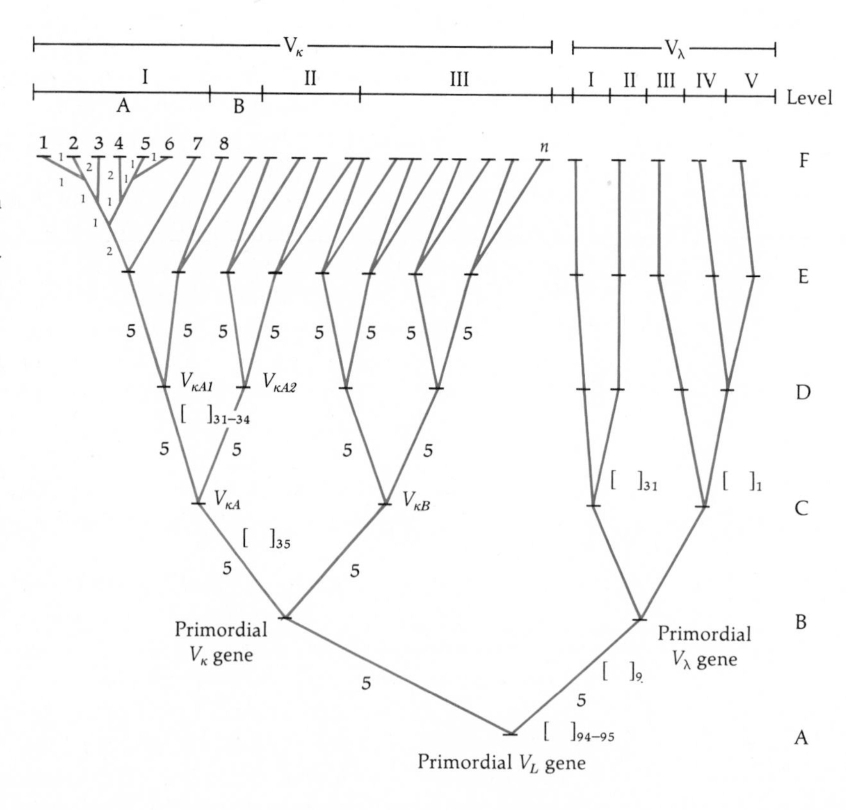

Figure 14.29
A possible genealogic tree for the variable-region genes coding for human light chains, deduced from amino acid sequence homologies between myeloma proteins. Numbers on the branches indicate the number of amino acid substitutions between branch points. Brackets indicate deletions of amino acids at specific positions. [After L. Hood, J. H. Campbell, and S. C. R. Elgin, *Annu. Rev. Genet.* 9:305 (1975).]

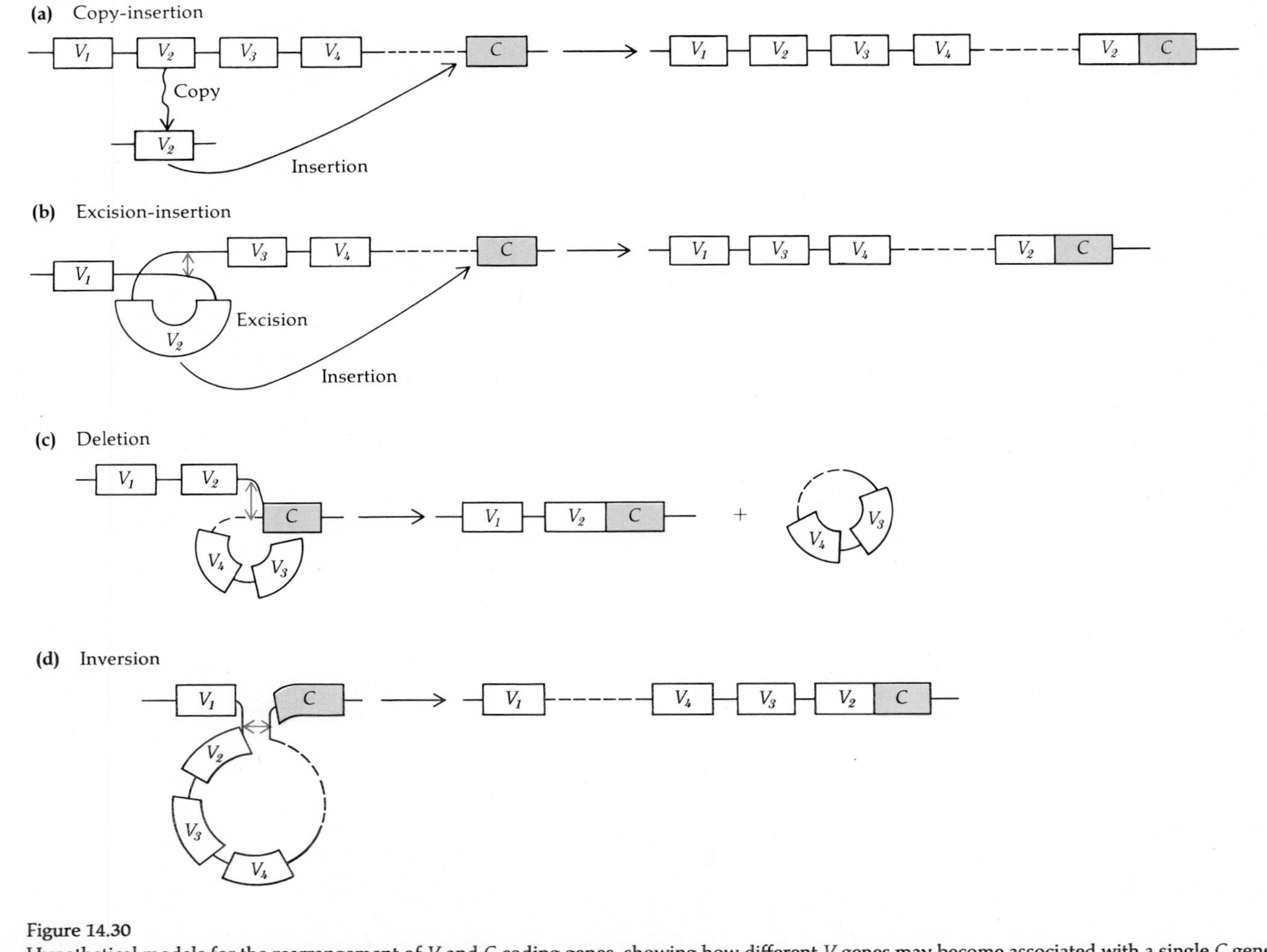

Figure 14.30
Hypothetical models for the rearrangement of *V* and *C* coding genes, showing how different *V* genes may become associated with a single *C* gene. [After S. Tonegawa et al., *Cold Spring Harbor Symp. Quant. Biol.* 41:877 (1976).]

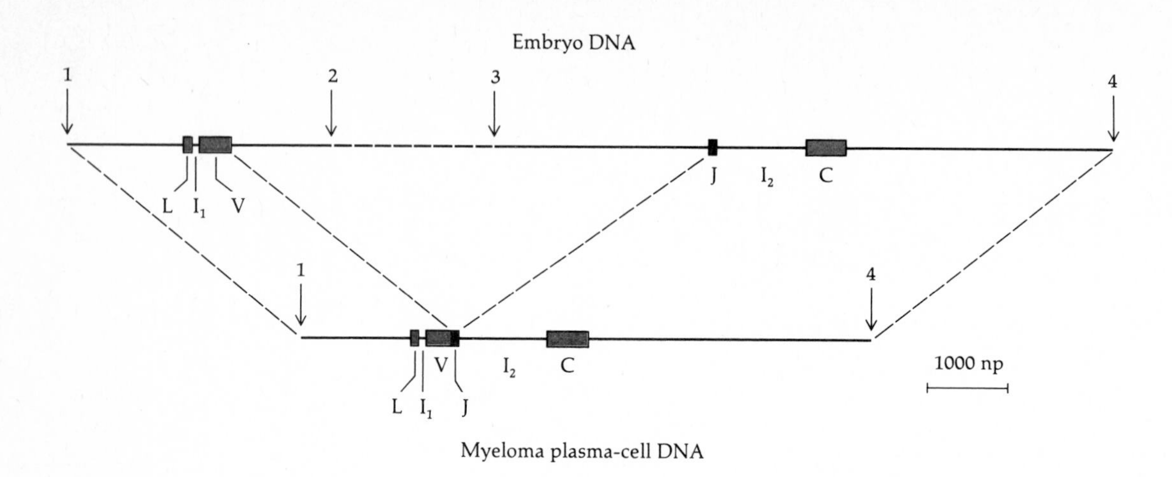

Figure 14.31
Arrangement of mouse λ_L light-chain DNA sequences in embryos and differentiated myeloma cells. The numbered arrows indicate *Eco* RI sites. The light-chain mRNA contains a leader sequence (L), a variable-region sequence (V and J), and a constant-region sequence (C). Two intervening sequences, I_1 and I_2, are present in the DNA separating the leader from the variable region and the variable region from the constant region. In the differentiated myeloma-cell DNA, these sequences reside on a single *Eco* RI restriction fragment produced by cleavage at sites 1 and 4. In embryo DNA, the sequences L, I_1, and V are found on one *Eco* RI fragment produced by cleavage at sites 1 and 2; the sequences J (coding for 13 amino acids of the variable region), I_2, and C are found on a second *Eco* RI fragment produced by cleavage at sites 3 and 4. [After C. Brack et al., *Cell* 15:1 (1978).]

The way in which different antibodies are synthesized, each containing identical constant regions but different variable regions, was a mystery for many years. Different genes were postulated to code for the constant and variable regions of heavy and light chains. The mechanism by which constant and variable regions of the polypeptide join to form the whole chain prompted different hypotheses. In 1976, S. Tonegawa and his colleagues showed that the V_κ and C_κ genes coding for a mouse myeloma light chain are adjacent in the DNA of the myeloma cells but are more distantly situated in the DNA from mouse embryos or mouse sperm. They postulated that, during lymphocyte differentiation, the selection for synthesis of a particular light chain is accompanied by rearrangement of the chosen V_κ and C_κ genes, so that both can be coordinately transcribed to yield an mRNA coding for that light chain (Figure 14.30).

Rearrangement of V_λ and C_λ light-chain genes has also been observed in other myeloma tumors. Commitment to produce a particular immunoglobin chain, which is accompanied by rearrangement of the V_λ and C_λ genes, excludes rearrangement of the V_κ and C_κ genes, and vice versa. Even after rearrangement, the V_λ and C_λ coding regions are still separated by two intervening sequences, which are removed by splicing of the RNA transcript prior to its transport to the cytoplasm (see Chapter 11).

The evidence for the rearrangement of variable- and constant-region genes is obtained using the tools of recombinant DNA research. An mRNA molecule coding for a myeloma light chain is purified from polysomes, transcribed into cDNA by reverse transcriptase, and amplified by cloning in a plasmid vector. The cloned light-chain cDNA is cleaved with a restriction enzyme that cuts the V portion from the C portion, and the two DNA fragments are separated by electrophoresis. The DNA fragments are radioactively labeled to high specific activity by nick translation, denatured, and used as probes to identify DNA restriction fragments derived from whole myeloma-cell DNA or whole embryo DNA, which contain sequences complementary to the probe sequence. As diagrammed in Figure 14.31, the crucial observation is that in myeloma-cell DNA a single *Eco* RI fragment contains both *V* and *C* coding regions, whereas in embryo DNA these regions are contained in separate *Eco* RI fragments. This observation indicates that, at some point between embryo and differentiated lymphocyte, a specific rearrangement of DNA occurred to eliminate an *Eco* RI restriction site, which was included in a large piece of DNA, thus bringing the *V* and *C* coding regions closer together.

Once this rearrangement has occurred in the differentiation of a lymphocyte, it is passed on to its mitotic progeny as a heritable property. This type of hereditary commitment is analogous to the acts of cell determination that occur during embryogenesis. Whether the molecular mechanisms are the same remains to be discovered. If directed DNA rearrangements do occur during development, then the totipotency exhibited by differentiated somatic-cell nuclei would suggest that the rearrangements are reversible. Conversely, the low frequency of success attained in Gurdon's nuclear transplantation experiments might indicate that many of the nuclei tested were no longer totipotent, owing to irreversible changes occurring during differentiation.

Problems

1. Some women who are carriers of red-green color blindness suffer from visual defects. How might you account for this?

2. When homozygous *yellow, vermilion, singed* female *Drosophila* are crossed to wild-type males, gynandromorphs occur among the progeny at a low frequency (about 2 per 10^4 XX zygotes). Some gynandromorphs have heads in which all the tissue is XO, as evidenced by *yellow* cuticle and *singed* bristles, and yet the eye color is wild type rather than *vermilion*. How might you explain this observation?

3. When *D. melanogaster* embryos that are homozygous for the homeotic mutation *ophthalmoptera* are raised at 17°C, the adults have normal eyes, but when they are raised at 29°C, wing tissue replaces most of the eye tissue (Figure 14.11). If eggs are collected at 17°C and the developing

creatures are shifted to 29°C at different stages of development, the effect on the adult phenotype is that shown in Figure 14.32. The reciprocal experiment—eggs collected at 29°C and later shifted to 17°C—is also shown in the figure. What interpretation of these data can you offer?

4. The data in the following table were compiled from A. H. Sturtevant's original drawings of *D. simulans* gynandromorphs. They record the number of disagreements of sex between the humerus (Hu), wing (W), abdominal tergites (t), abdominal sternites (s), and genitalia (G) in 758 gynandromorphic half-flies. (Because flies are bilaterally symmetrical, each fly provides two bits of data on the frequency with which the sex-dividing line falls between two primordia.)

	Hu	W	2t	2s	3t	3s	4t	4s	5t	5s	G
Hu	–	111	176	202	182	216	200	229	227	241	322
W		–	171	171	178	178	196	197	217	216	304
2t			–	60	22	78	48	99	77	120	209
2s				–	64	20	72	47	95	76	210
3t					–	70	33	89	69	111	232
3s						–	72	35	95	66	187
4t							–	79	43	96	220
4s								–	96	39	180
5t									–	97	213
5s										–	165
G											–

Data from A. Garcia-Bellido and J. Merriam, *J. Exp. Zool.* *170*:61 (1969).

Using a compass and ruler, construct the most precise fate map possible from these data. Give distances in sturt units. Remember that, as in genetic maps of chromosomes, additivity is best over small intervals.

5. A traveler who returned from the distant land of Sind reported visiting a family in which the mother and two of her three daughters suffered from a congenital absence of teeth in various portions of their jaws. The third daughter, however, was not missing any of her teeth. Also living with the family were an older married son and his wife. The son shared the congenital defect of his mother and two younger sisters; his wife was

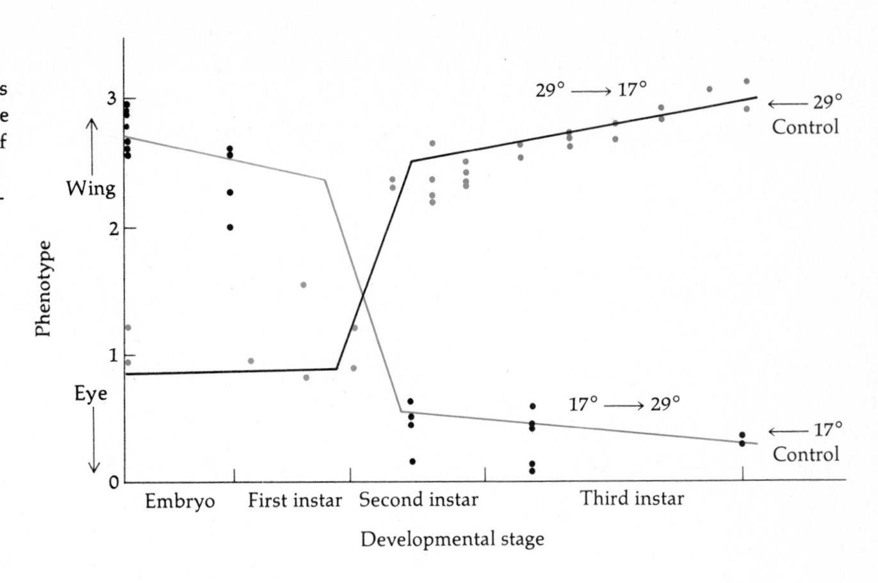

Figure 14.32
The effect of temperature shift, at the indicated stages during development, on the homeotic transformation of eye to wing in *D. melanogaster.* The average phenotype of adult flies raised throughout their development at 17° or 29°C is indicated by the level of the arrows labeled "Control." Each point represents the average phenotype of a population of flies shifted at the indicated stage of development. [After J. H. Postlethwait, *Develop. Biol.* 36:212 (1974).]

much mistreated by her in-laws because of her failure to have conceived any children after five years of marriage. Explain the failure of the son's wife to bear children.

6. It has been claimed that a human clone was produced several years ago in a secret laboratory, by transplantation of a diploid nucleus from a man's somatic cell into an enucleated human egg cell. Soon after development had begun, the embryo was said to have been inserted into the uterus of a surrogate mother and delivered nine months later as a healthy baby clone. The clone is said to be doing well and receiving a fine upbringing. In view of what you know about development, what is the possibility that a wealthy egoist could have an *exact* replica of himself produced by cloning?

7. Propose experiments employing recombinant DNA techniques and restriction enzyme analysis that would distinguish between models a, b, and c in Figure 14.30 for the rearrangement of the *V* and *C* genes during myeloma-cell differentiation.

8. *Sex-reversed* (*Sxr*) is an autosome-linked dominant mutation that causes XX mice to develop as phenotypic males, expressing H-Y antigen on their cells. What might be the origin of the *Sxr* mutation? Predict the sex phenotype of $X^{Tfm}X^{+}$, *Sxr* individuals.

9. Chimeric mice can be produced by uniting two or more eight-cell embryos of different genotypes. The cells rearrange themselves to form an abnormally large single morula, which develops into a normal-sized fetus and is

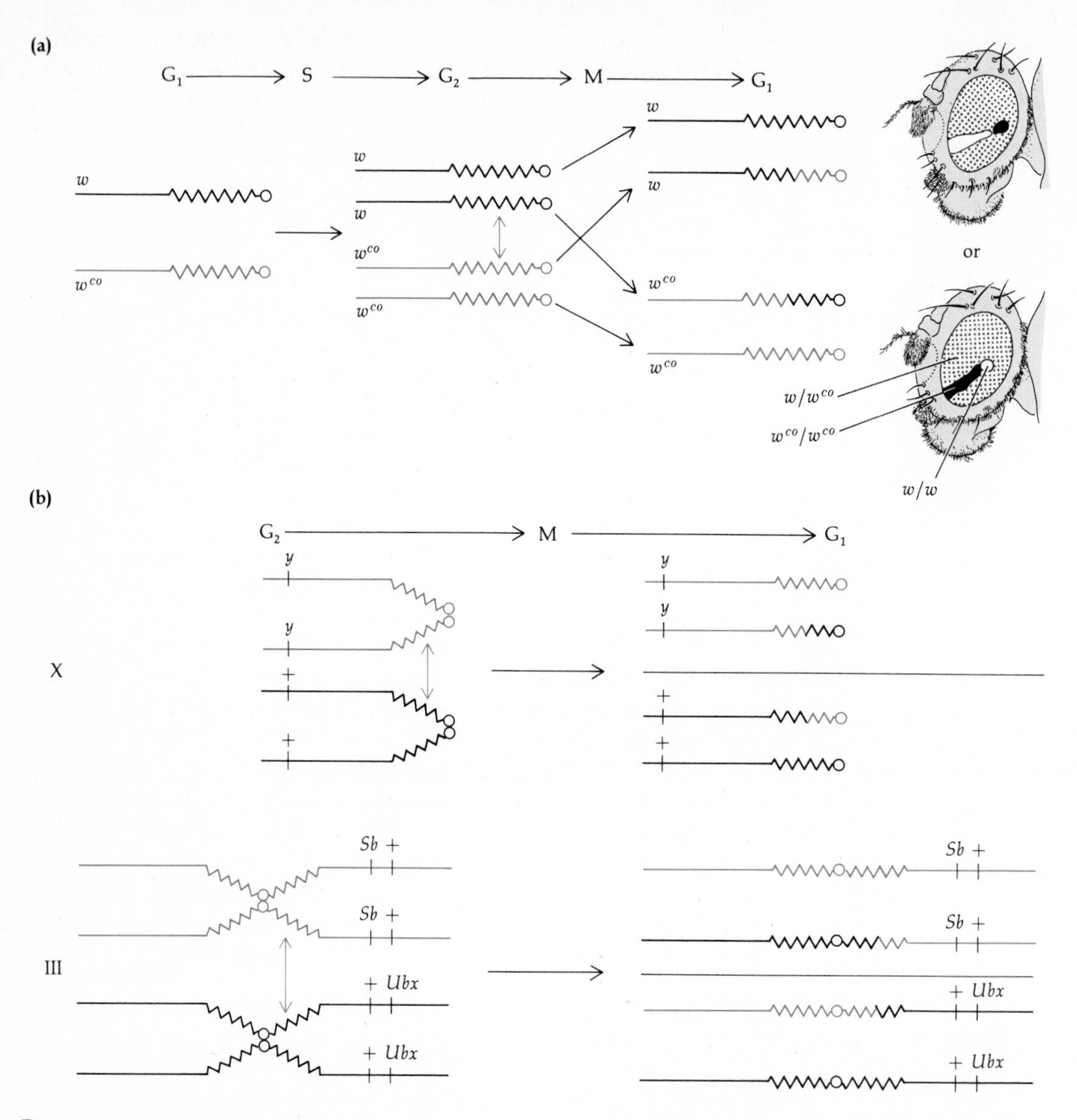

Figure 14.33
(a) Mitotic recombination in the X chromosome of *Drosophila*. When it occurs in a cell of the eye imaginal disk, proper segregation at mitosis leads to the development of an eye exhibiting "twin spots." Alleles w and w^{co} are different alleles of the *white* locus; w/w^{co} is dark ruby in color. Note the shape of the clones and the difference in size between clones in the anterior eye region and the posterior eye region, demonstrating regional effects on cell proliferation during development. (After R. Nöthiger, in *The Biology of Imaginal Disks*, ed. by H. Ursprung and R. Nöthiger, Springer-Verlag, New York, 1972.) **(b)** Mitotic recombination in cells of genotype y/y^+; $Sb\ Ubx^+/Sb^+\ Ubx$ can be detected in both the X chromosome and the right arm of the 3rd chromosome. At a low X-ray dose, recombination will occur in only a few cells of a larva, and the probability that both events will occur in a single cell is very low.

finally born as a normal-sized baby with cells of more than one genotype. Experiments uniting four genetically marked embryos have been carried out in which the four markers were the adult coat colors white, black, yellow, and light brown. Chimeric mice exhibiting two or three of these coat colors in patches have been obtained, but chimeras exhibiting all four coat colors have never been observed. How might these observations be explained?

10. Mitotic recombination can be induced by X irradiation of cells in a developing organism. If chromosomes carry appropriate genetic markers, the clonal descendants of a cell that has undergone mitotic recombination can be identified, as diagrammed in Figure 14.33a. Possible mitotic recombinations that can be induced in *Drosophila* cells of genotype y/y^+; Sb Ubx^+/Sb^+ Ubx are diagrammed in Figure 14.33b. The X chromosome is heterozygous for the *yellow* (*y*) mutant, causing yellow cuticle and bristles; the third chromosome is heterozygous for the dominant mutant *Stubble* (*Sb*), causing short bristles, and for *Ultrabithorax* (*Ubx*). When homozygous, *Ubx* causes the transformation of metathorax to mesothorax (Figure 14.12), as well as making other transformations that cause death before emergence of the adult. Clones of *Ubx*/*Ubx* cells can be identified by the appearance of normal, rather than short, bristles in the adult and, in the metathorax of the adult, by the appearance of mesothoracic structures such as wing cells and large bristles (the normal metathorax lacks any large bristles). How could you use this genetic system and X irradiation to ascertain at what period during development the Ubx^+ gene functions to cause cells to follow the metathoracic developmental pathway rather than the mesothoracic pathway?

15

Quantitative Characters

Genes and Phenotypes

Mendel's discovery of the basic laws of heredity was made possible by choosing contrasting characters that were easily distinguishable from one another. The peas were yellow or green, round or wrinkled; the flowers were axial or terminal; the plants were dwarf or tall, and so on. There were two alternative forms for each trait, determined by different alleles of the same gene. However, not all traits appear in such clearly distinguishable alternative forms. Humans, or pine trees, do not come in only two height classes, tall and short; rather, they vary continuously over a large range of heights. Height, weight, fertility, and longevity are a few examples of many traits that exhibit more or less *continuous variation*. The occurrence of continuous variation is due to (1) interactions between different genes, and (2) interactions between genes and the environment.

The interaction between the genes and the environment was already pointed out in Chapter 2, where we introduced the distinction between genotype and phenotype. The genotype of an organism is the genetic information it has inherited; the phenotype is its appearance, which we can observe (page 48). In Chapters 10–14 we reviewed the mode of action of genes—how the genetic information present in a zygote directs the development of an organism and determines its phenotype. A long sequence of processes usually intervenes between the synthesis of a polypeptide encoded by a structural gene and the manifestation of the organism's phenotype. The ultimate effect of a gene on the phenotype depends on the environmental conditions but also on the actions of other genes. The action of a gene is modulated by regulatory genes, but depends on other structural genes as well. For example, the gene determining the synthesis of indole in the tryptophan pathway of *Salmonella typhimurium* may fail to lead to the synthesis of tryptophan because of a mutation in some previous gene in the pathway, such as the one determining the synthesis of anthranilic acid (see page 346 and Figure 10.8). Tryptophan will, however, be synthesized if anthranilic acid is available from the environment.

In this chapter we shall introduce certain concepts describing various relationships that may exist between genes and phenotypes, precisely as a consequence of interactions among genes or between genes and the environment. Much of the chapter will deal with characters exhibiting continuous variation, in an effort to disentangle the environmental and genetic factors that account for such variation.

Penetrance and Expressivity

Some genes are rather invariant with respect to the phenotypes of their carriers, whereas others exhibit a great degree of variation. There may be variation both because not all individuals having a certain genotype manifest the expected phenotype and because the phenotype may be expressed to different degrees in different individuals. The *penetrance* of a gene is the *proportion of individuals* showing the expected phenotype. *Expressivity* is the *degree* to which the phenotype is manifested (in penetrant individuals). The dominant mutation *Lobe* (*L*) in *Drosophila melanogaster* is characterized by a reduction of the size of the eye, but the penetrance of the gene is 75%—only 75% of the individuals carrying the *L* gene have reduced eyes, while the other 25% have normal eyes. Moreover, the *L* gene has *variable* expressivity, because the 75% of individuals having small eyes may have them reduced to different degrees (Figure 15.1).

Many genes usually have complete penetrance and full expressivity. In Mendel's experiments, all peas having the dominant yellow allele (either in homozygous or heterozygous condition) were yellow, whereas all peas homozygous for the green allele were green. In humans, all individuals having the I^AI^A or I^Ai genotypes belong to the A blood group, those having the I^BI^B or I^Bi genotypes belong to the B blood group, those with the I^AI^B genotype belong to the AB blood group, and those with the *ii* genotype belong to the O blood group (Chapter 2, page 47).

An example of incomplete penetrance and variable expressivity is the dominant gene for Huntington's chorea in humans. Persons who carry this dominant gene may enjoy good health for most of their lives, but become ill at variable age. The disease starts with involuntary twitchings of the head, limbs, and body and progresses to degenerative changes of the nervous system, loss of physical and mental powers, and death. The age at which the disease first becomes noticeable ranges from infancy to old age

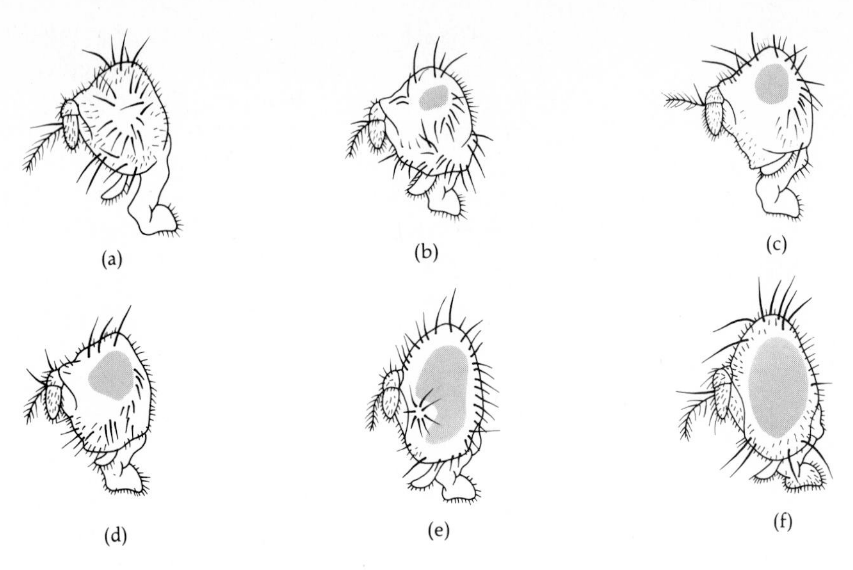

Figure 15.1
Penetrance and expressivity of the gene *Lobe* in *Drosophila melanogaster*. This dominant gene has variable expressivity, the size of the eye varying from zero (a) to normal (f). The gene is penetrant in only about 75% of the carriers (a–e).

(Table 15.1). The gene has incomplete penetrance, because it is virtually certain that the disease is never manifested in some carriers of the gene, i.e., they die of old age or other causes without having suffered from the chorea. Moreover, the gene has variable expressivity, because the disease appears at different ages and consequently has different effects on the lives of its victims.

The reasons why a gene may be penetrant in some individuals but not in others, and why it may have variable expressivity, are environmental and/or genetic. The environmental role in penetrance is obvious in the case of auxotrophic mutants and other conditional lethals. *Salmonella* carrying a mutation at the indole gene may be unable to survive in minimal medium, but do well if indole is added to the medium. *Drosophila* flies having some temperature-sensitive lethals do well at 20°C but become paralyzed or die at 29°C. Environmental effects also affect the expressivity of morphological traits (Figure 15.2). All *D. melanogaster* flies homozygous for the vestigial (*vg*) allele exhibit a vestigial phenotype, but the size of the wings depends on the temperature at which the flies were raised. The penetrance and expressivity of a gene may also depend on other genes carried by an individual, as is apparent with modifier genes and epistatic genes, discussed in the following section.

Modifier Genes and Epistatic Genes

The most basic gene interactions are those between alleles at the same locus: dominance, recessivity, and codominance. At the ABO locus in humans, alleles I^A and I^B are dominant over i, which is recessive to them; I^A

Table 15.1
Age of onset of Huntington's chorea.

Age	Number of Cases	Percent	Cumulative Percent
0–4	4	0.9	0.9
5–9	5	1.1	2.0
10–14	15	3.3	5.2
15–19	24	5.2	10.4
20–24	38	8.3	18.7
25–29	57	12.4	31.1
30–34	83	18.0	49.1
35–39	80	17.4	66.5
40–44	57	12.4	78.9
45–49	42	9.1	88.0
50–54	28	6.1	94.1
55–59	12	2.6	96.7
60–64	7	1.5	98.3
65–69	7	1.5	99.8
70–74	1	0.2	100.0
Total:	460	100.0	100.0

and I^B are codominant with respect to each other. Genes with more or less additive effects on the phenotype will be discussed below in connection with continuous variation. We shall now consider two other kinds of interactions, those due to modifier genes and epistatic genes.

A *modifier gene* is one that affects the phenotypic expression of a gene at a different locus. In mice the piebald spotting (*s*) gene determines the presence of white fur on the belly, but the amount of this may range from a small spot to the entire coat (Figure 15.3). By selective breeding one can produce strains in which all individuals have small white spots ("low" lines) as well as strains in which all individuals have nearly completely white fur ("high" lines). The results of crossing a low strain with a high strain are shown in Table 15.2. The F_1 and F_2 are intermediate between the two parental lines, with the F_2 exhibiting wider variation than the F_1. In the

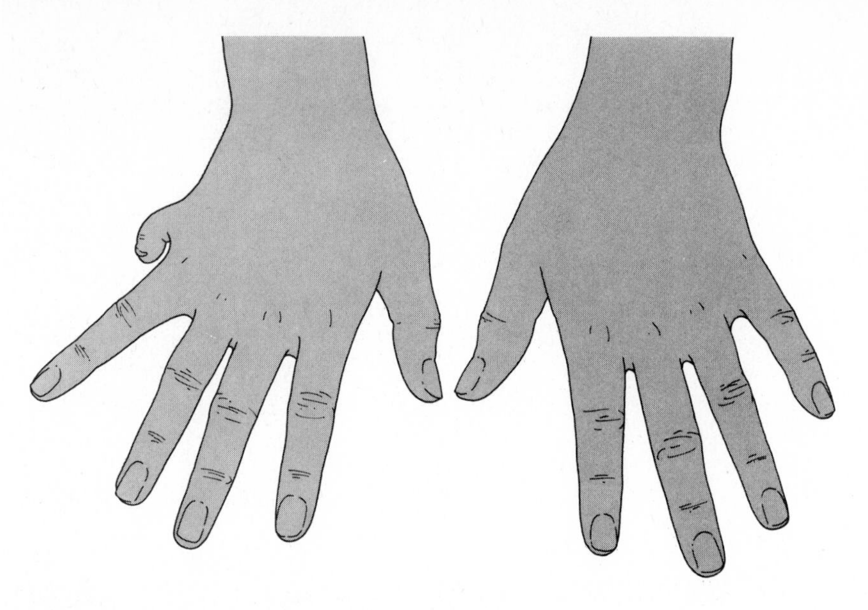

Figure 15.2
Variable expressivity of polydactyly. There are many forms of polydactyly: an extra digit sometimes grows next to the thumb, other times next to the little finger. Polydactyly, at least in some families, is due to a dominant gene (*D*) that controls the numbers of bony rays formed in the embryonic buds of hands and feet. The *Dd* genotype varies in its expressivity, at times even within the same individual, who may show five fingers on one hand but six on the other. In such cases the internal environment of the developing individual is responsible for the variable expressivity of the trait.

Table 15.2
Inheritance of piebald spotting in mice.

	Number of Individuals with Following Percentages of Dorsal White Fur:																			
Strain	5	10	15	20	25	30	35	40	45	50	55	60	65	70	75	80	85	90	95	100
Low line	7	89	35	2	1															
High line																	1	9	29	134
Low × High, F_1				1	1	4	3	3	10	7	9	9	7	1	1					
Low × High, F_2		1	2	6	15	10	12	14	19	27	29	28	19	34	21	14	2	3	2	
F_1 × High													5	16	33	24	15	10	9	12

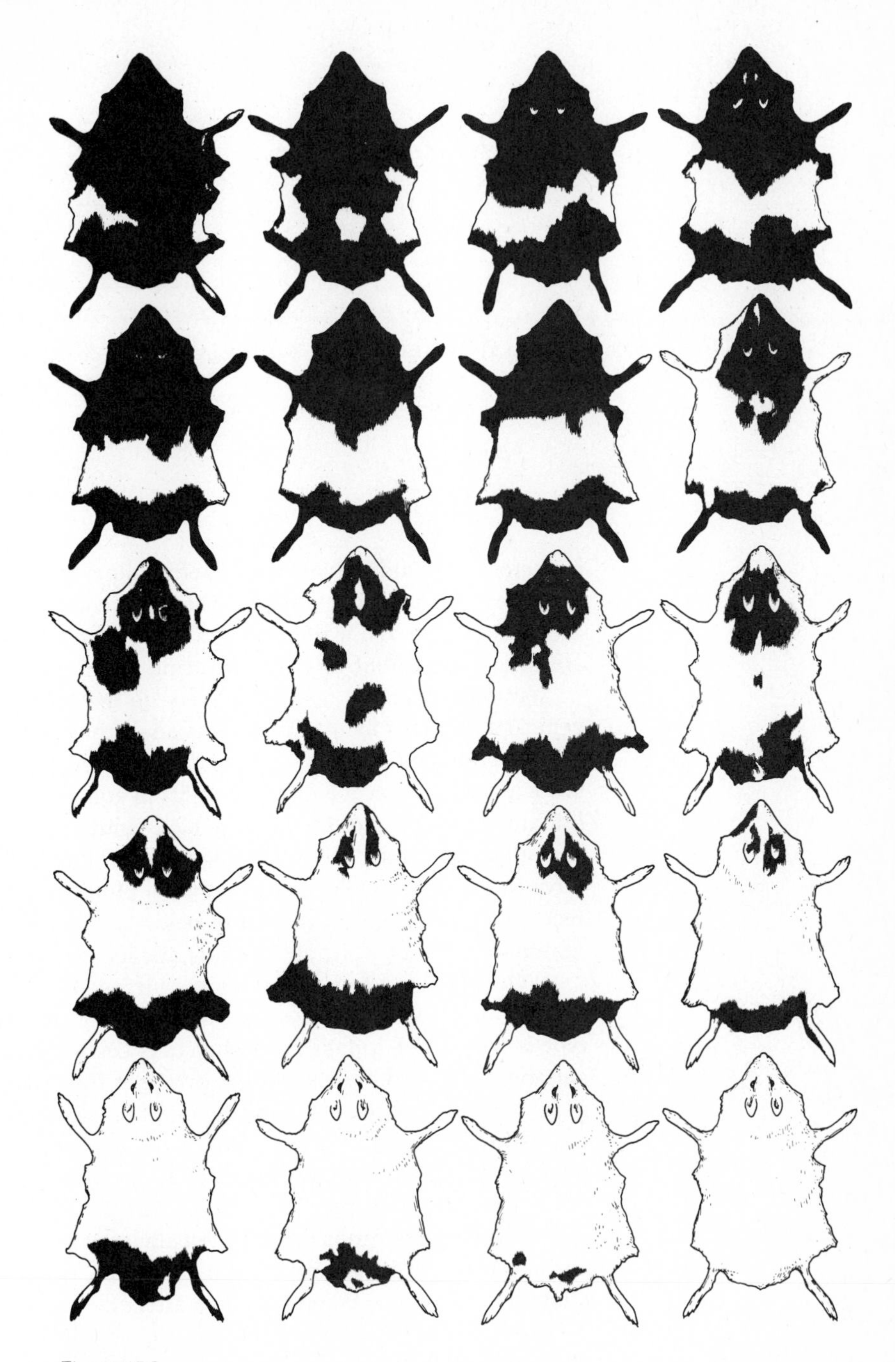

Figure 15.3
Variation in piebald spotting in strains of mice homozygous for the *s* gene. The amount of white fur is determined by other genes interacting with *ss*.

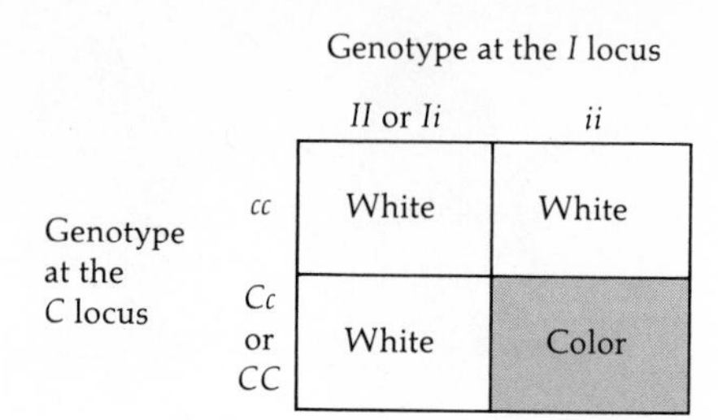

Figure 15.4
Epistasis in fowl. Colored plumage requires that the dominant allele *C* be present at the color locus, but this is not a sufficient condition. Color is suppressed in individuals carrying a dominant allele at the epistatic locus *I*.

backcross of F_1 individuals with the high line, the parental high-line types are recovered at approximately the frequency that would be expected (between 1/8 and 1/16) if there were three or four unlinked modifier genes affecting the amount of white spotting in the mice.

An *epistatic gene* is one that covers up the phenotypic expression of a gene at a different locus. Epistasis is analogous to dominance, because in both phenomena one gene covers the expression of another—at the same locus in dominance, at a different locus in epistasis. In fowl there is a gene, *I*, that is epistatic over the color gene *C*. Individuals that carry the dominant allele *I* will have white plumage even if they have the dominant allele *C* for color. Thus, fowls may have white plumage either because they are recessive homozygotes, *cc*, at the color gene or because they have the dominant allele *I* (*II* or *Ii*) at the epistatic locus (Figure 15.4).

White Leghorn fowls have white plumage because their genotype is *II CC*. White Wyandotte fowls also have white plumage because their genotype is *ii cc*. The F_1 offspring of the cross Leghorn X Wyandotte have the genotype *Ii Cc* and are also white (owing to the presence of the dominant gene *I*). The F_2 progeny, however, consist of white and colored chickens in the ratio 13:3 (Figure 15.5). This ratio departs from the ratio 9:3:3:1 [(double dominant):(single dominant at one locus):(single dominant at the other locus):(double recessive)] expected in the F_2 of a cross between individuals homozygous for different alleles at each of two loci in separate chromosomes. The standard ratio, however, occurs when the two loci affect separate traits; when both loci affect the same trait, other ratios can be expected. In the cross shown in Figure 15.5, the double dominants (9/16), the single dominants at the *I* locus (3/16), and the double recessive (1/16) are all white; only the dominants at the *C* locus but homozygous recessive (*ii*) at the other locus (3/16) are colored. Hence the 13:3 ratio of white to colored. The cross between two doubly homozygous white-flowered peas shown in Figure 15.6 gives a 9:7 ratio in the F_2. Other ratios are possible, depending on the mode of interaction between the alleles at the two loci.

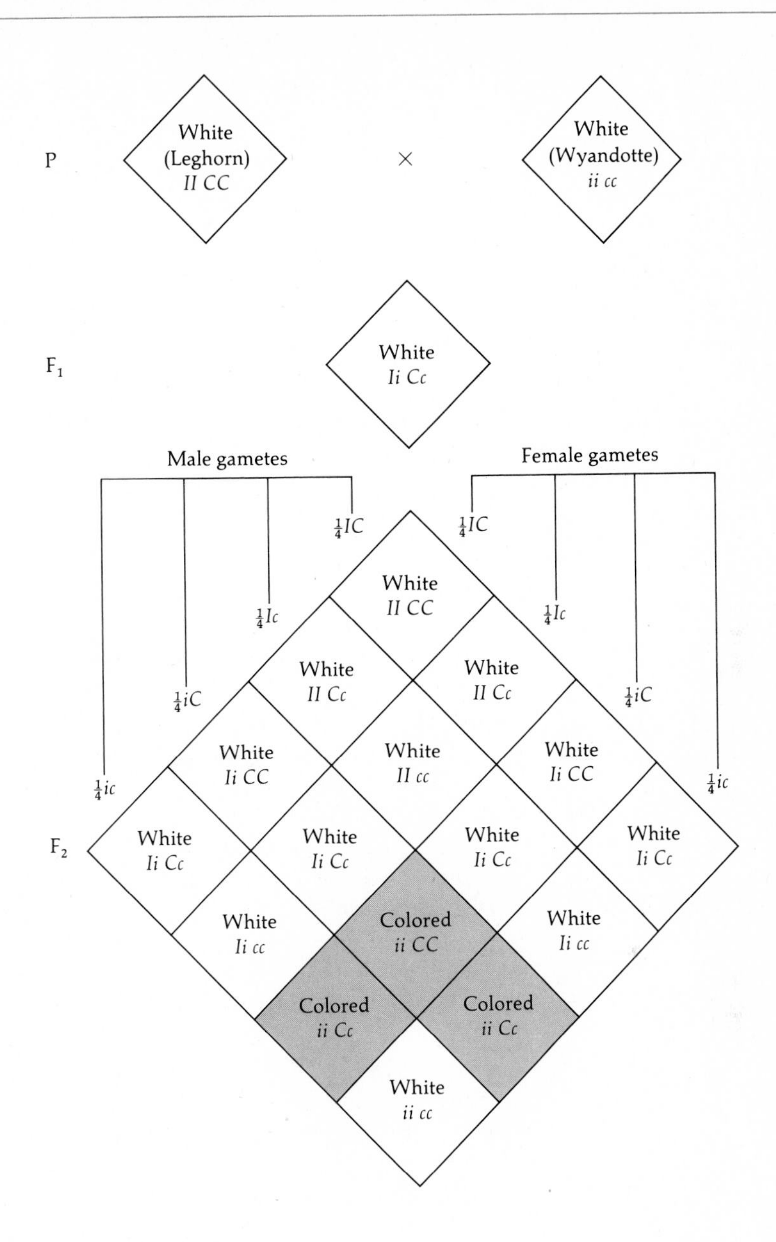

Figure 15.5
A 13:3 ratio obtained in the F_2 of a cross between two breeds of white fowl. The ratio is due to interaction between the color gene *C* and the *I* gene epistatic to it.

Pleiotropy

In the previous section we have considered interactions between loci affecting a single trait. The reverse of that situation, i.e., when a gene affects several traits, is known as *pleiotropy*. Mendel noted that one of the genes he studied affected simultaneously the color of the flowers (red or white), the color of the seeds (gray or brown), and the reddish spots (presence or absence) in the axils of the leaves (Figure 15.7).

Flies homozygous for the *vg* allele in *Drosophila* have vestigial wings, which is the most obvious effect of that gene. But they differ from wild-

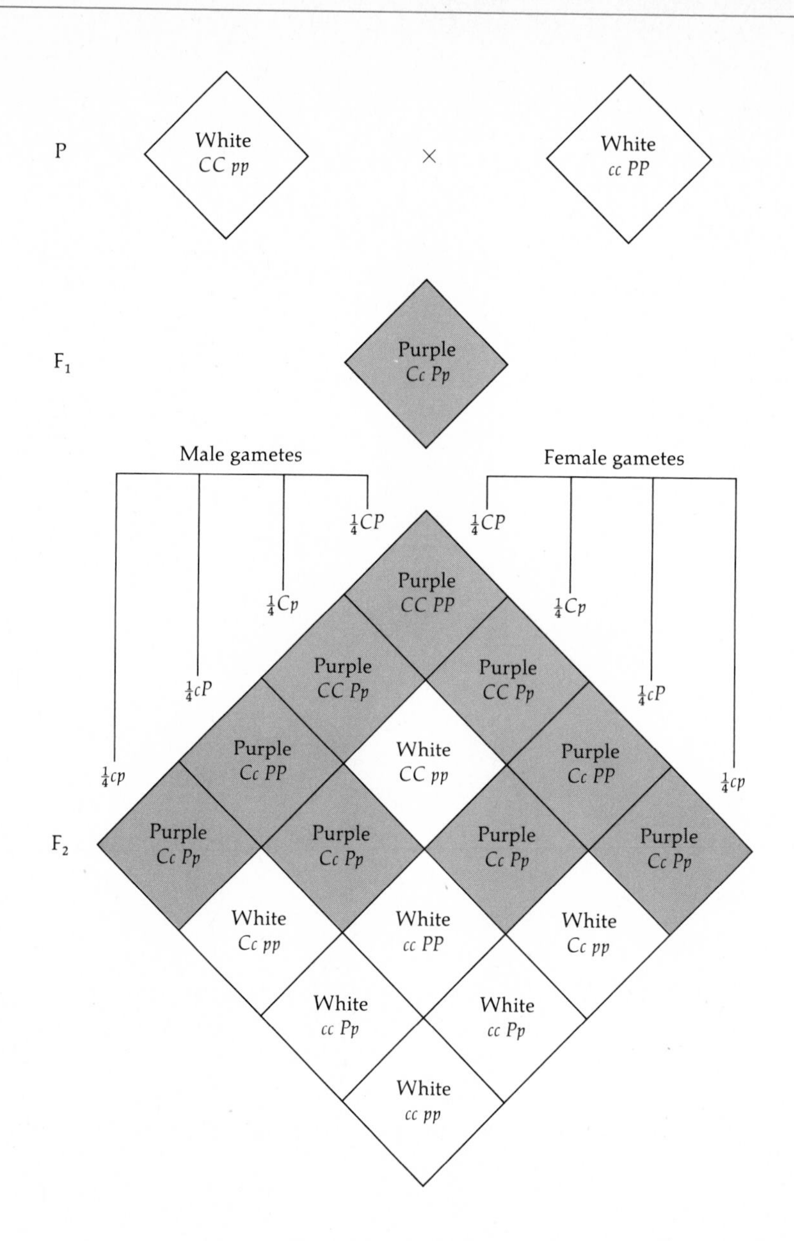

Figure 15.6
A 9:7 ratio obtained in the F_2 of a cross between two white-flowered sweet peas. Purple flowers require the presence of the dominant *P* allele, but the color is not expressed in individuals homozygous for the recessive allele at the epistatic locus *C*.

type flies in many other traits: they have modified balancers, a certain pair of dorsal bristles are erect instead of horizontal, the reproductive organs are somewhat different, longevity and fecundity are reduced, and so on.

An example of pleiotropy in humans is the recessive gene responsible for phenylketonuria (PKU), a condition resulting in severe mental retardation. Untreated homozygotes differ from normal individuals in the amount of phenylalanine in the blood, and in IQ, head size, and hair color (Figure 15.8). PKU also serves as an example of the effects of the environment on the expressivity of a gene. With simple biochemical tests, newborn babies can be screened for PKU, and this is now done routinely

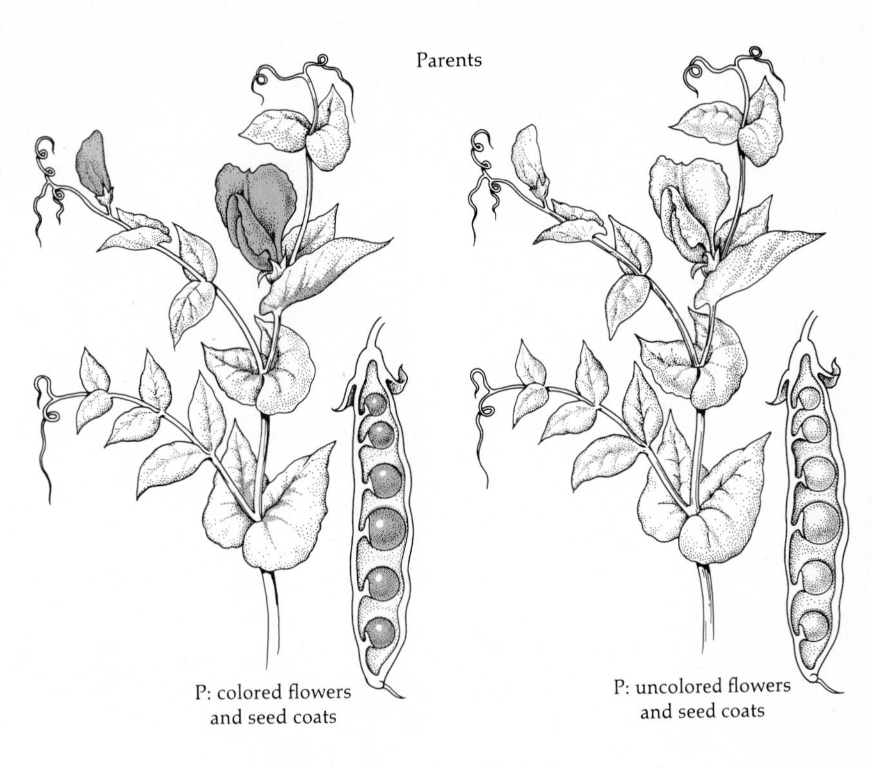

Figure 15.7
Pleiotropy in one of Mendel's pea experiments. The gene for flower color also affects the color of the seeds and of the leaf axils.

in the United States and Great Britain. PKU patients who are provided from birth with a diet low in phenylalanine develop into effectively normal individuals.

In Chapter 10 we presented the one gene–one polypeptide hypothesis. At the level of primary gene action, each gene has a single function: coding for one polypeptide. Pleiotropic effects therefore reflect the integration of developmental processes, in that the products of a single biochemical pathway may ultimately affect many developmental pathways. In the example of PKU, the normal gene codes for the enzyme that synthesizes tyrosine from phenylalanine. PKU victims lack that enzyme; consequently, phenylalanine accumulates in the blood. This also results in brain damage, which accounts for the reduction in IQ and head size. Moreover, hair pigment is formed from tyrosine; the light hair color of PKU victims is due to their inability to synthesize tyrosine from phenylalanine.

Continuous Variation

There are two kinds of variation among organisms of the same species. With respect to some traits, variation is discrete, or *discontinuous*: organisms fall into one or another of a few clearly distinguishable classes. The traits studied by Mendel are of this kind, and so are many other ex-

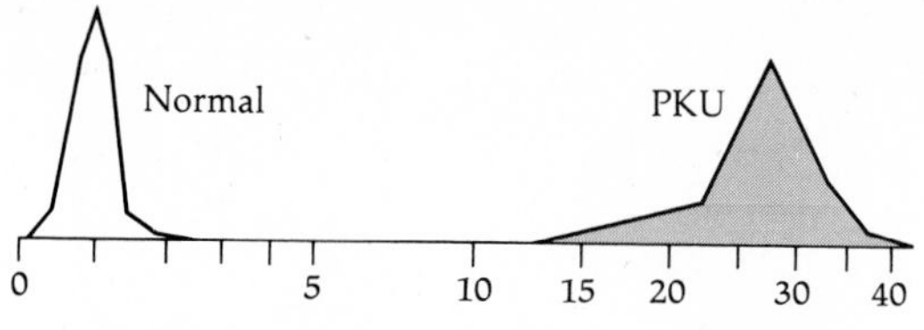

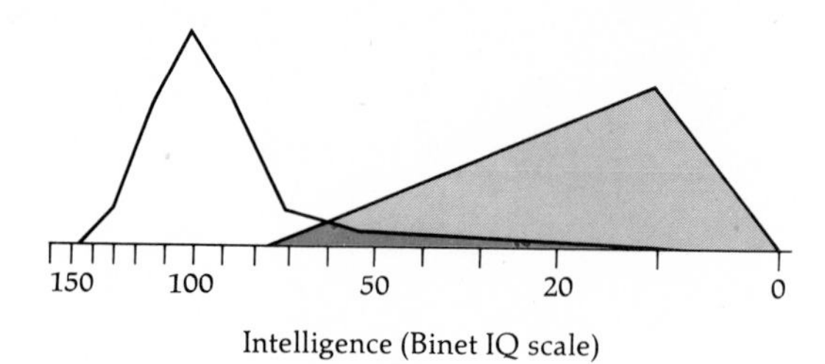

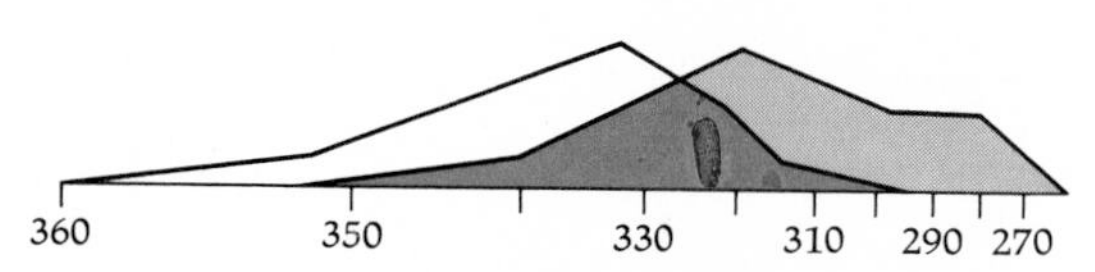

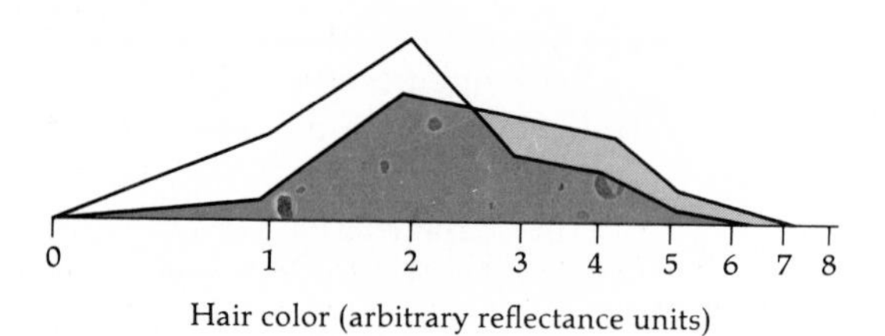

Figure 15.8
Pleiotropy of the PKU gene in humans. A single gene locus affects several phenotypic traits. In each case, variation among normal individuals is shown by the curve on the left, and variation among PKU patients, by the curve extending to the right.

amples of variation used so far in this book: humans have A, B, AB, or O blood groups; *Drosophila* flies have either normal or vestigial wings, and either red or white eyes; snapdragon flowers can be red or white; *Salmonella* bacteria are able or unable to synthesize indole. Yet with respect to other traits, variation is *continuous*: organisms vary more or less continuously over a range. Most people are between 145 and 185 cm tall; although they can be divided into height classes differing, for example, by 5 cm, there are people with all intermediate heights. *Drosophila* females may lay from a few to several hundred eggs; corn ears may have from a few score to several hundred seeds. Traits exhibiting continuous variation are sometimes called *quantitative*, or *metric*, characters because differences between individuals are quantitative, or small (requiring precise measurement), rather than qualitative, or large (requiring only simple observation, without precise measurement).

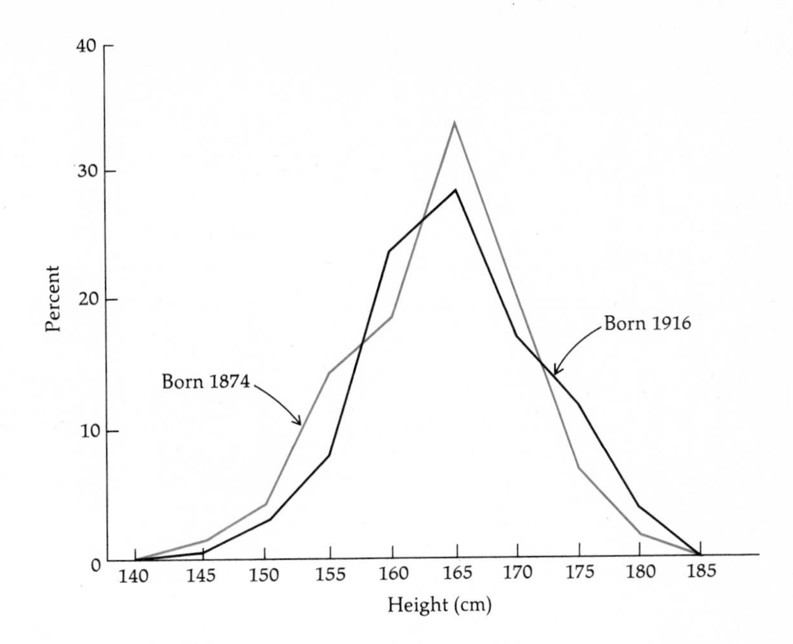

Figure 15.9
Distribution of height in 20-year-old Italian males. The men have been divided into classes differing by 5 cm. The mean height is 163.7 cm for men born in 1874, and 166.1 cm for men born in 1916. More than 200,000 men were included in each sample. A few individuals were excluded because they fell outside the height range shown in the figure.

One characteristic of quantitative traits is that their numerical values are usually distributed on a characteristic bell-shaped curve called a *normal distribution* (Figures 15.9 and 15.10). Some interesting properties of normal distributions are discussed in Appendix A.V.

In the early 1900s, geneticists raised the questions of whether quantitative variation is hereditary and, if so, whether it is inherited according to Mendelian laws. The questions were soon settled—quantitative variation is due in part to environmental influences and in part to genetic differences that are inherited like other genes. The Danish geneticist Wilhelm Johannsen showed in 1903 that continuous variation was partly environmental and partly genetic (he subsequently formulated the distinction between genotype and phenotype). In 1906 the mathematician George Udny Yule suggested that several gene loci each having a small effect could account for quantitative variation. This was experimentally confirmed in the following years by two geneticists, the Swede Herman Nilsson-Ehle and the American Edward M. East.

Johannsen found that the weight of beans (*Phaseolus vulgaris*) in a commercial seed lot ranged from "light" (15 centigrams) to "heavy" (90 cg). By allowing self-pollination of the beans over several generations, he established several lines, each highly homozygous. He planted seeds of different weights but *all from a single line* and weighed the seeds produced by each plant. Although the original seeds differed in size, the beans produced by all plants were of the same mean weight (Table 15.3). For example, the mean weight of beans produced by 57.5-cg seeds was 45.8 cg,

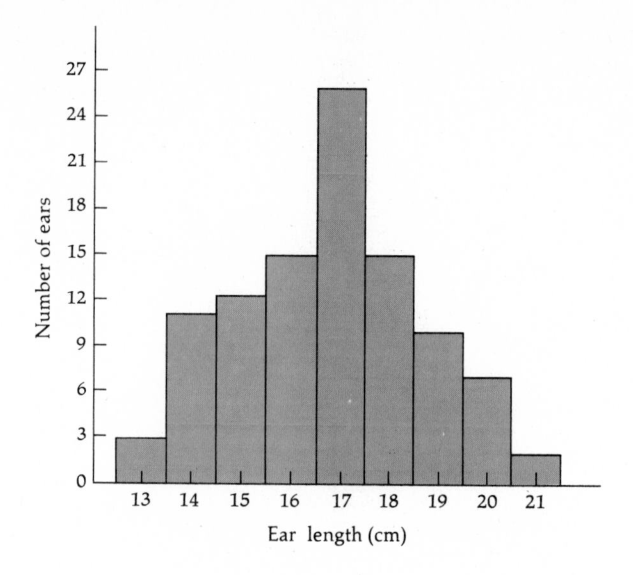

Figure 15.10
Distribution of ear length in a long-eared variety of corn. This distribution, like that shown in Figure 15.9, is roughly bell-shaped, with frequencies being highest at some intermediate value and tapering off toward both extremes.

not significantly greater than 44.5 cg, the mean weight of beans produced by seeds weighing only 27.5 cg. Thus he demonstrated that weight variation among beans of a single plant was environmental—as it had to be, since each line was homozygous and thus lacked genetic variation that could segregate in the progenies.

Johannsen demonstrated, moreover, that genetic differences contribute to bean weight by showing that beans from different lines had different average sizes. Table 15.4 shows the mean weight of seeds from

Table 15.3
Number of beans of different weights in the progeny of seeds of different weights from Johannsen's pure line 13.

Weight of Parental Seed (cg)	Number of Progeny Beans of Weight (cg):*									Mean Weight of Progeny (cg)
	22.5	27.5	32.5	37.5	42.5	47.5	52.5	57.5	62.5	
27.5		1	5	6	11	4	8	5		44.5
37.5	1	2	6	27	43	45	27	11	2	45.3
47.5		5	9	18	28	19	21	3		43.4
57.5		1	7	17	16	26	17	8	3	45.8

*Beans in the 22.5-cg class include all beans between 20 and 25 cg; those in the 27.5-cg class include all beans between 25 and 30 cg, etc.

Table 15.4
Mean weight of beans from different pure lines of Johannsen's.

Weight of Parental Seed (cg)	Mean Weight (cg) in Line Number:					
	19	18	13	7	2	1
20		41.0		45.9		
30	35.8	40.7	47.5			
40	34.8	40.8	45.0	49.5	57.2	
50			45.1		54.9	
60			45.8	48.2	56.5	63.1
70					55.5	64.9
Mean weight of all progeny of a single line:	35.1	40.8	45.4	49.2	55.8	64.2

different lines. From each line he used seeds of different weights: for example, he used seeds weighing 20 cg, 40 cg, and 60 cg in line 7. All progeny of a given line had the same mean weight, regardless of how heavy the parental seeds were, thus confirming the results shown in Table 15.3. But the mean weights of beans from *different* lines (last row in Table 15.4) were consistently different; for example, the mean weights of all beans from lines 19 and 1 were 35.1 cg and 64.2 cg, respectively.

Seed Color in Wheat

The experiments of Johannsen demonstrated that both the environment and the heredity contribute to continuous variation. However, the experiments did *not* show that the hereditary differences between pure lines were due to Mendelian genes. It was conceivable that quantitative characters were inherited in a different fashion than discrete characters, as some biologists were still arguing. In 1909 Nilsson-Ehle showed that continuous variation occurs when a trait is determined by several genes, each having a small effect, that behave according to Mendelian laws.

Nilsson-Ehle had several true-breeding lines of wheat with kernels ranging from white through various shades of red to dark red. Crosses between white and light red varieties produced F_1 offspring that were

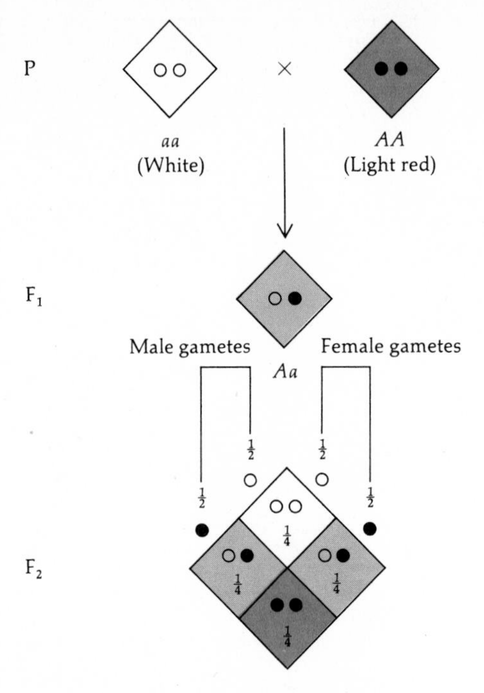

Figure 15.11
Kernel color in the F_2 of a cross between a white and a light red variety of wheat. Capital letters represent alleles for redness, so that two *A* alleles produce light-red kernels, but one *A* and one *a* produce kernels intermediate between white and light red.

intermediate in color between the parents; the F_2 consisted of white, intermediate, and light red kernels in the approximate proportions 1:2:1. When white and red varieties were crossed, the F_1 was light red, and in the F_2 there were five kinds of kernels, ranging from white to red; the proportion of white kernels was approximately 1/16 in each of several crosses. Crosses between white and dark red varieties produced F_1 offspring intermediate between the parents; the F_2 consisted of seven classes, ranging from white to dark red, with kernels of intermediate color being the most common, and white kernels representing about 1/64 of the total number of kernels. Nilsson-Ehle correctly explained his results as being due to three pairs of genes with two alleles each, one for white and one for red, in such a way that the alleles for redness each contributed a small amount of color to the phenotype.

Let us denote the alleles at the three loci as *A a*, *B b*, and *C c*, with the capital letters representing alleles for redness and the lower-case letters representing alleles for whiteness. The genotypes of the varieties can be represented as *aa bb cc* for white, *AA bb cc* for light red, *AA BB cc* for red,

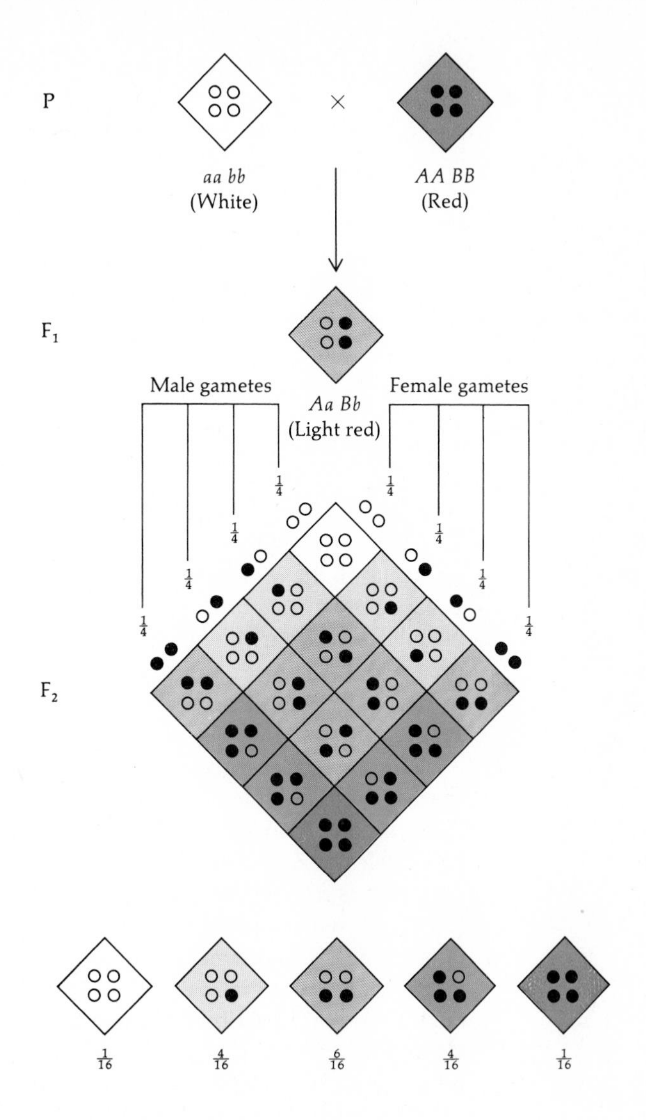

Figure 15.12
Kernel color in the F_2 of a cross between a white and a red variety of wheat. The color differences are assumed to be due to two pairs of genes. Capital letters represent alleles for redness.

and *AA BB CC* for dark red. The cross between the white and light red varieties is *aa bb cc* × *AA bb cc*. Since the two varieties are identical at the *B* and *C* loci, we need be concerned only with the *A* locus. The results of the cross are shown in Figure 15-11. The F_1 is intermediate in color between the two parents because it has one gene for red and one for white. In the F_2 the familiar ratio 1:2:1 appears.

The cross between the white and red varieties (*aa bb cc* × *AA BB cc*) is shown in Figure 15.12, where the *C* locus has been ignored because the two varieties are identical with respect to it. If, as assumed, each capital-letter allele (dark circles in the figure) contributes an equal amount of redness, color is determined by the number of such alleles, independently of the locus to which they belong. Thus, for example, the genotypes *AA bb*, *aa BB*, and *Aa Bb* all give light red kernels.

Finally, the cross between the white and dark red varieties (*aa bb cc* × *AA BB CC*) is shown in Figure 15.13. The F_1 is, as in the two previous crosses, intermediate between the parents. In the F_2 there are seven kinds of offspring in the proportions 1:6:15:20:15:6:1, approximately as observed by Nilsson-Ehle.

Polygenic Inheritance

Genes that each contribute a small amount to the variation in a quantitative trait are called *multiple factors* or *polygenes* ("many genes"). In the example of the wheat kernels, the effects of the polygenes are *additive* because the effects of the alleles are cumulative. It is not always true that all alleles have an identical effect—some may have greater effect than others at the same or at different loci. Moreover, the effects of polygenes are not always additive, because dominance or interlocus interactions may be present.

The results of increasing the number of gene loci that affect a quantitative trait are shown in Figure 15.14 and Table 15.5. We assume that there are only two alleles at each locus, that gene effects are additive, and that alleles at different loci have equal effects. It is, moreover, assumed that there is a certain amount of variation due to environmental influences (bottom row in Figure 15.14). Environmental effects could be ignored in Nilsson-Ehle's experiment because, indeed, they may contribute little to kernel-color variation in wheat. But most quantitative traits are affected by environmental variations. Human weight, the amount of milk produced by a cow, and the size of an ear of corn, for example, are affected by nongenetic factors such as nutrition, climate, and disease.

The effects of polygenic inheritance are clear. The greater the number of gene loci that cumulatively affect a trait, the more nearly continuous the variation in the trait will be. Figure 15.14 shows that the number of genetic classes in the F_2 is the number of alleles plus 1: three classes with one locus (two alleles), five with two loci, seven with three loci, thirteen with six loci. Environmental factors contribute to make the distribution more nearly continuous.

Quantitative traits tend to be normally distributed because the proportion of genotypes is greater in the intermediate classes than in the extreme classes. This tendency becomes stronger as the number of loci affecting the trait increases. Consequently the *variance* (which measures the dispersion of the distribution—see Appendix A.III) decreases as the number of genes increases. It can be seen in Figure 15.14 (and in Table 15.5) that the proportion of individuals falling in each extreme class is $(\frac{1}{2})^2 = \frac{1}{4}$ for one locus, $(\frac{1}{2})^4 = \frac{1}{16}$ for two loci, $(\frac{1}{2})^6 = \frac{1}{64}$ for three loci, etc.

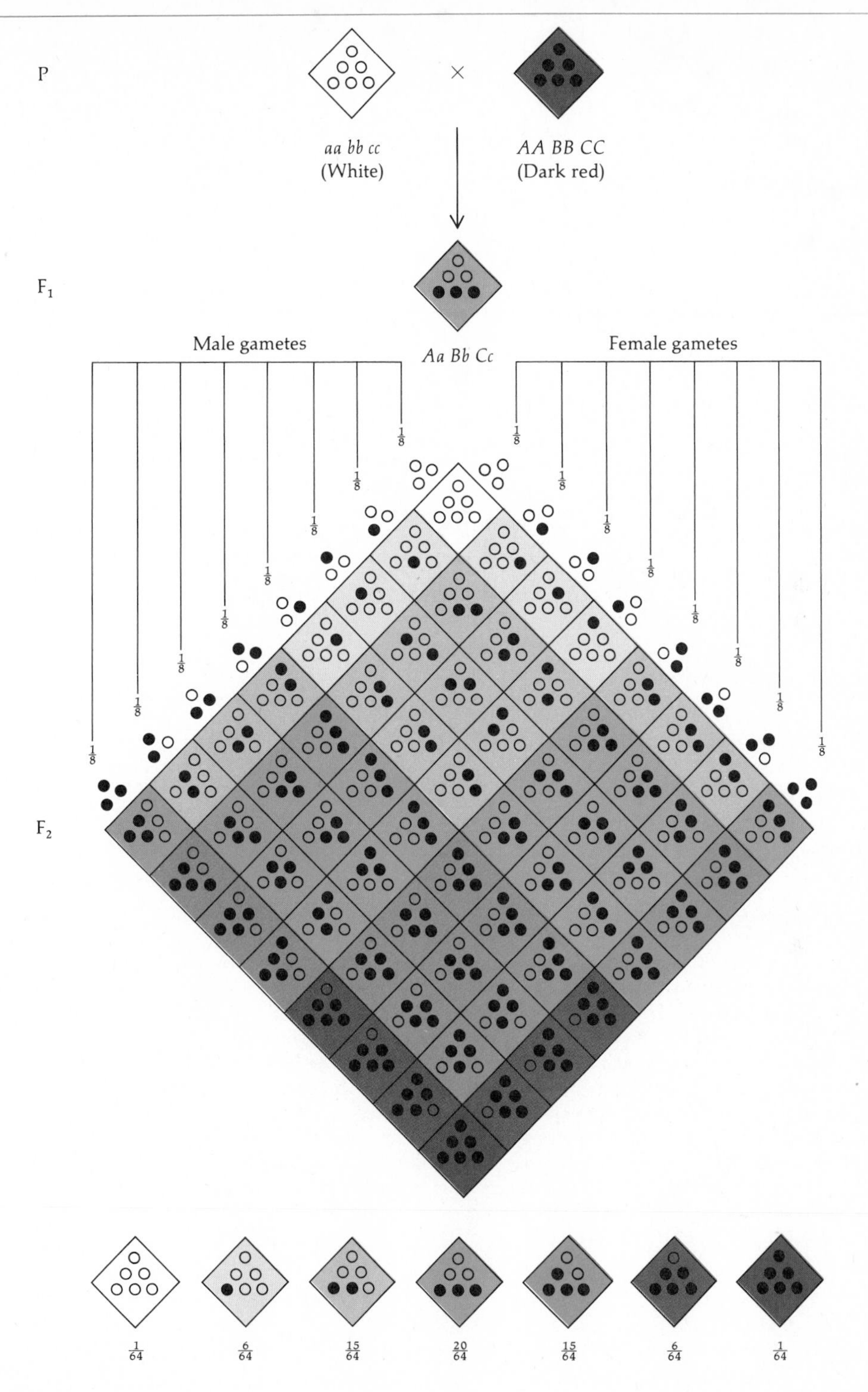

Figure 15.13
Kernel color in the F_2 of a cross between a white and a dark red variety of wheat. The color differences are assumed to be due to three pairs of genes. Six different genotypes appear in the F_2 in the proportions indicated at the bottom of the figure. Capital letters represent alleles for redness.

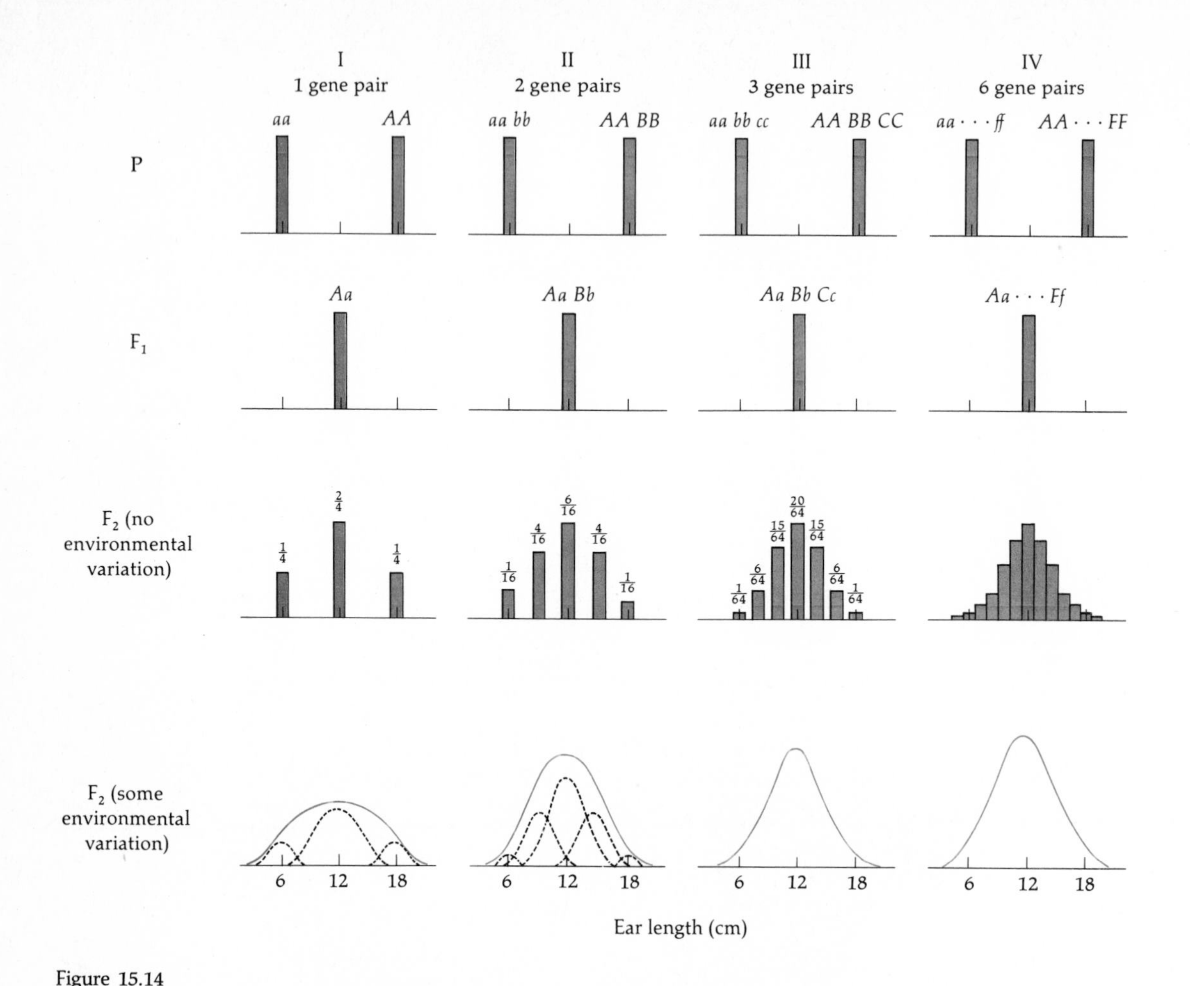

Figure 15.14
Crosses between two lines of plants differing at one, two, three, and six gene loci. The first three rows show the expected distributions for the P, F_1, and F_2 generations in the absence of environmental variation. The bottom row shows the F_2 distributions as they would appear when the environment contributes to the phenotypic variation. The parents in this hypothetical example are corn plants with ears that are 6 cm long in the short-eared parent and 18 cm long in the long-eared parent. In case I, the difference is assumed to be due to a single pair of genes, so that allele *A* causes an increase of 6 cm each time it occurs over the length of 6 cm characteristic of the short-eared parent; in case II, alleles *A* and *B* add 3 cm each; in case III, alleles *A*, *B*, and *C* add 2 cm each; and in case IV, alleles *A*, *B*, . . . , *F* add 1 cm each. The proportions of individuals falling in each of the F_2 genotypic classes of case IV are given in Table 15.5. Note that an increase in the number of gene pairs determining the same difference between two parental strains leads to a decrease in the variance of the F_2 distribution, because a greater proportion of individuals fall in the intermediate classes.

Table 15.5
Frequency distribution expected in the F_2 from hypothetical crosses between two lines of corn differing in ear length by 12 cm, when this difference is determined by one, two, three, and six pairs of genes with equal additive effects.

Gene	Frequency of Ear of Length (cm):*													
Pairs	6	7	8	9	10	11	12	13	14	15	16	17	18	Total
1	$\frac{1}{4}$						$\frac{2}{4}$						$\frac{1}{4}$	1
2	$\frac{1}{16}$			$\frac{4}{16}$			$\frac{6}{16}$			$\frac{4}{16}$			$\frac{1}{16}$	1
3	$\frac{1}{64}$		$\frac{6}{64}$		$\frac{15}{64}$		$\frac{20}{64}$		$\frac{15}{64}$		$\frac{6}{64}$		$\frac{1}{64}$	1
6	$\frac{1}{4096}$	$\frac{12}{4096}$	$\frac{66}{4096}$	$\frac{220}{4096}$	$\frac{495}{4096}$	$\frac{792}{4096}$	$\frac{924}{4096}$	$\frac{792}{4096}$	$\frac{495}{4096}$	$\frac{220}{4096}$	$\frac{66}{4096}$	$\frac{12}{4096}$	$\frac{1}{4096}$	1

*The fractions are the terms in the binomial expansion of $(\frac{1}{2} + \frac{1}{2})^n$, where n is the number of alleles (twice the number of loci, or gene pairs). The denominator is simply 2^n.

Owing to environmental variation, quantitative traits do not usually fall into classes that precisely reflect the genotype. This often makes it impossible to ascertain the number of genes affecting a quantitative trait by the kind of analysis performed by Nilsson-Ehle. The number of genes involved can, however, be estimated from the proportion of F_2 individuals falling in the parental classes.

East used the polygenic model of inheritance to explain variation in the length of flowers in *Nicotiana longiflora*. He crossed two varieties whose flowers had average lengths of 40.5 mm and 93.3 mm. The F_1 was intermediate in length, as expected. The F_2, consisting of 444 plants, had a broader distribution than the F_1, also as expected, but none of the F_2 flowers was either as short or as long as the average lengths of the parental flowers. With four pairs of genes, the expected proportion of F_2 individuals falling in each parental class is $(\frac{1}{2})^8 = \frac{1}{256}$. Because none of the 444 F_2 plants fell in the parental classes, it can be concluded that more than four gene pairs are involved in the flower-length differences between the two parental varieties (Figure 15.15).

Skin pigmentation in humans varies from light to dark. The most extreme differences occur between Caucasians and African Blacks. Although differences in skin color exist within each group, they are small relative to the differences between the groups. Matings between Blacks and Caucasians produce children of intermediate skin color; matings between F_1 individuals and their backcrosses to Blacks or Caucasians produce children with a range of skin pigmentations. The distribution of these F_2 and backcross individuals is approximately as expected if the difference in skin pigmentation between Blacks and Caucasians were due to three or four pairs of genes, each contributing equally to skin color (Figure 15.16).

Genetic and Environmental Variation

Francis Galton (1822–1911) used the terms *nature* and *nurture* to refer to the roles played by heredity and environment in determining quantitative traits. Both types of influence—genetic and environmental—are usually present. A well-fed mouse will be larger than a starved one, and a well-schooled person will have a higher IQ than one raised in a deprived environment. But variation among individuals can also be due to genetic differences. It is interesting—and practical, for example, in plant and animal breeding—to separate the genetic from the environmental effects on quantitative traits.

We might ask the question: Is it possible to ascertain to what extent a given trait is determined by the genotype and to what extent by the environment? Reflecting on this question makes us realize that it is not well formulated. Any trait depends completely for its development on both heredity *and* environment. For an individual to develop, it must have a genotype—the genetic constitution of the zygote—but development can

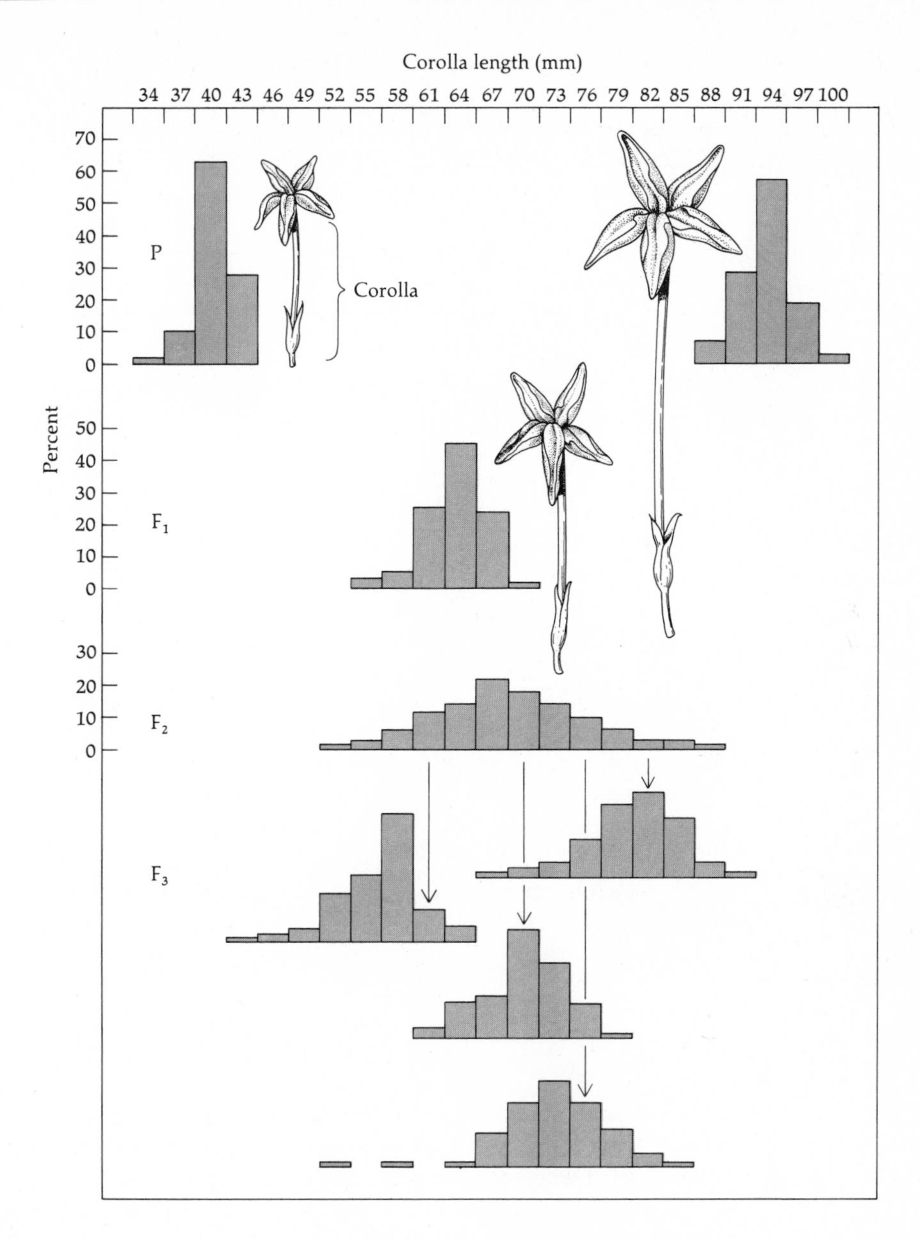

Figure 15.15
Corolla length in *Nicotiana longiflora*. The distribution of the proportion of individuals falling in each class is shown for the P, F_1, F_2, and F_3 generations. Corolla length varies continuously, but the flowers are grouped in classes each covering a range of 3 mm. The distributions of the F_1 and F_2 have means intermediate between the two parental strains, but the F_2 has a greater variance. The F_3 progenies demonstrate that part of the variance observed in the F_2 is genetic, because F_2 plants with different corolla lengths give rise to different distributions whose mean corolla lengths correspond approximately to the lengths of the F_2 plants from which they are produced.

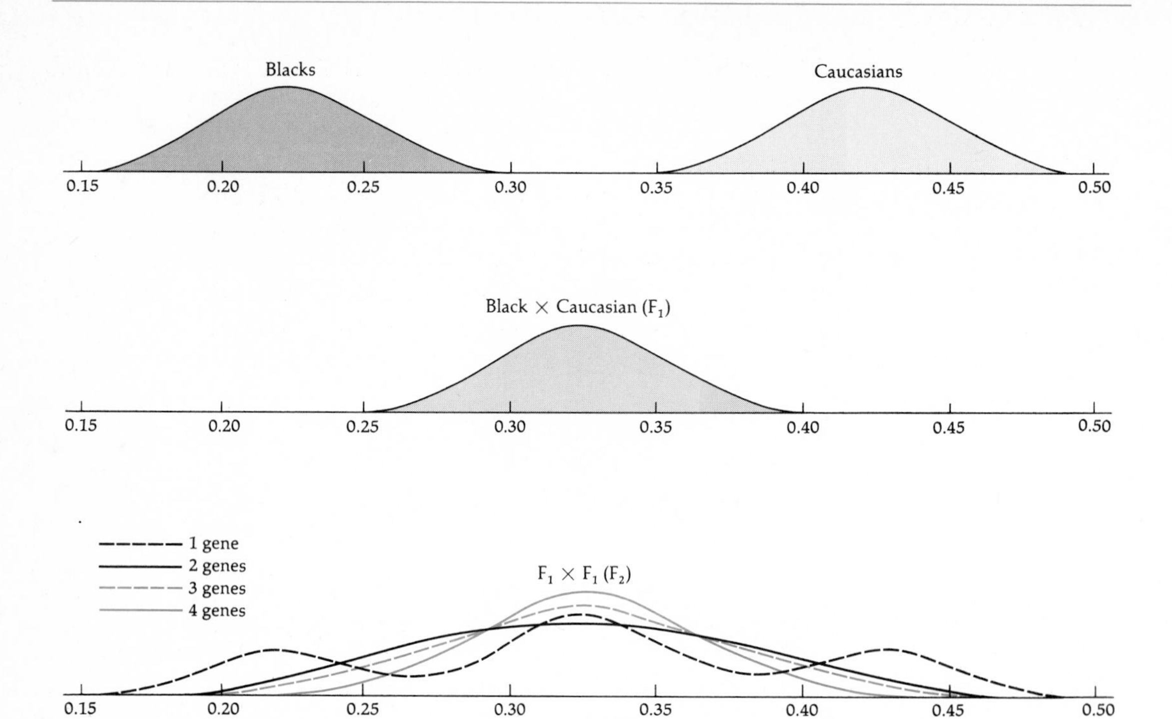

Figure 15.16
Distribution of skin color in Blacks and Caucasians. Skin color is measured by the reflectance of skin to red light of 685 nm wavelength. The F_2 curves are theoretical expectations based on various hypotheses about the number of genes involved in the skin-color differences between Blacks and Caucasians. Studies of F_2 progenies show distributions resembling those expected if three or four gene pairs are involved. (After W. F. Bodmer and L. L. Cavalli-Sforza, *Genetics, Evolution, and Man,* W. H. Freeman, San Francisco, 1976.)

only occur in some environment. There is no way in which we could measure the IQ of a human genotype that has not been exposed to any environment whatsoever; such individuals cannot exist.

The question of the relative effects of heredity and environment, the question of "nature versus nurture," can be properly raised as follows: To what extent is the *variation* among individuals with respect to a trait due to *genetic* variation (i.e., to genetic differences among the individuals), and to what extent is it due to *environmental* variation (i.e., to environmental differences)? It will soon be apparent why it is important to realize that this question, rather than the one formulated above, is the one being answered when geneticists investigate the relative effects of heredity and environment.

The fraction of the phenotypic variation in a trait that is due to genetic differences can be measured by the *heritability* of the trait, a concept advanced by the American geneticist Jay L. Lush. Let us use the following symbols:

H = heritability
V_T = the total phenotypic variance observed in a trait
V_G = the fraction of the phenotypic variance that is due to genetic differences among individuals
V_E = the fraction of the phenotypic variance that is due to differences in the environmental conditions to which the individuals were exposed

We thus have $V_T = V_G + V_E$, and, by definition,

$$\text{Heritability} = \frac{\text{Genetic variance}}{\text{Phenotypic variance}} \qquad H = \frac{V_G}{V_T} = \frac{V_G}{V_G + V_E}$$

Measuring the total phenotypic variance of a trait in a group of individuals is usually not difficult. (First the mean value of the trait is calculated; then the differences between each value and the mean are obtained and squared; the average of these squared differences is the variance—see Appendix A.III.) However, partitioning the total variance into the environmental and genetic components is not a simple matter. Geneticists use a variety of methods that will not be reviewed here. The rationales of the *twin method* and the *mass selection method* are presented in Boxes 15.1 and 15.2. Here we shall use East's data on flower length to illustrate the concept (Figure 15.15).

The varieties of *Nicotiana longiflora* crossed by East were homozygous. Hence the variance within each parental group is all environmental. The variance among the F_1 offspring is all environmental as well; all F_1 individuals are genetically identical to each other (although not homozygous) because all gametes produced by each parental strain are identical. The average variance within each of the two parental varieties and the F_1 offspring is 8.76, which is, therefore, an estimate of the environmental variance (V_E) in the environment where the experiments were conducted. Thus we have $V_E = 8.76$.

The genes inherited from the two parental strains segregate in the F_2. Consequently, the phenotypic variance of the F_2 consists of both genetic and environmental variance. The total phenotypic variance (V_T) of the F_2 offspring is 40.96; thus $V_T = 40.96$. Because $V_T = V_G + V_E$, it follows that

$$V_G = V_T - V_E = 40.96 - 8.76 = 32.20$$

The heritability of flower length in East's experiment is therefore

$$H = \frac{V_G}{V_T} = \frac{32.20}{40.96} = 0.79$$

Box 15.1 Heritability by the Twin Method

One way of estimating heritability is to measure the phenotypic variance in groups of relatives with known degrees of relatedness, such as twins or sibs or first cousins. There are two kinds of twins: identical and fraternal. *Identical*, or monozygotic, twins arise from a single zygote that splits in two early in embryogenesis and develops into two genetically identical individuals. *Fraternal*, or dizygotic, twins arise from two independent zygotes, i.e., two eggs fertilized by two sperm. The genetic relationship between fraternal twins is the same as that between ordinary full-sibs, because in both instances zygotes form from different eggs from the same mother and different sperm from the same father. Fraternal twins, like full-sibs, share on the average half their genes (for each gene received by one sib from the mother, the probability that the other sib will receive the same gene is one-half, and the same for the paternal genes).

Identical twins are always of the same sex; fraternal twins may be both female, one female and one male, or both male, in the proportions 1:2:1. In order to estimate heritability, sets of identical twins of one sex are compared with sets of fraternal twins of the same sex as the identical twins. Let V_i = phenotypic variance between identical twins and V_f = phenotypic variance between fraternal twins. Because the identical twins are genetically identical, their variance is all environmental: $V_i = V_E$

The variance between fraternal twins is part genetic and part environmental. However, because they have half of their genes in common, their genetic variance will be half that of unrelated individuals:

$$V_f = \tfrac{1}{2}V_G + V_E$$

Therefore the difference between the two phenotypic variances estimates half the genetic variance:

$$V_f - V_i = \tfrac{1}{2}V_G + V_E - V_E = \tfrac{1}{2}V_G$$

Twice that difference divided by the total phenotypic variance is an estimate of heritability:

$$H = \frac{2(V_f - V_i)}{V_T} = \frac{2(\frac{1}{2}V_G)}{V_T} = \frac{V_G}{V_T}$$

There are, nevertheless, some problems. The environmental variance has been assumed to be the same for identical and for fraternal twins of the same sex, because in both cases the twins are born at the same time, are of the same sex, live together, go to the same schools, etc. But it is possible that identical twins are treated more similarly than fraternal twins, precisely because identical twins are genetically more alike and thus tend to react more similarly. If the V_E of fraternal twins is greater than the V_E of identical twins, the difference between the two phenotypic variances contains not only genetic variance but also some residual environmental variance. The formula given above will then overestimate heritability.

Another difficulty is that twins are treated more similarly than unrelated individuals because the twins are raised in the same family, cultural milieu, etc. This and the problem raised in the previous paragraph can be largely circumvented by studying twins who have been adopted by different families. Although the proportion of such cases is limited, a number of them have been studied by geneticists.

Yet other problems remain that may also affect estimates of heritability. These problems need not concern us here, however, because the purpose of this discussion is only to understand the rationale and basic methodology in estimating heritability using twin studies.

Box 15.2 Heritability by the Mass Selection Method

Plant and animal breeders improve domestic stocks by *mass selection*, one form of artificial selection. The procedure is simply to breed the next generation from those individuals that are best with respect to the trait being improved. Dairies breed the cows producing more milk, racers breed the swiftest horses, farmers use for seed wheat plants with the greatest yield, and so on. Artificial selection has been practiced for more than 10,000 years. If phenotypic variation is due in part to genetic differences among individuals, mass selection will improve the genetic characteristics of the stock.

Assume that the trait under consideration is normally distributed, as is often true for quantitative traits. For purposes of illustration, we might think of flower length in *Nicotiana longiflora* (Figure 15.15). We need to be concerned with two measures: the *selection differential*, which is the difference between the mean of the selected parents and the mean of the population, and the *selection gain*, which is the difference between the mean of the progeny of the selected parents and the mean of the parental generation (Figure 15.17).

If the variation is all due to environmental causes, the distribution of the selected parents' offspring will be the same as that of the parental populations (Figure 15.17, left). This was the situation observed by Johannsen when he bred beans of different sizes from the same inbred line (Table 15.3). If the phenotypic variation is all genetic, on the other hand, the mean of the offspring will be the same as that of the selected parents (Figure 15.17, center). This would be the case with respect to kernel color in wheat (Figure 15.13). The most common situation is that the variation is partly genetic and partly environmental. In such cases there will be a selection gain, but the mean of the offspring will not be as high as that of the selected parents. Heritability will then be estimated by the ratio between the selection gain and the selection differential. If we let G represent the selection gain and D represent the selection differential, we have

$$H = \frac{G}{D}$$

For example, the number of certain abdominal bristles in a strain of *Drosophila melanogaster* is 38. Flies with an average of 42.8 bristles were used to breed the next generation; therefore $D = 42.8 - 38 = 4.8$. The mean number of bristles in the offspring of the selected parents was 40.6; therefore $G = 40.6 - 38 = 2.6$. The heritability is then estimated as

$$H = \frac{G}{D} = \frac{2.6}{4.8} = 0.54$$

This estimate agrees very well with an estimate of $H = 0.52$ obtained in a separate experiment by comparing half-sibs (which share one-quarter of their genes) with full-sibs (which share one-half of their genes).

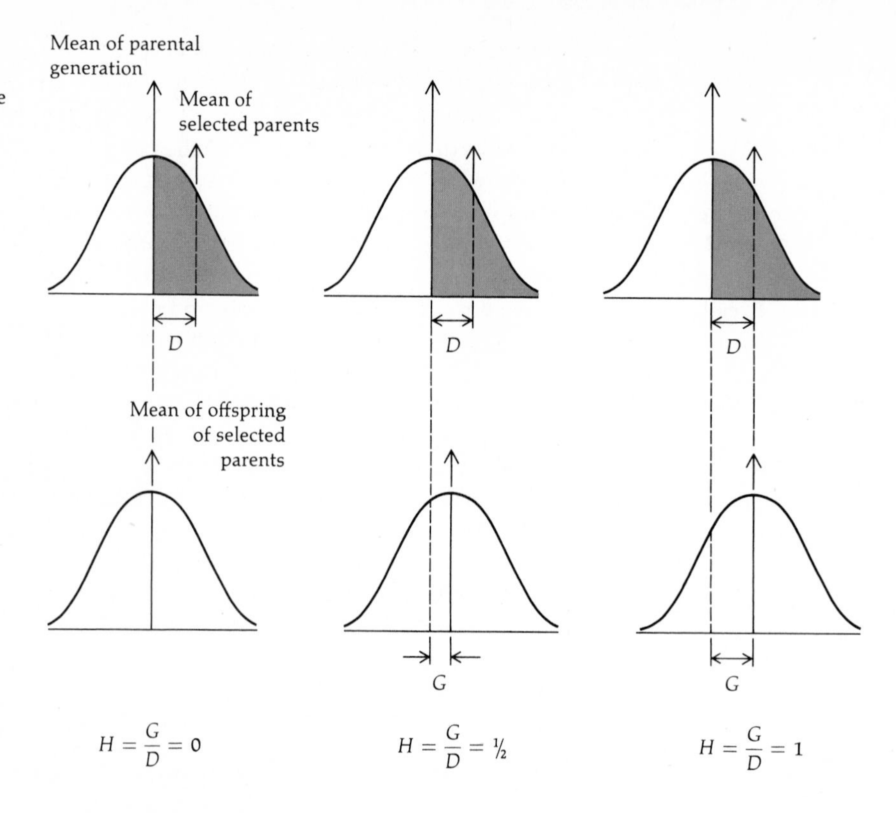

Figure 15.17
Selection differential (D) and selection gain (G). The selection gain divided by the selection differential estimates the heritability (H) in artificial selection experiments.

It would be incorrect, however, to say that flower length in *N. longiflora* is 79% determined by genes and 21% determined by the environment. What was said earlier must now be emphasized: Heritability measures *not the degree to which a trait is determined by genes,* but rather *the proportion of the phenotypic variation among individuals that is due to genetic variation.*

It must be pointed out that interactions between genes (dominance between alleles at the same locus as well as epistasis between alleles at different loci) affect heritability estimates, because such interactions are not additive. How to deal with these genetic interactions is a matter for texts more advanced than this one; awareness of their existence is sufficient here.

There is, however, another class of interactions affecting heritability estimates that is most important for understanding the significance of these estimates, namely, the interactions between genes and environments. Heritability estimates are valid only for the particular environment in which they are obtained. In other environments, they may be quite different. Consider, for example, the experiment illustrated in Figure 2.17. In the normal environment used for selecting bright and dull rats, the genetic differences between the two strains result in a considerable difference in phenotype: bright rats perform much better than dull rats. However, the difference between the two strains disappears when the rats are raised in a deprived environment. This means that the genetic differences between the strains are not expressed as phenotypic differences in the restricted environment; therefore, the heritability for brightness is greater in the normal environment than in the restrictive environment. Additional examples showing that heritability estimates are valid only for the environment in which they were obtained are given in the following section.

Despite their restricted significance, heritability estimates are quite useful in plant and animal breeding because they indicate the amount of response that can be expected in artificial selection of desirable traits. Heritability estimates also give us an idea of the role played by genetic differences in determining phenotypic variation among organisms living in the same environment. Some heritability estimates for various traits in animals and plants are given in Table 15.6, and for humans in Table 15.7. These values might, of course, be different if they were estimated under environmental conditions different from those under which they *were* obtained.

Heritability in Different Populations

Heritability is a *population-specific* measurement. It does not measure any invariant property of organisms, but only the relative contributions of genetic differences and environmental differences to phenotypic variation. If the genetic variation or the environmental variation is changed, heritability estimates will also change. Thus, measuring heritability for a group of organisms in two different environments, or for two different populations in the same environment, is likely to yield different results. These qualifications have already been stated, but they will be further developed here because of (1) their fundamental importance in understanding the concept of heritability and (2) their implications with respect to issues such as the much-debated question of whether IQ differences between human racial groups are genetic.

Consider the following "thought experiments," which we shall assume to have been done with *Potentilla glandulosa* (Figure 2.16). This plant can be reproduced by cuttings that make it possible to obtain a group of individuals genetically identical to each other when they are all derived from parts of a single plant.

EXPERIMENT 1. We cut one plant into many equal parts and place these parts in a very heterogeneous hillside where plants may be exposed to considerable environmental differences in the quality of the soil, the

Table 15.6
Examples of heritability for traits in various organisms (estimated by various methods).

Trait	Heritability
Amount of white spotting in Frisian cattle	0.95
Slaughter weight in cattle	0.85
Plant height in corn	0.70
Root length in radishes	0.65
Egg weight in poultry	0.60
Thickness of back fat in pigs	0.55
Fleece weight in sheep	0.40
Ovarian response to gonadotropic hormone in rats	0.35
Milk production in cattle	0.30
Yield in corn	0.25
Egg production in poultry	0.20
Egg production in *Drosophila*	0.20
Ear length in corn	0.17
Litter size in mice	0.15
Conception rate in cattle	0.05

amounts of sunlight and moisture, etc. We measure phenotypic variation in a certain trait, say, the total weight (biomass) of each plant. Because the plants are genetically identical, the variation is all environmental. The heritability of the trait as measured in these plants is therefore zero.

EXPERIMENT 2. We collect plants in different localities, so that they are genetically heterogeneous. We plant a small cutting from each plant in an experimental garden. We provide the plants with optimal conditions of soil, fertilizer, moisture, light, etc. We take great care that all plants receive uniform treatment. We estimate heritability in these plants for the same trait as in Experiment 1; we might obtain a very high value, say, 0.95, as we

Table 15.7
Heritabilities for certain traits in humans (estimated by twin studies).

Trait	Heritability
Stature	0.81
Sitting height	0.76
Weight	0.78
Cephalic index	0.75
Binet mental age	0.65
Binet IQ	0.68
Otis IQ	0.80
Verbal aptitude	0.68
Arithmetic aptitude	0.12
Aptitude for science	0.34
Aptitude for history and literature	0.45
Spelling ability	0.53
Foot-tapping speed	0.50

would expect, because the plants are genetically very heterogeneous and the environment quite uniform.

EXPERIMENT 3. We use cuttings from the same set of plants used for Experiment 2 and proceed similarly, except that we provide the plants with poor soil, no fertilizer, and marginal levels of moisture and light. Nevertheless, we make sure that all plants are treated uniformly. At the end of the growing season, the plants are very small owing to the poor environmental conditions. We estimate heritability and obtain a high value, as in Experiment 2, and for the same reasons: the plants are genetically heterogeneous but the environment is uniform.

EXPERIMENT 4. We use cuttings from the same set of plants as in Experiments 2 and 3 and plant them on a hillside where the plants grow

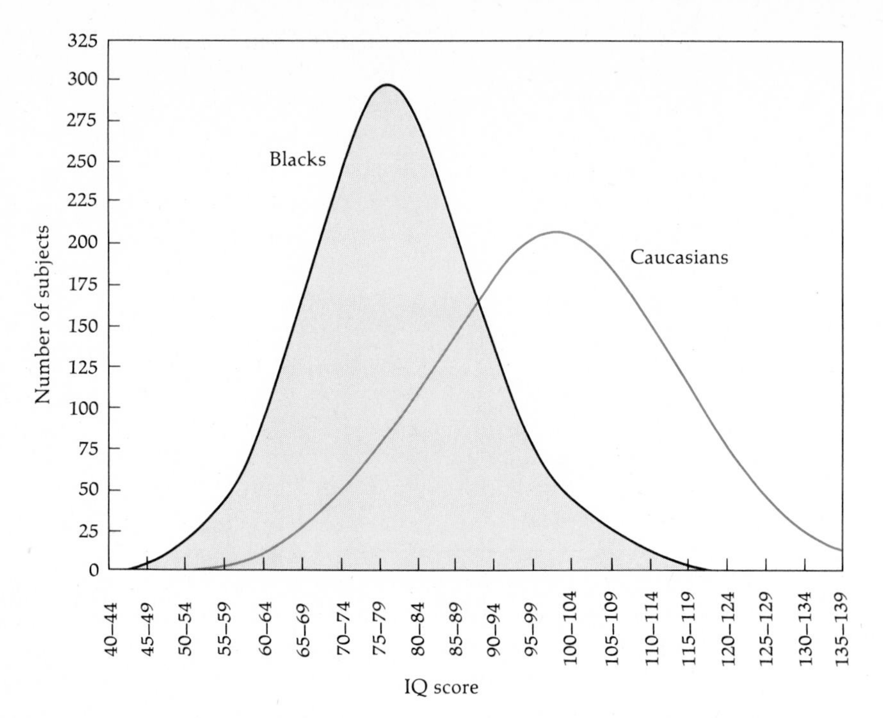

Figure 15.18
Distribution of IQ in Blacks and Caucasians. A sample of 1800 Black children from schools in Alabama, Florida, Georgia, Tennessee, and South Carolina is compared with a representative sample of Caucasians. The mean IQs in these distributions are 80.7 and 101.8. This difference of 21.1 points is larger than that observed in most other studies. The difference in mean IQ between U.S. Blacks and Caucasians is often around 15 points.

naturally. We estimate heritability at the end of the growing season and obtain an intermediate value, say, 0.60. This experiment corresponds better than any of the other three to the usual conditions found for animals and plants in nature, where there is variation in both the genotype and the environment.

These thought experiments illustrate, first, a point already made: heritability estimates do *not* measure the degree to which a trait is determined by genes, but rather the proportion of the *phenotypic variation* among individuals that is due to *genetic variation.* Estimates of heritability are population-specific, valid only for a given population in a given environment. In all four experiments we have estimated the heritability of the same trait, the total biomass of a plant. Yet the heritability of the trait is 0 in Experiment 1, 0.95 in Experiments 2 and 3, and 0.60 in Experiment 4. The heritability of the trait is different when estimated in different groups of individuals (as in Experiments 1 versus 2) or in different environments (as in Experiments 3 versus 4).

The experiments also illustrate a point of practical importance. From the fact that the heritability of a trait is high in each of two populations, *it does not follow* that average differences between the two populations are largely due to genetic differences. Assume that somebody would claim that the large differences in size between the plants of Experiment 2 and those of Experiment 3 are largely genetic, on the grounds that heritability is very high in both cases. The claim is clearly preposterous. Cuttings from the same set of plants were used in both experiments; the differences between the two populations are due to the two very different kinds of environments used in the two experiments.

A point that derives from the two previous ones is that differences in ranking order between populations or even between individuals do not necessarily imply that one population or one individual is *genetically* better than the other. The one that is best in one environment may not be best in another. The experiment shown in Figure 2.16 illustrates this point well: the coastal plant was the best in the sea-level garden but was the worst at 10,000 feet.

Numerous studies have shown that the average IQ score is higher in American Caucasians than in American Blacks (Figure 15.18). Other studies have also shown that the heritability of IQ is high in both populations; estimates range from 0.40 to 0.80, but the actual value is of no particular relevance here. We also need not be concerned with the extent to which IQ scores reflect intelligence. Some have argued that, because IQ is so highly heritable, the difference in average IQ between Caucasians and Blacks is largely genetic. The argument is not valid. High IQ heritability *within* Caucasian populations and *within* Black populations tells nothing about the cause of the difference *between* these populations. We may recall that heritability was very high in Experiment 2 as well as in Experiment 3, and that the plants of Experiment 2 performed much better than those of Experiment 3. Nevertheless, the difference was not genetic at all, but rather the two sets of plants were genetically identical and the difference between the two sets was entirely environmental.

Human populations are genetically different from each other. There is also considerable genetic heterogeneity within any one human population. These differences affect IQ as well as many other characters. But heritability studies do not allow us to conclude that one population is genetically "better" than another with respect to IQ. Moreover, IQ rankings might be quite different if cultural environments were significantly changed.

Problems

1. Identify the various kinds of gene interactions. What is the difference between epistasis and dominance?

2. The agouti coat color, characteristic of mice, rats, guinea pigs, squirrels, and other wild rodents, is due to the individual hairs' being for the most part black but having a narrow yellow band near the tip. The agouti pattern is due to dominant alleles at two loci, one (*C*) being necessary for the development of any color, and another (*A*) being required for the yellow banding of the black hairs. Individuals homozygous for the recessive *c* allele are albino. Individuals having the dominant *C* allele, but homozygous for the recessive *a* allele, are black. Assume that a black mouse, with genotype *CC aa*, is crossed with an albino, having the *cc AA* genotype. What will the phenotype of the F_1 be? What kinds of progeny, and in what proportions, will appear in the F_2?

3. In rats, two pairs of genes (*A a* and *R r*) interact, producing the following phenotypes:

   A- R-: gray
   A- rr: yellow
   aa R-: black
   aa rr: cream

   These phenotypes require, however, that a dominant color allele, *C*, be present, since *cc* individuals are albino. The three pairs of genes are inherited independently. Four different albino lines, each crossed with a true-breeding gray strain, produced gray F_1 progenies. The F_2 were as follows:

Albino line	F_2 progeny				
	Gray	Yellow	Black	Cream	Albino
1	48	0	0	0	16
2	104	33	0	0	44
3	174	0	65	0	80
4	292	87	88	32	171

   What is the probable genotype of each albino line? Use the chi-square method to test the agreement between the observed results and those predicted by your explanation.

4. In poultry the genes for rose comb, *R*, and pea comb, *P*, when present together, produce walnut comb. Individuals homozygous for the recessive alleles at both loci (*rr pp*) have single comb. The two pairs of genes are

inherited independently. Determine the kinds and proportions of the phenotypes expected in the progeny of the following crosses:

(a) *RR Pp* (walnut) × *rr Pp* (pea)
(b) *Rr Pp* (walnut) × *Rr Pp* (walnut)
(c) *Rr Pp* (walnut) × *Rr pp* (rose)
(d) *Rr pp* (rose) × *rr Pp* (pea)

What is the genotype of two parents, one walnut and the other single, that produced ¼ single, ¼ rose, ¼ pea, and ¼ walnut in the F_1 progeny?

5. Several families with deaf-mute individuals have been studied in Northern Ireland. One marriage between two normal individuals produced eight children, four of whom (two sons and two daughters) were deaf-mute. A second marriage between two deaf-mutes produced three daughters and one son, all deaf-mute. When a deaf-mute son from marriage 1 and a deaf-mute daughter from marriage 2 were married, all four of their children (one daughter and three sons) were also deaf-mute. However, when one of the deaf-mute sons from this marriage married a deaf-mute daughter from an unrelated family, all six children (all sons) were normal. What is the probable inheritance of deaf-mutism?

6. PKU victims (homozygous *pp*) may develop normally if maintained on a diet low in phenylalanine. When such individuals marry, they usually produce normal heterozygous offspring (*Pp*), since their spouses are usually normal homozygotes (*PP*). However, it has sometimes been found that a treated phenylketonuric woman (*pp*) married to a normal homozygous man (*PP*) produces children who are all mentally defective. How can this be explained?

7. In the Japanese morning glory (*Pharbitis nil*), crosses between two certain strains with purple flowers produce progenies all with blue flowers. Yet crosses between individuals of any one purple strain produce only purple-flowered progeny. How can these results be explained?

8. In maize, the development of scutellum color requires the presence of any two of the three genes S_2, S_3, and S_4. Give the ratios expected with respect to scutellum color in the F_2 of the following crosses:

(a) $S_2S_2s_3s_3s_4s_4$ (colorless) × $s_2s_2S_3S_3s_4s_4$ (colorless)
(b) $S_2S_2s_3s_3S_4S_4$ (colored) × $s_2s_2S_3S_3S_4S_4$ (colored)

9. Artificial selection often ceases to be effective after it is practiced for a number of generations. When two independently selected stocks in which selection is no longer effective are intercrossed, a selection response may be obtained in the following generations when selection is practiced in progenies from the crosses. How can this result be explained?

10. When pure parental strains differing with respect to a size character are intercrossed, the F_1 is usually no more variable than the parents, whereas the F_2 is considerably more variable. Why?

11. Assume that two highly inbred strains of oats consistently yield about 4 g and 10 g per plant, respectively. The two strains are intercrossed, and the F_1 is selfed. About 1/64 of the F_2 plants yield about 10 g per plant. How many genes are likely to be responsible for the difference between the two original inbred strains?

12. When the F_2 plants from the previous problem are selfed, it is observed that F_3 families differ markedly in their variability. Some have as little variation as the original parents, some have somewhat more variation, and some have as much variation as the F_2 itself. How do you explain this? Would you expect any F_3 family to exhibit more variability than the F_2? Why?

13. Assume that the difference between a corn plant with 6-cm-long ears and one with 18-cm-long ears is due to: (1) two pairs of genes; (2) three pairs of genes; (3) four pairs of genes. Assume also that in each case the genes have equal and cumulative effects on ear length and are inherited independently. The two plants are crossed, and the F_1 is backcrossed to the long-eared parent. What proportion of the progeny is expected to produce 18-cm-long ears in each of the three cases?

14. Nilsson-Ehle crossed two types of oats, one with white seeds, the other with black seeds. The F_1 between them had black seeds. The F_2 consisted of 560 plants as follows: 418 black, 106 gray, and 36 white. How can the inheritance of seed color be explained in this case?

15. In *Drosophila melanogaster*, crosses of Dichaete-winged flies × Dichaete give ⅔ Dichaete and ⅓ normal-winged flies. Dichaete × normal gives ½ Dichaete and ½ normal. Explain.

16. The following heritability estimates were obtained for a strain of cultivated strawberry: (1) yield, 0.48; (2) firmness of fruit, 0.46; (3) size of fruit, 0.20. The mean values and standard deviations in the parental generation were: (1) 380 ± 91 g; (2) 4.4 ± 0.6 (the method and units of measurement for firmness need not concern us here); (3) 11.3 ± 3.0 g. Assume that we produce a new generation, using as parents plants that are two standard deviations above the mean. What will be the expected gain for each of the three traits after one generation of selection?

17. The mean weight of 6-week-old mice in a laboratory population is 21.5 grams. Two sets of parents are used to produce the following generation: (1) heavy mice, with a mean weight of 27.5 g, and (2) light mice, with a

mean weight of 15.5 g. The mean weights of the progenies when 6 weeks old are: (1) 22.7 g and (2) 18.1 g. Calculate the heritability of weight in each of the two sets of progenies. Suggest at least one reason that might explain why the heritability is greater in one case than in the other.

18. The method for estimating heritability by the twin method, given in Box 15.1, requires knowing the total phenotypic variance for the trait in the population. When only the variance between twins is known, heritability can be estimated by the following formula:

$$H = \frac{V_f - V_i}{V_f}$$

where V_f and V_i are the phenotypic variances between fraternal twins and between identical twins, respectively. The values of V_f and V_i are calculated by obtaining the means of the squares of the differences between the two members of twin pairs.

The differences in IQ between the two twins of 10 identical pairs and 10 fraternal pairs are as follows (all 20 pairs of twins are males, and the two twins were raised together in every case):

Pair	Identical twins	Fraternal twins
1	4	12
2	7	4
3	5	9
4	3	7
5	6	7
6	1	11
7	9	13
8	7	10
9	3	9
10	7	9

Estimate the heritability of IQ based on this set of data.

III

Evolution of the Genetic Materials

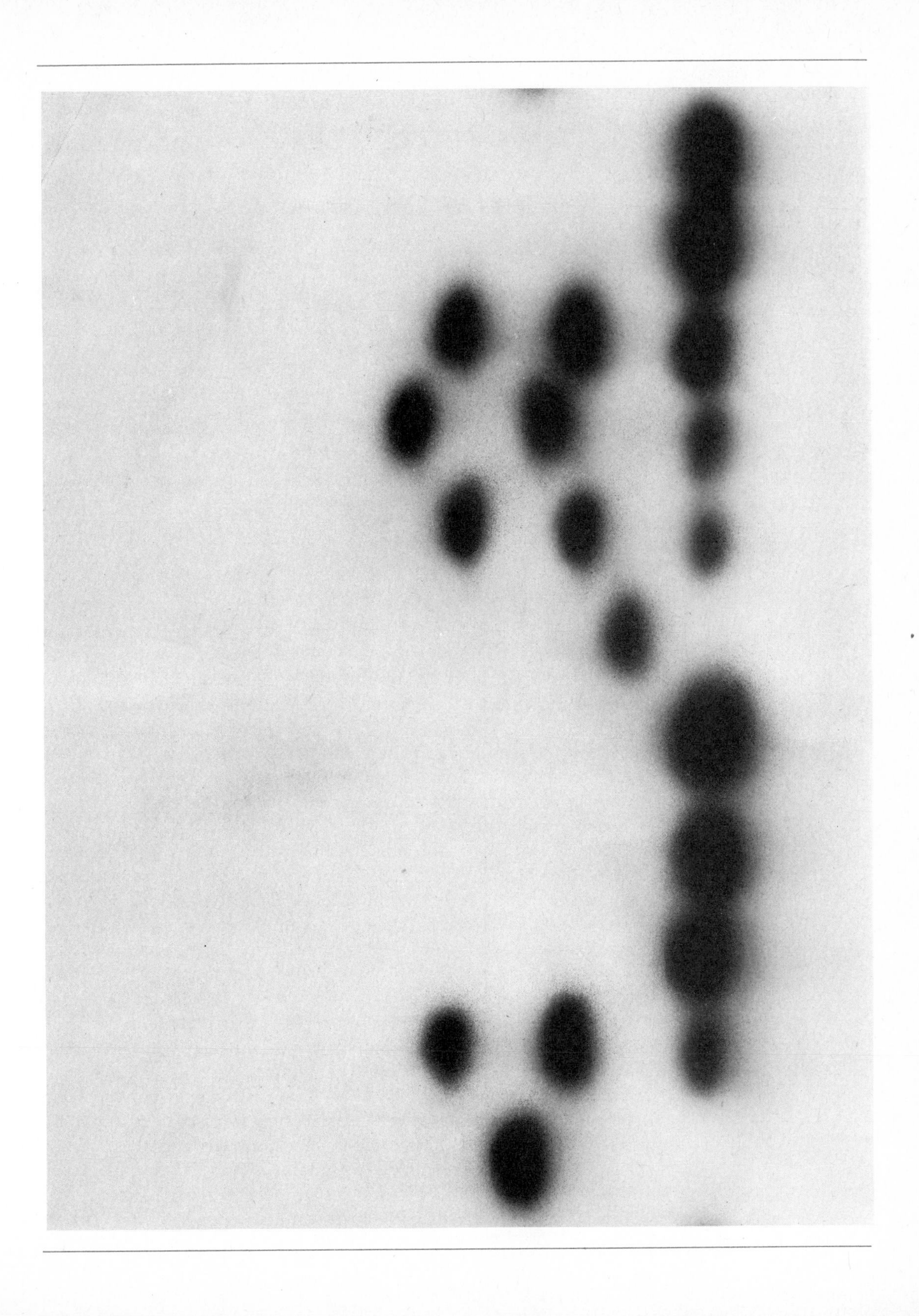

16

Gene Mutations

The largest living animals are the blue whales, which can reach up to 30 meters in length and weigh 13 metric tons; the giant sequoias of California are even larger, some weighing up to 1500 metric tons—about one hundred million trillion (10^{20}) times as much as a typical bacterium. Some bacteria divide every 30 minutes; giant sequoias and bristlecone pines may live several thousand years. Living beings are also very diverse in their shapes, the ways in which they obtain energy, the environments in which they live, and many other respects.

Diversity is, indeed, one obvious feature of the living world. The source of that diversity is the evolutionary process. The ancestors of human beings in the Cambrian geological period, about 600 million years ago, were wormlike creatures living in the seas; these also include among their descendants sea urchins, trouts, crocodiles, eagles, and cows. All cellular organisms derive from prokaryotic creatures that lived more than three billion years ago.

Morphological and functional changes are underlain by genetic changes—biological evolution occurs because the hereditary material, DNA, can change from generation to generation. Heredity is a conservative process, but not perfectly so. As we have seen in previous chapters, the hereditary information in cells is protected so as to be transmitted unaltered to future generations. Occasionally, however, "mistakes" occur, so that daughter cells differ from the parental cells in the DNA sequence or in the amount of DNA. These changes in the hereditary materials are called *mutations*. Here we shall discuss gene mutations, which affect one or a few nucleotides within a gene. Chromosomal mutations, which affect the number or structure of chromosomes, will be discussed in Chapter 17.

Molecular Basis of Gene Mutation

As discussed in Chapter 9, the hereditary information present in the nucleotide-pair sequence of DNA is maintained intact by a complex metabolism involving both replication and repair functions. Mutations may be the result of errors in these processes, at any one of many possible steps. Mutagenic agents may hinder proper function by altering the structure of the DNA itself or by interfering with enzymes that are involved, directly or indirectly, in its metabolism. Examples have been cited in Chapters 8, 9, 12, and elsewhere. Correlation of the actual changes in the DNA structure of known mutant genes with the action of particular mutagenic agents or with environmental conditions associated with "spontaneous" mutations requires a knowledge of the nucleotide sequence in both wild-type and mutant DNA, in order to deduce realistic mechanisms for the mutation. New DNA sequencing methods applied to cloned genes now make this possible. Until recently, most studies of the molecular basis of mutation have relied on amino acid substitutions present in proteins coded by mutant genes and interpreted through the genetic code table. Virtually all studies of the molecular nature of induced mutations have been done with bacteria. However, amino acid substitutions observed in human hemoglobin variants (see below) corroborate the conclusions drawn from bacterial studies.

Mutations can be divided into two general classes: *base-pair substitutions* and *frameshift mutations*. The latter entail the insertion or deletion of one to several nucleotide pairs. Fewer than 20% of spontaneous mutations are believed to be due to base-pair substitutions. Most of the rest are frameshift mutations of varying size. "Hot spots" (see Figure 8.8) are an exception. For example, nucleotide sequence studies of the cloned *lacI* gene of *E. coli* have revealed that two hot spots for spontaneous mutation occur at sites containing the methylated base 5-methylcytosine, created by a DNA modification enzyme (see Chapter 9).

Cytosine spontaneously deaminates at a significant rate to produce uracil (Figure 16.1). All cells possess a repair pathway for detecting dG/dU mispairs and correcting them by hydrolysis of the glycosidic bond to release free uracil (Chapter 9). Spontaneous deamination of 5-methylcytosine, however, creates thymine (Figure 16.1), a normal constituent of DNA, which escapes detection by this repair pathway. The dG/dT mispairs created in this way are repaired less efficiently; this gives rise to a high frequency of the mutation dG/dC→dA/dT, creating a hot spot.

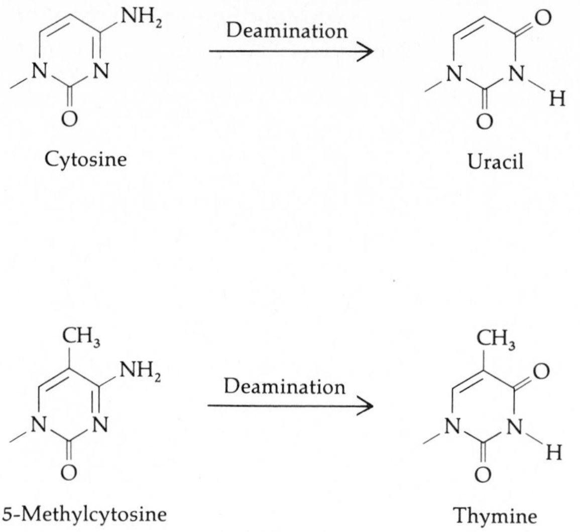

NH_2 Deamination → Cytosine Uracil

CH_3 NH_2 Deamination → 5-Methylcytosine Thymine

Figure 16.1
Spontaneous deamination of cytosine produces uracil; that of 5-methylcytosine produces thymine.

Base-Pair Substitutions

Two types of base-pair substitutions occur: transitions and transversions. *Transitions* are the substitution of one purine for another purine or of one pyrimidine for another pyrimidine (AT→GC, GC→AT, TA→CG, and CG→TA). *Transversions* are the substitution of a purine for a pyrimidine, or vice versa (AT→CG, AT→TA, etc). Base substitutions can occur in a number of ways. For example, mutations in the T4 DNA polymerase gene can produce a defective DNA polymerase that causes an increase in both transitions and transversions during DNA replication. Mutator genes, which increase mutation rates, are known in a variety of organisms, including both *E. coli* and *Drosophila*. In *E. coli* the *mutS* mutant causes both types of substitutions, while the *mutT* mutant specifically induces the transversion AT→CG (see Figure 9.5).

Spontaneous transitions may occur during DNA replication because of *tautomerism*—a shift in the position of a proton that changes the chemical properties of a molecule. Tautomeric shifts in the bases change their hydrogen bonding properties such that adenine assumes those of guanine, guanine those of adenine, cytosine those of thymine, and thymine those of cytosine (Figure 16.2). The base-analogue mutagen 5-bromouracil (5-BU), an analogue of thymine in which the methyl group is replaced by a bromine atom, exerts its mutagenic activity through tautomerism, which is enhanced by the great electron-withdrawing power of bromine compared to a methyl group (Figure 16.3). The mutations induced by 5-BU may be

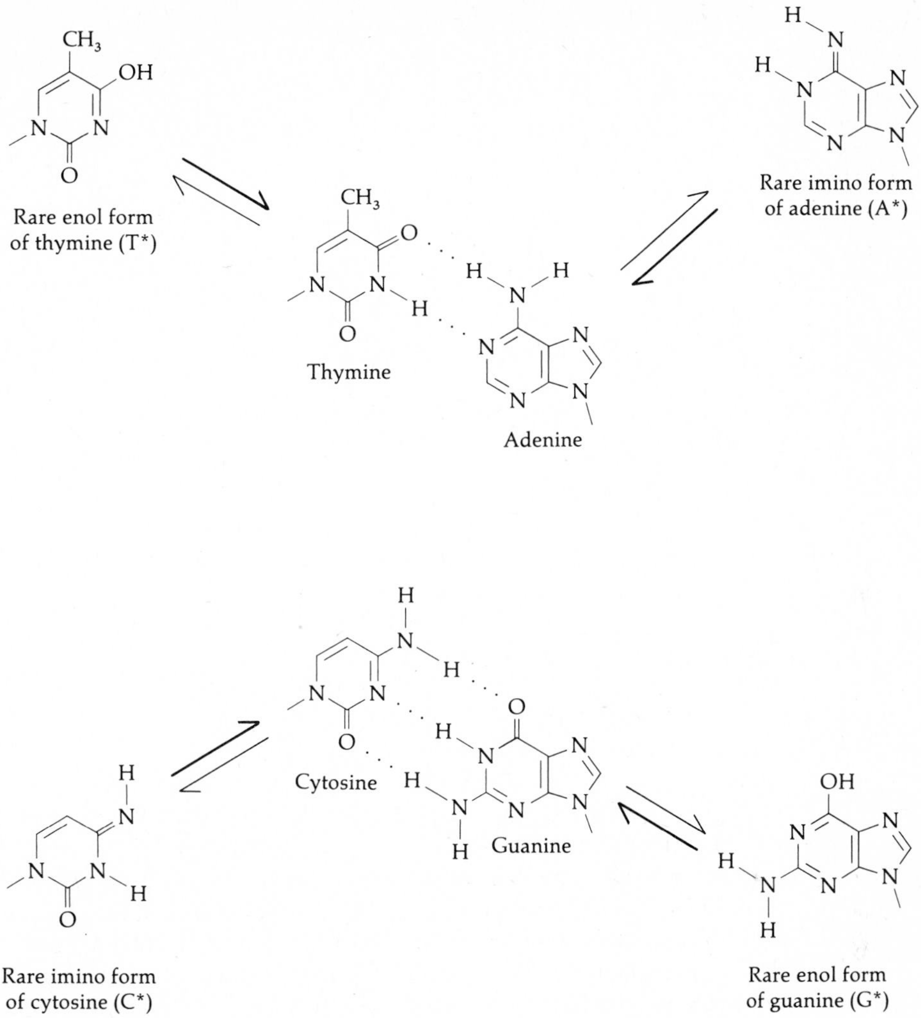

Rare enol form of thymine (T*)

Thymine

Adenine

Rare imino form of adenine (A*)

Cytosine

Guanine

Rare imino form of cytosine (C*)

Rare enol form of guanine (G*)

Figure 16.2
Tautomeric forms of the DNA bases. The most common forms, in which adenine hydrogen bonds with thymine and guanine hydrogen bonds with cytosine, are shown in the center. The relatively rare tautomers, with different hydrogen bonding properties, are indicated by arrows. C* will hydrogen bond with A, G* with T, T* with G, and A* with C.

caused by either incorporation errors or templating errors, resulting in the transitions GC→AT or AT→GC (Figure 16.4).

Another base-analogue mutagen is 2-aminopurine (2-AP), which can pair with either thymine or cytosine (Figure 16.5). Like 5-BU, 2-AP induces transition mutations that may be the result of either incorporation errors or templating errors. Mutations that are induced by base analogues are also induced to revert to wild type by base analogues. Induced reversion by base analogues is a useful means of identifying transition mutations (see Table 12.2).

The mutagen nitrous acid also causes the transitions GC→AT and AT→GC, by the deamination of cytosine to uracil and of adenine to hypoxanthine (which has the pairing properties of guanine). Induced re-

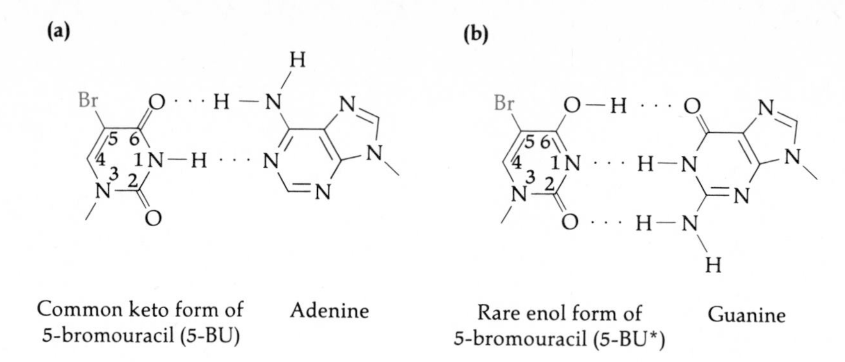

Figure 16.3
Tautomeric forms of 5-bromouracil, an analogue of thymine. **(a)** The most common tautomer, which hydrogen bonds with adenine. **(b)** The less common, but significant, tautomer, which hydrogen bonds with guanine.

version by nitrous acid is also a useful means of identifying transition mutations (Figure 16.6).

Since 2-AP, 5-BU, and nitrous acid are bidirectional in their action, they cannot provide evidence as to whether a transition mutation is GC →AT or AT→GC. The mutagen hydroxylamine, however, reacts specifically with cytosine to convert it to a form that pairs with adenine (Figure 16.7), and this action is unidirectional: GC→AT. Mutations induced by hydroxylamine are not revertible by hydroxylamine, but can be reverted by bidirectional agents. Evidence supporting the postulated action of 2-AP comes from observing the amino acid substitutions caused by 2-AP-induced reversions of specific mutations in the tryptophan synthetase A protein of *E. coli* (Figure 16.8).

Mechanisms by which transversions are induced are less well understood. It is believed that transversions are *not* induced by base mispairing during replication. Many mutations induced by ultraviolet radiation are transversions. Since pyrimidine dimers are a major lesion induced by UV radiation, they may be involved in initiating processes that cause transversions. Replication of DNA containing pyrimidine dimers may create gaps in the strand opposite a dimer. Gaps created in this way—or in other ways, such as cleavage of the glycosidic bond between the base and the sugar portion of a nucleotide—are believed to be necessary for the induction of transversions. An error-prone repair pathway may be involved. Transversions can be identified by their failure to revert following base-analogue mutagenesis. The transversion CG→AT can also be identified by reversion by *mutT* in *E. coli.*

Frameshift Mutations

Frameshift mutations were introduced in Chapter 12 and compared with base-substitution mutations. Frameshift mutations constitute a large

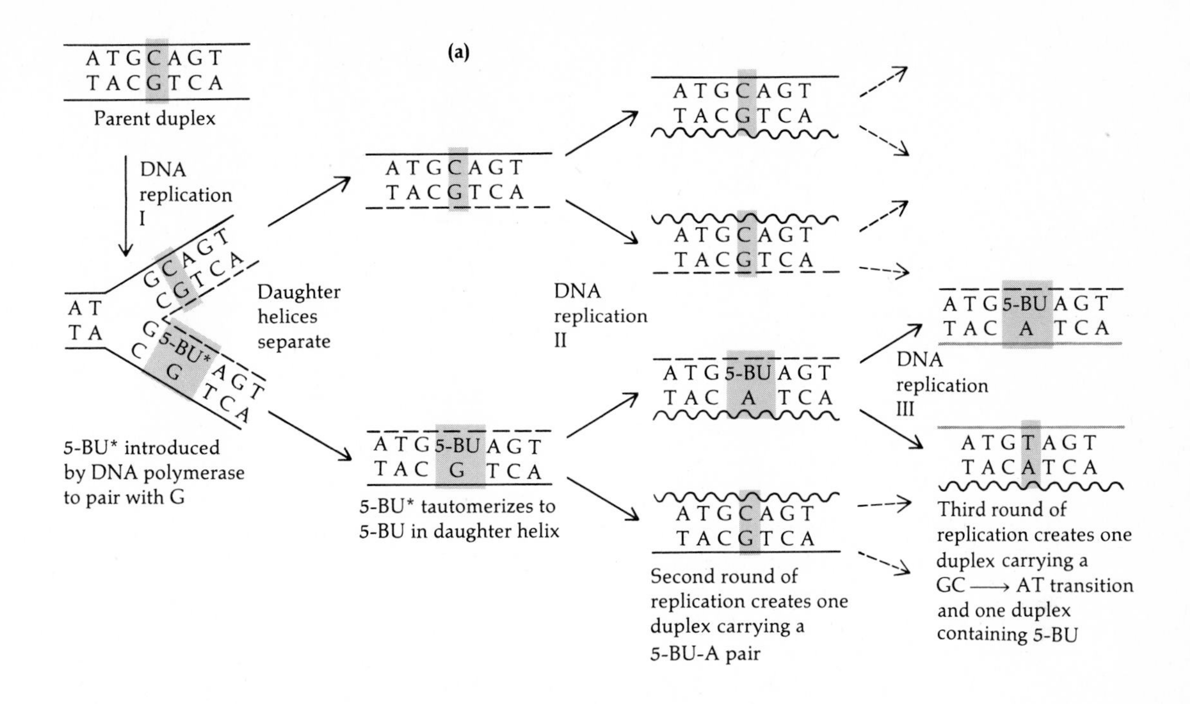

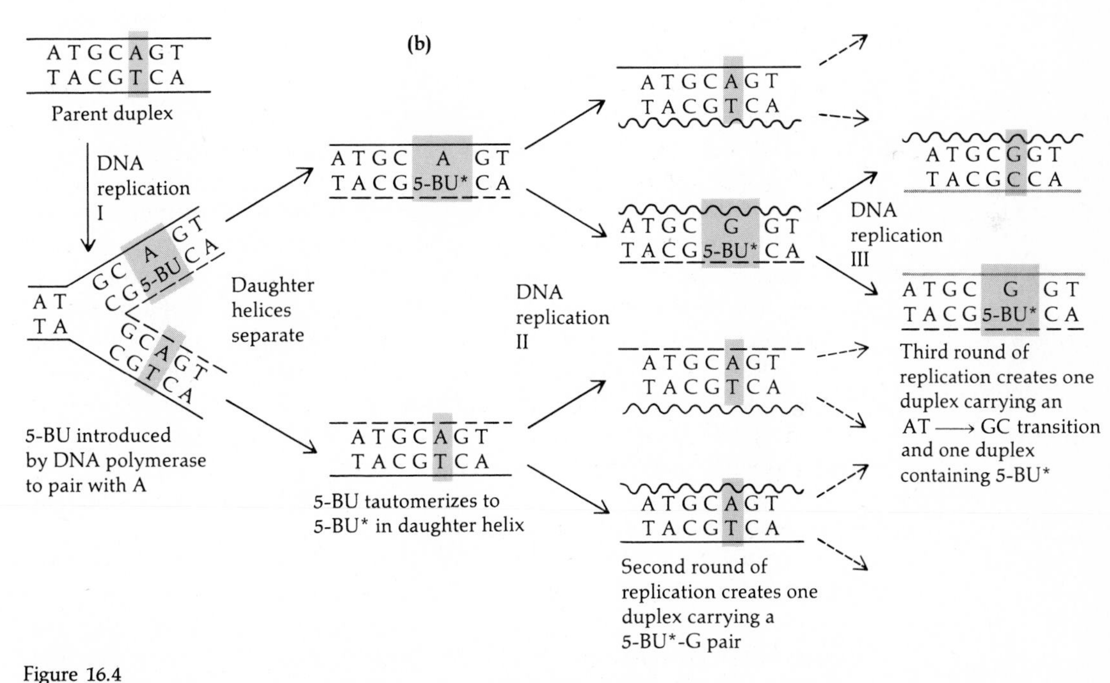

Figure 16.4
The induction of transitions by 5-bromouracil. **(a)** An incorporation error produces a GC→AT transition. **(b)** A templating error produces an AT→GC transition. (After U. Goodenough and R. P. Levine, *Genetics,* Holt, Rinehart and Winston, New York, 1974.)

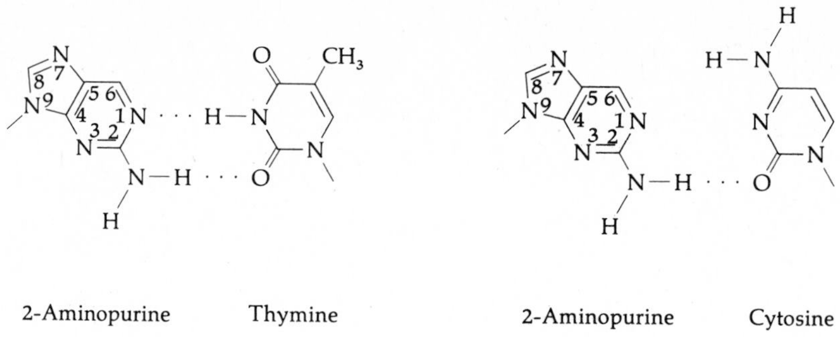

Figure 16.5
2-Aminopurine (2-AP), an analogue of adenine, hydrogen bonds with thymine and cytosine.

2-Aminopurine Thymine 2-Aminopurine Cytosine

fraction of all spontaneous mutations. Spontaneous T4 *rII* frameshift mutations arise during DNA replication in a host cell but not during storage of free phage particles. In contrast, most mutations that arise during storage of free phage particles are transitions, as judged by their induced reversion properties, and may be due to the spontaneous deamination of cytosine.

Evidence substantiating the postulated nature of frameshift mutations is provided by amino acid sequence analyses of proteins coded by genes containing mutually suppressing frameshift mutations (see Chapter 12). Three comparisons of wild-type T4 lysozyme protein with that produced by double frameshift mutants are shown in Figure 16.9. Using the genetic code table, the possible nucleotide sequences can be inferred for both wild-type and mutant mRNAs, which are found to differ by insertion and deletion of one or more nucleotides. Figure 16.9a is an example of the type of mutual suppression discussed previously: deletion of one A residue in a serine codon of the wild-type mRNA shifts the reading frame in one direction, while insertion of one G residue adjacent to the wild-type alanine codon shifts the frame in the opposite direction. Consequently, a sequence of five amino acids is changed in the protein of the double mutant. Figure 16.9b shows the effect of combining an insertion of two nucleotides, GU (following the same serine residue mentioned above), with the insertion of the G residue adjacent to the alanine codon. The result is that the reading frame is shifted by a whole codon, causing the insertion of an additional amino acid into the protein and the alteration of a sequence of four amino acids in the wild-type protein to a different sequence of five amino acids in the protein of the double mutant. Figure 16.9c shows a case in which the double mutant is the result of a combination of a two-nucleotide insertion and a two-nucleotide deletion. These cases show that single frameshift mutations may be the result of the insertion of two adjacent nucleotides rather than just one, as postulated in Chapter 12. Changes of as many as five contiguous nucleotides have been observed in some single-mutation events.

Most frameshift mutations have been observed to occur in sequences containing monotonous runs of one or a pair of adjacent bases (Figure

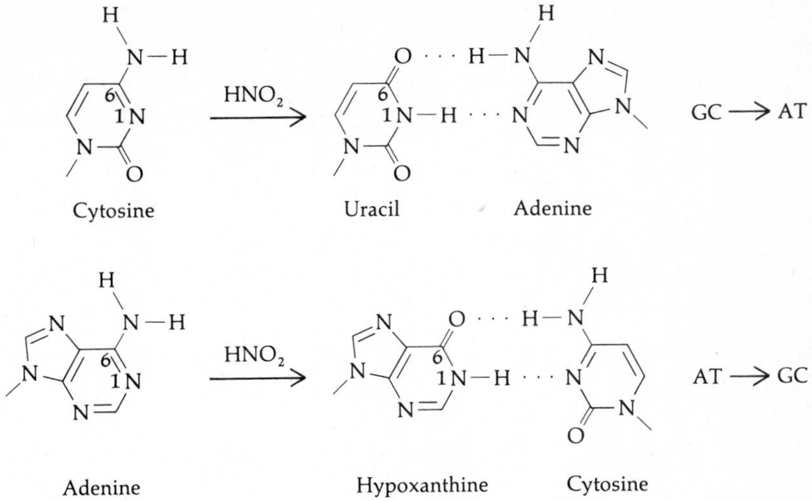

Figure 16.6
Nitrous acid induces GC →AT transitions by deamination of cytosine to uracil, and AT→GC transitions by deamination of adenine to hypoxanthine.

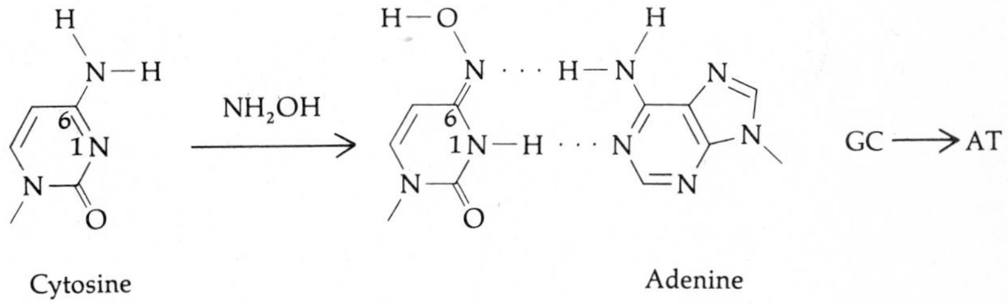

Cytosine —HNO_2→ Uracil · Adenine   GC → AT

Adenine —HNO_2→ Hypoxanthine · Cytosine   AT → GC

Figure 16.7
Hydroxylamine reacts specifically with cytosine, converting it to a form that hydrogen bonds with adenine. Hydroxylamine causes the unidirectional transition GC→AT.

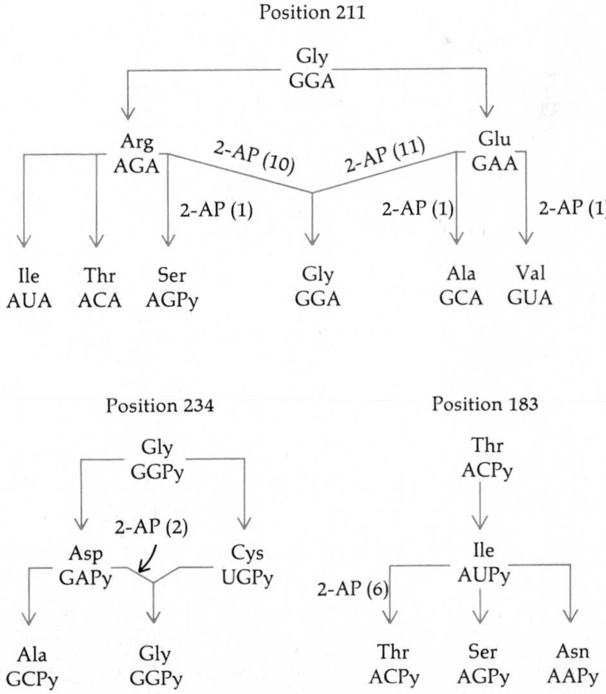

Cytosine —NH_2OH→ Adenine   GC → AT

Figure 16.8
Amino acid substitutions observed at three positions in the tryptophan synthetase A protein of *E. coli*. The numbers of independent substitutions observed are given in parentheses following 2-AP. Of the 32 reversions in the presence of 2-AP, 29 are transitions; the other 3 are transversions, but these may be the result of spontaneous reversion rather than that induced by 2-AP. The codons indicated for each amino acid are deduced from the substitutions observed; Py = pyrimidine (U or C).

Position 211
Gly GGA
Arg AGA
Glu GAA
2-AP (10)
2-AP (11)
2-AP (1)
2-AP (1)
2-AP (1)
Ile AUA
Thr ACA
Ser AGPy
Gly GGA
Ala GCA
Val GUA

Position 234
Gly GGPy
2-AP (2)
Asp GAPy
Cys UGPy
Ala GCPy
Gly GGPy

Position 183
Thr ACPy
Ile AUPy
2-AP (6)
Thr ACPy
Ser AGPy
Asn AAPy

(a) e+ · · · Thr Lys Ser Pro Ser Leu Asn Ala · · ·
· · · ACN AAPu AGU CCA UCA CUU AAU GCN · · ·
(−1, +1)
· · · ACN AAPu GUC CAU CAC UUA AUG GCN · · ·
eJ42 eJ44 · · · Thr Lys Val His His Leu Met Ala · · ·
(b) e+ · · · Lys Ser Pro Ser Leu Asn Ala Ala · · ·
· · · AAPu AGU CCA UCA CUU AAU GCN GCN · · ·
(+2, +1)
· · · AAPu AGU GUC CAU CAC UUA AUG GCN GCN · · ·
eJ17 eJ44 · · · Lys Ser Val His His Leu Met Ala Ala · · ·
(c) e+ · · · Asp Thr Glu Gly Tyr Tyr Thr Ile · · ·
· · · GAPy ACN GAA GGPy UAPy UAPy ACN AUN · · ·
(−2, +2)
· · · GAPy ACN AGG PyUA PyUN CAPy ACN AUN · · ·
eJD5 eJ201 · · · Asp Thr Arg Leu Leu His Thr Ile · · ·

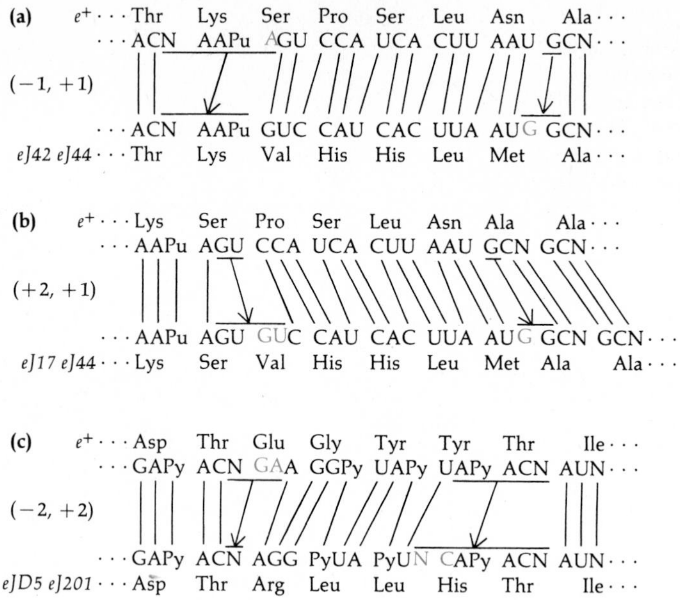

Figure 16.9
Amino acid sequences in the T4 lysozyme proteins coded by the wild-type e^+ gene and mutant *e* genes containing mutually suppressing frameshift mutations. The messenger RNA sequences are deduced from the observed replacements, the numbers of which are shown in parentheses in each case. N designates any base (or U, C, or A at the Ile positions in part c); Py = pyrimidine (U or C); Pu = purine (A or G).

trpA9813 ICR13 (+1)
CACGAGCAGG·GU·
CACGAGCAGGG·GU·
trpA21 ICR PR3 (−1)
AG U
AU·UCUGCCCCCGAC
AU·UCUGCCCCGAC
AG U
hisD3018 (+1)
A
GUCACCCCUGAGG
GUCACCCCCUGAGG
A
hisD2565 (+1)
GU·ACGCGUCC·GU·
GU·ACGCGUCCC·GU·
hisD3749 (+1)
A
CUGUAGCCC·GAG
CUGUAGCCCC·GAG
A
hisD2578 (+1)
GC·CU·
GC·CCU·
hisD3068 (+1)
CAGUAUGG·CCC·
CAGUAUGGG·CCC·
hisD3052 (−1)
A
GACACCGC·CGGCAG
GACACCGCCGGCAG
A
hisD3052 R14 (−2)
A
GAACUGCCGCGCGCGGACACCGCCG
UGCCGCGCGGACACC

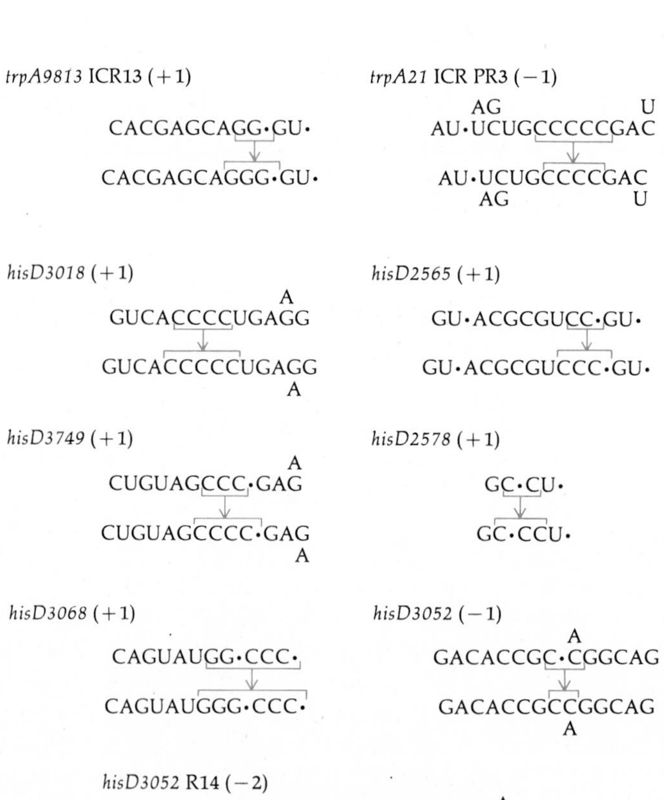

Figure 16.10
Mutagen-induced frameshifts in bacteria. Mutational changes inferred for a number of frameshift mutations in *E. coli* and *S. typhimurium* are shown. Changes are presented as they would appear in the mRNA. [After J. Roth, *Annu. Rev. Genet.* 8:319 (1974).]

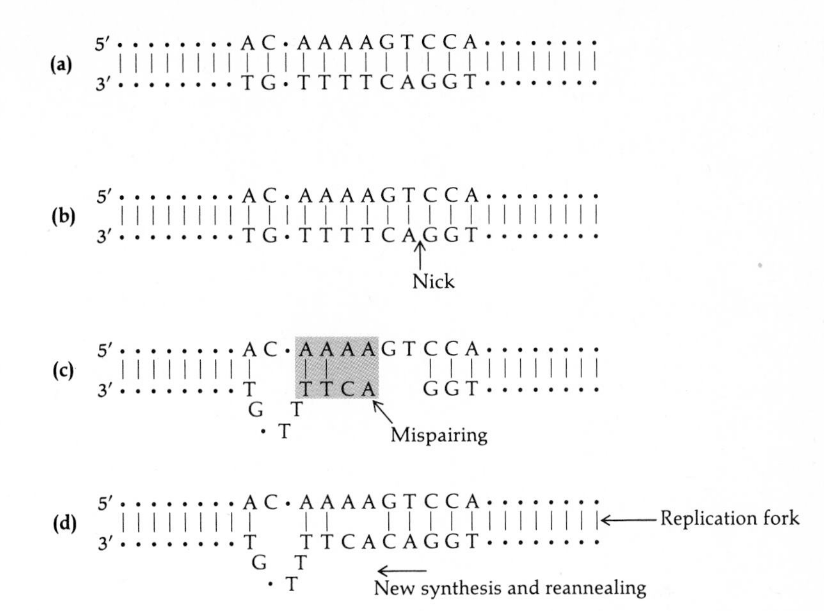

Figure 16.11
The Streisinger model for frameshift mutation, slightly modified. **(a)** Original DNA molecule. **(b)** A nick is introduced. **(c)** The DNA-unwinding protein induces local dissociation and incorrect reannealing. **(d)** Repair synthesis fills the gap created in part c just prior to the arrival of a replication fork before the looped-out strand is detected by other repair enzymes and corrected.

16.10). George Streisinger has proposed a model for the origin of frameshift mutations during DNA synthesis that postulates local dissociation and misannealing of base pairs in such runs (Figure 16.11). According to this model, the effect of frameshift mutagens is to facilitate or stabilize the formation of such misannealed sequences.

Polypeptide Termination in Human Hemoglobins

Two variant human hemoglobins in which the α globin chains are longer than normal have been sequenced. Hemoglobin Constant Spring (HbCS) has 31 extra amino acid residues appended to the C terminal of a normal α chain. Hemoglobin Wayne (HbW1) contains an α chain in which the three amino acids adjacent to the C terminal are altered and five extra amino acid residues are appended (Figure 16.12). (Constant Spring and Wayne are place names identifying the residences of the affected people.) Comparing the amino acid sequences of the normal, HbCS, and HbW1 α chains made it possible to infer the corresponding mRNA sequences by means of the genetic code table. Subsequent sequencing of the normal α-chain mRNA has produced a sequence largely in agreement with the inferred sequence. This sequence indicates that the normal termination codon in the α-chain mRNA is UAA, which has mutated to CAA in HbCS. This allows a normally untranslated portion of the mRNA to be read, resulting in the addition of 31 extra amino acids prior to termination. The

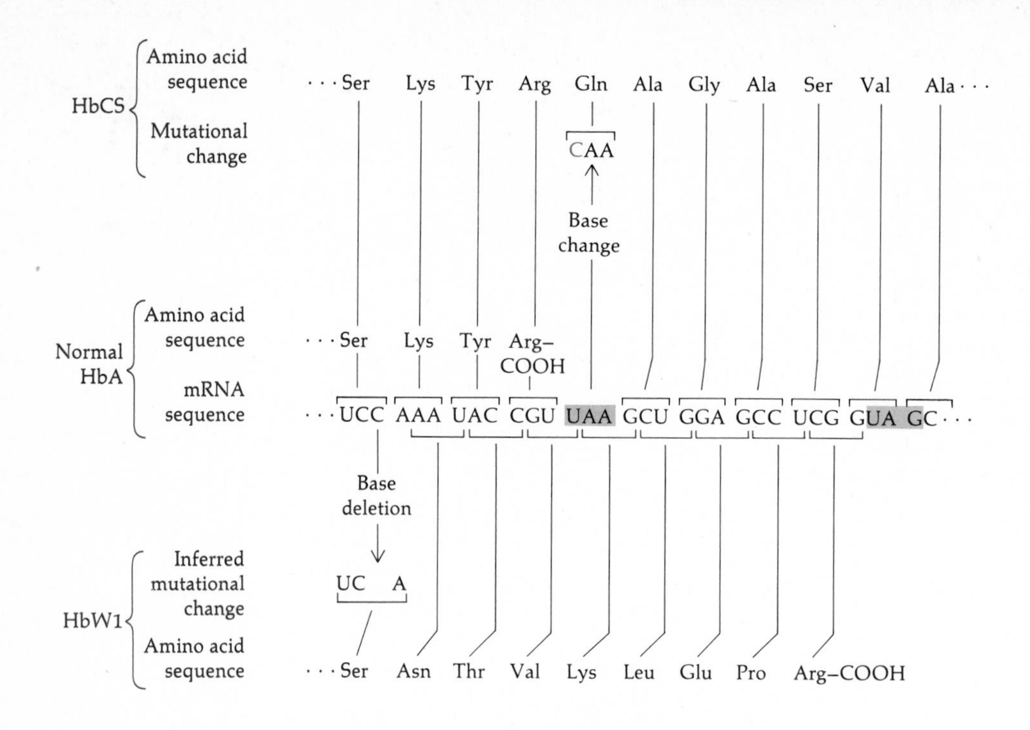

Figure 16.12
Mutations affecting termination of synthesis of the α chain of human hemoglobin. Amino acid sequences of normal (HbA) α chain and the altered α chains in Hb Constant Spring (HbCS) and Hb Wayne (HbW1) are shown with the nucleotide sequence of the α-chain mRNA. The HbCS chain extends another 24 amino acid residues to the right.

HbW1 mutation is due to a nucleotide deletion in the serine codon, resulting in a shift of the reading frame, as indicated in Figure 16.12. The shift of reading frame in HbW1 creates a UAG termination codon, which leads to termination after the addition of five extra amino acids to the α chain.

Two other variant human hemoglobins, Hb Cranston and Hb Tak, have abnormally long β chains. The amino acid sequences of these mutant chains and the normal β chain have been determined and have been compared with the nucleotide sequence determined for the mRNA by RNA sequencing studies. Both of these abnormal hemoglobins have been produced by frameshift mutations: the insertion of AG into the β-chain mRNA of Hb Cranston and the insertion of AC into that of Hb Tak, at slightly different positions (Figure 16.13).

Mutation Rates

Mutations are rare events. The probability that a given *E. coli* cell will mutate, for example, from $T1^S$ (sensitivity to infection by phage T1) to $T1^R$ (resistance to infection) is very small. When the probability of an event is very small and the number of trials in which the event may occur

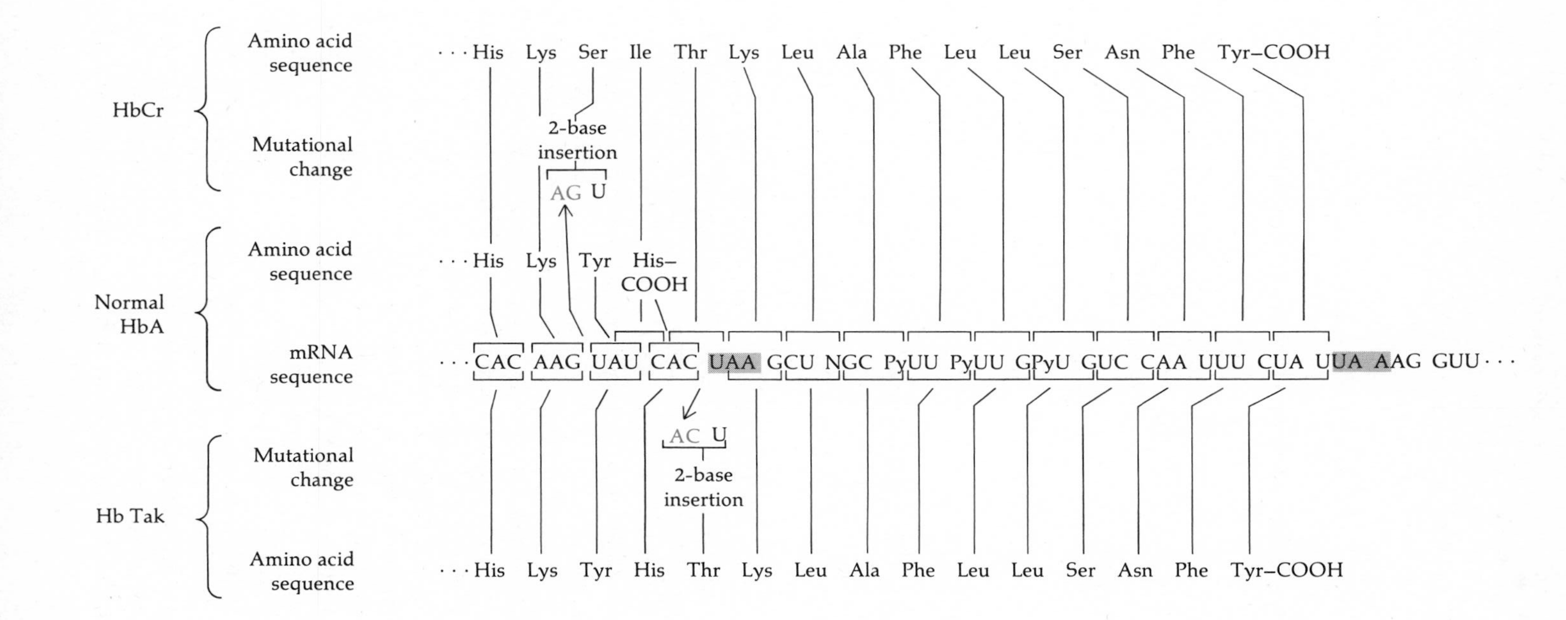

Figure 16.13
Mutations affecting termination of synthesis of the β chain of human hemoglobin. Amino acid sequences of normal (HbA) β chain and the altered β chains in Hb Cranston (HbCr) and Hb Tak are shown with the nucleotide sequence of the β-chain mRNA. N denotes any nucleotide; Py = pyrimidine (Uor C).

is very large, the frequency of events follows the *Poisson distribution* (see Appendix A.IV, page 793).

The Poisson distribution can be used to estimate the mutation rate, m, per generation from $T1^S$ to $T1^R$. Salvador Luria and Max Delbrück inoculated 10^3 *E. coli* $T1^S$ cells/ml in 20 cultures, each containing 0.2 ml of medium, and allowed the cells to multiply to a concentration of about 10^9 cells/ml (about 20 generations). Of the 20 cultures, 9 were found to contain variable numbers of $T1^R$ cells, while 11 had no $T1^R$ cells. The zero term of the Poisson distribution,

$$p_0 = e^{-mN}$$

where N is the number of cells per culture (here, $0.2 \times 10^9 = 2 \times 10^8$), gives the probability of cultures having no mutants. Taking logarithms, we obtain

$$\ln p_0 = -mN$$

$$m = \frac{-\ln p_0}{N}$$

Because 11 cultures had no $T1^R$ cells, $p_0 = 11/20 = 0.55$. Therefore

$$m = \frac{-\ln 0.55}{2 \times 10^8} = \frac{0.598}{2 \times 10^8} = 3 \times 10^{-9}$$

The measurement of spontaneous mutation rates in bacteria and phages is relatively easy, because large populations can be examined in the laboratory. The measurement of mutation rates in higher organisms is hampered by the smaller numbers of individuals that can be studied and by diploidy, which obscures the observation of recessive mutations. The most careful studies of single-gene mutation rates have been carried out with corn and *Drosophila*. In general, the observed mutation rates are low, but some genes are clearly more mutable than others (Table 16.1).

Most studies of mutation in higher organisms have not dealt with single-gene mutations, because of their rarity. Instead, whole-chromosome mutation rates have been studied. In 1927 H. J. Muller devised a rapid and easy way to study mutation, by screening for sex-linked lethal mutations occurring in the sperm of *Drosophila*. The *Muller-5* X chromosome developed for these studies is marked with the semidominant *Bar* (*B*) eye-shape mutation and the recessive *apricot* (w^a) eye-color mutation. It also contains inversions to suppress crossing over (see Chapter 17). Females that are homozygous for the Muller-5 chromosome are crossed to wild-type males, whose sperm is to be tested for the presence of recessive lethal mutations (Figure 16.14). Daughters from this cross carry one Muller-5 chromosome and one test chromosome and are mated individually in separate vials to Muller-5 males. The appearance of any

Table 16.1
Mutation rates of specific genes in various organisms.

Organism and Trait	Mutations per Genome per Generation
Bacteriophage T2 (virus)	
Host range	3×10^{-9}
Lysis inhibition	1×10^{-8}
Escherichia coli (bacterium)	
Streptomycin resistance	4×10^{-10}
Streptomycin dependence	1×10^{-9}
Resistance to phage T1	3×10^{-9}
Lactose fermentation	2×10^{-7}
Salmonella typhimurium (bacterium)	
Tryptophan independence	5×10^{-8}
Chlamydomonas reinhardi (alga)	
Streptomycin resistance	1×10^{-6}
Neurospora crassa (fungus)	
Adenine independence	4×10^{-8}
Inositol independence	8×10^{-8}
Zea mays (corn)	
Shrunken seeds	1×10^{-6}
Purple seeds	1×10^{-5}
Drosophila melanogaster (fruit fly)	
Electrophoretic variants	4×10^{-6}
White eye	4×10^{-5}
Yellow body	1×10^{-4}
Mus musculus (mouse)	
Brown coat	8×10^{-6}
Piebald coat	3×10^{-5}
Homo sapiens (human)	
Huntington's chorea	1×10^{-6}
Aniridia (absence of iris)	5×10^{-6}
Retinoblastoma (tumor of retina)	1×10^{-5}
Hemophilia A	3×10^{-5}
Achondroplasia (dwarfness)	$4\text{-}8 \times 10^{-5}$
Neurofibromatosis (tumor of nerve tissue)	2×10^{-4}

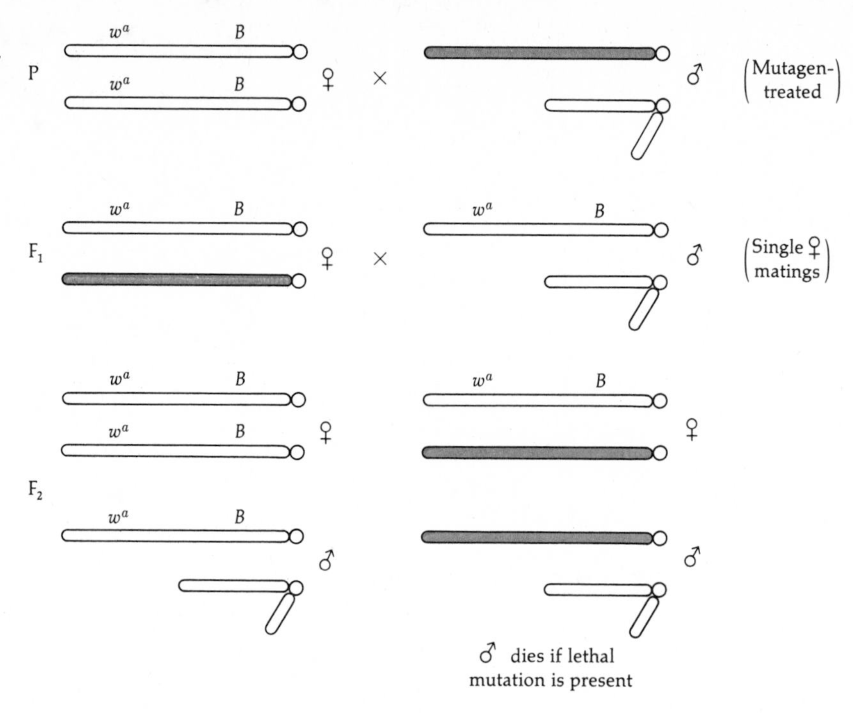

Figure 16.14
The Muller-5 technique for detecting sex-linked lethal mutations in *Drosophila*. The Muller-5 chromosome is marked with the recessive w^a mutation and the dominant *B* mutation and carries inversions to suppress crossing over.

wild-type males in the F_2 progeny indicates that the test chromosome does not carry any recessive, hemizygous lethal mutations. Conversely, the absence of wild-type male progeny indicates that the test chromosome carries at least one newly arisen lethal mutation. In one experiment, for example, 6346 F_1 daughters were mated individually, and 8 were found to carry newly arisen sex-linked lethals. This gives a spontaneous mutation rate per chromosome of 0.13%. Different strains of *D. melanogaster* exhibit spontaneous sex-linked lethal mutation rates ranging from about 0.08% to more than 1%.

The crosses in Figure 16.14 permit the recovery of X chromosomes carrying newly arisen lethal mutations for future study, since these chromosomes are carried in females in heterozygosis with Muller-5 and crossing over in the X chromosome is suppressed by the inversions mentioned above. This type of cross is also useful for recovering newly arisen recessive visible mutations, recessive female-sterile mutations, and male-sterile mutations.

The Muller-5 method for measuring mutation rates has been useful for detecting agents that induce mutations. Muller was the first to show that X rays greatly increase the mutation rate. He observed this in the offspring of wild-type males that were irradiated prior to mating them

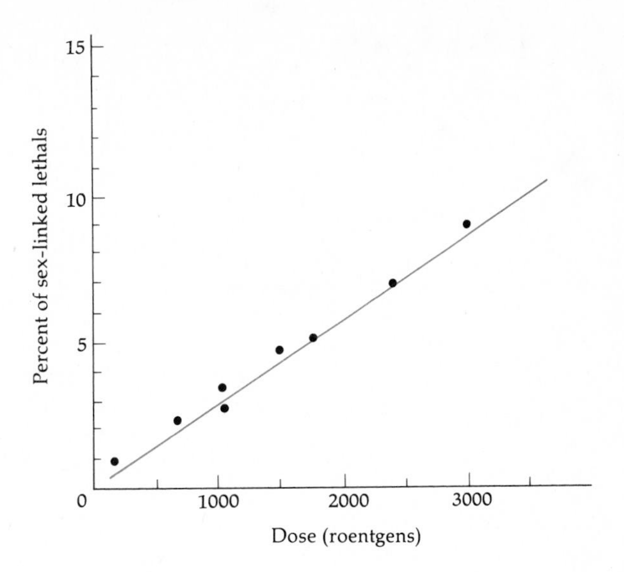

Figure 16.15
X rays increase the frequency of sex-linked lethal mutations in *Drosophila* in direct proportion to the amount of radiation.

with homozygous Muller-5 females (Figure 16.15). In general, the mutation frequency is directly proportional to the X-ray dose measured in *roentgen units* (2.08×10^9 ion pairs produced per cubic centimeter of air). The mechanism of X-ray mutagenesis is indirect. Ionizations created in or near the DNA cause chemical damage to the DNA, which may either be lethal or be repaired; if it is incorrectly repaired, mutations may result.

The first *chemical* agent proved to be mutagenic was also discovered using the Muller-5 method. Mustard gas was employed as a chemical weapon in World War I. During World War II it was shown that exposure of *Drosophila* males to a short, sublethal dose of mustard gas induced a high frequency of lethal mutations (7.3%) in the X chromosome. At present the Muller-5 test is an important component of the screening process used to detect environmental pollutants that are mutagenic.

Mutation Rates and Evolution

New mutations are the ultimate source of the genetic variation upon which biological evolution depends. We may raise the question whether the rates of mutation are sufficiently high to generate abundant genetic variation. The mutation rates of individual genes are low (Table 16.1), but each organism has many genes, and species consist of many individuals. When

Box 16.1 A System for Detecting Mutagens

Newly arisen mutations are much more likely to be deleterious than not. A new mutation may have an effect ranging from inconsequential to catastrophic on the viability of the individual affected. Some of the human hemoglobin variants described in this chapter have little or no effect on their carriers, most probably because the alterations are appended to a largely intact and functional protein. Other hemoglobin variants have more serious effects—for example, HbS, which is responsible for sickle-cell anemia. Deleterious mutations arising in the human population add their increments of misery to mankind.

It is now known that many chemical compounds that are synthesized and used for other purposes are mutagens. Human welfare would be improved if mutagenic compounds could be readily identified and their use restricted in order to avoid unnecessary exposure. Moreover, many chemical carcinogens (cancer-causing agents) have been found to be potent frameshift mutagens or to be metabolized in cells to forms that are frameshift mutagens. The hypothesis that carcinogens act by inducing mutations that in turn cause cancer is being much debated at present.

The existence of well-characterized frameshift mutations in the *hisD* gene of *Salmonella typhimurium* (Figure 16.10) has been used by Bruce Ames to develop an assay for potential mutagens and frameshift carcinogens that is both cheaper and quicker than assays involving mammals. Compounds are screened for their ability to revert a series of known frameshift mutations in the *hisD* gene to wild type, which is readily assayed because wild-mutated cells form colonies on medium that lacks histidine. The tester strains employed have been made more sensitive to mutagenesis by the incorporation of certain other mutations, which inactivate the excision repair system of the cells and make the cells more permeable to foreign organic molecules. Since many compounds require activation by mammalian enzymes to exhibit mutagenic or carcinogenic activity, the usefulness of the *Salmonella* assay has been extended by the incorporation of a rat-liver extract into the test medium.

Environmental health groups are currently using the Ames test and the Muller-5 method in order to screen for potentially hazardous chemicals used as food additives, cosmetic ingredients, flame retardants, etc.

whole individuals or species are taken into account, mutations are common events, not rare.

Assume that there are 100,000 pairs of genes in humans and 10,000 pairs in *Drosophila,* and that the average mutation rate per gene per generation is 10^{-5}. The average number of mutations arising per generation can then be estimated as $(2 \times 10^5$ genes$) \times (10^{-5}$ mutation per gene$) = 2$ mutations for a human zygote, and as $(2 \times 10^4) \times (10^{-5}) = 0.2$ mutation for a *Drosophila* zygote.

There are about 4×10^9 humans in the world; with two mutations per individual, the total number of mutations newly arisen in the present population of mankind becomes 8×10^9. The median number of individuals per insect species is estimated to be about 1.2×10^8; if we assume that 0.2 new mutation is acquired, on the average, by each individual, there would be 2.4×10^7 new mutations per generation in an insect species of median size. Species of other kinds of organisms, including vertebrates, may consist of fewer individuals than insect species, but even such species will acquire large numbers of mutations with each generation.

When whole species are considered, many mutations occur in every generation, even at a single locus. If the average mutation rate per gene per generation is 10^{-5}, about 80,000 new mutations arise, on the average, per generation at each locus in the human species (4×10^9 individuals $\times$ 2 genes per locus $\times$ 10^{-5}mutation per gene). In an insect species of median size, the average number of mutations per locus per generation may be about 2400. It is therefore not surprising that different species and different populations of the same species often become adapted to specific environmental challenges. For example, many species of insects have evolved resistance to DDT in regions of the world where spraying has been intense. We shall see in Chapter 18 that natural populations possess large stores of genetic variation, accumulated from mutations arisen in previous generations. But even if a (possible) allele necessary to meet a new environmental challenge were not present in a species, it would likely arise soon by mutation. The potential of the mutation process to generate new variation is, indeed, enormous.

Mutation as a Random Process

Mutations are often said to be accidental, undirected, random, or chance events. These terms are used as synonyms, but there are at least three different senses in which they can be predicated of the mutation process:

1. Mutations are accidental or chance events, first, in the sense that *they are rare exceptions to the regularity of the process of DNA replication,* which normally involves precise copying of the hereditary information encoded in the nucleotide sequences.
2. Mutations are accidental, random, or chance events also because *there is no way of knowing whether a given gene will mutate in a particular cell or in a particular generation.* We cannot predict, for a given gene locus, which individuals will have a new mutation and which ones will not, nor can we predict which gene will mutate in a given individual. However, this does not imply that no regularities exist in the mutation process. The regularities are those associated with stochastic processes, to which probabilities can be assigned. There is a definite probability (although it may not have been ascertained) that a given gene will mutate in any given individual

(see Table 16.1), and there is a definite probability that a new mutant will appear in a population of a given size.

It is not true, however, that any mutation is just as likely to occur as any other mutation. For example, transitions have different probabilities of occurrence than transversions, and base substitutions have different probabilities than frameshift mutations. Moreover, not any nucleotide sequence can arise from just any other sequence in a single step.

3. Mutations are accidental, undirected, random, or chance events in still another sense that is very important for evolution, namely, in the sense that *they are unoriented with respect to adaptation*. Mutations occur independently of whether or not they are adaptive in the environments where the organisms live.

   Mutations are known in humans at about 2500 gene loci; most of them result in various abnormalities, ranging from inconsequential to lethal. Mutant genes that cause lethality or sterility, or that decrease viability or fertility, are well known in many organisms. In natural populations of *Drosophila*, for example, about 20% or more of all chromosomes carry at least one mutation that is lethal in homozygous condition. Lethal and other mutations may express their effects at various stages of development, from early embryogenesis throughout the life of the individual. They may affect different tissues, organ systems, behavioral patterns, or metabolic processes.

Thus, even casual notice of the nature of most mutations observed in human beings or other familiar species, such as *Drosophila*, reveals their nonadaptive character. Mutations do *not* arise in order to produce better adaptation of individuals to their environment. This fact, long ago obvious to geneticists studying higher organisms, was not accepted by bacteriologists until the late 1940s. Most scientists studying bacterial mutations believed that mutants arose in bacterial populations in *response* to a new selective environment. For example, when penicillin-sensitive bacteria are spread on a Petri dish containing penicillin-supplemented medium, some resistant colonies appear that have the hereditary property of penicillin resistance. This observation was interpreted to mean that penicillin resistance was induced by the penicillin itself. The methodology employed by bacteriologists in using selective media for the isolation of mutant strains obscured the question of whether the mutants that were selected preexisted in the population or were induced by the selective agent. Indeed, some microbiologists questioned whether bacteria possessed genes at all! In their view, selected colonies might represent bacteria that had adopted a new physiological state that allowed survival in the new environment. In fact, such beliefs retarded the acceptance of the idea that DNA was the hereditary material, even though this was indicated by the transforming activity of *Pneumococcus* DNA (see Chapter 4).

Unity of understanding concerning the spontaneous and nonadaptive nature of mutations in all organisms ultimately came from the work of Luria and Delbrück, who studied the origin of *E. coli* mutations that confer resistance to phage T1. Their experiment, called a *fluctuation test*, compared the number of $T1^R$ mutants arising in small cultures of $T1^S$ cells with the number found in a much larger culture. Some of their data were employed above for estimating mutation rates using the Poisson distribution; more extensive data are given in Table 16.2. This work, published in 1943, marks the birth of bacterial genetics.

As described earlier, 20 0.2-ml cultures containing about 10^3 cells/ml of a $T1^S$ strain were grown to a concentration of about 10^9 cells/ml. At the same time, a large 10-ml culture of the same strain of cells was allowed to grow to the same density. Then, each 0.2-ml culture was spread on a plate inoculated with T1 phage, and the number of $T1^R$ colonies formed by each culture was determined. At the same time, 0.2-ml portions of the large culture were spread on T1-inoculated plates, and the number of $T1^R$ colonies formed by each of these was determined (Table 16.2). Each plate received 2×10^8 cells (0.2 ml $\times$ 10^9 cells/ml), yet the number of $T1^R$ colonies formed on each plate depended strongly on whether those cells came from the small cultures or the large one. Eleven of the 20 small cultures gave rise to no $T1^R$ colonies, and the 9 that did were very variable, ranging from 1 to 107 $T1^R$ colonies. In contrast, all of the portions of the large culture gave rise to a fairly similar number (mean: 16.7) of $T1^R$ colonies.

If $T1^R$ colonies are induced by the actual exposure of the sensitive cells to T1 phage (each cell having a small probability of being induced to $T1^R$), then all of the plates should contain some $T1^R$ colonies. The observation, however, that the history of the plated cells (i.e., the volume of the cultures from which they came) greatly affects the number of $T1^R$ colonies formed by a culture strongly suggests that $T1^R$ cells arise *prior* to plating and, therefore, independently of exposure to T1 phage.

The conclusion reached by Luria and Delbrück, that $T1^R$ cells arise spontaneously in the absence of the selective agent, T1 phage, was based on statistical arguments—the large variation among the small-culture plates compared with the small variation among the large-culture plates. Direct evidence for the spontaneous origin of $T1^R$ cells was provided by a replica-plating experiment (see Figure 7.1) devised by Joshua Lederberg and Esther Lederberg in 1952. About 10^7 cells from a culture of $T1^S$ *E. coli* were spread on a nutrient plate and allowed to grow for a few hours to produce 10^7 small colonies of bacteria. Inocula from these colonies were then transferred to a velvet surface, and from the velvet surface to each of three plates previously inoculated with T1 phage (Figure 16.16). The fact that $T1^R$ colonies appeared on all three T1-inoculated plates at identical positions indicated that these colonies owed their origin to particular colonies on the original master plate and that their resistance originated *before* they had ever been exposed to T1 phage. If the $T1^R$ colonies had been

Table 16.2
The fluctuation test of the spontaneous origin of T1 phage-resistant *E. coli* mutants.

Individual Cultures		Samples from Bulk Culture	
Culture number	$T1^R$ bacteria found	Sample number	$T1^R$ bacteria found
1	1	1	14
2	0	2	15
3	3	3	13
4	0	4	21
5	0	5	15
6	5	6	14
7	0	7	26
8	5	8	16
9	0	9	20
10	6	10	13
11	107		
12	0		
13	0		
14	0		
15	1		
16	0		
17	0		
18	64		
19	0		
20	35		
Mean ($\bar{n}$)	11.4		16.7
Variance	694		15
Variance/$\bar{n}$	61		0.9

From S. E. Luria and M. Delbrück, *Genetics* 28:491 (1943).

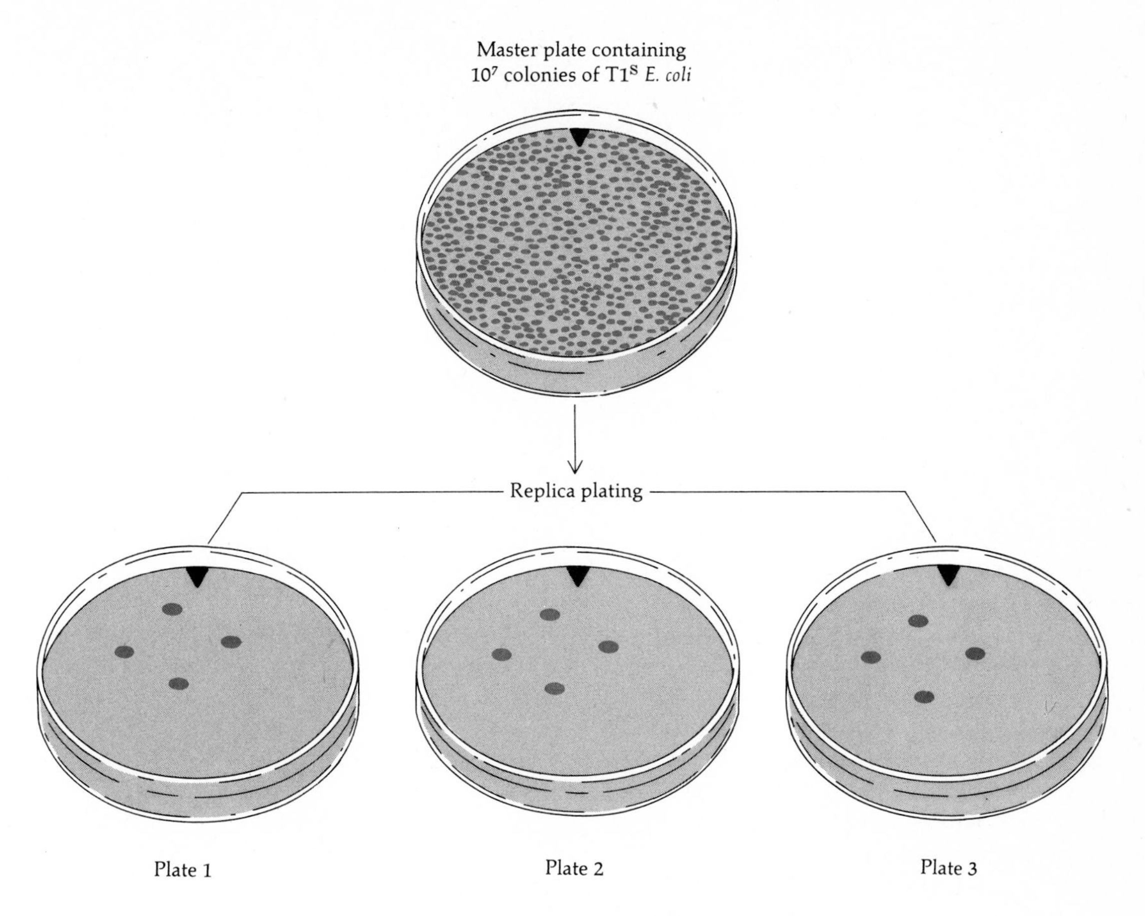

Figure 16.16
Proof of the spontaneous origin of $T1^R$ cells in a population of $T1^S$ cells. A culture of 10^7 *E. coli* $T1^S$ cells is spread on a nutrient plate and incubated for a short time to permit minute colonies to grow. This plate, which has never been exposed to T1 phage, is then replica-plated to three other plates, which have been inoculated with T1 phage. The $T1^R$ colonies that appear on all three selective plates appear at identical positions, indicating that they must have common origins in pre-existing $T1^R$ colonies on the original plate. (After G. S. Stent and R. Calendar, *Molecular Genetics*, 2nd ed., W. H. Freeman, San Francisco, 1978.)

induced by exposure to T1 on each replica plate, the resistant colonies would have appeared at different positions on different plates.

A pure culture of $T1^R$ *E. coli* that has never been exposed to T1 phage and therefore could not have been induced by T1 can be obtained from the master plate in Figure 16.16 by using the replica plates to identify the position of the spontaneous $T1^R$ colony. This experiment, and others like it, eventually convinced bacteriologists that bacteria possess spontaneously mutable genes just like those of other organisms.

Box 16.2 Mutation Randomness and Number of Genes

The random occurrence of both spontaneous and induced mutations in genes can give insight into the number of potentially mutable units in the genome, i.e., the number of genes in the genome (see Chapter 14). Estimates of the total number of potentially lethal complementation groups (essential functions) on the X chromosome of *D. melanogaster* are about 950 to 1000. This number is remarkably close to the approximately 1012 bands observed in the polytene X chromosome of this species. This seems to be only a coincidence, however, as a one-to-one correspondence of genes and bands has not been proved, and because there are certainly genes on the X chromosome that are nonessential and hence produce no lethal mutations.

Estimates can also be made of the number of genes having particular functions. Table 16.3 gives the distribution of X-linked female-sterile mutations induced in *D. melanogaster* by ethyl methane sulfonate (EMS); the number of mutations in each cistron is determined by complementation analysis. The mutants included in the table are those producing eggs that fail to hatch. They are therefore maternal-effect lethals—the maternal genotype rather than the embryonic genotype is responsible for lethality.

There are 30 cistrons with a single mutation each in Table 16.3, 12 cistrons with two mutations each, and so on. The distribution of mutations per cistron has the overall pattern expected for a Poisson distribution except that there are more cistrons with a high number of mutations than would be expected. Notably, there is one cistron with 15 different mutations. Such hot spots of mutability are exceptional (recall, however, that the fine-structure map of the T4 *rII* region shown in Figure 8.8 also exhibits hot spots). One possible cause of hot spots is discussed in this chapter.

If we exclude the hot spot with 15 mutations, there are a total of 131 female-sterile mutations in Table 16.3. The mean number, x, of mutations per mutable cistron is therefore

$$x = \frac{131}{N} \qquad (1)$$

where N represents the number of female-sterile (maternal-effect lethal) mutable cistrons in the X chromosome, the number we want to estimate.

The first and second terms of the Poisson distribution (see Appendix A.IV, page 793), i.e., the number of cistrons expected to have only one or only two mutations, are

$$p_1 = xe^{-x}$$

$$p_2 = \frac{x^2}{2}e^{-x}$$

Dividing p_2 by p_1, we obtain

$$\frac{p_2}{p_1} = \frac{x^2 \, e^{-x}}{2x \, e^{-x}} = \frac{x}{2}$$

Using the data given in Table 16.3, we have

$$\frac{p_2}{p_1} = \frac{12/N}{30/N} = 0.4$$

Therefore

$$\frac{x}{2} = 0.4 \quad \text{or} \quad x = 0.8$$

Substituting this value in the first equation for x, above, we have

$$0.8 = \frac{131}{N} \quad \text{or} \quad N = 164$$

Therefore, there are about 160 cistrons in the X chromosome that are essential for the production of viable eggs, although only 59 mutable cistrons have been identified in the table. The detection of all 160 cistrons by actual mutations would obviously have required screening many more potentially mutated X chromosomes than the approximately 6000 chromosomes that were examined in the experiments on which Table 16.3 is based.

Mutation and Adaptation

Mutations arise independently of whether they are beneficial or harmful to their carriers. Moreover, newly arisen mutations are more likely to be deleterious than beneficial. It is easy to see why this should be so. The genes occurring in a population have been subject to natural selection. Allelic variants that occur in substantial frequencies in a population are therefore adaptive; they are common in the population precisely because they are favored by natural selection. Any newly arising mutation is likely

Table 16.3
Frequency distribution of female-sterile mutations in the X chromosome of *Drosophila melanogaster.*

Number of Observed Mutations per Cistron	Number of Cistrons	Number of Mutations
1	30	30
2	12	24
3	5	15
4	3	12
5	4	20
6	1	6
7	1	7
8	1	8
9	1	9
15	1	15
Total:	59	146

Data modified from J. D. Mohler, *Genetics* 85:259 (1977).

also to have arisen earlier in the history of the population; if such a mutation does not already exist in substantial frequencies, it is because it has been eliminated or kept at low frequencies by natural selection, owing to its harmful effects on the organism.

It is nevertheless important to realize that mutations are not beneficial or harmful in the abstract, but rather with respect to some specific environment. A mutation increasing the density of hair may be adaptive in a population of mammals living in Alaska, but it is likely to be selected against in a population living in Florida. Increased melanin pigmentation may be beneficial to humans living in tropical Africa, where dark skin protects from the sun's ultraviolet radiation, but not in Scandinavia, where fair skin facilitates the sunlight-induced synthesis of vitamin D.

Auxotrophs and mutations to drug resistance in microorganisms are clear-cut examples of how the adaptive effects of mutants depend on the particular environments in which the organisms live. Auxotrophs will grow in supplemented medium, but not in minimal medium. A mutation conferring streptomycin resistance is useful to bacteria in the presence of streptomycin, but not in its absence. Well-studied examples of environmental dependence in higher organisms include the temperature-sensitive lethal mutations of *Drosophila*. Below certain temperatures, flies homozygous for these alleles survive and reproduce more or less normally, but at elevated temperatures, the flies become paralyzed or die, although wild-type flies can function normally.

Because of the environmental dependence of adaptation, the probability that a newly arisen mutation will increase the adaptability of an individual is greater when organisms colonize a new habitat or when environmental changes present a population with new challenges. In these cases, the adaptation of a population is less than optimal, and there is greater opportunity for new mutations to be adaptive. The evolutionary record shows that major evolutionary changes—such as the origin of terrestrial vertebrates—are often associated with the colonization of new habitats.

Problems

1. Suppose the Lederbergs had sought evidence for the spontaneous origin of phage-resistant bacterial cells by studying resistance to phage λ rather than to phage T1. What might they have concluded about mutations to phage resistance in this case?

2. The following sequence is part of a structural gene:

   3′ T A C A A G
   5′ A T G T T C

(a) What two possible mutations might hydroxylamine induce in this gene? Write the sequences of the two mutant genes.
(b) What mutagens might you choose to revert the mutant genes to wild type?
(c) If the top strand is used as a template by RNA polymerase, what will be the amino acid sequences coded by the wild-type and mutant genes? (Be sure to specify the amino terminal of each sequence.)

3. When newly induced mutations of specific genes are sought in mutagenesis experiments, it is frequently observed that mutational mosaics appear in the F_1 generation. When *Drosophila* males are fed EMS (ethyl methane sulfonate, an alkylating mutagen) and mated to attached-X females to screen for induced recessive visible mutations, part (but not all) of the bodies of some of the F_1 sons exhibit a mutant phenotype. For example, a newly induced *singed* bristle mutant may be observed in an individual with some *singed* and some normal bristles; moreover, the mutation is not always transmitted to the progeny of the mosaic male. Explain.

4. Penicillin kills only growing bacterial cells. It interferes with cell-wall synthesis, and as the bacteria grow, they develop weakened cell walls that eventually break, killing the bacteria. Penicillin is often used to increase the frequency of auxotrophic mutants in mutagenized populations of *E. coli* cells. Describe how this might be accomplished.

5. The mean number of mutations per mutable cistron for the data in Table 16.3 is calculated in Box 16.2 as $x = 0.8$. What is the mutation rate per mutable cistron induced by EMS?

6. Using the formula for the terms of the Poisson distribution (see Appendix A.IV, page 793), calculate the number of cistrons expected with 0, 1, 2, . . . , 9 mutations per cistron in Table 16.3, assuming $x = 0.8$. What do you infer from the discrepancy between the expected and observed values?

7. There are other methods, besides the one shown in Box 16.2, for estimating the number of mutable loci using the data in Table 16.3. One method consists of using only the p_1 term, as well as the total number (131) of mutations.

Since $x = 131/N$ and $p_1 = 30/N$,

$$p_1 = xe^{-x} = \frac{131}{N}e^{-131/N} = \frac{30}{N}$$

$$e^{-131/N} = \frac{30}{131} = 0.229$$

$$\frac{131}{N} = -\ln 0.229 = 1.474$$

$$N = \frac{131}{1.474} \approx 89$$

If the distribution of mutations were a good Poisson distribution, the best method for estimating the number of mutable loci would be the *maximum likelihood* method, which yields the following equation:

$$M(e^x - 1) - Cxe^x = 0$$

where M is the observed total number of mutations (131) and C is the number of cistrons at which mutations were observed (58). (Recall that the cistron with 15 mutations was excluded because it is a hot spot.) Solving the equation for the data in Table 16.3, one obtains

$$x = 1.931$$

and therefore

$$N \approx 68$$

Using each of these two new values of x (1.474 and 1.931) in turn, calculate the expected number of cistrons with 0, 1, 2, . . . , 9 mutations, and compare these expected numbers with the observed ones shown in Table 16.3. Do these comparisons confirm the inference you made in problem 6? In view of this conclusion, do you think that it is more reasonable in this case to use only the first two terms of the Poisson distribution (as done in Box 16.2), rather than the maximum likelihood method, in order to estimate the number of mutable cistrons?

8. Allen Shearn and his collaborators have selected lethal mutations of *Drosophila* that are stage-specific, namely, recessive lethals causing death at the late third-instar-larval/early pupal stage of development. Among 35 such mutations induced in the third autosome by the frameshift mutagen ICR-170, complementation tests show that 33 complementation groups are represented. Of these, 31 are identified by single mutations and two are identified by two mutations each. Use the Poisson distribution to estimate the total number of third-autosome complementation groups whose functions are required for development beyond the early pupal stage.

17

Chromosomal Mutations

Classification of Chromosomal Changes

Different cells of the same organism and different individuals of the same species have, as a rule, the same number of chromosomes, except that gametic cells have only half as many chromosomes as somatic cells. Homologous chromosomes are, also as a rule, uniform in the number and order of genes they carry. These rules have exceptions, known as *chromosomal mutations*, abnormalities, or aberrations. Chromosomal mutations can be subdivided as follows:

A. CHANGES IN THE STRUCTURE OF CHROMOSOMES. These may be due to changes in the *number of genes* in chromosomes (deletions and duplications; Figure 17.1) or in the *location of genes* on the chromosomes (inversions and translocations; Figure 17.2).

1. *Deletion,* or deficiency. A chromosome segment is lost from a chromosome.
2. *Duplication,* or repeat. A chromosome segment is present more than once in a set of chromosomes.
3. *Inversion.* A chromosome segment is reversed. If the inverted segment includes the centromere, the inversion is called *pericentric* ("around" the centromere); if not, the inversion is *paracentric.*
4. *Translocation.* The location of a chromosome segment is changed. The most common forms of translocations are *reciprocal,* involving the exchange of chromosome segments between two non-

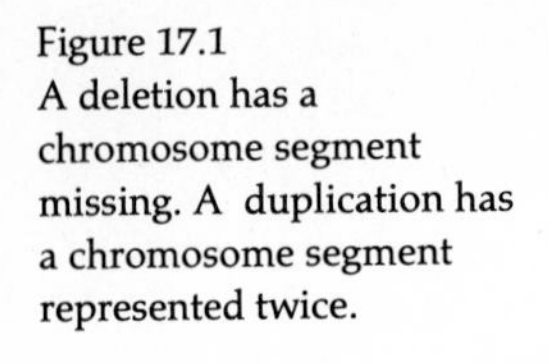

Figure 17.1
A deletion has a chromosome segment missing. A duplication has a chromosome segment represented twice.

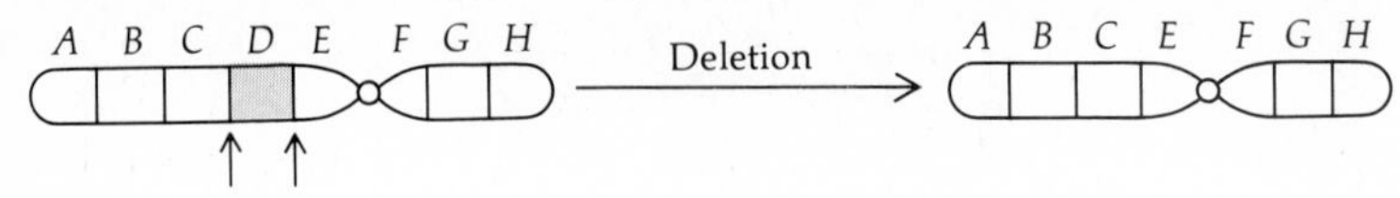

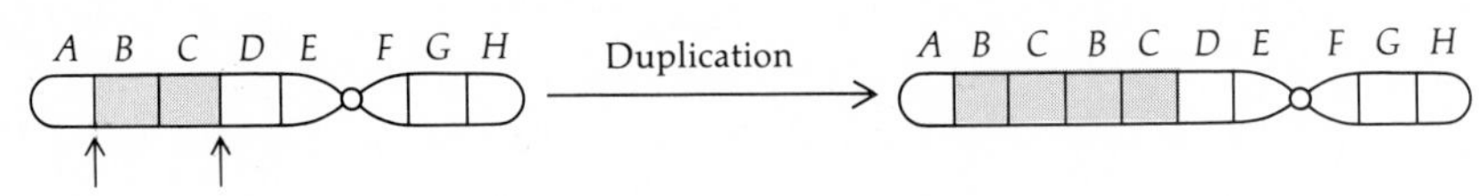

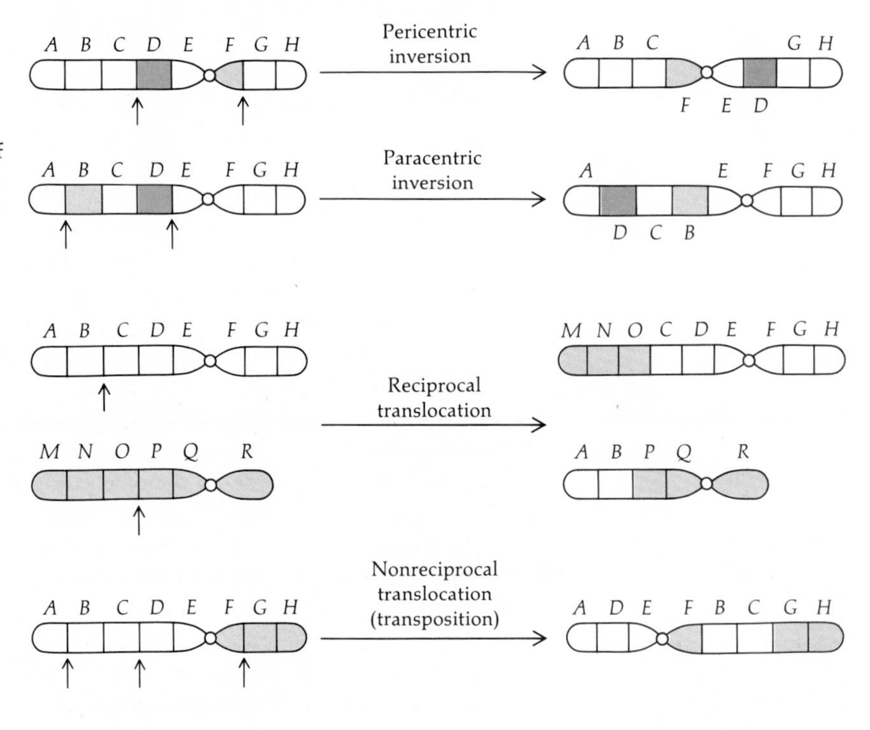

Figure 17.2
Inversions and translocations are chromosomal mutations that change the locations of genes in the chromosomes.

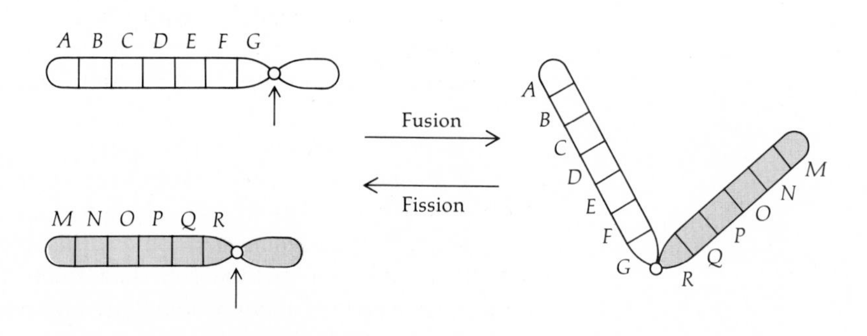

Figure 17.3
Centric fusions are the joining of two chromosomes at the centromere to become one single metacentric chromosome. Centric fissions, or dissociations, are the reciprocal of fusions: one metacentric chromosome splits into two telocentric chromosomes.

homologous chromosomes. A chromosomal segment may also move to a new location within the same chromosome, or in a different chromosome, without reciprocal exchange; these kinds of translocations are sometimes called *transpositions.*

B. CHANGES IN THE NUMBER OF CHROMOSOMES. Some changes do not alter the total amount of hereditary material (fusions and fissions; Figure 17.3); others do (aneuploidy, monoploidy, and polyploidy; Figure 17.4).

1. *Centric fusion.* Two nonhomologous chromosomes fuse into one. This entails the loss of a centromere.
2. *Centric fission.* One chromosome splits into two. A new centromere must be produced; otherwise the chromosome without a centromere would be lost when the cell divides.
3. *Aneuploidy.* One or more chromosomes of a normal set are lacking or are present in excess. The terms *nullisomic, monosomic, trisomic, tetrasomic,* etc., refer to the occurrence of a chromosome zero times, once, three times, four times, etc.
4. *Monoploidy* and *polyploidy.* The number of *sets* of chromosomes is other than two. Most eukaryotic organisms are *diploid,* i.e., they have two sets of chromosomes in their somatic cells, but only one set in their gametes. Some organisms are normally *monoploid,* i.e., they have only one set of chromosomes. Both monoploid and diploid individuals exist in certain social insects, such as the honeybee, in which the males are monoploid and develop from unfertilized eggs, while the females are diploid and develop from fertilized eggs. Monoploidy is sometimes also called *haploidy,* although this term is generally reserved for the number of chromosomes in the gametes, which in polyploids is greater than the monoploid number. *Polyploid* organisms have more than two sets of chromosomes; the organism is said to be *triploid* if it contains three sets of chromosomes, *tetraploid* if it contains four sets, and so on. The more common forms of polyploidy involve sets of chromosomes in multiples of two, i.e., tetraploids, hexaploids, and octoploids, which have four, six, and eight sets of chromosomes, respectively. Polyploidy is very common in some groups of plants, but is rare in animals.

Deletions

A deletion, or deficiency, is the loss of a chromosomal segment. The first chromosomal aberration manifested by genetic evidence was a deletion. It involved the mutant *Notch,* which produces an indented, or "notched,"

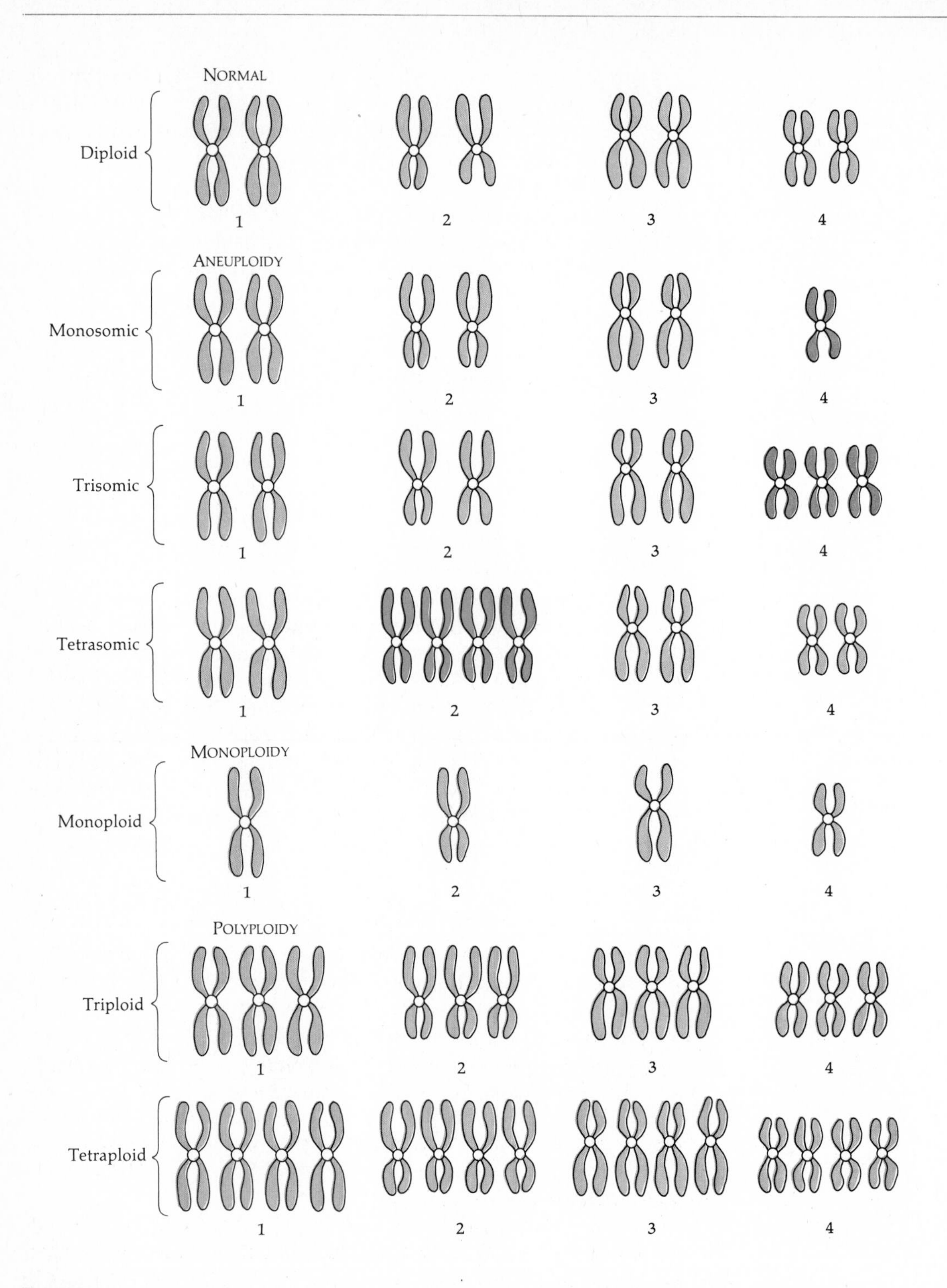

Figure 17.4
Aneuploidy occurs when one or more chromosomes are absent from, or added to, the normal complement of chromosomes. Monoploidy is the presence of only one set of chromosomes. Polyploidy is the presence of more than two sets of chromosomes.

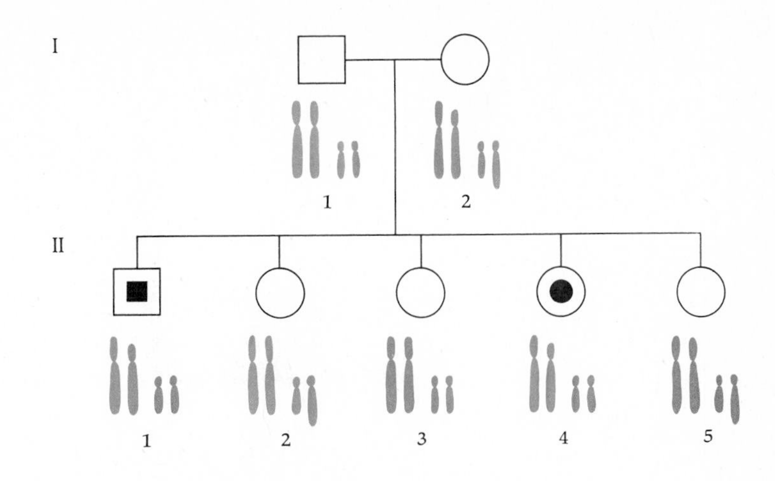

Figure 17.5
A pedigree with cases of the *cri-du-chat* syndrome. The mother (I-2) is phenotypically normal but has a reciprocal translocation involving chromosomes 5 and 13. One daughter (II-3) is chromosomally normal; two daughters (II-2 and II-5) have an extra chromosome segment (duplication); one son (II-1) and one daughter (II-4) have the *cri-du-chat* syndrome (deletion).

margin in the wings of *Drosophila melanogaster*, and was first studied by C. B. Bridges in 1917. *Notch* is a sex-linked trait that behaves as a dominant with respect to the notched phenotype, but also as a recessive lethal (see Chapter 5, page 160). Females heterozygous for *Notch* show the characteristic notched phenotype, but *Notch* is lethal in homozygous females and in hemizygous males. The *white* eye allele behaves as if it were dominant when *Notch* is in the homologous chromosome. Other recessive genes lying in the vicinity of *white* on the X chromosome also behave as dominant in the presence of *Notch*. This apparent dominance of recessive genes is called *pseudodominance* because it is due to the loss of a piece of chromosome in the homologous chromosome, which is therefore unable to complement the recessive mutation. Pseudodominance is one way to recognize deletions.

Deletions are usually lethal in homozygous condition (and in hemizygous condition if the deletion occurs in the X chromosome). This indicates that most genes are indispensable for the development of a viable organism. However, very small deletions that are not lethal in homozygous condition have been detected in corn, *Drosophila*, and other organisms. In *Escherichia coli*, nonlethal deletions of up to 1% of the genome are known. In *Drosophila*, the largest homozygous deficiency that allows survival to adulthood is about 0.1% of the genome (see Figure 5.24; the heterozygous female $Df(1)N^{64i16}/Df(1)dm^{75e19}$ survives to adulthood, although it completely lacks the bands 3C12 through 3D4). In heterozygous condition, deletions often have phenotypic effects, such as the wing indentations characteristic of *Notch*.

In humans, the *cri-du-chat* ("cry-of-the-cat") syndrome is associated with a heterozygous deficiency in the short arm of chromosome 5 (Figure 17.5). The name of this syndrome comes from the high-pitched, mewing cry made by babies with the disorder. The syndrome is also characterized by microcephaly (small head), severe growth abnormalities, and mental retardation—the IQ of *cri-du-chat* children is between 20 and 40. Death

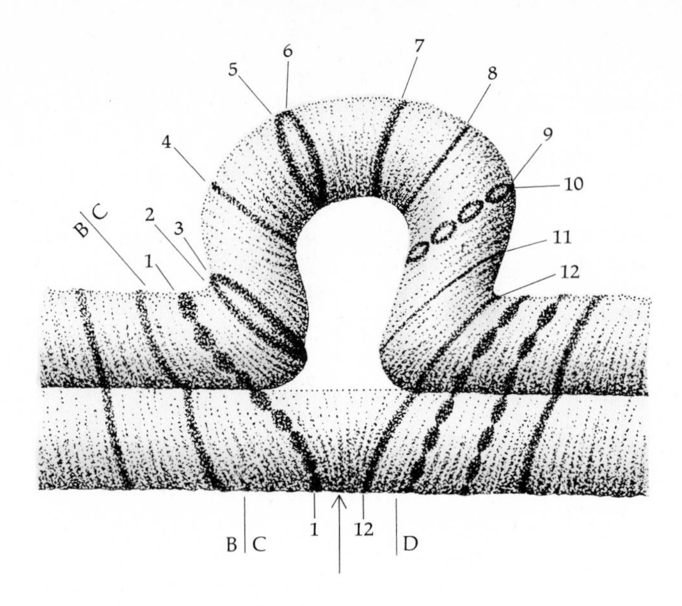

Figure 17.6
Buckling due to a heterozygous deficiency in the salivary gland chromosomes of *D. melanogaster*. Only a segment is shown of the paired X chromosomes of a larva heterozygous for *Notch*. Bands 3C2 through 3C11 are missing from the lower chromosome.

Figure 17.7
Banding pattern of metaphase chromosomes from a human male (see Figure 1.7). Chromosome banding techniques developed in the 1960s reveal alternating light and dark regions (*bands*) along the chromosomes. Two commonly used banding techniques are the giemsa method (used in the photograph shown) and the quinacrine mustard method. (Courtesy of Prof. W. Roy Breg, Yale University.)

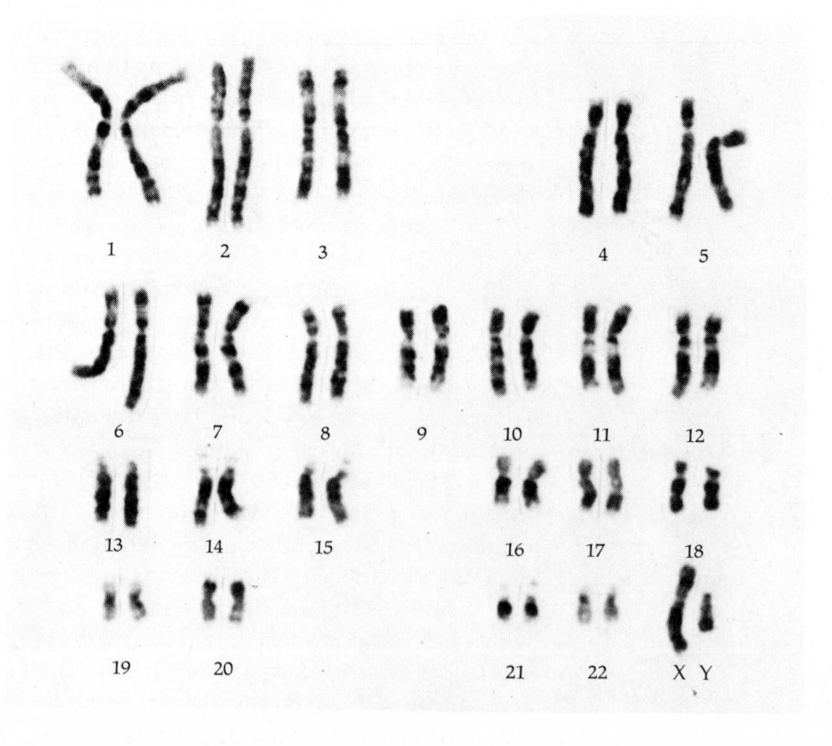

usually occurs in infancy or early childhood, although some survive to adulthood. The *Philadelphia* chromosome (a chromosome 22 with part of the long arm deleted) is associated with chronic myeloid leukemia; the name comes from the city where this chromosome deletion was first detected. Heterozygous deficiencies in other human chromosomes, such as 4, 13, and 18, are all associated with severe physical and mental defects.

Deficiencies may be recognized cytologically by the appearance in heterozygous individuals of a characteristic buckling when the two homologous chromosomes pair at meiosis, or in polytene chromosomes (Figure 17.6). The development of chromosome banding techniques (Figure 17.7) has facilitated the cytological detection of deficiencies in metaphase chromosomes. Deletions make possible the association of specific genes with specific sites on chromosomes (see Chapter 5).

Duplications

The presence of a chromosome segment more than once in the same chromosome or in a nonhomologous chromosome is known as a duplication, or repeat. Duplicated segments often occur in *tandem*, i.e., adjacent to each other. A *reverse tandem* duplication occurs when the order of genes in the duplicated segments is opposite. When the duplicated segment is at the end of the chromosome, the duplication is called *terminal* (Figure 17.8).

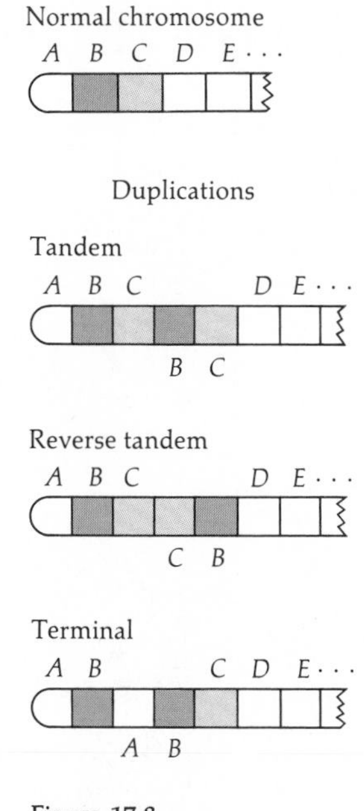

Figure 17.8
Common forms of chromosome duplications include tandem, reverse tandem, and terminal duplications.

Duplications may have phenotypic effects, such as *Bar* in the X chromosome of *Drosophila melanogaster*. *Bar* behaves as an incomplete dominant that reduces the number of facets in the eye. Females heterozygous for *Bar* have small, bar-shaped eyes. In homozygous or hemizygous condition, the eye is still smaller. *Bar* is due to a duplication of a small segment of the X chromosome. The segment may be present three times, a condition known as *Double Bar* or *Ultra Bar* (Figure 17.9).

Duplications are sometimes detected because individuals that are expected to be homozygous for a recessive allele fail to manifest the recessive phenotype, owing to the presence of a dominant allele in the duplicated segment. Cytologically, heterozygous duplications result in bucklings similar in appearance to those due to chromosome deficiencies (Figure 17.6).

Many duplications and deletions arise from chromosome breakage, which can be caused by a number of mechanisms, e.g., radiation, chemicals, and viruses. Chromosome breakage may also be induced by some genetic constitutions, such as the MR (male recombination) chromosomes common in *D. melanogaster* (20–50% of all second chromosomes sampled in a variety of populations are MR). MR chromosomes increase the mutation rate at specific gene loci (Table 17.1), as well as the frequency of mitotic recombination, in males and females. The mode of action of MR chromosomes involves chromosome breakage and, proba-

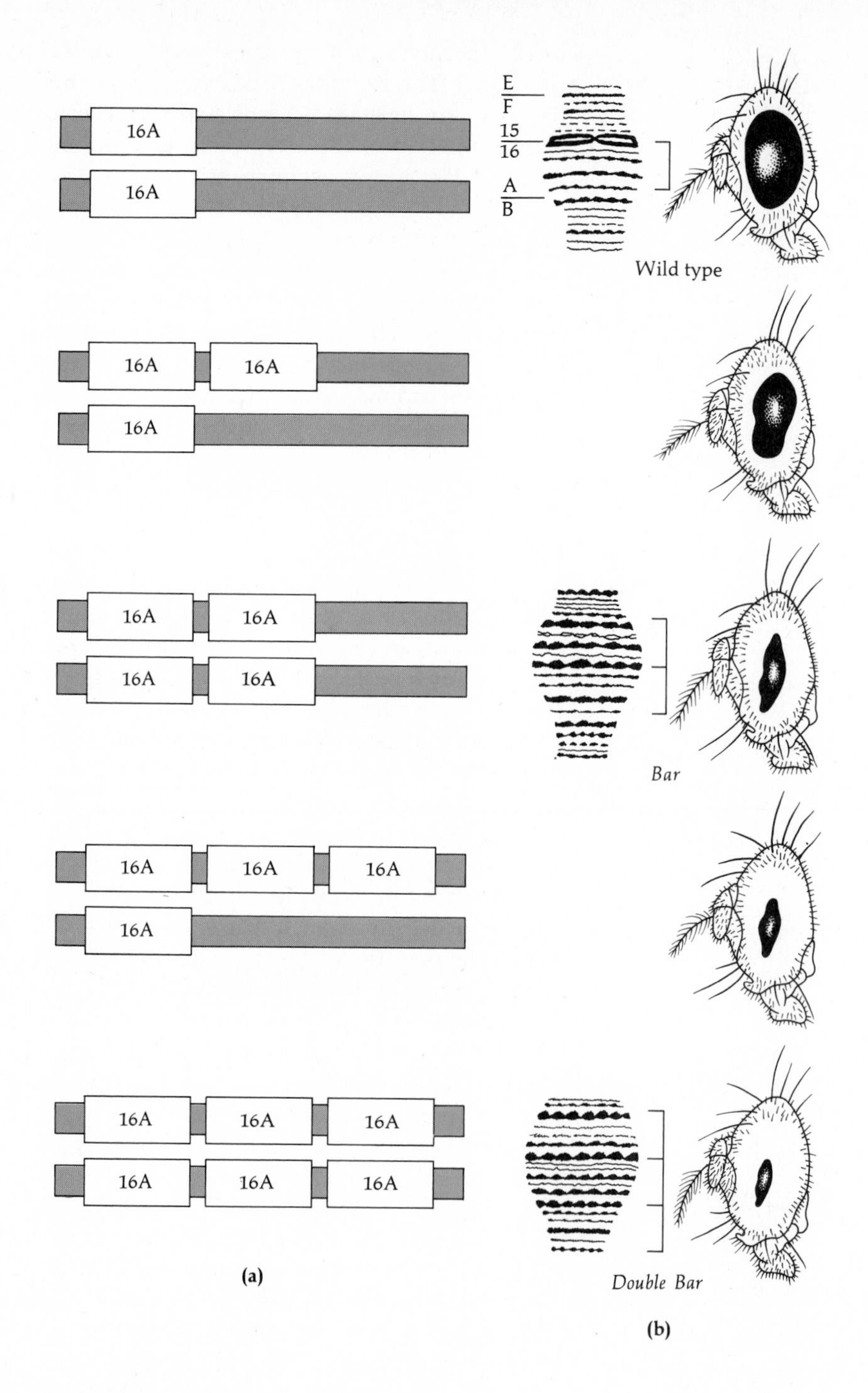

Figure 17.9
Duplications of section 16A of the X chromosome of *D. melanogaster* reduce the size of the eye (*Bar*).

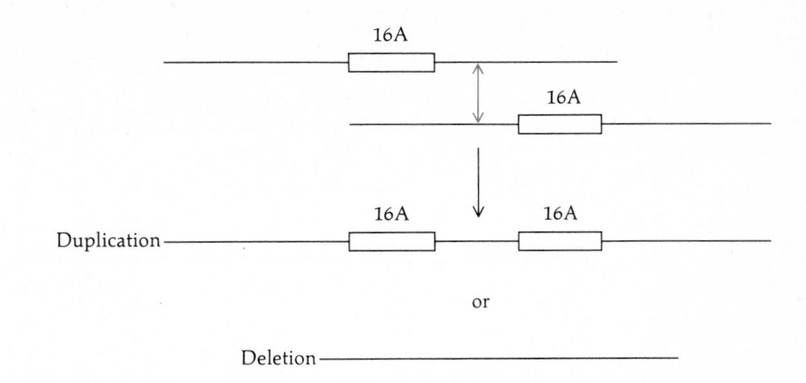

Figure 17.10
Duplications and deletions may arise by unequal crossing over. The figure shows the presumed origin of the duplicated 16A region of the X chromosome of *D. melanogaster* associated with *Bar*.

bly, the insertion of DNA sequences. A "mutator" gene that induces deletion mutations has been mapped on the third chromosome of *D. melanogaster*.

Duplications and deletions may sometimes arise by *unequal crossing over*. When similar DNA sequences occur in neighboring regions of a chromosome, the two homologues may pair inaccurately. Crossing over in the mispaired region will result in gametes with a duplication or a deletion (Figure 17.10). *Lepore* and *anti-Lepore* hemoglobins may have arisen by unequal crossing over (Figure 17.11). Duplications and deletions may also arise by crossing over in individuals heterozygous for inversions or for translocations.

Table 17.1
Mutation rate per locus per 10^5 X chromosomes at three loci of *D. melanogaster*.

	Mutant			
Chromosome*	*sn*	*ras*	*cm*	*y*
MR-h12	186	34	11	6
MR-n1	93	29	4	4
MR-s1	235	25	14	3.5
Control	0.4	0.2	0.8	1

*MR = male recombination.

After M. M. Green, *Stadler Symp. 10*:95 (1978).

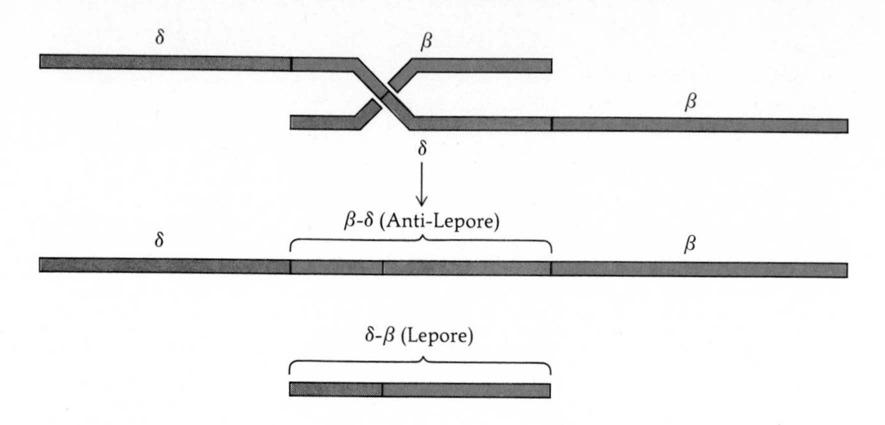

Figure 17.11
Possible origin of Lepore hemoglobins by unequal crossing over. Lepore hemoglobins (named after the family in which they were first found) have the amino end like that of a δ chain and the carboxyl end like that of a β chain. The β and δ chains each consist of 146 amino acids and are quite similar in sequence. The genes for the two chains are thought to be next to each other; their similarity might facilitate mispairing, so that the β gene in one chromosome pairs with the δ gene in the other chromosome rather than with its β homologue. Unequal crossing over would produce a chromosome with a deletion (Lepore hemoglobin) and a chromosome with a duplication (anti-Lepore hemoglobin). Anti-Lepore hemoglobins, which have the β amino end and the δ carboxyl end, have also been identified in humans.

Inversions

Inversions are 180° reversals of chromosomal segments; they do not change either the number of chromosomes or the number of genes in the chromosomes. If the gene sequence of a chromosome is represented as *ABCDEF*, inversion of the segment *BCD* will result in a chromosome with the sequence *ADCBEF*. Pericentric inversions include the centromere in the inverted segment; paracentric inversions do not (Figure 17.2).

The linkage order of genes becomes changed in individuals homozygous for a chromosomal inversion. A homozygote for the *ADC-BEF* chromosomal arrangement will show *A* closely linked to *D* (rather than to *B*, as in the original sequence, *ABCDEF*) and *E* closely linked to *B* (rather than to *D*). Pericentric inversions change the configuration of the chromosomes if the breakpoints are asymmetrically located relative to the centromere. In extreme cases, a metacentric chromosome may change into an acrocentric one, or vice versa. Pericentric inversions are responsible for some of the changes in chromosome configuration that take place during evolution. For example, human chromosome 17 is acrocentric, while the corresponding chimpanzee chromosome is metacentric (Figure 17.12). Several other chromosomes differ in humans and chimpanzees by pericentric inversions.

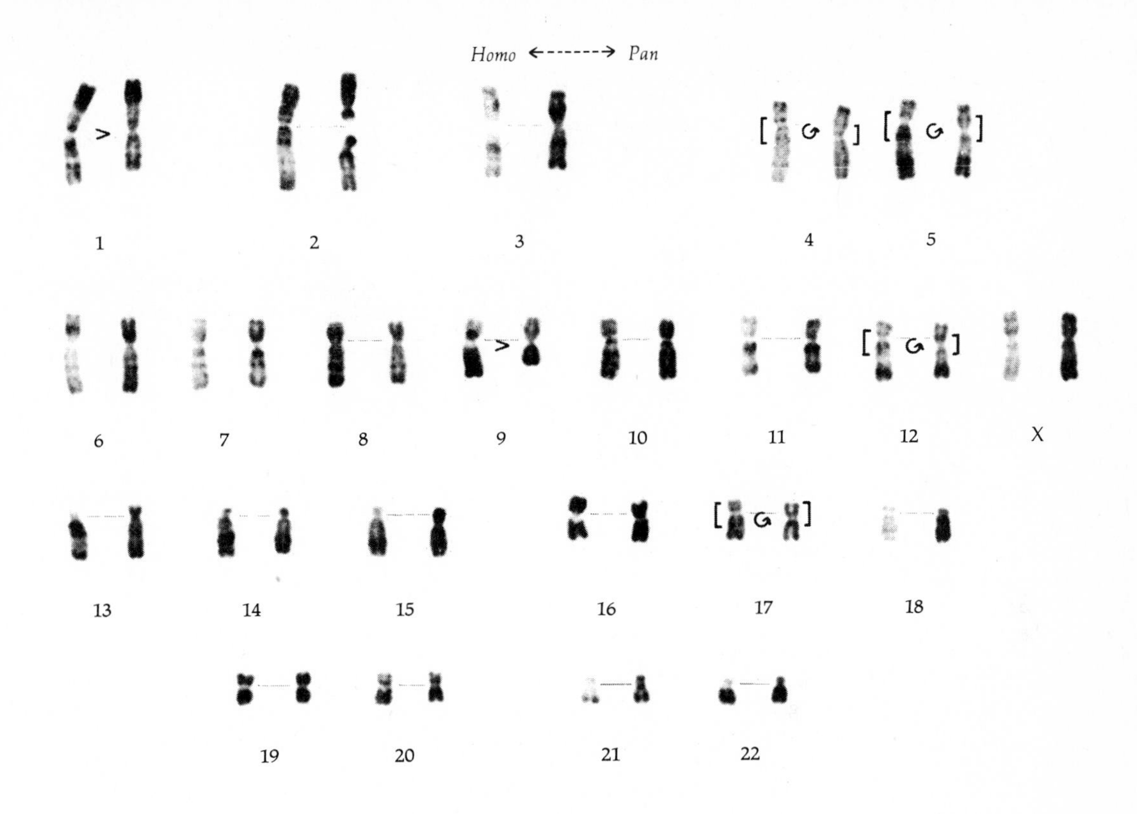

Figure 17.12
Comparison between human (left) and chimpanzee (right) chromosomes. The number of chromosomes is 46 in humans but 48 in chimpanzees. Human chromosome 2 contains most of the material of two chimpanzee chromosomes. Pericentric inversions have occurred in chromosomes 4, 5, 12, and 17. (Courtesy of Dr. Jean de Grouchy, Hôpital des Enfants Malades, Paris.)

In inversion heterozygotes, synapsis of the homologous chromosomes requires the formation of loops containing the inverted segments (Figures 17.13, 17.14, and 17.15). Heterozygous inversions can be recognized by the presence of such loops in preparations of cells at the pachytene stage of meiosis. Similar loops are also observed in polytene chromosomes (Figure 17.13).

The presence of inversions can also be detected genetically because they suppress, or considerably reduce, recombination in heterozygotes. The effects of crossing over in an individual heterozygous for a paracentric inversion are shown in Figure 17.14. Of the four chromosomes resulting from the meiotic divisions, one has two centromeres, one has none, and two are normal (noncrossover) chromosomes. Usually the only gametes that can give rise to viable progeny are those containing the noncrossover chromosomes.

The effects of crossing over in an individual heterozygous for a pericentric inversion are shown in Figure 17.15. Of the four chromosomes resulting from the meiotic divisions, two are the original, noncrossover

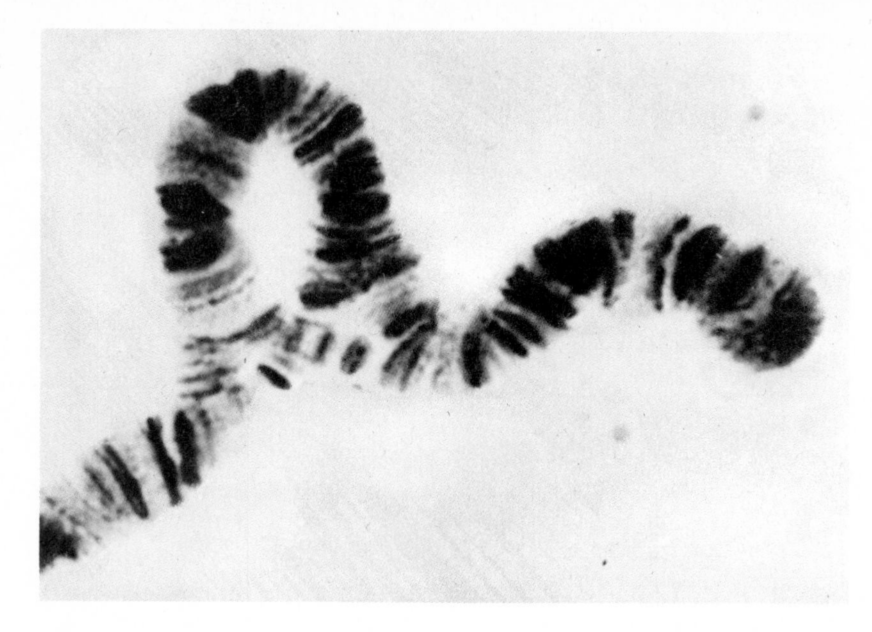

Figure 17.13
A segment of the polytene third chromosome of a *D. pseudoobscura* larva heterozygous for a paracentric inversion.

chromosomes, but the other two have some genes duplicated and others deleted. Usually only the gametes containing the two complete chromosomes can produce viable progeny and usually, therefore, no genetic recombination is found in the progeny of heterozygotes for either pericentric or paracentric inversions.

There are two notable exceptions to the rule that no viable recombinants appear in the offspring of inversion heterozygotes. The first exception concerns double crossing over. For example, the mutants *dark* eye (*d*) and *vestigial* wing (*vg*) are in the second chromosome of *D. melanogaster* and usually recombine with a frequency of about 8.5%. When females heterozygous for a normal chromosome carrying the dominant alleles for both loci (d^+vg^+) and an inverted chromosome carrying the recessive alleles (*d vg*) are crossed with males homozygous for both recessive alleles, they produce offspring with the following frequencies:

Eyes	Wings	Frequency
Normal	Normal	0.499
Normal	Vestigial	0.001
Dark	Normal	0.001
Dark	Vestigial	0.499

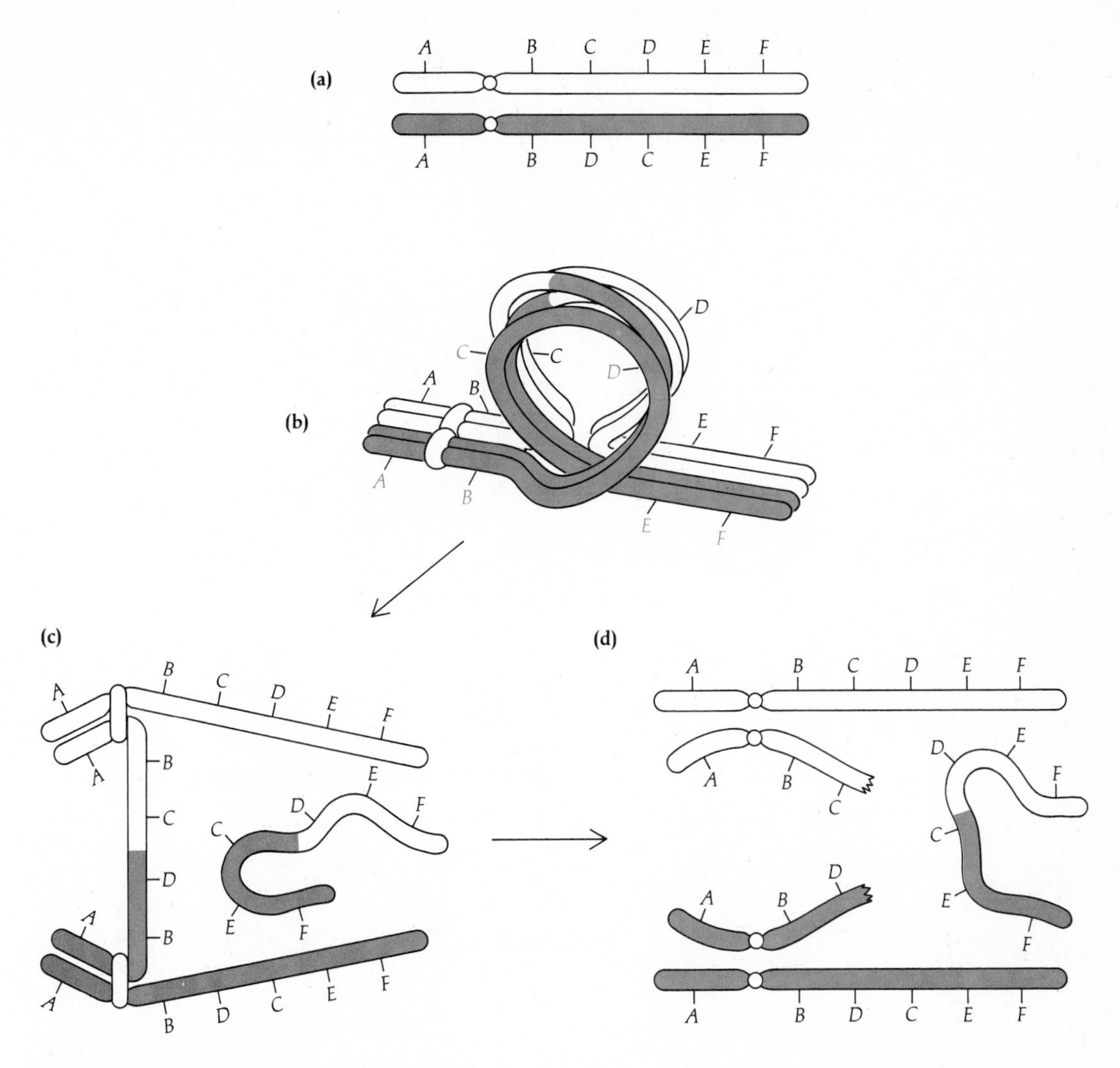

Figure 17.14
Crossing over in a heterozygote for a paracentric inversion. **(a)** The two homologous chromosomes. **(b)** Pairing at meiosis and crossing over between two nonsister chromatids. **(c)** Separation of the chromosomes during early anaphase of the first meiotic division. **(d)** The resulting chromosomes. The chromosomal segment without a centromere fails to move toward the poles during meiosis I and is usually lost; the chromosome segment with two centromeres breaks and, after meiosis II, yields gametes with deletions.

The 0.002 recombinants are all due to double crossing over, when one of the exchange points, but not the other, is between the two loci.

The other exception occurs when the duplicated and deleted segments in the crossover chromosomes do not interfere with the viability of the gametes or the zygotes formed by them. This is likely to occur only when the segments involved are small. This is a way in which duplications and deletions may come about. Duplicated and deleted chromosome segments appear in the crossover products not only in the case of pericentric het-

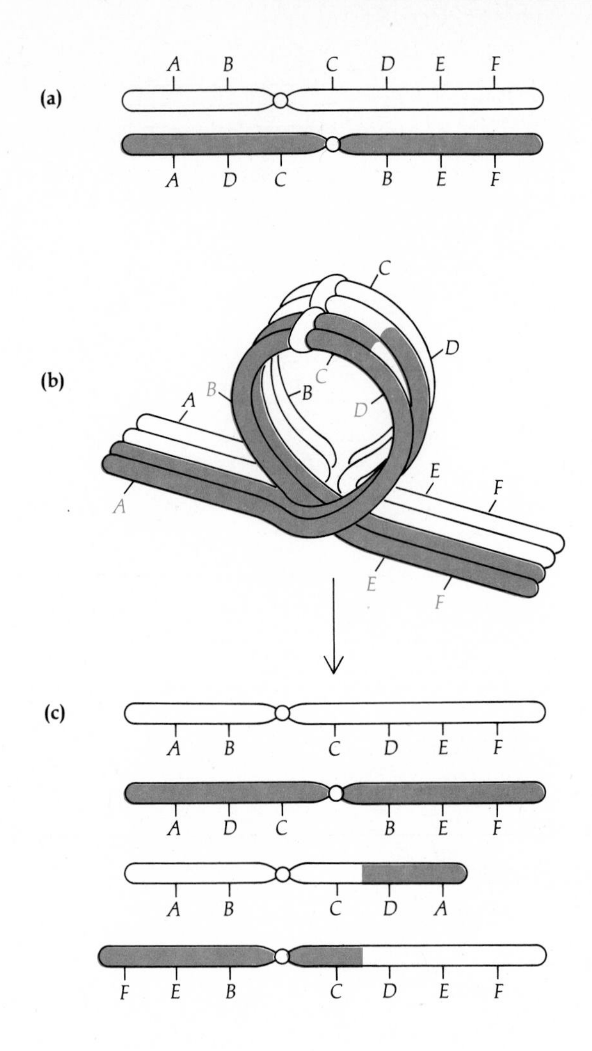

Figure 17.15
Crossing over in a heterozygote for a pericentric inversion. **(a)** The two homologous chromosomes. **(b)** Crossing over between two nonsister chromatids. **(c)** The four resulting chromosomes. Only the top two chromosomes have complete sets of genes; they are noncrossover chromosomes with the same gene sequences as the two original chromosomes.

erozygotes (Figure 17.15) but also in the case of heterozygotes for two paracentric inversions with overlapping breakpoints (Figure 17.16).

Inversion heterozygotes are often partially sterile because crossing over leads to gametes half of which cannot produce viable zygotes. There are exceptions, however. In drosophilids and related families of flies, crossing over does not occur during male meiosis; hence, males heterozygous for inversions do not exhibit reduced fertility. In females of these species that are heterozygous for paracentric inversions, one normal chromosome is always included in the egg nucleus, while the two abnormal chromosomes and the other normal chromosome are eliminated in the polar bodies; the females consequently have undiminished fertility. In other Diptera, such as the midge *Chironomus*, chiasmata are formed and crossing over takes place in male meiosis. Yet the number of progeny

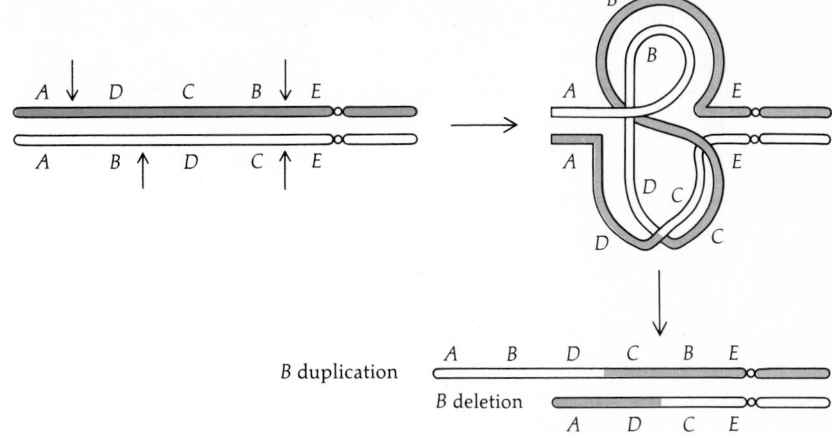

Figure 17.16
Crossing over in a heterozygote for two overlapping paracentric inversions. Overlapping inversions are those that include part of a previously inverted chromosome segment. In this figure, the chromosome on top has the segment *BCD* inverted; the bottom chromosome has *CD* inverted. For simplicity, only one chromatid is shown for each chromosome.

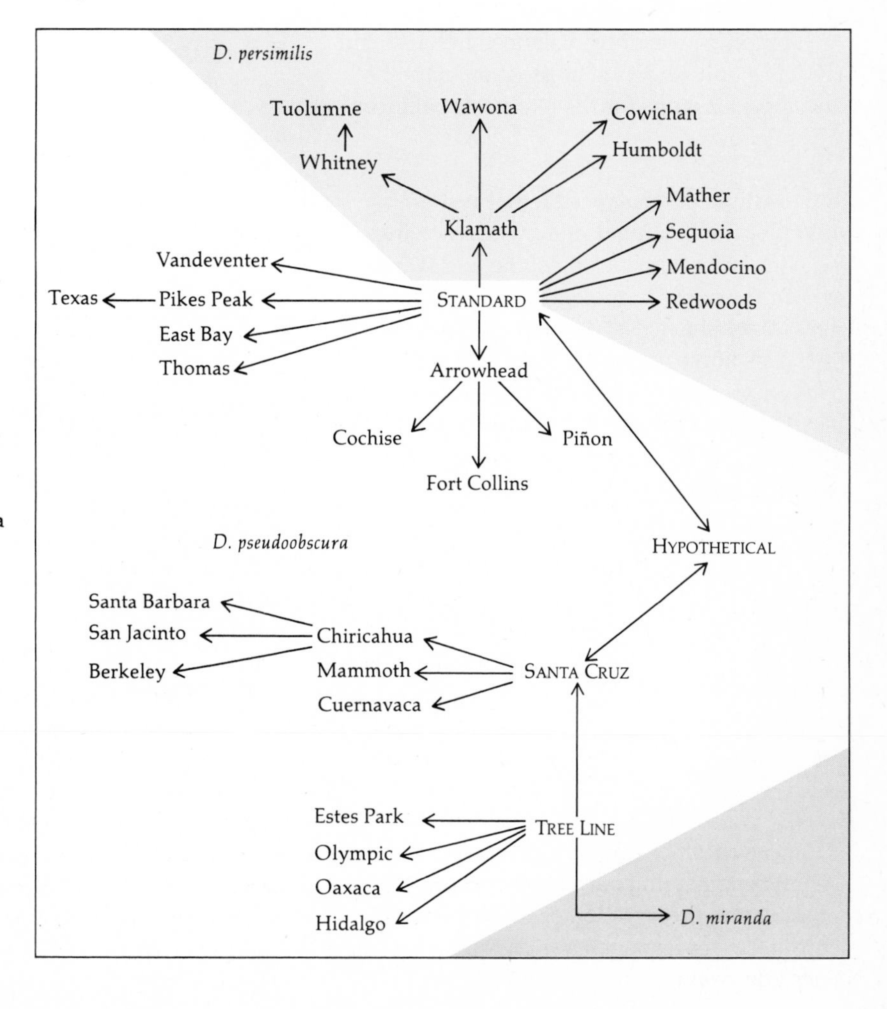

Figure 17.17
Phylogeny of the third-chromosome gene sequences found in natural populations of *Drosophila pseudoobscura, D. persimilis,* and *D. miranda*. The phylogeny is inferred from the overlapping inversions detected in polytene chromosomes. The Standard chromosomal sequence is found in both *D. pseudoobscura* and *D. persimilis*; all other sequences exist in only one species. The Hypothetical sequence has never been found, but is postulated as a "missing link," perhaps extinct, between Standard and Santa Cruz. The pattern suggests that the original ancestral arrangement is likely to be one of the four: Standard, Hypothetical, Santa Cruz, or Tree Line. (Most sequences are named after the places where they were first found.)

Box 17.1 Two Applications of Inversions

A. BALANCER CHROMOSOMES. The Muller-5 X chromosome of *D. melanogaster* was mentioned in Chapter 16 in connection with the study of mutation rates. A balancer is a chromosome having (1) several overlapping inversions, which impede recombination with the homologous chromosome, and (2) a dominant mutant, which allows identification of the progeny carrying the balancer chromosome. Balanced lethal stocks have been constructed in *Drosophila* that involve two homologous chromosomes, each carrying a different recessive lethal as well as different overlapping inversions. Balanced lethal stocks produce only one kind of progeny, namely, flies heterozygous for the two *whole* balancer chromosomes.

An example of a balanced lethal system involves the dominant mutants *Curly* wings (*Cy*) in one second chromosome and *Plum* eye color (*Pm*) in the homologue. A different recessive lethal exists on each chromosome, so that flies homozygous for either the *Curly* or the *Plum* chromosome are lethal. Therefore only heterozygous *Cy* +/+ *Pm* adult progeny are produced generation after generation by the balancer stock:

$$Cy\ +/+\ Pm \times Cy\ +/+\ Pm$$

Cy +/*Cy* +	*Cy* +/+ *Pm*	+ *Pm*/+ *Pm*
1 :	2 :	1
Dies	Balanced-lethal	Dies

Balancer chromosomes are early examples of genetic engineering because they make possible the preservation of intact whole chromosomes from generation to generation. Flies homozygous for an intact whole chromosome (or heterozygous for two given whole chromosomes) can then be obtained as follows ($+_1$ and $+_2$ are used here to represent any two wild chromosomes):

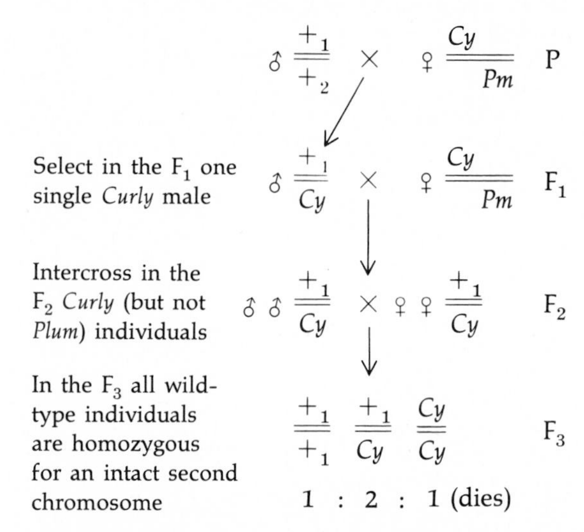

B. CHROMOSOMAL PHYLOGENIES. Many different inversions are found in some species of *Drosophila*, mosquitoes, grasshoppers, etc. In *Drosophila* and other dipterans, the breakpoints can be rather precisely identified by microscopic examination of the polytene chromosomes. If the inversions are overlapping, the ancestral relationships between inversions can sometimes be inferred.

Suppose that the sequences of bands in three different homologous chromosomes are ABCDEFGH, ABFEDCGH, and AEFBDCGH. The second sequence can arise from the first, or the first from the second, by a single inversion involving the segment CDEF. Similarly, the third sequence can arise from the second, or vice versa, by inversion of the segment BFE. However, the third sequence cannot arise from the first, nor the first from the third, by a single inversion. The second sequence represents a necessary intermediate step between the first and the third. If no other information is available, there is no way of knowing which of the three chromosome sequences was the original one, but only three ancestral relationships are possible: 1→2→3 or 3→2→1 or 1←2→3; any relationships involving the direct transitions 1→3 or 3→1 can be excluded. Figure 17.17 shows the ancestral relationships of several common inversions that occur in the third chromosome of two closely related American species, *Drosophila pseudoobscura* and *D. persimilis.*

produced by male inversion heterozygotes is not appreciably diminished, apparently because spermatozoa containing abnormal chromosomes fail to be involved in fertilization.

Translocations

Reciprocal translocations involve the interchange of blocks of genes between two nonhomologous chromosomes (Figure 17.2). If the gene sequences in two nonhomologous chromosomes are represented as *ABCDEF* and *GHIJKL,* the sequences *ABCDKL* and *GHIJEF* represent translocated chromosomes. In homozygotes for these translocations, the linkage relationships change: genes that were not linked in the original chromosomes are now linked, and vice versa. In the example given, the *KL* genes are linked to *ABCD* after the translocation, but are no longer linked to *GHIJ.*

In heterozygotes for reciprocal translocations, genes in *both* translocated chromosomes behave as if they were in the same linkage group, since only gametes containing the parental combinations of chromosomes can produce viable zygotes (Figure 17.18). Also, crossing over is reduced in translocation heterozygotes near the translocation breakpoints. This is because the cross-shaped configuration required for pairing of homologous parts at meiosis makes pairing imperfect near the breakpoints, and this reduces crossing over in such regions.

Cytologically, heterozygotes for reciprocal translocations exhibit characteristic crosslike configurations during meiotic prophase, owing to the requirements of pairing between homologous chromosome parts (Figure 17.18). Instead of *bivalents,* i.e., pairs of synapsed homologous chromosomes, there appear *quadrivalents* consisting of four associated chromosomes, each chromosome being partially homologous to two other chromosomes in the group. Segregation at the first meiotic anaphase may occur in three different ways. Of the six types of gametes that can be formed, only the two shown on the left in Figure 17.18—which result from "alternate" segregation—contain all the chromosomal parts once and only once. All other gametes have some chromosome segments duplicated and

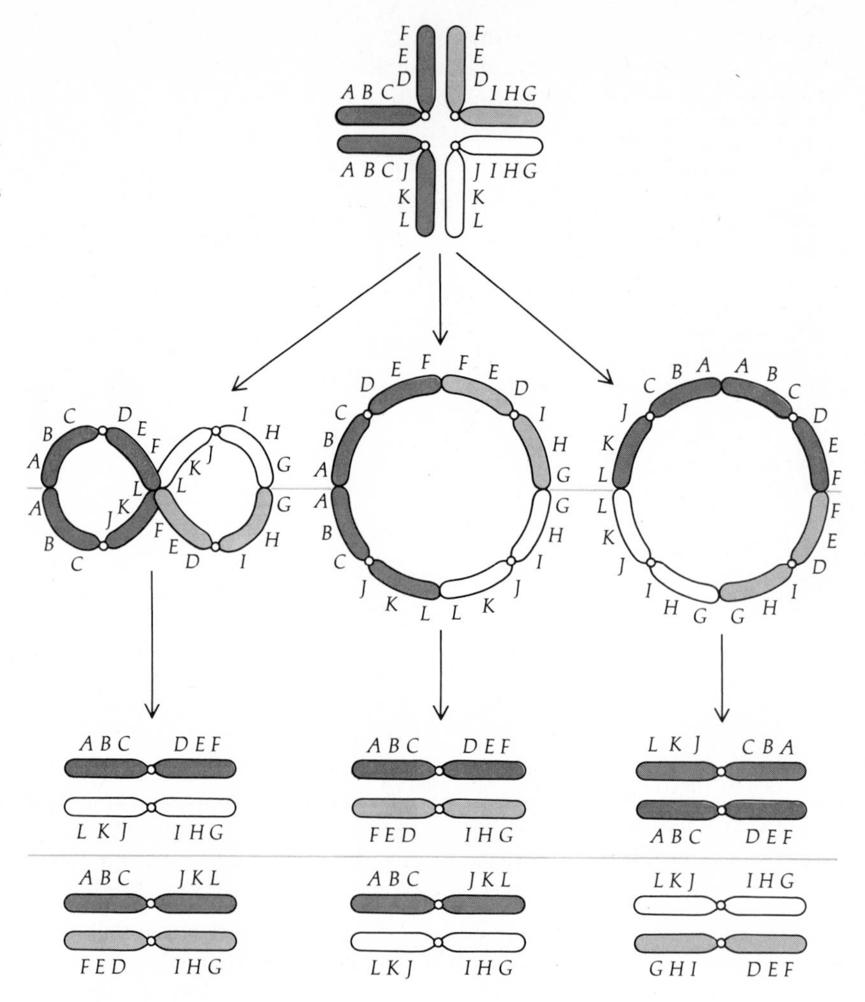

Figure 17.18
Meiosis in a translocation heterozygote. At the top is the cross-shaped configuration formed when the chromosomes of translocation heterozygotes pair during prophase of meiosis I. The second row shows the three configurations that may occur at metaphase I. The two lower rows show the six types of gametes that may be formed. Only the two types on the left contain complete sets of genes; the other four contain some duplicated, and some deleted, chromosome segments.

some deleted, and therefore cannot usually result in normal progeny. Since the only gametes producing normal progeny are those having either both nontranslocated chromosomes or both translocated chromosomes, it follows that all genes in both chromosomes will behave as a single linkage group, as pointed out in the previous paragraph.

Translocation heterozygotes are semisterile, owing to the production of abnormal gametes. In plants, pollen grains with duplicated or deleted chromosome segments are usually aborted. Animal gametes with duplicated or deleted chromosome parts may function, but the zygotes formed by such gametes usually die. However, if the duplicated and deleted chromosome segments are small, the gametes may function and produce viable offspring.

Heterozygotes for reciprocal translocations are relatively rare in animals, but have been found in natural populations of many plants, where

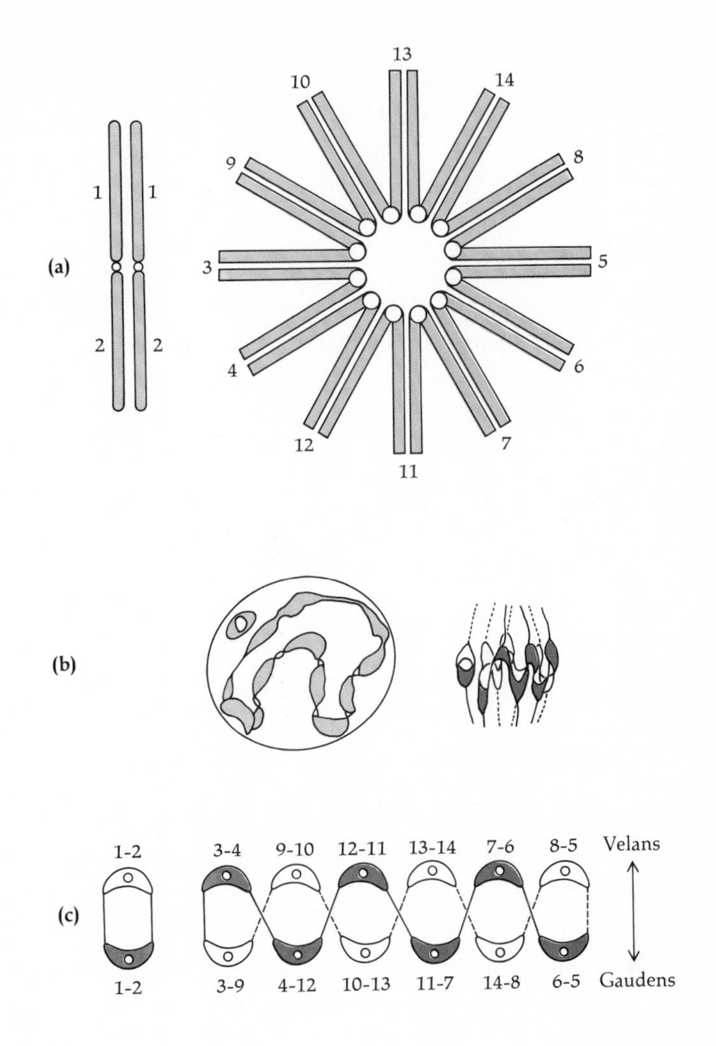

Figure 17.19
Meiosis in *Oenothera lamarckiana*. **(a)** Diagram of pairing between the nontranslocated pair of chromosomes and the six pairs of translocated chromosomes. **(b)** The chromosomes during prophase of meiosis I (left); alternate chromosome segregation during anaphase (right). **(c)** Chromosome segregation in terms of the chromosome arms of the two complexes velans and gaudens. (After M. W. Strickberger, *Genetics*, 2nd ed., Macmillan, New York, 1976.)

translocations sometimes involve more than two nonhomologous chromosomes. An extreme example is *Oenothera lamarckiana,* which is heterozygous for translocations involving 12 of its 14 chromosomes (Figure 17.19). If we represent the 14 *arms* of the seven nontranslocated chromosomes as 1-2, 3-4, 5-6, 7-8, 9-10, 11-12, and 13-14, all 14 *chromosomes* can be represented as follows: 1-2, 1-2, 3-4, 4-12, 12-11, 11-7, 7-6, 6-5, 5-8, 8-14, 14-13, 13-10, 10-9, and 9-3. With the exception of the first chromosome pair (1-2, 1-2), the homologues of the two arms of each chromosome appear one in each of two different chromosomes (e.g., the homologues of 3-4 are found one in 4-12, the other in 9-3). Pairing at meiosis produces one bivalent and one dodecavalent consisting of twelve chromosomes in a multiarmed, starlike configuration.

Segregation at meiosis produces only two kinds of functional gametes in *O. lamarckiana*: 1-2, 3-4, 9-10, 12-11, 13-14, 7-6, and 8-5 (called *velans*)

Box 17.2 Analysis of the *Drosophila* Genome by Means of Translocations

Dan L. Lindsley, Lawrence Sandler, and their colleagues have used reciprocal translocations between the Y chromosome and the autosomes for analyzing the viability and fertility effects of duplications and deletions for variable parts of the *Drosophila* genome. The procedure uses pairs of translocations with displaced breakpoints in the autosomes. By means of appropriate crosses, the autosomal segment can then be deleted or duplicated (Figure 17.20). The term *segmental aneuploidy* has been used to refer to the condition in which a chromosome

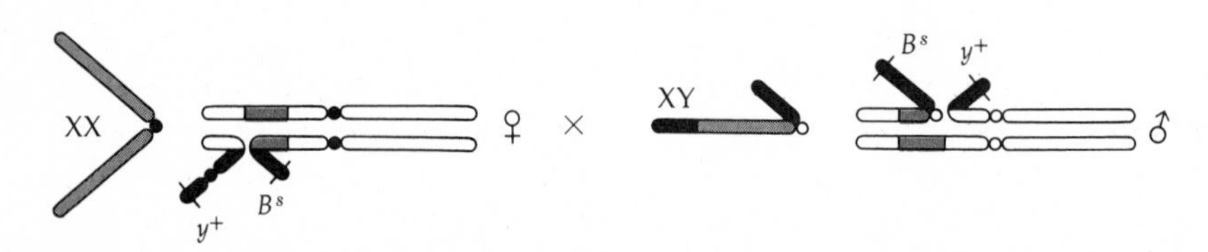

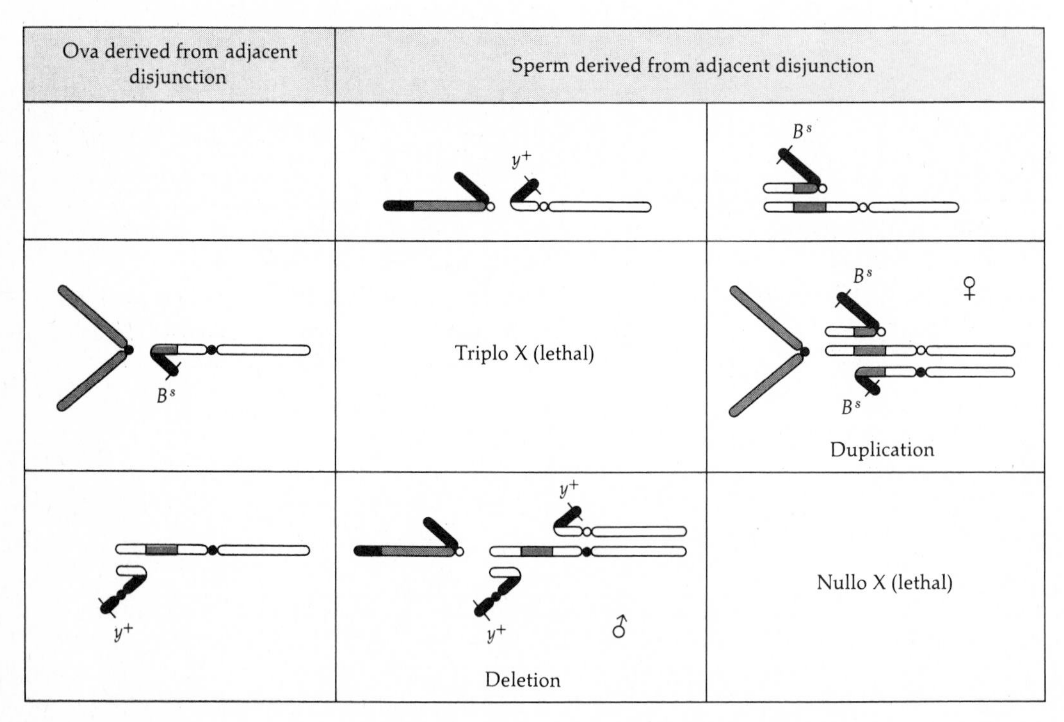

Figure 17.20
Production of segmental aneuploids by means of reciprocal translocations between the Y chromosome and the autosomes. The parents crossed are each heterozygous for a different reciprocal translocation. The breakpoints of these translocations are slightly displaced in the autosome. The autosomal segment between the two breakpoints (in color) is duplicated in some progeny and deleted in others. Y-chromosome material is represented by solid bars, X-chromosome material by shaded bars. The centromeres of maternal origin are solid; those of paternal origin are open. The two parts of the translocated Y chromosome are labeled with mutants (B^s and y^+). [After D. L. Lindsley, L. Sandler, et al., *Genetics* 71:157 (1972).]

segment is present in the genome some number of times different from two.

Virtually the whole genome of *D. melanogaster* has been analyzed by segmental aneuploidy. The results can be summarized as follows:

1. The third chromosome contains one locus that is lethal in haploid (monoploid) condition. No other haplo-lethal loci have been detected anywhere in the genome, indicating that few such loci exist (and perhaps only the one detected).
2. The third-chromosome locus that is haplo-lethal is also lethal in triploid condition. Since no other triplo-lethal loci have been found, such loci must be very rare throughout the genome.
3. Two classes of haplo-abnormal loci exist. The first class consists of *Minute* loci, which in haploid condition produce flies with short, thin bristles, reduced developmental rate, and low viability and fertility. There are at least 41 *Minute* loci in the genome—and probably not many more—distributed as follows: 7 in the X chromosome, 17 in the second, 16 in the third, and 1 in the fourth.
4. The second class of haplo-abnormal loci produces a variety of defective phenotypes in haploid condition. There are at least 11 such loci, distributed as follows: 1 in the X chromosome, 4 in the second, 6 in the third, and none in the fourth.
5. There are two loci in the X chromosome that cause abnormality when present in more than two copies; one of them is also haplo-abnormal.

Large deletions are lethal even in heterozygous condition; the largest deletion known that allows survival of the heterozygote represents about 3% of the *Drosophila* genome and is located on the second chromosome. The study of segmental aneuploids shows, however, that there are few haplo-lethal loci. It follows, therefore, that the lethality of heterozygous deficiencies is the cumulative result of many slightly deleterious effects, each due to haploidy for one gene locus.

Duplications are better tolerated than deletions. In general, however, the presence of more than 10% of the genome in triplicate causes lethality. But because there are few triplo-lethal loci, it follows that the lethal effects of segmental triploidy are due to the accumulation of deleterious effects produced by genes each of which has a small effect in triplicated condition.

and 1-2, 3-9, 4-12, 10-13, 11-7, 14-8, and 6-5 (called *gaudens*). The surprising thing, however, is that, except for the appearance of occasional "mutants," *O. lamarckiana* breeds true: only velans/gaudens heterozygotes, and no velans/velans or gaudens/gaudens homozygotes, are produced. This is due to the existence of recessive lethals in each of the two gametic complexes that make the homozygous combinations lethal.

Transpositions

Transpositions involve the transfer of a chromosome segment to another position, either in the same chromosome (Figure 17.2) or in a different chromosome. One interesting kind of transposition involves transposable elements, which were discussed in Chapter 7 (page 240) for prokaryotes.

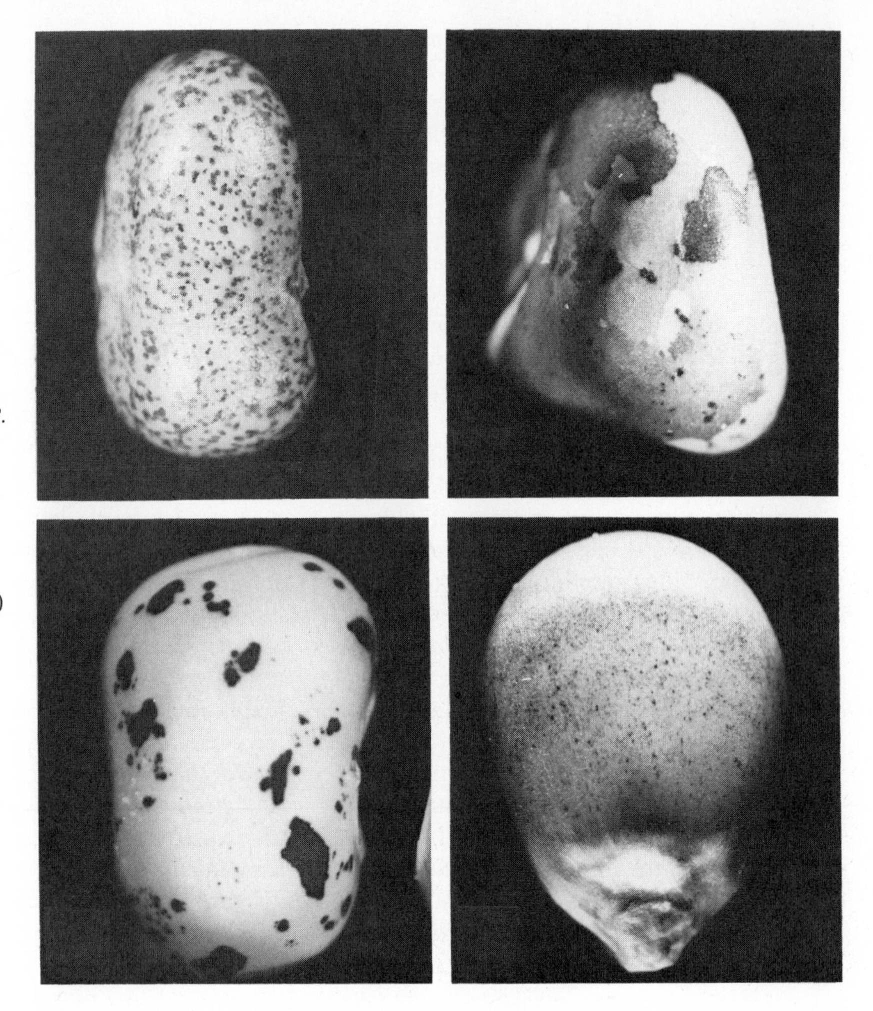

Figure 17.21
Corn kernels exhibiting spots of pigment produced by cells in which a controlling element has released the inhibition during development of a gene whose activity is required for pigment production. The cells within the white areas fail to produce pigment, owing to the continued inactivation of the pigment gene in those cells. (From P. A. Peterson, in *DNA Insertion Elements, Plasmids, and Episomes*, ed. by A. I. Bukhari, J. A. Shapiro, and S. L. Adhya, Cold Spring Harbor Laboratory, Cold Spring Harbor, N.Y., 1977.)

The first transposable genetic elements (originally called *controlling elements*) discovered in any organism were described by B. McClintock for maize. A number of genes function together to cause the red anthocyanin pigment to be synthesized in the kernels; inactivation of one of these genes causes a nonpigmented kernel. Controlling elements are a class of mutations that give rise to largely unpigmented kernels. These mutations are unstable in somatic kernel cells, causing occasional spots of pigment to occur on the kernel (Figure 17.21). They were called controlling elements because the activity of a pigment gene in a given kernel cell seems to be controlled by these mutations. They sometimes revert to wild type, and occasionally the reversion to wild type of one mutant gene is associated with the occurrence of mutation at another gene responsible for pigment

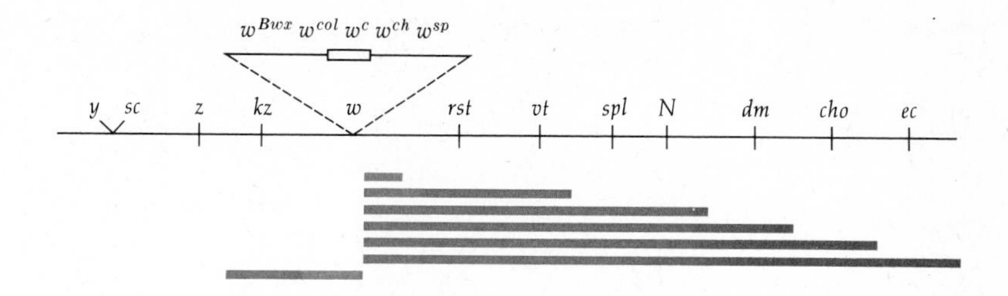

Figure 17.22
Deletions induced in the X chromosome of *D. melanogaster* by the transposable element *white-crimson*. The lengths of the deletions are shown by the bars. Each has one breakpoint located at the site of w^c within the *white* gene.

formation. Presumably a controlling element jumps from one anthocyanin locus to another.

Transposable elements have been identified in other eukaryotic organisms. The *white-crimson* (w^c) mutation in *Drosophila* has properties similar to those of IS1 of *E. coli* (see Chapter 7). It has been observed to cause transposition of the *white* gene to an autosome. Spontaneous deletions of adjacent X-chromosome genes originate at w^c (Figure 17.22) and extend to the left or right, similarly to the deletions caused by IS1.

Sex determination in the fly *Megaselia scalaris* is effected by a genetic element called the *sex realizer*. Males are hemizygous for the sex realizer, while females lack it. The sex realizer resides at the end of one of the three nonhomologous pairs of chromosomes of this species, converting that chromosome into a sex chromosome (Figure 17.23). Sperm are found (with a frequency of about 0.1%) in which the sex realizer has transposed to the end of another chromosome, converting it into a sex chromosome. Stocks can be constructed in each of which a different nonhomologous chromosome is the sex chromosome.

The discovery of transposable elements in both prokaryotes and eukaryotes suggests that they are a general feature of genomes. Transposable elements are responsible for a variety of chromosomal alterations and may be important in providing structural variability within species.

Robertsonian Changes

Chromosomal fusion and fission are sometimes called *Robertsonian changes*, after William R. Robertson, who was the first to postulate fusion as a mechanism for reduction in chromosome number. Chromosomal fusion occurs when two nonhomologous chromosomes fuse into one; chromo-

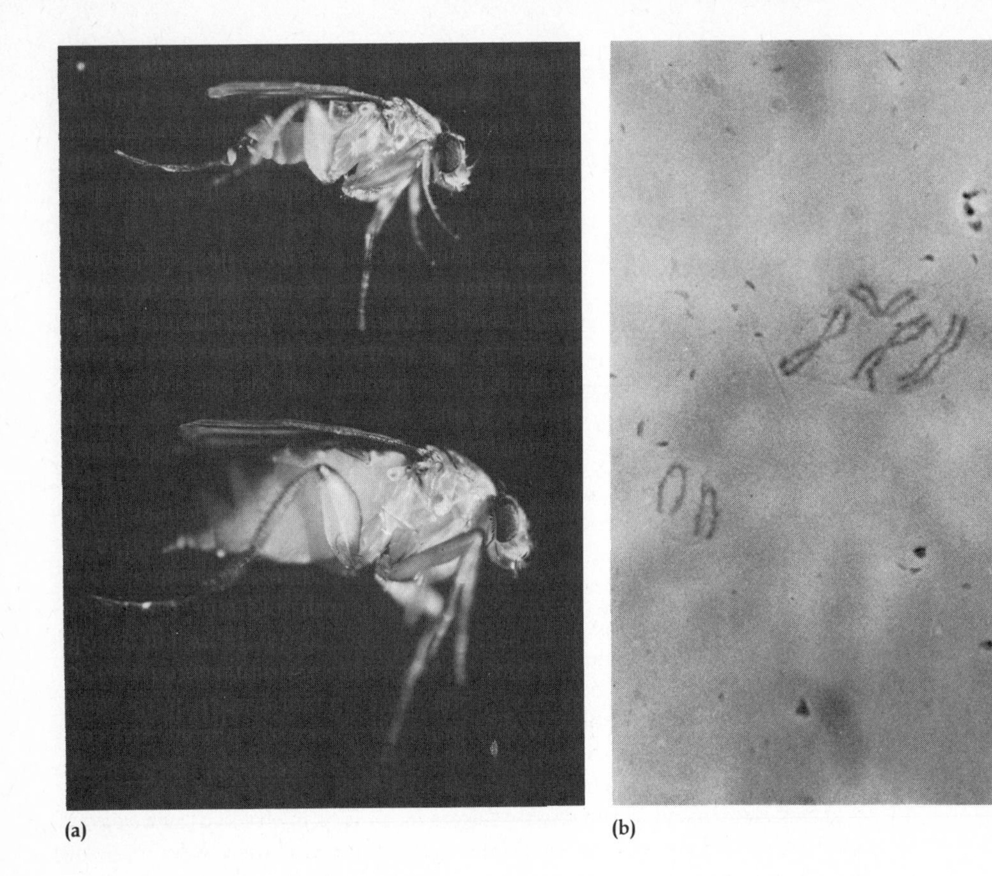

Figure 17.23
(a) Male (top) and female (bottom) *Megaselia scalaris.* **(b)** Mitotic chromosomes of *M. scalaris.* The karyotypes of males and females are identical. (Courtesy of Barbara Butler, University of California, Davis.)

somal fission, or dissociation, occurs when a chromosome splits into two (Figure 17.3). Fusions and fissions change the number of chromosomes but do not increase or reduce the amount of hereditary material.

Chromosomal fusions are thought to be much more common than fissions. Evidence of chromosomal fusion exists for virtually every major group of plants and animals. Increases in chromosome number by fission are well established in some cases, such as *Anolis* lizards. Robertsonian changes are common phenomena from the evolutionary point of view. The haploid chromosome numbers in most animals lie between 6 and 20, but the range extends from 1 (the variety *univalens* of the nematode worm *Parascaris equorum*) to about 220 (the butterfly *Lysandra atlantica*). In plants, the most common gametic numbers are 7, 8, 9, 11, 12, and 13, but they may be as high as 631 (the fern *Ophioglossum reticulatum*, which is almost certainly a polyploid). Even species of a single genus may differ in the gametic number of chromosomes, which in *Drosophila*, for example, ranges from 3 to 6 (Figure 17.24). Humans have 23 pairs of chromosomes, but chimpanzees and other apes have 24 pairs. Thus, at least one Robertsonian change, probably a fusion in the human lineage, has occurred in their evolution from a common ancestor.

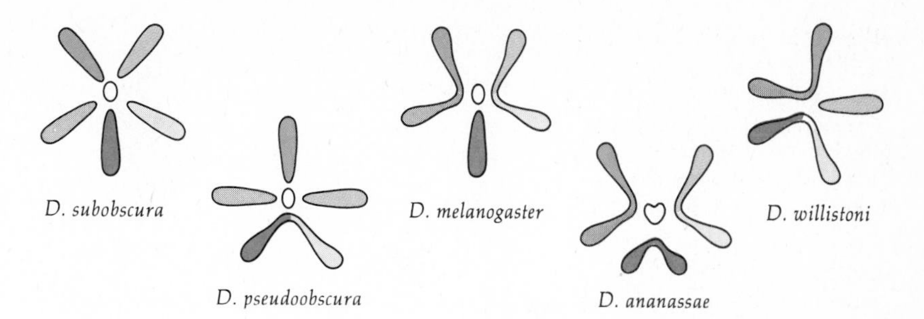

Figure 17.24
The haploid chromosome complements of five *Drosophila* species. Homologous chromosome arms are identified by shading. The ancestral condition for the genus appears to be five pairs of acrocentric chromosomes and one pair of dotlike chromosomes, as in *D. subobscura*. The other conditions can be derived from the primitive one through various chromosome fusions. The X chromosomes (solid color) of *D. melanogaster* and *D. ananassae* differ by a pericentric inversion that has changed the position of the centromere in *D. ananassae* from near the end of the chromosome to the middle.

Aneuploidy

Aneuploids have one chromosome present either fewer or more times than the other chromosomes (Figure 17.4). Nullisomics are organisms lacking both chromosomes of a pair, the total number of chromosomes being $2n - 2$. Monosomics lack one chromosome of a pair, the total number of chromosomes being $2n - 1$. Polysomics have one chromosome represented by more than two homologues: three times in trisomics (total number $2n + 1$), four times in tetrasomics (total number $2n + 2$), and so on. Aneuploidy changes the number of chromosomes as well as the total amount of genetic material.

Aneuploidy may result from abnormal segregation at meiosis. The first demonstration of aneuploidy was made by Bridges in 1916, when he discovered nondisjunction in *D. melanogaster*. Some females had three sex chromosomes—two X and one Y; some males had only one X and no Y chromosome.

Trisomics are known in many organisms, particularly plants, including such crop species as rice, corn, and wheat. Trisomics sometimes differ in appearance from normal individuals, as in the Jimson weed, *Datura stramonium*, in which all 12 possible trisomics are morphologically distinguishable (Figure 17.25). Often, particularly in animals, the presence of an extra chromosome has deleterious effects and may be lethal. In humans, trisomy for chromosome 21, or for the X chromosome, results in severe abnormalities. The reason why trisomics for some chromosomes are never found in humans, nor in other organisms, may be because they are lethal. Tetrasomics and even polysomics of higher order are known in some plants, but they are rarer than trisomics.

Figure 17.25
Capsule appearance in the Jimson weed, *Datura stramonium*. Each of the 12 mutant types is trisomic for one of the 12 pairs of chromosomes of this plant.

Monosomic, and more so nullisomic, individuals are often inviable, but have been found in polyploid plants. For example, the common tobacco plant, *Nicotiana tabacum*, is a tetraploid with 24 chromosome pairs; all 24 possible kinds of monosomics, differentiable by the appearance of the plants, have been identified. The bread wheat, *Triticum aestivum*, is hexaploid, having 21 pairs of chromosomes; all 21 possible kinds of nullisomics have been obtained.

Aneuploids are useful in genetic studies because they make it possible to associate specific genes with particular chromosomes. This is because unusual segregation ratios appear in their progenies. Consider a cross between a trisomic plant carrying only a dominant allele (*AAA*) and a normal plant, homozygous for the recessive allele (*aa*). One-half of the F_1

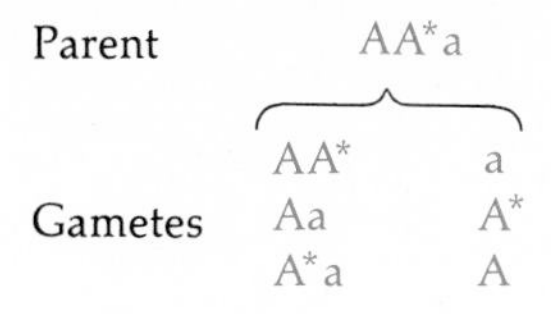

Figure 17.26
The six kinds of gametes produced by a trisomic individual. The two chromosomes carrying the dominant allele are distinguished by the asterisk added to one of them. Only one of the six gametes (a) will produce a homozygous recessive individual when joined with a gamete carrying the recessive allele *a*. Thus, the cross *AAa* × *aa* produces dominant and recessive phenotypes in the ratio 5:1.

offspring will be *AAa* trisomics. When these are crossed to the normal homozygous recessive, only one-sixth of the F_2 progeny will exhibit the recessive phenotype (Figure 17.26). However, if the gene is not in the tripled chromosome, the genotype of the trisomic will be *AA* at the locus; the F_1 will then be *Aa*, and the testcross will produce *Aa* and *aa* individuals in equal numbers.

Human Chromosomal Abnormalities

Suitable techniques for examining mammalian chromosomes were developed in the 1950s. These techniques spread the chromosomes of dividing cells in a way that permits the observation of each chromosome separately. Using these techniques, Joe-Hin Tjio and Albert Levan established that humans have 46 chromosomes; previously the number was thought to be 48. Shortly thereafter, in 1959, Jérôme Lejeune and Raymond Turpin demonstrated the first case of aneuploidy in humans: Down's syndrome patients were shown to be trisomics for chromosome 21.

Down's syndrome (mongolism) is a human congenital condition first described in the late nineteenth century. The syndrome is characterized by severe mental retardation, palmprint abnormalities, and typical facial features. Down's patients live to an average of 16 years; some live well into adulthood, but only rarely do they produce offspring. Most cases of Down's syndrome are due to the presence of 47 chromosomes, chromosome 21 being represented three times rather than only twice. A small proportion of the cases, however, have the regular 46 chromosomes. These individuals actually *do* carry three copies of chromosome 21, but one of them is fused with another chromosome (the long arm of chromosome 14; see Figure 17.27). The fusion of chromosomes 14 and 21 is inherited from one of the parents, who has only 45 chromosomes and is a *carrier*: the parent does not suffer the condition but passes the 14-21 fused chromosome to some of the children.

Down's syndrome is one of the most common severe diseases, appearing about once in every 700 live births. The frequency at conception is estimated to be about 7.3 per 1000, or about five times greater than at birth, but about four-fifths are lost by spontaneous abortion. The incidence of Down's syndrome increases with the mother's age, being about 40 times greater in the children of women in their forties than in the children of women in their early twenties (Figure 17.28). (This correlation is for the trisomic Down's syndrome, not the rarer cases involving the 14-21 chromosome fusion, which are not affected by the mother's age.) Other factors, such as the age of the father or the number of previous children, apparently have no effect on the incidence. Trisomy results from the union of a normal gamete with one having two homologous chromosomes as a

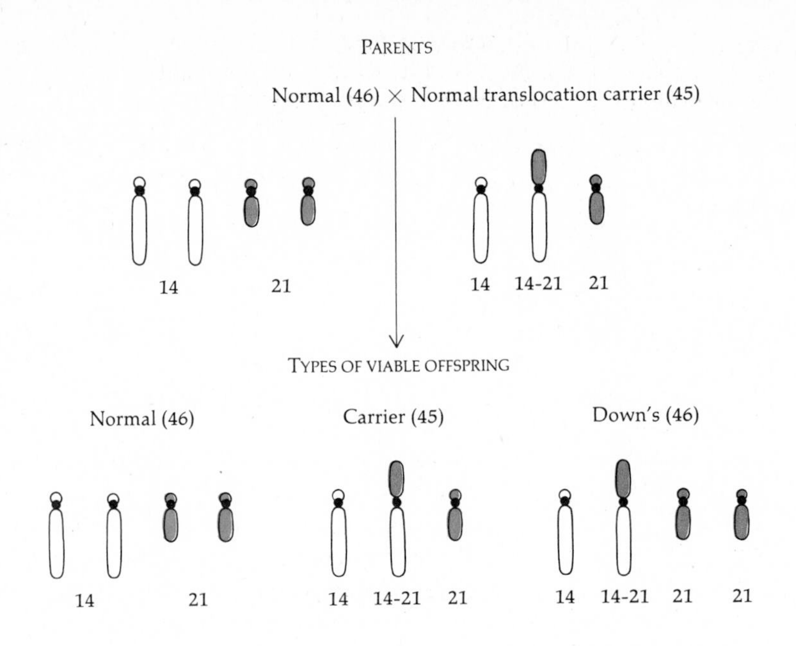

Figure 17.27 Down's syndrome associated with a 14-21 translocation. A normal carrier, having 45 chromosomes, can produce normal, carrier, and Down's syndrome children.

result of nondisjunction at meiosis. The increased incidence of Down's syndrome in children of older women is presumably due to an increase in the frequency of nondisjunction at meiosis.

Two other human infirmities due to autosomal aneuploidy are the Patau and Edwards syndromes. Patau's syndrome is due to trisomy for chromosome 13, has an incidence of about 1 per 5000 live births, and is characterized by harelip and cleft palate, as well as serious ocular, cerebral, and cardiovascular defects. Patau's patients usually die during the first three months of life, although a few have lived up to five years. Edwards' syndrome is due to trisomy for chromosome 18 and includes malformations in virtually every organ system. The incidence of Edwards' syndrome is not well known, but may be about 1 per 10,000 newborns. Patients live about six months on the average, but some have lived into their teen years.

Sex-chromosome aneuploidies also occur in humans. The Turner syndrome is due to the presence of one X and no Y chromosome (the only viable type of monosomy known in humans). Turner's syndrome patients are sterile—females have virtually no ovaries and only limited development of the secondary sexual characteristics. Other characteristics include short stature, abnormal jaws, webbed neck, and a shieldlike chest, but there is usually no mental deficiency. The Turner syndrome has an incidence of about 1 per 5000 live births. More common are *metafemales,* with a frequency of about 1 per 700, most of whom have three, but others more, X chromosomes. XXX females have underdeveloped genital organs, limited fertility, and as a rule are mentally retarded.

The Klinefelter syndrome has a frequency of 1 per 500 individuals and is usually due to XXY trisomy, although other karyotypes (XXYY,

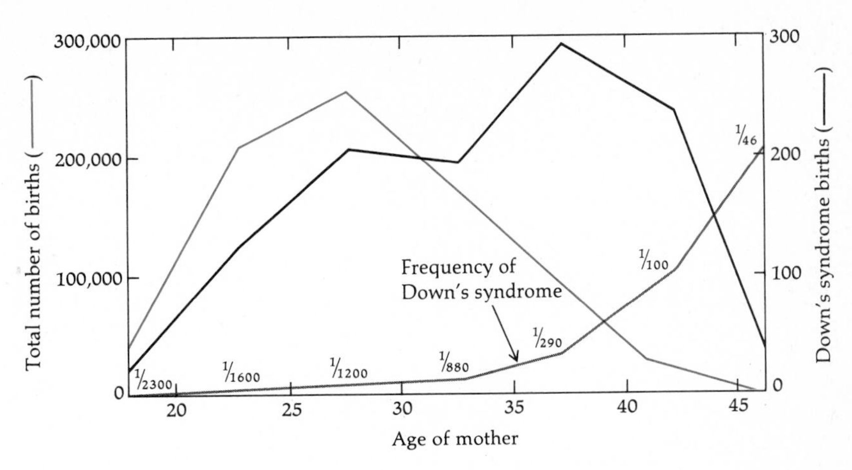

Figure 17.28
Correlation between the age of the mother and the incidence of Down's syndrome. The colored line plots the total number of births, the solid black line plots the number of babies born with Down's syndrome, and the gray line gives the incidence of Down's syndrome births per total number of births.

XXXY, XXXXY, and XXXXXY) are associated with the syndrome. Klinefelter's patients are sterile males with some tendency toward femaleness; they have underdeveloped testes and prostate, scanty body hair, and enlarged breasts. A few XXY individuals are mentally retarded, although most have IQs within the normal range. Those with higher numbers of sex chromosomes have a greater incidence of mental retardation. A summary of these aneuploid abnormalities is given in Table 17.2.

An interesting human aneuploidy is the XYY trisomy. These individuals are for the most part normal males, although they are somewhat

Box 17.3 Amniocentesis

Amniocentesis is a technique for the prenatal diagnosis of congenital abnormalities. A sample of 10–15 ml of the amniotic fluid surrounding the fetus is withdrawn with a surgical syringe. The fluid contains cells of fetal origin that can be cultured in a laboratory (Figure 17.29). The number of chromosomes in these cells can be examined, and the supernatant can also be subjected to a variety of biochemical tests that make possible the detection of metabolic defects. The sample is usually taken between the 14th and 16th weeks of pregnancy. If a severe abnormality is detected, consideration can be given to termination of the pregnancy by therapeutic abortion.

Amniocentesis is particularly advisable in high-risk pregnancies. Among these are pregnancies in which both parents are known carriers of a recessive gene responsible for a serious hereditary disease; those in which the mother is over 35 years of age and is hence a likely candidate for having a Down's syndrome child; and those in which one of the parents is a chromosomal translocation carrier. In one study involving 155 high-risk pregnancies, ten Down's syndrome trisomics were detected. In all ten cases the parents chose therapeutic abortion.

Table 17.2
Aneuploid abnormalities in the human population.

Chromosomes	Syndrome	Frequency at Birth
Autosomes		
Trisomic 21	Down	1/700
Trisomic 13	Patau	1/5000
Trisomic 18	Edwards	1/10,000
Sex Chromosomes—Females		
X0, monosomic	Turner	1/5000
XXX, trisomic XXXX, tetrasomic XXXXX, pentasomic	Metafemale	1/700
Sex Chromosomes—Males		
XYY, trisomic	Normal	1/1000
XXY, trisomic XXYY, tetrasomic XXXY, tetrasomic XXXXY, pentasomic XXXXXY, hexasomic	Klinefelter	1/500

taller than average. In the general population, the incidence of XYY males is about 1 per 1000; in prisons it is about 20 times greater, or 2 per 100. This suggests a greater tendency toward criminality in XYY males than in normal males, but it should be noted that only a small proportion of all XYY individuals are in penal or mental institutions. The proportion of the human population as a whole in such institutions is low, perhaps about 2 per 1000; if 2% of them are XYY, then (2/1000)(2/100), or about 4 per 100,000, are XYY institutionalized individuals. Since the incidence of XYY individuals in the general population is about 1 per 1000, it follows that (4/100,000)/(1/1000), or about 4%, of all XYY individuals are institutionalized, compared to 0.2% of the human population as a whole.

Like other aneuploidies, aneuploidy for the sex chromosomes is originally due to abnormal chromosome segregation at meiosis. A gamete having no sex chromosome and a normal gamete having an X chromosome will produce a Turner's female. The Klinefelter (XXY) syndrome may arise from one XX and one Y gamete, or from one XY and one X gamete.

Abnormal chromosome segregation may also occur in mitosis, resulting in the production of *mosaics*—individuals having cells with more than one genotype. Mitotic nondisjunction may occur in the first cell division of the zygotes or later, and this will determine the extent of the mosaicism (Figure 17.30). The most common type of human mosaics for

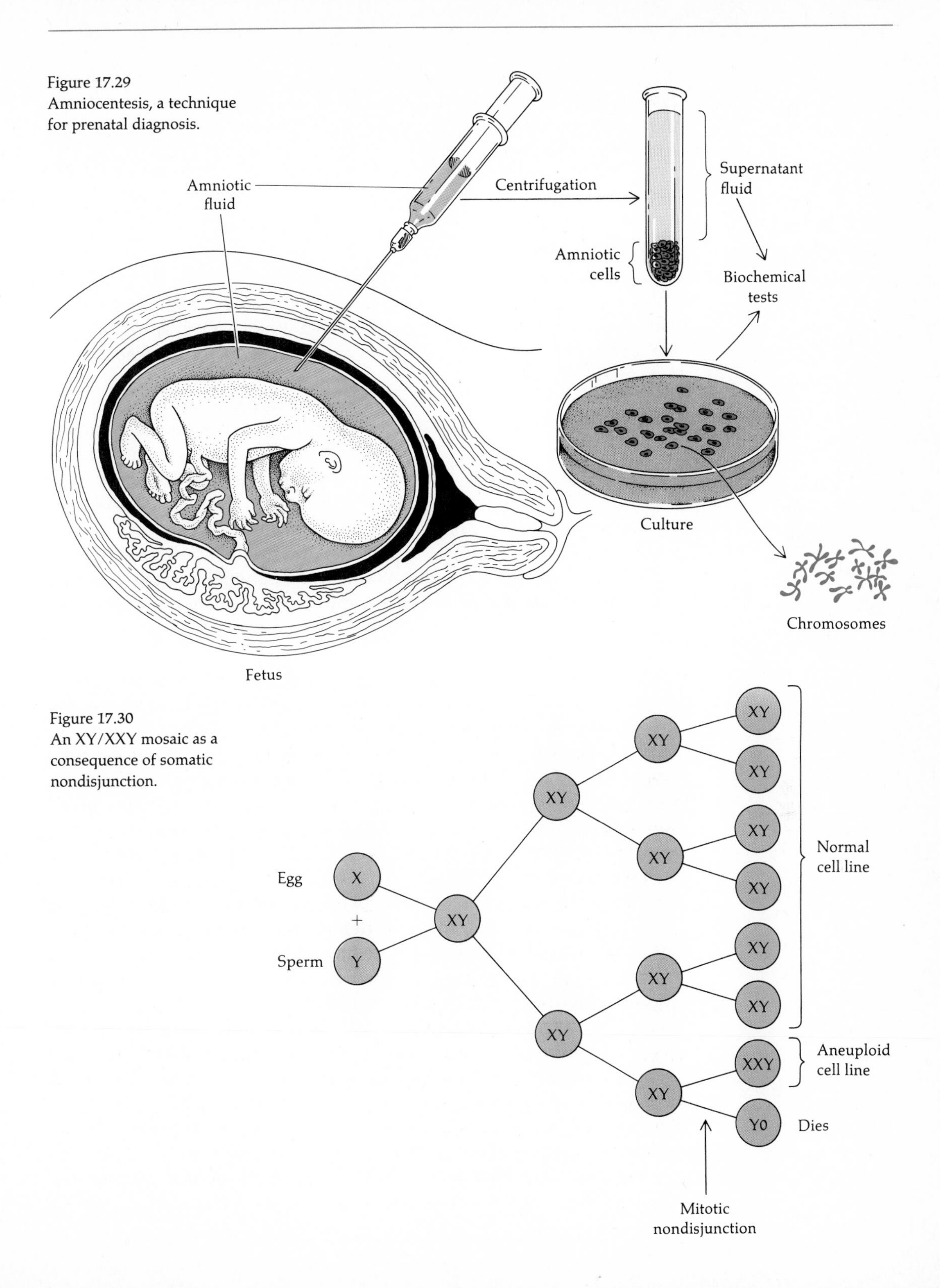

Figure 17.29
Amniocentesis, a technique for prenatal diagnosis.

Figure 17.30
An XY/XXY mosaic as a consequence of somatic nondisjunction.

the sex chromosomes involves XY and XXY cells. The extent of phenotypic abnormality in mosaics depends on the relative proportion of the different types of cells and, hence, on how early in development the somatic nondisjunction occurs.

Polyploidy

Polyploids are organisms with three or more *sets* of chromosomes. If we represent the number of chromosomes in a diploid organism as $2x$, then the number of chromosomes is $3x$ in a triploid, $4x$ in a tetraploid, $5x$ in a pentaploid, etc. (Figure 17.4).

Polyploidy is a rare phenomenon in animals but is relatively common in plants. Natural polyploid species occur in animals particularly among hermaphrodites (those having male and female organs), such as earthworms and planarians, or in forms with *parthenogenetic* females (females producing viable progeny without fertilization), such as some beetles, moths, sow bugs, shrimp, goldfish, and salamanders. Polyploid species occur in all major groups of plants. About 47% of all angiosperms (flowering plants) are polyploids (Table 17.3); polyploidy is also common among ferns, but is rare among gymnosperms (although the redwood, *Sequoia sempervirens*, is a polyploid). Some of the most important cultivated plants are polyploid (Table 17.4).

There are several reasons why polyploidy is rarer in animals than in plants. First, polyploidy disturbs the balance between the autosomes and the sex chromosomes required for sex determination. Second, most animals are cross-fertilized; a single newly arisen polyploid animal cannot reproduce by itself. Third, animals have more complex development, which may be affected, for example, by polyploidy-caused changes in cell size that distort the size of organs. Finally, plant polyploids often arise from chromosome duplication in hybrids, but animal hybrids are usually inviable or sterile.

Polyploidy may arise because irregularities at meiosis may yield unreduced gametes, i.e., gametes with the somatic rather than the gametic chromosome complement. If an unreduced gamete (chromosome number $2x$) from a diploid organism unites with a normal gamete ($1x$), the resulting zygote will be triploid ($3x$). The union of two unreduced gametes ($2x + 2x$) yields a tetraploid ($4x$) zygote.

Polyploids may also arise from the spontaneous doubling of the chromosomes in somatic cells, i.e., chromosome replication without cell division. This may result in tetraploid shoots that yield diploid gametes in their flowers. Self-fertilization will then result in an autotetraploid zygote; fertilization with a normal gamete will produce a triploid.

Polyploidy can be induced experimentally by various treatments that interfere with spindle formation during mitosis. One effective method

Table 17.3
Frequency of polyploid species among flowering plants in various regions of the world.

Region	Latitude (°N)	Percent Polyploids*
Sicily	37	37
Hungary	46–49	47
Denmark	54–58	53
Great Britain	50–61	57
Sweden	55–69	56
Norway	58–71	58
Finland	60–70	57
Iceland	63–66	64
South Greenland	60–71	72

*The frequency of polyploidy increases with latitude. This may be because northern lands have been relatively recently colonized by angiosperms, after becoming free of the continental ice sheets. Polyploids are thought to be better colonizers because they have greater genetic variability than diploids.

uses colchicine, an alkaloid drug obtained from the autumn crocus, *Colchicum autumnale.* After treatment with colchicine, the replicated chromosomes remain together in a single nucleus rather than going to opposite poles and forming two nuclei.

Two kinds of polyploids can be distinguished according to the origin of the chromosome sets. *Autopolyploids* have all their chromosomes derived from the same species. *Allopolyploids* have chromosome sets derived from different species. A diploid gamete (unreduced at meiosis or produced from tetraploid tissue) joined with a monoploid gamete from the same species produces an *autotriploid;* joined with a diploid gamete from the same species, it produces an *autotetraploid.* A diploid gamete joined with a monoploid from a different species yields an *allotriploid;* joined with a diploid gamete from a different species, it produces an *allotetraploid,* and so on. Allotetraploids may also arise from chromosome doubling caused by abnormal mitosis in a hybrid plant. The tissue and flower derived from such a cell will be allotetraploid; self-fertilization will yield an allotetraploid zygote.

Table 17.4
Examples of cultivated polyploid plants.

Plant	Ploidy	Somatic Chromosome Number	Gametic Chromosome Number
Banana	Triploid	27 (3 × 9)	Variable
Potato	Tetraploid	48 (4 × 12)	24
Bread wheat	Hexaploid	42 (6 × 7)	21
Boysenberry	Heptaploid	49 (7 × 7)	Variable
Strawberry	Octoploid	56 (8 × 7)	28

The presence of more than two homologous chromosomes in autopolyploids usually leads to the formation of gametes with incomplete sets of chromosomes. For example, in an autotetraploid there are four homologous chromosomes that may pair among themselves, forming a quadrivalent (all four chromosomes synapsed together). Unequal segregation may ensue, with three chromosomes going to one pole and one chromosome to the other pole, yielding inviable gametes. The production of gametes with incomplete sets of chromosomes makes autopolyploids partially sterile.

In allotetraploids, the different sets of chromosomes may be sufficiently different that only (or mostly) bivalents form, because only the homologous chromosomes of the same species pair with each other. Hence, allotetraploids are more often fertile than autotetraploids. This may be one reason why most polyploid species seem to be allopolyploids.

Polyploids with odd numbers of chromosome sets (triploids, pentaploids, etc.) are completely sterile, or nearly so, because they produce gametes with incomplete sets of chromosomes. This may be advantageous for people when the plants can be reproduced asexually, because the fruits will be seedless. This is true of cultivated bananas, which are triploid.

Polyploid plants can be produced experimentally by means of colchicine and by other methods. The first experimental polyploid was produced in 1928 by the Russian geneticist G. D. Karpechenko. Radish (*Raphanus sativus*), which has nine chromosome pairs, was crossed with cabbage (*Brassica oleracea*), which also has nine chromosome pairs. The hybrids were almost completely sterile because they had 18 chromosomes, which for the most part failed to pair at meiosis, yielding gametes with variable numbers of chromosomes. Occasionally, however, viable pollen and ovules were produced having all 18 chromosomes. Union of two of these gametes produced fully fertile F_2 plants with 36 chromosomes. The resulting allotetraploid plant is known as radocabbage, *Raphanobrassica*. Unfortunately for those who would eat it, radocabbage has foliage more like radish than like cabbage, and roots more like cabbage than like radish.

Problems

1. A plant homozygous for a dominant allele, *AA*, is treated with X rays and crossed to a homozygous recessive plant, *aa*. Three of the 400 F_1 plants show the recessive phenotype. How would you explain this observation?

2. An individual is heterozygous for the chromosome sequences *A•BCDEFG* and *A•BCFEDG* (the dot between *A* and *B* represents the centromere). What are the crossover products if a crossover takes place (a) between *B* and *C*, and (b) between *E* and *F*?

3. Diagram the pairing between two homologous chromosomes having the gene sequences *ABCDEFG•HI* and *ABFEDCG•HI*.

4. Diagram a double crossover and give the meiotic chromosome products in an individual heterozygous for (a) a paracentric inversion and (b) a pericentric inversion, assuming that both crossovers occur within the inverted segment.

5. The following gene sequences are found in the third chromosome in a natural population of *Drosophila*: *ABCDEFGHIJ*, *ABCGFEDHIJ*, *ABCGFIHDEJ*, *ABHIFGCDEJ*, *AFGCDEBHIJ*, and *AFGCBEDHIJ*. What are the likely phylogenetic relationships between these chromosomal arrangements?

6. In *Neurospora* the genes *biscuit* and *granular*, both of which affect growth, are normally located on the same arm of chromosome V. It is found in a certain stock, however, that these two genes assort independently. Which kind of chromosomal mutation would you suggest to explain this observation?

7. Plants heterozygous for a reciprocal translocation are often semisterile, but semisterility may also be due to heterozygosity for a recessive lethal allele. What methods would you use to distinguish between these two possibilities in a semisterile plant?

8. A semisterile plant heterozygous for a reciprocal translocation and for the alleles *P* (purple plant color) and *p* (green plant color), when crossed with a

fully fertile plant homozygous for p and having the standard chromosomal arrangement, produced 402 progeny plants as follows:

Semisterile, purple:	141
Fully fertile, green:	137
Fully fertile, purple:	69
Semisterile, green:	55

Determine the genetic distance between the locus P and the point of interchange of the translocation.

9. Assume that a *Datura* plant is trisomic for chromosome C and that association and segregation between the chromosomes occur at random (i.e., the gametic combinations CC and C are formed with equal probability). What kinds of progeny, and in what proportions, will be produced when such a plant is fertilized with pollen from a normal diploid plant?

10. Corn has 10 pairs of chromosomes, and all 10 different kinds of trisomics are available. The 10 kinds of trisomics are homozygotes for a certain dominant allele, A, but a normal diploid stock homozygous for the recessive allele, aa, is available. How would you ascertain on which chromosome the gene is located?

11. In culture media it is possible to obtain hybrid cells between human and mouse cells. The hybrid cells initially possess all the chromosomes of both species, but the human chromosomes are gradually lost as the cells replicate, so that cells may have only one or none of the homologues for a given human chromosome. The human and mouse forms of many enzymes can be distinguished by electrophoresis. This has made possible the discovery of the chromosomes in which enzyme genes are located, using human–mouse hybrid cells. What kinds of observations are necessary for such mapping?

12. Crosses between polyploid plants differing at one or more loci may give phenotypic segregation ratios different from those observed in diploids. Assume that a certain tetraploid forms only bivalents and that all possible combinations between pairs of the four homologous chromosomes form with equal probability. What kinds of progeny will be produced in the F_2 of the cross $AAAA \times aaaa$, and in what proportions are they expected?

18

Genetic Structure of Populations

Population Genetics

Genetics in general concerns the genetic constitution of organisms and the laws governing the transmission of this hereditary information from one generation to the next. *Population genetics* is that branch of genetics concerned with heredity in groups of individuals, i.e., in populations. Population geneticists study the genetic constitution of populations and how this genetic constitution changes from generation to generation.

Hereditary changes through the generations underlie the evolutionary process. Hence, population genetics may also be considered as *evolutionary genetics.* However, these two concepts can be distinguished. It is often understood that population genetics deals with populations of a given species, while evolutionary genetics deals with heredity in any populations, whether of the same or of different species. By these definitions, evolutionary genetics is a broader subject than population genetics—it includes population genetics as one of its parts. Evolutionary genetics is the subject of the rest of this book.

Populations and Gene Pools

The most obvious unit of living matter is the individual organism. In unicellular organisms, each cell is an individual; multicellular organisms consist of many interdependent cells, many of which die and are replaced by other cells throughout the life of an individual. In evolution, the relevant unit is not an individual but a population. A *population* is a community of individuals linked by bonds of mating and parenthood; in other words, a population is a community of individuals of the same species. The bonds of parenthood that link members of the same population are always present, but mating is absent in organisms that reproduce asexually. A *Mendelian population* is a community of interbreeding, sexually reproducing individuals, i.e., Mendelian populations are those in which reproduction involves mating.

The reason why the individual is not the relevant unit in evolution is that the genotype of an individual remains unchanged throughout its life; moreover, the individual is ephemeral (even though some organisms, such as conifer trees, may live up to several thousand years). A population, on the other hand, has continuity from generation to generation; moreover, the genetic constitution of a population may change—evolve—over the generations. The continuity of a population through time is provided by the mechanism of biological heredity.

The most inclusive Mendelian population is the *species* (Chapter 22). As a rule, the genetic discontinuities between species are absolute; sexually reproducing organisms of different species are kept from interbreeding by reproductive isolating mechanisms. Species are independent evolutionary units: genetic changes taking place in a local population can be extended to all members of the species, but are not ordinarily transmitted to members of a different species.

The individuals of a species are not usually homogeneously distributed in space; rather, they exist in more or less well-defined clusters, or local populations. A *local population* is a group of individuals of the same species living together in the same territory. The concept of local population may seem clear, but its application in practice entails difficulties because the boundaries between local populations are often fuzzy (Figure 18.1). Moreover, the organisms are not homogeneously distributed within a cluster, even when the clusters are quite discrete, as is true of organisms living in lakes or on islands; the lakes or islands may be sharply distinct, but individuals are not evenly distributed within a lake or on an island. Animals often migrate from one local population to another, and the

Figure 18.1 (*opposite*)
Geographic distribution of *Lacerta agilis*. **(a)** This lizard exists over a broad area encompassing large parts of Europe and western Asia, but its distribution is far from homogeneous. **(b)** *Lacerta agilis* has greater density along streams and rivers than in the intermediate areas. **(c)** The lizards occur in small family groups consisting of a few individuals each. Demes consist of about 20 to 40 family groups. Local populations consist of several demes each. Within a deme, matings occur rather freely between members of different family groups; however, fewer than 4% of all matings occur between individuals from different demes within the same local population. Fewer than 0.01% of all matings involve individuals from different local populations. (Data courtesy of Prof. Alexis B. Yablokov, Institute of Developmental Biology, USSR Academy of Sciences, Moscow.)

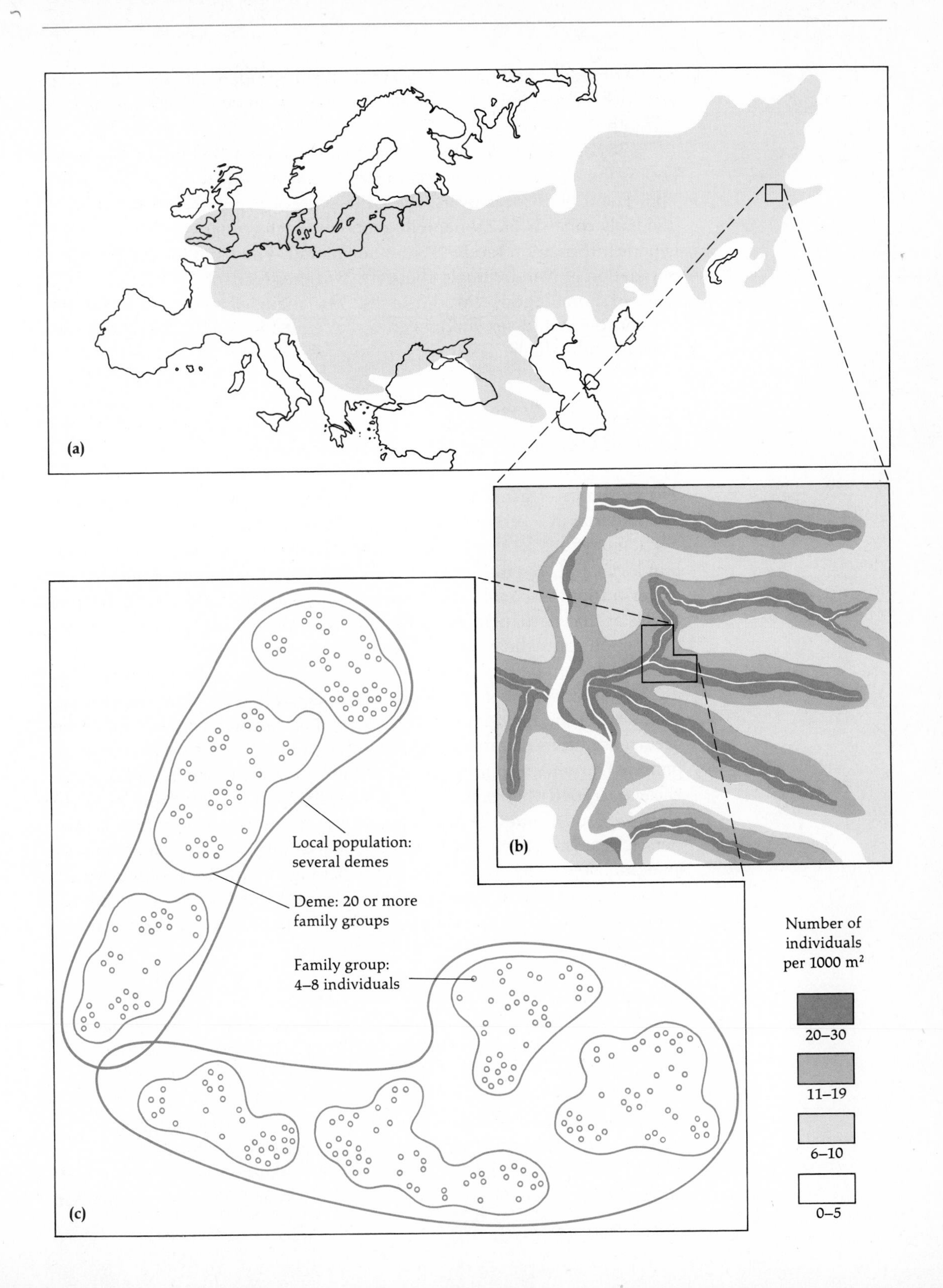
(a)
(b)
(c)
Local population: several demes
Deme: 20 or more family groups
Family group: 4–8 individuals
Number of individuals per 1000 m²
20–30
11–19
6–10
0–5

pollen or seeds of plants may also move from population to population, all of which makes local populations far from completely independent of each other.

The concept of a *gene pool* is useful in the study of evolution. The gene pool is the aggregate of the genotypes of all the individuals in a population. For diploid organisms, the gene pool of a population with N individuals consists of $2N$ haploid genomes. Each *genome* consists of all the genetic information received from one parent. Thus, in the gene pool of a population of N individuals, there are $2N$ genes for each gene locus, and N pairs of homologous chromosomes. The main exceptions are the sex chromosomes and sex-linked genes that exist in a single dose in heterogametic individuals.

Genetic Variation and Evolution

The existence of genetic variation is a necessary condition for evolution. Assume that at a certain gene locus all individuals of a given population are homozygous for exactly the same allele. Evolution cannot take place at that locus, because the allelic frequencies cannot change from generation to generation. Assume now that in a different population there are two alleles at that particular locus. Evolutionary change *can* take place in this population: one allele may increase in frequency at the expense of the other allele.

The modern theory of evolution derives from Charles Darwin (1809–1882) and his classic, *On the Origin of Species*, published in 1859. The occurrence of hereditary variation in natural populations was the starting point of Darwin's argument for evolution by a process of natural selection. Darwin argued that some natural hereditary variations may be more advantageous than others for the survival and reproduction of their carriers. Organisms having advantageous variations are more likely to survive and reproduce than organisms lacking them. As a consequence, useful variations will become more prevalent through the generations, while harmful or less useful ones will be eliminated. This is the process of *natural selection*, which plays a leading role in evolution.

A direct correlation between the amount of genetic variation in a population and the rate of evolutionary change by natural selection was demonstrated mathematically with respect to *fitness* by Sir Ronald A. Fisher in his Fundamental Theorem of Natural Selection (1930): *The rate of increase in fitness of a population at any time is equal to its genetic variance in fitness at that time.* (Fitness, in the technical sense used in the theorem, is a measure of relative reproductive rate; see Chapter 20, page 658).

The Fundamental Theorem applies strictly to allelic variation at a single gene locus, and only under particular environmental conditions. But the correlation between genetic variation and the opportunity for

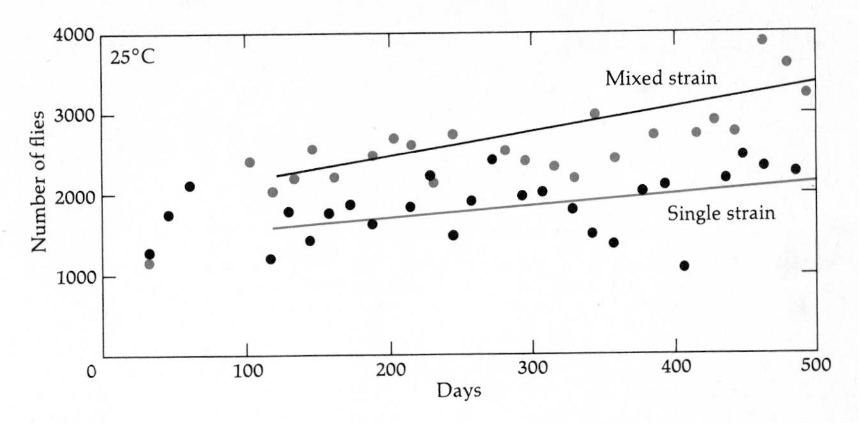

Figure 18.2
Correlation between amount of genetic variation and rate of evolution in laboratory populations of *Drosophila serrata* exposed to a new environment. The graph shows the change in number of flies during approximately 25 generations. The mixed-strain population initially had greater genetic variation than the single-strain population. Both populations increased in numbers throughout the experimental period, but the average rate of increase was substantially greater in the mixed-strain population than in the single-strain population. Increases in the number of flies over the generations reflect the increasing adaptation of the population to the experimental environment, which is promoted by evolution.

evolution is intuitively obvious. The greater the number of variable loci and the more alleles there are at each variable locus, the greater the possibility for change in the frequency of some alleles at the expense of others. This requires, of course, that there be selection favoring the change of some trait(s) and that the variation be relevant for the trait(s) being selected. Figure 18.2 and Table 18.1 give an experimental demonstration of the correlation between the amount of genetic variation and the rate of evolution when these conditions are met.

Genotypic and Genic Frequencies

We observe phenotypes directly, but not genotypes or genes. Variation in gene pools is expressed in terms of either genotype frequencies or gene frequencies. If we know the relationship between specific genotypes and the corresponding phenotypes, we are able to transform phenotypic frequencies into genotypic frequencies. Let us consider the M-N blood groups. There are three blood groups, M, N, and MN, which are determined by two alleles, L^M and L^N, at a single locus, according to the simple relationship shown in Table 18.2.

Table 18.1
Correlation between the amount of genetic variation and the rate of evolution in experimental populations of *Drosophila serrata* from Popondetta, New Guinea, and Sydney, Australia. The rate of evolution was measured by the rate at which the number of flies increased in the populations over about 25 generations. Figure 18.2 shows the data for the experiment at 25°C.

Population	Mean Number of Flies in Population	Mean Increase in Number of Flies per Generation
Experiment at 25°C		
Single strain (Popondetta)	1862 ± 79	31.5 ± 13.8
Mixed strain (Popondetta × Sydney)	2750 ± 112	58.5 ± 17.4
Experiment at 19°C		
Single strain (Popondetta)	1724 ± 58	25.2 ± 9.9
Mixed strain (Popondetta × Sydney)	2677 ± 102	61.2 ± 13.8

After F. J. Ayala, *Science 150*:903 (1965).

Examination of 730 Australian Aborigines produced the following results: 22 had blood group M, 216 had MN, and 492 had N. The frequencies of the blood groups and the corresponding genotypes are obtained by dividing the number of each kind observed by the total. For example, the frequency of blood group M is 22/730 = 0.030.

We can describe the variation at the M-N gene locus in this group of people by giving the frequencies of the three genotypes. If we assume that the 730 individuals are a *random sample* of Australian Aborigines, we may take the observed frequencies as characteristic of Australian Aborigines in general. A random sample is a representative, or unbiased, sample of a population.

It is convenient for some purposes to describe genetic variation at a locus using not the genotypic frequencies but the allelic frequencies. Allelic frequencies can be calculated either from the genotypic numbers observed or from the genotypic frequencies.

In order to calculate allelic frequencies directly from the genotype *numbers,* we simply count the number of times each allele is found and divide it by the total number of alleles in the sample. An $L^M L^M$ individual contains two L^M alleles; an $L^M L^N$ individual contains one L^M allele. Therefore, the number of L^M alleles in the sample described above is (2 × 22) + 216 = 260. The total number of alleles in the sample is twice the number of individuals, because each individual has two alleles: 2 × 730 = 1460. The frequency of the L^M allele in the sample is therefore 260/1460 = 0.178. Similarly, the frequency of the L^N allele is [(2 × 492) + 216]/1460 = 0.822.

Table 18.2
M-N blood group and genotypic frequencies in a population of Australian Aborigines.

Blood Group	Genotype	Number	Frequency
M	$L^M L^M$	22	0.030
MN	$L^M L^N$	216	0.296
N	$L^N L^N$	492	0.674
Total:		730	1.000

Allelic frequencies can also be calculated from the genotypic *frequencies*, by observing as before that all alleles in homozygotes are of a given kind, whereas only half the alleles of a heterozygote are of a given kind. Thus the frequency of an allele is the frequency of individuals homozygous for that allele plus half the frequency of heterozygotes for that allele. Among the Australian Aborigines, the frequency of L^M is 0.030 + ½(0.296) = 0.178; similarly, the frequency of L^N is 0.674 + ½(0.296) = 0.822. Table 18.3 gives genotypic and allelic frequencies for the M-N gene locus in three human populations. It is apparent that human populations are quite different from one another with respect to this locus.

The calculation of gene frequencies when the number of alleles at a locus is greater than two is based on the same rules that apply for two alleles: homozygotes carry two copies of one allele, heterozygotes carry one each of two alleles. For example, in a certain natural population of *Drosophila willistoni*, six different genotypes were found at the *Lap-5* locus

Table 18.3
Genotypic and allelic frequencies for the M-N gene locus in three human populations.

Population	Number Having Blood Group:			Total	Genotypic Frequency			Allelic Frequency	
	M	MN	N		$L^M L^M$	$L^M L^N$	$L^N L^N$	L^M	L^N
Australian Aborigines	22	216	492	730	0.030	0.296	0.674	0.178	0.822
Navaho Indians	305	52	4	361	0.845	0.144	0.011	0.917	0.083
U.S. Caucasians	1787	3039	1303	6129	0.292	0.496	0.213	0.539	0.461

Table 18.4
Genotypic frequencies observed at the *Lap-5* locus in a population of *Drosophila willistoni.*

Genotype	Number	Frequency
98/98	2	0.004
100/100	172	0.344
103/103	54	0.108
98/100	38	0.076
98/103	20	0.040
100/103	214	0.428
Total:	500	1.000

in the numbers shown in Table 18.4. (The *Lap-5* gene codes for a leucine aminopeptidase enzyme; each allele is identified by a number that refers to the mobility of the corresponding polypeptide under electrophoresis—see Box 18.1, page 612).

Genotypic frequencies are obtained by dividing the number of times each genotype is observed by the total number of genotypes. Thus the frequency of the *98/98* genotype is 2/500 = 0.004. The frequency of a given allele can be obtained from the genotypic frequencies by adding the frequency of the homozygotes for that allele and half the frequency of each of the heterozygotes for that allele. Thus the frequency of the allele *98* is the frequency of the homozygote *98/98* plus half the frequencies of the heterozygotes *98/100* and *98/103*, or $0.004 + \frac{1}{2}(0.076) + \frac{1}{2}(0.040) = 0.062$. Similarly, the frequencies of alleles *100* and *103* are calculated to be 0.596 and 0.342, respectively. The sum of these three frequencies is, of course, 1.000.

The allele frequencies can also be calculated by counting the number of times each allele appears and dividing it by the total number of alleles in the sample. Allele *98* appears twice in *98/98* homozygotes and once each in *98/100* and *98/103* heterozygotes, or $(2 \times 2) + 38 + 20 = 62$ times; because the number of alleles in the sample is $2 \times 500 = 1000$, the frequency of allele *98* is 0.062. The number of times each allele appears in the sample and the allelic frequencies for the data in Table 18.4 are shown in Table 18.5.

One reason why it is often preferable to describe genetic variation at a locus using allelic frequencies rather than genotypic frequencies is

Table 18.5
Allelic frequencies observed at the *Lap-5* locus in a population of *D. willistoni.*

Allele	Number	Frequency
98	62	0.062
100	596	0.596
103	342	0.342
Total:	1000	1.000

because usually there are fewer alleles than genotypes. With two alleles, the number of possible genotypes is three; with three alleles, it is six; with four alleles, it is ten. In general, if the number of different alleles is k, the number of different possible genotypes is $k(k + 1)/2$.

Two Models of Population Structure

Two conflicting hypotheses were advanced during the nineteen forties and fifties concerning the genetic structure of populations. The *classical* model argues that there is very little genetic variation, the *balance* model, that there is a great deal (Figure 18.3).

According to the classical model, the gene pool of a population consists, at the great majority of loci, of a wild-type allele with a frequency very close to 1, plus a few deleterious alleles arisen by mutation but kept at very low frequencies by natural selection. A typical individual would be homozygous for the wild-type allele at nearly every locus, but at a few loci it would be heterozygous for the wild allele and a mutant. The "normal," ideal genotype would be an individual homozygous for the wild-type allele at every locus. Evolution would occur because occasionally a beneficial allele arises by mutation. The beneficial mutant would gradually increase in frequency by natural selection and become the new wild-type allele, with the former wild-type allele being eliminated or reduced to a very low frequency.

According to the balance model, there is often no single wild-type allele. Rather, at many—perhaps most—loci, the gene pool consists of an array of alleles with various frequencies. Hence, individuals are heterozygous at a large proportion of these loci. There is no single "normal" or ideal genotype; instead, populations consist of arrays of genotypes that

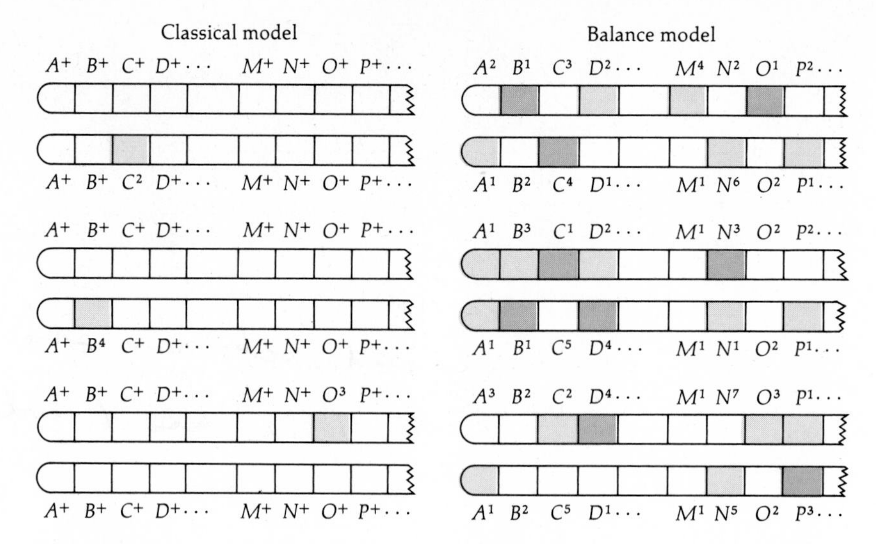

Figure 18.3
Two models of the genetic structure of populations. The hypothetical genotypes of three typical individuals are shown according to each model. Capital letters symbolize gene loci, and each number represents a different allele; the wild-type allele postulated by the classical model is represented by a + sign. According to the classical model, individuals are homozygous for the wild-type allele at nearly every locus, although they may be heterozygous for the wild allele and a mutant allele at an occasional locus (*C* in the first individual, *B* in the second, *O* in the third). According to the balance model, individuals are heterozygous at many gene loci.

differ from one another at many loci but are satisfactorily adapted to most environments encountered by the population.

The balance model sees evolution as a process of gradual change in the frequencies and kinds of alleles at many loci. Alleles do not act in isolation; rather, the fitness conferred by one allele depends on which other alleles exist in the genotype. The set of alleles present at any one locus is coadapted with the sets of alleles at other loci; hence, allelic changes at one locus are accompanied by allelic changes at other loci. However, like the classical model, the balance model accepts that many mutants are unconditionally harmful to their carriers; these deleterious alleles are eliminated or kept at low frequencies by natural selection, but play only a secondary, negative role in evolution.

Looking at Variation

It is now known that natural populations possess a great deal of genetic variation. Definitive evidence for this was not obtained until the late 1960s,

Figure 18.4
Variations in facial features, skin pigmentation, height, and other traits are apparent in human populations. (Owen Franken/Stock, Boston.)

however. It was known before that time that variation was a common phenomenon in nature. Whether the evidence suggested that allelic variation existed at many loci or only at a few was appraised differently by the balance school and the classical school. In any case, the evidence did not allow one to tell what proportion of the gene loci of an organism consisted of several alleles. The kind of evidence available before the 1960s was as follows.

Individual variation is a conspicuous phenomenon whenever organisms of the same species are carefully examined. Human populations, for example, exhibit variation in facial features, skin pigmentation, hair color and shape, body configuration, height and weight, blood groups, etc. (Figure 18.4). We notice human differences more readily than variation in other organisms, but morphological variation has been carefully recorded in many cases, e.g., with respect to color and pattern in snails, butterflies, grasshoppers, lady beetles, mice, and birds (Figure 18.5). Plants often differ in flower and seed color and in pattern, as well as in growth habit. A difficulty is that it is not immediately clear how much of this morphological variation is due to genetic variation and how much to environmental effects.

Geneticists have discovered that there is much more genetic variation than is apparent when organisms living in nature are observed. This has been accomplished by inbreeding, i.e., by mating close relatives, which increases the probability of homozygosis; recessive genes thus become expressed. Inbreeding has shown, for example, that virtually every *Drosophila* fly has allelic variants that in homozygous condition result in abnormal phenotypes, and that plants carry many alleles that when homozygous result in abnormal chlorophyll or none at all. Inbreeding has also shown that organisms carry alleles that in homozygous condition affect their fitness, i.e., that modify their fertility and survival probability (Table 18.6).

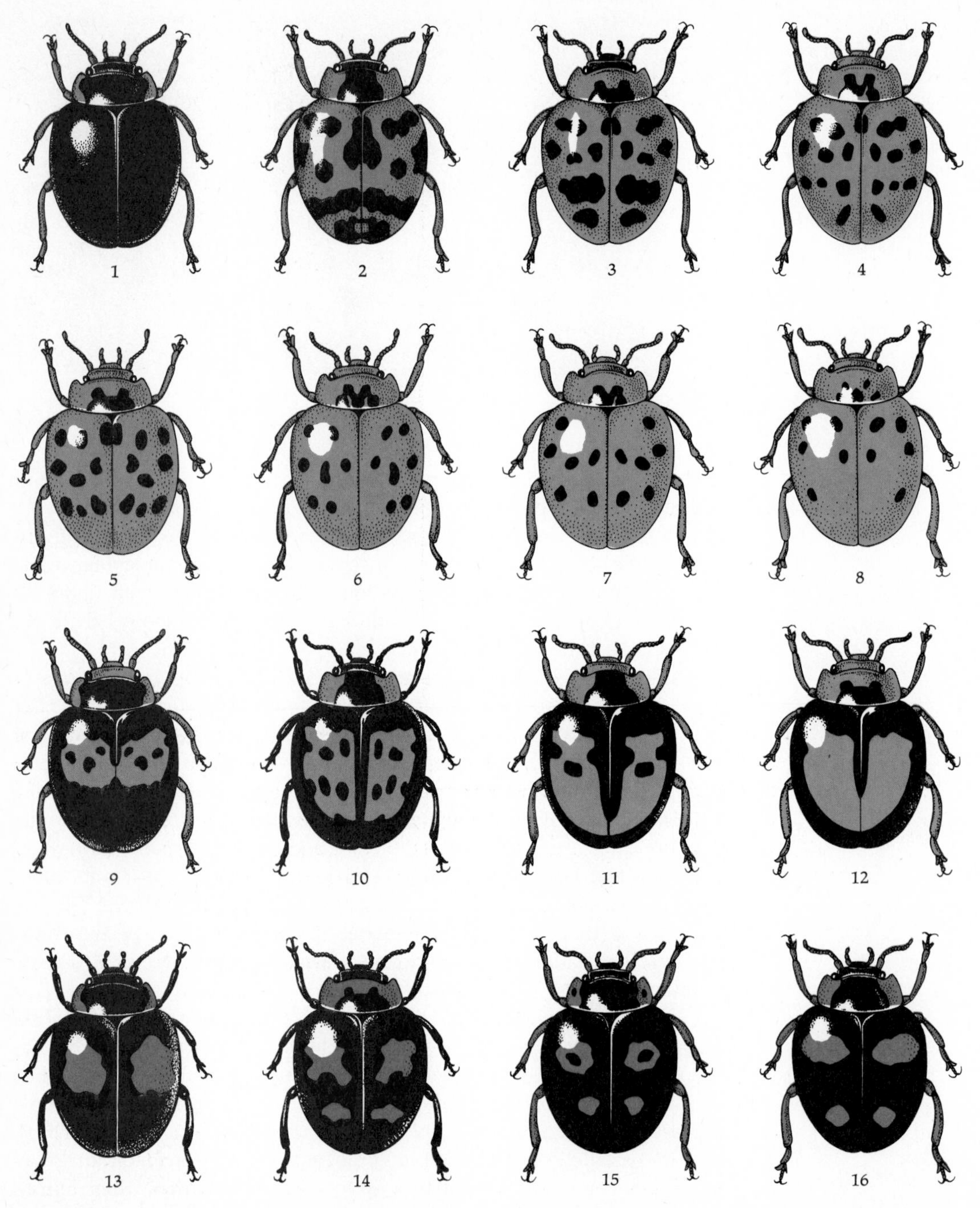

Figure 18.5
Morphological variation within a species is apparent in the color patterns of the wing covers (elytra) of the lady beetle *Harmonia axyridis*. The species occurs in Siberia, China, Korea, and Japan. The almost completely black phenotype (1) predominates in west-central Siberia, but farther eastward populations are more polymorphic, with increasing frequency of the black-spots-on-yellow-background phenotypes (2–8). The phenotypes 9–12 have yellow spots on a black background. Phenotypes 13–16 have red spots on a black background; these are found exclusively in the Far East.

Table 18.6
Frequencies of wild chromosomes of *Drosophila pseudoobscura* from the Sierra Nevada, California, that in homozygous condition have various effects on viability. A majority of the chromosomes have alleles that in homozygous condition reduce viability.

	Frequency (%) of Chromosome:		
Effect on Viability	Second	Third	Fourth
Lethal or semilethal	33.0	25.0	25.9
Subvitality (significantly lower viability than wild flies)	62.6	58.7	51.8
Normal (viability not significantly different from wild flies)	4.3	16.3	22.3
Supervitality (significantly higher viability than wild flies)	<0.1	<0.1	<0.1

After Th. Dobzhansky and B. Spassky, *Genetics 48*:1467 (1963).

A convincing source of evidence indicating that genetic variation is pervasive comes from artificial selection experiments. In artificial selection the individuals chosen to breed the next generation are those that exhibit the greatest expression of the desired characteristic (Chapter 15, page 521). For example, if we want to increase the yield of wheat, we choose in every generation the wheat plants with the greatest yield and use their seed to produce the next generation. If, over the generations, the selected population changes in the direction of the selection, it is clear that the original organisms had genetic variation with respect to the selected trait.

The changes obtained by artificial selection are often impressive. For example, the egg production in a flock of White Leghorn chickens increased from 125.6 eggs per hen per year in 1933 to 249.6 eggs per hen per year in 1965 (Figure 18.6). Artificial selection can also be practiced in opposite directions. Selection for high protein content in a variety of corn increased the protein content from 10.9 to 19.4%, while selection for *low* protein content reduced it from 10.9 to 4.9%. Artificial selection has been successful for innumerable commercially desirable traits in many domesticated species, including cattle, swine, sheep, poultry, corn, rice, and wheat, as well as in many experimental organisms, such as *Drosophila*, in which artificial selection has succeeded for more than 50 different traits. The fact that artificial selection succeeds virtually every time it is tried was taken by proponents of the balance model as showing that genetic variation exists in populations for virtually every characteristic of the organism.

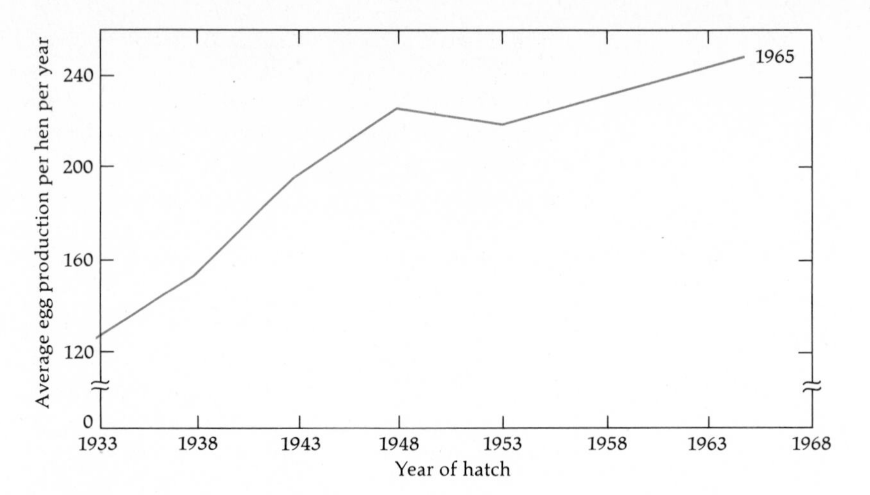

Figure 18.6
An example of artificial selection: egg production per hen per year in a flock of White Leghorn chickens. In the formation stock the average production was 125.6 eggs. Thirty-two years later, selection had increased productivity to 249.6 eggs, double the initial number. The success of selection indicates that the flock had considerable genetic variability with respect to egg production. The economic significance of doubling the number of eggs laid in a year is obvious. (After I. M. Lerner and W. J. Libby, *Heredity, Evolution, and Society*, 2nd ed., W. H. Freeman, San Francisco, 1976.)

The Problem of Measuring Genetic Variation

The evidence mentioned in the previous section may indicate that genetic variation is pervasive in natural populations, and hence that there is ample opportunity for evolutionary change. But in any case we would like to go one step further and find out precisely how much variation there is. For example, what proportion of all gene loci are polymorphic (i.e., variable) in a given population, and what proportion of all gene loci are heterozygous in a typical individual of the population? In trying to solve this problem, we find that the traditional methods of genetic analysis impose a methodological handicap.

Consider what we need to do in order to find out what proportion of the genes are polymorphic in a population. We cannot study every gene locus of an organism, because we do not even know how many loci there are, and because it would be an enormous task. The solution, then, is to look at only a sample of gene loci. If the sample is random, i.e., not biased and thus truly representative of the population, the values observed in the sample can be extrapolated to the whole population. Pollsters do quite well this way; for example, based on a sample of about 2000 individuals, they are able to predict with fair accuracy how millions of Americans will vote in a presidential election.

In order to ascertain how many gene loci are polymorphic in a population, we need to study a few genes that are an unbiased sample of all the gene loci. With the traditional methods of genetics this is impossible, because the existence of a gene is ascertained by examining the progenies of crosses between individuals showing different forms of a given character; from the proportions of individuals in the various classes, we infer whether one or more genes are involved. By such methods, therefore, the only genes known to exist are those that are variable. There is no way of obtaining an unbiased sample of the genome, because invariant genes cannot be included in the sample.

A way of out of this dilemma became possible with the discoveries in molecular genetics. It is now known that the genetic information encoded in the nucleotide sequences of the DNA of a structural gene is translated into a sequence of amino acids making up a polypeptide. We can select for study a series of proteins without previously knowing whether or not they are variable in a population—a series of proteins that, with respect to variation, represents an unbiased sample of all the structural genes in the organism. If a protein is found to be invariant among individuals, it is inferred that the gene coding for that protein is also invariant; if the protein is variable, we know that the gene is variable, and we can measure how variable it is, i.e., how many variant forms of the protein exist, and in what frequencies.

Quantifying Genetic Variation

Biochemists have known since the early 1950s how to obtain the amino acid sequences of proteins. One conceivable way to measure genetic variation in a natural population would therefore be to choose a fair number of proteins, say, 20, without knowing whether or not they were variable in the population, so that they would represent an unbiased sample. Then, each of the 20 proteins could be sequenced in a number of individuals, say, 100 (chosen at random), to find out how much variation, if any, existed for each of the proteins. The average amount of variation per protein found in the 100 individuals for the 20 proteins would be an estimate of the amount of variation in the genome of the population.

Unhappily, obtaining the amino acid sequence of a single protein is so demanding a task that several months, or even years, are usually required to do it. Hence it is hardly practical to sequence 2000 protein specimens in order to estimate the genetic variation in each population we want to study. Fortunately, there is a technique, *gel electrophoresis*, that makes possible the study of protein variation with only a moderate investment of time and money. Since the late 1960s, estimates of genetic variation have

Box 18.1 Gel Electrophoresis

The apparatus and procedures employed in gel electrophoresis for studying genetic variation in natural populations are shown in Figure 18.7. Tissue samples from organisms are individually *homogenized* (ground up) in order to release the enzymes and other proteins from the cells. The homogenate supernatants (liquid fractions) are placed in a gel made of starch, agar, polyacrylamide, or some other jelly-like substance. The gel is then subjected, usually for a few hours, to a direct electric current. Each protein in the gel migrates in a direction and at a rate that depend on the protein's net electric charge and molecular size. After the gel is removed from the electric field, it is treated with a chemical solution containing a substrate that is specific for the enzyme to be assayed, and a salt that reacts with the product of the reaction catalyzed by the enzyme. At the position in the gel to which the specific enzyme has migrated, a reaction takes place that can be written as follows:

$$\text{Substrate} \xrightarrow{\text{Enzyme}} \text{Product} + \text{Salt} \rightarrow \text{Colored spot}$$

The usefulness of the method lies in the fact that the genotype at the gene locus coding for the enzyme can be inferred for each individual in the sample from the number and positions of the spots observed in the gels. Figure 18.8 shows a gel that has been treated to reveal the position of the enzyme phosphoglucomutase; the gel contains the homogenates of 12 *Drosophila* flies. The gene locus coding for this enzyme can be represented as *Pgm*. The first and third individuals in the gel, starting from the left, have enzymes with different electrophoretic mobility, and thus different amino acid sequences; this in turn implies that they are coded by different alleles. Let us represent the alleles coding for the enzymes in the first and third individuals as Pgm^{100} and Pgm^{108}, respectively. (The superscripts indicate that the enzyme coded by allele Pgm^{108} migrates 8 mm farther in the gel than the enzyme coded by Pgm^{100}; this is a common way of representing alleles in electrophoretic studies, although letters—such as *S*, *M*, and *F* for slow, intermediate, and fast, or *a*, *b*, *c*, etc.—are sometimes used.)

Because the first and third individuals in Figure 18.8 each exhibit only one colored spot, we infer that they are homozygotes, with genotypes $Pgm^{100/100}$ and $Pgm^{108/108}$, respectively. The second individual exhibits two colored spots. One of these spots shows the same migration as that of the first individual and is thus coded by allele Pgm^{100}, while the other spot shows the same migration as that of the third individual and is thus coded by allele Pgm^{108}. We conclude that the second individual is heterozygous, with genotype $Pgm^{100/108}$.

Some proteins, such as the enzyme malate dehydrogenase, shown in Figure 18.9, consist of two polypeptides; heterozygotes will then exhibit three colored spots. Let us represent the locus coding for malate dehydrogenase as *Mdh*. The second individual in Figure 18.9 shows only one spot and is thus inferred to be homozygous, with genotype $Mdh^{94/94}$; the first individual is also homozygous, with genotype $Mdh^{104/104}$. A heterozygous individual has two kinds of polypeptides, which we can represent as A and B, coded by alleles Mdh^{94} and Mdh^{104}, respectively. Three associations of two different units are possible, namely, AA, AB, and BB. These correspond to the three colored spots that we see in the fourth individual of Figure 18.9.

There are proteins that consist of four or even more subunits; the electrophoretic patterns of heterozygous individuals will then show five or more colored spots, but the principles used to infer the genotypes from the patterns are similar to those just presented. The patterns shown in Figures 18.8 and 18.9 manifest the existence of two alleles at each locus. An invariant locus will be manifested by a colored spot that is the same for all individuals. On the other hand, more than two alleles are often found, as in Figure 18.10, which shows a pattern of the enzyme acid phosphatase in *Drosophila*. Protein variants controlled by allelic variants at a single gene locus and detectable by electrophoresis are called *allozymes*, or *electromorphs*. Electromorphs with identical migration in a gel may be the products of more than one allele, because (1) synonymous triplets code for the same amino acid and (2) some amino acid substitutions do not change the electrophoretic mobility of the proteins. Therefore, gel electrophoresis underestimates the amount of genetic variation, although at present it is not known by how much.

---

been obtained for natural populations of many organisms using gel electrophoresis (see Box 18.1).

Electrophoretic techniques show what the genotypes of the individuals in a sample are: how many are homozygous, how many are heterozygous, and for what alleles. In order to obtain an estimate of the amount of variation in a population, about 20 or more gene loci are usually studied. It is desirable to summarize the information obtained for all the loci in a simple way that would express the degree of variability of a population and that would permit comparing one population to another. This can be accomplished in a variety of ways, but two measures of genetic variation are commonly used: polymorphism and heterozygosity.

Polymorphism and Heterozygosity

One measure of genetic variation is the *proportion of polymorphic loci*, or simply the *polymorphism* (P), in a population. Assume that, using electrophoretic techniques, we examine 30 gene loci in *Phoronopsis viridis*, a kind of marine worm that lives on the coast of California, and assume that we find no variation whatsoever at 12 loci, but some variation at the other 18 loci. We can say that $18/30 = 0.60$ of the loci are polymorphic in that population, or that the degree of polymorphism in the population is 0.60. Assume that we examine three other populations of *P. viridis* and that the numbers of polymorphic loci, out of the 30 loci studied, are 15, 16, and 14. The degree of polymorphism in these three populations is 0.50, 0.53, and 0.47, respectively. We can then calculate the average polymorphism in the four populations of *P. viridis* as $(0.60 + 0.50 + 0.53 + 0.47)/4 = 0.525$ (Table 18.7).

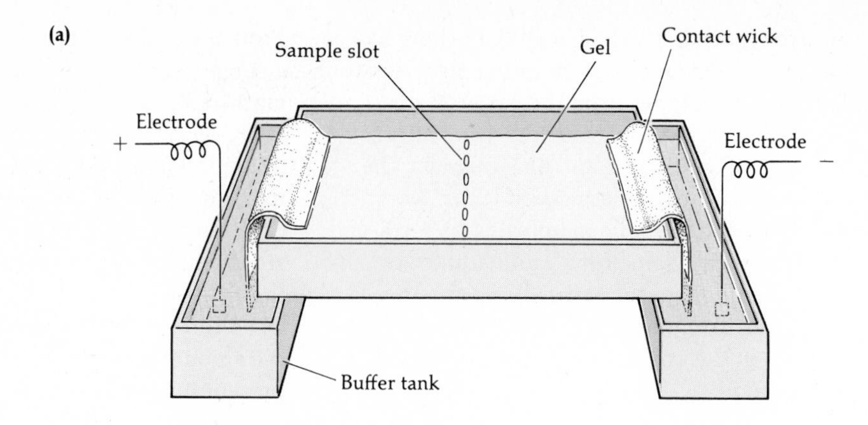

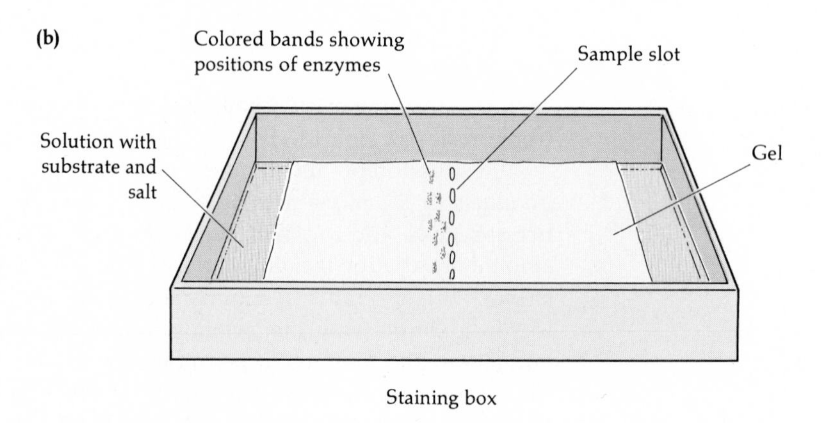

Figure 18.7
The techniques of gel electrophoresis and enzyme assay used to measure genetic variation in natural populations(the techniques are described in somewhat more detail in the text). **(a)**The liquid fractions from homogenized tissue samples are placed in a gel and subjected to a direct electric current. The enzymes and other proteins in the samples migrate to characteristic positions in the gel. **(b)** After the gel is removed from the electric field, it is treated with a specific chemical solution to reveal the positions to which the enzyme being assayed had migrated. The genotype at the gene locus coding for the enzyme can be determined for each individual from the pattern of the spots in the gels.

The amount of polymorphism is a useful measure of variation for certain purposes, but it suffers from two defects: arbitrariness and imprecision.

The number of variable loci observed depends on how many individuals are examined. Assume, for example, that we examined 100 individuals in the first *Phoronopsis* population. If we had examined more individuals, we might have found variation at some of the 12 loci that appeared invariant; if we had examined fewer individuals, some of the 18 polymorphic loci might have appeared invariant. In order to avoid the

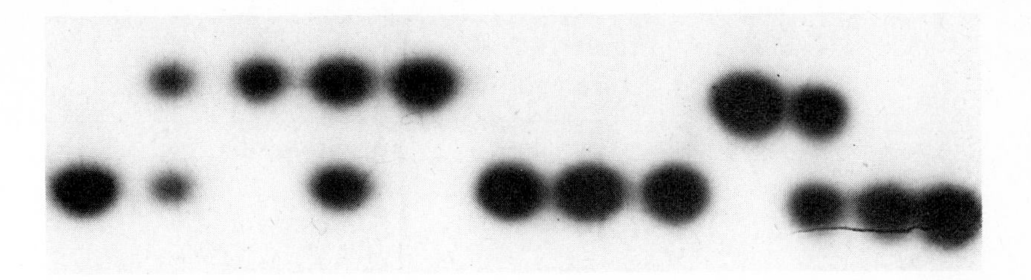

Figure 18.8
An electrophoretic gel stained for the enzyme phosphoglucomutase. The gel contains tissue samples from each of 12 females of *Drosophila pseudoobscura*. Flies with only one colored spot in the gel are inferred to be homozygotes; flies with two spots are heterozygotes. The genotypes of all 12 individuals are, from left to right: $Pgm^{100/100}$, $Pgm^{100/108}$, $Pgm^{108/108}$, $Pgm^{100/108}$, $Pgm^{108/108}$, $Pgm^{100/100}$, $Pgm^{100/100}$, $Pgm^{100/100}$, $Pgm^{108/108}$, $Pgm^{100/108}$, $Pgm^{100/100}$, and $Pgm^{100/100}$.

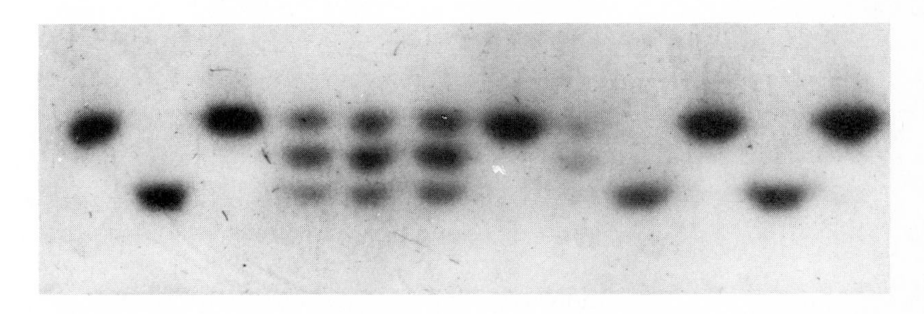

Figure 18.9
An electrophoretic gel stained for the enzyme malate dehydrogenase. The gel contains tissue samples from each of 12 flies of *Drosophila equinoxialis*. As in Figure 18.8, flies with only one colored spot in the gel are inferred to be homozygotes; but the heterozygotes exhibit three bands because malate dehydrogenase is a dimeric enzyme. The genotype of the second and ninth flies is inferred to be $Mdh^{94/94}$; the genotype of the first fly is inferred to be $Mdh^{104/104}$; the fourth, fifth, and sixth flies all have the heterozygous genotype $Mdh^{94/104}$, and so on.

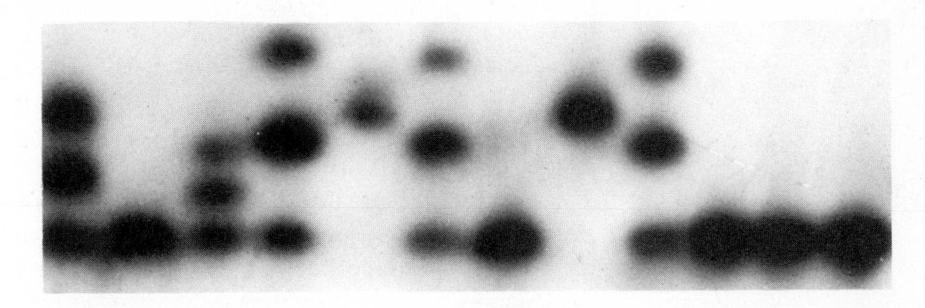

Figure 18.10
An electrophoretic gel stained for the enzyme acid phosphatase. The gel contains tissue samples from each of 12 flies of *Drosophila equinoxialis*. Acid phosphatase is a dimeric enzyme, and hence heterozygotes show three bands. Four different alleles (*88*, *96*, *100*, and *106*) are manifested in the gel. The first fly on the left has the genotype $Acph^{88/100}$, the second has $Acph^{88/88}$, the third has $Acph^{88/96}$, the fourth has $Acph^{88/106}$, the fifth has $Acph^{100/100}$, and so on.

Table 18.7
Calculation of the average polymorphism in four populations.

Population	Number of Loci		Polymorphism
	Polymorphic	Total	
1	18	30	18/30 = 0.60
2	15	30	15/30 = 0.50
3	16	30	16/30 = 0.53
4	14	30	14/30 = 0.47
			Average: 0.525

effect of sample size, it is necessary to adopt a *criterion of polymorphism*. One criterion often used is that a locus be considered polymorphic only when the most common allele has a frequency no greater than 0.95. Then, as more individuals are examined, additional variants may be found, but on the average the proportion of polymorphic loci will not change. It is, however, somewhat *arbitrary* to decide what criterion of polymorphism to use. Different values of polymorphism are obtained when different criteria are used. For example, if the criterion of polymorphism is that the frequency of the most common allele be no greater than 0.98, it is possible that some additional loci will be considered polymorphic that are not so with the 0.95 criterion (e.g., a locus with two alleles with the frequencies 0.97 and 0.03).

The polymorphism of a population is, moreover, an *imprecise* measure of genetic variation. This is because a slightly polymorphic locus counts as much as a very polymorphic one. Assume that at a certain locus there are two alleles with frequencies 0.95 and 0.05, while at another locus there are 20 alleles, each with a frequency of 0.05. It is obvious that more genetic variation exists at the second locus than at the first, yet both will count equally under the 0.95 criterion of polymorphism.

A better measure of genetic variation (because it is not arbitrary and is precise) is the *average frequency of heterozygous individuals* per locus, or simply the *heterozygosity* (H) of the population. This is calculated by first obtaining the frequency of heterozygous individuals at each locus and then averaging these frequencies over all loci. Assume that we study four loci in a population and find that the frequencies of heterozygotes at these loci are 0.25, 0.42, 0.09, and 0. The heterozygosity of the population, based on these four loci, is $(0.25 + 0.42 + 0.09 + 0)/4 = 0.19$ (Table 18.8). We

Table 18.8
Calculation of the average heterozygosity at four loci.

Locus	Number of Individuals: Heterozygotes	Total	Heterozygosity
1	25	100	25/100 = 0.25
2	42	100	42/100 = 0.42
3	9	100	9/100 = 0.09
4	0	100	0/100 = 0
			Average: 0.19

conclude that the heterozygosity of the population is 19%. Of course, in order for an estimate of heterozygosity to be valid, it must be based on a sample of more than four loci, but the procedure is the same. If several populations of the same species are examined, one can first calculate the heterozygosity in each population and then obtain the average over the various populations. If the heterozygosities in four populations are 0.19, 0.15, 0.13, and 0.17, the average heterozygosity for all four populations is $(0.19 + 0.15 + 0.13 + 0.17)/4 = 0.16$.

The heterozygosity of a population is the measure of genetic variation preferred by most population geneticists. It is a good measure of variation because it estimates the probability that two alleles taken at random from the population are different. (Each gamete from a different individual carries an allele at each locus that can be considered as randomly sampled from the population.) However, the observed heterozygosity does not reflect well the amount of genetic variation in populations of organisms that reproduce by self-fertilization, as some plants do, or organisms in which matings between relatives are common. In a population that always reproduces by self-fertilization, most individuals will be homozygous, even though different individuals may carry different alleles if the locus is variable in the population. There will also be more homozygotes in a population in which matings between relatives are common than in a population where they do not occur, even when the allelic frequencies are identical in both populations.

This difficulty can be overcome by calculating the *expected* heterozygosity, calculated from the allelic frequencies *as if* the individuals in the population were mating with each other at random. Assume that, at a locus, there are four alleles with frequencies f_1, f_2, f_3, and f_4. As we shall see in Chapter 19, the expected frequencies of the four homozygotes if there is random mating are f_1^2, f_2^2, f_3^2, and f_4^2. The expected heterozygosity at the locus will therefore be $H_{\text{expected}} = 1 - (f_1^2 + f_2^2 + f_3^2 + f_4^2)$. For example, if the allelic frequencies at a given locus are 0.50, 0.30, 0.10, and 0.10, the expected heterozygosity will be $H_{\text{expected}} = 1 - (0.50^2 + 0.30^2 + 0.10^2 + 0.10^2) = 1 - (0.25 + 0.09 + 0.01 + 0.01) = 0.64$.

Electrophoretic Estimates of Variation

Electrophoretic techniques were first applied to the estimation of genetic variation in natural populations in 1966, when three studies were published, one dealing with humans and the other two with *Drosophila* flies. Numerous populations of many organisms have been surveyed since that time, and many more are studied every year. Two studies will be reviewed here.

Table 18.9 lists 20 variable loci out of 71 loci sampled in a population of Europeans. The symbol used to represent the locus, the enzyme encoded by the locus, and the frequency of heterozygous individuals at the locus are given for each of the 20 variable loci. The heterozygosity of the population is the sum of the heterozygosities found at the 20 variable loci divided by the total number of loci sampled: 4.78/71 = 0.067.

A total of 39 gene loci coding for enzymes were studied in a population of the marine worm (phylum Phoronida) *Phoronopsis viridis,* from Bodega Bay, California. Table 18.10 gives the symbols used to represent the 27 loci in which at least two alleles were found. The table shows the observed and expected heterozygosities, as well as the loci that are polymorphic when the criterion of polymorphism is that the frequency of the most common allele be no greater than 0.95. By this criterion, 28.2% of the 39 loci studied are polymorphic. However, using the 0.99 criterion of polymorphism, 20 of the 39 loci, or 51.2%, are polymorphic. The observed heterozygosity is 7.2%, conspicuously less than the expected heterozygosity, 9.4%. This difference may be due to the occurrence of a certain amount of self-fertilization, since *P. viridis* is a hermaphroditic animal.

Table 18.10 also gives the allelic frequencies at the 27 variable loci. The number of alleles per locus ranges from only one (at the 12 invariant loci) to six (at the *Acph-2* and *G3pd-1* loci). Loci with a greater number of alleles do not necessarily have greater heterozygosities than loci with fewer alleles. For example, the observed and expected heterozygosities at the *Acph-2* locus are 0.160 and 0.217, respectively, while at the *Adk-1* locus, which has only two alleles, these heterozygosities are 0.224 and 0.496.

The importance of sampling a fairly large number of loci is apparent in Tables 18.9 and 18.10. For example, if only very few loci had been surveyed in the population of Europeans, the sample might have included a disproportionate number of highly variable loci (such as *Acph*, *Pgm-1*,

Table 18.9
Heterozygosity at 20 variable gene loci out of 71 loci sampled in a population of Europeans, as determined by electrophoresis.

Gene Locus*	Enzyme Encoded	Heterozygosity
Acph	Acid phosphatase	0.52
Pgm-1	Phosphoglucomutase-1	0.36
Pgm-2	Phosphoglucomutase-2	0.38
Adk	Adenylate kinase	0.09
Pept-A	Peptidase-A	0.37
Pept-C	Peptidase-C	0.02
Pept-D	Peptidase-D	0.02
Adn	Adenosine deaminase	0.11
6Pgdh	6-Phosphogluconate dehydrogenase	0.05
Aph	Alkaline phosphatase (placental)	0.53
Amy	Amylase (pancreatic)	0.09
Gpt	Glutamate-pyruvate transaminase	0.50
Got	Glutamate-oxaloacetate transaminase	0.03
Gput	Galactose-1-phosphate uridyl transferase	0.11
Adh-2	Alcohol dehydrogenase-2	0.07
Adh-3	Alcohol dehydrogenase-3	0.48
Peps	Pepsinogen	0.47
Ace	Acetylcholinesterase	0.23
Me	Malic enzyme	0.30
Hk	Hexokinase (white-cell)	0.05
	Average heterozygosity (including 51 invariant loci):	0.067

*Numbers or letters are used to distinguish several related enzymes and the loci coding for them; e.g., *Pgm-1* and *Pgm-2*, or *Pept-A*, *Pept-B*, and *Pept-C*.

After H. Harris and D. A. Hopkinson, *J. Human Genet.* *36*:9 (1972).

Table 18.10
Allelic frequencies at 27 variable loci in 120 individuals of the marine worm *Phoronopsis viridis.* The numbers used to represent alleles (*1, 2, 3,* etc.) indicate increasing mobility, in an electric field, of the proteins encoded by the alleles.

Gene Locus	Frequency of Allele: 1	2	3	4	5	6	Heterozygosity Observed	Expected	Is locus polymorphic by 0.95 criterion?
Acph-1	0.995	0.005					0.010	0.010	No
Acph-2	0.009	0.066	0.882	0.014	0.005	0.024	0.160	0.217	Yes
Adk-1	0.472	0.528					0.224	0.496	Yes
Est-2	0.008	0.992					0.017	0.017	No
Est-3	0.076	0.924					0.151	0.140	Yes
Est-5	0.483	0.396	0.122				0.443	0.596	Yes
Est-6	0.010	0.979	0.012				0.025	0.041	No
Est-7	0.010	0.990					0.021	0.021	No
Fum	0.986	0.014					0.028	0.028	No
αGpd	0.005	0.995					0.010	0.010	No
G3pd-1	0.040	0.915	0.017	0.011	0.011	0.006	0.159	0.161	Yes
G6pd	0.043	0.900	0.057				0.130	0.185	Yes
Hk-1	0.996	0.004					0.008	0.008	No
Hk-2	0.005	0.978	0.016				0.043	0.043	No
Idh	0.992	0.008					0.017	0.017	No
Lap-3	0.038	0.962					0.077	0.074	No
Lap-4	0.014	0.986					0.028	0.027	No
Lap-5	0.004	0.551	0.326	0.119			0.542	0.576	Yes
Mdh	0.008	0.987	0.004				0.025	0.025	No
Me-2	0.979	0.021					0.042	0.041	No
Me-3	0.017	0.824	0.159				0.125	0.296	Yes
Odh-1	0.992	0.008					0.017	0.017	No
Pgi	0.995	0.005					0.010	0.010	No
Pgm-1	0.159	0.827	0.013				0.221	0.290	Yes
Pgm-3	0.038	0.874	0.071	0.017			0.185	0.229	Yes
Tpi-1	0.929	0.071					0.000	0.133	Yes
Tpi-2	0.008	0.004	0.962	0.013	0.013		0.076	0.074	No
Averages (including 12 invariant loci)									
Heterozygosity:							0.072	0.094	
Polymorphism:									11/39 = 0.282

After F. J. Ayala et al., *Biochem. Genet. 18*:413 (1974).

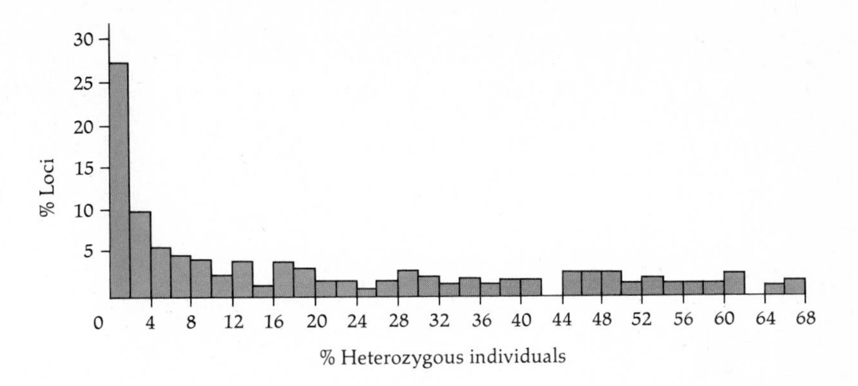

Figure 18.11 Distribution of heterozygosity among 180 gene loci studied by electrophoresis in six species of the *Drosophila willistoni* group. The average heterozygosity for all 180 loci is 0.177.

Pgm-2, and *Pept-A*). A distorted estimate of heterozygosity would then have been obtained. Figure 18.11 shows the distribution of heterozygosity among 180 loci studied in six closely related species of *Drosophila*. Characteristically, the distribution is widely spread (H ranges from 0 to 0.68) and is not at all like a normal distribution.

Experience with electrophoretic studies indicates that a sample of about 20 gene loci is usually sufficient; estimates of heterozygosity usually change little as the number of loci sampled exceeds 20. For example, a value of $H = 0.072$ was obtained for humans, using a sample of 26 gene loci. When the sample was extended to a total of 71 loci, the estimate became $H = 0.067$ (see Table 18.9).

Genetic Variation in Natural Populations

Considerable genetic variation exists in most natural populations. Table 18.11 summarizes the results of electrophoretic surveys obtained for 17 plant species and 125 animal species in which a fair number of loci have been sampled. Among animals it seems that, in general, invertebrates have more genetic variation than vertebrates, although there are exceptions. For the species in Table 18.11, the average heterozygosity is 13.4% for invertebrates and 6.0% for vertebrates. Plants show an average heterozygosity of 4.6%, with outcrossing plants exhibiting more genetic variation than those that reproduce primarily by selfing.

One way to appreciate the large amounts of genetic variation found in natural populations is the following. Consider humans, with a 6.7% heterozygosity detectable by electrophoresis. If we assume that there are 30,000 structural gene loci in a human being, which may be an underestimate, a person would be heterozygous at $30{,}000 \times 0.067 = 2010$ loci. Such an individual can theoretically produce $2^{2010} \approx 10^{605}$ different kinds of gametes. (An individual heterozygous at one locus can produce two

Table 18.11
Genic variation in natural populations of some major groups of animals and plants.

Organisms	Number of Species	Average Number of Loci per Species	Average Polymorphism*	Average Heterozygosity
Invertebrates				
Drosophila	28	24	0.529	0.150
Wasps	6	15	0.243	0.062
Other insects	4	18	0.531	0.151
Marine invertebrates	14	23	0.439	0.124
Land snails	5	18	0.437	0.150
Vertebrates				
Fishes	14	21	0.306	0.078
Amphibians	11	22	0.336	0.082
Reptiles	9	21	0.231	0.047
Birds	4	19	0.145	0.042
Mammals	30	28	0.206	0.051
Plants				
Self-pollinating	12	15	0.231	0.033
Outcrossing	5	17	0.344	0.078
Overall averages				
Invertebrates:	57	22	0.469	0.134
Vertebrates:	68	24	0.247	0.060
Plants:	17	16	0.264	0.046

*The criterion of polymorphism is not the same for all species.

From various sources.

different kinds of gametes, one with each allele; an individual heterozygous at n gene loci has the potential of producing 2^n different gametes.) This number of gametes will never be produced by any individual, however, nor by the whole of mankind; the estimated total number of protons and neutrons in the universe, 10^{76}, is infinitesimal by comparison.

Although not all possible gametic combinations are equally probable, calculations indicate that no two independent human gametes are likely to be identical and that no two human individuals (except those derived from the same zygote, such as identical twins) who exist now, have ever existed in the past, or will ever exist in the future are likely to be genetically identical. And the same can be said, in general, of organisms that reproduce sexually: no two individuals developed from separate zygotes are ever likely to be genetically identical.

Electrophoretic techniques have made it possible to obtain estimates of genetic variation in natural populations. How reliable are these estimates? Two conditions are required for making good estimates of genetic variation: (1) that a random sample of all gene loci be obtained and (2) that all alleles be detected at every locus.

The gene loci studied should represent a random sample of the genome with respect to variation because otherwise the estimates would be biased. The genes studied by gel electrophoresis code for enzymes and other soluble proteins. Such genes represent a considerable portion of the genome, but there are other kinds of gene loci, such as regulatory genes and genes coding for nonsoluble proteins. It is not known whether these genes are as variable as structural genes coding for soluble proteins. The estimates of heterozygosity might be biased for this reason, although we do not know whether they overestimate or underestimate genetic variation in this regard.

Electrophoresis separates proteins on the basis of their differential migration in an electric field. This differential migration is due to differences in molecular configuration and, primarily, to differences in net electric charge. But amino acid substitutions can occur that do not change the net electric charge of a protein or substantially modify its configuration. Electrophoresis therefore detects only a fraction of all differences in amino acid sequence. Several methods have been used to detect protein differences not distinguishable by electrophoresis, but at present we do not have reliable estimates of the proportion of the protein variants that remain undetected. We know that electrophoretic estimates of heterozygosity underestimate genetic variation at the protein level, although we do not know by how much.

Moreover, electrophoretic techniques do not detect nucleotide substitutions in the DNA that do not change the encoded amino acid. However, synonymous mutations that do not change the amino acid sequence of a protein are less likely to be important for evolution than those that do, since the latter usually have larger effects on the biology of the organism.

Problems

1. Three genotypes were observed at the *Pgm-1* locus in a human population. In a sample of 1110 individuals, the three genotypes occurred in the following numbers (*1* and *2* represent two different alleles):

Genotypes:	*1/1*	*1/2*	*2/2*
Numbers:	634	391	85

Calculate the genotypic and allelic frequencies.

2. Two human serum haptoglobins are determined by two alleles at a single locus. In a sample of 219 Egyptians, the three genotypes occurred in the following numbers (*1* and *3* represent the two alleles):

Genotypes:	*1/1*	*1/3*	*3/3*
Numbers:	9	135	75

What are the frequencies of the two alleles?

3. Calculate the expected frequency of heterozygotes, assuming random mating, from the data of problems 1 and 2. Use the chi-square test to determine whether the observed and expected *numbers* of heterozygous individuals are significantly different.

4. The following table gives the number of individuals for each of the M-N blood groups in samples from various human populations. Calculate the genotypic and allelic frequencies as well as the expected *number* of heterozygous individuals for each population. Test whether the observed and expected numbers of heterozygotes agree.

	Number having blood group:			
Population	M	MN	N	Total
Eskimos	475	89	5	569
Pueblo Indians	83	46	11	140
Russians	195	215	79	489
Swedes	433	564	203	1200
Chinese	342	500	187	1029
Japanese	356	519	225	1100
Belgians	896	1559	645	3100
English	121	200	101	422
Egyptians	140	245	117	502
Ainu	90	253	161	504
Fijians	22	89	89	200
Papuans	14	48	138	200

5. Samples from a population of *Drosophila pseudoobscura* taken in 11 successive months during 1973 gave the following numbers for each genotype at the *Hk-1* locus (*96, 100, 104,* and *108* are four different alleles):

Month	Number having genotype: *96/100*	*100/104*	*100/108*	*100/100*	*104/104*	Total
January	0	1	0	20	0	21
February	1	0	0	43	0	44
March	1	0	0	167	0	168
April	1	13	1	363	1	379
May	1	11	0	283	0	295
June	0	20	0	270	1	291
July	0	13	1	257	0	271
August	1	5	0	309	0	315
September	0	3	0	144	0	147
October	0	1	0	177	0	178
November	0	13	0	215	0	228
Total:	5	80	2	2248	2	2337

Calculate the allelic frequencies for each monthly sample. Use the chi-square test to determine whether the observed and expected numbers of heterozygotes are significantly different for the total of all monthly samples.

6. Several chromosomal arrangements, differing by a series of overlapping inversions, are known in the third chromosome of *Drosophila pseudoobscura*. Four arrangements (ST = Standard; AR = Arrowhead; CH = Chiricahua; TL = Tree Line) are found in three natural populations. The observed numbers of each genotype are as follows:

Locality	Number having genotype: *ST/AR*	*ST/CH*	*ST/TL*	*AR/CH*	*AR/TL*	*CH/TL*	*ST/ST*	*AR/AR*	*CH/CH*	Total
Keen Camp	53	66	3	48	3	6	30	11	44	264
Piñon Flat	40	53	5	37	3	7	31	11	21	208
Andreas Canyon	87	47	12	20	4	2	89	18	4	283

Calculate the frequency of each chromosomal arrangement and the expected frequency and number of heterozygotes in each of the three populations.

7. Twenty-two gene loci coding for blood proteins were studied in 23 chimpanzees (*Pan troglodytes*) and in 10 gorillas (*Gorilla gorilla*). The chimpanzees were all homozygotes at 21 loci; at the *Pgm-1* locus, 6 individuals were heterozygotes (*96/100*) and the other 17 were homozygotes (*100/100*). The gorillas were all homozygotes at 19 loci; at 3 loci, the following genotypes were observed (the number of individuals is given in parentheses after each genotype):

Ak:	*98/100* (4)	*100/100* (6)	
Dia:	*85/95* (5)	*85/85* (4)	*95/95* (1)
6-Pgdh:	*97/105* (3)	*105/105* (7)	

Calculate the observed and expected average heterozygosities for all 22 loci in the chimpanzees and in the gorillas. What proportion of the loci in each species are polymorphic by the 0.95 criterion?

8. Thirty-six gene loci coding for enzymes have been studied in the Antarctic krill, *Euphausia superba*. The allele frequencies and the observed frequency of heterozygotes at 21 polymorphic loci are given below; the other 15 loci were monomorphic. Calculate the observed and expected average heterozygosities for the 36 loci. What proportion of the loci are polymorphic by the 0.95 criterion of polymorphism? By the 0.99 criterion?

	Frequency of allele:						
Locus	*96*	*98*	*100*	*102*	*106*	*110*	Observed frequency of heterozygotes
Acph-1			0.996			0.004	0.008
Ao-1		0.012	0.960	0.028			0.081
Ald-1		0.012	0.988				0.024
Ald-2		0.169	0.831				0.274
Aph		0.004	0.996				0.008
Est-1	0.138		0.850	0.012			0.291
Est-4		0.012	0.988				0.024

(*continued*)

Locus	Frequency of allele: 96	98	100	102	106	110	Observed frequency of heterozygotes
Est-5		0.028	0.972				0.065
G6pdh-1		0.008	0.992				0.016
Got		0.402	0.594	0.004			0.449
Hk-1	0.028		0.969	0.004			0.063
Hk-2		0.004	0.996				0.008
Idh			0.996	0.004			0.009
Lap		0.004	0.996				0.008
Mdh-2	0.020		0.980				0.039
Mdh-3	0.004	0.123	0.874				0.236
Me-2		0.007	0.993				0.014
Odh		0.039	0.957	0.004			0.087
Pgi	0.020		0.787	0.178	0.016		0.323
To-2			0.988	0.012			0.024
Xdh	0.004	0.996					0.008

19

Processes of Evolutionary Change

Evolution, a Two-Step Process

Biological evolution is the process of change and diversification of organisms through time. Evolutionary change affects all aspects of living things—their morphology, physiology, behavior, and ecology. Underlying those changes, there are genetic changes, i.e., changes in the hereditary materials, which, in interaction with the environment, determine what the organisms are. At the genetic level, evolution consists of changes in the genetic constitution of populations.

Evolution at the genetic level may be seen as a two-step process. We have, first, mutation and recombination, the processes by which hereditary variation arises; then, we have genetic drift and natural selection, the processes by which genetic variants are differentially transmitted from generation to generation.

Evolution can occur only if there is hereditary variability. The ultimate source of all genetic variation is the process of mutation, but the variability is sorted out in new ways by the sexual process, i.e., by the independent assortment of chromosomes and by crossing over. The genetic variants arisen by mutation and recombination are not equally transmitted from one generation to another, but rather some variants may increase in frequency at the expense of others. Besides mutation, the processes by which allelic frequencies change in populations are natural selection, gene flow (migration) from one population to another, and random drift. Genotypic, though not allelic, frequencies may change by assortative mating, i.e., by deviations from random mating.

This chapter and the next two chapters will be dedicated to the processes by which allelic and genotypic frequencies change. Mutation, migration, and drift will be considered in this chapter, and selection and assortative mating in Chapters 20 and 21. Before studying these processes of change, we shall demonstrate that heredity by itself does not change gene frequencies, a principle known as the *Hardy-Weinberg law.*

Random Mating

It might at first appear that individuals exhibiting a dominant genotype should be more common than individuals with a recessive genotype. However, the 3:1 ratio applies to segregation in the offspring of two individuals heterozygous for the same two alleles. Different ratios appear in other mating combinations, and the frequencies of these combinations depend on the frequencies of the genotypes in a population. Mendel's laws do not tell us anything about the frequencies of genotypes in a population; the Hardy-Weinberg law does.

The main statement of the Hardy-Weinberg law is that, in the absence of the evolutionary processes—mutation, migration, drift, and selection—*gene frequencies remain constant from generation to generation.* The law also says that, if matings are random, *the genotypic frequencies are related to the gene frequencies by a simple formula* (the square expansion). A corollary of the Hardy-Weinberg law is that, if the allelic frequencies are initially the same in males and in females, *the equilibrium genotypic frequencies at any given locus are attained in one single generation of random mating.* If the allelic frequencies are initially different in the two sexes, they will become the same in one generation in the case of autosomal loci, because males and females inherit half of their genes from the males of the previous generation and half from the females; then the equilibrium genotypic frequencies will be reached in *two* generations. However, in the case of sex-linked loci, the equilibrium frequencies are attained only gradually (see Box 19.1, page 638). Before demonstrating the Hardy-Weinberg law, we must define random mating.

Random mating occurs when the probability of mating between individuals is independent of their genetic constitution. In a random mating population, therefore, matings between genotypes occur according to the proportions in which the genotypes exist.

The frequencies of the three genotypes for the M-N blood groups in U. S. Caucasians are given in Table 18.3 as $L^ML^M = 0.292$, $L^ML^N = 0.496$, and $L^NL^N = 0.213$. If U. S. Caucasians mate at random with respect to this trait, we expect the various kinds of marriages to occur with the frequencies given in Table 19.1. To obtain the probability of a given type of mating, we simply multiply the frequencies of the two genotypes involved. For example, matings between L^ML^M men and L^ML^N women will occur with a frequency of $0.292 \times 0.496 = 0.145$. We can verify that the frequencies of *all* types of matings add up to one: $0.085 + 0.145 + 0.062$

Table 19.1
Expected frequencies of the various types of marriages among U.S. Caucasians if matings are random with respect to the M-N blood groups.

Men	Women: 0.292 L^ML^M	0.496 L^ML^N	0.213 L^NL^N
0.292 L^ML^M	♂ *MM* × ♀ *MM* 0.292 × 0.292 = 0.085	♂ *MM* × ♀ *MN* 0.292 × 0.496 = 0.145	♂ *MM* × ♀ *NN* 0.292 × 0.213 = 0.062
0.496 L^ML^N	♂ *MN* × ♀ *MM* 0.496 × 0.292 = 0.145	♂ *MN* × ♀ *MN* 0.496 × 0.496 = 0.246	♂ *MN* × ♀ *NN* 0.496 × 0.213 = 0.106
0.213 L^NL^N	♂ *NN* × ♀ *MM* 0.213 × 0.292 = 0.062	♂ *NN* × ♀ *MN* 0.213 × 0.496 = 0.106	♂ *NN* × ♀ *NN* 0.213 × 0.213 = 0.045

+ 0.145 + 0.246 + 0.106 + 0.062 + 0.106 + 0.045 = 1.002. (The excess over 1.000 is due to rounded numbers.)

Random mating may occur with respect to a given gene locus or a trait, even though matings are not random with respect to some other loci or traits. Indeed, when choosing their mates, people have all sorts of preferences, and are influenced by circumstances such as socioeconomic status, schooling, neighborhood, and the like. But it seems likely that people choose their spouses without particular regard to their M-N blood groups; if so, mating *may* be random with respect to this trait.

Assortative mating occurs when choices of mates are affected by the genotype. For example, marriages in the U.S. are assortative with respect to racial features. Matings between two Caucasians or between two Blacks occur with higher frequency, and matings between one Caucasian and one Black with lesser frequency, than is expected from random mating. Assortative mating also occurs in organisms other than man. An extreme form of assortative mating is self-fertilization, which is the most common form of reproduction in many plants.

The Hardy-Weinberg Law

The Hardy-Weinberg law says that, by itself, the process of heredity does not change either allelic frequencies or (in a random mating population) genotypic frequencies at a given locus. Moreover, the equilibrium genotypic frequencies at any given locus are attained in one single generation of random mating whenever the allelic frequencies are the same in the two sexes.

The equilibrium genotypic frequencies are given by the square of the allelic frequencies. If there are only two alleles, *A* and *a*, with frequencies p and q, the frequencies of the three possible genotypes are:

$$(p + q)^2 = p^2 + 2pq + q^2$$

$$A \quad a \qquad AA \quad Aa \quad aa$$

where the alleles and genotypes to which the frequencies correspond are written in the second line.

If there are three alleles—say, A_1, A_2, and A_3—with frequencies p, q, and r, the genotypic frequencies are:

$$(p + q + r)^2 = p^2 + q^2 + r^2 + 2pq + 2pr + 2qr$$

$$A_1 \quad A_2 \quad A_3 \quad A_1A_1 \quad A_2A_2 \quad A_3A_3 \quad A_1A_2 \quad A_1A_3 \quad A_2A_3$$

The square expansion can be used similarly to obtain the equilibrium genotypic frequencies for any number of alleles. Note that the sum of all the allelic frequencies, and of all the genotypic frequencies, must always be 1. If there are only two alleles, with frequencies p and q, then $p + q = 1$, and therefore $p^2 + 2pq + q^2 = (p + q)^2 = 1$; if there are three alleles, with frequencies p, q, and r, then $p + q + r = 1$, and therefore also $(p + q + r)^2 = 1$; and so on.

The Hardy-Weinberg law was formulated in 1908 independently by the mathematician G. H. Hardy in England and by the physician Wilhelm Weinberg in Germany. A simple way of demonstrating the law is the following. Assume that at a given locus there are two alleles, *A* and *a*, and that their frequencies, in males as well as in females, are p for *A* and q for *a*. Assume also that males and females mate at random; this is equivalent to saying that the male and female *gametes* meet at random in the formation of the zygotes. Then the frequency of a given genotype will simply be the product of the frequencies of the two corresponding alleles (Table 19.2). The probability that one individual will have the *AA* genotype is the probability (p) of receiving the *A* allele from the mother multiplied by the probability (p) of receiving the *A* allele from the father, or $p \times p = p^2$. Similarly, the probability that an individual will have the *aa* genotype is q^2. The genotype *Aa* can arise in two ways: *A* from the mother and *a* from the father, which will occur with a frequency of pq, or *a* from the mother and *A* from the father, which will also occur with a frequency of pq; therefore the total frequency of *Aa* is $pq + pq = 2pq$. A geometric representation of the Hardy-Weinberg law for two alleles is shown in Figure 19.1, where the allelic frequencies are assumed to be 0.7 and 0.3.

Table 19.2
The Hardy-Weinberg law for two alleles.

Male Gametic Frequencies	Female Gametic Frequencies	
	p (A)	q (a)
p (A)	p^2 (AA)	pq (Aa)
q (a)	pq (Aa)	q^2 (aa)

We can now demonstrate the following three statements contained in the Hardy-Weinberg law:

1. The allelic frequencies do not change from generation to generation. This can easily be shown. The frequency of the A allele among the offspring in Table 19.2 is the frequency of AA plus half the frequency of Aa, or $p^2 + pq = p(p + q) = p$ (remember that $p + q = 1$).
2. The equilibrium genotypic frequencies are given by the square expansion and do not change. Since the allelic frequencies among the offspring are p and q, as they were among the parents, the genotypic frequencies in the following generation will again be p^2, $2pq$, and q^2.
3. The equilibrium genotypic frequencies are attained in one single generation. Notice that in Table 19.2 nothing is assumed with respect to the genotypic frequencies among the parents. Whatever these frequencies may be, if the allelic frequencies are p and q in males as well as in females, the genotypic frequencies among the progeny will be p^2, $2pq$, and q^2.

Table 19.3 uses the M-N blood groups among U.S. Caucasians as an example of the Hardy-Weinberg relationships. From the observed numbers of individuals shown in Table 18.3 for U.S. Caucasians, we can calculate the allelic frequencies. The frequency of L^M is the total number of L^M alleles observed (twice the number of $L^M L^M$ individuals plus the number of $L^M L^N$ individuals) divided by the total number of alleles in the sample (twice the number of individuals), or $[(1787 \times 2) + 3039]/(2 \times 6129) = 0.5395$. Similarly, we obtain the frequency of L^N as 0.4605. The expected equilibrium frequencies calculated according to the Hardy-Weinberg law [0.2911 ($L^M L^M$), 0.4968 ($L^M L^N$), and 0.2121 ($L^N L^N$)] are, in fact, very close to the genotypic frequencies observed in the population (0.292, 0.496, and 0.213).

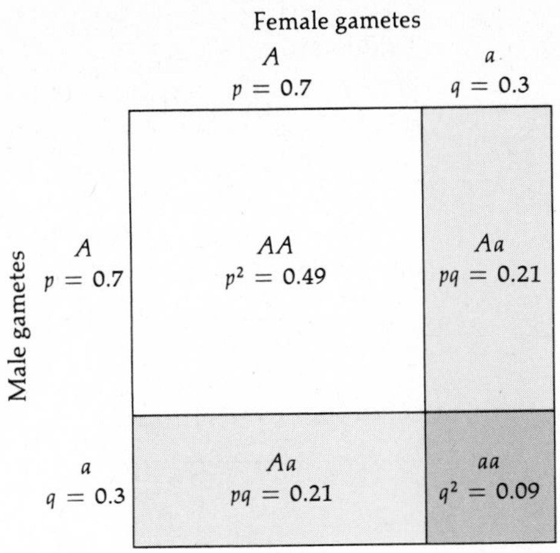

Male gametes \ Female gametes	A, $p = 0.7$	a, $q = 0.3$
A, $p = 0.7$	AA, $p^2 = 0.49$	Aa, $pq = 0.21$
a, $q = 0.3$	Aa, $pq = 0.21$	aa, $q^2 = 0.09$

Figure 19.1
Geometric representation of the relationship between allelic and genotypic frequencies according to the Hardy-Weinberg law.

The same procedure used for two alleles can be followed for demonstrating the Hardy-Weinberg law with any number of alleles. Table 19.4 demonstrates the equilibrium genotypic frequencies for a locus with three alleles, with frequencies p, q, and r, so that $p + q + r = 1$. The geometric representation shown in Figure 19.2 uses the ABO blood groups as an example of the three-allele case.

Applications of the Hardy-Weinberg Law

One application of the Hardy-Weinberg law is that it permits the computation of gene and genotypic frequencies in cases where not all genotypes can be distinguished, because of dominance. Albinism in humans is a relatively rare recessive condition. If the allele for normal pigmentation is represented as A and the allele for albinism as a, albinos have aa genotypes, while normally pigmented individuals are either AA or Aa. Assume that in a certain human population the frequency of albino individuals is 1 in 10,000. According to the Hardy-Weinberg law, the frequency of the homozygotes aa is q^2; thus $q^2 = 0.0001$ and $q = \sqrt{0.0001} = 0.01$. It follows that the frequency of the normal allele is 0.99; the frequencies of the other two genotypes are $p^2 = 0.99^2 = 0.98$ for AA, and $2pq = 2 \times 0.99 \times 0.01 \approx 0.02$ for Aa.

Table 19.3
Hardy-Weinberg equilibrium frequencies for the three genotypes of the M-N blood groups among U.S. Caucasians.

	Female Allelic Frequencies	
Male Allelic Frequencies	0.5395 (L^M)	0.4605 (L^N)
0.5395 (L^M)	0.2911 (L^ML^M)	0.2484 (L^ML^N)
0.4605 (L^N)	0.2484 (L^ML^N)	0.2121 (L^NL^N)

The ABO blood groups can serve as an example for three alleles. Assume that in a certain population the frequencies of the four blood groups are:

A (genotypes I^AI^A and I^Ai) = 0.45

B (genotypes I^BI^B and I^Bi) = 0.13

AB (genotype I^AI^B) = 0.06

O (genotype ii) = 0.36

Let us represent the frequencies of I^A, I^B, and i as p, q, and r, respectively. According to the Hardy-Weinberg law, the frequency of the genotype ii is r^2; therefore $r = \sqrt{0.36} = 0.60$, which is the frequency of the i allele. We now note that the joint frequency of blood groups B and O is $(q + r)^2$ (see Figure 19.2). Therefore, $(q + r)^2 = 0.13 + 0.36 = 0.49$, and $q + r = \sqrt{0.49} = 0.70$. Since we already know that $r = 0.60$, the frequency of allele I^B is $0.70 - 0.60 = 0.10$. Finally, the frequency of allele I^A is $p = 1 - (q + r) = 1 - 0.70 = 0.30$.

One interesting implication of the Hardy-Weinberg law is that rare alleles exist in populations, for the most part, in heterozygous genotypes, not in the homozygotes. Consider the albinism example given above. The frequency of albinos (aa) is 0.0001. The frequency of heterozygotes (Aa) is 0.02. Only half the alleles of the heterozygotes are a. Therefore the frequency of a in the population is 0.01 in heterozygous individuals and 0.0001 in homozygous individuals. There are about 100 times more a alleles in heterozygotes than in albinos.

In general, if the frequency of a recessive allele in a population is q, there will be pq recessive alleles (half the alleles in $2pq$ individuals) in the heterozygotes and q^2 recessive alleles in the homozygotes. The ratio of one to the other is $pq/q^2 = p/q$, which, if q is very small, will be approximately

Table 19.4
The Hardy-Weinberg law for three alleles.

Male Gametic Frequencies	Female Gametic Frequencies p (A_1)	q (A_2)	r (A_3)
p (A_1)	p^2 (A_1A_1)	pq (A_1A_2)	pr (A_1A_3)
q (A_2)	pq (A_1A_2)	q^2 (A_2A_2)	qr (A_2A_3)
r (A_3)	pr (A_1A_3)	qr (A_2A_3)	r^2 (A_3A_3)

$1/q$. Thus, the lower the frequency of an allele, the greater the proportion of that allele that exists in the heterozygotes. The frequency of the recessive gene for alkaptonuria is about 0.001. The frequency of alkaptonurics is $q^2 = 0.000001$, or 1 in 1 million, while the frequency of the heterozygotes is $2pq$, or about 0.002. The number of alkaptonuria genes is about 1000 times greater in heterozygotes than in homozygotes.

Imagine now that a misguided dictator, with eugenic ideals of "improving the race," wants to eliminate albinism from the population. If the heterozygotes cannot be identified, his program will be based on the elimination, or sterilization, of the recessive homozygotes. This will

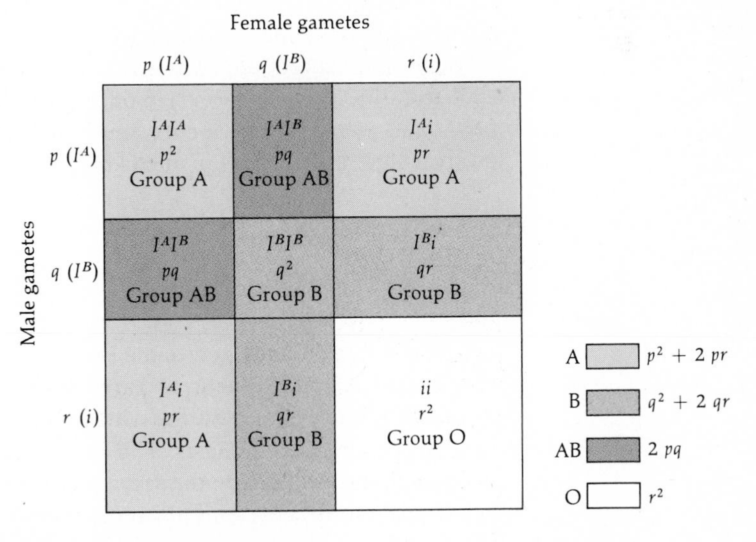

Figure 19.2
Geometric representation of the relationships between allelic and genotypic frequencies for the ABO blood groups.

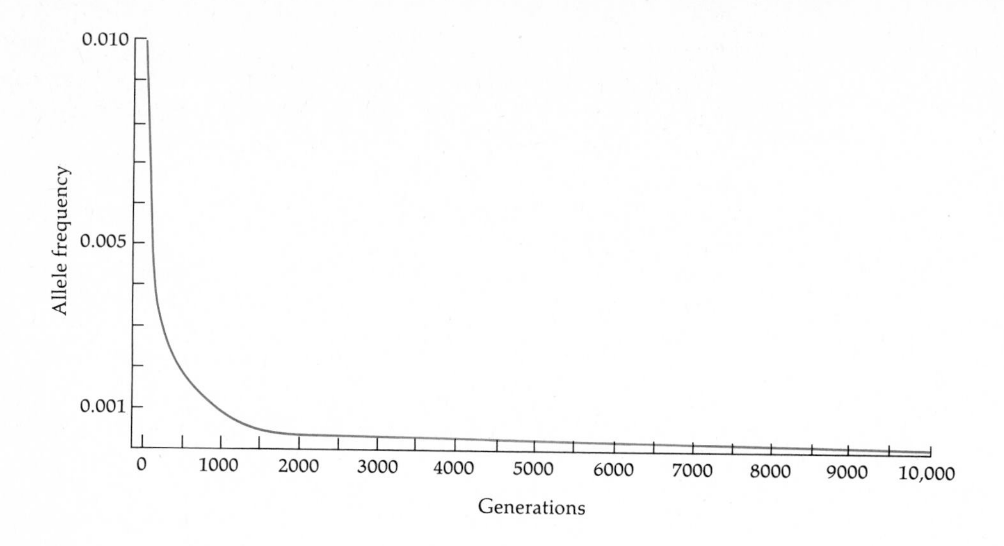

Figure 19.3
Change in allele frequency when the recessive homozygotes are eliminated from a population. If the initial allele frequency is 0.01, it takes 900 generations to reduce it to 0.001, and 9900 generations to reduce it to 0.0001. In general, the number of generations, t, required to change the allele frequency from q_0 to q_t is $t = 1/q_t - 1/q_0$ (see Chapter 20, page 669).

reduce the frequency of the allele in the population only very slightly, however, since most albino alleles go undetected in the heterozygotes. Albinos will appear in the next generation in very nearly the same frequency as they did previously. Many generations will be required to reduce greatly the frequency of the recessive allele (Figure 19.3).

The converse of this imaginary situation occurs in human populations for lethal recessive conditions that can now be cured. An example is phenylketonuria (PKU). The frequency of the allele is about 0.006. Even if all homozygotes were cured and they reproduced as effectively as normal individuals, the PKU gene would increase in frequency only very slowly, and the incidence of the condition even more slowly. If all PKU individuals were cured, the frequency of the gene would change only from 0.006 to 0.006036 ($q_1 = q + q^2$) in one generation. And, of course, if not all individuals were cured or if cured individuals retained some reproductive disadvantage, the increase in allelic frequency would be even slower.

Sex-Linked Genes

In the case of sex-linked genes, the equilibrium genotypic frequencies for females (or the homogametic sex) are the same as in the case of autosomal genes. If the alleles are A and a, with frequencies p and q, there will be p^2 AA, $2pq$ Aa, and q^2 aa females. The frequencies of the hemizygous males (heterogametic sex) will be the same as the allelic frequencies: p for the A males and q for the a males. This can be shown using the same reasoning as before. AA females receive one A gamete from the father and one A

gamete from the mother; if the frequency of A in males as well as in females is p, then AA females will be produced with a frequency p^2. Similarly, the frequency of aa females will be q^2 and the frequency of Aa females will be $2pq$. Males, however, receive their only X chromosome from the mother. Thus, the frequencies of the two hemizygous genotypes are the same as the frequencies of the two alleles among the females in the previous generation.

It follows that phenotypes determined by recessive genes will be more frequent among males than among females. If the frequency of a sex-linked recessive allele is q, the incidence of the phenotype will be q among males and q^2 among females. The ratio of one to the other will be $q/q^2 = 1/q$; the smaller the value of q, the greater the ratio of males to females exhibiting the recessive phenotype. The frequency of the allele for red-green color blindness is 0.08; therefore there are $1/0.08 = 12.5$ times more color-blind men than women. The frequency of the recessive gene for the most common form of hemophilia is 0.0001; according to the Hardy-Weinberg law, we expect $1/0.0001 = 10{,}000$ times more victims of this kind of hemophilia among men than among women (but very few in either case—1 in 10,000 men and 1 in 100 million women).

Mutation

The Hardy-Weinberg law in genetics is analogous to Newton's first law in mechanics, which says that a body remains at rest or maintains a constant velocity when not acted upon by a net external force. Bodies are always acted upon by external forces, but the first law is the point of departure for applying other laws. The Hardy-Weinberg law says that, in the absence of disturbing processes, gene frequencies do not change. But processes that change gene frequencies are always present, and without them there would be no evolution. The Hardy-Weinberg law is the point of departure from which we can calculate the effects of the processes of change.

The first process we shall consider is mutation. Although gene and chromosome mutations are the ultimate source of all genetic variability, they occur with very low frequency. Mutation is a very slow process that, by itself, changes the genetic constitution of populations at a very low rate. If mutation were the *only* process of genetic change, evolution would occur at an impossibly low rate. This is the main lesson to be learned from the calculations that follow.

Assume that there are two alleles, A_1 and A_2, at a locus, and that mutation from A_1 to A_2 occurs at a rate u per gamete per generation. Assume also that at a given time the frequency of A_1 is p_0. In the next generation, a fraction u of all A_1 alleles become A_2 by mutation. The

Box 19.1 Calculation of Allele Frequencies and Approach to Equilibrium for Sex-Linked Genes

In the case of sex-linked genes, the homogametic sex carries two-thirds of all genes in the population, while the heterogametic sex carries only one-third. Assume that there are two alleles, A and a, in a population and that the frequency of A is p_f among the females and p_m among the males. The frequency of A in the whole population will be

$$p = \tfrac{2}{3}p_f + \tfrac{1}{3}p_m$$

Similarly,

$$q = \tfrac{2}{3}q_f + \tfrac{1}{3}q_m$$

where q, q_f, and q_m are the frequencies of the a allele in the whole population, in females, and in males, respectively.

In humans, the *ma* (macroglobulin a) locus coding for the α_2 macroglobulin portion of blood serum is sex-linked. Presence of the antigen (ma^+) is dominant over absence (ma^-). The numbers of the two phenotypes in a sample of Norwegians were 57 ma^+ and 44 ma^- among females, but 23 ma^+ and 77 ma^- among males. Therefore the frequency of the ma^- allele among females was

$$q_f = \sqrt{44/101} = 0.66$$

and, among males,

$$q_m = 77/100 = 0.77$$

Hence the frequency of the ma^- allele in the population was

$$q = ⅔(0.66) + ⅓(0.77) = 0.70$$

When the allele frequencies at sex-linked loci are different between females and males, the population does not reach the equilibrium frequency in a single generation. Rather, the frequency of a given allele among the males in a given generation is the frequency of that allele among the females in the previous generation (because the males inherit their only X chromosome from their mothers), while the frequency among the females in a given generation is the average of its frequency in the females and the males of the previous generation (because the females inherit one X chromosome from their fathers and the other from their mothers). Consequently, the distribution of the allele frequencies of the two sexes oscillates, but the difference between the sexes is halved each generation; therefore the population rapidly approaches an equilibrium in which both sexes have the same frequency (Figure 19.4).

frequency (p_1) of A_1 after mutation will be the initial frequency (p_0) minus the frequency of mutated alleles (up_0), or

$$p_1 = p_0 - up_0 = p_0(1 - u)$$

In the following generation, a fraction u of the remaining A_1 alleles (p_1) will mutate to A_2, and the frequency of A_1 will become

$$p_2 = p_1 - up_1 = p_1(1 - u)$$

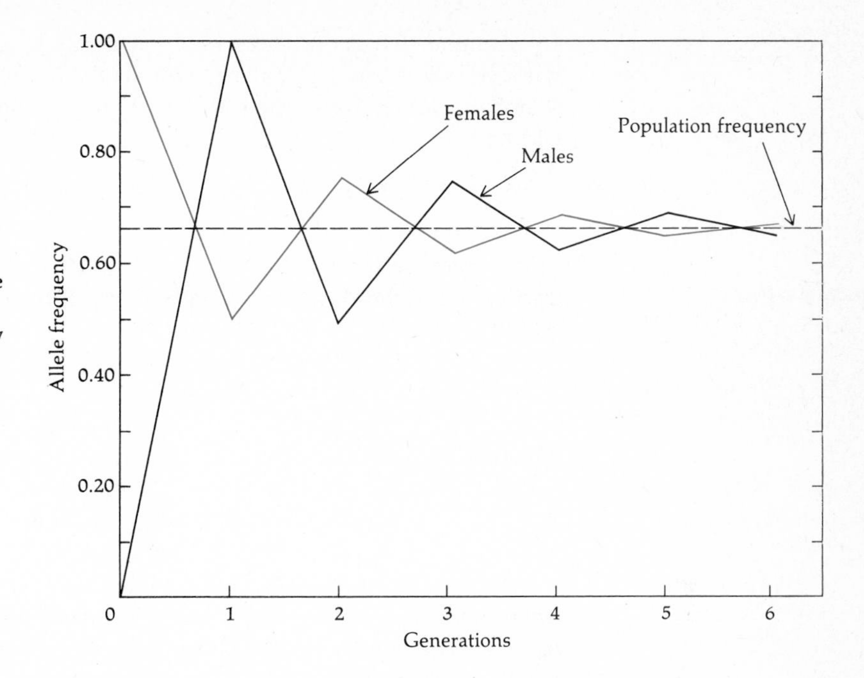

Figure 19.4
Changes in allele frequencies through the generations at a sex-linked locus when the two sexes have different allele frequencies. The case shown is the most extreme possible, namely, when the initial frequency of an allele is 1 in females and 0 in males. The initial frequency of the allele in the population (which is also the ultimate equilibrium frequency in males as well as in females) is therefore 0.67.

Substituting the value of p_1 obtained above, we have

$$p_2 = p_1(1 - u) = p_0(1 - u)(1 - u) = p_0(1 - u)^2$$

After t generations, the frequency of A_1 alleles will be

$$p_t = p_0(1 - u)^t$$

Because $1 - u$ is less than 1, it follows that, as t increases, p_t becomes ever smaller; if the process continues indefinitely, the frequency of A_1 will eventually decrease to zero. This result is intuitively obvious—the frequency of A_1 gradually decreases because a fraction of A_1 alleles changes to A_2 in every generation.

The rate of change is nevertheless very slow. For example, if the mutation rate is $u = 10^{-5}$ per gamete per generation, which is a typical gene mutation rate for eukaryotes, it will take 1000 generations to change the frequency of A_1 from 1.00 to 0.99, 2000 generations to change it from 0.50 to 0.49, and 10,000 generations to change it from 0.10 to 0.09. In general, as the frequency of A_1 decreases, it takes ever longer to accomplish a given amount of change (0.01 in the examples just given).

The model of mutation from one genetic variant to another without reverse mutation may apply to certain cases: for example, chromosomal inversions—a given chromosomal arrangement may yield inversions at a certain rate, but it is extremely unlikely that an inverted arrangement will revert to exactly the original sequence. Mutations, however, are often reversible: allele A_2 may mutate back to A_1.

Assume that A_1 mutates to A_2 at a rate u, as before, and that A_2 mutates to A_1 at a rate v. If at a given time the frequencies of A_1 and A_2 are p_0 and q_0, respectively, then after one generation the frequency of A_1 will be

$$p_1 = p_0 - up_0 + vq_0$$

This is because the fraction up_0 of the A_1 alleles changes to A_2, but the fraction vq_0 of the A_2 alleles changes to A_1. If the change in frequency of A_1 is represented by Δp, we have

$$\Delta p = p_1 - p_0$$

Substituting the value of p_1 obtained above, we have

$$\Delta p = (p_0 - up_0 + vq_0) - p_0 = vq_0 - up_0$$

An equilibrium between forward and backward mutations will exist when there is no net change in gene frequency, i.e., when $\Delta p = 0$. If we represent the equilibrium allele frequencies by $\hat{p}$ and $\hat{q}$, the requirement that $\Delta p = 0$ implies

$$u\hat{p} = v\hat{q}$$

This result says that the allelic frequencies will be at equilibrium when the number of A_1 alleles changing to A_2 alleles is the same as the number of A_2 changing to A_1. Since $p + q = 1$, and therefore $q = 1 - p$, there will be equilibrium when

$$u\hat{p} = v(1 - \hat{p})$$

$$u\hat{p} + v\hat{p} = v$$

$$\hat{p} = \frac{v}{u + v}$$

and, because $\hat{p} + \hat{q} = 1$,

$$\hat{q} = \frac{u}{u + v}$$

Assume that the mutation rates are $u = 10^{-5}$ and $v = 10^{-6}$; then

$$\hat{p} = \frac{10^{-6}}{10^{-5} + 10^{-6}} = \frac{1}{11} = 0.09$$

$$\hat{q} = \frac{10^{-5}}{10^{-5} + 10^{-6}} = \frac{10}{11} = 0.91$$

Two points need to be added. The first point is that allelic frequencies are usually not in mutational equilibrium, because other processes affect them. In particular, natural selection may favor one allele over the other; the equilibrium frequencies are then decided by the interaction between mutation and selection, as we shall see in Chapter 20. The second point is that allelic frequencies change at a slower rate when there is forward and backward mutation than when mutation occurs in only one direction, because backward mutation partially counteracts the effects of forward mutation. This emphasizes what has already been said—that mutation by itself takes a very long time in order to effect any substantial change in allele frequencies.

Migration

Migration, or *gene flow,* occurs when individuals move from one population to another and interbreed with the latter. Gene flow does not change allele frequencies for the whole species, but may change them locally when the allele frequencies in the migrants are different from those in resident individuals.

Assume that individuals from surrounding populations migrate at a certain rate into a local population and there interbreed with the residents. The proportion of migrants is m, so that in the next generation $(1 - m)$ of the genes are descendants of residents, and m are descendants of migrants. Assume that in the surrounding population a certain allele, A_1, has an average frequency P, while in the local population it has the frequency p_0. In the next generation, the frequency of A_1 in the local population will be

$$\begin{aligned} p_1 &= (1 - m)p_0 + mP \\ &= p_0 - m(p_0 - P) \end{aligned}$$

That is, the new allelic frequency will be the original allelic frequency (p_0) multiplied by the proportion of reproducing individuals that are residents $(1 - m)$, plus the proportion of reproducing migrant individuals (m) multiplied by their gene frequency (P). Reorganizing the terms as shown, we see that the new allelic frequency will be the original allelic frequency (p_0) minus the proportion of migrant individuals (m) multiplied by the difference in allelic frequency between residents and migrants $(p_0 - P)$.

The change Δp in allele frequency is

$$\Delta p = p_1 - p_0$$

Substituting the value of p_1 obtained above, we have

$$\Delta p = p_0 - m(p_0 - P) - p_0 = -m(p_0 - P)$$

That is, the greater the proportion of migrant individuals and the greater the difference between the two allele frequencies, the greater Δp becomes. Notice that Δp will be zero only when either m or $p_0 - P$ is zero. Therefore, unless migration stops ($m = 0$), the allelic frequency will continue to change until it becomes the same in the local population as in the surrounding populations ($p - P = 0$).

It is worthwhile to look at the difference in allele frequency between the local and the surrounding populations. After the first generation, it will be

$$p_1 - P = p_0 - m(p_0 - P) - P$$

$$= p_0 - mp_0 - P + mP$$

$$= (1 - m)p_0 - (1 - m)P$$

$$= (1 - m)(p_0 - P)$$

After the second generation, the difference in allele frequencies will be

$$p_2 - P = (1 - m)^2(p_0 - P)$$

and after t generations of migration, we have

$$p_t - P = (1 - m)^t(p_0 - P)$$

This formula allows one to calculate the effect of t generations of migration at a certain rate (m) if the initial allele frequencies (p_0 and P) are known:

$$p_t = (1 - m)^t(p_0 - P) + P$$

The formula is also helpful in investigating other interesting questions. For example, if we know the initial allelic frequencies (p_0 and P), the allelic frequency (p_t) in the resident population at a certain time, as well as the number of generations (t), we can calculate the rate of gene flow (m).

In the United States, people of mixed Caucasian and Black descent are considered members of the Black population. Racial admixture can hence be seen as a process of gene flow from the Caucasian to the Black population. The frequency of the R^0 allele at the locus determining the Rhesus blood groups is $P = 0.028$ in U.S. Caucasian populations. Among the African populations from which the ancestors of U.S. Blacks came, the frequency of R^0 is 0.630; this can be considered the initial allele frequency of the U.S. Black population, $p_0 = 0.630$. The African ancestors of the U.S. Blacks came to the United States about 300 years, or some 10 generations, ago; hence $t = 10$. The frequency of R^0 among U.S. Blacks is at present $p_t = 0.446$.

Rearranging the terms in the last equation shown above, we have

$$(1 - m)^t = \frac{p_t - P}{p_0 - P}$$

Substituting the values of the various parameters, we have

$$(1 - m)^{10} = \frac{0.446 - 0.028}{0.630 - 0.028} = 0.694$$

$$1 - m = \sqrt[10]{0.694} = 0.964$$

$$m = 0.036$$

Therefore the gene flow from U.S. Caucasians into the U.S. Black population has occurred at a rate equivalent to an average of 3.6% per generation. Ten generations of gene flow have left $(1 - m)^{10} = 0.694$ of all genes in U.S. Blacks derived from their African ancestors, and $1 - 0.694 = 0.306$, or somewhat more than 30%, of their genes derived from Caucasian ancestors.

The calculations above are only approximations, but they give us an idea of the extent of racial admixture in the United States. Using allele frequencies at other loci, calculations similar to the one just made lead to somewhat different results. Moreover, the degree of racial admixture may be different in various parts of the United States (Table 19.5). But it is clear that considerable gene flow has taken place.

Random Genetic Drift

Random genetic drift (*genetic drift,* for short, or simply *drift*) refers to changes in gene frequencies due to sampling variation from generation to generation. Assume that in a certain population two alleles, A and a, exist in frequencies 0.40 and 0.60; the frequency of A in the following generation may be less (or greater) than 0.40 simply because, by chance, allele A is present less often (or more often) than expected among the gametes that form the zygotes of that generation.

Genetic drift is a process of pure chance and represents a special case of the general phenomenon known as *sampling errors,* or *sampling variation.* The general principle is that the magnitude of the "errors" due to sampling is inversely related to the size of the sample—the smaller the sample, the larger the effects. With respect to organisms, this principle means that the smaller the number of breeding individuals in a population, the larger the allele frequency changes due to genetic drift are likely to be.

Table 19.5
Allele frequencies at several loci in African Blacks, U.S. Blacks, and U.S. Caucasians. The Africans are from a region from which slaves were shipped to the United States. The U.S. Blacks are from one southern city and one western city.

Allele	Blacks (Africa)	Blacks (Claxton, Georgia)	Blacks (Oakland, California)	Caucasians (Claxton, Georgia)
R^0	0.630	0.533	0.486	0.022
R^1	0.066	0.109	0.161	0.429
R^2	0.061	0.109	0.071	0.137
r	0.248	0.230	0.253	0.374
A	0.156	0.145	0.175	0.241
B	0.136	0.113	0.125	0.038
M	0.474	0.484	0.486	0.507
S	0.172	0.157	0.161	0.279
Fy^a	0.000	0.045	0.094	0.422
p	0.723	0.757	0.737	0.525
Jk^a	0.693	0.743		0.536
Js^a	0.117	0.123		0.002
T	0.631	0.670		0.527
Hp^1	0.684	0.518		0.413
$G6PD$	0.176	0.118		0.000
Hb^S	0.090	0.043		0.000

After J. Adams and R. H. Ward, *Science 180*:1137 (1973).

It is simple to see why there should be an inverse relation between sample size and sampling errors. Assume that we toss a coin and that the probability of getting heads in any one toss is 0.5. If we toss the coin only once, we can get heads or tails, but not both; although the probability of getting heads is 0.5, we get heads either once or not at all, but not half the time. If we toss the coin ten times, we are likely to get several heads and several tails; we would be surprised (and suspicious of the coin) if we got only heads, but not if we got, say, six heads and four tails. In the latter case the frequency of heads would be 0.6 rather than the expected 0.5, but we would attribute such deviation from expectation to chance. Assume now that we toss the coin 1000 times. We would be extremely suspicious of the coin if we got only heads, or even if we got 600 heads and 400 tails, although the frequency of heads in the latter case would be 0.6, the same frequency we observed without surprise when we threw the coin ten times. When the coin is tossed 1000 times, however, we would not be surprised to get 504 heads and 496 tails, which means a frequency of heads of 0.504, although the expected frequency is still 0.500.

The point of the coin example is that the larger the sample, the more likely it is to show a close agreement between the expected frequency of heads (0.5) and the observed frequency (1, 0.6, and 0.504 for one, ten, and 1000 throws in the example). With populations, we also expect that the larger the number of individuals producing the next generation, the closer the agreement between the expected allelic frequency (which is that in the parental generation) and the observed allelic frequency (which is that in the progeny).

Note that the relevant figure is not the total number of individuals in the population, but the *effective population size,* which is determined by the number of parents producing the following generation. This is so because the genes sampled to make up the following generation are from the individuals that become parents, not from the rest of the population.

There is an important difference between the coin examples and genetic drift. With a coin, the probability of obtaining heads in a run remains 0.5 even though in the previous run we may have obtained heads more or less often than expected. In populations, however, the frequency of an allele in a given sample (i.e., a generation) becomes the probability of obtaining the allele in the following sample (generation). If the frequency of an allele in one generation changes from, say, 0.5 to 0.6, the probability of obtaining the allele in the following generation becomes 0.6. Thus changes in allele frequencies become cumulative through the generations. However, because chance changes are random in direction, changes may be reversed, so long as the frequency of an allele does not become 0 or 1 (Figure 19.5). An allele that has increased in frequency in one generation may, in the next generation, increase again or decrease instead, with equal probability. If the allele becomes "lost" or "fixed" (i.e., if it reaches a frequency of either 0 or 1), the process stops. The allele cannot change in frequency again, unless a new allele arises by mutation.

Consider the following example. Assume that we have a large number of pea plants, *Pisum sativum,* like those used by Mendel, and that the frequency of the allele responsible for yellow peas, *Y*, is 0.5, the same as the frequency of the *y* allele, which produces green peas in homozygous condition. Assume also that the three genotypes occur in the expected frequencies of $\frac{1}{4}YY$, $\frac{1}{2}Yy$, and $\frac{1}{4}yy$. Suppose now that we pick up one seed (pea), without observing the phenotype, and grow a plant from it. What is the frequency of the *Y* allele among the peas produced by this plant through self-fertilization? Clearly, there are three possibilities: the frequency of *Y* will be 1, ½, or 0, depending on the genotype of the pea used

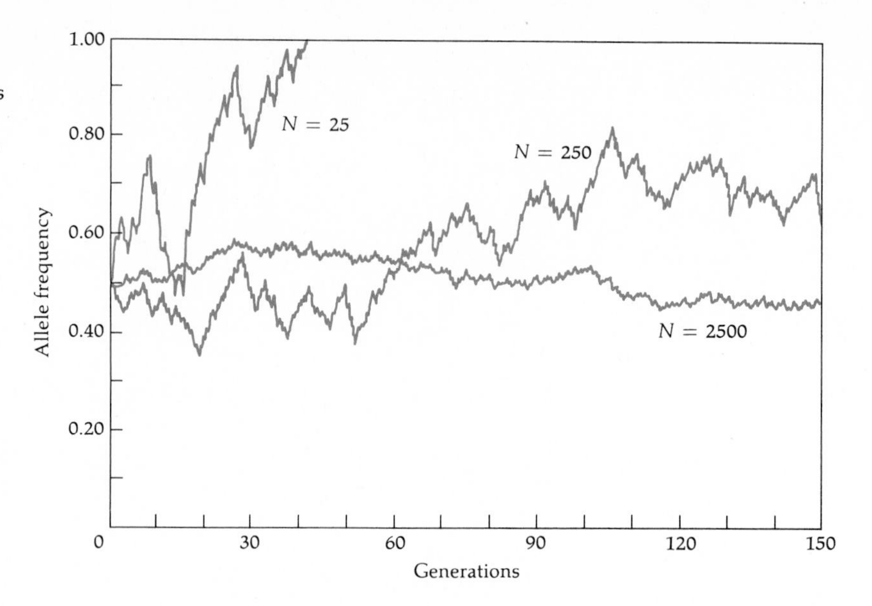

Figure 19.5
Population size and drift. The graphs show the results of computer experiments that simulate chance effects in three populations of different size, each starting with an allele frequency of 0.50. N is the effective population size. (After W. F. Bodmer and L. L. Cavalli-Sforza, *Genetics, Evolution, and Man*, W. H. Freeman, San Francisco, 1976.)

to produce the plant. The probability is ¼ that the pea was *YY*, and also ¼ that it was *yy*; thus the frequency of *Y* is likely to change to either 1 or 0 with a probability of ½. Now suppose that we collect 1000 peas from the original population and grow 1000 plants from them; the frequency of the *Y* allele among the peas produced by these plants is likely to be very nearly ½, although it may be slightly higher or lower.

Whenever we know the number of parents used to produce the following generation, and the allelic frequencies, as in the previous examples, it is possible to calculate the probability of obtaining a given allelic frequency in the following generation. In order to do this we need to know the *variance* of the allelic frequencies in the following generation, which is a measure of the amount of variation that would be found among different samples (see Appendix A.III). If there are two alleles, with frequencies p and q, and the number of parents is N (so that the number of genes in the sample used to produce the next generation is $2N$), the *variance* (s^2) of the allelic frequency in the following generation is

$$s^2 = \frac{pq}{2N}$$

and the *standard deviation* is

$$s = \sqrt{\frac{pq}{2N}}$$

These equations show the inverse relationship between sample size, $2N$, and the expected variation in allelic frequencies.

Table 19.6
Effects of random genetic drift from one generation to the next.

Population Size (N)	Gametes ($2N$)	Variance ($pq/2N$)	Standard Deviation ($\sqrt{pq/2N}$)	Expected Range in p in 95% of Populations ($p \pm 2$ s.d.)
Case 1: $p = q = 0.5$				
5	10	0.025	0.16	0.18–0.82
50	100	0.0025	0.05	0.40–0.60
500	1000	0.00025	0.016	0.468–0.532
Case 2: $p = 0.3$, $q = 0.7$				
5	10	0.021	0.145	0.01–0.59
50	100	0.0021	0.046	0.208–0.392
500	1000	0.00021	0.0145	0.271–0.329

Table 19.6 shows the likely effects of drift from one generation to the next in two cases: (1) when $p = q = 0.5$ and (2) when $p = 0.3$ and $q = 0.7$. In each case three different effective population sizes are considered: $N =$ 5, 50, and 500. The expected range of variation in p in 95% of the populations is given by $p \pm 2$ standard deviations (see Appendix A.III). In small populations with an effective size of 5 individuals, the range in the frequency p extends in one single generation from 0.18 to 0.82; in larger populations the range in allelic frequency is lower. Notice that the extent of the range decreases as the square root of the ratio of population sizes. For example, the range for the populations with 5 individuals is 0.64 (from 0.18 to 0.82), while in the populations 100 times greater (500 individuals), the range is $\sqrt{100} = 10$ times smaller ($0.532 - 0.468 = 0.064$).

Table 19.6 shows the range within which 95% of the populations (or the genes) are expected to fall. Within this range, the intermediate frequencies have a greater probability than the extreme frequencies. This is illustrated in Figures 19.6 and 19.7 for two cases with initial "frequencies" of 0.5 and 0.4, respectively, and each having two "population sizes," 10 and 100.

The *cumulative* effects of random genetic drift over the generations are shown in Figure 19.8. The experiment, performed by Peter F. Buri, consisted of 107 different populations, reproduced each generation by selecting at random 8 females and 8 males from the offspring of the preceding

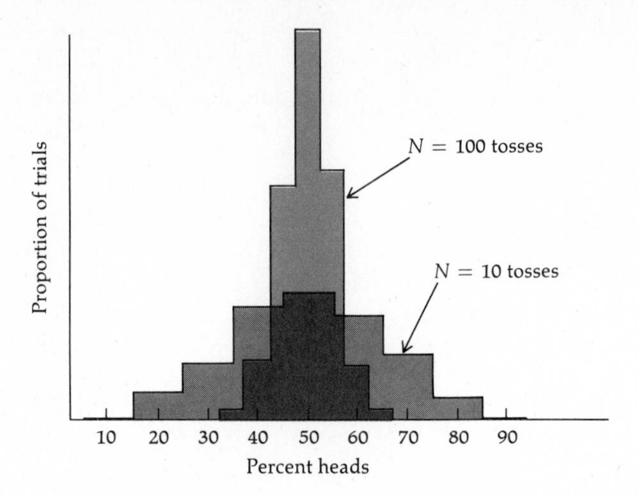

Figure 19.6
Experiments in coin tossing. Each of the two graphs represents 1000 independent trials of either 10 or 100 tosses. The ordinate represents the proportion of times a given number of heads was obtained. The distributions are approximately normal, with the mode at the expected probability of 0.50. The variance of the distribution is larger in the experiments with 10 tosses (equivalent to an effective population size of 5 diploid individuals) than in those with 100 tosses (effective population size of 50). (After W. F. Bodmer and L. L. Cavalli-Sforza, *Genetics, Evolution, and Man*, W. H. Freeman, San Francisco, 1976.)

generation (the approximate effective population size was 16 individuals, or 32 genes). The initial frequency of the two alleles, *bw* and bw^{75}, was 0.5 (all individuals were heterozygotes for these two alleles). In the first generation, the allele frequencies are spread around the mean value of 0.5, but with a greater number of populations in the intermediate frequencies. The frequencies obtained in the first generation become the initial frequencies for the second generation, etc. The spread in gene frequencies increases through the generations. The first "fixed" population appears in the fourth generation (frequency of allele $bw^{75} = 1$). The number of fixed populations gradually increases until generation 19, when the experiment was terminated (30 populations fixed for allele *bw* and 28 fixed for allele bw^{75}). If the experiment had continued indefinitely, all populations would have become fixed, approximately half for each allele.

Founder Effect and Bottlenecks

Unless a population is very small, changes in allelic frequencies due to genetic drift will be small from one generation to another, but the effects

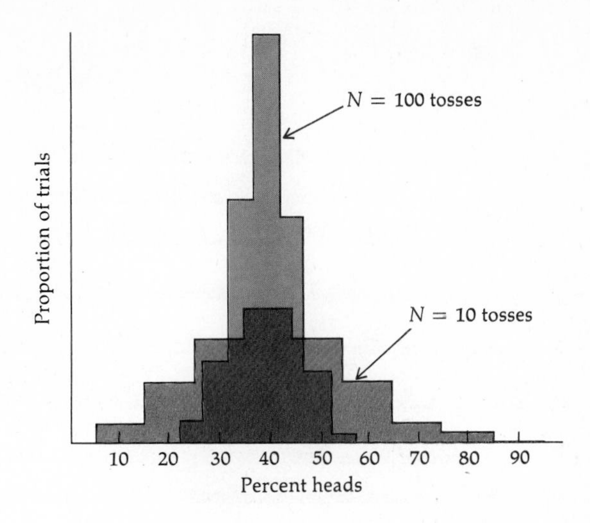

Figure 19.7
Computer-simulated experiments with coins that are "loaded" so as to give heads only 40% of the time. The experiments are analogous to those in Figure 19.6. The mode of the distribution is now at the expected probability of 0.40. (After W. F. Bodmer and L. L. Cavalli-Sforza, *Genetics, Evolution, and Man*, W. H. Freeman, San Francisco, 1976.)

over many generations may be large. If no other processes (mutation, migration, or selection) affect allelic frequencies at a gene locus, evolution will ultimately result in the fixation of one allele and the elimination of all others. When only drift is operating, the probability that a given allele will ultimately be fixed is precisely its frequency; e.g., if a certain allele has a frequency of 0.2 at a given time, it has a probability of 0.2 of ultimately being the only allele in the population. But this may require a very long time; the average number of generations required for fixation is about four times as large as the number of parents per generation.

Assume that a new allele arises by mutation in a population with effective size N. Since there are $2N$ alleles in the population, the frequency of the new mutant is $1/2N$, which is also the probability that the population will become fixed for that mutant by drift alone, since all alleles have the same probability of becoming fixed. The population will eventually be made up of genes all descended from that new mutant (or of one of the other $2N$ alleles in the population), but this can be shown to require, on the average, $4N$ generations when drift alone is operating. If the effective size of the population is 1 million individuals, the process will require about 4 million generations.

It is unlikely that random drift *alone* will affect allelic frequencies at any locus during long periods of time; mutation, migration, and selection are likely to take place at one time or another. These last three processes

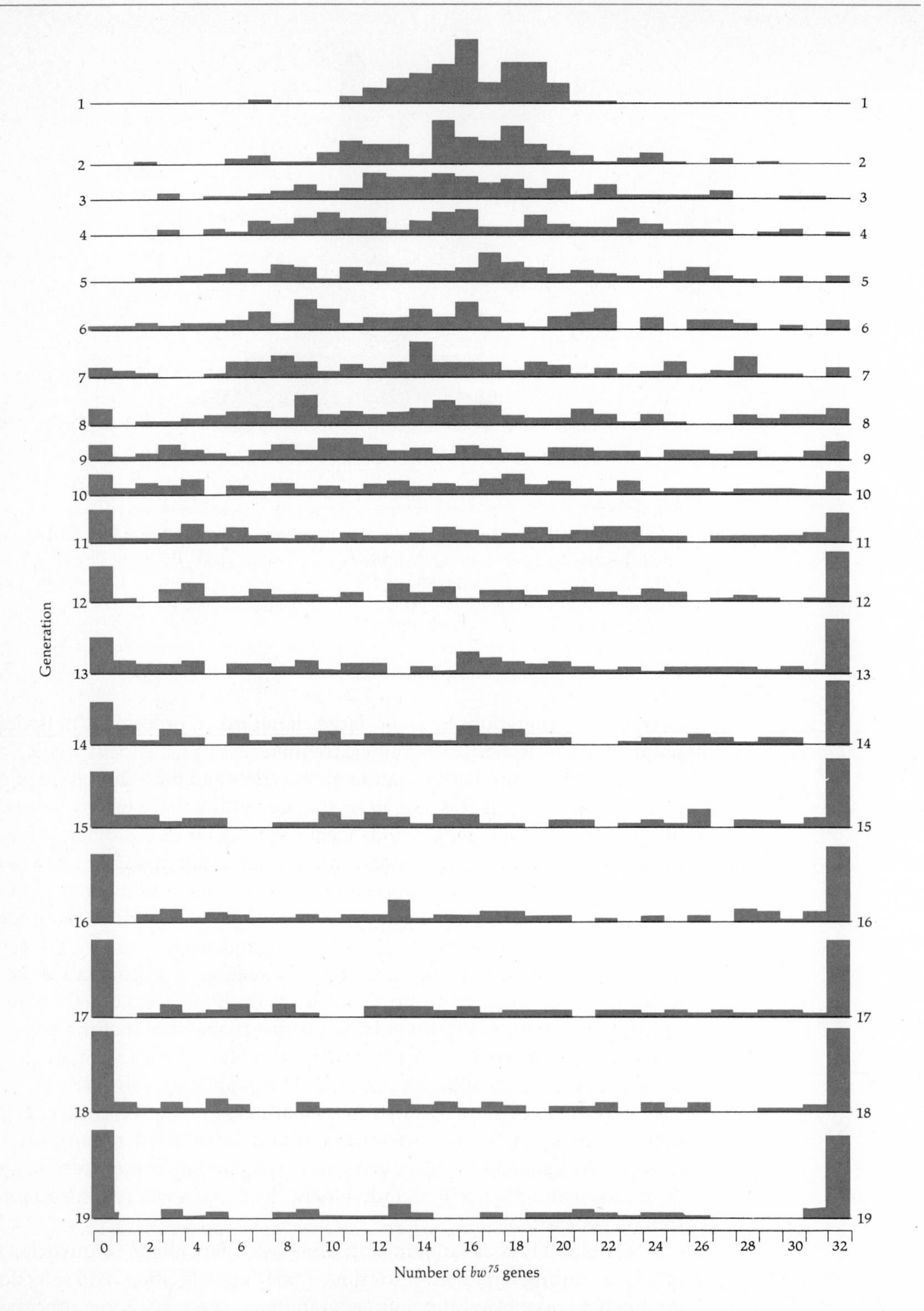

Figure 19.8
Distribution of allele frequencies in 19 consecutive generations among 107 lines of *Drosophila melanogaster*, each with 16 individuals. [After P. Buri, *Evolution 10*:367 (1956).]

are the *deterministic* processes of evolutionary change. Let us use x to represent the mutation rate (u) or the migration rate (m), or the selection coefficient (s, to be defined in Chapter 20); then gene frequency changes will be governed primarily by random genetic drift if, and only if,

$$4Nx << 1$$

where $<<$ means "much less than." If $4Nx \approx 1$ or greater, then gene frequency changes will be determined for the most part by the deterministic processes.

For example, assume that the mutation rate from allele A to allele a is $u = 10^{-5}$ (and assume there is no migration or selection). In a population of 100 breeding individuals, this rate will have little effect on gene frequencies relative to the effects of drift, because $4Nu = 4 \times 10^2 \times 10^{-5} = 4 \times 10^{-3} << 1$. In a population of one million breeding individuals, however, it would have a greater effect than drift, because then $4Nu = 4 \times 10^6 \times 10^{-5} = 40 > 1$. If the migration rate is 0.02 (or two individuals for every hundred) per generation and there is no mutation or selection, gene frequencies will change toward the frequencies in the population from which the migrants come, even in a small population with only 100 individuals, because in such case $4Nm = 4 \times 100 \times 0.02 = 8 > 1$.

Extreme cases of random genetic drift occur when a new population is established by only very few individuals; this has been called by Ernst Mayr the *founder effect.* Populations of many species living on oceanic islands, although they may now consist of millions of individuals, are descendants of one or very few colonizers arrived long ago by accidental dispersal. The situation is similar in lakes, isolated forests, and other ecological isolates. Because of sampling errors, gene frequencies at various loci are likely to be different in the few colonizers than in the population from which they came, which may have lasting effects on the evolution of such isolated populations.

An experimental demonstration of the founder effect is shown in Figure 19.9. Laboratory populations of *Drosophila pseudoobscura* were begun with samples from a population in which a certain chromosomal arrangement, represented as *PP*, had a frequency of 0.50. There were two types of populations; some ("large") were started with 5000 individuals, and the others ("small"), with 20 individuals. After 1½ years, or about 18 generations, the mean frequency of *PP* was about 0.30 in the large as well as in the small populations, but the range of frequencies was considerably greater in the small populations.

Chance variations in allelic frequencies similar to those due to the founder effect occur when populations go through *bottlenecks.* When climatic or other conditions are unfavorable, populations may be drastically reduced in numbers and run the risk of extinction. Such populations may later recover their typical size, but random drift may considerably alter their allelic frequencies during the bottleneck and, therefore, in the following generations. In primitive mankind, many tribes were decimated owing to various disasters; some of these tribes undoubtedly

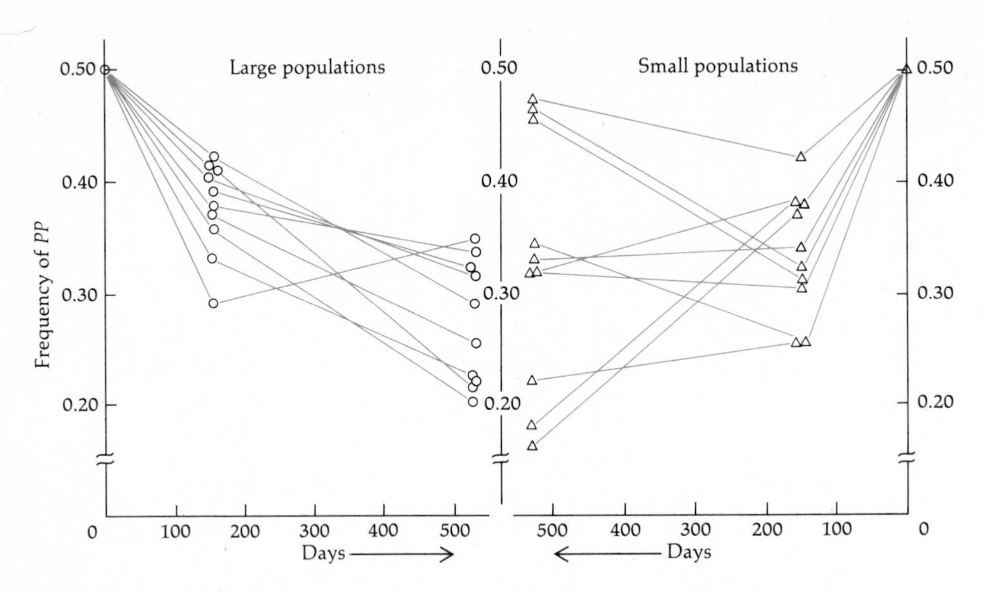

Figure 19.9
Founder effect in laboratory populations of *Drosophila pseudoobscura*. The graphs show the changes in the frequency of a chromosomal inversion, *PP*. Note that time proceeds from left to right for the 10 large populations, but from right to left for the 10 small populations. [After Th. Dobzhansky and O. Pavlovsky, *Evolution 11*:311 (1957).]

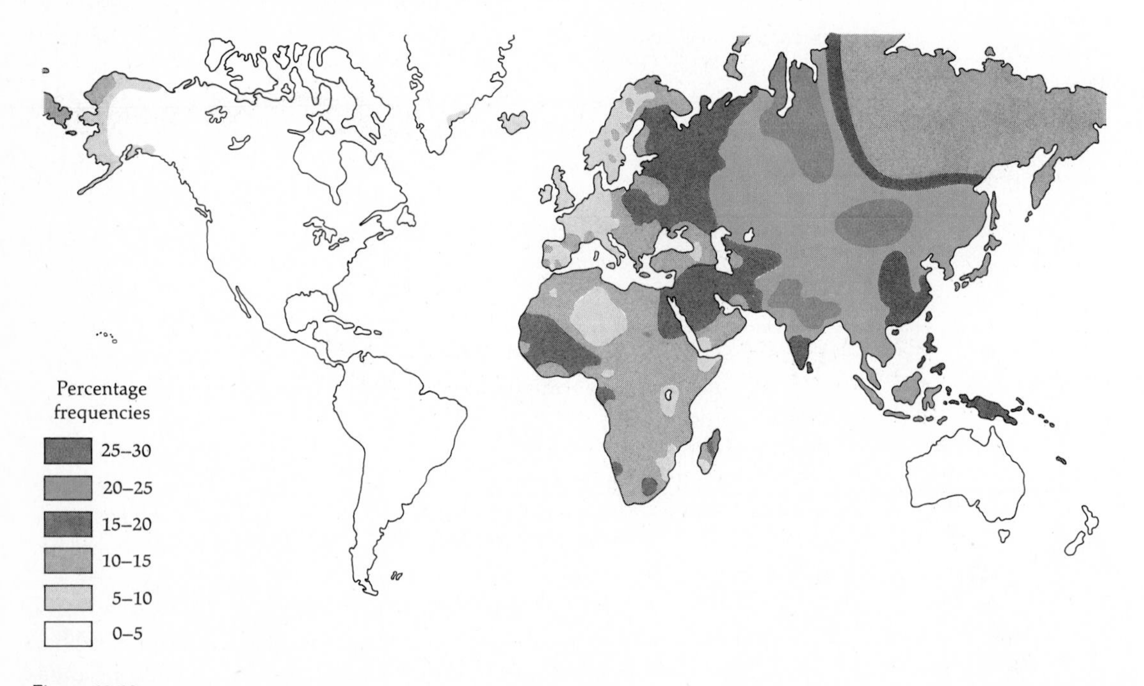

Figure 19.10
Frequency of the blood-group allele I^B in aboriginal populations of the world. The I^B allele is nearly or altogether absent among American Indians and aboriginal Australians unmixed with Europeans, but it is present in all Old World populations. The frequency of I^B is highest in northern India, Mongolia, central Asia, and in some aboriginal peoples from Siberia.

suffered rapid extinction as a result, but presumably most recovered from the survivors, possibly augmented by migrants from other tribes. Differences among human populations in the frequency of the ABO blood-group alleles may have resulted, at least in part, from population bottlenecks and founder effects (Figure 19.10).

A recent example of the founder effect in a human population is the Dunker communities in Pennsylvania. The Dunkers are a religious sect established by 27 families that emigrated from Germany in the mid-eighteenth century. Since then they have lived in small communities, intermarrying little with the surrounding populations. The effects of drift can be noted in several gene loci. The frequency of blood group A (genotypes I^AI^A and I^Ai) is 40–45% in German populations and in U.S. populations of German descent; among the Dunkers, however, it is 60%, while the I^B allele has gone nearly to extinction (frequency 2.5%). At the locus for the M-N blood groups, the frequency of the *M* allele is about 54% in both German and U.S. populations, but 65% among the Dunkers.

Problems

1. In a sample of 1100 Japanese from Tokyo, it was found that 356, 519, and 225 individuals had blood groups M, MN, and N, respectively. Calculate the allele frequencies and the expected Hardy-Weinberg genotypic frequencies. Use the chi-square test to determine whether the expected and observed numbers are in statistical agreement.

2. For the data in the previous problem, assume that the genotypic frequencies are the same in men and women. Assuming random mating, calculate the probabilities of all possible types of mating (see Table 19.1) and the probability of the progeny produced by each mating. When the progenies of all matings are taken into account, the expected genotypic frequencies should be the same as in the previous problem (i.e., those predicted by the Hardy-Weinberg law).

3. The frequency of red-green color blindness in the men in a certain population is 0.08. This form of color blindness is caused by a sex-linked recessive allele. What are the expected frequencies of the three genotypes among women?

4. The most common form of hemophilia is due to a sex-linked recessive allele with a frequency of 0.0001. What are the expected frequencies of the two male genotypes and the three female genotypes in the population?

5. Tay-Sachs disease is caused by an autosomal recessive allele. The disease is characterized by mental deficiency and blindness, with death occurring

by four years of age. The incidence of the disease among newborns is about 10 per million births. Assuming Hardy-Weinberg equilibrium, estimate the frequency of the allele and of the heterozygotes.

6. Cystic fibrosis is a recessive disease characterized by malabsorption of food and obstruction of the bronchial tubes and other tissues. Death usually occurs by the late teens. About 4 in 10,000 newborns will suffer from cystic fibrosis. Assuming Hardy-Weinberg equilibrium, what are the frequencies of the three genotypes among newborns?

7. Acatalasia is a recessive condition first discovered in Japan. Heterozygotes can be identified by the intermediate level of catalase in their blood. The frequency of heterozygotes is 0.09% in Hiroshima and Nagasaki, but 1.4% in the rest of Japan. Assuming Hardy-Weinberg equilibrium, calculate the allelic frequencies in (1) Hiroshima and Nagasaki and (2) the rest of Japan.

8. An experimental population of *Drosophila melanogaster* is started with 100 *bw*/*bw* females and 100 bw^+/bw^+ males. What will be the genotypic frequencies in the F_1, the F_2, and the following generations, assuming random mating and assuming that all genotypes reproduce equally effectively?

9. A population begins with the following composition at a sex-linked locus:

Males: 400 *A* 600 *a*

Females: 640 *AA* 320 *Aa* 40 *aa*

Assuming random mating, what will be the equilibrium genotypic frequencies?

10. The ratio between the recessive alleles (a) that exist in heterozygotes and in homozygotes is approximately $1/q$ when q is small. What proportion of *all a* alleles are found in homozygotes? (Note that in this case the approximation used in the text of Chapter 19 is not required.)

11. Calculate the proportion of all recessive alleles that are found in homozygotes for Tay-Sachs disease and for cystic fibrosis (see problems 5 and 6).

12. The procedure given on page 634 for calculating the allelic frequencies is not the most efficient, because it does not use the AB blood group for the calculations and therefore ignores part of the available information. The maximum likelihood method is the best for estimating the allelic frequencies, but it is quite complicated. A better method than the one used on page 634 starts by calculating p, q, and r, as is done there. Let $D = 1 -$

$p - q - r$. The corrected allelic frequencies are then given by $p^* = p(1 + D/2)$, $q^* = q(1 + D/2)$, and $r^* = (r + D/2)(1 + D/2)$. Use this method to estimate the allelic frequencies in the following populations:

	Number of individuals having blood group:				
Population	AB	B	A	O	Total
English	5,782	16,279	79,341	88,774	190,177
Chinese	606	1,626	1,920	1,848	6,000
Pygmies	103	300	313	316	1,032
Eskimos	7	17	260	200	484

(Note that $D = 0$ for the frequencies given on page 634, because those assumed frequencies exactly fit the Hardy-Weinberg expectations.)

13. In *Escherichia coli* the rate of mutation from histidine independence (his^+) to histidine requirement (his^-) and the rate of the reverse mutation have been estimated as

$$his^+ \rightarrow his^- \quad 2 \times 10^{-6}$$

$$his^- \rightarrow his^+ \quad 4 \times 10^{-8}$$

Assuming that no other processes are involved, what will be the equilibrium frequencies of the two alleles?

14. Assume that the forward and backward mutation rates at a certain locus of *Drosophila melanogaster* are

$$A \rightarrow a \quad 2 \times 10^{-5}$$

$$a \rightarrow A \quad 6 \times 10^{-7}$$

What are the expected allelic equilibrium frequencies if no other processes are involved? Is the calculation of mutation equilibrium frequencies of alleles different for diploid and haploid organisms?

15. Assume that at a certain locus the mutation rate of $A \rightarrow a$ is 10^{-6} and that there is no back mutation. What will be the frequency of A after 10, 1000, and 100,000 generations of mutation?

16. Assuming that ten generations have elapsed since the African ancestors came to the U.S., calculate the average rate of gene flow per generation between U.S. Blacks and Caucasians from Claxton, Georgia, using the frequencies of the Fy^a allele given in Table 19.5.

17. A population of *Drosophila melanogaster* is polymorphic for two alleles A_1 and A_2. One thousand populations are derived, and each is maintained by selecting at random ten females and ten males in each generation as parents of the following generation. After many generations, it is observed that 220 populations are fixed for allele A_1, and 780 for allele A_2. Estimate the allele frequencies in the original population, assuming that only genetic drift is involved.

18. What will be the range of allele frequencies (standard deviation) in the first generation among the 1000 populations of problem 17?

19. The effective size of a population, N_e, can be estimated by the equation

$$N_e = \frac{4N_m N_f}{N_m + N_f}$$

where N_m and N_f are the number of males and females producing the following generation. If $N_m = N_f$, then $N_e = N_m + N_f$; otherwise N_e is smaller than the sum of the two parents. Assume that in a cattle ranch 100 bulls and 400 cows are used to produce the following generation. What is the effective population size? A neighboring rancher maintains 500 cows, all artificially inseminated with the sperm of a single bull; what is the effective population size in this case?

20

Natural Selection

The Concept of Natural Selection

In Chapter 19 we considered three of the four processes that change gene frequencies—mutation, migration, and drift. Now we introduce the fourth and most critical one—natural selection. But first let us recall some essential features of the other three. We can predict the *direction* and *rate* of change in allele frequencies due to mutation or to migration whenever the appropriate parameters are known (i.e., the mutation or migration rate and the allele frequencies). With respect to drift, a knowledge of the relevant parameters (effective population size and allele frequencies) makes it possible to calculate the expected magnitude of allele frequency changes—i.e., the expected *rate* of change—but not the *direction* of change, because this is random.

There is, however, an important attribute that mutation, migration, and drift have in common: none of them is oriented with respect to adaptation. These processes change gene frequencies independently of whether or not such changes increase or decrease the adaptation of organisms to their environments. Therefore, because these processes are random with respect to adaptation, they would, by themselves, destroy the organization and adaptation characteristic of living beings. Natural selection is the process that promotes adaptation and keeps the disorganizing effects of the other processes in check. In this sense, natural selection is the most critical evolutionary process, because only natural selection accounts for the *adaptive* and highly organized nature of living creatures. Natural selection also explains the *diversity* of organisms because it promotes their adaptation to different ways of life.

The idea of natural selection as the fundamental process of evolutionary change was reached independently by Charles Darwin and Alfred Russel Wallace. In 1858 they made a joint presentation of their discovery to the Linnean Society of London. The argument for evolution by natural selection was developed fully, with considerable supporting evidence, in *The Origin of Species*, published by Darwin in 1859. He proposed that carriers of hereditary variations that are useful as adaptations to the environment are likely to survive better and produce more progeny than organisms possessing less useful variations. As a consequence, adaptive variations will gradually increase in frequency over the generations, at the expense of less adaptive ones. This process of differential multiplication of hereditary variations was called *natural selection.* The outcome of the process is organisms well adapted to their environments.

Natural selection can be defined simply as *the differential reproduction of alternative genetic variants*, determined by the fact that some variants increase the chances of survival and reproduction of their carriers relative to the carriers of other variants. Natural selection may be due to differential survival or differential fertility, or both.

Darwin emphasized that competition for limited resources, a common situation in nature, results in natural selection of the most effective competitors. He wrote, for example: "As more individuals are produced than can possibly survive, there must in every case be a struggle for existence, either one individual with another of the same species, or with the individuals of distinct species." But Darwin also noted that natural selection may occur without competition, as a result of inclement weather and other aspects of "the physical conditions of life." Populations of all sorts of organisms are often depleted during unfavorable seasons; some organisms are better equipped than others to withstand inclement weather. Moreover, natural selection may occur even if no organism would die before completing its reproductive period, simply because some organisms produce more offspring than others.

Darwinian Fitness

The parameters used to measure mutation, migration, and drift are the mutation rate, the migration rate, and the variance of allelic frequencies, respectively. The parameter commonly used to measure natural selection is *Darwinian fitness*, or *relative fitness* (also called *selective value* and *adaptive value*). Fitness is a measure of the reproductive efficiency of a genotype.

Natural selection operates by *differential* reproduction; accordingly, fitness is often expressed as a relative, not absolute, measure of reproductive efficiency. For mathematical convenience, geneticists usually

Table 20.1
Computation of the fitnesses of three genotypes when the number of progeny produced by each genotype is known.

	Genotype			
	A_1A_1	A_1A_2	A_2A_2	Total
(*a*) Number of zygotes in one generation	40	50	10	100
(*b*) Number of zygotes produced by each genotype in next generation	80	90	10	180
Computation				
1. Average number of progeny per individual in next generation (*b*/*a*)	80/40 = 2	90/50 = 1.8	10/10 = 1	
2. Fitness (*relative* reproductive efficiency)	2/2 = 1	1.8/2 = 0.9	1/2 = 0.5	

assign the fitness value 1 to the genotype with the highest reproductive efficiency. Assume that at a certain locus there are three genotypes and that, for every one progeny produced by an A_1A_1 homozygote, the heterozygote A_1A_2 produces, on the average, one progeny also, but the A_2A_2 homozygote produces only 0.8 progeny. Then the fitnesses of the three genotypes are 1, 1, and 0.8, respectively.

Table 20.1 shows how fitnesses can be computed when the number of progeny produced by each genotype is known. There are two steps. First, one calculates the average number of progeny produced per individual by each genotype in the next generation. Second, one divides the average number of progeny of each genotype by that of the *best* genotype.

If we know the genotypic fitnesses, we can predict the rate of change in the frequency of the genotypes. The converse is also true, and geneticists often compute fitnesses based on the changes in genotypic frequencies. A simple example is the following. Assume that in a haploid organism, such as *Escherichia coli*, the frequency of the two genotypes, A and a, in a large population is 0.50 at a given time, but the frequencies change to 0.667 A and 0.333 a in one generation; we infer that the fitnesses of A and a are 1 and 0.50, respectively. Note that a large population is assumed so that drift can be ignored; it is also assumed that there is neither mutation nor migration, or that they are so low that they can be ignored.

Table 20.2
Fitness differences due to differences in survival rate.

	Genotype		
Component	A_1A_1	A_1A_2	A_2A_2
Survival fitness	1	0.9	0.5
Fertility fitness	1	1	1
Net fitness	$1 \times 1 = 1$	$0.9 \times 1 = 0.9$	$0.5 \times 1 = 0.5$

Fitness is often represented by the letter w. A related measure is the *selection coefficient* (not to be confused with *selective value*, which is the same as fitness); it is usually represented by s, and is defined as $s = 1 - w$ (therefore, $w = 1 - s$). The selection coefficient measures the reduction in fitness of a genotype. In Table 20.1, the selection coefficient is 0 for the genotype A_1A_1, 0.1 for A_1A_2, and 0.5 for A_2A_2.

Relative fitness predicts the course of selection, i.e., it predicts how gene frequencies will change, but not how well the population will do. Because relative fitness values are relative measures, they do not say whether the population will increase or decrease in numbers. Assume, for example, that in Table 20.1 the numbers of zygotes produced by the three genotypes in the next generation are 40, 45, and 5. The relative fitnesses would be the same as those shown in the table now, although the total number of zygotes in the population would have decreased from 100 to 90 rather than increased from 100 to 180.

Aspects of the life of an individual that may affect its reproductive success contribute to natural selection and, therefore, to the fitness of genotypes. These various aspects—survival, rate of development, mating success, fertility, etc.—are called *fitness components.* The main ones are survival (often called *viability*) and fertility. Other components can be treated separately or incorporated into these two. For example, rate of development, mating success, and age of reproduction can be incorporated into fertility if this is expressed as a function of age.

Fitness differences may be due to one or several fitness components. From the point of view of natural selection, what counts is the overall, or net, fitness, not which components are involved (although this may be interesting in other regards). Tables 20.2 and 20.3 show three genotypes that have identical fitnesses in the two examples, although the components involved are different. Tay-Sachs disease is a human condition caused by the accumulation of complex fatty substances, called gangliosides, in the central nervous system, leading to complete mental degeneration, blind-

Table 20.3
Fitness differences due to differences in fertility.

	Genotype		
Component	A_1A_1	A_1A_2	A_2A_2
Survival fitness	1	1	1
Fertility fitness	1	0.9	0.5
Net fitness	$1 \times 1 = 1$	$1 \times 0.9 = 0.9$	$1 \times 0.5 = 0.5$

ness, and early death. Achondroplastic dwarfs reproduce, on the average, only 20% as efficiently as normal individuals. The fitness of Tay-Sachs patients is 0, owing to poor viability; that of achondroplastics is 0.20, owing to reduced fertility.

The ultimate outcome of natural selection may be either the elimination of one or another allele (although mutation may keep deleterious alleles at low frequencies, as shown below) or a stable polymorphism with two or more alleles. The effects of natural selection can be simply treated when there are only two alleles, and therefore three genotypes, at a single

Box 20.1 Kinds of Equilibrium

A physical system is in equilibrium when its state will not change if left undisturbed by external forces. The equilibrium may be stable, unstable, or neutral, depending on the behavior of the system when it is disturbed. A *stable* equilibrium exists when the system returns to its original state after the disturbance. An equilibrium is *unstable* when, after the disturbing force has ceased, the state of the system continues to change in the direction of the disturbance (until it finds a natural boundary or limit). An equilibrium is *neutral* when the state of the system does not change any more (does not react) after the disturbing force ceases.

Model examples of these three kinds of equilibria are (see Figure 20.1): a ball at the lowest point of a smooth concave surface (stable equilibrium), a ball at the highest point of a smooth convex surface (unstable equilibrium), and a ball on a perfectly horizontal surface (neutral equilibrium).

All three types of equilibrium may occur with respect to allele frequencies. The equilibrium may involve only one allele (*monomorphic* equilibrium) or more than one allele (*polymorphic* equilibrium).

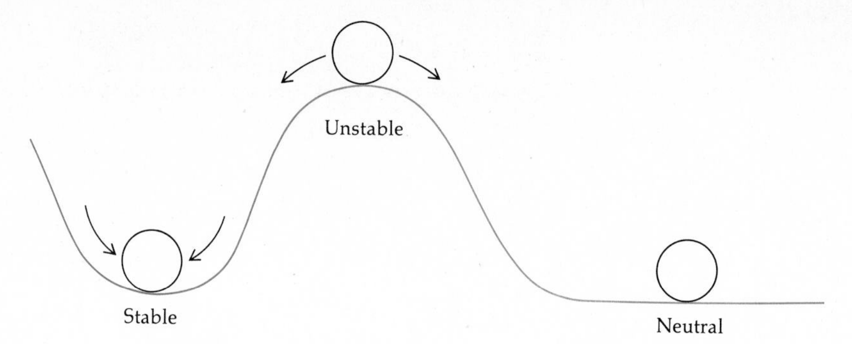

Figure 20.1
Three kinds of equilibrium. When displaced from a stable or unstable equilibrium position, the ball will move in the direction indicated by the arrows; a ball in neutral equilibrium, however, will remain at the position to which it has been displaced.

gene locus. In the following sections we will consider five cases: (1) selection against a recessive allele; (2) selection against a dominant allele; (3) selection against an allele without dominance; (4) selection in favor of the heterozygotes; and (5) selection against the heterozygotes. The first three cases lead to the eventual elimination of the disfavored allele. Case 4 leads to a stable polymorphism with both alleles present at frequencies determined by the selection coefficients against the two homozygotes. Case 5 has a point of polymorphic equilibrium, but the equilibrium is unstable, so that selection leads to fixation for one or the other allele. In the models considered, it is assumed that genotypic fitnesses are constant, i.e., independent of the frequency of the alleles themselves, of the density of the population, and of any other factors. Frequency-dependent selection—when the fitnesses are a function of the genotypic frequencies—will be considered later. Like selection favoring the heterozygotes, frequency-dependent selection may lead to stable polymorphic equilibrium.

Selection Against Recessive Homozygotes

Recessive alleles—such as those for colorless seeds (c) in corn, vestigial wings (vg) in *Drosophila,* or phenylketonuria in humans—produce heterozygotes that have phenotype and fitness identical to those of the dominant homozygotes. However, the recessive homozygotes may have considerably reduced fitness. Selection will, then, occur against the recessive homozygotes. We can study the effects of selection using the following general model:

Genotype:	AA	Aa	aa
Fitness (w):	1	1	$1 - s$

Table 20.4
Allele frequency changes after one generation of selection against recessive homozygotes.

	Genotype				
	AA	Aa	aa	Total	Frequency of a
1. Initial zygote frequency	p^2	$2pq$	q^2	1	q
2. Fitness (w)	1	1	$1 - s$		
3. Contribution of each genotype to next generation	p^2	$2pq$	$q^2(1 - s)$	$1 - sq^2$	
4. Normalized frequency	$\frac{p^2}{1 - sq^2}$	$\frac{2pq}{1 - sq^2}$	$\frac{q^2(1 - s)}{1 - sq^2}$	1	$q_1 = \frac{q - sq^2}{1 - sq^2}$
5. Change in allele frequency					$\Delta q = \frac{-spq^2}{1 - sq^2}$

The procedure used to calculate gene frequency changes from one generation to the next is shown in Table 20.4; additional details are shown in Box 20.2. The initial zygote frequencies are assumed to be in Hardy-Weinberg equilibrium, owing to the random combination of gametes from the previous generation. The basic step in the calculation is shown in the third row of Table 20.4: multiplication of the initial zygote frequencies (first row) by their relative fitnesses (second row), which gives the contribution of each genotype to the next generation. However, the values in the third row do not add up to 1. In order to convert them to frequencies that add up to 1, we must divide them by their sum total, an operation called *normalization*, which is done in the fourth row. From the frequencies of the progenies produced by each genotype, we can calculate the allele frequencies after selection (by the procedure explained in Chapter 18, page 602); the *change* in allele frequency due to selection is obtained by subtracting the original allele frequency from the frequency after selection. In the first, fourth, and fifth rows of Table 20.4 are shown the initial frequency q of the a allele, the frequency q_1 after one generation of selection, and the change in the frequency due to selection, $\Delta q = q_1 - q$.

The effect of selection against recessive homozygotes is a decrease in the frequency of the recessive allele, a result we would expect, because the recessive homozygotes are reproducing less effectively than the genotypes carrying the dominant allele.

Box 20.2 Selection Against Recessive Homozygotes

The model for selection against recessive homozygotes is given in the text. The basic steps are shown in Table 20.4. The frequencies of the two alleles at the beginning (zygote stage) of a certain generation are p and q, for A and a, respectively. The genotypes are assumed to be in Hardy-Weinberg equilibrium and will thus have the frequencies given in the first row. Since there are only two alleles, $p + q = 1$, and therefore $p^2 + 2pq + q^2 = (p + q)^2 = 1$. The basic step is shown in the third row of the table: multiplication of the initial zygote frequencies (first row) by their fitnesses (second row), which represent the relative rates of reproduction of the genotypes.

When we sum the values given in the third row, we see that they do not add up to 1:

$$p^2 + 2pq + q^2(1 - s) = p^2 + 2pq + q^2 - sq^2 = (p^2 + 2pq + q^2) - sq^2 = 1 - sq^2 \neq 1$$

In order to convert the third-row values to frequencies, we divide each value by their sum total, as shown in the fourth row; these now add up to 1:

$$\frac{p^2}{1 - sq^2} + \frac{2pq}{1 - sq^2} + \frac{q^2(1 - s)}{1 - sq^2} = \frac{p^2 + 2pq + q^2 - sq^2}{1 - sq^2} = \frac{1 - sq^2}{1 - sq^2} = 1$$

The frequency of the a allele after one generation of selection, q_1, is calculated by adding the frequency of the aa genotype and half the frequency of the Aa genotype:

$$q_1 = \frac{q^2(1 - s)}{1 - sq^2} + \frac{pq}{1 - sq^2} = \frac{pq + q^2 - sq^2}{1 - sq^2} = \frac{q(p + q) - sq^2}{1 - sq^2} = \frac{q - sq^2}{1 - sq^2}$$

The frequency of the A allele after one generation of selection, p_1, can be obtained by adding the frequency of the AA homozygotes and half the frequency of the Aa heterozygotes or, alternatively, by subtracting from 1 the frequency of a after selection. Using the first procedure, we obtain

$$p_1 = \frac{p^2}{1 - sq^2} + \frac{pq}{1 - sq^2} = \frac{p^2 + pq}{1 - sq^2} = \frac{p(p + q)}{1 - sq^2} = \frac{p}{1 - sq^2}$$

The allele frequency change is shown in the fifth row. The initial frequency of the a allele was q, and the frequency after selection is $q_1 = (q - sq^2)/(1 - sq^2)$. Therefore, the change in allele frequency per generation is

$$\Delta q = q_1 - q = \frac{q - sq^2}{1 - sq^2} - q = \frac{q - sq^2 - q(1 - sq^2)}{1 - sq^2} = \frac{q - sq^2 - q + sq^3}{1 - sq^2} = \frac{-sq^2(1 - q)}{1 - sq^2}$$

And, because $1 - q = p$, we obtain

$$\Delta q = \frac{-spq^2}{1 - sq^2}$$

Because the values of s, p, and q are either positive (but less than one) or zero, the numerator in the expression for Δq will be either negative or zero. And, because both s and q have values less than one, the denominator will be a positive number. Therefore, the value of Δq will be negative (unless it is zero), indicating that the value of q will decrease as a result of selection.

What will be the ultimate outcome of selection? By definition, there will be no further change in allele frequencies when

$$\Delta q = \frac{-spq^2}{1 - sq^2} = 0$$

Δq will be zero when the numerator is zero; this will happen only when $q = 0$ (or when s or p is zero, but this implies either that there is no selection or that the dominant allele is absent from the population). Therefore the value of q will gradually decrease (as it progresses through the values of q_1, q_2, q_3, etc.) to zero. The ultimate outcome of selection against recessive homozygotes is the elimination of the recessive allele.

Another interesting question concerns the size of Δq, i.e., the amount of change in allele frequency per generation. For a given value of s, the product pq^2 is greatest when $p = 0.33$ and $q = 0.67$; it becomes ever smaller as q decreases from that frequency to lower frequencies. This is so because, although p increases as q decreases, the decrease in q^2 is greater than the increase in p (the square of a number less than 1 is smaller than the number itself). Furthermore, the denominator increases as q^2 becomes smaller. Therefore, the rate of selection becomes extremely slow as q approaches zero (Figure 20.2 and Table 20.5).

The phenomenon of *industrial melanism* has been best studied in the moth *Biston betularia* in England. Until the middle of the nineteenth century, these moths were a uniformly peppered light gray color. Then, darkly pigmented variants started to appear in industrial regions, where the vegetation had gradually blackened owing to pollution from soot and other wastes. In some localities the dark varieties have almost completely replaced the light ones. The light gray moths are homozygous (*dd*) for a recessive allele; the dark moths are either heterozygous (*Dd*) or homozygous (*DD*) for the dominant allele.

The replacement of dark forms for light forms of *B. betularia* in industrial regions is due to differential predation by birds: on tree trunks darkened by pollution, the light gray forms are conspicuous, while the dark forms are well camouflaged. H. B. D. Kettlewell released marked dark and light moths near Birmingham, a heavily industrialized area. The proportions of recaptured moths were 53% for the dark form and 25% for the light form. Since the fertility of the two forms is about the same, we may assume that the relative fitnesses are determined exclusively by the difference in survival rate, mostly due to predation (Table 20.6). The procedure used to estimate fitnesses is essentially that shown in Table

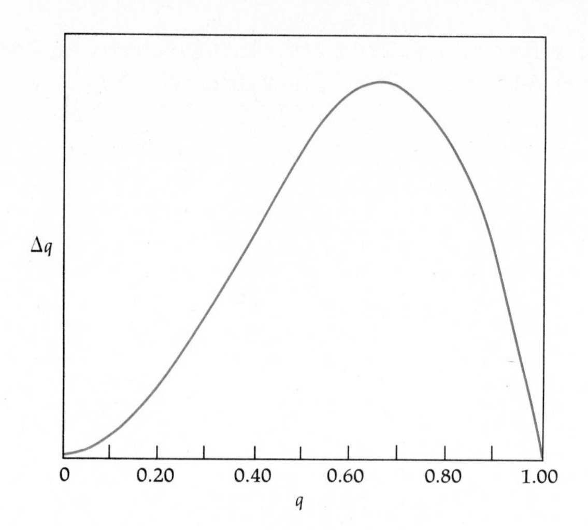

Figure 20.2
The change in allele frequency per generation (Δq) as a function of allele frequency (q) in the case of selection against recessive homozygotes.

20.1, except that the dominant homozygotes and the heterozygotes are treated jointly, since they are phenotypically indistinguishable.

Assume now that, at a certain time, the frequency of the *d* allele is $q = 0.50$ (frequency of the *dd* homozygotes $= 0.25$). Using the fitness values obtained in Table 20.6, we can calculate the selection coefficient against the recessive homozygotes (light form) as $s = 1 - w = 1 - 0.47 = 0.53$. With the formulas given in Table 20.4, we can calculate the changes that would occur owing to selection in a single generation. The frequency of the *d* allele is

$$q_1 = \frac{q - sq^2}{1 - sq^2} = \frac{0.50 - (0.53 \times 0.50^2)}{1 - (0.53 \times 0.50^2)} = \frac{0.3675}{0.8675} = 0.424$$

The change in allele frequency is

$$\Delta q = \frac{-spq^2}{1 - sq^2} = \frac{-0.53 \times 0.50 \times 0.50^2}{0.8675} = \frac{-0.06625}{0.8675} = -0.076$$

We can check that $0.424 + 0.076 = 0.50$, the initial frequency of the *d* allele. Note that the frequency of the *dd* homozygotes in the following generation will be $q^2 = 0.424^2 = 0.180$. This is *not* the same as the value for the zygotes *produced* by *dd* individuals, which is 0.135, as determined by the expression for *aa* in the fourth row of Table 20.4, namely, $q^2(1 - s)/(1 - sq^2)$. Recessive homozygotes are produced by matings involving the heterozygotes as well as the recessive homozygotes of the previous generation. Similarly, not all offspring produced by recessive homozygotes are

Table 20.5
Number of generations required to effect a given reduction in allele frequency (q) for various values of the selection coefficient (s) against recessive homozygotes.

	Generations Required				
Reduction in q	$s = 1$	$s = 0.50$	$s = 0.10$	$s = 0.01$	$s = 0.001$
0.99 to 0.50	1	11	56	559	5,585
0.50 to 0.10	8	20	102	1,020	10,198
0.10 to 0.01	90	185	924	9,240	92,398
0.01 to 0.001	900	1,805	9,023	90,231	902,314
0.001 to 0.0001	9,000	18,005	90,023	900,230	9,002,304

Table 20.6
Relative fitness of dark and light *Biston betularia* moths in Birmingham, England.

	Dark	Light
Genotype	*DD* and *Dd*	*dd*
(a) Number released	154	64
(b) Number recaptured	82	16
Survival rate (b/a)	0.53	0.25
Relative fitness (w)	$0.53/0.53 = 1$	$0.25/0.53 = 0.47$

also recessive homozygotes; matings with dominant homozygotes or with heterozygotes produce heterozygotes.

Biston betularia illustrates an important point concerning natural selection: the fitness of a genotype may be different in different environments. In unpolluted areas, the light-gray moths survive better than dark moths because, on lichen-covered trees, they are quite inconspicuous to bird predators. When light and dark moths were released in the unpolluted region of Dorset, a greater proportion of light moths than of dark moths was recovered. The data and the fitness calculations are shown in

Table 20.7
Relative fitness of dark and light *Biston betularia* moths in Dorset, England.

	Dark	Light
Genotype	*DD* and *Dd*	*dd*
(*a*) Number released	406	393
(*b*) Number recaptured	19	54
Survival rate (b/a)	0.047	0.137
Relative fitness (w)	0.047/0.137 = 0.343	0.137/0.137 = 1

Table 20.7. The reversal of fitness values between polluted and unpolluted areas is impressive. The fitness of light moths relative to dark moths is 0.47:1 in Birmingham, but 1:0.34 in Dorset.

Recessive Lethals

The limiting case of selection against recessive alleles occurs when the recessive homozygotes have zero fitness, i.e., when they die before attaining reproductive age or are sterile. A familiar example is phenylketonuria. The fitness of the *aa* homozygotes, if left untreated, is $w = 0$; therefore $s = 1$. Substituting 1 for s simplifies the formulas in Table 20.4. In particular, the frequency of allele *a* after one generation of selection becomes

$$q_1 = \frac{q - sq^2}{1 - sq^2} = \frac{q - q^2}{1 - q^2} = \frac{q(1 - q)}{(1 + q)(1 - q)} = \frac{q}{1 + q}$$

and the change in allele frequency becomes

$$\Delta q = \frac{-spq^2}{1 - sq^2} = \frac{-pq^2}{1 - q^2} = \frac{-(1 - q)q^2}{(1 + q)(1 - q)} = \frac{-q^2}{1 + q}$$

The amount of change in allele frequency after a given number of generations can now easily be calculated. Representing the frequency of *a* in the initial and following generations by $q_0, q_1, q_2, \ldots, q_t$, we have

$$q_1 = \frac{q_0}{1 + q_0} \quad \text{and} \quad q_2 = \frac{q_1}{1 + q_1}$$

Substituting the value of q_1 in the equation for q_2, we get

$$q_2 = \frac{\dfrac{q_0}{1 + q_0}}{1 + \dfrac{q_0}{1 + q_0}} = \frac{\dfrac{q_0}{1 + q_0}}{\dfrac{1 + q_0 + q_0}{1 + q_0}} = \frac{q_0}{1 + 2q_0}$$

Similarly, after t generations of selection,

$$q_t = \frac{q_0}{1 + tq_0}$$

The number of generations, t, required to change the allele frequency from a certain value, q_0, to another, q_t, can be derived from the last formula:

$$q_t(1 + tq_0) = q_0$$

$$q_t + tq_0q_t = q_0$$

$$t = \frac{q_0 - q_t}{q_0q_t} = \frac{1}{q_t} - \frac{1}{q_0}$$

This formula was used to calculate the values for $s = 1$ in Table 20.5. For the particular case when q_t is one-half the value of q_0, the formula becomes

$$t = \frac{1}{q_0/2} - \frac{1}{q_0} = \frac{2}{q_0} - \frac{1}{q_0} = \frac{1}{q_0}$$

That is, the number of generations required to reduce a certain gene frequency to half its initial value is 1 divided by the gene frequency. Thus it takes 10 generations to change q from 0.1 to 0.05, 100 generations to change it from 0.01 to 0.005, and 1000 generations to change it from 0.001 to 0.0005.

The frequency of the recessive allele for albinism in Norway is about 0.01. Assume that a eugenic goal is established to eliminate the allele from the population by sterilizing all albino individuals (see Chapter 19, page 635). It would take 100 generations to reduce the allele frequency to half its present value, and 9900 generations (see Table 20.5 and Figure 19.3) to reduce it to 0.0001. Eugenic measures are inefficient in the case of recessive alleles.

Box 20.3 Selection Against Dominants and Selection Without Dominance

The effects of one generation of selection against the dominant allele are shown in Table 20.8. The sum total of the relative proportions after selection (third row) is obtained as follows:

$$
\begin{aligned}
p^2(1-s) + 2pq(1-s) + q^2 &= p^2 - sp^2 + 2pq - 2spq + q^2 \\
&= p^2 + 2pq + q^2 - sp^2 - 2spq \\
&= (p+q)^2 - s(p^2 + 2pq) \\
&= 1 - s(p^2 + 2pq + q^2 - q^2) \\
&= 1 - s(1 - q^2) \\
&= 1 - s + sq^2
\end{aligned}
$$

The frequency of A after one generation of selection (fourth row) is the frequency of AA plus half the frequency of Aa, or

$$p_1 = \frac{p^2(1-s)}{1-s+sq^2} + \frac{pq(1-s)}{1-s+sq^2} = \frac{(p^2+pq)(1-s)}{1-s+sq^2} = \frac{p(p+q)(1-s)}{1-s+sq^2} = \frac{p(1-s)}{1-s+sq^2}$$

The change in the frequency of the A allele after selection (fifth row) is

$$\Delta p = \frac{p(1-s)}{1-s+sq^2} - p = \frac{p(1-s) - p(1-s+sq^2)}{1-s+sq^2} = \frac{p - sp - p + sp - spq^2}{1-s+sq^2} = \frac{-spq^2}{1-s+sq^2}$$

* * *

In the case of selection without dominance (Table 20.9), the sum total of the relative proportions after selection (third row) is

$$
\begin{aligned}
p^2 + 2pq(1 - s/2) + q^2(1-s) &= p^2 + 2pq + q^2 - spq - sq^2 \\
&= (p+q)^2 - sq(p+q) \\
&= 1 - sq
\end{aligned}
$$

The frequency of A_2 after selection (fourth row) is obtained by adding the frequency of A_2A_2 plus half the frequency of A_1A_2:

$$q_1 = \frac{q^2(1-s)}{1-sq} + \frac{pq(1-s/2)}{1-sq} = \frac{q^2 - sq^2 + pq - spq/2}{1-sq}$$

$$= \frac{q^2 + pq - (sq/2)(2q + p)}{1 - sq}$$

$$= \frac{q(q + p) - (sq/2)(q + q + p)}{1 - sq}$$

$$= \frac{q - sq(1 + q)/2}{1 - sq}$$

The change in the frequency of A_2 is

$$\Delta q = \frac{q - sq(1 + q)/2}{1 - sq} - q = \frac{q - (sq/2)(1 + q) - q + sq^2}{1 - sq}$$

$$= \frac{-(sq/2)(1 + q - 2q)}{1 - sq}$$

$$= \frac{-(sq/2)(p + q + q - 2q)}{1 - sq}$$

$$= \frac{-spq/2}{1 - sq}$$

Selection Against Dominants and Selection Without Dominance

Selection is more effective against dominant alleles than against recessive alleles because a dominant allele is expressed in the heterozygotes as well as in the homozygotes. Assume that dominance with respect to fitness is complete, so that dominant homozygotes and heterozygotes have equal fitness. The model is

Genotype:	AA	Aa	aa
Fitness (w):	$1 - s$	$1 - s$	1

The effects of one generation of selection are shown in Table 20.8; the calculations are detailed in Box 20.3. The procedure followed is similar to that used in the case of selection against recessives (Table 20.4).

Table 20.8
Allele frequency changes after one generation of selection against genotypes carrying the dominant allele.

	Genotype				
	AA	Aa	aa	Total	Frequency of A
1. Initial zygote frequency	p^2	$2pq$	q^2	1	p
2. Fitness (w)	$1 - s$	$1 - s$	1		
3. Contribution of each genotype to next generation	$p^2(1 - s)$	$2pq(1 - s)$	q^2	$1 - s + sq^2$	
4. Normalized frequency	$\frac{p^2(1 - s)}{1 - s + sq^2}$	$\frac{2pq(1 - s)}{1 - s + sq^2}$	$\frac{q^2}{1 - s + sq^2}$	1	$p_1 = \frac{p(1 - s)}{1 - s + sq^2}$
5. Change in allele frequency					$\Delta p = \frac{-spq^2}{1 - s + sq^2}$

The change in allele frequency due to selection is

$$\Delta p = \frac{-spq^2}{1 - s + sq^2}$$

As long as both alleles are in the population (p and q are positive) and there is selection (s is positive), the values of spq^2 and of $1 - s + sq^2$ will be positive. Therefore Δp will have a negative value and the frequency of A will gradually decrease to zero. If a dominant allele is sterile or lethal ($s = 1$), the change in allele frequency is

$$\Delta p = \frac{-pq^2}{1 - 1 + q^2} = \frac{-pq^2}{q^2} = -p$$

The frequency of the dominant allele then becomes zero in one single generation of selection—a result that is obvious, since neither the homozygotes nor the heterozygotes for the dominant allele leave any progeny.

The frequency of the allele selected against is represented by p in the case of selection against dominants (Table 20.8), and by q in the case of selection against recessives (Table 20.4). For any given frequency of the allele selected against, if s is the same, the change in allele frequency is greater in the case of selection against dominants than in the case of se-

Table 20.9
Allele frequency changes after one generation of selection when there is no dominance.

	Genotype				
	A_1A_1	A_1A_2	A_2A_2	Total	Frequency of A_2
1. Initial zygote frequency	p^2	$2pq$	q^2	1	q
2. Fitness (w)	1	$1 - s/2$	$1 - s$		
3. Contribution of each genotype to next generation	p^2	$2pq(1 - s/2)$	$q^2(1 - s)$	$1 - sq$	
4. Normalized frequency	$\frac{p^2}{1 - sq}$	$\frac{2pq(1 - s/2)}{1 - sq}$	$\frac{q^2(1 - s)}{1 - sq}$	1	$q_1 = \frac{q - sq(1 + q)/2}{1 - sq}$
5. Change in allele frequency					$\Delta q = \frac{-spq/2}{1 - sq}$

lection against recessives. This is as expected, because in the former case the heterozygotes are also selected against, but not in the latter case.

In some instances, the fitness of the heterozygotes is intermediate between the fitnesses of the two homozygotes. We shall consider only the general case when the selection coefficient against the heterozygotes is exactly half the selection coefficient against the disfavored homozygotes, i.e., when there is no dominance. The model is

Genotype:	A_1A_1	A_1A_2	A_2A_2
Fitness (w):	1	$1 - s/2$	$1 - s$

The effects of one generation of selection are shown in Table 20.9, and the detailed calculations are given in Box 20.3. The change in allele frequency is

$$\Delta q = \frac{-spq/2}{1 - sq}$$

It will be negative as long as both alleles persist in the population and there is selection. The equilibrium condition, $\Delta q = 0$, will be reached only when $q = 0$, i.e., when the allele selected against, A_2, is completely eliminated.

Selection and Mutation

The ultimate outcome of selection in all three cases considered so far (selection against recessive homozygotes, against dominants, and without dominance) is the elimination of the deleterious allele. Such alleles nevertheless remain in populations owing to mutation. Selection and mutation have opposite effects with respect to deleterious alleles. The net effect of the two processes will be nil when the number of deleterious alleles eliminated by selection is the same as that produced by mutation.

Let us first consider the case of recessive alleles. The frequency q of the recessive allele a will decrease, owing to selection, by the amount (Table 20.4)

$$\Delta q = \frac{-spq^2}{1 - sq^2}$$

Because the a allele will be at low frequency, the denominator will be nearly 1, and the approximate change in allele frequency due to selection will be

$$\Delta q \approx -spq^2$$

The a allele will, however, increase in frequency by the amount up, owing to mutation from A to a (Chapter 19, page 640). (We may ignore back mutation from a to A because the frequency of a is low.) There will be equilibrium between mutation and selection when

$$spq^2 \approx up$$

If p is nearly 1, we have

$$sq^2 \approx u$$

$$q \approx \sqrt{\frac{u}{s}}$$

When $s = 1$, this equation reduces to

$$q \approx \sqrt{u}$$

That is, the equilibrium frequency of an allele that causes lethality or sterility in homozygous condition is, approximately, the square root of the mutation rate. If $u = 10^{-5}$, the approximate equilibrium frequency of a recessive lethal allele will be $q = \sqrt{10^{-5}} = 0.003$. On the other hand, if $u = 10^{-5}$ but $s = 0.1$, the equilibrim frequency of a recessive deleterious allele will be $q = \sqrt{10^{-5}/10^{-1}} = \sqrt{10^{-4}} = 0.01$, or about three times higher than that of a lethal allele.

In the case of dominance, the frequency p of the dominant allele A will decrease, owing to selection, by the amount (Table 20.8)

$$\Delta p = \frac{-spq^2}{1 - s + sq^2}$$

But the A allele will increase in frequency by the amount uq, owing to mutation pressure. Making the same approximations as before (i.e., ignoring back mutation as well as the denominator in the expression for Δp), there will be an equilibrium when

$$spq^2 \approx uq$$

$$spq \approx u$$

Because p is small, q, will be nearly 1. Replacing q by 1, we obtain

$$sp \approx u$$

$$p \approx \frac{u}{s}$$

And, if $s = 1$ (i.e., the allele is lethal),

$$p \approx u$$

That is, in the case of a lethal dominant allele, the equilibrium frequency of the allele is simply the mutation rate. This we should expect. Individuals carrying a lethal dominant allele fail to reproduce; hence, the only such alleles found in a population will be those that have newly arisen by mutation in that generation.

Assuming the same mutation rates and the same selection coefficients, the equilibrium frequency is much higher for a recessive allele than for a dominant allele. (Note that the square root of a positive number less than one is greater than the number.) We expect this result because recessive alleles are hidden from selection in the heterozygotes.

If $u = 10^{-5}$, the equilibrium frequency of a lethal dominant allele is about 10^{-5}, 300 times smaller than 0.003, the approximate equilibrium frequency obtained for a lethal recessive allele. If $u = 10^{-5}$ and $s = 0.1$, the equilibrium frequency of a deleterious dominant allele is about $10^{-5}/10^{-1} = 10^{-4}$, 100 times smaller than the frequency obtained for a deleterious recessive allele. However, the number of *individuals* exhibiting the deleterious phenotype will be about twice as high for a dominant allele as for a recessive deleterious allele. In the former case, the frequency of individuals with the deleterious phenotype will be $2pq$ (the homozygotes with a frequency of p^2 will be very few if p is negligibly small). Because q will be nearly 1, we have $2pq \approx 2p$, which is approximately equal to $2u/s$.

Figure 20.3
An achondroplastic dwarf. Detail from *Las Meninas* (the ladies-in-waiting) by the Spanish painter Diego Velázquez. (Prado Museum, Madrid.)

In the case of a recessive allele, the deleterious phenotype is manifested only in the homozygotes, which have a frequency of $q^2 = (\sqrt{u/s})^2 = u/s$.

The allele equilibrium frequencies are, for dominant as well as recessive alleles, directly related to u and inversely related to s. Thus the equilibrium frequencies will be higher when u is greater or when s is smaller.

Achondroplasia is a deleterious condition caused by a dominant allele present at low frequencies in human populations. Because of abnormal growth of the long bones, achondroplastics have short, squat, and often deformed limbs, as well as bulging skulls (Figure 20.3). The mutation rate from the normal allele to the achondroplasia allele is 5×10^{-5}. Achondroplastics reproduce only 20% as efficiently as normal individuals; hence

$s = 0.8$. The equilibrium frequency of the allele can therefore be calculated as

$$p = \frac{u}{s} = \frac{5 \times 10^{-5}}{0.8} = 6.25 \times 10^{-5}$$

Because q is nearly 1, the heterozygote frequency, $2pq$, reduces approximately to $2p = 2 \times 6.25 \times 10^{-5} = 1.25 \times 10^{-4}$, or 1.25 individuals per 10,000 births, which is indeed the observed frequency of achondroplastic dwarfs in the population. The expected frequency of homozygotes is $(6.25 \times 10^{-5})^2 = 39 \times 10^{-10}$, or about 4 per billion individuals. Homozygotes have an extreme form of the syndrome, and the few known cases died in the fetal stage.

Chromosomal abnormalities may be treated as dominant mutations. Since people suffering from Down's syndrome do not reproduce, $s = 1$, and therefore $p \approx u/s \approx u$. The frequency of Down's syndrome trisomy is thus simply the frequency with which it arises in human populations owing to chromosomal nondisjunction. However, as with all dominant mutations, the frequency of individuals suffering from Down's syndrome is about twice the mutation rate (the frequency of heterozygotes is $2pq \approx 2p \approx 2u$). The incidence of Down's syndrome is about 1 per 700 births; the "mutation" rate to Down's trisomy is about 1 per 1400 gametes.

Estimation of Mutation Rates

Mutation rates from recessive to dominant alleles can be estimated directly by counting the number of dominant individuals born to recessive parents. In human populations, for example, the number of achondroplastics born to normal parents amounts to about 1 per 10,000 births. Hence the mutation rate to achondroplasia is 1 per 20,000 gametes, or 5×10^{-5} per gamete per generation.

In the case of recessive alleles, this simple method of estimation is not possible, because mutants in heterozygous condition are not expressed in the phenotype. The equations determining the allele equilibrium frequencies under mutation and selection can be used for estimating the mutation rates to recessive alleles (and, of course, to dominant alleles as well). If the selection coefficients and the equilibrium frequencies are known, the mutation rates can be calculated. For dominant alleles,

$$p = \frac{u}{s} \quad \text{or} \quad u = sp$$

For recessive alleles,

$$q = \sqrt{\frac{u}{s}} \quad \text{or} \quad u = sq^2$$

As pointed out in the previous section, these equations are only approximations. Moreover, selection coefficients may change in populations, and, consequently, the allelic frequencies observed may not always represent equilibrium frequencies. Despite these and other possible pitfalls, the equations for mutation-selection equilibrium are the best method for estimating recessive mutation rates in human populations, in which other methods (requiring, for example, inbreeding) cannot be used.

The frequency at birth of the recessive defect phenylketonuria (PKU) is approximately 4 per 100,000; therefore $q^2 = 4 \times 10^{-5}$. The reproductive efficiency of untreated PKU patients is zero, or $s = 1$. Then

$$u = sq^2 = 4 \times 10^{-5}$$

The frequency of this allele in human populations is

$$q = \sqrt{4 \times 10^{-5}} = 6.3 \times 10^{-3}$$

and the frequency of heterozygotes is

$$2pq \approx 2q = 2 \times 6.3 \times 10^{-3} = 1.26 \times 10^{-2}$$

That is, about 1.3 per 100 humans carry the allele, although only 4 per 100,000 suffer from PKU. The frequency of PKU alleles carried by heterozygotes is one-half of 1.26×10^{-2}, or 6.3×10^{-3}; the frequency of the allele carried by homozygotes is 4×10^{-5}. Therefore $(6.3 \times 10^{-3})/(4 \times 10^{-5}) = 158$ times more PKU alleles are present in PKU heterozygotes than in homozygotes. As pointed out earlier, rare alleles exist in populations mostly in heterozygous combinations.

Heterozygote Advantage

Selection in favor of the heterozygote over both homozygotes is known as *overdominance,* or *heterosis.* The model is

Genotype:	AA	Aa	aa
Fitness (w):	$1 - s$	1	$1 - t$

The effects of one generation of selection are summarized in Table 20.10 (some calculations are detailed in Box 20.4).

Selection in favor of the heterozygotes is different from the other modes of selection considered thus far, in a very significant way: overdominance leads to a stable polymorphic equilibrium, with frequencies determined by the coefficients of selection against the two homozygotes. The change in allele frequency due to selection is

$$\Delta q = \frac{pq(sp - tq)}{1 - sp^2 - tq^2}$$

Table 20.10
Allele frequency changes after one generation of selection when there is overdominance.

	Genotype				
	AA	Aa	aa	Total	Frequency of a
1. Initial zygote frequency	p^2	$2pq$	q^2	1	q
2. Fitness (w)	$1 - s$	1	$1 - t$		
3. Contribution of each genotype to next generation	$p^2(1 - s)$	$2pq$	$q^2(1 - t)$	$1 - sp^2 - tq^2$	
4. Normalized frequency	$\frac{p^2(1 - s)}{1 - sp^2 - tq^2}$	$\frac{2pq}{1 - sp^2 - tq^2}$	$\frac{q^2(1 - t)}{1 - sp^2 - tq^2}$	1	$q_1 = \frac{q - tq^2}{1 - sp^2 - tq^2}$
5. Change in allele frequency					$\Delta q = \frac{pq(sp - tq)}{1 - sp^2 - tq^2}$

The equilibrium condition, $\Delta q = 0$, will be satisfied only when the numerator is zero. If both alleles are present in a population (p and q are greater than zero), this will occur only when

$$sp = tq$$

$$s(1 - q) = tq$$

$$s = q(s + t)$$

$$q = \frac{s}{s + t}$$

Thus the equilibrium frequency of the a allele is the selection coefficient against AA divided by the sum of the two selection coefficients. Correspondingly, the equilibrium frequency of the A allele is

$$p = \frac{t}{s + t}$$

These two equilibrium frequencies are stable, because selection will change the two allele frequencies until the equilibrium values are reached. If p is greater than its equilibrium value, i.e., if $p > t/(s + t)$, then $sp > tq$, and Δq will be positive. Consequently, q will increase at the expense of p

Box 20.4 Overdominance and Selection Against Heterozygotes

The effects of one generation of selection in favor of the heterozygotes are shown in Table 20.10. The sum total of the relative proportions after selection (third row) is

$$p^2(1 - s) + 2pq + q^2(1 - t) = p^2 + 2pq + q^2 - sp^2 - tq^2 = 1 - sp^2 - tq^2$$

The frequency of *a* after selection (fourth row) is the frequency of *aa* plus half the frequency of *Aa*, or

$$q_1 = \frac{q^2(1 - t)}{1 - sp^2 - tq^2} + \frac{pq}{1 - sp^2 - tq^2} = \frac{q^2 + pq - tq^2}{1 - sp^2 - tq^2} = \frac{q(q + p) - tq^2}{1 - sp^2 - tq^2} = \frac{q - tq^2}{1 - sp^2 - tq^2}$$

The frequency of *A* after selection is

$$p_1 = \frac{p^2(1 - s)}{1 - sp^2 - tq^2} + \frac{pq}{1 - sp^2 - tq^2} = \frac{p^2 + pq - sp^2}{1 - sp^2 - tq^2} = \frac{p(p + q) - sp^2}{1 - sp^2 - tq^2} = \frac{p - sp^2}{1 - sp^2 - tq^2}$$

The change in the frequency of the *a* allele after selection (fifth row) is

$$\Delta q = \frac{q - tq^2}{1 - sp^2 - tq^2} - q = \frac{q - tq^2 - q + sp^2q + tq^3}{1 - sp^2 - tq^2}$$

$$= \frac{sp^2q - tq^2 + tq^3}{1 - sp^2 - tq^2}$$

$$= \frac{sp^2q - tq^2(1 - q)}{1 - sp^2 - tq^2}$$

$$= \frac{sp^2q - tpq^2}{1 - sp^2 - tq^2}$$

$$= \frac{pq(sp - tq)}{1 - sp^2 - tq^2}$$

* * *

The effects of one generation of selection against the heterozygotes are summarized in Table 20.13. The sum total of the relative proportions (third row) is

$$p^2 + 2pq(1 - s) + q^2 = 1 - 2spq$$

The frequency of *a* after selection (fourth row) is

$$q_1 = \frac{q^2}{1 - 2spq} + \frac{pq(1 - s)}{1 - 2spq} = \frac{q^2 + pq - spq}{1 - 2spq} = \frac{q(q + p) - spq}{1 - 2spq} = \frac{q - spq}{1 - 2spq}$$

The change in the frequency of *a* (fifth row) is

$$\Delta q = \frac{q - spq}{1 - 2spq} - q = \frac{q - spq - q + 2spq^2}{1 - 2spq}$$

$$= \frac{-spq + 2spq^2}{1 - 2spq}$$

$$= \frac{-spq(1 - 2q)}{1 - 2spq}$$

$$= \frac{-spq(1 - q - q)}{1 - 2spq}$$

$$= \frac{-spq(p - q)}{1 - 2spq}$$

$$= \frac{spq(q - p)}{1 - 2spq}$$

---

until $sp = qt$. On the other hand, if $p < t/(s + t)$, then $sp < tq$, and Δq will be negative, leading to a decrease in the value of q until the equilibrium frequencies are reached (Figure 20.4).

The equilibrium frequencies in overdominance depend on the relative magnitudes of the two selection coefficients, not on their actual values. Thus, an equilibrium frequency of $q = 0.25$ is obtained, for example, both when $s = 0.1$ and $t = 0.3$ and when $s = 0.02$ and $t = 0.06$. It follows that knowing the equilibrium allele frequencies does *not* allow one to calculate the actual values of s and t, but only their relative magnitudes.

An extreme form of heterosis occurs in *balanced lethal* systems, such as the *Curly-Plum* (*Cy-Pm*) system in *Drosophila melanogaster* (Box 17.1, page 576) and the velans-gaudens system in *Oenothera lamarckiana* (page 579 and Figure 17.19). In these cases, $s = t = 1$ and $p = q = 0.5$.

A well-known example of overdominance in human populations is sickle-cell anemia, a disease that is fairly common in some African and Asian populations (Chapter 10, page 348). The anemia is due to homozygosis for an allele, Hb^S, that produces an abnormal hemoglobin instead of the normal hemoglobin produced by allele Hb^A. Most of the

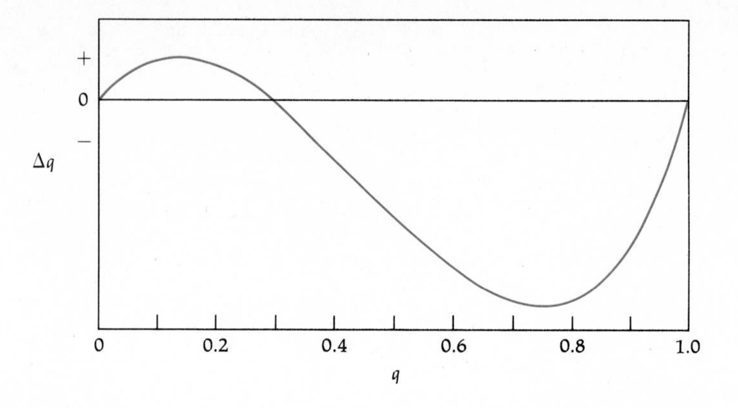

Figure 20.4
The change in allele frequency when selection favors the heterozygotes. The curve shown is the equation $\Delta q = pq(sp - tq) / (1 - sp^2 - tq^2)$, where the equilibrium frequency $q = s/(s + t) = 0.3$. The change in q is positive when $q < 0.3$ and negative when $q > 0.3$.

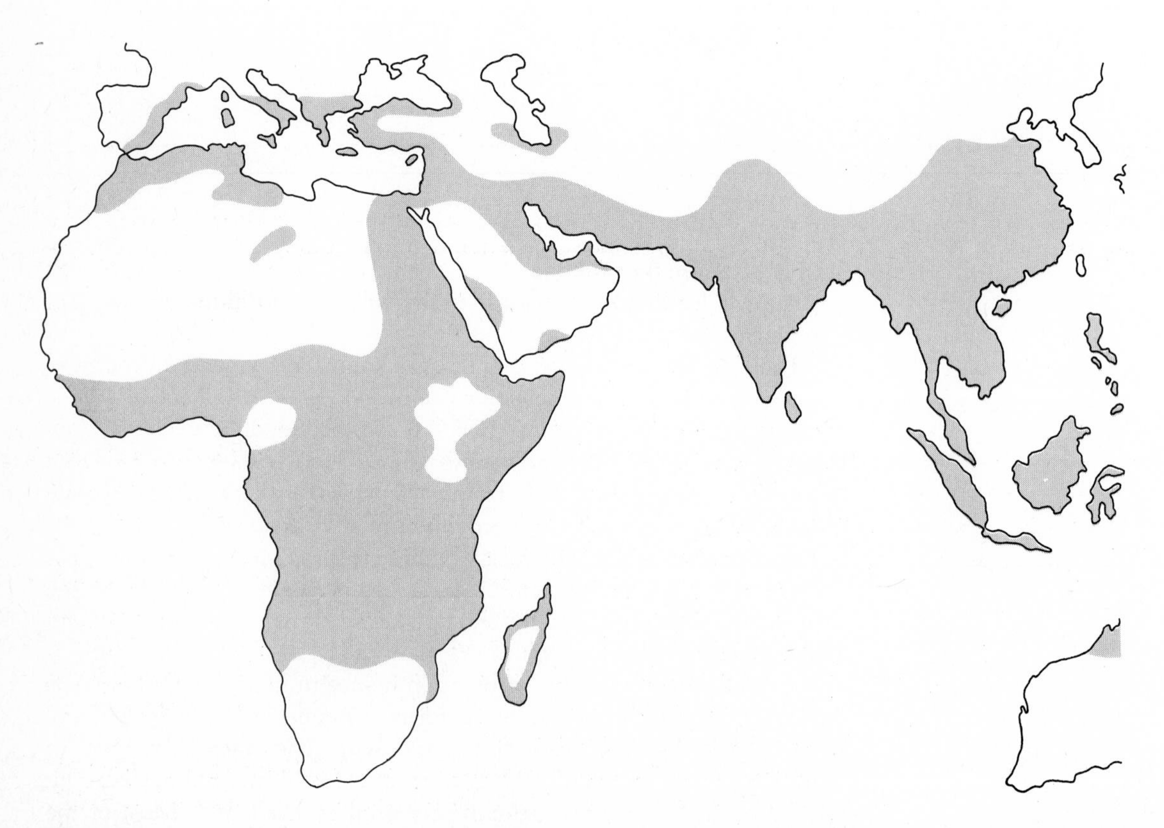

Figure 20.5
Distribution of malignant malaria caused by the parasite *Plasmodium falciparum* in the Old World.

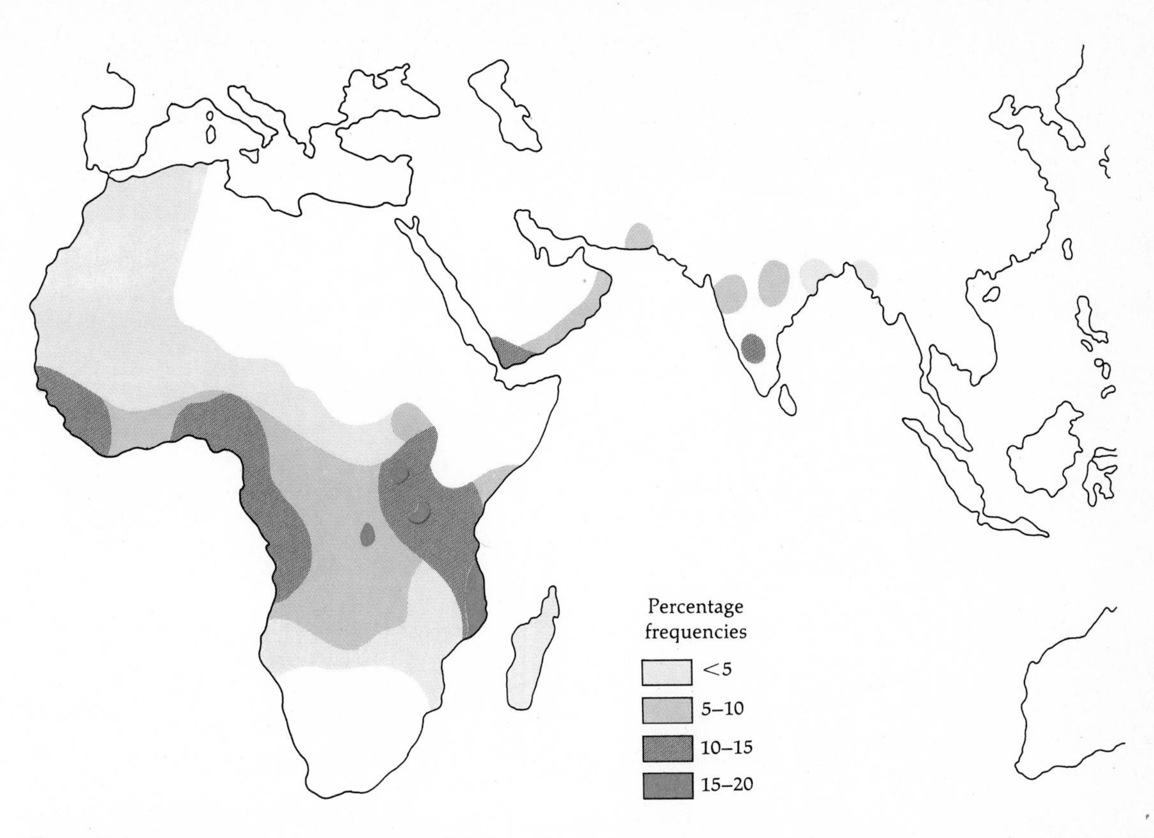

Figure 20.6
Distribution of the allele Hb^S, which in homozygous condition is responsible for sickle-cell anemia. The frequency of Hb^S is high in those regions of the world where *falciparum* malaria is endemic, because Hb^AHb^S individuals, heterozygous for the Hb^S and the "normal" allele, are highly resistant to malarial infection.

homozygotes Hb^SHb^S die before reaching sexual maturity, so their fitness is only slightly greater than zero. In spite of this, the Hb^S allele has fairly high frequencies in certain regions of the world, precisely in those regions where a certain form of malaria, caused by the parasite *Plasmodium falciparum*, is common (Figures 20.5 and 20.6).

The reason for the high frequency of the Hb^S allele in malarial regions is that heterozygotes Hb^AHb^S are resistant to malarial infections, whereas homozygotes Hb^AHb^A are not. Where *falciparum* malaria is rife, the heterozygotes have selective advantage over both homozygotes, which have a fairly high probability of dying from either anemia (Hb^SHb^S homozygotes) or malaria (Hb^AHb^A homozygotes).

Among 12,387 adult individuals examined in Nigeria, 29 were Hb^SHb^S homozygotes, 2993 were Hb^AHb^S heterozygotes, and 9365 were Hb^AHb^A homozygotes. The fitnesses of the three genotypes are computed in Table 20.11, using a slightly different procedure from that illustrated in Table

Table 20.11
Fitnesses of the three genotypes at the sickle-cell anemia locus in a population from Nigeria.

	Genotype				
	Hb^AHb^A	Hb^AHb^S	Hb^SHb^S	Total	Frequency of Hb^S (q)
1. Observed number	9365	2993	29	12,387	
2. Observed frequency	0.7560	0.2416	0.0023	1	0.1232
3. Expected frequency	0.7688	0.2160	0.0152	1	0.1232
4. Survival efficiency (observed/expected)	0.98	1.12	0.15		
5. Relative fitness (survival/1.12)	0.88	1	0.13		

20.1. First, the allele frequencies are calculated from the observed genotypic frequencies: the frequency of Hb^S is $q = 0.1232$. If we assume that the population is at equilibrium with respect to this locus, zygotes will be produced in the frequencies p^2, $2pq$, and q^2. If selection had operated completely by the time the adult genotypes were observed, then the ratios of the observed to the expected genotypic frequencies would give estimates of their relative survival efficiency (shown in the fourth row). These estimates are converted to relative fitnesses by dividing them by the largest value of the survival efficiency, 1.12, so that this becomes 1 (fifth row).

Using the formula developed earlier, we can estimate the allelic equilibrium frequency. The selection coefficients are $s = 1 - 0.88 = 0.12$ against the Hb^AHb^A homozygotes and $t = 1 - 0.13 = 0.87$ against the Hb^SHb^S homozygotes. The expected equilibrium frequency of the Hb^S allele is $0.12/(0.12 + 0.87) = 0.121$ (the frequency of Hb^S calculated from the observed genotypic frequencies is 0.123). The severity of the anemia is manifest in Table 20.11: the homozygotes for the sickle-cell allele survive only 13% as effectively as the Hb^AHb^S heterozygotes. On the other hand, owing to malarial mortality, the homozygotes for the "normal" allele survive only 88% as effectively as the heterozygotes.

Sickle-cell anemia is an additional example of how fitness depends on the environmental conditions. In places where malaria has been eradicated or never existed, the Hb^AHb^A homozygotes have the same fitness as the Hb^AHb^S heterozygotes. The selection mode is no longer overdominance, but selection against recessive homozygotes, which leads to the elimination of the recessive allele. A gradual reduction of Hb^S has occurred in U.S.

Table 20.12
Relative viabilities of flies homozygous for one of six chromosomes of *Drosophila pseudoobscura.*

Chromosome	Relative Viability (%) at Temperature: 25.5°C	21.0°C	16.5°C
1	99	98	100
2	95	89	87
3	92	109	109
4	0	43	89
5	28	73	106
6	3	39	0

Blacks, who have a much lower frequency of the sickle-cell allele than their African ancestors (even after taking into account the admixture with Caucasian ancestry, discussed in Chapter 19, page 642).

Many more examples of the dependence of fitness on the environmental conditions could be given; the example of industrial melanism was discussed earlier. Table 20.12 shows the relative viabilities of *Drosophila pseudoobscura* flies made homozygous for one of six different chromosomes found in nature. Viability is given as a percent of the viability associated with one chromosome, arbitrarily assumed to be "normal" (chromosome 1 at 16.5°C). The relative viabilities of the homozygous flies change from one temperature to another.

Selection Against Heterozygotes

There are situations in which the heterozygotes have lower fitness than either homozygote. Translocation polymorphisms are examples—the heterozygotes usually have lower fitness owing to reduced fertility. We shall consider the simplest case, when the two types of homozygotes have equal fitnesses. The model is

Genotype:	AA	Aa	aa
Fitness (w):	1	$1 - s$	1

Table 20.13
Allele frequency changes after one generation of selection against heterozygotes.

	Genotype				
	AA	*Aa*	*aa*	Total	Frequency of *a*
1. Initial zygote frequency	p^2	$2pq$	q^2	1	q
2. Fitness (w)	1	$1 - s$	1		
3. Contribution of each genotype to next generation	p^2	$2pq(1 - s)$	q^2	$1 - 2spq$	
4. Normalized frequency	$\frac{p^2}{1 - 2spq}$	$\frac{2pq(1 - s)}{1 - 2spq}$	$\frac{q^2}{1 - 2spq}$	1	$q_1 = \frac{q - spq}{1 - 2spq}$
5. Change in allele frequency					$\Delta q = \frac{spq(q - p)}{1 - 2spq}$

The effects of one generation of selection are shown in Table 20.13, and some detailed calculations are shown in Box 20.4.

The change in allele frequencies will be zero when $\Delta q = 0$. This will occur when $p = q$ (but the equilibrium values of p and q will be different when the two types of homozygotes have different fitnesses). However, the equilibrium frequencies are unstable. If $q > p$, then Δq is positive, and q will increase until the A allele is eliminated from the population. If $q < p$, then Δq is negative, and q will decrease further until the a allele is eliminated. Thus a population that is not at equilibrium will depart ever further from the equilibrium frequencies until the allele that initially had a higher-than-equilibrium frequency becomes fixed. The unlikely situation of a population that is initially at equilibrium will not persist; deviations from equilibrium caused by drift and other factors will tend to eliminate one or another allele (Figure 20.7).

The dynamics of selection against heterozygotes can be used for such practical purposes as pest control. Assume that a population has some undesirable property, e.g., mosquitoes carrying malarial parasites. In the laboratory we might be able to obtain a strain that lacks the undesirable trait, e.g., one in which a mutant allele makes the mosquitoes unsuitable hosts for the parasites. A translocation can be induced and the desirable strain made homozygous for the translocation. Mosquitoes of the desirable strain can then be released in sufficiently large numbers that their

Box 20.5 General Model of Selection at a Single Locus

Various modes of selection (against recessive alleles, against dominant alleles, without dominance, in favor of heterozygotes, and against heterozygotes) have been discussed in the text. They are all special cases of a more general model of selection at a single locus. The general model is

Genotype:	A_1A_1	A_1A_2	A_2A_2
Fitness:	w_1	w_2	w_3

The effects of one generation of selection are shown in Table 20.14. The frequency of A_2 after selection is

$$q_1 = \frac{pqw_2 + q^2w_3}{\overline{w}} = \frac{q(pw_2 + qw_3)}{\overline{w}}$$

The change in the frequency of A_2 is

$$\Delta q = \frac{q(pw_2 + qw_3)}{p^2w_1 + 2pqw_2 + q^2w_3} - q = \frac{pqw_2 + q^2w_3 - p^2qw_1 - 2pq^2w_2 - q^3w_3}{\overline{w}}$$

$$= \frac{pqw_2(1 - 2q) + q^2w_3(1 - q) - p^2qw_1}{\overline{w}}$$

$$= \frac{pqw_2(p - q) + pq^2w_3 - p^2qw_1}{\overline{w}}$$

$$= \frac{pq(w_2p - w_2q + qw_3 - pw_1)}{\overline{w}}$$

$$= pq\,\frac{p(w_2 - w_1) + q(w_3 - w_2)}{\overline{w}}$$

The allele frequency after selection and the change in allele frequency for the special cases presented in Tables 20.4, 20.8, 20.9, 20.10, and 20.13 can be obtained from the formulas for the general case by substituting the appropriate fitness values (w).

The change in allele frequency due to one generation of selection, Δq, can be analyzed in terms of three component parts—two in the numerator, and the third the denominator. The first component of the numerator is the product pq, which will always be positive (or zero) but will be small when either p or q is small, and larger when p and q have intermediate values.

Indeed, the effects of selection are generally greatest when the two alleles have intermediate frequencies.

The second component of the Δq expression is

$$p(w_2 - w_1) + q(w_3 - w_2)$$

which shows that the sign and magnitude of Δq are a function of the differences between the fitness values weighted by the allele frequencies.

The third component is the denominator,

$$p^2w_1 + 2pqw_2 + q^2w_3$$

often called the *mean fitness* of the population and represented by $\overline{w}$. This will always have a positive value; therefore, the sign of Δq is always the same as that of the second component of the expression for Δq. The magnitude of Δq is, of course, inversely related to that of the denominator, and it can be shown that natural selection tends to increase $\overline{w}$. Therefore, as the allele frequencies approach the equilibrium value determined by selection, the magnitude of Δq will tend to decrease, and the approach to equilibrium will be slower.

Table 20.14
General model of selection at a single gene locus.

	Genotype				
	A_1A_1	A_1A_2	A_2A_2	Total	Frequency of A_2
1. Initial zygote frequency	p^2	$2pq$	q^2	1	q
2. Fitness	w_1	w_2	w_3		
3. Contribution of each genotype to next generation	p^2w_1	$2pqw_2$	q^2w_3	$\overline{w} = p^2w_1 + 2pqw_2 + q^2w_3$	
4. Normalized frequency	$\frac{p^2w_1}{\overline{w}}$	$\frac{2pqw_2}{\overline{w}}$	$\frac{q^2w_3}{\overline{w}}$	1	$q_1 = \frac{q(pw_2 + qw_3)}{\overline{w}}$
5. Change in allele frequency					$\Delta q = pq\frac{p(w_2 - w_1) + q(w_3 - w_2)}{\overline{w}}$

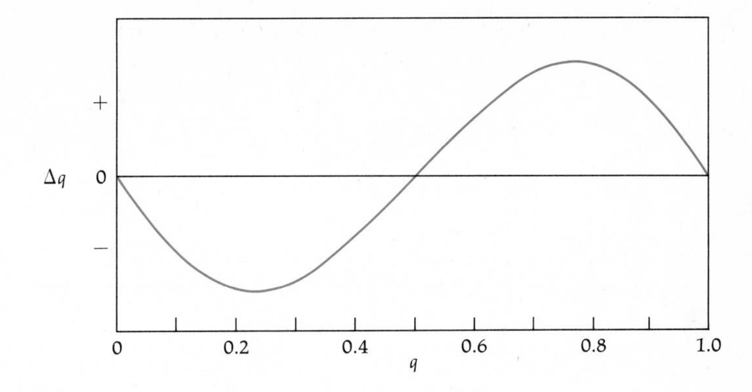

Figure 20.7
The change in allele frequency in the case of selection against the heterozygotes when the two homozygotes are assumed to have equal fitnesses. The equilibrium value ($\Delta q = 0$), when $q = 0.5$, is unstable: if the population is not at equilibrium, the allele frequencies will change until the allele with the lower-than-equilibrium frequency is eliminated from the population.

translocated genotype will occur in the population with greater frequency than the equilibrium value. The translocation genotype will then run to fixation, carrying with it the desirable allele.

Frequency-Dependent Selection

Other forms of selection besides heterozygote advantage may lead to balanced (i.e., stable-equilibrium) genetic polymorphisms. One is *frequency-dependent selection,* which is probably common in nature. Selection is frequency-dependent when the genotypic fitnesses vary with their frequency. In the examples of selection discussed so far, it was assumed that the fitnesses were constant, no matter what the frequencies of the genotypes. This simplifies the mathematical treatment of selection but is often unrealistic. Assume that the fitnesses of two genotypes, *AA* and *aa*, are inversely related to their frequencies—high fitness when a genotype is rare and low fitness when it is common. If a genotype is rare at a given time, natural selection will enhance its frequency; but as its frequency increases, its fitness diminishes, while the fitness of the alternative genotype increases. If there is a frequency at which the two genotypes have equal fitness, a stable polymorphic equilibrium will occur, even without heterosis.

In heterogeneous environments, a genotype may have high fitness when it is rare, because the subenvironments in which it is favored are relatively abundant. But when the genotype is common, its fitness may be

Table 20.15
Numbers of matings to two strains of *Drosophila pseudoobscura* placed together in variable proportions. Each line summarizes the results of several replicate experiments conducted in observation chambers.

Number of Flies in Chamber*	Males Mated			Females Mated		
	C	T	C:T	C	T	C:T
23C, 2T	77	24	3.2:1	93	8	11.6:1
20C, 5T	70	39	1.8:1	84	25	3.4:1
12C, 12T	55	49	1.1:1	50	54	1:1.1
5C, 20T	39	65	1:1.7	30	74	1:2.5
2C, 23T	30	70	1:2.3	12	88	1:7.3

*C = California; T = Texas.

After C. Petit and L. Ehrman, *Evol. Biol.* 3:177 (1969).

low, because its favorable subenvironments are saturated. Frequency-dependent selection has been extensively demonstrated in experimental populations of *Drosophila* and in cultivated plants. For example, in the lima bean, *Phaseolus lunatus*, the fitnesses of three genotypes, *SS*, *Ss*, and *ss*, change over the generations as their frequencies change. The fitness of the heterozygotes is equal to that of the homozygotes when the heterozygotes represent about 17% of the population, but is nearly three times as high when the heterozygotes represent only 2% of the population.

Frequency-dependent *sexual* selection occurs when the probability of mating of a genotype depends on its frequency. Often, the mates preferred are those that happen to be rare, a phenomenon not surprising, perhaps, to people who have experienced the exotic appeal of blondes in Mediterranean countries or of brunettes in Scandinavia. This phenomenon, known as the *rare-mate advantage*, has been thoroughly studied in *Drosophila*, where it commonly involves the males. The results of an experiment are shown in Table 20.15. *Drosophila pseudoobscura* males and females from California (C) and from Texas (T) were placed together in variable proportions. When flies from the C and T localities occur in equal frequencies (12C:12T), they mate with about equal frequencies (55:49 for males, 50:54 for females). But when the two localities are unequally represented, the less common males mate disproportionately more often than the more common males. For example, when the C and T flies exist in the ratio 23:2 (11.5:1), the ratio of matings in the males is 77:24 (3.2:1); that

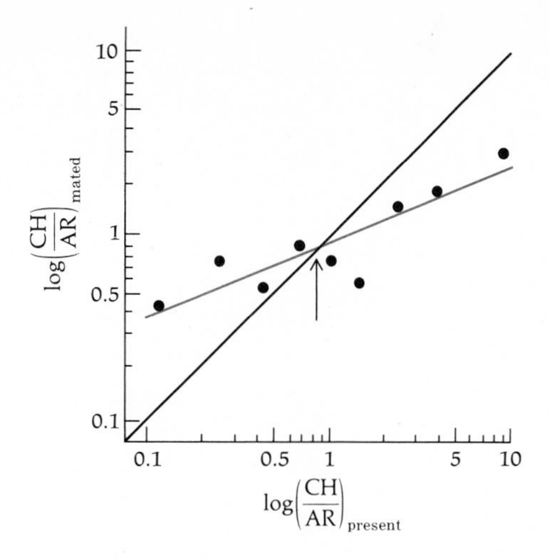

Figure 20.8
Frequency-dependent selection due to mating preferences. Two strains of *Drosophila pseudoobscura* (CH and AR) were combined in different ratios and the numbers of matings of each kind were recorded (in a way similar to that shown in Table 20.15). The graph plots the logarithm of the ratio (CH/AR) of the numbers of males that mated versus the logarithm of the ratio of the numbers of males that were present. Points above the diagonal indicate that males of the CH strain mated disproportionately more often than males of the AR strain; points below the diagonal indicate the opposite. It can be seen that the mating fitness of the CH males decreases as their frequency increases. If male mating differences are the only factor contributing to fitness, there will be a stable equilibrium between the CH and AR strains at the frequency determined by the point at which the diagonal is crossed by the curve representing mating success. [After F. J. Ayala, *Behav. Genet.* 2:85 (1972).]

is, each T male is mating nearly four times as often as each C male (11.5/3.2 = 3.6). When the ratio of flies is reversed (2C:23T), the now rare C males mate five times as often as the common T males (11.5/2.3 = 5.0).

Frequency-dependent selection in favor of rare genotypes is a mechanism contributing to the maintenance of genetic polymorphisms, since the fitness of a genotype increases as it becomes rarer (Figure 20.8). Frequency-dependent sexual selection may be particularly important in cases of migration. Immigrants may have a mating advantage because they are rare, thus making it more likely that their genes will become established in the population they have joined.

Problems

1. Using the generalized expression for $\triangle q$ given in Table 20.14 and the appropriate relative fitness values given in Tables 20.4, 20.8, 20.9, 20.10, and 20.13, derive the $\triangle q$ expressions given in these tables.

2. In an industrialized region, the fitness of *Biston betularia* moths is 1 for the dark form (DD and Dd) and 0.47 for the light form (dd). The allele frequencies at a certain time are $p = 0.40$ (D allele) and $q = 0.60$ (d allele). Place the appropriate values in the first and second rows of Table 20.4 and obtain the numerical values for all the expressions given in rows 3, 4, and 5 of that table. Assume now that the allele frequencies are (1) $p = 0.10$, $q = 0.90$ and (2) $p = 0.90$, $q = 0.10$. Calculate the corresponding values of Δq and compare them with each other and with the one obtained in the first part of this problem.

3. In a region where industrial pollution has been under control for a number of years, the fitness of *Biston betularia* moths is 0.47 for the dark form and 1 for the light form. Calculate the change in allele frequency, Δp, after one generation of selection when (1) $p = 0.40$, (2) $p = 0.10$, and (3) $p = 0.90$. Compare the results with those of the previous problem.

4. For a given gene locus, assume that the heterozygotes have fitness intermediate between the two homozygotes, but not exactly halfway between them, as it is in Table 20.9. That is, the fitness of the heterozygotes is $1 - hs$, where h is some positive number between 0 and 1. Derive the expression for Δq, the change in allele frequency after one generation of selection.

5. Individuals carrying the sickle-cell allele, Hb^S, can be identified through appropriate tests, because their red blood cells take on a characteristic sickle-like shape when exposed to low oxygen tension. This happens in heterozygotes, Hb^AHb^S, as well as in homozygotes, Hb^SHb^S, although to a lesser extent in the former. Is the Hb^S allele dominant with respect to sickling? In regions free of *falciparum* malaria, the fitness of the heterozygotes is similar to that of normal homozygotes, Hb^AHb^A, while homozygotes for the Hb^S allele have very low fitness. Is the Hb^S allele dominant with respect to fitness in malaria-free regions? What about those regions of the world infested with *falciparum* malaria? Do you know other examples of alleles that are recessive with respect to some aspect of the phenotype but not with respect to other aspects?

6. Retinoblastoma is a disease, due to a dominant allele, that leads to early death if left untreated. Assume that the mutation rate from the normal allele to the retinoblastoma allele is 10^{-5}. What is the equilibrium frequency of the allele in a population where the condition is not treated?

7. Assume that the mutation rate to a lethal recessive allele, such as that for Tay-Sachs disease, is 10^{-5}. What is the equilibrium frequency of the allele? Compare the answers to this problem and the previous one.

8. A certain allele in homozygous condition causes sterility in both male and female rats, but has no detectable effect in heterozygotes. The frequency of homozygotes in a wild population is 1 per 1000. Assuming Hardy-

Weinberg equilibrium, what is the frequency of the heterozygotes? If the mutation rate is doubled, what will be the *equilibrium* frequency of sterile individuals and of the heterozygotes?

9. The equilibrium frequency of a lethal recessive allele in a random-mating population of mice is 0.333. What are the fitnesses of the three genotypes?

10. The marine copepod *Tisbe reticulata* can be reared in seawater cultures in the laboratory, although larval mortality is considerable, particularly at high population densities. Heterozygotes $V^V V^M$ for two codominant alleles responsible for color differences were bred at low and high population densities. The numbers of adult F_1 progeny were as follows:

Density	$V^V V^V$	$V^V V^M$	$V^M V^M$	Total
Low	904	2023	912	2839
High	353	1069	329	1751

What are the relative fitnesses (viabilities) of the three genotypes at the two densities?

11. At a locus (*Est-6*) coding for an esterase, there are two alleles, *Est-6*^{F} and *Est-6*^{S}, in an experimental population of *Drosophila*. First-instar larvae of the three genotypes are placed in cultures, and the numbers of emerging adults are recorded. The results of two experiments are as follows:

	Number of larvae			Number of adults		
Experiment	*FF*	*FS*	*SS*	*FF*	*FS*	*SS*
1	160	480	360	80	240	90
2	360	480	160	90	240	80

Assuming that relative fitness depends only on larval viability, what are the fitnesses of the three genotypes in each experiment? Do you think that a stable polymorphic equilibrium might occur in the population and, if so, at what frequency?

21

Inbreeding, Coadaptation, and Geographic Differentiation

The Inbreeding Coefficient

The Hardy-Weinberg law applies only when mating is random, i.e., when the probability of mating between two genotypes is the product of their frequencies. Random mating was generally assumed throughout the two previous chapters. *Assortative mating* (see Chapter 19, page 630) prevails when matings do not occur at random: individuals with certain genotypes are more likely to mate with individuals of certain other genotypes than would be expected from their frequencies. Assortative mating does not, by itself, change gene frequencies, but it does change *genotypic* frequencies. If the probability of mating between like genotypes is greater than would be expected from randomness, the frequency of homozygotes will increase; if the probability is smaller, the frequency of homozygotes will decrease. In general, if the *mating system* is known (i.e., if we know the probabilities of the various types of matings), then the expected genotypic frequencies can be calculated from the genotypic frequencies in the previous generation.

A particularly interesting form of assortative mating is *inbreeding*—when matings between relatives are more frequent than would be expected from randomness. Because relatives are genetically more similar than unrelated individuals, inbreeding increases the frequency of homozygotes and decreases the frequency of heterozygotes relative to the expectations from random mating, although it does not change the allele frequencies. The most extreme kind of inbreeding is *self-fertilization*, or *selfing*, a common form of reproduction in some plant groups. Inbreeding is

Table 21.1
Results of selfing in a population started entirely with heterozygotes, *Aa*.

Generation	Genotypic Frequency *AA*	*Aa*	*aa*	F	Frequency of *a*
0	0	1	0	0	0.5
1	$1/4$	$1/2$	$1/4$	$1/2$	0.5
2	$3/8$	$1/4$	$3/8$	$3/4$	0.5
3	$7/16$	$1/8$	$7/16$	$7/8$	0.5
.	.	.	.	.	.
.	.	.	.	.	.
.	.	.	.	.	.
n	$\frac{1-(1/2)^n}{2}$	$(1/2)^n$	$\frac{1-(1/2)^n}{2}$	$1-(1/2)^n$	0.5
∞	$1/2$	0	$1/2$	1	0.5

often practiced in horticulture and animal husbandry. In human populations (as well as in those of other organisms), inbreeding increases the frequency of recessive hereditary infirmities.

The genetic consequences of inbreeding are measured by the *coefficient of inbreeding,* which is the probability that an individual receives, at a given locus, two alleles that are *identical by descent,* i.e., that are both copied from one single allele carried by an ancestor, belonging to a generation that must be specified. Two alleles with the same DNA sequence are identical *in structure* (or *in state*), but they will not be identical by descent if they have been inherited from unrelated ancestors. The coefficient of inbreeding is usually represented as F.

The results of inbreeding in the case of self-fertilization were worked out by Mendel, who calculated that, after n generations of selfing, the progeny of a heterozygote, *Aa*, consists of homozygotes and heterozygotes in the ratio $2^n - 1$ to 1 (Table 21.1). The progeny of a selfed heterozygote (*Aa*) consist of one-half heterozygotes (*Aa*) and one-half homozygotes (either *AA* or *aa*). The two alleles in each heterozygote are obviously not identical by descent. But in the homozygotes, both alleles are identical by descent because both are copies of the only allele of that type (either *A* or *a*) present in the selfed heterozygous parent. Thus, the proportion of individuals carrying two alleles identical by descent in the first generation of selfing is the same as the total frequency (one-half) of homozygotes, or $F = \frac{1}{2}$.

A selfed homozygote produces only homozygous offspring. These homozygotes all have two alleles that are identical by descent. Therefore the one-half inbreeding acquired in the first generation will remain thereafter and will accumulate with any additional inbreeding acquired in the following generations. In the second generation of selfing, one-half the progeny of the heterozygotes will again consist of homozygotes each having two alleles that are identical by descent. Thus the inbreeding coefficient in the progeny of the heterozygotes is again ½, and, since the heterozygotes represent one-half the population, the increment in the inbreeding coefficient is $\frac{1}{2} \times \frac{1}{2} = \frac{1}{4}$; this, added to the pre-existing ½, becomes $F = \frac{3}{4}$. In each of the following generations, the value of F will increase by one-half multiplied by the frequency of the heterozygotes in the previous generation.

Measuring Inbreeding

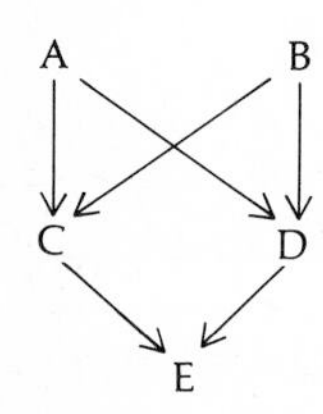

Figure 21.1
Pedigree of an offspring of a brother × sister mating.

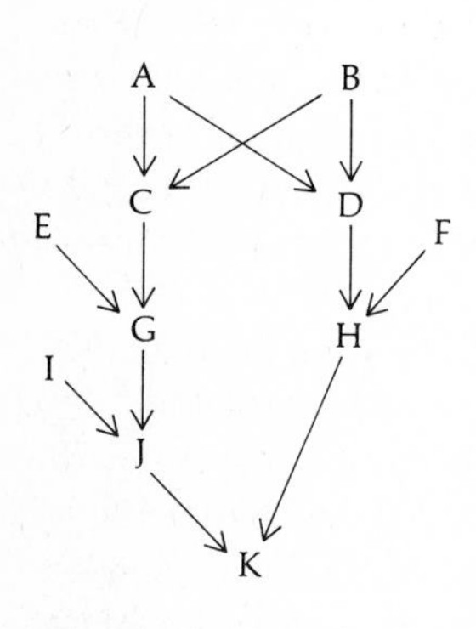

Figure 21.2
Pedigree of an offspring of a mating between first cousins once removed.

Let us calculate the F value in the offspring of full-sibs, i.e., of individuals having the same father and the same mother. Figure 21.1 represents the pedigree of a brother × sister mating; each arrow represents the transmission of one gamete. A and B are two unrelated parents, each of which contributes one gamete to C and one to D; E receives one gamete each from C and D, which are the full-sibs. Because A and B are unrelated, we assume that their alleles at a certain locus are not identical by descent. The two alleles in A can be represented as a_1a_2, and in B as a_3a_4. (Here, different subscripts indicate that the alleles are not identical by descent; however, this does not necessarily imply that the alleles are not identical in structure.) The probabilities of the four types of progeny from the mating A × B are $\frac{1}{4}(a_1a_3)$, $\frac{1}{4}(a_1a_4)$, $\frac{1}{4}(a_2a_3)$, and $\frac{1}{4}(a_2a_4)$. We are interested in the probability that offspring of the sibs are homozygous for any one allele, i.e., the probability that an offspring is homozygous a_1a_1 or a_2a_2 or a_3a_3 or a_4a_4. The answer is ¼.

Parent A produces two kinds of alleles, a_1 and a_2, each with probability ½. Therefore, the probability that C receives allele a_1 from A is ½, and the probability that C passes allele a_1 (if it carries it) to E is also ½. The probability that allele a_1 has passed from A to C to E is thus $\frac{1}{2} \times \frac{1}{2} = \frac{1}{4}$. The probability that A passes allele a_1 to D and that D passes it to E is also $\frac{1}{2} \times \frac{1}{2} = \frac{1}{4}$. Hence E has a probability of ¼ of receiving allele a_1 from C and a probability of ¼ of receiving a_1 from D; the probability that E receives a_1 from both C *and* D is $\frac{1}{4} \times \frac{1}{4} = \frac{1}{16}$.

We can repeat the reasoning of the previous paragraph for each of the other alleles. The probability that E is homozygous a_2a_2 is also $\frac{1}{16}$, and

Table 21.2
Coefficient of inbreeding, *F*, in the offspring of matings between various kinds of relatives.

Type of Mating	F
Selfing	1/2
Full-sibs	1/4
Uncle × niece, aunt × nephew, or double first cousins	1/8
First cousins	1/16
First cousins once removed	1/32
Second cousins	1/64
Second cousins once removed	1/128
Third cousins	1/256

the same is true for a_3a_3 and a_4a_4. The probability that E is homozygous for *any one* of the four alleles present in its two grandparents is therefore 1/16 + 1/16 + 1/16 + 1/16 = 1/4.

There is a simple method, called *path analysis,* based on the kind of reasoning just used for full-sib progeny, that allows one to compute the coefficient of inbreeding for an individual with known pedigree. The method consists of tracing the arrows in a pedigree from one individual back to itself through *each* ancestor common to *both* parents. Figure 21.2 shows the pedigree of an individual, K, whose parents are first cousins once removed. A and B are the two ancestors common to both parents, H and J. Hence, there are two paths: one is K–J–G–C–A–D–H–K, with seven steps; the other is K–J–G–C–B–D–H–K, also with seven steps. Because K appears twice in each path, the number of steps is reduced by one in each path. The contribution of each path to the inbreeding coefficient is $(1/2)^n$, where n is the number of steps in the path minus one (or just the number of steps, if the individual under consideration appears only once in each path). The value of F is obtained by adding the contributions of the various paths. In the pedigree in Figure 21.2, the contribution of each path is $(1/2)^6 = 1/64$, and the sum of the two paths is $F = 1/64 + 1/64 = 1/32$.

The coefficient of inbreeding in the progeny of matings between various kinds of relatives is given in Table 21.2. Systematic inbreeding is sometimes practiced in plant or animal breeding to obtain a certain degree of homozygosity. If the same type of inbred mating is practiced every generation, the coefficient of inbreeding will increase every generation (Figure 21.3).

The population consequence of inbreeding is to increase the frequency of homozygotes at the expense of heterozygotes. In a randomly mating population with two alleles having frequencies p and q, the frequency of heterozygotes is $2pq$. In a population with a coefficient of inbreeding F, the frequency of heterozygotes will be reduced by a fraction F of their total. The genotypic frequencies in an inbred population are

Genotype:	AA	Aa	aa
Frequency:	$p^2 + pqF$	$2pq - 2pqF$	$q^2 + pqF$

When there is no inbreeding, $F = 0$, and the genotypic frequencies reduce to the familiar Hardy-Weinberg equilibrium values.

The coefficient of inbreeding, F, measures the increase in the frequency of homozygous *individuals* at a locus; it also measures the increase in the proportion of homozygous *loci* per individual.

Inbreeding Depression and Heterosis

Plant and animal breeders try to improve their stocks with respect to certain traits (grain yield, egg production, etc.) by using the "best" individuals in each generation as parents of the next generation (artificial selection). Breeders also want homogeneity; they try to achieve this by systematic inbreeding, which increases homozygosity. However, breeders have long known that inbreeding usually leads to a reduction in fitness, owing to deterioration in important attributes, such as fertility, vigor, and resistance to disease. This phenomenon is known as *inbreeding depression.*

Inbreeding depression results from homozygosis for deleterious recessive alleles. Consider a recessive lethal allele, with mutation rate $u = 10^{-5}$. The equilibrium frequency of the allele is $q = \sqrt{u} = 0.0032$. In a random-mating population, the frequency of homozygotes is $q^2 = 10^{-5}$. Assume now that the coefficient of inbreeding in a certain stock is $F = 1/16$, similar to what is achieved in a single generation by matings between first cousins. Then the frequency of homozygotes for the allele will be

$$q^2 + pqF = 10^{-5} + (0.9968 \times 0.0032 \times 0.0625)$$

$$\approx 10^{-5} + (2 \times 10^{-4}) \approx 2 \times 10^{-4}$$

The frequency of homozygotes is thus about 20 times greater than in a random-mating population. A comparable increase in the frequency of

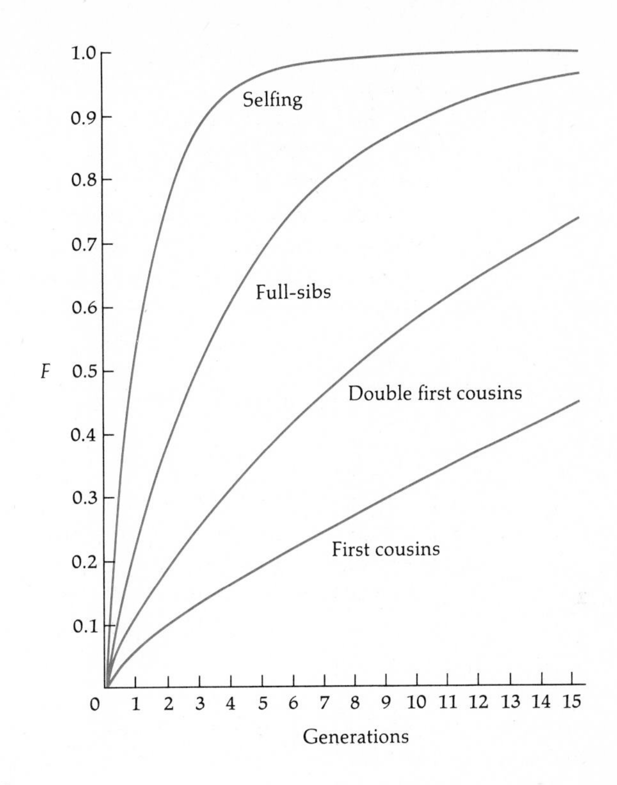

Figure 21.3
Increase in the inbreeding coefficient, F, when a given type of inbred mating is practiced in successive generations.

homozygotes will also occur with respect to other recessive deleterious alleles in the population.

Note that the increase in the proportion of homozygotes at any one locus is directly proportional to the value of F, since it is determined by pqF; thus if $F = 1/4$, the frequency of homozygotes in the previous example will be $10^{-5} + (8 \times 10^{-4})$, or about 80 times greater than in a random-mating population.

The effects of inbreeding depression can be counteracted by crossing independent inbred lines. The hybrids usually show a marked increase in fitness—size, fertility, vigor, etc. (Figure 21.4). This is called *hybrid vigor*, or *heterosis*. Independent inbred lines are likely to become homozygous for different deleterious recessive alleles. Intercrossing two inbred lines may retain homogeneity for the artificially selected traits while making the deleterious alleles heterozygous.

Hybrid vigor as a technique for crop improvement was first exploited in corn, with great success. The increase in productivity obtained with hybrid corn is indeed very large. The practice has been extended to other plants and to animals. It requires, however, that hybrid seed be obtained

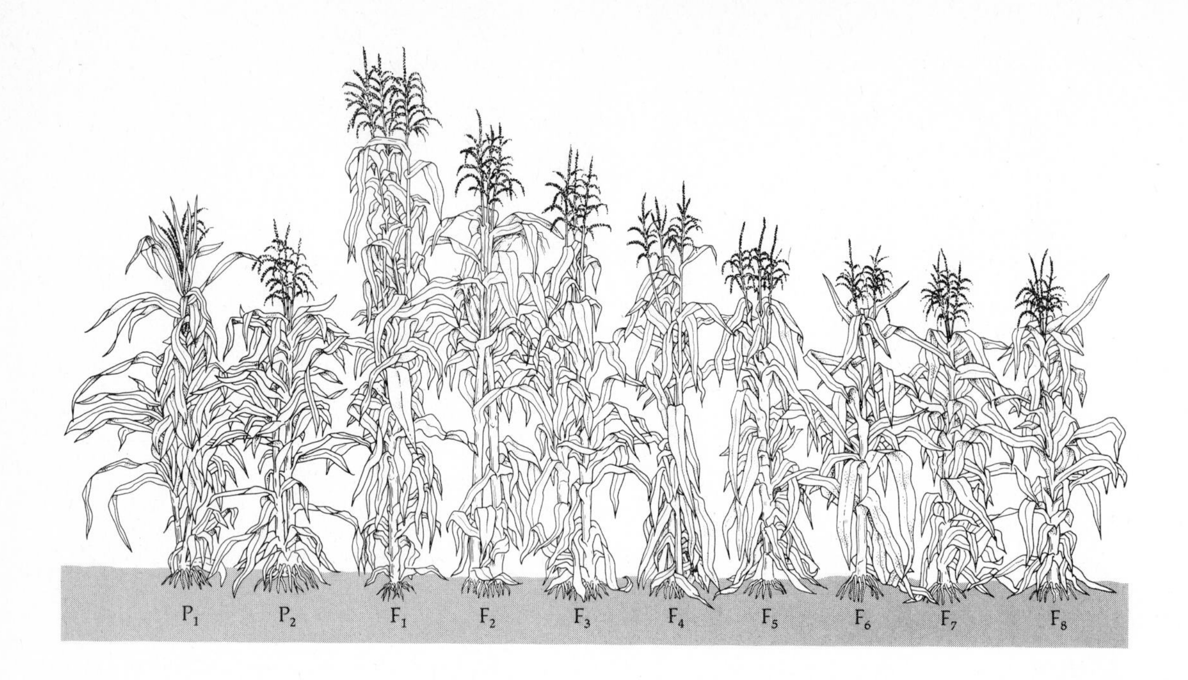

Figure 21.4
Inbreeding depression and hybrid vigor in corn. Heterosis arises in the F_1 of a cross between two inbred lines (P_1 and P_2). Selfing leads to increased inbreeding depression in the following generations (F_2 to F_8). [After D. F. Jones, *Genetics* 9:405 (1924).]

from supply houses, where it is produced using appropriate inbred stocks (Figure 21.5).

In nature, many plants normally reproduce by selfing. Inbreeding depression does not occur in these plants, because natural selection keeps deleterious recessive alleles at much lower frequencies than in random-mating populations. In normally self-fertilized organisms, homozygosis is very high; deleterious recessive alleles are eliminated by natural selection as they become homozygous. However, inbreeding depression follows the inbreeding of normally outbred animals and plants, because deleterious recessive alleles, mostly present in heterozygotes, become homozygous.

Inbreeding in Human Populations

Mating between parents and their children or between brothers and sisters is known as *incest.* There is an "incest taboo" in most human cultures, although many Egyptian pharaohs had incestuous marriages. Matings between close relatives, such as first cousins, are often forbidden by law or by religion. In the United States, for example, about half the states have laws prohibiting uncle-niece, aunt-nephew, and first-cousin marriages; in the other states, such marriages are legal.

Most religions and countries forbid marriages between close relatives, but exceptions are sometimes granted by the authorities. In the Roman

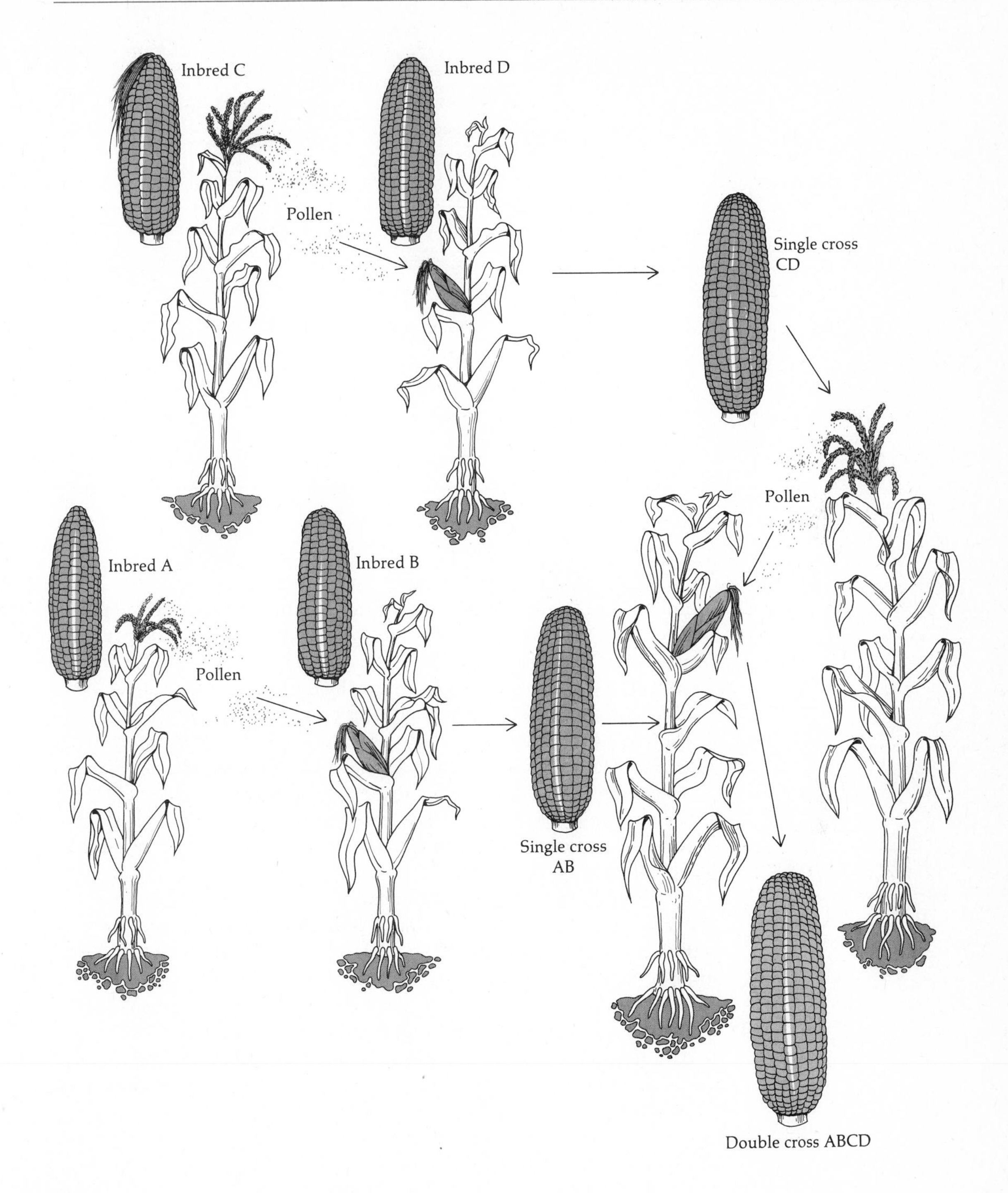

Figure 21.5
Production of hybrid corn from four inbred lines, A, B, C, and D. Paired crosses between the original inbred lines produce two vigorous hybrid plants, AB and CD, which are then intercrossed to yield the double-cross hybrid ABCD.

Table 21.3
Inbreeding depression in human populations. Frequencies of various diseases and of physical and mental defects among children of unrelated parents and of first cousins.

Population	Unrelated Parents		First Cousins	
	Sample size	Frequency (%)	Sample size	Frequency (%)
United States (1920–1956)	163	9.8	192	16.2
France (1919–1925)	833	3.5	144	12.8
Sweden (1947)	165	4	218	16
Japan (1948–1954)	3570	8.5	1817	11.7
Average:		6.5		14.2

After C. Stern, *Principles of Human Genetics,* 3rd ed., W. H. Freeman, San Francisco, 1973.

Catholic Church, marriages between uncle and niece, aunt and nephew, first cousins, first cousins once removed, and second cousins require dispensation. The parish records of these dispensations provide some of the best existing information concerning the frequency of consanguineous marriages—matings between relatives—in human populations.

Marriages between close relatives are not only allowed but considered desirable in some human societies. In Japan, for example, first-cousin marriages are encouraged, and up to 10% of the marriages in certain areas or social groups are between first cousins. In Andhra Pradesh (India), certain castes favor uncle-niece marriages, which account for more than 10% of all marriages.

The effects of inbreeding in human populations are shown in Table 21.3 and Figure 21.6. We calculated previously that the frequency of homozygotes for a *lethal* recessive allele with mutation rate $u = 10^{-5}$ is about 20 times greater in the offspring of first cousins than in the offspring of random-mating individuals. In the case of a *deleterious* recessive allele, with $s = 0.1$, the equilibrium frequency is $q = \sqrt{10^{-5}/10^{-1}} = \sqrt{10^{-4}} = 0.01$. The frequency of homozygotes among random-mating individuals is $q^2 = 10^{-4}$. In the offspring of first cousins, the frequency of homozygotes will be

$$q^2 + pqF = 10^{-4} + (0.99 \times 0.01 \times 0.0625)$$

$$\approx 10^{-4} + (6 \times 10^{-4}) = 7 \times 10^{-4}$$

or about seven times greater than in the offspring of random-mating individuals. If $s = 0.01$, the incidence of homozygotes is about three times

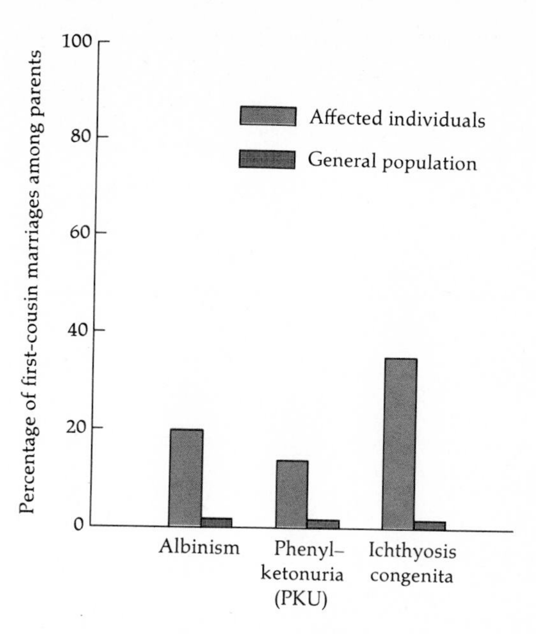

Figure 21.6
Consanguinity among parents of recessive homozygotes. The bars represent the frequencies, in European populations, of first-cousin marriages among the parents of affected individuals and in the general population. Ichthyosis congenita is a severe disease of the skin. (After W. F. Bodmer and L. L. Cavalli-Sforza, *Genetics, Evolution, and Man,* W. H. Freeman, San Francisco, 1976.)

greater in the offspring of first cousins than in the offspring of random-mating individuals.

Table 21.3 shows that, on the average, the incidence of defective newborn children is about twice as high when the parents are first cousins as when the parents are unrelated. This increase is considerably less than might be expected from the previous calculations. The calculations, however, apply to *recessive* infirmities. With respect to dominant alleles, the incidence of hereditary conditions is no greater, on the average, in consanguineous marriages than in unrelated marriages. Moreover, the data in Table 21.3 include nonhereditary defects. Figure 21.6 shows that deleterious recessive conditions increase by a large factor in the offspring of first cousins.

Cultural attitudes often have genetic consequences. Figure 21.7 shows the frequency of consanguineous marriages in some European populations. Until the year 1700, the Roman Catholic Church rarely granted dispensation from the prohibition against consanguineous marriages. Such marriages in Catholic Europe increased during the eighteenth century until the first half of the nineteenth century, when they started to decrease. The high frequency of consanguineous marriages during the first half of the nineteenth century seems to have been due in part to

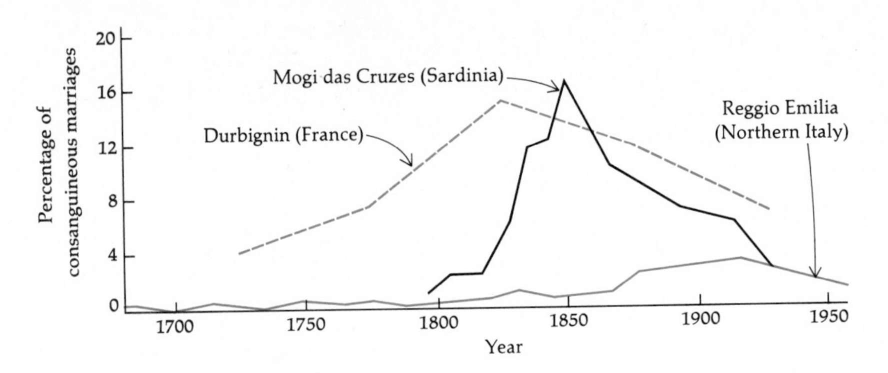

Figure 21.7
Frequency of consanguineous marriages in three European populations. (After A. Moroni, *Historical Demography, Human Ecology, and Consanguinity*, International Union for the Scientific Study of Population, Liège, 1969.)

Napoleon's abolition of the right of primogeniture, which caused the splitting of land property. This could be counteracted to some extent by marriages between close relatives. The Industrial Revolution, which greatly increased geographic mobility, may have been partially responsible for the decrease in consanguineous marriages observed since the nineteenth century. Whatever the reasons may have been for these changes, they had important genetic and health consequences because of their effects on the incidence of deleterious recessive infirmities.

Genetic Coadaptation

The mechanisms of evolutionary change—mutation, migration, drift, and selection—as well as inbreeding, have so far been considered primarily as they affect individual gene loci. However, genes survive and reproduce in whole organisms; a functional allele may fail to be passed on to the following generation if it occurs in an organism that fails to reproduce. Genes interact with the environment but also with other genes (Chapter 15). At any one locus, natural selection favors alleles that interact well with the alleles present at other loci. The term *genetic coadaptation* refers to the adaptive interaction between the genes that make up a gene pool.

Imagine a zygote made up of human genes, whale genes, and corn genes in equal numbers: such a chimera could not develop into a functional organism. Most living species cannot be intercrossed with each other; interspecific fertilization is sometimes possible between closely related species, but more often than not the hybrid zygotes fail to develop, or they develop into sterile organisms, such as the mule. The inviability or

Table 21.4
Allelic frequencies at the *Mdh-2* locus in two *Drosophila* species in Tame, Colombia.

Species	Frequency of Allele: *86*	*94*	Other*
D. equinoxialis	0.005	0.992	0.003
D. tropicalis	0.995	0.004	0.001

*Refers to several alleles with very low frequencies in both species.

sterility of interspecific hybrids is conspicuous evidence of genetic coadaptation. The horse and donkey genotypes are not mutually coadapted.

Whenever a new genic or chromosomal mutation arises that does not interact well with the rest of the genome, it is eliminated or kept at low frequency by natural selection, even though it might be functional in a different genetic background. The role of coadaptation between alleles at different loci can be illustrated by an analogy. A successful performance by a symphony orchestra requires not only that each player know *how* to play his instrument (a gene must be *able* to function), but also that he master his *part* in the piece being performed (a gene must *interact* well with the other genes). A violinist playing his part for Beethoven's Sixth Symphony while the rest of the orchestra was playing Ravel's *Bolero* would be cacophonic.

Owing to genetic coadaptation, a certain allele or set of alleles may be favorably selected in one species but be unfavorably selected in a different species. *Drosophila equinoxialis* has, at the *Mdh-2* locus (which codes for the cytoplasmic enzyme malate dehydrogenase), allele *94* with a frequency about 0.99, while *D. tropicalis* has allele *86* with a frequency about 0.99. Table 21.4 gives the allelic frequencies in both species in a certain South American locality. These allelic frequencies are relatively constant throughout the distributions of both species, which are abundant in the rain forests of Central and South America.

The polypeptides coded by alleles *86* and *94* differ by at least one amino acid, but they are functionally very similar, and flies of both species can function with either one of the two forms. It might seem that the two alleles are equivalent, their different frequencies being due to random drift. However, the experiment shown in Figure 21.8 indicates that genetic coadaptation is responsible for the different allelic frequencies in the two species. Laboratory populations of each species were set up, with the

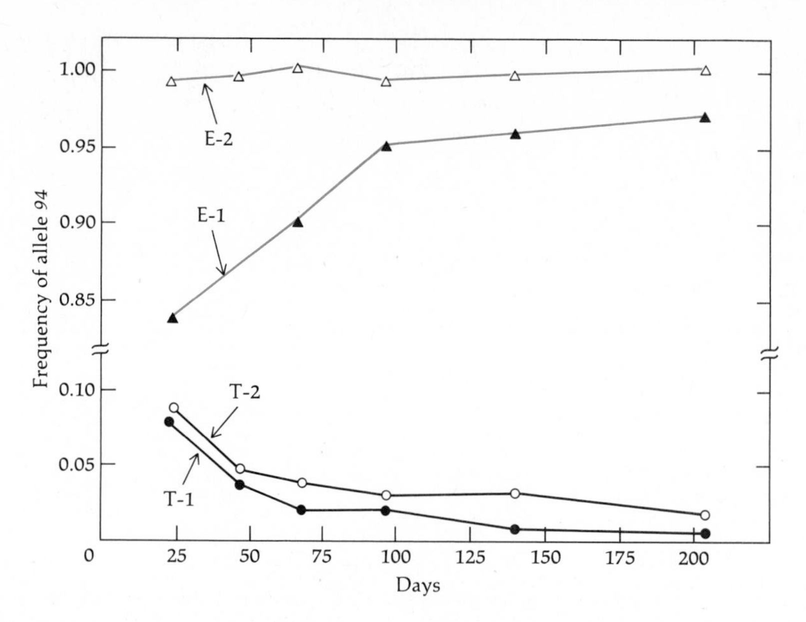

Figure 21.8
Natural selection in laboratory populations of *Drosophila equinoxialis* (E-1 and E-2) and *D. tropicalis* (T-1 and T-2). Two alleles, *86* and *94*, of the *Mdh-2* locus are present in all four populations. Natural selection tends to restore the frequencies occurring in nature: allele *94* increases in frequency in population E-1, where its initial frequency was lower than that in nature. In populations T-1 and T-2, however, where its initial frequency was greater than that in nature, allele *94* decreases in frequency.

frequency of the rare allele in a given species being artificially increased. Natural selection acts to restore the natural frequencies, although the experimental environment is the same for all populations. In *D. tropicalis*, allele *94*, which is rare in nature, decreases in frequency, while the same allele is favored in *D. equinoxialis*, a species in which allele *94* is common in nature. Alleles coding for polypeptides with a given electric charge are favored in one species but selected against in the other. The different directions of the process of natural selection in the two species must be due to the different genetic backgrounds in which the alleles are present (i.e., the different distributions of alleles present at the other loci), since the experimental environment is the same for all four laboratory populations.

Genetic coadaptation is a property of the gene pool of a species as a whole, but also of local populations. Alleles may be favored in one locality but not in another, because they interact well with other alleles in the first population but not in the second. An illustrative example is provided by the African swallowtail butterfly, *Papilio dardanus* (Figure 21.9). This species has several female phenotypes that mimic different butterfly species noxious to bird predators. The males, however, have a constant,

Figure 21.9
Mimicry and coadaptation in *Papilio dardanus* butterflies. (1) *P. dardanus* male. (2) *P. dardanus* female, nonmimetic. (3) and (4) *Amauris albimaculata* and its mimetic, *P. dardanus*, the "cenea" phenotype. (5) and (6) *Amauris niavius dominicanus* and its mimetic, *P. dardanus*, the "hippocoonides" phenotype. (7) and (8) F_1 offspring of a cross between cenea and hippocoonides from different African regions. When the butterflies come from the same region, the allele responsible for the cenea phenotype is dominant over that for hippocoonides, so that F_1 and F_2 progenies yield only one or the other phenotype. However, when cenea and hippocoonides from different regions are crossed, the progenies include intermediate, nonmimetic phenotypes, as exemplified by (7) and (8). [From C. A. Clarke and P. M. Sheppard, *J. Genet.* *56*:236 (1959); *Heredity* *14*:73, 163 (1960).]

nonmimetic phenotype. Birds find *P. dardanus* butterflies palatable, but avoid mimetic forms that they confuse with the noxious species. Several mimetic forms exist in some localities, while only one is found in other localities, depending on which noxious species happen to be present. Crosses can be made between two mimetic strains, which we may call A and B. What is interesting is that crosses between A and B give different results depending on whether or not the two strains come from the same locality. If both parents come from the same locality, only perfect female mimics are produced in the F_1, F_2, and backcross generations. When the two mimicking strains come from different regions, however, the F_1 female progenies are intermediate in appearance between the two female phenotypes of the parental strains, and the F_2 and backcross progenies also show intermediate phenotypes.

The mimetic patterns are determined for the most part by two major gene loci. At one locus there are two alleles, one determining the presence, the other the absence, of the "tails" that are typical of swallowtail butterflies. The other locus consists of several alleles, each determining the main color pattern of one mimetic form. There are, moreover, a number of modifier-gene loci that affect the expression of the major genes. At these modifier loci, alleles have been selected that maximize the mimetic characteristics of the butterflies. This, however, is accomplished by different sets of alleles in different local populations. Because in nature *P. dardanus* butterflies from different regions do not intercross, natural selection has not coadapted the sets of modifier alleles from separate regions. When mimetic forms from different regions are intercrossed, alleles that are not mutually coadapted are joined together, and imperfectly mimicking forms arise.

Linkage Disequilibrium

Genetic coadaptation may exist between some alleles but not others within the same population. In the case of polymorphic loci, certain alleles at one locus may be coadapted with some alleles at different loci, but not with others.

Assume that there are two loci, A and B, and that at each locus there are two alleles, A_1 A_2, and B_1 B_2. Assume further that alleles A_1 and B_1 interact well with each other so that they produce well-adapted phenotypes, and that the same is true for A_2 and B_2, but that the combinations A_1B_2 and A_2B_1 yield poorly adapted phenotypes. The adaptation of the population would be increased if the alleles would always (or most often) be transmitted in the combinations A_1B_1 and A_2B_2, and never (or rarely) in the combinations A_1B_2 and A_2B_1.

When alleles at different loci are not associated at random, they are said to be in *linkage disequilibrium*. When alleles at different loci *are* associated at random (i.e., in proportion to their frequencies), the loci are in linkage equilibrium.

Assume that the allelic frequencies at two loci are

First locus: $A_1 = p \qquad A_2 = q$

Second locus: $B_1 = r \qquad B_2 = s$

so that $p + q = 1$ and $r + s = 1$. If the alleles at the two loci are associated at random, we expect the four possible gametic classes to have frequencies that are the product of the frequencies of the alleles involved, that is,

$$A_1B_1 = pr$$

$$A_2B_2 = qs$$

$$A_1B_2 = ps$$

$$A_2B_1 = qr$$

Because these are the only possible kinds of gametes, the sum of their frequencies must be 1. Indeed,

$$pr + qs + ps + qr = p(r + s) + q(r + s) = p + q = 1$$

If the alleles are associated at random, the product of the frequencies of the two *coupling* gametes ($pr \times qs = pqrs$) is the same as the product of the frequencies of the two *repulsion* gametes ($ps \times qr = pqrs$). However, if the alleles are *not* randomly associated, the two products will be different; the extent of linkage disequilibrium, d, is measured by the difference between the two products:

$$d = (\text{freq. of } A_1B_1)(\text{freq. of } A_2B_2) - (\text{freq. of } A_1B_2)(\text{freq. of } A_2B_1)$$

The condition for linkage equilibrium is, therefore, $d = 0$.

Linkage disequilibrium is complete only when two gametic combinations exist—either only the two coupling gametes (A_1B_1 and A_2B_2) or only the two repulsion gametes (A_1B_2 and A_2B_1). The maximum absolute value that d can have is 0.25, namely, when linkage disequilibrium is complete and the allelic frequencies are 0.5 at both loci (Table 21.5). Note that, if the allelic frequencies at the two loci are different, complete linkage is not possible. For example, if the frequency of A_1 is 0.5 but the frequency of B_1 is 0.6, then not all B_1 alleles can be associated with either A_1 or A_2, but some must be associated with each of these two alleles.

According to the Hardy-Weinberg law, the equilibrium genetic frequencies at any one autosomal locus are reached in one single generation of random mating (or in two, if the allelic frequencies are different in the two sexes). This is not so when two loci are considered simultaneously (Box 21.1). However, linkage disequilibrium decreases with every generation of random mating—unless there is some process opposing the

Table 21.5
Maximum possible values of linkage disequilibrium, d, for three different cases. The frequency designated for each case is that of the most common allele and is the same at both loci.

Case	Frequency of Gametic Combination: A_1B_1	A_2B_2	A_1B_2	A_2B_1	d
1. Frequency 0.5					
Coupling	0.5	0.5	0	0	$(0.5)(0.5) - (0)(0) = 0.25$
Repulsion	0	0	0.5	0.5	$(0)(0) - (0.5)(0.5) = -0.25$
2. Frequency 0.6					
Coupling	0.6	0.4	0	0	$(0.6)(0.4) - (0)(0) = 0.24$
Repulsion	0	0	0.6	0.4	$(0)(0) - (0.6)(0.4) = -0.24$
3. Frequency 0.9					
Coupling	0.9	0.1	0	0	$(0.9)(0.1) - (0)(0) = 0.09$
Repulsion	0	0	0.9	0.1	$(0)(0) - (0.9)(0.1) = -0.09$

approach to linkage equilibrium. Permanent linkage disequilibrium may result from natural selection if some gametic combinations result in higher fitness than other combinations. Assume, for example, that the two coupling combinations produce, in either homozygous or heterozygous condition, viable zygotes, while the two repulsion combinations are lethal even in heterozygous combination; complete linkage disequilibrium would follow, even if the two loci were unlinked. Situations as extreme as this are not likely, however. Because the approach to linkage equilibrium is facilitated by the extent of recombination, the less closely linked two loci are, the greater the strength of natural selection required to maintain linkage disequilibrium. Consequently, in natural populations, linkage disequilibrium is more common between closely linked loci.

Supergenes

Linkage disequilibrium is decreased by recombination. The possibility of maintaining favorable allelic combinations in linkage disequilibrium is,

Box 21.1 Random Mating with Two Loci

It was shown in Chapter 19 that the genotypic equilibrium frequencies at any one autosomal locus are reached in one generation of random mating. This is not so when two loci are considered simultaneously. If the frequency of recombination between two loci is c, then the value of the linkage disequilibrium, d, decreases by the fraction cd in one generation of random mating (assuming that there is no selection). That is, if the value of d in one generation is d_0, its value in the following generation will be

$$d_1 = (1 - c)d_0$$

When two loci are unlinked, then $c = 0.5$, and the linkage disequilibrium value will be halved with every generation of random mating. If $c < 0.5$, then the approach to equilibrium will be slower. For example, if the frequency of recombination is 0.1, the value of d in one generation will be 90% of its value in the previous generation. The approach to equilibrium for various values of c, assuming random mating, is shown in Figure 21.10.

therefore, enhanced by reduction of the frequency of recombination between the loci involved. This may be accomplished through translocations and inversions. Assume that two loci, *A* and *B*, are located in different chromosomes; a translocation might bring them together in the same chromosome. Assume, now, that the two loci are separated within the same chromosome by a number of loci that we represent as *FG* · · · *MN,* so that the gene sequence along the chromosome is

· · · *AFG* · · · *MNB* · · ·

An inversion comprising the segment *FG* · · · *MNB* might bring together *A* and *B*; the new gene sequence would be

· · · *ABNM* · · · *GF* · · ·

Whenever linkage disequilibrium is favored by natural selection, chromosomal rearrangements increasing linkage between the loci will also be favored by natural selection. The term *supergene* is used to refer to several closely linked gene loci that affect a single trait or a series of interrelated traits.

A supergene is responsible for the expression of two flower phenotypes, known as "pin" and "thrum," found in the primrose and other species of the genus *Primula* (Figure 21.11). Pin-thrum polymorphisms were made famous by Darwin, who gave a detailed account of them in 1877. The pin phenotype is characterized by a long style above the ovary, which places the stigma at the same level as the mouth of the corolla; the pollen-bearing anthers are halfway down the corolla tube. The thrum phenotype has a short style, so that the stigma is halfway down the corolla tube, while the stamens are long, placing the anthers at the mouth of the

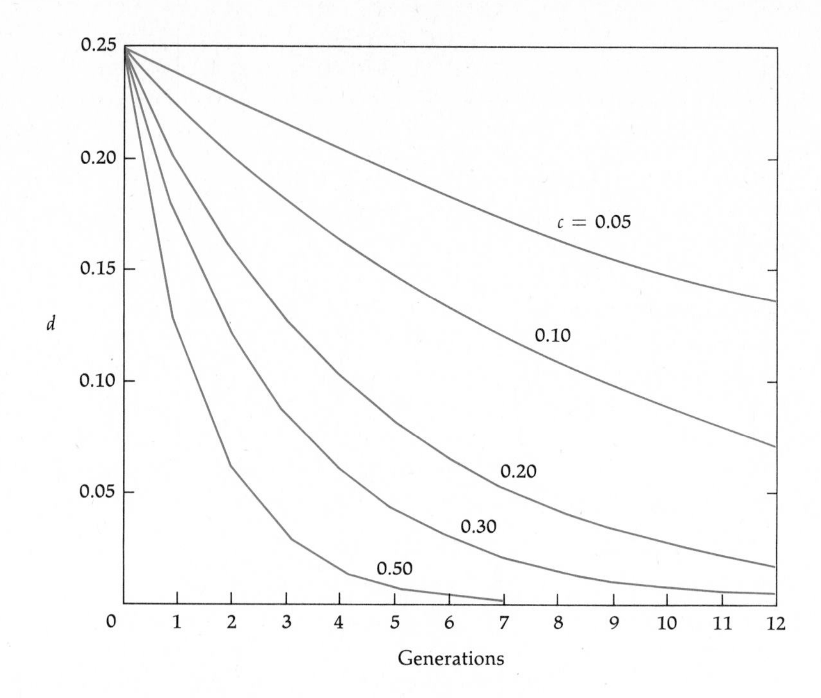

Figure 21.10
Decrease in linkage disequilibrium, d, over the generations for various levels of linkage (recombination frequency, c, from 0.05 to 0.50). The value of d after t generations of random mating is given by $d_t = (1 - c)^t d_0$.

corolla. Pin and thrum are phenotypically different in other ways also, such as the configuration of the stigma and the size of the pollen grains. Moreover, they differ physiologically: thrum pollen is more successful in fertilization when deposited on pin stigmas than on thrum stigmas; conversely, pin pollen fertilizes thrum flowers more successfully than it does pin flowers.

The pin-thrum phenomenon is known as *heterostyly* (meaning "different styles"). Heterostyly promotes cross-pollination. An insect that visits both pin and thrum flowers will receive pollen from one type of flower on parts of its body that get close to the stigma of the other type. The physiological differences reinforce the chances of cross-fertilization.

Pin and thrum phenotypes behave, as a rule, as if they were controlled by a single gene locus, with two alleles: S (for thrum) is dominant over s (for pin). Thrum plants, however, are generally heterozygous (Ss); when they are selfed or intercrossed, they produce pin and thrum plants in a typical 3:1 Mendelian ratio. Pin plants are homozygous (ss), and produce only pin types when selfed or intercrossed. In nature, most crosses are between thrum (Ss) and pin (ss) plants, producing thrum and pin progenies in about a 1:1 ratio. Pin and thrum flowers are found in approximately similar frequencies in natural populations.

The set of traits characteristic of the thrum or pin phenotypes are not determined by a single gene locus, however, but by several closely linked loci making up a supergene. The existence of multiple gene loci could be

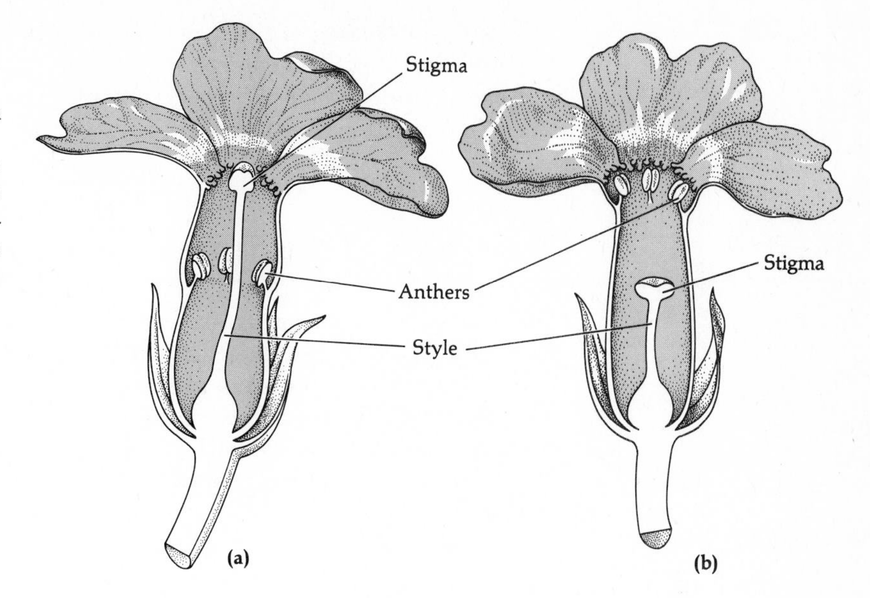

Figure 21.11
Two phenotypes in the primrose, *Primula officinalis.* **(a)** The pin phenotype has a highly placed stigma but low-placed anthers. **(b)** The thrum phenotype is just the opposite. This complementary arrangement facilitates cross-fertilization between the two phenotypes.

suspected, because the phenotypic and physiological differences between the thrum and pin types are multiple, and it has been confirmed by examination of large progenies from experimental thrum × pin crosses. Examples of mixtures between components of the two complex phenotypes are occasionally found, owing to recombination within the supergene. In nature, mixed phenotypes are occasionally found as well, but these remain rare, owing to their low fitness relative to that of the thrum and pin phenotypes. The supergene has become established in *Primula* because it makes possible the joint transmission of sets of alleles that produce adaptive phenotypes. The supergene control saves *Primula* populations from a high proportion of ill-adapted phenotypes.

A well-known example of a supergene is the set of gene loci controlling the color and the presence or absence of bands on the shell of the snail *Cepaea nemoralis,* studied by Arthur J. Cain and Philip M. Sheppard. The gradual process of formation of supergenes by means of successive translocations and inversions is evidenced by several species of grouse locusts. Robert K. Nabours has shown that the color patterns are determined by alleles at some 25 gene loci. In one species, *Acridium arenosum,* 13 of the genes are spread throughout one single chromosome and recombine fairly freely; in another species, *Apotettix eurycephalus,* the corresponding genes are combined into two groups (supergenes) of closely linked genes, the recombination between the groups being only 7%; and in a third species, *Paratettix texanus,* 24 of the 25 genes are all tightly linked, forming a single supergene. The formation of supergenes has advanced most in this last species.

Bacterial examples of supergenes are the operons. Genes involved in the same biochemical function, such as those controlling the synthesis of tryptophan, are often clustered together in close linkage within the genome (Chapter 10, page 346).

Inversion Polymorphisms

The formation of supergenes is a way of reducing crossing over, thereby facilitating the maintenance of linkage disequilibrium. Linkage disequilibrium can also be maintained by inversion polymorphisms, another mechanism for reducing genetic recombination. Assume, as we did above, that A_1B_1 and A_2B_2 are favorable allelic combinations, while A_1B_2 and A_2B_1 are unfavorable ones. Let us represent the gene sequence in the chromosome as

$$\cdots DEAF \cdots NBOP \cdots$$

Assume that an inversion of the segment from E to O takes place and that the alleles A_1 and B_1 are included in the segment. We would then have the following sequence:

$$\cdots DOB_1N \cdots FA_1EP \cdots$$

(Subscripts at loci other than A and B are not added, because we are not concerned with alleles at these other loci.)

Assume now that an individual heterozygous for the inversion and the original chromosome sequence carries alleles A_2 and B_2 in the original chromosome sequence, i.e., that this individual has the following genetic constitution:

$$\cdots DOB_1N \cdots FA_1EP \cdots / \cdots DEA_2F \cdots NB_2OP \cdots$$

As explained in Chapter 17 (page 571; see Figures 17.14 and 17.15), recombination is suppressed in the progenies of inversion heterozygotes. Thus, the individual described above will produce only two kinds of functional gametes, one containing the alleles A_1 and B_1, and the other containing A_2 and B_2. Natural selection may, then, favor original chromosome sequences that have alleles A_2 and B_2, and inverted chromosome sequences that have alleles A_1 and B_1. The population can then consist of only three types of individuals: (1) homozygotes for the chromosomal inversion, and therefore for alleles A_1 and B_1; (2) homozygotes for the original chromosome sequence, and therefore for alleles A_2 and B_2; and (3) heterozygotes for the inverted sequence and the original sequence. Only the two gametic combinations A_1B_1 and A_2B_2

Table 21.6
Relative frequencies of third chromosomes with different chromosomal arrangements in populations of *Drosophila pseudoobscura* in various localities.

Locality	Frequency (%) of Chromosomal Arrangement: *ST*	*AR*	*CH*	*PP*	*TL*	*SC*	*OL*	*EP*	*CU*
Methow, Washington	70.4	27.3	0.3	—	2.0	—	—	—	—
Mather, California	35.4	35.5	11.3	5.7	10.7	0.9	0.5	0.1	—
San Jacinto, California	41.5	25.6	29.2	—	3.4	0.3	—	—	—
Fort Collins, Colorado	4.3	39.9	0.2	32.9	12.3	—	2.1	7.2	—
Mesa Verde, Colorado	0.8	97.6	—	0.5	—	—	—	0.2	—
Chiricahua, Arizona	0.7	87.6	7.8	3.1	0.6	—	—	—	—
Central Texas	0.1	19.3	—	70.7	7.7	—	2.4	—	—
Chihuahua, Mexico	—	4.6	68.5	20.4	1.0	3.1	0.7	—	—
Durango, Mexico	—	—	74.0	9.2	3.1	13.1	—	—	—
Hidalgo, Mexico	—	—	—	0.9	31.4	1.7	13.5	1.7	48.3
Tehuacán, Mexico	—	—	—	—	20.2	1.1	—	3.2	74.5
Oaxaca, Mexico	—	—	10.3	—	7.9	—	0.9	1.6	71.4

After J. R. Powell, H. Levene, and Th. Dobzhansky, *Evolution 26*:553 (1973).

exist in all three kinds of individuals.

Inversion polymorphisms have been studied in many species of *Drosophila*. Some species have polymorphisms in all chromosomes—e.g., the European species *D. subobscura* and the American tropical species *D. willistoni*—whereas others have inverted segments concentrated mostly in one chromosome—e.g., the North American species *D. pseudoobscura*, which exhibits extensive polymorphism in only one of the five chromosomes, the third (Table 21.6).

As shown in Table 21.6, the frequencies of the various chromosomal arrangements in *Drosophila pseudoobscura* vary from one locality to another. Moreover, the frequencies may change from month to month throughout

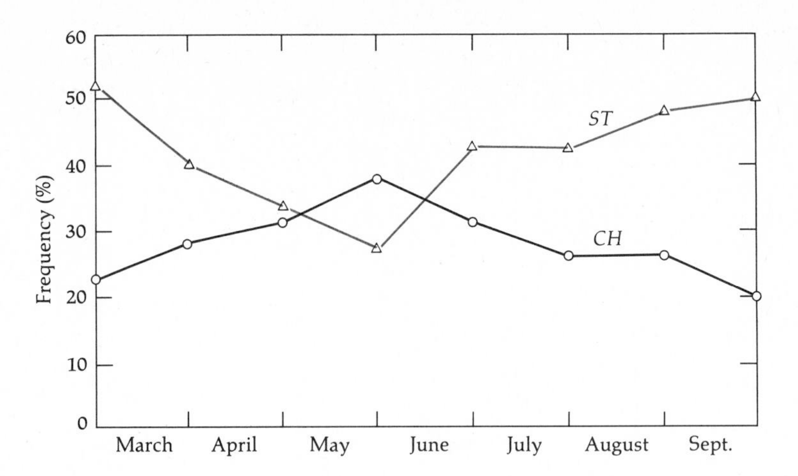

Figure 21.12
Frequencies of the *ST* and *CH* chromosomal arrangements of *Drosophila pseudoobscura* in San Jacinto, California. The frequencies of the two arrangements change throughout the year. *CH* reaches its highest frequency at the beginning of summer, when the frequency of *ST* is lowest. (The two frequencies do not add up to 100 because other arrangements also exist in the population.)

the year (Figure 21.12). These changes are seasonal, and are thus repeated in successive years. This suggests that the chromosomal arrangements differ in the sets of alleles they carry, and that these differences are adaptive: one arrangement is adaptively superior to the other during some period of the year but inferior during some other period. This hypothesis was tested with laboratory populations that were started with known frequencies of the chromosomal arrangements and allowed to breed freely within the laboratory cage. Typical results are shown in Figure 21.13. The frequencies of the inversions change rapidly in the early generations and more slowly later on, eventually reaching an equilibrium with both chromosomal arrangements present. From the rate of change and the equilibrium frequencies, the fitnesses of the three genotypes can be estimated; these are 1 for the heterozygote (*ST*/*CH*), 0.89 for one homozygote (*ST*/*ST*), and 0.41 for the other homozygote (*CH*/*CH*).

These results show that heterozygote superiority contributes to the maintenance of chromosomal polymorphism. Other laboratory experiments have shown that the fitnesses of the two homozygous genotypes depend on the temperature and on the density of the population, which may account for the seasonal oscillations observed in nature.

Direct evidence that chromosomal inversions differ in their allelic content has been obtained in *D. pseudoobscura* by Satya Prakash and Richard C. Lewontin. Two gene loci, *Pt-10* and *α-Amy*, coding for two proteins, were examined by gel electrophoresis. It was discovered that the allelic frequencies were quite different in different chromosomal arrangements (Table 21.7).

Inversion polymorphisms have been observed in natural populations of mosquitoes, black flies, midges, and other dipterans. It is uncertain how widespread such polymorphisms are in other organisms, since inversions are difficult to detect in the absence of polytene chromosomes. Neverthe-

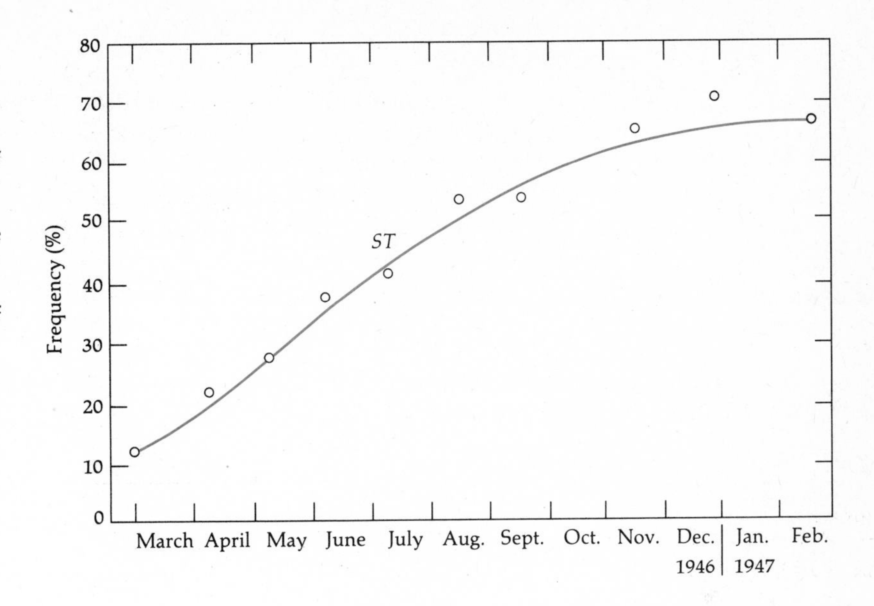

Figure 21.13
Change in the frequency of the *ST* chromosomal arrangement of *Drosophila pseudoobscura* in a laboratory population. Two chromosomal arrangments, *ST* and CH, are present in the population. The frequency of *ST* gradually increases from its initial frequency of 12% to an equilibrium frequency of about 70%. Correspondingly, *CH* decreases from its initial frequency of about 88% to an equilibrium frequency of about 30%.

less, inversion heterozygotes are known in many animals, such as grasshoppers, and in some plants.

Geographic Differentiation

Environmental conditions are infinitely variable. The weather and many physical and chemical conditions, as well as the foods, competitors, parasites, and predators may all vary to a greater or lesser extent. Natural selection promotes adaptation to local conditions, and this results in genetic differentiation between geographically separated populations.

The adaptive character of local differentiation is apparent in the experiments with the cinquefoil, *Potentilla glandulosa* (Figure 2.16). Plants collected at different altitudes are genetically different, as demonstrated by their different growth habits in identical environments. Moreover, the genetic differences are adaptive—plants grow best in the environment most similar to their natural habitat.

Table 21.8 shows the result of an experiment with two populations of *Drosophila serrata*, one collected in a temperate habitat (near Sydney, Australia, at latitude 34°S) and the other in a tropical habitat (Popondetta, New Guinea, at latitude 9°S). The populations were kept for about one year (18 generations) at two temperatures. At 19°C, a temperature considerably lower than those experienced in the tropics, the Sydney flies

Table 21.7
Allelic frequencies at two gene loci in four chromosomal arrangements of *Drosophila pseudoobscura.* The locus *Pt-10* codes for a larval protein, and *α-Amy* codes for the enzyme α-amylase. The Pikes Peak chromosomal arrangement is evolutionarily closely related to Standard, and Santa Cruz is closely related to Tree Line.

Chromosomal Arrangement	Allelic Frequency at Locus: *Pt-10*		*α-Amy*	
	104	*106*	*84*	*100*
Standard	1.00	0.00	0.15	0.85
Pikes Peak	1.00	0.00	0.00	1.00
Santa Cruz	0.00	1.00	1.00	0.00
Tree Line	<0.01	>0.99	>0.90	0.05

After S. Prakash and R. C. Lewontin, *Proc. Natl. Acad. Sci. USA* 59:398 (1968).

performed better, maintaining a greater population size than the flies from Popondetta. At 25°C, a temperature more common in the tropics, the differences disappeared. (See also Table 18.1 and Figure 18.2.)

The variation in skin pigmentation of different human populations is probably the result of past adaptation to local conditions. Humans, like other mammals, require vitamin D for calcium fixation and bone growth; an insufficient supply of vitamin D causes the bone disease called rickets. Vitamin D is produced in the deep layers of the skin under the stimulus of ultraviolet radiation from the sun. If the skin is heavily pigmented, the amount of UV radiation at high latitudes may not produce enough vitamin D, because the radiation is absorbed by the skin pigment. Consequently, natural selection has favored light skin pigmentations at higher latitudes. In tropical regions, too much vitamin D may be produced (as well as some skin damage) in lightly pigmented skins; hence, natural selection has favored deeper pigmentation. Figures 21.14 and 21.15 give two additional examples of adaptive geographic differentiation in human populations.

As a result of genetic differences among geographically separate populations of the same species, groupings may arise that include populations genetically more similar to each other than they are to populations placed in different groups. The genetic differences on which the geographic groupings are based may or may not be expressed in the

Table 21.8
Number of flies of *Drosophila serrata* populations from two different localities, maintained at two temperatures in population cages of the same size and with the same amount of food. The values given are the means and standard errors for 18 generations.

Locality	Population Size at Temperature: 19°C	25°C
Sydney, Australia	1803 ± 87	1782 ± 76
Popondetta, New Guinea	1580 ± 52	1828 ± 90

After F. J. Ayala, *Genetics* 51:527 (1965).

visible phenotype. The differences in human skin pigmentation or in beetle color and pattern (Figure 18.5) are examples of conspicuously expressed genetic differences. Differences in the frequencies of human blood groups or of chromosomal arrangements in *Drosophila* (Table 21.6) are examples of "hidden" genetic differences.

The Concept of Race

Geographically separate groups of populations are sometimes called *races,* which may be defined as *genetically distinct populations of the same species*. The concept of race, particularly when applied to human populations, has been much misunderstood—even abused—and deserves clarification.

Racial classification may be useful in order to recognize that geographic populations are genetically differentiated to some extent (as a consequence of drift and of adaptation to local conditions). Sometimes races are identified by a single trait (e.g., wing pattern in butterflies, and skin pigmentation or blood group in humans), but races are populations that have somewhat differentiated *gene pools*. The differences among races must involve the gene pool as a whole, and therefore allelic frequencies at many loci. Differences in one locus or trait may serve as indicators of overall genetic differentiation, but alone they do not constitute a sufficient fundament for distinguishing among races. Indeed, parents and their offspring may differ at a locus whenever this is polymorphic; for example, two A blood-group parents ($I^A i$) may have O blood-group children (*ii*).

Races are populations of the same species and are thus not reproductively isolated from one another. The formation of new species

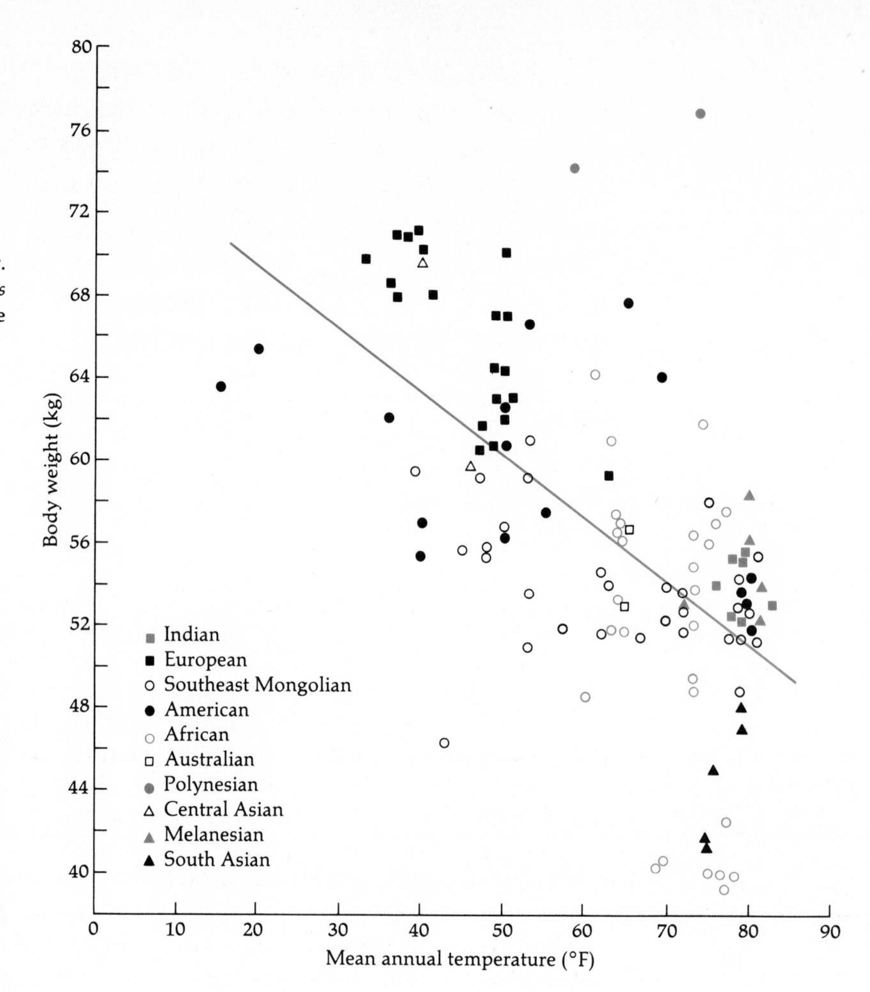

Figure 21.14
Negative correlation between body weight and mean annual temperature of the habitat in human populations. Although there is great variation, people living in colder climates tend to be heavier. This agrees with *Bergmann's rule,* according to which the members of local populations of many species from colder climates tend to be larger than the members of local populations of the same species from warmer climates. This is almost certainly connected with heat conservation and dissipation. The volume (and hence the weight) of the body increases as the cube of its linear dimensions, while the surface area increases only as the square. Therefore, other things being equal, larger bodies are advantageous in cold climates, whereas smaller bodies are advantageous in warm climates. [After D. F. Roberts, *Am. J. Phys. Anthropol.* 11:533 (1953).]

often involves transitional stages of racial differentiation. But races are not necessarily incipient species, because the process of racial differentiation is reversible. Racial differences may, and often do, decrease with time or even become obliterated. In humans, for example, racial differentiation has decreased in the last few centuries through migration and intermixture.

The formation and preservation of races requires that gene flow be limited; otherwise, races will fuse into a single gene pool. Usually, gene flow is curtailed by geographic separation. (Exceptions to this rule are possible through human choice. The differentiation of human races may be retained even where they are sympatric, because people choose their mates predominantly within their own race. As another example, people

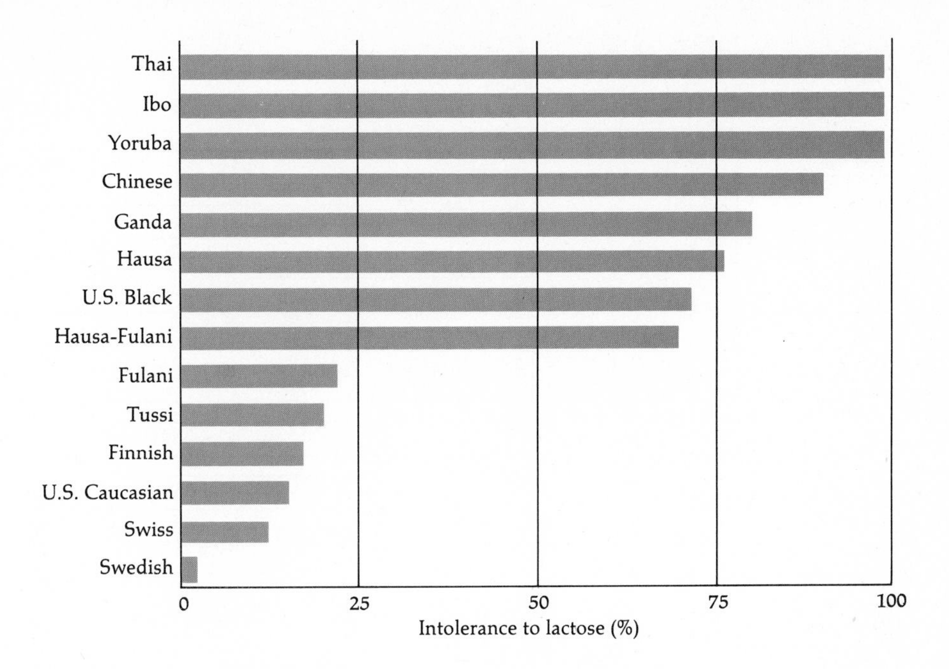

Figure 21.15
Intolerance to lactose in various human populations. The proportion of adult individuals tolerant to this milk sugar is highest in peoples that have used milk and dairy products in their diets for millennia. Populations that traditionally have not used dairy products in their diets are nearly 100% intolerant to lactose as adults. (After N. Kretchmer, *Scientific American*, October, 1972, p. 70.)

keep dog breeds separate, even when the breeds live in the same locality, by preventing interbreeding.) Sometimes, abrupt geographic barriers exist that facilitate the formation and identification of races, e.g., terrestrial organisms living on separate islands or aquatic ones living in separate lakes. Nevertheless, different degrees of gene flow and of genetic differentiation exist around various boundaries, some of which may include further subdivisions. The data in Table 21.6 serve to illustrate this point.

The geographic locations of the populations in Table 21.6 range from the North (Washington) toward the South (California), then to the East (Colorado, Arizona, and Texas), and then to the South again (Mexico). There is considerable differentiation in the frequency of chromosomal arrangements throughout the range. The frequency of *ST* is high in Washington, intermediate in California, and low or zero in the other localities. The frequency of *AR* is intermediate in Washington, California, and Fort Collins, high in Mesa Verde and Chiricahua, and becomes low and eventually zero as we move farther south; and so on for the other chromosomal arrangements. Some transitions in the frequency of chromosomal arrangement would be smoother if data for intermediate populations were added to the table.

The genetic differences reflected in the chromosomal frequencies can be the basis for racial differentiation in *D. pseudoobscura*. But how many races are there? One possible classification would distinguish four races: (1) a northern and central race, from Methow through Fort Collins, characterized by the presence of *AR* in intermediate frequencies; (2) a second

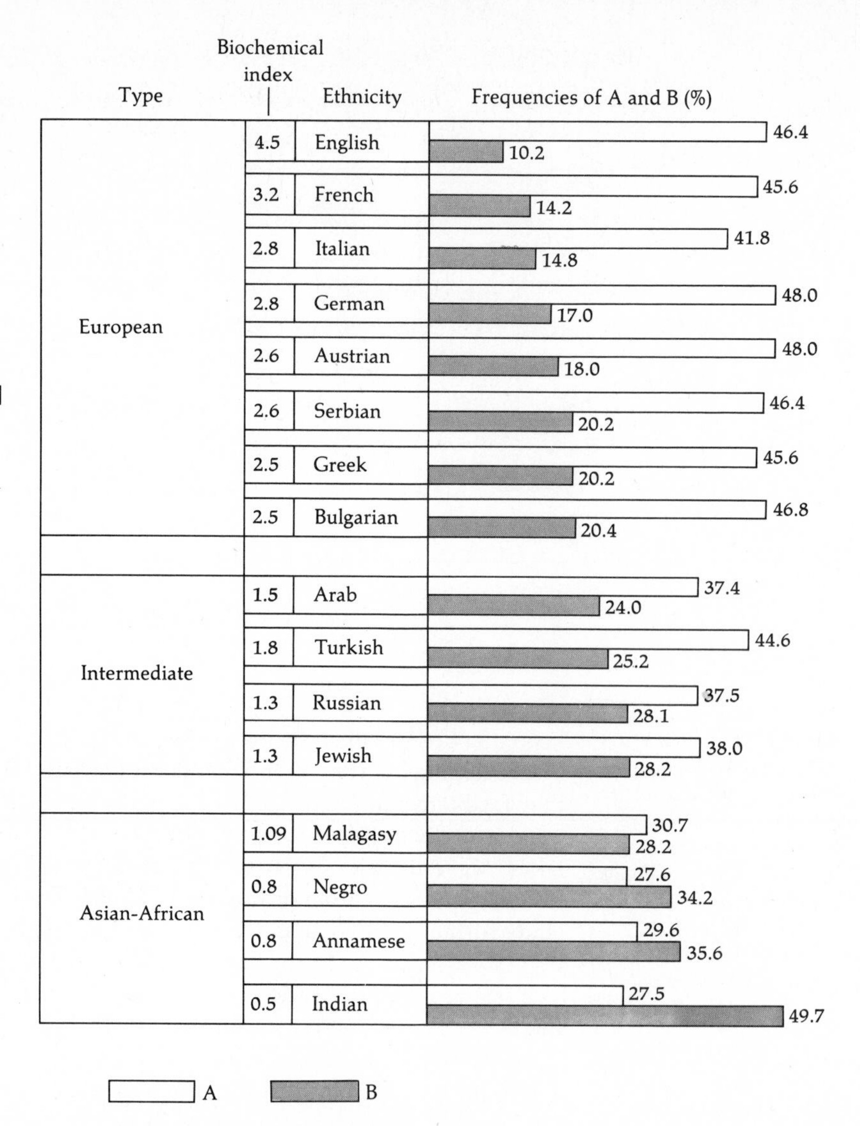

Figure 21.16
Ethnic differences manifested in the ABO blood groups. This graph represents the first use of genetic markers to identify racial differences. The numbers given are the frequencies of two blood groups, A and B. The biochemical index is the ratio of A to B. [After L. Hirszfeld and H. Hirszfeld, *Anthropologie* 29:505 (1919).]

race, from Mesa Verde through Chiricahua, with *AR* in high frequency; (3) a third race, from Central Texas through Durango, characterized by the presence of either *CH* or *PP* in high frequencies; and (4) a fourth race, from Hidalgo through Oaxaca, identified by the occurrence of the *CU* arrangement. But we might choose to divide the third race into two, distinguished by the high frequency of *CH* in one and of *PP* in the other. Or we might choose to separate the first two races not between Fort Collins and Mesa Verde, but between San Jacinto and Fort Collins; we would then have a northwestern race, characterized by relatively high frequencies of *ST*, and a central race, characterized by high frequencies of *AR*. This exercise illustrates a very important point: *the amount of genetic differentiation* required to distinguish among races and, consequently, the *number of races* and the *position of the boundaries* are largely arbitrary matters. Racial

Table 21.9
Frequencies, in percent, of various blood-group alleles in five human racial groups.

Allele	Caucasoids	Negroids	Mongoloids	Amerindians	Australoids
I^A	24–38	15–25	15–25	0–55	20–45
I^B	5–20	10–20	15–30	0	0
r	30–40	10–20	0–7	0	0
r'	0–2	0–6	0	0–17	13
r''	0–2	0–1	0–3	0–3	0
R^0	1–5	40–70	0–5	0–30	9
R^1	30–50	5–15	60–76	30–45	56
R^2	10–15	6–20	20–30	30–60	20
Fy^a	40	<10	90	0–90	?
Di^a	<1	0	1–12	0–25	0

After C. Stern, *Principles of Human Genetics*, 3rd ed., W. H. Freeman, San Francisco, 1973.

classification recognizes the existence of genetic differentiation within a species, but, rather than sharp differences, there is often a gradual transition (a *cline*) from less to more genetic differentiation along geographic lines.

Human Races

Once the concept of race is understood, it comes as no surprise to learn that there are multiple classifications of human races. Some classifications recognize only three races, others, more than fifty.

The ethnic diversity of mankind was recognized by Carolus Linnaeus, who identified four human varieties: African, American, Asiatic, and European. The familiar five "color" races were established by Johann Friederich Blumenbach in 1775: white, or Caucasian; yellow, or Mongolian; black, or Ethiopian; red, or American; and brown, or Malayan. Although the identifying characteristic was skin color, it is clear that ethnic groups differ in many other characteristics, such as facial features, hair, body build, and so on. The correspondence between the various characteristics is far from precise. In parts of India, for example, Caucasian facial features are found in people with black skin.

L. Hirszfeld and H. Hirszfeld suggested in 1918 that the ABO blood groups could be used for analysis of ethnic origins (Figure 21.16). The data

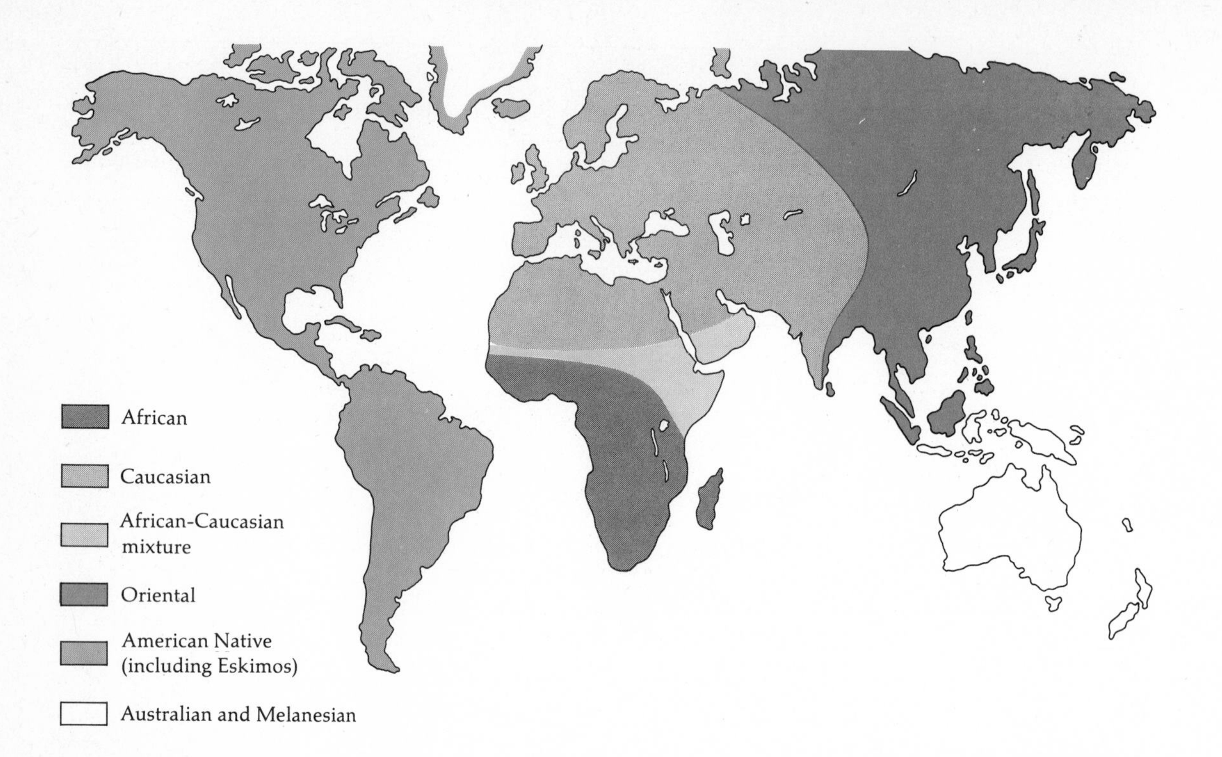

Figure 21.17
Geographic distribution of the main human ethnic groups. Five major groups can be distinguished: African, Caucasian, Oriental, American Native, and Australian and Melanesian. Admixtures between ethnic groups occur in various places, notably in Africa between Africans and Caucasians. (After W. F. Bodmer and L. L. Cavalli-Sforza, *Genetics, Evolution, and Man*, W. H. Freeman, San Francisco, 1976.)

then available for Old World peoples showed that the B blood group (genotypes I^BI^B and I^Bi) increases in frequency from about 10% in England to 50% in India; the A blood group (I^AI^A and I^Ai) has about the same frequency throughout Europe, a somewhat lower frequency in Russia and the Middle East, and a still lower frequency in Africa and India. A "biochemical index" (the ratio of the frequencies of blood groups A and B) was used to distinguish three racial groups: European, Intermediate, and Asian-African.

The frequency of the I^B gene throughout the world is shown in Figure 19.10. A racial classification based on blood-group gene frequencies does not, of course, imply that people with different blood groups belong to different races, but rather that allele frequency differences at the blood-group locus reflect overall differentiation in the gene pool. It has turned out, however, that the variations in ABO frequencies are not as great as variations in other blood groups, such as the Rhesus (*R*), Duffy (*Fy*), and Diego (*Di*), which are therefore more informative about ethnic groups (Table 21.9).

Geographic boundaries help to identify three major racial groups: Africans, Caucasians, and a highly heterogeneous group called Easterners. The Easterners can be subdivided into three groups: American Natives, Orientals, and Australians and Melanesians. The resulting five groups largely overlap the five "color" races of Blumenbach. Caucasians are a fairly homogeneous group ranging from western Europe to western Russia and through the Middle East to India, where there is a gradual transition to

Table 21.10
Nine geographical races and 34 local races in human populations.

Geographical Races		
1. European	4. Amerindian	7. Australian
2. Indian	5. African	8. Micronesian
3. Asiatic	6. Melanesian-Papuan	9. Polynesian

Local Races		
1. Northwest European	13. Lapp	25. Negrito
2. Northeast European	14. North American Indian	26. Melanesian-Papuan
3. Alpine	15. Central American Indian	27. Murrayian
4. Mediterranean	16. South American Indian	28. Carpentarian
5. Hindu	17. Fuegian	29. Micronesian
6. Turkic	18. East African	30. Polynesian
7. Tibetan	19. Sudanese	31. Neo-Hawaiian
8. North Chinese	20. Forest Negro	32. Ladino
9. Classic Mongoloid	21. Bantu	33. North American Black
10. Eskimo	22. Bushman and Hottentot	34. South African Black
11. Southeast Asiatic	23. African Pygmy	
12. Ainu	24. Dravidian	

After S. M. Garn, *Human Races*, C. C. Thomas, Springfield, Ill., 1961.

Easterners. In northern and central Africa, however, populations show various degrees of admixture between Caucasians and Africans (Figure 21.17).

Racial classifications recognize genetic heterogeneity among populations. The question is how much diversity to recognize. If only a few subdivisions are recognized, as in Figure 21.17, then there will be too much heterogeneity within some of them. On the other hand, with ever finer splitting, the differences become less sharp and the boundaries more blurred. An eclectic solution proposed by Stanley M. Garn distinguishes 9 "geographical races" and 34 "local races" in human populations (Table 21.10).

How genetically different are human races? In a variety of human populations, some 25 gene loci have been studied that are polymorphic in at least one racial group. The average heterozygosity of an individual gives a measure of the amount of genetic variation in a population, because it estimates the probability that two genes picked up at random at a given locus are different from each other. For any given human group, the average heterozygosity per individual at the 25 polymorphic loci ranges from 28 to 30%. The probability that two genes taken at random, each from a different racial group, are different (which is also the probability of heterozygosity in the progeny of an interracial cross) is about 35–40%. This is a small increase (from about 29% to about 37%) over the heterozygosity within groups. That is, the additional genetic differentiation *between* human races is relatively small in comparison to the genetic differentiation *within* groups. As was stated in the previous section, knowing the racial group of an individual provides little information about its genetic makeup. Each person has a unique genotype—it is different from that of every other person, whether or not they belong to the same race.

Problems

1. A cattle rancher breeds a bull (A) with its own daughter (C). What is the inbreeding coefficient of the offspring (D), assuming that A and B are unrelated?

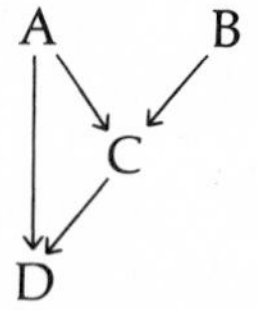

A B
C
D

2. Calculate the inbreeding coefficient of the offspring (G) of an uncle-niece mating:

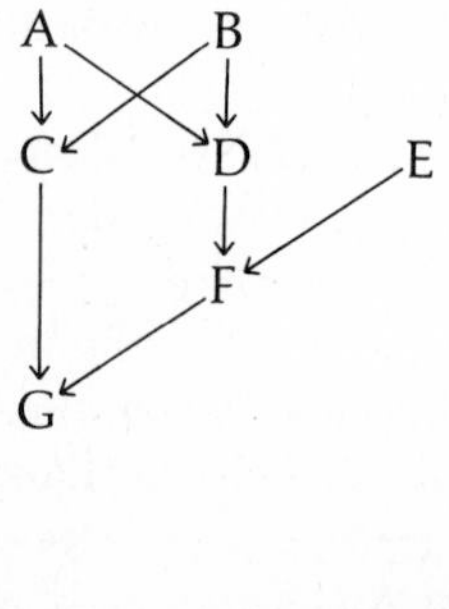

A B
C D E
F
G

3. Calculate the inbreeding coefficient of the offspring (K) of double first cousins:

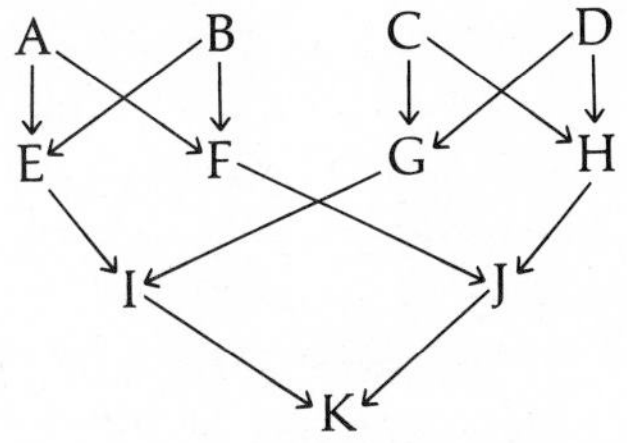

4. A rancher trying to obtain uniformity propagates his herd by brother-sister matings for six generations. Calculate the coefficient of inbreeding for each generation, assuming that he started with unrelated parents.

5. The frequencies of two autosomal alleles, A and a, in each of three plant populations are 0.80 and 0.20, respectively. The inbreeding coefficients in the three populations are 0, 0.40, and 0.80. What is the frequency of heterozygotes in each population?

6. The frequencies of two autosomal alleles, A and a, in a large, random-mating population of a monoecious plant are 0.60 and 0.40, respectively. A large sample is taken, in which each plant is reproduced by selfing, and the progeny are again selfed for five additional generations. Calculate the inbreeding coefficient and the frequency of heterozygotes of the resulting progeny.

7. In a small population, the numbers of individuals of the three possible genotypes at a locus are 28 AA, 24 Aa, and 48 aa. Calculate the inbreeding coefficient, assuming that inbreeding alone is responsible for any deviations of the genotypic frequencies from Hardy-Weinberg expectations.

8. Assume that the mutation rate to the recessive lethal allele responsible for cystic fibrosis is $u = 4 \times 10^{-4}$. Assume also that a population is at equilibrium with respect to this allele. What is the expected frequency of the disease in the offspring of first-cousin marriages? Assume now that in another population the mutation rate is twice as high ($u = 8 \times 10^{-4}$), owing to long-term exposure to background radiation, and that this population is also at equilibrium with respect to this allele. What is the expected frequency of the disease in the offspring of first-cousin marriages in this second population?

9. In Japan, the frequency of $L^M L^N$ heterozygotes for the M-N blood groups in large populations, in which random mating may be assumed, is 0.4928. However, in a town where matings between relatives are common, the frequency of MN individuals is 0.4435. Calculate the inbreeding coefficient in this town, assuming that the allele frequencies are the same as in the large populations.

10. In an African region, the frequency of the Hb^S allele, responsible for sickle-cell anemia, is 0.123. Calculate the frequency of Hb^SHb^S homozygotes among newborns: (1) in a random-mating population and (2) in the offspring of first cousins.

11. A population of fish in a lake is fixed at each of two unlinked loci for the dominant allele (*AA BB*). A canal is built that connects the lake with a smaller one, where the same fish species is fixed for the recessive alleles (*aa bb*). Assume that mating is random thereafter, that the original population was ten times greater in the large lake than in the small lake, and that the two loci are not affected by natural selection. What is the linkage disequilibrium, d, immediately after the populations of the two lakes become mixed, and after five generations of random mating?

12. In the previous problem, assume that the two loci are in the same chromosome and have a recombination frequency $c = 0.10$. What will be the value of d after five generations of random mating? How many generations will be necessary to reduce d to the value achieved for unlinked loci in five generations of random mating?

13. Assume now that the population in the large lake of problem 11 had allele frequencies $p = 0.80$ (A allele), $q = 0.20$ (a), $r = 0.60$ (B), and $s = 0.40$ (b), and that it was at equilibrium with respect to the two loci considered simultaneously. Assume, as before, that the population in the small lake is ten times smaller than in the large lake and is fixed for the a and b alleles. What is the initial value of d after the fish in the two lakes become thoroughly mixed? What will be the value of d after five generations of random mating: (1) if the two loci are unlinked and (2) if the loci are linked having a recombination frequency of $c = 0.10$?

14. Two linked esterase loci were examined in a population of *Drosophila montana*. At each locus there were two alleles, one exhibiting activity, the other being inactive ("null"). The observed numbers of each two-allele combination in a sample of 474 gametes were as follows:

	Locus 1	
Locus 2	Active	Null
Active	31	273
Null	97	73

Calculate the linkage disequilibrium value, d.

15. An experimental population of barley (*Hordeum vulgare*) was established by intercrossing 30 barley varieties from various parts of the world. The population was thereafter maintained by spontaneous self-fertilization. Two loci (A and B) coding for esterases were examined at various times; two alleles (*1* and *2*) were present at each locus. The gametic frequencies in three generations were as follows (several thousand gametes were examined in each generation):

Generation	A_1B_1	A_2B_2	A_1B_2	A_2B_1
4	0.453	0.019	0.076	0.452
14	0.407	0.004	0.098	0.491
26	0.354	0.003	0.256	0.387

Calculate the value of d for these three generations. What process(es) is (are) likely to be responsible for changing the linkage disequilibrium as the generations proceed?

22

Speciation and Macroevolution

Anagenesis and Cladogenesis

The process of evolution has two dimensions: (1) *anagenesis,* or evolution within a lineage, and (2) *cladogenesis,* or diversification. Changes occurring in a lineage as time passes are anagenetic evolution. They are often due to natural selection, which promotes adaptation to physical or biotic changes in the environment. Cladogenetic evolution occurs when a lineage splits into two or more lineages. The great diversity of the living world is the result of cladogenetic evolution, which results in adaptation to a greater variety of niches, or ways of life. The most fundamental cladogenetic process is *speciation,* the process by which a species splits into two or more species.

The last few chapters have been concerned with evolution as it occurs within a species; this is sometimes called *microevolution* (small-scale evolution). Correspondingly, evolution above the species level is called *macroevolution* (large-scale evolution). The genetic study of macroevolution has been made possible by the advances in molecular biology. The classical methods of Mendelian genetics ascertain the presence of genes by observing segregation in the progenies of crosses between individuals differing in some trait. But interspecific crosses are usually impossible and, when they do take place, the hybrid offspring are usually inviable or sterile. Genetic comparisons between different species can now be made by direct examination of their DNA or of the proteins encoded by the DNA.

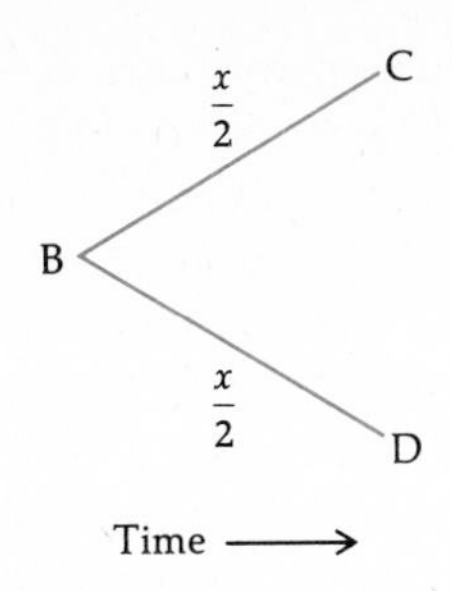

Figure 22.1
Inference of anagenetic evolution from cladogenetic data. C and D are two contemporary species having B as a common ancestral species. If the amount of genetic differentiation between C and D is x, we can assume, as a first approximation, that half of the change occurred in each of the two lineages.

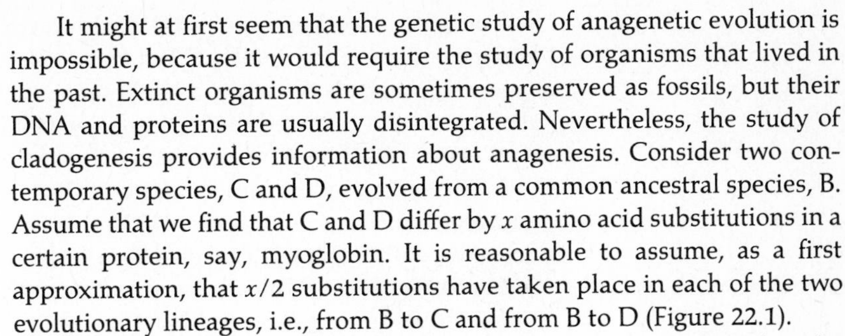

It might at first seem that the genetic study of anagenetic evolution is impossible, because it would require the study of organisms that lived in the past. Extinct organisms are sometimes preserved as fossils, but their DNA and proteins are usually disintegrated. Nevertheless, the study of cladogenesis provides information about anagenesis. Consider two contemporary species, C and D, evolved from a common ancestral species, B. Assume that we find that C and D differ by x amino acid substitutions in a certain protein, say, myoglobin. It is reasonable to assume, as a first approximation, that $x/2$ substitutions have taken place in each of the two evolutionary lineages, i.e., from B to C and from B to D (Figure 22.1).

The assumption that equal amounts of change have occurred in the two lineages can be dropped. Suppose that a third contemporary species, E, is compared with C and D and that the numbers of amino acid differences between the myoglobin molecules of the three species are as follows:

C and D: 4

C and E: 11

D and E: 9

If the *phylogeny* (evolutionary history) of the three species is as shown in Figure 22.2, we can estimate the number of substitutions that occurred in each of its branches. Let us use x and y to denote the number of amino acid differences between B and C and between B and D, respectively, and z to denote the number of differences between A and B *plus* those between A and E. We have, then, the following three equations:

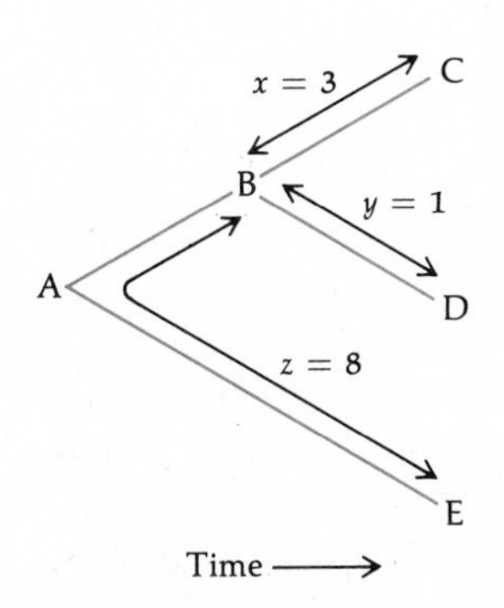

Figure 22.2
Estimated amounts of anagenetic change in the phylogeny of three contemporary species.

$$x + y = 4$$

$$x + z = 11$$

$$y + z = 9$$

Subtracting the third equation from the second, we get

$$x - y = 2$$

Adding this equation to the first, we get

$$2x = 6 \quad \text{or} \quad x = 3$$

Therefore

$$y = 4 - x = 1$$

$$z = 11 - x = 8$$

The procedure becomes more complicated when many more contemporary species are involved, but the conceptual basis for estimating anagenetic change is the same. An inherent problem is that an amino acid substitution that occurred in the past—say, from leucine to proline—may subsequently have been canceled by the reciprocal substitution—from proline to leucine—at the same position in the polypeptide, and may hence go undetected. The same problem exists at the level of the DNA, when a nucleotide substitution is canceled by the reciprocal one. The methods used to correct for such hidden substitutions need not be discussed here.

We have assumed above that the configuration of the phylogeny shown in Figure 22.2 was previously known. In fact, however, protein or DNA information may serve to reconstruct the phylogeny when this is *not* known from other sources or when it is doubtful. Indeed, because the number of substitutions between C and D is much smaller than that between either one of these species and E, we would infer that C and D have diverged from each other more recently than they diverged from E. We would thus arrive at the same phylogeny as shown in the figure. The reconstruction of phylogenetic history may be risky when it is based on the analysis of a single protein or DNA molecule, because a greater number of substitutions may have taken place in some lineages than in others, or at different times. However, the cumulative evidence obtained from many proteins studied in many species tends to converge upon phylogenies that are usually in good agreement with phylogenies deduced from morphological and paleontological evidence.

The Concept of Species

In sexually reproducing organisms, a *species* is a group of interbreeding natural populations that are reproductively isolated from other such groups. Species are natural systems, defined by the possibility of interbreeding between their members. The ability to interbreed is of great evolutionary import, because it establishes species as discrete and independent *evolutionary units*. Consider an adaptive mutation or some other genetic change originating in a single individual. Over the generations, this may spread by natural selection to all members of the species, but not to individuals of other species. This can be stated differently: individuals of a species share in a common gene pool, which is not, however, shared in by individuals of other species. Owing to reproductive isolation, different species have independently evolving gene pools.

Reproductive isolation is the criterion of speciation in sexual organisms. An ancestral species becomes transformed into two descendant species when an array of populations able to interbreed becomes segregated into two reproductively isolated arrays. It should not be surprising that reproductive isolation is used as the fundamental criterion to

Table 22.1
Classification of reproductive isolating mechanisms (RIMs).

1. *Prezygotic RIMs,* which prevent the formation of hybrid zygotes.
   a. *Ecological isolation:* populations occupy the same territory but live in different habitats, and thus do not meet.
   b. *Temporal isolation:* mating or flowering occur at different times, whether in different seasons or at different times of day.
   c. *Behavioral isolation* (also called *ethological isolation,* from the Greek *ethos,* "behavior," or *sexual isolation*): sexual attraction between females and males is weak or absent.
   d. *Mechanical isolation:* copulation or pollen transfer is forestalled by the different size or shape of genitalia, or the different structure of flowers.
   e. *Gametic isolation:* female and male gametes fail to attract each other, or the spermatozoa or pollen are inviable in the sexual ducts of animals or in the stigmas of flowers.

2. *Postzygotic RIMs,* which reduce the viability or fertility of hybrids.
   a. *Hybrid inviability:* hybrid zygotes fail to develop or at least to reach sexual maturity.
   b. *Hybrid sterility:* hybrids fail to produce functional gametes.
   c. *Hybrid breakdown:* the progenies of hybrids (F_2 or backcross generations) have reduced viability or fertility.

define species—reproductive isolation allows gene pools to evolve independently.

The biological properties of organisms that prevent interbreeding are called *reproductive isolating mechanisms* (RIMs). A classification of RIMs is shown in Table 22.1. Reproductive isolating mechanisms may be classified as prezygotic and postzygotic. *Prezygotic* RIMs impede hybridization between members of different populations, and thus prevent the formation of hybrid zygotes. *Postzygotic* RIMs reduce the viability or fertility of hybrids. Prezygotic and postzygotic RIMs all serve the same purpose: they forestall gene exchange between populations. But there is an important difference between them: the waste of reproductive effort is greater for postzygotic RIMs than for prezygotic RIMs. If a hybrid zygote is produced but is inviable (*hybrid inviability*—see Table 22.1), two gametes that could have been used in nonhybrid reproduction have been wasted. If the hybrid is viable but sterile (*hybrid sterility*), the waste includes not only the gametes but also the resources used by the hybrid in its development. The waste is even greater in *hybrid breakdown,* because it involves the resources used not only by the hybrids but also by their progenies. One prezygotic RIM, *gametic isolation,* may also involve reproductive waste when gametes fail to form viable zygotes. The other prezygotic RIMs avoid gametic waste, but some energy may be wasted in unsuccessful courtship (*behavioral isolation*) or in unsuccessful attempts to copulate (*mechanical isolation*). Natural selection promotes the development of prezygotic RIMs between populations already isolated by postzygotic RIMs, whenever the populations coexist in the same territory and there is opportunity for the formation of hybrid zygotes. This occurs precisely because reproductive waste is reduced or altogether eliminated when prezygotic RIMs exist.

The reproductive isolating mechanisms listed in Table 22.1 do not all occur between any two species, but it is usually true that two or more mechanisms, not just a single one, are involved in the reproductive isolation between species. Some RIMs are more common in plants (e.g., temporal isolation), others, in animals (e.g., behavioral isolation); but even among closely related species, different sets of RIMs are often involved when different pairs of species are compared. This is an example of the opportunism of natural selection: the evolutionary function of RIMs is to prevent interbreeding, but how this is accomplished depends on environmental circumstances as well as on the genetic variability available.

The Process of Speciation

Species are reproductively isolated groups of populations. The question of how species come about is, therefore, equivalent to the question of how reproductive isolation arises between groups of populations. In general, reproductive isolation usually starts as an incidental byproduct of genetic divergence, but is completed when it becomes directly promoted by natural selection. Speciation may occur in a variety of ways, but two main stages can be recognized in the process (Figure 22.3).

STAGE I. The onset of the speciation process requires, first, that gene flow between two populations of the same species be somehow interrupted—completely, or nearly so. Absence of gene flow enables the two populations to become genetically differentiated as a consequence of their adaptation to different local conditions or to different ways of life (and also as a consequence of genetic drift, which may play a lesser or greater role, depending on the circumstances). The interruption of gene flow is necessary, because otherwise the two populations would in fact share in a common gene pool and fail to become genetically different. As populations become ever more genetically different, reproductive isolating mechanisms appear, because different gene pools are not mutually coadapted; hybrid individuals will have disharmonious genetic constitutions and will have reduced fitness in the form of reduced viability or fertility.

Thus, the two characteristics of the first stage of speciation are that: (1) reproductive isolation appears primarily in the form of postzygotic RIMs, and (2) these RIMs are a byproduct of genetic differentiation—reproductive isolation is not directly promoted by natural selection at this stage.

Genetic differentiation, and the consequent appearance of postzygotic RIMs, is usually a gradual process. As such, it makes the decision as to whether or not the process of speciation has already begun between two

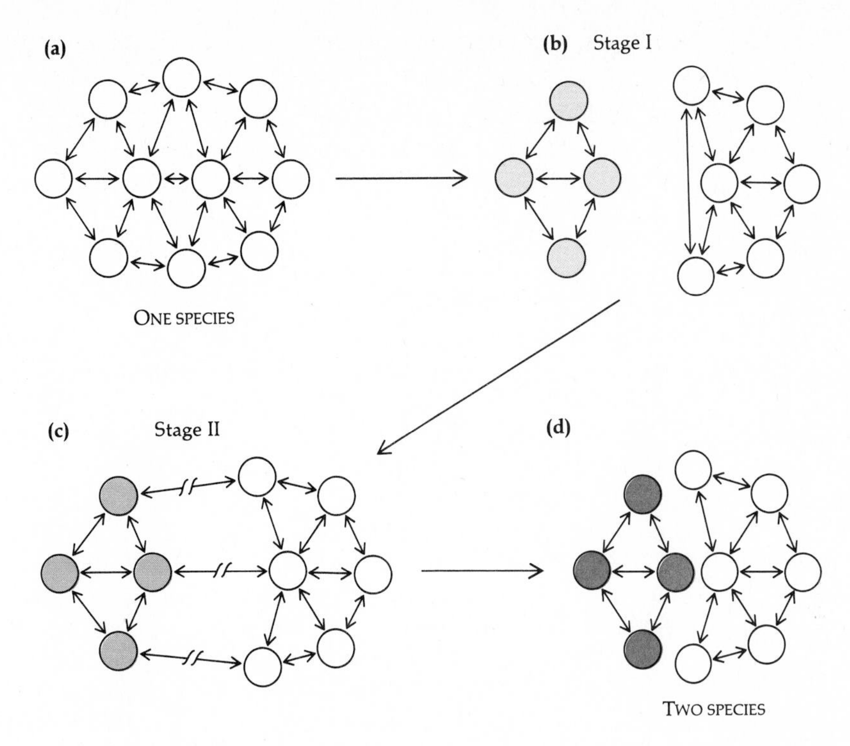

Figure 22.3
Generalized model of the process of speciation. **(a)** Local populations of a single species are represented by circles; the arrows indicate that gene flow occurs between populations. **(b)** The populations have become separated into two groups between which there is no gene flow. These groups gradually become genetically different, as indicated by the shading of the populations on the left. As a consequence of this genetic differentiation, reproductive isolating mechanisms arise between the two groups. This is the first stage of speciation. **(c)** Individuals from different population groups are able to intermate. However, owing to the pre-existing reproductive isolating mechanisms, little if any gene flow takes place, as indicated by the broken arrows. Natural selection favors the development of additional reproductive isolating mechanisms, particularly prezygotic ones, which prevent matings between individuals from different population groups. This is the second stage of speciation. **(d)** Speciation has been completed because the two groups of populations are fully reproductively isolated. There are now two species that can coexist without gene exchange.

populations a somewhat arbitrary matter. Populations may be considered to be in the first stage of speciation if RIMs have appeared between them. Local populations of a given species are often genetically somewhat different, but are not thought to be in the first stage of speciation if their differentiation is small and does not result in the appearance of RIMs.

STAGE II. This stage encompasses the completion of reproductive isolation. Assume that the external conditions that interfered with gene flow between two populations in the first stage of speciation disappear.

This might occur, for example, if two previously geographically separated populations expand and come to occupy, at least in part, the same territory. Two outcomes are possible: (1) a single gene pool arises, because the loss of fitness in the hybrids is not very great and the two populations fuse; (2) two species ultimately arise, because natural selection favors the further development of reproductive isolation.

The first stage of speciation is reversible: if it has not gone far enough, it is possible for two previously differentiated populations to fuse into a single gene pool. However, if matings between individuals from different populations leave progenies with reduced viability or fertility, natural selection will favor genetic variants promoting matings between individuals of the same population. Consider the following simplified situation. Assume that there are two alleles, A_1 and A_2, at a locus; A_1 favors matings between individuals of the same population, while A_2 favors interpopulational matings. Then A_1 will be present more often in progenies from intrapopulational crosses, that is, in individuals with good viability and fertility, while A_2 will be present more often in interpopulational hybrids; because the latter have low fitness, A_2 will decrease in frequency from generation to generation. Natural selection will result in the multiplication of alleles that favor intrapopulation matings and in the elimination of alleles that favor interpopulation matings; that is, natural selection will favor the development of prezygotic RIMs, which prevent the formation of hybrid zygotes.

The two characteristics of the second stage of speciation, then, are that: (1) reproductive isolation develops mostly in the form of prezygotic RIMs, and (2) the development of prezygotic RIMs is directly promoted by natural selection. These two characteristics of Stage II stand in contrast to the two characteristics of Stage I, pointed out above.

Nevertheless, speciation may take place without the occurrence of Stage II. In the absence of gene exchange, populations may develop complete reproductive isolation if the process of genetic differentiation continues long enough—for example, when the populations remain separated indefinitely on two islands. However, the speciation process is accelerated by Stage II because natural selection directly promotes the development of reproductive isolation.

Geographic Speciation

The general model of speciation just outlined may be realized in different ways, or modes, which can be classified as geographic speciation or quantum speciation. In *geographic speciation,* Stage I begins owing to geographic separation between populations. Terrestrial organisms may be separated by water (such as rivers, lakes, and oceans), mountains, deserts, or any kind of territory uninhabitable by the populations; freshwater organisms may be kept separate if they live in different river systems or unconnected lakes; marine organisms may be separated by land, by water of greater or lesser depth than the organisms can tolerate, or by water of different salinity.

As a result of natural selection, geographically separate populations become adapted to local conditions and thus become genetically differentiated. Random genetic drift may also contribute to genetic differentiation, particularly when populations are small or are derived from only a few individuals. If geographic separation continues for some time, incipient reproductive isolation may appear, particularly in the form of postzygotic RIMs; the populations will then be in the first stage of speciation.

The second stage of speciation begins when previously separated populations come into contact, at least over part of their geographic distributions. This may happen, for example, by topographic changes on the earth's surface, by ecological changes in the intervening territory that make it habitable by the populations, or by migration of members of one population into the territory of the other. Matings between individuals from different populations may then take place. Depending on the strength of the pre-existing RIMs and on the extent of hybridization, the two populations may fuse into a single gene pool or may develop additional (prezygotic) RIMs and become separate species.

The two stages of the process of geographic speciation can be illustrated with a group of closely related species of *Drosophila* that live in the American tropics (Figure 22.4). This group, collectively called the *Drosophila willistoni* group, consists of 15 species, 6 of which are *sibling species* (i.e., they are morphologically virtually indistinguishable). One of these siblings is *D. willistoni* itself, which consists of two subspecies: *D. w. quechua* lives in continental South America west of the Andes, and *D. w. willistoni* lives east of the Andes. There is incipient reproductive isolation between them, particularly in the form of hybrid sterility: when males and females from the two subspecies are intercrossed in the laboratory, the results depend on the direction of the mating:

♀ *D. w. willistoni* × ♂ *D. w. quechua* → fertile female and male progeny

♀ *D. w. quechua* × ♂ *D. w. willistoni* → fertile females but sterile males

If these two subspecies were to come into contact in nature and intercross, natural selection would favor the development of prezygotic RIMs, because the male progenies of all crosses between *quechua* females and *willistoni* males are sterile. In other words, these subspecies are two groups of populations in the first stage of geographic speciation.

The first stage of speciation is also found in another species of the group. *Drosophila equinoxialis* consists of two geographically separated subspecies: *D. e. equinoxialis* inhabits continental South America, and *D. e.*

Figure 22.4
Geographic distribution of six closely related species of the *Drosophila willistoni* group. *D. willistoni* and *D. equinoxialis* each consist of two subspecies. These subspecies represent populations in the first stage of geographic speciation.

caribbensis inhabits Central America and the Caribbean islands. Laboratory crosses between these two subspecies always produce fertile females but sterile males, regardless of the direction of the cross. Thus, there is somewhat greater reproductive isolation between the two subspecies of *D. equinoxialis* than between the two subspecies of *D. willistoni*. Natural selection in favor of prezygotic RIMs would be stronger in *D. equinoxialis,* because all hybrid males are sterile.

It is worth noting that prezygotic RIMs do *not* exist between the subspecies of *D. willistoni* or of *D. equinoxialis*. Reproductive isolation

Figure 22.5
Geographic distribution of the six semispecies of *Drosophila paulistorum*. These semispecies represent populations in the second stage of geographic speciation. Speciation has been virtually completed in places where two or three semispecies coexist without interbreeding.

between the subspecies is, therefore, far from complete, and they are not considered different species.

Stage II of the speciation process can be found within yet another species of the *D. willistoni* group. *Drosophila paulistorum* is a species consisting of six *semispecies*, or incipient species, two or three of which are sympatric in many localities (Figure 22.5). The semispecies exhibit hybrid sterility similar to that found in *D. equinoxialis*: crosses between males and females of two different semispecies yield fertile females but sterile males. But two or three semispecies have come into geographic contact in many places, and there the second stage of speciation has advanced to the point that ethological isolation is complete, or nearly so. When females and males from two different semispecies are placed together in the laboratory, the results depend on the geographic origin of the flies. When both semispecies are from the same locality, only *homogamic* matings (matings between members of the same semispecies) occur; when they are from different localities, however, *heterogamic* matings (matings between members of different semispecies) as well as homogamic matings occur, indicating that ethological isolation is not yet complete. The semispecies of *D. paulistorum* thus provide a remarkable example of the action of natural selection during the second stage of speciation: reproductive isolation has been completed where the semispecies are sympatric, but not elsewhere, because the genes involved have not yet spread fully throughout each semispecies.

Quantum Speciation

In geographic speciation, Stage I entails the gradual genetic divergence of geographically separated populations. The development of postzygotic RIMs as byproducts of genetic divergence usually requires a long period of time: thousands, perhaps millions, of generations. However, there are other modes of speciation in which the first stage, and the appearance of postzygotic RIMs, may require only relatively short periods of time. *Quantum speciation* (also called *rapid speciation* and *saltational speciation*) refers to these accelerated modes of speciation, particularly in the first stage.

One form of quantum speciation is polyploidy, the multiplication of entire chromosome complements (Chapter 17, page 592). Polyploid individuals may arise in just one or a few generations. Polyploid populations are reproductively isolated from their ancestral species and are thus new species. In polyploidy, the suppression of gene flow that is required for the onset of the first stage of speciation is due not to geographic separation but to cytological irregularities. Reproductive isolation in the form of hybrid sterility does not require many generations, but follows immediately, owing to chromosomal imbalance. If diploids and their fertile polyploid derivatives exist near each other and hybridization occurs, natural selection will favor the development of prezygotic isolating mechanisms (Stage II), which prevent interfertilization and the waste of gametes. Examples of polyploid species were given in Chapter 17.

Modes of quantum speciation other than polyploidy are known to occur in plants. One instance of quantum speciation, studied by Harlan Lewis, involves the two diploid species *Clarkia biloba* and *C. lingulata*. Both species are native to California, but *C. lingulata* has a narrow distribution, being known from only two sites in the central Sierra Nevada, at the southern periphery of the distribution of *C. biloba*. The two species are outcrossers, although capable of self-fertilization, and are similar in external morphology, although there are differences in petal shape. How-

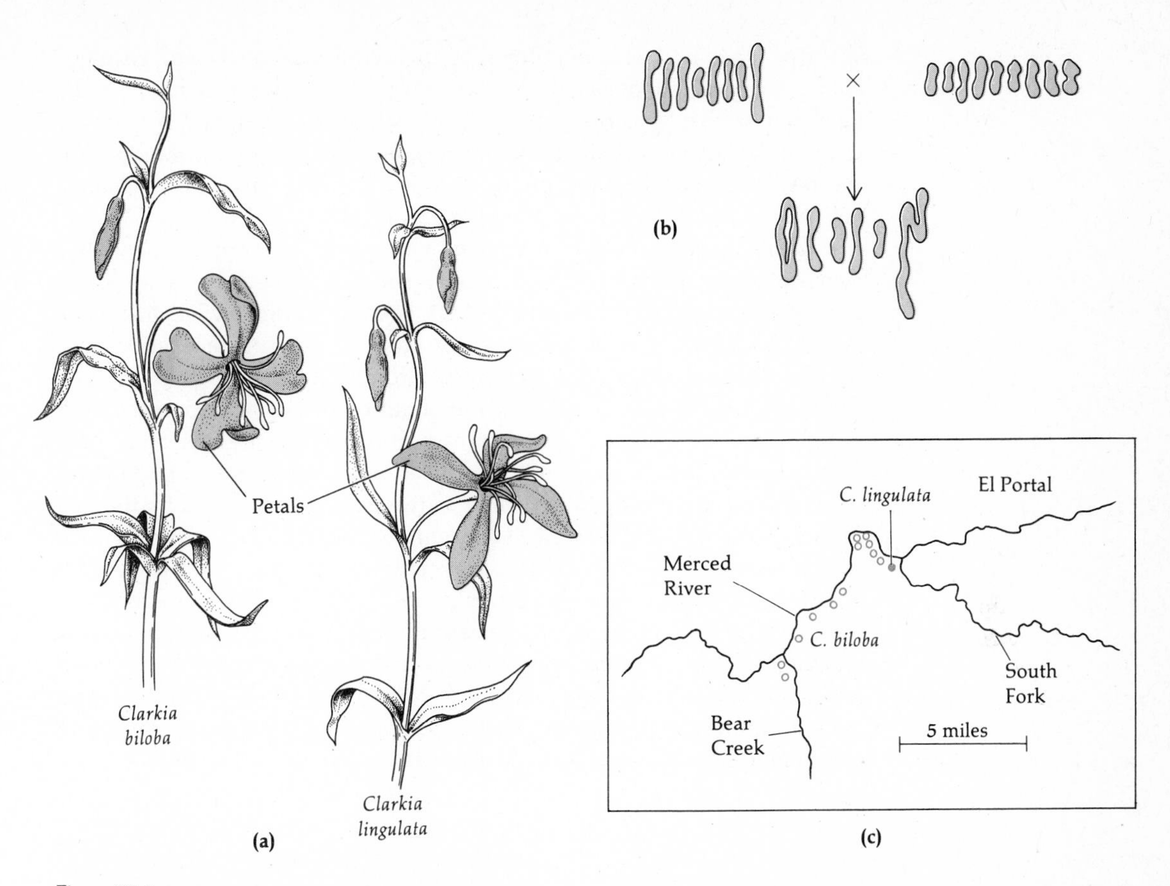

Figure 22.6
Two species of annual plants: *Clarkia lingulata* has arisen from *C. biloba* by quantum speciation. **(a)** Flowering branches of the two species, showing a difference in petal shape, which is bilobed in *C. biloba* but not in *C. lingulata*. **(b)** Paired chromosomes, at meiotic metaphase, of *C. biloba* (eight pairs, left), and of *C. lingulata* (nine pairs, right), and of the F_1 hybrid between the two. The chromosomes differ by at least two reciprocal translocations and a fission (or fusion). **(c)** A small portion of the Merced River canyon in the Sierra Nevada of California, west of Yosemite Valley, showing the southernmost populations of *C. biloba* (circles) and the two known populations of *C. lingulata* (dot).

ever, their chromosomal configurations differ by a translocation, several paracentric inversions, and an extra chromosome in *C. lingulata*, homologous to parts of two chromosomes of *C. biloba* (Figure 22.6). The narrowly distributed species, *C. lingulata*, has arisen from *C. biloba* by a rapid series of events involving extensive chromosomal reorganization. Chromosomal rearrangements, such as translocations, fusions, and fissions, reduce the fertility of individuals heterozygous for the new arrangements.

The first stage of speciation may thus be accomplished through chromosomal rearrangements without extensive allelic differentiation. Self-fertilization facilitates the propagation of the rearrangements; once there is a population of individuals exhibiting some reproductive isolation from the rest of the population, owing to the chromosomal rearrangements, natural selection favors the development of additional RIMs.

Rapid speciation initiated through chromosomal rearrangements has also occurred in animals, for example, in some flightless Australian grasshoppers, such as *Moraba scurra* and *M. viatica,* studied by Michael J. D. White. Incipient species differing by chromosomal translocations are found in adjacent territories. A translocation establishes itself at first in a small colony by genetic drift. If members of this colony possess high fitness, they may subsequently spread and displace the ancestral form from a certain area. The ancestral population and the derived population may then coexist contiguously, their individuality maintained by the low fitness of the hybrids formed in the contact zones, since the hybrids are translocation heterozygotes. The first stage of speciation is thus rapidly accomplished, and natural selection favors the development of additional RIMs (Stage II). This mode of speciation seems to be common in several animal groups, particularly in rodents living underground and having little mobility, such as mole rats of the group *Spalax ehrenbergi* in Israel and pocket gophers of the group *Thomomys talpoides* in the southern Rocky Mountains of the United States (Figure 22.7).

Genetic Differentiation During Speciation

The discovery that genes code for proteins and the development of the techniques of gel electrophoresis have made it possible to estimate the amount of genetic change during the speciation process. Before these techniques became available, there was already evidence suggesting that a fair amount of allelic substitution might be involved in speciation, since it was known that even closely related species are genetically quite different. For example, Erwin Baur had crossed two species of snapdragons, *Antirrhinum majus* and *A. molle,* which produce fertile hybrids. Considerable phenotypic variability appeared in the F_2. Most individuals showed various combinations of the parental traits, but some had characteristics present in neither parent, although they were found in other species of *Antirrhinum* or related genera. Baur estimated that more than 100 genetic differences exist between *A. majus* and *A. molle*. But it was not possible to estimate what proportion of the genes are different—invariant genes cannot be detected by Mendelian methods.

Figure 22.7
Distribution of pocket gophers of the *Thomomys talpoides* complex in the southern Rocky Mountains. The number of chromosomes in different populations ranges from 40 to 60, although individuals from different populations are morphologically indistinguishable. Differences in chromosome numbers prevent individuals from different populations from interbreeding in the regions where they are adjacent to one another.

Box 22.1 Genetic Identity and Genetic Distance

Electrophoretic studies provide data in the form of genotypic frequencies that can be readily converted to allelic frequencies. Assume that A and B are two different populations and K is a given gene locus, and that i different alleles are observed in the two populations. Let us represent the frequencies of the alleles in population A as a_1, a_2, a_3, etc., and in population B as b_1, b_2, b_3, etc. The genetic similarity between the two populations at this locus can be measured by I_K, defined as follows:

$$I_K = \frac{\sum a_i b_i}{\sqrt{\sum a_i^2 \sum b_i^2}}$$

where the symbol Σ means "summation of"; $a_i b_i$ represents the products $a_1 b_1$, $a_2 b_2$, $a_3 b_3$, etc.; a_i^2 means a_1^2, a_2^2, a_3^2, etc.; and b_i^2 means b_1^2, b_2^2, b_3^2, etc. The formula for I_K calculates the (normalized) probability that two alleles, one taken from each population, are identical.

Let us consider some simple examples. First assume that only one allele is observed, with frequency 1 in both populations. Then $a_1 = 1$, $b_1 = 1$, and therefore

$$I_K = \frac{1 \times 1}{\sqrt{1^2 \times 1^2}} = \frac{1}{1} = 1$$

Not surprisingly, the value of I_K is 1, indicating that the two populations are identical at this locus.

Assume now that we observe two different alleles, the first having a frequency of 1 in population A, and the second, a frequency of 1 in population B. Then $a_1 = 1$, $b_1 = 0$, $a_2 = 0$, $b_2 = 1$, and therefore

$$I_K = \frac{(1 \times 0) + (0 \times 1)}{\sqrt{(1^2 + 0^2)(0^2 + 1^2)}} = \frac{0 + 0}{\sqrt{1 \times 1}} = \frac{0}{1} = 0$$

The value of I_K is 0, indicating that the two populations are genetically completely different at this locus.

Now consider the case where two alleles exist in both populations, with frequencies $a_1 = 0.2$, $a_2 = 0.8$ $(a_1 + a_2 = 1)$, and $b_1 = 0.7$, $b_2 = 0.3$ $(b_1 + b_2 = 1)$. We then obtain

$$I_K = \frac{(0.2 \times 0.7) + (0.8 \times 0.3)}{\sqrt{(0.2^2 + 0.8^2)(0.7^2 + 0.3^2)}}$$

$$= \frac{0.14 + 0.24}{\sqrt{0.68 \times 0.58}}$$

$$= 0.605$$

The value of I_K lies between 1 and 0, as we would expect, since the two populations share common alleles, although not in identical frequencies.

In order to estimate the genetic differentiation between two populations, several loci need to be studied. Let I_{ab}, I_a, and I_b be the arithmetic means (the averages), over all loci, of $\Sigma a_i b_i$, Σa_i^2, and Σb_i^2, respectively. Then the *genetic identity, I*, between the two populations can be measured, as proposed by M. Nei, by

$$I = \frac{I_{ab}}{\sqrt{I_a I_b}}$$

and the *genetic distance, D*, between the populations can be measured by

$$D = -\ln I$$

Assume that the three examples of differentiation at single loci given above actually correspond to three different loci studied in two populations. Then we have

$$I_{ab} = \frac{1 + 0 + 0.38}{3} = 0.460$$

$$I_a = \frac{1 + 1 + 0.68}{3} = 0.893$$

$$I_b = \frac{1 + 1 + 0.58}{3} = 0.860$$

Therefore

$$I = \frac{0.460}{\sqrt{0.893 \times 0.860}} = 0.525$$

and

$$D = -\ln 0.525 = 0.644$$

That is, it is estimated that 0.644 allelic substitutions per gene locus (or 64.4 allelic substitutions per 100 loci) have occurred in the separate evolution of the two populations. More than three gene loci need to be studied in order to obtain an acceptable estimate of genetic differentiation between any two populations, but the three loci are sufficient to show how genetic identity and genetic distance are calculated.

Estimates of genetic differentiation between two populations can be obtained by studying, in both populations, a sample of proteins, chosen without knowing whether or not they are different in the populations. The genes coding for the proteins then represent a random sample of all the structural genes with respect to the differentiation between the populations. The results obtained from the study of a moderate number of gene loci can therefore be extrapolated to the whole genome.

An efficient technique for studying protein variation in natural populations is gel electrophoresis (Box 18.1), which provides estimates of genotypic and allelic frequencies in populations. A useful method, proposed by Masatoshi Nei, of estimating genetic differentiation between populations, using electrophoretic data, is shown in Box 22.1. Two parameters are used: (1) *genetic identity*, I, which estimates the proportion of genes that are identical in structure in two populations, and (2) *genetic distance*, D, which estimates the number of allelic substitutions per locus that have occurred in the separate evolution of two populations. There is an allelic substitution when one allele is replaced by a different allele, or when a set of alleles is replaced by a different set. The method takes into account the fact that not all observed allelic substitutions are complete: an allele may have been partly replaced by a different one, but the original allele may still exist in greater or lesser frequency.

Genetic identity, I, may range in value from zero (no alleles in common) to one (the same alleles and in the same frequencies are found in both populations). Genetic distance, D, may range in value from zero (no allelic change at all) to infinity; D can be greater than one because each locus may experience complete allelic substitution more than once as evolution goes on for long periods of time.

The statistics I and D can be used to measure genetic differentiation during the speciation process. We shall consider geographic speciation

Table 22.2
Genetic differentiation between populations of the *Drosophila willistoni* group at various levels of evolutionary divergence. Levels 2 and 3 represent Stage I and Stage II, respectively, of the process of geographic speciation. *I* estimates the degree of genetic similarity, *D*, the degree of genetic differentiation. The values given are the means and standard errors for several comparisons.

Level of Comparison	*I*	*D*
1. Local populations	0.970 ± 0.006	0.031 ± 0.007
2. Subspecies	0.795 ± 0.013	0.230 ± 0.016
3. Incipient species	0.798 ± 0.026	0.226 ± 0.033
4. Sibling species	0.563 ± 0.023	0.581 ± 0.039
5. Morphologically different species	0.352 ± 0.023	1.056 ± 0.068

From F. J. Ayala, *Evol. Biol.* 8:1 (1975).

first. The *Drosophila willistoni* group of species was used as a model of geographic speciation, because both stages of the process can be identified therein. This group of species has been extensively studied using electrophoretic techniques. The results are summarized in Table 22.2, in which five levels of evolutionary divergence are represented. The first level involves comparisons between populations living in different localities but without any reproductive isolation between them; the genetic identity is 0.970, indicating a very high degree of genetic similarity.

The second level involves comparisons between different subspecies (such as *D. w. willistoni* and *D. w. quechua*, and *D. e. equinoxialis* and *D. e. caribbensis*). These populations are in the first stage of speciation and exhibit postzygotic RIMs in the form of hybrid sterility. They also exhibit a fair amount of genetic differentiation, $I = 0.795$ and $D = 0.230$; complete allelic substitutions have occurred, on the average, in 23 of every 100 gene loci.

The third level of evolutionary divergence in Table 22.2 involves comparisons between the incipient species of the *D. paulistorum* complex. These are populations in the second stage of speciation, exhibiting some prezygotic as well as postzygotic RIMs. Apparently these populations are not genetically more differentiated than those in the first stage of speciation. This means that the second stage of speciation has not required much genetic change, which is perhaps not surprising. During the first stage of speciation, reproductive isolation comes about as a byproduct of genetic change: a fair amount of genetic change needs to take place over the whole genome before postzygotic RIMs develop. However, during the second stage of speciation, natural selection directly favors the development of prezygotic RIMs; only a few genes—those affecting courtship and mating behavior, for example—need to be changed to accomplish this.

The fourth level in Table 22.2 involves comparisons between sibling species (such as *D. willistoni* and *D. equinoxialis*). Despite their morphological similarity, these species are genetically quite different; about 58 allelic substitutions have occurred, on the average, per 100 loci. Species are independently evolving groups of populations. Once the process of speciation is completed, species will continue to diverge genetically. The results of this gradual process of divergence are also apparent in the comparisons between morphologically different species of the *D. willistoni* group (fifth level in Table 22.2). On the average, somewhat more than one allelic substitution per gene locus has occurred in the evolution of these nonsibling species.

Using the techniques of gel electrophoresis, comparisons between populations at various levels of evolutionary divergence have been carried out in many kinds of organisms during the past few years. Evolution is a complex process determined by the environmental conditions as well as by the nature of the organisms, and thus the amount of genetic change corresponding to a given level of evolutionary divergence is likely to vary from organism to organism, from place to place, and from time to time. The results of electrophoretic studies confirm this variation, but also show some general patterns (Table 22.3). With few exceptions, the genetic distance between populations in either the first or the second stage of speciation is about 0.20 (most comparisons fall in the range 0.16 to 0.30), for organisms as diverse as insects, fishes, amphibians, reptiles, and mammals. These results are consistent with the conclusions derived from the study of the *Drosophila willistoni* group: the first stage of the geographic speciation process requires a fair amount of genetic change (of the order of 20 allelic substitutions per 100 gene loci), whereas little additional genetic change is required during the second stage.

How much genetic change takes place in the quantum mode of speciation? It is clear that when a new species arises by polyploidy, no genetic changes other than the chromosome duplications are required; the new species has the alleles present in the parental species, and no others. However, because most polyploid species start from only one individual of each parental species, they possess at the beginning less genetic variation than the parental species (the founder effect; see Chapter 19, page 648).

Other modes of quantum speciation start with chromosomal rearrangements that cause either partial or total hybrid sterility. As in polyploidy, such rearrangements do not necessarily involve changes in allelic constitution, although there is often a reduction of genetic variation because the derivative population starts from only one or a few in-

Table 22.3
Genetic differentiation at various stages of evolutionary divergence in several groups of organisms. The average genetic identity is given first, followed by the average genetic distance (in parentheses).

	I (*D*)			
Organisms	Local populations	Subspecies	Incipient species	Species and closely related genera
Drosophila	0.987 (0.013)	0.851 (0.163)	0.788 (0.239)	0.381 (1.066)
Other invertebrates	0.985 (0.016)	–	–	0.465 (0.878)
Fishes	0.980 (0.020)	0.850 (0.163)	–	0.531 (0.760)
Salamanders	0.984 (0.017)	0.836 (0.181)	–	0.520 (0.742)
Reptiles	0.949 (0.053)	0.738 (0.306)		0.437 (0.988)
Mammals	0.944 (0.058)	0.793 (0.232)	0.769 (0.263)	0.620 (0.559)
Plants	0.966 (0.035)	–	–	0.510 (0.808)

Calculated from data in F. J. Ayala, *Evol. Biol.* 8:1 (1975).

dividuals. The first stage of speciation is therefore accomplished with little or no genetic change at the level of the individual genes.

What about genetic change in the second stage of quantum speciation? The second stage of speciation is similar in both geographic and quantum speciation. In both cases, the populations already exhibit postzygotic RIMs and are developing prezygotic isolation by natural selection. If, in geographic speciation, the second stage requires genetic changes in only a small fraction of the genes, the same should be true in quantum speciation. Experimental results confirm this prediction (Table 22.4). The first comparison is between the two annual plant species, *Clarkia biloba* and *C. lingulata*, discussed earlier as examples of quantum speciation. These species remain genetically quite similar: $I = 0.880$ and $D = 0.128$, indicating that only about 13 allelic substitutions per 100 gene loci have occurred in their separate evolution.

The second comparison is also between two annual plants, *Stephanomeria exigua* and *S. malheurensis*; the latter was derived from the former only very recently. Leslie Gottlieb has shown that the original and derivative populations differ by one chromosomal translocation and by their mode of reproduction: the original species reproduces by outcrossing,

Table 22.4
Genetic differentiation in quantum speciation. Little differentiation is observed between species or incipient species arisen by quantum speciation.

Populations Compared	*I*	*D*
Plants		
Clarkia biloba vs. *C. lingulata**	0.880	0.128
Stephanomeria exigua vs. *S. malheurensis*†	0.945	0.057
Rodents		
Spalax ehrenbergi†	0.978	0.022
Thomomys talpoides†	0.925	0.078

*Comparison between two recently arisen species.
†Comparison between incipient species, i.e., populations completing the second stage of speciation.

After F. J. Ayala, *Evol. Biol.* 8:1 (1975).

while the derivative species reproduces by selfing. As expected, the two species are genetically very similar (about 6 allelic substitutions per 100 loci).

The third and fourth comparisons in Table 22.4 involve rodents. *Spalax ehrenbergi* is a species of mole rat consisting of four groups of populations differing in their number of chromosomes (52, 54, 58, and 60). The populations are largely allopatric, although they are in contact with one another in narrow zones at the edges of their distributions, and some hybridization takes place there. The differences in chromosome number due to chromosomal fusions or fissions provide effective postzygotic RIMs; moreover, some ethological isolation has developed: laboratory tests show greater preference for matings between individuals of the same chromosomal type, although they appear morphologically indistinguishable. These four populations in the second stage of quantum speciation are, on the average, genetically very similar: only about 2 allelic substitutions per 100 gene loci have taken place in their separate evolution.

Thomomys talpoides is a species of pocket gopher consisting of more than eight populations differing in their chromosomal arrangements; they live in the north central and northwest United States and neighboring areas of southern Canada (Figure 22.7). As with *Spalax*, the populations of *Thomomys* are mostly allopatric, but are in contact at the edges of their distributions. The chromosomal rearrangements keep the populations from interbreeding in the zones of contact. Nevertheless, the average genetic distance between these populations is quite small (about 8 allelic substitutions per 100 loci).

In conclusion, quantum speciation can occur with little change at the level of the genes; that is, neither Stage I nor Stage II requires substantial allelic evolution in this mode of speciation. This result, in turn, confirms the conclusion reached with respect to geographic speciation, namely, that Stage II—when natural selection directly promotes prezygotic RIMs—does not require major genetic changes.

Genetic Change and Phylogeny: DNA Hybridization

Species are independent evolutionary units because they are reproductively isolated. Species evolve independently of each other and are likely to become genetically more and more different as time proceeds. It was pointed out at the beginning of this chapter that the amount of genetic change that has occurred in the branches of a phylogeny can be inferred by measuring the amount of genetic differentiation between living species. This depends on having previous knowledge of the phylogeny of the species. But it was also argued that the configuration of a phylogeny can also be inferred from the amounts of genetic differentiation between species. This is because evolution is a gradual process; hence, species that are genetically quite similar to each other are more likely to share a recent common ancestor with each other than with species from which they are genetically more different.

The degree of genetic differentiation between species can be measured directly by examining the DNA sequence of their genes or, indirectly, by examining the proteins encoded by structural genes. The methods used to estimate genetic change during evolution include DNA hybridization, protein sequencing, immunology, and electrophoresis.

Simple techniques have recently been developed for determining the nucleotide sequence of DNA molecules (or of the messenger RNA, which is transcribed from DNA sequences coding for polypeptides). These techniques have not yet been extensively applied to phylogenetic studies, because specific genes (or messenger RNA molecules) must first be obtained from the organisms to be compared. The isolation of specific genes is still a laborious enterprise.

A technique that estimates the overall similarity between the DNA of various organisms is *DNA hybridization.* Radioactively labeled DNA that has been "melted" (i.e., dissociated) and fractionated can be reacted with various amounts of melted DNA from a different species. Homologous sequences will hybridize to form duplexes; the extent of the reaction gives an estimate of the proportion of DNA sequences that are homologous (Figure 22.8). The repetitive DNA sequences are usually first removed so that only single-copy DNA is employed in the tests.

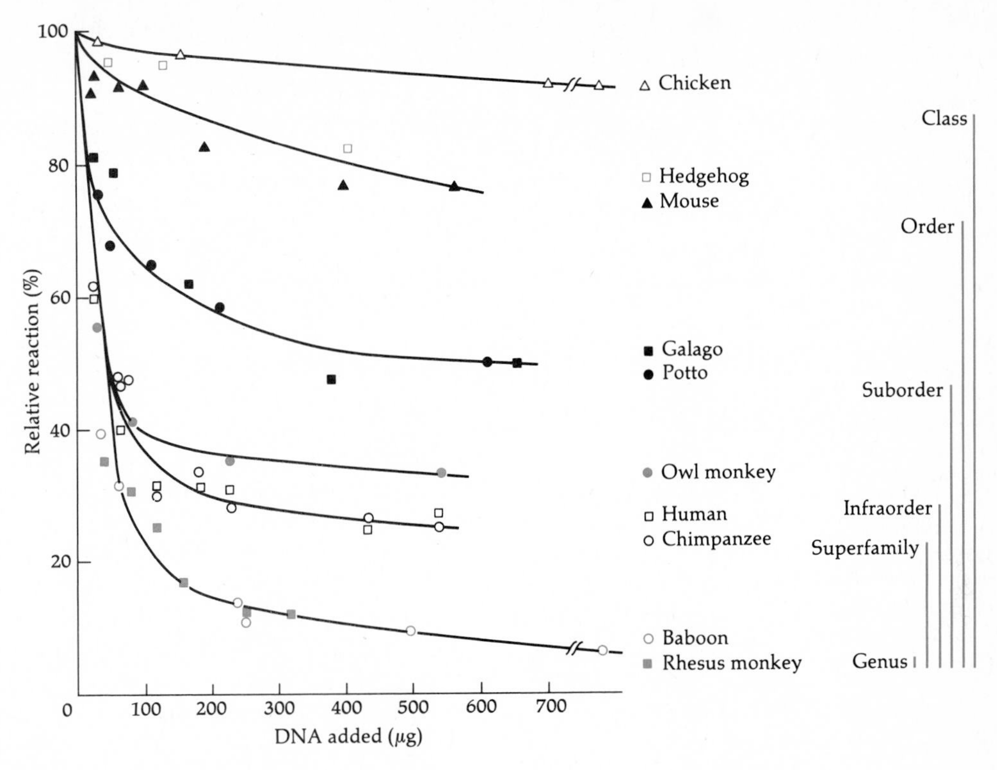

Figure 22.8
Homology between DNA sequences of rhesus monkeys and those of various other species. The experiments are carried out by hybridizing 130 μg of filter-bound DNA from rhesus monkeys with a solution containing 0.5 μg of radioactively labeled rhesus DNA together with increasing amounts of DNA from a second species. As the amount of DNA from the second species is increased, a smaller fraction of the radioactive DNA hybridizes with its homologous filter-bound DNA. The minimum amount of hybridizing radioactive DNA determines the proportion of DNA sequences that are *not* homologous in the two species. The proportions of nonhomologous DNA for various levels of taxonomic differentiation are indicated by the vertical bars. (After B. H. Hoyer and R. B. Roberts, in *Molecular Genetics*, Part II, ed. by J. H. Taylor, Academic Press, New York, 1967, p. 425.)

Sequences forming duplexes need not be complementary for every nucleotide. The proportion of noncomplementary nucleotides in interspecific DNA duplexes can be estimated by the rate at which the DNA strands separate at increasing temperatures. The critical parameter, called *thermal stability* (T_s), is the temperature at which 50% of the duplex DNA has dissociated (Figure 22.9). The difference (ΔT_s) between the T_s values of hybrid and control DNA molecules is approximately proportional to the proportion of unpaired nucleotides in the hybrid DNA, so that a ΔT_s value of 1°C corresponds approximately to 1% mismatched nucleotides. The results of comparing the DNA of various primates first with human DNA and then with green-monkey DNA (Table 22.5) serve to

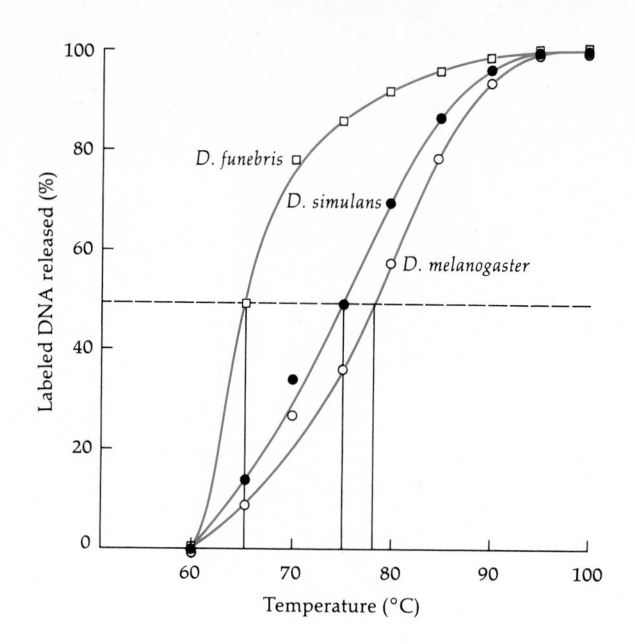

Figure 22.9
Thermal stability profiles of DNA duplexes having one strand from *Drosophila melanogaster* and the other from the species indicated. The thermal stability (T_s) value for the nonhybrid duplex DNA is 78°C; for the *D. melanogaster/D. simulans* duplex it is 75°C; and for the *D. melanogaster/D. funebris* duplex it is 65°C. Because $\Delta T_s = 1°C$ corresponds approximately to 1% mismatched nucleotides, the proportion of nucleotide pairs different from *D. melanogaster* is estimated as 3% for *D. simulans* and 13% for *D. funebris*. [After C. D. Laird and B. J. McCarthy, *Genetics* 60:303 (1968).]

estimate the percentage of nucleotide-pair substitutions that have occurred during primate evolution (Figure 22.10).

Protein-Sequence Phylogenies

Cytochrome *c* is a protein involved in cellular respiration; it is found in the mitochondria of animals and plants. The amino acid sequences of cytochrome *c* in humans, rhesus monkeys, and horses are shown in Figure 22.11. The amino acid at position 66 is isoleucine in humans, but threonine in rhesus monkeys and horses. Humans and rhesus monkeys have identical amino acids at the other 103 positions, but differ from horses in 11 additional amino acids (Table 22.6). It is known that the evolutionary divergence between the human and rhesus monkey lineages occurred well after these diverged from the horse lineage. The number of amino acid substitutions that have occurred in the various branches of the phylogeny are shown in Figure 22.12.

The genetic code (Table 12.1) makes it possible to calculate the minimum number of nucleotide differences required to change from a codon for one amino acid to a codon for another. At position 19 of cytochrome *c*, humans and rhesus monkeys have isoleucine, but horses have valine. Isoleucine may be encoded by any one of the three codons AUU, AUC, and AUA, and valine, by any one of the four codons GUU,

Table 22.5
Percent nucleotide differences between the DNA of various primates and the DNA of humans and green monkeys.

Species Tested	Tester DNA from: Human	Green monkey
Human	0	9.6
Chimpanzee	2.4	9.6
Gibbon	5.3	9.6
Green monkey	9.5	0
Rhesus monkey	–	3.5
Capuchin	15.8	16.5
Galago	42.0	42.0

After D. E. Kohne, J. A. Chiscon, and B. H. Hoyer, *J. Human Evol.* 1:627 (1972).

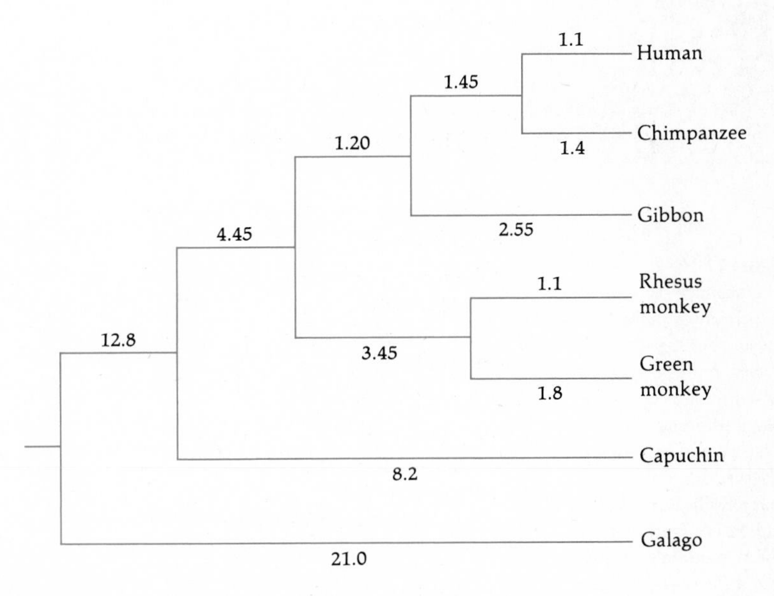

Figure 22.10
Phylogeny of primate species, based on thermal stability profiles of DNA hybrid duplexes. The numbers on the branches are estimated percent nucleotide-pair substitutions. [After D. E. Kohne, J. A. Chiscon, and B. H. Hoyer, *J. Human Evol.* 1:627 (1972).]

	1–8	9–20
Human	——	Gly–Asp–Val–Glu–Lys–Gly–Lys–Lys–Ile–Phe–Ile–Met–
Rhesus monkey	——	Gly–Asp–Val–Glu–Lys–Gly–Lys–Lys–Ile–Phe–Ile–Met–
Horse	——	Gly–Asp–Val–Glu–Lys–Gly–Lys–Lys–Ile–Phe–**Val–Gln**–

21–40
Lys–Cys–Ser–Gln–Cys–His–Thr–Val–Glu–Lys–Gly–Gly–Lys–His–Lys–Thr–Gly–Pro–Asn–Leu–
Lys–Cys–Ser–Gln–Cys–His–Thr–Val–Glu–Lys–Gly–Gly–Lys–His–Lys–Thr–Gly–Pro–Asn–Leu–
Lys–Cys–**Ala**–Gln–Cys–His–Thr–Val–Glu–Lys–Gly–Gly–Lys–His–Lys–Thr–Gly–Pro–Asn–Leu–

41–60
His–Gly–Leu–Phe–Gly–Arg–Lys–Thr–Gly–Gln–Ala–Pro–Gly–Tyr–Ser–Tyr–Thr–Ala–Ala–Asn–
His–Gly–Leu–Phe–Gly–Arg–Lys–Thr–Gly–Gln–Ala–Pro–Gly–Tyr–Ser–Tyr–Thr–Ala–Ala–Asn–
His–Gly–Leu–Phe–Gly–Arg–Lys–Thr–Gly–Gln–Ala–Pro–Gly–**Phe–Thr**–Tyr–Thr–**Asp**–Ala–Asn–

61–80
Lys–Asn–Lys–Gly–Ile–**Ile**–Trp–Gly–Glu–Asp–Thr–Leu–Met–Glu–Tyr–Leu–Glu–Asn–Pro–Lys–
Lys–Asn–Lys–Gly–Ile–Thr–Trp–Gly–Glu–Asp–Thr–Leu–Met–Glu–Tyr–Leu–Glu–Asn–Pro–Lys–
Lys–Asn–Lys–Gly–Ile–Thr–Trp–**Lys**–Glu–**Glu**–Thr–Leu–Met–Glu–Tyr–Leu–Glu–Asn–Pro–Lys–

81–100
Lys–Tyr–Ile–Pro–Gly–Thr–Lys–Met–Ile–Phe–Val–Gly–Ile–Lys–Lys–Lys–Glu–Glu–Arg–Ala–
Lys–Tyr–Ile–Pro–Gly–Thr–Lys–Met–Ile–Phe–Val–Gly–Ile–Lys–Lys–Lys–Glu–Glu–Arg–Ala–
Lys–Tyr–Ile–Pro–Gly–Thr–Lys–Met–Ile–Phe–**Ala**–Gly–Ile–Lys–Lys–Lys–**Thr**–Glu–Arg–**Glu**–

101–112
Asp–Leu–Ile–Ala–Tyr–Leu–Lys–Lys–Ala–Thr–Asn–Glu
Asp–Leu–Ile–Ala–Tyr–Leu–Lys–Lys–Ala–Thr–Asn–Glu
Asp–Leu–Ile–Ala–Tyr–Leu–Lys–Lys–Ala–Thr–Asn–Glu

Figure 22.11
The primary structures of cytochrome *c* in humans, rhesus monkeys, and horses. In these organisms, cytochrome *c* consists of 104 amino acids (positions 9–112; positions 1–8 are reserved for amino acids that exist in bacteria, wheat, and other organisms, but not in mammals). Amino acid differences between the sequences (see Table 22.6) are highlighted with color. (See Figure 10.13 for the standard amino acid abbreviations.)

GUC, GUA, and GUG. Thus, one single nucleotide-pair substitution (from A to G in the first position) is sufficient to change a codon for isoleucine to a codon for valine. At position 20, humans and rhesus monkeys have methionine (AUG), while horses have glutamine (CAA or CAG); therefore, at least two nucleotide-pair substitutions (in the first and second positions) must have occurred to change the methionine codon to a codon for glutamine. The minimum numbers of nucleotide-pair substitutions required to account for the amino acid differences between the cytochrome *c* molecules of humans, rhesus monkeys, and horses are shown (below the diagonal) in Table 22.6.

Assume, now, that we knew nothing about the phylogeny of humans, rhesus monkeys, and horses. The data in Table 22.6 suggest that the configuration shown in Figure 22.12 is the most likely. Evolution is, on the whole, a gradual process of change. Thus, species that are genetically more similar are likely to have diverged from each other more recently than they

Table 22.6
Number of amino acid differences (above the diagonal) and minimum number of nucleotide differences (below the diagonal) between the cytochrome *c* molecules of humans, rhesus monkeys, and horses. The cytochrome *c* in these organisms has 104 amino acids (see Figure 22.11).

	Human	Rhesus Monkey	Horse
Human	–	1	12
Rhesus monkey	1	–	11
Horse	15	14	–

did from species that are genetically less similar. Figure 22.13 shows two possible phylogenies of humans, rhesus monkeys, and horses, as well as the nucleotide substitutions required in each branch. The configurations of these phylogenies would seem very unlikely even if no information other than the amino acid sequences of the cytochrome *c* molecules were available.

The minimum numbers of nucleotide differences necessary to account for the amino acid differences in the cytochrome *c* molecules of 20 organisms are given in Table 22.7. A phylogeny based on that data matrix, as well as the minimum numbers of nucleotide changes required in each

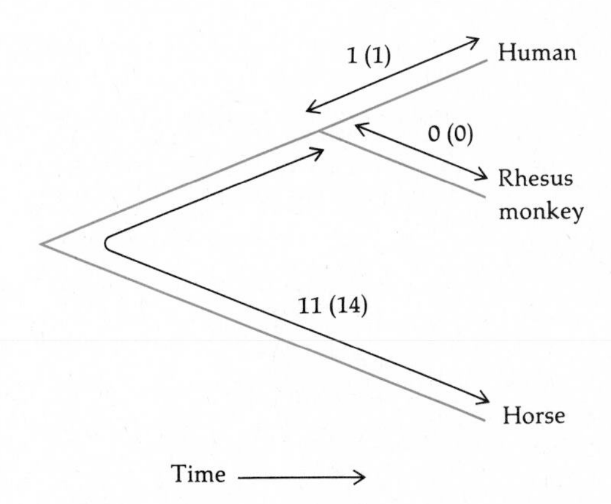

Figure 22.12
Anagenetic change in the evolution of cytochrome *c* from humans, rhesus monkeys, and horses. The numbers indicate the amino acid substitutions (and, in parentheses, the minimum number of nucleotide substitutions) that have taken place in each branch of the phylogeny.

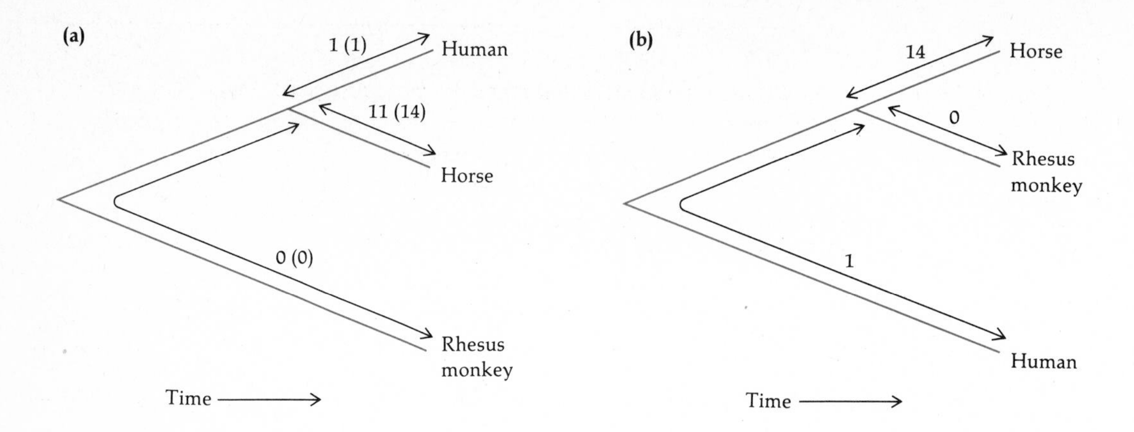

Figure 22.13
Two theoretically possible phylogenies of humans, rhesus monkeys, and horses. The numbers of amino acid (and nucleotide) substitutions required in each branch to account for the cytochrome *c* sequences indicate that neither of these two phylogenies is likely to be correct.

branch, is shown in Figure 22.14. Most of these changes are shown as fractions. It is obvious that a nucleotide change may or may not have taken place, but fractional nucleotide changes cannot occur. However, the values given in Figure 22.14 are those that best satisfy the data in Table 22.7.

The phylogenetic relationships shown in Figure 22.14 correspond fairly well, on the whole, with the phylogeny of the organisms as determined from the fossil record and other sources. There are disagreements, however. For example, chickens appear more closely related to penguins than to ducks and pigeons, and men and monkeys diverge from the other mammals before the marsupial kangaroo separates from the nonprimate placentals. Despite these erroneous relationships, it is remarkable that the study of a single protein should yield such an accurate representation of the phylogeny of 20 organisms as diverse as those in the figure. The amino acid sequences of proteins (and the genetic information contained therein) store considerable evolutionary information.

The reconstruction of phylogenies and the estimation of amounts of genetic change from protein sequence data are based on the assumption that the genes coding for the proteins are *homologous*, i.e., descended from a common ancestor. There are two kinds of homologous relationships among genes: orthologous and paralogous. *Orthologous* genes are descendants of an ancestral gene that was present in the ancestral species from which the species in question have evolved. The evolution of orthologous genes therefore reflects the evolution of the *species* in which they are found. The cytochrome *c* molecules of the 20 organisms shown in Figure 22.14 are orthologous, because they derive from a single ancestral gene present in a species ancestral to all 20 organisms.

Paralogous genes are descendants of a *duplicated* ancestral gene. Paralogous genes, therefore, evolve within the same species (as well as in different species). The genes coding for the α, β, γ, and δ hemoglobin chains in humans are paralogous. The evolution of paralogous genes reflects differences that have accumulated since the genes duplicated.

Table 22.7
Minimum number of nucleotide differences in the genes coding for cytochrome *c* molecules in 20 organisms.

Organism	1	2	3	4	5	6	7	8	9	10	11	12	13	14	15	16	17	18	19	20
1. Human	—	1	13	17*	16	13	12	12	17	16	18	18	19	20	31	33	36	63	56	66
2. Monkey		—	12	16*	15	12	11	13	16	15	17	17	18	21	32	32	35	62	57	65
3. Dog			—	10	8	4	6	7	12	12	14	14	13	30	29	24	28	64	61	66
4. Horse				—	1	5	11	11	16	16	16	17	16	32	27	24	33	64	60	68
5. Donkey					—	4	10	12	15	15	15	16	15	31	26	25	32	64	59	67
6. Pig						—	6	7	13	13	13	14	13	30	25	26	31	64	59	67
7. Rabbit							—	7	10	8	11	11	11	25	26	23	29	62	59	67
8. Kangaroo								—	14	14	15	13	14	30	27	26	31	66	58	68
9. Duck									—	3	3	3	7	24	26	25	29	61	62	66
10. Pigeon										—	4	4	8	24	27	26	30	59	62	66
11. Chicken											—	2	8	28	26	26	31	61	62	66
12. Penguin												—	8	28	27	28	30	62	61	65
13. Turtle													—	30	27	30	33	65	64	67
14. Rattlesnake														—	38	40	41	61	61	69
15. Tuna															—	34	41	72	66	69
16. Screwworm fly																—	16	58	63	65
17. Moth																	—	59	60	61
18. *Neurospora*																		—	57	61
19. *Saccharomyces*																			—	41
20. *Candida*																				—

*The differences between the horse and either human or rhesus monkey given here (17 and 16) are greater than those given in Table 22.6 (15 and 14). The two additional nucleotide substitutions are required when all the organisms included in this table are taken into account.

After W. M. Fitch and E. Margoliash, *Science* 155:279 (1967).

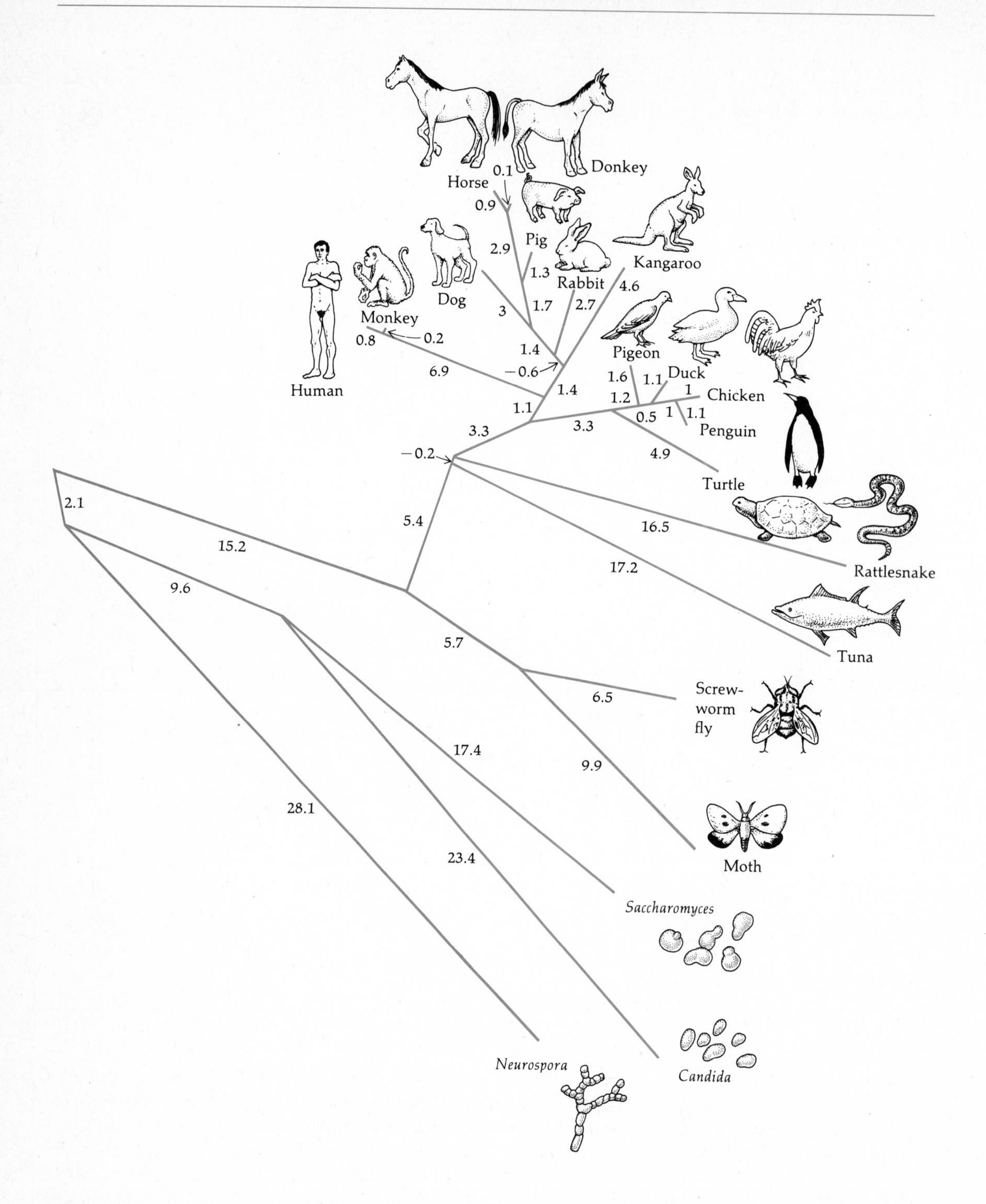

Figure 22.14
Phylogeny of 20 organisms, based on differences in the amino acid sequence of cytochrome *c*. The phylogeny agrees fairly well with evolutionary relationships inferred from the fossil record and other sources. The minimum number of nucleotide substitutions required for each branch is shown. [After W. M. Fitch and E. Margoliash, *Science* 155:279 (1967).]

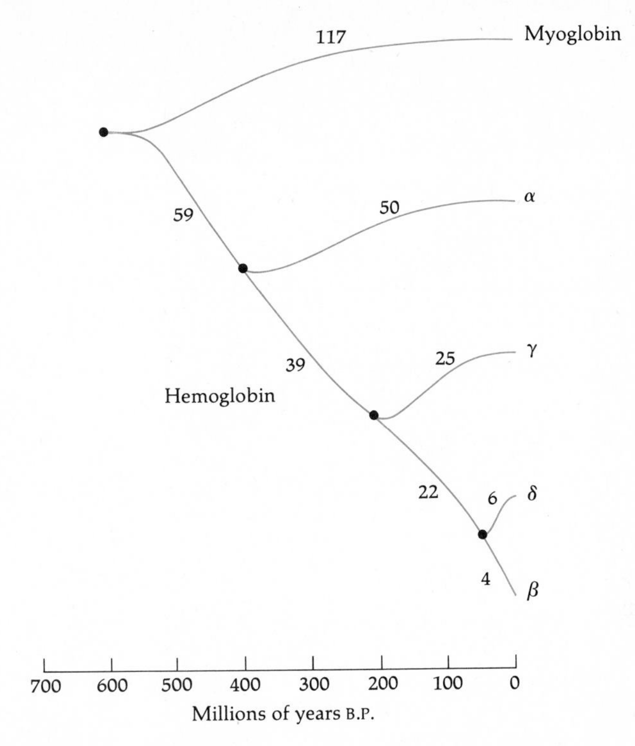

Figure 22.15
Evolutionary history of the globin genes. The dots indicate where the ancestral genes were duplicated, giving rise to a new gene line. The minimum number of nucleotide substitutions required to account for the amino acid differences between the proteins is shown for each branch. The first gene duplication occurred about 600 million years B.P. (before Present), one gene coding for myoglobin and the other being the ancestor of the various hemoglobin genes. About 400 million years ago, the hemoglobin gene became duplicated into one leading to the modern α gene and another that would duplicate again about 200 million years ago into the γ and β genes. The latter duplicated again some 40 million years ago in the ancestral lineage of the higher primates, giving rise to one new gene coding for the δ hemoglobin chain. The time estimates are based on paleontological and morphological studies of the organisms.

Homologies between paralogous genes serve to establish *gene* phylogenies, i.e., the evolutionary history of duplicated genes within a given lineage. Figure 22.15 is a phylogeny of the gene duplications giving rise to the myoglobin and hemoglobin genes found in modern humans. The evolutionary sequence of the duplications is shown, as well as the minimum number of nucleotide changes in each branch.

Cytochrome *c* molecules are slowly evolving proteins. Organisms as different as humans, silkworm moths, and *Neurospora* have a large proportion of amino acids in their cytochrome *c* molecules in common. The evolutionary conservation of this cytochrome makes possible the study of genetic differences among organisms that are only remotely related. However, this same conservation makes cytochrome *c* useless for

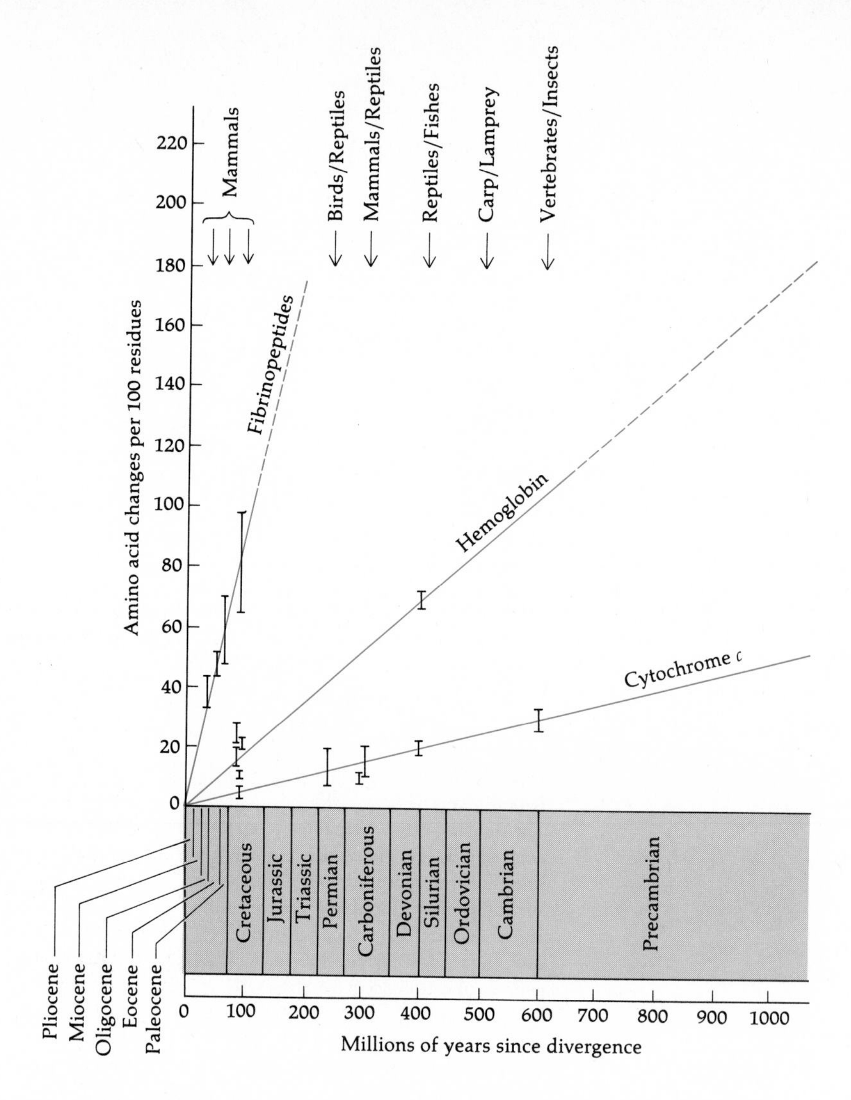

Figure 22.16
Rates of molecular evolution of different proteins. [After R. E. Dickerson, *J. Mol. Evol.* 1:26 (1971).]

determining evolutionary change in closely related organisms, since these may have cytochrome *c* molecules that are completely or nearly identical. For example, the primary structure of cytochrome *c* is identical in humans and chimpanzees, which diverged 10 to 15 million years ago; it differs by only one amino acid between humans and rhesus monkeys, whose most recent common ancestor lived 40 to 50 million years ago.

Fortunately, different proteins evolve at different rates. Phylogenetic relationships among closely related organisms can be inferred by studying the primary sequences of rapidly evolving proteins, such as fibrinopeptides in mammals (Figure 22.16). Carbonic anhydrases are rapidly evolving proteins that are physiologically important in the reversible hydration of CO_2 and in certain secretory processes. A phylogeny of

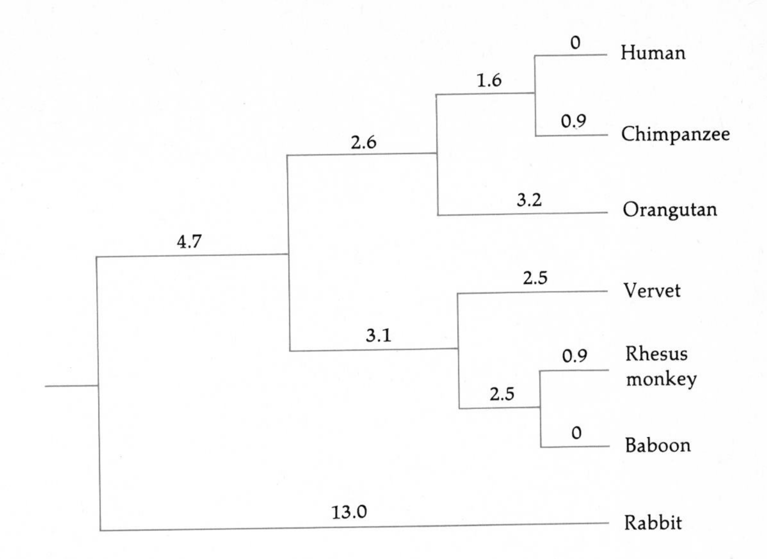

Figure 22.17
Phylogeny of various primates, based on differences in the sequence of 115 amino acids in carbonic anhydrase I. The numbers on the branches are the estimated numbers of nucleotide substitutions that have occurred in evolution. (After R. E. Tashian et al., in *Molecular Anthropology*, ed. by M. Goodman and R. E. Tashian, Plenum Press, New York, 1976, p. 301.)

various primates based on the amino acid sequence of carbonic anhydrase I, as well as the minimum number of nucleotide changes in each branch, is shown in Figure 22.17. Genetic changes in the evolution of closely related species can also be studied by other methods, such as DNA hybridization, immunology, and gel electrophoresis.

Immunology and Electrophoresis

More than 500 sequences or partial sequences of proteins, containing much evolutionary genetic information, are presently known. More sequences are obtained every year, although the procedures for determining the primary sequences of proteins are extremely laborious. Other methods, such as immunological techniques, permit estimating the degree of similarity among proteins with much less work than amino acid sequencing.

The immunological comparison of proteins is performed, in outline, as follows. A protein, say, albumin, is purified from the tissue of an animal, say, a chimpanzee. The protein is injected into a rabbit or some other mammal, which develops an immunological reaction and produces

Table 22.8
Immunological distances between albumins of various Old World primates.

Species Tested	Antiserum to: *Homo*	*Pan*	*Hylobates*
Homo sapiens (human)	0	3.7	11.1
Pan troglodytes (chimpanzee)	5.7	0	14.6
Pan paniscus (pygmy chimpanzee)	5.7	0	14.6
Gorilla gorilla (gorilla)	3.7	6.8	11.7
Pongo pygmaeus (orangutan)	8.6	9.3	11.7
Symphalangus syndactylus (siamang)	11.4	9.7	2.9
Hylobates lar (gibbon)	10.7	9.7	0
Old World monkeys (average of 6 species)	38.6	34.6	36.0

Calculated from data in V. M. Sarich and A. C. Wilson, *Science* 158:1200 (1967).

antibodies against the foreign protein, or *antigen*. The antibodies can be collected by bleeding the rabbit; thereafter, they will react not only against the specific antigen (chimpanzee albumin in the present example), but also against other related proteins (such as albumins from other primates). The greater the similarity between the protein used to immunize the rabbit and the protein being tested, the greater the extent of the immunological reaction. The degree of similarity between the specific antigen and other proteins is expressed as *immunological distance*. Such distances can be converted, if desired, into approximate numbers of amino acid differences.

Table 22.8 gives the immunological distances among humans, apes, and Old World monkeys. Antibodies against albumin obtained from a human (*Homo sapiens*), a chimpanzee (*Pan troglodytes*), and a gibbon (*Hylobates lar*) were prepared independently. The antibodies were then reacted against albumins obtained from humans, six species of apes, and six species of Old World monkeys. The phylogeny obtained from the data in the table is shown in Figure 22.18.

Electrophoresis is another relatively inexpensive method used to estimate protein differences among organisms. With electrophoresis, the number of amino acid differences between two species is not known, but only whether or not two proteins are electrophoretically identical. The simplicity of the method makes feasible the comparison of many proteins.

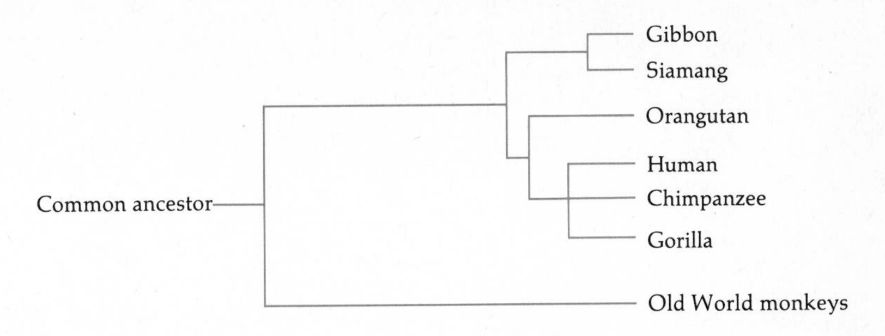

Figure 22.18
Phylogeny of humans, apes, and Old World monkeys, based on immunological differences between their albumins. Humans, chimpanzees, and gorillas appear more closely related to one another than any one of them is to the orangutan—a result confirmed by additional molecular studies. [After V. M. Sarich and A. C. Wilson, *Science 158*:1200 (1967).]

The overall results can be expressed as genetic distance, D, using the procedure shown in Box 22.1.

Electrophoresis is ineffective for comparing organisms that are evolutionarily very distant. These are likely to be electrophoretically different at all, or most, loci. Since the number of amino acid differences involved cannot be determined (but only whether or not the two proteins compared have identical electrophoretic migrations), the method fails to determine the degree of differentiation among various species when these differ at all, or nearly all, loci. On the other hand, electrophoretic distances have the advantage of being based on many loci; therefore, unequal rates of evolution in different lineages with respect to one locus may be compensated by other loci. Electrophoresis is, in general, an appropriate method for measuring genetic change among closely related organisms, in which the amino acid sequence of a single protein may fail to show any differences or give misleading results because of the small numbers of substitutions involved.

The phylogeny of the *Drosophila willistoni* group based on a matrix of genetic distances is shown in Figure 22.19. The numbers shown on the branches are expressed in units of genetic distance, D, and therefore estimate the average number of electrophoretically detectable allelic substitutions per locus that have occurred in each branch.

Neutrality Theory of Molecular Evolution

The reconstruction of a phylogeny from genetic similarities depends on the assumption that degrees of similarity reflect degrees of phylogenetic propinquity. On the whole, this is a reasonable assumption because

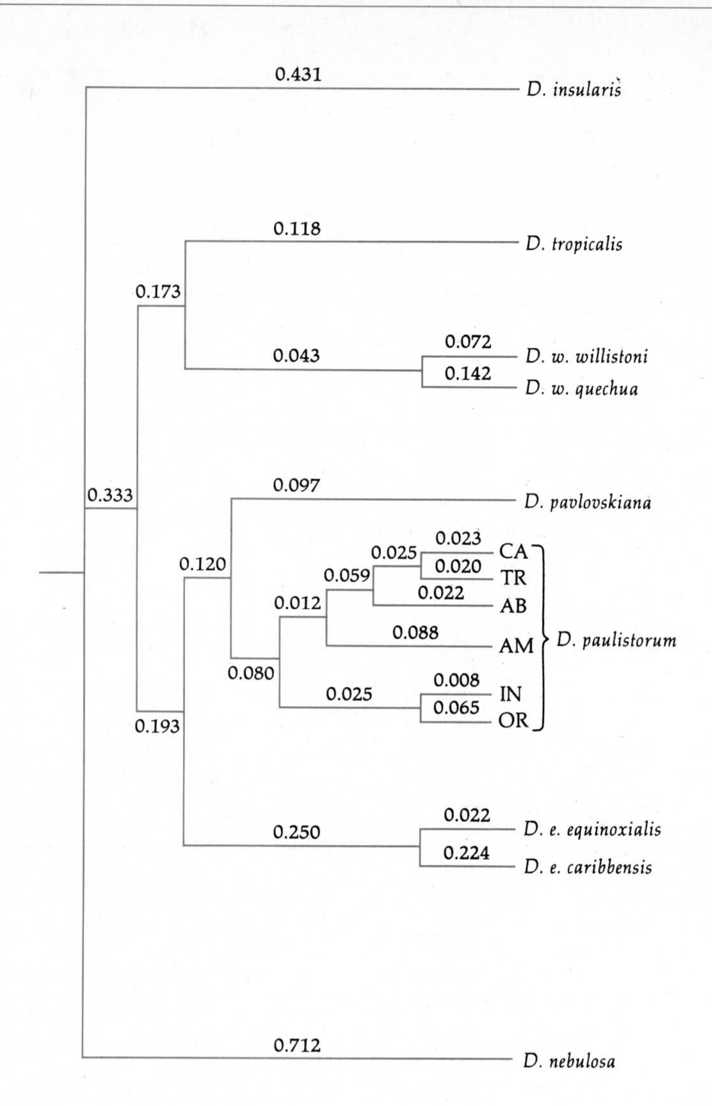

Figure 22.19
Phylogeny of species related to *Drosophila willistoni*, based on electrophoretic differences at 36 gene loci coding for enzymes. The numbers on the branches are the estimated allelic substitutions (detectable by electrophoresis) per locus that have taken place in evolution. There are seven species in the phylogenetic tree. Two species, *D. willistoni* and *D. equinoxialis*, are represented by two subspecies each. *D. paulistorum* is a complex of six semispecies (incipient species), called Central American (CA), Transitional (TR), Andean-Brazilian (AB), Amazonian (AM), Interior (IN), and Orinocan (OR). [After F. J. Ayala et al., *Evolution* 28:576 (1974).]

evolution is a process of gradual change. However, differences in rates of genetic change among lineages may be a source of error. Assume that a given species, A, diverged from the common lineage of two other species, B and C, before these diverged from each other. Assume also that a certain protein has evolved at a much faster rate in the lineage leading to C than in the other two lineages. It might be that the amino acid sequences of the protein would be more similar between A and B than between B and C. The phylogeny inferred from the amino acid sequences of the protein might then be erroneous.

The hypothesis has recently been advanced by Motoo Kimura and others that rates of amino acid substitutions in proteins and of nucleotide substitutions in DNA may be approximately constant because the vast majority of such changes are selectively neutral. New alleles appear in a population by mutation. If alternative alleles have identical fitness, changes in allelic frequencies from generation to generation will occur only by accidental sampling errors from generation to generation, i.e., by genetic drift (Chapter 19). Rates of allelic substitution would be stochastically constant, i.e., they would occur with a constant probability for a given protein. That probability can be shown to be simply the mutation rate for neutral alleles.

The neutrality theory of molecular evolution recognizes that, for any gene, a large proportion of all possible mutants are harmful to their carriers; these mutants are eliminated or kept at very low frequency by natural selection. The evolution of morphological, behavioral, and ecological traits is governed largely by natural selection, because it is determined by the selection of favorable mutants against deleterious ones. It is assumed, however, that a number of favorable mutants, adaptively equivalent to each other, can occur at each locus. These mutants are not subject to selection relative to one another because they do not affect the fitness of their carriers (nor do they modify their morphological, physiological, or behavioral properties). According to the neutrality theory, evolution at the molecular level consists for the most part of the gradual, random replacement of one neutral allele by another that is functionally equivalent to the first. The theory assumes that although favorable mutations occur, they are so rare that they have little effect on the overall evolutionary rate of nucleotide and amino acid substitutions.

Neutral alleles are not defined as having fitnesses that are identical in the mathematical sense. Operationally, neutral alleles are those whose differential contributions to fitness are so small that their frequencies change more owing to drift than to natural selection. Assume that two alleles, A_1 and A_2, have fitnesses 1 and $1 - s$ (where s is a positive number smaller than 1). The two alleles are effectively neutral if, and only if,

$$4N_e s << 1$$

where N_e is the effective size of the population (Chapter 19, page 645).

We now want to find the rate of substitution of neutral alleles, k, per unit time in the course of evolution. The time units can be years or generations. In a random mating with N diploid individuals,

$$k = 2Nux$$

where u is the neutral mutation rate per gamete per unit time (time measured in the same units as for k) and x is the probability of ultimate fixation of a neutral mutant. The derivation of this equation is straightforward: there are $2Nu$ mutants per unit time, each with a probability x of becoming fixed.

A population of N individuals has $2N$ genes at each autosomal locus. If the alleles are neutral, all genes have the identical probability of becoming fixed, which is simply

$$x = \frac{1}{2N}$$

Substituting this value of x in the previous equation, we obtain

$$k = 2Nu\frac{1}{2N} = u$$

That is, the rate of substitution of neutral alleles is precisely the rate at which the neutral alleles arise by mutation, independently of the size of the population and any other parameters. This is not only a remarkably simple result, but also one with momentous implications if it indeed applies to molecular evolution.

The Molecular Clock of Evolution

If the neutrality theory of molecular evolution were correct for a large number of gene loci, protein and DNA evolution would serve as evolutionary "clocks." The degree of genetic differentiation among species would be a measure of their phylogenetic relatedness; it would thus be justifiable to reconstruct phylogenies on the basis of genetic differences. Moreover, the actual chronological time of the various phylogenetic events could be roughly estimated. Assume that we have a phylogeny such as the one shown in Figure 22.14. If the rate of evolution of cytochrome *c* were constant through time, the number of nucleotide substitutions that have occurred in each branch of the phylogeny would be directly proportional to the time elapsed. Knowing from an outside source (such as the paleontological record) the actual geological time of any one event in the phylogeny would make it possible to determine the times of all the other events by simple proportions. That is, once it is "calibrated" by reference to a single event, the molecular clock can be used to measure the time of occurrence of all the other events in the phylogeny.

The molecular clock postulated by the neutrality theory is, of course, not a metronomic clock, like timepieces in ordinary life, which measure time exactly. The neutrality theory predicts, instead, that molecular evolution is a "stochastic clock," like radioactive decay. The *probability* of change is constant, although some variation occurs. Over fairly long periods of time, a stochastic clock is nevertheless quite accurate. Moreover, each gene or protein would represent a separate clock, providing an independent estimate of phylogenetic events and their time of occurrence.

Table 22.9
Statistical tests of the constancy of evolutionary rates of 7 proteins in 17 species of mammals.

Rates Tested	Chi-Square	Degrees of Freedom	Probability
Overall rates (comparisons among branches over all 7 proteins)	82.4	31	4×10^{-6}
Relative rates (comparisons among proteins within branches)	166.3	123	6×10^{-2}
Total	248.7	154	6×10^{-6}

After W. M. Fitch, in *Molecular Evolution,* ed. by F. J. Ayala, Sinauer, Sunderland, Mass., 1976, p. 160.

Each gene or protein would "tick" at a different rate (the mutation rate to neutral alleles, u, of the gene—see Figure 22.16), but all of them would be timing the same evolutionary events. The joint results of several genes or proteins would provide a fairly precise evolutionary clock.

Is there a molecular clock of evolution? This question can be investigated by examining whether or not the variation in the number of molecular changes that have occurred during equal evolutionary periods is greater than that expected by chance. This would also be a test of the neutrality theory of molecular evolution. Two kinds of test are possible. One kind consists of examining the number of molecular changes between phylogenetic events whose timing is known from the paleontological record and other sources. The other kind does not use actual times, but rather looks at parallel lineages derived from a common ancestor and tests whether the variation in the number of molecular changes along the branches is greater than that expected by chance.

Whether or not the neutrality theory is correct, and how accurate the molecular clock is, are at present controversial matters. The existing evidence suggests that the variation in the rate of molecular evolution is greater than that predicted by the neutrality theory. Nevertheless, molecular evolution appears to occur with sufficient regularity to serve as an evolutionary clock, although not as accurately as if the rate of evolution were stochastically constant, and would have only the variation expected from a Poisson distribution. The results of one test devised by Charles H. Langley and Walter M. Fitch are given in Table 22.9.

The test uses 7 proteins sequenced in 17 mammals, and starts by adding up the proteins one after another and treating them as if they were one single sequence. The minimum number of nucleotide substitutions

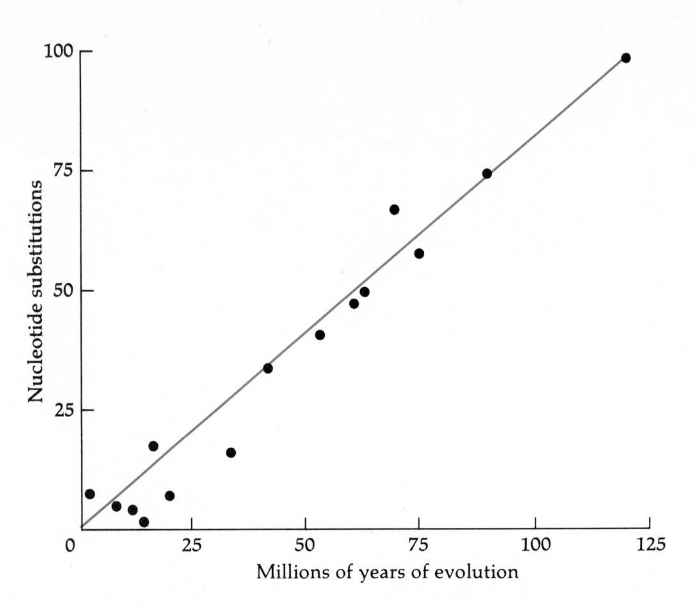

Figure 22.20
Nucleotide substitutions versus paleontological time. The minimum numbers of nucleotide substitutions for seven proteins (cytochrome *c*, fibrinopeptides A and B, hemoglobins α and β, myoglobin, and insulin C-peptide), sequenced in 17 species of mammals, have been calculated for comparisons between pairs of species whose ancestors diverged at the time indicated on the abscissa. The line has been drawn from the origin to the outermost point and corresponds to a rate of 0.41 nucleotide substitution per million years for all seven proteins together. Most points fall near the line, except for some representing comparisons between primates (points below the line at lower left), in which protein evolution seems to have occurred at a lower than average rate. (After W. M. Fitch, in *Molecular Evolution*, ed. by F. J. Ayala, Sinauer, Sunderland, Mass., 1976, p. 160.)

that accounts for the descent of the amino acid sequences from a common ancestor is found; the numbers of substitutions are then assigned to the various branches in the phylogeny. Two independent tests are made. First, the total number of substitutions per unit time is examined for different times; the hypothesis tested is whether the *overall* rate of change is uniform over time. The probability that the variation observed is due to chance is 4×10^{-6}, which is statistically highly significant. The conclusion follows that the proteins have not evolved at a constant rate with a Poisson variance. It is possible, however, that the proteins have all changed their rates *proportionately*—for example, because the rate of molecular evolution is constant per generation rather than per year; variations in generation length might have occurred through time. This possibility is examined by testing whether the rates of evolution of one protein *relative* to those of another are uniform through time. There is a marginally significant deviation from expectation (probability ≈ 0.06). The probability that all the variation observed (*total*) is due to chance is extremely small, 6×10^{-6}. The test is particularly valid because it makes no use of paleontological dates. The phylogeny is constructed using the protein data alone; this maximizes the probability of agreement between the data and the hypothesis of stochastically constant rates of molecular evolution. Even so, the data do not fit the hypothesis of constant probability of change with a Poisson variance. However, John H. Gillespie and C. H. Langley have recently shown that the data used for Table 22.9 *are* consistent with the hypothesis that molecular evolution occurs with a constant probability, if it is assumed that the variance of this probability is greater than that expected from a Poisson distribution.

Whether or not molecular evolution is stochastically constant, it appears that the heterogeneity of evolutionary rates is not excessively great. It is, therefore, possible to use genetic data as an approximate evolutionary clock; but in order to avoid large errors, it is necessary to use *average* rates obtained for many proteins and for long periods of time. Figure 22.20 plots the cumulative number of nucleotide substitutions required in 7 proteins against the paleontological dates of divergence in the evolution of 17 mammalian species. The overall correlation is fairly good for all phylogenetic events except those involving some primates, which appear to have evolved at a substantially lower rate than the average. This deviation from the average, observed at the lower left of the graph, illustrates an important point: the more recent the divergence of any two species, the more likely it is that the genetic changes observed will depart from the average evolutionary rate. This is simply because, as time increases, periods of rapid evolution and periods of slow evolution in any one lineage will tend to cancel each other out.

Structural Versus Regulatory Evolution

Our closest living relatives are the great apes—the African chimpanzee and gorilla and the Asian orangutan. Humans are classified in the family Hominidae; chimpanzees, gorillas, and orangutans are classified in the family Pongidae. The lesser Asian apes, gibbons and siamangs, are classified in the family Hylobatidae. The classification of humans and the great apes in different families is justified on biological grounds. As George Gaylord Simpson has written, "*Homo* is both anatomically and adaptively the most radically distinctive of all hominoids, divergent to a degree considered familial by all primatologists." Yet electrophoretic studies indicate that humans and apes are genetically as similar as are closely related species in other groups of organisms (Table 22.10 and Figure 22.21). The average genetic distance between humans and the great apes is $D = 0.354$, or about 35 electrophoretically detectable substitutions per 100 loci. Marie-Claire King and Allan C. Wilson have calculated that

Table 22.10
Genetic differentiation between humans and apes. Genetic identity, *I*, and genetic distance, *D*, are based on 23 gene loci studied electrophoretically.

Species Compared to Humans	*I*	*D*
1. *Pan troglodytes* (chimpanzee)	0.680	0.386
2. *Pan paniscus* (pygmy chimpanzee)	0.732	0.312
3. *Gorilla gorilla* (gorilla)	0.689	0.373
4. *Pongo pygmaeus abelii* (Sumatra orangutan)	0.710	0.347
5. *Pongo pygmaeus pygmaeus* (Borneo orangutan)	0.705	0.350
6. *Hylobates lar* (lar gibbon)	0.489	0.716
7. *Hylobates concolor* (concolor gibbon)	0.429	0.847
8. *Symphalangus syndactylus* (siamang)	0.333	1.099
Averages		
Great apes (1–5)	0.702 ± 0.009	0.354 ± 0.013
All apes (1–8)	0.595 ± 0.055	0.554 ± 0.105

After E. J. Bruce and F. J. Ayala, *Evolution* 33:1040 (1979).

humans and chimpanzees differ in only about 1% of all amino acids in their proteins.

These results lead to a paradox. We perceive ourselves as being considerably different in morphology and ways of life from apes, and this can hardly be attributed exclusively to the fact that humans more readily notice differences involving themselves. However, humans and apes appear to be no more genetically different than are morphologically indistinguishable (sibling) species of *Drosophila* (see Table 22.2). One possible resolution of this paradox is to assume that the estimates of genetic differentiation are biased; there are thousands of structural gene loci, and the few that have been surveyed might not represent a random sample of the whole genome. But the proteins surveyed by electrophoresis, as well as those sequenced or studied with immunological methods, *were* chosen at random (with respect to differentiation between species) and are more or less the same proteins studied in other animal groups. Therefore, it seems that a real discrepancy exists—the rate of organismal evolution appears to

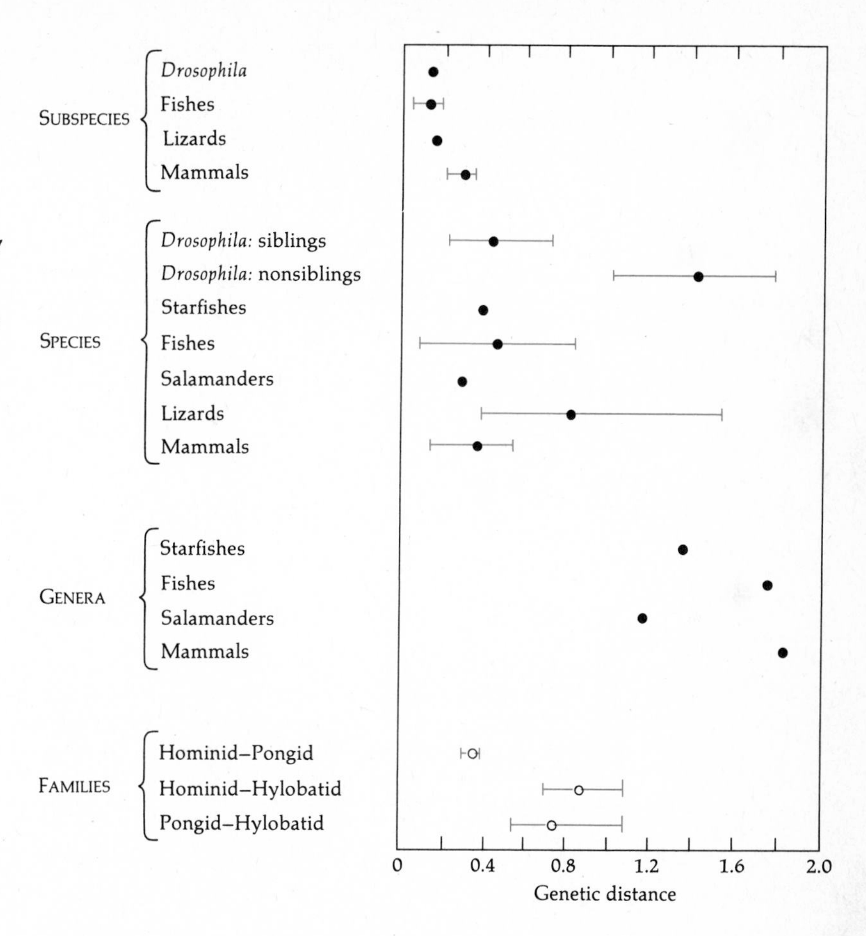

Figure 22.21
Average genetic distance, based on electrophoretic data, for organisms at various levels of evolutionary divergence. Comparisons involving primates are represented by open circles, all others, by closed circles. The bars indicate the range of values observed. [After E. J. Bruce and F. J. Ayala, *Evolution* 33:1040 (1979).]

be greater than the rate of protein evolution in the human lineage (Figure 22.22).

A different resolution of the paradox is possible. This is to postulate that organismal evolution is not determined primarily by changes in structural genes, but by changes in gene regulation. Then, organismal evolution would not necessarily proceed at the same rate as the evolution of structural genes. This hypothesis is supported by other indirect evidence, including the following. (1) Two African toad species, *Xenopus laevis* and *Xenopus borealis*, are morphologically very similar, but their difference in DNA sequence ($\Delta T_s = 12°C$) is greater than that between humans and New World monkeys ($\Delta T_s = 10°C$). (2) Protein evolution proceeds in mammals and in anurans (frogs and toads) at similar rates. Yet there is relatively little morphological differentiation among the more than 3000 known anuran species, while there is considerable morphological divergence among placental mammals (think of an armadillo, a mouse, a whale, and a human). (3) Moreover, frogs—but not mammals—that are

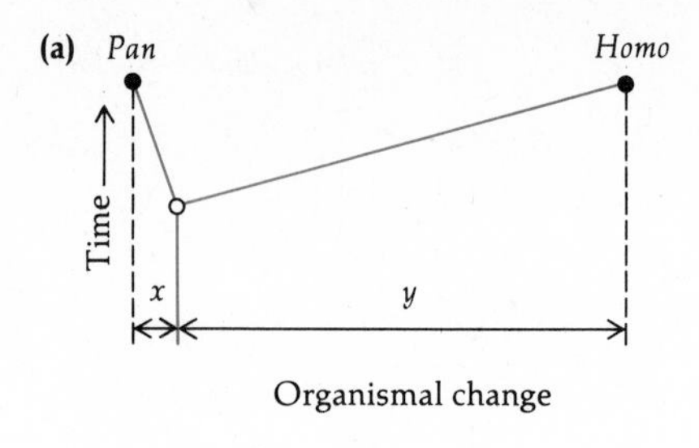

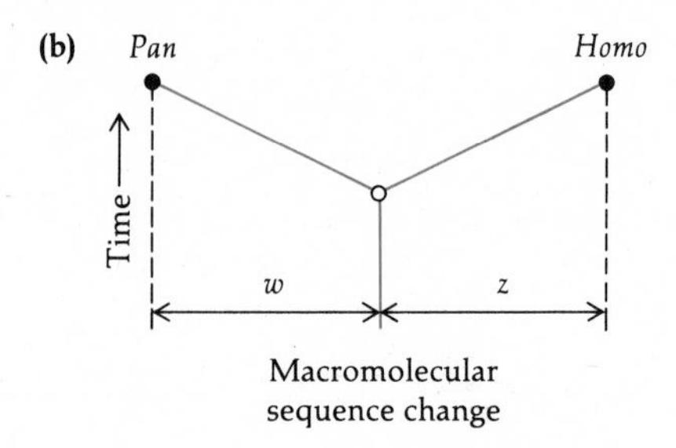

Figure 22.22
Contrast between morphological and molecular evolution since the divergence of the human and chimpanzee lineages. **(a)** More organismal change has taken place in the human lineage (y) than in the chimpanzee lineage (x). **(b)** However, at the protein and nucleic acid levels, the rate of change has been approximately the same in both lineages ($w \approx z$). [After M.-C. King and A. C. Wilson, *Science 188*:107 (1975).]

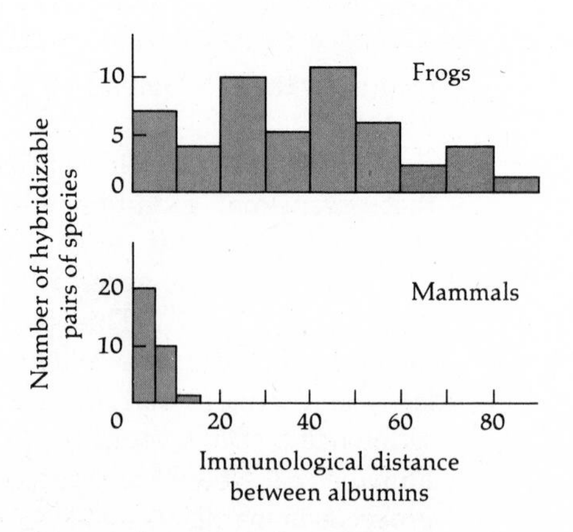

Figure 22.23
Interspecific hybridization as a function of immunological distance between the albumins of pairs of species capable of producing viable hybrids. In this study, 31 such pairs of placental mammals and 50 pairs of frogs were investigated. Immunologically quite different species of frogs are able to hybridize, but not mammals. [After A. C. Wilson et al., *Proc. Natl. Acad. Sci. USA 71*:2843 (1974).]

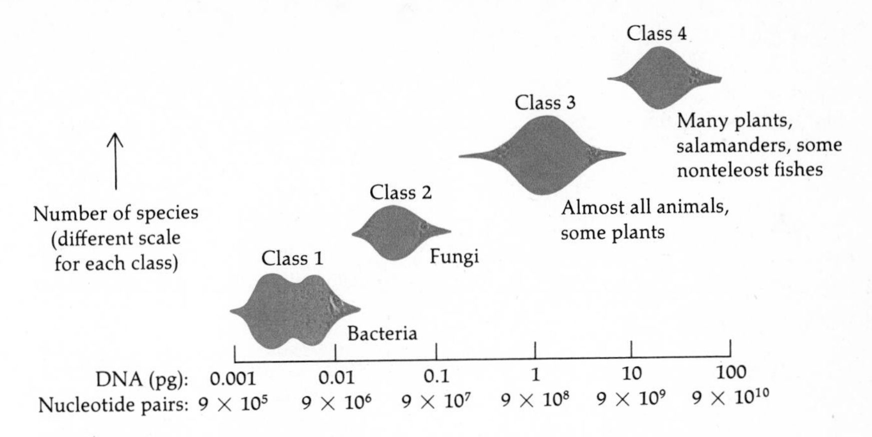

Figure 22.24
Organisms classified according to their amounts of DNA per cell. The amount of DNA is given by weight (1 pg $= 10^{-12}$g) and by the number of nucleotide pairs. It ranges within one order of magnitude for most organisms within each group. For all organisms, from bacteria to plants and animals, the amount of DNA varies by more than five orders of magnitude. (After R. Hinegardner, in *Molecular Evolution*, ed. by F. J. Ayala, Sinauer, Sunderland, Mass., 1976, p. 179.)

very different at the protein level are able to produce interspecific hybrids (Figure 22.23).

The role of gene regulation in adaptive evolution remains one of the major unsolved issues in evolutionary genetics. The evidence just mentioned suggests that changes in gene regulation may be most important in adaptive evolution, i.e., in the evolution of morphology, behavior, and reproductive isolation. Moreover, recent experimental studies with bacteria, yeast, and *Drosophila* have shown that adaptation to new environmental conditions often occurs by changes in gene regulation, although these may later be followed by changes in structural genes. However, little is known about gene regulation in higher organisms.

Evolution of Genome Size

Evolution consists not only of changes in the nucleotide sequence of the DNA, but also of changes in the *amount* of DNA. The early organisms ancestral to all DNA-containing living beings probably had only a few genes. Today, considerable variation in the amount of DNA per cell exists between organisms of different species. Organisms can be grouped into four broad classes, according to the amount of DNA they carry in each cell (Figures 22.24 and 22.25). The lowest amounts of DNA are found in some viruses, with about 10^4 nucleotide pairs per virus. Bacteria have, on the

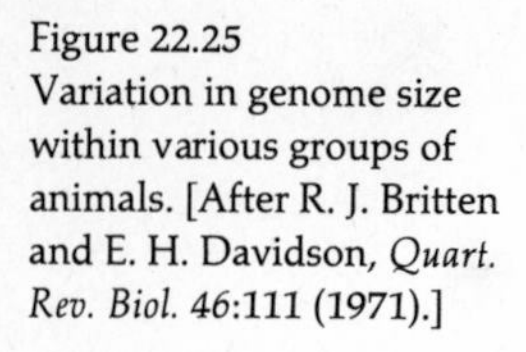

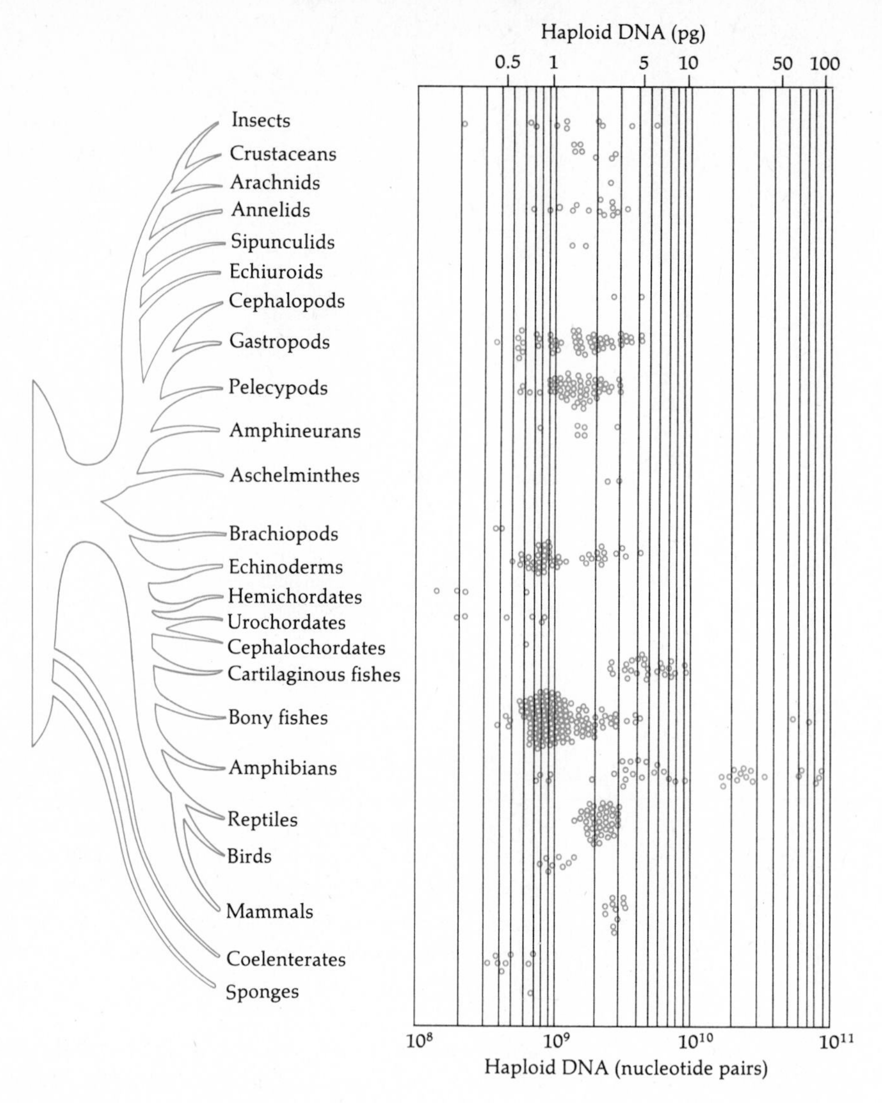

Figure 22.25
Variation in genome size within various groups of animals. [After R. J. Britten and E. H. Davidson, *Quart. Rev. Biol.* 46:111 (1971).]

average, about 4×10^6 nucleotide pairs per cell, and fungi have about ten times as much, or 4×10^7 nucleotide pairs per cell. Most animals and many plants have about 2×10^9 nucleotide pairs per cell, on the average. The most advanced plants, gymnosperms and angiosperms, often have 10^{10} or even more nucleotide pairs per cell. The animals with the largest amounts of DNA are salamanders and some primitive fishes, with more than 10^{10} nucleotide pairs per cell.

A substantial evolutionary increase in the amount of DNA per cell has occurred from bacteria to fungi, to animals, and to plants. More complex

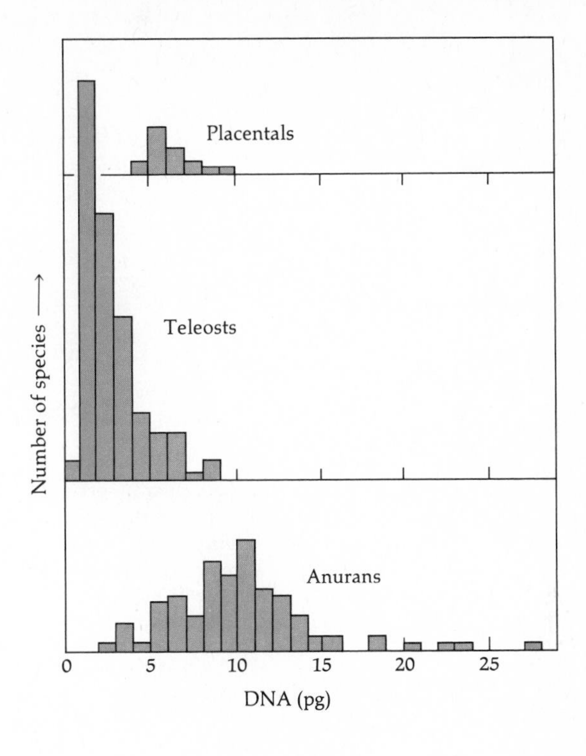

Figure 22.26
Distribution of DNA per cell in certain classes of mammals, fishes, and amphibians. The distributions vary around an intermediate mode. This suggests that evolutionary changes have been numerous and small. [After K. Bachmann, O. B. Goin, and C. J. Goin, in *Evolution of Genetic Systems*, ed. by H. H. Smith, *Brookhaven Symp. Biol.* *23*:419 (1972).]

organisms may need more DNA than a bacterium or a mold, but there seems to be no consistent relationship between the amount of DNA in an organism and its complexity of organization. For example, salamanders and flowering plants are not ten times more complex than mammals or birds, although some of the former have ten times more DNA than the latter.

How does the amount of DNA in the nucleus increase during evolution? Polyploidy is a process by which the amount of DNA can increase; when the number of chromosomes per cell is doubled, the amount of DNA is also doubled. Some organisms with very large amounts of DNA, such as some primitive vascular plants (*Psilopsida*), are polyploid. But polyploidy is a rare phenomenon in animals.

Deletions and duplications (Chapter 17) of relatively small segments of DNA appear to be the most general processes by which evolutionary changes in the amount of DNA have taken place. When the genome sizes of many fish, frog, and mammalian species are arranged in a frequency diagram, they are seen to vary around an intermediate mode (Figure 22.26). This indicates that evolutionary changes in the genome size of animals are numerous and individually small, as would be true with duplications and deletions. If changes in the amount of DNA had occurred primarily by

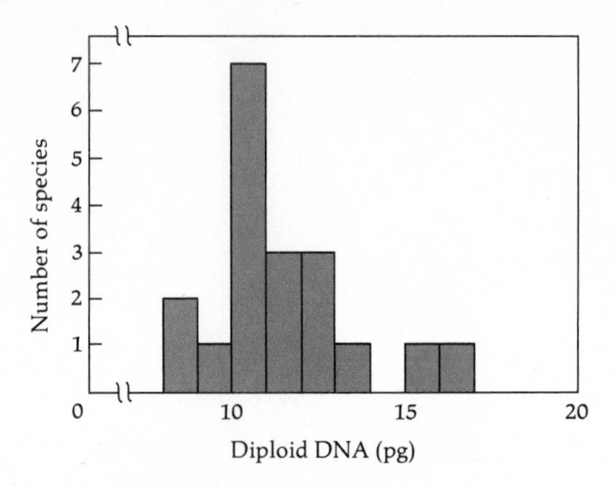

Figure 22.27
Distribution of DNA per cell in 19 species of toads of the genus *Bufo*. The shape of the distribution suggests that changes in the amounts of DNA within this genus have occurred in small increments and decrements rather than by polyploidy, which would have produced a distribution in which the amounts of DNA differed by exact multiples. [After K. Bachmann, O. B. Goin, and C. J. Goin, in *Evolution of Genetic Systems*, ed. by H. H. Smith, *Brookhaven Symp. Biol. 23*:419 (1972).]

polyploidy, organisms would differ in the amount of their DNA by exact multiples (double, quadruple, etc.).

The gradual change in the amount of DNA per cell can also be observed within a single genus, such as the toad *Bufo*. The DNA content has been determined in 19 of the 250 known species, and ranges from about 7 to 15 $\times$ 10^9 nucleotide pairs per cell, with a mode around 10 $\times$ 10^9 (Figure 22.27). Thus, the transition from about 7 $\times$ 10^9 nucleotide pairs per cell to about twice as much has occurred not by polyploidy, but by the cumulative addition of small amounts of DNA, resulting in a fairly smooth and continuous distribution.

Duplications of chromosomal segments often involve only one or a few genes. In recent years it has been discovered that many DNA sequences have originated by duplication, followed in some cases by evolutionary divergence of the duplicated sequences (e.g., the globin genes—see Figure 22.15). Of course, if the duplicated DNA sequences have diverged very substantially, it may no longer be possible to identify them as originally identical; as stated earlier, all genes must have originated by the duplication of a single or a few original genes. In other cases, however, such as the genes coding for ribosomal RNA or transfer RNA, genes exist in multiple copies that have remained essentially identical to each other in structure and in function. Finally, there are DNA sequences that are highly repetitive, each sequence being present from a few thousand to more than a million times (see Chapter 4). The role of these highly repetitive sequences remains unknown.

Problems

1. DNA hybridization experiments indicate that the estimated DNA nucleotide differences are 3% between *Drosophila melanogaster* and *D. simulans* and 13% between *D. melanogaster* and *D. funebris* (see Figure 22.9). Draw the most likely configuration of the phylogeny of these three species, indicating the percent nucleotide changes for each segment. Can you determine whether or not the percent change from the last common ancestor of *D. melanogaster* and *D. simulans* to each of these two species has been the same?

2. According to Table 22.5, the estimated DNA nucleotide differences are 9.6% between humans and green monkeys, 15.8% between humans and capuchins, and 16.5% between green monkeys and capuchins. Reconstruct the phylogeny of these three species and indicate the probable percent nucleotide changes for each segment. Why is it possible in this case to determine the percent change from the last common ancestor of humans and green monkeys to each of these two species, whereas you could not make a similar determination in problem 1?

3. The minimum numbers of nucleotide differences in the gene coding for cytochrome *c* between pairs of four species (see Table 22.7) and the configuration of the phylogeny are given below:

	Monkey	Dog	Tuna
Human	1	13	31
Monkey		12	32
Dog			29

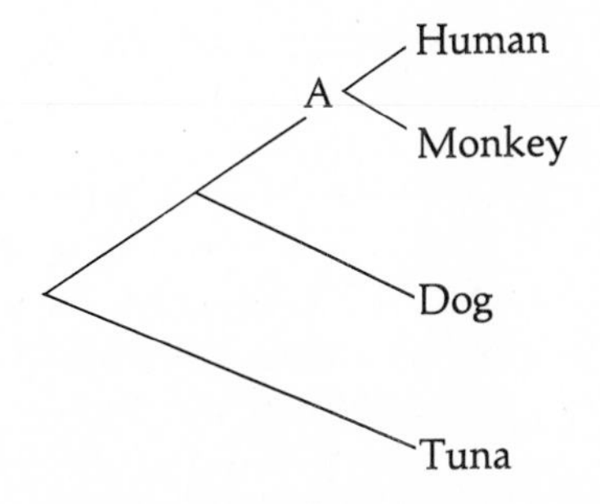

Calculate the minimum numbers of nucleotide differences from species A to human and monkey cytochrome *c* using: (1) only the data for human, monkey, and dog, and (2) only the data for human, monkey, and tuna. Are the two results identical? The inconsistency that you observe is, in fact, a common situation when a problem is "overdetermined" (there are more data than are necessary to find each unknown) and the data are not fully consistent. Evolutionists deal with these inconsistencies by finding out the changes for each segment of a phylogeny that minimize the total amount of discrepancy between the data and the proposed changes. Electronic computers are needed to find such "best solutions" whenever more than a few species are involved.

4. Compare the results obtained in problem 2 with the values shown in Figure 22.10. What is the reason for the discrepancy?

5. It has been known since the beginning of this century that *Anopheles maculipennis* mosquitoes are involved in the transmission of malaria. The situation was puzzling because *A. maculipennis* existed in many places where malaria did not occur. It was eventually discovered that at least six species that are morphologically identical as adults were confused under the name *A. maculipennis*; only some of them are able to transmit malaria. Several species may exist in the same territory, but some live in brackish water, some in running fresh water, and some in stagnant fresh water. What kind of reproductive isolating mechanism is involved?

6. Males and females of two closely related species of *Drosophila* were placed together in equal numbers. The following matings of each kind were observed:

♀ *D. bifasciata* × ♂ *D. bifasciata*	229
♀ *D. imaii* × ♂ *D. imaii*	375
♀ *D. bifasciata* × ♂ *D. imaii*	13
♀ *D. imaii* × ♂ *D. bifasciata*	9

Calculate the coefficient of sexual isolation, C_i, between these two species, defined as

$$C_i = p_{11} + p_{22} - q_{12} - q_{21}$$

where p_{11} and p_{22} are the frequencies of the two kinds of homogamic matings (♀A × ♂A and ♀B × ♂B), and q_{12} and q_{21} are the two kinds of heterogamic matings (♀A × ♂B and ♀B × ♂A).

7. Three species of the orchid *Dendrobium* are found in the same territories. Their flowers open at dawn and wither by nightfall, so that fertilization

can occur only within a period of less than one day. Flowering is brought about by a meteorological stimulus—such as a sudden storm on a hot day—that affects all three species. However, the lapse between the stimulus and flowering is 8 days in one species, 9 in another, and 10 or 11 in the third. What kind of reproductive isolating mechanism is preventing hybridization between these species?

8. Hybrids between the cotton species *Gossypium barbadense*, *G. hirsutum*, and *G. tomentosum* are vigorous and fertile, but F_2 progenies die in seed or early in development, or they develop into weak plants. What kind of reproductive isolating mechanism is keeping the three gene pools apart?

9. The last 20 amino acids of the alpha and beta hemoglobin chains in humans are as follows:

   Alpha: His–Ala–Ser–Leu–Asp–Lys–Phe–Leu–Ala–Ser–

   Beta: Gln–Ala–Ala–Tyr–Gln–Lys–Val–Val–Ala–Gly–

   Val–Ser–Thr–Val–Leu–Thr–Ser–Lys–Tyr–Arg

   Val–Ala–Asn–Ala–Leu–Ala–His–Lys–Tyr–His

   Using the genetic code (Table 12.1), calculate the minimum number of nucleotide changes that have occurred in the evolution of the two segments of DNA coding for the 20 amino acids since the duplication of the alpha and beta genes.

10. The rate of substitution of alleles, $k = 2Nux$ (page 765), may apply to selected alleles as well as to neutral alleles. If a newly arisen allele, A_2, is advantageous relative to the pre-existing allele, A_1, so that the fitness for A_1A_1 is 1, for A_1A_2 is $1 + s$, and for A_2A_2 is $1 + 2s$, the probability of fixation of the new allele is $x = 2N_es/N$, where N_e is the effective population size and N is the total number of individuals in the population. Assume that the rate of mutation to a new allele is $u = 10^{-5}$ per generation and that the effective population size is 10,000. Calculate the rate of substitution of alleles at three different loci in which (1) $s = 0$, (2) $s = 0.0001$, and (3) $s = 0.01$.

11. Genetic variation at 19 gene loci coding for blood proteins was studied in four primate species: chimpanzee (*Pan troglodytes*), gorilla (*Gorilla gorilla*), gibbon (*Hylobates lar*), and baboon (*Papio cynocephalus*). The results are given below. When only one allele was found, only the number identifying the allele is given; when there was more than one allele, the frequency of each is given in parentheses.

Locus	*Pan*	*Gorilla*	*Hylobates*	*Papio*
1. *Ak*	96	98 (0.20) 100 (0.80)	92	96
2. *Alb*	100	100	100	99
3. *Aph*	100	100	100	100
4. *Cer*	100	98	98	102
5. *Dia*	100	85 (0.67) 95 (0.33)	100 (0.67) 108 (0.33)	95 (0.88) 100 (0.12)
6. *Est-A*	100	101	102	96
7. *Est-B*	100	100	102	95 (0.17) 96 (0.08) 103 (0.75)
8. *G6pd*	100	100	102	102
9. *Got*	100	100	96	96 (0.14) 100 (0.86)
10. *Hb*	100	100	100	100 (0.92) 102 (0.08)
11. *Hpt*	105	107	107	107
12. *Icd*	96	100	100	100 (0.94) 107 (0.06)
13. *Lap*	100	100	100	100
14. *Ldh-A*	96	96	96	96
15. *Ldh-B*	100	100	100	100
16. *Mdh*	100	100	93 (0.62) 100 (0.38)	106
17. *6-Pgd*	97	97 (0.15) 105 (0.85)	94	94
18. *Pgm-1*	96 (0.12) 100 (0.88)	100	100	94
19. *Pgm-2*	100	96	100	102

Calculate the genetic identity, I, and genetic distance, D, and draw the likely configuration of the phylogeny.

Appendix

Probability and Statistics

Modern science, genetics included, is largely quantitative. Probability theory and many other branches of mathematics are used for developing models of natural phenomena and for predicting results. Mathematical statistics is used for summarizing and analyzing data. The results of experiments must often be counted or measured in some way; large bodies of data must be reduced to a simple, intelligible form. We may measure, for example, the height of all the students in a class and summarize the the information by calculating the average height. On a more sophisticated level, statistics is used for testing hypotheses. Scientific hypotheses often lead to quantitative predictions. Mendel's hypothesis, for example, led him to predict that, in a cross between two individuals differing with respect to a given trait, the F_2 should consist of dominant and recessive individuals in the ratio 3:1. Mendel crossed a tall plant and a short plant and obtained 1064 F_2 plants. 798 tall plants and 266 short plants would have given a 3:1 ratio; instead, he observed 787 tall plants and 277 short plants. Was this result compatible with the hypothesis? Suppose the agreement between prediction and result had not been that close—at what point would Mendel have been forced to reject his hypothesis?

In this appendix we present certain concepts and methods of probability theory and statistics that are necessary for an understanding of genetics. The treatment is quite elementary and will be superfluous for students with college courses in mathematics and statistics, but it will be useful to others. The following topics are introduced: (I) basic concepts of *probability* theory necessary for calculating the numerical expectations from certain kinds of hypotheses; (II) the *chi-square method* of testing hypotheses; (III) the *mean* and the *variance*—two measures frequently used in summarizing data; (IV) the *Poisson distribution*—a kind of distribution often found in genetics when measuring rare events, such as mutations; and (V) the *normal distribution*—another kind of distribution commonly encountered, particularly in population genetics.

A.I Probability

Probability can be defined as the number of favorable outcomes of an event divided by the total number of possible outcomes. For example, in a pea plant heterozygous (*Rr*) for the round-pea allele (*R*) and the wrinkled-pea allele (*r*), the probability that a gamete will carry the *R* allele is ½. In this case, the possible outcomes are *R* and *r*, and we designate *R* the "favorable" one.

Sometimes the probability of an outcome is not known *a priori;* it can then be determined empirically by observing the frequency with which it occurs. For example, in a sample of 6346 X chromosomes of *Drosophila,* 8 were found to carry a newly arisen recessive lethal (Chapter 16, page 548). The probability of mutation to a recessive lethal in one X chromosome is then estimated as $8/6346 = 0.0013$ per generation.

All probabilities must lie between zero and one. A probability of one indicates that the outcome is certain to occur; a probability of zero indicates that the outcome cannot occur.

There are certain laws that apply to the combination of probabilities. The most fundamental ones are called the law of the sum of probabilities and the law of the product of probabilities.

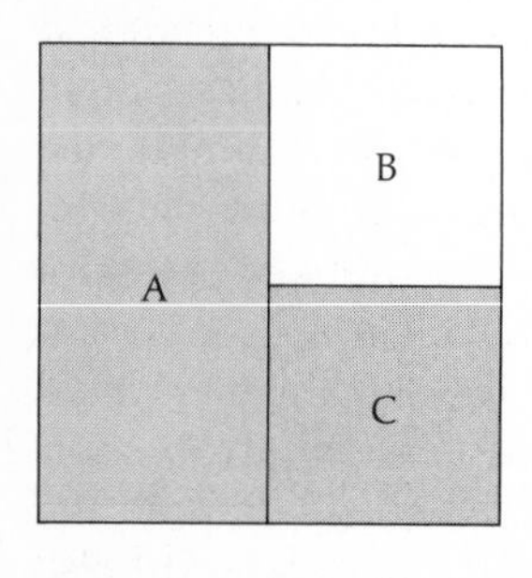

Figure A.1
Additivity of alternative probabilities. If A, B, and C represent three outcomes of a given event, with probabilities measured by their relative areas, the probability that either A or B will occur is the sum of the two corresponding areas.

LAW OF THE SUM OF PROBABILITIES The probability that one *or* another of several mutually *exclusive* outcomes of an event will occur is the sum of their individual probabilities. For example, suppose that we represent the probabilities of the only three possible outcomes of an event as areas in a square, as in Figure A.1, where the probability of A is ½, the probability of B is ¼, and the probability of C is ¼. Then the probability that, say, either A *or* B will occur is $\frac{1}{2} + \frac{1}{4} = \frac{3}{4}$, as is apparent from the figure. In the progeny of a cross between two *Rr* heterozygous pea plants, the probabilities of the three kinds of progeny are: ¼ homozygous *RR*, ¼ homozygous *rr*, and ½ heterozygous *Rr.* Then the probability that a progeny will exhibit the dominant phenotype, i.e., that it will be either homozygous *RR* or heterozygous *Rr*, is $\frac{1}{4} + \frac{1}{2} = \frac{3}{4}$.

LAW OF THE PRODUCT OF PROBABILITIES The probability that several mutually *independent* outcomes will *all* occur is the product of their individual probabilities. Consider, for example, the cross *Rr* × *Rr* mentioned in the previous section. The probability that an individual offspring will receive the *r* allele from one of the parents is ½; the probability that it will receive the *r* allele from the other parent is also ½. Hence the probability that an individual will receive an *r* allele from one parent *and* an *r* allele from the other parent is ½ × ½ = ¼.

It is important that the outcomes be independent. Consider, for example, the cross shown in Figure 3.5. The F_1 is a cross between a female, heterozygous (w^+/w) for the red-eye allele (w^+) and the white-eye allele (w), with a red-eyed male (w^+/Y). In the F_2, the probability of a white-eyed individual is ¼, because this requires that an individual receive the *w* allele from the mother (an outcome with a probability of ½) and the Y chromosome from the father (an outcome that also has a probability of ½). Now, the probability that an F_2 individual is a male is ½. But it would be erroneous to conclude that the probability that an F_2 individual is both white-eyed *and* male is (¼ × ½) = ⅛. This is because the two outcomes are clearly not independent. For an F_2 individual to have white eyes, it must inherit the Y chromosome from its father; thus, *all* F_2 white-eyed individuals are males.

A caveat. When calculating the probability of successive outcomes, it is important to distinguish the probability of all the successive outcomes *together* from the probability of a given outcome *alone.* Consider, for example, the following question: What is the probability that the first two children born in a family will both be males? For the answer, we need only apply the law of the product of probabilities. Assuming that the probability of having a male child is ½, the answer is ½ × ½ = ¼. Consider, now, the following, slightly different question: What is the probability that the second child in a family in which the first child is a male will also be a male? The answer in this case is ½. Independently of the sex of any previous children, the probability that the next child will be a male is always ½.

A second caveat. Sometimes, outcomes are samples from a "universe" that is limited (this situation is called by statisticians "sampling without replacement"). In such cases one must take into account that the probability of a given outcome depends on the number of previous trials and on the nature of the outcomes of those trials. As an illustration, consider a pack of 52 cards, with 4 aces. We deal the cards successively and show them face up. What is the probability that the first card will be an ace? The answer is 4/52. Assume that the first card was *not* an ace. What is the probability that the second card will be an ace? There are only 51 cards left; therefore this probability is 4/51. Assume, instead, that the first card *was* an ace. Now what is the probability that the second card will be an ace? There are only 3 aces and 51 cards left; the answer is therefore 3/51.

A.II The Chi-Square Method

One useful method for testing whether the results obtained in an experiment are consistent with a hypothesis is the *chi-square* (χ^2) *method.* The χ^2 function is

Table A.1
Calculation of χ^2 for Mendel's experiment with tall and short pea plants.

Operation	Tall Plants	Short Plants	Total
Observed (O)	787	277	1064
Expected (E)	$1064 \times 3/4 = 798$	$1064 \times 1/4 = 266$	1064
$O - E$	-11	$+11$	0
$(O - E)^2$	121	121	
$(O - E)^2/E$	0.15	0.44	$\chi^2 = 0.59$

$$\chi^2 = \sum \frac{(\text{observed} - \text{expected})^2}{\text{expected}}$$

where Σ stands for "summation" and applies to all classes in the experiment.

Consider the experiment in which Mendel crossed a tall plant (*TT*) and a short plant (*tt*). The F_1 cross was $Tt \times Tt$. According to Mendel's hypothesis, the F_2 should consist of tall (*TT* and *Tt*) and short (*tt*) plants in the ratio 3:1. He observed 787 tall plants and 277 short plants. The calculation of chi-square for this experiment is shown in Table A.1. The value of χ^2 is 0.59. Is this value consistent with the hypothesis? That is, can the difference between the observed and expected values be attributed to chance alone? In order to answer this question, we must introduce two concepts: the number of degrees of freedom and the level of significance.

The *number of degrees of freedom* can be simply calculated as the number of classes whose value must be known in order to determine the values of all the classes (once we know the total). In the example given, the number of degrees of freedom is 1, because if we are given the number in one class (e.g., 787 tall plants), we can determine the number in the other class by subtracting the first class from the total: $1064 - 787 = 277$. In general, in experiments of this type, the number of degrees of freedom is the number of classes minus one, or $k - 1$, because the last class can be calculated by subtracting the sum of all the other classes from the total. (We shall see below that in other types of experiments the number of degrees of freedom is not $k - 1$.)

The *level of significance* refers to the risk that we are willing to take of rejecting a hypothesis even if it is true. Differences between observed and expected values may vary as a result of chance, but if the probability of

Table A.2
Values of χ^2 corresponding to various levels of significance and degrees of freedom.

	Level of Significance		
Degrees of Freedom	0.05	0.01	0.001
1	3.84	6.64	10.83
2	5.99	9.21	13.82
3	7.82	11.34	16.27
4	9.49	13.28	18.47
5	11.07	15.09	20.52
6	12.59	16.81	22.46
7	14.07	18.48	24.32
8	15.51	20.09	26.13
9	16.92	21.67	27.88
10	18.31	23.21	29.59

observing a certain degree of discrepancy is too small, we may want to reject the hypothesis, even though it might be true. The level of significance most commonly used is 5%; that is, it is decided that the outcome is not consistent with the hypothesis whenever the probability of observing a certain discrepancy between the observed and expected values owing to chance alone is 5% or less. The chi-square values for various degrees of freedom are given in Table A.2 for levels of significance of 5%, 1%, and 0.1%.

We now return to the question of whether the results of Mendel's experiment are consistent with his hypothesis. The chi-square is 0.59, with one degree of freedom. This is an acceptable amount of discrepancy, because it is smaller than the chi-square value for one degree of freedom at the 5% level of significance, which is 3.84 (Table A.2). Hence, we are willing to accept that the results are consistent with Mendel's hypothesis and that the difference between the observed and expected values is due to chance.

In *Drosophila,* the allele for *dumpy* wings (*dp*) is recessive to that for normal wings (dp^+) and *cinnabar* eye (*cn*) is recessive to *red* eye (cn^+). If the

Table A.3
Calculation of χ^2 for the hypothesis of independent assortment of two loci.

Operation	Wild Type	*dumpy*	*cinnabar*	*dumpy* and *cinnabar*	Total
Observed (O)	268	50	54	28	400
Expected (E)	225	75	75	25	400
O − E	43	−25	−21	3	0
$(O - E)^2$	1849	625	441	9	
$(O - E)^2/E$	8.22	8.33	5.88	0.36	$\chi^2 = 22.79$

two loci assort independently, the cross dp^+/dp, $cn^+/cn \times dp^+/dp$, cn^+/cn should produce four kinds of progeny (wild type, *dumpy*, *cinnabar*, and *dumpy* and *cinnabar*) in the ratios 9:3:3:1. In a sample of 400 flies, the following numbers were observed:

Wild type	268
dumpy	50
cinnabar	54
dumpy and *cinnabar*	28

We want to test whether these results are consistent with the hypothesis that the two loci assort independently. The chi-square calculation is shown in Table A.3. The number of degrees of freedom is $k - 1 = 3$, because once we know the number of flies in three classes, we can determine the number in the fourth class by subtracting from the total. The chi-square for three degrees of freedom at the 0.05 level of significance is 7.82 (Table A.2). Therefore, we reject the hypothesis that the two loci assort independently. In this case, we would reject the hypothesis even at the 0.001 level of significance, because the χ^2 with three degrees of freedom at that level is 16.27.

TEST OF INDEPENDENCE It is sometimes desirable to test whether the results of two sets of observations made on the same individuals are independent or not. For example, 256 pairs of twins are classified as monozygotic or dizygotic and also as concordant or discordant with respect to bronchial asthma. A pair of twins is concordant if both suffer from the condition, and it is discordant if one does but the other does not; cases in which neither of the twins exhibits the trait are not included. The results are as follows:

Monozygotic concordant	30
Monozygotic discordant	34
Dizygotic concordant	46
Dizygotic discordant	146

In this case we do not have any hypothesis to tell us what frequency to expect in each class, but we can test whether the two characteristics are independent by means of a 2 × 2 *contingency table.* First we set the table with the *observed* results:

	Concordant	Discordant	Total
Monozygotic	30	34	64
Dizygotic	46	146	192
Total	76	180	256

We can now calculate the *expected* results (if type of twinning and concordance are independent) for each of the four classes by multiplying the corresponding marginal totals and dividing by the grand total. For example, the expected frequency of monozygotic concordant twins is $(64 \times 76)/256 = 19.00$. The 2 × 2 table of expected results is:

	Concordant	Discordant	Total
Monozygotic	19	45	64
Dizygotic	57	135	192
Total	76	180	256

The χ^2, calculated as in previous examples, is 12.08. Although there are four classes, the number of degrees of freedom in this case is 1, not 3. This can be seen by noting that it is enough to know one of the four values in the 2 × 2 contingency table in order to calculate the other three, by subtraction from the marginal totals (e.g., if the number of monozygotic concordant twins is 30, the number of monozygotic discordant twins must be 34, in order for the total number of monozygotic twins to be 64, and so on). Contingency tables may have any number of rows (r) and columns (c), not counting the "Totals," of course. In general, the number of degrees of freedom is $(r - 1)(c - 1)$.

Because the chi-square, 12.08, is larger than the chi-square for one degree of freedom at the 5% level of significance, we conclude that type of twinning and concordance or discordance with respect to bronchial asthma are not independent. (Lack of independence might be due to the existence of a genetic component for susceptibility to bronchial asthma.)

TESTING THE HYPOTHESIS OF HARDY-WEINBERG EQUILIBRIUM Two alleles exist at the *Pgm* locus in human populations. The number of individuals of each genotype in a sample of 200 Caucasians is

Pgm^1/Pgm^1	108
Pgm^1/Pgm^2	86
Pgm^2/Pgm^2	6
Total	200

We want to find out whether the observed numbers occur with the frequencies predicted by the Hardy-Weinberg law. First we calculate the frequency, p, of the Pgm^1 allele:

$$p = \frac{(108 \times 2) + 86}{400} = \frac{302}{400} = 0.755$$

The expected frequencies and the expected numbers of the three genotypes are

Genotype	Frequency	Number
Pgm^1/Pgm^1	$p^2 = 0.570$	114
Pgm^1/Pgm^2	$2pq = 0.370$	74
Pgm^2/Pgm^2	$q^2 = 0.060$	12

Calculating the chi-square as above, we obtain $\chi^2 = 5.26$. What is the number of degrees of freedom? It is 1—not 2, as it might seem by analogy with the cases of Mendelian segregation considered above. The reason is that we have used the data to calculate the allele frequency, $p = 0.755$.

Given this value and the total size of the sample, it is sufficient to know the number of individuals in one of the three genotypic classes in order to calculate the numbers in the other two classes.

This leads to another rule (equivalent to the one given earlier) for calculating the number of degrees of freedom: The number of degrees of freedom is the number of classes minus the number of independent values obtained from the data that are used for calculating the expected numbers. In the cases of Mendelian segregation considered earlier, the total number of individuals was the only value obtained from the data; with it and the Mendelian laws, we were able to calculate the expected numbers for each phenotypic class. In the Hardy-Weinberg problem, however, we calculate *two* values from the data: the total number of individuals and p. Note that $\chi^2 = 5.26$ is statistically significant at the 5% level of significance with one degree of freedom, but not with two degrees of freedom. If we had erroneously assumed that there were two degrees of freedom, we would not have rejected the hypothesis that the three genotypes are in Hardy-Weinberg equilibrium.

A caveat. The chi-square method is an approximate method that is good only if the total sample and the *expected* numbers in each class are large, but not if they are small. The practical rules are as follows: (1) If there is only one degree of freedom, the expected numbers in each class should be at least 5. (2) If the number of degrees of freedom is greater than 1, the expected numbers in each class should be at least 1. There are however, procedures that can be followed whenever these rules are broken.

If the number of degrees of freedom is 1 and one of the classes is smaller than 5, one must apply *Yates's correction.* This consists in making the differences (Observed − Expected) one-half unit closer to zero and then calculating the chi-square as before. This has been done in Table A.4 for the results of a backcross between a wild-type heterozygous rabbit (c^+/c^a) and an albino rabbit (c^a/c^a). Without Yates's correction, $\chi^2 = 4$, which is statistically significant at the 0.05 level of significance. *With* Yates's correction, $\chi^2 = 3.06$, which is *not* significant. We conclude that the results are consistent with the expectations.

If the number of degrees of freedom is greater than 1, but there are classes with expected values less than 1, we can combine the smaller classes with each other, so that the expected values in all new classes are 1 or larger. However, we must take into account that the number of classes after combining them is the number to be used for calculating the number of degrees of freedom. Table A.5 gives the results of a study in which the chromosome arrangements were ascertained in a sample of 50 *Drosophila pseudoobscura* larvae. First we calculate the frequency of each arrangement in the population:

$$p(AR) = \frac{(16 \times 2) + 22 + 6}{100} = 0.60$$

Table A.4
Calculation of χ^2, with and without Yates's correction, for the results of a backcross between a heterozygous (c^+/c^a) and an albino (c^a/c^a) rabbit.

Operation	Wild Type	Albino	Total
Observed (O)	12	4	16
Expected (E)	8	8	16
O − E	4	−4	0
$(O - E)^2$	16	16	
$(O - E)^2/E$	2	2	$\chi^2 = 4$
With Yates's correction:			
O − E	3.5	−3.5	0
$(O - E)^2$	12.25	12.25	
$(O - E)^2/E$	1.53	1.53	$\chi^2 = 3.06$

$$q(CH) = \frac{(4 \times 2) + 22 + 0}{100} = 0.30$$

$$r(TL) = 1 - 0.60 - 0.30 = 0.10$$

The expected frequencies of the genotypes can be calculated using the square expansion, $(p + q + r)^2$; the expected number of individuals is obtained by multiplying the total number in the sample (50) by the expected frequencies. This has been done in Table A.5. Three independent values are calculated from the data in order to obtain the expected values: p, q, and the total number of individuals (r is not independently obtained from the data, but is simply calculated as the difference $1 - p - q$). Since there are six classes in the data, the number of degrees of freedom is therefore 6 − 3 independent values = 3. The chi-square is 8.67, which is statistically significant at the 0.05 level with three degrees of freedom. In the lower part of Table A.5, the two classes with the lower expected numbers have been combined. Now we have five classes and, therefore, 5 − 3 = 2 degrees of freedom. The $\chi^2 = 1.81$ is, however, not statistically significant at the 0.05 level.

Table A.5
Calculation of χ^2, with and without combination of small classes, for a Hardy-Weinberg test.

Operation	*AR/AR*	*AR/CH*	*CH/CH*	*AR/TL*	*CH/TL*	*TL/TL*	Total
Observed (O)	16	22	4	6	0	2	50
Expected (E)	18	18	4.5	6	3	0.5	50
O − E	−2	+4	−0.5	0	−3	+1.5	0
$(O - E)^2$	4.00	16.00	0.25	0	9	2.25	
$(O - E)^2/E$	0.22	0.89	0.06	0	3	4.50	$\chi^2 = 8.67$
Combining small classes:							
Observed	16	22	4	6	2		50
Expected	18	18	4.5	6	3.5		50
O − E	−2	+4	−0.5	0	−1.5		0
$(O - E)^2$	4	16	0.25	0	2.25		
$(O - E)^2/E$	0.22	0.89	0.06	0	0.64		$\chi^2 = 1.81$

A.III Mean and Variance

Assume that we have a sample of individuals measured with respect to some trait, such as height. Two parameters that summarize the information are the arithmetic mean and the variance. The mean is said to be a measure of "central tendency," the variance, a measure of "dispersion."

The *arithmetic mean,* or simply the *mean,* of a distribution is the average of the value in all individuals of the sample:

$$\overline{X} = \frac{\Sigma X}{N}$$

where $\overline{X}$ is the mean, ΣX represents the sum of all the values observed, and N is the number of individuals.

Assume that we measure the height, to the nearest centimeter, of 10 students and obtain the following values: 170, 174, 177, 178, 178, 179, 179, 180, 181, and 184 cm. The mean height of the individuals in this sample is

$$\overline{X} = \frac{170 + 174 + 177 + 178 + 178 + 179 + 179 + 180 + 181 + 184}{10}$$

$$= 178.0 \text{ cm}$$

The *variance* is the sum of the squares of the differences between each value and the mean, divided by the number of individuals minus one, or

$$s^2 = \frac{\Sigma(X - \overline{X})^2}{N - 1}$$

where s^2 is the variance and the other symbols are as in the previous formula: X represents each one of the values, $\overline{X}$ is the mean, and N is the number of individuals. The variance in height among the 10 students in our sample is

$$s^2 = \frac{(-8)^2 + (-4)^2 + (-1)^2 + 0^2 + 0^2 + 1^2 + 1^2 + 2^2 + 3^2 + 6^2}{9}$$

$$= \frac{132}{9} = 14.67 \text{ cm}^2$$

It is often convenient to calculate the variance using the following formula, which is mathematically equivalent to the previous one:

$$s^2 = \frac{\Sigma X^2 - N\overline{X}^2}{N - 1}$$

For the example in question,

$$s^2 = \frac{170^2 + 174^2 + 177^2 + 178^2 + 178^2 + 179^2 + 179^2 + 180^2 + 181^2 + 184^2 - 10(178^2)}{9}$$

$$= \frac{316{,}972 - 316{,}840}{9} = 14.67 \text{ cm}^2$$

The variance is measured in squared units because it is based on the sum of *squared* differences. A related measure of dispersion which, however, is given in the same units as the mean is the *standard deviation*, which is simply the square root of the variance. Hence

$$s = \sqrt{\frac{\Sigma(X - \overline{X})^2}{N - 1}}$$

Table A.6
Observed and expected results (assuming a Poisson distribution) in a bacterial experiment.

Number of Colonies per Plate	Number of Plates	Number of Colonies	Expected Frequency of Plates	Expected Number of Plates
0	22	0	0.311	18.7
1	19	19	0.363	21.8
2	10	20	0.212	12.7
3	6	18	0.082	4.9
4	2	8	0.024	1.44
5	1	5	0.006	0.34
Total	60	70	0.998	59.9

where s is the standard deviation. For the height example discussed above, the standard deviation is $s = 3.83$ cm.

A.IV The Poisson Distribution

Consider the following experiment, similar to the one described in Chapter 16 (page 553), carried out by S. E. Luria and M. Delbrück [*Genetics 28:* 491 (1943)]. A large culture with liquid medium is inoculated with *Escherichia coli* carrying the mutant ton^S (sensitivity to infection by phage T1). The bacteria are allowed to grow until they reach maximum titer. Samples of 0.2 ml from this culture are spread on Petri plates having nutrient agar inoculated with the T1 virus. Mutations from ton^S to ton^R (resistance to T1) occur at a certain rate. On the plates, ton^R bacteria are able to grow and form colonies, but ton^S bacteria are not. The numbers of colonies found in each of 60 plates are given in Table A.6 (third column). The total number of colonies found in the 60 plates is 70, or 1.17 colonies per plate.

When the probability of an outcome (a mutant, in this case) is very small, but the number of trials (bacteria) is very large, the frequency of the outcomes follows a *Poisson distribution.* (It is also assumed that the outcomes are independent; in the example, the occurrence of one mutation in one bacterium must not affect the probability that another mutation will also occur.) Another example of a Poisson distribution is the number of achondroplastics born to normal parents for each 10,000 births in hospitals throughout the world.

The terms of the Poisson distribution are given by the following general formula:

$$p(k) = \frac{x^k}{k!}e^{-x}$$

where $p(k)$ is the probability that k outcomes will be observed in a given sample, x is the mean number of outcomes, and $k!$ (k factorial) represents the product $1 \times 2 \times 3 \times \cdots \times k$. Thus, according to the Poisson distribution, the expected frequency of samples having a given number of outcomes is:

Number of outcomes	0	1	2	3	4	. . .	k
Frequency	e^{-x}	xe^{-x}	$\frac{x^2}{2}e^{-x}$	$\frac{x^3}{2 \times 3}e^{-x}$	$\frac{x^4}{2 \times 3 \times 4}e^{-x}$	. . .	$\frac{x^k}{k!}e^{-x}$

In the bacterial example, the mean number of outcomes (mutants) per sample (plate) is $x = 1.17$. The expected *frequency* of plates with 0, 1, 2, etc., colonies can be calculated from the Poisson distribution (fourth column in Table A.6). The expected *number* of plates is obtained by multiplying the expected frequency by 60, which is the total number of plates in the study (fifth column in Table A.6). We could now test, for example, by means of a χ^2 test whether the observed results are consistent with the expectations of a Poisson distribution.

One useful property of the Poisson distribution is that the variance is equal to the mean. The observed variance for the data in Table A.6 is 1.50, which is not very different from the mean, 1.17.

The Poisson distribution has many applications in genetics. In Chapter 16 we have illustrated its use for the estimation of mutation rates (page 546) and of the number of genes (Box 16.2). The formula for calculating genetic distance from electrophoretic data is another application of the Poisson distribution (Box 22.1). Proteins with different electrophoretic mobility are different, but we do not know whether the difference involves just one or more than one amino acid. If the number of differences between proteins encoded by the same gene locus has a Poisson distribution (which is reasonable, because there are many amino acids in each protein, but the mean number of amino acid differences is small between closely related species), then the frequency of *identical* proteins represents the zero term (no difference) of a Poisson distribution. Thus,

Number of individuals:	1	0	0	1	5	7	7	22	25	26	27	17	11	17	4	4	1
Height in inches:	58	59	60	61	62	63	64	65	66	67	68	69	70	71	72	73	74

Figure A.2
Height distribution in 175 men recruited for the Army around the turn of the century. [From A. F. Blakeslee, *J. Hered.* 5:511 (1914).]

if the frequency of identical proteins is I and the mean frequency of differences is D, then

$$I = e^{-D}$$

Taking logarithms, we obtain

$$\ln I = -D \quad \text{or} \quad D = -\ln I$$

which is the formula for genetic distance given in Box 22.1.

A.V The Normal Distribution

For many quantitative traits, such as height, weight, and number of eggs laid, the frequency distributions among individuals are bell-shaped. The number of individuals is greater at the intermediate values and gradually decreases toward the extremes. An example is shown in Figure A.2; several others were given in Chapter 15. The *normal distribution* is a mathematical curve that corresponds to such bell-shaped distributions.

The normal distribution has some interesting properties based on its mean and its standard deviation. The most useful application concerns the number of observations that are expected to fall within certain values

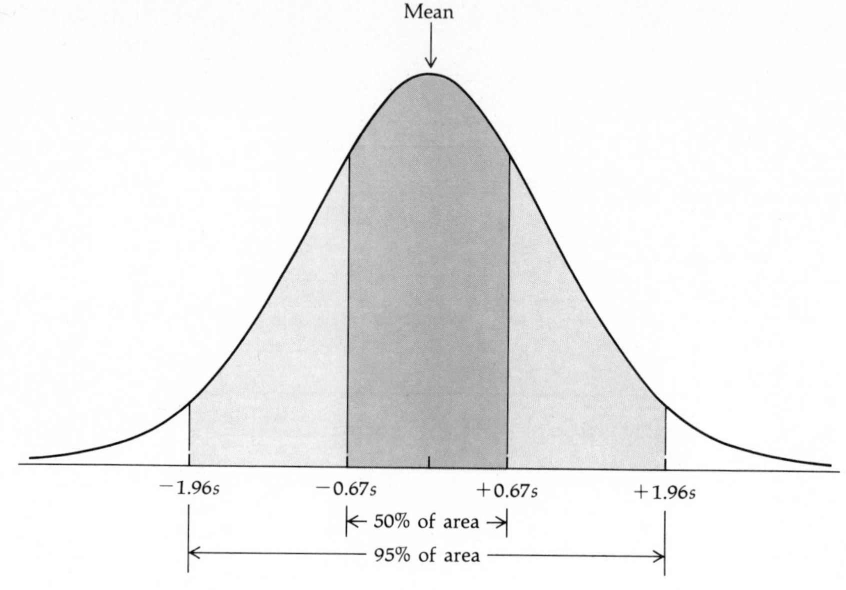

Figure A.3
The normal distribution, showing the fraction of the area of the curve that falls within certain ranges of values expressed as departures from the mean.

(Figure A.3). In a normal distribution, 50% of all observations fall between values separated by 0.67 standard deviation on each side of the mean ($\overline{X} \pm 0.67s$; dark-shaded area), 67% of the observations fall within $\overline{X} \pm s$, and 95% of the observations fall within $\overline{X} \pm 1.96s$ (light- and dark-shaded areas).

As an example, consider the height of the army recruits shown in Figure A.2. The mean and the standard deviation for that sample of 175 men are $\overline{X} = 67.3$ inches and $s = 2.7$ inches. The heights corresponding to $\overline{X} \pm s$ are 64.6 and 70.0 inches. The number of individuals taller than 64.6 inches but shorter than 70.0 inches is 117, which turns out to be exactly 67% of the 175 individuals in the sample. The heights corresponding to $\overline{X} \pm 1.96s$ are 62.0 and 72.6 inches. The number of individuals within this range is 163, or 93% of the total; 95% of the individuals should theoretically fall within this range. Although a sample of 175 individuals is not very large, the agreement between the observed and expected numbers is quite good.

Amniocentesis A procedure for diagnosing genetic abnormalities in an embryo or fetus.

Anabolic Pertaining to an enzymatic reaction leading to the synthesis of a more complex biological molecule from a less complex one (cf. *catabolic*).

Anagenesis The evolutionary change of a single lineage in the course of time (cf. *cladogenesis*).

Anaphase The third stage of mitosis or of meiosis (I or II), during which the chromosomes migrate toward opposite poles of the cell.

Aneuploidy The condition of a cell, tissue, or organism in which one or more whole chromosomes of a normal set either are absent or are present more than once (cf. *euploidy*).

Anlage The group of cells from which a given part of the organism develops.

Annealing The process also called nucleic acid hybridization by which two single-stranded polynucleotides form a double-stranded molecule, with hydrogen bonding between the complementary nucleotides of the two strands. Annealing can take place between complementary strands of either DNA or RNA to produce double-stranded DNA molecules, double-stranded RNA molecules, or RNA-DNA hybrid molecules.

Antibody A protein, synthesized by the immune system of a higher organism, that binds specifically to the foreign molecule (*antigen*) that induced its synthesis.

Anticodon The three adjacent nucleotides in a tRNA molecule that are complementary to, and pair with, the three nucleotides of a codon in an mRNA molecule during protein synthesis.

Antigen See *antibody*.

Artificial selection The process of choosing the parents of the following generation on the basis of one or more genetic traits (cf. *natural selection*).

Ascospore See *ascus*.

Ascus A sac containing *ascospores*, the products of meiosis, in the perithecia of actinomycete fungi such as *Neurospora* and *Sordaria*. Four or eight ascospores are found in each ascus.

Asexual reproduction The development of an organism from one or more cells in the absence of any sexual process; also called *vegetative reproduction*.

Assortative mating Nonrandom selection of mates with respect to one or more traits; it is positive (negative) when individuals with the same form of a trait mate more (less) often than would be predicted by chance (cf. *random mating*).

ATP Adenosine triphosphate, the primary repository of chemical energy in cells.

Attached-X chromosome A chromosome consisting of two homologous X chromosomes that are joined and that share a single centromere.

Attenuator A nucleotide sequence, following the promoter and preceding the structural genes of some operons, that causes RNA transcriptase to cease transcription of the operon prior to transcribing the structural genes into mRNA.

Autopolyploidy Polyploidy due to the presence of more than two chromosome sets of the same species.

Autosome A chromosome other than a sex chromosome.

Auxotroph A strain of organisms unable to synthesize a given organic molecule required for their own growth; growth can occur when the required compound is supplied in the food (cf. *prototroph*).

Glossary

Acrocentric chromosome A chromosome with the centromere near one end, so that one arm of the chromosome is short and the other is long.

Adaptation A structural or functional characteristic of an organism that allows it to cope better with its environment; the evolutionary process by which organisms become adapted to their environment.

Adaptive value A measure of the reproductive efficiency of an organism (or genotype) compared with other organisms (or genotypes); also called *selective value*.

Additive genes Genes that interact but that show no dominance, if they are alleles, or no epistasis, if they are not alleles.

Additive variance Genetic variance due to additive genes.

Allele One of two or more alternative forms of a gene, each possessing a unique nucleotide sequence; different alleles of a given gene are usually recognized, however, by the phenotypes rather than by comparison of their nucleotide sequences (cf. *heteroallele*).

Allopatric Of populations or species that inhabit separate geographic regions (cf. *sympatric*).

Allopolyploidy Polyploidy due to the addition of chromosome sets from two different species.

Allosteric transition A change from one conformation of a protein to another conformation (cf. *effector molecule*).

Allozymes Alternative enzyme forms encoded by different alleles at the same locus.

Amino acids The building blocks of proteins; several hundred are known, but only 20 are normally found in proteins.

Aminoacyl-tRNA A tRNA molecule covalently bound to an amino acid via an acyl bond between the carboxyl group of the amino acid and the 3′-OH of the tRNA.

Aminoacyl-tRNA ligase An enzyme that synthesizes a specific aminoacyl-tRNA molecule, employing a specific amino acid (e.g., alanine), its cognate tRNA (e.g., $tRNA^{Ala}$), and ATP to form, for example, alanyl-$tRNA^{Ala}$.

Backcross The cross of a heterozygote with one of its parents.

Back mutation A mutation that causes a mutant gene to regain its wild-type function (cf. *forward mutation*).

Bacteriophage A virus whose host is a bacterial cell; also called *phage.*

Balanced lethals Recessive lethals at different loci, so that each homologous chromosome carries at least one lethal, and associated with inversions, so that no recombination occurs between the homologous chromosomes.

Base pair Two nitrogenous bases that pair by hydrogen bonding in double-stranded DNA or RNA.

Bivalent Two homologous chromosomes that pair during the first meiotic division.

Blastocyst The embryonic stage of mammals that consists of about 64 cells (the outer, trophoblast, cells and the cells of the inner cell mass) and that implants itself into the uterine wall.

Blastoderm A multinuclear embryonic stage that results from nuclear divisions, without division of the cytoplasm of the zygotic cell in the process, and migration of nuclei to the periphery of the oocyte. A cellular blastoderm is established when cell membranes form around each of the nuclei at the periphery of the oocyte.

Blastula A multicellular embryonic stage resulting from complete cleavage divisions that apportion the cytoplasm of the zygotic cell into that of a number of smaller cells. The blastular stage preceeds gastrulation and organogenesis.

Bottleneck A period when a population becomes reduced to only a few individuals.

Catabolic Pertaining to an enzymatic reaction leading to the breakdown of a complex biological molecule into less complex components, which may either yield energy in the form of ATP or be used in subsequent anabolic reactions.

Cell cycle The growth cycle of an individual cell.

Centric fission The formation of two chromosomes by the splitting of one chromosome at (or near) the centromere.

Centric fusion The fusion of two acrocentric or telocentric chromosomes into one metacentric chromosome.

Centromere A chromosomal region that becomes associated with the spindle fibers during mitosis and meiosis.

Chiasmata The places at which pairs of homologous chromatids are in contact (from late prophase of meiosis to the beginning of first anaphase) and at which exchanges of homologous parts between nonsister chromatids have taken place by crossing over (singular: chiasma).

Chloroplast See *plastid.*

Chromatid In all duplicated chromosomes, each of the two longitudinal subunits that become visible during mitosis or meiosis.

Chromatin Material comprised of DNA, histone proteins, nonhistone proteins, and some RNA, identifiable in the nuclei of cells by its special staining properties.

Chromosomal aberration See *chromosomal mutation.*

Chromosomal abnormality See *chromosomal mutation.*

Chromosomal mutation A change in the structure or number of the chromosomes; also called *chromosomal aberration* or *chromosomal abnormality.*

Chromosomal polymorphism The presence in a population of more than one gene sequence for a given chromosome.

Chromosome A threadlike structure, found in the nuclei of cells, that contains the genes arranged in linear sequence; a whole DNA molecule comprising the genome of a prokaryotic cell; a DNA molecule complexed with histones and other proteins in eukaryotic cells.

Chromosome complement The group of chromosomes in a normal gametic or zygotic nucleus; it may consist of one (monoploid nucleus), two (diploid nucleus), or more (polyploid nucleus) chromosome sets (cf. *karyotype*).

Chromosome set The normal gametic complement of chromosomes of a diploid individual (cf. *chromosome complement*).

Cis-trans test See *complementation test*.

Cistron A nucleotide sequence in DNA specifying a single genetic function as defined by the complementation test; a nucleotide sequence coding for a single polypeptide; a gene.

Cladogenesis The splitting of an evolutionary lineage into two or more lineages (cf. *anagenesis*).

Cline A gradient in the frequencies of genotypes or phenotypes along a stretch of territory.

Coadaptation The harmonious interaction of genes; the selection process by which harmoniously interacting genes become established in a population.

Code A set of rules for transferring information from one alphabet or language to another.

Codominant Of alleles whose gene products are both manifest phenotypically.

Codon A group of three adjacent nucleotides in an mRNA molecule that code either for a specific amino acid or for polypeptide chain termination during protein synthesis.

Coefficient of selection The intensity of selection, as measured by the proportional reduction in the gametic contribution of one genotype compared with another.

Colinearity The linear correspondence between the order of amino acids in a polypeptide chain and the corresponding nucleotide sequence in the DNA molecule.

Complementation group A set of mutant alleles exhibiting mutant phenotypic behavior in the heterozygous state with each other.

Complementation test A genetic test to ascertain whether two gene mutations occur in the same functional gene and to establish the limits of the functional gene; also called *cis-trans test*.

Complexity The number of units in a nonrepeating sequence of nucleotide pairs in a prokaryotic genome or a haploid complement of chromosomes.

Conditional lethal mutation A mutation that kills the affected organism under one set of environmental conditions (*restrictive condition*) but that is not lethal under another set of conditions (*permissive condition*).

Conidium The haploid vegetative spore of an actinomycete fungus such as *Neurospora*. When conidia of one mating type come in contact with the protoperithecia of the opposite mating type, cell fusion followed by nuclear fusion occurs to produce diploid cells that develop into asci as meiosis takes place.

Conjugation The process by which DNA is transferred from bacteria of one mating type to bacteria of another during cell-to-cell contact.

Consanguinity The sharing of at least one recent common ancestor.

Continuous variation Variation, with respect to a certain trait, among phenotypes that cannot be classified into a few clearly distinct classes, but rather that differ very little one from the next.

Controlling element A eukaryotic transposable element, detectable through the abnormal activity of standard genes that it affects.

Conversion The process by which, during recombination, one allele of a gene participating in a cross is lost and is replaced by the other allele. Conversion results from the formation of heteroduplex DNA during recombination and the subsequent repair of noncomplementary bases in the heteroduplex DNA.

Crossing over The exchange of chromatid segments between homologous chromatids during meiosis; if different alleles are present on the chromatids, crossing over can be detected by the formation of genetically recombinant chromatids.

Dalton The unit of atomic mass; one dalton is exactly $\frac{1}{12}$ the mass of a single neutral atom of ^{12}C, the most common isotope of carbon.

Darwinian fitness The relative fitness of one genotype compared with another, as determined by its relative contribution to the following generations.

Deficiency A chromosomal mutation characterized by the loss of a chromosome segment; also called *deletion* (cf. *duplication*).

Degenerate code A code in which a single element in one language is specified by more than one element in the second language; e.g., the single amino acid isoleucine is specified by three different codons.

Deletion See *deficiency*.

Denatured DNA DNA that has been converted from double-stranded to single-stranded form by breaking the hydrogen bonds joining the two complementary strands (cf. *native DNA*).

Denatured protein A protein that has lost its natural configuration by exposure to a destabilizing agent such as heat.

Deoxyribonucleic acid (DNA) A polynucleotide in which the sugar residue is deoxyribose and which is the primary genetic material of all cells.

Dihybrid cross A cross between individuals that have different alleles at two gene loci.

Dioecious Of organisms (usually plants) having either male or female sex organs but not both (cf. *monoecious*).

Diploid A cell, tissue, or organism having two chromosome sets (cf. *haploid, polyploid*).

Dizygotic twins Twins developed from two separate fertilized ova; also called *fraternal twins*.

DNA See *deoxyribonucleic acid*.

DNA ligase An enzyme that creates a phosphodiester bond between the 5′-PO_4 end of one polynucleotide and the 3′-OH end of another, thereby producing a single, larger polynucleotide; also called *polynucleotide ligase*.

DNA polymerase The enzyme responsible for synthesizing DNA from deoxyribonucleoside triphosphates under the direction of a template DNA strand.

Dominant An allele, or the corresponding trait, that is manifest in all heterozygotes (cf. *recessive*).

Down's syndrome A syndrome characterized by physiological, behavioral, and mental defects that are due to the presence of an extra copy of the genetic material contained in chromosome 21.

Drift See *random genetic drift.*

Duplication A chromosomal mutation characterized by the presence of two copies of a chromosome segment in the haploid genome (cf. *deficiency*).

Effective population size The number of reproducing individuals in a population.

Effector molecule A (generally) small molecule whose concentration regulates the activity of a protein molecule by interacting with a specific binding site on the protein, causing an allosteric transition to occur in the structure and function of the protein molecule.

Electromorphs Allozymes that can be distinguished by electrophoresis.

Electrophoresis A technique for separating molecules based on their differential mobility in an electric field.

Endonuclease An enzyme that hydrolyzes internal phosphodiester bonds in a polynucleotide (cf. *exonuclease*).

Endosperm In flowering plants, a tissue specialized for nourishing the developing embryo.

Episome A genetic element (DNA molecule) that may exist either as an integrated part of a chromosomal DNA molecule of the host or as an independently replicating DNA molecule (plasmid) free of the host chromosome.

Epistasis Interaction between two nonallelic genes so that one of them (epistatic gene) interferes with, or even inhibits, the phenotypic expression of the other (hypostatic gene).

Euchromatic Of chromosome regions or whole chromosomes that have normal staining properties and that undergo the normal cycle of chromosome coiling (cf. *heterochromatic*).

Eugenics The study of methods under social control directed toward improving the hereditary constitution of future human generations.

Eukaryote A cell or organism that has a nucleus.

Euploidy The condition of a cell, tissue, or organism that has one or more exact multiples of a chromosome set (cf. *aneuploidy*).

Exonuclease An enzyme that hydrolyzes terminal phosphodiester bonds (at 3′ or 5′ ends) in a polynucleotide (cf. *endonuclease*).

Expressivity The degree of phenotypic expression of a penetrant gene.

Fate map A map of an embryo, identifying the cells whose mitotic progeny will give rise to specific adult tissues and organs.

Fertility factor (F factor) An episome capable of transferring a copy of itself from its host bacterial cell (an F^+ cell) to a bacterial cell not harboring an F factor (an F^- cell). When the F factor is integrated into the host chromosome (an Hfr cell), it is capable of mobilizing the bacterial chromosome for transfer to an F^- cell.

Fertilization The fusion of two gametes of opposite sex to form a zygote.

F factor See *fertility factor.*

Fission See *centric fission.*

Fitness The reproductive contribution of an organism or genotype to the following generations (cf. *Darwinian fitness*).

Forward mutation A mutation from the wild type to the mutant condition (cf. *back mutation*).

Founder effect Genetic drift due to the founding of a population by a small number of individuals.

Frameshift A mutation caused by either the insertion or deletion of a number of nucleotide pairs in DNA, whose effect is to change the reading frame of codons in an mRNA molecule during protein synthesis, causing an abnormal amino acid sequence to be synthesized from the site of the mutation on.

Fraternal twins See *dizygotic twins.*

Frequency-dependent selection Natural selection whose effects depend on the frequency of the genotypes or the phenotypes.

Fusion See *centric fusion.*

Gamete A mature reproductive cell capable of fusing with a similar cell of opposite sex to give a zygote; also called *sex cell.*

Gametophyte The haploid, sexual generation that produces the gametes in plants in which a sexual (haploid) generation alternates with an asexual (diploid) generation (cf. *sporophyte*).

Gastrula The embryonic stage marked by the beginning of cell movement and the initiation of organogenesis.

Gene In the genome of an organism, a sequence of nucleotides to which a specific function can be assigned; e.g., a nucleotide sequence coding for a polypeptide, a nucleotide sequence specifying a tRNA, or a nucleotide sequence required for the proper transcription of another gene.

Gene flow The exchange of genes (in one or both directions) at a low rate between two populations, due to the dispersal of gametes or of individuals from one population to another; also called *migration.*

Gene pool The sum total of the genes in a breeding population.

Genetic drift See *random genetic drift.*

Genetic locus See *locus.*

Genetic variance The fraction of the phenotypic variance that is due to differences in the genetic constitution of individuals in a population.

Genome The genetic content of a cell or virus; in eukaryotes, it sometimes refers to only one complete (haploid) chromosome set.

Genotype The sum total of the genetic information contained in an organism; the genetic constitution of an organism with respect to one or a few gene loci under consideration (cf. *phenotype*).

Germ cell An animal cell set aside early in embryogenesis that may multiply by mitosis or produce, by meiosis, cells that develop into either eggs or sperm (cf. *somatic cells*).

Gynandromorph An individual exhibiting both male and female sexual differentiation.

Haploid Of cells, such as gametes, having half as many chromosome sets as the somatic cells; sometimes used as a synonym for monoploid, i.e., a cell, tissue, or organism having only one chromosome set (cf. *diploid, polyploid*).

Hardy-Weinberg law A principle by which genotypic frequencies can be predicted on the basis of gene frequencies, under the assumption of random mating.

Hemizygous Of genes present only once in the genotype.

Heritability In the broad sense, the fraction of the total phenotypic variance that remains after exclusion of the variance due to environmental effects. In the narrow sense, the ratio of the additive genetic variance to the total phenotypic variance.

Hermaphrodite An individual that produces mature male and female gametes.

Heteroallele An allele that differs from other alleles of the same gene by nucleotide differences at *different* sites within the gene; contrast with "true" alleles, of which only four are possible at each site within the gene (AT, TA, GC, and CG).

Heterochromatic Of chromosome regions or whole chromosomes that have a dense, compact structure in telophase, interphase, and early prophase (cf. *euchromatic*).

Heteroduplex DNA Double-stranded DNA in which the two single strands have different hereditary origins; it may be produced either as an intermediate in recombination between two different DNA molecules or by the annealing *in vitro* of complementary single strands from different DNA molecules.

Heterogametic Of the sex whose gametes differ with respect to the sex chromosomes.

Heterogamic Of matings between individuals from different populations or species.

Heterokaryon A cell, or tissue composed of such cells, that has two or more nuclei of more than one genetic type.

Heterosis Superiority of the heterozygote over the homozygotes with respect to one or more characters; also called *hybrid vigor.*

Heterozygosity The condition of being heterozygous; the proportion of heterozygous individuals at a locus, or of heterozygous loci in an individual.

Heterozygote A cell or organism having two different alleles at a given locus on homologous chromosomes.

Hfr cell See *fertility factor.*

Histone Any of the basic proteins complexed with DNA in the chromosomes of eukaryotes.

Homeotic mutation A mutation that causes one body structure to be replaced by a different body structure during development.

Homogametic Of the sex whose gametes do not differ with respect to the sex chromosomes.

Homogamic Of matings between individuals from the same population or species.

Homologous Of chromosomes or chromosome segments that are identical with respect to their constituent genetic loci and their visible structure; in evolution, of genes and structures that are similar in different organisms owing to their having inherited them from a common ancestor.

Homozygosity The condition of being homozygous; the proportion of homozygous individuals at a locus, or of homozygous loci in an individual.

Homozygote A cell or organism having the same allele at a given locus on homologous chromosomes.

Hot spot A region of a DNA molecule much more susceptible to mutation than other regions of similar size.

Hybrid An offspring of a cross between two genetically unlike individuals.

Hybrid inviability Reduction of somatic vigor or survival rate in hybrid organisms.

Hybrid sterility Reduction or suppression of the reproductive capacity in hybrid organisms.

Hybrid vigor See *heterosis.*

Hypha The filamentous cellular structure composing the body of a fungus.

Identical by descent Of two genes that are identical in nucleotide sequence because they are both derived from a common ancestor.

Identical in structure Of two genes that are identical in nucleotide sequence, regardless of whether or not they are both derived from a common ancestor.

Identical twins See *monozygotic twins.*

Imaginal cell See *imago.*

Imaginal disk See *imago.*

Imago The adult form of an insect arising during pupation from the *imaginal cells* and *imaginal disks* of the larva.

Inbreeding Mating between relatives.

Inbreeding coefficient The probability that the two genes (alleles) at a locus are identical by descent.

Inbreeding depression Reduction in fitness or vigor due to the inbreeding of normally outbreeding organisms.

Incipient species Populations that are too distinct to be considered as subspecies of the same species, but not sufficiently differentiated to be regarded as different species; also called *semispecies.*

Inducer An effector molecule responsible for the induction of enzyme synthesis.

Induction The synthesis of new enzyme molecules in response to an environmental stimulus.

Insertion sequence One of a number of different nucleotide sequences found in bacteria and capable of moving from one chromosomal location to another. The spontaneous movement of such sequences may cause mutation at the original site of insertion or at the new site of insertion. These sequences may carry active promoters or terminators of mRNA synthesis; they may also provide the location of site-specific episome integration.

Interference A measure of the degree to which one crossover by a chromatid affects the probability of a second crossover by that same chromatid. Positive (negative) interference indicates that a crossover decreases (increases) the probability of a second crossover.

Intergenic suppressor A mutation that suppresses the phenotype of another mutation in a gene other than that in which the suppressor mutation resides.

Interphase The stage of the cell cycle during which metabolism and synthesis occur without any visible evidence of cell division.

Intersex An individual of a normally dioecious species whose reproductive organs or secondary sex characters are partly of one sex and partly of the other.

Intervening sequence A noncoding nucleotide sequence in eukaryotic DNA, separating two portions of nucleotide sequence found to be contiguous in cytoplasmic mRNA.

Intragenic suppressor A mutation that suppresses the phenotype of another mutation in the same gene as that in which the suppressor mutation resides.

Inversion A chromosomal mutation characterized by the reversal of a chromosome segment.

Inversion polymorphism The presence of two or more chromosome sequences, differing by inversions, in the homologous chromosomes of a population.

Isolating mechanism See *reproductive isolating mechanism.*

Karyotype The chromosome complement of a cell or organism, characterized by the number, size, and configuration of the chromosomes.

Klinefelter's syndrome A human syndrome that is due to the presence of one extra X chromosome in the male karyotype (XXY).

Leader The region of an mRNA molecule extending from the 5′ end to the beginning of the coding region of the first structural gene; it may contain ribosomal binding sites or an attenuator as well.

Lethal A gene or chromosomal mutation that causes death (in all carriers if it is dominant, but only in homozygotes if it is recessive) before reproductive age.

Ligase See *DNA ligase* and *aminoacyl-tRNA ligase.*

Linkage A measure of the degree to which alleles of two genes assort independently at meiosis or in genetic crosses.

Linkage disequilibrium Nonrandom association of alleles at different loci in a population.

Linkage group A set of gene loci that can be placed in a linear order representing the different degrees of linkage between the loci.

Linkage map A chromosome map showing the linear order of the genes associated with the chromosome.

Locus The place at which a particular mutation or a gene resides in a genetic map; often used interchangeably with *mutation* or *gene.*

Lymphocyte A cell capable of being stimulated by an antigen to produce a specific antibody to that antigen and to proliferate to produce a population of such antibody-producing cells.

Lysogen A strain of bacteria carrying a prophage.

Lysogeny One of two outcomes of the infection of a host cell by a temperate phage. One outcome is that the phage genome becomes repressed and the phage DNA replicates as part of the host DNA, forming a lysogen; infrequently, the lysogen may become induced and the host cell may burst, releasing a number of phage particles. The other outcome is the lytic cycle that produces progeny phage particles.

Macroevolution Evolution above the species level, leading to the formation of genera, families, and other higher taxa; also called *trans-specific evolution.*

Marker An allele whose inheritance is under observation in a cross.

Mass selection Artificial selection practiced by choosing, in each generation, individuals with maximum (minimum) expression of a given trait as parents of the following generation.

Mating system The pattern of mating in sexually reproducing organisms; two types of mating systems are *random mating* and *assortative mating.*

Mating type The mating compatibilities of an organism, which are usually genetically controlled.

Megaspore The larger of the two kinds of haploid spores produced by vascular plants; the smaller kind is called a *microspore.* In seed plants, the megaspore develops into the embryo sac (the female gametophyte), while the microspore gives rise to the pollen grain (the male gametophyte).

Meiosis Two successive divisions of a nucleus following one single replication of the chromosomes, so that the resulting four nuclei are haploid.

Mendelian population An interbreeding group of organisms sharing in a common gene pool.

Merozygote A partially diploid bacterial cell arising from conjugation, transduction, or transformation.

Messenger RNA (mRNA) An RNA molecule whose nucleotide sequence is translated into an amino acid sequence on ribosomes during polypeptide synthesis.

Metacentric chromosome A chromosome with the centromere near the middle.

Metafemale An individual that has a greater dose of female determiners than do normal females; also called *superfemale.*

Metaphase The second stage of mitosis or of meiosis (I or II), during which the condensed chromosomes line up on a plane between the two poles of the cell.

Metric character A trait that varies more or less continuously among individuals, which are therefore placed into classes according to measured values of the trait; also called *quantitative character.*

Microspore See *megaspore.*

Migration See *gene flow.*

Mitosis The division of a nucleus following replication of the chromosomes, so that the resulting daughter nuclei have the same number of chromosomes as the parent nucleus.

Modification enzyme See *restriction enzyme.*

Modifier gene A gene that interacts with other genes by modifying their phenotypic expression (cf. *regulatory gene, structural gene*).

Monoecious Of organisms (usually plants) having male and female sex organs on the same individual and producing male and female gametes (cf. *dioecious*).

Monoploid A cell, tissue, or organism having only one chromosome set (cf. *haploid, diploid, polyploid*).

Monosomic An aneuploid cell, tissue, or organism with one of two homologous chromosomes missing.

Monozygotic twins Twins developed from a single fertilized ovum that gives rise to two embryos at an early developmental stage; also called *identical twins.*

Mosaic An individual composed of different groups of cells, each expressing a different phenotype.

mRNA See *messenger RNA.*

Multiple alleles The occurrence in a population of more than two alleles at a locus.

Multiple-factor inheritance The determination of a phenotypic trait by genes at more than one locus.

Mutator gene A gene that increases the mutation rate of other genes in the same organism.

Muton The smallest unit of the gene capable of mutation; a base pair in DNA.

Native DNA Double-stranded DNA isolated from a cell with its hydrogen bonds between strands intact (cf. *denatured DNA*).

Natural selection The differential reproduction of alternative genotypes due to variable fitness (cf. *artificial selection*).

Nick translation A means of introducing radioactively labeled nucleotide triphosphates into native DNA *in vitro*, employing *E. coli* DNA polymerase I. Using a 5′-PO_4/3′-OH single-strand nick in a native DNA molecule, DNA polymerase I catalyzes the addition of an exogenous radioactive nucleotide to the 3′-OH side of the nick and hydrolyzes a nucleotide from the 5′-PO_4 side of the nick. This moves the nick along the native DNA molecule by one nucleotide. Subsequent polymerization and concomitant hydrolysis translates the nick along the DNA molecule while introducing more radioactive nucleotides into the DNA.

Nondisjunction The failure of two sister chromatids or homologous chromosomes to separate during cell division, so that both go to the same pole, thus producing aneuploid nuclei.

Nonhomologous Of chromosomes or chromosome segments that contain dissimilar genes and that do not pair during meiosis.

Nonrandom mating A mating system in which the frequencies of the various kinds of matings with respect to some trait or traits are different from those expected according to chance.

Norm of reaction See *range of reaction.*

Nucleolus A nuclear organelle of eukaryotes, associated with the chromosomal site of the genes coding for rRNA.

Nucleus A membrane-enclosed organelle of eukaryotes that contains the chromosomes.

Nullisomic An aneuploid cell, tissue, or organism in which both homologous chromosomes of a pair are missing.

Oligomer A protein composed of two or a few identical polypeptide subunits.

Oocyte See *oogonium.*

Oogenesis The process of differentiation of a mature egg cell from an undifferentiated germline cell, including the process of meiosis.

Oogonium A primordial germ cell that gives rise, by mitosis, to *oocytes*, from which the ovum and polar bodies develop by meiosis.

Operator In the DNA of an operon, a nucleotide sequence that is recognized and bound by a repressor protein, which, in turn, inhibits transcription of the operon (cf. *promoter*).

Operon Two or more adjacent structural genes, transcribed into a single mRNA, and the adjacent transcriptional control sites (promoter and operator). This organization places the expression of the structural genes in an operon under the coordinate control of a single promoter and operator.

Organogenesis The period during embryogenesis when the major organs of the body form.

Orthologous genes Homologous genes that have become differentiated in different species derived from a common ancestral species (cf. *paralogous genes*).

Outbreeding A mating system in which matings between close relatives do not usually occur.

Overdominance The condition when the heterozygote exhibits a more extreme manifestation of the trait (usually, fitness) than does either of the homozygotes.

Ovum A female gamete.

Palindrome A sequence of symbols that reads identically in both directions.

Panmixia See *random mating.*

Paracentric inversion A chromosomal inversion that does not include the centromere.

Paralogous genes Homologous genes that have arisen through a gene duplication and that have evolved in parallel within the same organism (cf. *orthologous genes*).

Parental type An association of genetic markers, found among the progeny of a cross, that is identical to an association of markers present in a parent (cf. *recombinant type*).

Parthenogenesis The production of an embryo from a female gamete without participation of a male gamete.

Pedigree A diagram showing the ancestral relationships among individuals of a family over two or more generations.

Penetrance The frequency with which a dominant or homozygous recessive gene manifests itself in the phenotype of its carriers.

Peptide bond A covalent bond formed between the NH_2 group of one amino acid and the COOH group of another, with the elimination of H_2O.

Pericentric inversion A chromosomal inversion that includes the centromere.

Perithecium A fruiting body, containing a number of asci, that develops from a *protoperithecium* of an actinomycete fungus.

Permissive condition See *conditional lethal mutation.*

Phage See *bacteriophage.*

Phagocyte A white blood cell that engulfs and destroys bacterial and other cells bearing antigen-antibody complexes on their surfaces.

Phenocopy A nonhereditary phenotypic modification that mimics a similar phenotype due to a gene mutation.

Phenotype The observable characteristics of an individual, resulting from the interaction between the genotype and the environment in which development occurs.

Phenotypic variance The variance among individuals with respect to some phenotypic trait or traits (cf. *genetic variance*).

Plaque A hole in a lawn of bacterial cells created by the growth of a bacteriophage and its concomitant killing of cells.

Plasmid A genetic element (DNA molecule), harbored within a host cell, that replicates independently of the host chromosomes.

Plastid In plants, a self-replicating organelle that can differentiate into a *chloroplast.*

Pleiotropy An effect whereby a single mutant gene affects two or more apparently otherwise unrelated aspects of the phenotype of an organism.

Polar bodies The smaller cells that are produced during meiosis in oogenesis and that do not develop into functional ova (cf. *spermatids*).

Polar mutation A mutation of one gene that affects the expression of the adjacent nonmutant gene on one side, but not of that on the other side.

Polygenic Of traits determined by many genes, each having only a slight effect on the expression of the trait.

Polymerase An enzyme that assembles a number of similar or identical subunits into a larger unit, or polymer; e.g., DNA polymerase and RNA polymerase.

Polymorphism The presence of several forms (of a trait or of a gene) in a population; the proportion of polymorphic gene loci in a population.

Polynucleotide ligase See *DNA ligase.*

Polypeptide A chain of amino acids covalently bound by peptide linkages.

Polyploid A cell, tissue, or organism having three or more complete chromosome sets (cf. *haploid, monoploid, diploid*).

Polyribosome See *polysome.*

Polysome A polyribosome, consisting of two or more ribosomes bound together by their simultaneous translation of a single mRNA molecule.

Polysomic A cell, tissue, or organism having one chromosome represented three or more times.

Polytene chromosome An interphase chromosome that has undergone a number of rounds of DNA replication, without accompanying nuclear divisions, and in which the resulting chromosome strands are paired lengthwise to create a ropelike giant chromosome revealing a specific banding pattern of the chromatin.

Position effect A change in the phenotypic effect of one or more genes, due to a change in their positions within the genome.

Primary sex ratio See *sex ratio*.

Primer A substrate that is required for a polymerization reaction (e.g., DNA synthesis) and that is structurally similar to the product of the reaction itself.

Prokaryote A cell or organism that lacks a membrane-bound nucleus and does not undergo mitosis or meiosis. The viruses, bacteria, and blue-green algae are prokaryotes.

Promoter In the DNA of an operon, a nucleotide sequence that is recognized by RNA transcriptase as a site at which to begin transcription (cf. *operator*).

Prophage The repressed form of a phage genome present in a lysogen.

Prophase The first stage of mitosis or of meiosis (I or II) during which the chromosomes condense and become visible as distinct bodies.

Protein A polymer composed of one or more polypeptide subunits and possessing a characteristic three-dimensional shape imposed by the sequence of its component amino acid residues.

Protoperithecium See *perithecium*.

Prototroph A strain of organisms capable of growth on a defined minimal medium from which they can synthesize all of the more complex biological molecules they require (cf. *auxotroph*).

Pseudoalleles Mutations that are allelic to one another on the basis of complementation, but that are separable from one another by recombination.

Pseudodominance The apparent dominance of a recessive gene (allele), owing to a deletion of the corresponding gene in the homologous chromosome.

Quadrivalent Four completely or partially homologous chromosomes that are associated by pairing from prophase to metaphase of the first meiotic division.

Quantitative character See *metric character*.

Quantum speciation The rapid rise of a new species, usually in small isolates, with the founder effect and random genetic drift playing important roles; also called *saltational speciation*.

Race A population or group of populations distinguishable from other such populations of the same species by the frequencies of genes, chromosomal arrangements, or hereditary phenotypic characteristics. A race that has received a taxonomic name is a subspecies.

Random genetic drift Variation in gene frequency from one generation to another due to chance fluctuations.

Random mating Random selection of mates with respect to one or more traits; also called *panmixia* (cf. *assortative mating*).

Random sample A sample obtained in such a way that each individual in the population, or each gene in the genome, has the same chance of being selected.

Range of reaction The range of all possible phenotypes that may develop, by interaction with various environments, from a given genotype; also called *norm of reaction*.

Recessive An allele, or the corresponding trait, that is manifest only in homozygotes (cf. *dominant*).

Reciprocal translocation A translocation that involves an exchange of chromosome segments between two nonhomologous chromosomes.

Recombinant type An association of genetic markers, found among the progeny of a cross, that is different from any association of markers present in the parents (cf. *parental type*).

Recombination The creation of a new association of DNA molecules (chromosomes) or parts of DNA molecules (chromosomes).

Recon The unit of genetic recombination; the individual nucleotide pair in a DNA molecule.

Regulatory gene In the broad sense, any gene that regulates or modifies the activity of other genes. In the narrow sense, a gene that codes for an allosteric protein that (alone or in combination with a corepressor) regulates the genetic transcription of the structural genes in an operon by binding to the operator (cf. *modifier gene, structural gene*).

Repeat A chromosomal duplication in which the duplicated segments are adjacent (in tandem), inverted or not; also called *tandem duplication*.

Replicon A self-replicating genetic element possessing a site for the initiation of DNA replication and genes specifying the necessary functions for controlling replication.

Repressor A protein that binds to an operator sequence in DNA and thereby inhibits the transcription of adjacent genes by blocking RNA transcriptase from the promoter for those genes.

Reproductive isolating mechanism (RIM) Any biological property of an organism that interferes with its interbreeding with organisms of other species.

Reproductive isolation The inability to interbreed due to biological differences.

Restriction enzyme An endonuclease that recognizes specific nucleotide sequences in DNA and then makes a double-strand cleavage of the DNA molecule. The DNA of an organism possessing such an enzyme is usually modified at the recognition sites, to prevent self-cleavage, by a *modification enzyme* that recognizes the sites and methylates specific nucleotides at each site.

Restrictive condition See *conditional lethal mutation.*

Reversion A second mutation that restores the genetic information altered by a first mutation.

Ribonucleic acid (RNA) A polynucleotide in which the sugar residue is ribose and which has uracil rather than the thymine found in DNA.

Ribosomal RNA (rRNA) The RNA molecules that are structural parts of ribosomes, i.e., 5S, 16S, and 23S RNAs in prokaryotes and 5S, 18S, and 28S RNAs in eukaryotes.

Ribosome An organelle, consisting of two subunits composed of RNA and proteins, that synthesizes polypeptides, whose amino acid sequences are specified by the nucleotide sequences of mRNA molecules.

RIM See *reproductive isolating mechanism.*

RNA See *ribonucleic acid.*

RNA polymerase See *RNA transcriptase.*

RNA transcriptase The enzyme responsible for transcribing the information encoded in DNA into RNA; also called *transcriptase* or *RNA polymerase.*

Robertsonian change A chromosomal mutation due to centric fusion or centric fission.

rRNA See *ribosomal RNA.*

S See *svedberg.*

Saltational speciation See *quantum speciation.*

Secondary sex ratio See *sex ratio.*

Selection See *natural selection* and *aritficial selection.*

Selection coefficient See *coefficient of selection.*

Selection differential In artificial selection, the difference in mean phenotypic value between the individuals selected as parents of the following generation and the whole population.

Selection gain In artificial selection, the difference in mean phenotypic value between the progeny of the selected parents and the parental generation.

Selective value See *adaptive value.*

Self-fertilization The union of male and female gametes produced by the same individual.

Selfing Breeding by self-fertilization.

Semispecies See *incipient species.*

Sex cell See *gamete.*

Sex chromosomes Chromosomes that are different in the two sexes and that are involved in sex determination (cf. *autosome*).

Sex-limited Pertaining to genetically controlled characters that are phenotypically expressed in only one sex.

Sex linkage Linkage of genes that are located in the sex chromosomes.

Sex ratio The number of males divided by the number of females (sometimes expressed in percent) at fertilization (primary sex ratio), at birth (secondary sex ratio), or at sexual maturity (tertiary sex ratio).

Sickle-cell anemia A human disease characterized by defective hemoglobin molecules and due to homozygosity for an allele coding for the beta chain of hemoglobin.

Sickle-cell trait A phenotype recognizable by the sickling of red blood cells exposed to low oxygen tension, and determined by heterozygosity for the allele responsible for sickle-cell anemia.

Somatic cells All body cells except the gametes and the cells from which these develop.

Speciation The process of species formation.

Species Groups of interbreeding natural populations that are reproductively isolated from other such groups.

Spermatids The cells that are produced during meiosis in spermatogenesis and that eventually develop into functional spermatozoa (cf. *polar bodies*).

Spermatocyte See *spermatogonium.*

Spermatogenesis The process of differentiation of a mature sperm cell from an undifferentiated germ-line cell, including the process of meiosis.

Spermatogonium A primordial germ cell that gives rise, by mitosis, to *spermatocytes*, from which the spermatozoa develop by meiosis.

Spermatozoon In animals, a male gamete.

Spindle In eukaryotes, an ellipsoidal collection of fibers visible during mitosis and meiosis and involved in the separation of homologous chromosomes or sister chromatids toward opposite poles of the cell.

Sporophyte The diploid, asexual generation that produces the spores in plants in which a sexual (haploid) generation alternates with an asexual (diploid) generation (cf. *gametophyte*).

Structural gene A gene that codes for a polypeptide (cf. *modifier gene, regulatory gene*).

Subspecies A population or group of populations distinguishable from other such populations of the same species by the frequencies of genes, chromosomal arrangements, or hereditary phenotypic characteristics. Subspecies sometimes exhibit incipient reproductive isolation, although not sufficiently to make them different species.

Supercoil A double-stranded DNA molecule containing extra twists in the helix that cause the helix to coil upon itself. In order for these extra twists to be maintained, it is necessary that the ends of the double-stranded helix be constrained from rotating freely, e.g., in a covalently closed circular DNA molecule.

Superfemale See *metafemale*.

Supergene A DNA segment that contains a number of closely linked genes affecting a single trait or an array of interrelated traits.

Suppressor See *intergenic suppressor* and *intragenic suppressor*.

Suppressor-sensitive mutation A mutation whose phenotype is suppressed in a genotype that also carries an intergenic suppressor of that mutation, e.g., *amber, ochre*, or *opal*.

Svedberg (S) A unit of measure for the rate at which a particle sediments in a centrifugal field; $1\ s = 10^{-13}$ sec. In general, the greater the mass of a particle, the greater the observed sedimentation rate.

Sympatric Of populations or species that inhabit, at least in part, the same geographic region (cf. *allopatric*).

Synapsis The pairing of chromosomes at meiosis.

Synaptinemal complex An organelle, present during meiosis, that mediates close pairing between homologous regions of chromatids.

Tandem duplication See *repeat*.

Telocentric chromosome A chromosome with a terminal centromere.

Telophase The fourth and final stage of mitosis or of meiosis (I or II).

Temperate phage A bacteriophage capable of conferring lysogeny upon a host cell (cf. *virulent phage*).

Template The DNA single strand, complementary to a nascent RNA or DNA strand, that serves to specify the nucleotide sequence of the nascent strand.

Terminator A nucleotide sequence in DNA that causes RNA transcriptase to cease transcription.

Tertiary sex ratio See *sex ratio*.

Testcross A cross between a heterozygote (at one or more loci) and the corresponding recessive homozygote.

Tetraploid A cell, tissue, or organism having four chromosome sets.

Tetrasomic A cell, tissue, or organism having one chromosome represented four times.

Transcriptase See *RNA transcriptase.*

Transcription The transfer of genetic information encoded in the nucleotide sequence of DNA into a nucleotide sequence of an RNA molecule.

Transduction The transfer of DNA from one cell to another, effected by a virus.

Transfer RNA (tRNA) Special RNA molecules that are associated with specific amino acids to form aminoacyl-tRNAs and that transfer their amino acids to growing polypeptides during association with ribosomes.

Transformation The direct assimilation of exogenous DNA by a cell, leading to recombination between that DNA and the DNA of the cell.

Transition A base-pair substitution mutation resulting in the replacement of one purine by another purine, or of one pyrimidine by another pyrimidine (cf. *transversion*).

Translocation A chromosomal mutation characterized by a change in position of a chromosome segment.

Transposition A translocation of a chromosome segment from one position to another without a reciprocal exchange (cf. *reciprocal translocation*).

Transposon A transposable DNA sequence carrying one to many genes bounded at each end by identical insertion sequences, which confer the ability to move from one location to another.

Transspecific evolution See *macroevolution.*

Transversion A base-pair substitution mutation resulting in the replacement of a purine by a pyrimidine, or vice versa (cf. *transition*).

Trihybrid cross A cross between individuals that have different alleles at three gene loci.

Triploid A cell, tissue, or organism having three chromosome sets.

Trisomic A cell, tissue, or organism having one chromosome represented three times.

tRNA See *transfer RNA.*

Turner's syndrome A human syndrome that is due to monosomy for the X chromosome with absence of a Y chromosome (X0); affected individuals are phenotypically female but usually have underdeveloped gonads.

Univalent An unpaired chromosome at the first meiotic division.

Variance A measure of variation, calculated as the sum of the squares of the differences between the value of each individual and the mean of the population.

Vegetative reproduction See *asexual reproduction.*

Virulent phage A bacteriophage whose infection invariably kills the host cell rather than conferring lysogeny (cf. *temperate phage*).

Zygote The diploid cell formed by the union of egg and sperm nuclei within the cell.

Bibliography

Chapter 1

Cairns, J., G. S. Stent, and J. D. Watson, eds., *Phage and the Origins of Molecular Biology*, Cold Spring Harbor Laboratory, Cold Spring Harbor, N.Y., 1966.

Dunn, L. C., *A Short History of Genetics*, McGraw-Hill, New York, 1965.

Stubbe, H., *History of Genetics*, MIT Press, Cambridge, Mass., 1972.

Sturtevant, A. H., *A History of Genetics*, Harper & Row, New York, 1965.

Wilson, E. O., T. Eisner, W. R. Briggs, R. E. Dickerson, R. L. Metzenberg, R. D. O'Brien, M. Susman, and W. E. Boggs, *Life on Earth*, 2nd ed., Sinauer, Sunderland, Mass., 1978.

Wolfe, S. L., *Biology of the Cell*, Wadsworth, Belmont, Cal., 1972.

Chapter 2

Clausen, J., D. D. Keck, and W. M. Hiesey, Experimental studies on the nature of species. I. Effects of varied environments on western North American plants, Carnegie Institution of Washington Publ. No. 520, Washington, D.C., 1940, pp. 1–452.

Cooper, R. M., and J. P. Zubek, Effects of enriched and restricted early environments on the learning ability of bright and dull rats, *Can. J. Psych. 12*:159–164 (1958).

Dunn, L. C., ed., *Genetics in the 20th Century*, Macmillan, New York, 1951.

Mendel, G., Experiments in plant hybridization. *English translation of Mendel's classic work. Reprinted, for example, in the following collections by Peters and by Stern and Sherwood.*

Peters, J. A., ed., *Classic Papers in Genetics*, Prentice-Hall, Englewood Cliffs, N.J., 1959.

Stern, C., and E. R. Sherwood, eds., *The Origin of Genetics*, W. H. Freeman, San Francisco, 1966.

Chapter 3

Bridges, C. B., Nondisjunction as proof of the chromosome theory of heredity, *Genetics 1*:1–52, 107–163 (1916).

Bridges, C. B., Sex in relation to chromosomes and genes, *Amer. Nat. 59*:127–137 (1925).

Galán, F., Teoría genética del sexo zigótico en el caso de *Ecballium elaterium, Revista de Biología 4*:187–220 (1964).

Morgan, L. V., Non criss-cross inheritance in *Drosophila melanogaster, Biol. Bull. 42*:267–274 (1922).

Morgan, T. H., Sex-limited inheritance in *Drosophila, Science 32*:120–122 (1910).

Sutton, W. S., The chromosomes in heredity, *Biol. Bull. 4*:231–251 (1903).

Zulueta, A. de, La herencia ligada al sexo en el coleóptero *Phytodecta variabilis* (Ol.), *EOS, Revista Española de Entomología, 1*:203–229 (1925).

Chapter 4

Avery, O. T., C. M. MacLeod, and M. McCarty, Studies on the chemical nature of the substance inducing transformation of Pneumococcal types, *J. Exp. Med. 79*:137–158 (1944).

Britten, R., and D. E. Kohne, Repeated sequences in DNA, *Science 161*:529–540 (1968).

Cairns, J., The bacterial chromosome and its manner of replication as seen by autoradiography, *J. Mol. Biol. 6*:208–213 (1963).

Hershey, A. D., and M. Chase, Independent functions of viral protein and nucleic acid in growth of bacteriophage, *J. Gen. Physiol. 36*:39–56 (1952).

Kavenoff, R., and B. Zimm, Chromosome-sized DNA molecules from *Drosophila, Chromosoma 41*:1–27 (1973).

Laird, C. D., and B. J. McCarthy, Molecular characterization of the *Drosophila* genome, *Genetics 63*:865–882 (1969).

Meselson, M., and F. Stahl, The replication of DNA in *Escherichia coli, Proc. Natl. Acad. Sci. USA* 44:671–682 (1958).

Watson, J. D., and F. H. C. Crick, Molecular structure of nucleic acids. A structure for deoxyribose nucleic acid, *Nature 171*:737–738 (1953).

Watson, J. D., and F. H. C. Crick, Genetical implications of the structure of deoxyribonucleic acid, *Nature 171*:964–967 (1953).

Wetmur, J., and N. Davidson, Kinetics of renaturation of DNA, *J. Mol. Biol. 31*:349–370 (1968).

Wilkins, M. H. F., et al., Molecular structure of deoxypentose nucleic acids, *Nature 171*:738–740 (1953).

Chromatin, Cold Spring Harbor Symposia on Quantitative Biology XLII, Cold Spring Harbor Laboratory, Cold Spring Harbor, N.Y., 1978.

Chromosome Structure and Function, Cold Spring Harbor Symposia on Quantitative Biology XXXVIII, Cold Spring Harbor Laboratory, Cold Spring Harbor, N.Y., 1973.

Chapter 5

Burton, W. G., et al., Role of methylation in the modification and restriction of chloroplast DNA in *Chlamydomonas, Proc. Natl. Acad. Sci. USA 76*:1390–1394 (1979).

Creighton, H. B., and B. McClintock, A correlation of cytological and genetical crossing-over in *Zea mays, Proc. Natl. Acad. Sci. USA 17*:492–497 (1931).

Gillham, N. W., *Organelle Heredity*, Raven Press, New York, 1977.

Lewis, E. B., ed., *Selected Papers of A. H. Sturtevant: Genetics and Evolution*, W. H. Freeman, San Francisco, 1961.

Muller, H. J., *Studies in Genetics: The Selected Papers of H. J. Muller*, Indiana University Press, Bloomington, 1962.

Sturtevant, A. H., The linear arrangement of six sex-linked factors in *Drosophila*, as shown by their mode of association, *J. Exp. Zool. 14*:43–59 (1913).

The following two papers should be compared with that of Sturtevant listed above. Great discoveries are often not accepted immediately by everyone. Castle offers another interpretation of Sturtevant's data.

Castle, W. E., Is the arrangement of the genes in the chromosome linear? *Proc. Natl. Acad. Sci. USA* 5:25–32 (1919).

Muller, H. J., Are the factors of heredity arranged in a line? *Amer. Nat.* 54:97–121 (1920).

Chapter 6

Benbow, R. M., et al., Genetic map of bacteriophage ϕX174, *J. Virol. 7*:549–558 (1971).

Benbow, R. M., et al., Genetic recombination in bacteriophage ϕX174, *J. Virol. 13*:898–907 (1974).

Cairns, J., G. S. Stent, and J. D. Watson, eds., *Phage and the Origins of Molecular Biology*, Cold Spring Harbor Laboratory, Cold Spring Harbor, N.Y., 1966.

Hershey, A. D., ed., *The Bacteriophage Lambda*, Cold Spring Harbor Laboratory, Cold Spring Harbor, N.Y., 1971.

Kaiser, A. D., Mutations in a temperate bacteriophage affecting its ability to lysogenize *Escherichia coli, Virology* 3:42–61 (1957).

Sanger, F., et al., The nucleotide sequence of bacteriophage ϕX174, *J. Mol. Biol. 125*:225–246 (1978).

Chapter 7

Bukhari, A. I., J. A. Shapiro, and S. L. Adhya, eds., *DNA Insertion Elements, Plasmids, and Episomes*, Cold Spring Harbor Laboratory, Cold Spring Harbor, N.Y., 1977.

Jacob, F., and E. Wollman, *Sexuality and the Genetics of Bacteria*, Academic Press, New York, 1961.

Miller, J. H., ed., *Experiments in Molecular Genetics*, Cold Spring Harbor Laboratory, Cold Spring Harbor, N.Y., 1972.

Shapiro, J. A., DNA insertion elements and the evolution of chromosome primary structure, *Trends in Biochemical Sciences*, August, 1977, pp. 176–180.

Yanofsky, C., and E. Lennox, Transduction and recombination study of linkage relationships among the genes controlling tryptophan synthesis in *Escherichia coli, Virology* 8:425–447 (1959).

Chapter 8

Benzer, S., Genetic fine structure, in *Harvey Lectures*, Vol. 56, Academic Press, New York, 1961.

Fogel, S., and R. Mortimer, Informational transfer in meiotic gene conversion, *Proc. Natl. Acad. Sci. USA 62*:96–103 (1969).

Fogel, S., et al., Gene conversion in unselected tetrads from multipoint crosses, *Stadler Symp. 1–2*:89–110 (1971).

Gelbart, W., et al., Extension of the limits of the XDH structural element in *Drosophila melanogaster, Genetics 84*:211–232 (1976).

Hurst, D., et al., Conversion-associated recombination in yeast, *Proc. Natl. Acad. Sci. USA 69*:101–105 (1972).

Lefevre, G., Salivary chromosome bands and the frequency of crossing over in *Drosophila melanogaster, Genetics 67*:497–513 (1971).

Miller, J. H., The promoter-operator region of the *lac* operon of *Escherichia coli, J. Mol. Biol. 38*:413–420 (1968).

Chapter 9

Abelson, J., Recombinant DNA: examples of present-day research, *Science 196*:159–160 (1977). *This same issue also contains many other important papers in the field of recombinant DNA.*

Alberts, B., and R. Sternglanz, Recent excitement in the DNA replication problem, *Nature 269*:655–661 (1977).

Chang, S., and S. N. Cohen, *In vivo* site-specific genetic recombination promoted by the *Eco RI* restriction endonuclease, *Proc. Natl. Acad. Sci. USA* 74:4811–4815 (1977).

Grell, R. F., ed., *Mechanisms in Recombination,* Plenum Press, New York, 1974.

Itakura, K., et al., Expression in *Escherichia coli* of a chemically synthesized gene for the hormone somatostatin, *Science 198*:1056–1063 (1977).

Kornberg, A., *DNA Replication*, W. H. Freeman, San Francisco, 1980.

Landy, A., and W. Ross, Viral integration and excision: structure of the lambda *att* sites, *Science 197*:1147–1160 (1977).

Lehman, I. R., DNA ligase: structure, mechanism, and function, *Science 186*:790–797 (1974).

Lehman, I. R., and D. G. Uyemura, DNA polymerase I: essential replication enzyme, *Science 193*:963–969 (1976).

Maniatis, T., et al., The isolation of structural genes from libraries of eucaryotic DNA, *Cell 15*:687–701 (1978).

Roberts, R. J., Restriction endonucleases, *CRC Crit. Rev. Biochem.* 4:123–164 (1976).

Ullrich, A., et al., Rat insulin genes: construction of plasmids containing the coding sequences, *Science 196*:1313–1319 (1977).

Wechsler, J. A., and J. D. Gross, *Escherichia coli* mutants temperature-sensitive for DNA synthesis, *Mol. Gen. Genet. 113*:273–284 (1971).

Chapter 10

Beadle, G. W., and B. Ephrussi, Development of eye colors in *Drosophila:* diffusable substances and their interrelations, *Genetics 22*:76–86 (1937).

Beadle, G. W., and E. L. Tatum, Genetic control of biochemical reactions in *Neurospora, Proc. Natl. Acad. Sci. USA 27*:499–506 (1941).

Beet, E. A., The genetics of the sickle-cell trait in a Bantu tribe, *Ann. Eugen. 14*:279–284 (1949).

Demerec, M., and Z. Hartman, Tryptophan mutants in *Salmonella typhimurium,* Carnegie Institution of Washington Publ. No. 612, Washington, D.C., 1956, pp. 5–33.

Dickerson, R. E., and I. Geis, *The Structure and Action of Proteins,* W. A. Benjamin, Menlo Park, Cal., 1969.

Fincham, J. R. S., *Genetic Complementation*, W. A. Benjamin, New York, 1966.

Harris, H., *The Principles of Human Biochemical Genetics*, 2nd ed., Elsevier, New York, 1975.

Kan, Y. W., and A. M. Dozy, Polymorphism of DNA sequence adjacent to human beta-globin structural gene: relationship to sickle mutation, *Proc. Natl. Acad. Sci. USA 75*:5631–5635 (1978).

Kühn, A., Über die Determination der Form-Struktur-und Pigment der Schuppen bei *Ephestia kühniella, Zeit. Arch. Entwick.-Mech. Org. 143*:408–487 (1948).

Neel, J. V., The inheritance of sickle-cell anemia, *Science 110*:64–66 (1949).

Pauling, L., H. A. Itano, S. J. Singer, and I. C. Wells, Sickle-cell anemia, a molecular disease, *Science 110*:543–548 (1949).

Watson, J. D., *Molecular Biology of the Gene*, 3rd ed., W. A. Benjamin, Menlo Park, Cal., 1976.

Weatherall, D. J., and J. B. Clegg, Recent developments in the molecular genetics of human hemoglobins, *Cell 16*:467–479 (1979).

Chapter 11

Dugaiczyk, A., et al., The natural ovalbumin gene contains seven intervening sequences, *Nature 274*:328–333 (1978).

Lai, E., et al., The ovalbumin gene: structural sequences in native chicken DNA are not contiguous, *Proc. Natl. Acad. Sci. USA 75*:2205–2209 (1978).

Losick, R., and M. Chamberlin, eds., *RNA Polymerase*, Cold Spring Harbor Laboratory, Cold Spring Harbor, N.Y., 1976.

McReynolds, L., et al., Sequence of chicken ovalbumin mRNA, *Nature 273*:723–728 (1978).

Nomura, M., A. Tissieres, and P. Lengyel, eds., *Ribosomes*, Cold Spring Harbor Laboratory, Cold Spring Harbor, N.Y., 1974.

O'Farrell, P. Z., et al., Structure and processing of yeast precursor tRNAs containing intervening sequences, *Nature 274*:438–445 (1978).

Tilghman, S. M., et al., The intervening sequence of a mouse beta-globin gene is transcribed within the 15S beta-globin mRNA precursor, *Proc. Natl. Acad. Sci. USA 75*:1309–1313 (1978).

Yanofsky, C., G. R. Drapeau, J. R. Guest, and B. C. Carlton, On the colinearity of gene structure and protein structure, *Proc. Natl. Acad. Sci. USA 51*:266–272 (1964).

The Mechanism of Protein Synthesis, Cold Spring Harbor Symposia on Quantitative Biology XXXIV, Cold Spring Harbor Laboratory, Cold Spring Harbor, N.Y., 1969.

Transcription of the Genetic Material, Cold Spring Harbor Symposia on Quantitative Biology XXXV, Cold Spring Harbor Laboratory, Cold Spring Harbor, N.Y., 1970.

Chapter 12

Crick, F. H. C., et al., General nature of the genetic code for proteins, *Nature 192*:1227–1232 (1961).

Garen, A., Sense and nonsense in the genetic code, *Science 160*:149–159 (1968).

Goodman, H., et al., Amber suppression: a nucleotide change in the anticodon of a tyrosine transfer RNA, *Nature 217*:1019–1024 (1968).

Hirsh, D., Tryptophan transfer RNA as the UGA suppressor, *J. Mol. Biol. 58*:439–458 (1971).

Nirenberg, M. W., and J. H. Matthaei, The dependence of cell-free protein synthesis in *E. coli* upon naturally occurring or synthetic polyribonucleotides, *Proc. Natl. Acad. Sci. USA 47*:1588–1602 (1961).

Piper, P. W., et al., Nonsense suppressors of *Saccharomyces cerevisiae* can be generated by mutation of the tyrosine tRNA anticodon, *Nature 262*:757–761 (1977).

The Genetic Code, Cold Spring Harbor Symposia on Quantitative Biology XXXI, Cold Spring Harbor Laboratory, Cold Spring Harbor, N.Y., 1966.

Genetics is a rapidly advancing science. As this text is about to go to press, exceptions to the universality of the genetic code have just been discovered and are reported in the following references.

Barrell, B. G., A. T. Bankier, and J. Drouin, A different genetic code in human mitochondria, *Nature 282*:189–194 (1979).

Hall, B. D., Mitochondria spring surprises, *Nature 282*: 129–130 (1979).

Macino, G., et al., Use of the UGA terminator as a tryptophan codon in yeast mitochondria, *Proc. Natl. Acad. Sci. USA 76*:3784–3785 (1979).

Chapter 13

Bertrand, K., et al., New features of the regulation of the tryptophan operon, *Science 189*:22–26 (1975).

Dickson, R. C., et al., Genetic regulation: the *lac* control region, *Science 187*:27–35 (1975).

Farabaugh, P. J., Sequence of the *lac* I gene, *Nature 274*:765–769 (1978).

Gilbert, W., and A. Maxam, The nucleotide sequence of the *lac* operator, *Proc. Natl. Acad. Sci. USA 70*:3581–3584 (1973).

Herskowitz, I., Control of gene expression in bacteriophage lambda, *Annu. Rev. Genet. 7*:289–324 (1973).

Johnson, A., et al., Mechanism of action of the *cro* protein of bacteriophage lambda, *Proc. Natl. Acad. Sci. USA 75*:1783–1787 (1978).

Lee, F., and C. Yanofsky, Transcription termination at the *trp* operon attenuators of *Escherichia coli* and *Salmonella typhimurium*: RNA secondary structure and regulation of termination, *Proc. Natl. Acad. Sci. USA 74*:4365–4369 (1977).

Ogata, R., and W. Gilbert, An amino-terminal fragment of *lac* repressor binds specifically to *lac* operator, *Proc. Natl. Acad. Sci. USA 75*:5851–5854 (1978).

Pabo, C. O., et al., The lambda repressor contains two domains, *Proc. Natl. Acad. Sci. USA 76*:1608–1612 (1979).

Ptashne, M., et al., Autoregulation and function of a repressor in bacteriophage lambda, *Science 194*:156–161 (1976).

Wu, A., and T. Platt, Transcription termination: nucleotide sequence at 3′ end of tryptophan operon in *Escherichia coli*, *Proc. Natl. Acad. Sci. USA 75*:5442–5446 (1978).

Zieg, J., et al., Regulation of gene expression by site-specific inversion, *Cell 15*:237–244 (1978).

Zurawski, G., et al., Nucleotide sequence of the leader region of the phenylalanine operon of *Escherichia coli*, *Proc. Natl. Acad. Sci. USA 75*:4271–4275 (1978).

Chapter 14

Bennett, D., The T-locus of the mouse, *Cell 6*:441–454 (1975).

Brack, C., et al., A complete immunoglobin gene is created by somatic recombination, *Cell 15*:1–14 (1978).

Davidson, E. H., *Genetic Activity in Early Development*, 2nd ed., Academic Press, New York, 1976.

Galau, G., et al., Structural gene sets active in embryo and adult tissues of the sea urchin, *Cell 7*:487–505 (1976).

Gehring, W., Control of determination in the *Drosophila* embryo, in *Genetic Mechanisms of Development*, ed. by F. H. Ruddle, Academic Press, New York, 1973, pp. 103–128.

Gurdon, J. B., *The Control of Gene Expression in Animal Development*, Clarendon Press, Oxford, 1974.

Hood, L. E., I. L. Weissman, and W. B. Wood, *Immunology*, Benjamin/Cummings, Menlo Park, Cal., 1978.

Hotta, Y., and S. Benzer, Mapping of behavior in *Drosophila* mosaics, in *Genetic Mechanisms of Development*, ed. by F. H. Ruddle, Academic Press, New York, 1973, pp. 129–167.

Lewis, E. B., A gene complex controlling segmentation in *Drosophila*, *Nature 276*:565–570 (1978).

Lucchesi, J. C., Dosage compensation, *Amer. Zool. 17*:685–693 (1977).

Postlethwait, J. H., and H. A. Schneiderman, Developmental genetics of *Drosophila* imaginal discs, *Annu. Rev. Genet.* 7:381–433 (1973).

Wachtel, S. S., Immunogenetic aspects of abnormal sexual differentiation, *Cell 16*:691–695 (1979).

Chapter 15

Bodmer, W. F., and L. L. Cavalli-Sforza, Intelligence and race, *Scientific American*, October, 1970, pp. 19–29.

Bodmer, W. F., and L. L. Cavalli-Sforza, *Genetics, Evolution, and Man*, W. H. Freeman, San Francisco, 1976.

East, E. M., Studies on size inheritance in *Nicotiana*, *Genetics 1*:164–176 (1916).

Falconer, D. S., *Introduction to Quantitative Genetics*, Ronald Press, New York, 1961.

Johannsen, W., *Über Erblichkeit in Populationen und in reinen Linien*, G. Fischer, Jena, 1903. *An English translation of the summary and conclusions can be found in pp. 21–26 of the collection by Peters, cited for Chapter 2.*

Lewontin, R. C., Race and intelligence, *Bulletin of the Atomic Scientists*, March, 1970, pp. 2–8.

Lush, J. L., *Animal Breeding Plans*, Iowa State University Press, Ames, 1945.

Mather, K., Polygenic inheritance and natural selection, *Biol. Rev. 18*:32–64 (1943).

Nilsson-Ehle, H., Kreuzungsuntersuchungen an Hafer und Weizen, *Lunds Univ. Aarskr. N. F. Atd.*, Ser. 2, Vol. 5, No. 2, 1909, pp. 1–122.

Chapter 16

Ames, B. N., W. E. Durston, E. Yamasaki, and F. D. Lee, Carcinogens are mutagens: a simple test system combining liver homogenates for activation and bacteria for detection, *Proc. Natl. Acad. Sci. USA 70*:2281–2285 (1973).

Ames, B. N., J. McCann, and E. Yamasaki, Methods for detecting carcinogens and mutagens with the *Salmonella*/mammalian-microsome mutagenicity test, *Mut. Res. 31*: 347–364 (1975).

Coulondre, C., J. H. Miller, P. J. Faraburgh, and W. Gilbert, Molecular basis of base substitution hotspots in *Escherichia coli*, *Nature 274*:775–780 (1978).

Drake, J. W., *The Molecular Basis of Mutation*, Holden-Day, San Francisco, 1970.

Lindahl, T., An N-glycosidase from *Escherichia coli* that releases free uracil from DNA containing deaminated cytosine residues, *Proc. Natl. Acad. Sci. USA 71*:3649–3653 (1974).

Livneh, Z., D. Elad, and J. Sperling, Enzymatic insertion of purine bases into depurinated DNA *in vitro*, *Proc. Natl. Acad. Sci. USA 76*:1089–1093 (1979).

Luria, S. E., and M. Delbrück, Mutations of bacteria from virus sensitivity to virus resistance, *Genetics 28*:491–511 (1943).

Miller, J. H., C. Coulondre, and P. J. Faraburgh, Correlation of nonsense sites in the *lac* I gene with specific codons in the nucleotide sequence, *Nature 274*:770–775 (1978).

Mohler, J. D., Developmental genetics of the *Drosophila* egg. I. Identification of 59 sex-linked cistrons with maternal effects on embryonic development, *Genetics 85*:259–272 (1977).

Tye, B.-K., and I. R. Lehman, Excision repair of uracil incorporated in DNA as a result of a defect in dUTPase, *J. Mol. Biol. 117*:293–306 (1977).

Chapter 17

Blakeslee, A. F., New Jimson weeds from old chromosomes, *J. Hered. 25*:80–108 (1934).

Bridges, C. B., Deficiency, *Genetics 2*:445–465 (1917).

Bridges, C. B., The Bar "gene," a duplication, *Science 83*:210–211 (1936).

Carson, H. L., Chromosome tracers of the origin of species, *Science 168*:1414–1418 (1970).

Cleland, R. E., *Oenothera: Cytogenetics and Evolution*, Academic Press, London, 1972.

Dobzhansky, Th., Chromosomal races in *Drosophila pseudoobscura* and *D. persimilis*, Carnegie Institution of Washington Publ. No. 554, Washington, D.C., 1944, pp. 47–114.

Garber, E. D., *Cytogenetics: An Introduction*, McGraw-Hill, New York, 1972.

Green, M. M., The genetic control of mutation in *Drosophila*, *Stadler Symp. 10*:95–104 (1978).

Liapunova, E. A., and N. N. Vorontsov, Chromosomes and some issues of the evolution of the ground squirrel genus *Citellus* (Rodentia: Sciuridae), *Experientia 26*:1033–1038 (1970).

Lindsley, D. L., L. Sandler, et al., Segmental aneuploidy and the genetic gross structure of the *Drosophila* genome, *Genetics 71*:157–184 (1972).

Mainx, F., The genetics of *Megaselia scalaris* Loew (Phoridae): a new type of sex determination, *Amer. Nat. 98*:415–430 (1964).

McClintock, B., Some parallels between gene control systems in maize and in bacteria, *Amer. Nat. 95*:265–277 (1961).

Lejeune, J., B. Dutrillaux, M. O. Rethoré, and M. Prieur, Comparison de la structure fine des chromatides d'*Homo sapiens* et de *Pan troglodytes*, *Chromosoma 43*:423–444 (1973).

Stebbins, G. L., *Chromosomal Evolution in Higher Plants*, E. Arnold, London, 1971.

Stern, C., *Principles of Human Genetics*, 3rd ed., W. H. Freeman, San Francisco, 1973.

Stewart, B., and J. Merriam, Regulation of gene activity by dosage compensation at the chromosomal level in *Drosophila*, *Genetics 79*:635–647 (1975).

Sturtevant, A. H., The effects of unequal crossing over at the Bar locus in *Drosophila*, *Genetics 10*:117–147 (1925).

White, M. J. D., *Animal Cytology and Evolution*, 3rd ed., Cambridge University Press, Cambridge, 1973.

Chapter 18

Ayala, F. J., Evolution of fitness in experimental populations of *Drosophila serrata*, *Science 150*:903–905 (1965).

Ayala, F. J., J. W. Valentine, L. G. Barr, and G. S. Zumwalt, Genetic variability in a temperate intertidal phoronid, *Phoronopsis viridis*, *Biochem, Genet. 18*:413–427 (1974).

Dobzhansky, Th., and B. Spassky, Genetics of natural populations. XXXIV. Adaptive norm, genetic load and genetic elite in *D. pseudoobscura*, *Genetics 48*:1467–1485 (1963).

Fisher, R. A., *The Genetical Theory of Natural Selection*, Clarendon Press, Oxford, 1930.

Gottlieb, L. D., Electrophoretic evidence and plant systematics, *Ann. Missouri Bot. Gard. 64*:161–180 (1977).

Harris, H., and D. A. Hopkinson, Average heterozygosity in man, *J. Human Genet. 36*:9–20 (1972).

Lerner, I. M., and W. I. Libby, *Heredity, Evolution, and Society*, 2nd ed., W. H. Freeman, San Francisco, 1976.

Lewontin, R. C., *The Genetic Basis of Evolutionary Change*, Columbia University Press, New York, 1974.

Selander, R. K., Genic variation in natural populations, in *Molecular Evolution*, ed. by F. J. Ayala, Sinauer, Sunderland, Mass., 1976, pp. 21–45.

Chapter 19

Adams, J., and R. H. Ward, Admixture studies and the detection of selection, *Science 180*:1137–1143 (1973).

Bodmer, W., and L. L. Cavalli-Sforza, *Genetics, Evolution, and Man*, W. H. Freeman, San Francisco, 1976.

Buri, P., Gene frequency in small populations of mutant *Drosophila*, *Evolution 10*:367–402 (1956).

Crow, J. F., and M. Kimura, *An Introduction to Population Genetics Theory*, Harper & Row, New York, 1970.

Dobzhansky, Th., and O. Pavlovsky, An experimental study of interaction between genetic drift and natural selection, *Evolution 7*:198–210 (1953).

Glass, H. B., and C. C. Li, The dynamics of racial intermixture: an analysis based on the American Negro, *Amer. J. Human Genet. 5*:1–20 (1953).

Hardy, G. H., Mendelian proportions in a mixed population, *Science 28*:49–50 (1908).

Li, C. C., *Population Genetics*, University of Chicago Press, Chicago, 1955.

Mayr, E., *Populations, Species, and Evolution*, Harvard University Press, Cambridge, Mass., 1970.

Mettler, L. E., and T. G. Gregg, *Population Genetics and Evolution*, Prentice-Hall, Englewood Cliffs, N.J., 1969.

Mourant, A. E., *The Distribution of the Human Blood Groups*, Blackwell, Oxford, 1954.

Spiess, E., *Genes in Populations*, J. Wiley, New York, 1977.

Stern, C., Wilhelm Weinberg (1862–1937) biography, *Genetics 47*:1–5 (1962).

Chapter 20

Allison, A. C., Polymorphism and natural selection in human populations, *Cold Spring Harbor Symp. Quant. Biol. 29*:139–149 (1964).

Ayala, F. J., Frequency-dependent mating advantage in *Dro-*

sophila, Behav. Genet. 2:85–91 (1972).

Ayala, F. J., and C. A. Campbell, Frequency-dependent selection, *Annu. Rev. Ecol. Syst.* 5:115–138 (1974).

Bajema, C. J., ed., *Natural Selection in Human Populations*, J. Wiley, New York, 1971.

Battaglia, B., Balanced polymorphism in *Tisbe reticulata*, a marine copepod, *Evolution 12*:358–364 (1958).

Kettlewell, H. B. D., The phenomenon of industrial melanism in the Lepidoptera, *Annu. Rev. Entom. 6*:245–262 (1961).

Kojima, K., and K. M. Yarbrough, Frequency-dependent selection at the esterase-6 locus in *Drosophila melanogaster, Proc. Natl. Acad. Sci. USA 57*:645–649 (1967).

Petit, C., and L. Ehrman, Sexual selection in *Drosophila, Evol. Biol. 3*:177–223 (1969).

Schull, W. J., ed., *Genetic Selection in Man*, University of Michigan Press, Ann Arbor, 1963.

Spiess, E. B., Low frequency advantage in mating of *Drosophila pseudoobscura* karyotypes, *Amer. Nat. 102*:363–379 (1968).

Wallace, B., *Topics in Population Genetics*, W. W. Norton, New York, 1968.

Chapter 21

Ayala, F. J., Relative fitness of populations of *Drosophila serrata* and *Drosophila birchii, Genetics 51*:527–544 (1965).

Ayala, F. J., and W. W. Anderson, Evidence of natural selection in molecular evolution, *Nature New Biology 241*:274–276 (1973).

Bodmer, W. F., and L. L. Cavalli-Sforza, *Genetics, Evolution, and Man*, W. H. Freeman, San Francisco, 1976.

Clarke, C. A., and P. M. Sheppard, The evolution of mimicry in the butterfly *Papilio dardanus, Heredity 14*:163–173 (1960).

Ford, E. B., *Ecological Genetics*, 3rd ed., Chapman & Hall, London, 1971.

Garn, S. M., *Human Races*, C. C. Thomas, Springfield, Ill., 1961.

Hirszfeld, L., and H. Hirszfeld, Essai d'application des méthodes sérologiques au problème des races, *Anthropologie 29*:505–537 (1919).

Jones, D. F., The attainment of homozygosity in inbred strains of maize, *Genetics 9*:405–418 (1924).

Kretchmer, N., Lactose and lactase, *Scientific American*, October, 1972, pp. 70–78.

Moroni, A., *Historical Demography, Human Ecology, and Consanguinity*, International Union for the Scientific Study of Population, Liège, 1969.

Nabours, R. K., I. Larson, and N. Hartwit, Inheritance of color patterns in the grouse locust, *Acridium arenosum* Burmeister (Tettigidae), *Genetics 18*:159–171 (1933).

Powell, J. R., H. Levene, and Th. Dobzhansky, Chromosomal polymorphism in *Drosophila pseudoobscura* used for diagnosis of geographic origin, *Evolution 26*:553–559 (1973).

Prakash, S., and R. C. Lewontin, A molecular approach to the study of genic heterozygosity in natural populations. III. Direct evidence of coadaptation in gene arrangements of *Drosophila, Proc. Natl. Acad. Sci. USA 59*:398–405 (1968).

Roberts, D. F., Body weight, race, and climate, *Amer. J. Phys. Anthro. 11*:533–558 (1953).

Stern, C., *Principles of Human Genetics*, 3rd ed., W. H. Freeman, San Francisco, 1973.

Weir, B. S., R. W. Allard, and A. L. Kahler, Analysis of complex allozyme polymorphisms in a barley population, *Genetics 72*:505–523 (1972).

Chapter 22

Ayala, F. J., Genetic differentiation during the speciation process, *Evol. Biol. 8*:1–78 (1975).

Ayala, F. J., ed., *Molecular Evolution*, Sinauer, Sunderland, Mass., 1976.

Ayala, F. J., M. L. Tracey, D. Hedgecock, and R. C. Richmond, Genetic differentiation during the speciation process in *Drosophila, Evolution 28*:576–592 (1974).

Bachmann, K., O. B. Goin, and C. J. Goin, Nuclear DNA amounts in vertebrates, *Brookhaven Symp. Biol. 23*:419–450 (1972).

Britten, R. J., and E. H. Davidson, Repetitive and non-repetitive DNA sequences and a speculation on the origin of evolutionary novelty, *Quart. Rev. Biol. 46*:111–138 (1971).

Bruce, E. J., and F. J. Ayala, Phylogenetic relationships between man and the apes: electrophoretic evidence, *Evolution 33*:1040–1056 (1979).

Carson, H. L., Speciation and the founder principle, *Stadler Symp. 3*:51–70 (1971).

Dickerson, R. E., The structure of cytochrome *c* and the rates of molecular evolution, *J. Mol. Evol. 1*:26–45 (1971).

Dobzhansky, Th., *Genetics of the Evolutionary Process*, Columbia University Press, New York, 1970.

Dobzhansky, Th., F. J. Ayala, G. L. Stebbins, and J. W. Valentine, *Evolution*, W. H. Freeman, San Francisco, 1977.

Fitch, W. M., Molecular evolutionary clocks, in *Molecular Evolution*, ed. by F. J. Ayala, Sinauer, Sunderland, Mass., 1976, pp. 160–178.

Fitch, W. M., and E. Margoliash, Construction of phylogenetic trees, *Science 155*:279–284 (1967).

Gillespie, J. H., and C. H. Langley, Are evolutionary rates really variable? *J. Mol. Evol. 13*:27–34 (1979).

Gottlieb, L. D., Genetic confirmation of the origin of *Clarkia lingulata, Evolution 28*:244–250 (1974).

Hinegardner, R., Evolution of genome size, in *Molecular Evolution*, ed. by F. J. Ayala, Sinauer, Sunderland, Mass., 1976, pp. 179–199.

Hoyer, B. H., and R. B. Roberts, Studies on nucleic acid interactions using DNA-agar, in *Molecular Genetics*, Part II, ed. by J. H. Taylor, Academic Press, New York, 1967, pp. 425–479.

Kimura, M., Evolutionary rate at the molecular level, *Nature 217*:624–626 (1968).

King, J. L., and T. H. Jukes, Non-Darwinian evolution, *Science 164*:788–798 (1969).

King, M.-C., and A. C. Wilson, Evolution at two levels: molecular similarities and biological differences between humans and chimpanzees, *Science 188*:107–116 (1975).

Kohne, D. E., J. A. Chiscon, and B. H. Hoyer, Evolution of primate DNA sequences, *J. Human Evol. 1*:627–644 (1972).

Laird, C. D., and B. J. McCarthy, Magnitude of interspecific nucleotide sequence variability in *Drosophila, Genetics 60*:303–322 (1968).

Lewis, H., Speciation in flowering plants, *Science 152*:167–172 (1966).

Mayr, E., *Animal Species and Evolution*, Harvard University Press, Cambridge, Mass., 1963.

Nei, M., Genetic distance between populations, *Amer. Nat. 106*:283–291 (1972).

Nevo, E., Y. J. Kim, C. R. Shaw, and C. S. Thaeler, Genetic variation, selection, and speciation in *Thomomys talpoides* pocket gophers, *Evolution 28*:1–23 (1974).

Sarich, V. M., and A. C. Wilson, Immunological time scale for hominid evolution, *Science 158*:1200–1203 (1967).

Tashian, R. E., M. Goodman, R. E. Ferrell, and R. J. Tanis, Evolution of carbonic anhydrase in primates and other mammals, in *Molecular Anthropology*, ed. by M. Goodman and R. E. Tashian, Plenum Press, New York, 1976, pp. 301–319.

White, M. J. D., *Modes of Speciation*, W. H. Freeman, San Francisco, 1977.

Wilson, A. C., C. R. Maxson, and V. M. Sarich, Two types of molecular evolution: evidence from studies of interspecific hybridization, *Proc. Natl. Acad. Sci. USA 71*:2843–2847 (1974).

Index